T0132987

Insights and Manipulations

Other Books of Interest from St. Augustine's Press

Insights and Manipulations: What Classical Geometry Looked Like at Its Peak, and How It Was Transformed— A Guidebook

Harvey Flaumenhaft

Volume One: Apollonian Vistas

Volume Two: The Way to Cartesian Mastery

St. Augustine's Press
South Bend, Indiana

Manufactured in the United States of America.

2 3 4 5 6 28 27 26 25 24 23

Library of Congress Control Number: 2020940560

∞ The paper used in this publication meets the minimum requirements of the American National Standard for Information Sciences - Permanence of Paper for Printed Materials, ANSI Z39.48-1984.

St. Augustine's Press
www.staugustine.net

This work is dedicated to the memory of
Morty Traum and Joe Oxenhorn—
men of science struck down too soon for me to share it with them.

CONTENTS

Volume One: Apollonian Vistas

Volume Two: The Way to Cartesian Mastery 631

ACKNOWLEDGEMENTS

This work derives from the legacy of Jacob Klein. Klein's book on Greek logistic and the origins of algebra, since its publication more than seven decades ago, has gradually been achieving wider recognition as an extraordinary achievement in uncovering the buried foundations upon which the modern world is built. To understand and judge that book is hard, however, for readers who have not invested the forbidding amount of time and effort required to attain familiarity with and clarity about the technical content of Klein's mathematical sources—and not only the content but the form as well, for Klein insists that what others treat as a mere change in form is rather a fundamental transformation of content. To speed along the modern highway of the mind cannot be a substitute for the geometrical journey required by the absence of a royal road. But a suitable guidebook full of maps can shorten the time and ease the effort of those who, without it, would not be inclined to traverse the winding tangle of difficult pathways that enable one to re-think the thoughts that have constituted the world we live in now.

To provide such a guidebook was my aim in producing the work before you. While my interest in its theme was first excited early in my high-school years, what enabled me to write it was my career as a member of the faculty at St. John's College in Annapolis. There I learned what I needed to know, and the college was very generous in frequently allowing me partial leave to think through what I had learned and to put it down on paper. Much of that time on leave was funded by generous grants from the National Endowment for the Humanities and from the Lynde and Harry Bradley Foundation. Especially helpful to me at the former institution were John Agresto and Daniel Jones, as was Hillel Fradkin at the latter.

The transition from private study to a project for publication owed much to the inspiring example and effective good offices of Curtis Wilson, a former Dean of the College then still teaching at St. John's. While I myself was serving as Dean of the College, a generous grant from the Andrew W. Mellon Foundation was arranged by Mary Patterson McPherson. It funded three summers of participation by members of the St. John's faculty in study groups taught by me using this book in draft. That grant also funded a conference on the topic, as well as several years of editorial assistance and other expenses incurred in putting the manuscript into final form. Under that Mellon funding, I was helped by a number of student interns—Rachel Roccia Sullivan, Wyatt Dowling, Sirrobert Burbridge, William Whittaker, and Robert Abbott—as well as by my colleague Marilyn Higuera, who, giving the manuscript close editorial scrutiny, rooted out errors and suggested improvements in style and substance.

Some of the manuscript was also used in honors courses at the University of Delaware, where I was visiting professor at the invitation of Frank Murray, the Dean of the College of Education, with funding from the Exxon Education Foundation. Jan Blits, who arranged for me to teach there and was my partner in team-teaching the courses, has provided countless hours of critical conversation and encouraging companionship over the decades during which this work has been in process.

Leon Kass, during more than sixty years of friendship, has provided more than I have words to thank him for. Here, I will only note that besides helping me obtain the wherewithal that provided time for writing the book's first draft, he also supplied encouragement and applied pressure for me to get it done, while suggesting improvements as he took several successive groups of University of Chicago students through an early version of the entire manuscript.

And it is hard to imagine how I could have completed so demanding a piece of work as this without the companionship, the devotion, and the support of my late wife Mera.

Several people were particularly helpful in moving the work from manuscript to publication. Charles Butterworth introduced me to the publisher of St. Augustine's Press, who responded to my presentation by saying, "You've done something important, but I've never published anything like it, and I can think of many problems there would be in producing it; still, if I can't figure out how to solve them, maybe I'm in the wrong business." Well, Bruce Fingerhut is certainly in the right business. And the editor to whom he assigned my manuscript, Cathy Bruckbauer, was a pleasure to work with—as was the typesetter, Benjamin Fingerhut, who devoted considerate and strenuous effort to getting the book's intricate layout as close as possible to my conception of it. I am truly grateful to them all.

Harvey Flaumenhaft
St. John's College, Annapolis
Summer 2020

INTRODUCTION

We often take for granted the terms, the premises, and the methods that prevail in our time and place. We take for granted, as the starting-points for our own thinking, the outcomes of a process of thinking by our predecessors.

What happens is something like this: Questions are asked, and answers are given. These answers in turn provoke new questions, with their own answers. The new questions are built from the answers that were given to the old questions, but the old questions are now no longer asked. Foundations get covered over by what is built upon them.

Progress thus can lead to a kind of forgetfulness, making us less thoughtful in some ways than the people whom we go beyond. We can become more thoughtful, though, by attending to the originating thinking that while out of sight is still at work in the achievements it has generated. To be thoughtful human beings—to be thoughtful about what it is that makes us human—we need to read the record of the thinking that has shaped the world around us, and continues still to shape our minds.

Scientific thinking is a fundamental part of that record, but it is a part that is read even less than the rest. It was not always so. Only recently has the prevalent division between "the humanities" and "science" come to be taken for granted. At one time, educated people read Euclid and Ptolemy along with Homer and Plato, whereas nowadays readers of Shakespeare and Rousseau rarely read Copernicus or Newton.

That is often held to be because books in science, unlike those in the humanities, simply become outdated: in science the past is held to be *passé*. Now science is indeed progressive, and progress is a good thing, but so is preservation of what is good. Progress even requires preservation: unless there is keeping, our getting is also great losing—and keeping takes plenty of work.

Precisely if science is essentially progressive, we can truly understand it only by seeing its progress *as* progress. This means that our minds must move through its progressive stages. We ourselves must think through the process of thought that has given us what we otherwise would thoughtlessly accept as given. By refusing to be the passive recipients of ready-made presuppositions and approaches, we can avoid becoming their prisoners. Only by actively taking part in scientific discovery—only through engaging in re-discovery ourselves—can we avoid both blind reaction against the scientific enterprise and blind submission to it.

We and our world are products of a process of thinking, and truly thoughtful thinking is peculiar: it cannot simply outgrow the process of thinking it grows out of. When we utter deceptively simple phrases that in fact are the outcome of a complex development of thought—phrases like "the square root of two"—we may work wonders as we use them to take part in building vast and intricate structures from the labors of millions of people, but we do not truly know what it is that we are doing unless we at some time ask the questions which the words employed so casually now were once an attempt to answer.

When we combine the scientific quest for the roots of things with the humanistic endeavor to make the dead letter come alive in a thoughtful mind, then the past becomes a living source of wisdom that prepares us for the future—a more solid source of wisdom than vague attempts at being "interdisciplinary" which all too often merely provide an excuse for avoiding the study of scientific thought itself. The love of wisdom in its wholeness requires exploration of the sources of the things we take for granted—and this includes the thinking that has sorted out the various disciplines, making demarcations between fields as well as envisioning what is to be done within them.

To foster the reading of classic texts in science, guidebooks are needed. In 1987, as a contribution to supplying the need, I developed a series of such guidebooks with the support of the National Endowment for the Humanities. The purpose of that series, called *Masterworks of Discovery*, was to help non-specialists gain access to formative scientific writings, ancient and modern.

The volumes in the series were *not* to be books *about* thinkers and their thoughts—not to be histories or synopses to take the place of the original works. They were to be guides to help non-specialists read for themselves the thinkers' own expressions of their thinking. While addressed to an audience including scientists as well as scholars in the humanities, the volumes were meant to be readable by any intelligent person who had been exposed to the rudiments of high-school mathematics.

Several volumes in the series were published by the Rutgers University Press, while some books that were expected to form part of the series eventually found a place elsewhere, most notably at the Green Lion Press. Indeed, the present guidebook, on which I was at work when I undertook to develop the series, ended up in need of too much time for preparation, and of too much space for elaboration, to fit within requirements at Rutgers. In the years ahead, many more such guidebooks may well appear in various places as the need for them gradually comes to be more widely appreciated.

Readers of such guidebooks will be unlikely to succumb to notions that reduce science to a mere up-to-date body of concepts and facts, and the humanities to mere frills left over in the world of learning after scientists have done the solid work. By their study of classic texts in science, readers of these guidebooks will be taking part in continuing education at the highest level. The education of a human being requires learning about the process by which the human race obtains its education—and there is no better way to do this than to read the writings of those master-students who have been the master-teachers.

The thoughts to which we are accustomed become most truly our own only when the thinking that went into them becomes our own, so that we can see them as answers to questions that we may not at first have thought to ask—answers to questions that themselves arise from answers to still prior questions. That is the principle behind this guidebook. Its particular purpose is as follows.

Later scientific works are built upon the thinking present in the classic writings of the Greeks; and in much of that thinking, the study of mathematics is central. Mathematical writing, however, is the kind that is most forbidding to non-specialists; even more so are Greek mathematical classics for readers of today, who are accustomed to mathematics of a somewhat different sort. And yet, in studying the mathematical writings of the ancient masters together with those of the founders of the modern world, we consider something fundamental to the understanding of our world—and of ourselves. This guidebook therefore presents ancient mathematics at its peak, along with the critique that transformed it into modern mathematics. The texts chiefly examined are the beginning of the *Conics* of Apollonius of Perga (who wrote in Greek in the third century before the Christian era) and the beginning of the *Geometry* of René Descartes (who wrote in French in the seventeenth century). Attention is also paid to some selections from writings by other ancients—Euclid, Pappus, and Diophantus—and to the book by Viète that has some claim to being the founding text of mathematical modernity.

Through these works, readers gain access to sources of the tremendous transformation in thought whose outcome has been the mathematicization of the world around us and the primacy of mathematical physics in the life of the mind. Scientific technology and technological science have depended upon a transformation in mathematics which made it possible for the sciences as such to be mathematicized, so that the exact sciences became knowledge par excellence. The modern project for mastering nature has relied upon the use of equations, often represented by graphs, to solve problems. When the equation replaced the proportion as the heart of mathematics, and geometric theorem-demonstration lost its primacy to algebraic problem-solving, an immense power was generated. Because of this, Descartes' *Geometry* has been called the greatest single step in the progress of the exact sciences. To determine whether it was indeed such a step, we need to know what it was a step *from* as well as what it was a step *toward*. We cannot understand what Descartes did to transform mathematics unless we understand what it was that underwent the transformation. By studying classical mathematics on its own terms, we prepare ourselves to consider Descartes' critique of classical mathematics and his transformation not only of mathematics but of the world of learning generally—and therewith his work in transforming the whole wide world.

In Volume One of this study, readers will gain a knowledge of the character of classical mathematics through an introductory treatment of the conic sections. But why the conic sections—why not simply look at Euclid's *Elements*? Well, much of classical geometry may seem to many readers just too familiar to think very hard about. The figures that Euclid studies in his *Elements* are made of the simplest lines—lines that are either straight or circular. But classical geometry is not easily taken for granted when reading Apollonius. His *Conics* treats the simplest of the lines that are neither straight nor circular. These are the conic sections. Now parabolas, hyperbolas, and ellipses may be familiar to the modern reader as the graphs of certain equations in high-school mathematics, but the lines are no longer studied *as* conic sections—that is to say, as cuttings of a cone. It is in the classical study of the conic sections that the modern reader can most easily see both the achievement of classical mathematics and the difficulty that led Descartes and his followers to turn away from classical mathematics.

The First Part of this guidebook examines the beginning of the First Book of Apollonius's *Conics*. That beginning, which ends with Apollonius's 10th proposition, prepares the way for his definition of the conic sections. The Third Part of the guidebook examines the conic sections as characterized in Apollonius's 11th through 14th propositions. Between the First Part and the Third, the Second Part treats some material from Euclid's *Elements* that is fundamental for classical mathematics but strange for modern students. It has to do with ratio, and with the notions of number and of magnitude. For Apollonius, as for Euclid in the century before him, the handling of ratios is founded upon a certain view of the relation between numbers and magnitudes. When Descartes made his new beginning, almost two millennia later, he said that the ancients were handicapped by their having a scruple against using the terms of arithmetic in geometry. Descartes attributed this to their not seeing clearly enough the relation between the two mathematical sciences. Before modern readers can appreciate why Descartes wanted to overcome the scruple, and what he saw that enabled him to do it, they must be clear about just what that scruple was.

Readers must, at least for a while, make themselves at home in a world where how-much and how-many are kept distinct—a world which gives an account of shapes in terms of geometric proportions rather than in terms of the equations of algebra. For a while, readers must stop saying "AB-squared," and must speak instead of "the square arising from the line AB"; they must learn to compound ratios instead of multiplying fractions; they must not speak of "the square root of 2." Some study of Euclid will convey an awareness of the foundation covered over by the later modern quantitative superstructure which was built upon it, and in which modern readers are so used to dwelling.

With this Euclidean foundation, and having only so much familiarity with the elementary properties of fairly simple figures as is acquired early in high-school mathematics, readers of this guidebook can move on through the First Book of Apollonius's *Conics*. High-school mathematics may have accustomed us to the presentation of theorems and their demonstrations, but Apollonius presents his theorems and demonstrations in a way that is very different from what students may be used to now. His way of doing things may seem needlessly indirect and cumbersome. The mere enunciation of a theorem is often very long and hard to follow; and the demonstration, in addition to its being long, often provides few of the signposts that would allow the reader easily to discern its parts and their relation to each other. Even the reasons for the selection and the order of the propositions are not easy to discern.

But the way will be much clearer if the traveler takes a guided tour, especially one that hands out maps. This guidebook therefore supplies help of various kinds: previews and flashbacks; diagrams that present matters stage-by-stage instead of all-at-once; tables, outlines, and flowcharts that give summaries of enunciations and also give overviews of the structures of demonstrations, as well as of the relations among propositions. Much effort has been expended to make the two-page spread the unit of composition, thus concentrating readers' efforts on those difficulties that cannot be avoided. Drawings appear in profusion, but almost all of them are placed within the same two-page spread as the words that they are meant to illuminate. The placement of the drawings thus enables the reader to move easily back and forth between words and pictures. To bring the experience of the reader closer to that of someone in the presence of another human being acting as a friendly guide through difficult terrain, the drawings are all produced by hand. Although the drawings for the most part explicate the statements they are placed among or opposite, they also often suggest much more than they show at first glance. They are visual aids, but they are not a substitute for the exertions of the reader. What is depicted in the drawings needs to be studied as carefully as what is said in the text.

With such help, what might otherwise appear to be merely one thing coming after another will more easily be seen as an elegantly and even dramatically arranged journey. This guidebook, then, can help its readers come to see why Apollonius presents things in the way that he does. How one chooses to do things depends on what one thinks one is doing. What is strange to us in classical mathematics is strange because in modern times we may have different things on our minds when we study mathematics. Awareness of such differences can help us think more deeply about what knowledge is and how to get it.

In the first three parts of this guidebook, then, the reader learns some fundamentals of classical geometry, culminating in Apollonius's presentation of what distinguishes the conic sections. The next two parts go on to exhibit more fully the character of classical mathematics at its peak. They show how Apollonius went on to develop his account of the conic sections, coming to a provisional conclusion at the end of the First Book of his *Conics*. This is also the conclusion to Volume One of this guidebook. The guidebook as a whole is entitled *Insights and Manipulations*. Volume One is entitled *Apollonian Vistas*.

Volume Two, containing the last three parts of the guidebook, is entitled *The Road to Cartesian Mastery*. It exhibits the character of the transformation of classical into modern mathematics under the influence of Descartes, and helps the reader understand what Descartes was doing when he criticized his ancient predecessors and suggested another way to proceed. The closing part of the guidebook examines the *Geometry*, where Descartes denigrated classical mathematics as obtuse, having earlier (in his *Rules for the Direction of the Mind*) denigrated it as desultory fooling-around and as disingenuous showing-off.

Descartes' *Geometry* is not a book of theorems and their demonstration. By homogenizing what is studied, and by making the central activity the manipulative working of the mind, rather than the mind's visualizing of form and its insight into what informs the act of vision, Descartes transformed mathematics into a tool of physics for the mastery of nature. He went public with his project in a cunning discourse about the method of well conducting one's reason and seeking truth in the sciences; and this discourse introduced a collection of try-outs of this methodical science, the third and last of which "essays" was his *Geometry*.

To help readers understand the movement from the Apollonian mathematics of Volume One to the Cartesian mathematics with which Volume Two concludes, Volume Two begins with two parts that present some propositions from later books of Apollonius's *Conics*, along with some material from other authors, both ancient and modern. These authors turn the mind from considering theorems to considering the solving of problems. Thus Part Six of this guidebook, treating more of the *Conics* along with Pappus on loci, leads to the geometrical problem-solving that is the point of departure for Descartes' *Geometry*; and then Part Seven, treating the classical numerical problem-solving of Diophantus and the innovative algebraic art of Viète, presents the problem-solving matrix of mathematical modernity. After that, Part Eight treats Descartes' *Geometry* and thus concludes this guidebook.

Among the questions that readers will consider, as they move through this guidebook, are the following: What is the relation between the demonstration of theorems and the solving of problems? What separates the notions of "how-much" and "how-many"? Why try to overcome that separation by the notion of quantity as represented by a number-line? What is the difference between a mathematics of proportions which provides images for viewing being, and a mathematics of equations which provides tools for mastering nature? How does mathematics get transformed into what lends itself to being interpreted as a system of signs referring to signs—as a symbolism which takes on meaning when applied, becoming a source of immense power? What is mathematics, and why study it? What is learning, and what promotes it?

With minds shaped by the thinking of yesterday and of the day before it, we struggle to answer the questions of today, in a world transformed by the minds that did the thinking. We shall proceed more thoughtfully in the days ahead if we have thought through that thinking for ourselves. A means to that end is a guidebook like this. Written for serious amateurs by a serious amateur, it contains hand-drawn pictures in profusion which will help to ease the way of readers who give them as much attention as they give the words.

NOTE ON THE TRANSLATION OF APOLLONIUS AND ITS USE

This guidebook contains a new English translation of the Greek of Apollonius. It strives for consistency—that is, to render any two instances of the same word or phrase or grammatical construction in Greek by the same word or phrase or grammatical construction in English. When the differences between the languages make the result merely awkward, this fault has been accepted as a small enough price to pay for getting the reader as close as possible to awareness of what the original text is like. But when that price seems too high, readable inconsistency has been preferred to an unreadable consistency. This is sometimes indicated by words inserted in square brackets. Square brackets also surround explanatory asides inserted when needed for readability. While words in square brackets are always additions to the text, words in parentheses are not additions, but rather are part of the attempt to put into English the Greek words of Apollonius, or to point out just which Greek words he is using. That is not to say that the English of this translation conveys the Greek of Apollonius word-for-word. Prolixity or repetitiousness has been preferred to letting the reader become more confused or overwhelmed than is unavoidable in beginning to read the text.

Here are some examples of how departures from consistency or from thrift result from differences between the kinds of language to which ancient Greek and modern English belong: Greek is rich in little words (called "particles") that indicate a connection between thoughts, or an attitude of the speaker toward the statement, which an English speaker would indicate by tone of voice or by an entire phrase, and which a translator must convey instead by inserting various punctuation marks or by being prolix. Greek, being richly inflected, can indicate emphasis or sequence of thoughts by word order, which indication the translator can roughly imitate by adding phrases—as when, in the sentence "the circle (direct object) the straight line (subject) touches," the translator may keep the direct object in the emphatic initial position by saying "as for the circle, the straight line touches it," adding the words "as for" and "it," to which no words in the Greek correspond. Inflection makes the Greek verb by itself grammatically sufficient without including a personal pronoun to supply an unnamed subject, so that the emphasis which is given when the Greek includes the personal pronoun is lost in translation when the English needs the pronoun merely to be grammatical. And in English extra words are often needed to clarify what in Greek is clear because inflection sorts out adjectives and nouns by gender. The straight line with endpoints A and B is in Greek referred to as "the AB," the feminine gender of the definite article in Greek suggesting what sort of item is being named—and so this translation speaks of "the [line] AB"; but sometimes the straightness of the line is emphasized in the Greek by including the adjective—so the translation speaks of "the straight [line] AB."

Concern for clarity has also led not only to employment of some punctuation not found in the Greek text that has come down to us from Apollonius, but also to insertion of paragraph breaks (indicated by a new line begun with a double dash), as well as of some additional sentence breaks.

It would have been traditional to include a single complete diagram in each box that contains the translation of an item from the Greek of Apollonius. But any pictures that Apollonius may have produced have been handed down much less reliably than his words. The diagrams differ in the different editions of his work. At any rate, a single complicated diagram is more difficult to follow than a sequence of simpler ones. This guidebook therefore, rather than providing (as is customary) a single complete diagram for each translated item, will present in the explication (not in the translation) a sequence of generously labeled depictions of parts or aspects of the single translated item that is contained in the shaded box with heavy borders. (In the limited number of cases of a translated item where the explication does not contain such depictions, a single diagram of the traditional sort has been included in an appendix to the volume.) Readers, in studying a chapter of the guidebook, or a section of a chapter, should begin by taking a merely cursory first look at each item of text translated from Apollonius—after which, upon reaching the end of the chapter (or of the section of the chapter) that explicates, with depictions in profusion, the translated text, they can go back and, as they carefully examine each translated item, produce for themselves a diagram to accompany it.

I should add that my work with Apollonius for years relied upon the translation of *Conics* (Books I–III complete) that was prepared by R. Catesby Taliaferro for the program of instruction at St. John's College and was published by Encyclopedia Britannica. While Britannica let it go out of print, Taliaferro's translation has been available for some time, in a fine improved edition, from Green Lion Press, where it has been edited by Dana Densmore and supplied with diagrams by William H. Donahue and an introduction by me.

Consider, dear reader, as you set out, Johannes Kepler's words of caution in Chapter 59 of his *New Astronomy*, as translated by W. H. Donahue:

"If anyone thinks that the obscurity of this presentation arises from the perplexity of my mind, ... I urge any such person to read the *Conics* of Apollonius. He will see that there are some matters which no mind, however gifted, can present in such way as to be understood in a cursory reading. There is need of meditation, and close thinking through of what is said."

VOLUME ONE: APOLLONIAN VISTAS

PART I
CUTTING OF CONES

Apollonius's Conics, Book I: through proposition 10

CHAPTER 1. WHAT'S WHAT

The study of Apollonius presupposes an elementary Euclidean study of geometry—a study of configurations involving lines which are all either straight or circular.

In the definitions with which Euclid opens the First Book of his *Elements*, a circle is defined as a plane figure contained by one line such that all the straight lines falling upon the containing line from one point within the figure (called its center) are equal to each other. Right after the definitions at the beginning of that First Book come five "things required" for going on. (These are usually called "postulates" in English, after the Latin word for "demand.") The first three are these: to draw a straight line from any point to any other point; to extend a bounded straight line continuously straight ahead; and to carve out a circle around any center at any distance from that center. The First Book of Euclid's *Elements* opens not only with definitions and postulates but also with common notions (that is, notions common to figures and numbers), and the Eleventh Book opens with definitions for solid things. You would do well to have at hand a copy of the *Elements* of Euclid as you read the *Conics* of Apollonius.

Apollonius opens the First Book of his *Conics* with definitions. They are identified not simply as definitions, but as "definitions—first ones." These begin with the conic surface and go on to deal with certain relations between straight lines and curves of an unspecified sort.

First Definitions: conic surfaces and cones

Definitions—first ones

—If, from some point, joining it to a circle's periphery, which [circle] is not in the same plane with the point, a straight [line] is extended both ways, and, while the point remains where it is, the straight [line], having been taken around the circle's periphery, returns to the same place from which it began to be carried, then the surface that is carved out by the straight [line]—namely, the surface that is composed of two surfaces lying vertically opposite one another, each of which boundlessly grows as the straight [line] that carves it out is boundlessly extended—I call a "conic surface"; and the point that remains where it is, [I call] its "vertex"; and its "axis" [is what I call] the straight [line] that is drawn through that point and the center of the circle;

[In this translation, "draw" will render *agô* (which can also be rendered as "lead" or "drive"), and *graphô* will be rendered as "carve out," although it too might be rendered as "draw." And rather than using the latinate "circumference," although that would perfectly convey the Greek word's components (by using Latin *circum* for Greek *peri*, and *ferentia* for *phereia*), this translation will use the English cognate "periphery," since the reference will often be to an arc on the circle's periphery rather than to the entire line that surrounds it.]

Apollonius defines the conic surface by generating it. He starts with a circle, as well as with a fixed point that is not in the same plane with the circle. Connecting that fixed point with any point on the circle's circumference, he draws a straight line. He prolongs the straight line in both directions—that is, he prolongs it on through the fixed point as well as on through the circle's circumference. Then, keeping the fixed point still (the fixed point being one of the straight line's points), he moves the straight line all the way around the circle, so that it passes in turn through every point of the circle's circumference (fig. 1).

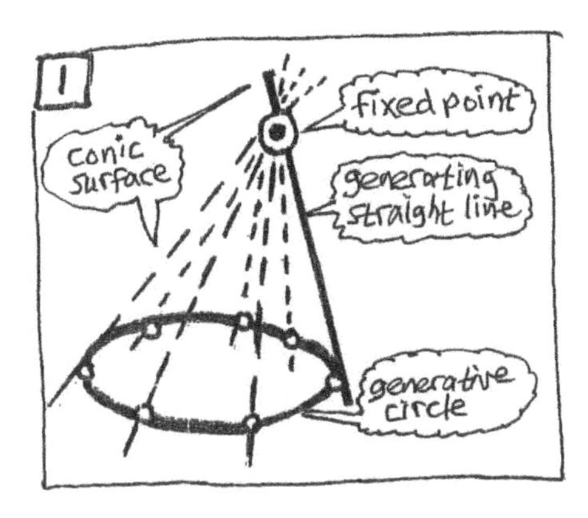

On each side of the fixed point, a surface thus is generated by the straight line. The two surfaces together make up a conic surface. The more that the generating straight line has been prolonged, the greater is the conic surface that it generates.

The conic surface will, we are told, grow indefinitely long as we prolong the generating straight line. The term translated "extended" is, more literally, "thrown out forward" (*pros+ek+ballô*). The word *ballô* ("throw"), from which we get our word "ballistics," will assume some prominence later.

The generating line need not come to an end at the fixed point or at the circle—or anywhere in particular beyond them. It is, the Greek says, prolongable *eis apeiron*. The definition speaks of the conic surface as growing *eis apeiron*, and of the generating straight line as being prolonged *eis apeiron*. The expression *eis apeiron* ("boundlessly") could be translated as "indefinitely," or even as "infinitely," but not as "to infinity." The word *eis* means "to" or "into," but the phrase does not refer to some place to which the line goes. It is adverbial.

The terms *ekballô* and *eis apeiron* are used also by Euclid in his *Elements*. He says in his opening definitions that parallels are lines that do not meet when thrown outward boundlessly (*ekballomenai eis apeiron*). And, after the first postulate, which requires that we be able to conduct (*agô*, usually translated "draw") a straight line from any point to any other point, the second postulate requires that we be able to throw outward (*ekballein*), continuously and straight, any bounded (*peperasmenên*) straight line. What is unlimited is thus not some extent, but rather the power to extend limits.

Nowadays, by contrast, it is usual to say that a straight line is infinite—that what is finite is the line segment made when we pick two definite endpoints that cut it off from an infinite straight line. For Euclid and for his successors, however, every straight line is finite, and is named by its endpoints: "the AB" line is the straight line that leads from the endpoint named "A" to the endpoint named "B." What is infinite is the capability of prolonging the finite straight line. Every act of the unlimited power of extending the stretch between its endpoints results in another straight line which itself also has bounds, its endpoints. Although the *power* of a line to be prolonged does *not* have limits, what is thrown out by every *act* of that power *does* have limits. Our minds have the power to extend the bounds of what we visualize, and then to visualize what has thus been extended; but though this power of our minds is unbounded, whatever we visualize is bounded. What we "com-prehend," we *take hold* of all *together*. That is why men of Euclid's mind did not speak of "infinity." (Moreover, just as a magnitude is a definite so-much, so a multitude is a definite so-many. While nowadays it is usual to speak of the infinity of prime numbers, Euclid speaks rather of there being more prime numbers than any given number.) Aristotle puts the matter this way: there isn't anything that is *actually* infinite; whatever is infinite is infinite *potentially*.

Not only is every straight line finite, so is every plane. A plane, like a straight line, can be extended infinitely; but, as with a straight line, any act of extending it leaves it definitely bounded. But whereas a straight line is bounded by two endpoints, a plane is bounded by a line (or by several lines that share endpoints—as in the case of a triangle, which is sometimes called by Apollonius a "plane" rather than a "figure"). The figure that is contained by the closed circular line called a "circle" is usually called by Apollonius a "circle," but sometimes he just calls it a "plane." It is a circular plane.

> and a "cone" [is what I call] the figure (*skhêma*) which—contained by the circle and the conic surface—is between the vertex and the circle's periphery; and "the cone's vertex" [is what I call] the point which is also the surface's vertex; and its "axis" [is what I call] the straight [line] that is drawn through its vertex and the center of the circle, and its "base" [is what I call] the circle;

Apollonius began the book with a circle—with what might be called a perfectly thin disk. Now he takes this circular plane figure, around whose circumference moved the straight line that generated the conic surface, and he puts it together with the part of the conic surface that is between the circle's circumference and the original fixed point. The conic surface and the plane surface together will contain a solid figure—a cone.

A *conic* surface is not the same thing as a *cone's* surface. A *cone's* surface is heterogeneous: a cone has two kinds of surface—one of them being conic, while the other one (the base) is planar. A *conic* surface has two parts, located on opposite sides of the vertex, each one of which is itself a conic surface. One of the two surfaces is generated by the part of the moving straight line that extends above the fixed point; and the other one of the two surfaces is generated by the part of the moving straight line that extends below the fixed point. But although the movement of the straight line about the circumference of the circle generates a conic surface in which there are two surfaces, these surfaces are of the same kind—both are conic.

While the conic surface has two parts that have the same look, its nature is not homogeneous. Its heterogeneous nature is discovered by considering its genealogy. It is a kind of hybrid: it is fathered by a straight line out of a circle. By considering its dual nature, we shall better understand what it is and what it in turn can generate.

Coming-into-being, however, is not what we are studying here; this is not physics but mathematics. Verbs of generation are used by Apollonius and by Euclid in a form which indicates not the ongoing process, but rather the accomplishment of an act—as, for example, when Euclid in his first postulate requires a straight line to be drawn, the infinitive is in a form called the "aorist." That aorist infinitive means not "*to engage in the process* of drawing a straight line," but rather "*to accomplish the act* of drawing a straight line." (This is something like the distinction that we make in English when we say "Apollonius *wrote* a book" in contrast to saying "Apollonius *was writing* a book." Consider the difference in the sentences "Apollonius wrote a book on conics before he died" and "Apollonius for several years was writing a book on conics.")

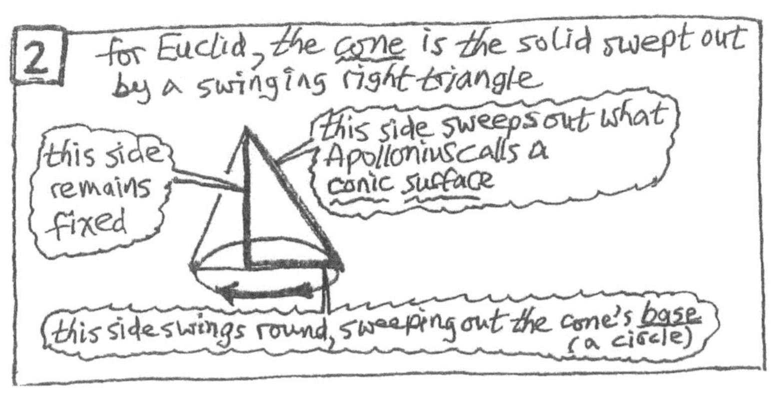

Cones had been defined in Euclid's work prior to the work of Apollonius. Euclid's definitions are, however, different from those of Apollonius. The two authors had different things on their minds.

When Euclid gives definitions of solid figures at the beginning of the Eleventh Book of the *Elements*, he says that a cone is the figure comprehended when, taking one of the sides about the right angle in a right-angled triangle, you keep the side fixed and carry the triangle all the way around to the same position from which you began to move it (fig. 2). The cone's axis is the straight line which remains fixed, about which the triangle is turned; and the base is the circle swept out by the straight line which is carried round. Euclid, like Apollonius, defines a cone by generating it.

Apollonius, defining a conic surface by generating it, gives the conic surface before he gives the cone, and then he puts the cone together—out of a conic surface and a circular planar surface that serves as the base of the cone. Euclid gives the cone without giving the conic surface at all—thus leaving a conic surface to be merely what is left from the surface of a cone after you take away the part that is planar (the circular base). Euclid in his *Elements* was interested in a definite solid, whereas Apollonius here will be interested in cones and in conic surfaces only insofar as they are means for getting certain lines.

The handling of solid things especially seems to call for generative definitions. Euclid's definition of a *circle*, as we have seen, does not generate a circle. What the definition gives is not a circle but a test for circularity. A postulate is what gives the generation of a circle. The postulates also give the generation of a straight line. The postulates give the generation of no other lines than circles and straight lines. Euclid could easily have defined a *sphere* in the way that he defines a circle. He could have said that a sphere is a solid figure contained by one surface such that all the straight lines falling upon it, from one point among those lying within the figure, are equal to one another. But instead, Euclid defines a sphere by generating it.

Euclid's definition of a sphere uses terms defined at the beginning of the First Book of his *Elements*. The First Book gives definitions of a circle's diameter and of a semicircle: a circle's *diameter* is any straight line drawn through the center and terminated in both directions by the circle's circumference (the circle's diameter thus being a bisector of the circle); and a *semicircle* is the figure contained by the diameter and the circumference cut off by it (the semicircle's center being the same as the circle's). In the Eleventh Book, just before Euclid defines a cone, he says that a *sphere* is the figure comprehended when, while keeping a semicircle's diameter fixed, you carry the semicircle all the way around to the same position from which you began to move it. (The sphere's axis is the straight line which remains fixed and about which the semicircle is turned; the sphere's center is the same as the semicircle's; and the sphere's diameter is any straight line drawn through the center and terminated in both directions by the surface of the sphere.) Thus Euclid defines a sphere by generating it, even though he does not have to.

To get a good grip on some kinds of mathematical things, it seems best to refer to some kind of coming-into-being. Still, coming-into-being is not at the heart of classical geometry. Classical geometry directs our attention to what always stays the same. Our minds may change as we consider what does not change, but the movement of our minds is not what we attend to. We might ask whether, in studying mathematics, we *should* be thinking primarily of the movement of our minds. This question will recur as we proceed. In the end, indeed, it will come to be the center of our attention. But we must attend to other things first.

Having generated cones, we now need to consider their kinds.

and, of the cones, I call those "right" that have their axes at right [angles] to their bases, and [I call] those "uneven" (*skalênous*) that do not have their axes at right [angles] to their bases.

[In this translation, "right" renders the Greek *orthos*, and "at right [angles]" renders the Greek *pros orthas*; and when the Greek cognate *orthios* appears later, it will be rendered as "upright." When *orthos* is followed by the dative, it will be rendered by "upright to...." Traditionally, *skalênos* is rendered as "oblique," which in this translation will be reserved for rendering *plagia*.]

Not all cones here spoken of by Apollonius are *right* cones. Consider, by contrast, the cones with which Euclid is concerned in the definitions at the beginning of the Eleventh Book of his *Elements*. There, as we have seen, Euclid generates a cone by rotating a right triangle (as he generates a sphere by rotating a semicircle). He differentiates one kind of cone from another by the angle at the vertex, which depends upon the relative sizes of the legs of the generating right triangle. If the straight line that remains fixed be equal to the remaining side about the right angle that is carried round, then the cone will be right-angled; if less, then it will be obtuse-angled; and if greater, then acute-angled. Euclid's right-angled cone, and his obtuse-angled cone, and his acute-angled cone, are all of them *right* cones (fig. 3).

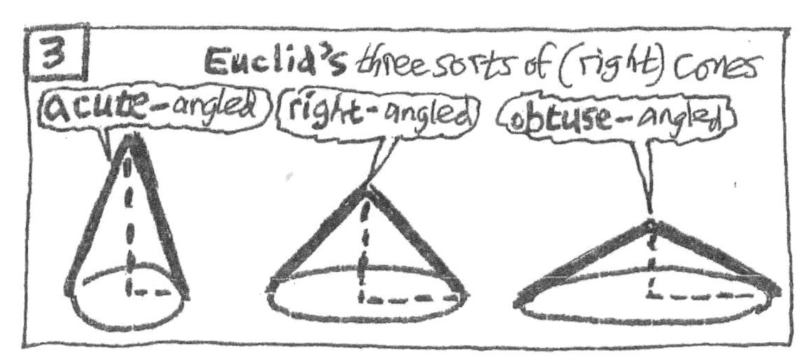

From Apollonius you can get right cones—when the axis is set up right. That is, the line drawn from the fixed point on the generating straight line, straight to the center of the generative circle, must be upright with respect to the plane of that circle. When the axis is upright, the cone is right. "Upright" and "right," with all their connotations, translate the same Greek root, *orth-*, from which we get such words as "orthopedic" and "orthodox." In Apollonius, by contrast with Euclid, we encounter not only cones that are right cones but also cones that are not right—ones that stand unevenly rather than upright (fig. 4). Although our translation will reserve "oblique" for the line that will be called *plagia*, we shall for convenience in our comments and drawings use "oblique" also for the cone that is called *skalênos*.

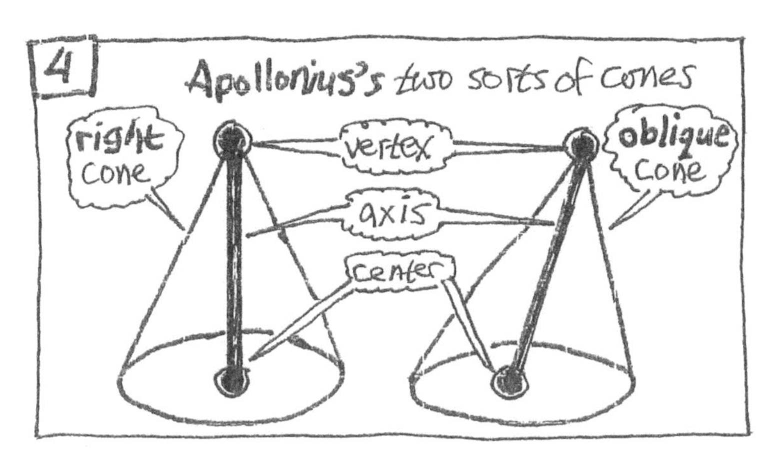

Before Apollonius, right cones were all that mathematicians made use of to obtain curved lines that were conic sections. They obtained such curved lines by cutting into a right cone with a plane perpendicular to the hypotenuse of the generating right triangle in one of its positions. When the right cone was *right*-angled, then the plane would cut into the cone a line of *one* sort (later called the "parabola"); but when the right cone was *obtuse*, then the line of section was of *another* sort (this one the "hyperbola"); and it was of *yet another* sort when the right cone was *acute* (this one the "ellipse"). (Fig. 5.)

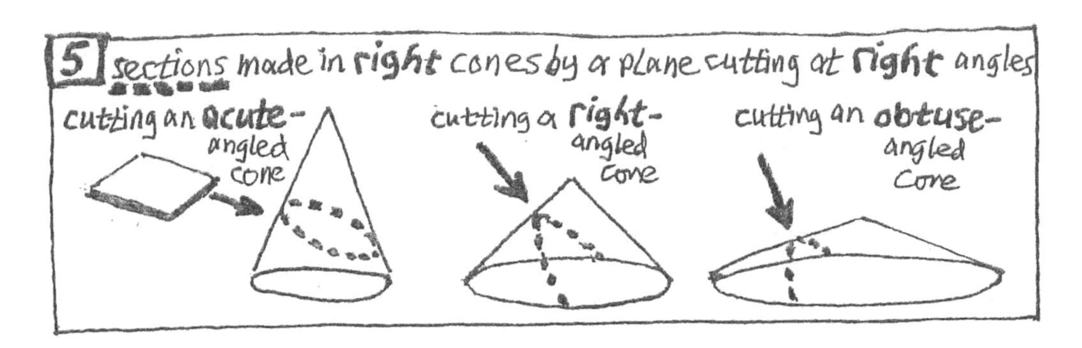

It was Apollonius who introduced cuts at any angle (cuts made obliquely as well as at right angles) in any cone (oblique as well as right). It was also Apollonius who introduced the names "parabola" and "hyperbola" and "ellipse" for the sections. In the course of the *Conics*, he will show why he did so.

Note that a right cone is radially symmetrical, but an oblique cone is not. That is to say: a right cone, unlike an oblique cone, is the same all the way around. An oblique cone slants; it tilts to the side. Even an oblique cone, however, won't tilt to the side when it is viewed from a certain direction. When you view it from that particular direction, it tilts either directly toward you or directly away from you, rather than to the side. And so, we can say that—with respect to that direction—an oblique cone is bilaterally symmetrical, since the right-hand side of its appearance is the same as its left-hand side. A right cone, however, being radially symmetrical, could be said to be bilaterally symmetrical viewed from *every* direction.

Whether or not a cone is a right cone, its non-planar surface is the surface with which Apollonius will deal—and in dealing with that conic surface, he will make use of the fact that even if its cone is oblique, and therefore is not symmetrical radially, that cone is at least symmetrical bilaterally.

But although he will be dealing with the conic surface, he will not be studying it as such. He will be dealing with the conic surface in order to study certain curved lines—as is suggested by the later items in the set of first definitions. These deal with various sorts of diameters. Later, the 7th proposition, when read in the light of these first definitions, shows that if Apollonius had restricted himself to right cones as Euclid does, then the only diameters that his conic sections could have had to begin with would have been those that he will soon call "axes."

First Definitions: diameters and ordinatewise lines

—Of each curved line that is in one plane, a "diameter" is what I call any straight [line] which, having been drawn from the curved line, divides in two [halves] all straight lines that, drawn to this [curved] line, are parallel to some straight [line]; and a "vertex" of the [curved] line [is what I call] the straight [line]'s bound [in other words, the diameter's endpoint] on that [curved] line; and "ordinatewise" [that is, orderly] (*tetagmenôs*) is what I call how it is that the parallels are each drawn down onto the diameter.

For a given curve, is there some straight line that will bisect all the straight lines that have their endpoints on the curve and are drawn parallel to some straight line of reference? If there is, then we call the bisecting straight line a diameter. And we say, of the bisected straight lines which are set in order, all drawn parallel to a given straight line of reference, that they are drawn in an orderly manner—that they are "ordinatewise." Note that the definition does not generate what it defines; it does not say that there is a diameter for any given curve. Nor does it say, if there is one, that there is only one. The definition merely tells you what a line must be to be a diameter (*a*, not *the* diameter).

In figure 6 there is a curve, and at its left is a straight line to which many straight lines have been drawn parallel. The midpoints of these parallel straight lines have been marked. (The endpoints of the straight lines are on the curve, for otherwise the lines would have no determinate lengths, and would therefore have no determinate midpoints.) What the definition says is this: if these midpoints lie along a straight line, then this straight line is a diameter. In the curve shown here, the midpoints do not lie along a straight line. Hence there is no diameter for this curve—at least, not with *these* parallel lines as the lines drawn ordinatewise. It might be, that if the lines were drawn parallel to some *other* given straight line, then they *would* have their midpoints lying on a straight line. Or perhaps for this curve there just isn't any diameter, no matter what straight line is used to ordain a way to draw the requisite parallel lines.

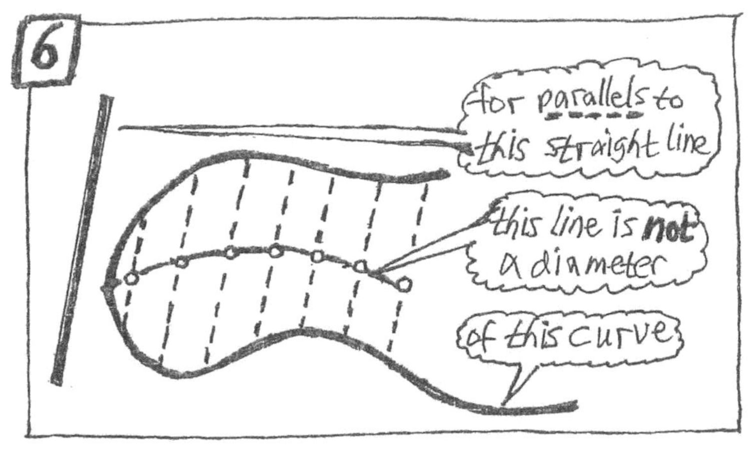

In figure 7, there is another curve, and at its left is a straight line to which many parallel straight lines have been drawn, with their midpoints marked. These midpoints do lie on a straight line. *This* straight line therefore *is* a diameter.

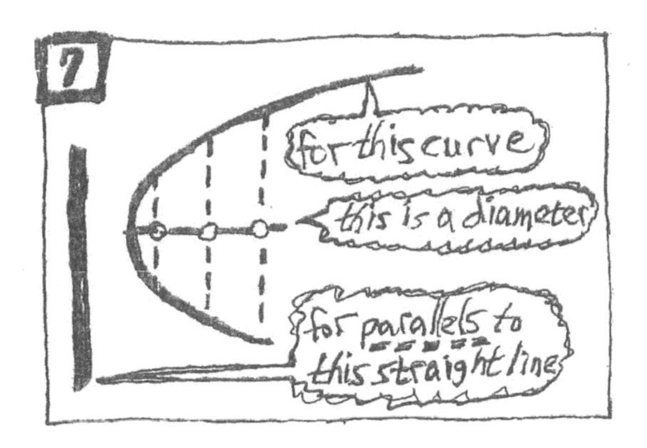

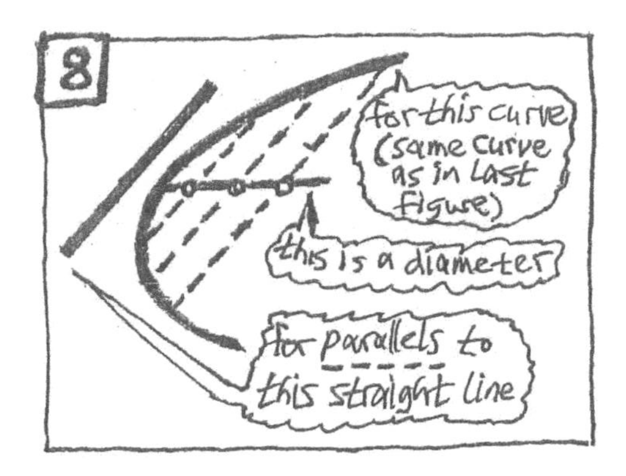

In figure 8, there is the same curve, but now at its left is a different straight line to which many parallel straight lines have been drawn, with their midpoints marked. These midpoints, like the midpoints of the parallel lines drawn in the previous figure for the same curve, lie on a straight line. This straight line too is therefore a diameter. It is *another* diameter of the *same* curve.

The definition suggests what the rest of these First Definitions will make clearer—that attention is going to be directed not to cones and conic surfaces as such, but to curved lines that they yield for study. A key to this study will be the relation between those curved lines called "conic sections" and certain other lines that are straight.

Cones may be interesting three-dimensional figures, and there may be various reasons for studying them, but the very title of the book indicates that the subject of the book is "conics"—"things conic"—rather than cones as such. The cone seems to be of interest to Apollonius because its generation combines the straight line and the circular line, which are the only sorts of lines in Euclid's *Elements*, so that the cone, when cut by planes which are oriented in various ways, will yield as intersections various non-straight lines which are also non-circular. These lines are the simplest of the non-straight lines that are also non-circular; after the circle, they are the simplest of the curved lines.

The only lines whose study is presupposed by Apollonius in the First Book of his *Conics* are those that are either straight or circular. Starting from these, which are the only sorts of lines in Euclid's *Elements*, Apollonius engages the reader in a study of those simplest lines which—as we shall see—

either are *not straight*	or else are *not circular*
but nonetheless	but nonetheless
are indefinitely prolongable	are closed
(like lines that *are* straight),	(like lines that *are* circular).

In approaching the study of such curves obtained by intersection, readers should consider the cone in cross-section. They should do so not merely because two dimensions at a time are easier to visualize than three, but also because a planar view of the cone from above and a planar view of the cone from beside, taken together, will exhibit the two aspects of interest in exploring the sections that can be cut in a cone. From above, the cone in outline will come to view as the simplest figure that is bounded by a curved line—a circle. From beside, it will come to view as a triangle—the simplest figure bounded by straight lines. These two views will reveal the ratios that lurk within and can illuminate the character of the conic sections.

And likewise, also of two curved lines that lie in one plane: I call an "oblique" (*plagian*) diameter [traditionally rendered as "transverse" diameter] any straight [line] which, cutting the two [curved] lines, cuts in two [halves] all the [straight lines] that are drawn, to either of the [curved] lines, parallel to some straight [line]; and its "vertices" [are what I call] the diameter's bounds, [that is, endpoints] on the [curved] lines; and [I call] an "upright" (*orthian*) diameter any [straight line] which, lying between the two [curved lines], cuts in two [halves] all the [straight lines] that are drawn parallel to some straight [line] and that are taken off between the [curved] lines; and "ordinatewise" (*tetagmenôs*) is what I call how it is that the parallels are each drawn down onto the diameter [whether, that is, the diameter be oblique or upright].

A *single* diameter of a *single* curved line was defined previously. Now *two* diameters are defined, each of which belongs to the same *pair* of curved lines as does the other. There is, on the one hand, a diameter called *plagios* and, on the other hand, a diameter called *orthios*. In translating the latter term, there is no alternative to "upright"; but to translate the other term traditionally, as "transverse," is less faithful and less suggestive than to translate it as "oblique." "Oblique diameter" will be a more suggestive translation in examining the work of Apollonius, but, since the more traditional translation is what became the standard technical term, "transverse diameter" will be more helpful for dealing with his successors. For convenience, we shall render the Greek in both ways, often saying "oblique (transverse)," but using the briefer "transverse" in the figures. More about that later.

Now, as previously when a *single* curved line's diameter was defined, we are not told that any two curved lines lying in a plane do have diameters of the sort defined. Rather, we are told merely what two straight lines must be in order to be called such diameters of two curved lines that lie in a plane.

The pair of curved lines now being defined, like the single curve defined previously, is located in a plane. We are not going to leave the plane behind in our dealings with the conic surface. The curves that we are considering are curves upon a plane. We do not get them with a plane alone, but in getting them on a conic surface we will get them also on a plane (the plane that cuts them onto the conic surface).

Now see figure 9 for a depiction of the situation of that pair of curved lines.

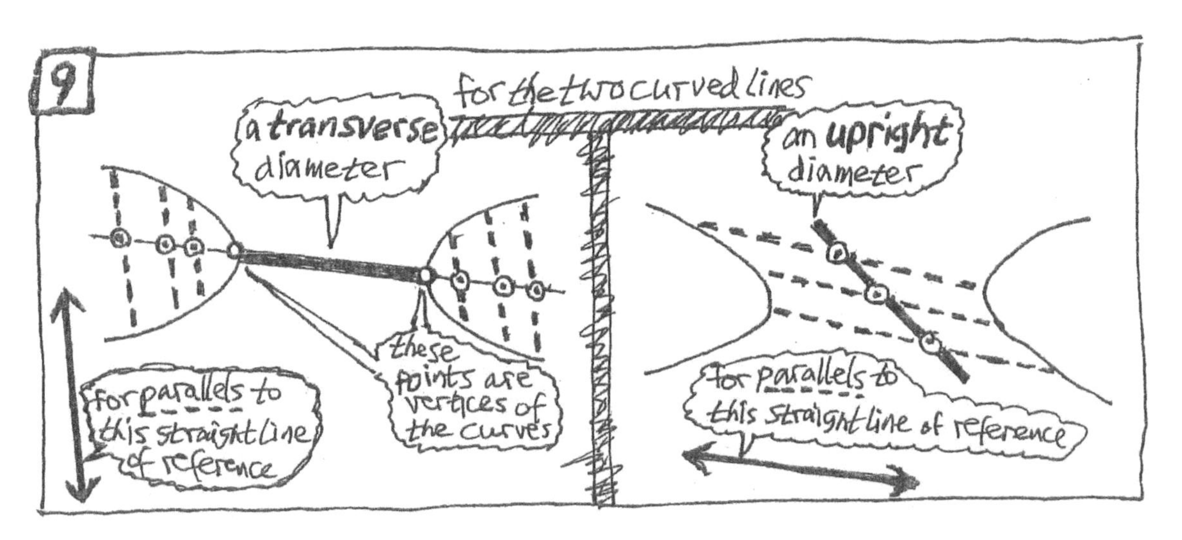

For a pair of curved lines in a single plane, the situation is as follows:

Their *transverse* diameter is a	Their *upright* diameter is a
straight line that *cuts* them	straight line that *lies between* them
and bisects all straight lines that	and bisects all straight lines that
are drawn *to either* of them	are *intercepted between* them drawn
parallel to some straight line.	parallel to some straight line.

The vertices of the curved lines are the ends of the transverse diameter on those curves, one on each curve. Nothing is said about any vertices of the other diameter—presumably because this other diameter, the upright diameter, has no determinate endpoints on the curved lines. And why does it have no endpoints on curves? Because of how it lies.

Thus, a *transverse* diameter	By contrast, an *upright* diameter
runs *across*	lies *upright* in
the space between the two curves,	the space between the two curves,
hitting *each* of them.	hitting *neither* of them
And the straight lines	And the straight lines
(drawn parallel to each other)	(drawn parallel to each other)
which this diameter	which this diameter
goes on to bisect	bisects
are *within*	run *across* the space
each of the curves.	between the two curves.

A *transverse* diameter of two curved lines is a prolongation of a diameter of that single curved line which is on one or the other side. A transverse diameter belongs to the *pair* of curves because it is a prolongation of *both* of the diameters that belong to the curves in the pair when each curve is taken *singly*. Both of these single-curve diameters are of the sort which was presented previously. A further step is taken in now presenting an *upright* diameter of two curved lines. Such a diameter has nothing to do with *either* of the curves in the pair when each curve is taken *singly*; it is *only* when the curves are taken together as a *pair* that there is an upright diameter of the curves. The lines which it bisects run from a point on one curve, not to another point on the same curve, but rather to a point on the other curve. The space they cross is not in any sense inside a single curve but it could be said to be in some sense inside a pair of curves taken as a single thing.

Finally, each of the parallels is said to be drawn ordinatewise to the diameter which bisects them.

So far, these definitions about diameters have been *either* (earlier) for a single curved line, *or* (just now) for a pair of curved lines. The definitions that remain to be considered will all apply *both* to a single curve *and* to a pair of curves.

First Definitions: conjugate diameters and axes

—"Conjugate (*suzugeis*) diameters" of a curved line—and of two curved lines—[are what I call] straight [lines], each of which, being a diameter, divides in two [halves] the parallels to the other diameter; and "axis" of a curved line—and of two curved lines—[is what I call] any straight [line] which, being a diameter of the [curved] line or of the [curved] lines, cuts at right [angles] the parallels.
—"Conjugate axes" of a curved line—and of two curved lines—are what I call any straight [lines] which, being conjugate diameters, cut at right [angles] the parallels to each other.

Now we are told, first, that when two straight lines are diameters of the same single curve or pair of curves, and the lines that each diameter bisects are parallel to the other diameter, then the two diameters are *conjugate* diameters. That is to say, two diameters are conjugate when each diameter is parallel to the lines drawn ordinatewise to the other. (The *con +jug* in Latin means "yoked together," as does the *sy(n) + zyg* in the Greek.) In figure 10, there is an example of conjugate diameters for a single curve, and also an example for a pair of curves.

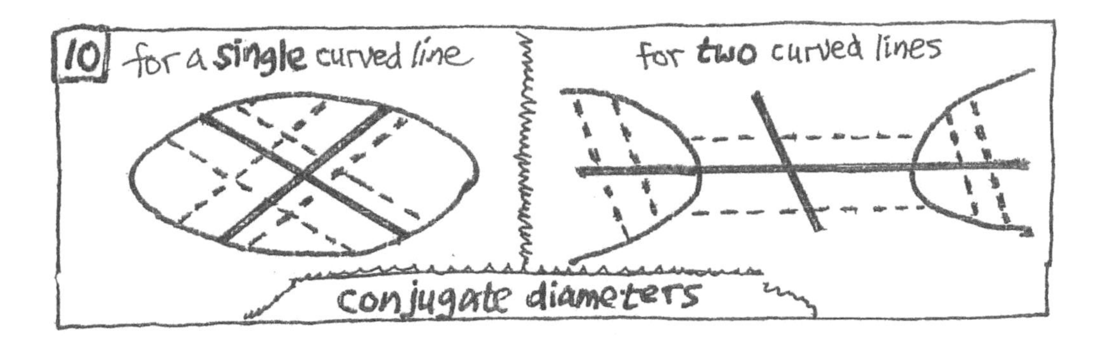

Next, we are told that when a diameter of a single curve or of a pair of curves is perpendicular to the lines drawn ordinatewise, then it is an *axis* of the curve(s). That is to say, a straight line is not only a diameter, but is also an axis, when it not only slices each of the parallel lines into two equal lines, but also makes two equal angles on the same side of each of the lines (fig. 11).

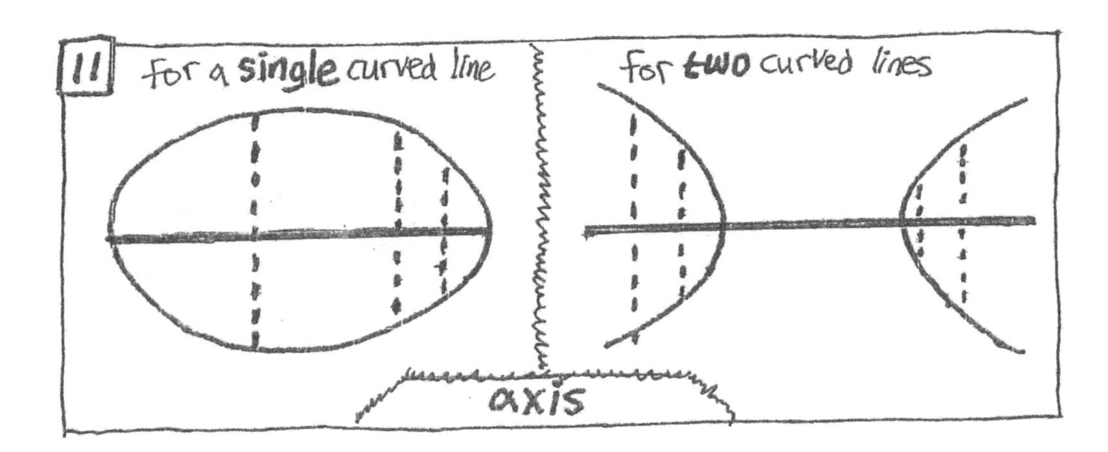

Having learned, first, when to call a pair of diameters *conjugate*, and, next, when to call a diameter an *axis*, we learn, finally, when to call conjugate diameters also *conjugate axes.*

We are told that when each of the two conjugate diameters of a single curve or of a pair of curves is perpendicular to its lines drawn ordinatewise, then they are not only conjugate diameters but also conjugate axes. This means that when conjugate diameters are perpendicular to each other, they are also conjugate axes (fig. 12).

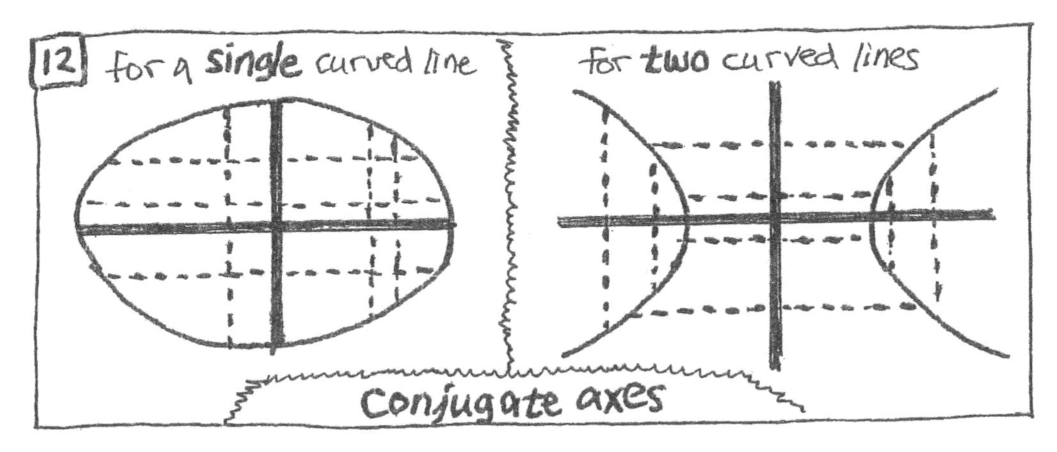

The only curve a knowledge of which is presupposed here is circular. But a circle's every diameter is an axis (since a circle's every diameter must be perpendicular to its chords drawn ordinatewise); moreover, a circle's every diameter has a conjugate diameter, and all the conjugate diameters are also conjugate axes. The First Definitions thus conclude by suggesting that we are going to be looking at non-circular curves. These definitions conclude by showing us that when we examine those new curved lines, we shall be looking at special straight lines in them: various diameters with their lines drawn ordinatewise.

Overview of *First Definitions*, and transition to the propositions

Below, you will find an overview of the First Definitions. In the Greek text the statements are grouped by means of the Greek particle *de*, meaning "and" or "but." In the overview below, this is indicated by "&" linking the items within the groupings, as well as by a double-line separator between the groupings.

	conic surface	vertex	axis
&			
	cone	vertex	axis and base
&			
	cones: right and oblique		
==================			
of a curved line:	diameter	vertex	
&			
of two curved lines:	transverse diameter upright diameter ordinatewise	vertices	
==================			
of a curved line			
& of two:	conjugate diameters		
&			
of a curved line			
& of two:			axis
==================			
of a curved line			
& of two:			conjugate axes

Many of these "definitions—first ones" will not be used until much later in the First Book of the *Conics*, yet Apollonius places them here at the beginning of the book. Here at the beginning, however, is not where he places all of the definitions that he will use in the First Book. A few definitions, which he identifies as "second definitions," are withheld by him until after he has completed sixteen of the First Book's sixty propositions.

What are labeled as *horoi* (traditionally translated as "definitions") appear at the beginning of Euclid's *Elements* too (as well as later in the text). The central item in those opening *horoi* of Euclid, which (like those of Apollonius) are grouped by the particle *de*, is this one: "A *horos* is anything's extremity." The Greek word *horos* primarily means "boundary." The word was eventually put to technical use in matters of terminology. *Terminus*, the Latin word for "boundary," underwent a similar extension. *Horoi* might well have been translated as "terms" rather than, in accordance with tradition, as "definitions."

The definitions that precede the propositions of Euclid involve no surface but a plane and no lines but those that are either straight or circular. As we have seen, the definitions that precede the propositions of Apollonius begin with the conic surface and go on to deal with certain relations between lines that are straight and curves that are of an unspecified sort.

After the First Definitions come the propositions. As has been said, their study presupposes that we have completed an elementary Euclidean study of figures involving lines that are all either straight or circular. With that knowledge, we are prepared to study other lines—the simplest lines that are neither straight nor circular. A consideration of Apollonius's opening definitions suggests how we might begin our advanced studies in geometry: by making use of what we know already—the straight line and the circle. (To adhere strictly to Apollonian parlance, we should not say "circle" when we mean a line that is circular. A "circle" is a disc-like figure, whereas the line that contains a circle, or a stretch of such a line, is called by Apollonius a "periphery of a circle"—*peri+phereia* being, in Latin, *circum+ferentia*. To avoid being cumbersome, however, we shall speak more loosely in what follows.)

By putting *together* the straight line and the circle, we can generate a conic surface—a surface that is neither flat nor spherical; and if we cut that curvy surface with a plane surface, we shall get (as the intersection of the surfaces) some kind of a curvy line. By cutting the conic surface with a plane, we cut a conic section into the plane. We can begin to study this new kind of line in the plane by considering first the conic surface from which it was generated. And we can study this surface by considering the old kinds of lines which generated it. Both of the old kinds of lines (the straight lines and the circles) have an easy simplicity about them. There are no bumps or dips in either of them; neither of them has a ripple. But what about the conic surface generated by the two of them together: what is it like? Apollonius will tell us in the first two propositions.

It is common to refer to the items in a book of classical geometry as "propositions." The term is Latin for "things put forward." The Greek texts themselves, however, do not employ that term. (Its Greek counterpart—*prothesis* or *protasis*—is part of the terminology of grammar, poetry, and politics, but it is only sometimes used by ancient mathematicians, some of whom use it to refer merely to an introductory enunciation.) Apollonius in a letter to a friend, writing to introduce his *Conics*, speaks of its "theorems" and also of its "problems," but the headings in his text (like those in Euclid's *Elements*) are mere numbers devoid of any other identification. Most of the numbered items in the First Book of Apollonius's *Conics* are theorems; the problems appear late in the book. This guidebook will refer to the numbered items, whether they are theorems or are problems, as "propositions." But we do so only for convenience—so that we can, without any complication, use the numbers of the items in sequence. In the course of our study, the relation between theorems and problems will eventually emerge as a very important question. For now, let us say that a theorem is something to be looked at, whereas a problem is something to be done or made. In Greek, a *theos* is a divinity, and *horaô* is to see; a *theôros* is someone sent as an ambassador to view another community's religious procession—and thus *theôria* came to mean a certain kind of leisurely viewing, and *theôrêma* to mean something so viewed. The Greek words *pro* and *ballô* we have encountered already, and so it will not be difficult to think of a *problêma* as something thrown out before one. "A figure of such-and-such a sort has this-or-that property"—that would be the enunciation of a theorem. "Given this-or-that, to construct such-and-such a figure"—that would be the statement of a problem.

Classical geometers present their theorems in the following form:

First comes the general *enunciation* of what's to be seen.

Then comes the *setting out* of an exemplary figure,

followed by the *specification* of what's to be shown from what's set out in the figure.

Finally, comes the *demonstration* proper:
the showing—from what's been set out—of what's been specified,
and therewith of what's put forward in general by the enunciation.

After reading each theorem, try to write down answers to these two questions:

What is the point of the *enunciation*?

What is the gist of the *demonstration*?

Take your time, trying different formulations, until you are satisfied that you have put your answers into words as *clearly*, and as *concisely*, as you can. It will be helpful, in moving through the text, to:

re-state the enunciation in the form "*if* this, *then* that";

re-state the demonstration—
adding emphasis and
laying out the statements schematically to exhibit the logical structure,
sometimes by means of a map or a flow-chart,
accompanied by a drawing of
that part of the diagram which is immediately relevant;

state the order and the structure of the argument
not only within but also among theorems.

However, to avoid being misled while easing your way in untangling complexity, do keep this in mind: the enunciation of a proposition in classical geometry characteristically states the general truth of the matter about a being of some kind, whereas its restatement as an "if ..., then ..." has the form of one statement's following from another.

The rest of Part One of this guidebook will examine the first ten propositions of the *Conics* of Apollonius. A chart in the last chapter of Part One will provide an overview of the steps taken in the first ten propositions. These constitute the first leg of our journey through the First Book of the *Conics*. The next part of the guidebook will examine certain Euclidean foundations on which our further work will rest; and the next part after that will examine the propositions in which Apollonius characterizes various sorts of conic sections, beginning with the eleventh proposition.

Part of the difficulty in reading texts of classical geometry is that everything is put into words. The text could be spoken aloud without loss. Pictures, typographical devices, and notations are not employed as substitutes for words. (Bare letters of the alphabet are, to be sure, used to name points, but there are no signs that replace words for operations, for relations, or for technical entities.) Diagrams which may help in comprehending the text can be visualized or drawn by its readers for themselves (albeit with very great effort) by carefully attending to the specifications given in the words of the text.

The fact that the author of the text does not replace words and letters with drawings or other notations makes for a kind of prolixity. More words are necessary than would be needed if some of the message were conveyed in drawings or other notations. That is not at all to say, however, that the spirit of the text rules out its being accompanied by diagrams. But, amid uncertainty about just what diagrams might have accompanied the original text of Apollonius, different modern editions do not all provide the very same diagrams. In this guidebook, the diagrams are placed in the explication rather than in the translation of the original text of Apollonius.

The text is also more prolix than it would be if the statements which it contains omitted the numerous qualifications (subordinate clauses especially) that are necessary in order to avoid making statements that are not quite true. The requirements of accuracy often make it hard to survey at a glance even the enunciation of the proposition, let alone its lengthy demonstration—to say nothing of the even longer sequences of propositions.

This brings us to another source of difficulty, one of an opposite sort: the conciseness of the text. Accuracy of speech, as we just remarked, requires a certain prolixity—and yet, given the demands of accuracy, no more is said by Apollonius than he must. Just how much he must say is, of course, a matter of his purpose. He certainly does little to supply his readers with aids to seeing at a glance the point of much of what he says. It does seem from what he does supply that he could have supplied more had he wished to. As we proceed, we shall need to consider why he does not.

Our study of the mathematics of Apollonius involves considering Apollonius as a teacher, as a helper in learning. "Mathematics" is a term borrowed from Greek. *Mathêmatikê* is Greek for "the art that has to do with the *mathêmata*." The *mathêmata* are the "things that are learned or learnable." Mathematical study thus is somehow understood as learning par excellence. What about Apollonius's teaching—is it teaching par excellence?

CHAPTER 2. THE SURFACE

1st proposition

> \# 1 The straight [lines] that are drawn from the vertex of the conic surface to the points on the surface are on the surface.
> —Let there be a conic surface whose vertex is what the point A is, and let there be taken some point on the conic surface, namely, the [point] B, and let there be joined some straight [line], namely the [line] ACB. I say that the straight [line] ACB is on the surface.
> —For—if possible (*dunaton*)—let it not be [on the surface]. And let the [line] which has carved out the surface be what the straight [line] DE is; and let the circle along which the straight [line] ED is carried be what the circle EF is. Then if, while the [point] A remains where it is, the straight [line] DE is carried along the periphery of the circle EF, it will also go through the point B, and there will be, of two straight [lines], the same bounds—which very thing is absurd (*atopos*). ["Nowhere," that is to say, can two lines have the same endpoints if those lines are straight.]
> —Not so, therefore, that the straight [line] joined from the [point] A to the [point] B is not on the surface. On the surface it therefore is.
> **Porism** Also, it is evident that if, from the vertex to some point of those inside the surface, there be joined a straight [line], it will fall inside the conic surface; and if it be joined to some [point] of those outside, [the straight line] will be outside the surface.

(A prefatory word: As you proceed through this guidebook, you will find it helpful to keep in mind the advice contained in the "Note on the translation of Apollonius and its use" which follows the Introduction.)

Note the parts of the proposition. The *enunciation* here is the first sentence. The *setting out* is the next sentence after the enunciation; it begins with the words "Let there be . . . ." Then comes the *specification*; it begins with the words "I say that . . . ." After that, comes the *demonstration*.

Added on at the end of this proposition is something called a *porism*. We shall consider it at the end of our consideration of the proposition itself.

1st proposition: enunciation

The enunciation of this 1st proposition may be simple, but it is easy to misunderstand. It does *not* claim that if a line lying entirely on the conic surface goes from the vertex down to any other point on the surface, then the line must be straight. Such a line, for all we know, could be squiggly (fig. 1). Nor does it claim that such a line, even though it does not have to be straight, *can* be straight. While that assertion is true (fig. 2), it is not what the enunciation says. What it does say is that if a line from the vertex down to another point on the conic surface is straight, then it must lie entirely on the surface. If a straight line has its endpoints on a conic surface, and one of the endpoints is the vertex of the conic surface, then every other point of the straight line must also be on that surface.

In other words, a straight line from the vertex down to another point on the conic surface nowhere leaves the surface. As we go straight down along it, we do not go over any dips in the surface, nor under any bumps on it. The surface nowhere ripples going down (fig. 3).

For convenience, I say "*down* from the vertex." Strictly speaking, I could as well say *up* or *sideways*—depending on how we imagine the conic surface to be oriented. It does seem easiest to imagine the vertex as being at the top of what we are visualizing—so I shall keep saying "down from the vertex." Of course, the vertex at the top of what we are visualizing is also at the bottom of the conic surface's other part, which we are disregarding.

1
Here are two ways to go down from the vertex to another point on the surface, but not go straight.

2 But we can, if we wish, go down straight from the vertex to another point on the surface, while staying on the surface.

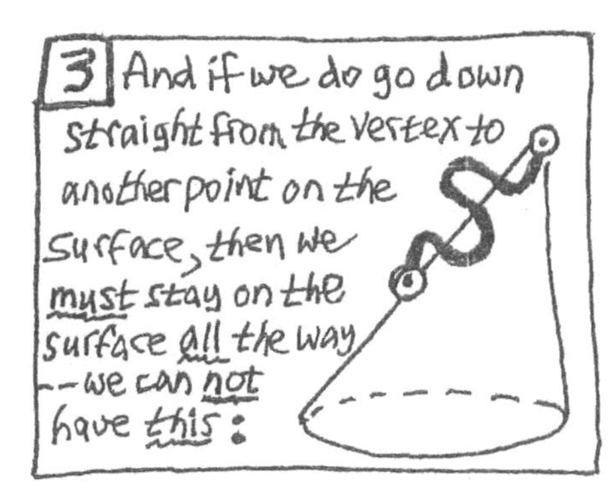

3
And if we do go down straight from the vertex to another point on the surface, then we must stay on the surface all the way—we can not have this:

1st proposition: demonstration

We *set out* as our figure a conic surface having on it point A as vertex, and also some other point B; and we join those two points to make straight line A(C)B—with point C being a point between A and B, on the line. We *specify* that from what has been set out, it must be shown that straight line A(C)B is on that conic surface which has those points A and B (fig. 4).

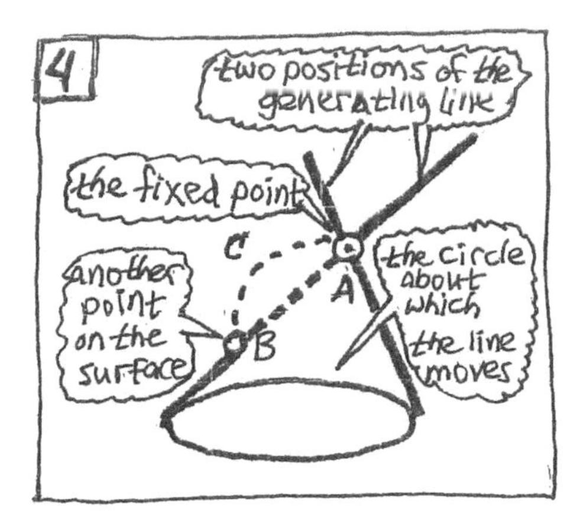

It makes no difference what choice we make among the three possible cases for placement of the non-vertex point B. It can be placed as follows (fig. 5):

—between the vertex and the generative circle, or
—on the other side of the vertex from the circle (i.e., above the vertex), or
—on the other side of the circle from the vertex (i.e., below the circle).

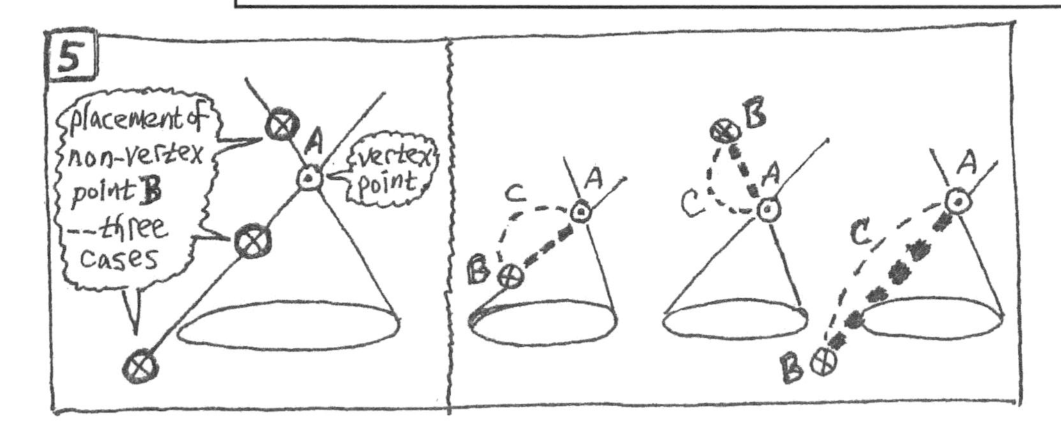

Now let us examine the demonstration proper. So quickly does the demonstration follow from the definition of a conic surface, that it might not seem to be much of a demonstration at all. But it is indeed a demonstration, for it does go beyond the definition. The definition of a conic surface is not the whole of the demonstration, but is only what it starts from.

Because a conic surface is, by its definition, a surface generated in a particular way by a straight line, there must be—*on the surface*—some straight line that joins the vertex and any other point on the surface. The demonstration takes that fact about *this* surface and *this* straight line, and puts it together with a property of *any* straight line that joins two points—namely, its uniqueness. Its uniqueness means that two points can be joined by *only one* straight line.

The demonstration is indirect—it is what is often called a *reductio*. This is a shortening of the Latin phrase *reductio ad absurdum*, which means "reduction to what's absurd." Such a demonstration gets its validity from the following principle: if a statement is such that its denial *must* lead to an absurdity—that is, to a statement which *must* be false—then the original statement itself *must* be true.

What then is the absurdity here? Before you read further, try to put an answer down in writing. Some possibilities follow.

It is absurd to assert the existence of two distinct straight lines, each having the same endpoints as the other. That absurdity follows necessarily from denying the enunciation. It follows, that is to say, from supposing that there is a straight line which has as endpoints the vertex and another point on the conic surface, but is itself *not* entirely on the surface. The supposition which denies the enunciation therefore must be false; and therefore the enunciation itself must be true. Every straight line which has as its ends the two points given (namely, the vertex and any other point on the conic surface) must itself be entirely on the conic surface.

Consider the following two statements:

The definition of a conic surface	The character of a straight line
guarantees this:	guarantees this:
on the surface there's at least one	nowhere does more than one
straight line joining	straight line join
the vertex to any other point on the surface.	any point to any other.

Taken together, those two statements guarantee this: any straight line that joins the vertex and any other point on a conic surface will itself be on that surface. In short:

those two points are the ends of a straight line which is *entirely on the surface*—but
any two points can be the ends of *only* one straight line—hence
any straight line whose ends are *those* two points must be *entirely on the surface.*

That is one way to state the demonstration, but perhaps you will prefer the following formulation.

Between *those* two points (namely,
the vertex and any other point on the conic surface),
there *must* be *at least* one straight line that is on the surface,
but between *any* two points
there *cannot* be *more than* one straight line.
—The existence of the *one* straight line (the one
between the vertex and the other point) is *necessary*,
and the existence of *another* such straight line is *impossible*.
—But we imply that there *is* such another *if* we suppose that
a straight line which joins those two points (namely,
the vertex of a conic surface and another point on the surface)
is, itself, *not* entirely on the conic surface.
—Therefore that straight line *must* be *entirely* on the surface.

Are the different ways of saying the same thing just as good as one another? As we proceed in our study, you will find it useful to try out different ways to say what you see. Some formulations will help you to see more clearly than others will. Your reformulations can make things clearer for you, or can help you to see more things with the same clarity.

When you study a proposition, you may find it useful to write out the statements of the demonstration, each on a separate line, and so arranged as to display how they are related (with whatever supplementary words, phrases, and sentences help to make the argument clearer). Here, for example, is what might be done with the demonstration of the 1st proposition:

Let the straight line AB *not* be on the surface (this will be a *reductio*)

[i.e., suppose that there is some point C which
is on the straight line between its endpoints, but is *not* on the surface].

The straight line which generated the surface—let it be DE;
and the circle along which this straight line moves—let it be FE.

Straight line DE is such that [by the definition of a conic surface]
if, with point A fixed, it moves along circumference of circle EF,
then it will go through the point B.

But then,
[since DE is a straight line that
is on the surface,
and goes through both this point B and the vertex A,
and ACB is a straight line that,
like DE, goes through both points B and A,
but unlike DE, is *not* on the surface]
two [distinct] straight lines [DE and ACB] will have the same ends—
which is an absurdity.

Hence, it *cannot* be the case that straight line AB is *not* on the surface;
and so, it *must* be the case that it *is* on the surface

1st proposition: porism

This 1st proposition has said that a straight line will itself lie entirely *on* the surface if it joins the vertex to a point which is *on* the surface. But what if this other point is *not* on the surface—what about a straight line joining the vertex to *it*?

Such a straight line will *not* lie entirely on the surface. Indeed, except for the vertex at its end, it cannot pierce the surface, but will be entirely *off* the surface. In particular, if a line goes straight from the vertex to a point that is *not* on the surface, then the straight line itself will be:

either entirely *inside* the surface if that point is *inside*	or entirely *outside* the surface if that point is *outside*.

This is stated in the *porism* given immediately after the demonstration of the 1st proposition.

What is a porism? A porism is a helpful bonus provided by the demonstration of a proposition. It falls out from the proposition without any need at all for further demonstration. The one that we are given here will be needed in the very next proposition, Apollonius's 2nd proposition.

2nd proposition

#2 If, on either of the vertically opposite surfaces, two points be taken, and the straight [line] joining the points does not verge towards the vertex, it will fall inside the surface; and its straight prolongation will fall outside.
—Let there be a conic surface such that its vertex is what the point A is, and the circle along which is carried the straight [line] that carves out the surface is what the circle BC is, and let there be taken on either of the vertically opposite surfaces two points, namely the [points] D, E; and let the joined [line] DE not verge towards the point A. I say that the [line] DE will be inside the surface, and its straight prolongation will be outside.
—Let there be joined the [lines] AE, AD; and let them be extended. They will fall, then, on the circle's circumference. Let them fall to the [points] B, C; and let there be joined the [line] BC. The [line] BC will therefore be inside the circle, and so will be inside the conic surface also.
—Let there be taken, then, on the [line] DE a chance point, namely the [point] F, and let the [line] joining AF be extended. It will fall then on the straight [line] BC; for the triangle BCA is in one plane. Let it fall down to the [point] G. Since then the [point] G is inside the conic surface, therefore also the [line] AG is inside the conic surface, and so too the [point] F is inside the conic surface; likewise, then, will it be shown that also all the points on the line DE are inside the surface. The [line] DE, therefore, is inside the surface.
—Let there be extended, then, the [line] DE to the [point] H. I say, then, that it will fall outside the conic surface.
—For—if possible—let there be some [point] of it, say, the [point] H, not outside the conic surface; and let the joined [line] AH be extended. Then it will fall either on the circle's periphery or inside—which very thing is impossible (*adunaton*), for it falls on the [line] BC extended, as on the [point] K, say. The [line] EH is, therefore, outside the surface.
—The [line] DE, therefore, is inside the conic surface; and its straight prolongation is outside.

2nd proposition: enunciation

The 1st proposition showed this: next to us on a conic surface there will be no *ripples* (no bumps or dips) if we go *down* along a straight line from the vertex to another point which is on the surface. But being flat in a certain way is only one aspect of the conic surface. The surface would be a plane if it were simply flat. It is flat only if we go from the vertex directly down to another point. But what if we go, not down, but *about* the surface? Suppose, that is, we go from one point on the surface to another point on it when the situation is like this: neither of the points is the vertex, and also the pair of them is so aligned that we would not be repeating the situation of the previous proposition. (In order to avoid such a repetition, the points must be aligned so that if we were to go from one point to the other along a straight line, we would not be going directly down away from the vertex, or directly up toward it.) What then? Well, if we lay it down in advance that we must stay on the surface, how many ways are there for us to go? There are as many as we please, and they can be as squiggly as we please (fig. 6).

We had better take hold of the situation by considering our relation to the surface if we go from one point to the other by the route that is most direct, the one route that stands out from all others. That is to say: let us suppose that we go from the one point on the surface *straight* to the other point on the surface. (And let us suppose also that the two points are not in a straight line with the vertex. Why not? Because such a line would put us right back again in the situation of the previous proposition.) Now what would happen to the relationship between us and the surface just as soon as we left the starting-point of our trip going straight to the other point?

No matter how close together the two endpoints may be located on the conic surface, and no matter how far apart they may be, we would go *inside* the surface right after leaving one end, and we would *stay* inside until we reached the other end, at which point we would *pierce* the surface. And if we then continued straight on, we would go *outside* the conic surface; and from then on, however far we might go, we would *stay* outside. We would *not* go in and out and in and out. What we *would* do, is go *in* and *stay* inside—until we go *out* at the end and *stay* outside after that. An example is shown in figure 7.

But figure 7 shows only a special case: the case when two points are all the way across from each other. It must be emphasized, however, that no matter how close together the two points may be upon the surface, the inside/outside property is exactly the same (fig. 8).

If we take two points upon a conic surface, when neither of them is the vertex and the two of them are not in a straight line with the vertex, and if we go along the straight line between them—then what must be true about the straight line that we go along is this: *in-between* the endpoints, it is entirely *inside* the surface, and, extended *beyond* the points, it is entirely *outside* the surface—and that must be so however small or however large may be the distance between the two points upon the surface.

Note that the two points that are under consideration in the 2nd proposition must be located on the same one of the two surfaces (the surfaces lying on opposite sides of the vertex from each other) which together comprise the entire conic surface that was generated in the very first definition. This is because unless both points are situated on the same one of the two vertically opposite surfaces, the straight line *between* them will be *outside*, and its *prolongation* will be *inside*—rather than vice versa (fig. 9).

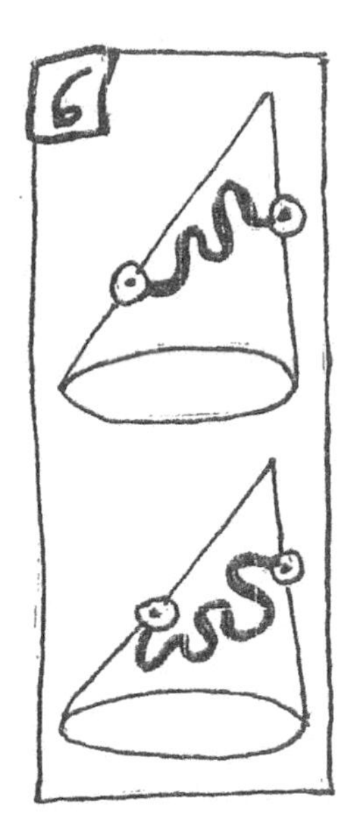

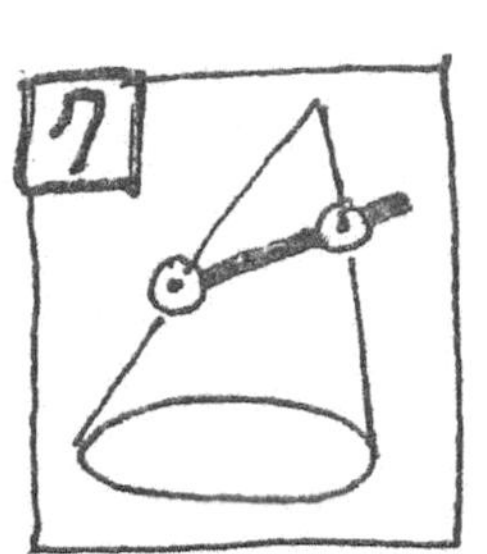

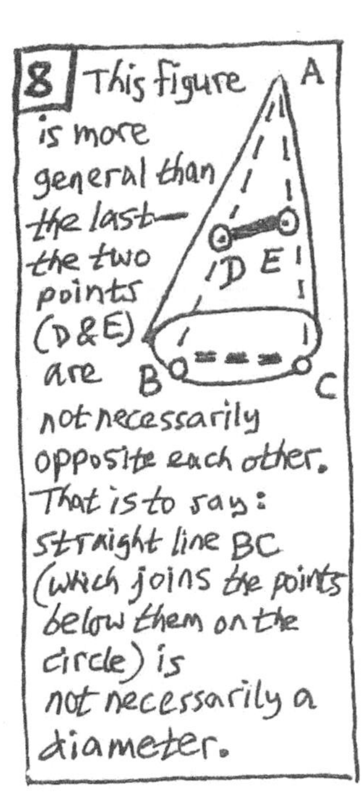

8
This figure is more general than the last—the two points (D&E) are not necessarily opposite each other. That is to say: straight line BC (which joins the points below them on the circle) is not necessarily a diameter.
A
D
E
B
C

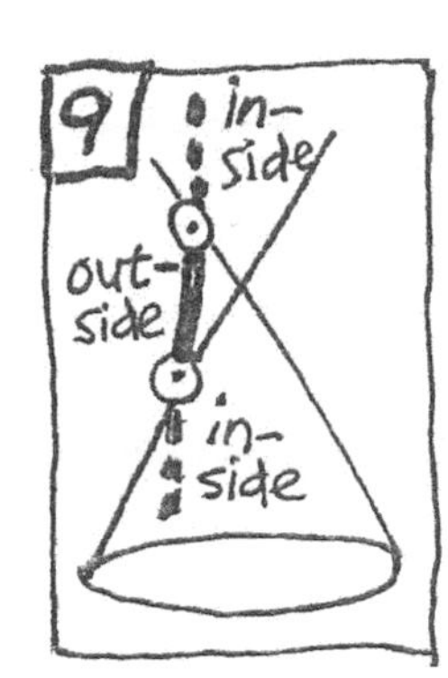

Now what is the significance of what the 2nd proposition tells us about our straight line? This: the conic surface, as it goes about alongside our straight line, will not have a ripple. There will not be the tiniest bump or dip no matter how tiny is the distance between the two points on the surface. And if the two points are all the way across the surface from each other, then when we go about the surface alongside the straight line, our path will everywhere bulge outward as we go all the way around (fig. 10).

If we walk around it with the vertex on our left sometime, then the vertex will be on our left for the whole trip (and if it is sometime on our right, then it will also be so for the whole trip). We will, that is to say, be bearing always leftward (or always rightward) all the way around—whereas if there *were* any bumps or dips, then for some of the trip we would have to bear leftward and for some of it rightward. Our path will look nothing like the path in figure 11. Rather, everywhere the path will be curved convexly—call it "roundish." Maybe it will be a perfectly round path (fig. 12). Or maybe a somewhat flattened roundish one (as was fig. 10). Or maybe even, for all we know, a lopsided flattened roundish one (fig. 13). (Indeed, for all we know now, our roundish path might not even take us all the way back to where we started from; it might not close.) If the points are closer together than the ones we have just been talking about, then alongside the straight line between them there is, perhaps, a path that is a part of some one of the paths we have just been looking at, each of which has this property: it everywhere bulges outward; so that if any two points are taken on it, and are joined by a straight line, then—however close together the points may be—the straight line will not cross the curve between the points. The curve in the next figure bulges outward (fig. 14). Take a point on it (call it P), and then take another one (call it A), and see how the straight line drawn between them will not cross the curve no matter how close you bring the second point to the first one (taking—after A—the points B, and C, and D, and so on).

Now take a curve with bumps and dips. Do the very same thing with it that you did with the previous curve: take a point P and then another one A; join P to A by a straight line; then come closer to P with B, and C, and D, and so on (fig. 15). The curve does bulge outward everywhere between P and E, but not everywhere between E and A.

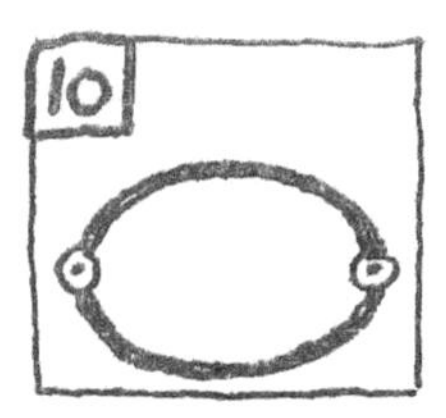

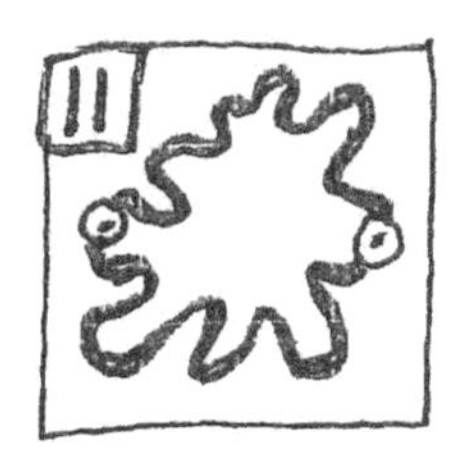

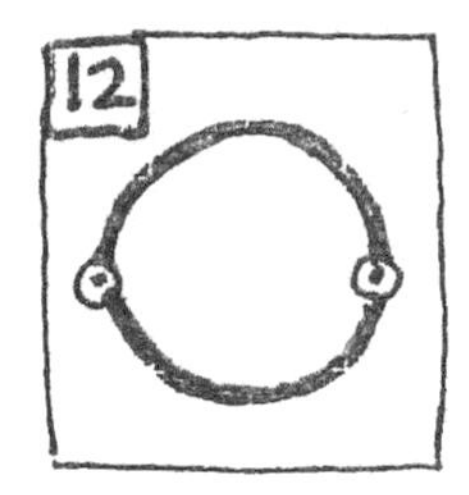

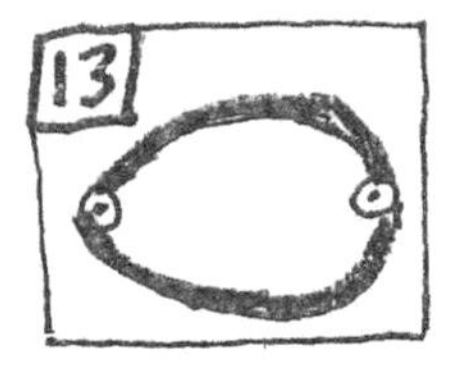

Let us sum up what we have seen so far:

> *Previously*, the *1st proposition* showed that a line stretching straight down
> from the vertex to another point on the surface
> will everywhere be *on* the surface.
>
> *Now*, the *2nd proposition* shows that a line stretching straight across
> between two non-vertex points on a conic surface
> (that is, when the points are
> not in a straight line with the vertex,
> nor on vertically opposite surfaces),
> will, everywhere between its endpoints, be *inside* the surface,
> and will, anywhere it is prolonged to, be *outside* the surface.

In other words: *around* the conic surface (as well as *down* it) there is not a ripple.

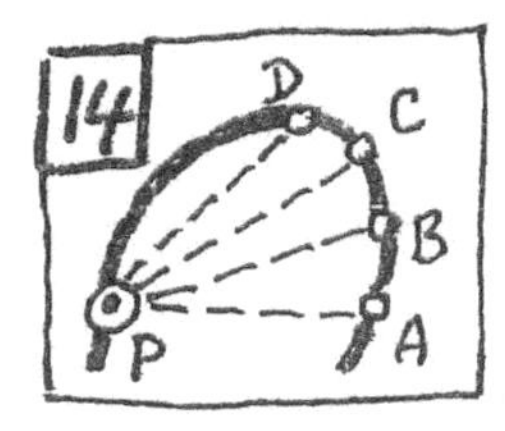

2nd proposition: demonstration introduced

This 2nd proposition is concerned with a straight line between two points: neither one of them is the vertex; and the pair of them do not lie on a straight line with the vertex, nor on opposite sides of the vertex. This straight line shows that there are no bumps or dips upon the surface going *around*. It shows this in a way very much like the way in which we earlier demonstrated that there are no bumps or dips upon the surface going *down*.

How did that *previous* (1st) proposition show the *straightness* of the surface alongside a straight line that goes down from the vertex to another point on the surface?	How can this *present* (2nd) proposition show the *roundness* of the surface alongside a straight line that goes across from one point on the surface to another one?
It showed it by referring to the perfect straightness of the *straight line* that generated the conic surface.	It can show it by referring to the perfect roundness of the *circle* that served for generating the conic surface.

Now in the 2nd proposition, the demonstration takes the fact that the conic surface is generated by the motion of a straight line (when one of its points is fixed) around a circle, and adds a fact about any circle—namely, that a straight line between any two points on a circle is entirely inside the circle, and has its prolongation entirely outside. This property belongs not to circles alone, but to any curve that bulges everywhere outward. In all such curves—curves that are convex (like the curves in figure 16)—every chord of the curve (that is, every straight line between two points on the curve) is inside the curve, and has its prolongation outside.

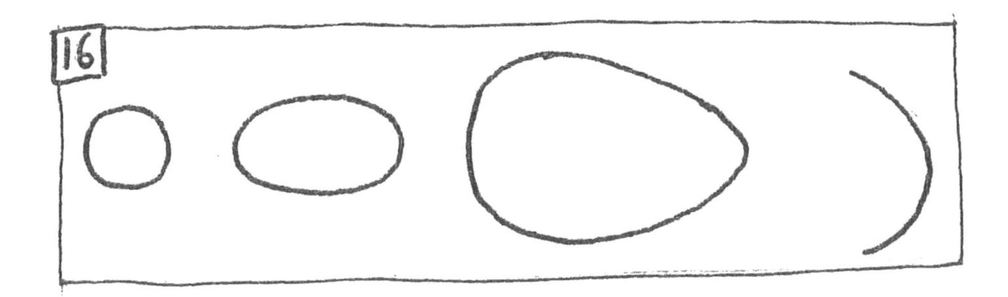

In the 1st proposition, the demonstration showed that the straight line between the endpoints was entirely on the surface because the straight line that generated the surface went, in one of its positions, through the endpoints: that straight line was entirely on the surface—and there can be only one straight line through any two points. The 1st proposition, in other words, was demonstrated by taking the fact that the conic surface is generated by a straight line, and adding a fact about a straight line between any two points—its uniqueness.

All that we had to work with in the 1st proposition was the definition that says how to generate a conic surface. All that we have to work with now in the 2nd proposition is that very same definition and, in addition, the 1st proposition with its porism.

Here is a side-by-side comparison of the 2nd proposition's demonstration with that of the 1st one:

Before, the 1st proposition	*Now*, the 2nd proposition
took the generating straight line in the position where it goes through the non-vertex point—	first puts the generating straight line in each of the two positions where it goes through one of the points on the surface,
	then finds the points where the two straight lines meet the circle around which the generating straight line moves,
	and then takes a straight line drawn between these two points on that generative circle—
and showed that because the *generating straight line* did not leave the surface,	and shows that because this *chord of the generative circle* does not leave the inside of the circle (unless it is prolonged, and then it never leaves the outside),
therefore neither did the straight line whose endpoints are the two points which must have been on the generating straight line in one of its positions.	therefore neither does the straight line whose endpoints are the two points which were located on the surface at the beginning.

The demonstration of the 2nd proposition shows, first, that the straight line in-between the points falls entirely inside the surface; and, second, that the prolongation beyond the points falls entirely outside the surface. This latter part, like the whole demonstration of the 1st proposition, is a *reductio*. A *reductio* is an indirect demonstration. Are such demonstrations less persuasive than direct demonstrations?

As we have seen, the principle of demonstration by *reductio* is this: something *must* be so if it is *impossible* for it *not* to be so. Its validity depends upon the certainty that the alternatives have been exhausted. Exhaustion of the alternatives can be guaranteed by employing dichotomy: it must be that something either *is* so or else is *not* so—and no third alternative remains.

When should a *reductio* be used? Can a *reductio* be more persuasive than a direct demonstration would be? Are simpler things likelier than more complex things to need a demonstration that is indirect?

2nd proposition: the demonstration

Like the non-vertex point in the previous proposition, in this proposition the pair of points on the surface (call them D and E) can be placed as follows:

—between the vertex and the generative circle, or
—on the other side of the vertex from the circle, or
—on the other side of the circle from the vertex. (fig. 17)

17 three cases for placement of the pair of points
I
II
III

(It is also possible that while the pair is on the same side of vertex, the members of the pair are separated by the generative circle.) The demonstration first provides an apparatus (fig. 18):

Join the vertex (A) to each point (D and E),
thus getting straight lines (AD and AE).
Since they are straight lines that
(by the 1st proposition) lie entirely on the surface,
prolong them and they will fall upon the
circumference of the generative circle.
Take the points where they do so fall (B and C),
and join them by a straight line.
This straight line (BC) will be within the circle—
since a circle has the property that
it everywhere bulges outward:
a circle, that is to say, is met by any of its chords
nowhere else but at the chord's endpoints.

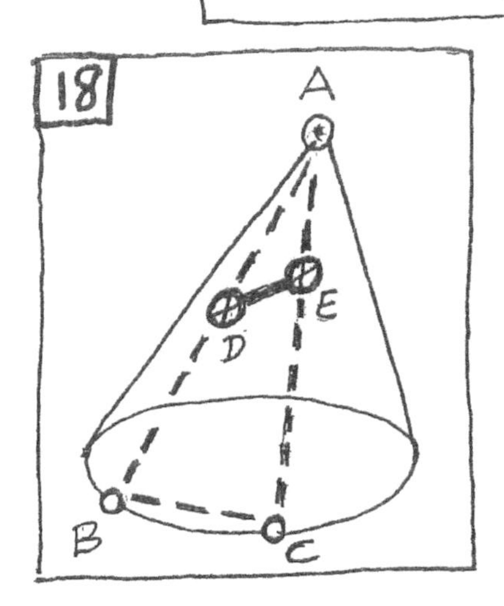

This chord BC of the generative circle is used to show that the straight line between points D and E is entirely inside the conic surface, and has its prolongation entirely outside. DE *itself* is shown to be *inside* the conic surface, as follows (in box with figure 19): DE *prolonged* is shown to be *outside* the conic surface, as follows (in box with figure 20):

On DE, take a random point (call it F);
and to this point, join the vertex (A).
This straight line (AF) that you get—prolong it.
It will fall on chord BC at some point (call it G)
(since if any two straight lines cut one another,
they are in one plane;
and [by Euc XI.2] triangle BCA is in one plane).
And now, since BC, being a chord, is inside the circle,
therefore any point on BC
is inside the surface which was generated by
moving the generating straight line around the circle.

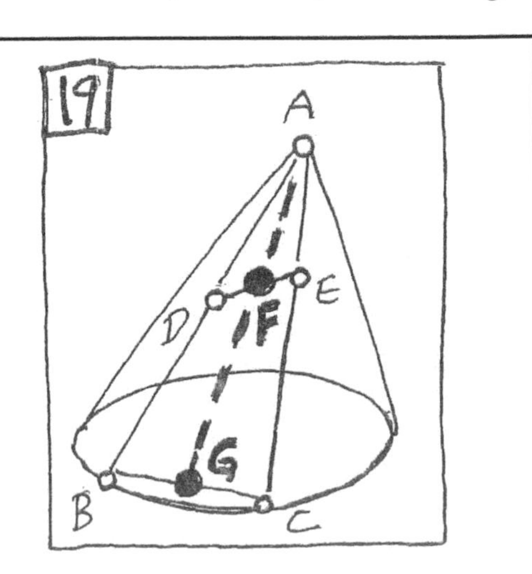

Point G is therefore inside the conic surface,
and therefore (by the porism to the 1st proposition)
point F is likewise inside,
and since F is any point in DE, so all the points in DE are likewise inside.
Hence DE is inside the conic surface.

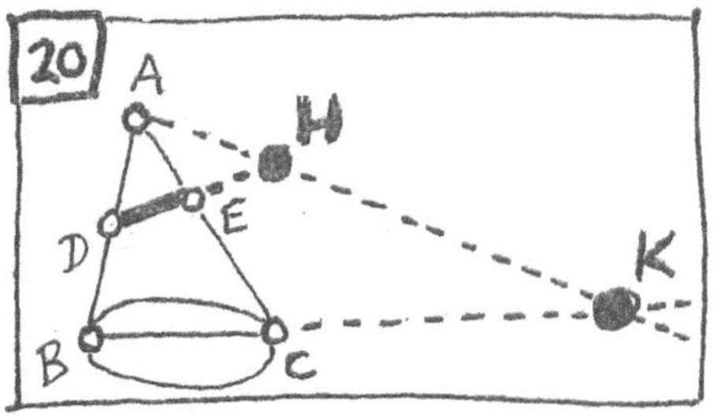

Prolong DE—say, to point H.
H must be outside the surface—by *reductio*,
for suppose that H is *not* outside the surface:
join the vertex (A) to H, and extend this straight line AH;
and now, since H is supposed to be not outside the surface,
either H is on the surface
(in which case, AH extended will meet chord BC at C),
or else H is inside the surface
(in which case, AH extended will meet chord BC between B and C)—
but in either case, if H is supposed to be not outside the surface,
then AH extended will meet chord BC.
But that is absurd: AH extended cannot meet chord BC. Here is why:
Since H is not on DE itself but on DE's extension,
AH extended will not meet BC itself but BC's extension (say, at K).
And now, since K must be on the extension of chord BC,
and the extension of any circle's chord is outside the circle,
therefore K is outside chord BC's circle—which is the generative circle—
and so every point on AK is outside the generated surface.
So point H, being on AK, is outside the surface.

20

A H D E K B C

transition to the next proposition

In these first two propositions of the book, we see that there are only two possibilities for a straight line joining two points on a conic surface: such a line

either goes directly to the vertex, and is entirely *on* the surface,	or else does not end in the vertex, and is entirely *inside* the surface.

Thus we have established something about the shape of the conic surface:

going *down*, it is *straight*;	going *about*, it is *roundish* (curvily ever bulging outward).

But we have not yet used the conic surface to get new lines for us to study.

CHAPTER 3. GETTING OLD FIGURES AS SECTIONS

3rd proposition

#3 If a cone be cut by a plane through its vertex, the cut (*tomê*) is a triangle. [Hereafter, *tomê* will be rendered not by "cut" but rather by the Latinate technical term that is traditional—namely, "section."]
—Let there be a cone whose vertex is what the point A is; and whose base is what the circle BC is; and let it be cut by some plane through the point A; and let it make as sections on the surface the lines AB, AC, and in the base the straight [line] BC. I say that the [figure] ABC is a triangle.
—For since the line joined from the [point] A to the [point] B is the common section of the cutting plane and of the cone's surface, therefore straight is the [line] AB—and likewise also [straight] is the [line] AC. And also the [line] BC is straight. A triangle, therefore, is what the [figure] ABC is.
—If, therefore, a cone is cut by any plane through its vertex, the section is a triangle.

The 3rd proposition now gives us our first conic *section*. How do we get it? By cutting a cone with a plane *through the vertex*. Note that this is the section of a cone, not of a conic surface. If it were a *conic surface* that we cut with a plane through the vertex, then what we would get would be a *pair* of straight lines meeting at the vertex; but since a cone (unlike a conic surface) has a base, the section that we get now has a *third* straight line—making it a *triangle*.

To get a triangle, however, we do not need to cut a cone. This cut is of merely instrumental interest—useful for getting hold of sections of a conic surface that will be of interest in themselves. To get hold of these sections, we shall continue to cut cones rather than conic surfaces, except in the next proposition, the 4th, where we shall cut a conic surface in order to get a cone.

Note that triangle ABC is not necessarily an *axial* triangle. (The axis, as the definition said, is the straight line joining the vertex to the center of the circle; so an axial triangle is produced by passing the plane not merely through the vertex, which is the topmost point of the axis, but through the entire axis. The significance of doing so will appear later.) The figure could therefore be drawn more generally, as in figure 1.

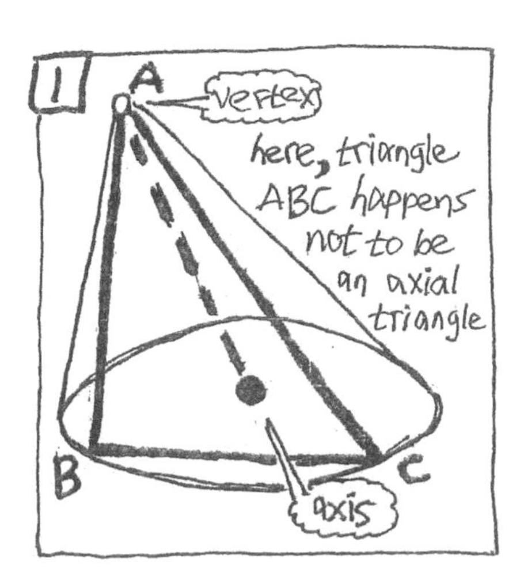

It is easy to demonstrate that if a cone is cut by a plane through the vertex, then the section is a triangle.

> We let there be (as set out) a cone with point A as vertex and circle BC as base.
> This cone is cut by some plane through vertex A,
> making the following sections:
>
> | on the surface, | in the base, |
> | --- | --- |
> | lines AB and AC; | line BC. |
>
> From that, we specify, it is to be shown that ABC is a triangle.
>
> Now the line joining A to B is a line that contains the vertex A and is
> on the cutting plane as well as on the cone's surface—
> and hence it is straight.
> Also straight, for the same reason, is the line from vertex A to point C.
> Finally, the line joining the two non-vertex points B and C is
> the intersection of two planes (the cutting plane and the base) —
> and hence it too is straight.
> Hence, ABC is a triangle (fig. 2).

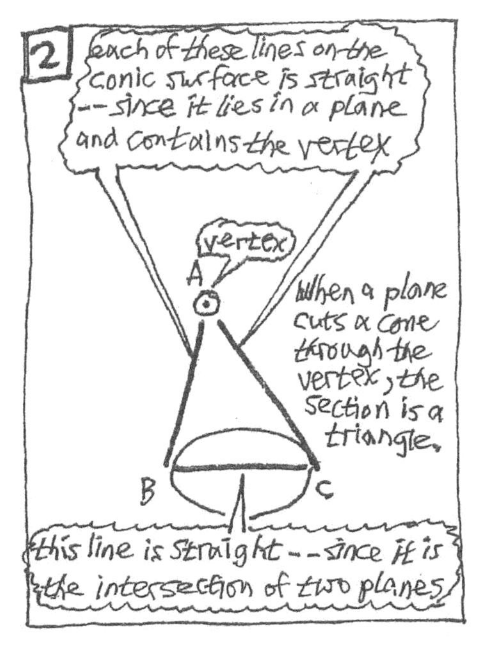

2
each of these lines on the conic surface is straight --since it lies in a plane and contains the vertex
vertex
A
When a plane cuts a cone through the vertex, the section is a triangle.
B
C
this line is straight --since it is the intersection of two planes

It seems obvious that the lines AB and AC are straight. What was asserted in the enunciations of the first two propositions may have seemed equally obvious, yet Apollonius used a *reductio* to demonstrate them. We could use a *reductio* here in the 3rd proposition too; we could show this: if a line from the vertex of a conic surface to another point on it has been cut into the conic surface by some other surface, then the cutting surface cannot be a plane if that line of intersection of the two surfaces is not straight. That is not what Apollonius does. Why, then, did he use a *reductio* to demonstrate the 2nd proposition? He could have demonstrated it directly from the proposition that now comes 3rd, which does not seem to depend upon the one that just came 2nd. Why did he not present the propositions in the order 1-3-2?

It seems that while the order of Apollonius's propositions does not violate logic, logical dependence alone is not sufficient to explain the order. It would have suited mere logic just as well if the proposition that now comes 3rd had come before the one that came 2nd. But if the present 3rd proposition had indeed come earlier, then it would have separated the 1st one from the present 2nd one—and we have seen that the propositions that now stand as the first two do point out features of the conic surface that belong together. Moreover, the one that now comes 3rd would (if it were placed after the 1st) also then be separated (by the present 2nd one) from the ones that now come 4th and 5th—and we shall see that it rather belongs immediately before what comes in the next two propositions.

In the 3rd proposition, we have cut a cone and the section that results is a *straight-line* figure. Now what would most appropriately come immediately after that? Would it not be the cutting of a cone that results in a *circular* section?

transition to the next propositions

Circles are the sections that we shall get from cutting in both of the next two propositions. In the 4th proposition, by cutting a *conic surface* in a certain way, we shall get a circle, and along with it a cone; and then, in the 5th proposition, by cutting a certain kind of *cone* in a certain way, we shall get a circle too.

To study the circle, however, we do not need to cut a conic surface or a cone, any more than we need such cutting to study a figure (like the triangle) that is made up of lines that are all straight. But the circle is a curve, and by getting circles from cutting cones we prepare ourselves to study other curves gotten from such cutting—curves that are *non-circular*.

Circles have certain properties that will be useful for us to remember as we proceed. So now, before we proceed to the next two propositions, let us consider those useful properties of circles.

The 4th proposition will use the property by which Euclid defines a circle. That property is this: a plane figure which is contained by one line is a circle if, and only if, there is some one point from which all of that line's points are equally distant. We might note here that in Greek the circle is named by the word for wheel—*kyklos* (from which comes our word "cycle"). From the hub of a wheel radiate the spokes (in Latin, the *radii*), every one of which is equal to every other one (fig. 3).

The other proposition, the 5th, will use another property—one which, while it is not the property that is used to define the circle, is nonetheless a property that is distinctive of the circle. This property follows from putting together two propositions that Euclid demonstrates in his *Elements*. In the next part of this guidebook, we shall examine some passages selected from Euclid's *Elements*. For now, so as not to interrupt our train of thought, let us just accept some things for which we shall get a firmer foundation later on.

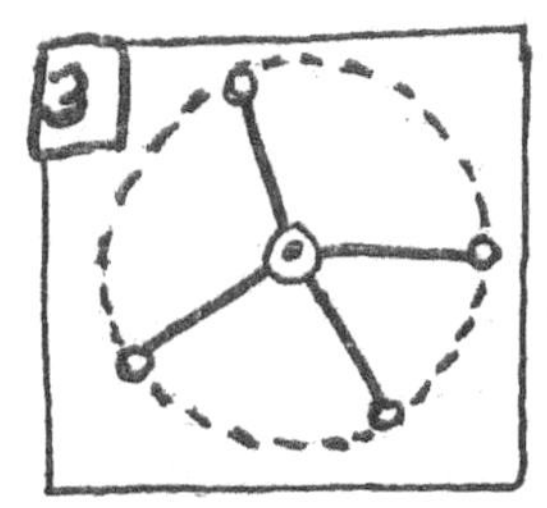

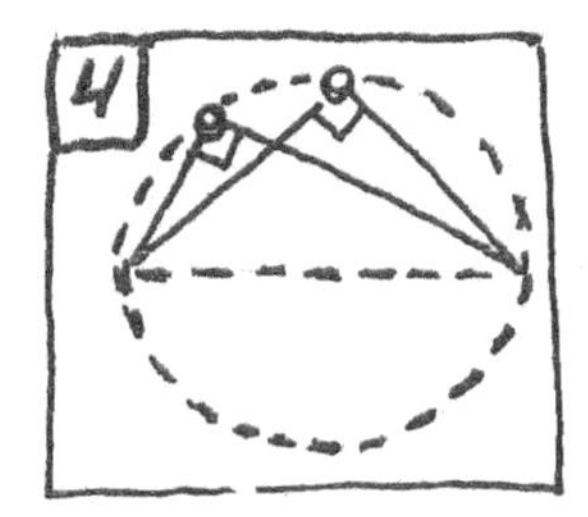

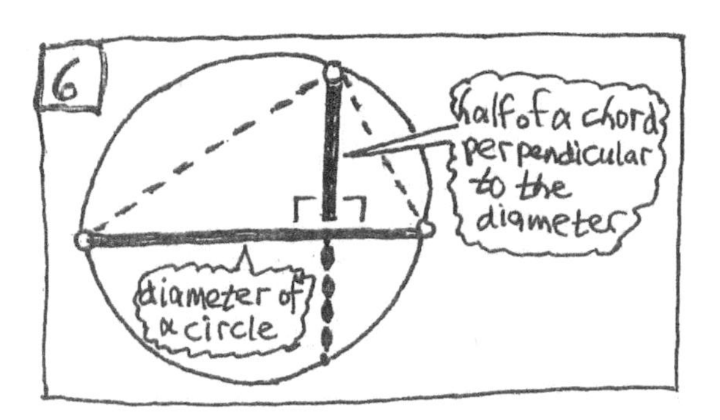

Euclid demonstrates (*Elements*, III.31) that if an angle is drawn in a semicircle, then it must be a right angle. This Euclidean proposition implies that when an angle in a semicircle is put together with the diameter that stretches under it, what we shall get is a right triangle (fig. 4).

In any right triangle, as Euclid also demonstrates (VI. 8), if a perpendicular is dropped from the vertex of the right angle to the hypotenuse, then it will make two triangles, each of which is similar to the other, as well as being similar to the whole original triangle. Let us label the legs of the two triangles as in figure 5; and then we can say that the right triangle with legs A and M is similar to the one with legs M and B (as well as being similar to the whole original triangle). Now, since the corresponding sides of similar triangles are proportional,

as A is to M, so also M is to B.

If (for example) A is double M, then M is double B; or if A is triple M, then M is triple B. (The now common notation for this proportionality, which the ancients, by contrast, wrote out in words, not abbreviations—along with letters as names of points and lines—is "A : M :: M : B.") Of these proportional terms, M is the one in the middle; in other words, it is the "mean." Hence, as Euclid says (in his porism to this proposition),

M is the mean proportional between A and B.

The comparison between M and the other lines, which is here stated in a proportion, can be restated in an equality. From a later proposition of Euclid's *Elements* (VI. 17, following from VI.16), it can be seen that the preceding proportion (the mean-proportionality established in the porism to Euclid's VI.8) implies the following equality:

if M is made the side of a square,
and A and B are made the sides of a rectangle,
then the square and the rectangle will be equal.

Now we take the inference we drew from Euclid's III.31 a few moments ago, and we put it together with the inference we just drew from the porism to his VI.8—and then (remembering that a chord of a curve is merely a straight line joining two points of the curve) we arrive at the following distinctive property of a circle:

Draw a chord perpendicular to a diameter,
and take half of the chord;
and take as well the two segments into which
the diameter is split by that half-chord (fig. 6).

Then a square which
has as its side that perpendicular half-chord
will be equal to a rectangle which
has as its sides the segments into which
the diameter is split by that perpendicular half-chord.

That property of square-rectangle equality does not immediately come to mind when we look at a circle. But then neither does the property that we use to *define* a circle. A child is able to recognize a circle without ever having heard about equidistance from a center. A circle is contained by a perfectly round line, immediately recognizable as such from its own looks—from its very shape without any added marking. Yet we define the circle, not in terms of its own perfect roundness, but rather in terms of the equality of certain straight lines which we can attach to it. In order to be able to *say* very much about a circle, or to be able more easily to *handle* the circle in order to say something about *other* figures, we *mark* the circle's hub and spokes. So, in order to enable us to manage the articulation of various properties of figures, Euclid even in his definition of the circle departs from what immediately comes to mind in looking at a circle. Now, even further from what immediately comes to mind is the circle's property that we have drawn out of Euclid's propositions. Whereas the definition speaks of the equality of straight lines, this other property speaks of the equality of a square and a rectangle that we put together out of straight lines. But this will enable us to put into words an even more complex handling of figures.

And now a word of caution. When you encounter the statement that the square whose side is line M is equal to the rectangle whose sides are lines A and B, it would be easy to assume that if only the author had known better, he would have spoken of equal *areas*, as we do now, and would have written something more familiar to us and easier to handle. It would be a mistake, however, to lapse into familiar ways by writing the statement as saying this: $M^2 = A*B$. Of course it is true that if, for example, A is 3 meters long and B is 12 meters long, then you will find the length of M in meters by finding that number which, when multiplied by itself, equals the product of multiplying 3 and 12 together. But to apply to M the term "square root," would be to presuppose Descartes' transformation of the work of writers like Apollonius—a transformation that cannot be appreciated without seeing what it was that was transformed, and why.

We should not merely assume that classical geometry is saying, in an unfamiliar and somewhat cumbersome way, what our algebra says more crisply. The fact that algebra can illuminate the study of geometry does not by itself make classical geometry a form of algebra. Much can be learned about curved lines without reading Euclid, or Apollonius, or Descartes; but by reading them, in their own terms, we can learn other things about curved lines—and also about learning itself.

When classical geometry is put into writing, then speech may, without change of meaning, be represented by signs that are abbreviations. We might, for example, write this: sqr [M] = rct [A, B]. But something new will be introduced if we go beyond mere abbreviation, to use symbols that get their meaning from a system that does not refer directly to countings and comparisons in figures which we can imagine drawing.

"The square whose side is line-M" does not mean what "M^2" means. As we shall see, the signs do not even mean in the same way. If the sign "M^2" is intended merely as an abbreviation for "the rectangle whose 2 sides are M" (that is, for "the square with side M"), then it is a misleading abbreviation. It is no accident that while the sign "M^2" is read "M-squared," and "M^3" is read "M-cubed" (thus indicating what that sign grew out of), nonetheless "M^4" and "M^n" are read "M to the 4th power" and "M to the nth power" (thus indicating that with the exponential notation any simply visualizable signification was outgrown). Likewise, "the rectangle contained by lines A and B" should not be turned into "A times B"—even though the rectangle contained by lines that are 3 meters long and 12 meters long, respectively, will be covered by 36 squares, each with a side as long as a meter-stick. In our study we shall be trying to see clearly what the difference is and why it makes a difference. But that will take a while.

Now let us turn to the next proposition, which uses the simpler of those two properties of the circle that we have just discussed—the property that Euclid uses to define the circle—namely, the equality of a circle's several radii.

As you first read the words of the proposition, do not be discouraged by difficulty in trying to visualize what they describe. In the explication that follows the proposition, you will see pictures that show what the words of the proposition say.

4th proposition

#4 If either of the vertically opposite surfaces be cut by some plane that is parallel to the circle along which is carried the straight [line] that carves out the surface, the plane that is taken off amid the surface will be a circle having its center on the axis; and the figure (*skhêma*) which is contained, by both the circle and the conic surface that is taken off by the cutting plane, on the side toward the vertex, will be a cone.

—Let there be a conic surface whose vertex is what the point A is; and whose circle along which is carried the straight [line] that carves out the surface is what the [circle] BC is; and let the surface have been cut by some plane that is parallel to the circle BC, and let the plane make as a section on the surface the line DE. I say that the line DE is a circle having its center on the axis.

—For let there be taken, as the center of the circle BC, the [point] F, and let there be joined the [line] AF; it therefore is the axis—and it meets with the cutting plane. Let it meet with it at the [point] G, and let some plane have been extended through the [line] AF. The section will be, then, the triangle ABC. And since the points D, G, E are in the cutting plane, and are also in the plane ABC, therefore straight is the [line] DGE.

—Let there be taken, then, some point on the line DE—the [point] H, say—and let the joined [line] AH be extended. It will then meet with the circumference BC. Let it meet with it at the [point] K, and let there be joined the [lines] GH, FK. And since two parallel planes—the [planes] DE, BC—are cut by a plane—the [plane] ABC—their common sections are parallel. Parallel, therefore, is the [line] DE to the [line] BC. On account of the same things, then, also the [line] GH is parallel to the [line] KF. Therefore as is the [line] FA to the [line] AG, so is the [line] FB to DG, and the [line] FC to GE, and the [line] FK to GH. And the three [lines] that are the [lines] BF, KF, FC are equal to each other. Also, therefore, the three [lines] that are the [lines] DG, GH, GE are equal to each other.

—Then likewise will we show (*deixomen*) that also all the straight [lines] falling forth from the point G onto the line DE are equal to each other.

—A circle, therefore, is what the line DE is—one having its center on the axis.

—Also evident is that the figure (*skhêma*) that is contained by both the circle DE and the conic surface that is taken off by it, on the side toward the conic surface's point A, is a cone.

—Also together with that is it demonstrated (*sunapodedeiktai*) that the common section of the cutting plane and of the triangle through the axis is a diameter of the circle. [The root of both *deixomen* ("we will show") and *sunapodedeiktai* ("is demonstrated") is the same: *deik-*.]

In this, as in subsequent propositions, you should first quickly look over the words of the proposition, then study the explication, and then, returning to the proposition, draw for yourself, as you study the words, a picture of what you saw in your study of the explication.

4th proposition: enunciation

In this proposition, it is not a cone that we cut (although a cone is what we did cut in the previous proposition, and what we shall cut in subsequent propositions). Here what we cut is, rather, a conic surface; a cone is merely the by-product of our cutting. Here we do not need to begin with the circular plane that a cone would have as its base. What we do need is the circular line that the straight line moves along as it generates the conic surface. What we get as we go is a planar region that is demarcated in the plane with which we cut the conic surface, and this region of the cutting plane becomes the base of the cone that is the by-product of our cutting of the conic surface.

How do we cut the conic surface? With a plane that is *parallel to the generative circle*. And what do we then get as the section? We get a line of the sort that we know already: we get a *circular* line.

This circle of section has its center on the axis. If we made another such cut—and another, and another—then we would see that the cone is like a stack of circles pierced through their centers by the axis. (In a *right* cone, the axis is *upright*; but in an *oblique* cone the axis *tilts*, thus skewing the stack of circles.)

So, the enunciation tells us this: if either one of the vertically opposite surfaces is cut by a plane, and this cutting plane is *parallel to the generative circle*—then the plane that is cut off within the surface will be a *circle* with its *center* on the axis; and when this circle is put together with the conic surface extending from the section to the vertex, the figure that is contained will be a *cone*.

4th proposition: demonstration introduced

We let there be (as set out) a conic surface which has point A as vertex, and which has been generated by a straight line moving along circle BC. This surface is cut by a plane which is parallel to circle BC. The cutting makes, as section on the surface, line DE (fig. 7). (As before, the figure is *simplest* when the thing of interest—here, it is the cutting plane—is placed *between* the vertex and the generative circle; but it could be placed, instead, either on the other side of the vertex from the generative circle, or else on the other side of the generative circle from the vertex.) We specify that from this, it is to be shown that DE is a circle whose center is on the axis.

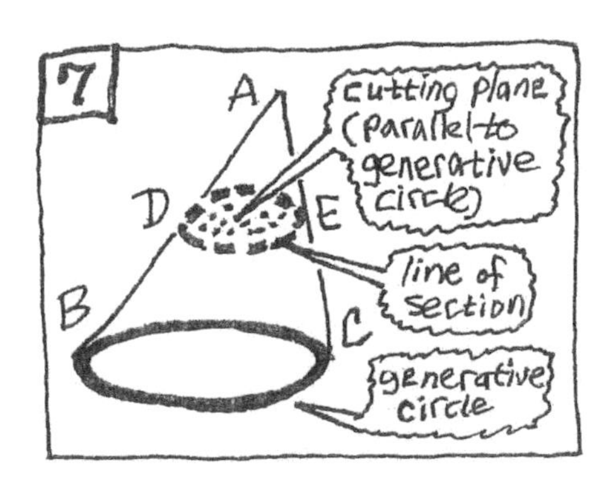

How shall we demonstrate that the line thus cut into the surface is circular? By showing that if straight lines are drawn from any point on the line to a certain single point, then they will all be equal to each other.

Where would we expect that single point to be? Nowhere else but where the cutting plane is intersected by the axis of the surface.

And how shall we show that the straight lines are equal? Well, if magnitudes are in the same ratio to equal magnitudes, then they will be equal to each other. And there are some straight lines here that we already know to be all equal, every one to every other—namely, the radii of the generative circle. What if each one of the straight lines that we want to get equal to each other had—to some one of those radii of the generative circle—the same ratio that any other one had to one of those radii? Then, since each radius of the generative circle is equal to every other one, all the straight lines that might be drawn (in the cutting plane) from the section line to the axis would have to be equal to each other. The question, then, is this: how can all those straight lines be shown to be proportional (that is, in the same ratio)?

Well, we can take a *top*-view—of the generative *circle*, and also of the section line which we would like to see as circular (fig. 8).

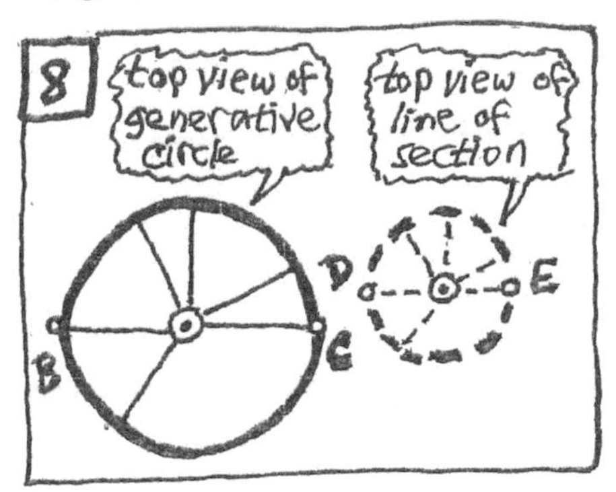

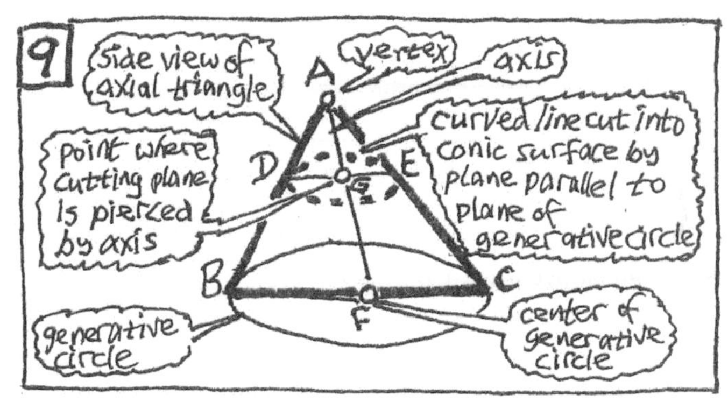

Now let us take a *side*-view. It will show us a figure made up entirely of *straight lines*: axial triangle ABC (fig. 9). Point A is the vertex; BC is the generative circle; curved line DE is cut into the conic surface by a plane that is parallel to the circular plane BC; point F is the center of circle BC; and the axis AF pierces the cutting plane at G.

In the plane of this axial triangle (which is depicted as the plane of the page), FB and FC are two radii of the generative circle; and GD and GE are two straight lines from the line of section to that point in the cutting plane where the plane is met by the axis (point G). The axial triangle's intersection with the two planes makes two straight lines—one of them (straight line DE) in the cutting plane, and the other (straight line BC) in the generative circle's plane. Since the cutting plane and the generative circle's plane are parallel to each other, the two straight lines (DE and BC) are also parallel to each other—and hence we have equiangular triangles (fig. 10).

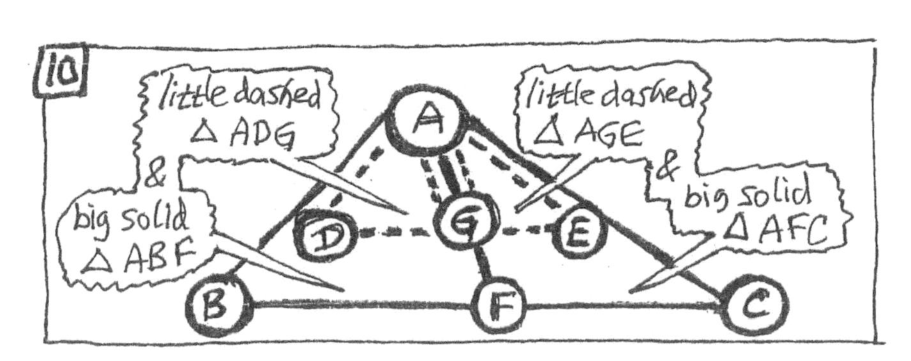

Moreover, the two pairs of similar triangles have sides in common. Side FA is common to the big triangle on the left and to the other big one on the right; and side GA is common to the small triangle on the left and to the other small one on the right (fig. 11). By means of the ratio of big common side FA to small common side GA, we shall relate the other sides.

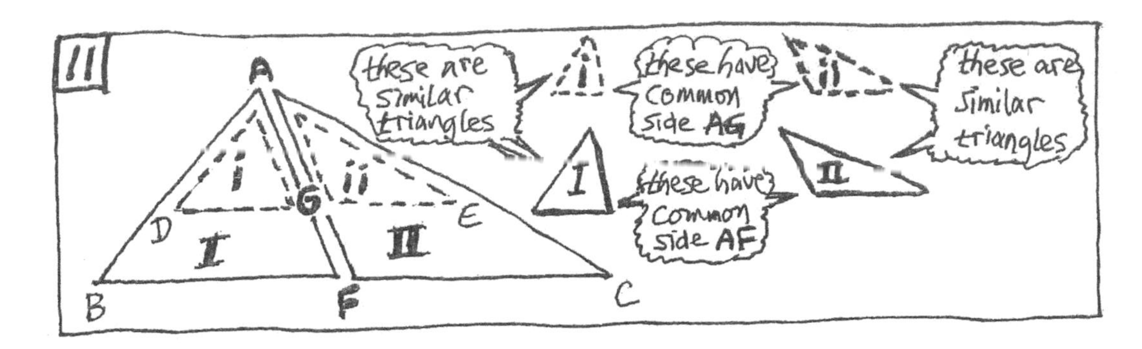

Since corresponding sides of equiangular triangles are proportional (Euclid VI.4), the ratio of the lower left-hand base to the one above it (of FB to GD) and the ratio of the lower right-hand base to the one above it (of FC to GE) are both ratios that are the same as the ratio of the big common side to the small one (that is, of FA to GA)—and therefore the ratios (of FB to GD, and of FC to GE) are the same as each other. The lower bases (FB on the left, and FC on the right), being both radii of the generating circle, are equal to each other. Since the lower bases not only have the same ratio to the upper bases, but also are equal to each other, therefore the upper bases too (GD on the left, and GE on the right) are equal to each other.

But D and E are special points on the line of section (they are the points where the line of section is intersected by the axial triangle); and in order to show that the line of section makes a circle having center G, we must show that the straight line drawn from G to *any* point on the line of section (call it the point H) is a radius—that is, that GH is equal both to GD and to GE. How, then, shall we show this?

Well, *first* take any point H at random on the section. *Now* we can use it to get another axial triangle: let this random point H be joined to the vertex A, and (as the 1st proposition guarantees) our straight line AH will lie upon the surface, and its prolongation will meet the generative circle at some point K. Let this K be joined to point F, the center of the generating circle. And *now*, in this triangle (AFK), we shall get a small similar triangle (AGH), by letting our random point H be joined to G, the point where the axis pierces the cutting plane (fig. 12).

Now, a side in the one triangle of *this* pair has, to its correspondent in the other, the same ratio that we saw in the two pairs we considered a few moments ago—since all three pairs have the same common sides (namely, big FA and small GA) along the axis (fig. 13).

But FK is a radius of the generative circle below, and hence it is equal to each of the generative circle's other radii FB and FC. Therefore the lines above as well are equal to each other: that is, GH is equal both to GD and GE. The relationship is depicted in figure 14.

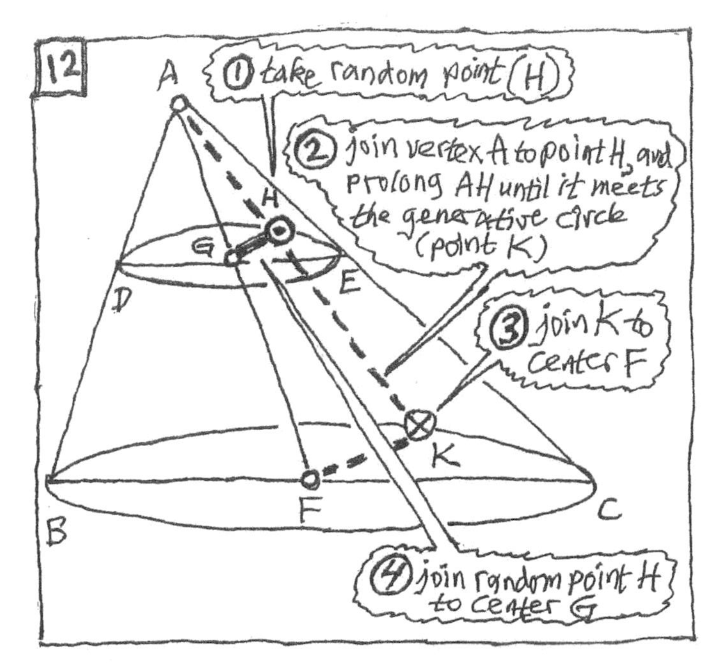
12
A
① take random point (H)
② join vertex A to point H, and prolong AH until it meets the generative circle (point K)
H
G
D
E
③ join K to center F
K
F
B
C
④ join random point H to center G

13
A
G H
III
now, those triangles are similar
D
E
F K
K
B
F
C
and these have common side AG
D G G E
and these have common side AF
I
II
B F F C

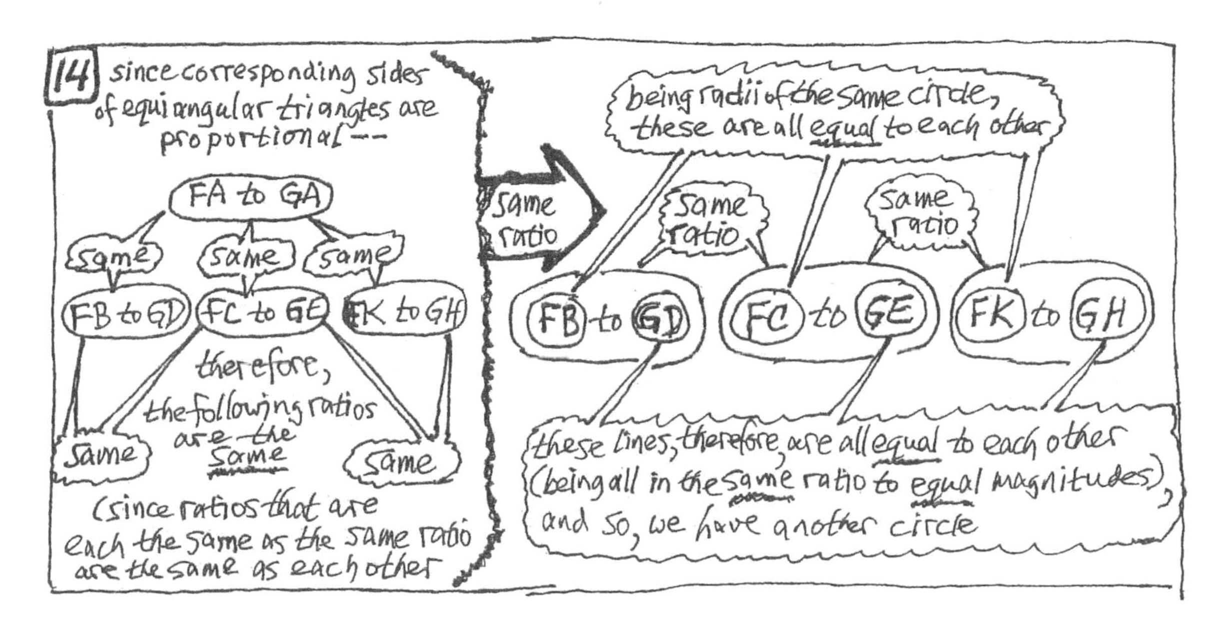
14
since corresponding sides of equiangular triangles are proportional --
FA to GA
same
same
same
FB to GD
FC to GE
FK to GH
therefore, the following ratios are the same
same
same
(since ratios that are each the same as the same ratio are the same as each other
same ratio
being radii of the same circle, these are all equal to each other
same ratio
same ratio
FB to GD
FC to GE
FK to GH
these lines, therefore, are all equal to each other (being all in the same ratio to equal magnitudes), and so, we have another circle

Thus, we begin with a *top*-view that is *circular*: in the generative circle, as in any circle, the straight lines from its circumference to its center are equal. Then, we take *side*-views that are *rectilineal*: from the equality of the straight lines that were the generative circle's radii, we get—by means of axial triangles—the equality of the straight lines that were drawn from the curved line of section to a certain single point. And so, in a concluding view from above, we are able to see the line of section as itself making a *circle*, whose center is the point where the axis meets the cutting plane.

The line of section's shape is shown from putting together what is circular and what is rectilineal. The rectilineal apparatus is the complex part of the demonstration. In it, we must remember this: the intersection of two planes is a straight line; and if two planes are parallel to each other, then their intersections with a third plane will be two parallel straight lines. The two parallel planes in the present proposition are these: the generating circle's plane and the cutting plane. There are *two* "third" planes that intersect with them: one is *any* plane passing through the axis, and the other is the one plane passing through the axis that passes also through the straight line joining the vertex to any point that is taken at random on the line of section. In the first of these "third" planes, we make two pairs of similar triangles; and in the second of these "third" planes, we make one pair of similar triangles. All three pairs of triangles have common sides along the axis, so their other sides are shown to be proportional. The proportionality is a link between equalities: namely, the equality with which we begin, and the equality with which we conclude.

4th proposition: demonstration

Now let us lay out the demonstration without so much explanatory elaboration (fig. 15).

First we need to get the axis.
Take circle BC's center (F), and join it with the vertex, getting AF.
This AF is the axis (by the 1st definition),
and it meets the cutting plane at some point (G).

Now we can get an axial triangle.
Through axis AF, produce some plane, and (by the 3rd proposition)
the section that this plane cuts into the surface will be a triangle (ABC).
In the 3rd proposition, the triangle did not have to be an axial triangle;
but here it is taken in such a way that it *will* be an axial triangle.

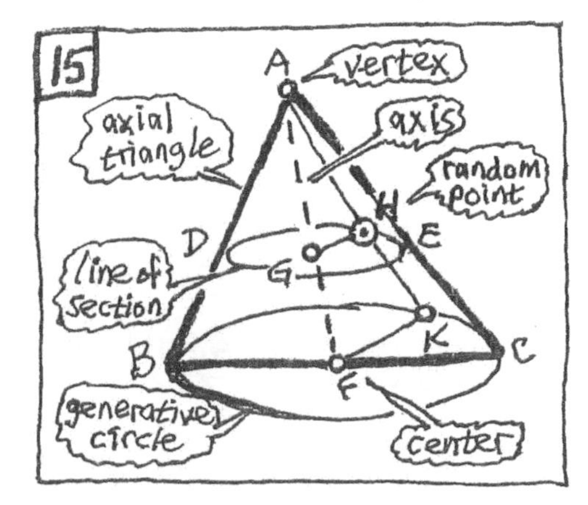

—Points D, G, E are in both the cutting plane and the plane of axial triangle ABC—and hence (since the intersection of two planes is a straight line [Euc XI.3])
DGE is a straight line.
—On line of section DE (the line where cutting plane meets conic surface),
take a random point H, and now get the following two straight lines on which
this H (the random point on the section) is also located:

one line from H to A (the vertex);
and one from H to G (the point where the axis meets the cutting plane).
So, join AH; it, when produced, will
(by the 1st proposition, since A is the vertex and H is on the surface)
fall on circumference BC at some point K;
and to this K, join F (the center of circle BC), getting FK.
Now we have GH and FK, as well as DE and BC.
Since cutting plane DE and generative circle's plane BC
are (by supposition) two parallel planes cut by axial triangle's plane ABC,
their common sections are therefore [Euc XI.16] parallel:
that is, straight lines DE and BC are parallel.
For the same reason (having to do with the planes mentioned above):
straight lines GH and FK are parallel
—Hence (since in equiangular triangles,
sides about equal angles are proportional)
each base of the big triangles has to each base of the little triangles
the same ratio that
the common side of the big triangles has
to the common side of the little triangles;
that is: **as FA is to GA,** so
FB is to GD, & FC to GE, & FK to GH.
Hence we conclude
(since ratios that are the same as the same ratios are the same as each other) that
as **FB is to GD,** so **FC to GE**; and that
as **FC to GE,** so **FK to GH.**
—These conclusions from what we saw by taking a side-view of similar triangles will be combined with what we shall see next by taking a top-view of circles;
and from that we shall draw a further conclusion.
—Since each big triangle's base is a radius of the generative circle, therefore
FB is equal to FC, equal to FK.
—Hence we can conclude (since magnitudes to which
equal magnitudes have the same ratio are equal to each other) that [Euc V.9]
GD is equal to GE, equal to GH;
which is to say: all the small triangles' bases are equal to each other.
—And likewise we could show that *all* straight lines
from point G falling on the line of section DE are equal to each other;
hence they are radii with center G,
and line DE makes a *circle* having its *center* on the axis.
—And it is evident that the figure contained by
this circle DE and the conic surface that it cuts off on the side of point A
is a *cone*,
and therewith that the common section of
the cutting plane and the axial triangle (the triangle through the axis)
is a *diameter* of the circle.

A reminder as you proceed: you'll want first merely to skim the proposition and then to study it only after reading the explication that follows it, which contains pictures to help you visualize what the words say.

5th proposition

#5 If an uneven cone [in traditional parlance an oblique (*skalênos*) cone] be cut by a plane through its axis and at right [angles] to its base, and be cut also by another plane, a plane which is at right [angles] to the triangle that goes through its axis, and which takes off, on the side toward the vertex, a triangle—a triangle which is similar to the triangle that goes through the axis, and which lies subcontrariwise [*hupenantiôs*]—if so, then the section is a circle, and let such a section be called "subcontrary" (*hupenantia*).

—Let there be an uneven cone whose vertex is what the point A is, and whose base is what the circle BC is; and let it have been cut by a plane that goes through the axis and is upright (*orthos*) [in other words, is perpendicular] to the circle BC, and let it make as section the triangle ABC. Let the cone, then, have been cut also by another plane, this one being a plane that is at right [angles] to the triangle ABC and that takes off, on the side toward the point A, the triangle AKG which is similar to the triangle ABC and which lies subcontrariwise—lies, that is, so that the angle by AKG [in other words, the angle contained by the lines AK and KG] is equal to the [angle] by ABC [in other words, the angle contained by the lines AB and BC]. And let it make as section on the surface the line GHK. I say that a circle is what the line GHK is.

—For let there be taken any points on the lines GHK, BC—say, the [points] H, L—and from the points H, L let there be drawn perpendiculars to the plane through the triangle ABC. They will fall, then, to the common sections of the planes. Let them fall as, say, the [lines] FH, LM. Parallel, therefore, is the [line] FH to the [line] LM.

—Let there be drawn, then, through the [point] F and parallel to the [line] BC, the [line] DFE. And the [line] FH is parallel to the [line] LM. The plane through the [lines] FH, DE is therefore parallel to the base of the cone. A circle, therefore, is what it [namely, that parallel plane] is—one whose diameter is what the [line] DE is. Equal, therefore, is the [rectangle] by the [lines] DF, FE to the [square] from the [line] FH. [In other words: a rectangle so built that it could be contained by the straight lines DF and FE as its two adjacent sides would be equal to a square so built that it could be raised from the straight line FH as its side. Note that as a phrase like "the angle by ABC" is an Apollonian abbreviation for "the angle contained by the straight lines AB and BC," so a phrase like "the rectangle by the lines ABC" is an abbreviation for "the rectangle that could be contained by the straight lines AB and BC," and "the square from FH" is an abbreviation for "the square that could arise from the straight line FH."]

—And since parallel is the [line] ED to the [line] BC, the angle by ADE is equal to the [angle] by ABC; and the [angle] by AKG is supposed equal to the [angle] by ABC. Also the [angle] by AKG is therefore equal to the [angle] by ADE. And also the vertical angles at the point F are equal. Similar, therefore, is the triangle DFG to the triangle KFE; therefore as is the [line] EF to the [line] FK, so is the [line] GF to FD. The [rectangle] by the [lines] EFD, therefore, is equal to the [rectangle] by the [lines] KFG. But equal to the [rectangle] by the [lines] EFD there has been shown to be the [square] from the [line] FH; therefore also the [rectangle] by the [lines] KFG is equal to the [square] from the [line] FH.

—Likewise, then, will it be shown that also all the perpendiculars drawn from the line GHK to the [line] GK are [lines] having capability (*dunamenai*) equal to the [rectangle contained] by the cuttings (*tmêmata*) [rendered hereafter traditionally, as "segments"] of the [line] GK.

[The capability that a line is said to have is the power of the line, taken as a side, to give rise to a square of a certain size. Here, each of the straight lines that has been drawn perpendicularly, from some point H on the curve GHK to some point F on the straight line GK, has the following power: the square to which HF can, as a side, give rise is equal to the rectangle that would be contained by the segments KF and FG into which the point F will cut that straight line GK.]
—A circle, therefore, is what the section is—a circle whose diameter is what the [line] GK is.

5th proposition: enunciation

Now we see another way to get a circle by conic cutting. How to do the cutting—that is different here; but so is *what* is cut:

The previous circle (the circle of the 4th proposition) resulted from cutting *by* a plane having the orientation that *is* parallel to the generative circle.	This "subcontrary" circle (this circle of the 5th proposition) results from cutting *by* a plane having a certain orientation that is *not* parallel to the generative circle.
The previous circle resulted from cutting *into* a *conic surface*.	Now the 5th proposition specifies that what is cut *into* be an *oblique cone*.

Is there some connection between these two requirements—that what is cut be an oblique cone, and that the plane which cuts it have the specified orientation? It will take us a while to prepare for answering that question.

Consider a *right* cone. As we have seen, it is symmetrical all around; it is *radially* symmetrical. Now consider a cone whose vertex-point is *not* located right over its base's center-point. Its axis will not be perpendicular to the base. Such a cone, one that stands "unevenly," is an *oblique* cone. Since it leans to one side, it is not radially symmetrical; but if we stand on the extended plane of its base, and go walking around the cone, it *will* look upright from a *single* direction (as well as from the directly opposite direction). Directing our gaze from there, we see the cone leaning neither to the right nor to the left, but only forward (directly toward us) or backward (directly away from us). So, although the oblique cone is not radially symmetrical like the right cone, it is nonetheless *bilaterally* symmetrical.

It follows that *any* plane that is passed through the axis of a *right* cone will be perpendicular to the base, and will slice the cone into identical halves; but *only one* of the planes that's passed through the axis of an *oblique* cone will do so (fig.16).

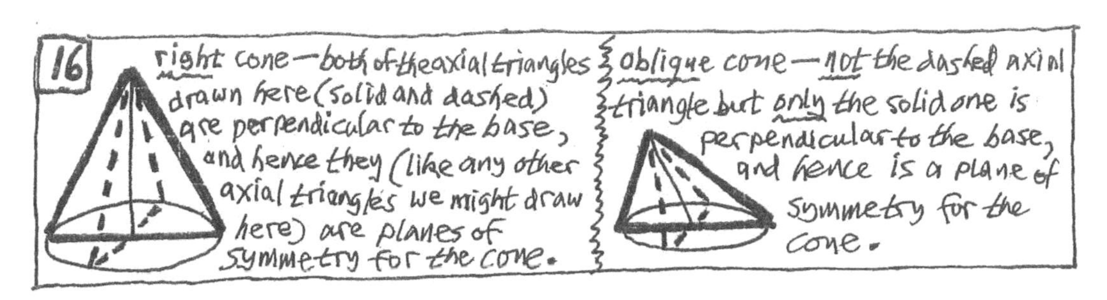

16
right cone—both of the axial triangles drawn here (solid and dashed) are perpendicular to the base, and hence they (like any other axial triangles we might draw here) are planes of symmetry for the cone.
oblique cone—not the dashed axial triangle but only the solid one is perpendicular to the base, and hence is a plane of symmetry for the cone.

Let us take that unique plane through the axis—that *plane of symmetry* which slices the oblique cone into symmetrical halves, and which thus makes upon the surface of the cone an axial triangle that is perpendicular to the base. Now, if we cut the cone with a plane that is perpendicular to *that* axial triangle (to the axial triangle which is itself perpendicular to the base), then the intersection of the cone and the cutting plane will be a curved line that is bilaterally symmetrical about the straight line that is the intersection of the cutting plane and the axial triangle.

We now have *part* of the specification about how to orient the plane that cuts the oblique cone: the cutting plane must be perpendicular to that unique axial triangle which is perpendicular to the base. Such an orientation of the cutting plane will result in a symmetry between what is on the one side of the axial triangle and what is on the other side. It also will allow us to deal with the leaning of the cone by viewing the cone from the side, in cross-section, facing the perpendicular axial triangle, thus making the cutting plane appear edge-on (that is, as a straight line). (Fig. 17.)

17
side view of oblique cone
edge-on view of cutting-plane
The unique axial triangle that is perpendicular to the oblique cone's base is in the plane of the page — so, the cutting-plane, which is perpendicular to the page (i.e. is perpendicular to the plane of the axial-triangle) is viewed edge-on — and so, the only thing that remains to be specified is this: the angle that the dashed straight line makes with the triangle's side.

And now we shall *complete* the specification about how to orient the cutting plane if we merely specify how to orient, in the plane of the axial triangle, a single straight line—namely, the intersection that is made with the axial triangle's plane by a cutting plane that is perpendicular to it.

Not only is the cutting plane going to be taken perpendicular to this axial triangle (the one that is perpendicular to the base), but its intersection with this axial triangle is going to make another triangle in it that is similar to the axial triangle and lies "subcontrariwise" in it. This smaller triangle has the *same shape* as a triangle that would be made by cutting parallel to the base, but it is *differently situated*; it is flipped around, so that the *opposite* angles, one *under* the other, are equal (fig. 18). (From that, we get the term we use. "Under" in Greek is *hypo*, and "against, or opposite" is *anti*—the corresponding Latin being "sub" and "contra." Hence the term *hypenantia*—translated "subcontrary.")

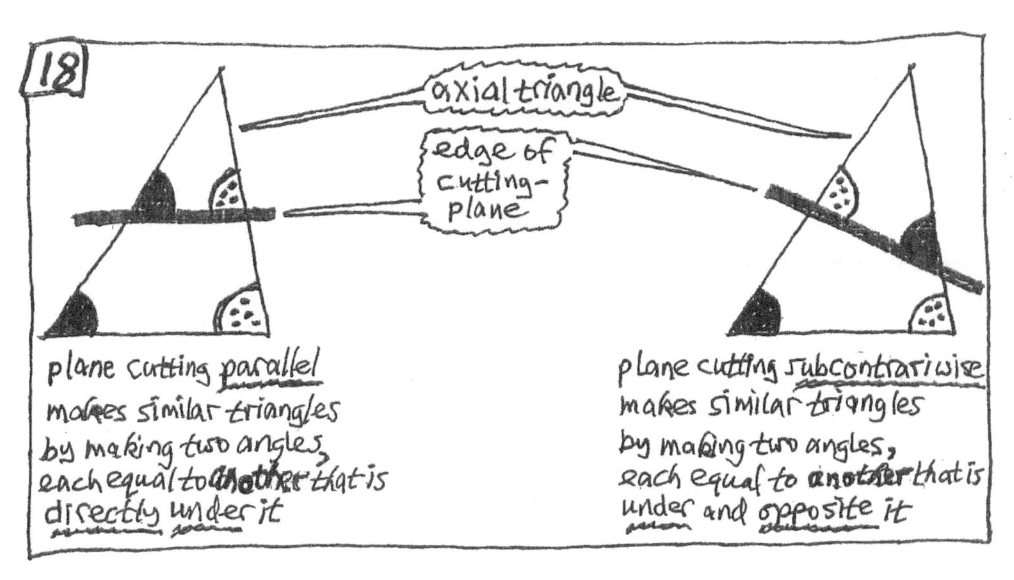

It may not seem strange that when an oblique cone is cut by a plane parallel to the base, then the section is circular like the base; but it may very well seem strange for the section to be circular even when the plane does not cut parallel to the base. And unless there were some reason to believe, prior to a demonstration, that a circle can also be cut subcontrariwise, there would be no reason to seek to demonstrate it. What might lead us to believe that there is this second way to get a circle in an oblique cone?

To see that, consider the side-view of an oblique cone. It shows, in the plane of the page, the single axial triangle which is perpendicular to the base, and which thus constitutes the plane of symmetry of the cone.

Now, take the triangle's two sides that make an angle whose vertex is the vertex of the cone. Extend the shorter side enough to make it equal to the longer side, and shorten the longer side enough to make it equal to the once shorter side; thus you'll get the sides of another triangle—one with the same vertex and vertex angle, but with the sides reversed. Now connect the endpoints, thus making the third side of this other triangle. The base of *either* of these two triangles could be the diameter of a circle for generating a cone that has the *same* upper conic surface.

Thus, every oblique cone will have a *twin*. That is to say, there belongs to it another oblique cone with identically the same upper conic surface, and with its circular base the same size but oriented in reverse. The twins have the same shape and size, but they are not in the same place, though their places do overlap. A circular section will therefore be cut into an oblique cone not only by a plane that is parallel to the cone's base, but *also* by a plane that is *not* parallel to the cone's *own* base—provided that the cutting plane *is* parallel to the *twin's* base. That is because, *either* way, the cutting plane will be given the same orientation in the *common* upper conic surface (fig. 19).

Since, in the side view of the twin cones, both triangles have the same upper portion, a straight line that cuts through that upper portion from side to side will,

if it is *parallel* to	if it is *subcontrariwise* to
the *original* triangle's base,	the *original* triangle's base,
be *subcontrariwise* to	be *parallel* to
the *other* triangle's base,	the *other* triangle's base.

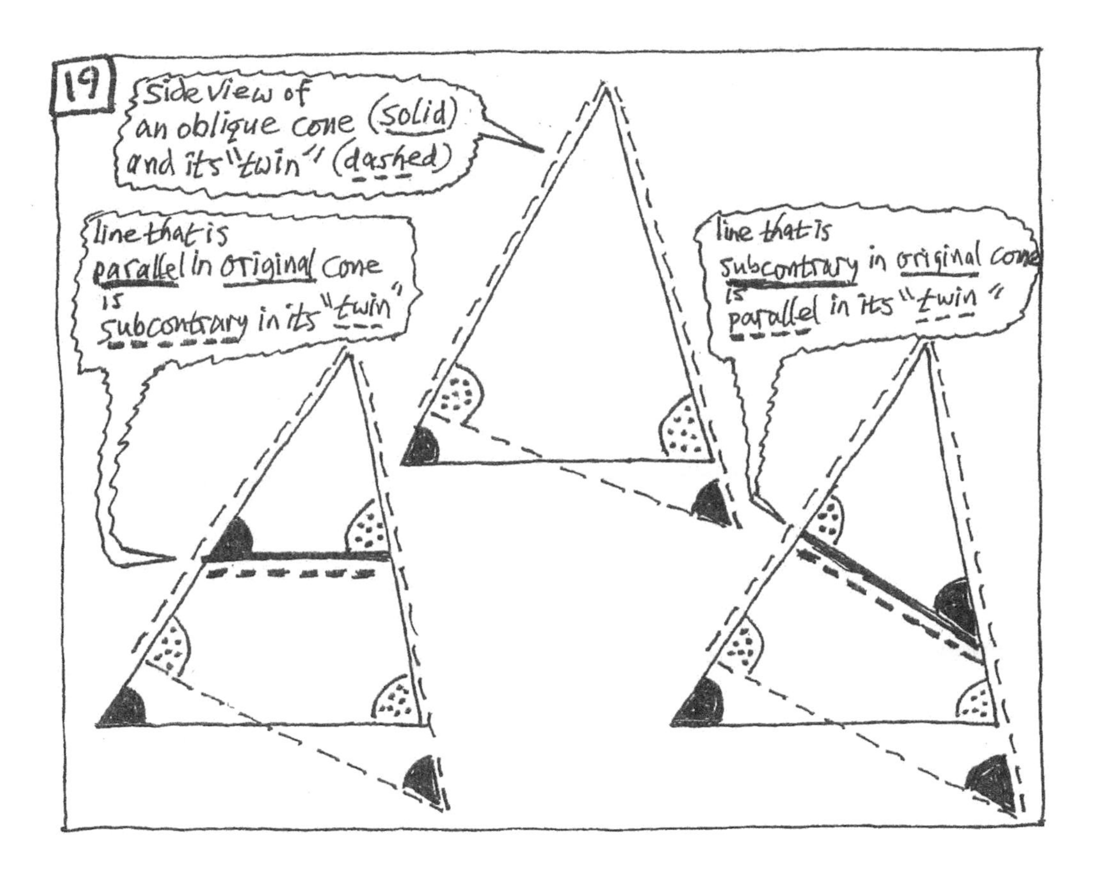
19
Side view of an oblique cone (solid) and its "twin" (dashed)
line that is parallel in original cone is subcontrary in its "twin"
line that is subcontrary in original cone is parallel in its "twin"

In a *right* cone, because of the radial symmetry, *every* axial triangle is *isosceles*—hence the little triangle that is made in the axial triangle by the cutting plane in a right cone can*not* be *similar* to the axial triangle *unless* the cutting plane is *parallel* to the base (fig. 20).

But in an *oblique* cone, we *can* get a similar triangle *even if* we do *not* cut parallel to the base. This is because the only axial triangle that is isosceles in an oblique cone is the one whose plane is perpendicular to the plane of symmetry. In other words, it is the one whose plane is perpendicular to the single axial plane that is perpendicular to the base (fig. 21). By taking the axial triangle that is in the plane of symmetry (in other words, by taking the one that is perpendicular to the base), we guarantee that the axial triangle is not isosceles; and by taking a cutting plane that is perpendicular to that axial triangle, we guarantee (because of the symmetry between things on the two sides of the axial triangle that is perpendicular to the base) that if the *two*-dimensional *triangles* are similar, then so also will be the *three*-dimensional *cones*.

In this 5th proposition, the cut cone is *oblique* and the cutting plane is oriented *subcontrariwise*. Again let us ask: what is the connection between these two features of the enunciation? Now we can say what it is.

If a cone is *right*, then a plane cutting it *subcontrariwise* must be *parallel* to the base. But that situation—a right cone cut parallel to the base—would be a situation already dealt with in the previous proposition. If we are to move beyond the situation of the 4th proposition, and deal in this 5th proposition with an entirely new situation (rather than presenting a situation that will turn out to be only partly new, because it covers in addition a situation that has already been covered), we must take the subcontrariwise cutting of an *oblique* cone *only*. We must leave out the subcontrariwise cutting of a *right* cone, because this turns out to be the same as the already-treated situation of the *parallel* cutting of a conic surface.

So, what the enunciation tells us is this:

if, through the axis of an oblique cone,
you pass a plane that is perpendicular to the base
(thus getting an axial triangle that is
a plane of symmetry for that leaning cone),

and you cut the cone with another plane—
this cutting plane being
perpendicular to the axial triangle that was made by the first plane,
and also being
oriented so that it cuts off (on its vertex side)
a triangle that is similar to the axial triangle and lies subcontrariwise—

then the section is a circle (called "subcontrary").

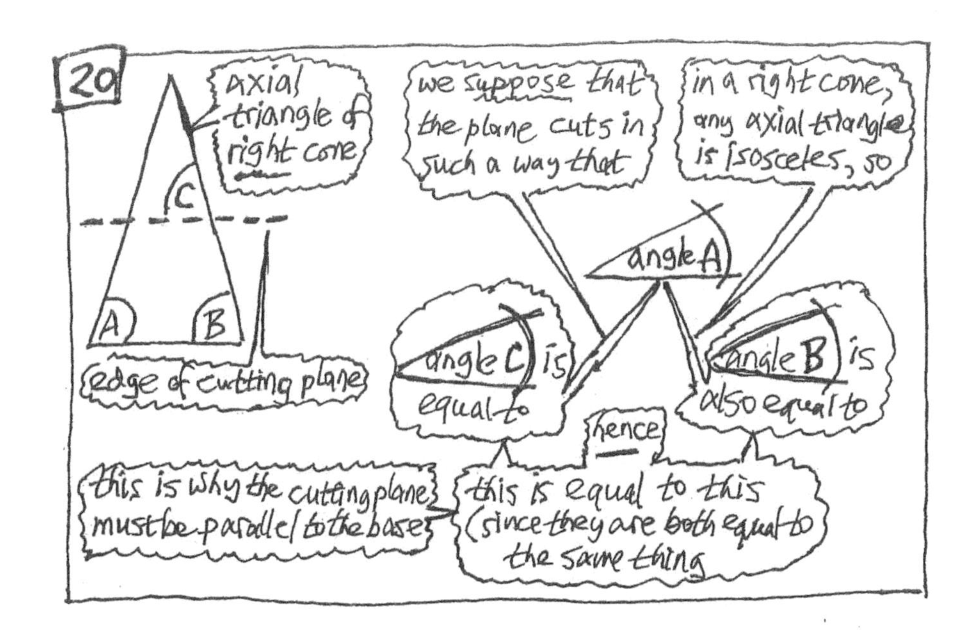
20
axial triangle of right cone
C
A
B
edge of cutting plane
we suppose that the plane cuts in such a way that
in a right cone, any axial triangle is isosceles, so
angle A
angle C is equal to
angle B is also equal to
hence
this is why the cutting plane must be parallel to the base
this is equal to this (since they are both equal to the same thing

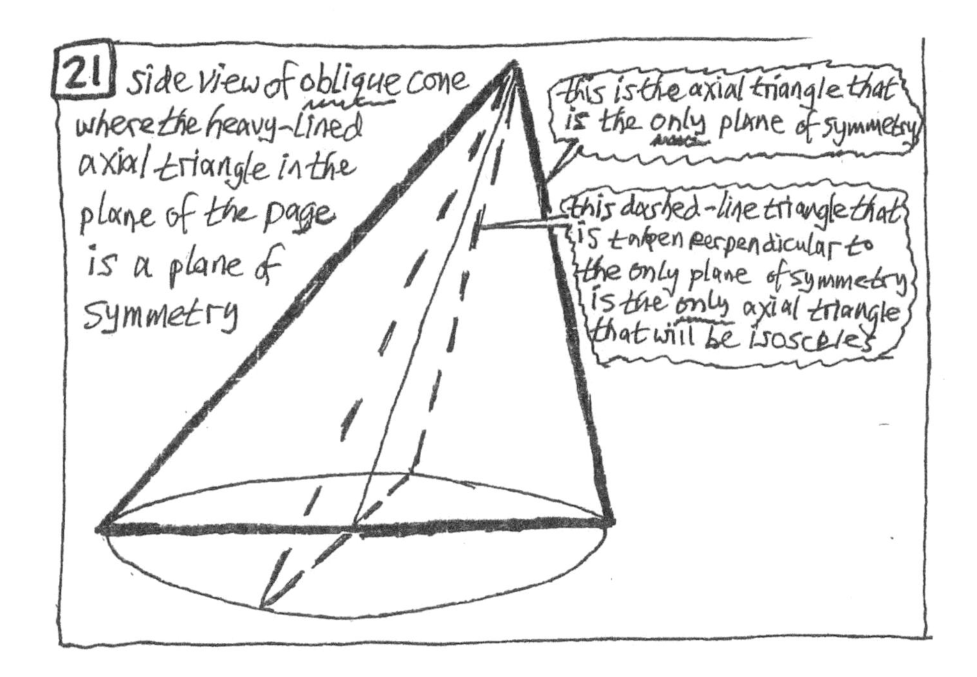
21
side view of oblique cone where the heavy-lined axial triangle in the plane of the page is a plane of symmetry
this is the axial triangle that is the only plane of symmetry
this dashed-line triangle that is taken perpendicular to the only plane of symmetry is the only axial triangle that will be isosceles

5th proposition: demonstration

We let there be (as set out) an oblique cone having point A as vertex and circle BC as base. This oblique cone is cut by two planes. One plane goes through the axis and is perpendicular to circle BC, making the section that the 3rd proposition says it will: a triangle (here, it is the axial triangle ABC). The other plane is perpendicular to this axial triangle ABC, and cuts off (on the side toward vertex point A) the triangle AKG, which is similar to triangle ABC and lies subcontrariwise (that is to say, angle AKG is equal to angle ABC); this plane makes on the conic surface the line of section GHK. We specify that from this, it be shown that line GHK makes a circle. How shall we demonstrate this?

The axial triangle that is perpendicular to the base will be a plane of symmetry for an oblique cone. Would it not therefore bisect any lines (for example, HF) that we draw perpendicular to the plane of the triangle from the line of section, if we prolonged them to the other side of the surface (fig. 22)? That is to say: would not the intersection of the cutting plane and axial triangle (namely, GK) be a diameter of the section?

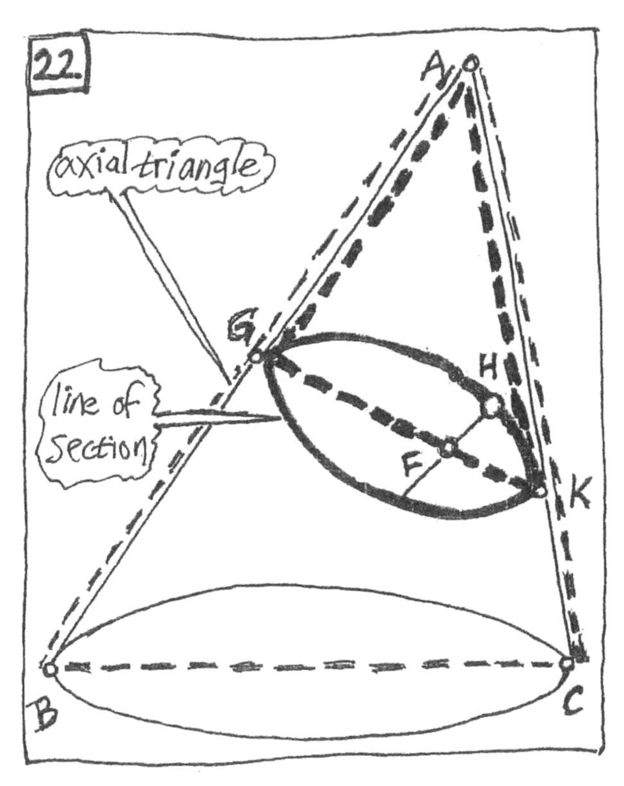

Consider GK. If it were parallel to the base, then (by the 4th proposition) the section would be a circle, and then line HF (which is a perpendicular half-chord of the section) would be capable of being the side giving rise to a square equal to the rectangle that would be contained by the segments into which F splits the diameter. (This follows from what Euclid shows in the *Elements*, for in circle DHE, if lines DH and HE are joined, then [III.31] they contain a right angle; hence [VI.8, porism] triangle DFH is similar to triangle HFE, and so, as DF is to FH so is HF to FE; hence [VI.17] the square arising from FH is equal to the rectangle contained by DF and FE.)

But there is more to say about GK. We have previously taken a cutting plane that is parallel to the base. Now, we take one that is not parallel to the base—not parallel, however, in a particular way. That is, GK, instead of making similar triangles by being parallel, is to be oriented so as to make angles like this: the axial triangle is similar to a smaller triangle that is flipped around so that the angles that are equal to those in the bigger triangle are opposite them.

Now suppose we draw through F a line that is parallel to the base (call it DE, and let it go all the way across the axial triangle). Then the straight line DE does two things:

first, with HF it determines a plane parallel to the base—
which will make (by intersecting with the cone) a circle,
of which HF is a perpendicular half-chord and DE a diameter;

second, with GK it will make similar triangles in
the plane of the axial triangle.

DE will thus be central in both situations. The segments into which DE is split by F (namely, DF and FE) are (*first*, in the circular top-view) related to HF, and are (*second*, in the rectilineal side-view) related to the segments into which GK is split by F (namely, GF and FK). (Fig. 23.) Since DE is central in both situations, it can be central in relating the elements specific to each of the two situations. That is to say: it is by means of DE that the line we catch in the top-view (HF) can be related to the lines we catch in the side-view (GK's segments—namely, GF and FK).

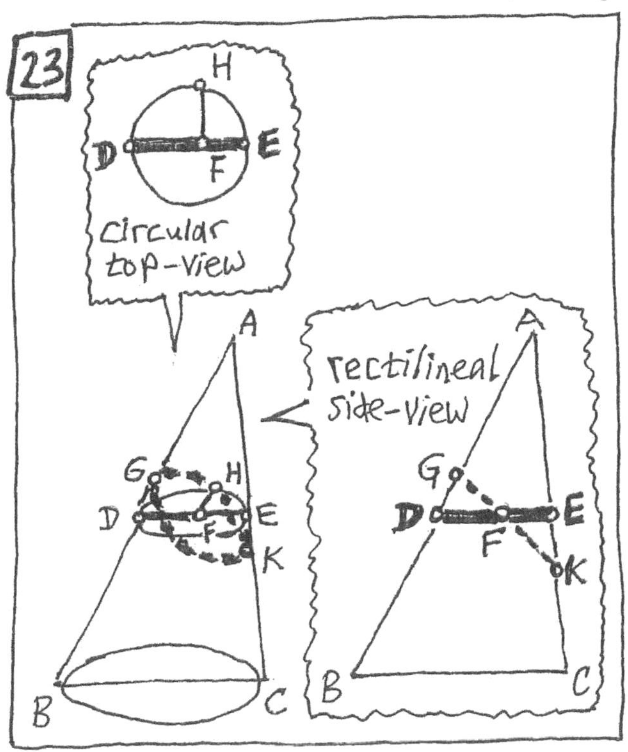

Now let us complete the demonstration. For this, see the next two boxes, with the accompanying figures—figure 24 for the first, and figure 25 for the second.

On the line of section GHK, having taken any point (say, H),
we now, on line BC (the generative circle), take any point (say, L).
From these points, take HF and LM, both dropped perpendicular to
the plane of the axial triangle ABC
(we already have perpendicular HF, but must draw LM).
These perpendiculars fall to the intersections of the planes—
which are (for HF) straight line GK, and (for LM) straight line BC—
and so (since a straight line that is perpendicular to a plane
will be perpendicular to every straight line that it meets in that plane)
HF will be perpendicular to GK, and LM perpendicular to BC.
HF and LM (since they are perpendicular to the same plane)
are [Euc XI.6] parallel to each other.

Now HF and DE are two straight lines which meet each other,
and each of them is parallel to one of
two straight lines (LM and BC) which meet each other;
hence the two planes through the two pairs of lines are parallel [Euc XI.15]:
the plane through HF and DE is parallel to the cone's base—
and hence (by the 4th proposition) it makes as section
a circle with diameter DE.
Hence (by the square-rectangle property which we saw before)
rectangle [EF, FD] is equal to square [HF].

The parallelism of planes has just given us a circle that establishes a certain square's equality to a certain rectangle. The similarity of triangles will next give us a proportion that establishes another rectangle's equality to that same rectangle.

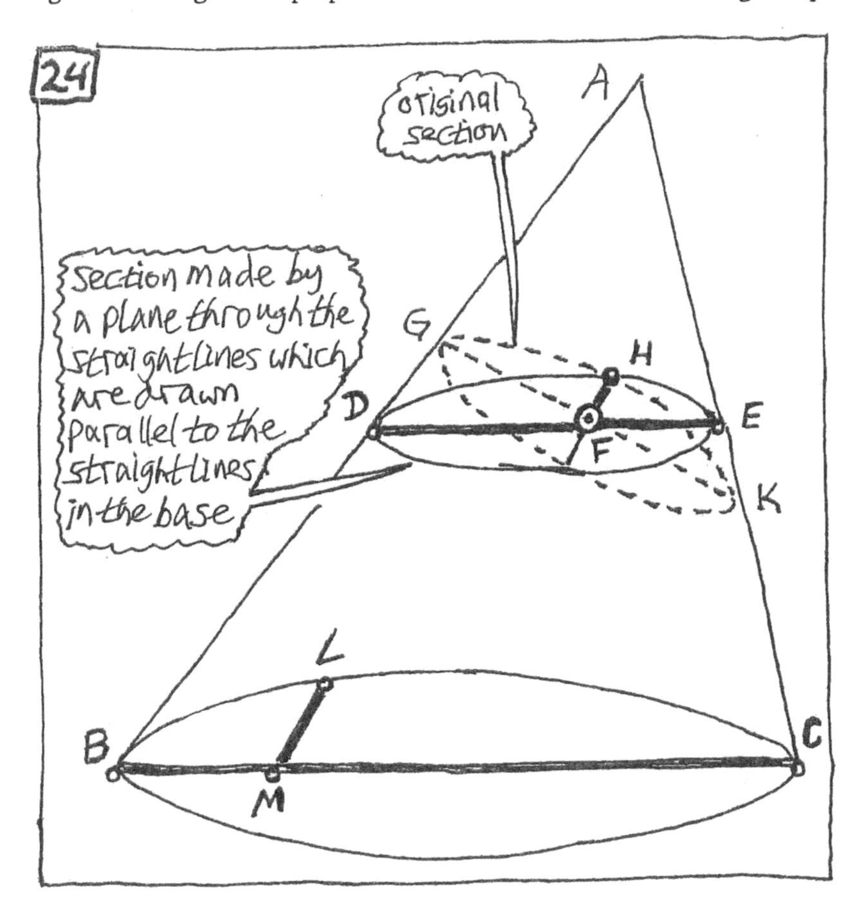

DE was drawn parallel to BC;
hence angle ADE is equal to angle ABC.
And we supposed the cut made in such a way
that angle AKG is equal to angle ABC.
Therefore, angle AKG and angle ADE
(since they are both equal to angle ABC)
are equal to each other.
But we have another pair of equal angles:
at point F (since vertical angles are equal)
angle DFG is equal to angle KFE; and so
(since a pair of angles in one triangle is
equal to a pair of angles in the other)
triangle DFG is similar to triangle KFE; and so
(since corresponding sides of equiangular triangles are proportional [Euc VI.4])
GF is to FD as EF is to FK.
But if four straight lines be proportional,
then a rectangle contained by the extremes
will be equal to a rectangle contained by the means (Euc VI.16);
hence: rectangle [EF, FD] is equal to rectangle [KF, FG].

Since we now have equal to the same rectangle [EF, FD]
both another rectangle and also a square,
that other rectangle and that square must be equal to each other:
square [HF] is equal to rectangle [KF, FG].
Line HF is capable of giving rise to a square that
will be equal to the rectangle that would be contained by
segments KF and FG.

Likewise (since H is any random point on the section)
it could be shown that every perpendicular drawn
from section line GHK to straight line GK
is capable of giving rise to a square that
will be equal to the rectangle that would be contained by
the segments (in each case) of straight line GK.
The section is therefore a circle whose diameter is GK (fig. 25).

proportionals and box-building in some propositions of Euclid

Now, having encountered, in our transition to the last two propositions, a connection between proportionality and box-building, you may feel the need for a better grasp on the elementary knowledge of geometry that Apollonius presupposes. To have examined a few propositions from Euclid's *Elements* (some of them mentioned earlier) can help as we proceed. The Sixth Book shows how to find a fourth proportional if three straight lines are given (12th proposition) and a mean proportional if two straight lines are given (13th proposition); the propositions upon which these propositions most directly depend are the 1st and 2nd. The 16th and 17th propositions connect the proportionality of lines with the equality of the rectangles which they contain; the proposition upon which these propositions most directly depend is the 14th. The propositions from which we get the square-rectangle property for the circle are the 31st proposition of the Third Book and the 8th proposition of the Sixth Book.

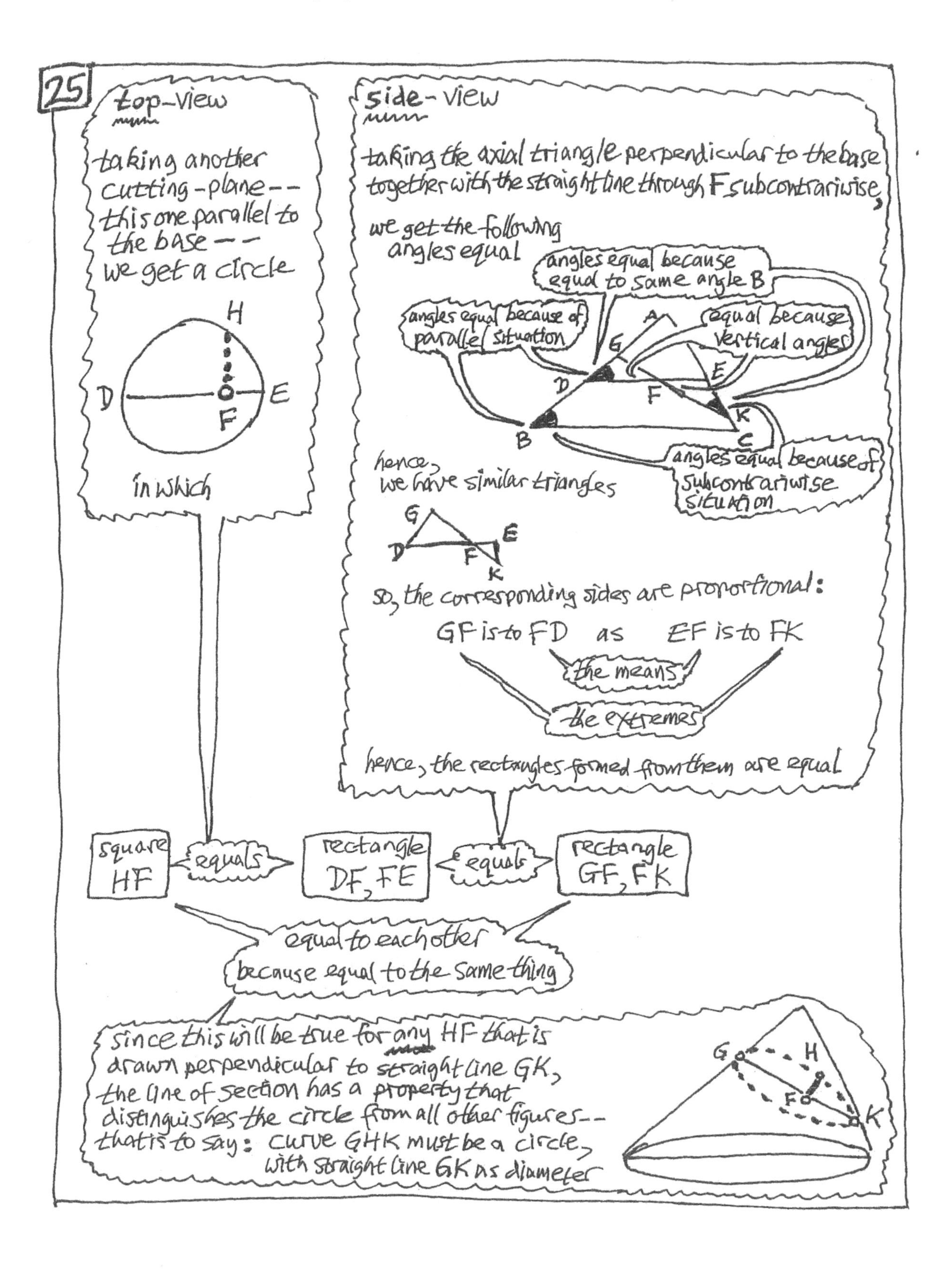

transition to the next propositions

This last (5th) proposition is the very first whose demonstration shows how to characterize a curved line of section in the following way:

> the sizes of straight lines in the section are related by equating *rectangles that those straight lines were used to build.*

The previous proposition (the 4th) characterized the curved line of section in the following way:

> the sizes of straight lines in the section were related by equating *those straight lines themselves.*

The gist of the demonstrations of these two propositions is as follows:

5th proposition (*just* done)—	4th proposition (done *previously*)—
First, we take a *circular* view from above: and since we have a circle, therefore a square and a rectangle are equal.	First, we took a *circular* view from above and since we have a circle, therefore straight lines that radiate from the center are equal.
Then, we take a *rectilineal* view from beside: and since we have similar triangles, and hence also have sides that are proportional, therefore another rectangle is equal to that first rectangle.	Then we took *rectilineal* views from beside: and since we have similar triangles, hence those straight lines are in proportion with other straight lines that radiate from a point inside the section.
Then, we put them *together:* since things that are equal to the same thing are equal to each other, therefore the square and the second rectangle are equal to each other.	Then, we put them *together:* since straight lines that are in proportion with equal straight lines are equal to each other, therefore the straight lines that radiate from the point inside the section are equal to each other.
And we conclude: since that square and that rectangle are equal to each other, therefore the section is a circle.	And we conclude: since all the straight lines that radiate to the section from a point inside it are equal to each other, therefore the section is a circle.

The 4th proposition might be described as fairly straightforward elementary Euclidean work:

> in it, we were using the proportionality of straight lines to establish
> an equality of *straight lines*
> which characterizes the circularity of a line.

But the 5th proposition exemplifies the more advanced work that is to come; it is the first step on our long, winding road of exploration:

> in it, we shall be using the proportionality of straight lines to establish
> equality or inequality of *rectangles*
> (indeed, what will be rectangles *with certain shapes*),
> so as to characterize certain kinds of curved lines that are non-circular.

In the 5th proposition the curved line of interest may not be very exciting—it is merely that old familiar curve, the circle. But this is only the beginning.

In the conclusion of the 5th proposition, as also in the conclusion of the previous (4th) proposition, the intersection of the cutting plane and the perpendicular axial triangle is identified as a diameter of the circular section. In the earlier (4th) proposition, this was said to have been demonstrated along with what was enunciated at the beginning of the proposition, but the enunciation of that earlier 4th proposition did not mention any diameter—and neither does the enunciation of the 5th proposition. Diameters are not yet up front in the propositions.

The first definitions that introduced the whole book did not present any new curves, but they did present a novel discussion of diameters—kinds of diameters unheard of in the elementary study of the circle. Now, after five propositions, we do not yet have a new curve (the only curved line so far has been a circular curve), but we do have a new way to get a diameter for a circular curve. Before we go on to characterize new curves, we need to know about diameters for sections that may not be circular. Such a diameter is what we shall get in the 7th proposition.

But before we can have this diameter of a new curved-line section, we shall need to know something about the relation of two surfaces, one flat and one not. So, the 6th proposition will show us that there is an at least partial symmetry of the conic surface—of that surface into which the cutting plane of the 7th proposition will cut a new curve as it slices this curve's diameter into the axial triangle. What the 6th proposition will tell us is this: straight lines from one side of a conic *surface* to the other side (if they are drawn parallel to a straight line that is perpendicular to an axial triangle's base) will be bisected by the axial triangle (and this is so, by the way, whether or not the axial triangle itself is perpendicular to the cone's base). The 6th proposition thus will make an assertion for the conic *surface* that will enable the 7th proposition to make a corresponding assertion for the conic *section*.

CHAPTER 4. A BRIDGE

From the 6th and 7th propositions, we get no sections at all. They constitute a bridge. It enables us to go from the cutting that gives old lines to the cutting that gives new ones.

The 6th proposition will prepare for the 7th by giving us the axial triangle as a bisecting plane; and the 7th in turn will prepare for the 8th and 9th by giving us, as a bisecting straight line, that axial plane's intersection with the cutting plane. The porism will indicate that there are new lines to be gotten by cutting a cone, but it will not show us what they are—although it will lead us to see something about them.

6th proposition: preparation

In the previous (5th) proposition, the circularity of the section was established by using a half-chord that was perpendicular to the axial triangle. We obtained that half-chord by drawing a straight line from one side of the cone's surface, perpendicularly across to the axial triangle. Had we then continued across to the other side of the cone's surface, we would have obtained the other half of the whole chord. In the present (6th) proposition, we shall again give our attention to a straight line that is drawn from one side of the cone's surface across to the axial triangle—but this one will continue on through the axial triangle, all the way across to the other side of the cone's surface. What the proposition will show, is that the part of the drawn straight line that is taken only as far as the axial triangle is exactly half of the whole. For that to be so, however, the straight line that is drawn from one side of the cone's surface across to the other side must be drawn in a certain way: it must be drawn parallel to a certain reference line. This reference line is a straight line that is in the plane of the cone's base, and is drawn perpendicular to the straight line that is the axial triangle's base (fig. 1).

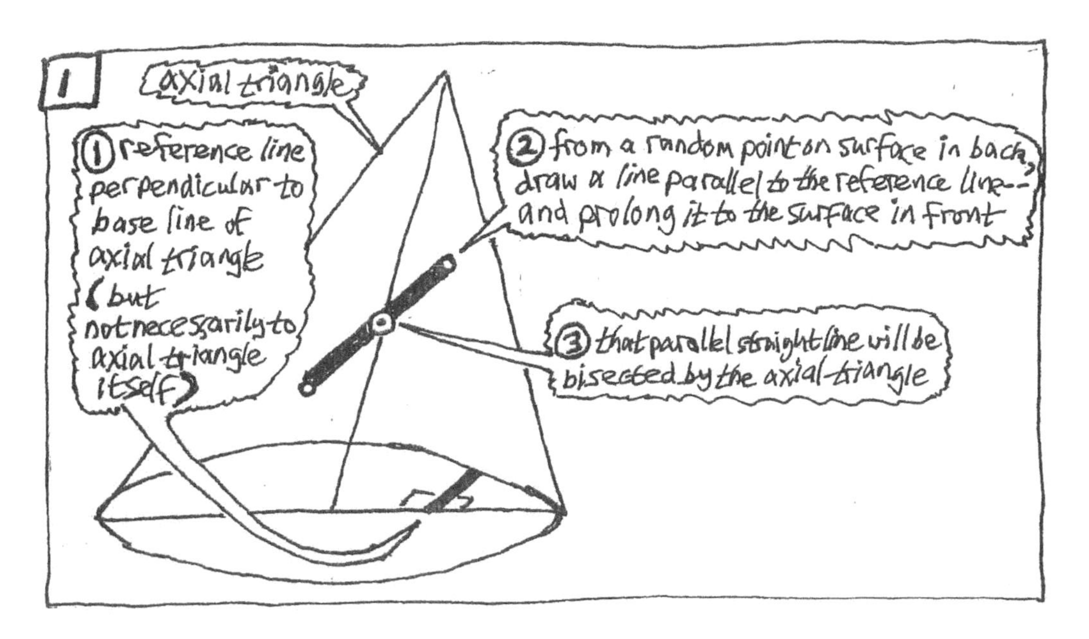

The straight line of interest (the one that is bisected by the axial triangle) may be, although it does not have to be, perpendicular to the axial triangle. It will be perpendicular to the axial triangle only if the reference line for parallelism is perpendicular to that triangle. The reference line, however, does not have to be perpendicular to that triangle. Although the reference line must be perpendicular to the axial triangle's base (which is a straight line), it may or may not be perpendicular to the axial triangle's plane.

That is an instance of the following general situation: when a straight line lies in a plane, a straight line that is perpendicular to that first straight line does not also have to be perpendicular to the plane in which the first straight line lies.

To make the situation easier to visualize, consider a very thin book—so thin that its spine may be regarded as the straight-line intersection of two planes (the back cover and the front cover). Suppose that on the front cover there is a narrow stripe that is perpendicular to the spine of the book. When the book is half-open—when it is open just enough for the front cover to be perpendicular to the back cover—then the straight-line stripe on the front cover will be perpendicular to the plane of the back cover. When, however, the book is open only a little, then the stripe on the front cover (though it will still make a right angle with the line of the spine) will now make an acute angle with the plane of the back cover (fig. 2).

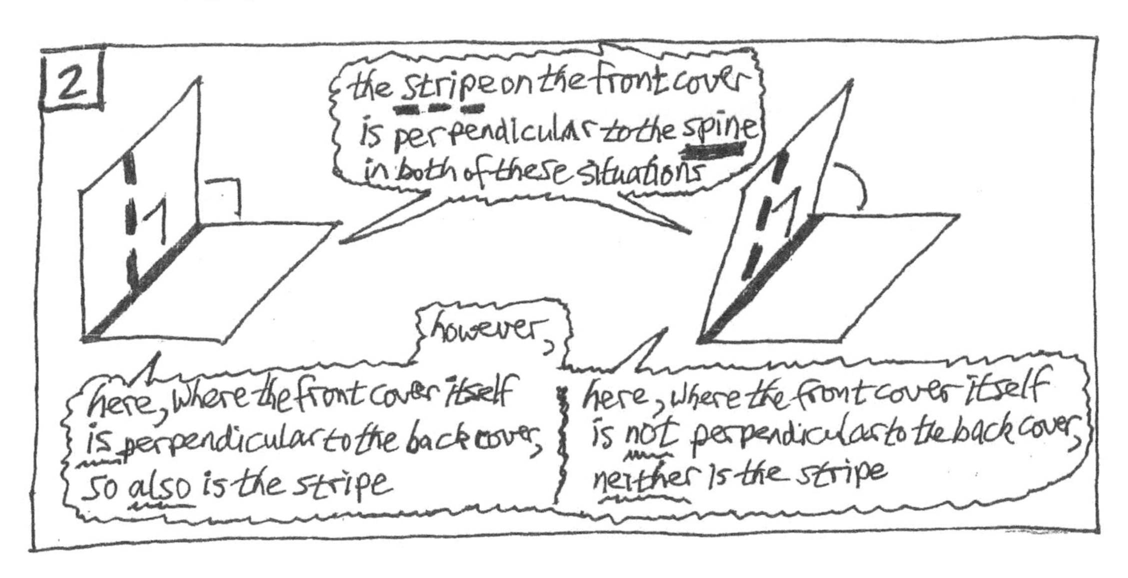

Thus (to return to the cone) the line of reference, which lies in the plane of the cone's base and is perpendicular to the straight-line base of the axial triangle, will be perpendicular also to the plane of the axial triangle if—and only if—the plane of the axial triangle is perpendicular to the plane of the cone's base. As we have seen, in a right cone every axial triangle is perpendicular to the base, but in an oblique cone there is only one that is. The present proposition does not restrict itself to the case of a right cone, nor to any particular axial triangle in the case of an oblique cone. While the axial triangle may be perpendicular to the base, nothing can depend upon its being so, since it just as well may not be so. In either situation, the axial triangle will mark the half-way point of a straight line drawn all the way across the inside of the cone in that certain way described (that is to say, parallel to the straight line of reference—which reference line, lying in the base of the cone, is perpendicular to the base of the axial triangle).

We are ready now for the presentation of the 6th proposition.

6th proposition

Again: patience as you give this a first skim-through! Helpful pictures follow in the explication.

#6 If a cone be cut by a plane through its axis, and there be taken on the cone's surface some point which is not on a side of the triangle through the axis, and from the point be drawn a straight [line] parallel to some [line] which is a perpendicular from the circumference of the circle to the base of the triangle, then [that straight line drawn parallel] will meet with the triangle through the axis, and it will, on being extended to the other side of the surface, be cut in two [halves] by the triangle.

—Let there be a cone whose vertex is what the point A is, and whose base is what the circle BC is, and let the cone have been cut by a plane through the axis, and let it make as common section the triangle ABC; and from some point of those on the periphery BC—say the [point] M—let there be a perpendicular drawn to the [line] BC—namely the [line] MN. Let there have been taken, then, on the surface of the cone some point—the [point] D, say—and, through the [point] D and parallel to the [line] MN, let there be drawn the [line] DE. I say that the line [DE] extended will meet with the plane of the triangle ABC, and, further extended toward the other side of the cone so far that it may fall together with its surface, will be cut in two [halves] by the plane of the triangle ABC.

—Let there be joined the [line] AD, and let it be extended. It will therefore fall together with the periphery of the circle BC. Let it fall together with it at the [point] K, and from the [point] K to the [line] BC let a perpendicular be drawn—namely the [line] KHL. Parallel, therefore, is the [line] KH to the [line] MN—and therefore also to the [line] DE.

—Let there be joined, from the [point] A to the [point] H, the [line] AH. Since then, in the triangle AHK, a parallel to the [line] HK is what the [line] DE is, therefore the [line] DE extended will fall together with the [line] AH. But the [line] AH is in the plane of the triangle ABC. The [line] DE will, therefore, fall together with the plane of the triangle ABC.

—On account of the same things, it [the line DE] falls together with the [line] AH too. Let it fall together with it at the [point] F, and let the [line] DF be extended in a straight prolongation so far that it may fall together with the cone's surface. Let it fall together with it at the [point] G. I say that equal is the [line] DF to the [line] FG.

—For since the points A, G, L are on the cone's surface, but are also in the plane extended through the [lines] AH, AK, DG, KL—which very thing is a triangle that goes through the vertex of the cone—therefore the points A, G, L are on the common section of the cone's surface and of the triangle. Straight, therefore, is the [line] through the [points] A, G, L. Since, then, in the triangle ALK there has been drawn parallel to the base KHL the [line] DG, and also there has been drawn through, [there has been drawn, that is, across those parallel lines DG and KHL] some [line] from the [point] A, namely, the [line] AFH—therefore as is the [line] KH to HL, so is the [line] DF to FG. But equal is the [line] KH to the [line] HL—since, in the circle BC, a perpendicular to the diameter is what the line [KL] is—therefore equal also is the [line] DF to the [line] FG.

6th proposition: enunciation

What the enunciation tells us, is this:

—*if*, in a cone through whose axis a plane has been passed
(thus making an axial triangle—which
may be, or may not be, perpendicular to the cone's base),
a straight line of reference is drawn
perpendicular to the axial triangle's straight-line base,
from some point on the circumference of the cone's circular base,

and a straight line is drawn parallel to the reference line,
from any point on the cone's surface that is
not on a side of the axial triangle,

—*then* this straight line that is parallel to the reference line
will meet the axial triangle,
and when it is prolonged through the axial triangle
all the way across to the other side of the surface,
it will be bisected by the axial triangle.

But why is it, that when we pick points on the conic surface from which to draw the parallels, we must not pick a point that is on a side of the axial triangle? Because a parallel through such a point would not go inside the surface, and hence it would not stretch some definite distance across to the other side of the surface; and therefore it would not be bisected by the axial triangle.

6th proposition: demonstration

We let there be a cone, having point A as vertex and circle BC as base. We cut the cone with a plane through its axis, thus making as common section the triangle ABC. We draw, from some point M on the circumference, a straight line of reference MN that is perpendicular to the triangle's base BC; and we also draw, parallel to that line MN, and through some random point D on the surface of the cone, another straight line DE. From that (we specify) this is to be shown: straight line DE will, if prolonged, meet the plane of triangle ABC, and will, if prolonged yet further, until it meets the surface, be bisected by triangle ABC. To show that, we need to draw some more lines (fig. 3). (It will be simpler to draw the figure with the axial triangle lying flat in the plane of the page, so that it is perpendicular to the base of the cone. But remember: nothing here depends on whether or not the axial triangle is perpendicular to the base. It may be perpendicular—but it does not have to be.)

First, we show that the straight line, if prolonged, *meets* the axial triangle.

Join A (the vertex) and D (the point on the surface). This line AD, if prolonged, will (by the 1st proposition) meet the circumference of circle BC at some point (say, K). From this point K, draw the straight line KHL that is perpendicular to BC (which is itself perpendicular to MN). KH will be parallel to MN. But DE too is parallel to MN, and hence (since lines parallel to the same line will be parallel to each other [Euc XI.9]) KH will be parallel to DE. Join AH (join, that is to say, the point that is the vertex and the point where BC is hit by the perpendicular line from K); and now AHK is a triangle in which DE (being parallel to KH) will, when prolonged, meet that line AH. But since DE meets a line (AH) which is in the plane ABC, DE will meet the plane of triangle ABC.

And now, we show that the straight line, if prolonged yet further, is *bisected by* the axial triangle.

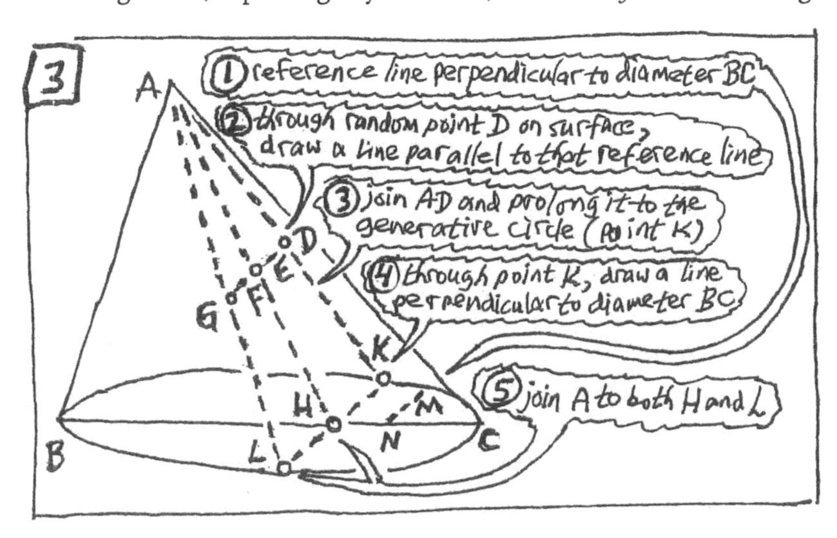

DE, if prolonged, meets AH at some point (say, at F).
DF, if prolonged, meets the cone's surface at some point (say, at G).
Points A, G, L are on the surface of the cone, and are also in
the plane extended through straight lines AH, AK, DG, KL—
a plane that (by the 3rd proposition) is a triangle through the cone's vertex.
Hence those points A, G, L are on the common section of
the cone's surface and the triangle ALK—
which intersection is a line that is straight.
In triangle ALK, line DG is parallel to base KL,
and both lines (DG and KL) are met by line AFH drawn from point A;
hence (since DG is a straight line parallel to a side of a triangle)
as KH is to HL, so also DF is to FG.
But KH is equal to HL (since straight line KL is—in circle BC—
a chord perpendicular to the diameter).
Also, therefore, DF is equal to FG.

Much argument is not needed to establish that in any triangle (such as AHK in figure 4) a straight line DE drawn parallel to one side (HK) from a point on another side (AK) must, when prolonged, meet the third side (AH). Why? Because of the following manifest fact: if a straight line goes into a triangle, then it must eventually—if we prolong it far enough—come out. That must be so in any situation that we can visualize, regardless of the relative sizes of the lines and the angles in our visualization. Since Euclid does not state it explicitly beforehand, a man named Pasch suggested in the nineteenth century that it be made an axiom, thus avoiding implicit appeal to visualization. In any case, whether that principle is made explicit or remains implicit, it does guarantee the following assertion: line DE (since it enters through side AK and, when prolonged, cannot meet side HK because of parallelism) must exit through the remaining side, AH.

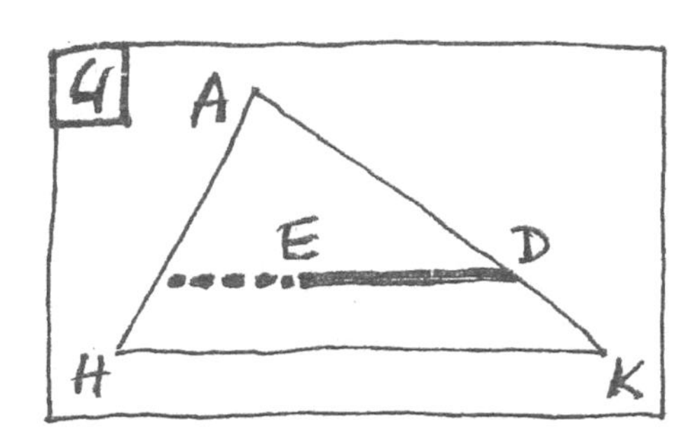

After showing that the straight line meets the axial triangle, the demonstration shows that the straight line is bisected by the axial triangle. The bisection is demonstrated by combining two views of the cone: a view from above, and a view from the side. The top-view shows a symmetry that belongs to chords in circles; the side-view shows a proportionality that belongs to sides of similar triangles.

The circular top-view exhibits this: a chord that is perpendicular to a diameter is bisected by that diameter (Euc III.3). Here (fig. 5), chord KL, since it is perpendicular to the diameter of generative circle BC, is bisected by the diameter at H—which is to say that KH is equal to HL.

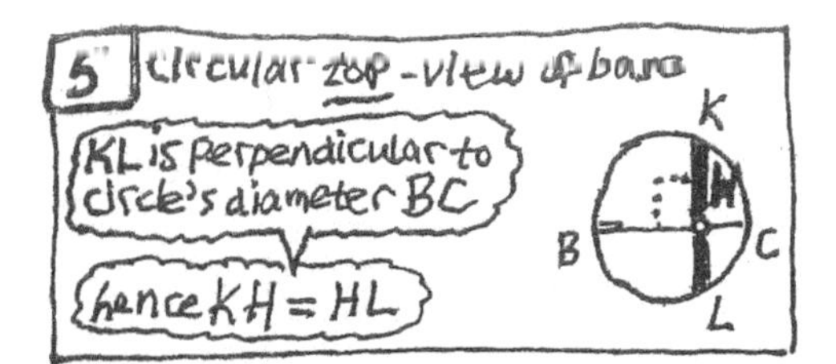

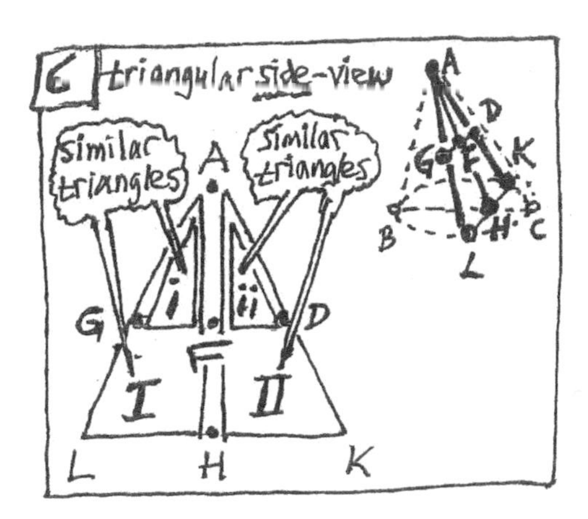

The triangular side-view exhibits this: a line that goes across a triangle and is parallel to its side makes equiangular triangles, which therefore have their corresponding sides proportional (Euc VI.4). Here (fig. 6), since line GD is parallel to side LK of triangle ALK, we have that the little triangle on the left (i) is similar to the big triangle on the left (I), and that the little triangle on the right (ii) is similar to the big triangle on the right (II). In the left-hand triangles [(i) and (I)], therefore, we have that GF is to LH as AF is to AH; and in the right-hand triangles [(ii) and (II)], we have that as AF is to AH, so is FD to HK. But if two ratios are each the same as the same ratio (which here is the ratio of AF to AH), the two are therefore the same as each other; hence the ratio of GF to LH, on the one hand, and the ratio of FD to HK, on the other, are the same.

The core of the demonstration thus is this (fig. 7):

a *side-view* yields	ratios that are the same, and
a *top-view* yields	the equality of the consequents in these ratios;
those two results *together* yield	the equality of the antecedents.

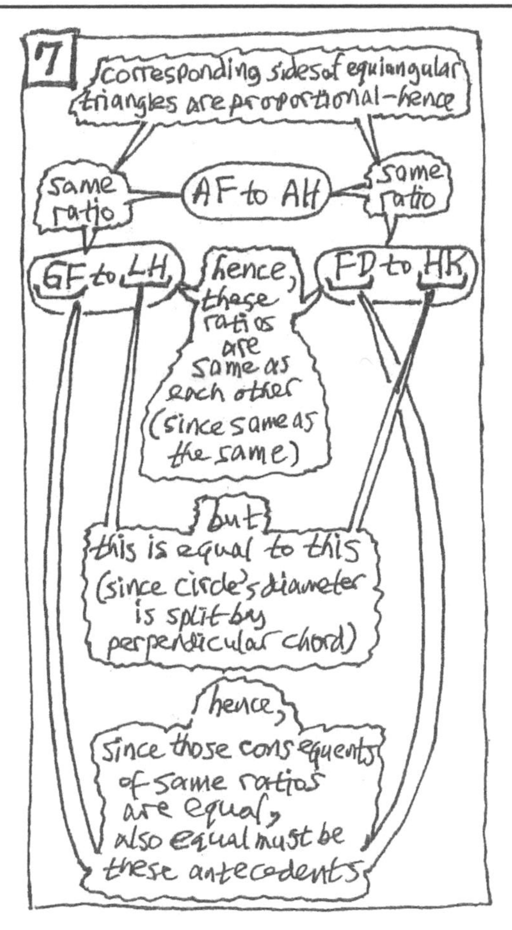

That looks much like what was done in the demonstrations of the last two propositions of the previous chapter—the 4th and 5th propositions. Those are the only propositions so far that have given us, as conic sections, curves of some kind—but the curves were merely of the circular kind. Each of those two previous propositions in its demonstration took a *side*-view in order to use a property of similar triangles, as well as a *top*-view in order to use a property of circles, and then it put the two results *together*. Now in the 6th proposition, again putting two such views together, we have just taken our first step upon the road that will lead us from curved sections that are circular to curved sections that are not those old familiar lines.

7th proposition: preparation

What the 6th proposition has just said generally about the cone's conic *surface*, the 7th proposition now applies to that particular place on the conic surface where a plane cuts into the surface a curved line of *section*. The 7th proposition brings in the cutting plane not only to slice such a curved line on the conic surface, but also at the same time to slice a straight line in the axial triangle. The proposition now says, about those two lines (the one curved, the other straight), what the preceding proposition said about the two surfaces (the one curved, the other flat) into which those two lines are now sliced by the cutting plane (fig. 8).

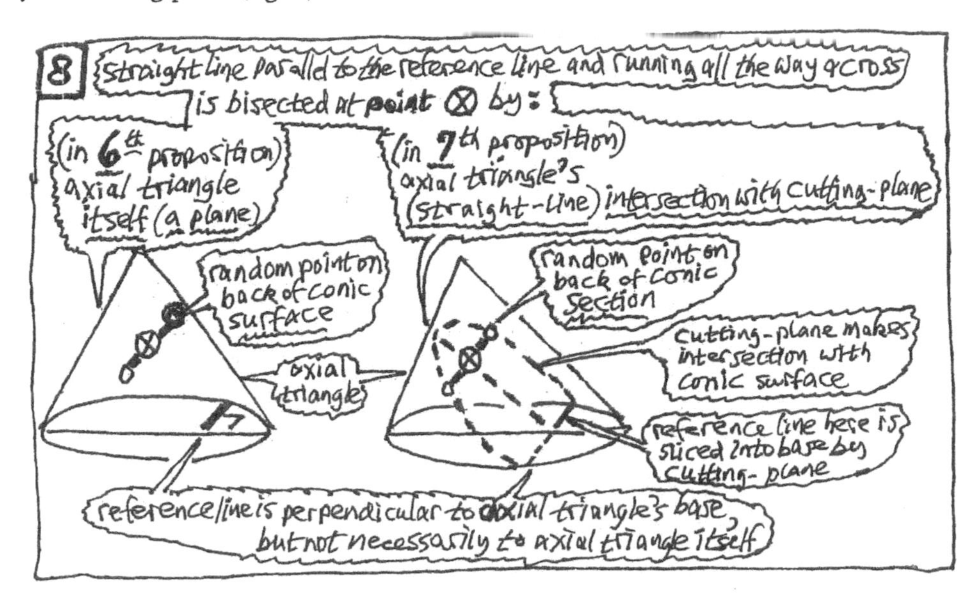

The *preceding* proposition (the 6th) said this:	The *present* proposition (the 7th) says this:
—*if*, inside the conic surface,	—*if*, inside the curved-line intersection of the conic surface and the cutting plane,
straight lines are drawn parallel to a reference line which lies in the cone's base (as its intersection with the cutting plane) and is perpendicular to an axial triangle's base,	straight lines are drawn parallel to a reference line which lies in the cone's base (as its intersection with the cutting plane) and is perpendicular to an axial triangle's base or to its extension
—*then*, when the parallel lines are prolonged to the other side, so that they stretch all the way across inside *the conic surface,*	—*then*, when the parallel lines are prolonged to the other side, so that they stretch all the way across inside *the curved-line intersection of the conic surface and the cutting plane,*
they will be bisected by the *plane which is the axial triangle.*	they will be bisected by the *straight-line intersection of the axial triangle and the cutting plane.*

Since the straight-line bisector is in the same plane as the curved-line section that is cut onto the surface of the cone, what the 7th proposition gives us will be identified afterward as a *diameter* of the curvilinear *conic section*.

We were given diameters earlier, in the 4th and 5th propositions. But the conic sections then were circular. Is there any reason for suspecting now, at the very beginning of the 7th proposition, that we are going to get something new?

Yes, there is reason to suspect it. To begin with, the 7th proposition now opens new possibilities for the orientation of the cutting plane—the plane that cuts into the conic surface to make the curvilinear section. In the situation of the 4th proposition, the cutting plane—because it was parallel to the cone's base plane—could not have any intersection with that plane. And in the situation of the 5th proposition, although the cutting plane would (if extended) have had an intersection with the cone's base plane, this straight-line intersection of the planes would have met the extension of the axial triangle's base, rather than meeting the axial triangle's base itself. But now, in the 7th proposition, the cutting plane may intersect the cone's base plane in a straight line which meets the axial triangle's base itself (fig. 9).

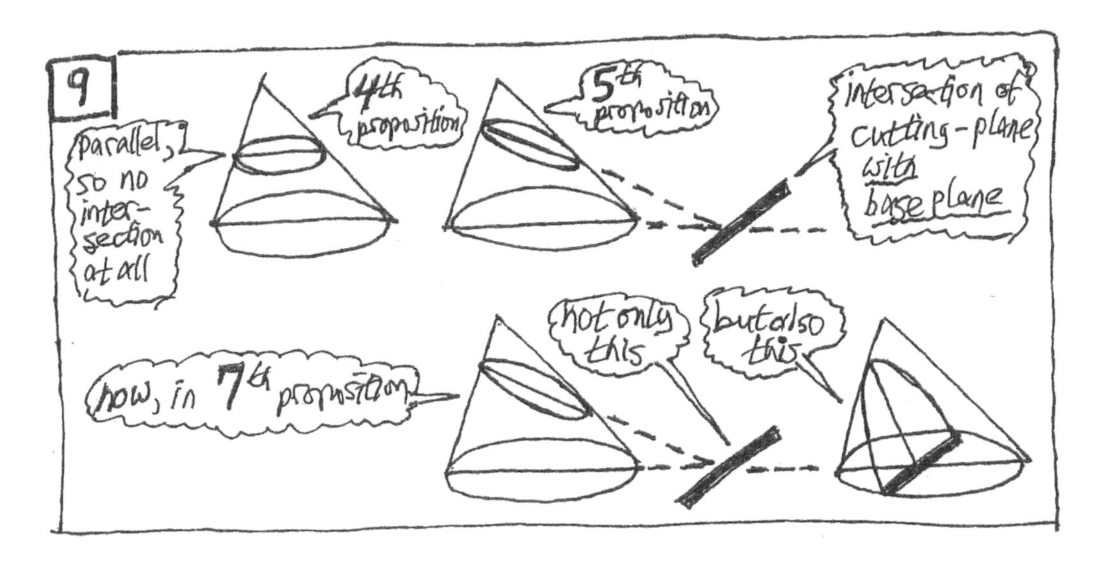

But the 7th proposition goes on to do more than suggest new possibilities for the orientation of the cutting plane, while giving us (for whatever section it might make upon the surface of the cone) a diameter—a straight-line bisector of straight lines that are drawn in an orderly array of parallelism. The 7th proposition goes on as well to open new possibilities for the curve that the plane cuts onto the cone. These new possibilities for the curve, as we shall see, have to do with the angle at which the curve's diameter meets the ordinatewise straight lines that it bisects. Having shown us the ordinatewise chords' symmetry of *length* (on the one side of the bisector and on the other), the proposition will go on to show us the conditions for their symmetry of *angle* (fig. 10).

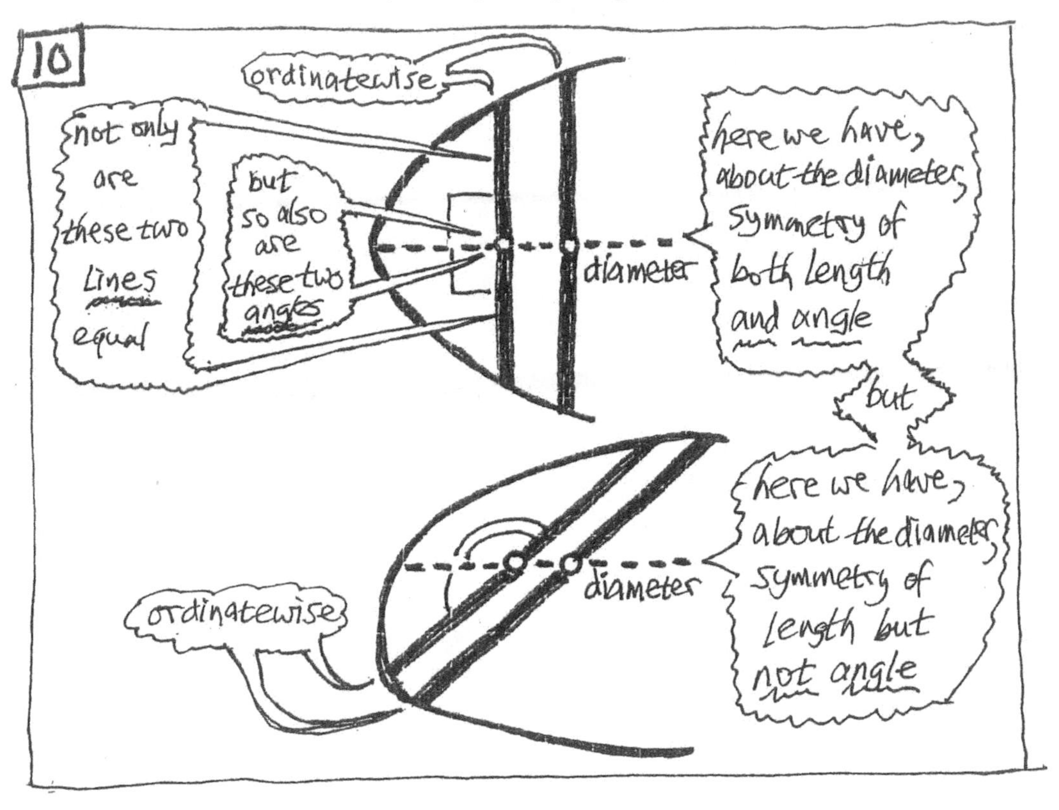

We are now ready for the presentation of the 7th proposition.

7th proposition (fig. 11)

#7 If a cone be cut by a plane through its axis, and be cut also by another plane, one that cuts the plane in which is the base of the cone—cuts it, that is to say, along a straight [line] which is at right [angles] either to the base of the triangle through the axis or to its straight prolongation— and if the straight [lines] that are drawn from the section which has been generated on the cone's surface (the section, that is to say, which has been made by the cutting plane) be parallel to the straight [line] that is at right [angles] to the base of the triangle, then [these straight lines] will fall upon the common section of the cutting plane and of the triangle through the axis, and [these straight lines], further extended to the other side of the section, will be cut in two [halves] by it [that is, by the straight-line common section]; and if (on the one hand) a right one be what the cone is, the straight [line] in the base will be at right [angles] to the common section of the cutting plane and of the triangle through the axis, but if (on the other hand) an uneven [that is, oblique] one be what the cone is, [the straight line in the base] will not always be at right [angles] [to that common section], but will [only] be at right [angles] to it whenever the plane through the axis be at right [angles] to the base of the cone.

—Let there be a cone whose vertex is what the point A is, and whose base is what the circle BC is, and let the cone be cut by a plane through the axis, and let [this axial plane] make as section the triangle ABC. And let the cone be cut also by another plane, this one cutting the plane in which is the circle BC, cutting it, that is to say, along the straight [line] DE which is at right [angles] either to the [line] BC or to its straight prolongation, and let [this other plane] make as section on the surface of the cone the [line] DFE. The common section, then, of the cutting plane and of the triangle ABC is what the [line] FG is. And let there be taken any point on the section DFE—the [point] H, say—and through the [point] H let there be drawn parallel to the [line] DE the [line] HK. I say that the [line] HK will meet with the [line] FG, and, extended to the other side of the section DFE, will be cut in two [halves] by the straight [line] FG.

—For since a cone whose vertex is what the point A is, and whose base is what the circle BC is, has been cut by a plane through its axis, and [the plane] makes as section the triangle ABC, and since there has been taken on the surface some point that is not on a side of the triangle ABC, namely the [point] H, and perpendicular is the [line] DG to the [line] BC—therefore the [line] through the [point] H drawn parallel to the [line] DG (that is, the [line] HK) will meet with the triangle ABC, and, further extended to the other side of the surface, will be cut in two [halves] by the triangle.

—Since, then, the [line] through the [point] H drawn parallel to the [line] DE meets with the triangle ABC and is in the plane through the section DFE, therefore on the common section will it fall, that is, on the common section of the cutting plane and of the triangle ABC. But the common section of the planes is what the [line] FG is. Therefore the [line] through the point H drawn parallel to the [line] DE will fall on the [line] FG; and, further extended to the other side of the section DFE, will be cut in two [halves] by the straight [line] FG.

—Then either the cone is one that is right (*orthos*), or the triangle ABC through its axis is upright (*orthon*) to [in other words, is perpendicular to] the circle BC, or neither.

—First, let the cone be one that is right (*orthos*). Also, then, the triangle ABC would be upright (*orthon*) to the circle BC. Since, then, the [triangular] plane ABC is upright to the [circular] plane BC, and in one of the planes (namely, the plane BC) there has been drawn, at right [angles] to their common section (namely to the [line] BC), the [line] DE—therefore the [line] DE is at right [angles] to the triangle ABC; and also, therefore, [it is upright] to all the straight [lines] that touch it and are in the triangle ABC; and so also it is at right [angles] to the [line] FG.

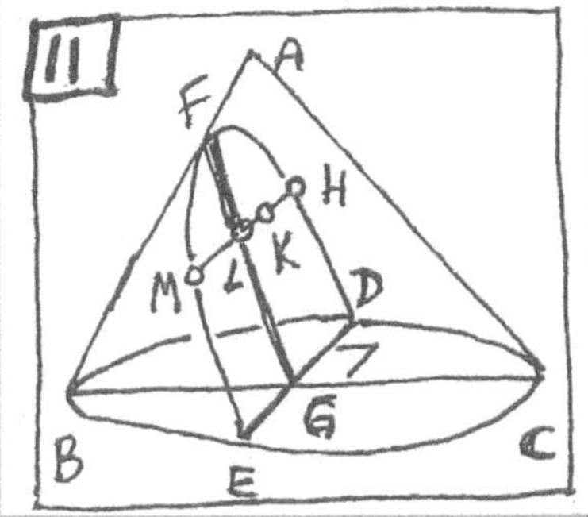

—Then, let not the cone be one that is right. Now then, if the triangle through the axis is upright to the circle BC, likewise will we show that also the [line] DE is at right [angles] to the [line] FG. Then let not the triangle ABC through the axis be upright to the circle BC. I say that neither is the [line] DE at right [angles] to the line FG, for—if possible—let it be. But it is also at right [angles] to the [line] BC. Therefore the [line] DE is at right [angles] to each of the lines BC, FG; and therefore it will be at right [angles] also to the plane through the [lines] BC, FG. But the plane through the [lines] BC, GF is what the triangle ABC is. And therefore the [line] DE is at right [angles] also to the triangle ABC. And also all the planes through it [namely, through the line DE] are at right [angles] to the triangle ABC. But a certain one of the planes that go through the [line] DE is what the circle BC is. The circle BC, therefore, is at right [angles] to the triangle ABC. And so also the triangle ABC will be upright to the circle BC—which very thing is supposed not to be so. Not so, therefore, is it that the [line] DE is at right [angles] to the [line] FG.

Porism: From this it is evident, then, that a diameter of the section DFE is what the [line] FG is, since indeed it cuts in two [halves] the [lines] that are drawn parallel to a certain straight [line], namely to the [line] DE; and it is evident also that it is possible for some parallels to be cut in two [halves] by the diameter (the [line] FG) and yet not be at right [angles] [to it].

7th proposition: enunciation

What the enunciation tells us, is this:

—*if*, through a cone's axis a plane has been passed (thus making an axial triangle),

and in the cone's base plane there is made, by a cutting plane,
a straight line of reference that is
perpendicular to the axial triangle's base or to its base's extension,

and from the section that is made on the cone's surface by that cutting plane
straight lines are drawn parallel to the straight line of reference

—*then* those parallel straight lines will meet the
straight-line intersection of the axial triangle and cutting plane,

and they will, if further prolonged to the conic section's other side, be bisected by
that straight-line intersection of the axial triangle and cutting plane;

and the straight line of reference in the base will,
if the cone is right,
always be perpendicular to
the straight-line intersection of the axial triangle and cutting plane,
but if the cone is oblique,
then it will be perpendicular only when
the plane through the axis is perpendicular to the cone's base plane.

7th proposition: demonstration

We let there be a cone that has point A as vertex and circle BC as base. It is cut by a plane through the axis—making as section (according to the 3rd proposition) a triangle ABC. The cone is cut also by another plane; this other plane cuts the plane of circle BC in a straight line DE that is perpendicular either to BC or to BC's prolongation—and thus makes as section on the cone's surface the line DFE. The intersection of the axial triangle ABC and the cutting plane will be a straight line (call it FG). On the conic section DFE, we take any point (say, H) and through it we draw a straight line that is parallel to DE (say, HK). From that (we specify) this is to be shown: HK will (in the first place) meet FG (the straight line that is the intersection of the axial triangle and cutting plane); and will (in the second place), if prolonged to the other side of conic section DFE, be bisected by that same intersection FG (fig. 11). (Note that here in this 7th proposition, as in the 6th proposition before it, the axial triangle need not be perpendicular to the cone's base: it may be—but it does not have to be.)

The demonstration follows easily from what was demonstrated just before this (in the 6th proposition). The first thing demonstrated here is the bisection of the straight lines drawn ordinatewise across the inside of the conic section. The demonstration then goes on to something which the previous proposition did not treat at all: the angle that the bisected parallel lines make with what bisects them. Let us now look over the whole demonstration, after which we shall examine part of it, and then consider the significance of what is shown in that part.

First, the *bisection* of the ordinatewise chords:

Since the cone with A as vertex and circle BC as base
is cut by a plane through the axis,
making as intersection (by the 3rd proposition) the triangle ABC;
and some point H is on the surface and not on triangle ABC;
and DG is perpendicular to BC—

it therefore follows (by the 6th proposition) that
straight line HK, drawn through H, parallel to DG, will meet triangle ABC,
and will, if prolonged to the other side of the surface,
be bisected by that triangle.

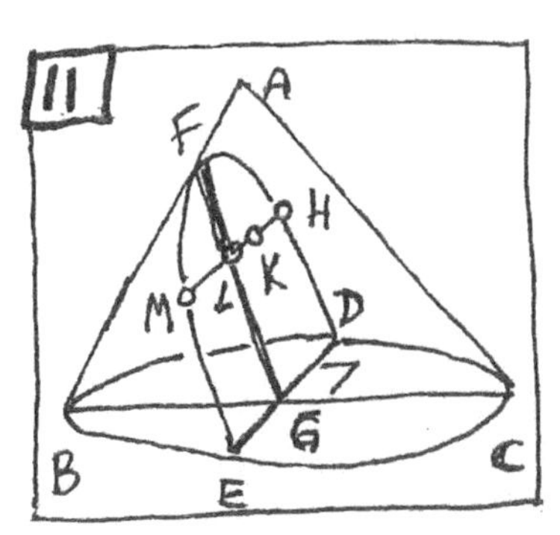

Since HK not only meets triangle ABC,
but also is in the plane of section (plane DEF)—

it therefore will fall on
the intersection of triangle ABC and the cutting plane
(which intersection is straight line FG),
and will, if prolonged to the other side of section DFE,
be bisected by that straight line FG.

And now, the *angle* made by the ordinatewise chords with their bisector. This shows that the ordinatewise chords will be perpendicular to their bisector if—and only if—the axial triangle is perpendicular to the cone's base:

If the given cone is *right*, then DE is perpendicular
not only to the axial triangle's straight-line base BC
but also to the axial triangle's whole plane ABC,
and hence to every straight line in it that meets DE—
and such a line is FG
(which is the axial triangle's intersection with the cutting plane);
hence DE *must* be perpendicular to FG.
But why must DE be perpendicular to the axial triangle if the cone is right?
For the following reason.
If a straight line is perpendicular to a plane,
then all planes through that straight line will be perpendicular to that plane—
and so, since triangle ABC is axial,
and the axis of a right cone is a straight line perpendicular to the cone's base,
it follows that
plane ABC (the axial triangle) is perpendicular to plane BC (the circular base).
But DE, which is drawn in one of the planes (the cone's base), is
perpendicular to straight line BC
(which is the intersection of the axial triangle and the cone's base).
Hence (by the definition of the perpendicularity of a plane to a plane)
straight line DE is perpendicular to axial triangle ABC.
So (by the definition of the perpendicularity of a straight line to a plane)
DE is perpendicular to all straight lines in triangle ABC that meet it,
and hence DE is perpendicular to FG (which is such a straight line).

If, on the other hand, the given cone is *not* right, then we have the following:
when the axial triangle is perpendicular to circle BC,
then DE is perpendicular to FG, which could be shown as before;
but when the axial triangle is not perpendicular to circle BC,
then DE is not perpendicular to FG, which is shown as follows (by *reductio*).
If possible, let DE be perpendicular to FG.
Since DE is perpendicular not only to FG but also to straight line BC,
it is perpendicular to the plane through FG and BC—
which is triangle ABC.
Then all planes through DE are perpendicular to triangle ABC.
And circle BC, since it is a plane through DE, is then
perpendicular to triangle ABC.
But this is contrary to the supposition that
the circle and axial triangle are not perpendicular.
Hence DE *cannot* be perpendicular to FG.

Some people find it easy to visualize the three-dimensional relationships upon which depends this demonstration (that the lines drawn ordinatewise will be perpendicular to their bisector if—and only if—the axial triangle is perpendicular to the cone's base). But other people find it hard to visualize. If you would like a more thorough account of what is involved in establishing the perpendicularity, you will find it in the appendix at the end of this chapter.

7th proposition: the perpendicularity—its significance

It does not matter whether the cutting plane meets the cone's base plane in the axial triangle's base itself (that is, inside the cone's circular base) or in the extension of the axial triangle's base (that is, outside the cone's circular base). The reference line—which is the intersection of the cutting plane and the base plane—will still be in the cone's base plane, and it will still be perpendicular to the axial triangle's straight-line base or to its extension (fig. 12).

Nor does it matter at what angle the cutting plane meets the axial triangle's base. The ordinatewise lines' bisector—which is the intersection of the cutting plane and the plane of the axial triangle—will still be in the plane of the axial triangle; and it will still be perpendicular to the reference line if and only if the reference line itself is perpendicular to the axial triangle's plane (and not merely to the axial triangle's base, or to its extension) (fig. 13).

The perpendicularity of the axial triangle to the cone's base is the necessary and sufficient condition of the ordinatewise lines' perpendicularity to their bisector—since, as we have seen:

the reference line *must* be perpendicular to that bisector *if* the axial triangle is perpendicular to the cone's base;	the reference line *cannot* be perpendicular to that bisector *unless* the axial triangle is perpendicular to the cone's base.

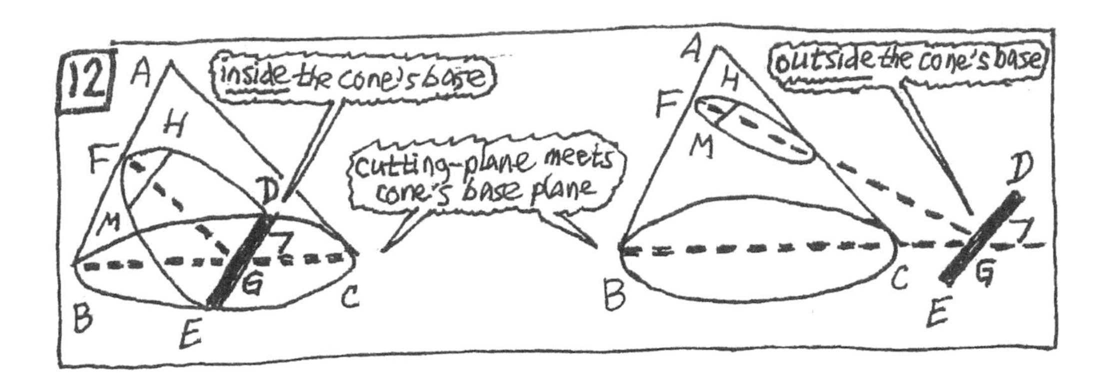

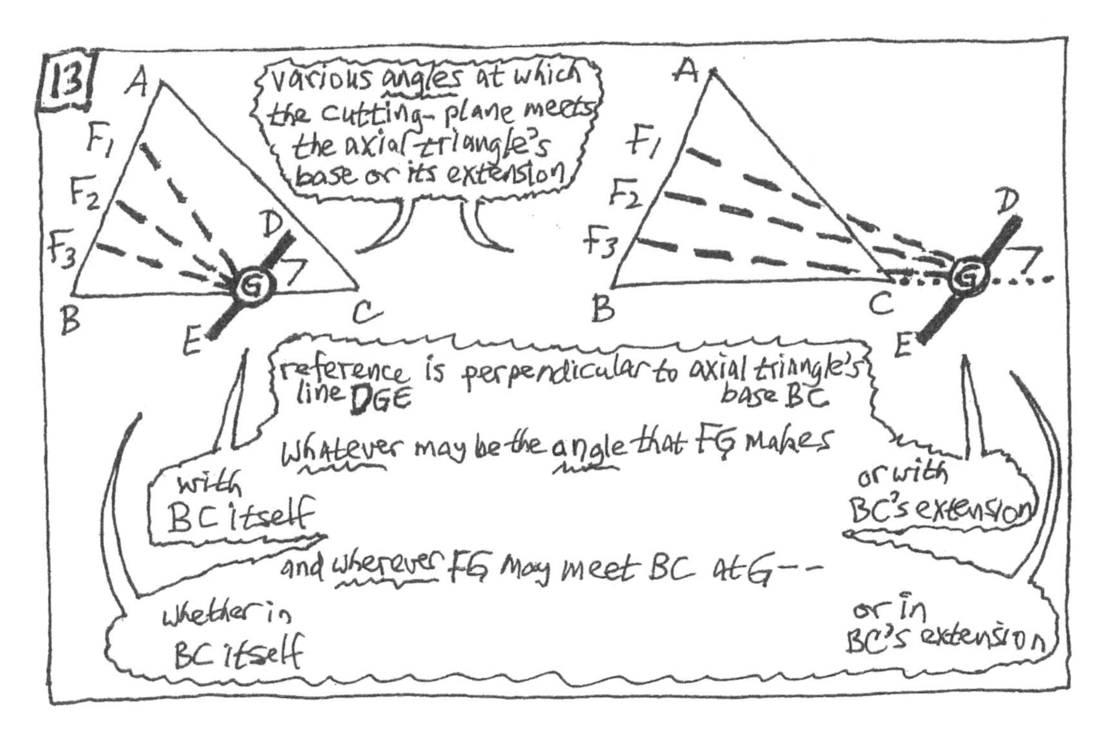

Although the bisector as such hits the ordinatewise chords so as to make equal *lengths* of line on either side of itself, it does not always make equal *angles* on either side of itself. The symmetry of angles, unlike the symmetry of lengths, is conditional. The ordinatewise lines will make equal angles (that is, be perpendicular to their bisector) if—and only if—the straight-line intersection of the cutting plane with the cone's base plane is perpendicular not merely to the straight-line base of the axial triangle, but to the whole plane of the axial triangle. This condition will be met if—and only if—the axial triangle is perpendicular to the cone's base. The condition is necessary and sufficient.

So, to find out when the ordinatewise chords will be perpendicular to the diameter, we need only ask this question: when will the axial triangle be perpendicular to the cone's base?

In a right cone, the answer is: always. In a right cone, every axial triangle is perpendicular to the cone's base, since a right cone is one whose axis is perpendicular to the cone's base. But the answer is different in an oblique cone. Since an oblique cone is one whose axis is tilted, there is only one axial triangle whose axis is perpendicular to the cone's base. This unique triangle is the one whose plane gives us what we are calling the side-view of the cone, the view in which the cone does not at all lean up out of the page or down into the page, but leans only to the right or to the left. That is to say, we slice a cross-section into the cone so that it leans to neither side but only toward or away from us, and then we view that cross-section from the side. In an oblique cone, only that single one of all the axial triangles is perpendicular to the base.

7th proposition: transition to succeeding propositions

The 6th proposition was needed for this 7th proposition. This 7th proposition will be needed for the 8th and 9th propositions, which will present us new lines—and which, with the 10th, will prepare for the 11th proposition. Only then will the new lines begin to be characterized.

The 4th and 5th propositions got circles by cutting the cone with a plane that, in direct side-view, made an angle equal to one of the base angles—either equal to the one directly below it (as when the plane in the 4th proposition cut parallel to the base) or equal to the one below and opposite it (as when the plane in the 5th proposition cut subcontrariwise in an oblique cone). (Fig. 14.)

The 6th proposition did not give us a section of any sort. Rather, it told us something about the relation between an axial triangle and the conic surface. It told us that an axial triangle, even if it is not simply a plane of symmetry for a cone, is nonetheless a plane that provides a kind of symmetry for the cone. Straight lines bounded by a cone's conic *surface*, if they are drawn parallel to a straight line that is perpendicular to the base of an axial triangle, are bisected by the axial triangle (whether or not the axial triangle, by being perpendicular to the cone's base, is a plane of symmetry for the cone).

The 6th proposition provided what we need in order to speak, here in the 7th proposition, about a straight line that provides a kind of symmetry for a conic *section.* Such a straight line is needed in order to discuss the new lines that cutting a cone will give us.

That straight line which is the bisector of the straight lines drawn ordinatewise—when will it be perpendicular to the lines that it bisects? That is, when will the angles, and not merely the lengths of line, be equal on both sides of the bisector? The angles will be equal when the plane of the axial triangle is perpendicular to the plane of the cone's base. That is, only when the plane of the cone's base is perpendicular to the plane of the axial triangle—and not merely when one straight line in the plane of the cone's base is perpendicular to the one straight line which is the axial triangle's base. In a right cone, the two planes must always be perpendicular; but in an oblique cone, since they need not be perpendicular, there need not be the symmetry of the angles astride the bisector.

7th proposition: the porism

We encountered a porism earlier—at the end of the very first proposition. It helpfully provided the means for demonstrating the proposition that came immediately after it, the 2nd proposition.

Now we have another porism. This porism to the 7th proposition identifies the bisecting line as a diameter (note that again the Greek does not use the article, and should therefore be translated "*a* diameter," not "*the* diameter"); and the porism asserts that something is possible for the ordinatewise chords and their diameter: the parallels can be bisected by this diameter and yet not be perpendicular to it.

The porism thus gives us something new—by saying, in effect, that a cone can be cut to yield a curved line that is not a circle.

Let us see why that novelty is implicit in the porism. A *straight* line of course cannot have a diameter, since there cannot be any chords in a straight line. In a *circular* curve, on the other hand, all the chords that are drawn parallel to some straight line of reference can be bisected by a single straight line only if the chords are perpendicular to their bisector. In other words, every diameter of a circular line must be perpendicular to the straight lines drawn from it ordinatewise. So, if we find a line that *does* have a diameter, but this diameter is *not* perpendicular to the lines drawn from it ordinatewise, then we have found a *curved* line that is *not a circle*. And that is what we find in the porism. When we cut an oblique cone with a plane that is not parallel to the base or subcontrariwise—having taken an axial triangle that is not perpendicular to the base—then (since the lines drawn ordinatewise are not perpendicular to the diameter) the line of section is something new. It must be a non-circular curve.

Depending on whether or not the axial triangle is a plane of symmetry for the curve, the diameter will be or will not be simply a line of symmetry for the conic section. In every *circle*, every diameter is a line of symmetry. However, if some diameter in some curve *is* a line of symmetry, then, even so, the curve, while it *may* be a circle, nonetheless *may not* be a circle. But if some diameter in some curve is *not* a line of symmetry, then we can be sure of this: the curve *cannot* be a circle. We are surely on the track of new lines.

Thus, we now know that we can get a line that has a diameter that is not simply a line of symmetry for the curve. When we get such a line with such a diameter, that line cannot be straight or circular. What we know about it so far is only what it is not. We do not yet know what it is. We do know, however, how we may get to know what we do not yet know about it—namely, by way of the relation between what we do not know yet and what we do know already. What we know already are the straight line and the circle. Prior knowledge about the straight line and the circle, which has enabled us to get to know *that* there is this new line, will enable us also to get to know *what* it is.

how to draw the cutting of a cone

In drawing the cutting of a cone to get a non-circular section, you may find figure 15 helpful. It suggests that lines be drawn in a certain order.

> After drawing the *cone* (steps 1 and 2) and the *axial triangle* (step 3),
> draw, in the following order, the following intersections of the cutting plane:
>
> first (step 4) draw its straight-line intersection with the cone's base plane—
> that will be the *reference line* for the lines drawn ordinatewise;
>
> then (step 5) draw its straight-line intersection with the axial triangle—
> that will be the *diameter*;
>
> and only then (step 6) draw its curved-line intersection with the conic surface—
> that will be the conic *section*.

Even though the axial triangle need not be perpendicular to the base of the cone, you can simply draw the special case in which it is. That way, the axial triangle will lie flat on the page. The tilt of the axis will not at all be up out of the page toward you, or back down out of the page away from you, but only in the page toward the right or left. But remember, as you consider the figure, that nothing depends upon the axial triangle's being perpendicular to the base of the cone.

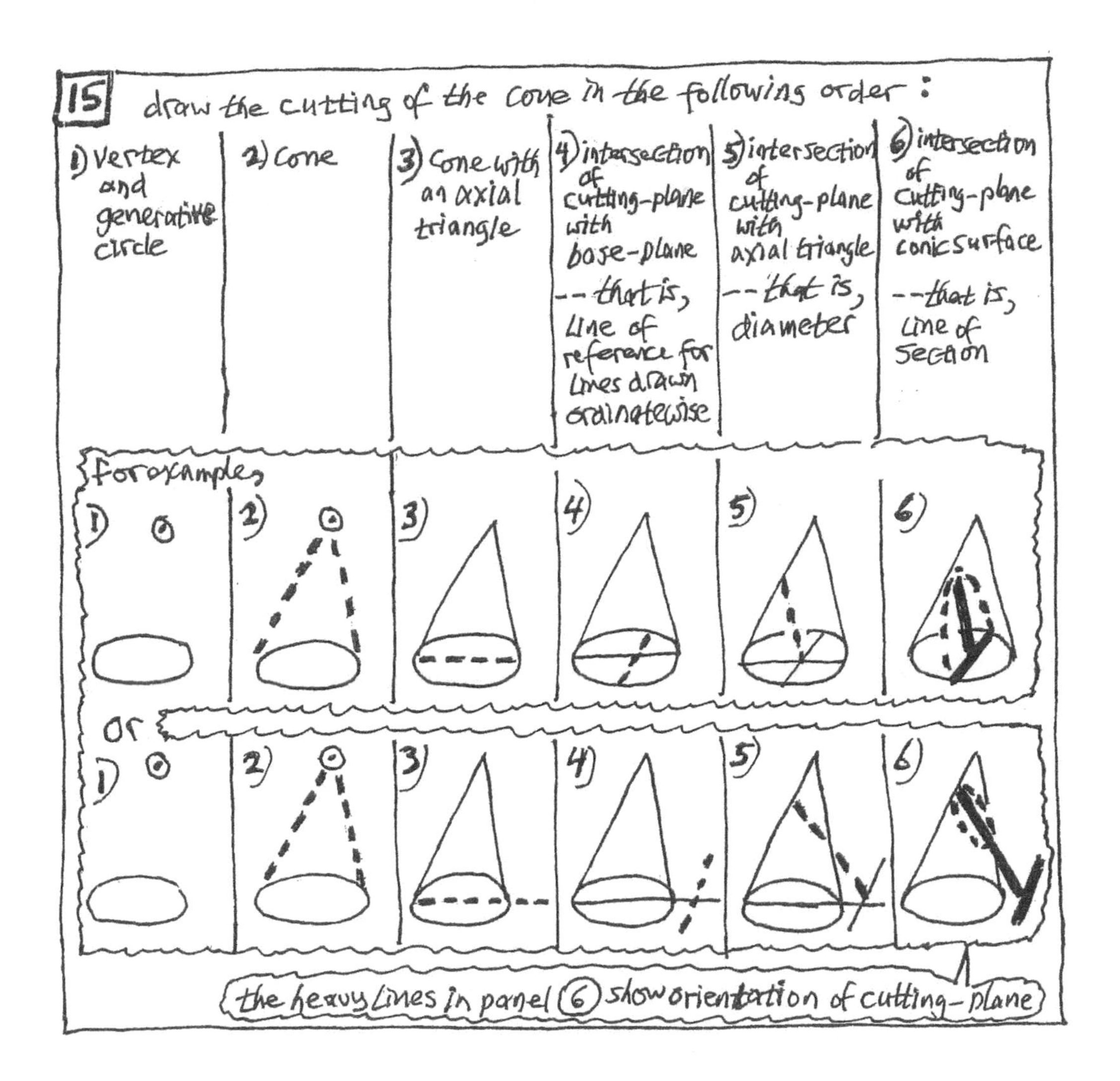
15
draw the cutting of the cone in the following order:
1) vertex and generative circle
2) cone
3) cone with an axial triangle
4) intersection of cutting-plane with base-plane --that is, line of reference for lines drawn ordinatewise
5) intersection of cutting-plane with axial triangle --that is, diameter
6) intersection of cutting-plane with conic surface --that is, line of section
for examples,
1)
2)
3)
4)
5)
6)
or
1)
2)
3)
4)
5)
6)
the heavy lines in panel 6 show orientation of cutting-plane

appendix on demonstrating the perpendicularity in the 7th proposition

Suppose that the axial triangle is perpendicular to the cone's base (either because the cone is a right cone, and hence every one of the cone's axial triangles is perpendicular to the cone's base—or because, although the cone is an oblique cone, the particular axial triangle taken is that unique one which is perpendicular to the cone's base.) If, either way, the axial triangle thus is perpendicular to the cone's base, then the reference line is perpendicular to the bisector of the lines drawn ordinatewise. Why?

Imagine a thin book that has, on its cover, thin stripes perpendicular to its spine, and you hold the book with its spine on a tabletop. The book is a plane inclined to another plane, which is the tabletop. The book's spine is the straight-line intersection of the two planes. The stripes are straight lines which are in the plane of the book, and are perpendicular to the intersection of the plane of the book with the plane of the tabletop. If the book's stripes are also perpendicular to the tabletop, then the plane of the book is perpendicular to the plane of the tabletop. Thus, figure 16 illustrates the Euclidean definition of the perpendicularity of a plane to a plane:

one plane is perpendicular to another plane
if every straight line which is drawn, in the one plane,
perpendicular to the intersection of the two planes,
is perpendicular to the other plane. (*Elements* XI, defs.)

(The perpendicularity of a *plane* to a plane is thus defined in terms of the perpendicularity of a *straight line* to a plane.)

Here in the demonstration of the 7th proposition, what we have is this: the reference line is in the plane of the cone's base, and is perpendicular to this plane's intersection with the plane of the axial triangle; and the plane of the cone's base is perpendicular to the plane of the axial triangle; it follows, by the definition, that the reference line is perpendicular, not merely to the axial triangle's base (which is the line where the cone's base plane intersects the plane of the axial triangle) but to the whole plane of the axial triangle.

From this in turn it follows (by another Euclidean definition—the definition of the perpendicularity of a straight line to a plane) that the reference line is perpendicular as well to the bisector of the lines drawn ordinatewise. How?

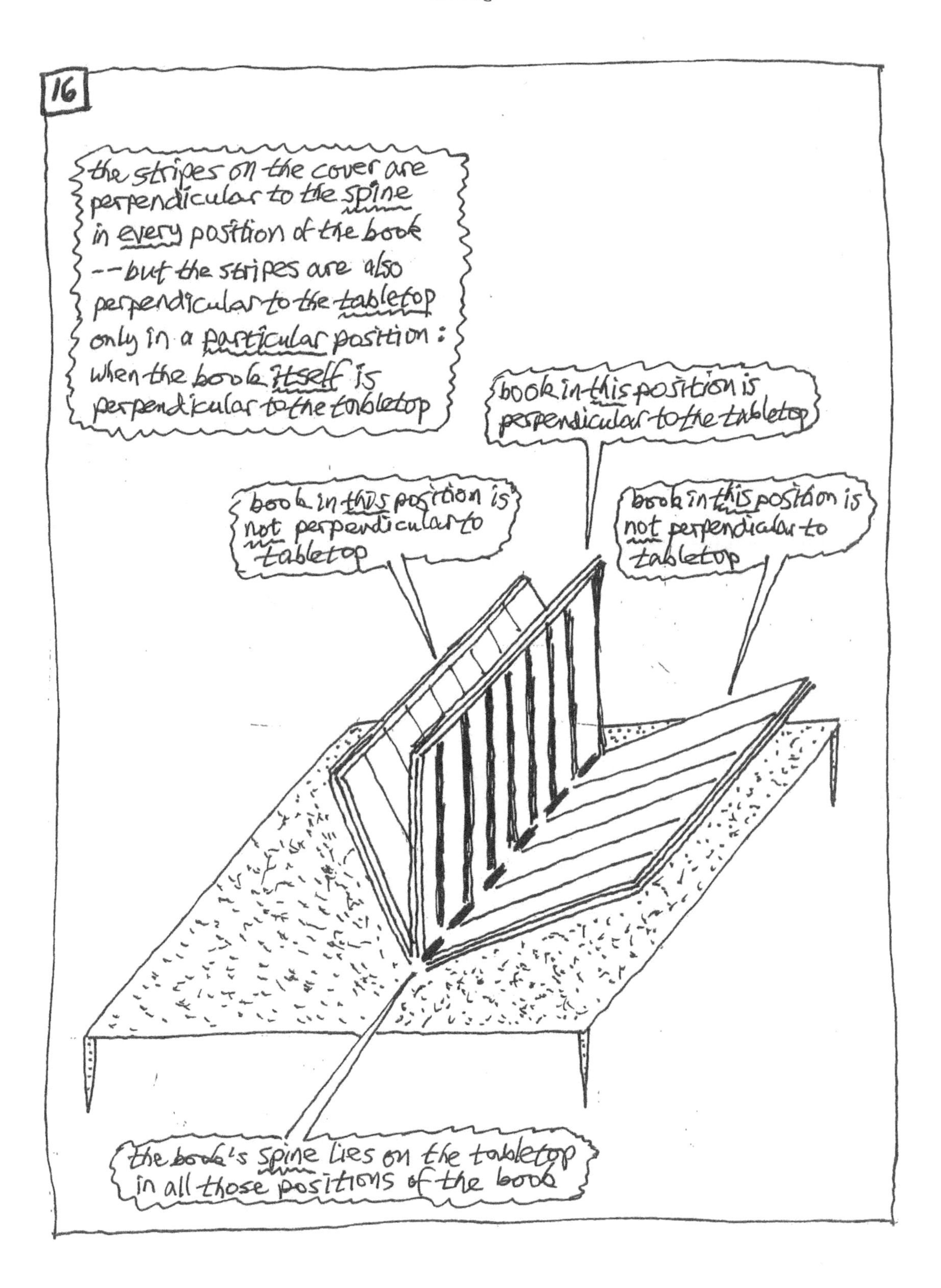
16
the stripes on the cover are perpendicular to the spine in every position of the book --but the stripes are also perpendicular to the tabletop only in a particular position: when the book itself is perpendicular to the tabletop
book in this position is perpendicular to the tabletop
book in this position is not perpendicular to tabletop
book in this position is not perpendicular to tabletop
the book's spine lies on the tabletop in all those positions of the book

Imagine that you hold a pencil so that it stands with its tip on a tabletop—and that you now take other pencils and lay them on the tabletop so that they all meet the standing pencil at its tip. If the standing pencil is perpendicular to every one of the other pencils, then it must be standing perfectly upright: it must be perpendicular to the tabletop. This illustrates the Euclidean definition (*Elements* XI):

> A straight line is perpendicular to a plane
> if it is perpendicular to every straight line which
> meets it and is in that plane.

In figure 17, for example,

> FG (diameter) and BC (axial triangle's base) are pencils that
> lie on a tabletop (plane of axial triangle), where they meet at G.
> Perpendicular to BC (axial triangle's base) is pencil DG (reference line).
> If DG is also perpendicular to *any* pencil that meets G and *lies* on the tabletop,
> then it will be perpendicular to the tabletop (which is
> the whole plane on which FG and BC lie—the plane of the axial triangle).

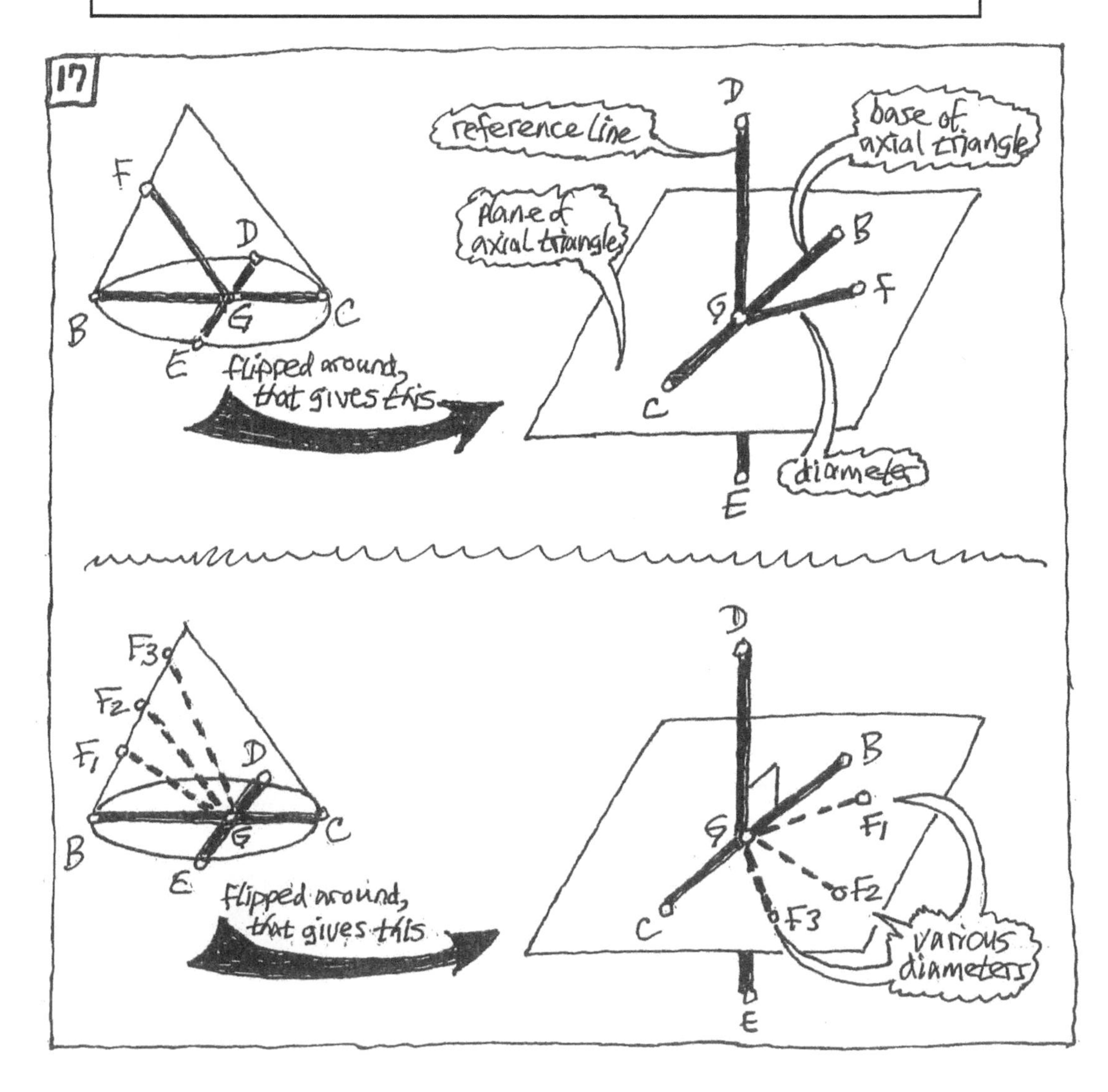

Note that point G need not be inside BC—it can be on the prolongation of BC, as in figure 18.

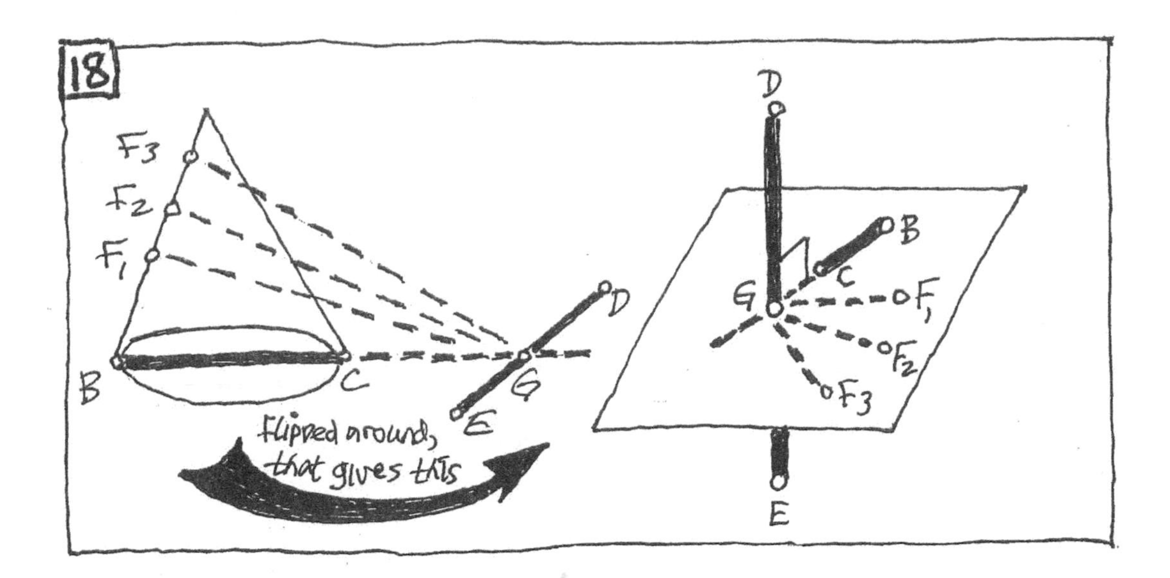

So far in the demonstration of the 7th proposition, what we have is this: the reference line is perpendicular to the plane of the axial triangle. In the plane of the axial triangle, however, is located the intersection of the axial triangle with the cutting plane; and this straight-line intersection, which is the bisector of the lines drawn ordinatewise, meets the reference straight line. Hence it follows (by the definition of the perpendicularity of a straight line to a plane) that the reference line is perpendicular to the bisector of the lines drawn ordinatewise—when, that is, the axial triangle is perpendicular to the cone's base.

So far, we have shown this:	Now, we must show this:
if the axial triangle is	if the axial triangle is
perpendicular to the cone's base	*not* perpendicular to the cone's base
(as *must* be the case in a right cone,	(as *may* be the case—
and as *may* be in an oblique cone),	but *only* in an oblique cone),
then the reference line is	then the reference line is
perpendicular to the bisector of	*not* perpendicular to the bisector of
the lines drawn ordinatewise.	the lines drawn ordinatewise.

This last thing will be shown by a *reductio* proof—one that involves the following absurdity:

> If we suppose that
> that the axial triangle is *not* perpendicular to the cone's base,
> and we suppose *also*
> that the reference line is perpendicular to the ordinatewise lines,
> then it will follow
> that the axial triangle *is* perpendicular to the cone's base.

To show this, we need a Euclidean proposition. Imagine that a book is opened so that it stands upright on a tabletop. Consider its covers to be planes, ;and its spine to be a straight line perpendicular to the plane of the tabletop. If, while keeping the spine upright on the same spot, you make the angle between the covers larger or smaller, you will be taking different planes through the straight line that is perpendicular to the tabletop—yet every one of those planes will be perpendicular to the plane of the tabletop. Thus, figure 19 illustrates the Euclidean proposition (*Elements* XI.18):

> if a straight line is perpendicular to a plane,
> then every plane through that straight line
> is perpendicular to that plane.

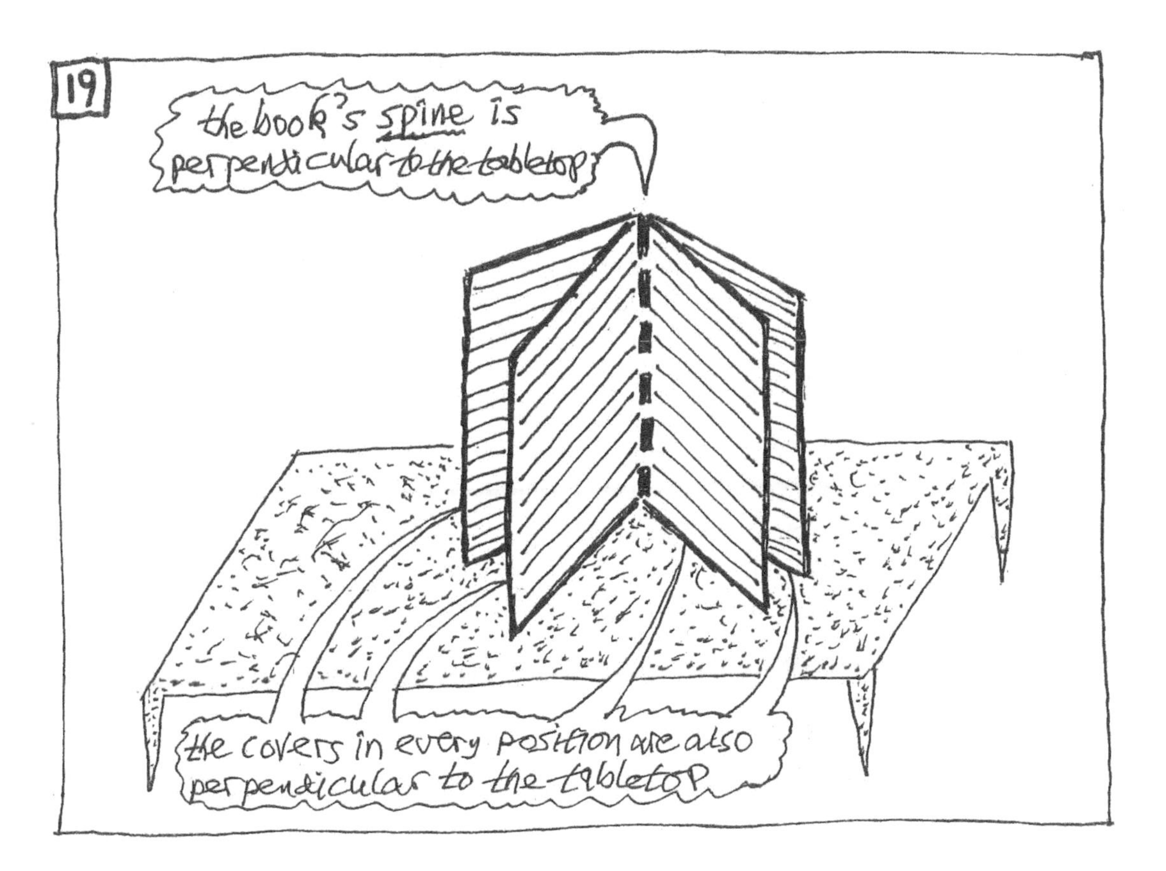

Now we are equipped for the proof, the reductio ad absurdum:

The reference line is supposed to be perpendicular to
the bisector of the lines drawn ordinatewise
(which is the intersection of the cutting plane and the axial triangle).
But it was originally supposed to be drawn perpendicular to
the axial triangle's base
(which is the intersection of the cone's base plane and the axial triangle).
Being perpendicular to both straight lines of intersection made by the axial triangle
(one of them made with the cutting plane, and the other with the cone's base plane), the reference line is
perpendicular to the plane through those two straight lines—
which is the plane of the axial triangle.
However, not only is the reference line
thus perpendicular to the plane of the axial triangle,
it is also located in the plane of the cone's base—
and hence
the plane of the cone's base is
perpendicular to the plane of the axial triangle.
This conclusion contradicts the supposition that it is not so perpendicular.
The conclusion follows by way of the Euclidean proposition (*Elements* XI.18)
with which this account of the demonstration began:
if a straight line is perpendicular to a plane,
then every plane through that straight line is perpendicular to that first plane.

CHAPTER 5. GETTING NEW LINES AS SECTIONS

where we are

The Latin word *sectio(n-)* and the Greek word *tomê* signify a "cut" or a "cutting." (That Greek word is the source of *a-tom*—which was something conceived of as "*un*able to be *cut*.") Conic sections are the lines that are the inter-sections when a plane cuts (and thus is cut by) a conic surface. When there is such a cutting, there is a line called a "cut."

The first lines that Apollonius got by cutting a cone were the straight lines that constituted a triangle in the 3rd proposition; he then got circles when he cut a cone in the 4th proposition, and again in the 5th proposition. Then, for two propositions after getting those old lines, he did not try to get new kinds of lines by cutting a cone. Instead, in preparation for describing conic sections, he got a straight line inside a nondescript conic section. The 6th proposition enabled him to get, in the 7th, a diameter.

Diameters now help to show, in the 8th and 9th propositions, how cutting cones with planes can yield new sorts of lines. These new lines of section differ from the straight line and the circle in the following ways.

The line of section that is	The line of section that is
cut in the 8th proposition,	cut in the 9th proposition
since it has a diameter,	also
must be a curve;	is a curve;
but although it is	but although it is
not straight,	not circular,
it nonetheless	it nonetheless
is indefinitely extendable	is closed
like a straight line	like a circle
(rather than	(rather than
being closed	being indefinitely extendable
like a circle).	like a straight line).

Apollonius, here introducing lines that are neither straight nor circular, thus presents them in two varieties: a line of the one variety is a curve that is somehow like a straight line and unlike a circle; a line of the other variety is a curve that is somehow like a circle and unlike a straight line.

In the 10th proposition (which will be treated in the next chapter, where it will be the last proposition to be treated here in Part One of this guidebook), Apollonius will present a property that belongs to all the curved lines that a plane cuts into a conic surface; he will show that all of the curved conic sections—like the curved conic surface into which they are cut by the plane—are perfectly free of wiggles. Then, in the four propositions that come right after that, Apollonius will present the differences and the similarities to be found among the curves of conic section—by speaking about relations of size.

We are now ready for the presentation of the 8th proposition.

8th proposition

(Again, here and hereafter: the translated proposition should first be merely skimmed. Then read it closely after reading the subsequent explication. The drawings in the explication will help you visualize what is being said in the proposition, and the words in the explication will help you understand the significance of what is being said in the words of the proposition.)

#8 If a cone be cut by a plane through its axis, and be cut also by another plane, one that cuts the base of the cone along a straight [line] that is at right [angles] to the base of the triangle through the axis, and the diameter of the section that is generated on the surface either be parallel to one of the sides of the triangle or fall together with one of the sides continued outside beyond the vertex of the cone, and both the surface of the cone and the cutting plane be extended boundlessly—then also the section boundlessly will grow; and, from the diameter of the section, on the side toward the vertex, some straight [line] that is drawn from the section of the cone, parallel to the straight [line] in the base of the cone, will take off a [line] equal to any given straight [line].

—Let there be a cone whose vertex is what the point A is, and whose base is what the circle BC is, and let it have been cut by a plane through its axis, and let this plane make as section the triangle ABC. And let the cone have been cut also by another plane which cuts the circle BC along the straight [line] DE that is at right [angles] to the [line] BC, and let this plane make as section on the surface the line DFE. And the diameter of the section DFE—namely the [line] FG—either let it be parallel to the [line] AC or let it, extended, fall together with the [line] AC outside beyond the point A. I say that if both the surface of the cone and the cutting plane be extended boundlessly, also the section DFE boundlessly will grow.

—For let there have been extended both the surface of the cone and the cutting plane. Evident it is, then, that also the [lines] AB, AC, FG will, together, be extended. Since the [line] FG either is parallel to the [line] AC or, extended, falls together with it outside beyond the point A, therefore the [lines] FG, AC when extended toward the side of the points C, G will never fall together. Let them have been extended, then, and let there be taken some chance point on the [line] FG—the [point] H, say—and, through the point H, parallel to the [line] BC, let there be drawn the [line] KHL, and, parallel to the [line] DE, [let there be drawn] the [line] MHN. The plane through the [lines] KL, MN is therefore parallel to the plane through the [lines] BC, DE. A circle, therefore, is what the plane KLMN is. And since the points D, E, M, N are in the cutting plane and are also on the surface of the cone, therefore they are on the common section. Therefore it has grown—the [line] DFE has—as far as the points M, N. With the surface of the cone and the cutting plane having grown as far as the circle KLMN, therefore also the section DFE has grown as far as the points M, N; likewise, then, will we show that also if there be boundlessly extended both the surface of the cone and the cutting plane, also the section MDFEN boundlessly will grow.

—And also evident is it that some [line] will take off, from the straight [line] FH on the side toward the point F, a [line] equal to any given straight [line]. For if, equal to the given [line], we lay down the [line] FX and we draw, through the point X, a parallel to the [line] DE, it will fall together with the section, just as also the line through the [point] H was demonstrated to fall together with the section at the points M, N. And so there is drawn some straight [line]—falling together with the section and being parallel to the [line] DE—that takes off, from the [line] FG, a straight [line] equal to the given one, on the side toward the point F.

8th proposition: enunciation

The 8th proposition, like the first definition of the book, uses the expression *eis apeiron*, translated "indefinitely." The if-clause, speaking of the conic surface and the cutting plane, says: *prosekballêtai eis apeiron* ("is hurled out forward indefinitely"). And the then-clause, speaking of the cut, says: *eis apeiron auxêthêsetai* ("indefinitely will be augmented"). It should be noted that a new curve of the sort presented here is not of infinite extent. As was the case with the infinitely prolongable straight line, what is infinite is not its extent but rather the prolongability of its extent.

How do we orient the plane with which we cut a cone so that we get a line of section which is indefinitely extendable? We do so by cutting the cone so that the cutting plane, after hitting the conic surface on one side of the axial triangle, does not hit the surface on the axial triangle's other side below the vertex—however far we may extend the cutting plane and the conic surface. However much they are extended (and now, for the sake of brevity, I shall again speak—albeit loosely—of the conic surface as having "sides") the cutting plane hits only one side of the surface on the same side of the cone's vertex. It may be parallel to the other side of the surface; but if it is not parallel, then it will get farther and farther away from the other side of the surface on the same side of the cone's vertex, and its extension in the opposite direction will hit the extension of the other side of the cone on the other side of the cone's vertex (fig. 1).

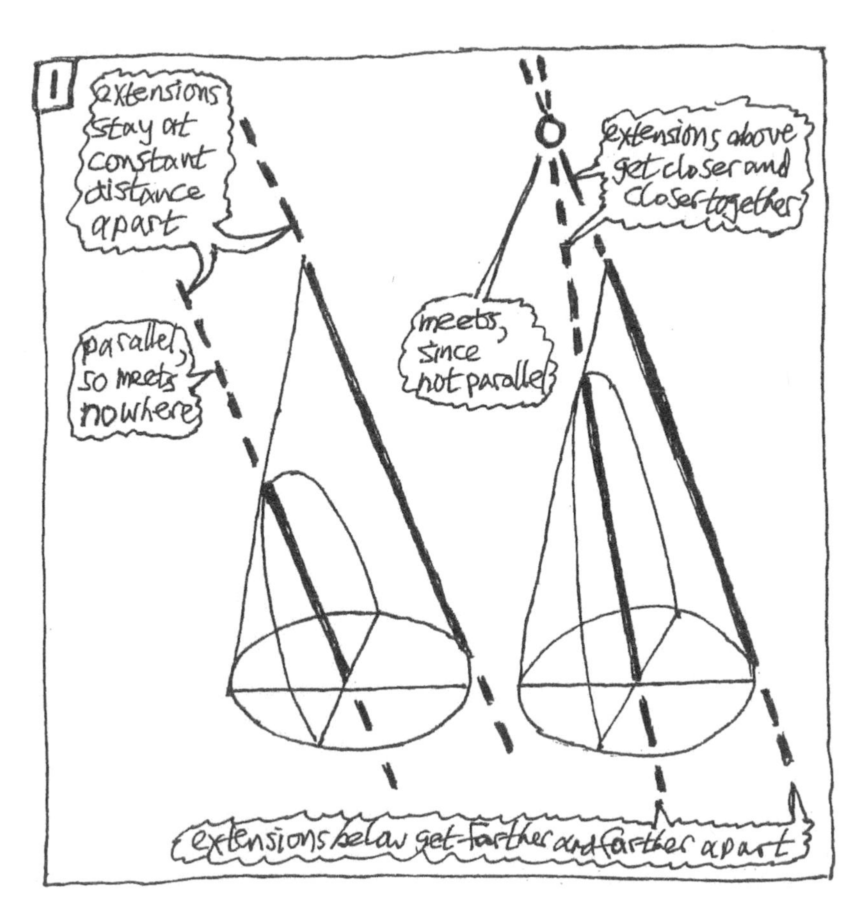

But the proposition shows more than that the resulting conic section can be extended as much as we please. It shows also that the diameter's straight-line prolongation will be met, at any distance along it that we please, by a chord (a chord in the section's extension) which is parallel to the ordinatewise chords bisected by the unprolonged diameter. Nothing is said here, however, about whether that chord which is parallel to the ordinatewise chords bisected by the unprolonged diameter (and which thus meets the diameter's prolongation) will be bisected by the diameter's prolongation. That is: Apollonius does not say that the diameter's prolongation will be a diameter for the ordinatewise chords of the section's extension. He says only that the diameter is part of a straight line such that an ordinatewise straight line drawn through any point of it will lie along a chord of the augmented section (fig. 2).

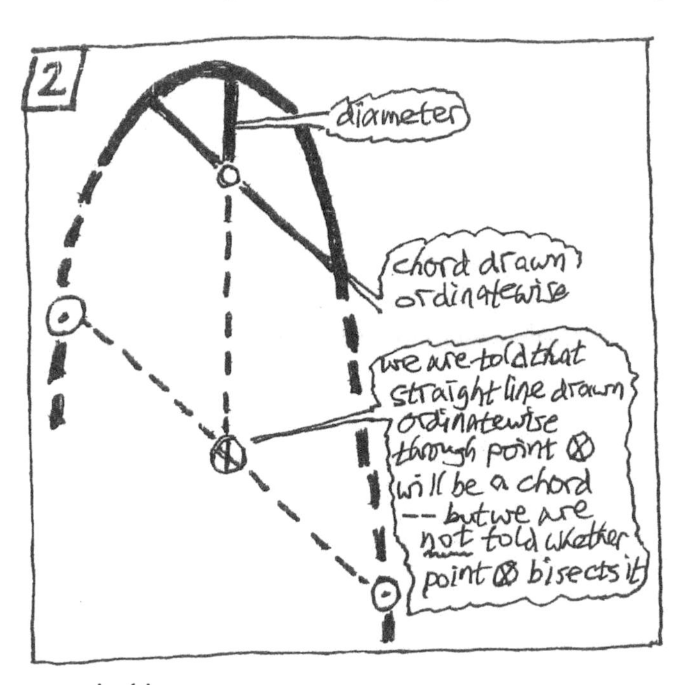

So, what the enunciation says, is this:

—*if*, in a cone through whose axis
a plane has been passed (making an axial triangle),
another plane cuts the cone so as to
slice into the cone's base a straight line of reference
perpendicular to the axial triangle's base,

and the diameter (of the section that results on the surface)
either is parallel to one of the axial triangle's sides
or else meets one of the axial triangle's sides extended beyond the vertex,

—*then* if both the cone's surface and the cutting plane are
extended indefinitely,
the section too will increase indefinitely;

and from some point on the conic section a straight line can be drawn that
is parallel to the straight line of reference
and cuts off from the prolonged diameter (on the vertex-side)
a straight line equal to any given straight line.

8th proposition: demonstration

We let there be a cone with point A as vertex and circle BC as base. It is cut by a plane through its axis, making as section (by the 3rd proposition) triangle ABC. That is not, however, the section in which we are interested. The cone is cut also by another plane, which, while slicing the cone's base (circle BC) in straight line DE that is perpendicular to the axial triangle's base BC, makes, as section on the conic surface, line DFE—whose diameter (see the 7th proposition and its porism) either is parallel to AC, or else, when prolonged, meets AC beyond vertex A. From that (we specify) what is to be shown is this: if both the cone's surface and the cutting plane are produced indefinitely, then conic section DFE will also increase indefinitely, as will the system of ordinatewise chords that cut off lines along the prolonged diameter (fig. 3).

First, we show that the cone's surface and the cutting plane can be increased as far as a circle that is parallel to the cone's base and has its center at any point on the prolonged diameter. That is shown as follows:

Extend the cone's surface and the cutting plane. It is evident that this also prolongs straight lines AB, AC, FG. FG either is parallel to AC— or else, if it does meet AC when they are prolonged, it does so beyond vertex A. Hence FG never meets AC if they are prolonged in the direction of G and C. Take, on the prolonged FG, a point at random (say, H). Through this point H, draw two straight lines— one of them (KHL) parallel to the axial triangle's base BC, and the other one (MHN) parallel to DE (that other straight line in the cone's base). Hence (Euc XI.15) the plane through the two lines so drawn (KL and MN) is parallel to the plane through the lines (BC and DE) to which they are parallel. Hence, by the 4th proposition (since a plane cuts the cone parallel to its base), plane KMLN is a circle.

Next, we show that the conic section too is increased as far as points on that circle. That is shown as follows:

Points D, E, M, N are all points that are both in the cutting plane and on the cone's surface. Hence they are all on the intersection. Hence section DFE has increased as far as points M and N. Hence, with both the cone's surface and the cutting plane having been increased as far as circle KMLN, section DFE has increased as far as points (on the circle) M and N.

Likewise it could be shown that if both the cone's surface and the cutting plane are increased *indefinitely*, then that conic section (MDFEN) too is increased indefinitely.

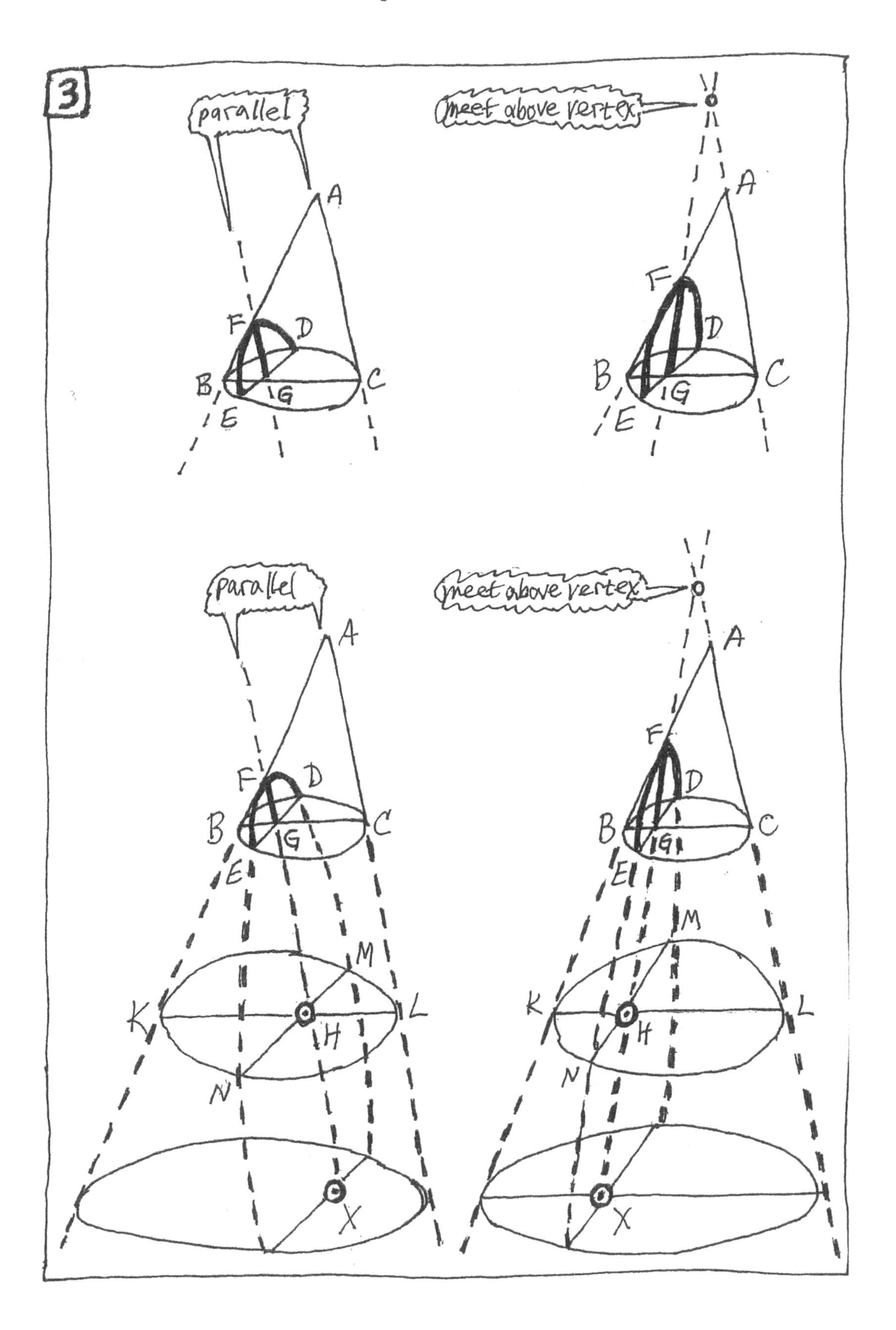
3
parallel
meet above vertex
A
F
D
B
C
G
E
parallel
meet above vertex
A
F
D
B
C
G
E
K
L
M
H
N
X

Finally, it is evident that some straight line will cut off on FH (the diameter's prolongation), on the side of point H, a straight line equal to any given straight line.

> *If* we lay down FX equal to the given straight line
> ("X" will not seem so strange a letter to use after "N" and before "O" if you take note that the Greek alphabet puts the letter Chi, which is written as X, after Nu and before Omicron),
> and we draw through X a parallel to DE,
>
> *then* that parallel will meet the section—
> just as the straight line through H also proved to meet the section
> in points M and N—
> and so some straight line is drawn,
> parallel to DE, and meeting the section,
> and cutting off on FG, on the side of point H,
> a straight line equal to a given straight line.

We could not have shown the indefinite extendability of the conic section unless we had already had a diameter for any non-circular conic section. That is to say, we could not have done the 8th proposition unless we had already done the 7th. And since we could not have done the 7th without having done the 6th before it, we could not have moved on from the circular curves of section of the 4th and 5th propositions, to the non-circular curves of section of the 8th and 9th, without moving through the 6th and 7th propositions—which did not give us any kinds of conic sections.

The line through X is said to meet the section for the same reason that the line through H meets it in points M and N (fig. 4). Since the only difference between the case of X and that of H seems to be that X is at a *given* distance from F, why not just leave out the talk about H—and proceed directly to doing for X alone what the demonstration does first for H and then again for X? To see why not, consider *first* the step involving F*H*. That step establishes this assertion: drawing a line through any extension of the diameter (here, as far as H) will give us points (here called M and N) on an extension of the *curve of section*. Consider *next* the step involving F*X*. That step establishes this assertion: drawing a line through any extension of the diameter—however long that extension may be (here, as far as X)—will extend *the system of ordinatewise chords*. A parallel through a point at any distance along the diameter's extension will be a chord of the section; in other words, the section farther out will *not* curve in such a way that lines that are drawn parallel to the ordinatewise chords cease to form chords of the section. (For an example of what the demonstration rules out, see figure 5.) Thus, the demonstration first establishes (by treating FH) that the curve is indefinitely extendable, and then establishes (by treating FX) that any definite extension of the diameter will also extend the system of ordinatewise chords.

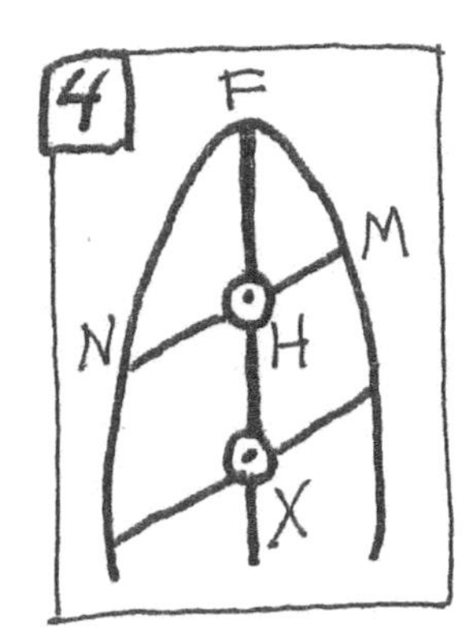

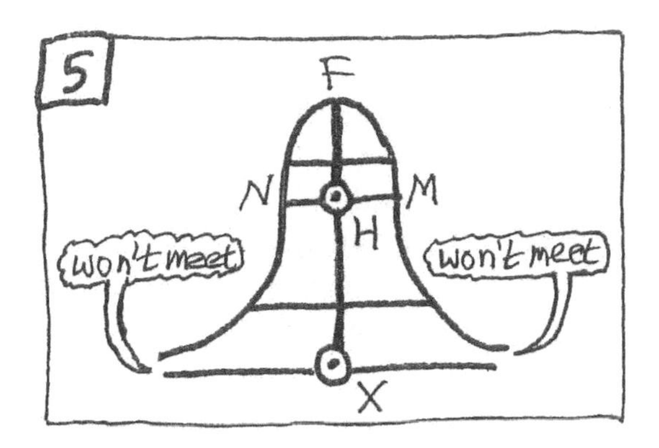

Now we can consider why the demonstration ends by repeating what was said before, thus placing such emphasis upon the straight lines cut off along the diameter of the extended section—when what would seem to be interesting is the extendability of the curve itself. Our attention is being drawn to something. Although we would like to get non-circular curves, what we would like is not merely to *get* them but rather to get them so that we are able to *speak about* them. About circles Euclid does not say: "They are perfectly round—and that is all we care to say about the lovely things"; rather, he says much about them in terms of various configurations of straight lines of various sorts—radii, diameters, chords, tangents, right angles, inscribed angles, and what-not else. So here we need to be alert to what may help us to be articulate about these lines that are now coming into view. It is true that we might have saved the step at the end of the present demonstration—if, near the beginning, the point on the prolongation of the diameter had been picked, not at random (at H), but rather at X, so as to cut off along the prolonged diameter a straight line equal to the given straight line. But then the line cut off along the diameter, and the ordinatewise line that meets it at the cut-off point, would have been lost as indicators of how we shall get hold of these conic sections. For now, let it suffice to point out that those who have heard talk of "ordinates" (albeit without ever cutting a cone) will also have heard of "abscissas"—and "ab-scissa" is merely the Latin word for "off-cut." Those who have heard talk of "parabolas" and "hyperbolas" and "ellipses"—without ever cutting a cone—will be discovering, after two more propositions, just what those names (which are Greek) have to do with lines cut off on diameters that are met, at the points of cut-off, by lines drawn ordinatewise. To say more now might spoil the enjoyment of reading an author like Apollonius. We must be patient.

An author does not always tell us the significance of the things that he tells us. Instead, he may leave us to discover—on re-reading—the importance of the little things that, although they did not seem at first to tell us much, did not go smoothly past us either. Sometimes an oblique indication is sufficient for the author's purpose, and is even better than showing or telling something outright. "The lord whose oracle is at Delphi neither speaks out nor stays silent, but indicates," said Heraclitus with reference to Apollo. Considering that saying, we might ask ourselves the following questions as we proceed in our study of Apollonius: If mere indication is the way that a master of discovery has chosen to foster learning, will explication become an impediment to the learning that he fosters? How much of Apollonius's art of teaching depends upon re-reading, and how much re-reading is it reasonable for a teacher to require?

9th proposition: preparation

In the proposition at which we have just looked, the 8th, Apollonius did not cut through the cone from one side to the other; he hit only one side. From these cuts—whether he avoided hitting the other side by cutting parallel to the axial triangle's other side (since parallels do not meet, however far extended), or by cutting so as to diverge from the axial triangle's other side, thus hitting the axial triangle's other side only on the other side of the vertex (that is, by extending the cutting plane in the opposite direction)—either way, the section that he got was a line of a new kind, an infinitely extendable line that was nonetheless not straight. It could not be straight because it had chords bisected by a diameter.

Earlier, he had cut through the cone from one side to the other—not, however, at just any angle. The way that he did it was to hit both sides so as to make angles there that were exactly the same size as the base angles of the axial triangle. From those cuts—whether, as in the 4th proposition, the base angles of the axial triangle are directly underneath their equals (which is the parallel situation), or, as in the 5th proposition, the base angles of the axial triangle are underneath and opposite their equals (which is the subcontrariwise situation)—either way, the section that he got was a circle.

Now, in the 9th proposition, he again cuts through the cone from one side all the way across to the other (making extensions if necessary). But this time he takes the remaining possibility: he cuts all the way across, but he does it so as not to make the angles the same size as the base angles of that axial triangle by which the cone presents itself to us sideways. That is to say, he now cuts all the way across but neither parallel nor subcontrariwise (fig. 6).

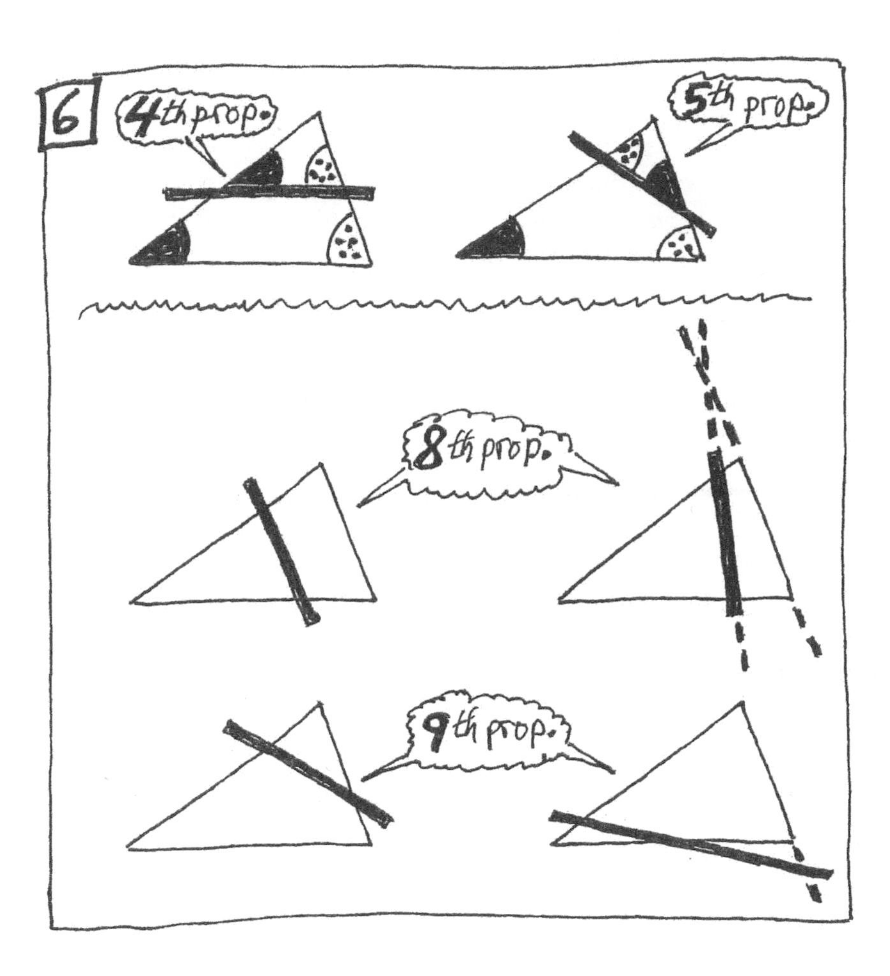

From such a cut, the section that he gets is a line of another new kind—namely, a line that is closed but is nonetheless not circular. He shows that it cannot be a circle—because its perpendicular half-chord cannot be the side of a square that is equal to a rectangle contained by the segments into which that half-chord splits its diameter.

We now are ready for the presentation of the 9th proposition.

9th proposition

#9 If a cone be cut by a plane—a plane which falls together with each side of the triangle through its axis, and which has been drawn neither parallel to the base nor subcontrariwise—the section will not be a circle.
—Let there be a cone whose vertex is what the point A is, and whose base is what the circle BC is, and let it have been cut by some plane that is neither parallel to the base nor subcontrariwise, and let the cutting plane make as section on the surface the line DKE. I say that the line DKE will not be a circle.
—For—if possible—let it be one. And let the cutting plane fall together with the base, and let the common section of the planes be what the [line] FG is, and let the center of the circle BC be what the [point] H is, and from it, [that is, from the point H] let a perpendicular be drawn to the [line] FG—namely the [line] HG. And let there have been extended through the [line] GH and the axis a plane, and let it make as sections on the conic surface the straight lines BA, AC. Since then the points D, E, G are in the plane through the [line] DKE, and are also in the plane through the points A, B, C—therefore the points D, E, G are on the common section of the planes. Straight, therefore is the [line] GED.
—Let there be taken, then, some point on the line DKE—the [point] K, say—and, through the [point] K let there be drawn parallel to the line FG the [line] KL; then equal will be the [line] KM to the [line] ML. Therefore the [line] DE is a diameter of the [so-called] circle DKLE. Then let there be drawn, through the [point] M parallel to the [line] BC, the [line] NMX. But also the [line] KL is parallel to the [line] FG. And so the plane through the [lines] NX, KM is parallel to the [plane] through the [lines] BC, FG—it is parallel, that is, to the base—and the section will be a circle. Let it be the [circle] NKX.
—And since the [line] FG is at right [angles] to the [line] BG, also the [line] KM is at right [angles] to the [line] NX. And so the [rectangle] by the [lines] NMX is equal to the [square] from the [line] KM. But the [rectangle] by the [lines] DME is equal to the [square] from the [line] KM—for a circle is what it is supposed that the line DKEL is, and its diameter what the [line] DE is. The [rectangle] by the [lines] NMX, therefore, is equal to the [rectangle] by the [lines] DME. Therefore as is the [line] MN to MD, so is the [line] EM to MX. Similar, therefore, is the triangle DMN to the triangle XME, and the angle by DNM is equal to the one by MEX. But the angle by DNM is equal to the one by ABC—for parallel is the [line] NX to the [line] BC. Therefore also the [angle] by ABC is equal to the [angle] by MEX. Subcontrary, therefore, is what the section is—which very thing is what it is supposed not to be. Not a circle, therefore, is the line DKE.

[A reminder: we were shown back in #5 what it means to say, for example, that the rectangle by the lines NMX is equal to the square from the line KM; it means that the rectangle which would be contained by the lines NM and MX, as sides, is equal to the square which could arise from the line KM, as a side.]

9th proposition: enunciation

In the 8th proposition we got a new line that was like a circle in being a curve with a diameter. The new line that we get now in the 9th proposition is also like a circle in being curved, but it is even more like a circle—for not only is it curved, it is also closed. Not only does it, like a circle, have a diameter, but its diameter is like a circle's diameter: the diameter here is bounded by the curve. That is why the enunciation does not say anything about doing something that was done in the previous proposition—namely, prolonging the diameter to any length.

So, what the enunciation tells us is this: if a cone is cut by a plane that meets both sides of the axial triangle, but is neither parallel to the base nor situated subcontrariwise, then the section is not a circle. In other words, when the cutting plane meets both sides of the conic surface on the same side of the cone's vertex, but does so at angles that are not the same size as the base angles of the axial triangle, then, although the section is both curved and closed, nonetheless it is not circular.

Do not, by the way, let it bother you that we would not call a square a curve just because a square can be said to have a diameter (namely, its diagonal). And why not be bothered by that? Because a square is not contained by a single line; it is a figure that is contained by several straight lines placed end-to-end at right angles.

9th proposition: demonstration

We let there be a cone with point A as its vertex and circle BC as its base. The cone is cut by a plane that is neither parallel to the base nor situated subcontrariwise. The cutting plane makes on the surface the line of section DKE. From that (we specify) what is to be shown is this: line DKE is not a circle (fig. 7).

We employ a *reductio* argument, showing the absurdity into which we are led by supposing that what we wish to demonstrate to be so, is not so. We begin, that is to say, by letting DKE be a circle.

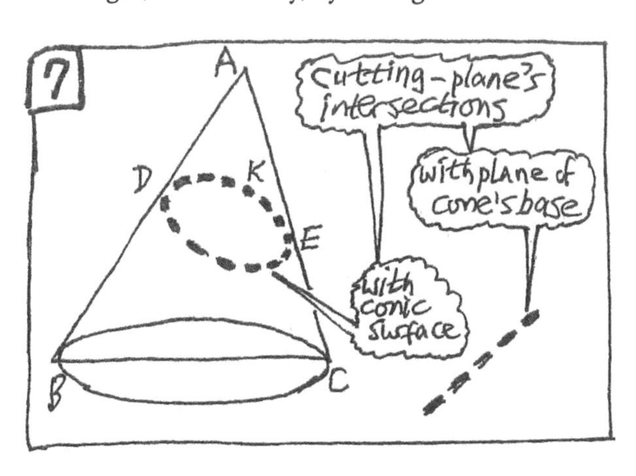

First, we get the supposed circle's diameter, and a half-chord which is perpendicular to it, as follows (fig. 8).

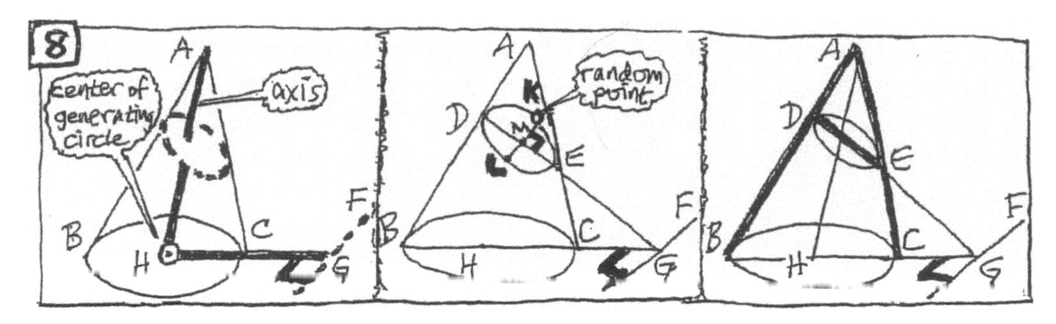

Since the cutting plane
is not parallel to the base plane,
these two planes intersect—in a straight line (say, FG).
From the center (H) of circle BC,
and perpendicular to
that straight-line intersection of
cutting plane and base plane,
draw straight line HG;
and through this straight line HG and the axis,
extend a plane.

This plane (by the 1st proposition) makes,
as sections on the conic surface,
the axial triangle's sides BA and AC.

Located in the axial triangle's plane
(the plane through points A, B, C)
are points D, E, G—which
are also located in the cutting plane (through DKE);
hence those points D, E, G are
on the intersection of two planes,
which means (Euc XI.3) that GED is a straight line.

Through some point on the section line DKE (say, K), and
parallel to FG (the intersection of cutting plane and base plane),
draw KL—which is thus drawn ordinatewise.
This KL is (by the 7th proposition) bisected by
the intersection (DE) of axial triangle and cutting plane.
In other words, KM is equal to ML.
Hence (by a definition from the beginning)
line DE is the circle DKEL's diameter
(supposing, as we are, for purposes of argument,
that the section is indeed a circle).

Next, we get a circle that intersects the section, and then we get its property, as follows (fig. 9).

Through point M, and parallel to line BC, draw line NMX.
Since line KL is parallel to line FG, and line NMX is parallel to line BC,
it follows
(since [Euc XI.15] there's parallelism of two planes if
two straight lines in one of the planes meet, and
each is parallel to one of two lines in another plane that meet)
that the plane through lines KM and NX is
parallel to the plane through lines FG and BC.
But the latter plane is the base plane,
and therefore (by the 4th proposition) section NKXL must be a circle.

And (since line KM is parallel to line FG, which is
perpendicular to line BG, which is parallel to line NX)
therefore (Euc XI.10) line KM is perpendicular to line NX.

Hence (Euc III.31, VI.8, VI.17) in NKXL (the circle that is parallel to the base),
rectangle [NM, MX] is equal to square [KM].

Then, by supposing that the original section is a circle, we get its property too (fig. 10).

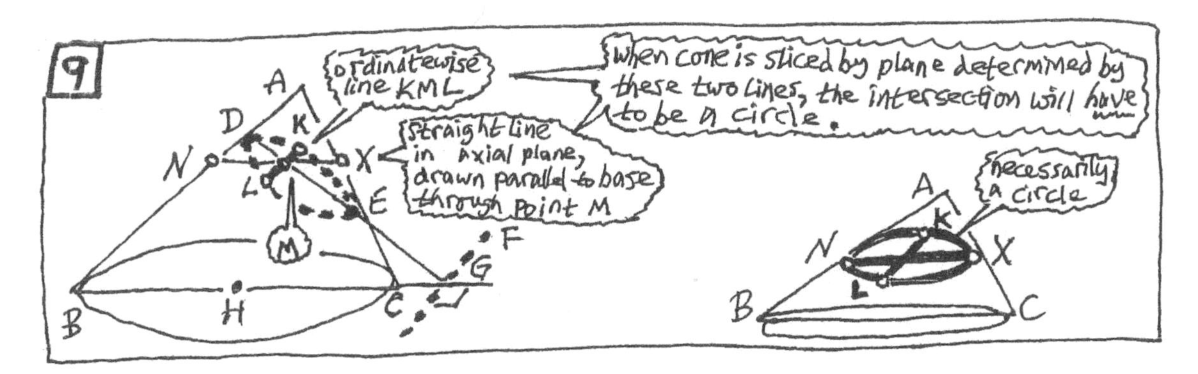

DKEL (by supposition) is also a circle, and DE is its diameter.
Hence in that supposed circle that is not parallel to the base,
rectangle [DM, ME] is equal to square [KM].

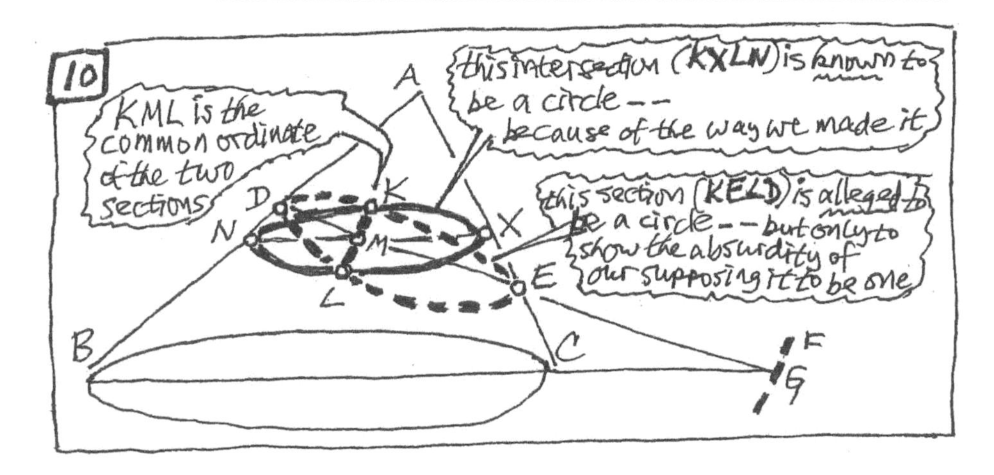

Finally, we use those top-views to get a proportionality of lines, from which it follows that we have equiangular triangles in the side-view.

KM is an ordinatewise half-chord for both diameters, each of which
it splits into two parts.
The rectangle contained by the two segments of the diameter of
the line that is *known* to be a circle,
and the rectangle contained by the two segments of the diameter of
the line that is *alleged* to be a circle,
are both equal to the square whose side is
the ordinatewise half-chord KM.
These two rectangles ([NM, MX] and [DM, ME]),
being equal to the same thing, are therefore equal to each other (fig. 11).

But the rectangles can be equal to each other
only if their sides are proportional
(so that the rectangles contained by
the lines that are mean terms in the proportion
will be equal to the rectangles contained by
the terms that are at the extremes).
Hence (Euc VI.16), as MN is to MD, so EM is to MX.
These lines are sides of triangles DMN and XME, which (Euc VI.6) are similar
(since they have proportional sides,
and these make—on opposite sides of M—
vertical angles, hence equal angles) (fig. 12).

Since the triangles DMN and XME are similar,
therefore angle DNM is equal to angle MEX.
But angle DNM (since NX was drawn parallel to BC)
is equal to angle ABC too.

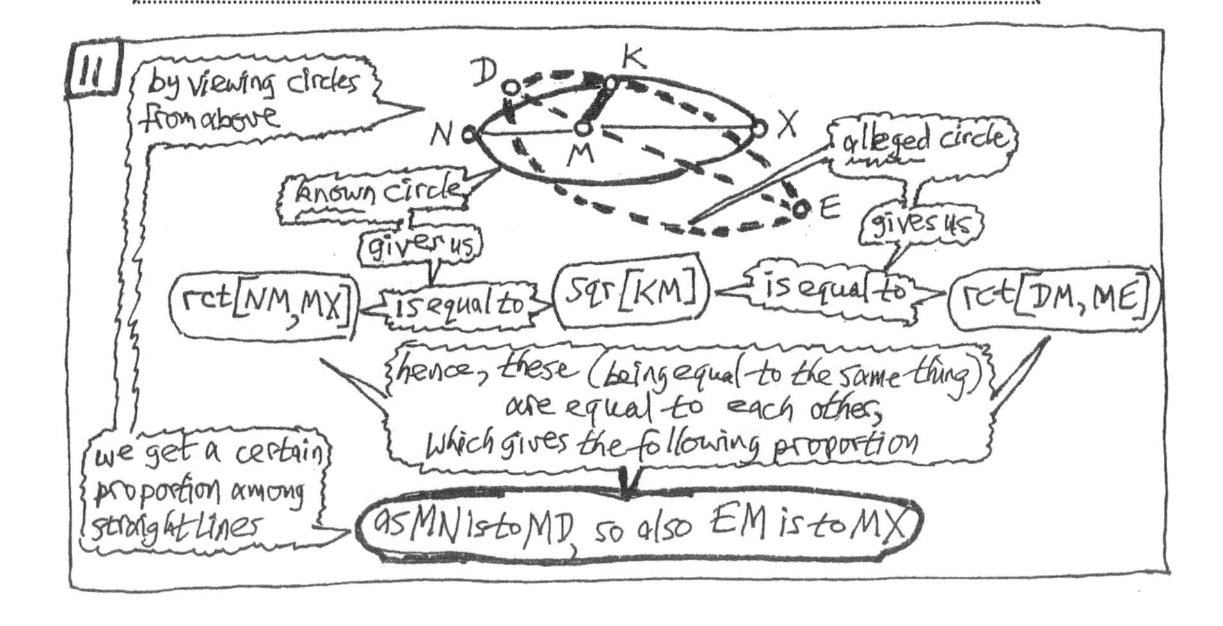

Hence (since the same angle DNM is equal to both angles MEX and ABC)
angle ABC is equal to angle MEX.
And therefore we have the situation of the 5th proposition:
the angle that is made by the cutting is equal to
the one that is under and opposite it in the axial triangle,
and so the section would have to have been cut subcontrary,
even though we *supposed* that the cutting plane was *not* subcontrary (fig. 13).

12
there can be equality between these rectangles
rct[MN, MX] & rct[MD, ME]
extreme terms
mean terms
only if there is proportionality among the lines that are their sides
So, since there is that proportionality
as MN is to MD, so ME is to MX
there must be similarity of these triangles
Why similarity? Because the proportional sides surround vertical angles-- and vertical angles are equal.
D
M
N
X
E

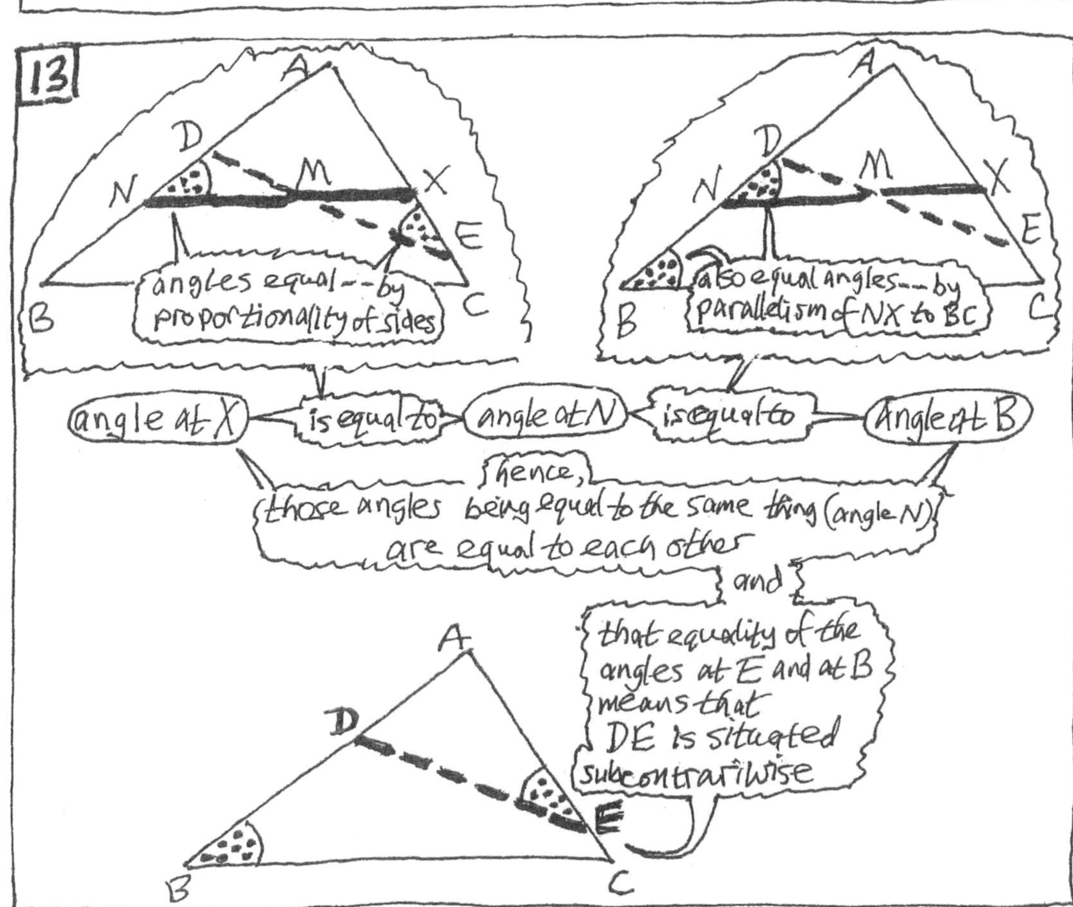

angle at X

angle at N

Angle at B

that equality of the angles at E and at B means that DE is situated subcontrariwise

This 9th proposition is, in effect, a converse of the 4th and 5th propositions in combination. In the earlier propositions, we saw that if our cutting is parallel (4th) or subcontrariwise (5th), then we get a circle as section. Now (9th) we see that if our cutting is not parallel or subcontrariwise, then we do *not* get a circle as section—which is to say: *only if* our cutting is parallel or subcontrariwise, do we get a circle as section.

We suppose that we cut neither way, and we ask: what if we did get the same result? We show that if we did get the same result (a circle) when we did not cut one way (parallel), then we must have cut the other way (subcontrariwise). But we did not cut the other way. So, we could not have gotten the same result. That demonstration is by *reductio ad absurdum*.

If a cut gives a circle as the section, and the cut is not parallel, then the cut must be subcontrary—that has been shown in the way that many things have been shown here before: by viewing *circles* from *above* the cone, and *triangles* from *beside* it. From above, we here surveyed the section that is cut or alleged to be cut, and from the side we assessed the inclination of the cutting plane. We took a top-view of a true circle from directly over the base, and also a top-view of an alleged circle from above but aslant the base, and we attributed to both of them the square-rectangle property that we have used before (the property that follows from the fact that if a half-chord is perpendicular to the diameter of a circle, then it is the mean proportional between the segments into which it splits the diameter). Putting together the two results from above, we got a proportionality that we used to establish the equiangularity of triangles in the side-view, and thus to establish the inclination of the cutting plane. Thus, by viewing circles from above, we got a proportion among certain straight lines, which we then used in viewing triangles from beside.

Many things in the demonstration of this 9th proposition are similar to things in the demonstration of the 5th. But we could not have gone directly from the demonstration of the 5th to the demonstration of the 9th proposition. Here in the 9th proposition, as in the 8th before it, the demonstration needs the diameter which was provided (for any conic section) by the 6th and 7th propositions. Here, on the basis of that earlier work, the intersection of the cutting plane with the axial triangle is said to be a diameter of the section that was made by the cutting plane, and hence to be a diameter of the alleged circle.

As with the 7th proposition, so also here with the 9th, the demonstration depends upon three-dimensional relationships that some people find easy to visualize but others find hard. If you would like a more thorough account of why the line KM (which is the half-chord of the true circle, as well as of the alleged circle) is perpendicular to the diameter NX of the true circle, you will find it in an appendix at the end of this chapter.

For this 9th proposition, the way in which Apollonius proceeds is different from the way in which he proceeded before. Before, he oriented the cutting plane by characterizing its straight-line intersections with two other planes—namely, the plane of the axial triangle, and the plane of the cone's base. The straight line in the axial triangle's plane was variously oriented with respect to the axial triangle's sides, and the straight line in the plane of the cone's base was always perpendicular to the axial triangle's base (fig. 14). For orientation here in the 9th proposition, we might expect Apollonius to use the line DG and the line FG (the latter, however, being perpendicular to the extension, BG, of the axial triangle's base, rather than to the axial triangle's base itself, BC). But Apollonius here does not proceed in that order which we might expect.

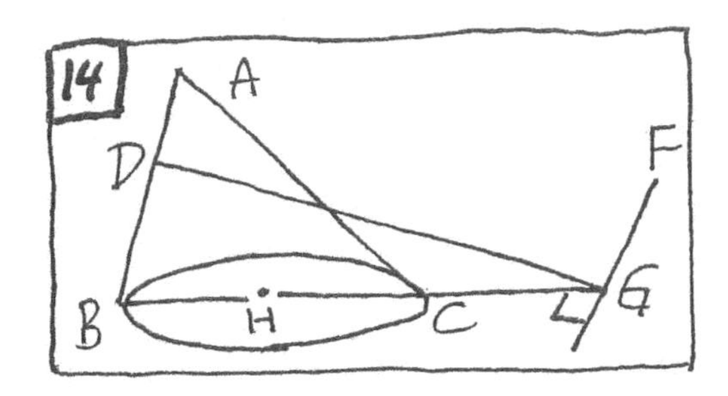

Previously, he first got an axial triangle, and only then did he cut—so that the intersection of cutting plane and cone's base is a straight line that is perpendicular to the axial triangle's base. Here, however, he first cuts the cone in a certain way that he has not used before—thus getting the straight line that is the intersection of the cutting plane and the cone's base extended—and *afterwards* he draws, perpendicular to this straight line, a straight line HG from the center of the cone's base; and after that, through this straight line HG and also through the axis, he passes a plane, getting straight lines AB and AC. Thus, from the cutting plane he gets straight line FG, from which he gets BC, from which he gets AB and AC—all the while never calling ABC the axial triangle.

Along with that difference in procedure, goes this difference in result: the 9th proposition, in contrast to previous propositions, shows not what you get from cutting the cone but rather what you do *not* get from doing so. The section that results from cutting thus is nondescript. Of course, the section that results in the previous 8th proposition is also somehow nondescript, since the 8th proposition does not specify a single way of cutting. Rather, it specifies two ways of cutting, even though both ways could have been given in a single specification. Instead of that simple specification—namely, cut so that the other side (even if extended) is not hit below the vertex—Apollonius gives a two-fold specification: he specifies cutting so as *either* to be parallel to the other side *or* to hit its extension above the vertex. The two sorts of sections that result from those two ways of cutting will later be characterized in the 11th and the 12th propositions.

The 8th proposition comes before the 9th, not because the latter is logically dependent on the former, but rather for the following reason: what is not a straight line but is like a straight line belongs before what is not a circle but is like a circle. Note that while the 8th is positive, the 9th is negative. That is: the 8th points out, not that the cutting gives a section that is *not* a straight line, but rather that the cutting gives a curved section that is *like* a straight line; whereas the 9th points out, not that the cutting gives a section that is *like* a circle, but rather that the cutting gives a closed curved section that is *not* a circle. How the 8th proposition's sections are each like a straight line can easily be stated positively without saying what each is; how the 9th proposition's section is not like a circle cannot be stated positively without stating what it is.

appendix on demonstrating the perpendicularity in the 9th proposition

In order to use the rectangle-square property with the diameter of the other line of section (that is, of the line which we cut in such a way as to guarantee its being a true circle), we need to show that the line which is the half-chord of the true circle (as well as of the alleged circle) is perpendicular to the diameter of this true circle. How is this perpendicularity shown? We use the following Euclidean proposition (*Elements* XI.10):

> A pair of straight lines will contain angles equal to those contained by
> another pair of straight lines, ones that meet in another plane than the first pair,
> if the ones in the first pair meet one another and
> each of them is parallel to one in the other pair.

In the proposition we have (fig. 15) that

> in the plane of the base, straight lines FG and BG are perpendicular to each other;
> and in another plane, two straight lines which meet are
> KM (which is parallel to the former) and line NX (which is parallel to the latter);
> hence these straight lines KM and NX are also perpendicular to each other.

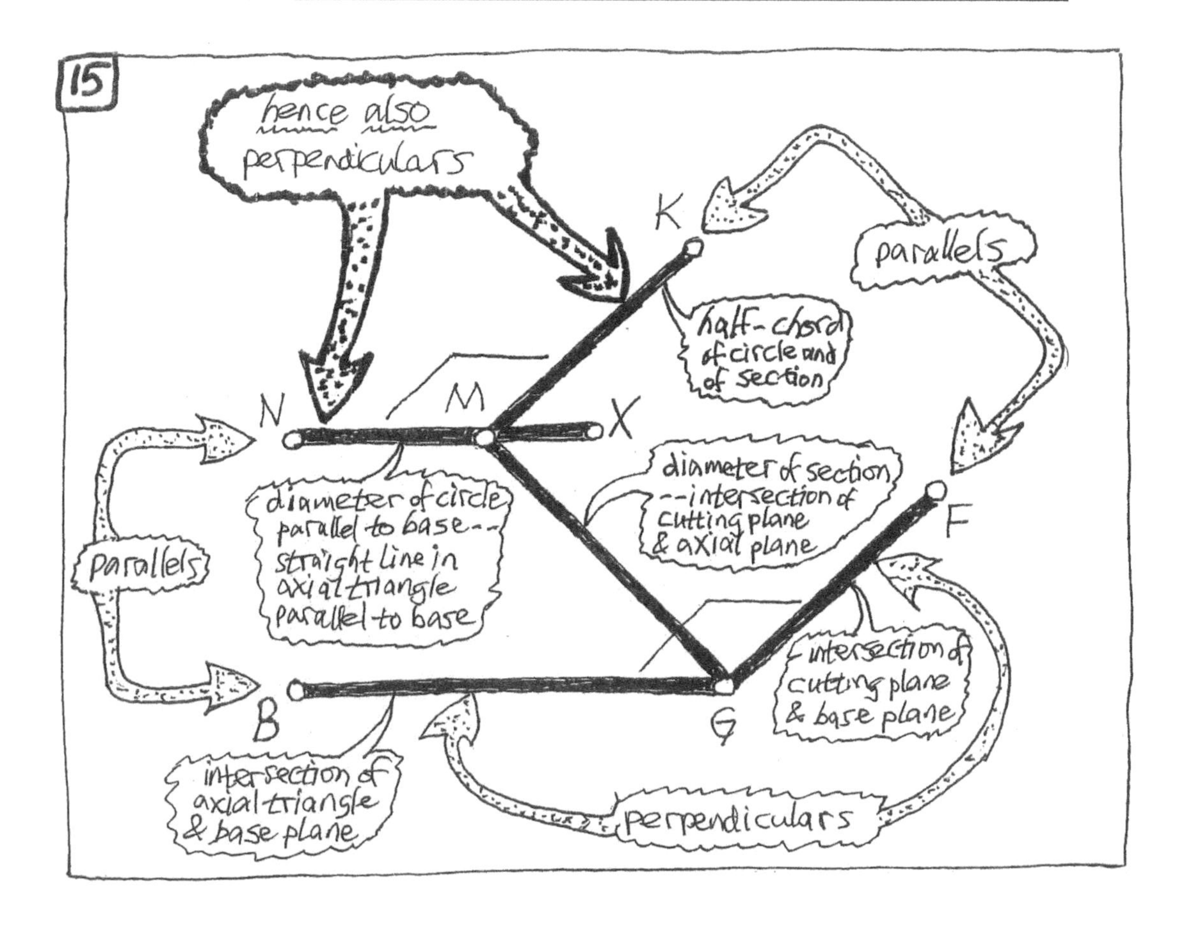

The situation might be clearer if we illustrate it with a book, a tabletop, and a pencil. Imagine (fig. 16) that a book is lying on a tabletop with its front cover facing up, but turned around so that its top edge is near you. Open the front cover a little and hold it there. Now hold the pencil at the higher corner of the top edge of the book's slightly opened front cover, and make sure of two things: that the pencil is parallel to the top edge of the back cover of the book (that is, the edge near you), and that neither end of the pencil tilts toward or away from you (that is, that the pencil is in the plane determined by the top edge of the front cover and the top edge of the back cover). By using a book, we have guaranteed that the side edge of the front cover is parallel to the book's spine, and that the book's spine is perpendicular to the top edge of the back cover. These, together with the other things, will guarantee that the pencil will be perpendicular to the side edge of the book's slightly opened front cover.

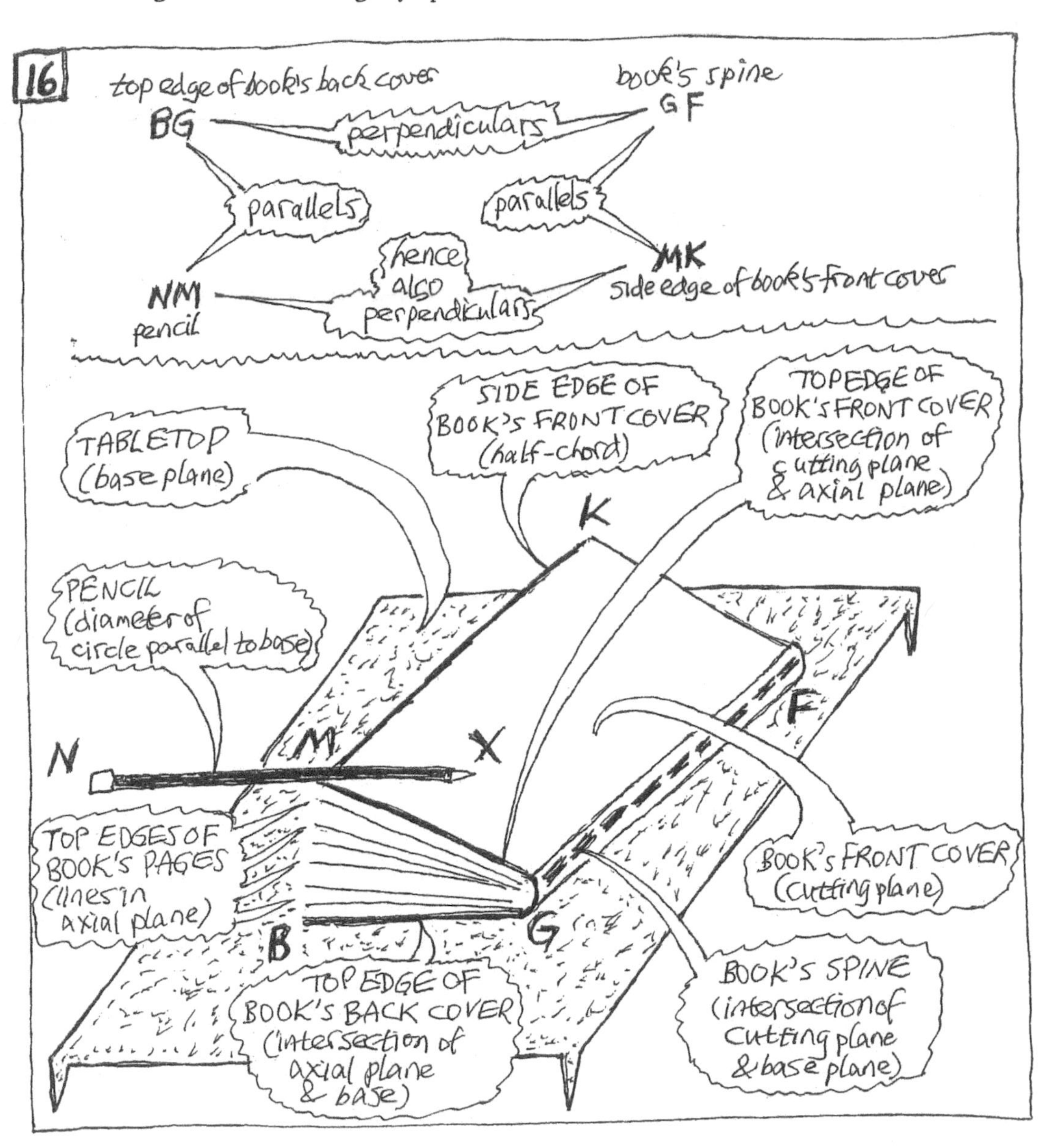

CHAPTER 6. LIKE SURFACE, LIKE SECTION

10th proposition

#10 If on a section of a cone there be taken two points, the straight [line] joining the two points will fall inside the section, and its straight prolongation will fall outside.

—Let there be a cone whose vertex is what the point A is, and whose base is what the circle BC is, and let [the cone] be cut by a plane through its axis, and let [the plane] make as section the triangle ABC. Let the cone have been cut, then, also by another plane [this one not through the vertex], and let [this plane] make as section on the cone's surface the line DEF, and let there be taken, on the [line] DEF, two points—namely the [points] G, H. I say that the straight [line] joining the [points] G, H will fall inside the line DEF, and its straight prolongation will fall outside.

—For since a cone whose vertex is what the point A is, and whose base is what the circle BC is, has been cut by a plane through its axis, and there have been taken some points on its surface—namely, the [points] G, H—which are not on the side of the triangle through the axis, and since the straight [line] joining the [point] G to the [point] H does not verge toward the [point] A, therefore the straight [line] joining the [points] G, H will fall inside the cone, and its straight prolongation will fall outside [the cone], and so [will fall] outside the section DFE also.

10th proposition: enunciation

This 10th proposition shows, in effect, that a conic section is nowhere wiggly. Conic sections are not straight, but they curve without a bump or dip. They bulge in the same way all the way, curving only rightward—or only leftward—as we go along them. That is to say: if any two points on the line of section are joined by a straight line, then—however close together, or however far apart those two points may be—no other point on the line of section is also on the straight line that joins the two points. Any chord of a conic section is entirely inside the section; and if the chord is prolonged in a straight line beyond its endpoint on the section line, it will go outside the section, and after that it will nowhere again be inside. This 10th proposition thus says about the conic *section* what the 2nd proposition said earlier about the conic *surface*.

The 2nd proposition showed that if any two points on the conic surface are joined by a straight line, then—if the straight line does not verge to the vertex—no other point on the surface will be on that straight line whose endpoints are on the surface. It was easy to see what the 1st proposition showed—that the surface does not wiggle going straight down. What was more in need of being shown was what the 2nd proposition showed—that even if we do not go straight down, but rather curve about the surface, even so, the conic surface does not wiggle. The 10th proposition now says that there is not a wiggle in the conic section either.

But since the conic section is a line in the conic surface, and the 2nd proposition showed that the conic surface does not have a wiggle in it, we might ask: could not what is said now by the 10th proposition have been said way back then, at the end of the 2nd proposition? Why the wait?

It is not that there is any need for a long and complicated demonstration. The demonstration here in the 10th proposition does not seem to be much. Would it not have been sufficient to present this proposition at the end of the 2nd proposition—as a porism like the two porisms that have already appeared in this book, without a demonstration? Why, then, should seven propositions have had to intervene before this property that was demonstrated for the conic surface is finally presented for the conic section?

In answering that question, asking this question will help: when did Apollonius get conic sections to speak of?

Well, way back at the end of the 2nd proposition he did not yet have any sections at all. At the end of the 3rd proposition he did not yet have any curved sections. At the end of the 4th and 5th propositions, he did not yet have any noncircular sections: the only curved sections that he had were circles, and it needs no showing here that there are no wiggles in circles (since it is shown, in effect, in III.2 of Euclid's *Elements*). The 6th proposition was merely preparatory; and at the end of the 7th proposition, while what was said about any section did suggest that there are lines of section that are non-circular curves, he still had not yet actually gotten any non-circular curves by cutting a cone.

Now that he has gone through the 8th and 9th propositions, however, he has shown how a conic surface can be cut by a plane in order to get non-circular curves. As we approach the 10th proposition, something can be said that is true of all those curves—something that was true also of the old lines—something with which the book began. As soon as we leave this 10th proposition, Apollonius will employ his knowledge of ratios to distinguish the various new lines.

Consider the enunciation of this 10th proposition: if two points be taken on the section of a cone, then the straight line joining them will fall inside the section, and will, if prolonged, fall outside. Had we encountered that assertion immediately after the 2nd proposition, we would have had some trouble just as soon as we had read the enunciation of the very next proposition—the 3rd proposition—which was the first to give us a section. The 3rd proposition produced a triangle by cutting a cone through the vertex, and it called this triangle a section. But if two points be taken on a single side of the triangle—that is, on the same side of such a section—then the straight line joining them will fall not inside the section but rather on it; and, if the straight line joining them is produced in a straight line, it can *remain* on the section for a stretch. Thus, in such a case the enunciation of this 10th proposition would be false (fig. 1).

The 10th proposition thus tacitly excludes from consideration, among the lines that are of interest as sections of a cone, the first section that we obtained—the triangular section of the 3rd proposition. That rectilineal cut—the triangle that is constituted by the straight lines that are cut into the surface of a cone by a plane that passes through the cone's vertex—is of interest here only as a means of discussing the sections that are curvilinear. Apollonius here is interested not so much in what lines he can get by cutting cones, as in what kinds of curves are conic sections.

10th proposition: demonstration

The remarks just made about the enunciation of the 10th proposition are helpful for understanding some strange features of the rest of the proposition.

We let there be a cone with point A as vertex and circle BC as base. It is cut by a plane through its axis, making as section (according to the 3rd proposition) a triangle, ABC. But that is not the section in which we are interested. The cone is cut also by another plane, making as section on the cone's surface the line DEF. On this line of section DEF, we take two points (say, G and H). From that (we specify) what is to be shown is this: the straight line that joins GH will fall inside section line DEF, and will, if prolonged, fall outside it (fig. 2).

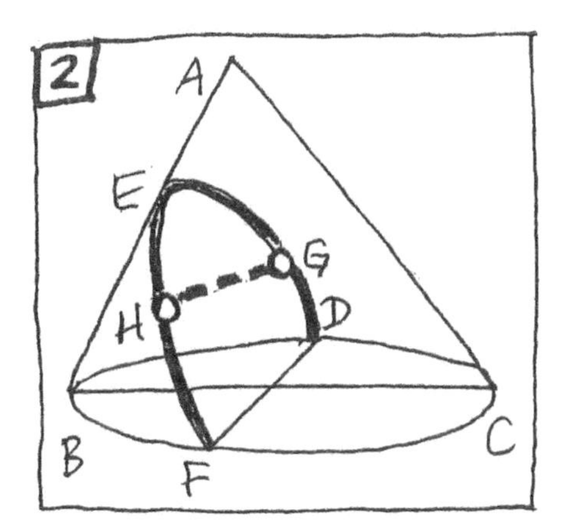

The demonstration is this:

Points G and H taken on the cone's surface
are not on a side of its axial triangle ABC,
and the straight line GH that joins them
does not verge to vertex point A;
hence (by the 2nd proposition)
it will fall inside the cone,
and will, if prolonged, fall outside the cone—
and hence also outside the section DFE.

In reading that brief statement, a number of questions arise: Even if the enunciation does have to wait until after the 9th proposition, why would it not suffice, as a demonstration, to make a one-line reference to the 2nd proposition, instead of dragging in that axial triangle? Why is that triangle required to be axial? And why is it there at all? If it is needed, why must mention of it wait until the setting-out of the figure, instead of being presented (as with earlier propositions) in the enunciation; and why, in the setting-out of the figure, can mention of it not wait until the plane has cut the surface and thus has given us the section? A proper answer to any one of these questions will be an answer to all of them. They all are aspects of the same thing.

An implication of the 2nd proposition was this: any line that is cut into a conic surface by a plane will not wiggle if it is a curve, and it will be a curve if a straight line connecting two points on it does not verge to the cone's vertex. An implication of the 3rd proposition was this: the line cut into a cone by a plane will not be a curve but rather will be a triangle if the plane passes through the cone's vertex—if (that is to say) the straight line that connects two points on the line verges to the vertex. (The 3rd proposition is something like a converse of part of the 2nd proposition: the 2nd proposition implies that the line is curved if it does not verge to the vertex; the 3rd proposition implies that if the line does go through the vertex, then it is not curved but rectilineal.)

To demonstrate what the *10th* proposition enunciates—that is, to apply to the conic *section* what the *2nd* proposition says about the conic *surface*—we must exclude the possibility that the section that is cut by the plane is not curved but rather is made of straight lines. What must be excluded in order to exclude this possibility? The 3rd proposition has shown us one way to cut a cone with a plane and get a rectilineal section instead of a curve—namely, pass the cutting plane through the cone's vertex, thus getting a triangle whose sides constitute the sides of an angle whose vertex is the cone's vertex. What other ways are there? None. The 2nd proposition implied that no other way will do it. All we need to do, therefore, is arrange things so that our cutting plane does not pass through the vertex. How can we do that? As follows.

If, before we cut the cone, we set up a triangle that is sliced into the cone by a plane that passes through the cone's vertex, and, only *after* having done that, do we cut the cone with a plane that first hits the triangle within a side (that is: not at the endpoint which is the vertex of the triangle), then the cutting plane is guaranteed not to pass through the vertex of the cone. Thus, no two points on the resultant section can be joined by a straight line that verges to the vertex of the cone, and so it will be possible to take what the 2nd proposition said about any two points that are on the conic surface and whose connecting straight line does not verge to the vertex, and apply that statement to any two points on the conic section, as in the 10th proposition.

In short, the reason why the axial triangle is brought in, and is brought in before the cutting plane itself, is this: in order to guarantee that the section will be a curve (rather than having straight-line sides that verge to the vertex). The axial triangle provides the guarantee by allowing us to ensure that the cutting plane does not pass through the vertex.

But why not merely make explicit at the very outset that the section under consideration is one which—because the plane that cuts the cone does not pass through the vertex—is nowhere straight? Well, that would deprive the reader of one inducement to ask what is being done for what. When (as here) the exclusion of the straight line section is left implicit, the reader is led to consider what it is about cuts in a cone that makes them worth examining, and what thus governs how the examination is presented.

Still, even if it is better to leave the restriction to curves implicit, why is there any need for the explicit restriction that the plane that passes through the vertex, which is the topmost point of the axis, has to pass also through the entire axis of the cone?

Well, a plane that passes through the vertex might nonetheless not cut the cone at all—it might pass *only* through the vertex. In any case, therefore, we would need to make some additional restriction besides its merely passing through the vertex. What restriction should we add then? A plane that passes through the vertex would have to cut the cone if it passed not only through the topmost point of the axis but through the rest of the axis as well—thereby freeing us from any need to mention passing through the vertex. Passing through the axis includes passing through the vertex, and also includes cutting the cone—while passing through the vertex does not as such include passing through any other point in or on the cone (fig. 3).

Besides: the triangle to which we have referred the cutting plane in the earlier propositions was always an axial triangle—and so it seems best to use a triangle of that sort, even though others could be used. If any one of infinitely many triangles could be used, it seems best to use one of the sort that stands out, and of the sort that has been used before. The difference is only that previously we took the cutting plane so as to slice, into the cone's base plane, a straight line that is perpendicular to the axial triangle's base—whereas now we are taking the cutting plane so as to hit the axial triangle's side below the cone's vertex.

We had already been shown, in the 1st and 2nd propositions, that a conic *surface* is nowhere wiggly. It therefore doesn't take much work here in the 10th propositions to show that a conic *section* too is nowhere wiggly. But it does take *some* work—since the section in the surface is not the same thing as the surface itself, and it is not an instance of it either. All the points on the conic section are points on the conic surface, but the conic section is not a conic surface. It is true that Socrates is mortal because Socrates is a man and every man is mortal; but although every student enrolled in a class on conic sections is thereby also enrolled in the college in which the class is given, the class will not occupy forty acres even though the college does. The class is a part of the college, not an instance of it.

Still, the argument is not a long one. It almost goes without saying. Why, then, say it? What is its effect? It shows us that what we are being shown is something about the special lines gotten from cutting a conic surface by a plane.

Apollonius here is not interested in cones as such, nor is he interested in cutting as such. He is not someone who sees a surface or a solid and is driven to cut it.

The title says that this is a book "of things conic" (*kônikôn*). Although among the other books of Apollonius were ones whose titles identified them as being on the cutting of a ratio, and on the cutting of an area, and on determinate cutting, the title of the book that we are reading now says nothing of cutting—and nothing of cones either. The cutting of cones is of interest as a source of lines of a new shape, ones that are neither straight nor circular. These lines are gotten by cutting, with a perfectly flat surface, a curved surface of a certain sort—one that has been generated by a straight line, when one of its points is fixed and it moves around a circle that lies in a perfectly flat surface that does not contain the fixed point. A surface so generated is conic. The lines in which we are interested are curved lines lying in a plane while also lying on a surface that is conic.

CHAPTER 7. LOOKING BACK AND LOOKING FORWARD

Let us now look back at the road along which we readers of this book of conic things have been led by Apollonius. The road has taken us from the elementary Euclidean study of straight things and circular things, into an advanced study that, by putting together straight things and circular things, generates the simplest things that are neither straight nor circular.

First, Apollonius showed that a conic surface is not wiggly in any direction—that it is like a straight line in being flat (if you go from a point on it toward its vertex) [Prop. 1]; and that it is like a circle in being everywhere curvy bulging outward (if you go from a point on it toward another point on it, but without going straight toward its vertex) [Prop 2].

Then (by cutting through the vertex, as well as through the cone) he obtained a conic section that was straight-lined (being a triangle) [Prop. 3], and another conic section that was circular (by cutting so as to make an angle equal to one of the base angles: either parallel [Prop. 4] or subcontrariwise [Prop. 5]).

Then came propositions that acted as a bridge from old lines (straight and circular) to new ones; the first of them [Prop. 6] was preparatory—it gave a bisecting surface; and the second [Prop. 7] gave the bisecting line that is prerequisite for presenting lines of section that are neither straight nor circular—and its porism told us that there were in fact new ones, and identified what it is that will tell us about them.

Then he obtained (by cutting only one side of a single surface, however much extended) a conic section that was like a straight line in being indefinitely extendable, but was unlike it in having a diameter (being curved in a certain way) [Prop. 8]; and he obtained (by cutting both sides of a single surface—extended if necessary—but so as not to make an angle equal to one of the base angles) another conic section that was like a circle in being closed, but was unlike it in not having its perpendicular half-chord be a mean proportional between the segments into which it splits the diameter [Prop. 9].

Finally, he showed that a conic section is not wiggly [Prop. 10].

In the next few propositions, Apollonius will present several varieties of the conic sections that are lines of a new sort. The first ten propositions have thus constituted a prologue to the exhibition of these new lines.

The prologue begins by showing that the conic surface is nowhere wiggly, whether we view it from beside (1st proposition) or view it from above (2nd proposition); and immediately after that we are shown (3rd proposition) that the lines cut into a cone's surface by a plane that passes through the vertex, far from wiggling, will not even curve. The prologue ends by showing (10th proposition) that the lines cut into a cone's surface by a plane that does not pass through the vertex will, like the conic surface into which they are cut, curve without wiggling. Thus, the prologue's closing proposition shows for the conic section what the opening two propositions showed for the conic surface—that it is nowhere wiggly. Why is it important thus to emphasize the lack of wiggliness? Because from it we learn why it is a plane that we employ for cutting the conic surface to get our first non-circular curves. Cutting the conic surface with a plane will guarantee that the non-circular curves that we get are simpler than the non-circular curves that we might get upon a conic surface in some other way. A line that is free of wiggles is simpler than a wiggly one. If a line that is drawn upon a conic surface is wiggly, then it cannot be cut into the conic surface with a plane—that is to say, we cannot slide it out for examination upon a flat surface.

In between the propositions about non-wiggliness that open and close the prologue, Apollonius places propositions from which we learn what lines of section can be yielded by the conic surface when cut by a plane: not just old familiar lines that are perfectly straight or perfectly round, but interesting hybrid lines that have properties like those that belong to those elemental lines the straight and the circular. The conic surface gives us interesting hybrid lines because it itself is a hybrid surface—flat in one way and round in the other. Viewed from above, it is like a stack of larger and larger circles; viewed from beside, it is like a ladder of larger and larger similar triangles (fig. 1).

To say much more about the various new sorts of curved lines that we can get by cutting a cone, we must compare the sizes of straight lines in them. How so?

Well, the cone when viewed from *above* is like a stack of circles. If we slice ourselves a circle from the stack, we can (by using the perpendicular half-chord's mean-proportionality with respect to the segments into which it splits the diameter) get the equality of the squares and rectangles that we saw before.

And the cone when viewed from *beside* is like a ladder of similar triangles. If we pull out a couple of triangles from the ladder, we can (by using the proportionality of the straight lines that are the sides, and putting together the same ratios of different straight lines) get the equality of different rectangles.

By putting together the results of those two kinds of views—the straight-line aspect and the circular—it is possible to give an account of the lines that are formed in a plane when the plane intersects a cone so that the lines that are cut are neither straight nor circular.

To do that, however, we shall need to have considered some elementary Euclidean matters. Among them are these: the use of the equality of two rectangles to present in brief the proportionality of four straight lines, and the use of the ratio of two rectangles to present in brief the ratio put together out of two ratios of straight lines. We shall now turn to considering all that, in the next part of this guidebook.

PART II
EUCLIDEAN FOUNDATIONS

Elements: selections

CHAPTER 1. RATIOS, LINES, AND RECTANGLES

We have seen that if a half-chord in a circle is perpendicular to a diameter, which it thus splits into segments, then that half-chord has a certain power: the half-chord can be taken as a side from which to raise up a square—and this square on the perpendicular half-chord will be equal to the rectangle contained by the segments into which the half-chord splits the diameter (fig. 1).

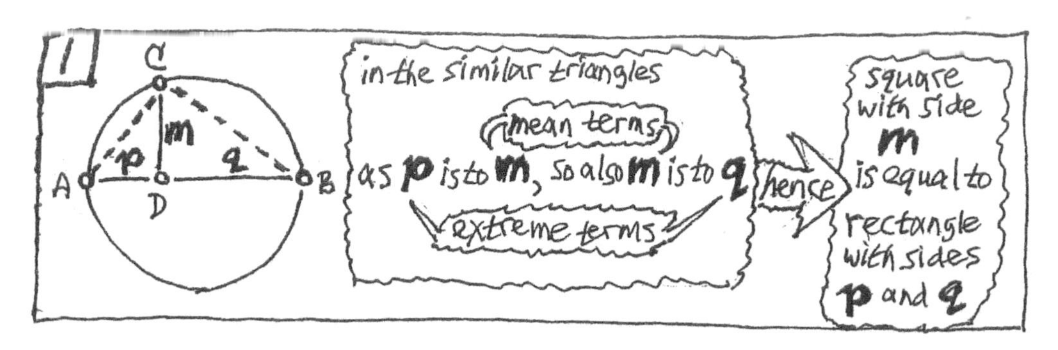

A circle associated with three straight lines of this sort (namely, such a perpendicular half-chord and the two segments into which it splits the diameter) will be the indispensable link in each of the demonstrations in the next few Apollonian propositions, which will give us lines that are neither straight nor circular.

In these new curved lines, the straight lines that will be neatly related to each other will be called *ordinates* and *abscissas*. We shall use these terms, departing slightly from the terminology of Apollonius for the sake of brevity, as follows. Apollonius has spoken of straight lines drawn "ordinatewise"—that is, of chords ordered in parallel to a certain straight line of reference. Now, a line that is *set in some order* can be called, in Latin, "ordinata," and we shall use the name "ordinate" for an ordinatewise half-chord—that is, for a certain part of a chord drawn ordinatewise, namely, that part which stretches from the curve only up to the diameter. Now, a line that is *cut off* can be called, in Latin, "abscissa," and we shall use the name "abscissa" for the straight line which is cut off from the diameter by the ordinate—that is, we shall use it for a certain part of the diameter, namely, that part which stretches from the vertex-point (where the diameter meets the curve) down to the cutoff point (where the diameter meets the ordinate). Thus, the ordinate. lying along the ordinatewise chord, stretches from the curve to the diameter; and the abscissa, lying along the diameter, stretches from the curve to the ordinatewise chord. (Fig. 2.)

We shall see that in the new curves there is a relation between those two straight lines, the ordinate and abscissa. This relation is similar to the relation that we earlier saw in that old familiar figure the circle; but is not the same as it. In the new curves, the straight line called the ordinate, taken as a side, will indeed make a square that is equal to the rectangle contained by two other straight lines, and one of these will indeed be the diameter's segment called the abscissa (the straight line cut off from the diameter by the ordinate); the other one, however, will be, not another segment of the diameter, but rather a straight line which Apollonius will have to contrive to be of just the right size to do the job.

Thus, the ordinate has the power to be the side of a square that will be of interest; and a certain rectangle will be of interest because of two things: it has the abscissa as a side, and it is equal to the square of interest. Apollonius will demonstrate that the square of interest and the rectangle of interest are equal to each other because they are both equal to the same mediating rectangle. This mediating rectangle to which both are equal will be the rectangle whose sides are the segments into which the ordinate splits the diameter of a certain circle. This circle is made by cutting parallel to the cone's base.

To show for each curve that the square on the ordinate is equal to this mediating rectangle—that will be quick work. We have seen it before. The new, more difficult work will be to show that this mediating rectangle, in its turn, is equal to the rectangle of interest. To show that this follows from the very definition of the specially contrived line, we shall have to put ratios together—"compound" them (and we shall also have to undo such "compounding")—and we shall have to make use of the fact that in equiangular triangles corresponding sides are in the same ratio (fig. 3).

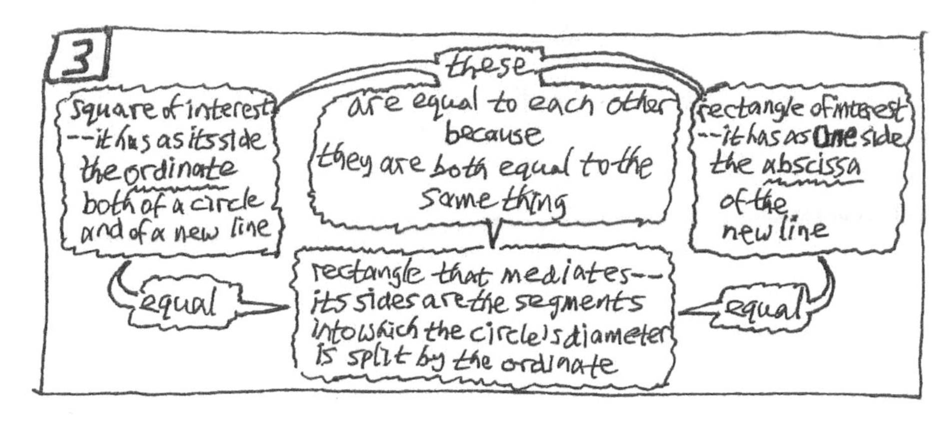

The work that Apollonius does with ratios presupposes a knowledge of their compounding and its use. It also presupposes a knowledge of when straight lines can be compared indirectly without resort to ratios. Both sorts of things will be presented in this part of this guidebook. The presentation will be helpful for understanding classical geometry in its own terms, and also for preparing us to understand the modern transformation which sought to supersede it.

In the rest of this part's first chapter, we shall examine how straight lines can be related even without speaking of ratios—by building squares and other rectangles out of them and then saying which rectangles are bigger than, or equal to, or smaller than which. The classic presentation of this is to be found in the Second Book of Euclid's *Elements*, which some readers have taken to be a form of algebra. "Geometrical algebra," they call it. It will be important for us to try to see how what is done by Euclid and Apollonius *differs* from algebra. A consideration of what makes for the difference (which is suggested in this chapter and is discussed in the next) will lead us to an examination (in this part's chapters 3 to 5) of the classical notions of number and ratio, after which we shall be ready to consider the compounding of ratios (in chapter 6) and its use (in chapter 7).

After that, this guidebook's next part will go on to present the propositions of Apollonius that characterize the various kinds of conic sections.

This treatment of Euclidean foundations, in this second part of the guidebook, is an interruption in our reading of Apollonius. Apollonius himself does not ever say that his 10th proposition is the completion of anything. To a new reader, it may not even very clearly appear that the first ten propositions constitute a distinct part of Apollonius's presentation.

Why, then, not wait to introduce the material included in this part of the guidebook? Because the work of Apollonius presupposes the work of Euclid, and it is in the 11th proposition of Apollonius's *Conics* that what is presupposed from Euclid's *Elements* will begin to seem especially strange to modern readers. Without some careful preparation, modern readers may assimilate what is strange to what is familiar, misinterpreting Apollonius as saying something like "$y^2 = p*x$." Or they may be utterly confused by what is strange, not knowing what to make of it when Apollonius says that a ratio is compounded of two other ratios.

Why, then, not require the reader to study Euclid before even beginning to read Apollonius—why not present Euclid first, that is, instead of interrupting Apollonius to bring in Euclid? Because if readers had to wait—for a study of Apollonius and Descartes—until they had worked through Euclid, few would stay the course. To many readers, much of Euclid's *Elements* may seem like nothing very new, and may therefore seem not worth the time. What is most relevant to a modern appreciation of classical mathematics is unlikely to be reached soon enough in Euclid without a great deal of skipping. But a study of classical mathematics that begins with a great deal of skipping around in the text under consideration will predispose readers to an uncritical acceptance of Descartes' disparagement of ancient books of mathematics as mere grab-bags of whatever random discoveries the ancient mathematicians just happened to stumble upon. Moreover, even if there were no objection to beginning the study of classical mathematics by skipping around in the middle of a text without concern for its integrity, there would still be reason for postponement of considering fundamental notions that are strange—until readers can see their use, as they will be able to do in the next proposition of Apollonius, to which we shall turn at the beginning of the next part of this guidebook.

A study of Apollonius's tightly constructed text, with selective help from Euclid *along the way*, provides an excellent way to begin to appreciate the significance of classical mathematics and its transformation. Very young students nowadays learn to characterize lines as graphs in what is called a "Cartesian" coordinate system. By supposing that there is a one-to-one correspondence between the points on a straight line and "the real numbers," and then laying down two such number-lines as axes, students are taught to establish a one-to-one correspondence between points on the plane and pairs of "real numbers." In this way, a line is associated with an equation. The equation relates the two items in any pair of "real numbers" that corresponds to a point on the line. As we shall see, it is no accident that the "real number" obtained from one axis is called the "abscissa" and that the "real number" obtained from the other axis is called the "ordinate." In the study of classical mathematics, however, when you read (for example) that the square arising from a line called the "ordinate" is equal to a rectangle contained by one straight line called the "abscissa" and another straight line of constant size, what this will mean is not what is now meant by writing down the string of signs "$y^2 = p*x$" and calling it the equation of a "parabola." Not that it is an accident that readers of Apollonius nowadays may be inclined to write something like "$y^2 = p*x$." An effort is required for us to see such writing as re-writing, and to understand just why it was, that such writing came to replace the writing that preceded it.

Now then, let us consider this matter of relating the sizes of straight lines in classical geometry. So far, the most important such relationship that we have seen is the one mentioned a few moments ago, which we first examined a while back when we were preparing for Apollonius's 5th proposition. That was the relation among a certain trio of straight lines in a circle. Straight lines in ratios were related to each other by the equality of the square or oblong rectangles which they can make.

In order to see how also some straight lines which we have *not* got involved in ratios are related to each other by the equality of the square or oblong rectangles which they can make, we need to consider the beginning of the Second Book of Euclid's *Elements*. It is preceded by an exploration of equality in the First Book of the *Elements*, beginning with things that are equal because they can fit upon each other. The concluding quarter of the First Book—after considering how figures that do not have the same shape, and therefore do not fit upon each other, nonetheless can have the same size—culminates in the Book's next-to-last proposition (whose converse is the last): the celebrated "Pythagorean theorem," which shows that in a triangle two sides of which contain a right angle, the remaining side (the "hypotenuse," which "stretches under" the right angle) serves as a side from which is raised a square that is equal to the squares, taken together, that are raised from those two sides which contain the right angle. This 47th proposition of the First Book is the first instance of Euclid's relating the sizes of straight lines indirectly by relating the sizes of boxes built from them. After that special case in which the lines whose sizes are related constitute a right triangle, providing a spectacular conclusion to the First Book, the Second Book goes on to relate the sizes of lines that are various segments of straight lines. (This Book finally returns to lines that constitute triangles—triangles that are obtuse-angled in the next-to-next-to-last proposition, and that are acute-angled in the next-to-last proposition. The very last proposition returns to what was the theme right before the conclusion of the First Book. On the 47th proposition of Euclid's First Book depend the propositions that come in the Second Book long before its end, and the propositions that come at the end of the Second Book depend on propositions that come earlier in the Second Book.)

So now let us look at the beginning of the Second Book of the *Elements*. It opens with the following definitions, depicted in figure 4.

(II) Definitions
Every rectangular parallelogram is said to be "contained by" the two straight [lines] containing the right angle; and, of the parallelograms about its diameter, let any one whatever—[when taken along] with the two complements—be called a "gnomon."

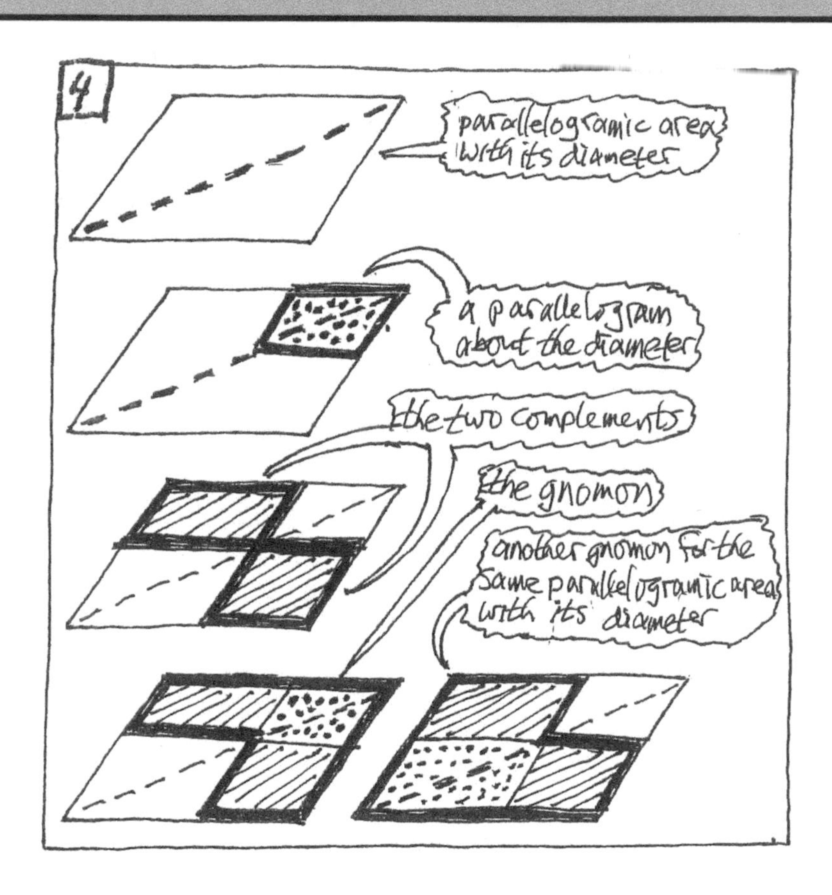

Here are the first four propositions of the Second Book of Euclid's *Elements*. You should examine the enunciations and look over the demonstrations, but you need not tarry over them on a first reading. Instead, move on briskly to the figure that follows them (namely, figure 5).

(II) #1 If there be two straight [lines], and one of them be cut into however many segments, then the rectangle contained by the two straight [lines] is equal to, [taken together,] the rectangles contained by the uncut [line] and each of the segments.
—Let there be two straight [lines], namely the [lines] A, BC; and also let the [line] BC be cut as chanced, namely at the points D, E. I say that the rectangle contained by the [lines] A, BC is equal to—[taken together]—the rectangle contained by the [lines] A, BD, and also the [rectangle contained] by the [lines] A, DE, and moreover the [rectangle contained] by the [lines] A, EC.
—For let there be drawn, from the [point] B, at right angles to the [line] BC, the [line] BF; and also let there be put equal to the [line] A the [line] BG; and also through the [point] G, parallel to the [line] BC, let there be drawn the [line] GH, and through the [points] D, E, C let there be drawn parallel to the [line] BG the [lines] DK, EL, CH. Then equal is the [rectangle] BH to the [rectangles] BK, DL, EH [taken together].
—And (on the one hand) the [rectangle] BH is the [rectangle] contained by the [lines] A, BC—for (on the one hand) it is contained by the [lines] GB, BC and (on the other hand) equal is the [line] BG to the [line] A.
—And (on the other hand) the [rectangle] BK is the [rectangle contained] by the [lines] A, BD—for (on the one hand) it is contained by the [lines] GB, BD, and (on the other hand) equal is the [line] BG to the [line] A. And the [rectangle] DL is the [rectangle contained] by the [lines] A, DE—for equal is the [line] DK, that is, the [line] BG, to the [line] A. And moreover, similarly the [rectangle] EH is the [rectangle contained] by the [lines] A, EC.
—Therefore the [rectangle contained] by the [lines] A, BC is equal to, [taken together,] the [rectangle contained] by the [lines] A, BD, and also the [rectangle contained] by the [lines] A, DE, and moreover the [rectangle contained] by the [lines] A, EC.
—Therefore, if there be two straight [lines], and one of them be cut into however many segments, then the rectangle contained by the two straight [lines] is equal to the rectangles contained by the uncut [line] and each of the segments—the very thing which was to be shown.

(II) #2 If a straight line be cut as chanced, then the rectangles contained by the whole [line] and each of the segments are, [taken together,] equal to the square from the whole [line].
—For let a straight [line] be cut as chanced, namely the [line] AB at the point C; I say that the rectangle contained by the [lines] AB, BC [taken] along with (*meta*) the rectangle contained by the [lines] BA, AC is equal to the square from the [line] AB.
—For let there be drawn up from the [line] AB the square ADEB, and let there be drawn through the [point] C, parallel to either of the [lines] AD, BE, the [line] CF.
—Equal, then, is the [square] AE to the [rectangles] AF, CE [taken together]; and also the [square] AE is the square from the [line] AB. And the [rectangle] AF is the rectangle contained by the [lines] BA, AC—for [the rectangle AF] is contained by the [lines] DA, AC, but equal is the [line] AD to the [line] AB. And the [rectangle] CE is the [rectangle contained] by the [lines] AB, BC—for equal is the [line] BE to the [line] AB.
—Therefore the [rectangle contained] by the [lines] BA, AC, [taken] along with (*meta*) the [rectangle contained] by the [lines] AB, BC, is equal to the square from the [line] AB.
—Therefore, if a straight line be cut as chanced, then the rectangles contained by the whole [line] and each of the segments are, [taken together,] equal to the square from the whole [line]—the very thing which was to be shown.

(II) #3 If a straight line be cut as chanced, then the rectangle contained by the whole [line] and one of the segments is equal to, [taken together,] the rectangle contained by the segments and the square from the aforesaid segment.
—For let a straight [line] be cut as chanced, namely the [line] AB at the [point] C. I say that the rectangle contained by the [lines] AB, BC is equal to the rectangle contained by the [lines] AC, CB along with the square from the [line] BC.
—For let there be drawn up from the [line] CB a square, namely the [square] CDEB; and also let the [line] ED be drawn through to the [point] F; and also, through the [point] A, let there be drawn parallel to either of the [lines] CD, BE the [line] AF.
—Equal, then, is the [rectangle] AE to, [taken together,] the [rectangles] AD, CE.
—Now the [rectangle] AE is the rectangle contained by the [lines] AB, BC, for it is contained by the [lines] AB, BE—but equal is the [line] BE to the [line] BC.
—But the [rectangle] AD is the [rectangle contained] by the [lines] AC, CB, for equal is [line] DC to the [line] CB—and the [square] DB is the square from the [line] CB.
—Therefore the rectangle contained by the [lines] AB, BC is equal to the rectangle contained by the [lines] AC, CB along with the square from the [line] BC.
—Therefore if a straight line be cut as chanced, then the rectangle contained by the whole [line] and one of the segments is equal to, [taken together,] the rectangle contained by the segments and the square from the aforesaid segment—the very thing which was to be shown.

(II) #4 If a straight line be cut as chanced, then the square from the whole [line] is equal to, [taken together,] the squares from the segments and twice the rectangle contained by the segments.
—For let a straight line, namely the [line] AB, be cut as chanced, namely at the [point] C. I say that the square from the [line] AB is equal to, [taken together,] the squares from the [lines] AC, CB and twice the rectangle contained by the [lines] AC, CB.
—For let there be drawn up from the [line] AB a square, namely the [square] ADEB; and also let there be joined the [line] BD; and also, through the [point] C, let there be drawn parallel to either of the [lines] AD, EB the [line] CF, and, through the [point] G, let there be drawn parallel to either of the [lines] AB, DE the [line] HK.
—And, since parallel is the [line] CF to the [line] AD, and upon them has fallen the [line] BD, the exterior angle, namely the [one contained] by CGB—is equal to the interior and opposite [angle], namely the [one contained] by ADB. But the [angle contained] by ADB is equal to the [one contained] by ABD—since a side is equal [to a side], namely the [line] BA to the [line] AD—therefore also the angle [contained] by CGB is equal to the [angle contained] by GBC; and so also a side, the [line] BC, is equal to a side, the [line] CG. But the [line] CB is equal to the [line] GK, and the [line] CG to the [line] KB; and also, therefore, the [line] GK is equal to the [line] KB; therefore the [quadrilateral] CGKB is equilateral.
—I say next that it is also right-angled. For, since parallel is the [line] CG to the [line] BK, therefore the angles [contained] by KBC, GCB are equal to two right [angles]; but a right [angle] is what the [angle contained] by KBC is; therefore also the [angle] by BCG is a right [angle], so that also the opposite [angles contained] by CGK, GKB are right [angles].
—Rectangular, therefore, is the [quadrilateral] CGKB; and it was shown also to be equilateral. A square, therefore, is what it is; and it is [drawn up] from the [line] CB.
—On account of the same things, then, also the [figure] HF is a square; and it is [drawn up] from the [line] HG—that is [from] the [line] AC. Therefore the [figures] HF, KC are squares [drawn up] from the [lines] AC, CB.
—And since equal is the [figure] AG to the [figure] GE, and [the figure] AG is the [rectangle contained] by the [lines] AC, CB—for equal is the [line] GC to the [line] CB—therefore also the [figure] GE is equal to the [rectangle contained] by the [lines] AC, CB. Therefore the [figures] AG, GE, [taken together,] are equal to twice the [rectangle contained] by the [lines] AC, CB. But the squares HF, CK are also the squares from [the lines] AC, CB. Therefore the four [figures] HF, CK, AG, GE, [taken together,] are equal to, [taken together,] the squares from the [lines] AC, CB and twice the rectangle contained by the [lines] AC, CB. But the [figures] HF, CK, AG, GE, [taken together,] are the whole [figure] ADEB, which is a square from the [line] AB.

Therefore the square from the [line] AB is equal to, [taken together,] the squares from the [lines] AC, CB and twice the rectangle contained by the [lines] AC, CB.

Therefore if a straight line be cut as chanced, then the square from the whole [line] is equal to, [taken together,] the squares from the segments and twice the rectangle contained by the segments—the very thing which was to be shown

Figure 5 shows at a glance what Euclid states in the enunciations of these propositions. (The figure is misleading, however. It might lead you to believe that what the propositions say is so trivial as to go without saying. That's because what the demonstrations amount to is merely setting out the enunciations in drawings like those in the figure—drawings justifiable by what Book One has done, and so devised that the various boxes can be displayed simply as component parts and as a whole that they make up.)

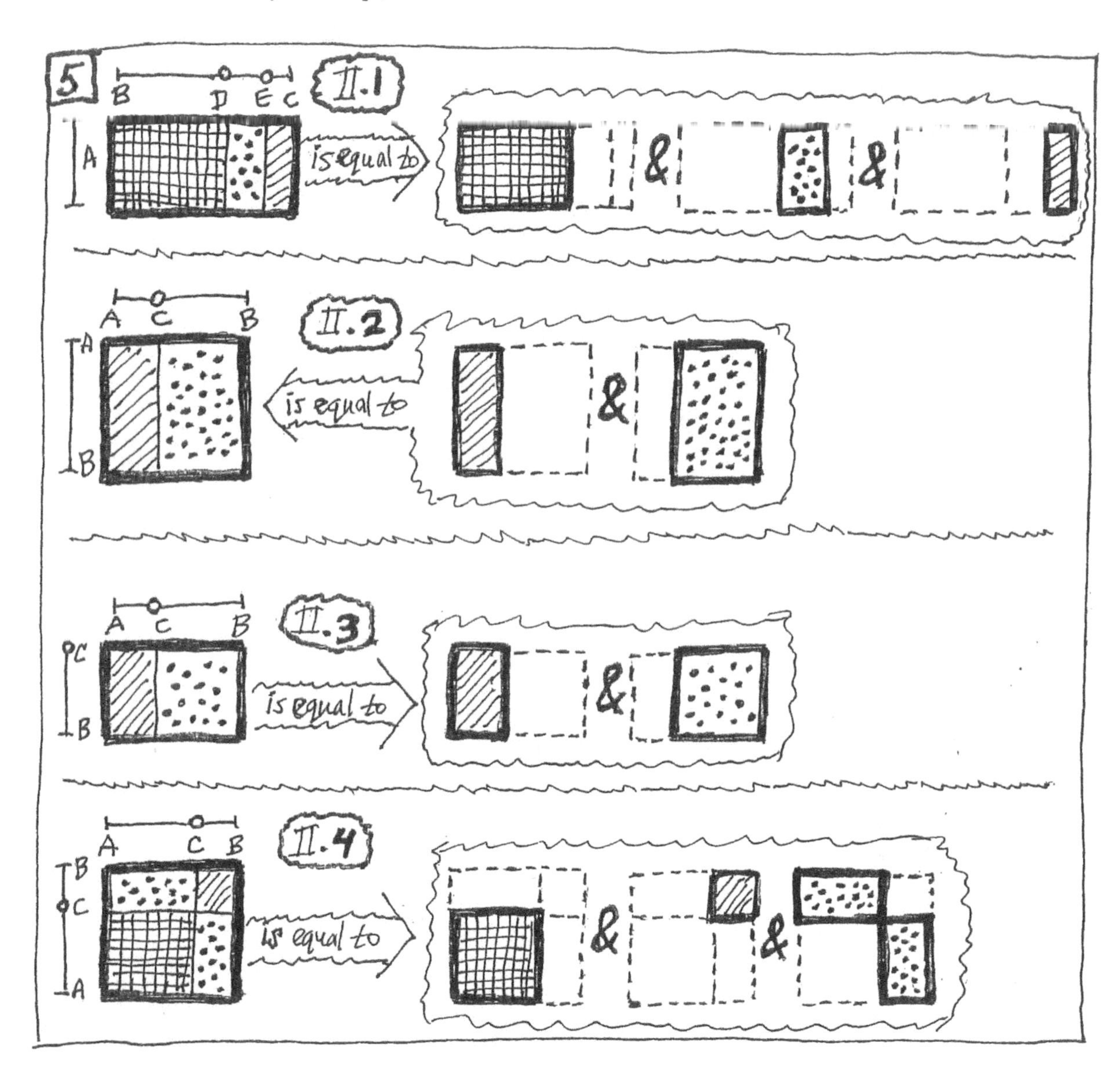

If we use some abbreviations, we can sum up these enunciations as follows:

> if there be a straight line A, and also a straight line BC that is cut at D and at E, then:
>
> (II. 1) rct [A, BC] = rct [A, BD] + rct [A, DE] + rct [A, EC]
>
> and if there be a straight line AB that is cut at C, then:
>
> (II.2) rct [AB, AC] + rct [AB, BC] = sqr [AB]
>
> (II.3) rct [AB, BC] = ret [AC, CB] + sqr [CB]
>
> (II.4) sqr [AB] = sqr [AC] + sqr [CB] + 2 rct [AC, CB].

Those statements somehow correspond to, but are not the same as the following equations, which will be familiar to someone who has studied basic algebra:

(1)	$a(b+c+d)$	$= ab+ac+ad$
(2)	$a(a+b)+b(a+b)$	$= (a+b)^2$
(3)	$b(a+b)$	$= ab+b^2$
(4)	$(a+b)^2$	$= a^2+2ab+b^2$.

Euclid's enunciations can be transformed into these equations, but the equations do not mean what those enunciations do. We shall be examining things that indicate how the statements differ. For now, let the following suffice. In these equations from algebra, the letters represent "real numbers"—which are the numbers on the number-line familiar to students in school nowadays. But in the statements from Euclid (restated using abbreviations) that appear above that list of equations which are their algebraic counterparts, the letters are merely names for endpoints of lines (except for the letter "A" that appears in the first of the statements—that "A" being the name for a line whose endpoints do not need to be named). The only number that appears in these statements (namely, the "2" in the last of them) simply gives a count, telling how many times the rectangle is to be taken.

Even though you may find the algebraic statements more familiar than Euclid's enunciations, and also easier to keep in mind, it is nonetheless easy to see what Euclid presents. Relationships that are a bit more complex are presented in his next two propositions, *Elements* II.5 and II.6. Apollonius in his *Conics* will first use *Elements* II.5 in the 15th proposition—which is just as soon as he has characterized the new, non-circular curves. When, in the 30th proposition of the *Conics*, Apollonius first uses *Elements* II.6, he will use II.5 along with it.

By way of preview for *Elements* II.5 and II.6, suppose first that you cut a straight line and use the two pieces as sides to contain a rectangle (as in figure 6). When will you get the biggest possible rectangle from so cutting a given straight line? You will get it when you cut the line in half, thus making equal sides—from which you will get a square. A square has the largest area of any rectangle with a given perimeter—or, as we have here, with a given half-perimeter (fig. 7). (*Elements* VI.27 supplies a reason for believing that the square indeed is largest.)

Suppose now that, having cut the straight line AB equally (at its mid-point M), you then cut it also unequally (at any other point K). What is the difference in size between the square and the rectangle which these segments can make—what is the difference, that is to say, between the square whose sides are the equal halves of the given straight line, and the rectangle which is contained by the unequal segments of the line? It turns out that the difference in area is equal to the square whose side is the line between the two points of section, M and K. That is (again using abbreviations):

sqr [MA, MB] — rct [KA, KB] = sqr [MK]. (Fig. 8.)

Now, however, suppose that, having cut the straight line AB equally, you again take some point that is unequally distant from the endpoints A and B of the given line, but this time you do so not internally but externally. That is, take some point K' which is on the prolongation of the given line. Before, when the unequal segments were lines KA and KB, drawn from the *internal* point of section K to the endpoints A and B of the given line, the rectangle contained by those unequal segments had to be smaller than the square contained by the equal segments (which are the halves, AM and MB, made by midpoint M). Now, however, when the unequal lines are K'A and K'B, drawn from the *external* point K' to the given line's endpoints A and B, the rectangle contained by those lines does not have to be smaller than the square contained by the equal segments. But it does again turn out to be equal to the difference in area between the square contained by the equal segments and the square whose side is the line between the two points of section. That is:

rct [K'A, K'B] + sqr [MA, MB] = sqr [K'M]. (Fig. 9.)

Thus, when you take a point that is equally distant from the endpoints of a given straight line and you also take another point, one that is unequally distant from the endpoints (whether inside the line, or outside it but in line with it), then the rectangle contained by the unequal sides is the difference between two squares—the square whose side is the half, and the square whose side is the line between the points. The square on the half is the larger square in the case of the internal point, and the smaller square in the case of the external point.

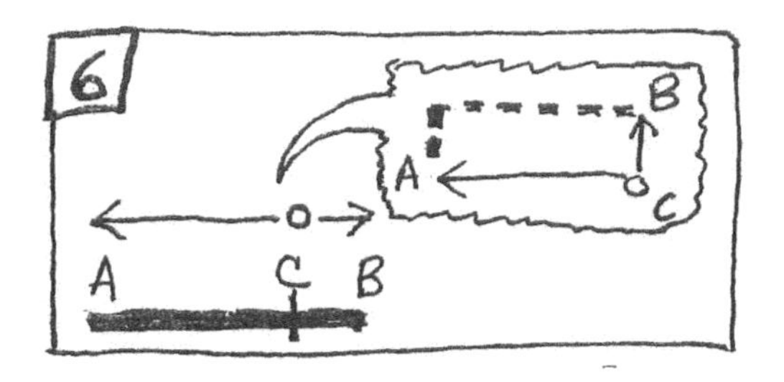
6
A
B
C
A C B

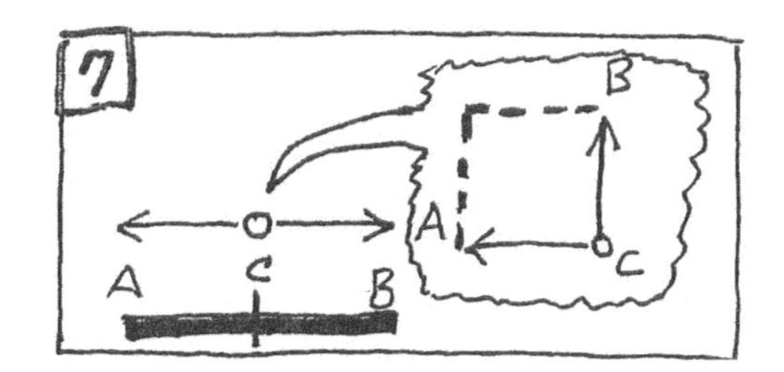
7
A
B
C
A C B

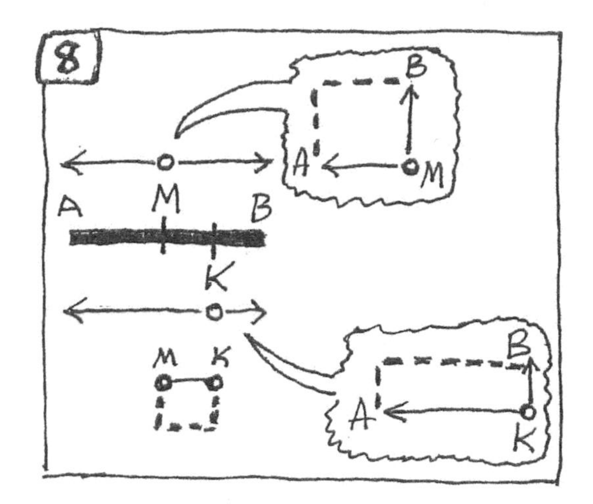
8
A
B
M
A M B
K
M K
A
B
K

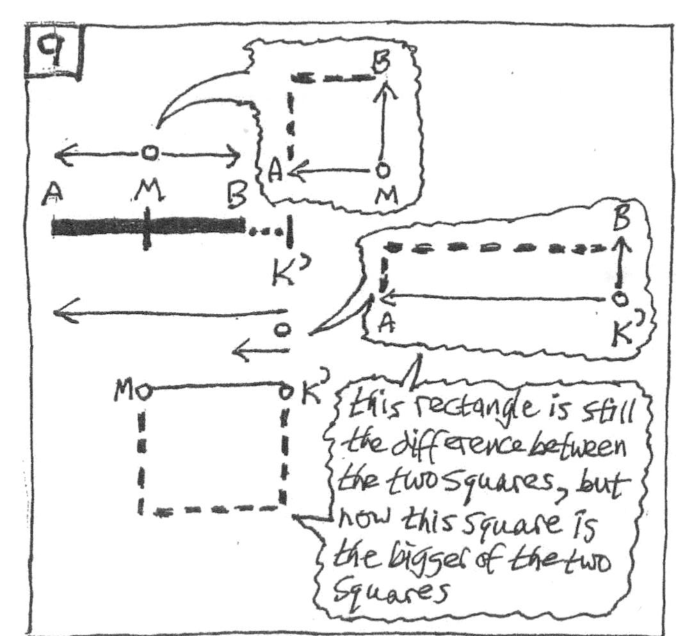
9
A
B
M
A M B
K'
A
B
K'
M K'
this rectangle is still the difference between the two squares, but now this square is the bigger of the two squares

In the 5th proposition, Euclid treats the internal point; and in the 6th, the one that is external. Here are the two propositions. In the demonstrations, blocks are drawn on the straight lines—smaller blocks within larger blocks—and then the blocks are rearranged. The presentation of these two propositions is followed by help in sorting out the demonstrations: figures 10, 11, 12, and 13.

(II) #5 If a straight line be cut into equal and unequal [segments], then the rectangle contained by the unequal segments of the whole, [taken] along with the square from the [line] between the cut[point]s, is equal to the square from the half.

—For let some straight [line], say the [line] AB, be cut into equal [segments] at the [point] C, and into unequal [segments] at the [point] D. I say that the rectangle contained by the [lines] AD, DB, [taken] along with the square from the [line] CD, is equal to the square from the [line] CB.

—For let there be drawn up from the [line] CB a square, namely the [square] CEFB; and also let there be joined the [line] BE; and also, through the [point] D, let there be drawn parallel to either of the [lines] CE,BF the [line] DG, and, again, through the [point] H, let there be drawn parallel to either of the [lines] AB, EF the [line] KM, and, again, through the [point] A, let there be drawn parallel to either of the [lines] CL, BM the [line] AK.

—And, since equal is the complement CH to the complement HF, let a common [figure] be put to, [that is, added to,] each complement—namely the [common figure] DM. The whole [figure] CM, therefore, is equal to the whole [figure] DF.

—But the [figure] CM is equal to the [figure] AL—since also the [line] AC is equal to the [line] CB—therefore also the [figure] AL is equal to the [figure] DF. Let a common [figure] be put to, [that is, added to each]—namely the [common figure] CH. The whole [figure] AH, therefore, is equal to the gnomon NOP.

—But the [figure] AH is the [rectangle contained] by the [lines] AD, DB—for DH is equal to DB—therefore also the gnomon NOP is equal to the [rectangle contained] by AD, DB. Let a common [figure] be put to, [that is, added to each] —namely the [common figure] LG, which is equal to the square from the [line] CD.

—Therefore the gnomon NOP and the [figure] LG, [taken together,] are equal to, [taken together,] the rectangle contained by the [lines] AD, DB and the square from the [line] CD. But the gnomon NOP and the [figure] LG, [taken together,] are the whole square CEFB which is [drawn up] from the [line] CB.

—Therefore the rectangle contained by the [lines] AD, DB, [taken] along with the square from the [line] CD, is equal to the square from the [line] CB.

—Therefore if a straight line be cut into equal and unequal [segments], the rectangle contained by the unequal segments of the whole, [taken] along with the square from the [line] between the cut[point]s, is equal to the square from the half—the very thing which was to be shown.

(II) #6 If a straight line be cut in two [equal parts], and there be put to it some straight [line] [which is added on to it] in a straight line, then the rectangle contained by [as the one side] the whole [line taken together] with the added [line] and [as the other side] the added line [itself], then [that rectangle, taken] along with the square from the half, is equal to the square on the straight line [made] out of the half and the added [line].
—For let some straight [line], say the [line] AB, be cut in two [equal parts] at the point C, and let there be put to it some straight [line] in a straight [line], namely the [line] BD. I say that the rectangle contained by the [lines] AD, DB, [taken] along with the square from the [line] CB, is equal to the square from the [line] CD.
—For let there be drawn up from the [line] CD the square CEFD; and also, let there be joined the [line] DE; and also, through the point B, let there be drawn parallel to either of the [lines] EC, DF the [line] BG, and, through the point H, let there be drawn parallel to either of the [lines] AB, EF the [line] KM, and also, again, through the [point] A, let there be drawn parallel to either of the [lines] CL, DM the [line] AK.
—Then, since equal is the [line] AC to the [line] CB, equal is also the [figure] AL to the [figure] CH—but the [figure] CH is equal to the [figure] HF—therefore the [figure] AL also is equal to the [figure] HF. Let a common [figure] be put to, [that is, added to each]—namely the [common figure] CM. The whole [figure] AM, therefore, is equal to the gnomon NOP.
—But the [figure] AM is the [rectangle contained] by the [lines] AD, DB—for equal is the [line] DM to the [line] DB—therefore also the gnomon NOP is equal to the [rectangle contained] by the [lines] AD, DB. Let a common [figure] be put to, [that is, added to each]—namely the [common figure] LG, which is equal to the square from the [line] BC.
—Therefore the rectangle contained by the [lines] AD, DB, [taken] along with the square from the [line] CB, is equal to, [taken together,] the gnomon NOP and the [figure] LG. But the gnomon NOP and the [figure] LG, [taken together,] are the whole square CEFD which is [drawn up] from the [line] CD.
—Therefore the rectangle contained by the [lines] AD, DB, [taken] along with the square from the [line] CB, is equal to the square from the [line] CD.
—Therefore if a straight line be cut in two [equal parts], and there be put to it some straight [line] [which is added on to it] in a straight line, then the rectangle contained by [as the one side] the whole [line taken together] with the added [line] and [as the other side] the added line [itself], then [that rectangle, taken] along with the square from the half, is equal to the square on the straight line [made] out of the half and the added [line]—the very thing which was to be shown.

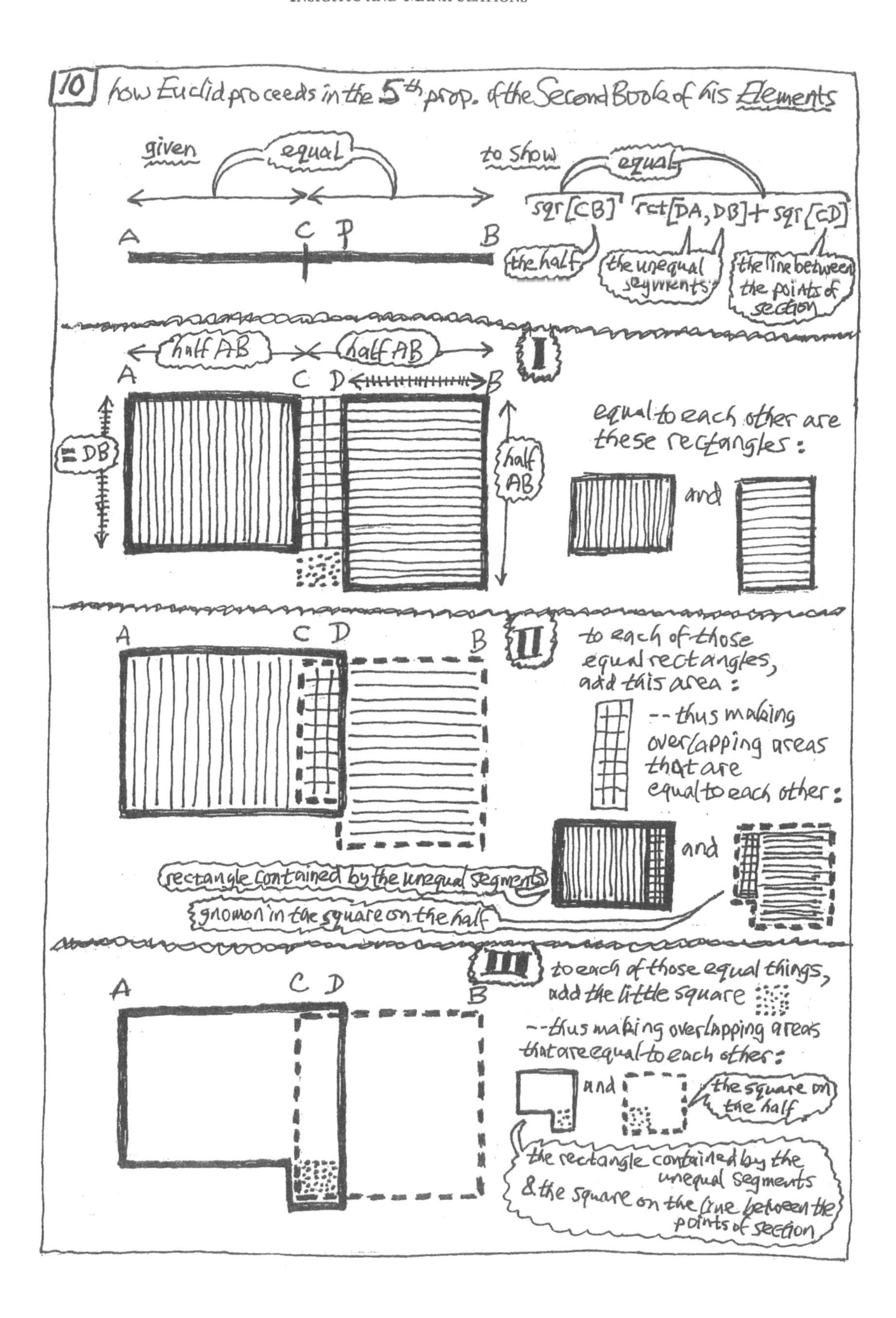
10 how Euclid proceeds in the 5th prop. of the Second Book of his Elements
given
equal
to show
equal
sqr[CB]
rct[DA, DB] + sqr[CD]
A
C
D
B
the half
the unequal segments
the line between the points of section
half AB
half AB
I
A
C
D
B
= DB
half AB
equal to each other are these rectangles:
and
A
C
D
B
II
to each of those equal rectangles, add this area:
-- thus making overlapping areas that are equal to each other:
and
rectangle contained by the unequal segments
gnomon in the square on the half
III
to each of those equal things, add the little square
-- thus making overlapping areas that are equal to each other:
A
C
D
B
and
the square on the half
the rectangle contained by the unequal segments & the square on the line between the points of section

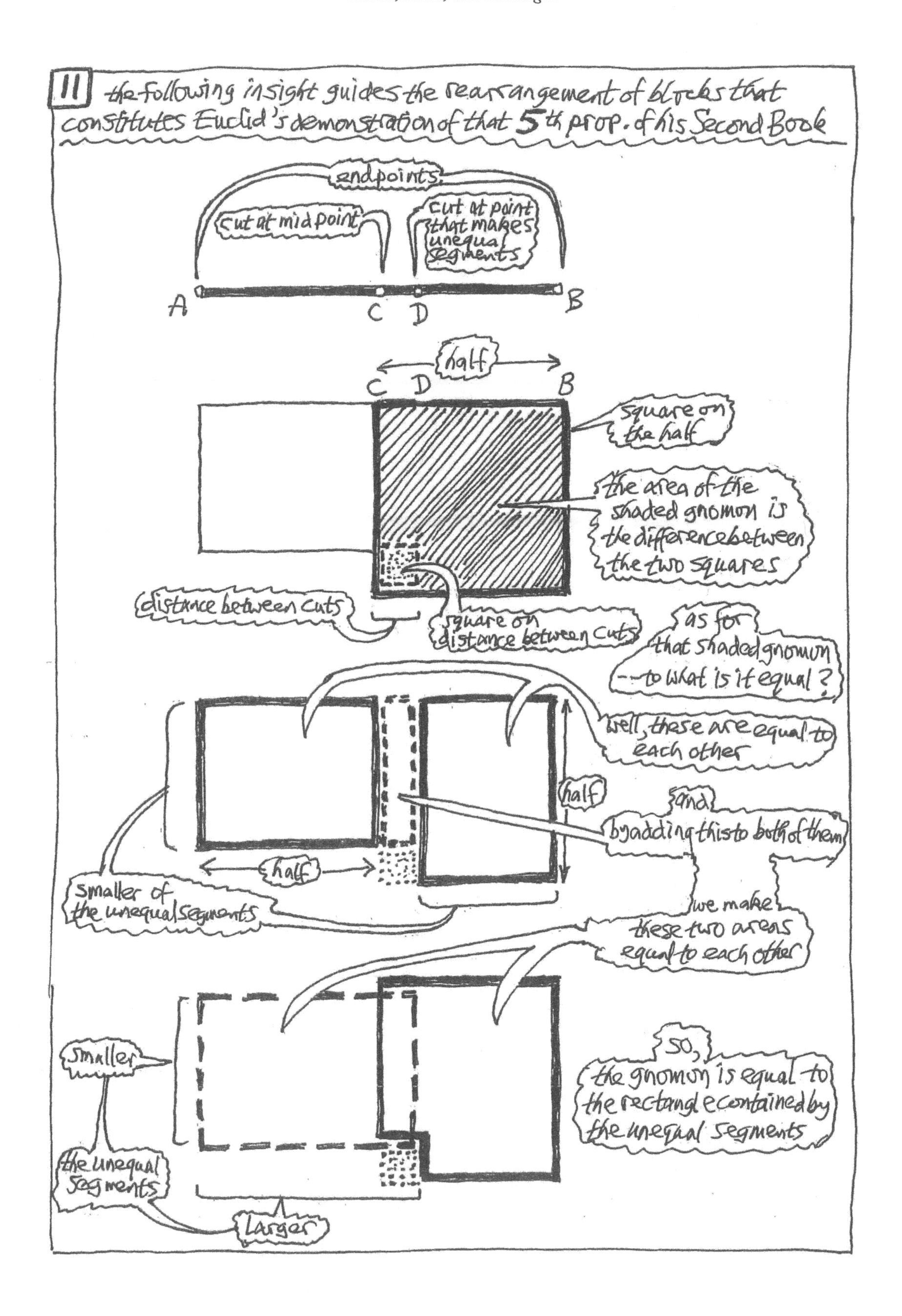
11 the following insight guides the rearrangement of blocks that constitutes Euclid's demonstration of that 5th prop. of his Second Book
endpoints
cut at mid point
cut at point that makes unequal segments
A
C
D
B
half
C D B
square on the half
the area of the shaded gnomon is the difference between the two squares
distance between cuts
square on distance between cuts
as for that shaded gnomon -- to what is it equal?
well, these are equal to each other
half
and by adding this to both of them
half
smaller of the unequal segments
we make these two areas equal to each other
smaller
the unequal segments
larger
so, the gnomon is equal to the rectangle contained by the unequal segments

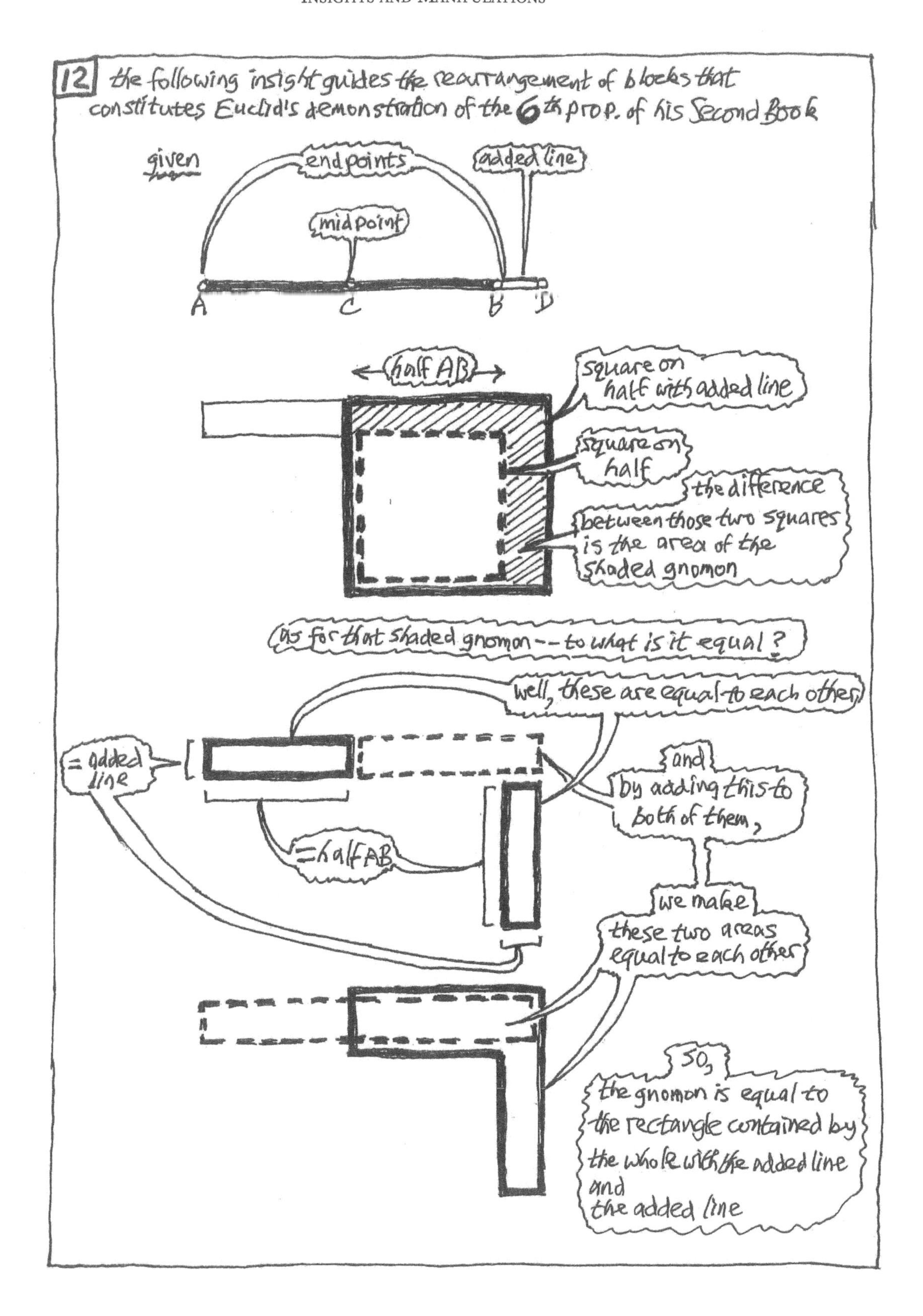
12 the following insight guides the rearrangement of blocks that constitutes Euclid's demonstration of the 6th prop. of his Second Book
given
endpoints
added line
midpoint
A
C
B
D
half AB
square on half with added line
square on half
the difference between those two squares is the area of the shaded gnomon
as for that shaded gnomon -- to what is it equal?
well, these are equal to each other,
= added line
and by adding this to both of them,
= half AB
we make these two areas equal to each other
so, the gnomon is equal to the rectangle contained by the whole with the added line and the added line

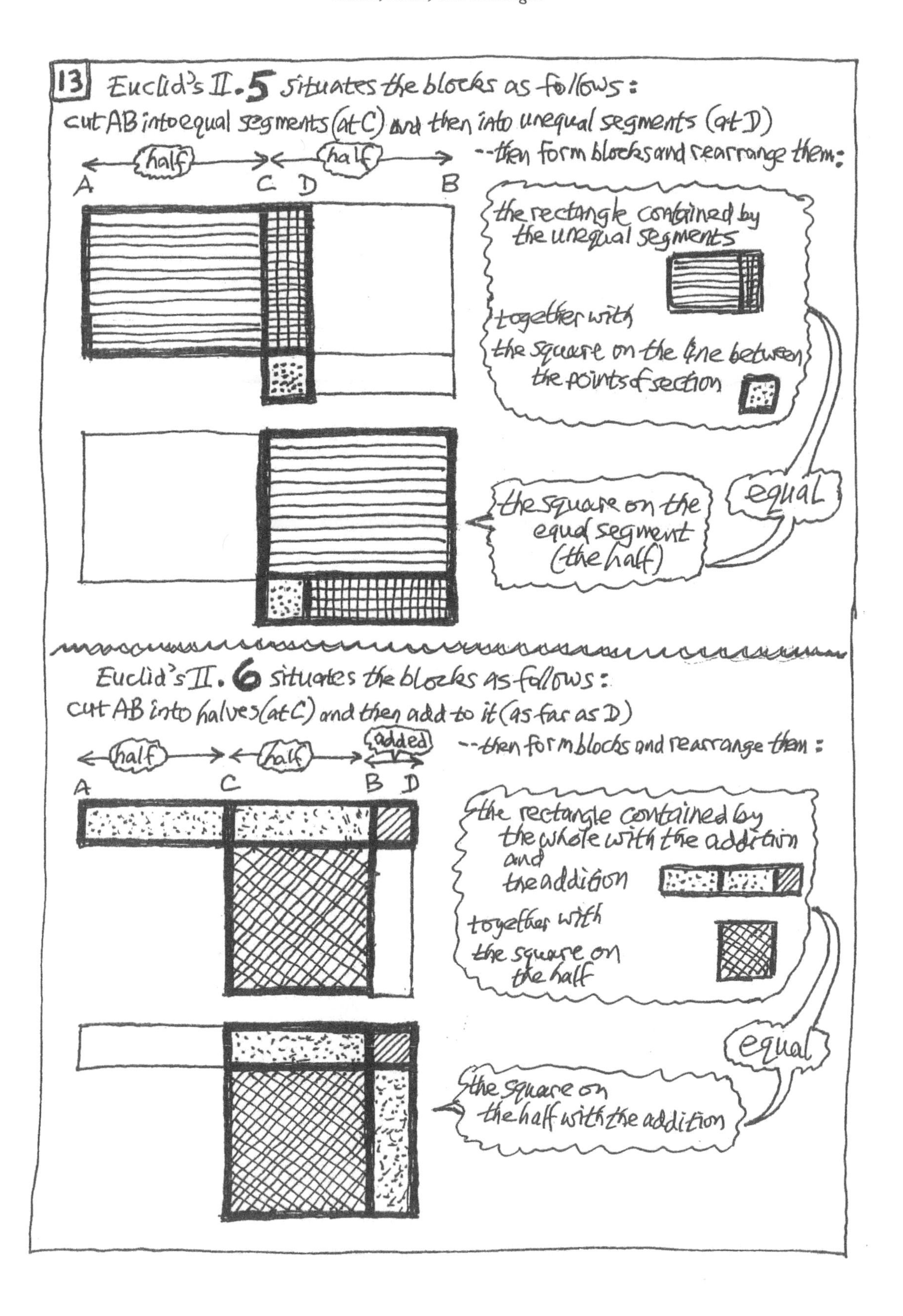
13
Euclid's II.5 situates the blocks as follows:
cut AB into equal segments (at C) and then into unequal segments (at D)
half
half
--then form blocks and rearrange them:
A
C
D
B
the rectangle contained by the unequal segments
together with
the square on the line between the points of section
equal
the square on the equal segment (the half)
Euclid's II.6 situates the blocks as follows:
cut AB into halves (at C) and then add to it (as far as D)
half
half
added
--then form blocks and rearrange them:
A
C
B
D
the rectangle contained by the whole with the addition and the addition
together with
the square on the half
equal
the square on the half with the addition

As is shown by the figures that we have just seen, the insight guiding the rearrangement of blocks in the demonstration of Euclid's II.6 is similar to the insight guiding the rearrangement in II.5. That insight suggests paying attention to the difference of the squares in each of the propositions. In doing so, we are led to treat the point that is external to the line as a counterpart of the cutting-point that is internal to it. Now it would be very convenient if we could treat Euclid's two propositions as merely two cases of the same thing—and we could indeed do so if we would just allow ourselves to speak of a line as being "cut *externally*."

Euclid himself does not speak of an external cut. He speaks, rather, of a line added on to a given straight line. But if we treated both of his propositions as involving the unequal cutting of a line, we then would give ourselves less to remember. We would only need to say this:

> When a straight line is cut equally and also unequally—
> regardless of whether the cut that is unequal is *internal* or is *external*—
>
> then the rectangle contained by the unequal segments is equal to
> the difference between two squares, namely,
> the square on the half, on the one hand, and, on the other,
> the square on the line between the two cutting-points
> (one of these points cutting the line equally, and the other cutting it unequally).
> (Fig. 14.)

Does it make sense, however, to speak of cutting a line *outside* the line—after all, must not a cut be *inside* what is being cut? What, then, have we just done by introducing the term "an *external* cut"—have we, for the sake of great convenience, abandoned common sense? Or would it be better to say that we have changed the meaning of our terms, stretching them in order to increase our power? Is something like that what happened (we shall, eventually, want to ask) when the practitioners of classical geometry were repudiated algebraically by Descartes?

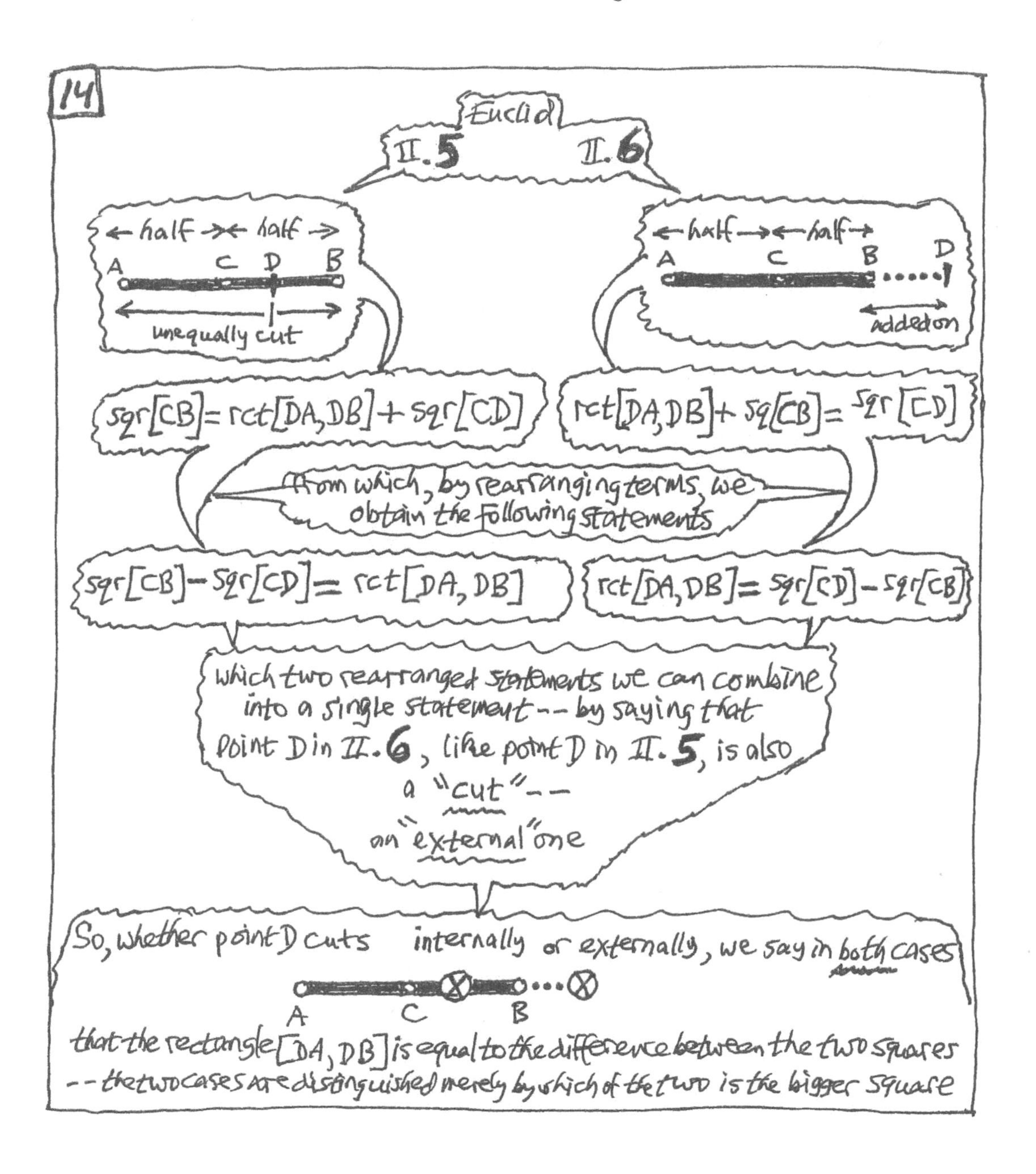
14
Euclid
II.5
II.6
← half →← half →
A C D B
unequally cut
← half →← half →
A C B D
added on
sqr[CB] = rct[DA, DB] + sqr[CD]
rct[DA, DB] + sq[CB] = sqr[CD]
From which, by rearranging terms, we obtain the following statements
sqr[CB] − sqr[CD] = rct[DA, DB]
rct[DA, DB] = sqr[CD] − sqr[CB]
which two rearranged statements we can combine into a single statement -- by saying that point D in II.6, like point D in II.5, is also a "cut" -- an "external" one
So, whether point D cuts internally or externally, we say in both cases
A C B
that the rectangle [DA, DB] is equal to the difference between the two squares -- the two cases are distinguished merely by which of the two is the bigger square

CHAPTER 2. ALGEBRA

We have just seen how the sizes of certain straight lines can be related without any reference to ratios—by building rectangles out of them and comparing those rectangles. We shall soon see another way to relate the sizes of certain straight lines—by compounding their ratios so as to get rectangles that are then compared. This can, however, get complicated and difficult.

It is true that the complications and difficulties can be avoided if we submit without resistance to the algebraic temptation to ease our way by operating with signs that do not keep our eyes on what our signs refer to. Eventually, with Descartes, we shall indeed turn our attention to our signs and the workings of our minds; but by postponing that turn, we shall be better able to appreciate the motive for it. Only after we have looked with care at older forms shall we take up their transformation—and then, in the spirit of Descartes, we shall ask whether our learning has been only incidentally a learning about shape, and been more essentially a learning about learning to achieve mastery. We shall ask whether learning is a removal of what gets in the way of our seeing all things as "images" of beings that we "see with the mind's eye," or is rather a "hands-on" acquisition of a power of "mental manipulation."

For a while, however, we shall be talking not about configurations of signs that we manipulate, but about figures that we look at.

What we shall be talking about is not so much visual as visualizable. Much of our difficulty in reading texts of classical geometry is that the author makes his presentation not in drawings but in sentences. The sentences must therefore be long and complex. The reader can see the point only by visualizing in a diagram the succession of things said by the author. The diagram that is drawn need not be exact. For determining the sizes of things, what is relied upon is what is said, not what is seen. Appeal is made to the diagram, not for determining the sizes of things, but rather for keeping in mind the arrangements of things. So, in the very first proposition of Euclid's *Elements* (which is this: to construct an equilateral triangle on a straight line with given bounds), when two large circles are drawn from the straight line's endpoints, it is not necessary to demonstrate that these circles have a point in common. It goes without saying, that a circle cannot pass from being outside another circle to being inside it without somewhere crossing it (fig. 1). No matter how lumpy a circle you may draw, you will not visualize the circles as non-intersecting. You would not need an axiom of continuity here to show what follows from the visualization of form; you would, however, need it if you wanted instead to engage in a merely formal manipulation of strings of signs.

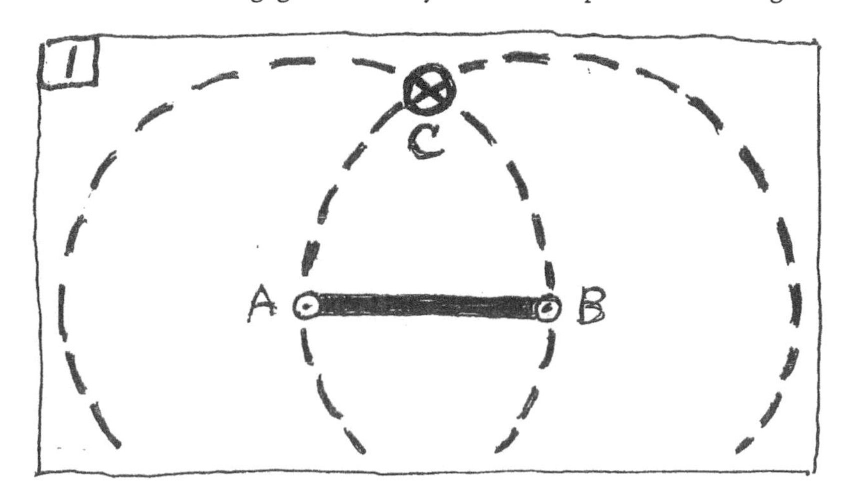

The very first proposition of Euclid's *Elements* makes it manifest that classical geometry is *speech* about what is *visualizable*. Here is Euclid's first proposition:

(I) #1 On a given delimited straight [line], to construct an equilateral triangle.
—Let the given delimited straight [line] be what the [line] AB is. Required, then, on the [line] AB to construct an equilateral triangle.
—With center the [point] A and distance the [line] AB let a circle be carved out, namely the [circle] BCD; and also, again, with the center [point] B and distance the [line] BA let a circle be carved out, namely the [circle] ACE; and also, from the point C at which the circles cut each other, to the points A, B let there be joined straight [lines], namely the [lines] CA, CB.
—And, since the point A is center of the circle CDB, equal is the [line] AC to the [line] AB. Again, since the point B is center of the circle CAE, equal is the [line] BC to the [line] BA. But it was shown also that the [line] CA is equal to the [line] AB. Therefore each of the [lines] CA, CB is equal to the [line] AB. But the equals to the same thing are also equals to each other; therefore also the [line] CA is equal to the [line] CB. Therefore the three [lines] CA, AB, BC are equal to each other. Equilateral, therefore, is the triangle ABC—and it is also constructed on the given delimited straight [line] AB.
—On a given delimited straight [line], therefore, an equilateral triangle has been constructed—the very thing which was required to be done.

Note, by the way, that the given straight line is here called *peperasmenê* not to distinguish it from any straight lines that might nowadays be considered infinite, but rather to indicate that besides having a certain size the straight line has definite endpoints or bounds or limits (each rendered in Greek as a *peras*) that will figure in the construction. That's why the usual translation "finite" is here replaced by "delimited."

As this proposition shows us, what Euclid does in his *Elements* differs greatly from what, for example, Henry George Forder proposes to do in his book entitled *The Foundations of Euclidean Geometry* (New York, 1958, Dover re-publication of the Cambridge University Press edition of 1927). Let us consider Forder's characterization of his enterprise.

On the first page of his book, Forder says: "The object of this work is to show that all the propositions of Euclidean geometry follow logically from a small number of Axioms explicitly laid down, and to discuss to some extent the relations between these Axioms. The most famous work whose aim was a logical deduction of the propositions of Geometry is of course that of Euclid, but many flaws have been noticed in his treatment during the two thousand years that have elapsed since his work was written. In particular, Euclid almost completely ignored the relations of order, suggested by such words as 'between,' 'inside,' though such relations are, in fact, of great importance in a deductive treatment. They are fundamental in much of the present work."

"*Any* set of entities satisfying the Axioms will give a representation of the Geometry," Forder later says. "If we illustrate our argument by figures, nothing save what is explicitly stated and deduced may be used from these figures. Theoretically, figures are unnecessary; actually they are needed as a prop to human infirmity. Their sole function is to help the reader to follow the reasoning; in the reasoning itself they must play no part" (pp. 42–43).

Forder then goes on to say: "Most of our Axioms, if regarded as treating of the Physical Space in which we live, would be considered trivial, but the fact that all Euclidean Geometry follows from them is most surprising and anything but trivial and it is the main purpose of this book to show just this fact. Our Axioms and Definitions could also be regarded as a final analysis of the properties of Euclidean Space." Although Euclidean geometry may, as a matter of fact, apply to our physical world, Forder urges us to ignore that fact: "Whenever possible, it is best to take the formalist view of our deductions and to regard the investigation as a game played in accordance with our rules, the Axioms, and starting from the fixed positions of the pieces given in the hypotheses of the theorems. And in order that no unstated assumption may creep in, we must move slowly and warily ... Our Geometry is an abstract Geometry. The reasoning could be followed by a disembodied spirit who had no idea of a physical point; just as a man blind from birth could understand the Electromagnetic Theory of Light."

Now such "foundations of Euclidean geometry" are not Euclid's *Elements*. Euclid, and those (like Apollonius) who followed him, were engaged in a different sort of study. We have already seen Apollonius rely upon visualizability in a matter that did not involve the determination of size. That was in the *Conics*' 6th proposition, where a line from D when extended had to meet line AH (fig. 2). There, since DE is parallel to the base, and hence (by the definition of parallelism) cannot meet line KH, it must hit the other side. Why? Because (as goes without saying for classical geometry) if a straight line goes into a triangle through one side, it—or its extension—must eventually come out through another side. None of Euclid's postulates or axioms asserts this, nor do any of his propositions demonstrate it. But if we merely visualize a triangle, we can "see" that it must be true. Modern students of geometry who refuse to rely on any visualizable properties, even properties that do not involve any determination of size, have made this an axiom. It is called Pasch's axiom, after Moritz Pasch, who led the movement to axiomatize mathematics.

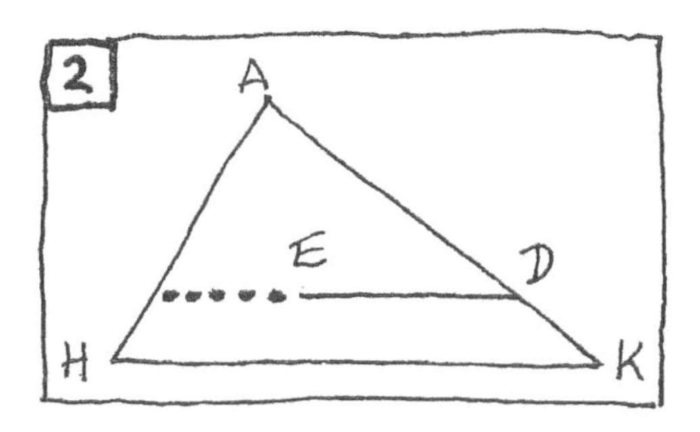

On ground prepared by Pasch, following the development of geometries that were non-Euclidean, Hilbert laid down his *Foundations of Geometry*, from which is derived Forder's own project to go beyond Euclid. For a brief, clear, introductory presentation of the attitude toward geometry exemplified by Forder's book, you might look at "Chapter VII: The Nature of a Logical or Mathematical System," in Morris R. Cohen and Ernest Nagel, *An Introduction to Logic and Scientific Method* (New York, 1934).

We might say (shifting metaphors) that the step taken by Hilbert came far down the path along which the study of form was transformed into a formalist study. An earlier step, one as important as any other along the way, had to do with disregarding the distinction between how-many and how-much in the handling of relationships of size. This involved the transformation of abbreviations into symbols.

If we turn from looking at figures, to considering the configurations of terms that we have used in talking about the figures that we look at, then we may be tempted to take the *abbreviations* that we use for easing the strain of seeing things, and transform them into *symbols* that are instruments in a system for promoting manipulability.

The language of abbreviations is a language in which the things signified are not themselves signs. We merely use simpler signs to continue referring to whatever things we have been referring to by using more complex signs. Symbols, however, are signs that have no *immediate* reference to things outside of the symbol system of which they are a part, though the language of symbols may be usefully *applied* to things outside the system of symbols.

For example, when we are making inferences and we speak of "substitution," we are referring to the manipulation of signs in a system of signs, rather than referring to any things signified outside the system of signs. A proposition of Euclid or of Apollonius does not speak of substitution. Instead of speaking of substitution, classical writers say (for example) that "two things are equal to each other if each of them is equal to the same third thing," or that "two ratios are the same if each of them is the same as some third ratio." Only by appealing to statements like these do we know when it is permissible to rewrite a string of signs with some substitution, and when it is not. If A and B are lines, for example, that permit us to write "A = B" then

permission to	but permission to
write this: B + C = D	write this: B ⊥ C
will permit us to	will *not* permit us to
write this as well: A + C = D	write this as well: A ⊥ C.

Thus, when we employ strings of signs to help us speak about equality, we rather freely engage in the act of substituting one sign for another; but we are more restrained in making substitutions when what we wish to speak about involves perpendicularity.

Even if you are stepping into another frame of mind when you adopt the language of substitution, it is easy to see why you did it, and hence it is easy to retrace your path. But it is much harder to find your way if, rather than compounding ratios (as the classical geometers do), you instead adopt, for linear quantities, the numerical language of multiplication and division—and that is what you would be doing if you were to write "P = A * B," where "A" and "B" and "P" do not stand for the outcome of some act of counting but rather are "real numbers." These so-called "real numbers"—what the student of basic algebra gets used to thinking of as all of the numbers, including negatives, fractions, irrationals like the square root of two, and even those irrationals called transcendentals, like pi—these in their totality result from making a complicated connection between straight lines and the numbers with which we count.

Students nowadays get used to hearing something like the following. Take a straight line of infinite extent; call some point on it the origin, and, taking some other point on the line, call the distance between the two points a unit distance; then you will have "the number-line," because every "real number" will correspond to one and only one point on the line, and every point on the line will correspond to one and only one "real number."

Of course, the only points as yet that you have labeled are zero and one. But once you have located the point for zero and the point for one, you can easily find the point that will have to correspond to any positive or negative integer or fraction that you choose. Without tremendous difficulty you can also find the points that will have to correspond to the roots of any non-negative integer. Thus you will see that for every positive or negative integer, fraction, or root of a non-negative integer or fraction that you might choose, there is a point (and only one point) to associate with such a "real number" in this enterprise (fig. 3).

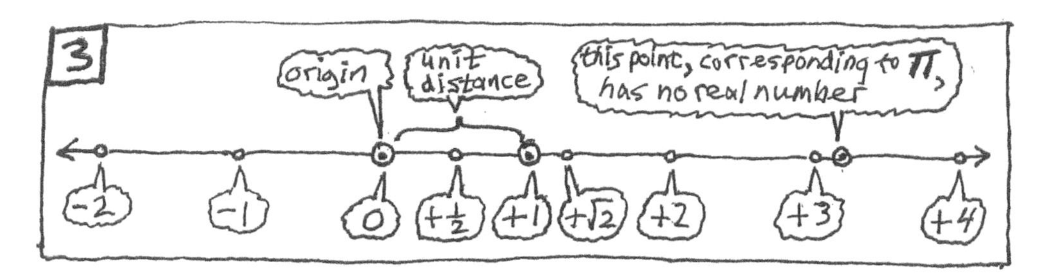

But even after every such "real number" has been in that way associated with a point, some points will still be left without any number. Indeed, it can be shown that there are infinitely many of them that would still lack a number. To suppose "the real number system," however, is to suppose that for every point on the line there is some number—though it would take some doing to say what that means. In this guidebook, we shall not try to do that. But we shall try to understand what led to the attempt to fuse the study of lines and numbers. We shall be considering classical mathematics and the transformation that led to the prevalence of non-classical mathematics in our world. Until we have examined that transformation—that is to say, as long as we are still doing classical geometry—we shall be saying things like "the square arising from line C is equal to the rectangle contained by lines A and B," or "the square on C is equal to the rectangle in A and B," rather than saying "C-squared equals A-times-B." We shall *eventually* say things like "C-squared equals A-times-B," but only after we have gone along the road that shows the reasons *why*. If we have grown up using "the number-line," it does not take much effort to manipulate "A * B," whatever A and B may be; but to say just what we mean by doing so is not very easy.

We can pretty easily say what we mean when we speak of taking some number of units another number of times, or of taking some magnitude some number of times. If three units be taken four times, for example, the product will be the same number of units as if two units be taken six times. Or if, for example, a line be taken three times as long as another line, then it will be the side of a square nine times as great.

But it would be difficult to say what it would mean to take line A, not (for example) *three* times, but rather "*line-B* times."

We cannot avoid the difficulty by saying that "line-A times line-B" merely means "the product of multiplying the number of units long that A is, by the number of units long that B is." Consider the strange results to which that would lead. Suppose that line B is to line A as 3 is to 2, and that we take "A times B" and call it C. How does the size of C compare with the size of A? Surely it ought *not* to depend on the size of the units in which we choose to measure the lines—and yet, by that definition of multiplication which was just given, it *will* depend on the size of the units. If line A is 400 inches long, and line B is 600 inches long, then line C will be 240,000 inches long—600 times as long as A. But if we measure the lines in centimeters, then line A will be 1,016 centimeters long, and line B will be 1,524 centimeters long, and so line C will be 1,548,384 centimeters long—which is 1,524 times as long as line A. Thus line C, the product of line A and line B, will be six hundred times as long as line A if we measure the lines in inches; whereas if we measure in centimeters, then it will be more than fifteen hundred times as long.

This indefiniteness of the result of using the definition is not the only difficulty. There is also the following difficulty: the definition cannot be used at all when lines A and B are incommensurable.

For example, what exactly would be the product of multiplying together the numbers that we would get by measuring *together* a square's side and its diagonal? We cannot say. Why not? Because the side of a square and its diagonal are incommensurable. Whenever some unit does exactly measure one of those two lengths—either the side or the diagonal—then, no matter how small may be the unit that measures the one, it will not also exactly measure the other. In other words, there cannot be any pair of numbers such that the ratio which one of the numbers has to the other is the same as the ratio which a square's diagonal has to its side.

To see why that is so, we need to be clear about what is meant when we speak of *numbers* here. Let us therefore now examine the notion of number that is set out in Euclid's *Elements*.

CHAPTER 3. NUMBERS

In classical mathematics, a number is a number of *things of some kind*. When there are more than one of something, their number is the "multitude" of them. It tells how many of them there are. Suppose, for example, that in a field there are seven cows, four goats, and one dog. If we count cows, then the cow is our unit, the cow being that according to which any being in the field will count as one item; and there is a multitude of seven such units in the field. If we count animals, however, then the animal is our unit; and now there is a multitude of twelve units. Of cows, there are seven; but of animals, there are twelve.

Of dogs as dogs, there is no reason to make a count (as distinguished from including them in the count of animals). There is no reason to count dogs, for the field does not contain dogs. It contains only one dog, and one dog alone does not constitute any multitude of dogs. If what we were thinking was "There is *only* one dog in the field," we might say "There is *one* dog in the field"; but if we were not thinking of that single dog in relation to some possible multitude of dogs, we would simply say "There is *a* dog in the field."

Of pigs, there is no reason to make a count, for the field does not contain a single pig, let alone a multitude of pigs.

If there had earlier been six horses in the field, which were then lent out, the field might now be said to lack six horses; but while six is the multitude of horses that are lacking *from* the field, there is not any multitude of horses *in* the field.

If we chop up one of the cows, arranging to divide its remains into four equal pieces, and we take three of them, then we have not taken any multitude of cows. We have taken merely three pieces that are each a fourth part in the equal division of what is a heap of the makings for beef stew.

And so, when we have several items, each of which counts as one of what we are counting, then what we obtain by counting is a number. The numbers, in order, are these: two units, three units, four units, and so on. The numbers, we might say are these: a duo, a trio, a quartet, and so on. To us, nowadays, that seems strange: a lot seems to be lacking.

"One" does not name a multitude of units: it is true that a unit can be combined with, or compared with, any number of units, but a unit is not itself a number—so "one" does not name a number.

"Zero" also does not name a number, for there is no multitude of units when there are no units for a multitude to be composed of.

Neither does "negative-six" name any multitude, although "six" does. Six is a number, but there is no number named "negative-six"—and hence there is no number named "positive-six" either.

"Three-fourths" does not name a number of units, although "three" does, and so does "four." We can, if we wish, break things apart to get a *new* unit—say, a fourth part of a cow's equally divided carcass—and then take three of these new units; but the old unit that has been broken up, being no longer unitary, has ceased to be the unit. The same Latin word that gives us "fracture" gives us "fraction." A fraction is not a number.

As for "the square root of two," it is not a number; indeed, as we shall soon see, it is not even a fraction. And neither is "pi" a number.

To say it again, the numbers are: two units, three units, four units, and so on—taking as many units as we please. That is what Euclid tells us as he begins the Seventh Book of his *Elements* with the following definitions:

(VII) Definitions

—An "unit" (*monas*) is that according to which each of the beings is said [to be] one (*hen*); and a "number" (*arithmos*) is a multitude (*plêthos*) put together (*sugkeimenon*) out of (*ek*) units.

After thus telling us what the numbers are, Euclid goes on to speak of the relations among numbers of different sizes, in the following definitions:

—A "part" (*meros*) is what a number is of a number, the lesser of the greater, whenever [the lesser] may measure (*katametrêi*) the greater; and "parts" *(merê)* whenever it may not measure it; and a "multiple" (*pollaplasios*) is the greater of the lesser, whenever it may be measured by the lesser.

We learn, for example, that because fifteen "measures" sixty—that is to say, because a multitude of fifteen units taken four *times* is equal to a multitude of sixty units—fifteen is called "part" of sixty, and sixty is called a "multiple" of fifteen.

But fifteen is not "part" of forty. That is because you cannot get a multitude of forty units by taking a multitude of fifteen units some number of times—which is to say that the multitude of fifteen units does not "measure" forty units. Both fifteen units and forty units are, however, measured by five units; and five units, which is the eighth part of forty units (the part obtained by separating forty units into eight equal parts), is also the third part of fifteen units, so fifteen units is three of the eighth parts of forty units. This is to say that fifteen units, which is not a "part" of forty units, is, however, "parts" of it. Indeed, a smaller number, if it is not part of a larger number, must be parts of it, since both numbers must—just because they are numbers—be measured by the unit.

Euclid goes on to tell us about sorts of numbers—"even" and "odd" and "even-times-even" and "even-times-odd" and "odd-times-odd" and "prime" and "composite"—and about the relation of numbers that are "prime to one another" or "composite to one another":

—An "even number" is that which is separable into two [equals]; and an "odd number" is that which is not separable into two [equals], or that which differs by a unit from an even number.

—An "even [number-taken-an]-even-[number-of]-times" is a number measured by an even number according to an even number; and an "odd [number-taken-an]-even-[number-of]-times" is a number measured by an even number according to an odd number; and an "odd [number-taken-an]-odd-[number-of]-times" is a number measured by an odd number according to an odd number.

—A "prime (*prôtos*) number" is that which is measured by a unit only.

—"Prime-to-one-another numbers" are those which are measured by—as a common measure—a unit only.

—A "composite (*sunthetoi*) number" is that which is measured by some number.

—"Composite-to-one-another numbers" are those which are measured by—as a common measure—some number.

Numbers are thus of interest not only with respect to their relations of size but also with respect to their relations of kind. Even and odd are distinguished by separability into two equal parts (or by odd's differing from even by a unit), and the more complex terms here are distinguished by "measuring by a number according to another number"—which nowadays would be stated as "dividing by an integer so as to yield another such integer as quotient without a remainder." Thus Euclid defines some kinds of numbers by employing notions that have to do with relations of size. He also, however, sorts numbers into kinds by referring to shape. How so?

Well, suppose we use a dot to represent a unit, and then represent a number by dots in a line. For example, four units might be represented as in figure 1.

Suppose we now represent the number produced by taking a number some number of times. Take, for example, the number produced when four is taken three times. The arrangement of dots might most easily take shape as in figure 2. In the figure, the numbers four and three are "sides" of the "plane" number twelve.

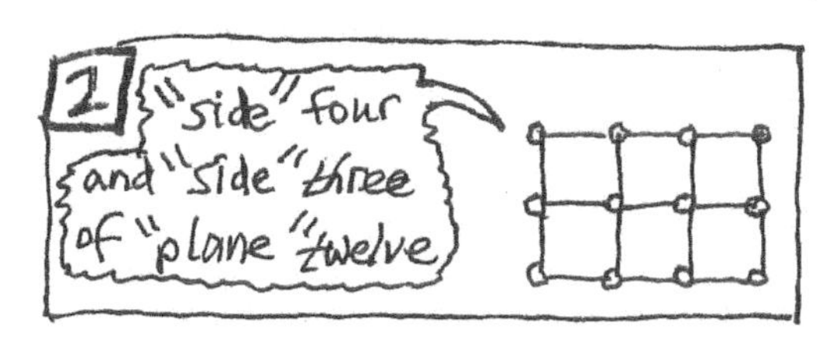

Likewise, six and four are "sides" of the "plane" number twenty-four. But the number twenty-four can also be represented as a solid, as shown in figure 3. Two and three and four are "sides" of the "solid" number twenty-four.

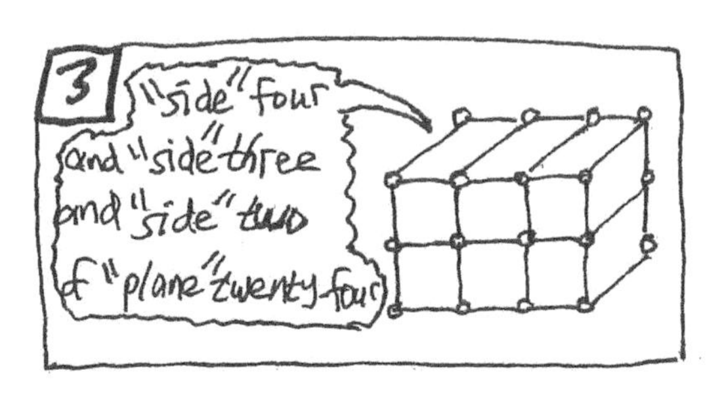

Having spoken of numbers that are "sides," "planes," and "solids," Euclid goes on to speak of the multiplication of a number by a number, and then of the sorts of numerical products of that multiplication—such as the numbers called "square" and "cube," whose factors are called "sides":

—A number is said to "multiply" a number whenever as many units as are in it, [in, that is, the number which multiplies], so many times has been put together (*suntethêi*) the [number] which is multiplied, and some [number thus] be generated; and whenever two numbers, having multiplied one another, may make some [number], the [number] generated is called "plane," and "sides of it" are what the numbers that have multiplied one another are called; and whenever three numbers, having multiplied one another, make some [number], then the [number] generated is called "solid," and "sides of it" are what the numbers that have multiplied one another are called.
—A "square" number is what the equal [number] [taken-an]-equal-[number-of]-times is called, or a [number] contained by two equal numbers is called; and a "cube" [number] is what the equal [number] [taken-an]-equal-[number-of]-times [taken-an]-equal-[number-of]-times is called, or what a [number] contained by three equal numbers is called.
—Numbers are "proportional" (*analogon*) whenever the first is, of the second, and the third is, of the fourth, a multiple [taken]-an-equal-[number-of]-times, or whenever they are, [the first of the second, and the third of the fourth,] the same part, or the same parts,.
—"Similar" plane—and solid—numbers are those having their "sides" proportional.

But Euclid does *not* represent numbers by dots in lines. To use dots would force him to pick this or that number. To signify not this number or that, but rather any number at all, it is more convenient to use bare lines, without indicating how many dots are carried on each line, although it might be confusing that we then have to represent by a line the unit also.

On the other hand, since the unit does have a ratio to any number, just as any number has a ratio to any other number, and since the product of multiplying any number by any other number is also a number, it might make everything more convenient if we represented as lines not just the numbers that are "sides," as well as any "side" that is a unit, but also all the figurate numbers that we can produce. After all, though the sort of number called a "square" number may be somehow different from the sort of number called a "cube," it is nonetheless true that any "square" number has a ratio to any "cube" number (since any number has a ratio to any other number), whereas a square cannot have a ratio to a cube (no more than a length can have a ratio to a weight, since they are not magnitudes of the same kind). If we go along with speaking in this way, however, we must take care to keep in mind that when it is numbers that are called "sides" or "squares" or "cubes" we are not engaged in speech about figures; we are using figures of speech. Several of the books of Euclid's *Elements* deal with what would nowadays be called "number theory" rather than "geometry."

The number definitions at the beginning of this central Seventh Book of the thirteen books of Euclid's *Elements* are completed by defining a "perfect," that is, a "complete" number:

—A "complete" (*teleios* [traditionally rendered as "perfect"]) number is that which is equal to its own parts [taken all together to make one number].

Thus, the number six is "complete," because six itself is equal to the whole obtained by adding up *all* the parts of six—the numbers which measure it (two and three), as well as the unit, which also measures it.

Some of the numbers are thus said to be "complete"; but, we might ask, do all the numbers—the ones that are called complete together with the ones that are not—constitute a *system* that is (in a more usual sense) complete? The numbers in the classical sense—the multitudes two, three, four, and so on—tell us the result of a count, and nowadays we often speak as if those numbers are merely some of the items contained in an expanded system that we call "the *real* numbers." Are they indeed that? To think about the matter, let us begin again.

To tell how many things of some kind there are, we number them. Because there are horses rather than only a single horse, we count horses. Farmer Brown has a herd, say, and so does Farmer Gray. To compare the manyness of the one herd with that of the other herd, we can say: Farmer Brown has a horse … and another … and another … and another; whereas Farmer Gray has a horse … and another … and another … and another … and another … and another. That way of doing things makes it hard to tell just how the herds compare. It would be better to say: Farmer Brown has a horse … and a second … and a third … and a fourth. Or he has one horse … two of them … three … and the fourth one is the last—so, he has four of them; and likewise Farmer Gray has six of them. In Brown's herd there are four horses, as compared with six horses in Gray's.

We might find it convenient to speak not of numbers of horses, or of numbers of saddles. We might speak rather of numbers of pure units, letting the unit be whatever will make something count as one of whatever we are counting. Then, to save words, we might say this: just as we answer "four" rather than "four horses" when asked how many horses belong to Farmer Brown, so we might say "seven and five make twelve" instead of saying "seven units and five units make twelve units."

Whenever we put numbers of things together we get some number of things, and whenever we take a number of things a number of times we get some number of things; but only sometimes can we take a given number of things away from a given number of things, and even when we can, only sometimes can we get some number of things by doing so. For example, we cannot take seven things away from five things, and although we can take six things away from seven things, we shall not have a *number* of things left if we do, but only a single thing. It is also only sometimes that we can separate a given number of things into a given number of equal parts. A multitude of ten things can be divided into five equal parts, but a multitude of eight things cannot. In dividing a manyness there is constraint from which we are free in dividing a muchness.

We often measure a muchness to say just how much of it there is compared with some other muchness: having chosen some unit of muchness, we count how many times such a unit must be taken in order to be equal to the muchness that we are measuring. A line, for example, is a kind of a muchness: there is more line in a longer line than in a shorter line. If two lines are not of equal size, one of them is greater than the other. Lines are magnitudes. If we measure them, we obtain multitudes that tell their sizes—their lengths. But magnitudes are not multitudes. Whereas the size of a magnitude has to do with how *much* there is of it, the size of a multitude has to do with how *many* are the units which constitute it.

Not only magnitudes but *multitudes* as well are often measured, and what is used to measure a multitude is another multitude. Thus, we say that twelve is six taken two times, or is twice six. We say, moreover, that eight is twice the third part of twelve, or—to abbreviate—that eight is two-thirds of twelve. "Two-thirds" does not name a multitude (unless we say that we are merely treating a third part of twelve as a new unit) but "two-thirds" is nonetheless a numerical expression. If we therefore begin to treat it like the name of a number, we are on the road to devising a system of fractions. "Two-thirds" of something is *less* than one such something, but it is not *fewer*. Two thirds *are* fewer than three thirds—but to say this, we must have broken the unit into three new and smaller units. The numerator of a fraction is a number or else it is "one." The denominator is also a number: it is the number of parts that result from a division (although we will allow, to simplify, a denominator as well as a numerator to be "one"). With a fraction, we make a division into equal parts and then we count them. A fraction is not itself a number (a multitude), but it may be the counterpart of a number. "Twelve-sixths" is not the number "two"; but it *is* a counterpart, among the fractions, of the item "two" among the numbers. Another counterpart of the number "two" is the fraction "eight-fourths." Those counterparts are really the same counterpart differently expressed, since they are both equal to the fraction whose numerator is the number "two" and whose denominator is "one."

As with division, so with subtraction. In taking things away, we may get a numerical expression that we can treat like a number in certain respects. When we have twelve horses and eight of them are taken away, there is a remainder of four horses. But if eleven are taken away, then there are not horses left; only a single horse remains. And if twelve horses are taken away, then not even a single horse is left: none remain. But twelve horses can be compared with a single horse: the twelve are to the one as twenty-four are to two. "Twenty-four" and "two" are numbers, and "twelve" is also a number—so "one" is like a number in being comparable with a number as a number is. Moreover, "one" can be added to a number as a number can, and "one" is sometimes what we reply when asked how many of some kind of thing there are. "One" is therefore a numerical word even if it is not a number. But in that case, so is "none": if twelve horses are taken away, leaving not a single horse, and we are asked how many remain, we can then say "none." Now suppose that Farmer Brown owes fourteen horses to Farmer Gray, but has in his possession only ten horses. Farmer Gray takes away the ten. How many horses does Farmer Brown now own? None. But he might be said in some sense to own even less than none, for he still owes four. If we did say that, then we would have to say that he owns four *less* than no horses—that fourteen less than ten is four less than none. But then we would be counting, not horses *owned*, but horses *either-owned-or-owed*. We would be letting the payment of four-horses-owed count as wealth equal to no-horses-owed-or-owned. Again, as with the fractions, we have numerical expressions that we are beginning to treat like numbers. We are on the road to devising a system of so-called "rational numbers," which includes "negative" items as well as such non-negative items as "zero" and "one" in addition to fractions (which include the counterparts of the multitudes that were first called "numbers"—namely, two, three, four, and so on).

We must say "counterparts" because the multitude "three" (for example), while it corresponds to, is not the same as "the positive fraction nine-thirds" or "the positive integer three." We have not brought about an *expansion* of a number system by merely introducing new *additional* items alongside old ones; rather, we have made an *entirely* new system. This new system does contain some items that *correspond* to the items in the old system, but they *differ* from the old items in being items of the very same *new* sort as are the new items that do *not* have correspondents in the old system. For example: although the number "three" is a multitude, "positive three" is not a multitude; it is an item defined by its place in a system where, among other things, "positive six" takes the place belonging to the item that is the outcome of such operations as multiplying "negative two" by "negative three."

We have just taken a look at part of the road that leads from numbers in the classical sense to what we nowadays are used to thinking of as numbers. We have looked at some steps on the road to the so-called "rational numbers," the so-called numbers that have something to do with ratios. But the system of what we nowadays are used to thinking of as numbers is the system of "*real* numbers," only some of which are items that have counterparts among the items in the system of "rational numbers." The system of what we nowadays are used to thinking of as numbers is replete with "irrational numbers," some of them "algebraic" (such as "the square root of two") and some of them "transcendental" (such as "pi").

Classically, however, numbers are multitudes, and although for a particular relation in size of one magnitude to another there may be some number ratio that is the same, there does not have to be any number ratio that is the same.

If there did have to be such a number ratio for every ratio of magnitudes, then there would not have been a reason for devising a system of "real numbers." The reason for taking the step from "rational numbers" to "real numbers" has to do with the difference between *multitudes* and *magnitudes*.

The difference between how we can speak about multitudes (that is, numbers in the classical sense) and how we can speak about magnitudes classically—this manifests itself in the contrast between the definitions with which Euclid begins his treatment of magnitudes in the Fifth Book of the *Elements* and those with which he begins his treatment of numbers afterwards in the Seventh Book.

Although Euclid says what a number is, he does not define what a magnitude is. Examples of magnitudes can be found in his propositions, however. After the Fifth Book of the *Elements* demonstrates many propositions about the ratios of magnitudes as such, these are used by later books of the *Elements* to demonstrate propositions about such magnitudes as straight lines, triangles, rectangles, circles, pyramids, cubes, and spheres. Such magnitudes correspond to what we nowadays call lengths, areas, and volumes. Weight is yet another sort of magnitude. Although weights are not mentioned in the *Elements*, what Euclid there says generally about magnitudes is applied, for example, by Archimedes (in his work on equilibrium) to weights in particular.

At the beginning of the Fifth Book of the *Elements*, Euclid defines ratio for magnitudes, but at the beginning of the Seventh Book he does not define ratio for numbers. There in the Fifth Book (after saying when it is that magnitudes are said to have or hold a ratio to one another) he also defines magnitudes' being in the same ratio, and only after doing that does he define magnitudes' being proportional. But here in the Seventh Book he does not define numbers' being in the same ratio: he goes directly into defining numbers' being proportional, which he needs to do in order to define the similarity of numbers that are of the sorts called "plane" or "solid."

The Greek term translated as "proportional" is *analogon*. After a prefix meaning "up; again" (*ana*), the term contains a form of the word *logos*. This is the Greek term translated as "ratio." *Logos* is derived from the same root from which we get "collect" (which is what the root means); and in most contexts, it can be translated as "speech" or as "reason." Proportionality is, in Greek, *analogia* (from which we get "analogy")—a condition in which things that are different may be said to bear or carry or hold up or have again the same articulable relationship.

Later in the *Elements*, in the enunciation of the 5th proposition of the Tenth Book, for example, Euclid does use the term "same ratio" in speaking of numerical ratios. As we shall soon see, Euclid says, in the definitions with which the Tenth Book begins, that lines which have no numerical ratios to a given straight line are said to be "without a *logos*"—that is, *alogoi*. This Greek term for lines lacking any articulable ratios to a given line is translated (through the Latin) as "irrational."

Such a line and the given line to which it is referred cannot both be measured by the same unit, no matter how small a unit we may use to try to measure them together. They are said to be "without a measurement together"—that is, they are *asymmetra*. This Greek term is translated (through the Latin) as "incommensurable." Because magnitudes of the same kind can be incommensurable, magnitudes are radically different from multitudes, and so we must speak of them differently.

Let us now turn to the classic example of incommensurability: let us consider the relation between the side of a square and its diagonal.

CHAPTER 4. INCOMMENSURABILITY

If a square's diagonal were in fact commensurable with its side, then the ratio of the diagonal to the side would be the same as the ratio of some number to some other number. But it cannot be. Why not? Because *that* ratio of a number to a number would have to have two properties that are incompatible—one property following from its being *any* of the ratios which *a number has to a number*, and the other property following from its being the same as that *particular* geometrical ratio which a *square's diagonal has to its side*. Self-contradiction thus results from asserting the commensurability of a square's diagonal and its side. Let us examine the reasoning that shows the contradiction.

First, let us look at that property which belongs to any ratio whatever that is numerical. It is this: any numerical ratio whatever must either be in lowest terms already, or else be reducible to them eventually. Consider, for example, the ratio that thirty has to seventy. Those two numbers have in common the factors two and five; so, dividing each of that ratio's terms by ten, we see that the ratio which thirty has to seventy is the same as the ratio which three has to seven. The ratio of three to seven is the ratio of thirty to seventy reduced to lowest terms. If a ratio of one number to another number were not reducible to lowest terms, then the two original numbers would have to contain an endless supply of common factors, which is impossible; any number must sooner or later run out of factors if you keep canceling them out by division.

That property of every ratio which a number has to a number (namely, reducibility to lowest terms) cannot belong to the special ratio which a square's diagonal has to its side. To see why this is so, consider any given square. If you make a new square using as the *side* of the *new* square the *diagonal* of the *original* square, then the new square will be double the size of the original square.

Figure 1 shows that doubling. In the left-hand portion of the figure, the diagonal of a square divides it into two triangles, and it takes *four* of these triangles to fill up a new square which has as its own side that diagonal. In the right-hand portion, the figure also shows what happens if we now make another new square, using as this *newest* square's own side a *half* of the diagonal of the original square. The original square, now being itself a square on the newest square's diagonal, will be double the size of this newest square—since this newest square is made up of two triangles, and it takes four of these triangles to make up the original square.

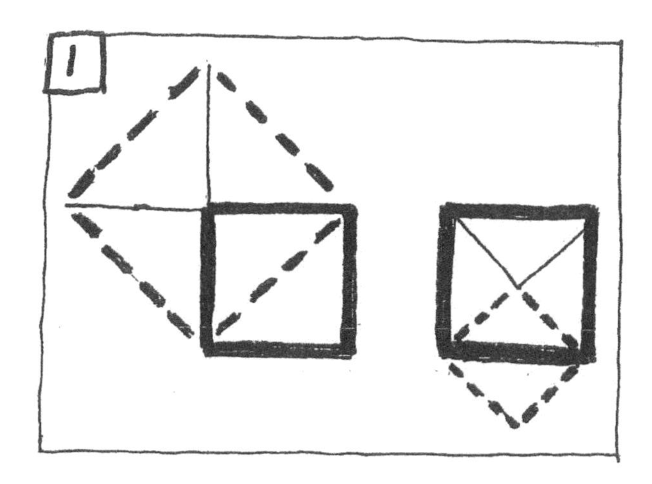

Let us suppose that it were in fact possible to divide a square's diagonal as well as its side into pieces that are all of the same size. Let us count the pieces and say that K is the number of pieces into which we have divided the diagonal, and that L is the number of pieces into which we have divided the side. By taking as a unit any of those equal pieces into which we have supposedly divided the diagonal and the side, we have measured the two lines together: the ratio which the number K has to the number L would be the same as the ratio which the square's diagonal has to its side.

Now, disregarding for a moment just what ratio the number K has to the number L, but considering only that it is supposed to be a ratio of numbers—which, as such, must be reducible to lowest terms—we can say that there would have to be two numbers (let us call them P and Q) such that both of the following statements are true: (first) the ratio which P has to Q is the same ratio which K has to L and (second) P and Q do not have a single factor in common. Since the ratio that K has to L is supposed to be a ratio of numbers, there cannot be any such pair of numbers as K and L unless there is also such a pair of numbers as P and Q. (If the ratio of K to L is already in lowest terms, then we will just let K and L themselves be called by the names P and Q.) So now we have P being the number of equal pieces into which the square's diagonal is divided, and Q being the number of such pieces into which the square's side is divided (fig. 2).

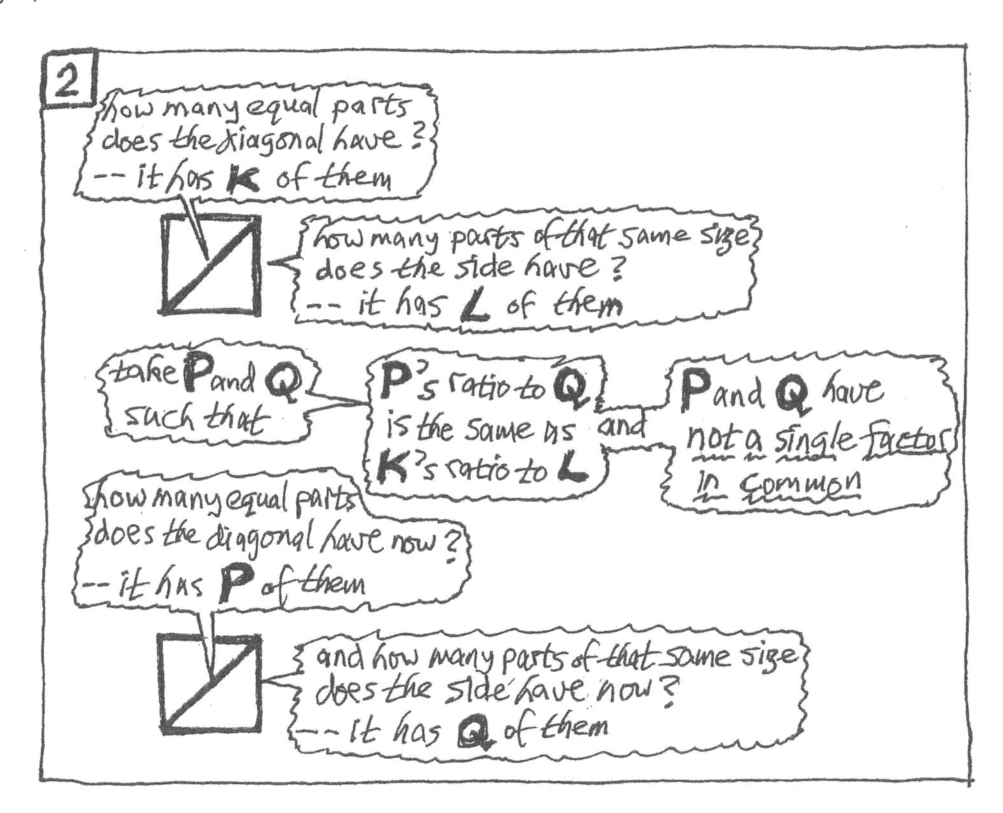

And now we shall see that there just cannot be any such pair of numbers as P and Q. That is because this pair of numbers would have to satisfy contradictory requirements. The first requirement results from the necessities of *numbers*; the second, from the consequences of *configuring lines*. Because (as has been said) the ratio of P to Q is a ratio of numbers reduced to lowest terms, P and Q are numbers that cannot have a single common factor; but (as will be shown) because the ratio of P to Q is the same ratio that a square's diagonal has to its side, P and Q must both be numbers divisible by the number two. That is to say, P and Q are such that they cannot both be divisible by the same number, and yet they also must both be divisible by the number two—a direct contradiction.

Now we must see just why the latter claim is true. We ask: why is it, that if any pair of numbers are in that special ratio which a square's diagonal has to its side, then both numbers must be divisible by the number two?

In order to see why, we must first take note of another fact about numbers—namely, this: if some number has been multiplied by itself and the product is an even number, then the number that was multiplied by itself must *itself* also have been a number that is even. (The reason is simple: a number cannot be both odd and even—it must be the one or the other—and when two even numbers are multiplied together, then the number that is the product is also even; however, when two odd numbers are multiplied together, then the number that is the product is not even but odd.) That said, we are now ready to see why it is, that if any two numbers are in that special ratio which a square's diagonal has to its side, then they must have the factor two in common.

The numbers P and Q have the factor two in common because both of them must be even. They must both be even because each of them when multiplied by itself will give a product that must be double another number—and therefore each must itself be double some number. Why?

Look at the left-hand portion of figure 3, and consider why P must be an even number. Because the square on the diagonal is double the square on the side, the number produced when P is taken P times is double the number produced when Q is taken Q times. Hence (since any number which is double some other number must itself be even) the multiplication of P by itself produces an even number, and hence the number P itself must be even.

And now look at the right-hand portion of the figure, and consider why Q also must be an even number. Since it has already been shown that P must be an even number, it follows that there must be another number that is half of P—call this number H. But because the square on the side is double the square on the half-diagonal, it follows that the number produced when Q is taken Q times is double the number produced when H is taken H times. Hence (since, as we said before, any number which is double some other number must itself be even) the multiplication of Q by itself produces a number that is even, and hence the number Q itself must be even.

Since both P and Q must therefore be even numbers, they must have the number two as a common factor, even though they cannot have any number as a common factor. It is therefore absurd to say that there is a pair of numbers like P and Q. And hence there cannot be a pair of numbers like that K and L which we supposed that there could be.

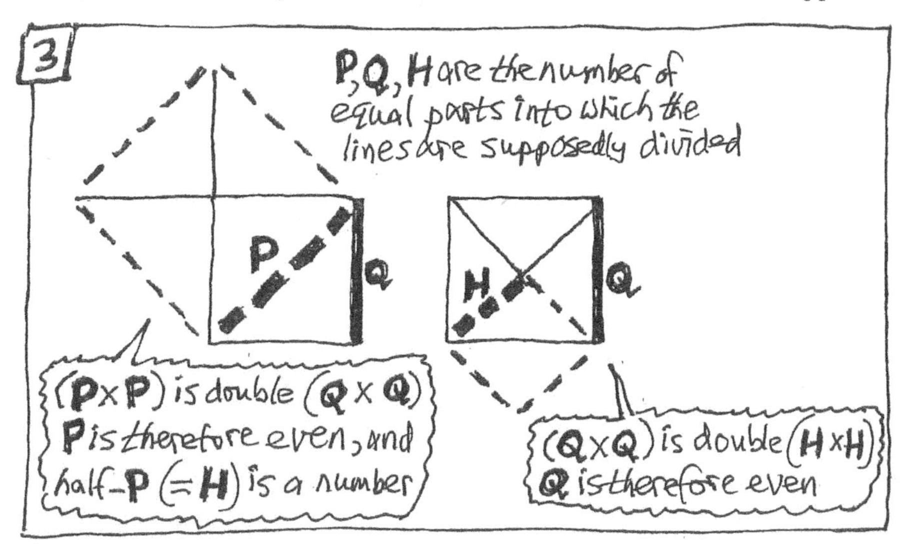

If a pair of numbers did have the same ratio that the square's diagonal has to its side, then no matter how many times we halved both numerical terms in the ratio, both terms that we got would have to remain even. But no pair of numbers can be like that. Each number of the pair would have to contain the number two as a factor in endless supply. Such a ratio would be irreducible to lowest terms, because no matter how many common twos we were to strike as factors from the terms of the ratio, there would still have to be yet another two in each. No number, however, can contain an endless supply of factors.

It is therefore self-contradictory to say that any ratio can be a ratio of *one number to another* and *also* be the ratio of *a square's diagonal to its side*. To avoid being led into absurdity, we must say that a square's diagonal and its side are *incommensurable*.

The diagonal of a square and its side are not the only pair of lines that are incommensurable. From Euclid we learn that there are many others. The Tenth Book of his *Elements* begins with the following definitions:

(X) Definitions—first ones

—"Commensurable" (*symmetra*)—that is what magnitudes are said to be which are measured by the same measure; and "incommensurable" (*asymmetra*)—[that is what magnitudes are said to be which are] those of which no common measure admits of being generated.

—Straight [lines] are "commensurable in power" (*dunamei*) [traditionally rendered "in square"] whenever the squares [raised] from them may be measured by the same piece of territory (*khôriô*) [traditionally rendered "area"]; and "incommensurable [in power]" whenever, for the squares from them, no common measure may admit of being generated.

—These two things being supposed, it is shown that there belong to the straight [line] laid down beforehand both commensurable and incommensurable [lines which are] boundless in multitude, some in length only and also others in power. Then let the straight [line] that is laid down beforehand be called "rational" (*rhêtê*), and those [straight lines which are] commensurable with this [line], either in length and in power or in power only, be called "rational," and those [straight lines which are] incommensurable with this [line] be called "irrational" (*alogoi*). And also let the square [raised] from the straight [line] laid down beforehand be called "rational" (*rhêton*); and also the [figures that are] commensurable with this [square] be called "rational" (*rhêta*), but the [figures which are] incommensurable with this [square] be called "irrational" (*aloga*); and also [let] the [straight lines] that have the power (*dunamenai*) [to give rise to] them, [to give rise, that is, to those squares,] be called "irrational" (*alogoi*)—if, [that is, those figures are] squares, then the sides themselves, but if [they are] some other rectilineal figures,] then the [straight lines which] draw up [from themselves] squares which are equal to them [equal, that is, to those non-square rectilineal figures].

In the tenth proposition of the Tenth Book, Euclid first begins to show us how to find incommensurable lines—many of them. After presenting thirteen sorts of them, he concludes the Tenth Book by saying that from what he has presented, there arise innumerably many others.

That makes it hard to say what we mean when we say that one pair of lines is in the same ratio as is another pair of lines. Even if we could somehow ignore the difficulty with the ratio of a square's diagonal to its side, other ones like it would still keep popping up all over the place, since the ratio of a square's diagonal to its side is only one of innumerably many ratios of incommensurable lines. And it is not only with lines that the difficulty arises. Whatever the kinds of magnitudes that are being compared, whatever it is with respect to which they are greater or smaller, comparisons of muchness are not as such reducible to comparisons of manyness. The sizes of magnitudes cannot always be compared by comparing counts obtained by measuring. Magnitudes of *any* kind can be incommensurable. It is true that one cube may have to another, or one weight may have to another, the same ratio that three has to five, for example; but the one cube may have to the other, or the one weight may have to the other, the same ratio that a square's diagonal has to its side.

Magnitudes are called "incommensurable"—incapable of being measured together—not when they are of different kinds, but rather when they are not measurable by the same unit even though they are of the very same kind. Two magnitudes may be of the very same kind and yet it may be true that whatever unit measures one of them will not measure the other too. By a unit's "measuring" a magnitude, we mean that when the unit is multiplied (that is, when it is taken some number of times), it can equal that magnitude. Therefore, when magnitudes of the same kind are incommensurable, one of them has to the other a ratio that is not the same as any numerical ratio.

The demonstrable existence of incommensurability means that comparisons of size can only be approximated by using numbers. We can, to be sure, be ever more exact—even as exact as we please—but if we wish to speak with absolute exactness, numbers fail us.

For the ancient Pythagoreans, who were the first to conceive of the world as thoroughly mathematical, knowledge of the world was knowledge of numerical relationships. In the world stretching out around us, the Pythagoreans saw correlations between shapes and numbers, such as we encountered when we considered kinds of numbers—"square" numbers and "cube" numbers, for example.

The Pythagoreans noted also that the movements of the heavenly bodies take place in cycles. Their changes in position, and their returns to the same configuration, have rhythms related by recurring numbers that we get by watching the skies and counting the times. With numbers, the world goes round. We are surrounded by a cosmos. (*Cosmos* is the Greek word for a "beautiful adornment.") The beauty on high appears to us down here in numbers.

Even the qualitative features of the world of nature show a wondrous correlation with numbers: numbers make the world sing. It is not merely that rhythm is numerical; it is that tone, or at least pitch, is numerical too. If a string stretched by a weight is plucked, it gives off some sound. The pitch of the sound will be lowered as the string is lengthened. If another string is plucked (a string of the same material and thickness, and stretched by the same weight) then the longer the string the lower will be the pitch of the sound produced by plucking it. Now, what set the Pythagoreans thinking was the relation between numbers and harmony. (*Harmonia* is the Greek word for "the condition in which one thing fits another"; a word from carpentry thus is used to describe music.) When the lengths of two strings of the sort just mentioned are adjusted so that one of them has to the other the same ratio that one small number has to another (or has to the unit), then there is music. With the Pythagoreans' mathematicization of music, mathematical physics begins.

When we listen to an interval between the tones of sounds that are produced by strings whose lengths are in small-number ratios, it sounds good. The tones "fit." Pluck a string, and then pluck another one that is a little longer. The longer string will give off a sound a little lower than the first string. Pluck a series of strings that are longer and longer, until you have a string that is exactly twice as long as the first string. The tone that is now produced will sound as if it is somehow the same as the tone produced by the first string—the same again, yet also different, for it will be lower. It will be lower by the interval that is called "the octave."

Suppose that we start with the tone produced by a string (call it A) that is 120 units long. If we pluck another string Z that is 240 units long, then we will get a tone that is an octave below our first tone. The numerical ratio for the string lengths that give tones separated by the interval of an octave is 2 to 1. Now let us raise the tone. If after plucking string Z, which is 240 units long, we pluck another string M which is 160 units long, the interval over which the tone rises will be a "perfect fifth." The numerical ratio for the string lengths that give tones separated by the interval of a perfect fifth is 3 to 2. If, after plucking string Z (the string that is 240 units long), we had plucked another string N that is 180 units long, the interval over which the tone would have risen would have been a "perfect fourth." The numerical ratio for the string lengths that give tones separated by the interval of a perfect fourth is 4 to 3.

Thus not only were the sights on high early seen to be an expression of mathematical relationships, so were the sounds down here that enter our souls and powerfully move what lies deep down within us. Thinking that nature is a display of numerical relationships, and that human souls are properly ordered by attending to those relationships, the Pythagoreans formed societies that sought to shape the thinking of the political societies of their time by being the givers of their laws. It is said that when the discovery of incommensurability was first revealed to outsiders, thus making public the insufficiency of number, the man who thus had undermined the Pythagorean enterprise was murdered.

But never mind the whole wide world; even the relationships of size in mere geometrical figures cannot be understood simply in terms of multitudes. If geometry could in fact be simply arithmetized, if we could just measure lines together, getting numbers which we could then just multiply together, and we could thus simply express as equations all the relationships that we have to handle, then much that is difficult for us in Apollonius and in other writers of antiquity could be handled much more easily—and Descartes would not have had to undertake a radical transformation of geometry in the seventeenth century. Instead of manipulating equations, however, we must learn to deal with non-numerical ratios of magnitudes, and with boxes that are built from lines devised to exhibit those ratios. And before we can freely deal with ratios, we must learn what is meant when two ratios are said to be the same even when they are not the same as some numerical ratio. We are told that by Euclid, in the Fifth Book of the *Elements*. That is long before Euclid's discussion of number, which does not take place until the Seventh Book of the *Elements*, thus raising a question about what kind of a teacher he is: after all, is it not a principle of good teaching that questions should be raised before answers are presented? In any case, incommensurability is responsible for the difficulty of Euclid's definition of sameness of ratio for magnitudes, as well as for the difficulty that modern readers encounter in the classical presentation of relationships of size generally. Having considered incommensurability, we are now ready to turn back to Euclid's *Elements* for the classic mode of coping with the difficulty.

CHAPTER 5. RATIOS OF MAGNITUDES

Even when two magnitudes are incommensurable, the ratio of one to the other can be the same as the ratio of one to another of two other incommensurable magnitudes. For example, the ratio which a small square's diagonal has to its side is the same ratio which a big square's diagonal has to its own side. Sameness of ratio for magnitudes is therefore not to be defined in terms of measuring magnitudes. We could so define it only if we could divide each of the magnitudes into pieces, each equal to some magnitude that is small enough to be a suitable unit, and then, when we counted up all of those small equal pieces, we found no left-over unaccounted-for even-smaller piece of either of the magnitudes; but we cannot do that with magnitudes that are incommensurable. We do say that two ratios of magnitudes are the same as each other *if* they are both the same as the same ratio of numbers, but we do not say it *only if* so. How, then, can we articulate when it is, that two ratios of magnitudes are the same as each other, without regard to whether or not both of the ratios are the same as the same ratio of numbers?

Euclid provides an answer to that question in the Fifth Book of his *Elements*, which gives a general account of ratios of magnitudes. The Fifth Book begins with definitions that say when one magnitude is a part or a multiple of another:

(V) Definitions

—A "part"—that is what a magnitude is of a magnitude, the lesser of the greater, whenever it measure the greater; and a "multiple"—[that is what a magnitude is of a magnitude,] the greater of the lesser, whenever it be measured by the lesser.

When one magnitude measures another, the smaller one is a "part" of the larger—which is a "multiple" of the smaller. In contrast to the definitions for numbers (given, as noted above, in the Seventh Book of the *Elements*), here in the Fifth Book nothing is said of one magnitude's being "parts" of another. Why not? Because magnitudes may be incommensurable. In the case of two numbers, even if one number is not a part or a multiple of the other number, it must still be true that the two numbers can both be measured by the same unit (even if not by the same number)—but this is not true of magnitudes. One magnitude may not be "parts" of another. In other words, magnitudes do not have any part that they must have by their very nature, whereas all numbers by their very nature have the unit as a part. Because of that, the parts of a magnitude are not of such interest in their own right as are the parts of a number. With magnitudes, multiples are of more interest than parts are.

Ratio is what Euclid defines next; he then tells us what sort of magnitudes are said to have a ratio to one another—namely, those magnitudes which, being multiplied, can exceed one other.

—A "ratio" (*logos*)—that is what, of two homogenous magnitudes, a certain sort of having (*skhesis*) in respect of size is.

—"To have a ratio to one another"—magnitudes are said [to have that,] which are capable, when multiplied, of exceeding (*huperekhein*) [literally, "having it over"] one another.

A magnitude cannot have a ratio to another magnitude unless the two magnitudes are of the same kind. A line, for example, can have a ratio to a line. Their sizes can be compared. The shorter line, if taken enough times, will be longer than the longer line; and the originally longer line, if taken enough times, will be longer than the line that was produced by multiplying the shorter line. Likewise, a weight can have a ratio to a weight. Their sizes can be compared. The lighter weight, if taken enough times, will be heavier than the heavier weight; and the originally heavier weight, if taken enough times, will be heavier than the weight that was produced by multiplying the lighter weight. But while any two lines can be compared with each other (and can, by being multiplied, each exceed the other), and so can any two weights, we cannot say the same of a line and a weight. A line is a magnitude of one kind, and a weight is a magnitude of another. A line cannot be longer or shorter than a weight, or equal to it; and a weight cannot be heavier or lighter than a line, or equal to it.

So also, a line and a square are magnitudes of different kinds—as are also a square and a cube, or a cube and a line. Lengths, areas, and volumes are different sorts of sizes. A small line, if multiplied enough times, will have a greater length than the side of a very large square, but it will not thereby get any area. A very small square, however, if multiplied enough times, will have a greater area than a very large triangle: a square and a triangle are comparable in size. A square and a triangle are magnitudes of the same kind, as are a cube and a pyramid.

But why not just say that one magnitude can be compared in size with another only if the two magnitudes are of the same kind? After all, if they are of the same kind, then either they are equal or one of them exceeds the other—and then there is no need to compare their multiples.

The reason for speaking of multiples of magnitudes, rather than merely of magnitudes themselves, will emerge if we examine the next definitions:

"In the same ratio"—that is what magnitudes are said to be, first [magnitude having the ratio] to second and third [having it] to fourth, whenever the multiples [which are taken-an]-equal-[number-of]-times, [that is, the equimultiples] of the first and third, [if considered with respect to] the multiples [which are taken-an]-equal-[number-of]-times, [that is, the equimultiples] of the second and fourth—whatever [the number of times the] multiplying [involves]—whenever, that is, each [being considered with respect to] each, it either at the same time exceeds or at the same time is equal or at the same time lacks, [if, that is, the magnitudes are] taken in corresponding order, [in other words, ratios are said to be the same whenever, if multiples of the first and third are equimultiples, and also multiples of the second and fourth are equimultiples, then, however many times the multiplyings may be done, those multiples of the first and second always compare with each other, in respect of which is greater, as do those multiples of the third and fourth]; and let the magnitudes that have the same ratio be called "proportional" (*analogon*); and whenever, of the equimultiples, the multiple of the first [magnitude] exceed the multiple of the second but the multiple of the third do not exceed the multiple of the fourth, then the first is, to the second, said "to have a greater ratio" than [has] the third to the fourth.

It is by a seemingly cumbersome procedure of examining whether *multiples* of one magnitude are equal to (or greater than, or less than) multiples of another that we are to determine whether one *ratio* of magnitudes is the same as (or greater than, or less than) another ratio. Why could Euclid not have proceeded in a simpler way?

To understand Euclid's definitions for magnitudes, we must consider what would be their counterparts for numbers. When one number (call it M) is the same part, parts, or multiple of a second number (N) that a third (P) is of a fourth (Q), then we would say that the ratio of number M to number N is the same as the ratio of number P to number Q. For example: we would say that 4 has to 12 the same ratio that 8 has to 24—since 4 is a third part of 12, as 8 is of 24.

But what if two number ratios are not the same? Then we would say that one of the ratios is greater than the other. We decide which of them is the greater ratio by considering that if one number R is greater than another S, then we would say that the ratio of the greater number R to any third number T is greater than the ratio of the smaller number S to that same third number T.

Thus, there is little difficulty in comparing two ratios when both of them are ratios of numbers. It is easy to say what we mean when we say, of any two ratios of numbers, that the ratios are the same—or, if they are not the same, then which is greater and which smaller.

Take, for example, the ratio that 3 has to 5 and the ratio that 7 has to 9. Are these ratios the same? If not, which is greater and which is smaller? The consequents of the ratios (the numbers 5 and 9) have 45 as their least common multiple (which is the smallest number that contains them both as factors). What number has to 45 the same ratio that 3 has to 5? The number 27 does. And what number has to 45 the same ratio that 7 has to 9? The number 35 does. But 27 is fewer parts of 45 than 35 is (since if we take a 9th part of 45, then 27 is 3 of them, whereas 35 is 7 of them), and therefore 27 has a smaller ratio to 45 than 35 has. Therefore 3 has to 5 a smaller ratio than 7 has to 9.

Consider now how two figures that differ in size can have the same shape. For this, the components of the one figure must have the same relationships to each other that the corresponding components of the second figure have to each other. For example, two rectangles have the same shape if you measure their sides and get the following result: the number of units in the *first* rectangle's *height* is the same part, parts, or multiple of the number of units in its *width* that the number of units in the *second* rectangle's height is of the number of units in *its* width. What makes it possible for us to say that the ratio of one side to the other in that first rectangle is the same as the ratio of one side to the other in the second rectangle? The fact that each of those ratios is the same as the same number ratio. How could we determine that each rectangle's height has to its width the same ratio that, for example 3 has to 4? By dividing the width into 4 equal parts, and then dividing the height into as many pieces of that same size (equal to one 4th part of the width) as we can. If the height, when divided in that way, ends up having 3 pieces of just that size—no more, no less—then we know that the height has to the width the same ratio that 3 has to 4. (If the height ends up having 3 pieces but only two of them are of the same size, while the last one is a left-over smaller piece, then the height has to the width a ratio greater than 3 has to 4.)

Even without measuring, however, we wish to say that the parts of a figure have to each other the same relationship of size that corresponding parts of a figure similar to it have to each other—whatever may be the size of the other figure. Thus, in those two rectangles we were just talking about, the one's diagonal must have to its width the same relationship in size that the other one's diagonal has to its own width. But how can we say this when there is no numerical relationship in size that is the same as the relationship in size of either rectangle's diagonal to its side? How can we say that the ratio of the diagonal to the side in the one rectangle is the same as the ratio of the diagonal to the side in the other rectangle if these two ratios are not the same as the same numerical ratio? Euclid does tell us, for ratios of magnitudes, what sameness is; but it is very difficult at first to understand just what he means. While his definition may be the proper place of departure on a road that we need to travel, for a beginner it seems to constitute a locked gate.

There is, however, a key to open the lock. It is this fact: while there can be *no* ratio of numbers that is *the same as* a ratio of incommensurable straight lines, *any* ratio of numbers that we may take must be *either greater or less than* such a ratio of straight lines.

For example, we might ask: which is greater—the ratio of some line to another line, or the ratio of seven to twelve? We would divide the second line into twelve equal parts, and then take one of those pieces in order to measure the first line. Suppose that this first line, although it may be longer than six of these pieces put together, turns out to be shorter than seven of them. We would conclude that the first line has to the other line a ratio that is less than the numerical ratio of seven to twelve.

Let us consider that ratio of lines which has been of such concern to us, the ratio which a square's diagonal has to its side. And let us take the following ratio of numbers: the ratio that three has to two. Since, as we have seen, that ratio of lines which we are considering cannot be the same as any ratio of numbers—any ratio of numbers whatever—it must therefore be either greater or else less than that ratio of numbers which we have taken. So, is it greater or is it less?

It must be less. Why? Because if it were greater—that is, if a square's diagonal were greater than three-halves of its side—then the diagonal taken two times would have to be greater than the side taken three times. If that were so, however, then (for the reason given in the box below) a new square whose side is the original square's diagonal would have to be more than double the original square.

<table>
<tr><td colspan="2">The original square's diagonal—call it D;</td><td>and its side—call it S.</td></tr>
<tr><td>Suppose it indeed true that</td><td>2D</td><td>were greater than 3S.</td></tr>
<tr><td>Then</td><td>a square with that greater side 2D</td><td>would be greater than a square with that lesser side 3S.</td></tr>
<tr><td></td><td>But a square with side 2D
is equal to
4 squares with side D</td><td>a square with side 3S
is equal to
9 squares with side S.</td></tr>
<tr><td>So,</td><td>4 squares with side D</td><td>would be greater than
9 squares with side S.</td></tr>
<tr><td>So,</td><td colspan="2">greater than the ratio of 9 to 4 would be</td></tr>
<tr><td></td><td>the ratio of a square with side D</td><td>to a square with side S—</td></tr>
<tr><td colspan="3">which would make the one square more than double the other.</td></tr>
</table>

But a new square whose side is the original square's diagonal cannot be more than double the original square (since, as we have seen, it must be exactly double); therefore a square's diagonal cannot have to its side a ratio that is greater than the ratio that three has to two. And so (since neither can it have a ratio that is the same as any numerical ratio, this one included) the ratio of lines must be less than the ratio of three to two.

Having thus shown that the ratio of a square's diagonal to its side is less than the ratio which three has to two, we could in like manner also show that the ratio of those two lines is greater than the ratio that, say, four has to three. And we could go on showing, for the ratio of *any* pair of numbers which we might choose, that the ratio of a square's diagonal to its side is greater—or is less—than the ratio of the pair of numbers chosen.

Indeed, although no ratio of numbers is the same as the ratio of a square's diagonal to its side, we can nonetheless confine this ratio of lines as closely as we please by using pairs of ratios of numbers, as follows. Let us divide the square's side into ten equal parts, and take such a tenth part as the unit; then the diagonal will be longer than fourteen of these units but shorter than fifteen of them. And then let us consider what we get when we take as the unit the side's hundredth part, and then its thousandth. We can go on and on in that way, eventually reaching numbers large enough to give us a pair of number-ratios that differ from each other as little as we please; we can show the ratio of diagonal to side to be *greater* than the ever-so-slightly-*smaller* number-ratio, and *smaller* than the ever-so-slightly-*greater* number-ratio. So, although the ratio of a square's diagonal to its side cannot be the same as any number-ratio, it can be confined between a pair of number-ratios that differ from each other as little as we please (fig. 1).

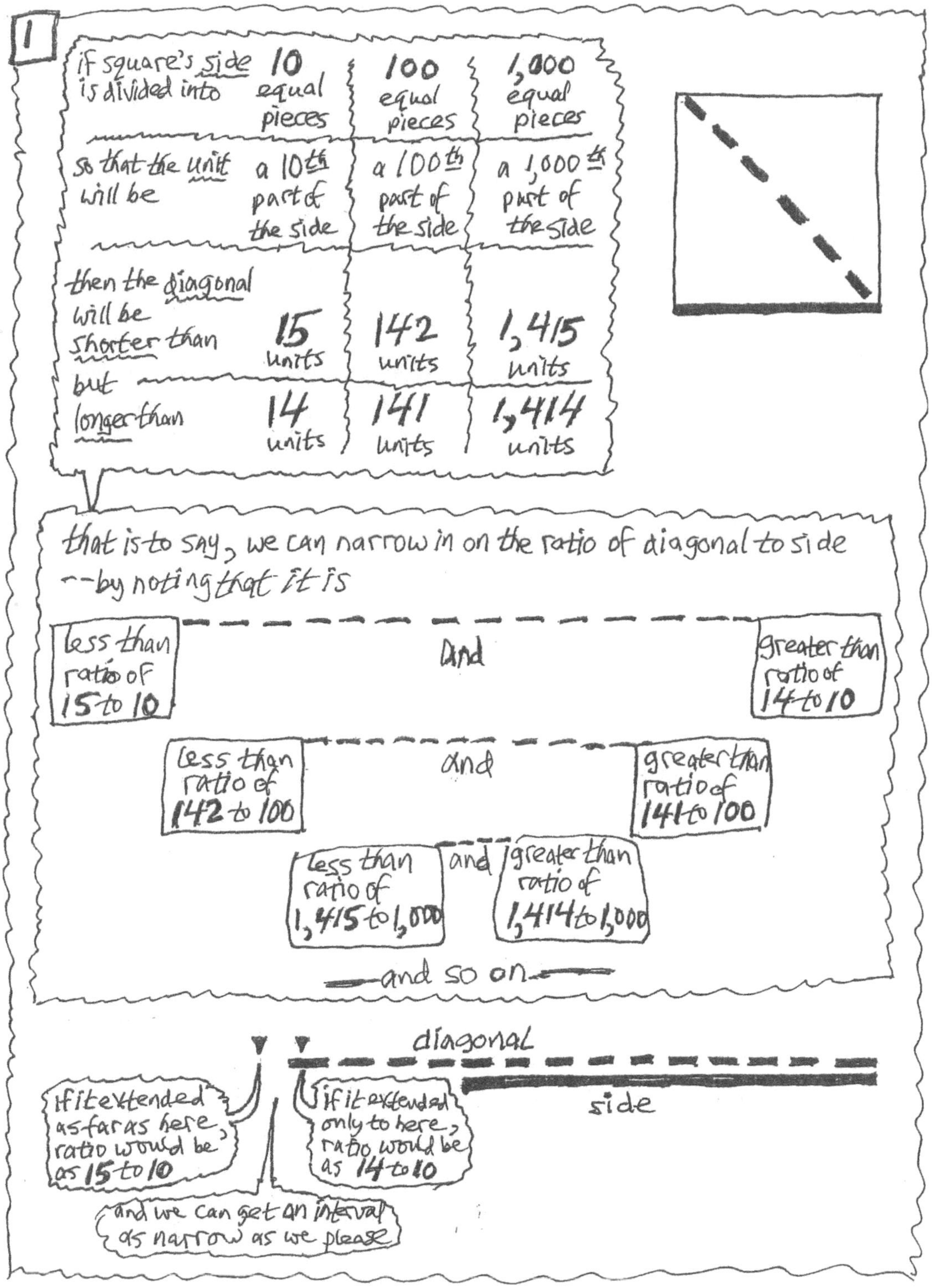

1

if square's side is divided into	10 equal pieces	100 equal pieces	1,000 equal pieces
so that the unit will be	a 10th part of the side	a 100th part of the side	a 1,000th part of the side
then the diagonal will be shorter than	15 units	142 units	1,415 units
but longer than	14 units	141 units	1,414 units

that is to say, we can narrow in on the ratio of diagonal to side
--by noting that it is

less than ratio of 15 to 10 — and — greater than ratio of 14 to 10

less than ratio of 142 to 100 — and — greater than ratio of 141 to 100

less than ratio of 1,415 to 1,000 — and — greater than ratio of 1,414 to 1,000

—and so on—

diagonal

side

If it extended as far as here, ratio would be as 15 to 10

If it extended only to here, ratio would be as 14 to 10

and we can get an interval as narrow as we please

Where nowadays we use a decimal expansion to approximate the *algebraic* irrational "number" called "the square root of two," classical geometers would use number ratios to approximate the ratio of a square's diagonal to its side. We say "the square root of two is equal to 1.414 ..." where the classical geometers would say that the ratio of the square's diagonal to its side is smaller than the ratio of 1415 to 1000 and greater than the ratio of 1414 to 1000. If we choose to go out another million decimal places, they, for their part, could choose to find two numbers, differing only by a unit, such that the greater number's ratio to one billion is greater than, and the smaller number's ratio to one billion is smaller than, the ratio of a square's diagonal to the square's side. By using a tiny enough unit to give two ratios of huge enough numbers, classical geometers can approximate a ratio of incommensurable magnitudes as closely as they please. The smaller the unit, and hence the larger the numbers in the ratios, the smaller will be the difference between the ratio that is a little greater and the ratio that is a little smaller than the one being approximated.

Things are similar with respect to any of our irrational "numbers" which is not algebraic but *transcendental* (that is, which transcends algebra in not being the solution to an equation in which a polynomial in one unknown is equal to zero). Consider our "pi," which we might say is approximately equal to "three and a seventh," or to "3.14." We cannot sneer at Archimedes, who showed, in the generation after Apollonius, that the ratio of a circle's circumference to its diameter is less than the ratio of 31,429 to 10,000 and is greater than the ratio of 31,409 to 10,000. The ability to get some good approximations does not distinguish us from the ancient geometers; what does so is rather what we think the approximations give us. It is one thing to use ratios of multitudes to approximate ratios of incommensurable magnitudes—as was done above for the ratio of the square's diagonal to its side, and as was done by Archimedes for the ratio of the circle's circumference to its diameter. It is something else to use real numbers that are rational, expressed as decimal fractions, to approximate real numbers that are irrational, such as the algebraic "square root of two" or the transcendental "pi."

The question that we want to answer, however, is not when it is that ratios of magnitudes are *almost* the same, but rather when it is that they are *exactly* the same—even when both of them are not the same as the same ratio of numbers. The key to the answer, restating what we said, is this: given any ratio of *magnitudes*, then any ratio of *numbers* that is *not* the *same* as that ratio of magnitudes must be *either greater* than it *or smaller*.

That gives us the following answer:

> Given *two* ratios of *magnitudes*,
> they will be the same as *each other*
> (whether or not they are the same as some ratio of *numbers*)
>
> whenever it is true that—
> for *any* ratio of *numbers* whatever that we might take—
>
> if that ratio of numbers is *greater* than the *one* ratio of *magnitudes*,
> then it is *also* greater than the *other*;
> and
> if it is *less* than the *one* ratio of *magnitudes*,
> then it is *also* less than the *other*.

Euclid's definition seems much more complicated than that, however, because it emphasizes "equimultiples." Why does it do so? Because that makes it simpler to compare ratios of magnitudes with ratios of numbers: taking multiples is a way to make the comparisons *without* performing any divisions. We could in fact get a definition by using division, but the definition would be clumsier if we did. How so?

Let us put the notion of *equi*-multiples aside for a moment, and consider multiples *simply*. That is, let us consider multiples in contradistinction to divisions and parts.

Suppose, for example, that the ratio of some magnitude (called A) to another magnitude (called B) is greater than the ratio of nine to seven but is less than the ratio of ten to seven. This means that if we divide B into seven equal parts and we use as a unit (for trying to measure A) one of those seventh parts of B, then A will turn out to be greater than nine of those parts of B, but less than ten of them. If A and B happen to be incommensurable, then, no matter how small we make the B-measuring unit with which we try to measure A, we shall find that we cannot divide A into pieces of that size without having a smaller piece left over.

So, in comparing the ratio of A to B with the numerical ratio of some multitude m to some multitude n, it is less awkward to speak of A-taken-n-times and of B-taken-m-times than to speak of the little piece of A that may be left over when we divide A into m pieces that are each equal to the nth part of B—or, in other words, to speak of the little piece that may be left over in A, no matter how small are the equal parts into which we divide B. To make the requisite comparisons, we *must* consider *multiples* of magnitudes, but we *need not* consider *parts* of them; we *have* to *multiply*, but we do *not also* have to *divide*. Instead of trying to measure magnitudes together, by dividing them and counting the parts, we can speak merely of *multiples of the magnitudes*.

What we have seen just now is this: to say that the ratio of magnitude A to magnitude B is either the same as the numerical ratio of m to n, or is greater or less than it, is to say that A-taken-n-times is either equal to B-taken-m-times, or is greater or less than it. Let us take that, and put it together with what we saw earlier—which was this: to say that the ratio of A to B is the same as the ratio of C to D is to say that whatever numerical ratio you may take (say, that of m to n) this ratio of numbers will not be the same as, or greater than, or less than *either* one of the ratios of magnitudes *unless* it is likewise so with respect to the *other* one.

All that remains is for multiples of a certain sort to be brought in—namely, "*equi*-multiples." Equimultiples of two magnitudes are two other magnitudes that are obtained by multiplying the two original magnitudes an equal number of times. Here—as figure 2 shows—both ratios' antecedent terms (namely, A and C) are each taken *n* times, and both their consequent terms (namely, B and D) are each taken *m* times. In other words, equimultiples (*n*A and *n*C) are taken of the *first* and *third* magnitudes (A and C); and *also* equimultiples (*m*B and *m*D) are taken of the *second* and *fourth* magnitudes (B and D). And *m* and *n* are any numbers at all: we are interested in all the equimultiples of the first and third magnitudes, and also of the second and fourth ones.

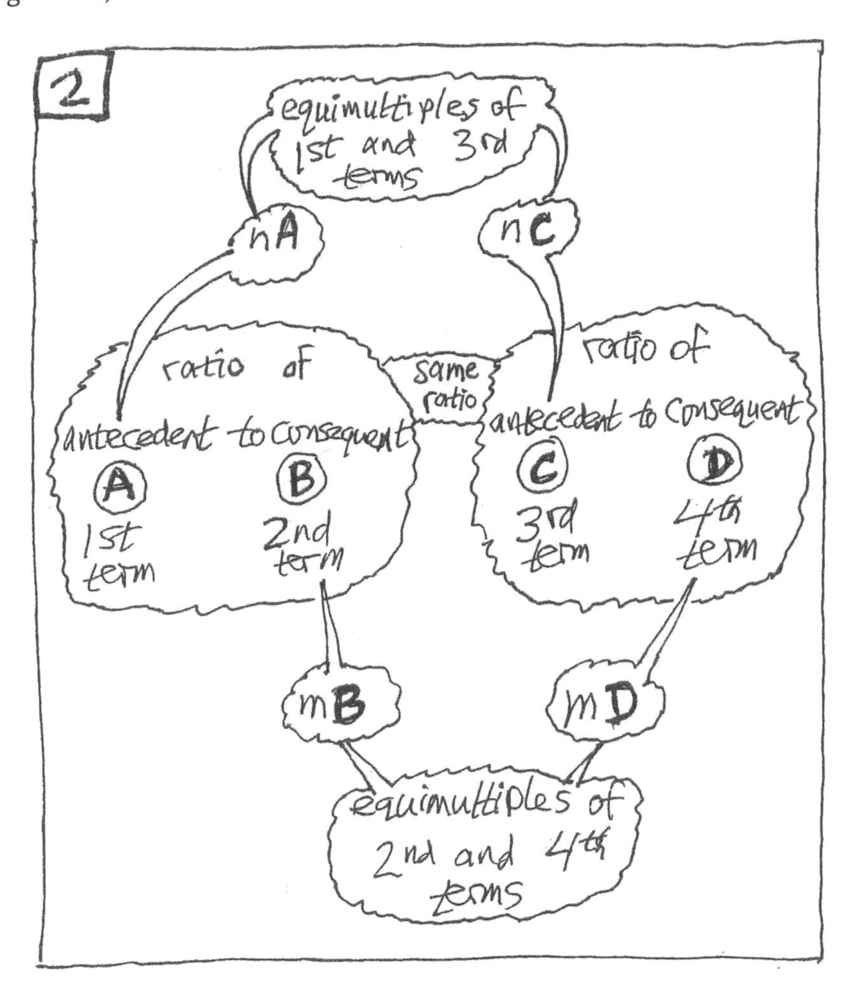

Now at last we are in a position to see that Euclid's definition is not so bewildering as it might have seemed at first. There are several ways of saying what we can do with what we have seen. In abbreviated form, they are exhibited in figure 3. In interpreting the figure, just remember that whenever we say "*if* <something>, *then* <something else>," that is equivalent to saying "*not* <something> *unless* <something else>." So, the ways laid out in the figure are as follows.

First way. Staying with ratios, we can say that two ratios of magnitudes (call them the ratios of A to B, and of C to D) are the same if, and only if, whatever numerical ratio we may take (say, the ratio that some number called *m* has to some other number called *n*) the following is true: this numerical ratio which *m* has to *n* will not be greater than, or the same as, or less than one of the two ratios of magnitudes, unless it is likewise so with respect to the other one.

Second way. We might want to get more explicit by making divisions into equal parts—that is, by dividing B into *n* equal parts and doing the same to D (these, B and D, being the consequent terms of the two ratios of magnitudes). Then the ratios would be the same if the following is true: antecedent term A will not be greater than, or equal to, or less than *m* of those *n*th parts of its consequent term B unless the other antecedent term C is likewise so with respect to that same number *m* of those *n*th parts of its own consequent term D.

Third way. Someone who was willing to do all that (namely, was willing to divide magnitudes into equal parts and count them up and compare sizes) might insist on using fractions to restate that as follows: magnitude A will not be greater than, or equal to, or less than *m*/*n* ths of B unless C is likewise so with respect to that same fraction (*m*/*n* ths) of D.

Fourth—and final—way. Rather than getting into the complicated business of dividing, and then counting divisions, and then comparing sizes—even though we can briefly restate it all by using fractions—we might simply take multiples alone; and then we would say that the ratios are the same if the following is true: A-taken-*n*-times will not be greater than, or equal to, or less than B-taken-*m*-times unless C-taken-exactly-as-many-times-as-A is likewise so with respect to D-taken-exactly-as-many-times-as-B.

Those are several ways to determine, despite incommensurability, when we may say the ratio of one magnitude to another is the same as the ratio of a third magnitude to a fourth one. The first way formulates our initial insight, and the final way formulates Euclid's definition. Like Euclid's definition, the other definitions of same ratio for magnitudes do cover all ratios of magnitudes, whether or not they are the same as ratios of multitudes; but Euclid's definition, unlike the others, manages to do the job by speaking simply of multitudes of magnitudes.

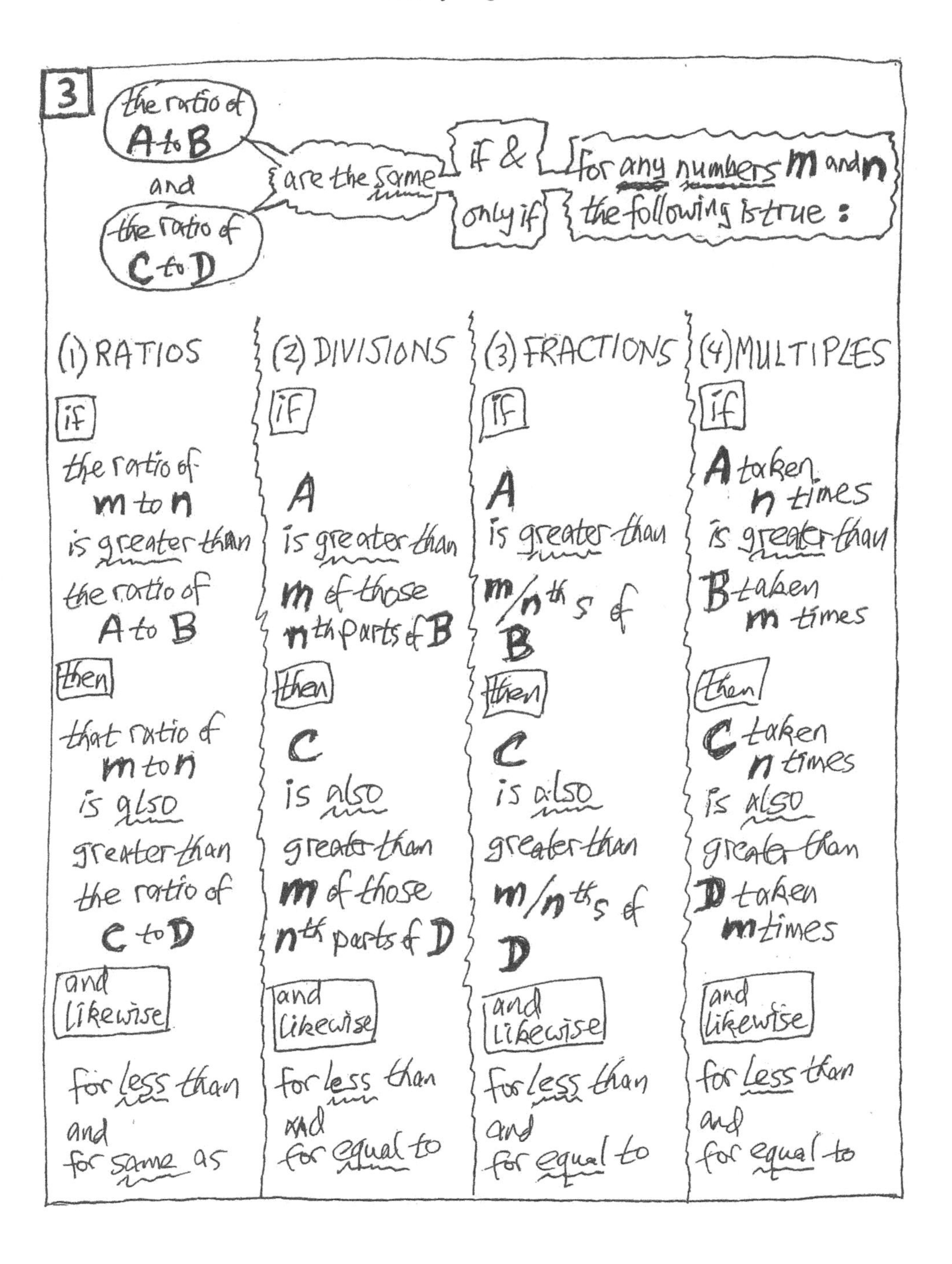
3
the ratio of A to B
and
the ratio of C to D
are the same
if & only if
for any numbers m and n the following is true:
(1) RATIOS
if
the ratio of m to n is greater than the ratio of A to B
then
that ratio of m to n is also greater than the ratio of C to D
and likewise
for less than and for same as
(2) DIVISIONS
if
A is greater than m of those nth parts of B
then
C is also greater than m of those nth parts of D
and likewise
for less than and for equal to
(3) FRACTIONS
if
A is greater than m/nths of B
then
C is also greater than m/nths of D
and likewise
for less than and for equal to
(4) MULTIPLES
if
A taken n times is greater than B taken m times
then
C taken n times is also greater than D taken m times
and likewise
for less than and for equal to

Magnitudes and multitudes are *not* the *same* as each other, nor are either of them the same as ratios of them, but magnitudes and multitudes and their ratios *are* all, in a way, *alike*. How?

In the first place, any two magnitudes of the same kind are equal, or else one of them is greater than the other; and likewise, any two multitudes are equal, or else one of them is greater than the other. Moreover, any magnitude has a ratio to any other magnitude of the same kind, just as any multitude has a ratio to another multitude. Finally, any two ratios (whatever it may be that they are ratios of, whether magnitudes or multitudes) are the same as each other, or else one of them is greater than the other. This last fact supplies the insight that enabled Euclid to define same ratio for magnitudes regardless of whether the magnitudes are commensurable.

Euclid's definition can help us to understand what enabled Dedekind several thousand years afterward to define "the real numbers" in terms of "the rational numbers." Dedekind found himself in the following situation. In modern times, proportions containing magnitudes and multitudes had given way to equations containing "real numbers," but there was still some difficulty in saying just what "real numbers" were. To say much about the matter, it seemed necessary to refer not only to the multitudes to which we come by counting (that is, numbers in the strict sense) but also to magnitudes that we visualize (namely, lines). If we place the counting numbers along a line as in figure 4, it is then clear not only where to put all the "rational numbers," including those which are fractional and those which are non-positive (fig. 5)—but also where to put such "irrational numbers" as "the square root of two." To put "the square root of two" in its place, for example, just raise up a square from a side that is a unit long, and swing its diagonal down (fig. 6). But if "the square root of two" is a number, it should be definable without appeal to visualization. After all, we need not visualize in order to count, or even to go on to think up fractions and negatives. That thought troubled Dedekind as he taught calculus in the nineteenth century, relying upon appeals to a number-line. He finally saw how to treat "real numbers" (the modern counterpart of Euclidean ratios of linear magnitudes) in terms of collections of "rational numbers" (the modern counterpart of Euclidean ratios of numbers), thus making it more plausible to speak of "the real numbers" as being really like numbers—grasped by thought without our having to imagine any line. You can find out how Dedekind did it if you read his paper "Continuity and Irrational Numbers."

Reading Dedekind is one way to think about what it means to rely on a system of "real numbers." Unless we do think through the meaning of relying on a system of real numbers, many of the things we take for granted in the modern world are without foundation. But, we should ask, does Dedekind's work simply represent progress beyond Euclid's, or are very important differences covered over by the important similarities between what Dedekind had on his mind and Euclid on his?

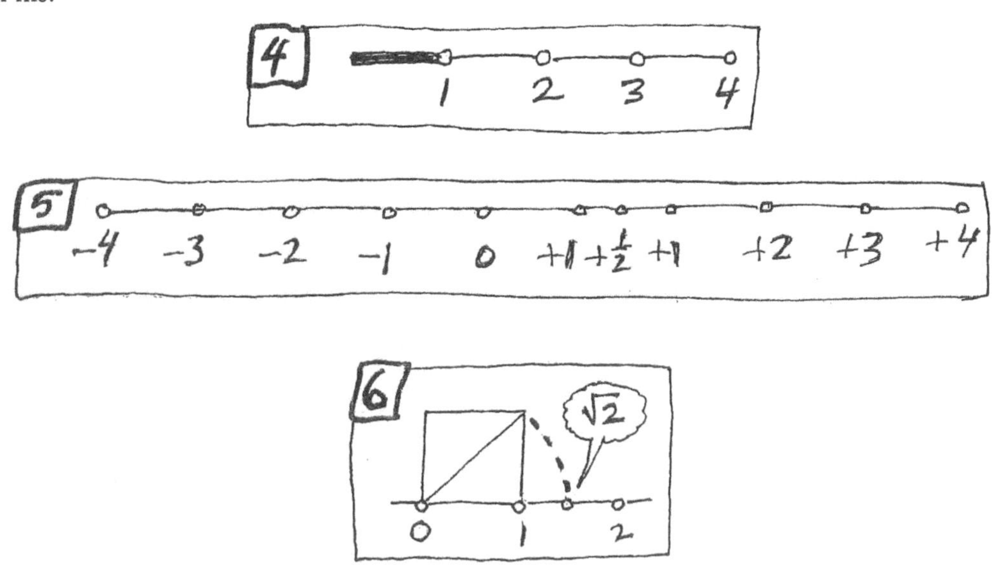

In this guidebook's treatment of classical mathematics, we shall not make use of modern replacements for the handling of ratios. We shall deal with proportions of magnitudes and multitudes, not with equations relating real-number quantities. Do not, as we go, radically transform what you read, by writing "A/B = C/D" or "A/C = B/D," as people often do, having seen "A:B :: C:D" used as an abbreviation for the statement of proportionality "as is A to B so is C to D." Since a magnitude can have a ratio to another magnitude only if the two magnitudes are of the same kind, we cannot say that a proportionality, which states the sameness of two ratios, means the same thing that the corresponding equality of two quotients does.

It is true that if line A is to line B as weight C is to weight D, and both the lines and the weights are as 2 is to 5, then line A is double the fifth part of line B, as weight C is double the fifth part of weight D. In other words, A is parts of B—it is two fifth-parts of B—as C is of D. But a ratio of magnitudes need not be the same as any numerical ratio: two lines (like two weights) may be incommensurable. And even if they are commensurable, what would it mean to take a quotient like "line/weight"? Nothing—unless we had assigned real numbers so as to turn line and weight into quantities. When we use quotients, we do so because we are dealing in quantities that are real numbers. Any two real numbers can be combined by the operations of addition, subtraction, multiplication, and division, as well as root extraction, to yield another real number. Even so, it is hard to see how the quotient "length/weight" could mean much to us. There are other quotients, however, that have come to mean much in modern times.

Suppose, for example, that one man runs 12 miles in 6 minutes and another man runs 8 miles in 4 minutes. Neither man is swifter than the other, since the distances traversed are as the times taken to traverse them: as a distance of 12 miles is to one of 8 miles, so also a time interval of 6 minutes is to one of 4 minutes. Nowadays we would say that velocity is the quotient of distance/time, and that since 12 miles/ 6 minutes = 8 miles/4 minutes, the (average) velocities of the two men are equal. Speaking this way is very useful to us: it pervades our lives. The example is relatively simple, but it presupposes the same great conceptual transformation that more complex examples would.

The following configuration of signs—S:S' :: T:T'—is a string of abbreviations for common-sense notions made systematic. (The language of proportions was the language in which Archimedes spoke, and Galileo too; even Newton still spoke that way in presenting mathematically his principles of natural science.) It compares spatial intervals with each other and temporal intervals with each other, and states that the spaces traversed are to each other as the times of traversal are to each other. Discourse that employs multitudes and magnitudes and speaks of ratios and proportions is closer to the world of ordinary pre-scientific experience than is discourse that employs real numbers and speaks of operations and equations.

The following configuration of signs—V = S/T =S'/T'—is a string of signs that get their meaning from the relations of signs to each other. This meaning is the outcome of reworking the signs borrowed from common-sense usage. Students of physics and other disciplines nowadays encounter many such important quantities that are defined by equating them with expressions composed of other quantities that are assigned real numbers. ("Kinetic energy" for example, is said to be a quantity equal to $\frac{1}{2}[mv^2]$.) We must be careful what we say when using quantitative terms of different sorts.

As we have seen, we can compare not only any two numbers, or any two magnitudes of the same kind, but also any two ratios, whether of numbers or of magnitudes. We need to be discriminating in doing so, however. One ratio of magnitudes or of numbers can be said to be greater or smaller than another, just as one magnitude (or number) can be said to be greater or smaller than another magnitude of the same kind (or number). But whereas two magnitudes of the same kind are said to be *equal* to each other when neither is greater than the other, two ratios are said to be the *same* when neither is greater than the other.

The reason for this distinction is that for Euclid and Apollonius equality is a sameness of a certain kind: it is a sameness, in size, of *different* things. Suppose, for example, that A is twice the size of B, and that C is twice the size of D. In both ratios, the first magnitude has to the second the relationship in size of being its double—and so, although the four magnitudes are different, the two ratios are the very same.

For the same reason, a magnitude is not said by Euclid or Apollonius to be equal to itself. A magnitude is not equal to itself, because it is the same as itself.

When a ratio is said to be "a ratio of equality," what is meant is not that there are *ratios* which are equal to each other, but rather that the *magnitudes* which have that ratio to each other are equal to each other. That is: when two magnitudes have the same size as one another, and are therefore equal to one another, then the magnitudes are said to be in a ratio of equality to one another.

Magnitudes, and multitudes, and ratios of either of them—all of these are alike in being quantitative (that is, in being a matter of greater and smaller), but they differ in the way that they are quantitative.

Any two ratios can be compared: they have an order of size—that is what makes them quantitative. In this way, ratios are like both multitudes and magnitudes (magnitudes of any one sort, that is to say). But in another way ratios are not like either multitudes or magnitudes.

Ratios are unlike magnitudes in the following way: magnitudes of different sorts cannot be compared with each other in respect to size, whereas any ratios can be compared with each other in respect to size. (Consider, for example, a ratio of squares and a ratio of weights. The ratio of one square to another square of which it is the triple is a greater ratio than the ratio of a weight to another weight of which it is the double.) In this homogeneity, ratios are like multitudes (numbers of units), since any number of units can be compared with any other number of units in respect to size. In another way, however, ratios are unlike numbers: there is a continuity among ratios, but not among numbers. Numbers are discrete. For example, there is no multitude greater than 14 units yet less than 15 units. There are, however, lines greater than 14 units long yet less than 15 units long, and any such line has a ratio to one that is 10 units long; this ratio is greater than the ratio of 14 to 10 yet less than that of 15 to 10. The lack of continuity among the multitudes will not be eliminated by calling fractions "numbers." If one line is the diagonal of a square whose side is the other line, then no numbers however large will have the same ratio that those lines have. Among the fractions there is no more continuity than there is among the multitudes. The difficulty is not there are not enough multitudes between any two given multitudes, say 4 and 8. The difficulty is that there are not any of the right sort.

What we have been saying is this. Ratios relate multitudes, and they relate magnitudes. Ratios (like *multitudes*) have *homogeneity*, and ratios also (like *magnitudes*) have *continuity*; but magnitudes do *not* have homogeneity, and multitudes do *not* have continuity. Ratios themselves are therefore *not* things of the same sort as the things that they relate.

Ratios are not things of the same sort as quotients either. To be sure, ratios are like quotients in that they are quantitative. Like any two magnitudes of the same sort, or any two multitudes—or any two quotients—any two ratios (whether of magnitudes or of multitudes) have an order of size. But quotients, unlike ratios, are themselves like what they are quotients *of*: a quotient is a "real number" obtained through the operation of dividing a "real number" by a "real number." If A and B are "real numbers," then the quotient A/B is a "real number" also.

Yet, although ratios and "real numbers" are things of different sorts, they do have a similarity—and not merely that they are both quantitative. Although the magnitudes are not homogeneous, and the multitudes are not continuous, the ratios are both homogeneous and continuous—and so also are "the real numbers."

CHAPTER 6. COMPOUNDING RATIOS

Two magnitudes of the same kind (or two numbers) are equal to each other or else one is greater and the other is smaller. Likewise, two ratios are the same or else one is greater and the other is smaller.

Magnitudes and numbers can be combined, in more ways than one: a magnitude can be added to another of the same kind, as a number can be added to any other number; and a magnitude, or a number, can be taken any number of times—it can be multiplied—the many magnitudes then being added together to get another magnitude of the same kind, or the many numbers then being added together to get another number. But can ratios—not magnitudes or numbers, but ratios—be put together? Euclid and Apollonius both speak of "compounding" ratios. The English word "com-pound" (a form of "com-pose") comes from the Latin for "put-together," which is the counterpart of the Greek term that was used by Euclid and Apollonius. What would it mean to "compound" ratios?

To begin thinking about that, let us think back to how ratios of string lengths were correlated with intervals of pitch when we were considering the Pythagorean attempt to render the world intelligible in terms of relationships of numbers. We took four strings: string A was 120 units long, M was 160, N was 180, and Z was 240. From Z's tone, the interval of pitch to M's tone was a perfect fifth, and to N's tone was a perfect fourth. A perfect fourth is the interval of pitch not only from Z's tone to N's, but also from M's tone to A's (which a Pythagorean might say is because the ratio of M's length to A's—160 units to 120 units—is the same as the ratio of Z's length to N's—240 units to 180 units). If, now, we *put together the intervals* of the perfect fourth and the perfect fifth (the intervals of pitch from Z's tone to M's and from that same M's tone to A's), we get the interval of the octave (the interval of pitch all the way from Z's tone to A's). The ratio of string lengths for the first of the component intervals is 4 to 3, and that for the second is 3 to 2. The ratio of string lengths for the composite interval is the same as the ratio obtained by putting together the two component ratios in such a way that we get this:

"4 to 3 *and* 3 to 2"
is compressed into
"4 to 3 to 2"
and finally into
"4 to 2"
or, in other words, into
"2 to 1."

We have correlated the putting together of certain intervals of tone with the putting together of the string-length ratios that give those intervals. By putting together string-length ratios of 4 to 3 and 3 to 2, we have thus obtained the string-length ratio of 2 to 1 (fig. 1).

1 string name	length in units	ratio of string lengths & interval of tones		
A	120	1 to 2 octave		3 to 4 perfect fourth
M	160		2 to 3 perfect fifth	
N	180			3 to 4 perfect fourth
Z	240			

When two lines are commensurable—that is, when we can use a common unit to measure them together—then the ratio of one of them to the other is the same as the ratio that the number of units counted up in one line has to the number of them counted up in the other. We are used to saying that one such line is some fraction of the other. If we have one line so related to a second, and we also have a third related to a fourth, then we have two numerical ratios—and we are used to saying that we have two fractions. A moment ago, when we took two ratios of strings and put them together to get another ratio of strings, we got something like what we are used to calling the product of multiplying two fractions together.

> String Z's length was 4/3 of string M's, which in turn was 3/2 of string A's
>
> and string Z's length was therefore 4/3 of 3/2 of string A's.
>
> We say string Z's length is equal to (4/3 * 3/2) of string A's.
>
> We get a product of fractions by multiplying numbers used in naming the fractions,
>
> that is, $(4*3)/(3*2) = 12/6 = 2/1 = 2$,
>
> and we say that string Z is therefore 2 times as long as string A.

No fraction, however, can express the relation of one line to another if the lines are incommensurable. If one line is not a part, or parts, or a multiple of another (and it may well not be), then we will not have fractions to multiply together. The association of number ratios with pitch intervals of plucked strings might lead us to the putting together of ratios, but we have to think further if we are to be able to compound ratios regardless of whether the ratios are numerical.

Compounding of ratios is not defined by Euclid. He begins to put ratios together by defining a ratio's duplicate, its triplicate, and so on. Let us therefore consider what it means to multiply something that has a size but is not a number.

We can multiply a line—we can take it a certain number of times. Numbers can be multiplied together. We speak of "A times B" when A and B are both numbers. We say for example, that 5 times 7 is equal to 35. By speaking of "5 times 7" we mean either "the number 5 taken 7 times" or "the number 7 taken 5 times." We can get along without always being clear about which we mean, however, since each is equal to 35. But we can get into trouble by speaking of "A times B" when A and B are not both numbers. We can say that line A taken 5 times is equal to line B, but what would it mean to speak of multiplying a line *by a line*—what would it mean to say "A times B" when A and B are *both* lines?

Suppose we have a line called A, and we triple it to get another line called B. Line B has to line A the ratio that is given by tripling. If we now double this line B (which is itself a magnitude equal to line A taken 3 times), then the line C that we get, which is equal to line B taken 2 times, will also be equal to line A taken 6 times. We have thus taken the triple line B twice, but we have not done the tripling twice. To duplicate the tripling rather than merely to double the triple, we must do the tripling twice—we have to triple the triple. Line B, which is equal to line A taken 3 times, must itself be taken 3 times—to yield a line (call it D) which is equal, not to line A taken 6 times, but rather to line A taken 9 times (fig. 2). To duplicate a tripling, what we take twice is not the triple but the tripling. What we then duplicate is not a magnitude, but a relationship of magnitudes—or is not a number, but a relationship of numbers.

So, if D has to B the same ratio which 3 has to 1, and B in turn has to A the same ratio which 3 has to 1, then D has to A the ratio which is the duplicate of the ratio which 3 has to 1. This duplicate ratio is the ratio which 9 has to 1. More generally: if the ratio which A has to B is the same as the ratio which B in its turn has to D, then the ratio which the first term (A) has to the last term (D) is the duplicate of the ratio which the first term (A) has to the middle term (B).

In the same way we can go on to triplicate a ratio. We have just seen when we can say that the ratio which A has to some K is the duplicate of the ratio which A has to B. It is the duplicate when K is such that as A is to B, so also B is to K. Now let us ask: when can we say that the ratio which A has to L is the triplicate of the ratio which A has to B? It is the triplicate when L is such that as A is to B, so also B is to some K, and so as well this K is to L. Suppose, for example, that B has to A the ratio of the double, and that the triplicate of this ratio is the ratio which some L has to the same A—what, then, is this ratio which this L has to A? It is the ratio of the octuple: the ratio of 8 to 1 is the triplicate of the ratio of 2 to 1 (fig. 3).

In the same way we could go on to quadruplicate a ratio. Indeed, we could go on and on, in this same way, replicating the ratio as many times as we please.

The relevant definitions as given by Euclid will be found in the box below figure 3. (The remainder of the definitions given by Euclid at the beginning of the Fifth Book of his *Elements* are presented at the end of this chapter.)

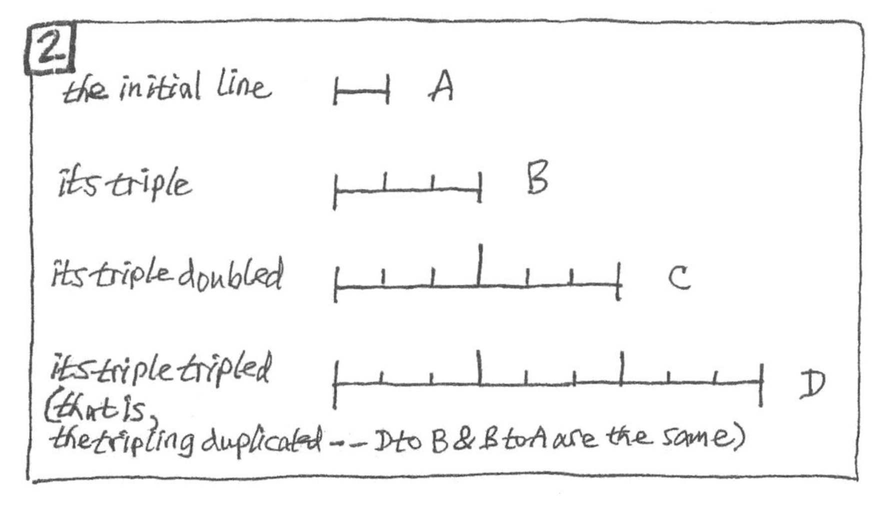

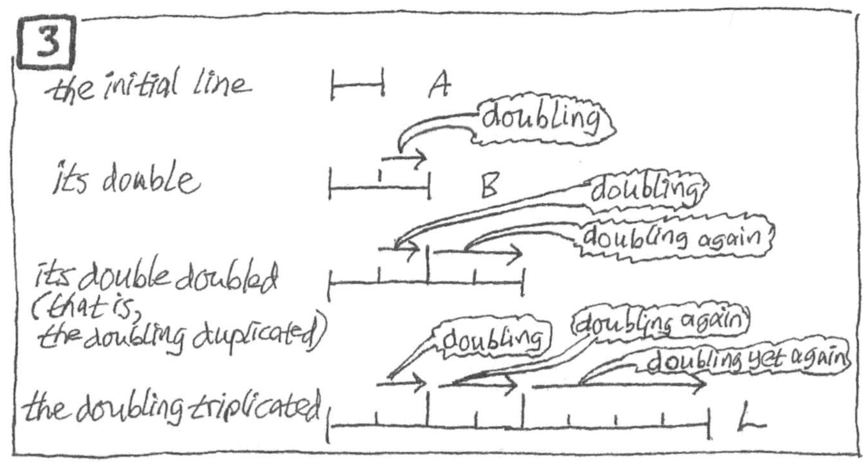

And a proportion (*analogia*) in three terms (*horoi*) is the least [possible]; and whenever three magnitudes be proportional, [that is, whenever as is the first magnitude to the second so is the second to the third,] the first is said to "have a ratio [that is] double" [what it has] to the second, [this being rendered traditionally by speaking of "the duplicate ratio of..."]; and whenever four magnitudes be [continuously] proportional, the first is, to the fourth, said to "have a ratio [that is] triple" [what it has] to the second, [this being rendered traditionally by speaking of "the triplicate ratio of..."], and ever on successively in the same way, as the proportion may arise. [The meaning of the editorially inserted corrective word "continuously" is this: four magnitudes—say A, B, C, D—are "continuously" proportional when as is A to B, so is B to C, and so is C to D. (In other words, each consequent is the same as the antecedent of the next ratio.)]

Duplicating a ratio is easily accomplished when the ratio's second term in one instance (that is, the term B in the ratio which A has to B) is the same as the ratio's first term in another instance (also the term B, in the ratio which B has to K). Then, since the interval in the *first* instance of the ratio comes to its *end* at B, which is also the *beginning* of the interval in the *second* instance, the adjoining intervals are easily put together to compose another interval. The duplicate ratio is the relation between the terms at the very beginning (A) and at the very end (K) of the composite interval (fig. 4).

If the terms are convenient, not only can we put a ratio together with itself, we can also put it together with another ratio. As long as the term that ends the interval in one ratio also begins the interval in the other ratio, the intervals can be joined together. By composing a new interval in this way and then taking the ratio which the term at that interval's very beginning has to the term at its very end, we will manage to put two ratios together. By going all the way from the beginning of the first interval to the end of the second interval, we do what we did in duplicating one-and-the-same ratio. Even when the ratio which A has to B is not the same as the ratio which B has to K, it is nonetheless still true that if the interval from A to B is joined to the interval from B to K, then the terms of the composite interval from A to K will give us the ratio which A has to K, a ratio which is compounded out of the ratios which A has to B and which B has to K (fig. 5).

But what if the terms are not so convenient as to afford such a "squoosh"? What if our other ratio (the one which is not the ratio which A has to B) is not the ratio which that same B has to some C, but is rather the ratio which some other D has to some E? How then could we put *these* two ratios together? We could do so by remembering what we had to do to duplicate a ratio in the first place. We did not then take the ratio of A to B and then take it again in the very same terms, trying to "squoosh" together the interval from A to B and the interval from A to B. Instead, we found *another term* K such that the ratio of B to K was the same as that of the original A to B, and then we were able to "squoosh" together the interval from A to B and the interval from B to K, since the term ending the first interval was the same as the term beginning the second interval (fig. 4). So now, we have to get magnitudes other than D and E which have the same ratio as they do but will nonetheless allow for "squooshing." That is, we need to find some K such that B (the second term of the first ratio) has to K (this term that we find) the same ratio which D has to E. Then, since K is such that as D is to E so also B is to K, it will follow that the product of putting together the ratio which A has to B and the ratio which D has to E will be the same as the product of putting together the ratio which A has to B and the ratio which B has to K—that is, it will be the same as the ratio which A has to K (fig. 6).

The magnitude requisite for doing this can always be found. Given two magnitudes (D and E) and a third magnitude (B), we can always find a fourth proportional (K). That is: we can always find a fourth magnitude (K) to which the third magnitude (B) will have the same ratio which the first magnitude (D) has to the second (E). Euclid's *Elements* (VI. 12) shows how to do this for straight lines. (At the end of the next chapter of this guidebook, we shall look at this way of finding a straight line which is a fourth proportional to any three straight lines that are given.)

For numerical examples of what we have done, see figures 7 to 11, following.

We have seen how to compound ratios that are not the same, by getting a "squooshing" situation—a situation in which there is a middle term, so that the end of the one interval is the same as the beginning of the other interval. In this respect, the compounding of ratios is like the duplicating of a ratio. But it is not duplication—since what it involves is not one-and-the-same ratio but rather two different ratios. We might think of the duplication of a ratio as a special case of compounding two ratios: we duplicate a ratio if we compound it with itself.

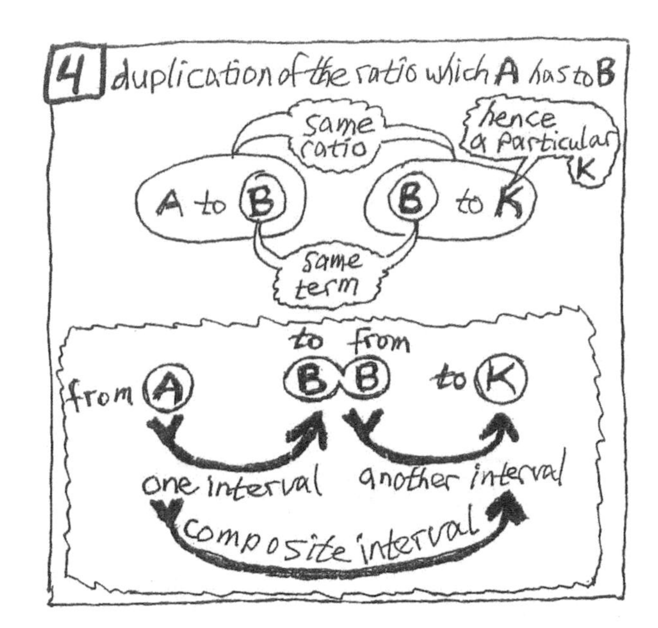
4
duplication of the ratio which A has to B
same ratio
hence a particular K
A to B
B to K
same term
to from
from A
B B
to K
one interval
another interval
composite interval

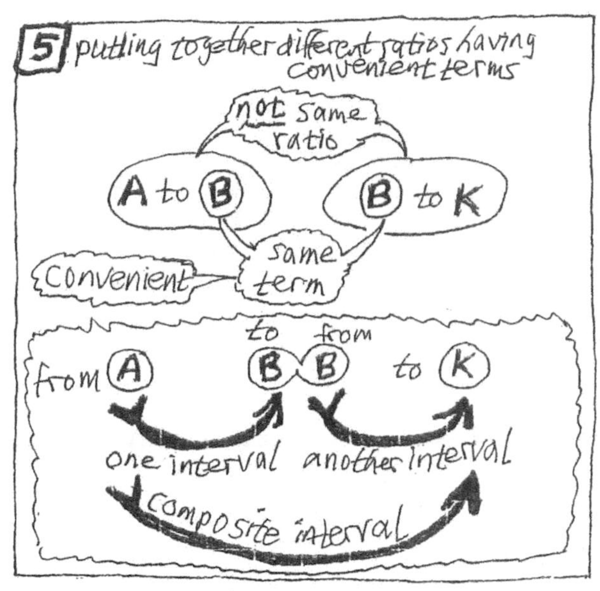
5
putting together different ratios having convenient terms
not same ratio
A to B
B to K
same term
convenient
to from
from A
B B
to K
one interval
another interval
composite interval

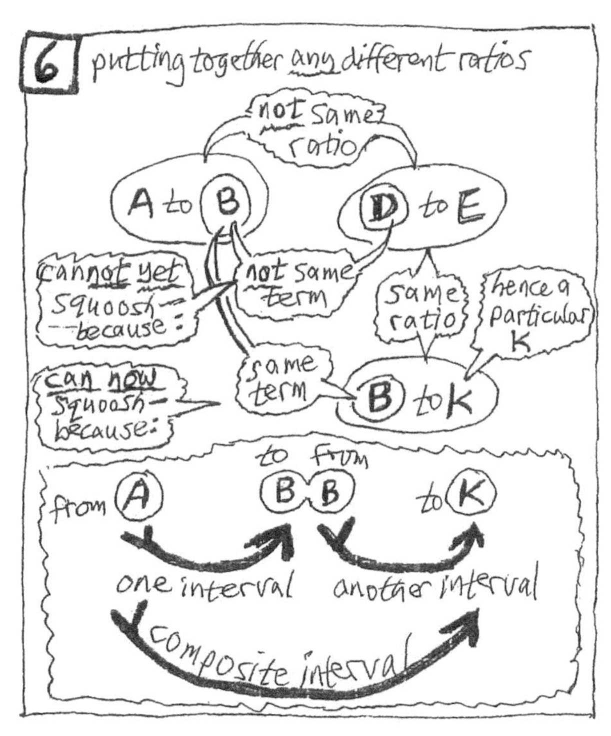
6
putting together any different ratios
not same ratio
A to B
D to E
cannot yet squoosh because:
not same term
same ratio
hence a particular K
same term
can now squoosh because:
B to K
to from
from A
B B
to K
one interval
another interval
composite interval

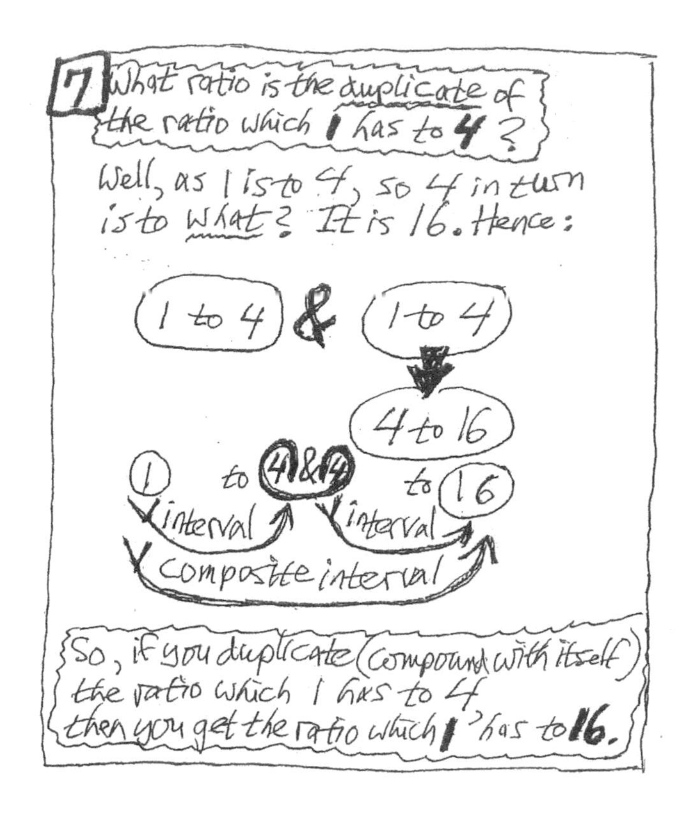
7
What ratio is the duplicate of the ratio which 1 has to 4?
Well, as 1 is to 4, so 4 in turn is to what? It is 16. Hence:
1 to 4
&
1 to 4
4 to 16
1
to
4&4
to
16
interval
interval
composite interval
So, if you duplicate (compound with itself) the ratio which 1 has to 4 then you get the ratio which 1 has to 16.

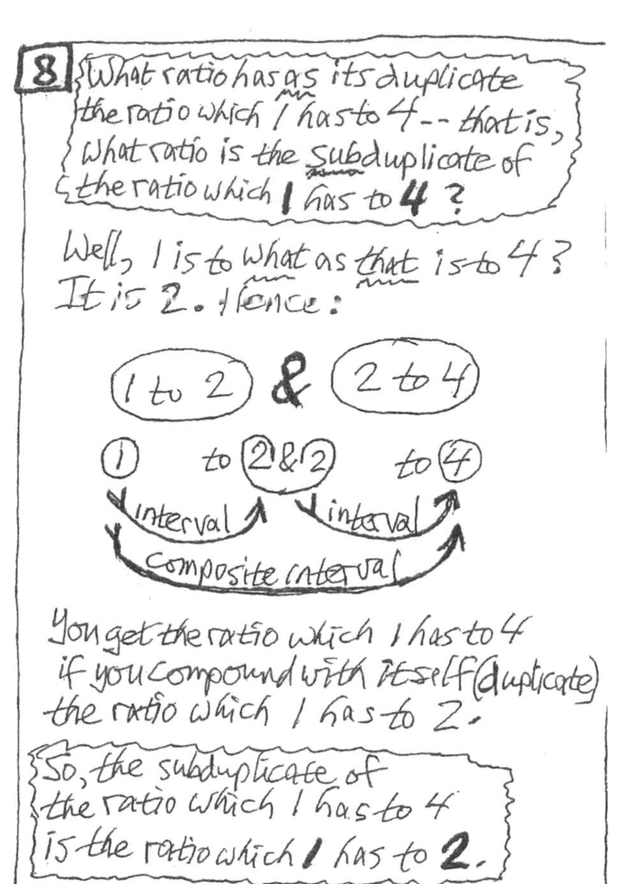
8
What ratio has as its duplicate the ratio which 1 has to 4 -- that is, What ratio is the subduplicate of the ratio which 1 has to 4?
Well, 1 is to what as that is to 4? It is 2. Hence:
1 to 2
&
2 to 4
1
to
2&2
to
4
interval
interval
composite interval
You get the ratio which 1 has to 4 if you compound with itself (duplicate) the ratio which 1 has to 2.
So, the subduplicate of the ratio which 1 has to 4 is the ratio which 1 has to 2.

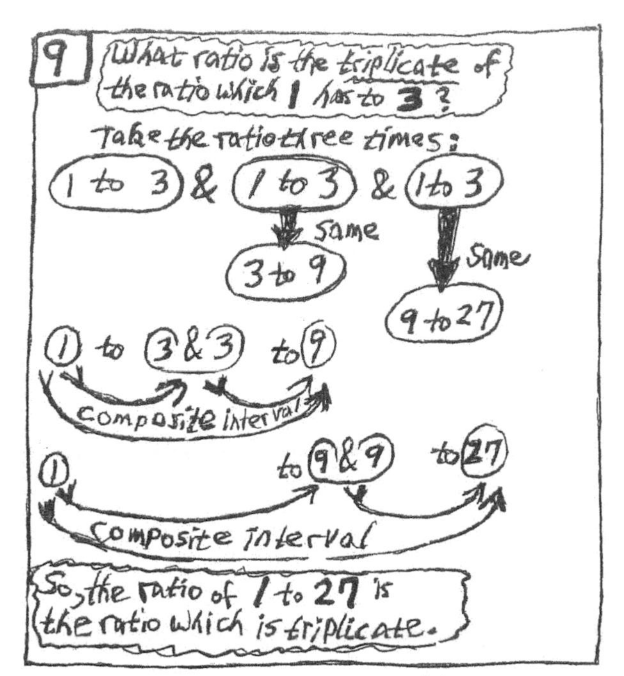
9
What ratio is the triplicate of the ratio which 1 has to 3?
Take the ratio three times:
1 to 3 & 1 to 3 & 1 to 3
same
3 to 9
same
9 to 27
1 to 3 & 3 to 9
composite interval
1 to 9 & 9 to 27
composite interval
So, the ratio of 1 to 27 is the ratio which is triplicate.

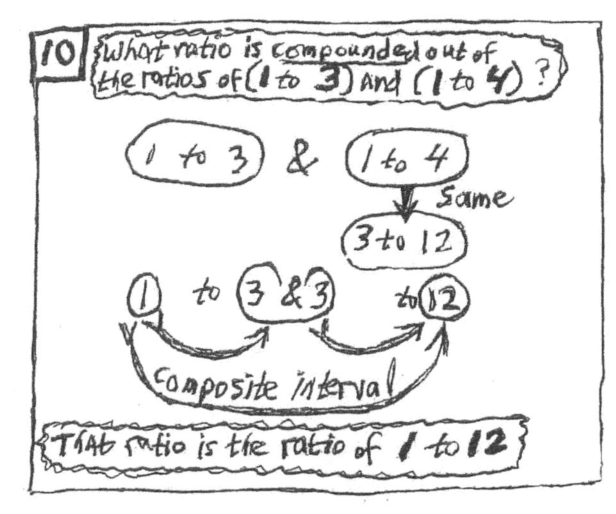
10
What ratio is compounded out of the ratios of (1 to 3) and (1 to 4)?
1 to 3 & 1 to 4
same
3 to 12
1 to 3 & 3 to 12
composite interval
That ratio is the ratio of 1 to 12

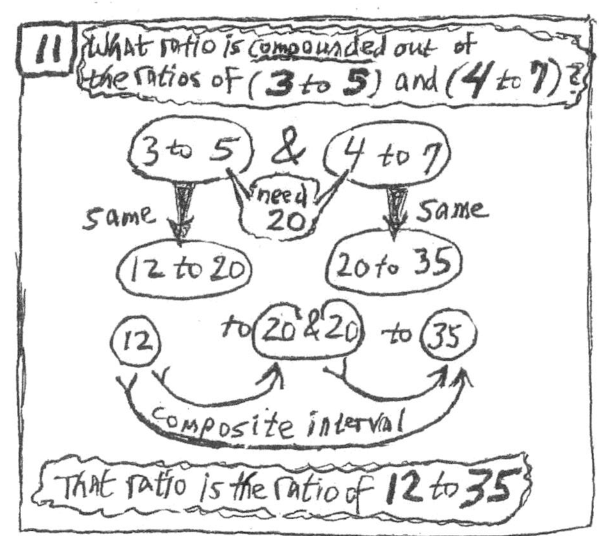
11
What ratio is compounded out of the ratios of (3 to 5) and (4 to 7)?
3 to 5 & 4 to 7
need 20
same
same
12 to 20
20 to 35
12 to 20 & 20 to 35
composite interval
That ratio is the ratio of 12 to 35

The definitions at the beginning of the Fifth Book of Euclid's *Elements* conclude as follows

—"Corresponding (*homologa*) magnitudes"—said to be such [in ratios] are the antecedents in respect to the antecedents, and the consequents in respect to the consequents.

—An "alternately[taken] (*enallax*) ratio" is a taking of the antecedent in relation to the antecedent and of the consequent in relation to the consequent.
—An "inversely[taken] (*anapalin*) ratio" is a taking of the consequent as antecedent in relation to the antecedent as consequent.

—"Composition (*sunthesis*) [that is, 'putting together'] of a ratio" is a taking of the antecedent along with (*meta*) the consequent as of one thing, in relation to the consequent by itself.
—"Separation (*diairesis*) of a ratio" is a taking of the excess by which the antecedent exceeds the consequent, in relation to the consequent by itself.
—"Conversion (*anastrophê*) of a ratio" is a taking of the antecedent in relation to the excess by which the antecedent exceeds the consequent.

—"Through an equal [number of terms] (*di isou*)] a ratio [goes]" [hereafter traditionally rendered by the traditional Latin term *ex aequali*] whenever—there being several magnitudes and also other magnitudes equal to them in multitude, which taken two by two are in the same ratio—as is, among the first magnitudes, the first to the last, so is, among the second magnitudes, the first to the last. [That is, to put it] otherwise, [it is] a taking of the extremes while pulling out the intermediates.
—But a "perturbed (*tetargmenê*) proportion" is whenever it may arise that—there being three magnitudes and also other magnitudes equal to them in multitude—as is, among the first magnitudes, antecedent to consequent, so is, among the second magnitudes, antecedent to consequent among the second magnitudes, while, as is, among the first magnitudes, the consequent to a certain other [third] magnitude, so is, among the second magnitudes, that certain other [third] magnitude to the antecedent.

When we have magnitudes in the proportion that as is A to B so is C to D, the definitions at the beginning of the Fifth Book of Euclid's *Elements* refer to the following proportions (here identified also by their traditional Latin labels):

(*alternando*) "alternately"	as A	to C	so B	to D
(*invertendo*) "inversely"	as B	to A	so D	to C
(*componendo*) "by composition"	as (A with B)	to B	so (C with D)	to D
(*separando*) "by separation"	as (A less B)	to B	so (C less D)	to D
(*convertendo*) "by conversion"	as A	to (A less B)	so C	to (C less D)

When we have magnitudes in the proportions
that as A to B.........so D to E
and that as B to C..........so E to F, then the *ex aequali* proportion is
that as A......to C....so D.......to F

—or when we have magnitudes in the three proportions
that as A to B..............so D to E
and that as B to C..............so E to F,
and that......as C to K..............so F to L, then the *ex aequali* proportion is
that as A............to K...so D............to L

—and so on, for any number of pairs of magnitudes, with intermediate terms disposed similarly.

But when the intermediate terms are disposed oppositely, as in the proportions
that as A to B.............so E to F
and that as B to C...so D to E, then the "perturbed" proportion is
that as A.......to C...so D......to F.

CHAPTER 7. RECTANGLES, LINES, AND RATIOS

Among the definitions which begin the Fifth Book of the *Elements*, Euclid defines the duplication of a ratio and its triplication, but nowhere in the *Elements* does he define the compounding of ratios. In the 23rd proposition of the Sixth Book, however, he does speak of a ratio as being compounded of other ratios (he uses the phrases *logos syngkeimenos* and *logos syngkeitai*). The proposition is not a definition of compounding—for if it were, then there would be no need for a demonstration. Rather, the proposition makes use of a notion of compounding which it presupposes we already somehow have. The enunciation says that equiangular parallelograms have to each other the ratio which is compounded of the ratios of their sides. A consideration of the proposition will show us how compounding ratios of magnitudes involves something that is similar to multiplying numbers, but is not the same as it.

For convenience, let us here restrict ourselves to a special case—that is, let us consider only parallelograms that are rectangles. First, let us consider rectangles all of whose sides are lines that are commensurable.

Suppose that we triple the side of a square and then construct a new square on the new side (fig. 1). The new side is 3 times as large as the old side, but how many times as large as the old square is the new square? We get the number of times larger if 3 times is taken 3 times. That is to say: the new square is 9 times as large as the old one. The new square has to the old square a ratio which is the duplicate of the ratio which the new side has to the old side. Squares have to each other the ratio that is the duplicate of the ratio that their sides have.

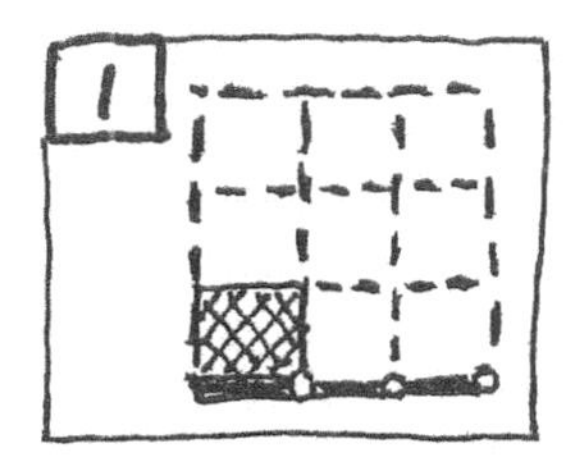

Suppose now that we have a rectangle and we double its height and also triple its base (fig. 2). The new height is 2 times as large as the old height, and the new base is 3 times as large as the old base—but how many times as large as the old rectangle is the new rectangle? It is 6 times as large.

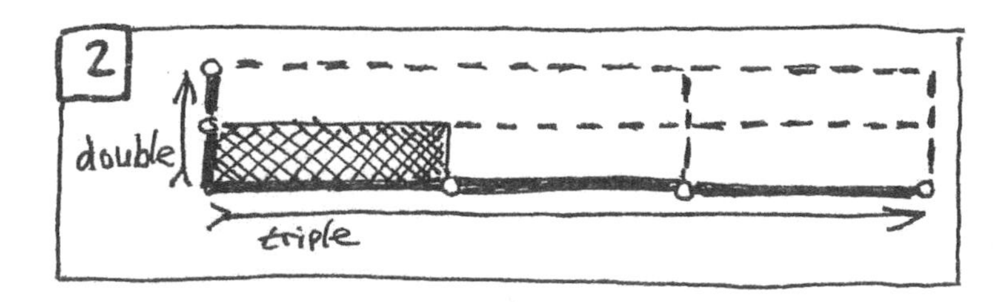

Suppose now that we have two rectangles, one of them having as its height a line 2 units long and as its base a line 3 units long, and the other having as its height a line 5 units long and as its base a line 7 units long (fig. 3). Their heights have to each other the same ratio which 2 has to 5, and their bases have to each other the same ratio which 3 has to 7; but what is the ratio which the rectangles themselves have to each other? This is the ratio which 6 has to 35. The ratio can be obtained by counting squares.

Instead of counting squares, we can save labor by taking the ratio of the numbers which are obtained by first multiplying 2 and 3 together and then multiplying 5 and 7. The number that is the product of the one multiplication (of 2 by 3) has to the number that is the product of the other multiplication (of 5 by 7) the ratio which is compounded of the ratio which 2 has to 5 and the ratio which 3 has to 7. This is the ratio which 6 has to 35 (fig. 4).

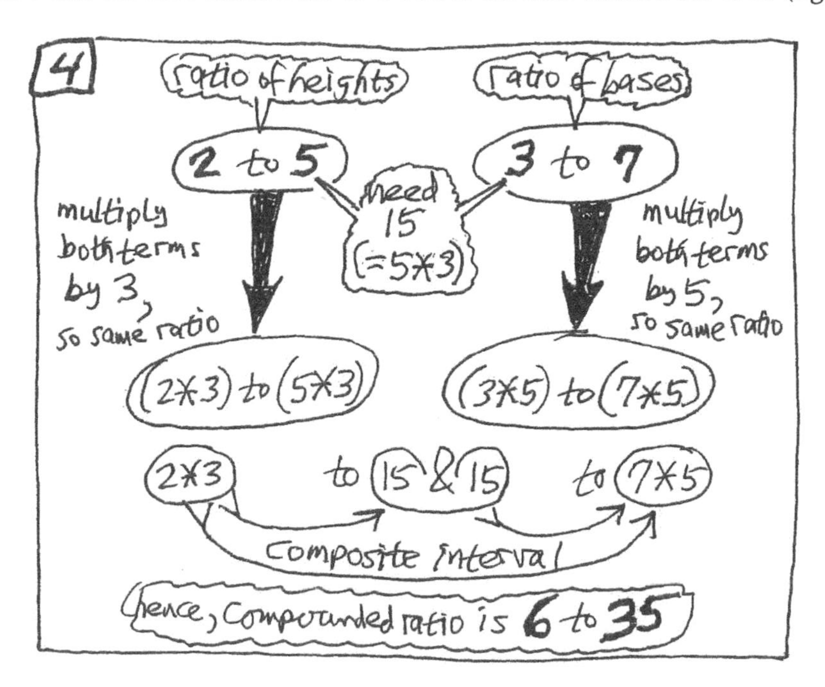

But the sides of two rectangles may not be commensurable. Suppose, for example, that we construct two rectangles in the following way. We take a straight line (call it S), and with S as side we construct a square, while with 2S as side we construct an equilateral triangle. As we have seen, the diagonal of the square (call it D) will be a straight line which is incommensurable with S. That is because a new square raised up from the diagonal of a square will be double this original square. Now let us also use a ratio of incommensurables other than a square's side and its diagonal. It can be shown (by the so-called "Pythagorean theorem"—Euclid's 1.47) that a square on the altitude of an equilateral triangle will be triple the square on half the side of that equilateral triangle. (The half-side and the altitude form the legs of a right triangle whose hypotenuse is the full side; and so, since a square on the full-side is quadruple a square in the half-side, a square on the altitude must be triple that square on the half side.) The triangle's altitude (call it A) is therefore incommensurable with S (since a square on A is triple a square on S). Now let us construct two rectangles—one with sides D and S, and the other with sides A and S (fig. 5). Since a square with side A will be half-again as large as a square with side D, the ratio of the heights (D and A) of the two rectangles that we constructed cannot be the same as any numerical ratio—hence, neither can the ratio of the rectangles themselves.

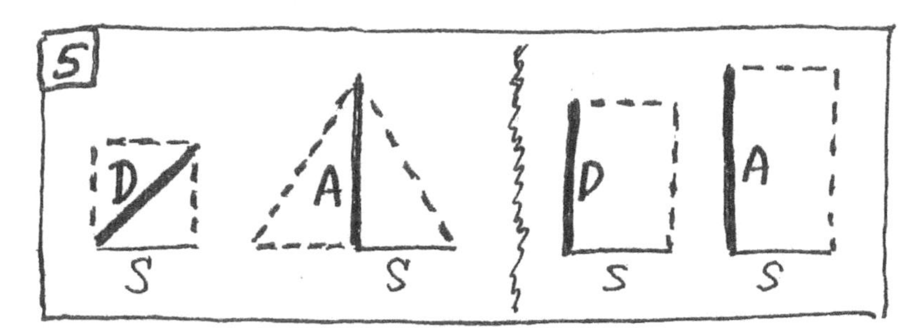

If, to overcome the distinction between how-much and how-many, we had devised a "number-line," we might have said that if S has unit length, then D's length is "the square-root of 2," and A's is "the square-root of 3," and the area "A*S" is the product "D*S*one-half the square-root of 6." Eventually we shall see what the devising of a "number-line" has to do with overcoming the distinction between how-much and how-many—but since we have not yet thought through the thinking that eventually leads to a "number-line," we cannot now have recourse to such "multiplication" of "real numbers." To demonstrate what Euclid presents in VI.23 (that equiangular parallelograms have to each other the ratio compounded of the ratios of their sides), we cannot proceed as if the ratios of the sides of the parallelograms are the same as some ratios of numbers.

However, rectangles whose sides happen to be measurable together can be compared by measuring their sides and then compounding the ratios of the numbers obtained. And ratios of numbers can be compounded by multiplying the numbers which are their terms. So although the compounding of ratios cannot be reduced to the multiplication of numbers, what we do in multiplying the numbers that are the terms of ratios being compounded is suggestive about the compounding of ratios of magnitudes.

Consider again how ratios of numbers can be compounded by multiplying the numbers which are their terms. The ratio compounded of the ratio which 7 has to 3 and the ratio which 8 has to 5, for example, is the ratio of one numerical product to another—one of these products comes from multiplying 7 and 8, the other from multiplying 3 and 5 (fig. 6).

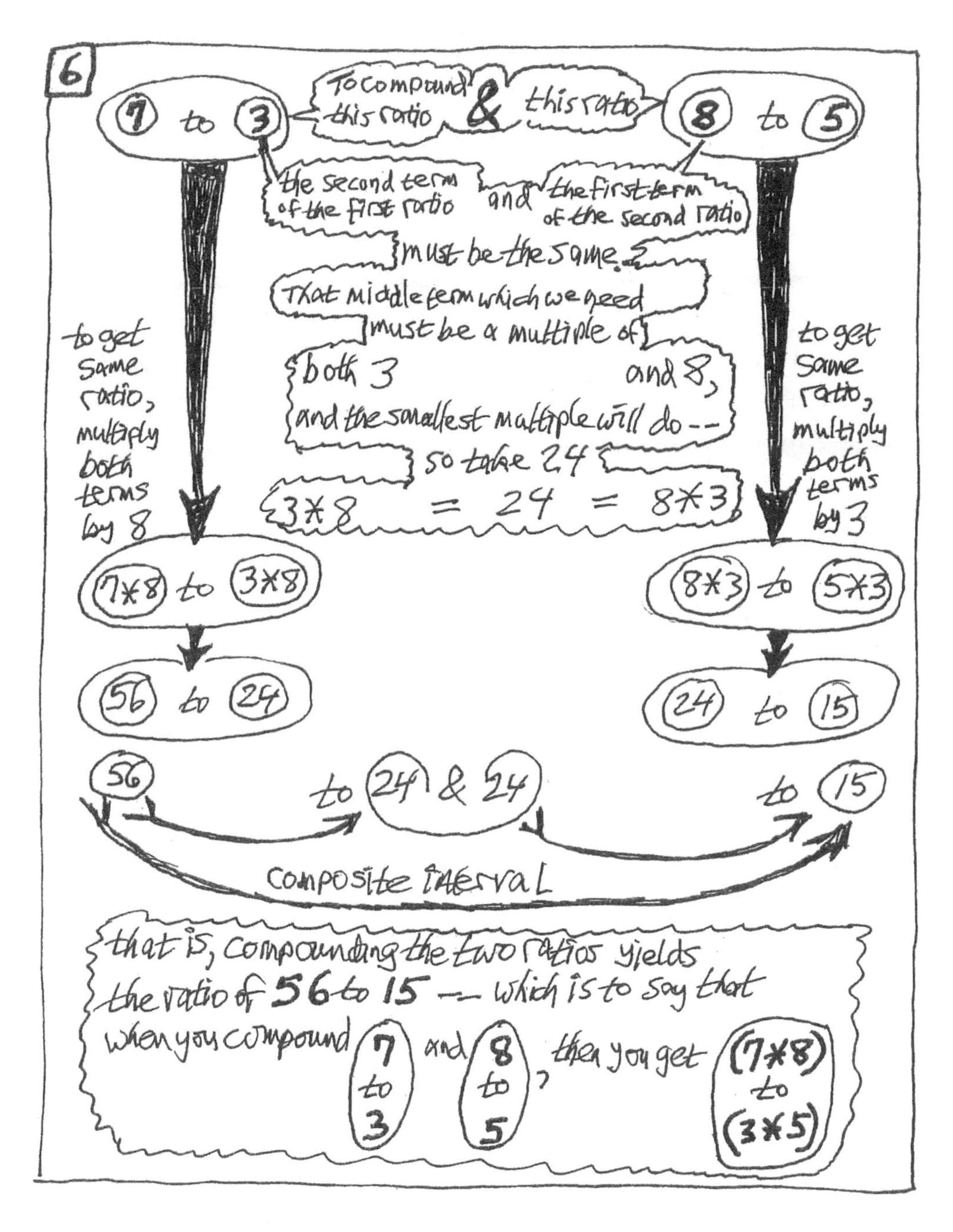

Now let us consider all that more generally—as Euclid presents it.

The Eighth Book of Euclid's *Elements* includes the numerical counterpart of the Sixth Book's proposition for the ratios of equiangular parallelogramic figures. The 5th proposition of the Eighth Book says that "plane" numbers have to one another the ratio which is compounded of the ratios of their "sides":

(VIII) 5. Plane numbers have to one another the ratio compounded from [the ratios of] their sides.

—Let there be as plane numbers the [numbers] A, B—and also as sides of the [number] A let there be the numbers C, D, and as sides of the [number] B the [numbers] E, F. I say that the [number] A, to the [number] B, has the ratio compounded from [the ratios of] their sides.

—For, the ratios being given which the [number] C has to the [number] E and also which the [number] D has to the [number] F, let there be taken least numbers G, H, K that are continuously (*hexês*) in the ratios C [to] E, D [to] E—and so, as is the [number] C to the [number] E so is the [number] G to the [number] H, and as is the [number] D to the [number] F so is the [number] H to the [number] K. And also, let the [number] D, by multiplying the [number] E, make the [number] L.

—And—since the [number] D has, on the one hand, by multiplying the [number] C, made the [number] A, and has, on the other hand, by multiplying the [number] E, made the [number] L—therefore, as is the [number] C to the [number] E so is the [number] A to the [number] L. And as is the [number] C to the [number] E so is the [number] G to the [number] H—therefore also, as is the [number] G to the [number] H so is the [number] A to the [number] L.

—Again—since the [number] E has, by multiplying the [number] D, made the [number] L, but has also, by multiplying the [number] F, made the [number] B—therefore, as is the [number] D to the [number] F so is the [number] L to the [number] B. But as is the [number] D to the [number] F so is the [number] H to the [number] K. And therefore also, as is the [number] H to the [number] K so is the [number] L to the [number] B.

—It was, however, shown also, that as is the [number] G to the [number] H so is the [number] A to the [number] L; therefore—*ex aequali*—as the [number] G is to the [number] K so is the [number] A to the [number] B. The [number] G to the [number] K, however, has the ratio compounded from [the ratios of] their sides; and also, therefore, the [number] A to the [number] B has the ratio compounded from [the ratios of] their sides—the very thing which was to be shown.

Having looked at Euclid's proposition VIII.5, let us now look at his earlier proposition VI.23, which shows how to compare two parallelograms when their angles are equal even if neither their heights nor their bases are equal.

The demonstration of the 23rd proposition in Book Six of Euclid's *Elements* depends upon some things shown in the Fifth Book, which contains his general treatment of the ratios and proportions of any magnitudes at all. The demonstration depends also upon the very first proposition of the Sixth Book, which relates ratios of parallelograms to ratios of lines which are their sides, as follows: equiangular parallelograms under the same height have to each other the same ratio which their bases do. If, therefore, one of the two sides that contain a rectangle is equal to a side of another rectangle, then the first rectangle will have to the other rectangle the same ratio which the other side of that first rectangle has to the other side of that other rectangle. This can be useful in comparing two rectangles in the more complex case when neither side of one rectangle is equal to a side of the other rectangle. It can be useful, that is, if we make use of our ability to put ratios together. Here is how Euclid does the job in VI.23:

(VI) 23. Equiangular parallelograms have to one another the ratio compounded from [the ratios of] their sides.
—Let there be as equiangular parallelograms the [parallelograms] AC, CF that have the angle [contained] by [the lines] BCD equal to the angle [contained] by [the lines] ECG. I say that the parallelogram AC has to the parallelogram CF the ratio compounded from [the ratios of] their sides.
—For let them be laid so that the [line] BC is in a straight [line] with the [line] CG; therefore in a straight [line] is also the [line] DC with the [line] CE. And also, let there be completed the parallelogram DG, and also let there be laid out a certain straight [line], namely the [line] K, and let it be generated that (on the one hand) as is the [line] BC to the [line] CG so is the [line] K to the [line] L, and (on the other hand) as is the [line] DC to the [line] CE so is the [line] L to the [line] M.
—Therefore the ratios of the [line] K to the [line] L and of the [line] L to the [line] M are the same as the ratios of the sides—namely [the ratios] of the [line] BC to the [line] CG and of the [line] DC to the [line] CE. But the ratio of the [line] K to the [line] M is compounded from the ratio of the [line] K to the [line] L and the [ratio] of the [line] L to the [line] M—so that also the [line] K has to the [line] M the ratio compounded from [the ratios of] the sides.
—And—since, as is the [line] BC to the [line] CG so is the parallelogram AC to the [parallelogram] CH, but as is the [line] BC to the [line] CG so is the [line] K to the [line] L—therefore also, as is the [line] K to the [line] L so is the [parallelogram] AC to the [parallelogram] CH.
—Again—since, as is the [line] DC to the [line] CE so is the parallelogram CH to the [parallelogram] CF, but as is the [line] DC to the [line] CE so is the [line] L to the [line] M—therefore also, as is the [line] L to the [line] M so is the parallelogram CH to the parallelogram CF.
—Since, then, it was shown that as is the [line] K to the [line] L so is the parallelogram AC to the parallelogram CH, and that as is the [line] L to the [line] M so is the parallelogram CH to the parallelogram CF, therefore—*ex aequali*—as is the [line] K to the [line] M so is the [parallelogram] AC to the parallelogram CF. The [line] K to the [line] M, however, has the ratio compounded from [the ratios of] the sides; and also, therefore, the [parallelogram] AC to the [parallelogram] CF has the ratio compounded from [the ratios of] the sides. Therefore equiangular parallelograms have to one another the ratio compounded from [the ratios of] their sides—the very thing which was to be shown.

In VI.23, we have two rectangles (call them K and L) which we would like to relate to each other, but which we are not able to relate directly. We therefore relate them indirectly. Let us relate the one rectangle (K) to something (call it M) that is related to the other rectangle (L). What we need to find is another rectangle (M) such that even though we do not have the ratio of K to L, we do have both the ratio of K to M and the ratio of M to L. If we had this M, then we would have the ratio of K to L—for if K and L and M are any three magnitudes at all, then the ratio which K has to L is compounded of the ratio of K to M and the ratio of M to L. What, then, is this rectangle M that we need? Well, we can compare any two rectangles with equal *heights* when their *bases* are unequal, and we can compare any two rectangles with equal *bases* when their *heights* are unequal (Euc VI.1). So, let rectangle M get one of its sides from rectangle K, and its other side from rectangle L. If rectangle M thus gets the height of rectangle K and the base of rectangle L, then rectangles K and M will have the same ratio as their bases, and rectangles M and L will have the same ratio as their heights. The compounding of these ratios will give us the ratio of rectangles K and L (fig. 7).

If you consider that a rectangle will remain the same size when you turn it around, letting its height become its base and its base become its height, you will see that either side of one rectangle can be related to either side of the other. Whichever way you form the pair of ratios of the sides, you will get the same result when you compound the ratios.

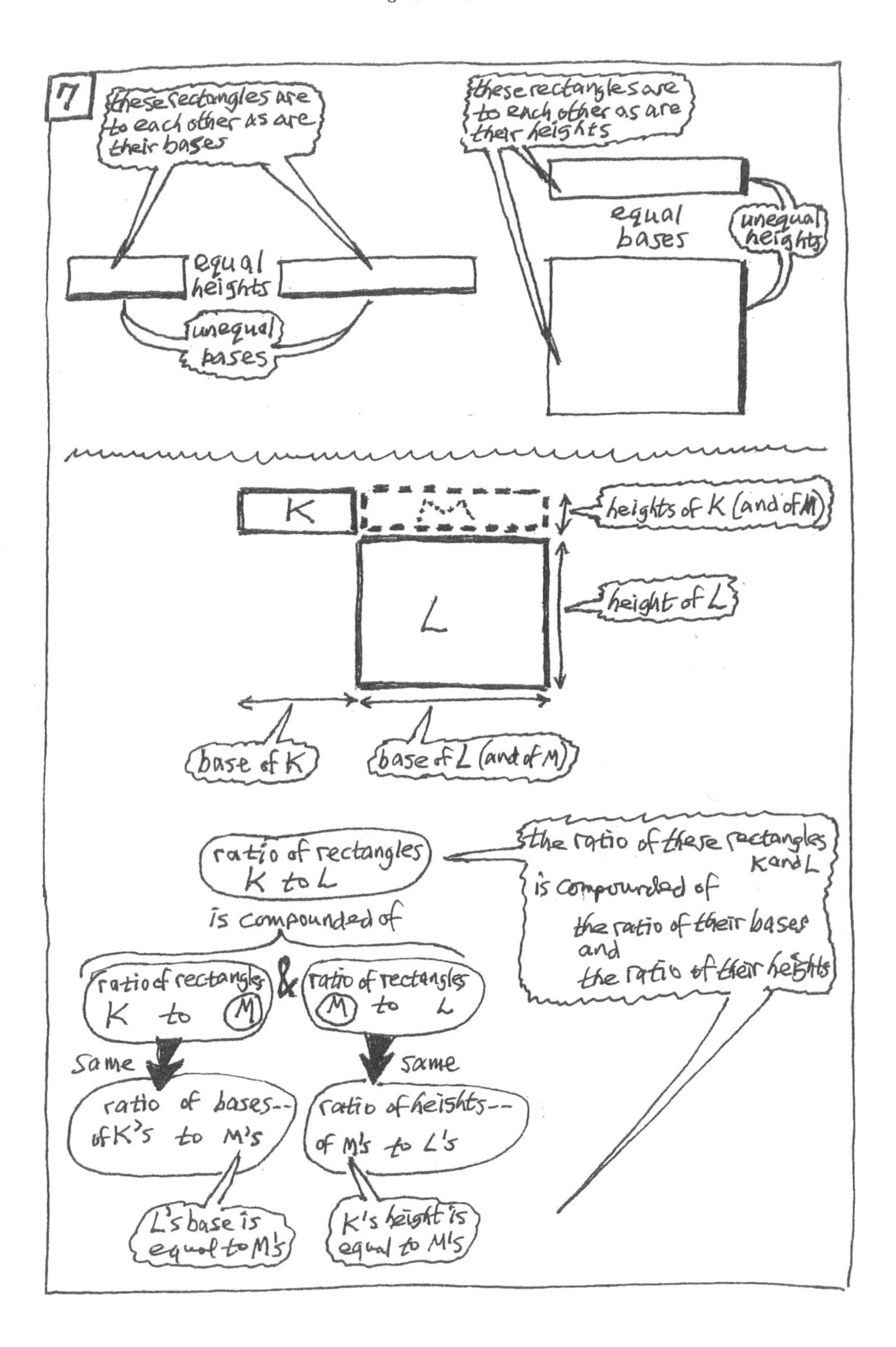
7
these rectangles are to each other as are their bases
equal heights
unequal bases
these rectangles are to each other as are their heights
equal bases
unequal heights
K
M
heights of K (and of M)
L
height of L
base of K
base of L (and of M)
ratio of rectangles K to L
is compounded of
ratio of rectangles K to M
&
ratio of rectangles M to L
same
same
ratio of bases-- of K's to M's
ratio of heights-- of M's to L's
L's base is equal to M's
K's height is equal to M's
the ratio of these rectangles K and L
is compounded of
the ratio of their bases
and
the ratio of their heights

Notice the similarity of structure in the demonstration of Euclid's VI.23 and in the demonstration of his VIII.5. The letters in the different demonstrations refer to different sorts of things, but they are related in similar ways (fig. 8).

Disregarding the meaning of signs can make them easier to manipulate. Statements written down, in abbreviations, about a number that is the product of multiplying some number P by some other number Q can look a lot like statements written down, in abbreviations, about the rectangle that is the product of containment by some line P and by some other line Q. When our statements get complicated enough, we may seek simpler ways to represent what we are talking about. To do so, we may talk about patterns in the signs that we use to represent what we are talking about. We may change the way we represent what we are talking about.

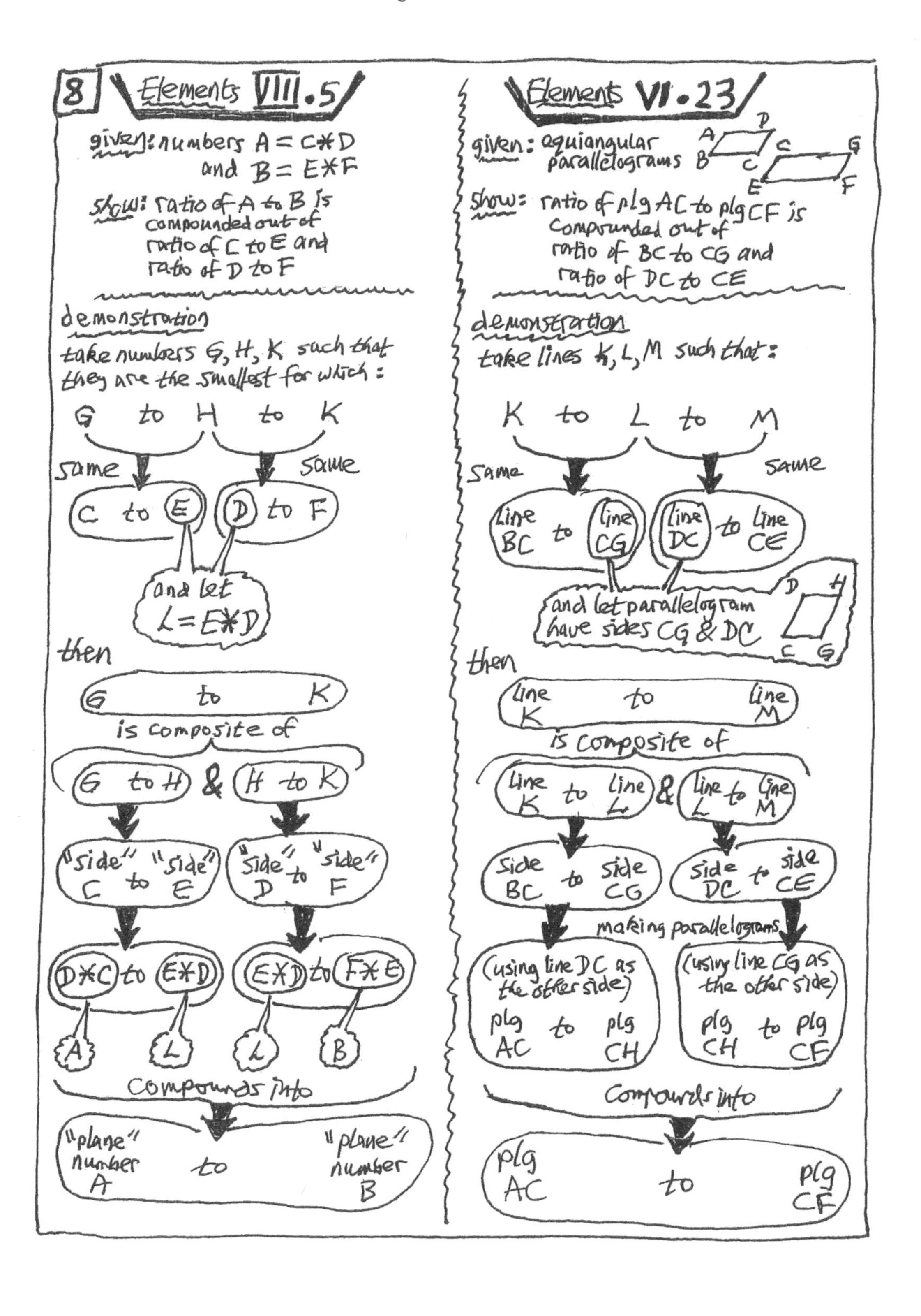
8
Elements VIII.5
given: numbers A = C*D
and B = E*F
show: ratio of A to B is
compounded out of
ratio of C to E and
ratio of D to F
demonstration
take numbers G, H, K such that
they are the smallest for which:
G to H to K
same
same
C to E
D to F
and let
L = E*D
then
G to K
is composite of
G to H
&
H to K
"side" C to "side" E
"side" D to "side" F
D*C to E*D
E*D to F*E
A
L
L
B
compounds into
"plane" number A
to
"plane" number B
Elements VI.23
given: equiangular
parallelograms
A
B
C
D
C
E
G
F
show: ratio of plg AC to plg CF is
compounded out of
ratio of BC to CG and
ratio of DC to CE
demonstration
take lines K, L, M such that:
K to L to M
same
same
line BC to line CG
line DC to line CE
and let parallelogram
have sides CG & DC
D
H
C
G
then
line K to line M
is composite of
line K to line L
&
line L to line M
side BC to side CG
side DC to side CE
making parallelograms
(using line DC as the other side)
plg AC to plg CH
(using line CG as the other side)
plg CH to plg CF
compounds into
plg AC
to
plg CF

Euclid shows that if we have two rectangles, then their ratio is the same as the ratio compounded of two ratios of straight lines. We can also turn the situation around. If we have a ratio that results from compounding two ratios of straight lines, then we can represent that relationship as a ratio of two rectangles. We can, that is to say, represent a relationship among four items by presenting a relationship between only two items. The relationship between two rectangles shows the relationship between four straight lines. In particular, the proportionality of four straight lines can be shown by the equality of two rectangles.

But suppose that we have two rectangles (call them A and B) and we get another rectangle (call it C) which not only is equal to rectangle B but also has the same height as rectangle A (fig. 9).

Then the ratio which rectangle A has to rectangle C will be the same as the ratio which rectangle A has to rectangle B, and this will be the same ratio which rectangle A's base has to the base of rectangle C. Thus, as rectangle A is to rectangle B, so will rectangle A's base be to rectangle C's base. And for these two straight lines (the base of A and the base of C) together with any given third straight line at all, we can always find a fourth proportional.

Thus, a complicated relationship among four straight lines can be compressed into a relationship between two items (rectangles) which still exhibit the four (since the straight lines still appear as sides containing the rectangles), and this can then further be compressed into a relationship between two items of the same sort as the original four (since straight lines—albeit new ones—are again the terms of the relationship).

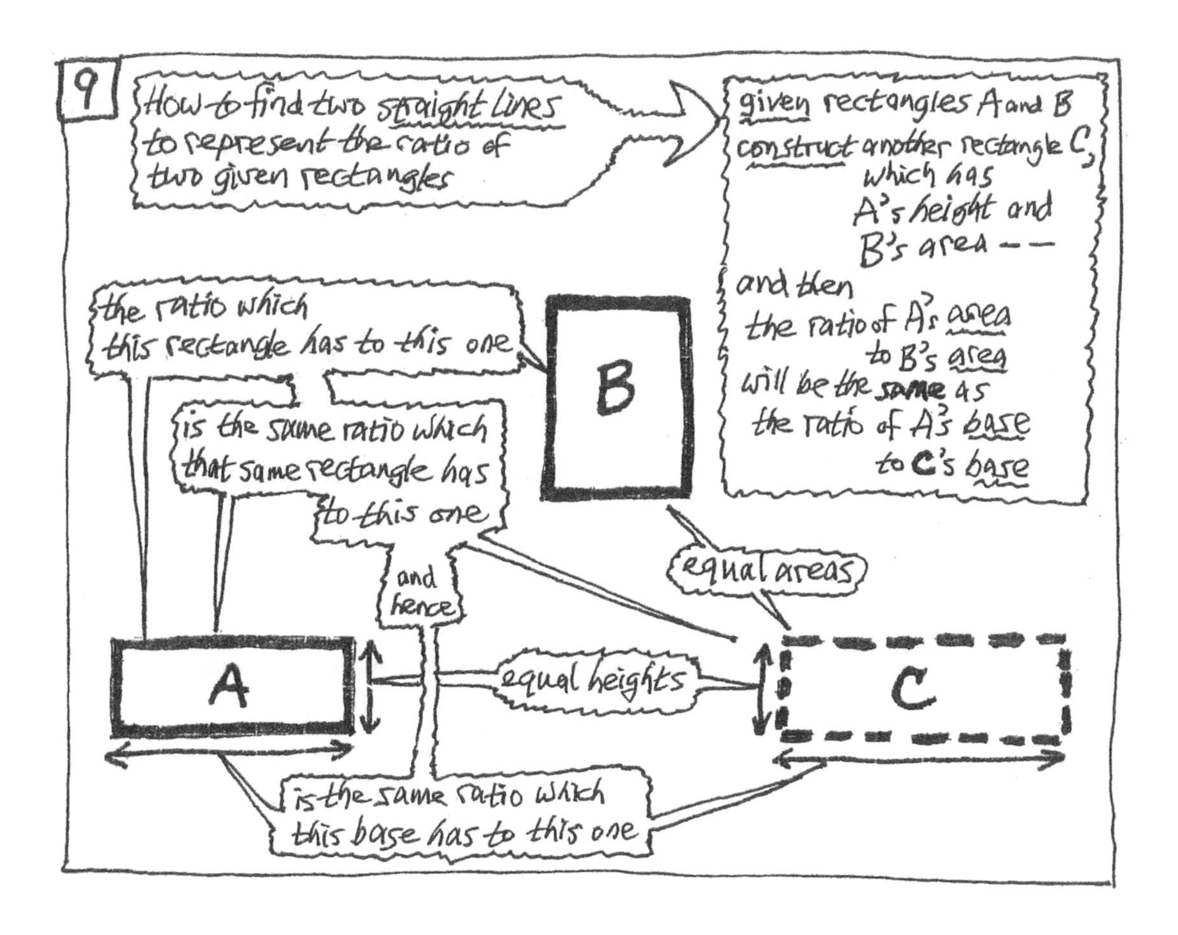
9
How to find two straight lines to represent the ratio of two given rectangles
given rectangles A and B construct another rectangle C, which has A's height and B's area --
and then the ratio of A's area to B's area will be the same as the ratio of A's base to C's base
the ratio which this rectangle has to this one
is the same ratio which that same rectangle has to this one
and hence
is the same ratio which this base has to this one
equal areas
equal heights
A
B
C

Indeed, rectangles need not be involved at all in the process of representing as another straight-line ratio the ratio compounded of two straight-line ratios. Suppose that we have one pair of straight lines (call them A and B) and also another pair (C and D). To "squoosh" these two ratios, pick *any* straight line you please (call it L). Now, there will be *some* straight line (call it P) which has to L the same ratio that A has to B; and there will also be *some* straight line (call it Q) to which L has the same ratio that C has to D. (We shall soon see just how to find fourth proportionals like P and Q.) Now:

<table>
<tr><td>since</td><td>as A is to B, so</td><td>P is to L,</td><td>and</td><td>as C is to D, so L is to Q,</td><td></td></tr>
<tr><td colspan="6">therefore the ratio compounded of</td></tr>
<tr><td></td><td>A's ratio to B</td><td></td><td>and</td><td>C's ratio to D</td><td></td></tr>
<tr><td colspan="6">will be the same as the ratio compounded of</td></tr>
<tr><td></td><td></td><td>P's ratio to L</td><td>and</td><td></td><td>L's ratio to Q,</td></tr>
<tr><td colspan="6">which is the same as the ratio whose terms are the two lines that we have found—
namely:</td></tr>
<tr><td></td><td></td><td>P's ratio to ....</td><td></td><td></td><td>Q</td></tr>
</table>

Thus, we have represented the ratio compounded of two line ratios (line A's ratio to line B, and line C's ratio to line D) as another line ratio (line P's ratio to line Q). If, however, we represented the ratio as a ratio of rectangles (the rectangle contained by line A and line C, to the one contained by line B and line D), we would not lose sight of the original lines.

By doing so, however, we would lose something else. If we represented a ratio compounded of straight-line ratios by another straight-line ratio, we could compound as many straight-line ratios as we please and still have a representation for the ratio that results. If, by contrast, we insist on keeping sight of the original lines when we compound *many* straight-line ratios together, we cannot continue to have a representation for the resultant ratio. That will become a matter of some importance, so let us pause to make sure that it is clear.

Note first, by the way, that although only two ratios are ever compounded at a time, there is no difficulty in speaking of *three* ratios as being compounded together, since you get the same final result whatever may be the order in which you perform the compounding of two ratios at a time. (If, for example, one ratio is compounded with a second, and the ratio resulting from that compounding is then compounded with a third ratio, you get the same ratio as if you had compounded the first ratio with the ratio that would result from compounding the second ratio with the third (fig. 10).)

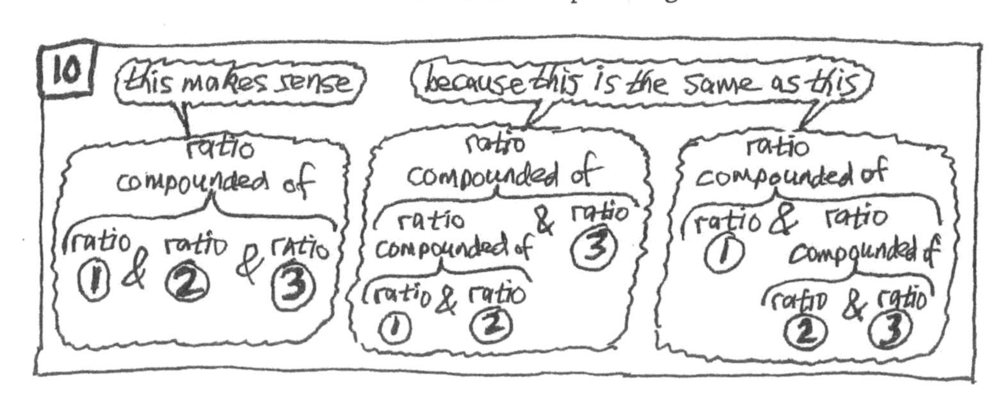

Now suppose that the ratios to be compounded are ratios of straight lines—the ratio of A to B, and the ratio of C to D, and the ratio of E to F. Three-dimensional boxes, like boxes that are two-dimensional, have to each other the same ratio as the ratio compounded of the ratios of their sides. That is to say: just as *rectangles* have to each other the ratio compounded of the ratio of their *heights* and the ratio of their *widths*, so *parallelepipeds* have to each other the ratio composed of the ratio of their *heights*, and the ratio of their *widths*, and the ratio of their *depths*. For example, if you take a box and you double its height and triple its width and quadruple its depth, you will make the box twenty-four times as large. (The ratio of 24 to 1 is compounded of the ratios of 2 to 1, and of 3 to 1, and of 4 to 1.) So, by making parallelepipeds, you can represent the ratio compounded of three straight-line ratios, while keeping in sight the original straight lines. The ratio compounded of the ratios of A to B, and of C to D, and of E to F, is the same as the ratio of a parallelepiped with sides A,C,E to a parallelepiped with sides B,D,F.

The ratio compounded of the three straight-line ratios does not have to be represented as a ratio of parallelepipeds; it could also be represented as another straight-line ratio. Figure 11 shows how. In this representation of the compound by the ratio of the one straight line (P) to the other (R), however, the original straight lines (A,B,C,D,E,F) are lost to sight, whereas they were kept in sight when the compound was represented as a ratio of parallelepipeds.

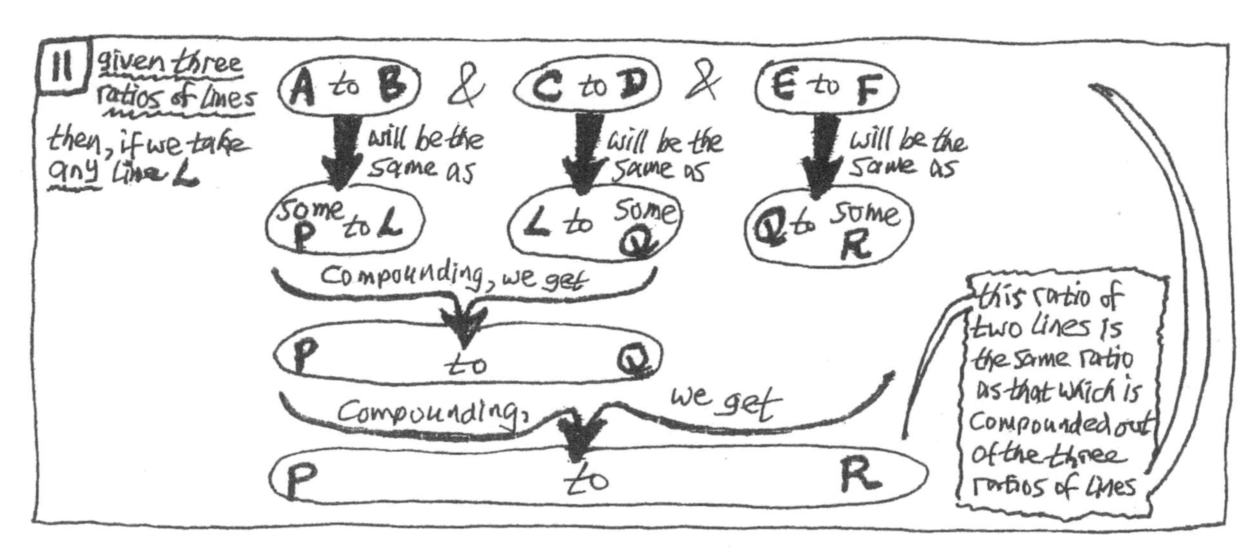

Now what happens if we move on to compound more than three ratios of straight lines? We can find straight lines whose ratio is the same as the ratio that results from compounding four or more ratios of straight line, but the "squooshing" by which we find those lines squeezes out the original straight lines. The original straight lines cannot be kept in sight by making them the sides of figures of several dimensions. With two ratios, we built two-dimensional boxes (rectangles); and with three ratios, we built three-dimensional boxes (parallelepipeds). Since we cannot visualize a box of more than three dimensions, we must eventually abandon representations in which the original lines appear as the sides of boxes for which the ratio of the boxes is compounded of the ratios of the original lines. But we do not have to lose sight of the original lines so long as we do not have our eye on compounding that involves more than three ratios.

We might wish to treat all compounding as the same, no matter how many ratios they involve—and then we would be concerned about the disadvantage of letting the ratio compounded of two ratios of straight lines be represented by the ratio of the two rectangles which those straight lines would contain if used as sides. For now, however, we shall let the ratio which is compounded of two ratios of straight lines be represented by the ratio of the rectangles which those straight lines would contain if used as sides. The advantage of using this representation will outweigh its disadvantage.

It came to be thought, a few centuries ago, that the disadvantage was too great. The study of classical geometry and its transformation prepares us to see why those who followed Descartes ceased to represent the proportionality of four straight lines by the equality of the two rectangles which those straight lines can contain, and chose instead to represent that proportionality by the equality of two straight lines—each of these straight lines being taken as the "product" of "multiplying" two straight lines together. It was this seemingly small change in representation that made it possible for geometrical proportions to be swept away by algebraic equations, which then transformed the world.

The geometrical key to Descartes' transformation of mathematics is an elementary Euclidean proposition which Apollonius will use in his next proposition. That requisite proposition from Euclid is the 12th proposition of the Sixth Book of the *Elements*. It shows how to find a fourth proportional if three straight lines are given. The procedure is presented in figure 12.

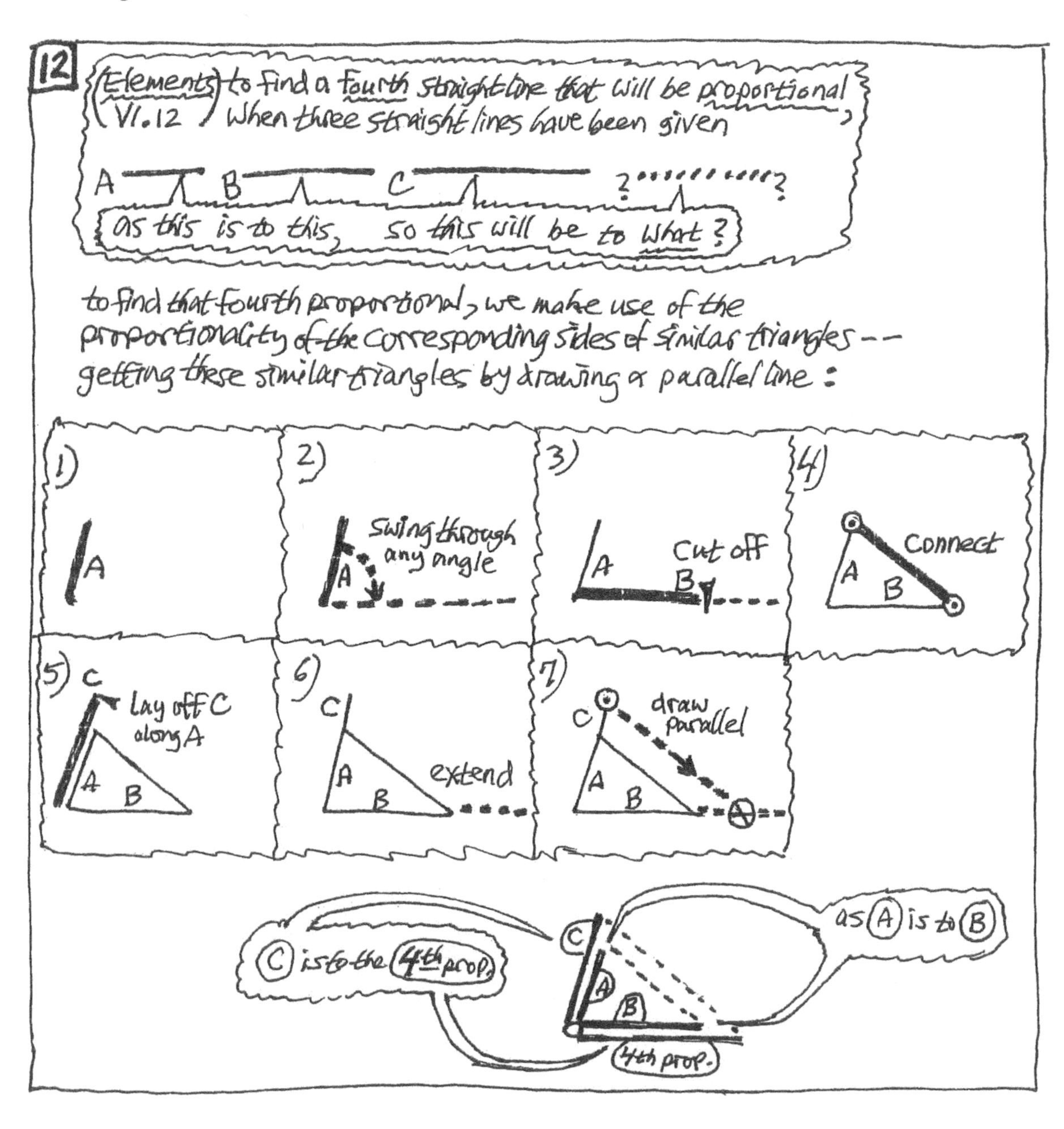

(And, for how Euclid finds a mean proportional if two straight lines are given, see the 13th proposition of that Sixth Book of the *Elements*. The 14th proposition then makes it possible to establish, in the 16th and 17th propositions, the connection between the proportionality of straight lines and the equality of the rectangles that those lines contain. The 31st proposition of the Third Book of the *Elements* and the 8th proposition of its Sixth Book give us the square-rectangle property for the circle.)

That procedure for finding a fourth proportional to three given straight lines may not seem like much to make a fuss about, but it proved to be highly explosive material when the match was struck. Some two thousand years after the time of Euclid and Apollonius, it provided Descartes with the means to transform statements involving ratios of straight lines or equalities of rectangles into equations involving "multiplications" of "linear" quantities that yield other "linear" quantities as "products." For us, this 12th proposition of the Sixth Book of Euclid's *Elements* is like a ticking time-bomb.

The reason why a square's diagonal and its side are incommensurable can now be restated as follows. Any square has to another square the ratio that is the duplicate of the ratio of the side of the one square to the side of the other square. But a ratio is said to be the subduplicate of a ratio that is its duplicate, so we can say that the side of any square has to the side of another square the ratio that is the subduplicate of the ratio of the one square to the other. Since a new square made on the diagonal of an old square will be 2 times the size of the old square, the old square's diagonal must have to its side the ratio that is the subduplicate of the ratio of 2 to 1. But although the ratio of 2 to 1 is numerical, the ratio that is its subduplicate is not. There are no two numbers such that the product of multiplying one of them by itself is double the product of multiplying the other by itself and at least one of those numbers is odd. A square's side therefore cannot be exactly any fraction of its diagonal. That is why a square's diagonal and its side are incommensurable.

Incommensurability is a fact of fundamental significance in classical geometry, a source of many difficulties. In the end we shall have to ask this question: when Descartes' bomb explodes, are those difficulties blown away—or are they merely covered over?

PART III
CHARACTERIZATION OF CONIC SECTIONS

Apollonius's *Conics*, Book I: propositions 11 through 14

CHAPTER 1. THE PARABOLA

transition

When Apollonius first obtained new sections (in the 8th and 9th propositions), he was not able to say much about them.

He showed first (in the 8th proposition) what happens if he cuts through one side of an axial triangle so that he does not hit the other side of the triangle. How does he avoid hitting the other side? Either by placing the cutting plane parallel to the other side, so that it will not hit it at all; or by placing it to hit the other side's extension above the cone's vertex, so that below the vertex it will get farther and farther away from the other side. Either way, the line that results is a curve, and it has a diameter. Since this diameter is indefinitely prolongable, and prolongable also is the curve itself along with its array of ordinates, the curve is one that is not closed. The line obtained in the 8th proposition is therefore surely new.

Immediately after getting it, Apollonius showed (in the 9th proposition) what happens if he cuts so that he hits the axial triangle again on the other side, this time below the vertex, but differently from the way that he cut when he got circles before (in the 4th and 5th propositions). The line that results is a curve that is closed but is not circular. The line obtained in the 9th proposition is therefore also surely new.

So, we can say that there are some lines which, although they are curved, never close when they are prolonged; and that there are also some curved lines which, although they are closed, are not circular. But we wish to go beyond that: we wish to characterize these new curves by saying more than merely whether they are prolongable or closed.

The circle was defined at the beginning of Euclid's *Elements* not merely by its being closed (by its being a planar figure contained by one line) but by something in addition: the equality of the straight lines drawn from the points of the containing line to a certain point inside it. Perhaps these new curves too can be characterized by the size of straight lines drawn from the points on them to something inside them.

Apollonius first examines new curves of the sort that he showed us first (in the 8th proposition). These are the ones least like a circle, because, rather than being closed, they can be prolonged indefinitely. The very first of these new curves are those cut by a plane which does not hit the axial triangle again on the other side or on the other side's prolongation. Apollonius calls such a curve a *parabola*. He does so in his 11th proposition.

In treating this proposition that presents the parabola, we shall consider the following:

—how the cone is cut to get the curve
—what it is that is shown about the curve
—how it is shown
—why we might want to show it.

Here first is the text of Apollonius's 11th proposition. As usual in this guidebook, diagrams follow in the subsequent explication.

#11 If a cone be cut by a plane through its axis, and be cut also by another plane which cuts the base of the cone along a straight [line] that is at right [angles] to the base of the triangle through the axis, and, moreover, the diameter of the section be parallel to one side of the triangle through the axis, and any [straight line] from the section of the cone be drawn parallel to the common section of the cutting plane and of the base of the cone, that is, be drawn as far as the diameter of the section, then [this straight line so drawn from the section to its diameter] will have the power (*dunêsetai*) [the power, that is, to give rise to a square that is equal to] the [rectangle] contained by [the following two lines:] [first,] the [line] which is taken off by it, [by, that is, that straight line which is drawn from the section to the diameter], [taken off, that is,] from the diameter, on the side toward the section's vertex—and, [second,] a certain other [straight line] which has, as its ratio to the [straight line] between the cone's angle and the vertex of the section, the ratio which the square [arising] from the base of the triangle through the axis has to the [rectangle] contained by the triangle's remaining two sides. And let such a section be called a parabola (*parabolê*). —Let there be a cone of which the point A is vertex and the circle BC is base. And let [the cone] be cut by a plane through the axis; and let [the plane] make as section the triangle ABC, and let the cone be cut also by another plane which cuts the base of the cone along a straight [line], namely the [line] DE, that is at right [angles] to the [line] BC; and let [this other plane] make as section on the surface of the cone the [line] DFE, and let the diameter of the section, namely the [line] FG, be parallel to one side of the triangle through the axis—[let it be parallel, that is to say,] to the [line] AC. And from the point F let there be drawn, at right [angles] to the straight [line] FG, the [line] FH. And let [line FH] have been so made that as is the [square] from BC to the [rectangle] by BAC, so is the [line] FH to FA. And let there be taken some chance point on the section—the [point] K, say—and through the [point] K [let there be drawn] parallel to the [line] DE the [line] KL. I say that the [square] from the [line] KL is equal to the [rectangle] by the [lines] HFL.

—For let there be drawn through the [point] L parallel to the [line] BC the [line] MN. And also the [line] KL is a parallel—to the [line] DE. The plane through the [lines] KL, MN is therefore parallel to the plane through the [lines] BC, DE—that is, it is parallel to the base of the cone. The plane through the lines KL, MN therefore is a circle, a diameter of which the [line] MN is. And perpendicular to the [line] MN is the line KL, since also perpendicular is the [line] DE to the [line] BC. The [rectangle] by the [lines] MLN, therefore, is equal to the [square] from the [line] KL. And since as is the [square] from the [line] BC to the [rectangle] by the [lines] BAC, so is the [line] HF to FA—and the [square] from the [line] BC has, to the [rectangle] by the [lines] BAC, the ratio put together out of (*sunkeimenon ek*) [hereafter, rendered traditionally as "compounded from"] that which the [line] BC has to CA and also that which the [line] BC has to BA—therefore the ratio of the [line] HF to FA is compounded from that of the [line] BC to CA and also that of the [line] CB to BA. But (on the one hand) as is the [line] BC to CA, so is the [line] MN to NA—that is, [so is] the [line] ML to LF; and (on the other hand) as is the [line] BC to BA, so is the [line] MN to MA—that is, [so is] the [line] LM to MF, and, as remainder, so is the [line] NL to FA. The ratio of the [line] HF to FA, therefore, is compounded from that of the [line] ML to LF and also that of the [line] NL to FA. But the ratio compounded from that of the [line] ML to LF and also that of the [line] LN to FA is that of the [rectangle] by MLN to the [rectangle] by LFA. As, therefore, is the line [HF] to FA, so is the [rectangle] by MLN to the [rectangle] by LFA. But as is the [line] HF to FA, so—with the [straight line] FL taken as a common height—is the [rectangle] by HFL to the [rectangle] by LFA. As, therefore, is the [rectangle] by MLN to the [rectangle] by LFA, so is the [rectangle] by HFL to the [rectangle] by LFA. Equal, therefore, is the [rectangle] by MLN to the [rectangle] by HFL. But the [rectangle] by MLN is equal to the [square] from the [line] KL. Therefore the [square] from the [line] KL also is equal to the [rectangle] by the [lines] HFL.

—And let such a section be called a "parabola" (*parabolê*); and also, the [line] HF—the one along (*para*) which the straight lines that are drawn down ordinatewise to the diameter FG have the power (*dunantai*)—let it be called an "upright" (*orthia*) [side].

[We need to remember what capability is meant when a straight line is said to "have the power." It means that a line of that size can, when taken as a side, give rise to a square of a certain size. Here, that square arising from the line HF will be equal to a certain rectangle—the rectangle, namely, which is "laid alongside" (*parakeitai para*, traditionally rendered as "is applied to") the line HF, as one of its own sides, and which has, as its other side, the piece taken off from the diameter, on the side toward the vertex, by one of those lines that is drawn down ordinatewise.]

11th proposition: the enunciation—how we cut

The first thing to consider is the orientation of the cutting plane. A plane is determined by any two intersecting straight lines that lie in it. In order to specify straight lines for the orientation of the cutting plane, we first pass another plane through the axis. This *axial* plane, by intersecting with the cone, makes an axial *triangle*; and it is with respect to this axial triangle that we orient the cutting plane, the plane which will cut into the cone the curve that is a conic *section*.

Two results are guaranteed by the various ways in which we shall orient the cutting plane to cut the various sections into the cone. *First*: whatever different sorts of sections we may obtain by cutting, every one of them will have a diameter that bisects all of its chords that are parallel to a reference line. And *second*: when we cut in different ways, we shall obtain sections of different kinds. That two-fold result is achieved by making—in each proposition that gives us a new kind of curve by cutting a cone (the 11th, 12th, and 13th propositions of Book One)—a two-fold stipulation about the orientation of the cutting plane: in all three propositions, there will be a sameness in how the cutting plane meets the base plane; and in each of the three propositions, there will be a difference in how the cutting plane meets the axial plane.

As we now examine the orientation of the cutting plane, you can see a depiction of it in figure 1.

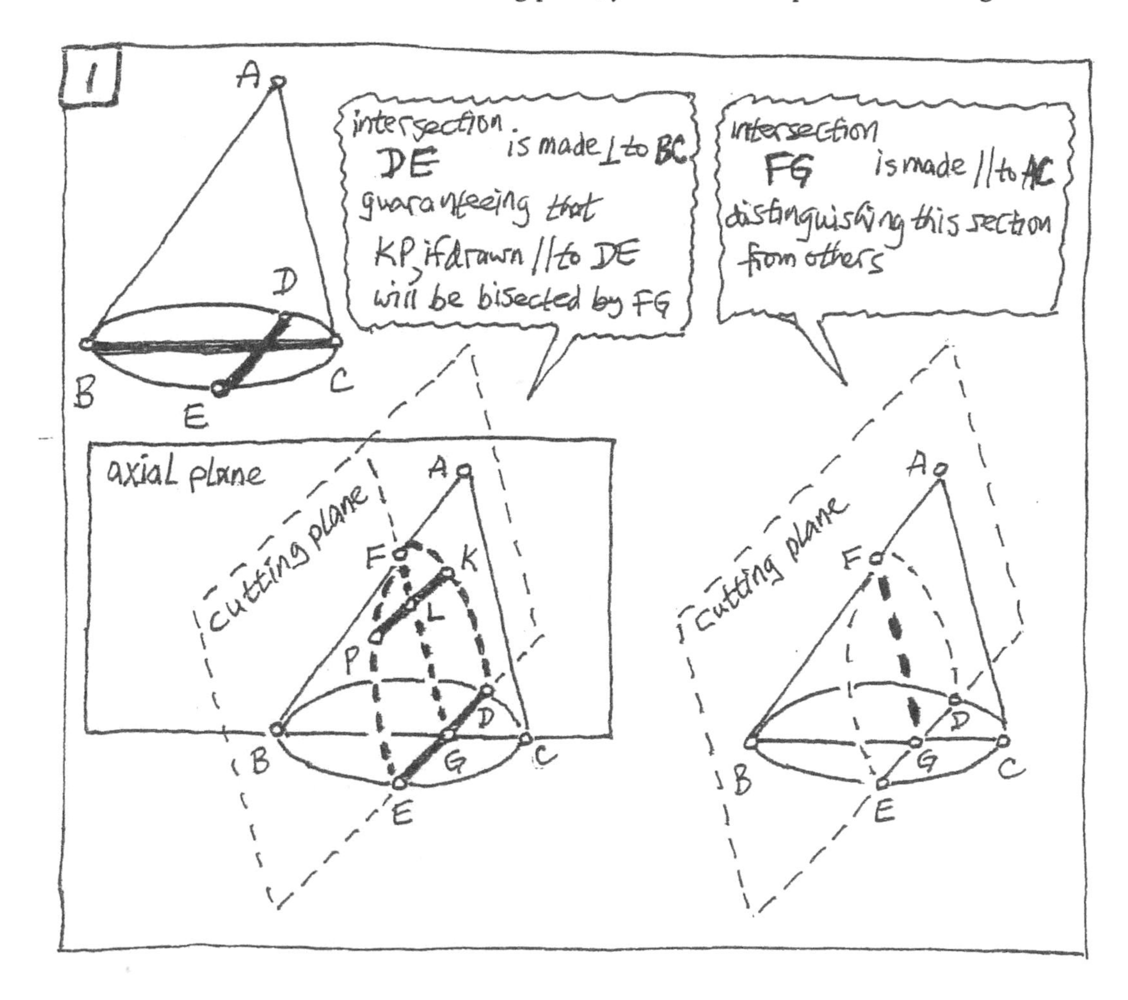

The *first* stipulation for the cutting plane's orientation guarantees that there will be a diameter. It does so by providing a reference line for straight lines drawn "ordinatewise," lines which the diameter will bisect. Here is how. The cutting plane makes straight line DE by intersecting with the cone's base plane, and it also makes straight line FG by intersecting with the axial plane; the axial plane makes the axial triangle's straight-line base BC by intersecting with the cone's base plane. Now consider that line DE, which is of interest because it can be a reference line for straight lines drawn in an orderly way across the section parallel to that reference line. The line DE will indeed be such a reference line if we orient the cutting plane so that DE is perpendicular to BC. That is something we learned earlier in Book One. (Keep in mind, however, that while DE is perpendicular to the *base-line* BC of the axial triangle, it is *not necessarily* perpendicular to the *whole plane* of the axial triangle ABC. Although the axial triangle may be perpendicular to the cone's base, it also may not be—but, even when it is not, the lines drawn ordinatewise will still be bisected by the diameter, which is what FG, the cutting plane's intersection with the axial triangle, turns out to be. Thus, the lines drawn ordinatewise may or may not be perpendicular to their diameter, the line that bisects them.) Figure 1 shows a plane so oriented that it cuts into the cone the curve EFD. By stipulating that DE be perpendicular to BC, we guarantee that the cutting plane's intersection FG with the axial triangle ABC will bisect each of the section's chords KP that is parallel to the line of reference DE. All chords drawn ordinatewise like KL will be bisected by diameter FG. The perpendicularity of the two straight lines that have been specified (the intersections made with the cone's base plane by the cutting plane and by the axial plane) is the guarantee that the cutting plane's intersection with the axial triangle will bisect each of the section's chords which is parallel to the line of reference—it is the guarantee, in other words, that there will be a diameter for lines drawn ordinatewise. The cutting plane's intersection with the cone's base plane can be the line of reference because it is perpendicular to the base plane's intersection with the axial plane, and that perpendicularity makes the cutting plane's intersection with the axial plane a diameter for the lines drawn ordinatewise. That perpendicularity will be a stipulation not only in the case of sections of the kind presented here in this proposition, but also in subsequent propositions presenting sections of other kinds. Diameters bisecting their lines drawn ordinatewise provide access to conic sections of *every* kind.

What distinguishes the parabola, the section of this *particular* kind presented in the 11th proposition, is the *second* stipulation in arranging for the cutting. It concerns the relation of the cutting plane, not to the axial triangle's *base*, but rather to its *sides*. For this kind of section (which is, in figure 1, the curve EFD), the cutting plane's straight-line intersection (FG) with the axial triangle (ABC) is parallel to the axial triangle's opposite side (AC).

11th proposition: the enunciation—what we show about the line we cut

The cutting plane has its orientation fixed by two lines (DE and FG) made by its intersection with two other planes—the cone's base and the axial triangle. Those lines are situated so that:

DE, in the base plane,	FG, in the axial plane,
is perpendicular to BC;	is parallel to AC.

When we cut that way, then (fig. 2) the intersection of plane and cone (the curve EFD) has the following property: any straight line (call it KL) drawn ordinatewise from the section to the diameter (FG) will be related in a certain way to the straight line (LF) which it cuts off on the diameter going away from the section's vertex. In speaking of this relationship, it will be convenient to use the terminology of ordinates and abscissas. The term "ordinate" comes from the Latin for something that has been "set in order," and the term "abscissa" comes from the Latin for something that is "cut off, or away from." A line drawn ordinatewise across the section is cut in half by the diameter, and it cuts off a portion of the diameter extending away from the vertex. Point L is the endpoint both for an *ordinate* (which is the line that extends ordinatewise half-way across the section from point K on the section) and also for an *abscissa* (which is the line that extends from the section's vertex F along the diameter down to where it is "cut off" by its meeting with the ordinate).

Now why are ordinate and abscissa of interest? Because their relationship tells us how the section widens as it lengthens. The section is shaped by the relationship between the size of the diameter's abscissa and the size of its ordinate.

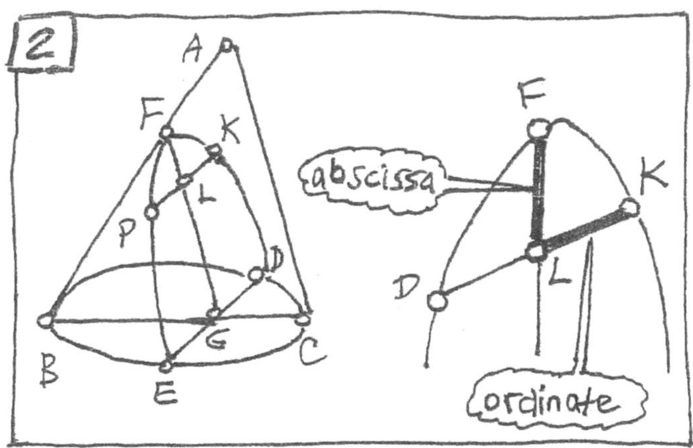

Shaping the curve, that size-relationship between the diameter's abscissa and its ordinate turns out to be the following (fig. 3).

> *If* we make a square whose side is the ordinate (KL),
> and, equal to that square, we make a rectangle whose *one* side is the *abscissa* (LF),
> *then* this rectangle's *other* side (FH) must be a line of a certain size;
> we must contrive it to be just the right length to fit the following proportion:
>
> the ratio which that line (FH) has to the line (FA) that stretches between
> the section's vertex (F) and the cone's vertex (A)
> must be the same as
> the ratio which a certain other square has to a certain other rectangle—
> a square whose side is the axial triangle's base (BC),
> and a rectangle whose sides are the axial triangle's sides (BA and AC).

Thus, what enables us to articulate the relationship between the ordinate and abscissa is a third straight line, one which we make just the right size to embody certain ratios. The line so contrived is called (at the end of the proposition) "the upright side."

In this 11th proposition, the ordinate and abscissa are related indirectly—by first contriving a straight line into which we build the relevant relationships, and by then building boxes out of those several straight lines (namely, the line that is the ordinate, the line that is the abscissa, and the line that is contrived to be the upright side).

It is because of how the contrived straight line can make a rectangle to exhibit the relevant relationships of lines, that Apollonius bestows the name "parabola" upon the curved line that will be cut into a cone by a plane that has been oriented as described. Bestowal of this name upon the curve is the last thing done by Apollonius in the enunciation. The reasons for his nomenclature will be clearer if we consider terminology that he inherited from Euclid—terminology that characterizes relationships between straight lines used to make rectangles.

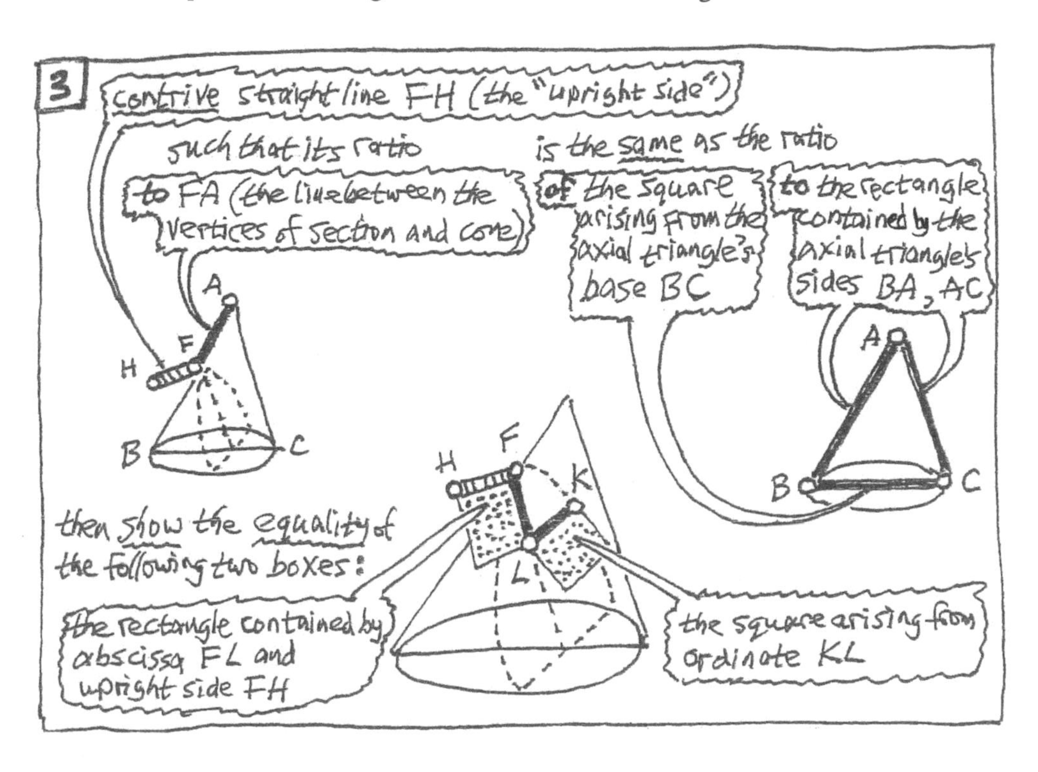

11th proposition: the enunciation–terminology

The situation here is similar to one that we saw earlier. What we saw then was that although we *cannot* by numerical means *directly* state the ratio that the diagonal of a square has to its side, we nonetheless can relate the lines *indirectly*. We can by numerical means relate the *squares* which they have the power to generate. As a *power* for square-generating, a square's *diagonal* is twice what the square's *side* is (fig. 4).

Traditionally, translations of this 11th proposition (and of subsequent propositions) speak of a line as being "equal in square" to a rectangle. A more literal translation would speak of a line as "having the power" with respect to a rectangle. The power that is meant is the power to be the side from which can arise a square equal to the rectangle. The lines which have such a power are often called simply "powers" (*dynameis*). (For an example, see Plato's *Theaetetus*, at 148A.) You might consider not lines and squares, but numbers. The number 3 is the "side" of the "square number" 9, and thus it would be the "power" from which the square number 9 arises. (Nowadays, however, terminology differs: we say that 3 is the "root"—"the square root"—of 9, and that 9 is 3 "squared" or "raised to the second power.")

The traditional translation of Apollonius's *Conics*, here and elsewhere, tries to follow the terminology of Heath's standard translation of Euclid's *Elements*. But perhaps it would be better here to translate both Euclid and Apollonius more literally. By translating Euclid more literally than Heath does, we can sometimes see where the work of Apollonius is coming from. Apollonius will soon speak, in the traditional translation, of a straight line's being equal in square to an area that is "applied to" some straight line. Euclid also speaks, in Heath's translation, of an area's being "applied to" a straight line. When the *Elements* in several places speaks of "throwing" a parallelogram "alongside" a given straight line under certain specified conditions, Heath's translation renders the verb *parabalein* (which is, literally, "alongside+to throw") as "to apply," and the preposition *para* ("alongside") as "to."

Toward the end of the First Book of the *Elements*, the 44th proposition performs the following task: along (*para*) a given straight line, to throw (*parabalein*), in a given rectilineal angle, a parallelogram equal to a given triangle (fig. 5). The very next proposition (I.45) performs a task which is similar but is more general: in a given angle, to construct a parallelogram equal to a given rectilineal figure. (In this more general proposition, the parallelogram to be constructed need not be thrown alongside a given straight line, and the given rectilineal figure to which it must be equal need not be a triangle.) Immediately after that, the 46th proposition performs the following task which is similar but is more specific: from (*apo*) a given straight line, to draw up (*anagrapsai*) a square. (Heath translates this as: to "describe" a square "on" the line. Later on, translating the qualification stated in the enunciation of VI.28, Heath renders it as having a parallelogram "described on" a straight line, rather than "drawn up from" it.) The next propositions (the 47th and 48th) provide the First Book of Euclid's *Elements* with its grand finale, which relates straight lines by way of the squares arising from them—these straight lines being the sides of a right triangle. These last two propositions, the famous "Pythagorean" theorem and its converse, give the First Book a spectacular conclusion—as well as setting the reader on the way to relating straight lines by way of squares and rectangles in the Second Book.

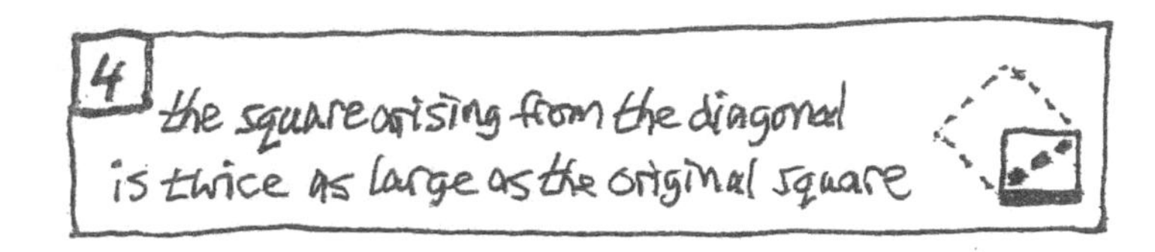

4
the square arising from the diagonal
is twice as large as the original square

In propositions toward the end of the Sixth Book of the *Elements* (in the 25th and 27th, and, most importantly, in the 28th and 29th propositions) Euclid again speaks of what Heath translates as the "application" of areas "to" straight lines. The first two of those propositions (and the proposition between them, the 26th) make possible the tasks of the last two, which are:

(in the 28th)	(in the 29th)
alongside (*para*) a given straight line	alongside (*para*) a given straight line,
and equal to a given rectilineal figure,	and equal to a given rectilineal figure,
to throw-alongside (*parabalein*)	to throw-alongside (*parabalein*)
a parallelogram that is	a parallelogram that is
lacking (*elleipon*)	thrown over-beyond (*hyperballon*)
by a parallelogramic figure that is	by a parallelogramic figure that is
similar to a given one;	similar to a given one. (Fig. 6.)

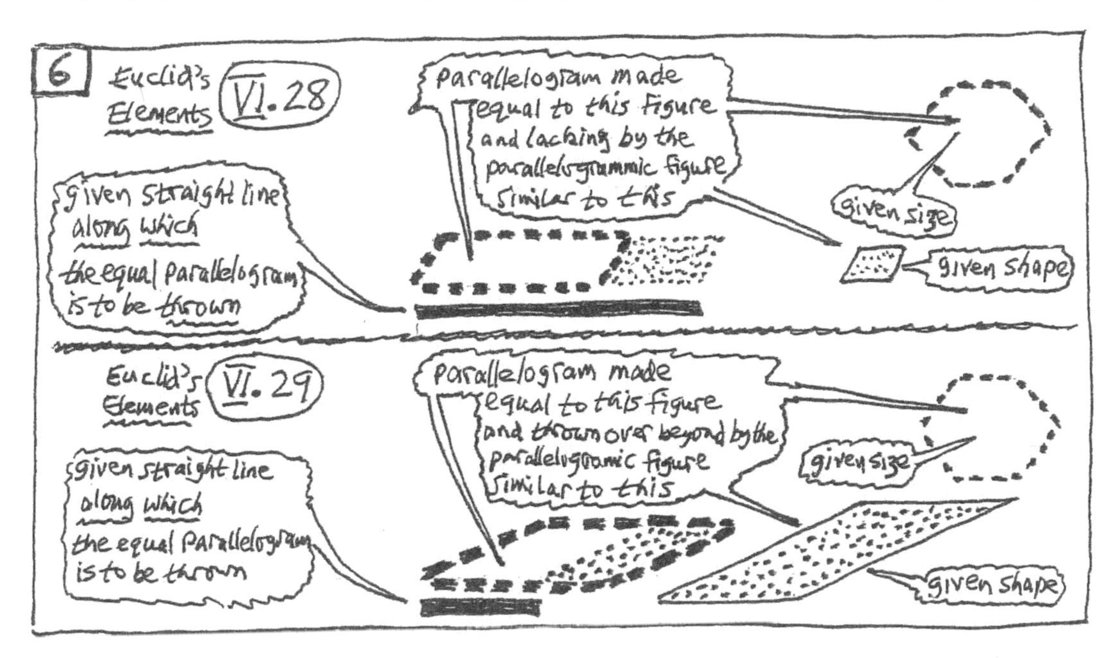

Euclid's Greek text, after the word *parabalein*, and immediately alongside it, has the word *elleipon* in the 28th proposition, and *hyperballon* in the 29th.

This terminology of Euclid is echoed by Apollonius. In the present proposition of the *Conics*, Apollonius deals with the conic section that he calls the *parabolê*. In the next proposition, he will deal with the one that he calls the *hyperbolê*; and in the next proposition after that, with the one that he calls the *elleipsis*.

The line that Apollonius contrives for the characterization of the conic sections (here and also in subsequent propositions) is sometimes called (not by him, however) "the parameter"—*para+metron* being Greek for an "alongside-measure." This name emphasizes the line's size rather than its position. By contrast, calling it "the upright side" emphasizes its placement in a configuration. The line so placed is of interest because of the difference that its size makes with respect to the rectangle of interest that is thrown alongside it.

In the present proposition, the upright side itself makes one side of a rectangle (since the rectangle's side is thrown alongside it just as far as its endpoint and no farther). In subsequent propositions, however, although the upright side will still be the line along which one side of an interesting rectangle is thrown, this rectangle's side will either go over beyond the endpoint of the upright side, or else will lack the length to reach so far.

Here in the 11th proposition, the contrivance is called (translating literally) "the [line] along which the [lines] drawn ordinatewise onto the diameter have the power"—which is later abbreviated to "the [line] along which the drawn [lines] have the power" and "the [line] along which they have the power." In defining the various conic sections, a square will be raised up from the ordinate, and a rectangle of equal area will be made, having as one of its sides that ordinate's abscissa. As this equal rectangular area thrown out (*bolê*) from the abscissa (1) falls just alongside (*para*) the parameter, or else (2) goes over beyond (*hyper*) it, or (3) is a lacking (*elleipsis*) of the length to reach the end of it, Apollonius will call the section (1) a "parabola," or else (2) a "hyperbola," or (3) an "ellipse."

At a different length cut off along the diameter by an ordinate, the ordinate itself has a different length—but the upright side is contrived by Apollonius to have the same length for every one of that diameter's ordinates in that section (fig. 7).

Here in the 11th proposition—where we get the parabola—the upright side is a line of constant length that is so contrived that it will relate an ordinate to its abscissa in the following way: a square raised up from the ordinate will be equal to a rectangle contained by the ordinate's abscissa and *the upright side*.

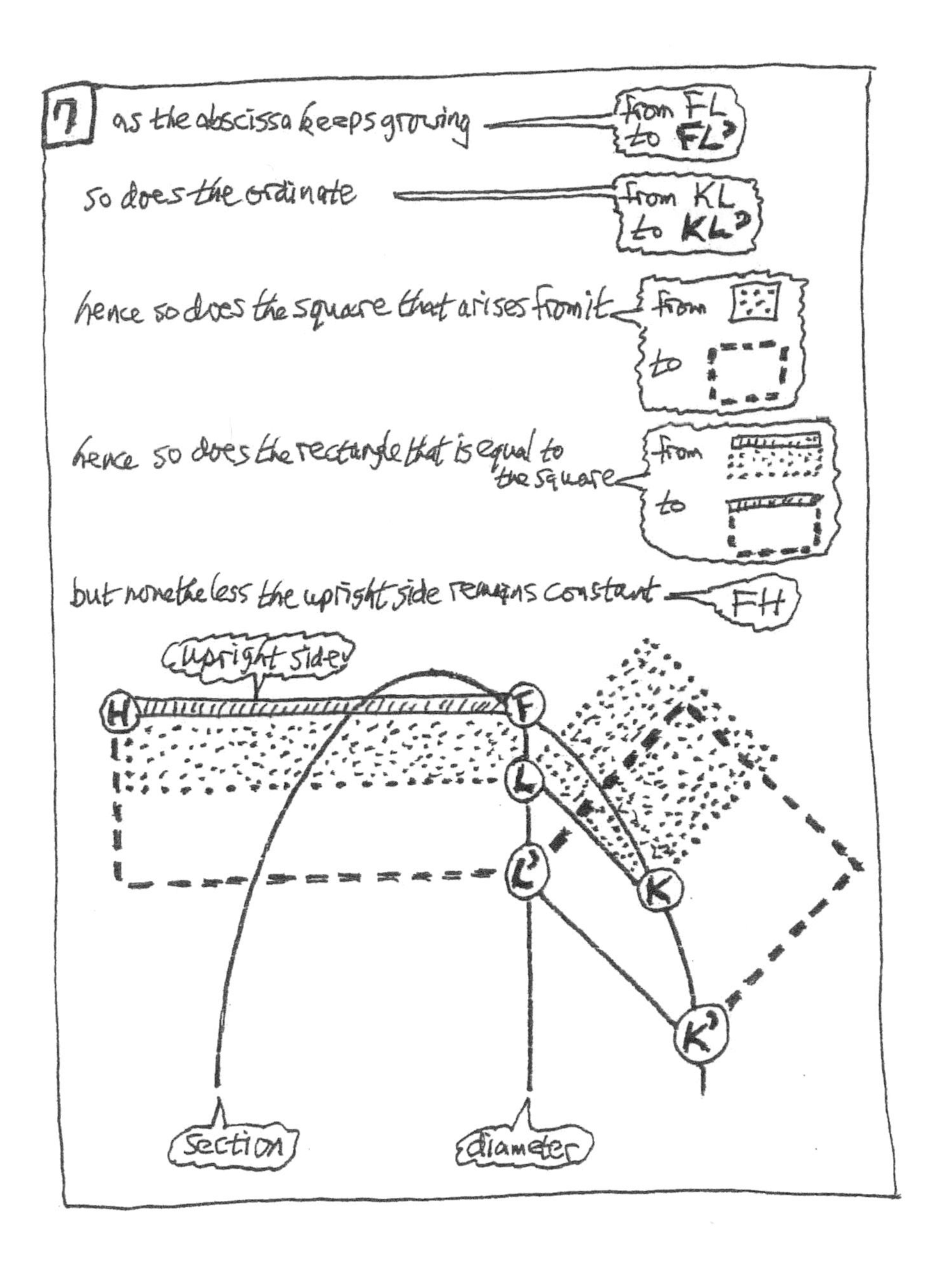
7
as the abscissa keeps growing
from FL to FL'
so does the ordinate
from KL to KL'
hence so does the square that arises from it
from
to
hence so does the rectangle that is equal to the square
from
to
but nonetheless the upright side remains constant
FH
upright side
H
F
L
L'
K
K'
section
diameter

Later in the book, Apollonius will easily show a more direct relationship between ordinates and abscissas in a parabola: he will demonstrate that in a parabola the squares on the ordinates will have the same ratio to each other as do the abscissas *themselves*. If, for example, we go 9 times as far along the diameter to reach the cutoff point, then the square arising from the ordinate will also be 9 times as large—and so the ordinate itself (which is the square's side) will be 3 times as large. From this recipe for growing a parabola, we can answer questions like this: how many times longer an ordinate will we get by taking an abscissa that is 25 times longer? (The answer is, of course, that the ordinate will be 5 times longer.)

This property is the counterpart of the equation ($y = ax^2$, for example) that nowadays is used to define a parabola, but it is not the property used by Apollonius to characterize it in his 11th proposition. Does the 11th proposition tell us what this curved line is—or does the proposition merely give us one of its many properties? This property justifies giving to the curve the name "parabola," but perhaps another name would be better. Would it be better to give it a name that has to do with how we get it—calling it, say, "conic parallelotoma" (since the Greek root for "cut" is *tom-*, as in *ana+tom+y*)? Or would it be better to find some name that emphasizes its shape rather than how it is generated? Or is it best to follow Apollonius, and use this rectangle-square property?

Consider the circle, which Euclid defines in terms of the equality to each other of the straight lines drawn to some one point from any of the points of the closed line that contains it. Would it be better to define a circle in terms of the rectangle-square property of its diameter and its perpendicular half-chord? This property, after all, is a characteristic that is distinctive of the circle. It is complicated, to be sure, but then so is Euclid's definition. After all, a child can recognize a circle without ever thinking about the equidistance of straight lines from a center. Does Euclid's definition come closer to capturing a circle's essence than would defining it as perfectly round? If so, why?

In any case, Apollonius in his treatment of the conic sections directs our attention to the contrivance called the parameter. Each parabola's parameter is a constant length of line. It is defined as having a certain constant ratio to a certain constant line (here FA). This line FA is the same for all of the abscissas and ordinates which belong to that diameter in that section; and that ratio also is the same for all of them.

Do keep in mind, by the way, that when we speak of the change and constancy of various items in the cone or conic section, we do so only to be brief. Do not be misled when we speak of constancy and change, rather than of sameness and difference. What will be changing is not any of the items in the cone, but rather our selection of them for attention. We shall be considering the sameness and the difference of the same sort of item taken at different places in a figure that does not change.

11th proposition: the demonstration

The task of the demonstration in the 11th proposition is to show that by cutting in the way described, we get a section with the following property (fig. 8):

if straight line FH is contrived so that
the ratio which it has to the straight line FA is the same as
the ratio which a square with side BC has to a rectangle with sides BA and AC—

then a square whose side is ordinate KL will be equal to
a rectangle whose sides are
abscissa FL and that line FH contrived to be the upright side.

How does Apollonius show that the square whose side is KL and the rectangle whose sides are LF and FH are equal to each other? By showing that they are both equal to the same thing—namely, the rectangle whose sides are ML and LN. The demonstration thus requires establishing two equalities. (Fig. 9.)

To establish the first equality is the easier part of the demonstration. It does not use the line that has been contrived. It merely takes the view from *above* that presents us with a circle—the best known of the curves obtainable as sections of a cone. What we do is the following. Through L, the point where the ordinate meets the diameter, we pass a plane parallel to the base of the cone, thus slicing into the cone a circle. The circle's half-chord is the original section's ordinate KL—which is the side of that square which concerns us. Because this half-chord KL is perpendicular to the circle's diameter (MLN), we can say (as we have repeatedly said before) that the square whose side is KL is equal to the rectangle whose sides are ML and LN.

What remains to be done now is to establish that this rectangle with sides ML and LN, which is equal to the square with side KL, is also equal to the rectangle with sides LF and FH.

How do we establish this second equality? Well, having taken the view from *above* in order to use the square-rectangle property that derives from the proportionality of certain straight lines in a *circle*, we now take the view from *beside* in order to use the proportionality of straight lines in similar *triangles*. The view from above and the view from beside are linked by the straight line MLN; for not only is MLN the diameter of the circle, it is also a parallel to the base of the axial triangle—the triangle whose sides are used to determine the size of the line that is contrived to be the upright side.

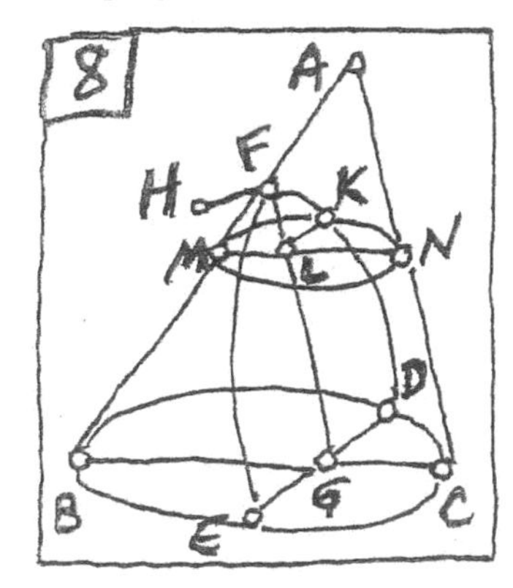

We need to show for those two rectangles—one whose sides are ML and LN, the other whose sides are LF and FH—that their equality follows from the sameness of the ratios which determine the size of the upright side (FH). The upright side is defined in terms of a relationship among lines in the axial triangle. We need to show how, from that relationship, another relationship follows. We need to relate the relationship that we have to start with (because we made things that way) to the relationship that we would like to have established (but do not yet have). Since the lines in the relationship that we already have (all of them) are lines in the axial triangle, we give ourselves a side view of the figure, and we take hold of what we need in the axial triangle.

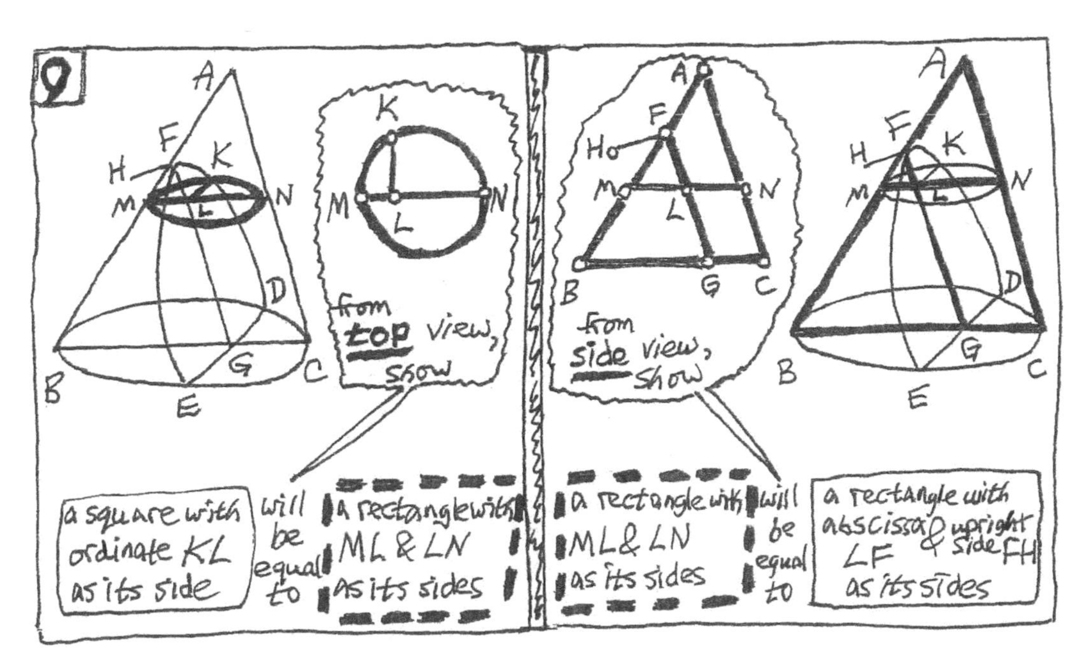

What we have already—by virtue of getting the section by cutting parallel to the side of the axial triangle, and slicing a circle whose diameter is parallel to the triangle's base—is the *similarity of triangles*. From that, we shall get the *equality of rectangles*.

A ratio of rectangles (a square, of course, being a kind of rectangle) was used in contriving the line that is to be the upright side. We take that ratio of rectangles, and we do as follows: first, we *take apart* the ratio of *rectangles* into component ratios of *straight lines*; then, we use corresponding sides of similar triangles to get *other* straight lines in the *same* ratios; and finally, we *put together* the ratios of these other straight lines into a ratio of other *rectangles*. This final ratio of rectangles is the same as the initial ratio of rectangles, which was used to contrive what turns out to be the upright side, but the straight lines involved are different.

We do all that in the following way (fig. 10).

(**Step I**) The ratio of the square (whose both sides are BC)
to the rectangle (whose sides are BA and AC)
is *taken apart* into its two component ratios, which are
the ratios of their sides (the ratio of BC to AC, and of BC to BA).

Now, taking *the one component* ratio,
we get same ratios by the similarity of triangles:
(**step II**) as BC is to AC, so is MN to NA;
and (**step III**) as MN is to NA, so is ML to LF.
Thus, the one component ratio (the ratio of BC to AC)
is the same as the ratio of ML to LF.

Now, taking *the other component* ratio,
we also get same ratios by the similarity of triangles:
(**step IV**) as BC is to BA, so is NM to MA; and
(**step V**) as NM is to MA, so is LM to MF;
and (**step VI**) as LM is to MF, so is NL to FA.
Thus, the other component ratio (the ratio of BC to BA)
is the same as the ratio of NL to FA.

And now (**step VII**), if we *put together* those two resulting ratios of lines
(namely, that of ML to LF and that of NL to FA), we get
the ratio of the rectangles that would be built by using those lines as their sides.
This is the ratio of the rectangle whose sides are ML and LN
to the rectangle whose sides are LF and FA.

(Note that in decompounding any ratio of rectangles, there is a choice among alternatives. Which alternative we choose will depend upon which lines we wish to relate. For more on this, see this chapter's appendix on decompounding.)

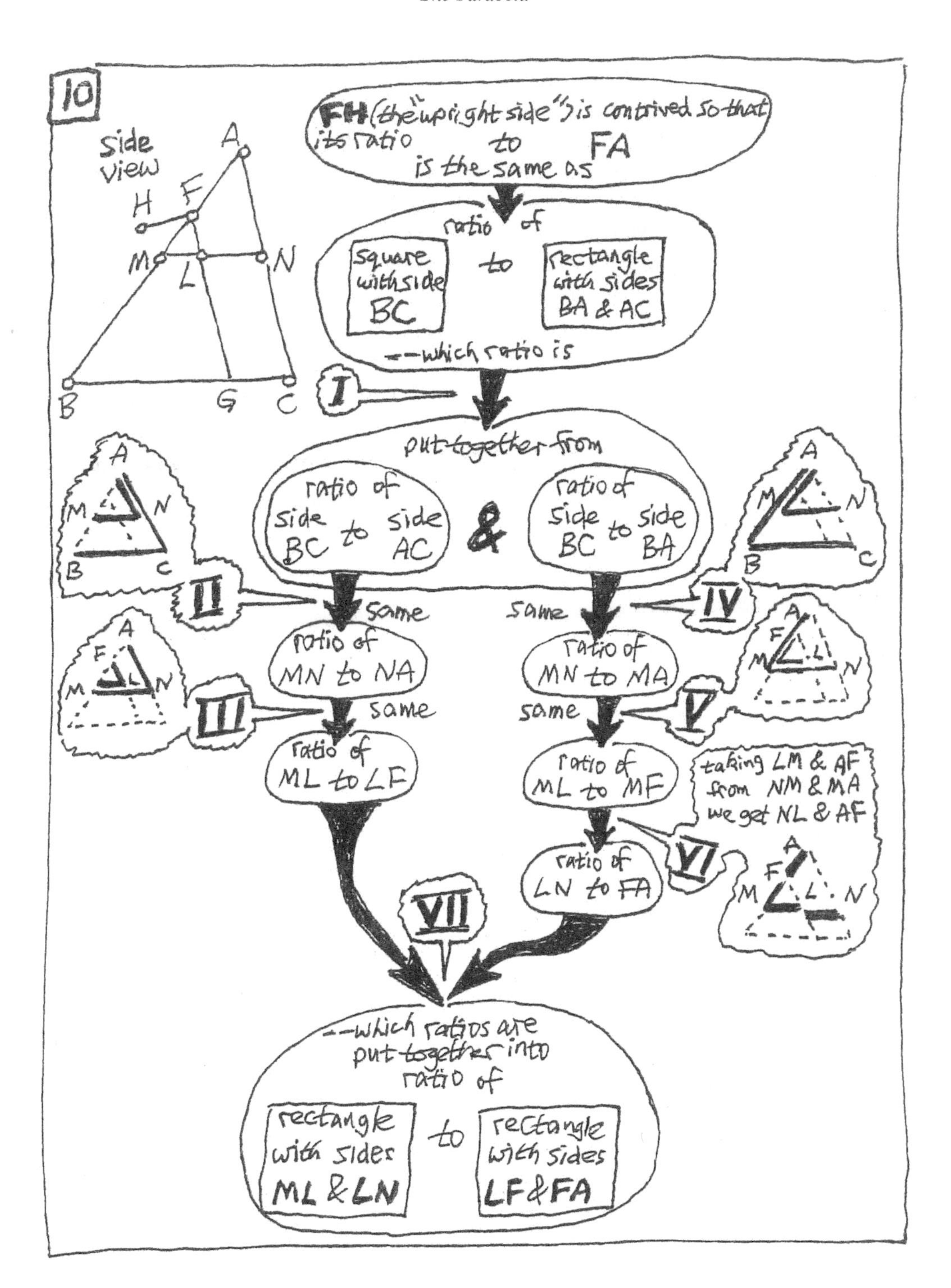
10
side view
A
H
F
M
L
N
B
G
C
FH (the "upright side") is contrived so that its ratio to FA is the same as
ratio of
square with side BC
to
rectangle with sides BA & AC
--which ratio is
I
put together from
ratio of side BC to side AC
&
ratio of side BC to side BA
II
same
ratio of MN to NA
III
same
ratio of ML to LF
IV
same
ratio of MN to MA
V
same
ratio of ML to MF
taking LM & AF from NM & MA we get NL & AF
VI
ratio of LN to FA
VII
--which ratios are put together into ratio of
rectangle with sides ML & LN
to
rectangle with sides LF & FA

All of that is put together in the 11th proposition of Apollonius. You have seen an English translation of the synthetic presentation by Apollonius himself. This guidebook has tried to show that understanding what Apollonius has done does not have to be a task as difficult as a glance at the proposition might have made it seem to be.

In studying the propositions of Apollonius, brute force of memorization will not suffice to do the work. To be sure, much must be committed to memory; but much less needs to be remembered once the proposition has been understood. As you read the propositions, you will understand them better if you write things down as follows.

List the statements, one-by-one, on paper. Arrange the statements so that their separation, juxtaposition, and subordination give you a structural overview. Use abbreviations to reduce how much there is for you to have to pay attention to. Use arrows and marks of emphasis to direct the flow of your attention. In short, make a map. And after you have a map, look over it until you see as much as you can in a glance or two. Set your map aside, and then try to reproduce it on paper without a look at it. Go over it with a view to making it simpler. Successive reproductions and remakings of the map will make the territory yours. Figure 11 gives an example of a simplified map for this 11th proposition. The map, while marking the order of the steps that are taken by Apollonius, depicts their logical structure, along with the aspects which we view, and by which we are guided, in taking those steps.

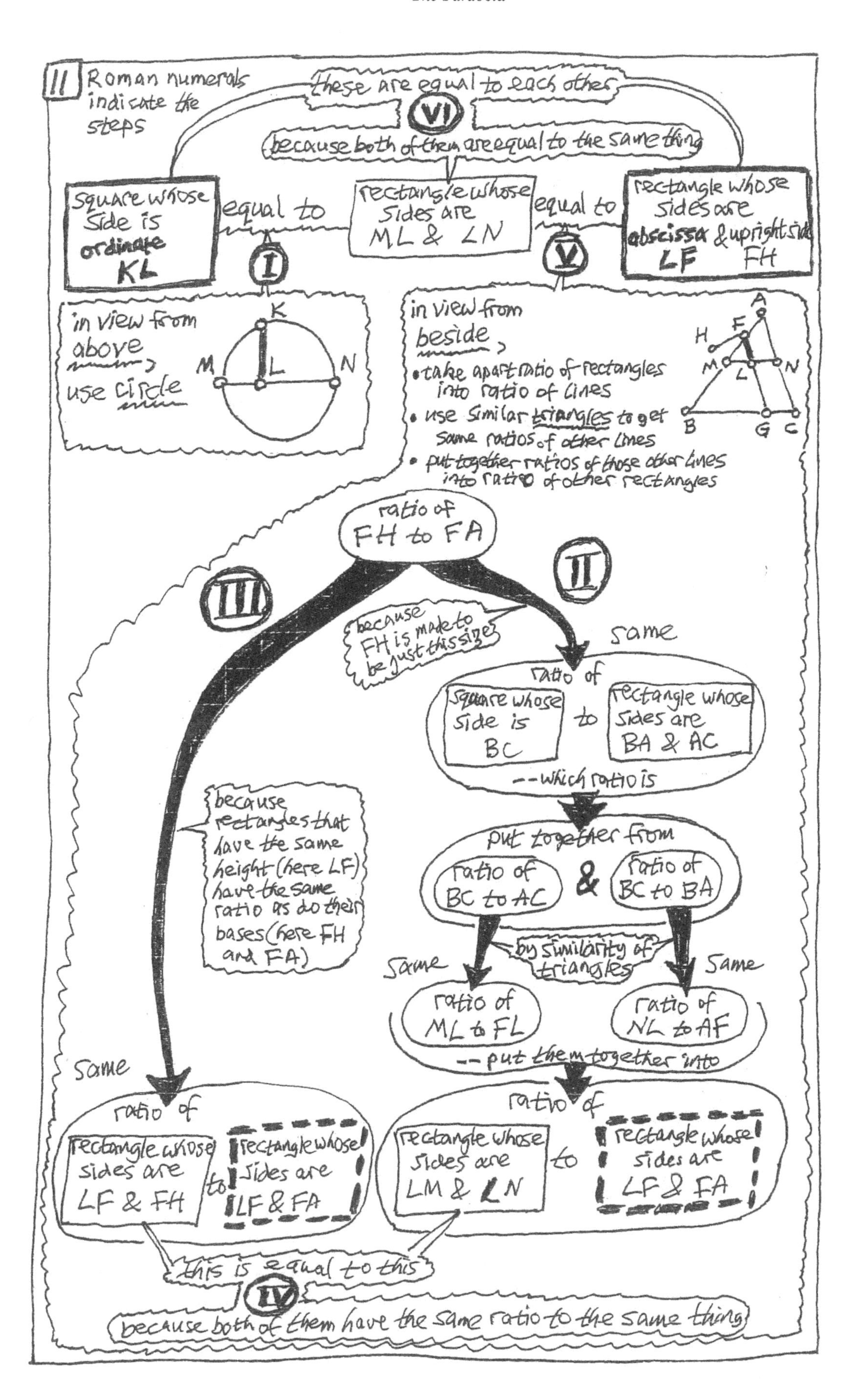
II
Roman numerals indicate the steps
these are equal to each other
VI
because both of them are equal to the same thing
square whose side is ordinate KL
equal to
rectangle whose sides are ML & LN
equal to
rectangle whose sides are abscissa & upright side LF FH
I
V
in view from above → use circle
K M L N
in view from beside →
• take apart ratio of rectangles into ratio of lines
• use similar triangles to get same ratios of other lines
• put together ratios of those other lines into ratio of other rectangles
A H F M L N B G C
ratio of FH to FA
III
II
because FH is made to be just this size
same
ratio of
square whose side is BC
to
rectangle whose sides are BA & AC
-- which ratio is
put together from
ratio of BC to AC
&
ratio of BC to BA
because rectangles that have the same height (here LF) have the same ratio as do their bases (here FH and FA)
same
by similarity of triangles
same
ratio of ML to FL
ratio of NL to AF
-- put them together into
same
ratio of
rectangle whose sides are LF & FH
to
rectangle whose sides are LF & FA
ratio of
rectangle whose sides are LM & LN
to
rectangle whose sides are LF & FA
this is equal to this
IV
because both of them have the same ratio to the same thing

an invitation to inquiry

Although the simplest of curves, the circle, can be defined by a simple equality of straight lines, a more complex relationship among straight lines that is distinctive to the circle suggests a way to define a variety of other curves. To make use of this relationship, however, we must take into account various ratios that determine the relative sizes of various straight lines; and, since each ratio has merely a part in the determination, the ratios need to be put together. Built into the line that is contrived to be the upright side—enabling it to do its work—is the compounding of certain ratios. We have seen how the upright side is shown to do its work; it does it as follows: in the proportion that determines the size of the upright side, one of the ratios—a ratio of certain rectangles—is taken apart into component ratios of straight lines, and these ratios are shown (by similar triangles) to be the same as ratios of other lines, which ratios are then compounded to give the same ratio as in the definition of the contrived line; this final ratio is the same, but the lines involved in it are different.

Now it is all well and good to figure out the demonstration if we already have the enunciation—that is, if we have been given the parametric definition—but suppose that we had been presented with nothing but a line of section cut into the cone by a plane oriented as prescribed, without any upright side already contrived. How then could we have *found our way to* the contrivance? In other words, how can we see the upright side as the *outcome* of a process of thought—as the *answer* to a *question*—rather than as something *given* as a *starting-point*?

In the first of curves, the circle, a distinctive relationship holds among certain straight lines in it—among, that is to say, any half-chord perpendicular to the diameter, and the segments into which it splits the diameter. The perpendicular half-chord is the mean proportional between the segments into which it splits the diameter. That is to say, the following ratios are the same: the ratio of one segment to the half-chord, and the ratio of the half-chord to the other segment. Or—to speak not of sameness of ratio but rather of sameness of size—the following things are equal: the half-chord in its square-generating power, and the segments in their rectangle-containing power.

It is this rectangle that mediates the parts of the demonstration here in the 11th proposition, and it will do so also in the 12th and 13th after it. In the demonstrations for all the different curves in the First Book of the *Conics*, a square erected on the ordinate and a certain interesting rectangle will be equal to each other because both of them are equal to the same other rectangle; and this other mediating rectangle is made from the segments into which the ordinate acting as a perpendicular half-chord splits the diameter of a circle sliced by a plane parallel to the base of the cone.

Showing the equality of this mediating rectangle and the square on the ordinate is the easy part of these demonstrations. The more difficult part is showing the equality of the two rectangles—of the mediating rectangle (the rectangle with sides ML and LN) and another rectangle (fig. 12). This more difficult work is done in each of the three demonstrations by contriving a line of just the right size to fit a certain proportion. Built into the straight line that is made to be the upright side, enabling it to do its work, is the putting together of certain ratios that are embodied in certain boxes (certain squares or other rectangles).

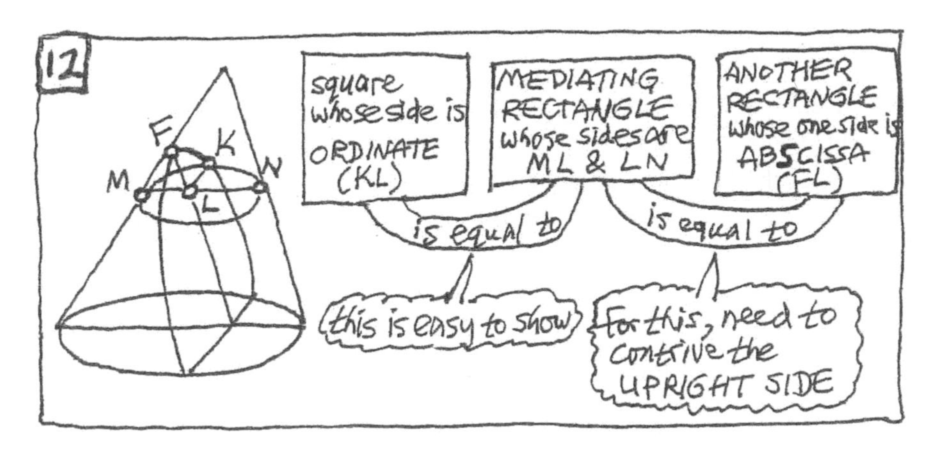

So, let us consider again what is said in the demonstration of the 11th proposition. Now, however, let us not consider how to get from the beginning to the end. That is to say: let us now *not* move from the strange "if" part of the enunciation (in which the manner of the cutting is combined with a contrived line which seems to come from nowhere) to the remarkable "then" part (which states the section's property). Rather, let us now consider how an alert examination of what we have done by cutting in that manner might have as its *outcome* the suggestion to us of that very contrivance.

Our question now is this (looking, in figure 13, at something we have seen before): how might we *find out* that we *should* make a line whose ratio to FA will be the same as that ratio which a square with side BC has to a rectangle with sides BA and AC—in order to be able to show that the rectangle whose one side is the line so made, and whose other side is the abscissa, is equal to a square whose side is the ordinate?

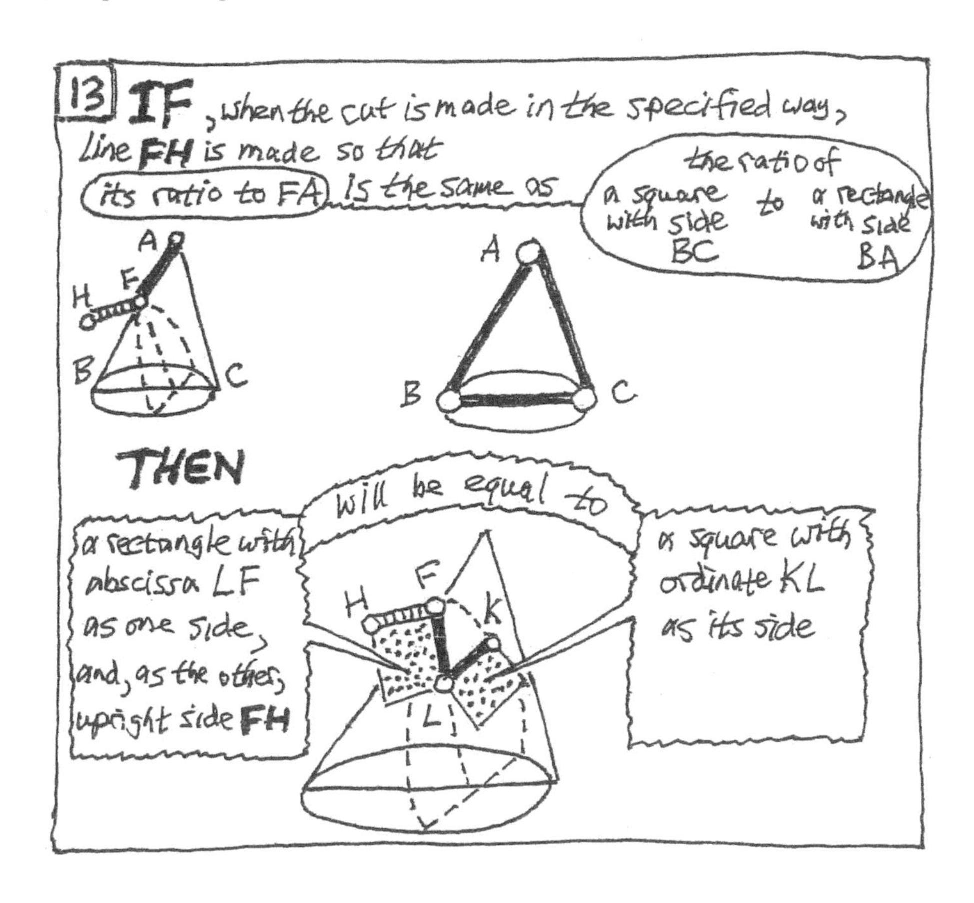

analysis to find contrivance for parabola

The upright side is contrived in order to characterize the conic section. How might we try to characterize this section without taking the upright side as our point of departure? An answer to that would help us to see the definition of the upright side as an outcome of thinking through the question of how to characterize the section.

So, let us begin by asking the following question (fig. 14): as we move down along the conic section's diameter from its vertex F, what stays the same and what becomes different at different points on the diameter? Let us consider things at some point—call it L.

First let us take the *top view* (fig. 15, on the left). It shows a circle. Even though the circle's diameter MN grows as we move down along the section's diameter, the square on the ordinate KL stays equal to the rectangle contained by ML and LN. The square changes size as we move down along the section's diameter, but its size is always the same as that of the rectangle.

Now let us take the *side view* (fig. 15, on the right). It shows what stays the same about what becomes different. As we go down along the conic section's diameter, cutting off a longer and longer abscissa LF, and getting a bigger and bigger diameter MLN for the circle, what happens to that rectangle with sides ML and LN which is equal to the square on the ordinate KL? That rectangle grows. Although its one side LN stays the same, its other side ML grows. But something also stays the same about this side (ML) that grows: what stays the same about it, is its ratio to the growing abscissa (LF). That ratio of ML to LF stays the same because (since these two lines are parallel to the sides of the axial triangle) the little triangle MLF keeps the same shape as the axial triangle. Now we have a *constant* ratio for the side (ML) which *grows* in the axial triangle. As for the side (ML) which stays the same, it will stay in some constant ratio to any constant line that we may choose. So, let us do as follows.

Let us consider (fig. 16) that little triangle MLF in relation to the rectangle whose sides are ML and LN. One of the rectangle's sides (ML) is also a side of the little triangle, which also has as a side the abscissa (LF). The rectangle's other side (LN) is the extension of that very side of the triangle at which we just looked (ML). This extension LN is constant, and the only other constant extension of the little triangle is FA—and FA, after all, is as good as any other constant line to give us a constant ratio with the constant line LN.

So, now we have a constant ratio for each of the sides (ML and LN) of that rectangle which is always equal to the square on ordinate KL. As for the one side, ML, it grows—but its ratio to the growing abscissa LF stays constant; and as for the other side, LN, it is constant—so its ratio to the constant line FA stretching between the vertices of section and of cone stays constant too.

Thus, the ratio of ML to LF and the ratio of LN to FA are both constant. Let us put these two constant ratios together, so that we shall have only one ratio to consider. If these two ratios of straight lines are compounded, they will yield a ratio that is the same as a ratio of rectangles made by taking those straight lines as sides—namely, the ratio of the rectangle with sides ML and LN to the rectangle with sides LF and FA. This ratio, since it is compounded out of constant ratios, will itself be constant. Let us see how this can help us.

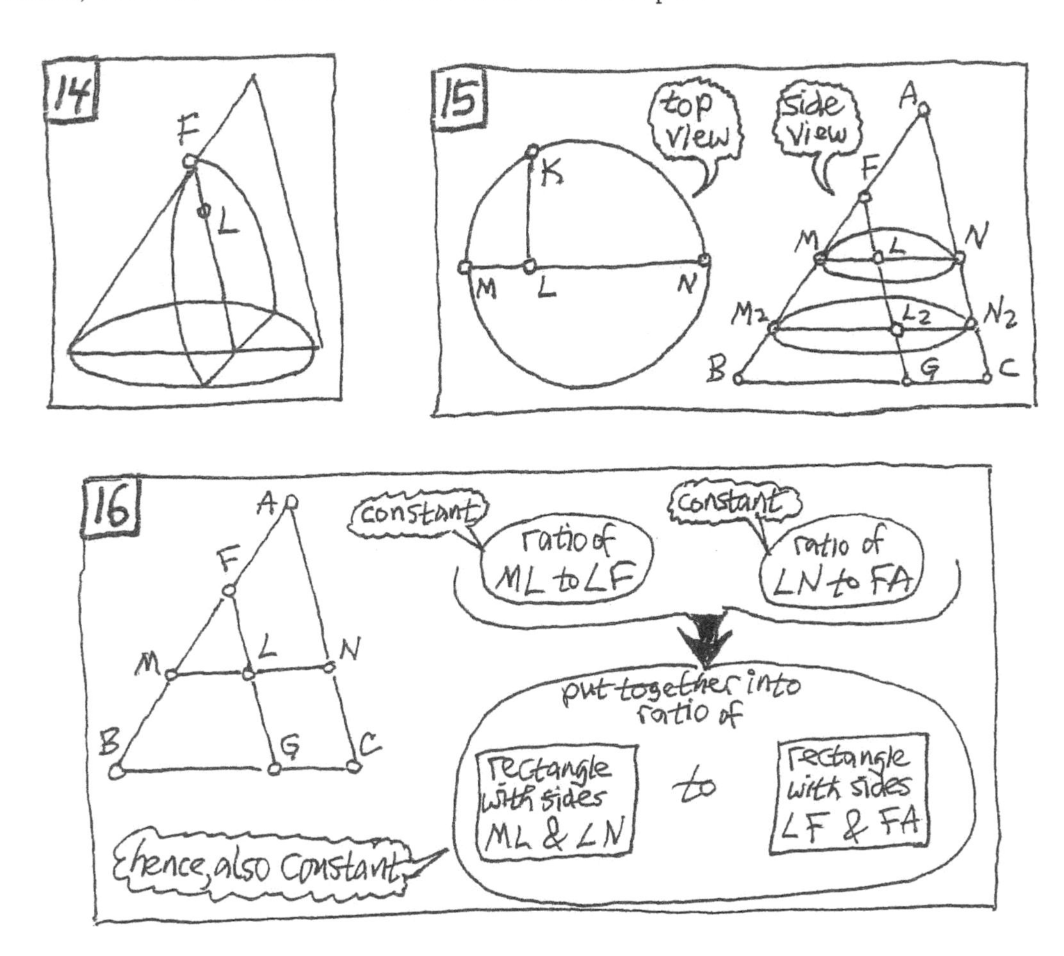

From this constancy of the ratio of the rectangle with sides ML and LN to the rectangle with sides LF and FA, what we now know is this—that although the ordinate KL grows as the abscissa FL grows, nonetheless a certain relationship between them stays the same. The relationship is this: the rectangle with sides ML and LN, which is equal to the square whose side is the growing *ordinate*, will have a constant ratio to a rectangle with the following sides—one side (FA) is the constant distance between the vertices of section and of cone, and the other side (FL) is the growing *abscissa*.

So now (fig. 17) we have the constancy of the ratio which the rectangle with sides ML and LN has to the rectangle with sides LF and FA. One term in this ratio is the rectangle with sides ML and LN (which are segments into which the circle's diameter is split by the ordinate). The other term is the rectangle with sides LF (which is the abscissa) and FA (which is a line of constant size). That is to say: we have the rectangle which concerns us (the rectangle with sides ML and LN) in a constant ratio to another rectangle, one of whose sides is the line which concerns us (abscissa LF).

But this ratio is not what we are after. It is only a step on the way. What we *have* is a rectangle that is in *constant ratio* to the rectangle with sides ML and LN, but what we would *like* to have is a rectangle that is *equal* to the rectangle with sides ML and LN—because then it would be equal to the square that has the ordinate as its side. Moreover, we want that equal rectangle (like the rectangle of constant ratio which we now have) to have the abscissa FL as one of its own sides. If we had the rectangle that we want, then we would have the ordinate and abscissa in relation to each other—not directly, to be sure, but at least through their power to make a square or other rectangle.

What do we need in order to have what we want? (See figure 18.) We already have the constancy of the ratio of the rectangle with sides ML and LN to the rectangle with sides LF and FA. We want to have a rectangle which is equal to the first rectangle in the ratio, and which has as *one* of its own sides one of the sides of the second rectangle—namely, the abscissa (LF). If we had the *other* side, then we would have that rectangle itself—which is want we want. Our problem, then, is this: to find out what should be its other side.

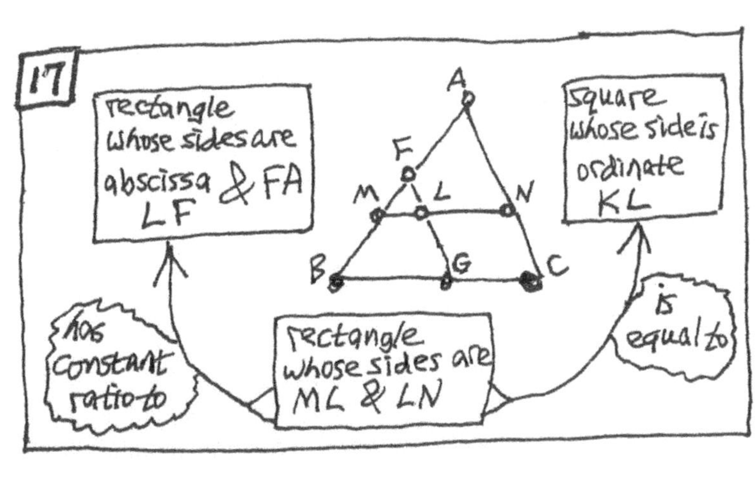

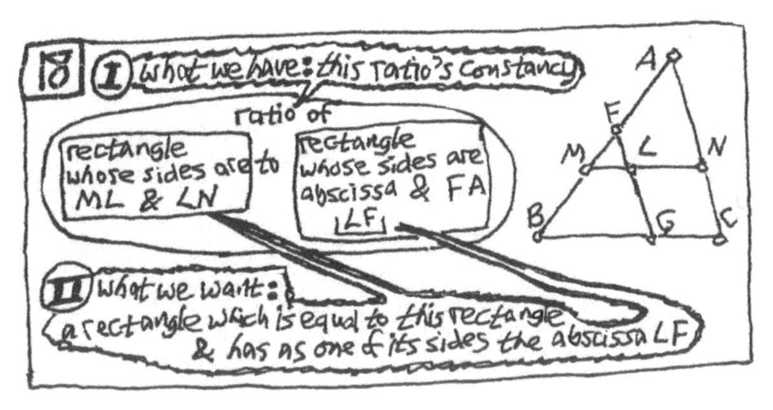

Let us call that unknown other side "(u)"—and, even though we do not have it, let us *suppose* that we do. That is, let us ask (step I): *what if* we *did* have it? (Fig. 19.)

Now (step II), two magnitudes can be equal to each other only if each of them has the same ratio that the other one also has to the same third magnitude. So if we *did* have that unknown side (u), then the rectangle that it makes with the abscissa—namely, the rectangle with sides FL and (u)—would have, to any rectangle (to any one at all), the same ratio that its *equal* (the rectangle with sides ML and LN) has to that same thing. And we already know something about the ratio that the rectangle with sides ML and LN has to one rectangle in particular—namely, its ratio to the rectangle with sides LF and FA. Hence, we know that the rectangle whose sides are ML and LN can be equal to the rectangle whose sides are FL and (u) only if each of these rectangles has the same ratio to the rectangle whose sides are LF and FA.

Now (step III), taking the ratio which involves the unknown line (u), we see that the two rectangles—one whose sides are FL and (u), the other whose sides are LF and FA—both have the same height LF; and so, the ratio of the one rectangle to the other is the same as the ratio of their bases, which is the ratio of line (u) to line FA.

Therefore (step IV), although we do not have the unknown line (u), we do know its relation to things we do have—we know that *if* we had it, then its ratio to FA (which is a constant line) would be the same as the ratio of the rectangle whose sides are ML and LN to the rectangle whose sides are LF and FA, which is a ratio that is constant.

But although this ratio is itself constant, it is *not* stated entirely in terms of *lines* that are constant—and so we have not yet determined the size of the unknown line (u).

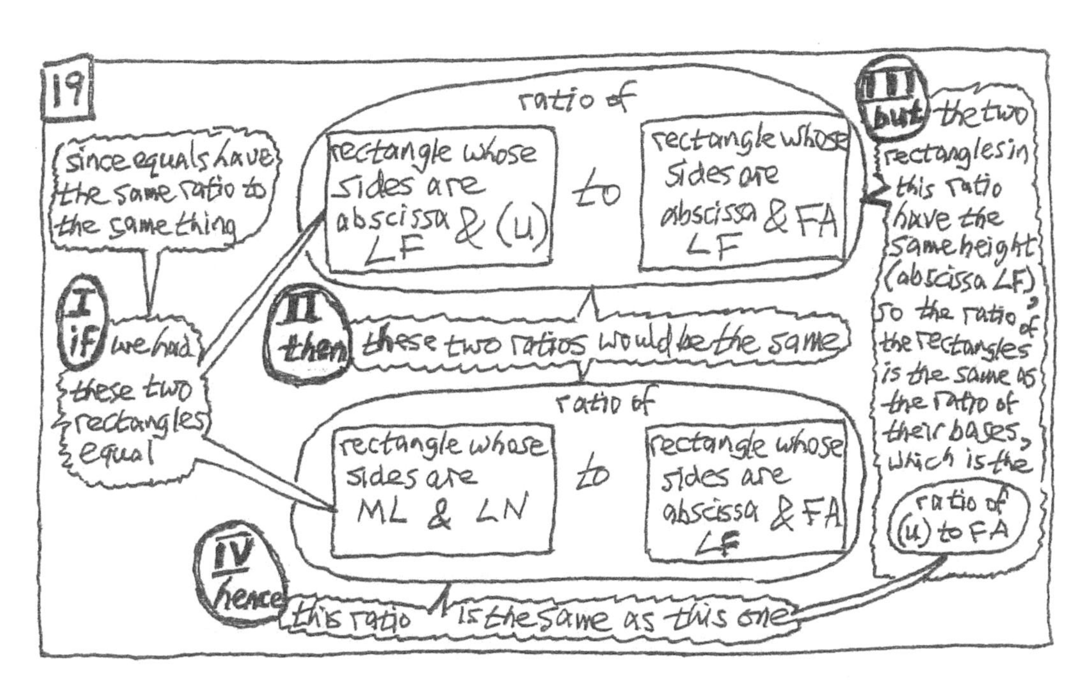

We need to find a ratio which involves only lines that are constant but which is the same as that ratio which we do have—a ratio involving four lines, only one of them constant—the ratio of the rectangle whose sides are ML and LN to the rectangle whose sides are LF and FA.

To find the ratio that we need (fig. 20), we (step I) take apart the ratio that we have. We take apart the single ratio of rectangles into its two component ratios of lines. Since the ratio of the rectangles is the same as the ratio compounded out of the ratios of their sides, the ratio of the rectangles is the same as the ratio compounded out of two ratios—those of ML to LF and of LN to FA.

Now (step II), by the similarity of triangles, we get those same ratios in terms of constant lines—namely, as ratios of BC to AC and of BC to AB.

And now (step III), we put the component ratios back together into one ratio of rectangles, which will be the same *ratio* that we had before, but now with different *terms*, terms now involving none but *constant* lines; this is the ratio that the square whose side is BC has to the rectangle whose sides are BA and AC.

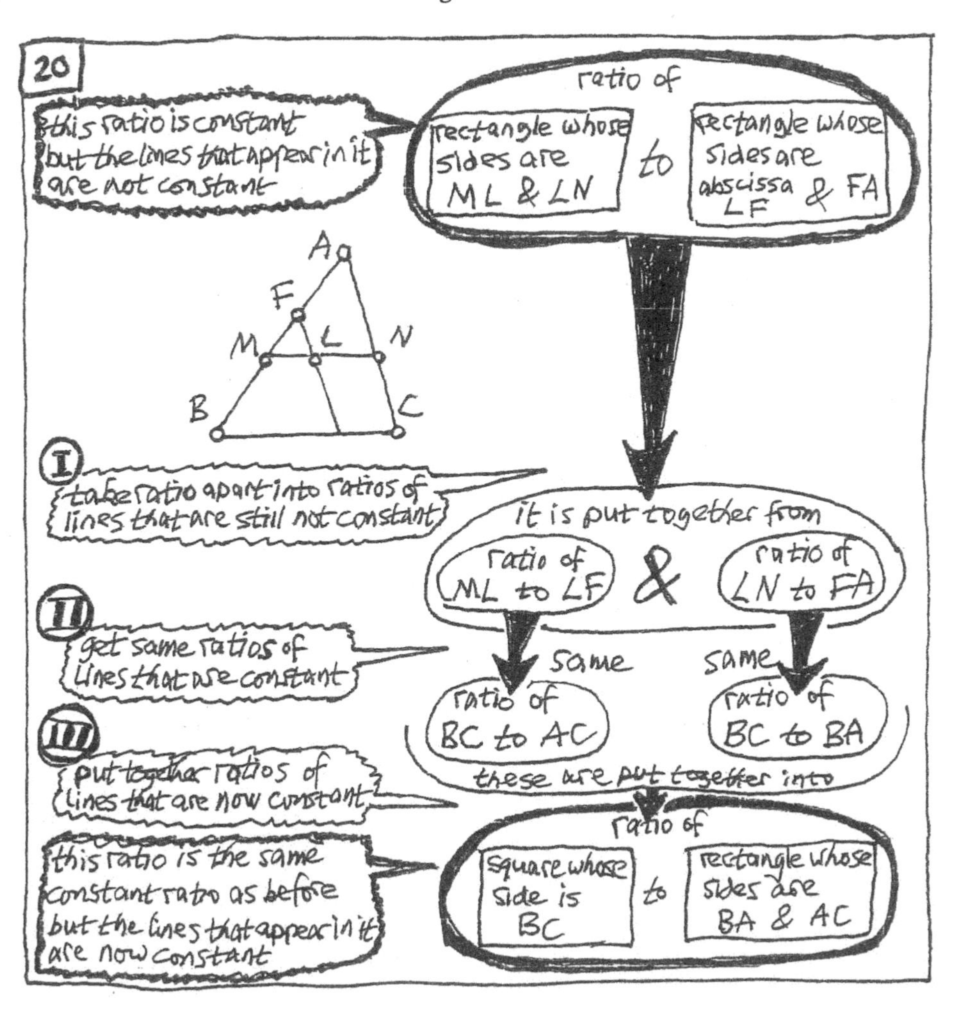

So, now we are ready to do the requisite contriving. Figure 21 shows why. We have seen (step I) that if we had the line that we called "(u)" then its ratio to line FA would be the same as the ratio of the rectangle whose sides are ML and LN to the rectangle whose sides are LF and FA, and (step II) that this ratio of rectangles is the same as the ratio of the square whose side is BC to the rectangle whose sides are BA and AC, and so (step III) that if we had line (u) then its ratio to line FA would be the same as this last ratio.

We therefore now contrive an upright side that is of just the right size to make the following two ratios the same: one of them is the ratio which this upright side has to the line (FA) that stretches between the vertices of section and cone; and the other one is the ratio which the square arising from the axial triangle's base (BC) has to the rectangle contained by the axial triangle's sides (BA and AC). This ratio of square to rectangle involves only constant lines (lines of the axial triangle), but it is the same ratio as that original constant ratio which involved lines not all of which were constant.

Thus, the *outcome* of *our* inquiry is the contriving of the line taken by Apollonius as the *starting-point* for *his* demonstration—the line that he calls "the upright side."

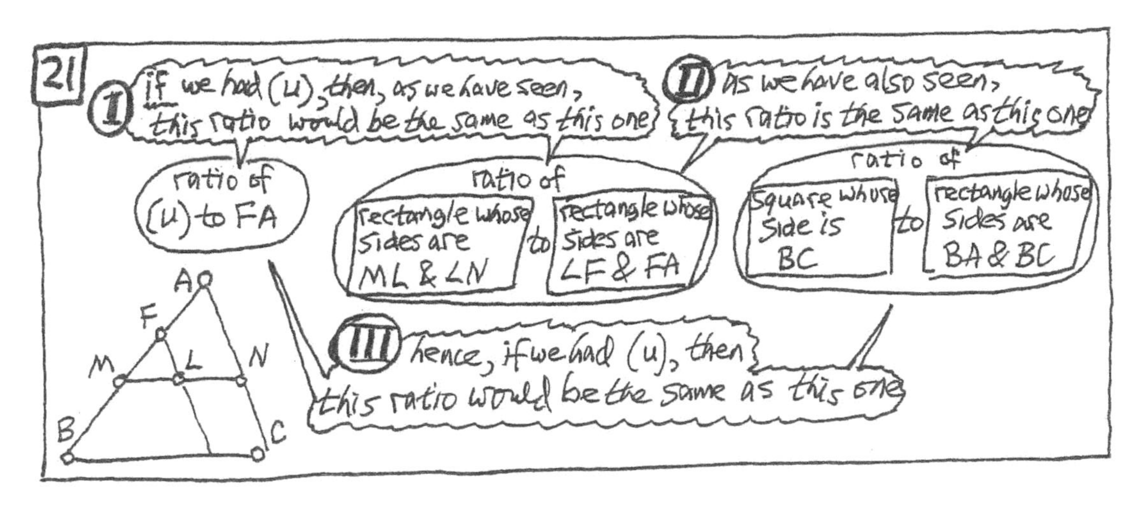

Although our inquiry followed upon a reading of Apollonius's proposition, it did not depend upon what his proposition demonstrates. So, we must ask: can our analysis—our way of inquiry whose outcome was our finding the contrivance—be a *substitute* for Apollonius's demonstration? No, it cannot. But *why* not?

analysis and synthesis

Suppose we say: "If it rains, then we play chess." By making that statement, we are not implying this one: "If we are playing chess, then it is raining." It may be that we play chess every day—come rain or come shine. To imply that if we are playing chess then it must be raining, we would have to say this: "We play chess *if and only if* it rains." More generally: The assertion "*If and only if* P, then Q" implies not only the assertion "If P, then Q" but also its converse "If Q, then P"—but the assertion "If P, then Q" by itself does *not* imply its converse "If Q, then P."

The converse of a true proposition is *possibly* true. Consider , for example, this proposition: if the two sides of a triangle are equal, then so are its two base angles. That proposition is true, and so is its converse: if a triangle's two base angles are equal, then so are its two sides. But the converse of a true proposition is *not necessarily* true. For example: if two triangles are congruent, then they must be similar; but if they are similar, then they *may* be congruent—*or* they may *not*. (Fig. 22.)

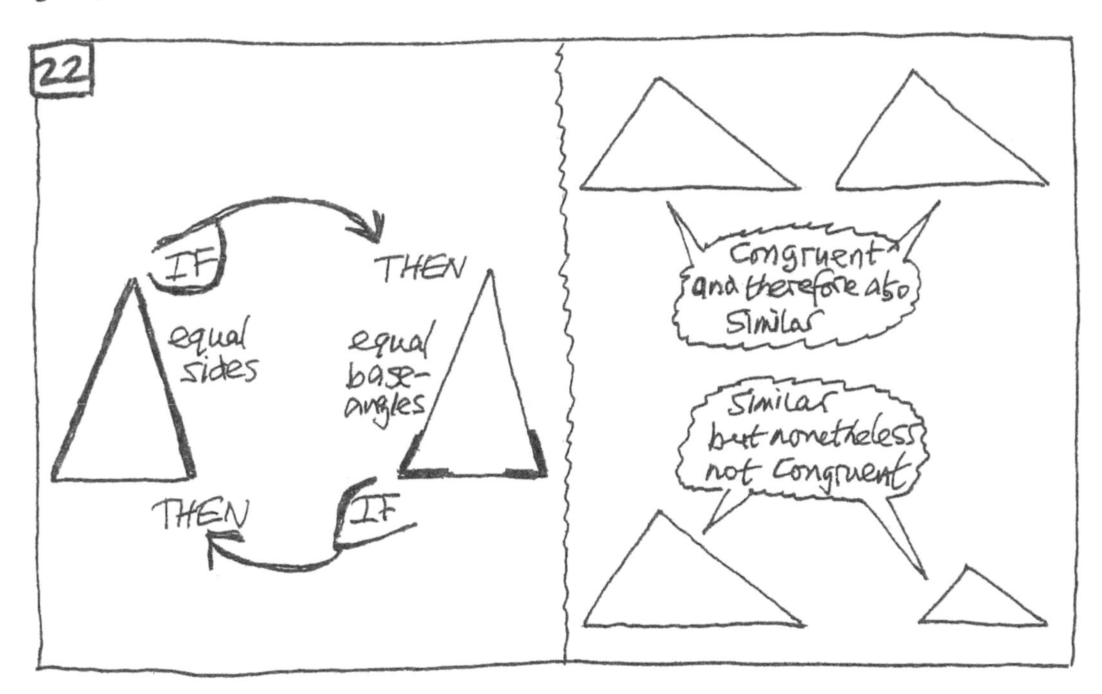

What we said in our analysis of the 11th proposition is something like this: "If we *did* have E , then we *would* have D, and then we would *also* have C and then we would *also* have B, and we would have B *easily* if we *got* A—but this last thing A is something that we *can* get easily." However, "if" does *not* mean "*only* if"; our analysis therefore is not a demonstration of the following statement: "If we *did* get A, then we would easily have B, and then we would have C, and then also D, and then we'd *finally* have E."

In order to *demonstrate* the equality of the rectangles with which we dealt a few moments ago in our *analysis* for the curve, we must establish that the statements can be put together to form a chain of consequence in the order that is the *reverse* of the order in the analysis. To demonstrate the equality of the rectangles, we must say what Apollonius says in the 11th proposition in the very *same* order in which *he* says it.

After we have taken something apart to see how it is put together, we must put it together. *Syn+the+sis* is Greek for "together-with+put+ing." *Ana+ly+sis* is Greek for "up-loosen-ing": an analysis is a loosening-up, an untying of what is knotted up. The process of synthesis is the opposite of the process of analysis. The difference lies not in the steps, but rather in the order in which they are taken.

We can *learn*—we can find out something that we don't know—by drawing *consequences* from *supposing* that we do know something about what we don't know. How do we *know* that we *don't* know something? Only by knowing something about its *relation* to what we *do* know. And if we know how what we *don't* know is *related* to what we *do* know, then we can *get* to know the *un*known by *thinking through* what we know about the relation of what is as *yet un*known to what is known *already*.

We can get what we don't have, if we *suppose* that we do have something of a hold on it—at least enough of a hold to answer this question: "*What if* we *already* had what we are still looking for: what could we *then* get—and what would follow from *that*?" But when, by such inquiring, we finally reach something that we *do* have, then we must reverse direction and show that we can get *from* what we *already* have, to what we *don't yet* have but just *supposed* ourselves to have *provisionally*. *De+monstra+tion* is Latin for "from+show+ing"; it is a translation of the Greek term, which is *apo+deik+sis*. To *figure out* the way to go to get where we want to be, we may have to be *analytic*; but to get there *from* where we are, we have to be *apodeictic*. The road that Euclid and Apollonius take us along is apodeictic.

We need to be familiar with analysis as well as with synthesis in order to understand the reasons why classical mathematics gave way to modern mathematics. What Descartes later did will be clearer if we consider, as we go along, what might lie behind what Apollonius puts before our eyes. Although this may seem to slow us down, it will in fact prepare us to travel farther faster.

In giving an analysis for the proposition that we have been considering, we have saved some time by not examining the blind alleys into which we might have wandered. But, although we have not examined any false births that might have occurred before our labor produced the outcome for which we wished, nonetheless, the analytic way that we have taken resembles Socratic dialectic. Our guide has been this thought: when we *know* that we *don't* know something, we *do* know how what we *don't* know is *related* to what we *do* know—and hence we know how to *get* to know what we don't already know.

Our situation was something like the following (fig. 23): (step 1) we are looking for a place (A), and we don't know where to find it; but (step 2) we know what's near it (B); and (step 3) we know what's near what's near it (C); and (step 4) we know what's near what's near what's near it (D); and (step 5) this is near where we in fact are (E)—a place we surely know. So now we know where to find place A—but *only if* we are sure that we can take the requisite steps in the direction that we want to go. Thinking our way from A to E was tricky—we didn't know which of the many possible nearby places to go to next, in order to get eventually to where we are in fact. From A, for example, we might have gone first to a dead end, J or K or L, and from B we might have gone to dead end M or N. But now that we have thought out a route, the question is not *which* steps to try to take from where we are in fact, to where we want to go. The only question is *whether* we can be sure that we can take each of those steps in the right *direction*—that is, *from* where we are in fact, place E.

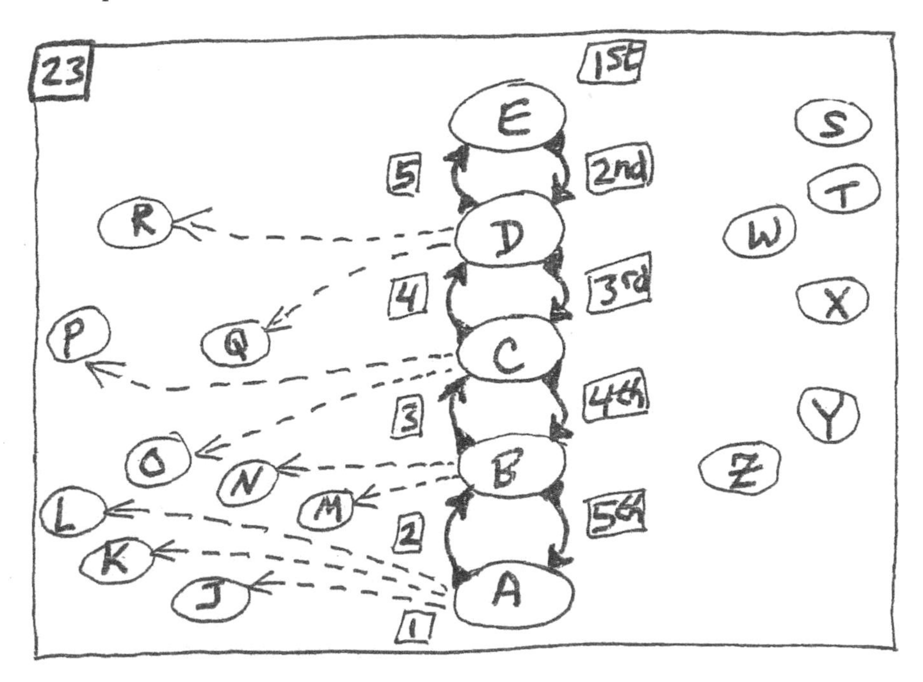

So, we try to take them in reverse—in the opposite order from that in which we found them. (1st) we start from E, the place we surely know; and then (2nd) we try to step to D to make sure that we can; and then (3rd) we try to step to C; and then (4th) we try to step to B; and then (5th) we try to step to A—so, finally, if we are sure of all our steps, we have found what we were looking for.

If we are given an analysis, then it is easy to supply a demonstrative synthesis; but if we are given a demonstrative synthesis alone, it takes some work to figure out what analysis could have gotten us to it. We need first to see the steps as steps, and then to state the steps in reverse, and then to see what questions guide the sequence.

The *analysis* has the following form (fig. 24): Find a way to get to E from A. Well, if we had E, then we would have D; then we would have C; then we would have B; but B is something that we can get—because we do have A or we know how to get it—so, we have found a way to get to E from A. The *demonstrative synthesis* has the following form: From A, get to E. Well, from A, we get B; and since we have B, we therefore have C; and therefore D; and therefore E—so, from A we have gotten to E.

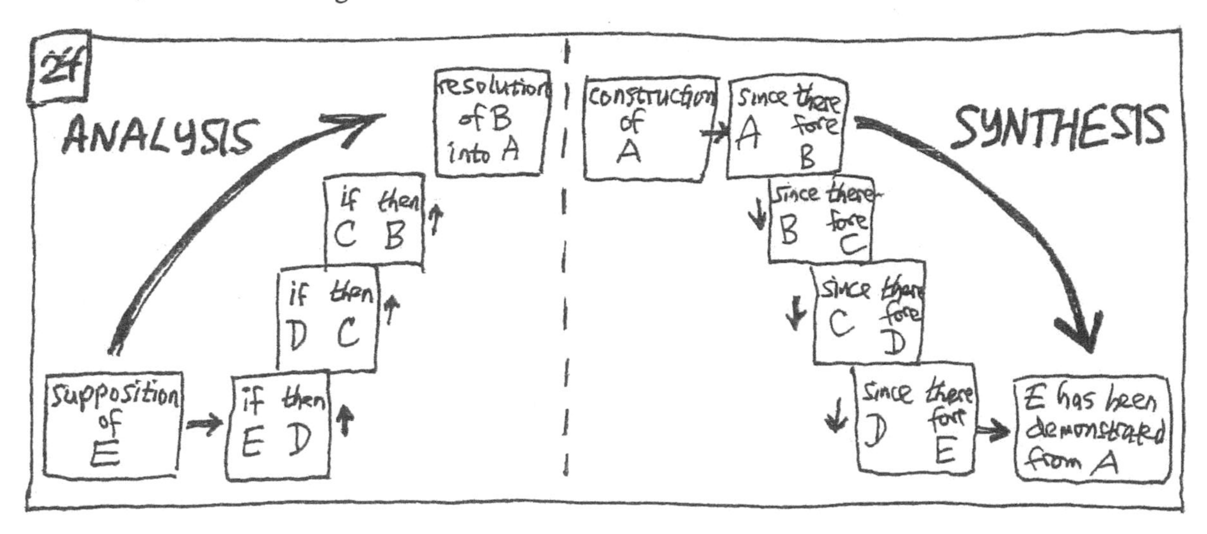

In the demonstrative synthesis, the very same steps are taken all over again (albeit this time in reverse). We must engage in this repetition because saying "If C, then B, then A" is not equivalent to saying "If A, then B, then C." By contrast, saying "If, and *only* if, C, then B" *does* imply saying "If B, then C"—but the *analysis* may not give us an "*only* if" at every step. Unless an "only if" can be attached to the "if," the road from the "if" to the "then" is not a two-way street. We cannot assume that the road will take us in the proper direction; we have to show that it will. An analysis does not suffice; there must also be a demonstrative synthesis. (Fig. 25.)

Apollonius *rarely* presents an *analysis*, but he *always* presents a *demonstrative synthesis*. He presents *very many* demonstrative syntheses *without* analyses, and he will present (what we have not yet encountered) *a few* demonstrative syntheses *with* analyses—but we shall encounter in his book *no analyses without demonstrative syntheses*, even though the demonstrative synthesis repeats in reverse the steps of the analysis.

This is not a personal peculiarity, for in Euclid's *Elements* there is also the same insistence on synthetic demonstration rather than analysis. (A dramatic example may be found in the appendix at the very end of this chapter.) Apollonius is not essentially different from Euclid in this respect.

A nineteenth-century German scholar thought it a Greek *ethnic* peculiarity, when in fact an analysis has been presented, to insist on then presenting all over again, in reverse order, the very same steps which have all just been presented in the analysis, so that a demonstration will take place in the synthesis. But it is not a mere ethnic peculiarity, nor is it a matter of mere style as distinguished from content. It has to do with what we think we are teaching and learning. Indeed, it has to do with what we think teaching and learning *are*.

The English word "mathematics" (as was mentioned earlier) comes from the Greek word for the kind of know-how that has to do with *ta mathêmata*, which is Greek for "the things that are learned, or that are learnable." Just what are they? And is there any way simply to separate the question of *how to present* mathematics from the question of *what it is*, or to separate the question of what mathematics is from the question of what *learning* and *teaching* are?

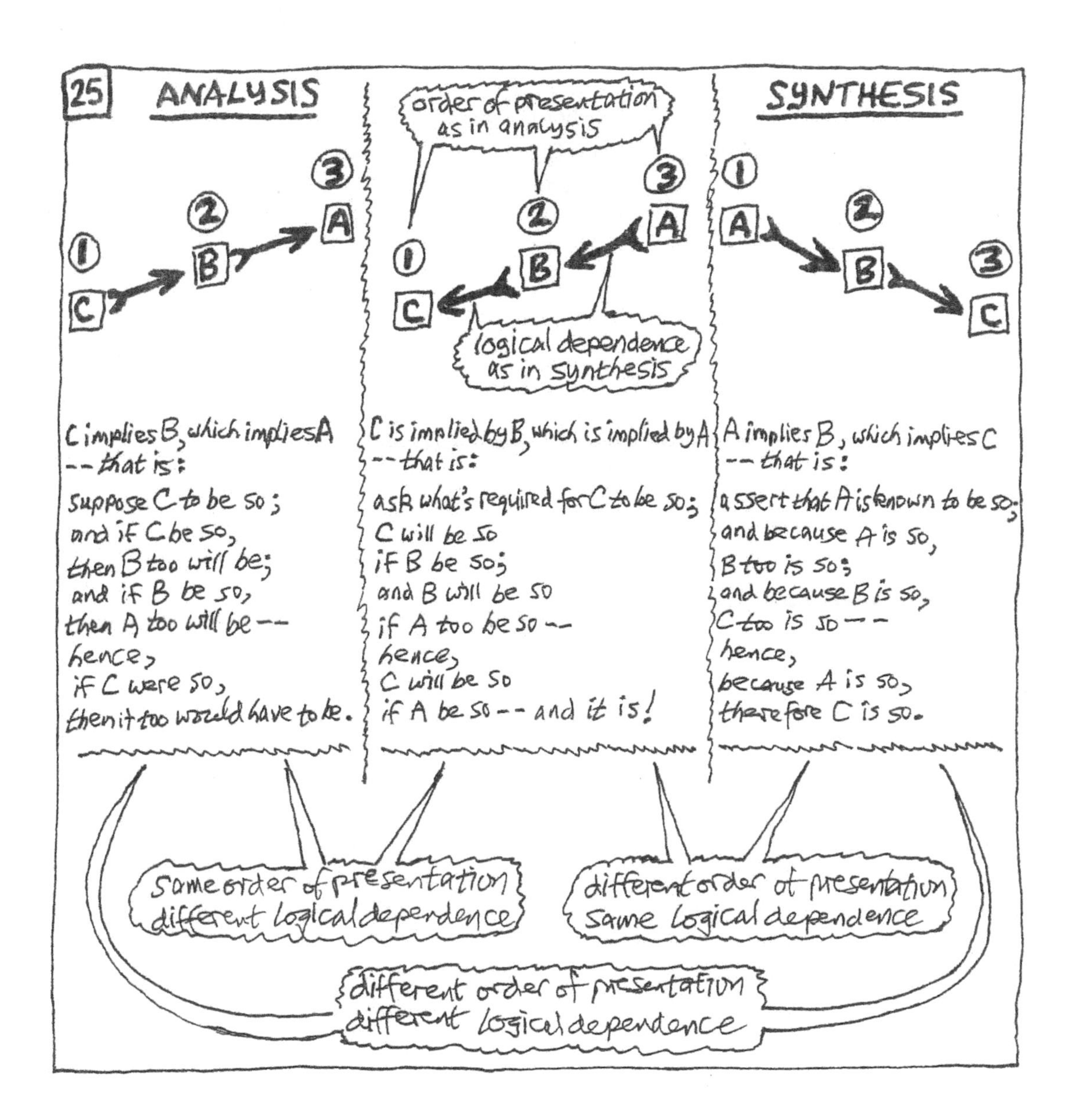
25
ANALYSIS
order of presentation as in analysis
SYNTHESIS
1
2
3
C
B
A
1
2
3
C
B
A
1
2
3
A
B
C
logical dependence as in synthesis
C implies B, which implies A
-- that is:
suppose C to be so;
and if C be so,
then B too will be;
and if B be so,
then A too will be --
hence,
if C were so,
then it too would have to be.
C is implied by B, which is implied by A
-- that is:
ask what's required for C to be so;
C will be so
if B be so;
and B will be so
if A too be so --
hence,
C will be so
if A be so -- and it is!
A implies B, which implies C
-- that is:
assert that A is known to be so;
and because A is so,
B too is so;
and because B is so,
C too is so --
hence,
because A is so,
therefore C is so.
same order of presentation
different logical dependence
different order of presentation
same logical dependence
different order of presentation
different logical dependence

appendix on decompounding

In decompounding any ratio of rectangles, there is a choice to be made between two alternatives. Which alternative you choose depends on which lines you wish to relate (fig. 26). You may want to refresh your memory by looking again at the presentation of the compounding of ratios, and at the presentation of ratios of rectangles, in figures 7 and 8 of chapter 7 of the Second Part of this guidebook.

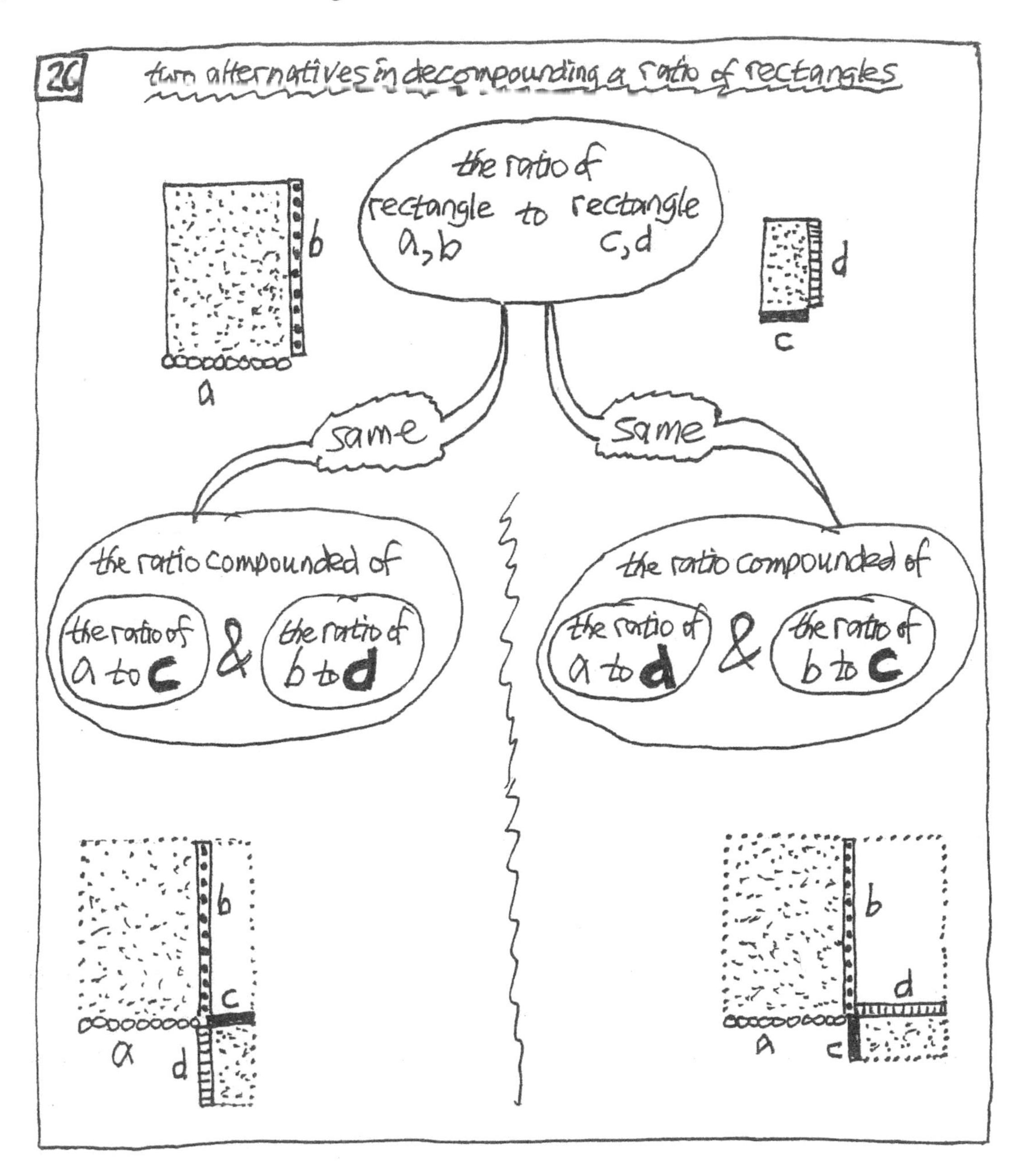

In remembering a demonstration which contains a decompounding of ratios, you will go astray if you choose the wrong alternative in decompounding. In order to get the right ratios, you need to keep in mind what it is for which you want the ratios. The purpose that guides the decompounding in the demonstration of the 11th proposition is schematized in figure 27.

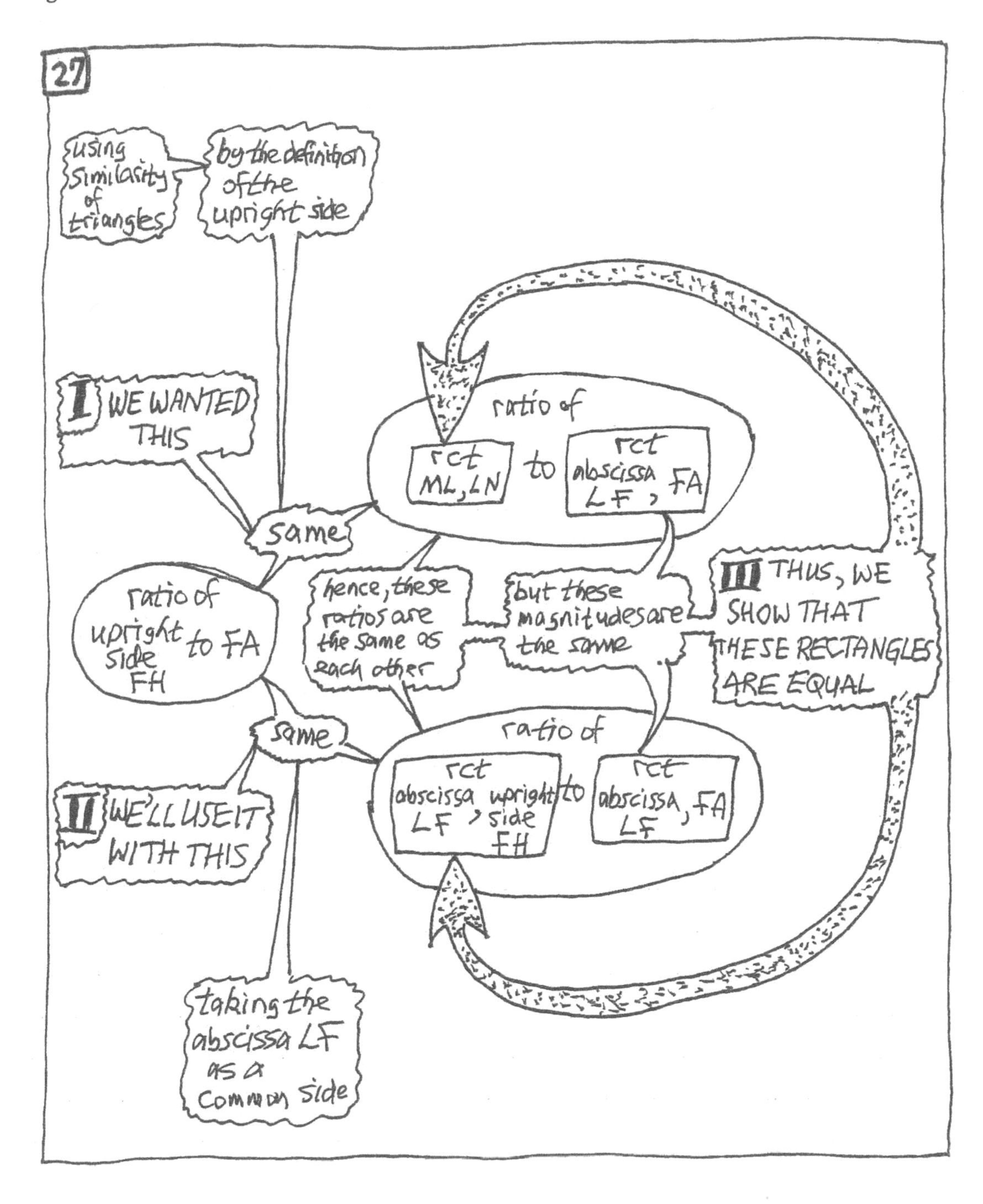

appendix on Euclid's synthetic presentation

Consider the 47th proposition of the First Book of Euclid's *Elements* (the demonstration of the so-called "Pythagorean" theorem).

Elements I.47 In the right-angled triangles, the [square] from the side stretching under (*hupoteinousês*) the right angle is equal to the squares from the sides containing the right angle.
—Let a right-angled triangle be what the [figure] ABC is, having the angle by ABC right. I say that the square from BC is equal to the squares from BA, AC.
—For let there be drawn up from the [line] BC the square BDEC, and from the [lines] BA, AC the [squares] GB, HC. And also, through the [point] A let a parallel to either of the [lines] BD, CE be drawn, namely the [line] AL. And also, let there be joined the [lines] AD, FC.
—And also, since right is what each of the angles by BAC, BAG is, then with a certain straight [line], namely the [line] AB, and at the same point A, the two straight lines AC, AG not lying on the same side make the adjacent angles equal to two right [angles]. Therefore in a straight {line] is the [line] CA with the [line] AG. Then on account of the same things is also the [line] BA with the [line] AH in a straight [line].
—And also—since equal is the angle by DBC to the [angle] by FBA, for a right [angle] is what each is—let a common [angle] be added, namely the angle by ABC. Therefore the whole angle by DBA is, to the whole angle by FBC, equal.
—And also—since equal is the [line] DB is to the [line] BC, and the [line] FB to the [line] BA—hence the two [lines] DB, BA to the two sides FB, BC are equal respectively;
and also the angle by ABD to the [angle] by FBC is equal. Therefore the base AD to the base FC is equal; and also, the triangle ABD to the triangle FBC is equal.
—And also, the triangle ABD's double is what the parallelogram BL is—for as base they have the same [line] BD, and are in the same parallels, namely the [lines] BD, AL—and the triangle FBC's double is what the square GB is—for, again, as base have the same [line] FB and are in the same parallels, namely the [lines] FB, GC. [But the doubles of equals are equal to one another.] Therefore also the parallelogram BL is equal to the square GB.
—Then, similarly, there being joined the [lines] AE, BK, there can be shown also the parallelogram CL equal to the square HC.
—The whole square BDEC, therefore, to the two squares GB, HC is equal.
—And the square BDEC is from the [line] BC drawn up, and the squares GB, HC from the [lines] BA, AC. Therefore the square from the side BC is equal to the squares from the sides BA, AC.
—In the right-angled triangles, therefore, the [square] from the side stretching under the right angle is equal to the squares from the sides containing the right angle—which very thing it was required to show.

Euclid's reasons for dropping the perpendicular in the demonstration of I.47 are unintelligible unless we do some analysis—that is, unless we say something like this: one non-right angle of a right triangle suffices to determine its shape, so the perpendicular dropped from the right angle's vertex to the hypotenuse will give similar triangles, and so on and so forth. But Euclid gives no analysis in this proposition, leaving the reader in the dark for a while.

By ending the First Book of his *Elements* with that proposition and its converse, Euclid provides the First Book with a conclusion which is spectacular—and which is somewhat inconclusive, since his readers do not quite know what has hit them, but would like to find out.

Having ended the first Book by relating the sizes of the straight lines that constitute a right triangle, Euclid goes on in the Second Book to show other ways to relate straight lines by means of squares and rectangles. What lies behind the spectacular conclusion of the First Book is not revealed until the 8th proposition of the Sixth Book—which demonstrates that if you drop a perpendicular from the vertex of the right angle to the base of the right triangle, then the two triangles into which the whole right triangle is split will be similar both to the whole and to one another. Thus it is evident (Euclid says in the porism) that if you drop a perpendicular from the vertex of the right angle to the base of a right triangle, then the straight line so drawn will be a mean proportional between the segments into which it splits the base. The demonstration of VI.8 does not depend upon I.47, whereas VI.8 clearly shows the reason for what is shown obscurely in I.47. (See figure 28.)

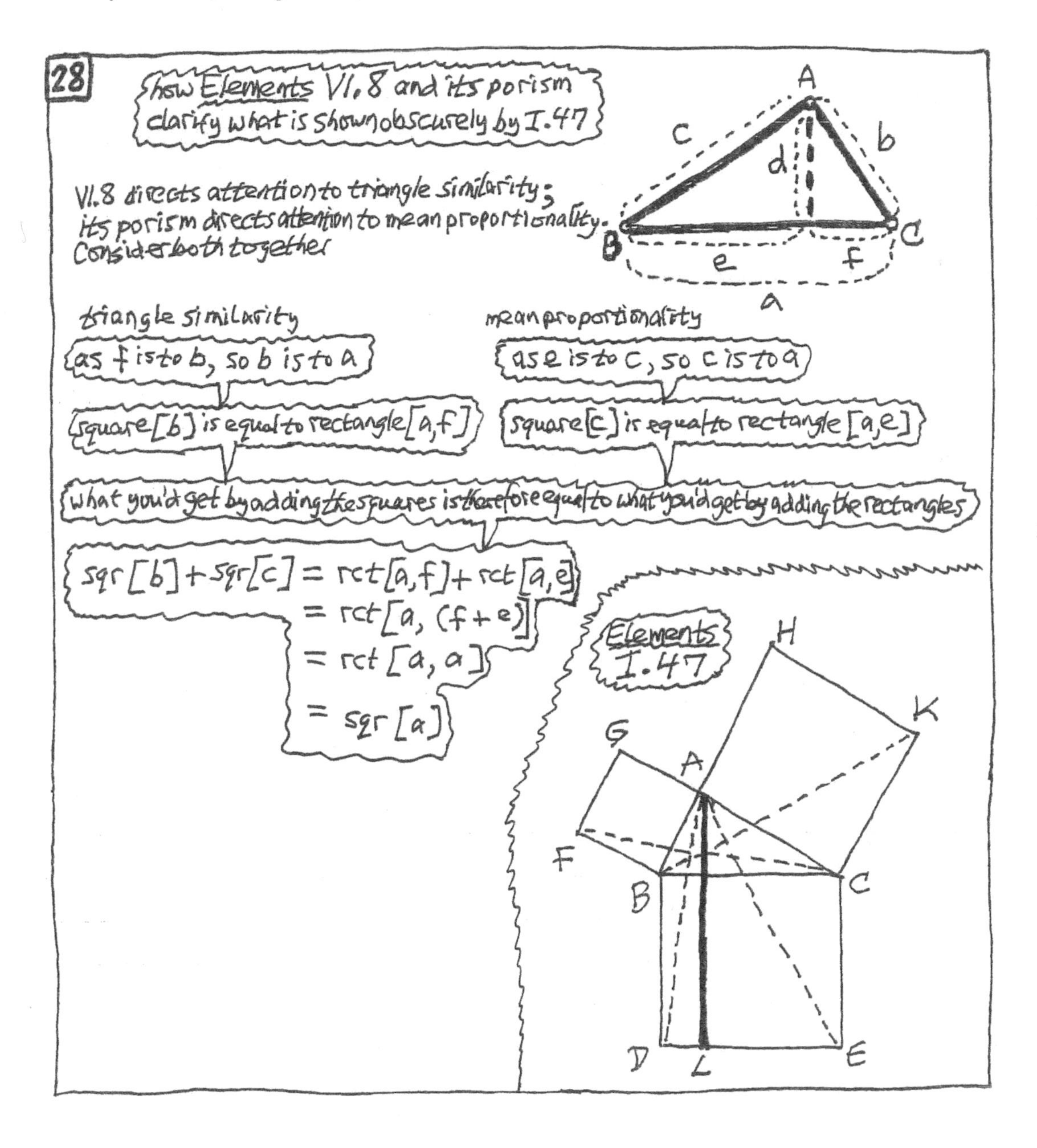

CHAPTER 2. THE HYPERBOLA

transition

The First Book of the *Conics* began by introducing new curves, in a prologue comprising ten propositions. The 11th proposition then characterized the first sort of the new curves that had been introduced in the 8th proposition.

Both sorts of new curves introduced in the 8th proposition were less like a circle than were the sort of new curves introduced in the 9th proposition. Whereas circles (like the new curves in the 9th proposition) are closed curves, the curves in the 8th proposition can be prolonged indefinitely as we prolong their diameters. All the curves in the 8th proposition were cut by a plane that does not hit the axial triangle's other side (or its extension) below the cone's vertex. Nonetheless, the enunciation of the 8th proposition did not speak generally. It could have said that the cutting plane does not hit the axial triangle's other side (or its extension) below the cone's vertex. But that is not what it said. It said, rather, that the cutting plane has *either* of *two* orientations—one of them being parallel to a side of the axial triangle, and the other one hitting an extension of a side of the axial triangle above the vertex of the cone.

When the 11th proposition took up these prolongable curves, it considered only the first of the two ways of getting them—namely, the way in which the cutting plane is parallel to a side of the axial triangle. In that way of cutting, the angle which the cutting plane makes with the axial triangle's side is equal to the axial triangle's vertex angle. (Fig. 1.)

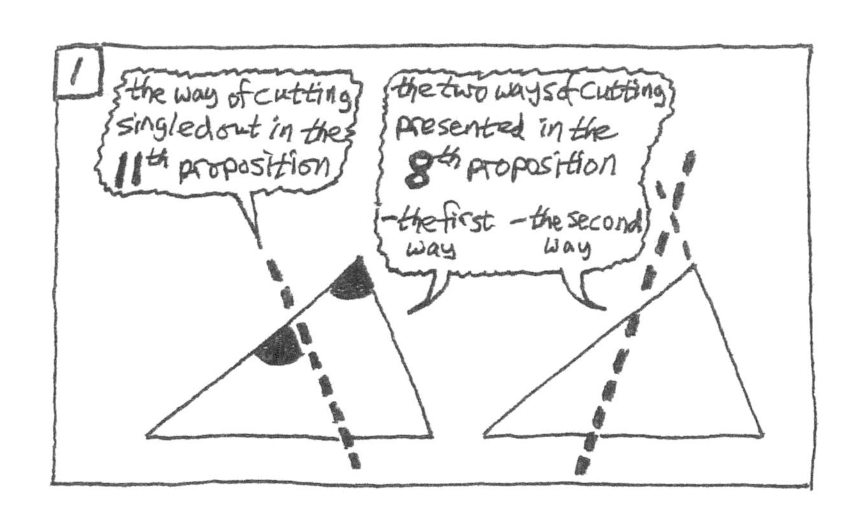

Cutting in the other way would allow for a range of angles, not just one angle. Of all the cuttings that are instances of the cutting in the 8th proposition, the one that was singled out in the 11th proposition is the one that stands out from the others.

It also stands out as a counterpart of the two cuttings which the 9th proposition demonstrated to be the only cuttings that give circles. Those circle-cuttings meet the axial triangle's two sides in angles which are equal to two angles of the axial triangle (its base angles). The cutting in the 11th proposition also meets the axial triangle's side and its base in angles which are equal to two angles of the axial triangle (namely, the vertex angle and one base angle). (Fig. 2.)

We might wonder why we do not now again in a single proposition consider both ways of cutting, as we did earlier in the 8th proposition. But back then we might have wondered why the enunciation of the 8th proposition did distinguish two ways which did not need to be distinguished in the demonstration of the proposition. That enunciation could have said merely that the cutting plane meets the axial triangle's base and only one of its sides below the cone's vertex. It seems that we need first to take separate note of that way of not hitting the other side which stands out—namely, cutting parallel to the other side. Only after the 11th proposition has taken separate note of this, does the 12th proposition consider the other way, which includes innumerable non-parallel ways of cutting that do not hit again—ways of cutting (that is to say) so that the second hit takes place on the other side's extension above the vertex of the cone.

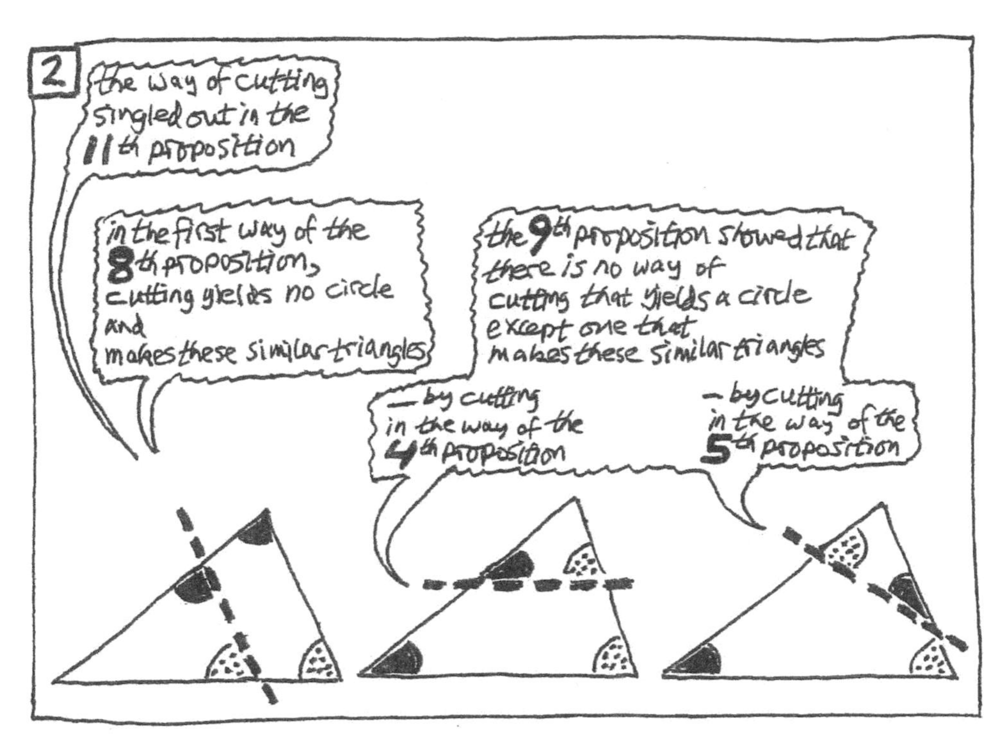

Here, then, is the text of the 12th proposition.

#12 If a cone be cut by a plane through its axis, and be cut also by another plane which cuts the base of the cone along a straight [line] that is at right [angles] to the base of the triangle through the axis, and, moreover, the section's diameter, being extended, fall together with one side of the triangle through the axis, [that is, if the diameter meet that axial triangle's side] out beyond the cone's vertex, and any [straight line] from the section of the cone be drawn parallel to the common section of the cutting plane and of the base of the cone, as far as the diameter of the section, then [this straight line so drawn to the diameter] will have the power (*dunêsetai*) [the power, that is, to give rise to a square that is equal to] a certain [rectangular] area (*khôrion*) ["area," we need to remember, meaning not a number of any sort, but an extent of territory,] laid alongside (*parakeimenon para*) [or, in traditional parlance, "applied to"] a certain straight [line], one to which the [line] that, being a straight prolongation for the diameter of the section, and subtending, [that is, "stretching under"] the exterior angle of the cone's vertex, has the ratio that the square from the [line] drawn from the cone's vertex, parallel to the section's diameter, as far as the base of the triangle, has to the [rectangle] contained by the base's segments which the [line] drawn [from the cone's vertex] makes—[with that equal area] having as breadth the line taken off by it, [taken off, namely, by that line drawn from the section to the diameter], [taken off, that is,] from the diameter, on the side toward the vertex of the section, [and also with that equal area] overshooting (*huperballon*) [or "going beyond," or "exceeding"] by a figure (*eidei*) that is similar to, and is also laid similarly to, the [rectangle] contained by [the following two straight lines]: the [line] subtending the exterior angle [at the vertex] of the axial triangle, and also the [line just characterized by a certain ratio, namely, the line] along which the [straight lines that are] drawn down ordinatewise have the power (*dunantai*), [that is, that power mentioned above]. And let such a section be called an hyperbola (*hyperbolê*).
—Let there be a cone whose vertex the point A is and whose base the circle BC is. And let [the cone] have been cut by a plane through the axis; and let it make as section the triangle ABC, and let the cone have been cut also by another plane which cuts the base of the cone along a straight [line], namely the [line] DE, which is at right [angles] to the base BC of the triangle ABC; and let [this other plane] make as section on the surface of the cone the line DFE, and let the diameter of the section, namely the [line] FG, being extended, fall together with one side of the triangle ABC, namely the [line] AC, out beyond the cone's vertex—at the [point] H, say—and, through the [point] A, parallel to the diameter of the section, namely the [line FG], let there be drawn the [line] AK. And let [this line AK] cut the [line] BC. And, from the [point] F let there be drawn, at right [angles] to the [line] FG, the [line] FL. And let [this straight line FL] be so made that as is the [square] from KA to the [rectangle] by BKC so is the [line] FH to FL. And let there be taken some chance point on the section—the [point] M, say—and, through the [point] M, let there be drawn parallel to the [line] DE the [line] MN, and, through the [point] N, let there be drawn parallel to the [line] FL the [line] NOX, and—there being joined the [line] HL—let it be extended, to the [point] X, say, and, through the points L, X, let there be drawn parallel to the [line] FN the [lines] LO, XP. I say that the [line] MN has the power (*dunatai*) [to give rise to a square equal to] the [area] FX that is laid along (*parakeitai para*) the [line] FL, while having as breadth the [line] FN, and overshooting (*huperballon*) by a figure (*eidei*), namely by the [area] LX, which is similar to the [rectangle contained] by the [lines] HFL.

—For let there be drawn through the [point] N parallel to the [line] BC, the [line] RNS. And also the [line] NM is a parallel to the [line] DE. The plane through the [lines] MN, RS therefore is parallel to the [plane] through the [lines] BC, DE—that is, [it is parallel] to the base of the cone. If, therefore, there be extended the plane through the [lines] MN, RS, the section will be a circle, a diameter of which the [line] RNS is. And perpendicular to it is the [line] MN. The [rectangle], therefore, by the [lines] RNS is equal to the [square] from the [line] MN. And since as is the [square] from AK to the [rectangle] by BKC so is the [line] FH to FL—and the ratio of the [square] from the [line] AK to the [rectangle] by BKC is compounded from that which the [line] AK has to KC and also that which the [line] AK has to KB—therefore also the ratio of the [line] FH to the [line] FL is compounded from that which the [line] AK has to KC and also that which the [line] AK has to KB. But (on the one hand) as is the [line] AK to KC so is the [line] HG to GC—that is, [so is] the [line] HN to NS; and (on the other hand) as is the [line] AK to KB so is the [line] FG to GB—that is, [so is] the [line] FN to NR. Therefore the ratio of the [line] HF to FL is compounded from that of the [line] HN to NS and also that of the [line] FN to NR. But the ratio compounded from that of the [line] HN to NS and also that of the [line] FN to NR is the ratio of the [rectangle] contained by the [lines] HNF to the [rectangle] by the [lines] SNR. Also, therefore, as is the [rectangle] by the [lines] HNF to the [rectangle] by the [lines] SNR so is the [line] HF to FL—that is, [so is] the [line] HN to NX. But as is the [line] HN to NX, so—with the [straight line] FN taken as common height—is the [rectangle] by the [lines] HNF to the [rectangle] by the [lines] FNX. Also, therefore, as is the [rectangle] contained by the [lines] HNF to the [rectangle] by the [lines] SNR so is the [rectangle] by the lines HNF to the [rectangle] by the [lines] XNF. Therefore, the [rectangle] by SNR is equal to the [rectangle] by XNF. But the [square] from MN was shown to be equal to the [rectangle] by SNR. Therefore also the [square] from the [line] MN is equal to the [rectangle] by the [lines] XNF. But the [rectangle] by XNF is the parallelogram XF.

—The [line] MN, therefore, has the power (*dunatai*) [to give rise to a square equal to] the [area] XF which is laid alongside (*parakeitai para*) [or, in traditional parlance, "is applied to"] the [line] FL—[that area] XF having as its breadth the [line] FN, and overshooting [or "exceeding"] (*huperballon*) [the rectangle contained by the lines FN and FL] by the [area] LX, which is similar to the [rectangle contained] by the [lines] HFL.

—And let such a section be called an "hyperbola" (*huperbolê*); and as for the [line] LF—the one along (*para*) which the lines drawn down ordinatewise to the [line] FG have the power (*dunantai*)—let that same [line LF] be called an "upright" (*orthia*) [side], and an "oblique" [or "slanting"] (*plagia*) [side] be what the [line] FH is called.

[To refer to a straight line which, like FH, stretches *across* from the section's vertex F to the point H, and also figures in the proportion that here obliquely characterizes the size of the overshoot in terms of a shape, the traditional albeit misleading translation has been "transverse" side. We shall render the Greek *plagia* as "oblique"; this better contrasts with the paired term "upright," which is the traditional rendering of the Greek *orthia*. The reader should bear in mind, however, that the traditional rendering of *plagia* into English as "transverse" merely transliterates the traditional Latin rendering, as well as the derived French usage in the *Geometry* of Descartes, which we shall be examining later.]

[Again, to be more explicit about the business of "having the power": A straight line drawn down ordinatewise, from the section to its diameter FG, "has the power," it is here said, "along LF." This means that such a line drawn down ordinatewise (say, MN), when taken as a side, gives rise to a square that will be equal to a certain rectangle which lies along (*para*) the "upright" LF, whose size has been characterized by a certain proportion. The equal rectangle shoots beyond (*huper*) the line FL which it lies along (*para*). That is, the equal rectangle's length FP is greater than FL. If the rectangle's length were not greater than the upright side, the rectangle would be too small to be equal to the square, since the rectangle's breadth is the piece (line FN) that is taken off from the diameter, on the side toward the section's vertex F, by the line MN that is drawn down ordinatewise.]

12th proposition: the enunciation—how we cut

As in the previous proposition, we first pass a plane through the axis, to get an axial triangle. It is with respect to the axial triangle that the cutting plane is oriented (fig. 3).

As before, the straight line (DE) that is the cutting plane's intersection with the cone's base plane is perpendicular to the straight line (BC) that is the axial triangle's base. By orienting that intersection (DE) in that way, we guarantee that every one of the section's chords which is parallel to it will be bisected by the cutting plane's other straight-line intersection—namely, its intersection (FG) with the axial triangle.

Straight line DE thus serves as the reference line for lines drawn ordinatewise across the section. These lines will be bisected by the axial triangle, and thus by straight line FG; and hence this line FG is the diameter of the half chords that they make, which are this diameter's ordinates.

Line DE, the cutting plane's intersection with the plane of the cone's base, thus has, in the present proposition, the same orientation which it had in the previous proposition. What is different here is the orientation of the cutting plane's other intersection with a plane—namely, its intersection FG with the plane of the axial triangle.

What is the *same*	What is *different*
is the orientation of the cutting plane	is the orientation of the cutting plane
viewed from *above* the cone	viewed from *beside* the cone
(that is to say, the orientation of	(that is to say, the orientation of
the *reference line*	the *diameter*, from which we cut
for the *ordinates*).	off the *abscissas*).

How is the diameter oriented now? The diameter is not parallel to the other side of the axial triangle, as it was in the previous proposition. But neither does the diameter now hit the other side or its extension below the vertex, as it will do in the next proposition. Rather, it hits the other side's extension above the vertex. (Below the vertex, therefore, it gets farther and farther away from the other side and its extension.)

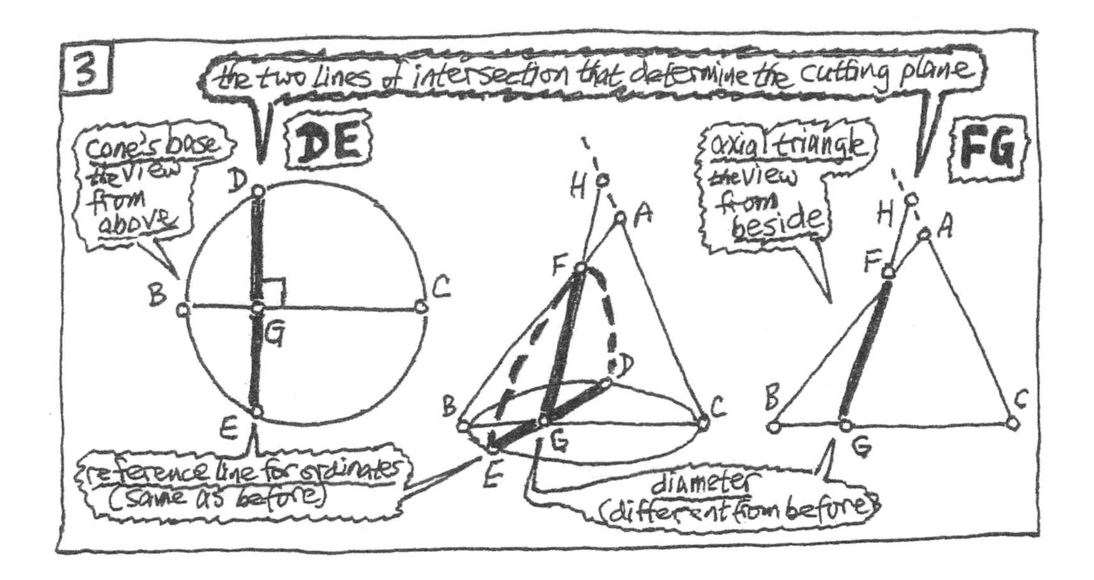

12th proposition: the enunciation–what we show about the line we cut

If our cutting is done in the way just described, then the intersection of cone by cutting plane is a conic section of a sort distinguished by a certain property of the straight lines that we draw in it.

As before, the ordinate and the abscissa are related by showing the equality of a square and a rectangle. (Fig. 4.) The square has as its side the ordinate (MN); and the rectangle has as one of its sides the abscissa (FN). Now, however, to determine the size of the rectangle's other side (NX) involves much greater complication than in the previous proposition. (By the way, the point X is so called because in the Greek alphabet the letter "Xi" comes after "Nu" and before "Omicron.")

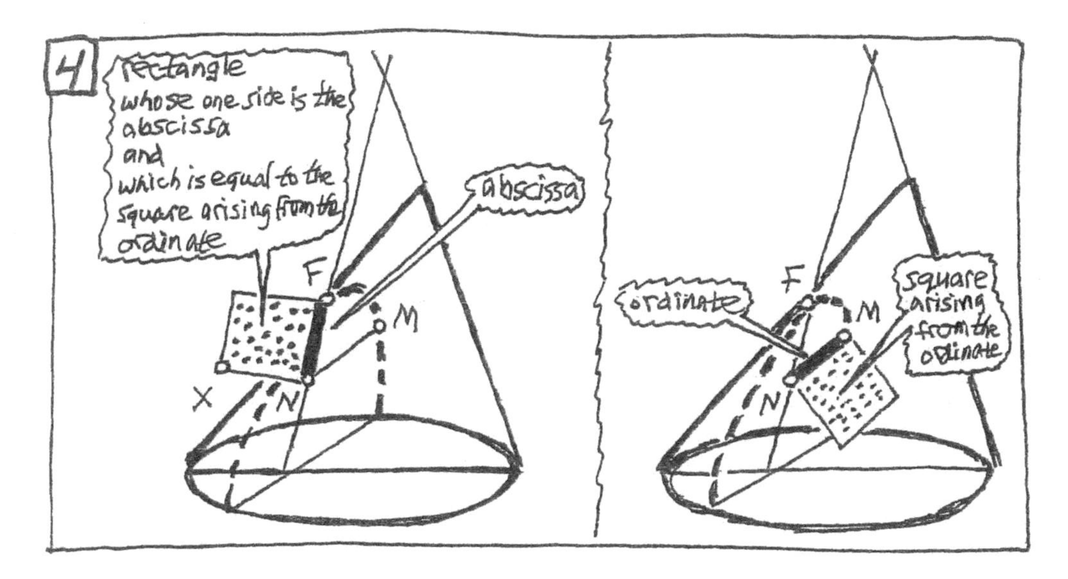

As before, a line must be contrived to be the upright side by being made just the right size to fit a certain proportion. In order to state the proportion for the present proposition, however, we first must name a line and we then must draw another line.

First, we give the name "oblique side" to a straight line which goes in line with the diameter that has been sliced into the axial triangle by the cutting plane. The oblique side goes from the section's vertex all the way across to the extension, above the cone's vertex, of the axial triangle's opposite side. (We shall be calling this side both "oblique," in accordance with the meaning of the Greek term *plagia*, and also "transverse," in accordance with the traditional parlance to be found later in Descartes.)

Then, from the cone's vertex, we draw a straight line parallel to the diameter, all the way until it meets the axial triangle's base. A square on that straight line has a certain ratio to a rectangle contained by the segments into which the axial triangle's base is split by that straight line. That same ratio is to be the ratio which the oblique (transverse) side has to the line that we contrive to be the upright side.

Figure 5 depicts those lines by which the size of the upright side is determined—namely, the oblique (transverse) side HF and the parallel line AK drawn in.

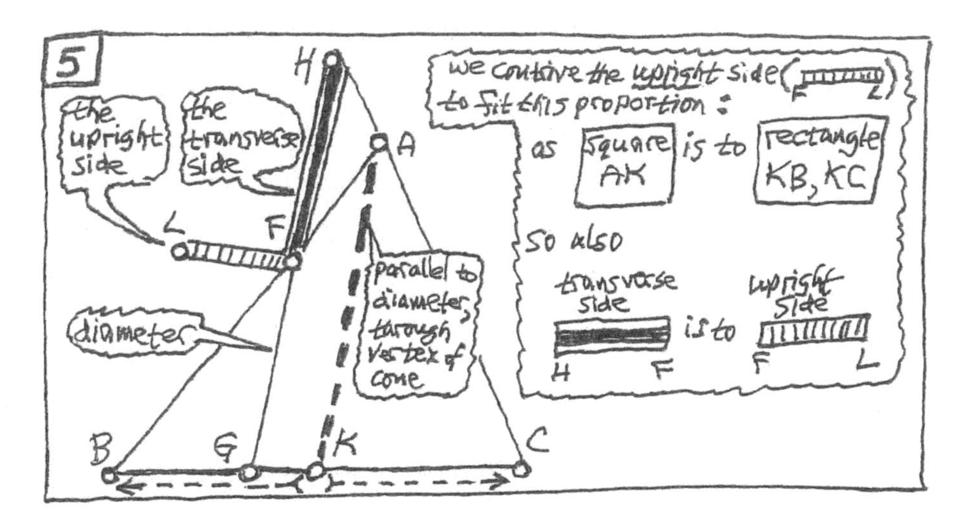

In this 12th proposition—by contrast with the previous one—the rectangle that is equal to the square on the ordinate, while it does have as one of its sides the abscissa, has as its other side not the upright side but rather a line that is longer than the upright side. The upright side, although it is not itself the equal rectangle's other side, is, however, the means of getting this other side. To get it, we look at another rectangle. The rectangle that will be equal to the square on the ordinate is not itself equal to—but rather goes beyond—the rectangle contained by the abscissa and the upright side. That is to say (fig. 6):

> the rectangle which
> is equal to the square on the ordinate and has the abscissa as one of its sides,
>
> overshoots
>
> the rectangle which
> also has the abscissa as one of its sides but has as its other side the upright side.

The rectangular overshoot—that is what we need to attend to.

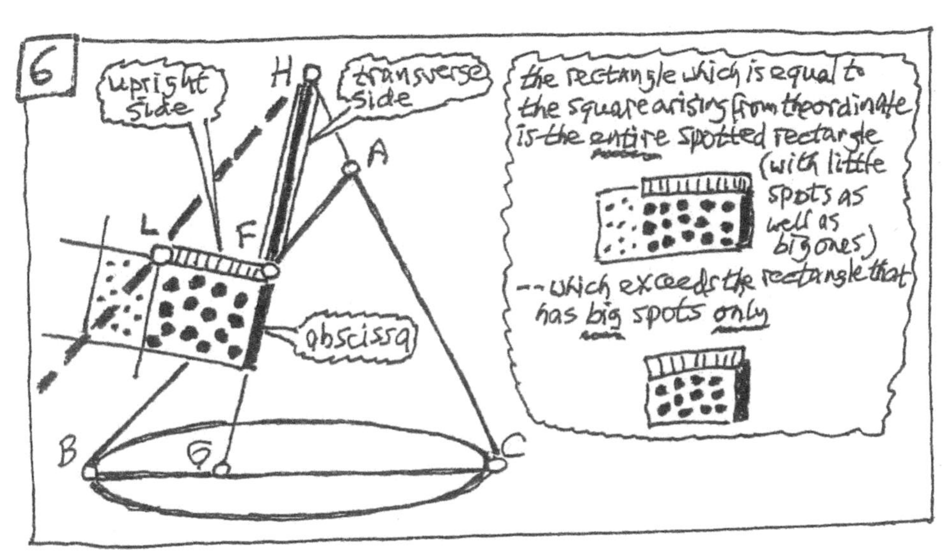

Just how much of an overshoot is there? Well, the excess is rectangular in form (*eidos*), and one of its sides has the same size as the abscissa. Of what size is its other side? Of just the right size to give it a certain shape—namely, this shape: the rectangular excess, though it may not be the same size as, still looks just like the rectangle contained by the oblique (transverse) side and the upright side. In other words: as the oblique (transverse) side is to the upright side, so is the abscissa (which is the rectangular excess's one side) to the rectangular excess's other side. The reference rectangle formed by the oblique (transverse) side and the upright side exhibits the form of the rectangular excess. (Fig.7.)

Thus, the size of the excess for each abscissa will be determined by these two items: the size of the abscissa, and the constant ratio of the constant oblique (transverse) side to the constant upright side.

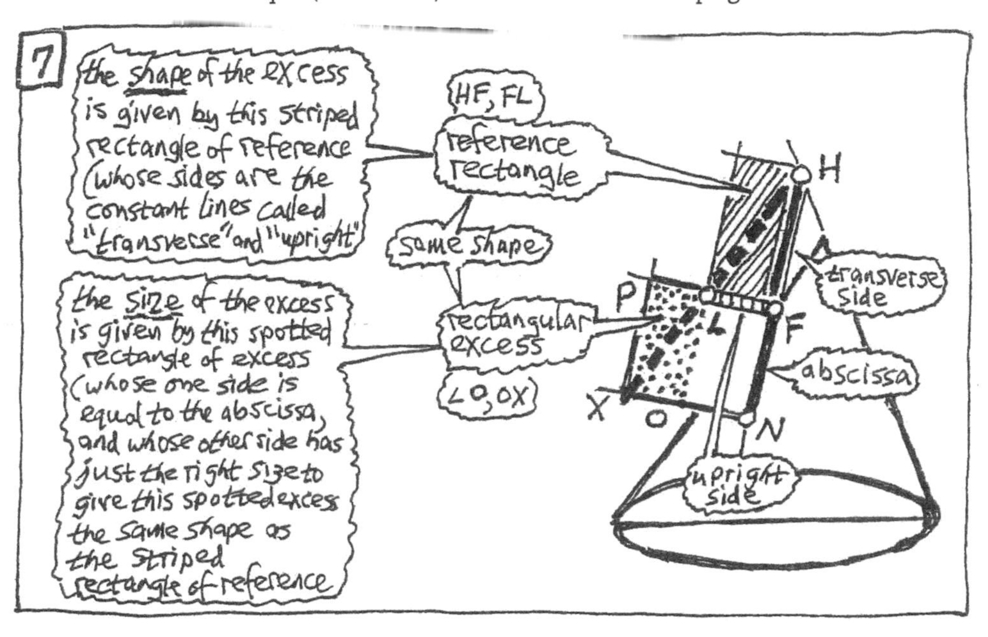

The changing size of the equal rectangle's non-abscissa side (its width) will therefore be determined by the constant upright side. The width, that is to say, will be of just the right size to fit this proportion: the ratio which the oblique (transverse) side has to it (the upright side) will be the same as the ratio which the line that is the sum of the oblique (transverse) side and the abscissa has to the width of the equal rectangle. (What we will get from the oblique (transverse) side by adding the abscissa on to it, is the line that stretches from the abscissa's non-vertex endpoint to the oblique (transverse) side's non-vertex endpoint.) That is (fig. 8), NF+FH (oblique [transverse] side+abscissa) will be to NX (width) as HF (abscissa) is to FL (upright side).

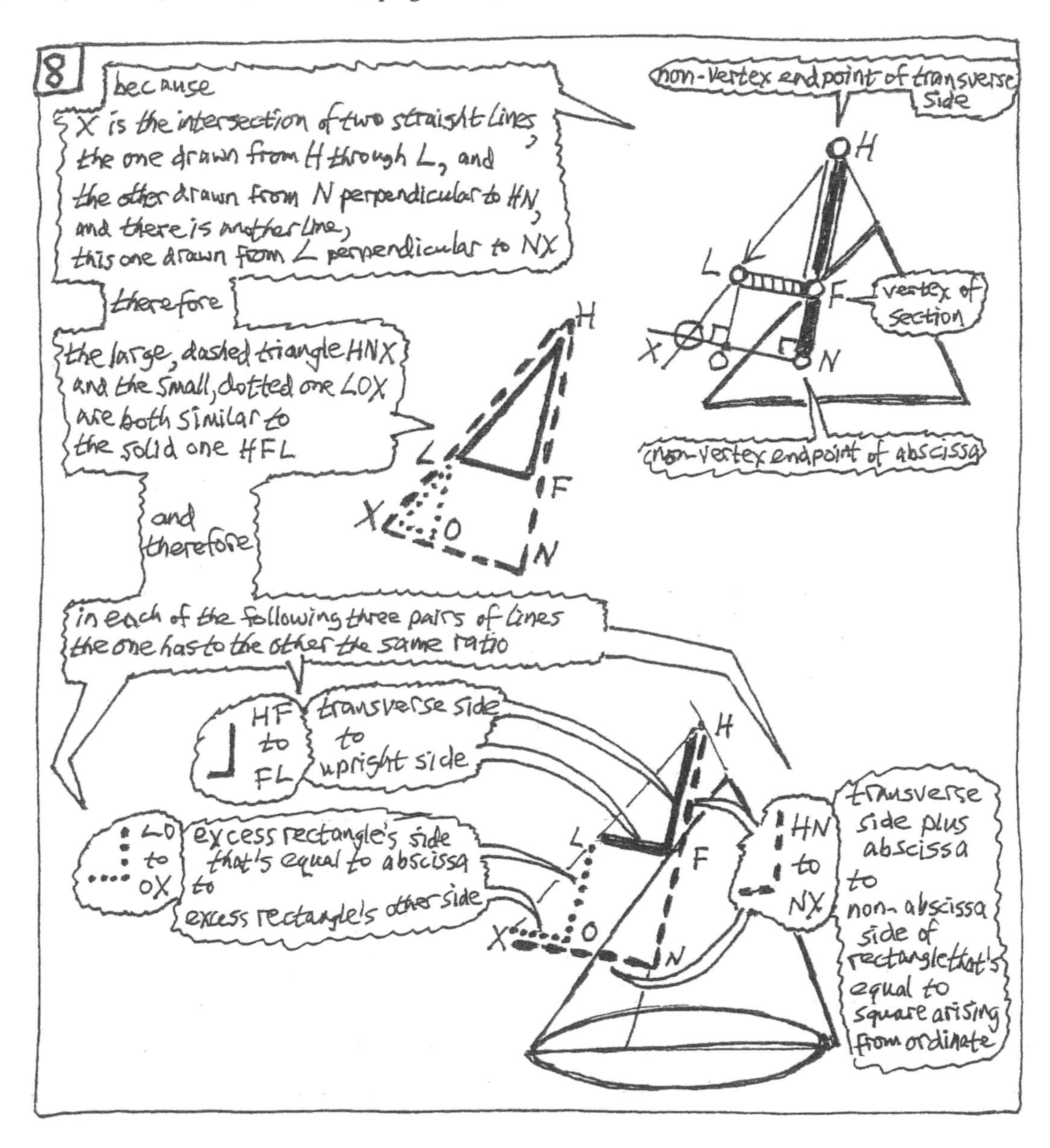

At the end of the enunciation, the section is given the name "hyperbola." In Greek, that is to say, the curve is called the "overshoot." *Hyper+bolê* is Greek for "over (or) beyond+throw." The rectangle that here is thrown along the upright side goes beyond or overshoots the upright side. In the parabola (*para+bolê* is Greek for "along+thrown"), the rectangle that is thrown alongside the upright side does not overshoot but stays along the upright side just until the end. These curves were not discovered by Apollonius, but he put them together in the way that they are presented here—and it was he who gave them the names used here, which stuck. Giving them these names was part of putting them together in the way that they are presented here.

The names that have been given out already might lead us to expect a third to come. Since, that is, the rectangle in one case is thrown exactly alongside the upright side, and in the next case overshoots it, we might therefore expect to see a case in which the rectangle lacks the length to stretch all the way along the upright side. Our expectation will be fulfilled in the next proposition. It presents a section which is given the name "ellipse," since *elleip+sis* is Greek for "lack+ing (or) being+deficient." In Greek, that is to say, the curve is called the "lacking."

The present proposition is the middle one of three that define a sort of conic section by stating a relationship between ordinate and abscissa. In each of these three propositions, the relationship is stated indirectly. A square is raised up, having the ordinate as its side. Equal to this square, a rectangle is built, having the abscissa as one of its sides. To get the other side is a complicated affair. We have seen it for the parabola in the 11th proposition. Now, in the 12th, we are seeing it for the hyperbola. Afterward, in the 13th, we shall see it for the ellipse.

For each of these conic sections, a certain line is first contrived as a fourth proportional. For the parabola, this line ("the upright side") is itself the other side of the equal rectangle. For the hyperbola and the ellipse, however, the upright side is not itself the other side of the equal rectangle. Instead, it is the means of getting that other side. How?

In the hyperbola, the rectangle that is equal to the square on the ordinate is greater than the rectangle contained by the abscissa and the upright side; in the ellipse, it is smaller. The linear overshoot in the hyperbola, or the linear lacking in the ellipse, is of just such a size as to give a certain *shape* to the rectangular excess or the rectangular deficiency by which the equal rectangle differs from the rectangle that is contained by the abscissa and the upright side. That rectangular excess or rectangular deficiency has the same shape as the rectangle contained by the oblique (transverse) side and the upright side. What is displayed before the eye (the *shape* of a rectangle) embodies a constant relationship apprehended with the mind's eye (the *ratio* of the constant oblique [transverse] side to the constant upright side).

The constant relationship between the varying ordinate and the varying abscissa is indicated by the equal or greater or smaller size of a square raised upon the ordinate, in comparison with a rectangle contained by the abscissa and the constant upright side. The constant relationship between ordinate and abscissa is not itself a ratio, but it can be found in the squares and rectangles that do embody certain constant ratios—ratios which, taken together, constitute the relationship between the ordinate and the abscissa.

In these propositions, figures displaying certain looks or shapes are brought to sight by the enunciations; it is the demonstrations, however, that can give us insight into ratios that lurk behind or underneath the display.

Note that although, for convenience, I shall speak of the non-abscissa side as the "width" of that rectangle which is equal to the square on the ordinate, and I shall therefore sometimes also speak of the abscissa as the rectangle's "height," those are nonetheless not names given by Apollonius. Apollonius does not provide such convenience of nomenclature—except in the case of the abscissa of the hyperbola and of the ellipse, which he calls the rectangle's *platos*. However, since a plane figure has length and *platos* (as can be seen in the definitions at the beginning of Euclid's *Elements*) and a solid has length, *platos*, and depth (as can be seen in the first definition of the Eleventh Book of the *Elements*), it would be closer to the terminology of Apollonius if I spoke of the rectangle's "breadth" and "length," instead of speaking of its "height" and "width." I have chosen to speak, rather, of "width" (and therefore also of "height") because then I can use, as an abbreviation, "line (w)." This will have mnemonic usefulness—since double-u is the initial letter of the interrogative pronoun in English, and is also a partner of the initial letter of the word "upright," which it will sometimes be useful to abbreviate as "(u)."

Speaking of mnemonics, we shall find it useful to think of KB and KC, not as the segments into which the axial triangle's base is split by the straight line AK (which we draw through the axial triangle's vertex and parallel to the diameter), but rather as the lines from point K to each end of the base of the axial triangle—then when we get to the ellipse in the next proposition, we shall have fewer different things to remember. Earlier, in speaking of propositions II.5 and II.6 of Euclid's *Elements*, we were able to unify two enunciations by speaking of a line as being "cut externally." So here we might think of point K (the foot of line AK) as cutting the axial triangle's base (line BC) "internally" (in the case of the hyperbola) and "externally" (in the case of the ellipse). (Fig. 9.)

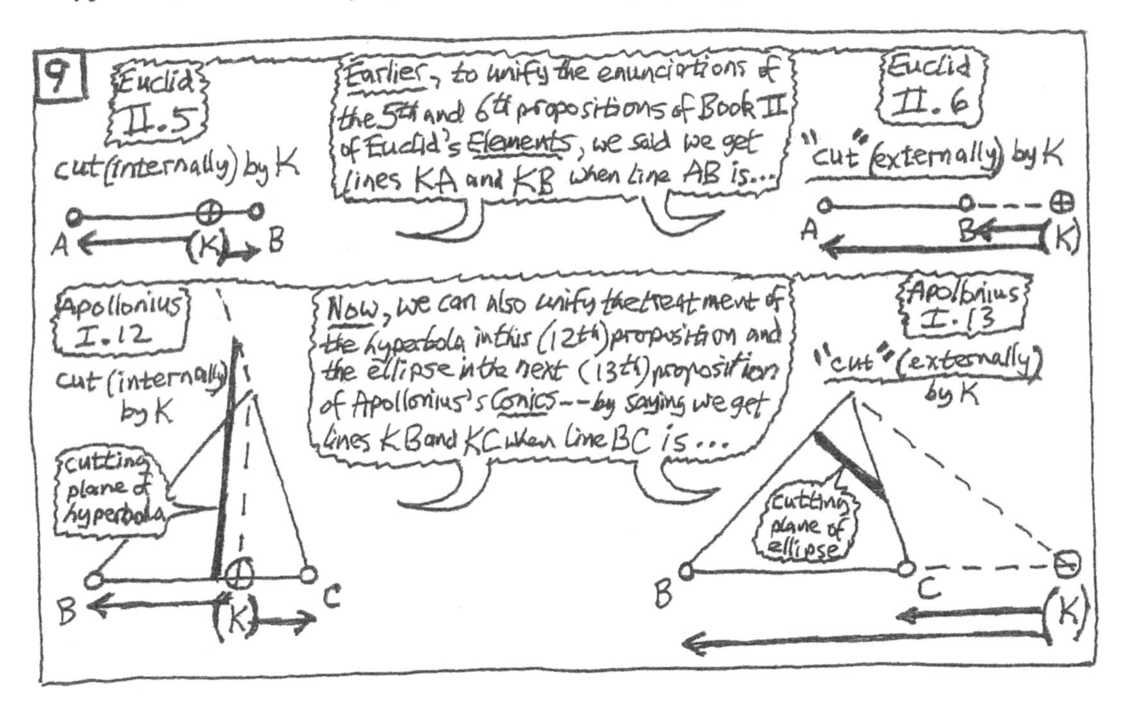

12th proposition: the demonstration

The task of the demonstration is to show that by cutting in the way described, we get a section with the following property (fig. 10):

> *if* straight line FL is contrived so that the ratio which
> the oblique (transverse) side HF has to it
> is the same ratio which
> a square with side AK has to a rectangle with sides KB and KC—
>
> and if FL is used to draw up from the abscissa-point a line NX which
> goes beyond the contrived upright side FL just far enough so that the ratio which
> the sum of the oblique (transverse) side and abscissa will have to it (NX)
> is the same ratio which
> the oblique (transverse) side has to the contrived line (FL),
>
> *then* a square whose side is ordinate MN will be equal to
> a rectangle whose sides are abscissa FN and the line NX.

That way of putting the matter may make things easier to grasp, but its emphasis differs from the emphasis in the text. What is emphasized in the text is not the equality of the square and the rectangle, but rather the equal rectangle's excess over another rectangle. The emphasis is two-fold—on size and on shape:

> The equal rectangle is greater than
> the rectangle whose sides are the abscissa and the upright side.
>
> The excess has a certain shape—
> it presents itself in the form of a certain rectangle—
> the rectangle whose sides are the oblique (transverse) side and the upright side.
> In other words,
> the rectangular excess's side that is equal to the abscissa
> has a certain ratio to the rectangular excess's other side—
> the same ratio which the oblique (transverse) side has to the upright side.

The work of the demonstration, as it was in the previous proposition, is to establish the equality of a rectangle and a square. The rectangle's sides are FN and NX, and the square's side is the ordinate MN. How does Apollonius show the equality? As in the previous proposition, square [MN] and rectangle [FN, NX] are shown to be equal to each other by showing that they are both equal to the same thing—namely, rectangle [RN, NS]. The demonstration, again, thus requires the establishment of two equalities (fig. 11).

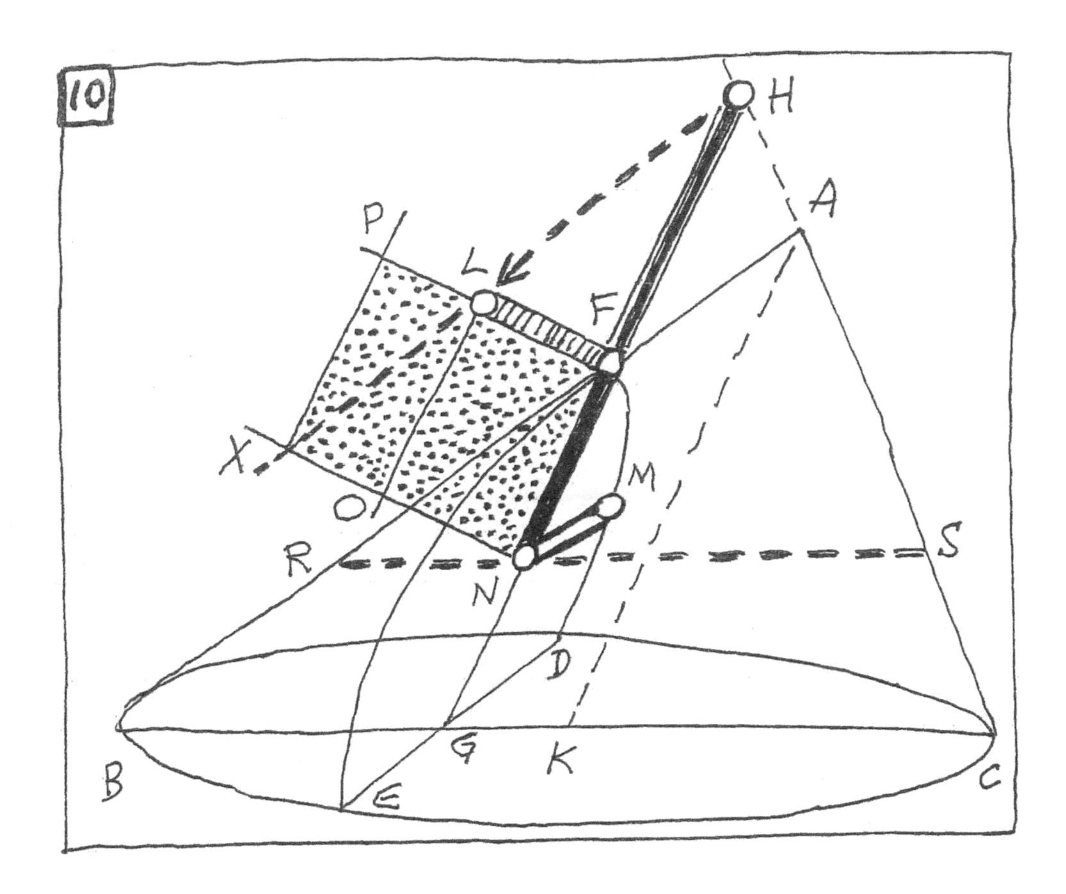
10
H
A
P
L
F
X
M
O
R
S
N
D
G
K
B
C
E

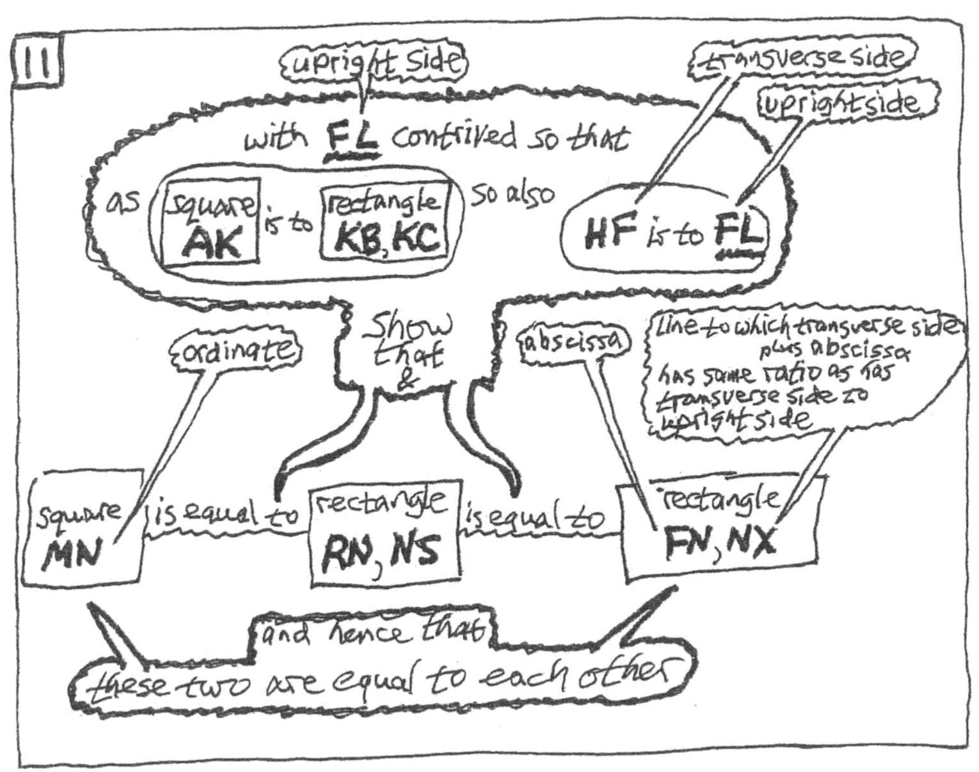
11
upright side
transverse side
upright side
with FL contrived so that
as square AK is to rectangle KB, KC so also HF is to FL
Show that &
ordinate
abscissa
line to which transverse side plus abscissa has same ratio as has transverse side to upright side
square MN is equal to rectangle RN, NS is equal to rectangle FN, NX
and hence that
these two are equal to each other

As in the previous proposition, the establishment of the first equality is the easier part of the demonstration. It does not involve using the line that has been contrived. It merely takes the view from *above* that presents us with a circle.

What we do is the following (fig. 12). Through N, the point where the ordinate meets the diameter, we pass a plane parallel to the base of the cone, thus slicing into the cone a circle. The circle's half-chord is the original section's ordinate MN—which is the side of the square that concerns us. Because this half-chord (MN) is perpendicular to the circle's diameter (RNS), we can say, as before, that the square whose side is MN is equal to the rectangle whose sides are RN and NS.

What remains to be done now is to establish that this rectangle with sides RN and NS, which is equal to the square with side MN, is *also* equal to the rectangle with sides FN and NX.

How do we establish this second equality? As in the previous proposition. Having taken the view from *above* in order to use the square-rectangle property that derives from the proportionality of certain straight lines in a *circle*, we take the view from *beside* in order to use the proportionality of straight lines in similar *triangles*. (Fig. 13.) The view from above and the view from beside are linked by the straight line RNS; for not only is RNS the diameter of the circle, it is also parallel to the base of the axial triangle—the triangle whose sides when extended are used to determine the size of the line that is contrived to be the upright side.

We need to show for those two rectangles—one whose sides are RN and NS, the other whose sides are FN and NX—that their equality follows from the sameness of the ratios which determine the size of the upright side (FL). The upright side is defined in terms of a relationship among lines that have to do with the axial triangle. We need to show how, from that relationship, another relationship follows. We need to relate the relationship that we have to start with (because we made things that way) to the relationship that we would like to have established (but do not yet have). Since the lines in the relationship that we already have (all of them) are lines that have to do with the axial triangle, we give ourselves a side view of the figure (fig. 13); and, from the axial triangle thus displayed, we take hold of what we need.

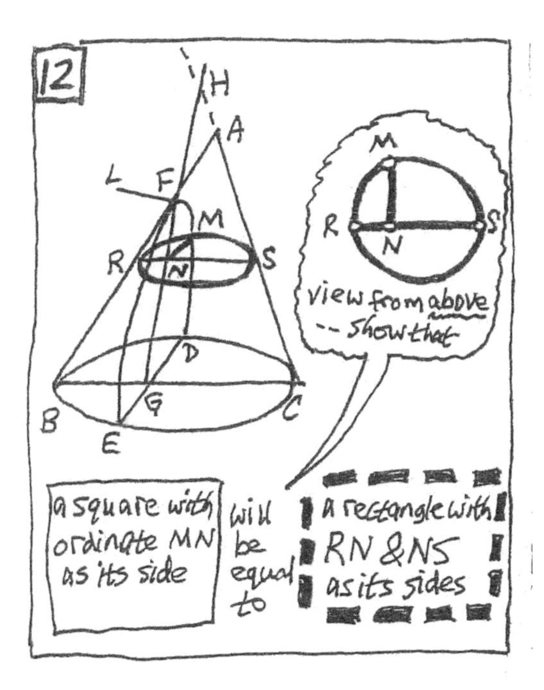
12
H
A
L
F
M
R
N
S
D
B
G
C
E
M
R
N
S
view from above
-- show that
a square with ordinate MN as its side
will be equal to
a rectangle with RN & NS as its sides

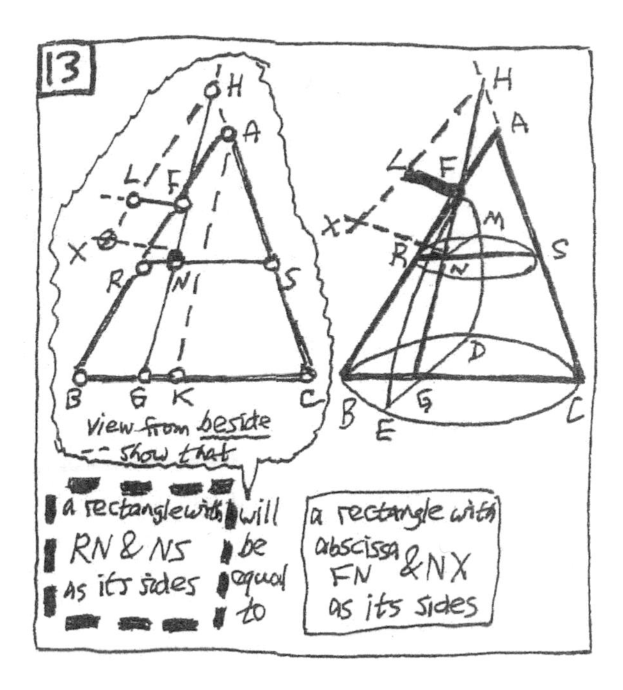
13
H
A
L
F
X
R
N
S
B
G
K
C
view from beside
-- show that
H
A
L
F
X
M
R
N
S
D
B
E
G
C
a rectangle with RN & NS as its sides
will be equal to
a rectangle with abscissa FN & NX as its sides

What we have already—by virtue of our slicing the diameter into the axial triangle, and our drawing a straight line in the axial triangle through the cone's vertex, parallel to the diameter, as well as our slicing a circle whose diameter is parallel to the triangle's base—is the *similarity of triangles*. From that, we shall get the *equality of rectangles*.

We move from what we already have, to what we do not have but want to have, in the way that is schematized in figure 14. We take the ratio of rectangles used to define the upright side, and we do this:

> (**Step I**) The ratio of the square (whose both sides are AK)
> to the rectangle (whose sides are KB and KC)
> is *taken apart* into its two component ratios, which are
> the ratios of their sides (the ratio of AK to KC, and of AK to KB)
>
> Now, taking *the one component* ratio,
> we get same ratios by the similarity of triangles:
> (**step II**) as AK is to KC, so is HG to GC;
> and (**step III**) as HG is to GC, so is HN to NS.
> Thus, the one component ratio (the ratio of AK to KC)
> is the same as the ratio of HN to NS.
>
> Now, taking *the other component* ratio,
> we also get same ratios by the similarity of triangles:
> (**step IV**) as AK is to KB, so is FG to GB;
> and (**step V**) as FG is to GB, so is FN to NR.
> Thus, the other component ratio (the ratio of AK to KB)
> is the same as the ratio of FN to NR.
>
> And now (**step VI**), if we *put together* those two resulting ratios of lines
> (namely, that of HN to NS and that of FN to NR), we get
> the ratio of the rectangles that would be built by using those lines as their sides.
> This is the ratio of the rectangle whose sides are FN and HN
> to the rectangle whose sides are RN and NS.

We do all that as depicted in figure 14. This figure for the hyperbola of Proposition 12 should be compared and contrasted with figure 10 of the previous chapter, which is the corresponding figure for the parabola of Proposition 11.

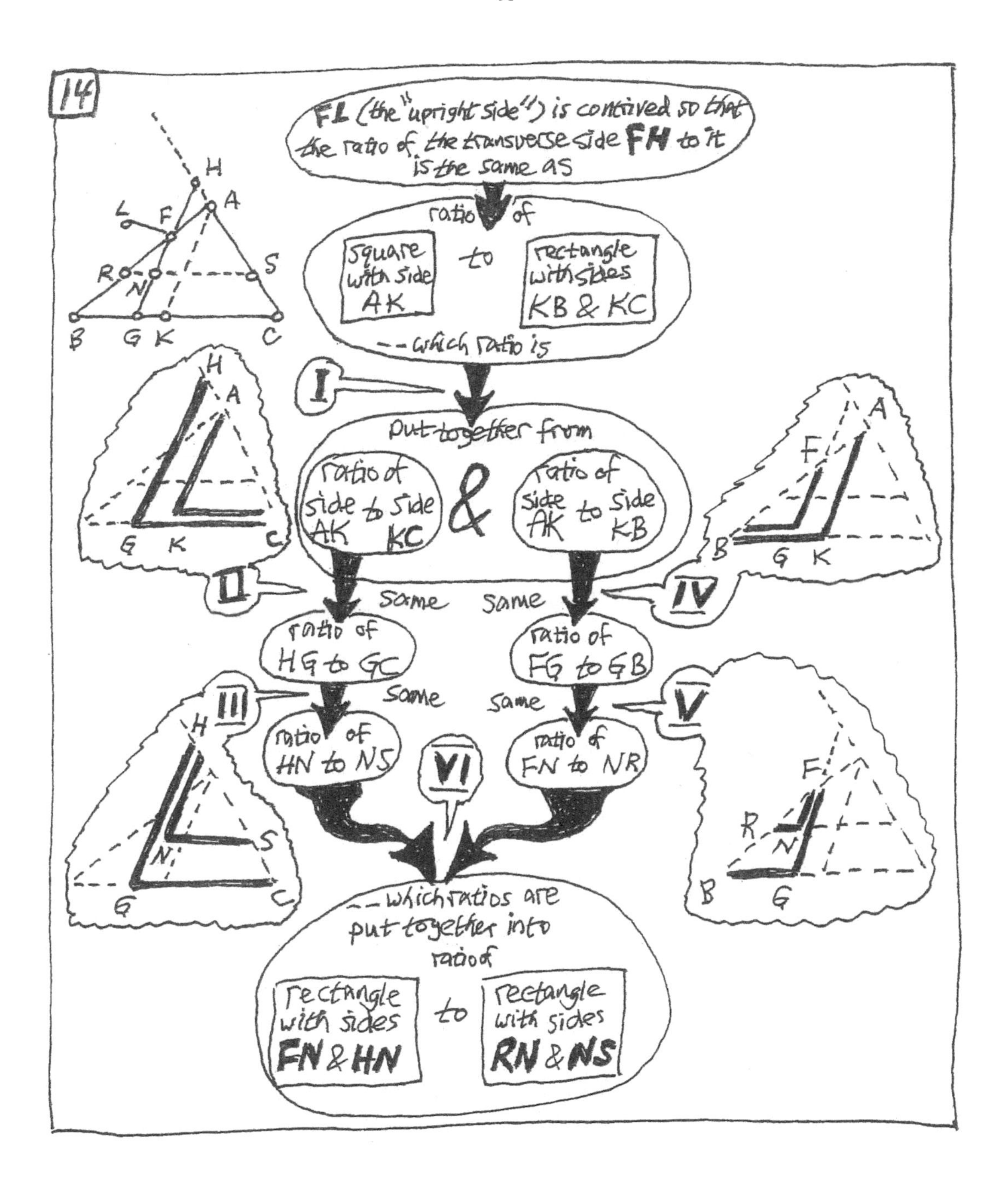
14
FL (the "upright side") is contrived so that the ratio of the transverse side FH to it is the same as
ratio of
square with side AK
to
rectangle with sides KB & KC
-- which ratio is
I
put together from
ratio of side AK to side KC
&
ratio of side AK to side KB
II
same
same
IV
ratio of HG to GC
ratio of FG to GB
III
same
same
V
ratio of HN to NS
VI
ratio of FN to NR
-- which ratios are put together into ratio of
rectangle with sides FN & HN
to
rectangle with sides RN & NS
H
A
L
F
R
N
S
B
G
K
C

Figure 15 presents a map that puts together the several pieces of the hyperbola demonstration of this 12th proposition. It should be compared and contrasted with figure 11 of the previous chapter, which is the corresponding figure for the parabola demonstration of Proposition 11.

In the case of the hyperbola as in the case of the parabola, what enables us to articulate the relationship between the ordinate and the abscissa is a third straight line, one which we make just the right size to embody certain ratios. Although we cannot relate the ordinate and abscissa directly, we can do so indirectly—by first contriving a line (the upright side) into which we build the relevant relationships, and by then building boxes out of those several straight lines (namely, out of the line that is the ordinate, the line that is the abscissa, and the line that is contrived to be the upright side).

But now, in the case of the hyperbola, the relationship is even more indirect than it was in the case of the parabola. Now there is a more complex relationship built into the articulation of the relationship between the ordinate and abscissa. This new complexity is the relationship between the upright side and another side that is not upright, but oblique. In the case of the hyperbola, the equal rectangle is larger than the rectangle whose sides are the abscissa and the line contrived to be the upright side. The form of the rectangular excess is given (as the form of any rectangle is given) by the ratio of its sides. The ratio here is the ratio which another line has to the upright side—this other line is the oblique (transverse) side.

It is not immediately obvious why one ingredient of complexity is introduced—namely, the straight line drawn from the axial triangle's vertex, parallel to the diameter, down to the axial triangle's base. That straight line AK does not seem necessary. Couldn't the job that it does be done instead by straight lines FG and HG, which do not have to be drawn? They are in the picture already, as are the segments (GB and GC) into which the axial triangle's base is split by the common bottom-point (G) of those straight lines which could do the job. They could do it because (since AK is drawn parallel to straight line HFG) the line FG has to GB, and the line HG has to GC, the same ratios as the drawn line AK has, respectively, to KB and to KC. The ratio Apollonius uses to define the hyperbola's upright side—the ratio of square [AK] to rectangle [KB, KC]—is the same as the ratio compounded out of the two ratios of AK to KB, and AK to KC. But these two ratios are the same, respectively, as the two other ratios—of FG to GB, and HG to GC. So, the ratio compounded out of these last two ratios—which is the ratio of rectangle [FG, HG] to rectangle [GB, GC]—could have been used to define the upright side just as well, it would seem, as the ratio that Apollonius used. The ratios are the same even though the lines are different. Why, then, draw AK rather than making do with what is already in the picture—FG and HG? After all, FG and HG do not have to be drawn in and yet are nonetheless as constant as is the line AK. So, the labor of drawing line AK does seem superfluous if we merely want to contrive an upright side that will enable us to demonstrate the square-rectangle property that relates the ordinate and the abscissa of the hyperbola. But since the next proposition, which presents the ellipse, also has this extra line drawn in, let us wait until then to consider further the question of why we draw this extra line. For now, let us ask instead about our contriving of the upright side.

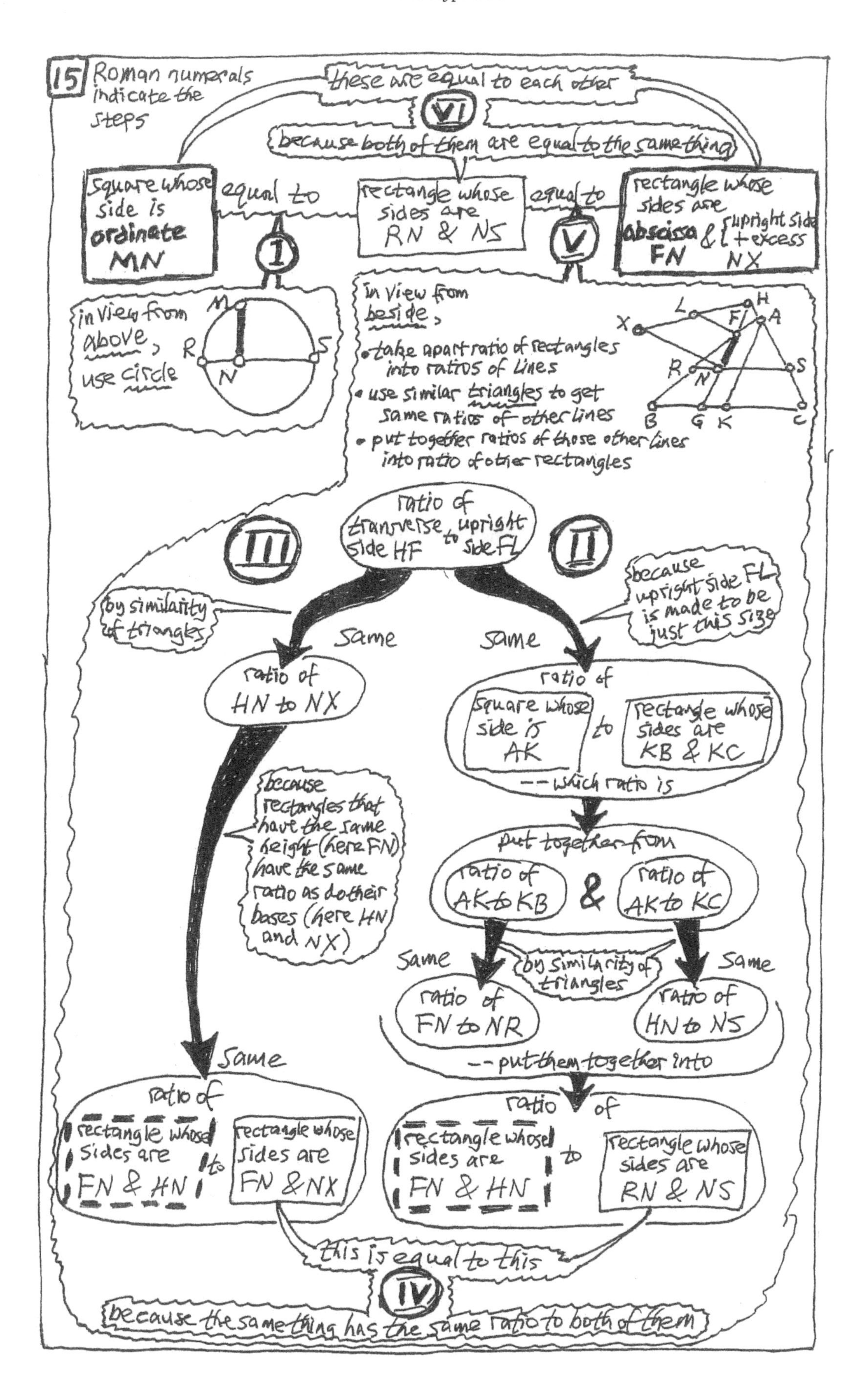
15
Roman numerals indicate the steps
these are equal to each other
VI
because both of them are equal to the same thing
square whose side is ordinate MN
equal to
rectangle whose sides are RN & NS
equal to
rectangle whose sides are abscissa FN & upright side + excess NX
I
V
in view from above, use circle
M R N S
in view from beside,
• take apart ratio of rectangles into ratios of lines
• use similar triangles to get same ratios of other lines
• put together ratios of those other lines into ratio of other rectangles
X L H F A R N S B G K C
ratio of transverse side HF to upright side FL
III
II
by similarity of triangles
because upright side FL is made to be just this size
same
same
ratio of HN to NX
ratio of square whose side is AK to rectangle whose sides are KB & KC
-- which ratio is
put together from
ratio of AK to KB & ratio of AK to KC
because rectangles that have the same height (here FN) have the same ratio as do their bases (here HN and NX)
same
by similarity of triangles
same
ratio of FN to NR
ratio of HN to NS
same
-- put them together into
ratio of
rectangle whose sides are FN & HN to rectangle whose sides are FN & NX
ratio of
rectangle whose sides are FN & HN to rectangle whose sides are RN & NS
this is equal to this
IV
because the same thing has the same ratio to both of them

In an hyperbola as in a parabola, at different lengths cut off along the diameter, the ordinates themselves have different lengths; the upright side, however, has the same length. That is to say: the parameter which we contrive is the same for every one of that diameter's ordinates in that section.

But we could have contrived any number of lines of constant length. Where could we have gotten the notion to contrive *that* one?

What we are interested in doing is articulating the relationship between the size of the ordinate and the size of the abscissa. In the case of this section, the hyperbola, the relationship turns out to be quite a mouthful (fig. 16):

Contrive a line FL of such a size that
the oblique (transverse) side HF will have to it
the same ratio which
a certain square has to a certain rectangle.
The square has as its side a line AK drawn
parallel to the section's diameter,
from the axial triangle's vertex all the way down to its base;
and the rectangle has as its sides
the segments (KB and KC) into which
the axial triangle's base is split by that drawn line AK.

And from the line FL, get a line NX which
goes beyond the contrived upright side FL just far enough so that
the sum of the oblique (transverse) side and the abscissa
will have to it (NX)
the same ratio which the oblique (transverse) side has to the upright side.

If, now, you take a square whose side is the ordinate MN
and, equal to that square, you make a rectangle whose one side is the abscissa FN,
then this rectangle's *other* side must be line NX.

In other words:

a rectangle which has as one of its sides the abscissa
and which is equal to a square whose side is the ordinate,
will be greater than
a rectangle which also has as one of its sides the abscissa
but which has as its other side
(its side standing upright upon the abscissa) the contrived line—

and it will exceed by just so much, that
the rectangular excess which has as one of its sides
a line of the same size as the abscissa
will (except for size) look just like the rectangle whose sides are
the oblique (transverse) side and the upright side.

But we could have contrived any number of lines of constant length. Where could we have gotten the notion to contrive *that* one?

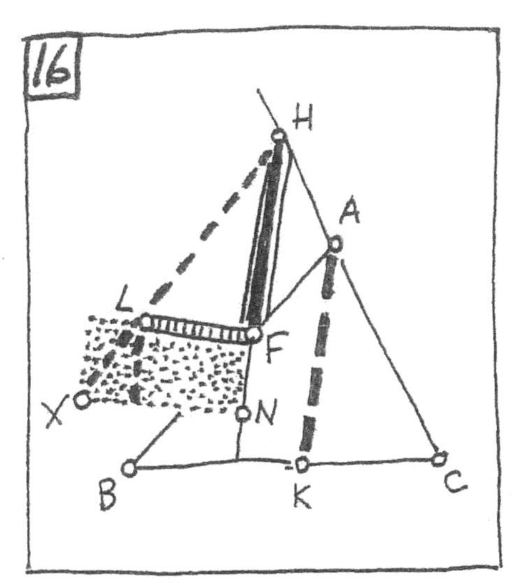

As before with the parabola, so also here with the hyperbola, we find a constant relationship between the changing abscissa and the changing ordinate. We relate these changing lines indirectly, by forming boxes out of them together with a constant line (the upright side).

In both the parabola and the hyperbola, the relationship is this: the square on the ordinate is equal to a rectangle having the abscissa as one of its sides.

In the parabola, the relationship is not very complicated, since the other side of the equal rectangle is the constant upright side. But in the hyperbola, the relationship is more complicated, since the other side of the equal rectangle is a width that changes, and so cannot be *equal to* any constant upright side. It is, however, *determined by* a constant upright side.

That line, if we do contrive it, will do the work of the demonstration. But what work will show *why* we should have contrived it that way in the first place? We have seen what comes from the contriving of the upright side. We would like to see what the contriving of the upright side itself comes from. That is the work of analysis. We can proceed somewhat as we did with the parabola—but the analysis for the hyperbola is longer and more complicated.

analysis to find contrivance for hyperbola

As we move down along the conic section's diameter from its vertex F, what stays the same and what becomes different at different points on the diameter? Let us consider things at some point—call it N.

Let us take the *top* view first (fig. 17). It shows us a circle. Even though the circle's diameter RS grows as we move down along the section's diameter, the square on the ordinate MN stays equal to the rectangle contained by RN and NS. The square changes size as we move down along the section's diameter, but its size is always the same as that of the rectangle.

Now let us take the *side* view (fig. 18). As we go down along the conic section's diameter, cutting off a longer and longer abscissa FN, and getting a bigger and bigger diameter RNS for the circle, what happens to that rectangle with sides RN and NS which is equal to the square on ordinate MN? That rectangle grows. As the rectangle grows, what happens to its sides (RN and NS)? Whereas earlier, in the parabola's rectangle, only one of the sides grew, now, by contrast, both sides do. Both RN and NS grow—but something stays the same about each one that grows. What stays the same about the side RN is its ratio to the growing abscissa (FN): the ratio of RN to FN stays the same because (since RNS stays parallel to the axial triangle's base BC) the little triangle RNF (on the left) keeps the same shape while RN descends. And what stays the same about the side NS is its ratio to the growing line HN (which is composed by adding the growing abscissa FN to the constant oblique [transverse] side FH): the ratio of NS to HN stays the same because (again, since RNS stays parallel to the axial triangle's base BC) the little triangle SNH (on the right) keeps the same shape while NS descends.

Thus, the ratio of FN to NR, and the ratio of HN to NS, are both constant. Let us put these two constant ratios together, so that we shall have only one ratio to consider. If these two ratios of straight lines are compounded, they will yield a ratio that is the same as a ratio of rectangles made by taking those straight lines as sides—namely, the ratio of rectangle [NH, NF] to rectangle [RN, NS]. This ratio, since it is compounded out of constant ratios, will itself be constant. Let us see how this can help us.

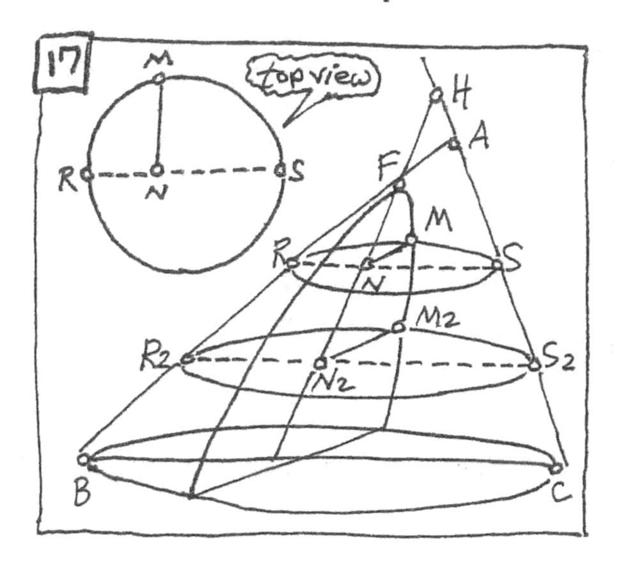

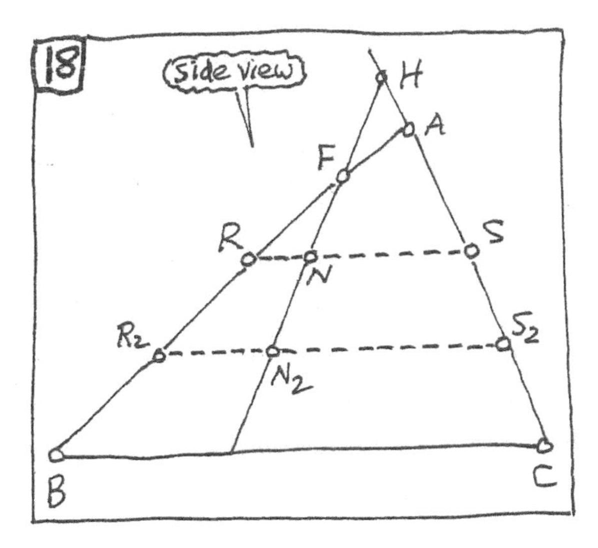

From this constancy of the ratio of rectangle [NH, NF] to rectangle [RN, NS], what we now know is this—that although the ordinate MN grows as the abscissa FN grows, nonetheless a certain relationship between them stays the same. The relationship is this: the rectangle with sides RN and NS, which is equal to the square whose side is the growing *ordinate*, will have a constant ratio to a rectangle with the following sides—one side (NH) has only a part of it constant (this constant part being the oblique [transverse] side FH), but the abscissa is its growing part, and this growing *abscissa* (NF) is the other side.

So, now (fig. 19) we have the constancy of the ratio of the rectangle with sides NH and NF to the rectangle with the sides RN and NS. The latter term in this ratio is the rectangle contained by the segments (RN and NS) into which the circle's diameter is split by the ordinate. The other term is the rectangle contained by the abscissa (NF) and the sum (NH) of the oblique (transverse) side and the abscissa. That is to say: we have the rectangle which concerns us (the rectangle with sides RN and NS) in a constant ratio to another rectangle, *one* of whose sides is the line which concerns us (abscissa FN), which line is also the non-constant part of this rectangle's *other* side (NH).

But this ratio is not what we are after. It is only a step on the way. What we *have* is a rectangle that has a *constant ratio* to the rectangle with sides RN and NS, but what we would *like* to have is a rectangle that is *equal* to the rectangle with sides RN and NS, because then it would be equal to the square that has the ordinate as its side. Moreover, we want that equal rectangle (like the rectangle of constant ratio which we now have) to have the abscissa FN as one of its own sides. If we had the rectangle that we want, then we would have the ordinate and abscissa in relation to each other—not directly, to be sure, but at least through their power to make a square or some other rectangle.

What do we need in order to have what we want? We already have the constancy of the ratio of the rectangle with sides NH and NF to the rectangle with sides RN and NS. We want to have a rectangle which is equal to the second rectangle in the ratio, and which has as *one* of its own sides one of the sides of the first rectangle in the ratio—namely, the abscissa NF. If we had the *other* side, then we would have that rectangle itself—a rectangle whose side is equal to the abscissa while the rectangle itself is equal to a square whose side is equal to the ordinate (fig. 20). Our problem, then, is this: to find out what should be that rectangle's other side.

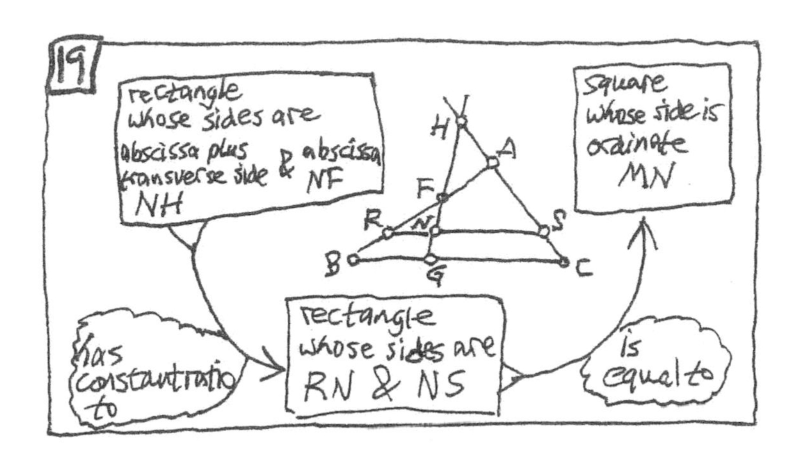
19
rectangle whose sides are abscissa plus transverse side NH & abscissa NF
square whose side is ordinate MN
H
A
F
R
N
S
B
G
C
has constant ratio to
rectangle whose sides are RN & NS
is equal to

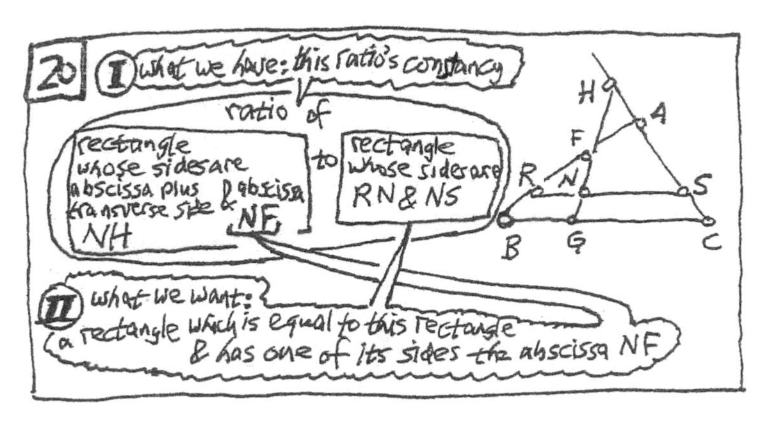
20
I what we have: this ratio's constancy
ratio of
rectangle whose sides are abscissa plus transverse side NH & abscissa NF
to
rectangle whose sides are RN & NS
H
A
F
R
N
S
B
G
C
II what we want:
a rectangle which is equal to this rectangle
& has one of its sides the abscissa NF

Let us call that unknown other side "width (w)"—and, even though we do not have it, let us *suppose* that we do. That is, let us ask (step I): *what if* we *did* have it? (Fig. 21.)

Now (step II) two magnitudes can be equal to each other only if a third magnitude has the same ratio to each of them that it has to the other. So if we *did* have that unknown side, width (w), then any rectangle (any one at all) would have, to the rectangle that the unknown width (w) makes with the abscissa—to the rectangle, namely, with the sides FN and (w)—the same ratio that it has to its *equal* (to the rectangle, namely, with sides RN and NS). And we already know something about the ratio, to this rectangle with sides RN and NS, of one rectangle in particular; we know the ratio to it of the rectangle, namely, with sides FN and HN. Hence, we know that the rectangle with sides RN and NS can be equal to the rectangle with sides FN and (w) only if the rectangle with sides FN and HN has the same ratio to each of them.

Now (step III), taking the ratio which involves the unknown line, width (w), we see that the two rectangles—one whose sides are FN and HN, and the other whose sides are FN and (w)—both have the same height FN; and so, the ratio of the one rectangle to the other is the same as the ratio of their bases, which is the ratio of line HN to width (w).

Therefore (step IV), although we do not have the unknown line, width (w), we do know its relation to things we do have—we know that *if* we had it, then the ratio to it of HN (which is a line with a constant part) would be the same as the ratio of the rectangle with sides FN and HN to the rectangle with sides RN and NS, which is a ratio that is constant.

But although this ratio is itself constant, it is *not* stated entirely in terms of *lines* that are constant, and so we have not yet determined the unknown line, width (w).

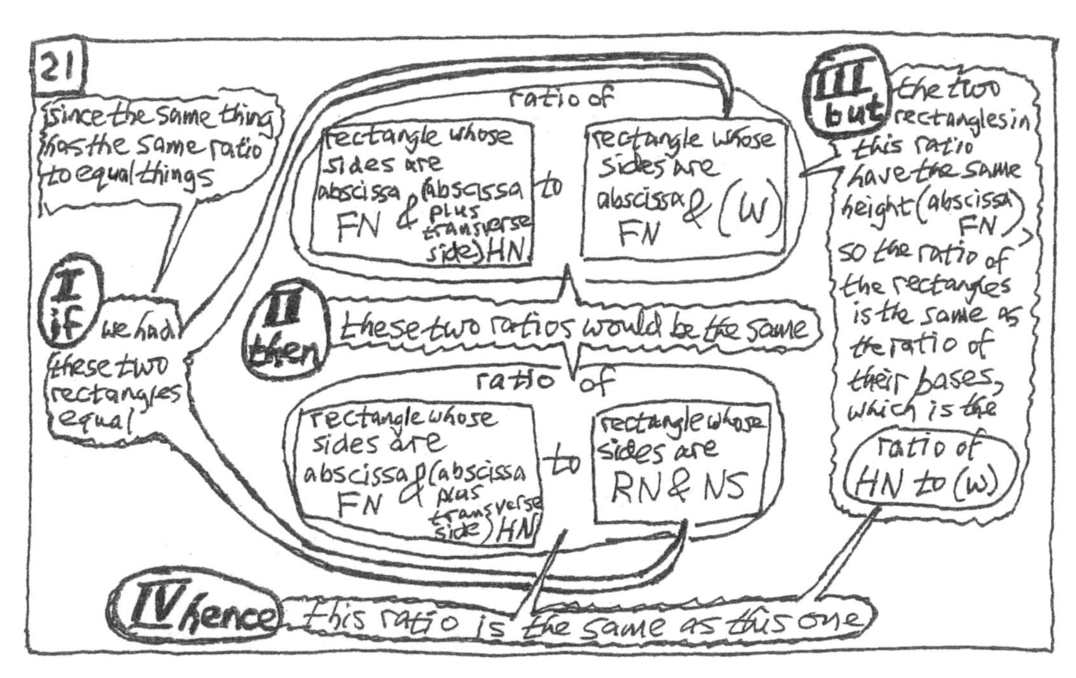

We need to find a ratio which involves only lines that are constant but which is the same as that ratio which we do have. Of the four lines involved in the ratio which we do have (the ratio of the rectangle whose sides are FN and HN to the rectangle whose sides are RN and NS), not one of them is constant.

To find the ratio that we need (fig. 22), we (step I) take apart the ratio that we have. We take apart the single ratio of rectangles into its two component ratios of lines. Since the ratio of the rectangles is the same as the ratio compounded out of the ratios of their sides, the ratio of the rectangles is the same as the ratio compounded out of two ratios—those of FN to RN and of HN to NS, which, by the similarity of triangles, are the same as the ratios of FG to GB and of HG to GC.

Now (step II), we use similar triangles whose sides will be constant lines. And we draw a straight line parallel to the diameter, from the vertex A, down to the axial triangle's base, so that this line AK splits the base into segments KB and KC. And so, by the similarity of triangles, those same ratios that we got in the first step we get now in terms of constant lines, as ratios of AK to KB and of AK to KC.

And now (step III), we put the component ratios back together into one ratio of rectangles, which will be the same *ratio* that we had before, but now with different *terms*, terms now involving none but *constant* lines; this is the ratio that the square whose side is AK has to the rectangle whose sides are KB and KC.

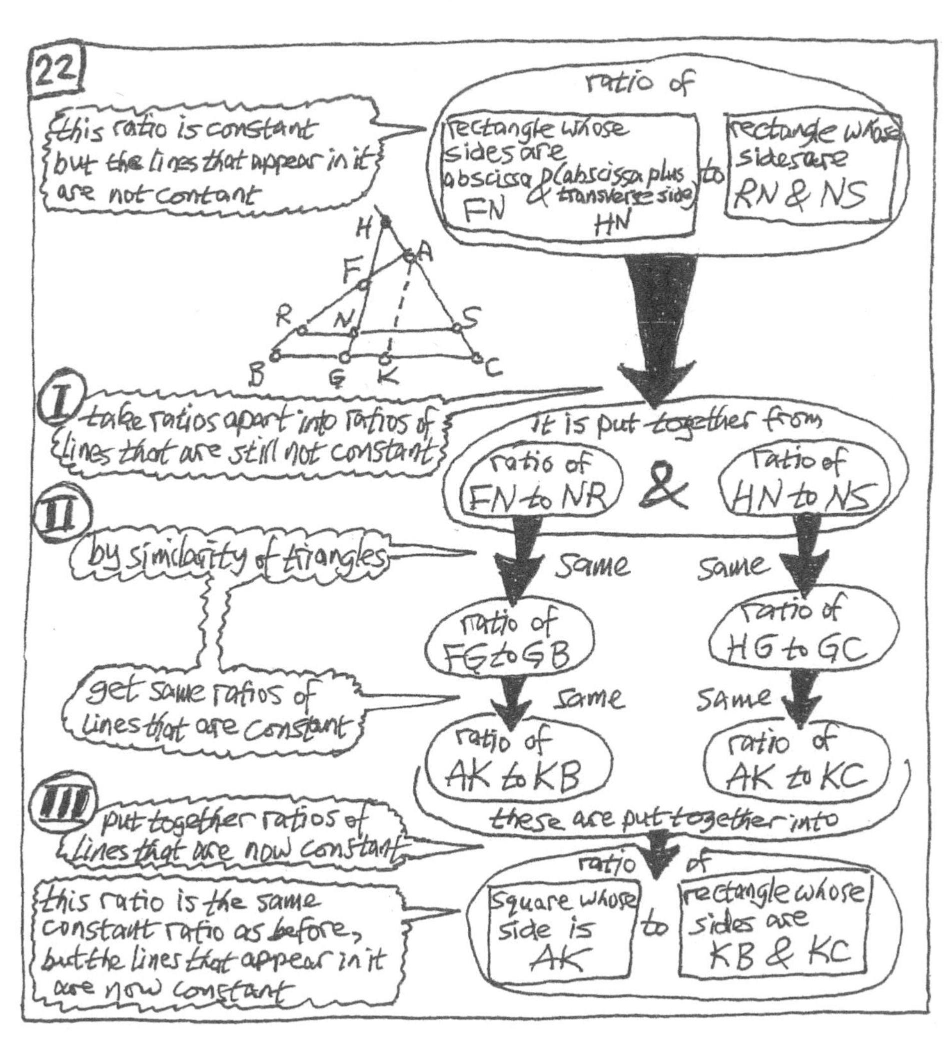

Now we are ready to do the requisite contriving. Figure 23 shows why. We have seen (step I) that if we had the width (w) then the ratio of the line HN to it would be the same as the ratio of the rectangle whose sides are NH and NF to the rectangle whose sides are RN and NS, and (step II) that this ratio of rectangles is the same as the ratio of the square whose side is AK to the rectangle whose sides are KB and KC, and so (step III) that if we had the width (w) then the ratio of line HN to it would be the same as this last ratio.

We therefore now contrive a line (call it NX) having just the right size to be the width (w). It (fig. 24) does so by having just the right size to make two ratios the same. One of them is the ratio, to this contrived line NX, of the line HN (this HN being the line that stretches from the abscissa's non-vertex endpoint to the oblique [transverse] side's non-vertex endpoint — which is to say: HN is the sum of the oblique [transverse] side and the abscissa). That ratio of HN to NX is (by our contriving) the same as a ratio set up by our drawing the line AK— namely, the ratio of the square arising from AK to the rectangle contained by the segments (KB and KC) into which the drawn line splits the axial triangle's base. This ratio of square to rectangle involves only constant lines, but it is the same ratio as that original constant ratio which involved lines not all of which were constant.

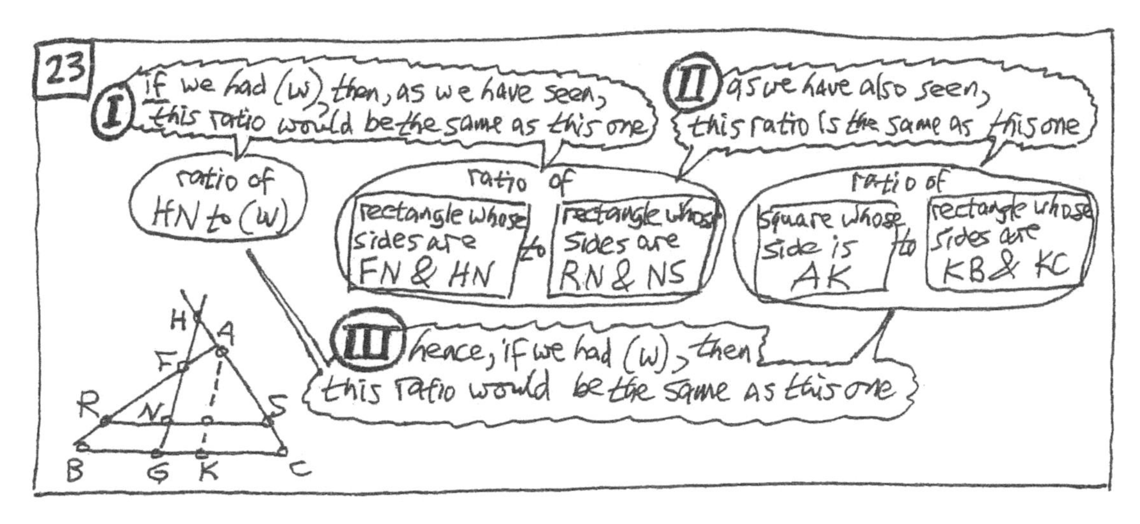

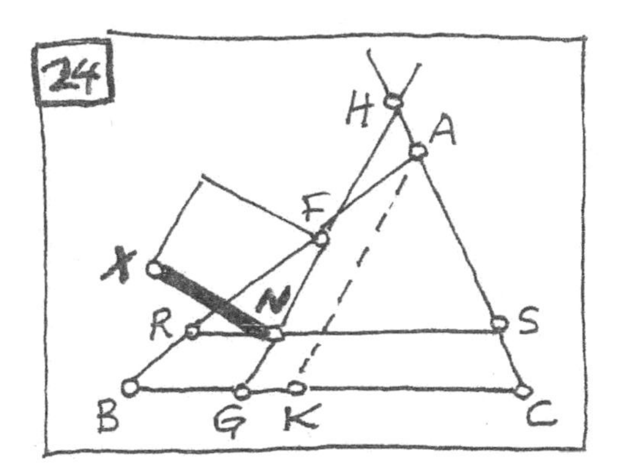

If we do contrive in that way, then we should get equality between the two rectangles—the one having as sides RN and NS, and the other having as sides the abscissa FN and the contrived line NX. This line NX which we have contrived is the width (w) which will make, with the abscissa, a rectangle equal to the square whose side is the ordinate. So, by means of our contrivance (line NX), we have related the abscissa and the ordinate.

Our analytic procedure so far for the hyperbola has been the same as it was for the parabola, as will be evident from comparing the figures that we have just seen for the analysis of the hyperbola with the figures that we saw for the analysis of the parabola in the previous chapter. The contrived line NX does now for the hyperbola the job that the upright side did earlier for the parabola.

But this line NX—by contrast with the parabola's upright side—does not have a size that is constant. This line NX which we have contrived for the hyperbola cannot be constant because the line HN, which has a ratio to it that is constant, does not itself have a size that is constant.

This line HN does, however, have a constant part, the oblique (transverse) side (HF), as well as having the abscissa (FN) as its growing part. Can we get some constant line that will, together with the constant oblique (transverse) side (HF), determine the growing width (NX) that is associated with the growing abscissa (FN)? If we can, then we shall have related the growing ordinate and the growing abscissa, indirectly, by means of a line that is constant. Let us therefore continue the analysis.

The equal rectangle's width (NX) is different for every different height (abscissa FN). (Fig. 25.) But while both those lines—height and width—grow, a certain ratio stays the same. Consider line HN, which is the sum of the oblique (transverse) side HF and the abscissa FN (the abscissa being the equal rectangle's height). This line HN has a constant ratio to the rectangle's width NX (fig 26). While both these lines (HN and width NX) grow as height FN grows, the ratio of the one to the other (the ratio of HN to NX) stays the same; this is always the same ratio that the constant oblique (transverse) side HF has to another constant line—call it FL (fig. 27). Thus the constant rectangle which the oblique (transverse) side HF makes with the constant line FL gives the constant shape of the rectangle by which the equal rectangle exceeds the rectangle which the abscissa FN makes with FL (fig. 28). If a rectangle is made with the abscissa FN as height and with this constant line FL as width, then the equal rectangle—whose height is the abscissa FN, and whose width (w) is NX—will exceed it by a certain amount. This excess is equal to a rectangle whose height is also the abscissa FN, but whose own width is just the right size to give it the same shape as a rectangle whose sides are the constant oblique (transverse) side HF and that constant line FL.

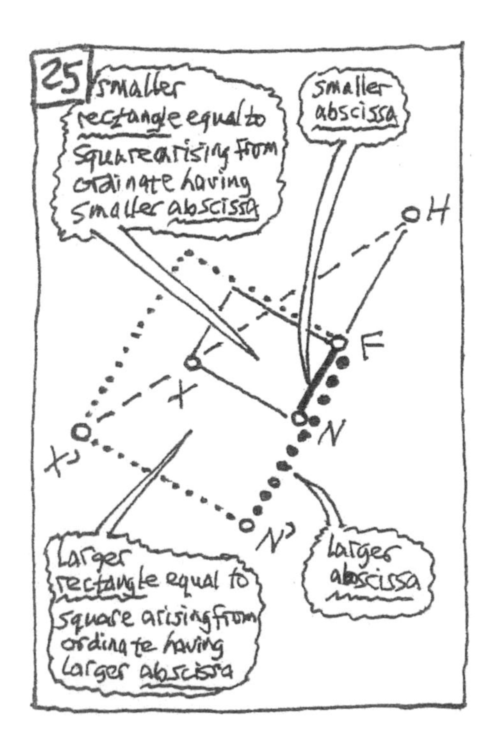

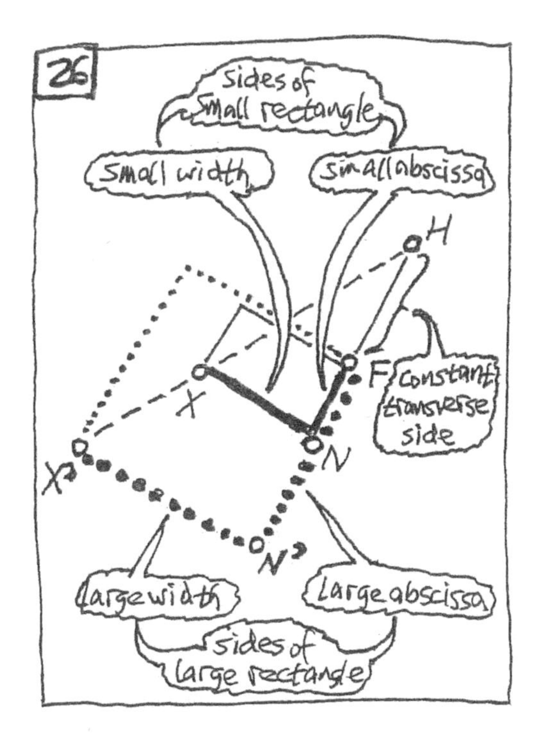

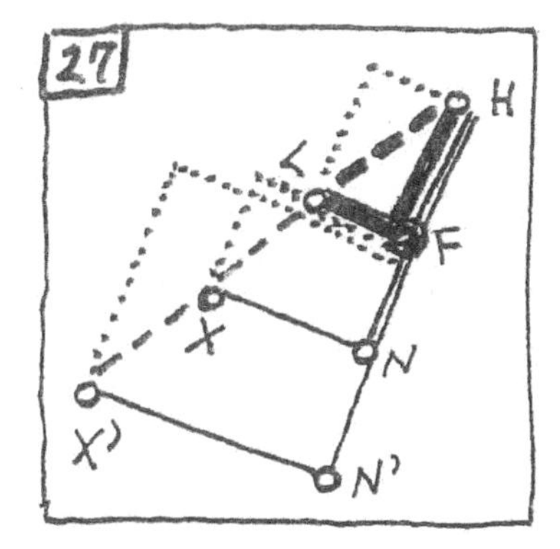

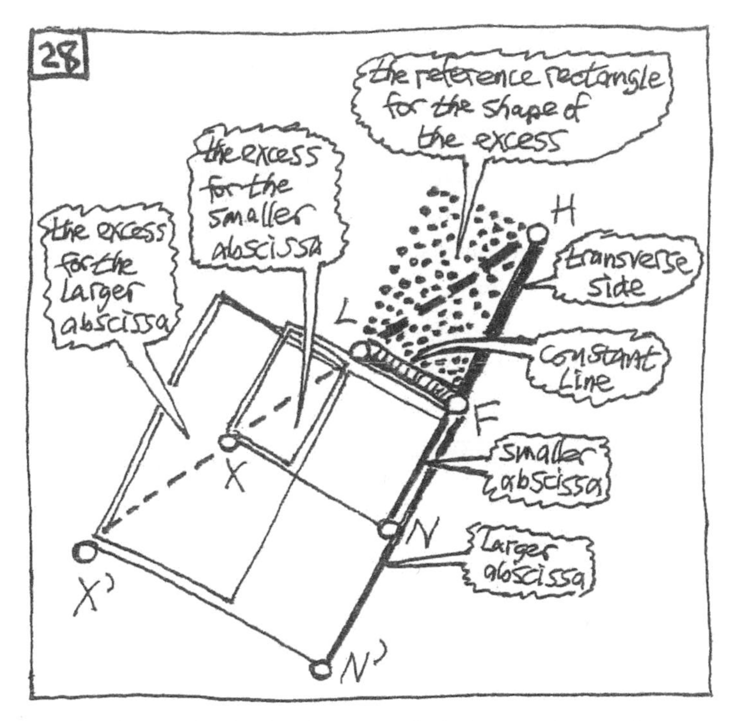

This rectangle to which the rectangular excess is always similar has its one side in line with the diameter; its other side is a line which sticks upright from the diameter. This other side (FL) will be called the *upright side* of this conic section. Earlier, with the parabola, the upright side was a constant line which stuck upright from the diameter and which gave us the width of the equal rectangle. Now, with the hyperbola, the upright side is also a constant line which sticks upright from the diameter and which gives us the width of the equal rectangle, but it gives it in a more complicated way than with the earlier section. Things are more complicated because the equal rectangle (equal to the square arising from the hyperbola's ordinate) now has a width that always, by some definite amount, overshoots the upright side.

So, this is what we do to get a constant upright side for determining the width NX of the rectangle that is equal to the square on the ordinate for the abscissa FN:

we contrive the upright side FL so that it will have,
to the constant oblique (transverse) side HF,
the following ratio—

the constant ratio which
the equal rectangle's growing width NX
has to the growing line HN
(HN being the sum of
the oblique [transverse] side HF and
the equal rectangle's growing height, abscissa FN).

Let us consider where we are, now that we have a proportion for contriving the upright side—namely, this: oblique (transverse) side HF will have to it (the upright side) the same ratio that square [AK] has to rectangle [KB, KC]. Not only are the ratios identified in this contriving proportion both constant, but the lines mentioned in the ratios are all constant. The ratios are the same as each other because each of them is the same as another ratio—namely, that of HN to NX. Although this ratio of HN to NX was useful for our purposes because it is a constant ratio, it was a mere step on the way—because its terms (HN and NX) are not constant. We have used the ratio of HN to NX as a step on the way to finding a proportion that defines a constant line that will serve as the upright side.

This upright side is *not, itself,* the other side of that rectangle which is equal to the square on the ordinate and has the abscissa as one side. This upright side is constant, whereas that rectangle's other side is not. The upright side is a line that *determines* the rectangle's other side for us. The rectangle's other side is *greater* than the upright side. The *rectangular excess* varies in size, but it has a *constant look* about it. What we can say now is the following.

If there is some width (w) which would make, with the abscissa as height,
a rectangle equal to the square whose side is the ordinate,
then that line (w) would be such that:

it would be greater than the constant upright side (u) which is
determined by the following proportion—namely,
as square [AK] is to rectangle [KB, KC],
so also the oblique (transverse) side is to the upright side (u);

and the excess of the width (w) with respect to the upright side (u)
would be determined by the following proportion—
as the oblique (transverse) side is to the upright side (u),
so also the sum of the oblique (transverse) side and the abscissa is to the width (w).
(Fig. 29.)

But can we be sure that a line contrived that way *will in fact* do the work for which we want it? Only if that is *demonstrable*. We must contrive the upright side *first* and then show that we can retrace the steps of the analysis in reverse. That is to say, our analysis cannot substitute for the synthesis in which Apollonius demonstrates the following:

if we cut with a plane properly oriented,
and we place upright a line which we contrive to be of such a size that
the oblique (transverse) side has the same ratio to it that
square [AK] has to rectangle [KB, KC],
then from this cutting and contriving, it will follow that

a square whose side is the ordinate
will be greater than
a rectangle whose sides are the abscissa and that line contrived as upright side,

and the excess will be equal to
a rectangle whose sides are the abscissa and a line to which the abscissa has
the same ratio that the oblique (transverse) side has to the upright side—
so that this rectangular excess has the same shape as a rectangle whose sides are
the oblique (transverse) side and the upright side.

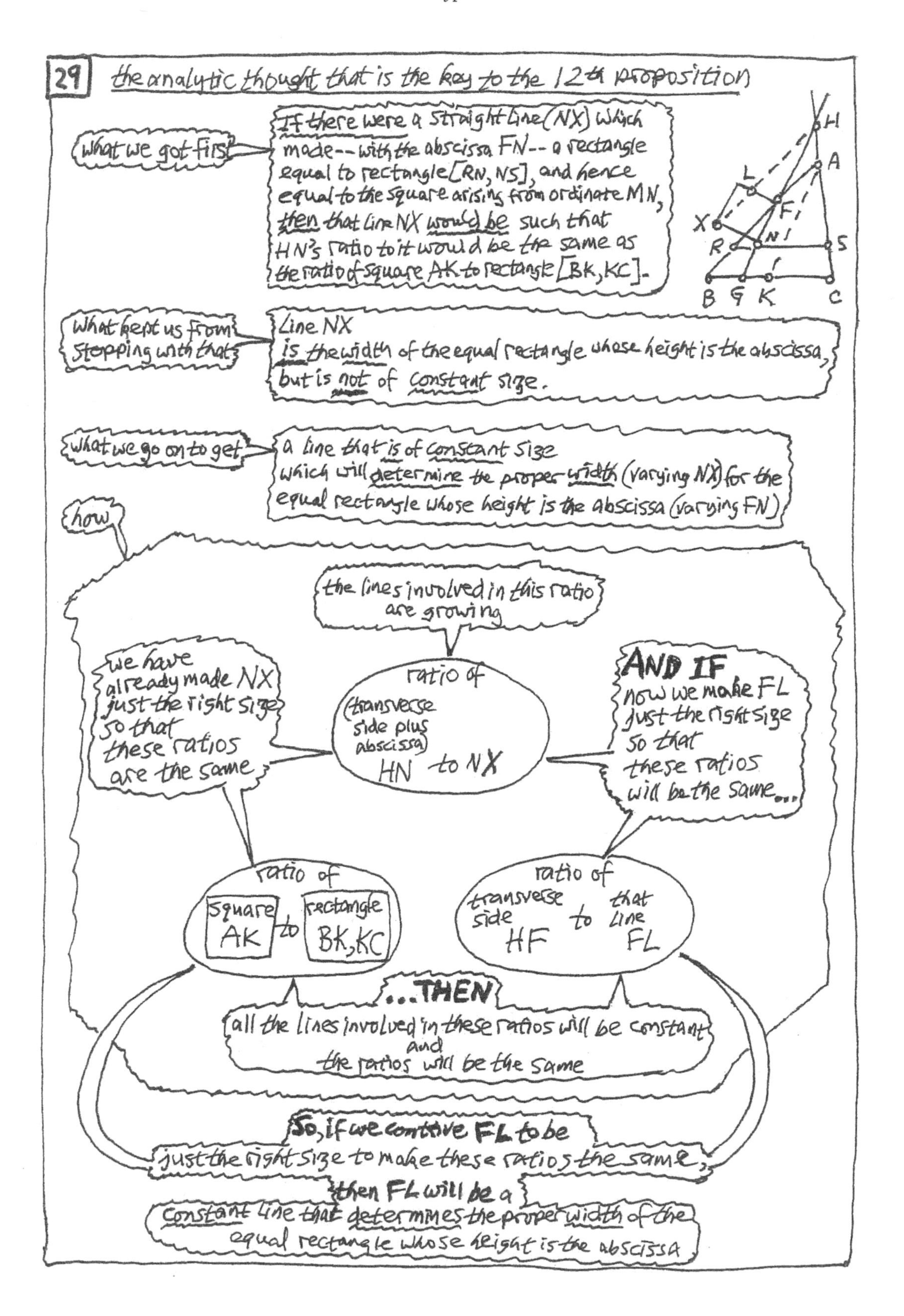
29
the analytic thought that is the key to the 12th proposition
what we got first
If there were a straight line (NX) which made-- with the abscissa FN-- a rectangle equal to rectangle [RN, NS], and hence equal to the square arising from ordinate MN, then that line NX would be such that HN's ratio to it would be the same as the ratio of square AK to rectangle [BK, KC].
H
A
L
F
X
R
N
S
B
G
K
C
What kept us from stopping with that
Line NX is the width of the equal rectangle whose height is the abscissa, but is not of constant size.
what we go on to get
a line that is of constant size which will determine the proper width (varying NX) for the equal rectangle whose height is the abscissa (varying FN)
how
the lines involved in this ratio are growing
we have already made NX just the right size so that these ratios are the same
ratio of (transverse side plus abscissa) HN to NX
AND IF now we make FL just the right size so that these ratios will be the same...
ratio of square AK to rectangle BK, KC
ratio of transverse side HF to that line FL
...THEN
all the lines involved in these ratios will be constant and the ratios will be the same
So, if we contrive FL to be
just the right size to make these ratios the same,
then FL will be a
constant line that determines the proper width of the equal rectangle whose height is the abscissa

Analytic inquiry cannot substitute for synthetic demonstration, but it can lead us towards a demonstration, or can clarify a demonstration that has been provided.

Why should we have even begun looking for a straight line that, with this curve's abscissa, will contain a rectangle equal to the square whose side is the curve's ordinate? Because we wanted to relate the abscissa and the ordinate, and it seemed from the previous proposition (on the parabola) that we could not relate them directly.

The simplest way to relate two straight lines if we cannot do so directly is to speak of their power to give rise to squares or to contain rectangles. That was how things had turned out in the parabola—when, that is to say, we cut in the only way which stood out as unique among the ways of getting a section which would not be straight or circular: when, in other words, we cut parallel to the axial triangle's side. The result for the parabola was the simplest way of all ways to get a rectangle having the abscissa as one side—namely, so that its other side will be a line of constant size.

It is not clear at the beginning of the 12th proposition whether this proposition's way to cut a cone will or will not yield a parabola again. The analysis therefore cannot assume that the other side of the rectangle will be a constant line. This other side turns out to be not itself a line that is constant, but nonetheless a line that can be related to a line that is constant. And although it cannot have a constant ratio to that constant line (since it is itself not constant), it does have a constant ratio to a line that is partly constant. As the oblique (transverse) side is to the upright side (namely, in a certain constant ratio), so the varying but partly constant sum of the oblique (transverse) side and the abscissa is to the varying side of the rectangle (fig. 30).

The constancy of the shape of the excess—its unvarying form—is an image of this constancy of the ratio which the varying side of the equal rectangle has to the varying but partly constant sum of the constant oblique (transverse) side and the growing abscissa.

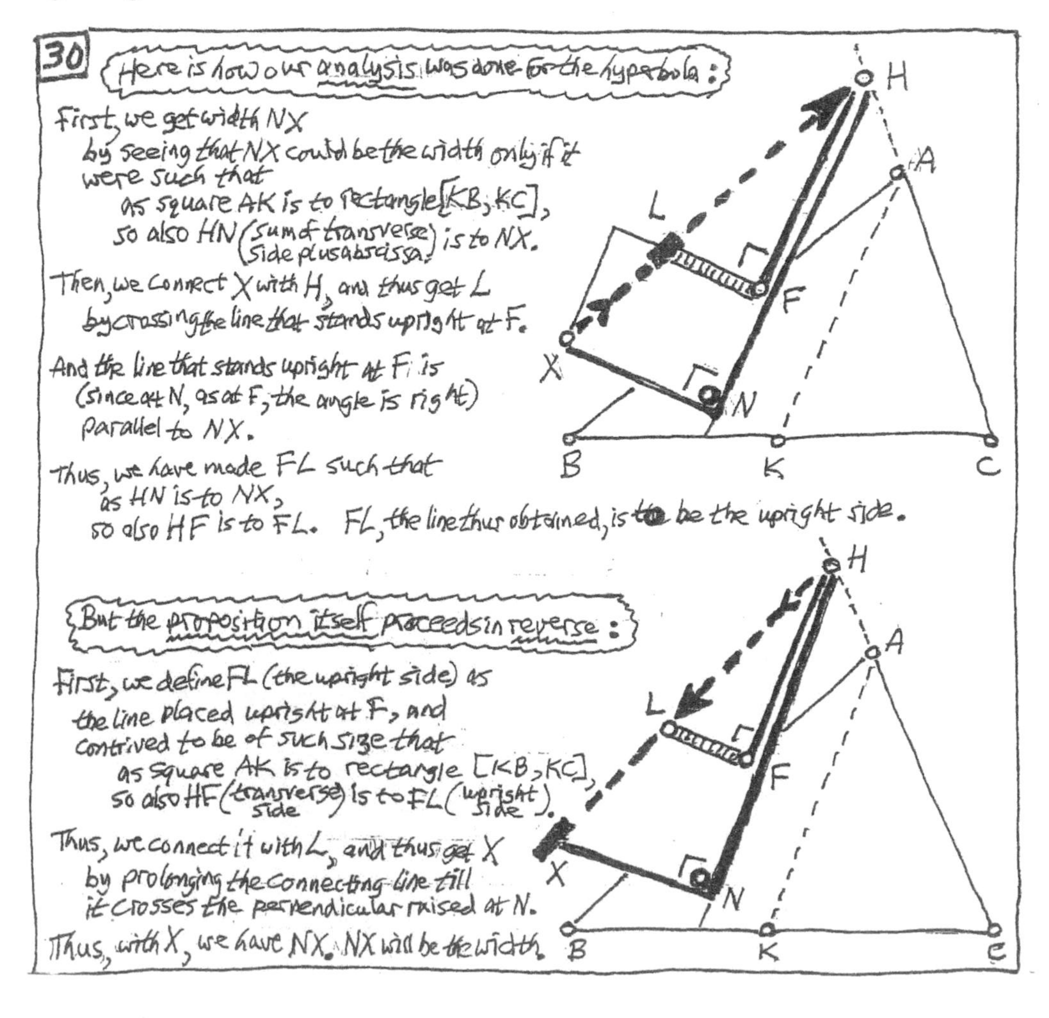

With the hyperbola as with the parabola, we look for the width of that rectangle which, having as its length the abscissa, is equal to the square on the ordinate. In the parabola, the equal rectangle's width is constant, so we contrive the upright side to be that rectangle's width itself; but in the hyperbola, the equal rectangle's width is not constant, so we do not stop there, but instead go on to contrive another line to be a constant upright side that will give us the equal rectangle's growing width.

The look of the rectangular excess is an image of the ratio of certain straight lines. The sameness of visible shape is an image of something unseeable but knowable that gives form to the rectangular excess and informs our seeing and our knowing.

In the enunciation of this proposition, Apollonius says that the equal area exceeds by an *eidos* which is similar to and similarly situated to the rectangle contained by two straight lines: along one of them lies the rectangle equal to the square on the ordinate, and the other one subtends the exterior angle of the axial triangle. At the end of the demonstration, he calls the one line "upright," and the other "oblique."

Eidos is Greek for the "look" or "form" of something; it comes from the same root that gives the Greek words for seeing and for knowing, which are cognate with the English words (derived from Latin) "vision" and "visible." It is the word used by Plato and his followers to refer to the unseen forms received by our intellects in knowing. *Eidê* is the plural. The *eidê* are the "looks" of things at which not our eye, but our "mind's eye" has, so to speak, looked when we know. They inform our knowing. Coming back to our text, we might say that here, in the upright and the oblique sides taken together, is the *eidos* which informs the various *eidê* (fig. 31).

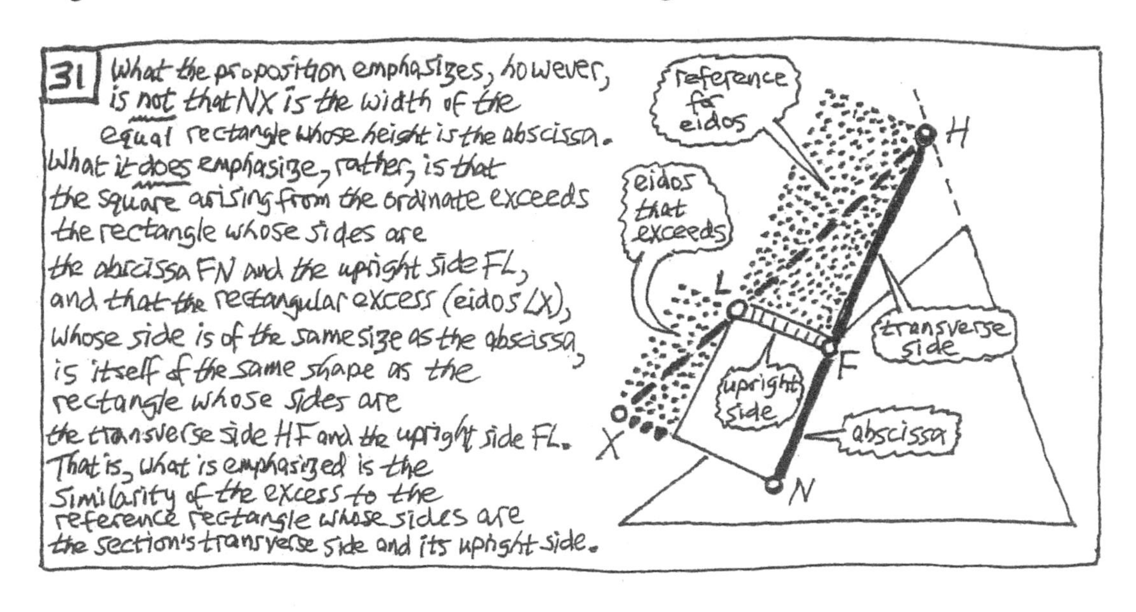

the hyperbola and the parabola

Now, before going on, let us compare the gist of these propositions, the 11th and 12th, in which Apollonius characterizes the parabola and the hyperbola. What is the point of each one's enunciation and demonstration, and where does the definition of each one's upright side come from?

The enunciations convey the following. In both the parabola and the hyperbola, any abscissa and its ordinate are related by making boxes: a certain rectangle that has the abscissa as its height is equal to a square that has the ordinate as its side. The parabola and the hyperbola differ with respect to that rectangle's width. For the parabola, the width is the constant upright side; for the hyperbola, however, the width is not constant, but it is determined by the constant upright side. The hyperbola's constant upright side determines the rectangle's varying width in the following way. The rectangle that is equal to the square on the ordinate exceeds the rectangle whose height is the abscissa and whose width is the upright side. The rectangular excess has a constant shape. Since the shape of any rectangle is constituted by the ratio of its sides, the shape of the rectangular excess is constituted by the ratio of the abscissa to the excess in width. This ratio is the same as the ratio of the oblique (transverse) side to the upright side. That is because both of those ratios are the same as another ratio—namely, the ratio of the sum of the oblique (transverse) side and the upright side to the equal rectangle's width. Thus, the shape of the rectangular excess is the same as that of a reference rectangle whose sides are the oblique (transverse) side and the upright side.

And here is the point of the demonstrations—

parabola (fig. 32):	*hyperbola* (fig. 33):
Sqr O and rct [A, (u)] are equal to	Sqr O and rct [A, (w)] are equal to
each other, because both are equal to	each other, because both are equal to
the same thing, namely, rct [L, R].	the same thing, namely, rct [L, R].
Why the equality with the square?	Why the equality with the square?
The top-view sameness of the	The top-view sameness of the
ratios of L to O and of O to R.	ratios of L to O and of O to R.
Why the equality with the rectangle?	Why the equality with the rectangle?
The side-view constancy of the	The side-view constancy of the
ratios of L to A and of R to F,	ratios of L to A and of R to (T+A),
which makes constant also the ratio	which makes constant also the ratio
compounded from them;	compounded from them;
and, since this ratio	and, since this ratio
of rct [L, R] to rct [A, F]	of rct [L, R] to rct [A, (T+A)]
will become bigger or smaller	will become bigger or smaller
if we replace F	if we replace (T+A)
by a line that is smaller or bigger,	by a line that is smaller or bigger,
a ratio of equality will result from	and a ratio of equality will result from
replacing that constant line F by a	replacing that varying line (T+A) by a
constant line (u) of just the right size.	varying line (w) of just the right size;
	and the ratio of (T+A) to (w),
	will be the ratio also
	of T to some constant line (u),
	and also of A to (w-u).

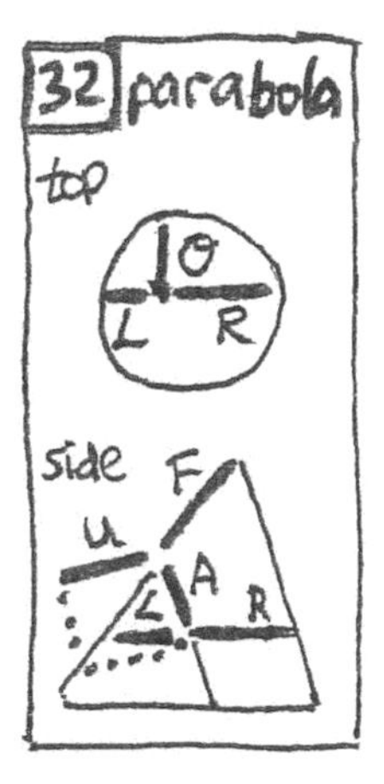

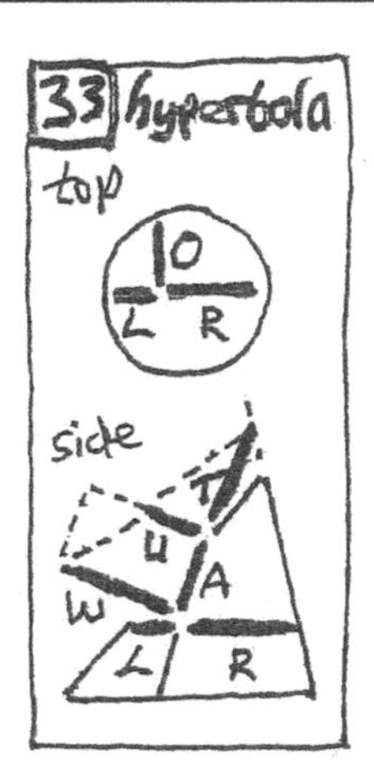

By looking at the gist of the demonstrations, we see that behind the display of squares and other rectangles there are ratios—that what matters is the sameness and the constancy of ratios. The upright side (u) is contrived by building into it the ratios needed for the display. The heart of the matter is the insight into ratios that makes possible the work of the upright side. In the propositions, this insight is found right in the middle of the demonstration, where its central importance is obscured by its central location. There is so much said before and after it that great effort is necessary in order to see it as the heart of the matter.

The point of the definition of the upright side for the parabola is shown in figure 34. For the hyperbola, the story is similar but more complicated; the point of it is shown in figure 35.

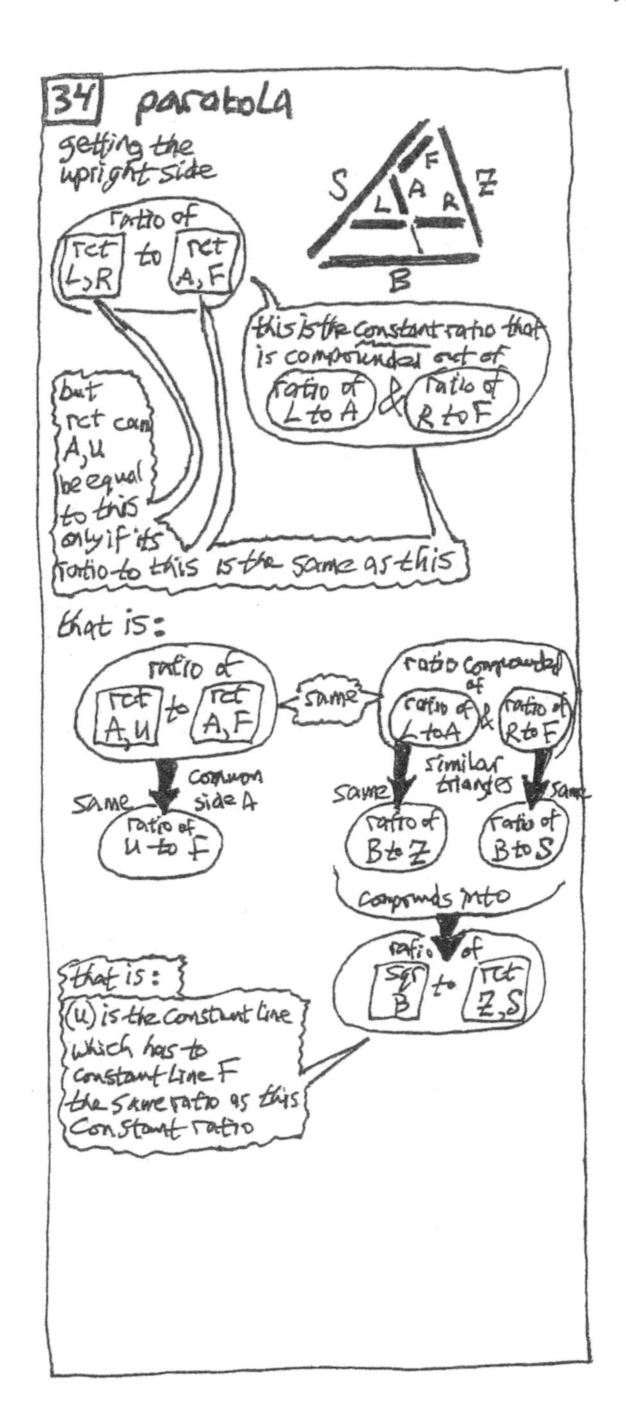
34 parabola
getting the upright side
S L A F R Z B
ratio of rct L, R to rct A, F
this is the constant ratio that is compounded out of ratio of L to A & ratio of R to F
but rct can A, U be equal to this only if its ratio to this is the same as this
that is:
ratio of rct A, U to rct A, F
same
ratio compounded of ratio of L to A & ratio of R to F
same
common side A
ratio of U to F
similar triangles
ratio of B to Z
ratio of B to S
compounds into
ratio of sqr B to rct Z, S
that is:
(U) is the constant line which has to constant line F the same ratio as this constant ratio

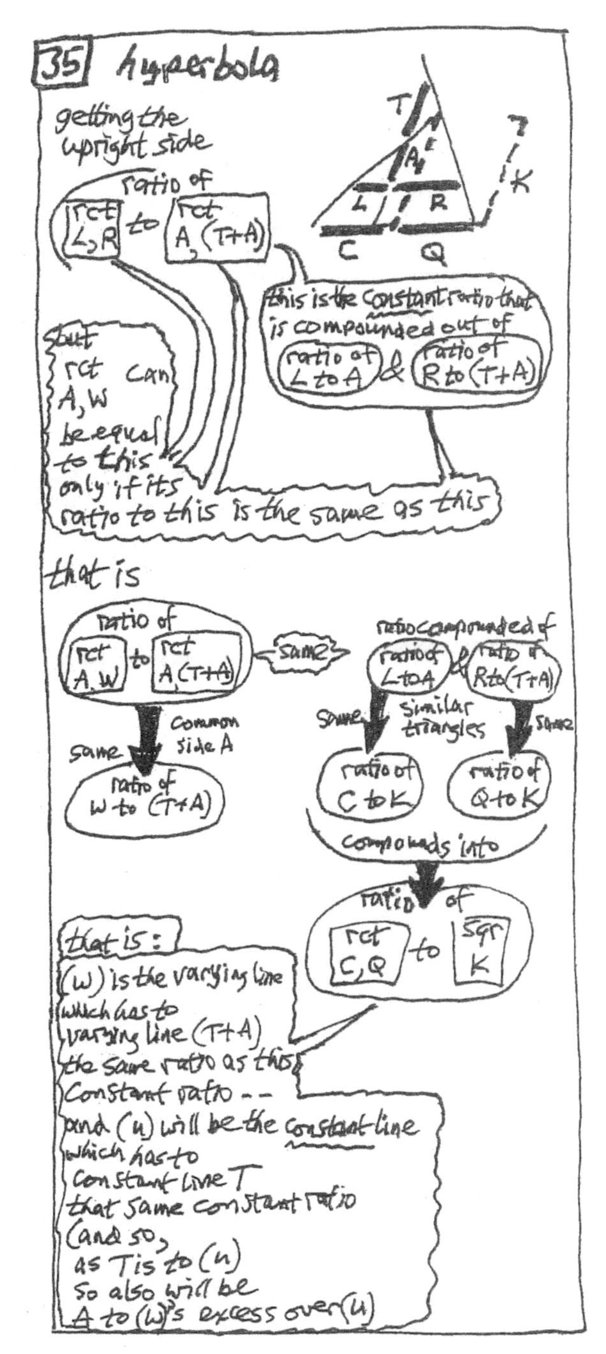
35 hyperbola
getting the upright side
T A E L R C Q K
ratio of rct L, R to rct A, (T+A)
this is the constant ratio that is compounded out of ratio of L to A & ratio of R to (T+A)
but rct can A, W be equal to this only if its ratio to this is the same as this
that is
ratio of rct A, W to rct A, (T+A)
same
ratio compounded of ratio of L to A & ratio of R to (T+A)
same
common side A
ratio of W to (T+A)
similar triangles
ratio of C to K
ratio of Q to K
compounds into
ratio of rct C, Q to sqr K
that is:
(W) is the varying line which has to varying line (T+A) the same ratio as this constant ratio - -
and (U) will be the constant line which has to constant line T that same constant ratio
(and so, as T is to (U) so also will be A to (W)'s excess over (U))

CHAPTER 3. THE ELLIPSE

transition

Now we consider the only remaining way to cut a cone and get a new curve. The new sorts of curves that have been characterized in the 11th and 12th propositions were first seen back in the 8th proposition. The new sort of curve now to be characterized in the 13th proposition was first seen back in the 9th proposition. Back then, Apollonius demonstrated that if we cut so as to hit the axial triangle's other side (or its extension) below the vertex of the cone, but, while doing so, did not cut either parallel to the base or subcontrariwise (fig. 1), then the section that would be cut into the cone is not a circle.

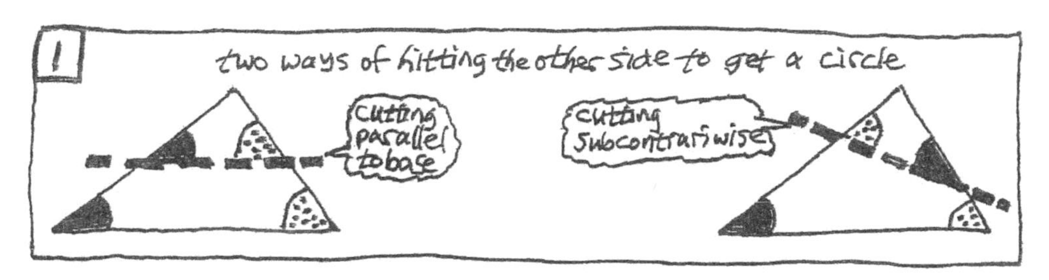

Consider the straight line that runs, in a line with the diameter, from the section's vertex to the other conic side. In the *hyperbola*, this line—this very important line called the oblique (transverse) side—met the other conic side on the other side of the cone's vertex; it met it (that is to say) in the opposite conic surface, on the extension of the triangle's opposite side (fig. 2). *Now*, by contrast, the line that we are considering will meet the other conic side *below* the cone's vertex. (Even if it does not meet the axial triangle's *side* below the vertex, nonetheless it will meet the *extension* of the axial triangle's side below the vertex rather than above.) (Fig. 3.)

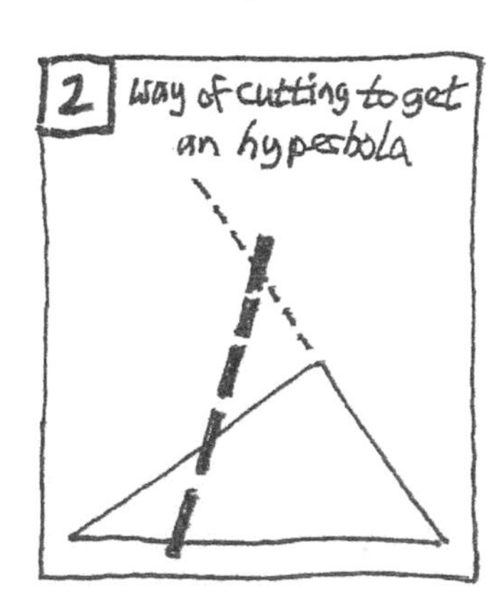

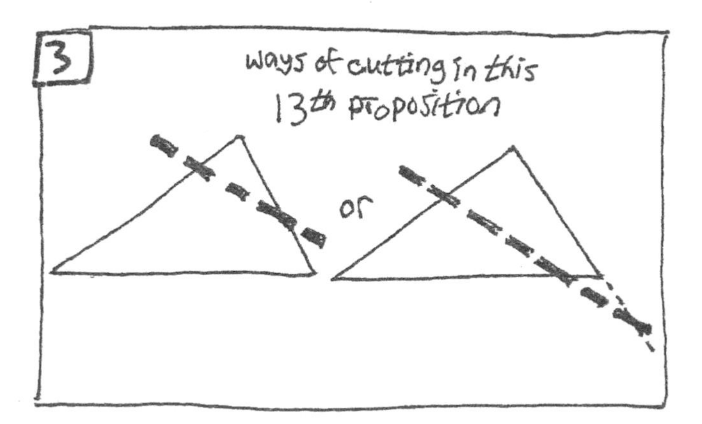

So now, when we cut this way, the diameter itself will be the oblique (transverse) side. What about that other line which was of great significance in the hyperbola—the one which went, in line with the diameter, from the non-vertex end of the abscissa, straight to the other side of the axial triangle, or rather to its extension? In other words, what about the line which stretches from the abscissa's non-vertex endpoint to the oblique (transverse) side's non-vertex endpoint? That other line, in the hyperbola, is what we got from the oblique (transverse) side when the abscissa was *added on to it*. Now, in the new curve that we shall call the ellipse, it is what we shall get from the oblique (transverse) side when the abscissa is *taken away from it*. (Fig. 4.)

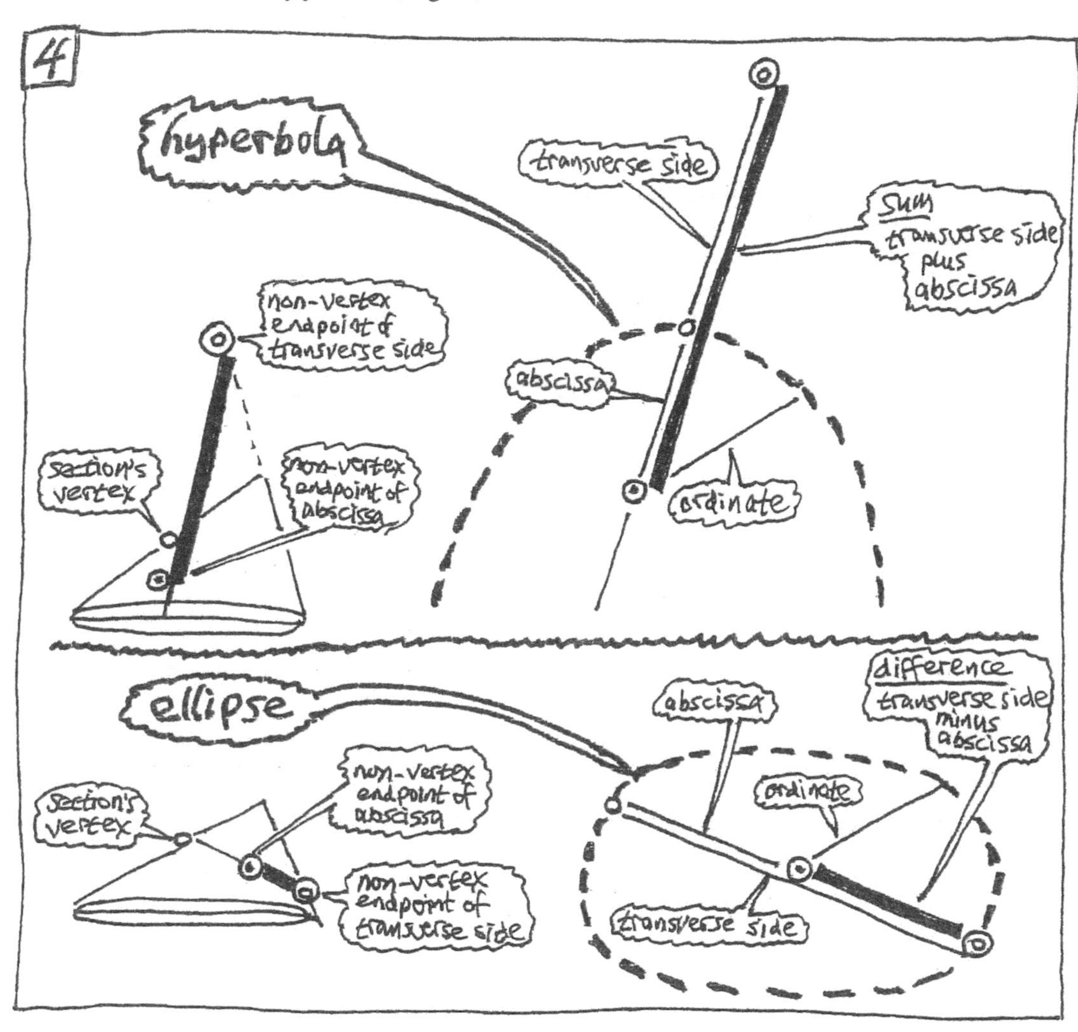

It has a function of the greatest significance—this line that in the hyperbola is longer than the oblique (transverse) side and in the ellipse is shorter. In the case of the hyperbola, the line was used to determine the width of a rectangle which had the abscissa for a side and was equal to the square on the ordinate. The determination was made in accordance with the following proportion:

> as the *hyperbola*'s upright side is to its oblique (transverse) side,
> so the equal rectangle's width is to the *sum* of
> the oblique (transverse) side and the abscissa.

In the present case, what would correspond to that would be the following proportion:

> as the *ellipse*'s upright side is to its oblique (transverse) side,
> so the equal rectangle's width is to the *difference* between
> the oblique (transverse) side and the abscissa.

It is true that what the enunciation of the hyperbola proposition gave was not quite that. But it did give its equivalent. It said the following:

> the equal rectangle which has the abscissa for a side
> *overshoots*
> that rectangle which has the abscissa for one side but has as its other side
> the upright side;
> and the rectangular *excess* has the same shape as the rectangle contained by
> the oblique (transverse) side and the upright side.

In the present case, that of the ellipse proposition, what will correspond is the following:

> the equal rectangle which has the abscissa for a side
> is *lacking* in comparison with
> that rectangle which has the abscissa for one side but has as its other side
> the upright side;
> and the rectangular *deficiency* has the same shape as the rectangle contained by
> the oblique (transverse) side and the upright side.

In both cases, there is a discrepancy between two rectangles—the equal rectangle whose side is the abscissa, and an unequal rectangle whose sides are the abscissa and the upright side. The discrepancy is a rectangular form shaped by the ratio of its sides—and this is the same ratio that the oblique (transverse) side has to the upright side. The two ratios are the same as each other because each is the same as another ratio: the ratio that a certain line has to the width of the equal rectangle. That certain line, in both cases, goes from the non-vertex side of the abscissa, in line with the diameter, straight to the other side of the axial triangle (or to its extension). In the case of the *hyperbola*, that line is the *sum* of the oblique (transverse) side and the abscissa; in the case of the *ellipse*, it is their *difference*. The present proposition and the previous one are variations on the same theme.

What we did in examining the propositions that characterize the parabola and the hyperbola, we shall do now again in examining the proposition that characterizes the ellipse: we shall consider how the cone is cut to get the curve, what it is that's shown about the curve, how it is shown, and why we might want to show it. The details of the work that we shall do with the 13th proposition will very much resemble the details of the work that we did with the 12th. What is repetitious will, however, serve to consolidate your grasp of things, and the points of contrast, arising as they will amid so much that is the same again, will prepare you for what you might not otherwise have thought to look for.

Here, then, is the text of Apollonius's 13th proposition. Before reading it, though, you may want to pause for a while, and think about what you might expect it to say....

#13 If a cone be cut by a plane through its axis, and be cut also by another plane which falls together with each side of the triangle through the axis, and which is drawn neither parallel to the base of the cone nor subcontrariwise, and if the plane in which the base of the cone is and the cutting plane fall together along a straight [line] that is at right [angles] either to the base of the triangle through the axis or to its straight prolongation, and any [straight] line from the section of the cone be drawn parallel to the common section of the planes, as far as the diameter of the section, then [this straight line so drawn to the diameter] will have the power (*dunêsetai*), [will have the power, that is, to give rise to a square that is equal to] a certain [rectangular] area (*khôrion*) laid along (*parakeimenon para*) [or "applied to"] a certain straight [line], one to which the diameter of the section has the ratio that the square from the [line] drawn from the cone's vertex parallel to the section's diameter, as far as the base of the triangle, has to the [rectangle] contained by the lines taken off on it [taken off, that is, on the base] by this same [straight line], on the side toward the straight [lines that are the sides] of the [axial] triangle—[with that equal area] having as breadth the line taken off by it, [taken off, namely, by that line drawn from the section to the diameter,] [and taken off] from the diameter, on the side toward the vertex of the section, and [also with that equal area] being lacking (*elleipon*)] [or "being deficient," or "falling short"] by a figure (*eidei*) that is similar to, and is also laid similarly to, the [rectangle] contained by [the following two straight lines]: the diameter, and also the [line just characterized by a certain ratio, namely the line] along which [the straight lines drawn down ordinatewise] have the power (*dunantai*), [namely that power mentioned above]. And let such a section be called an "ellipse" (*elleipsis*).
—Let there be a cone whose vertex the point A is and whose base the circle BC is. And let [the cone] be cut by a plane through the axis; and let it make as section the triangle ABC, and let the cone be cut also by another plane which falls together with each side of the triangle through the axis, and which is drawn neither parallel to the base of the cone nor subcontrariwise, and let it make as section on the surface of the cone the line DE. And the common section of the cutting plane and of the [plane] in which is the base of the cone, let the [line] FG, which is at right [angles] to the [line] BC, be that; and the diameter of the section, let the [line] ED be that; and from the [point] E let there be drawn, at right [angles] to the [line] ED, the [line] EH; and through the [point] A let there be drawn parallel to the [line] ED the [line] AK; and let [that straight line EH] be so made that as is the square from AK to the [rectangle] contained by BKC so is the [line] DE to the [line] EH. And let there be taken some point on the section—the [point] L, say—and through the [point] L let there be drawn parallel to the [line] FG the [line] LM. I say that the line LM has the power (*dunatai*) [to give rise to a square equal to] a certain area (*khôrion*) that is laid along (*parakeitai para*) the [line] EH, while having as breadth the [line] EM, and being lacking (*elleipon*) by a figure (*eidei*) that is similar to the [rectangle contained] by the [lines] DEH.

—For let there be joined the [line] DH. And through the [point] M, let there be drawn parallel to the [line] HE the [line] MXN; and through the [points] H, X, let there be drawn parallel to the [line] EM the [lines] HN, XO; and through the [point] M, let there be drawn parallel to the [line] BC the [line] PMR. Since, then, the [line] PR is parallel to the [line] BC, and also the [line] LM is parallel to the [line] FG, therefore the plane through the [lines] LM, PR is parallel to the plane through the [lines] FG, BC—is parallel, that is, to the base of the cone. If, therefore, through the [lines] LM, PR there be extended a plane, the section will be a circle, whose diameter is what the line PR is. And a perpendicular to it, [that is, to the line PR] is what the [line] LM is. Therefore the [rectangle] by the [lines] PMR is equal to the [square] from the [line] LM.
—And since as is the [square] from the [line] AK to the [rectangle] by the [lines] BKC so is the [line] ED to the [line] EH—and the ratio of the [square] from the [line] AK to the [rectangle] by the [lines] BKC is compounded from that which the [line] AK has to KB and also that which the [line] AK has to KC—but (on the one hand) as is the [line] AK to KB so is the [line] EG to GB, that is, [so is] the [line] EM to MP; and (on the other hand) as is the [line] AK to KC so is the [line] DG to GC, that is, [so is] the [line] DM to MR—therefore the ratio of the [line] DE to the [line] EH is compounded from that of the [line] EM to MP and also that of the [line] DM to MR. But the ratio compounded from that which the [line] EM has to MP and also that which the [line] DM has to MR is the ratio of the [rectangle] by the [lines] EMD to the [rectangle] by the [lines] PMR. Therefore as is the [rectangle] by the [lines] EMD to the [rectangle] by the [lines] PMR so is the [line] DE to the [line] EH—that is, [so is] the [line] DM to the [line] MX. And, as is the [line] DM to MX, so—with the [straight line] ME taken as common height—is the [rectangle] by DME to the [rectangle] by XME. Also, therefore, as is the [rectangle] by DME to the [rectangle] by PMR so is the [rectangle] by DME to the [rectangle] by XME. Equal, therefore, is the [rectangle] by PMR to the [rectangle] by XME. But the [rectangle] by PMR was shown equal to the [square] from the [line] LM; therefore also the [rectangle] by XME is equal to the [square] from the [line] LM.
—The [line] LM, therefore, has the power (*dunatai*) [to give rise to a square equal to] the [area] MO which is laid alongside (*parakeitai para*) [or "is applied to"] the [line] HE—[that area] MO having as its breadth the [line] EM, and being lacking (*elleipon*) [or "being deficient" or "falling short"] [with respect to the rectangle contained by the lines EM and EH], by a figure (*eidei*), namely the [area] ON, which is similar to the [rectangle contained] by DEH.
—And let such a section be called an "ellipse" (*elleipsis*); and as for the [line] EH—the one along (*para*) which the lines that are drawn down ordinatewise to the [line] DE have the power (*dunantai*)—let the same [line EH] be called an "upright" (*orthia*) [side], and an "oblique" [or "slanting"] (*plagia*) [side] be what the line ED is called.

[To repeat, since it will need to be remembered later, when we consider the work of Descartes: *plagia* has traditionally been rendered, not by our more literal "oblique," which suggests being aslant, but rather by "transverse," which suggests going across. (We might also note that the Greek *orthia/plagia* contrast can have a moral usage, suggesting the difference between being upright and not being straightforward.)]

[To be more explicit—in this situation of "deficiency"—about the business of "having the power": A straight line drawn down ordinatewise, from the section to its diameter ED, "has the power," it is here said, "along" EH. This means that such a line drawn down ordinatewise (say, the line LM), when taken as a side, gives rise to a square that will be equal to a certain rectangle which lies along (*para*) the "upright" EH, whose size has been characterized by a certain proportion. The equal rectangle is lacking (*elleipon*) in its stretch along (*para*) the line EH . That is, the equal rectangle's length EO is shorter than "upright side" EH. If the rectangle's length were not less than the upright side, the rectangle would be too large to be equal to the square, since the rectangle's breadth is the piece (line EM) that is taken off from the diameter, on the side toward the section's vertex E, by the line LM that is drawn down ordinatewise.]

13th proposition: the enunciation–how we cut

As in the previous proposition, we first pass a plane through the axis, to get an axial triangle. It is with respect to the axial triangle that the cutting plane is oriented (fig. 5).

As before, the straight line (FG) that is the cutting plane's intersection with the cone's base plane is perpendicular to the straight line (BC) that is the axial triangle's base. By orienting that intersection (FG) in that way, we guarantee that every one of the section's chords which is parallel to it will be bisected by the cutting plane's other straight-line intersection—namely, its intersection (ED) with the axial triangle.

Straight line FG thus serves as the reference line for lines drawn ordinatewise across the section. These lines will be bisected by the axial triangle, and thus by straight line ED; and hence this line ED is the diameter of the half-chords that they make, which are this diameter's ordinates.

Line FG, the cutting plane's intersection with the plane of the cone's base, thus has, in the present proposition, the same orientation which it had in the previous two propositions. What is different here is the orientation of the cutting plane's other intersection with a plane—namely, its intersection ED with the plane of the axial triangle.

What is the *same*	What is *different*
is the orientation of the cutting plane	is the orientation of the cutting plane
viewed from *above* the cone	viewed from *beside* the cone
(that is to say, the orientation of	(that is to say, the orientation of
the *reference line*	the *diameter*, from which we cut
for the *ordinates*).	off the *abscissas*).

How is the diameter oriented now? The diameter is not parallel to the other side of the axial triangle, as it was in the proposition before the previous proposition. But neither does the diameter's extension now hit the other side's extension above the vertex, as in the previous proposition. Rather, the diameter hits the other side or else it hits, below the vertex, the other side's extension.

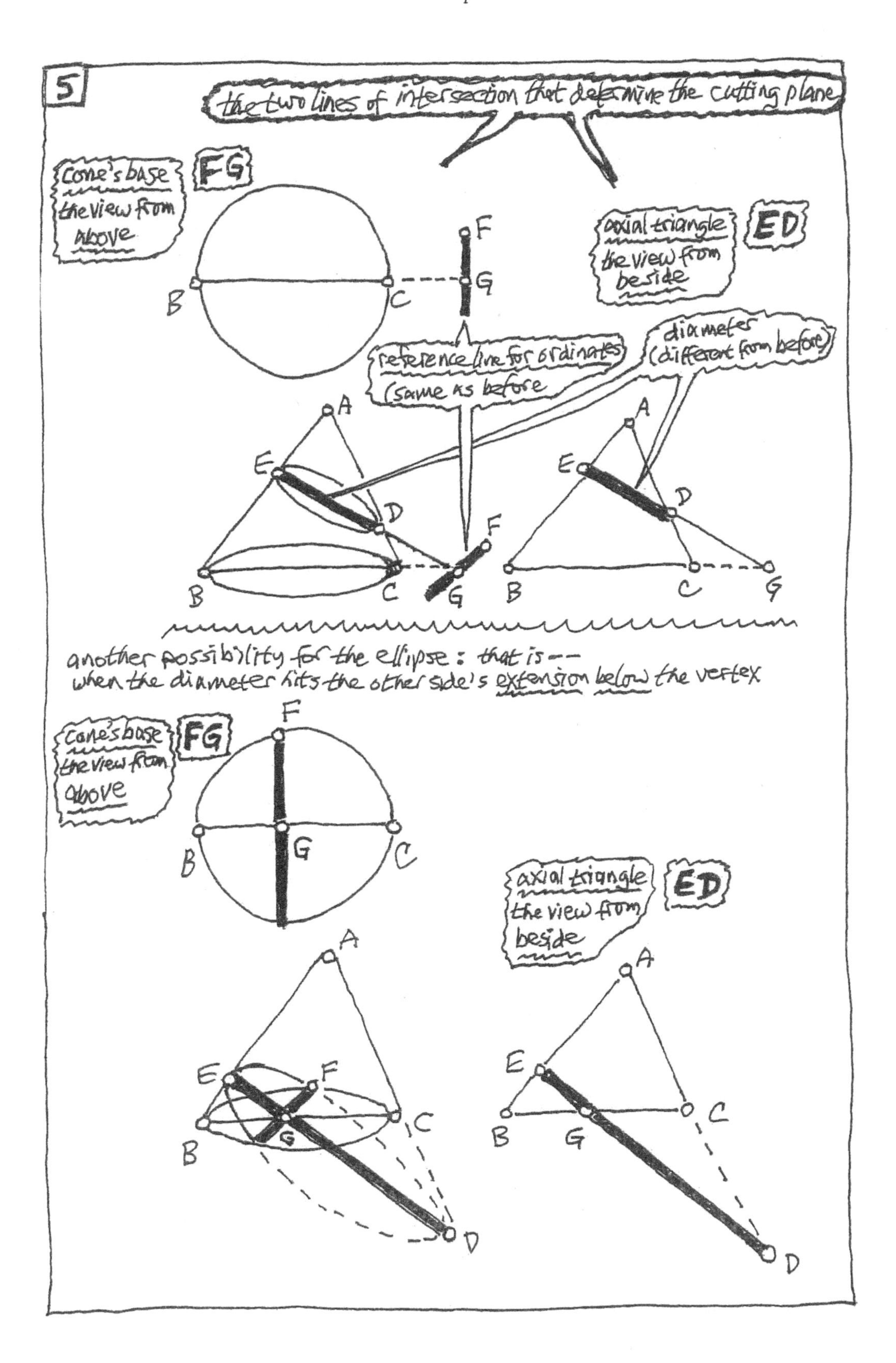
5
the two lines of intersection that determine the cutting plane
cone's base
the view from above
FG
axial triangle
the view from beside
ED
F
G
B
C
reference line for ordinates
(same as before
diameter
(different from before)
A
E
D
F
B
C
G
A
E
D
B
C
G
another possibility for the ellipse: that is--
when the diameter hits the other side's extension below the vertex
F
cone's base
the view from above
FG
G
B
C
axial triangle
the view from beside
ED
A
A
E
F
B
G
C
D
E
B
G
C
D

13th proposition: the enunciation—what we show about the line we cut

If our cutting is done in the way just described, then the intersection of cone by cutting plane is a conic section of a sort distinguished by a certain property of the straight lines that we draw in it.

As before, the ordinate and the abscissa are related by showing the equality of a square and a rectangle. (Fig. 6.) The square has as its side the ordinate (LM); and the rectangle has as one of its sides the abscissa (EM). To determine the size of the rectangle's other side (MX) in this proposition for the ellipse involves a complication like the one for the hyperbola that we saw in the previous proposition.

As with both parabola and hyperbola, a line must be contrived to be the upright side; it must be made just the right size to fit a certain proportion. As with the hyperbola, in order to state the proportion for the present proposition, we must name a line and draw another line.

With the hyperbola, the oblique (transverse) side was a straight line that went from the section's vertex across to the extension of the axial triangle's opposite side above the cone's vertex; but with the ellipse, the oblique (transverse) side will be a straight line that goes from the section's vertex across to the axial triangle's opposite side or to its extension below the cone's vertex. That is to say: whereas the oblique (transverse) side for the hyperbola was an extension of the diameter, the oblique (transverse) side for the ellipse will be the diameter itself.

With the hyperbola, we drew a straight line from the vertex of the cone, parallel to the diameter, all the way down until it met the axial triangle's base. We again draw such a line here; but here, with the ellipse, rather than meeting the axial triangle's base itself, it meets the extension of that base. As before, we are interested in a ratio given to us by this straight line that we draw—the ratio of a certain square to a certain rectangle. The square here, as in the case of the hyperbola, is the one that arises from the drawn straight line (AK). The rectangle here is contained by lines that may look different from the corresponding lines in the case of the hyperbola earlier, but the lines are in fact the same. That is to say: in this case as in the case of the hyperbola, they are the lines stretching from the bottom-point (K) of the drawn straight line, to each end (B and C) of the axial triangle's base. The difference is only whether K is inside (in the case of the hyperbola) or outside (in the case of the ellipse) the line joining B and C (the axial triangle's base). (In each case, you will find it helpful to call the lines "KB" and "KC"—thinking of them as two lines drawn from K, one of them to B and the other to C—rather than speaking of them without attending to the order in which you name their endpoints. You might even find it helpful to employ the labor-saving terminology suggested earlier, by speaking of K as "cutting" the axial triangle's base BC in both cases—"internally" in the case of the hyperbola and "externally" in the case of the ellipse.) The square whose side is KA has a certain ratio to the rectangle whose sides are KB and KC. This ratio is used to determine the size of the line that we contrive to be the upright side. We contrive the upright side to be of just the right size so that the oblique (transverse) side will have to it the very same ratio that the square whose side is KA has to the rectangle whose sides are KB and KC.

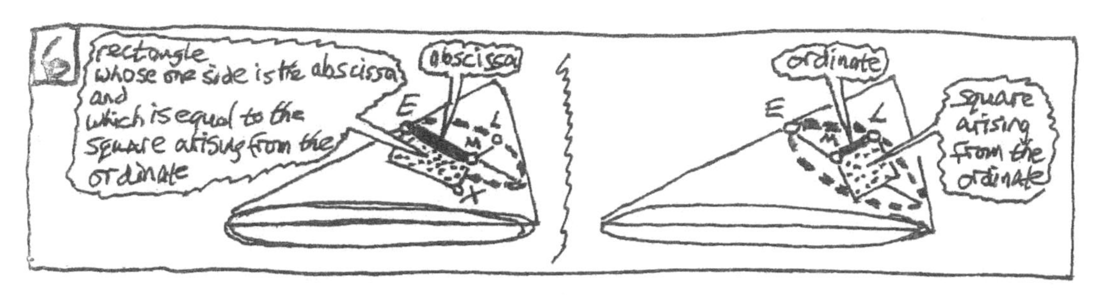

Figure 7 depicts those lines by which the size of the upright side is determined—namely, the oblique (transverse) side ED and the parallel line AK drawn in.

In this 13th proposition—as in the previous one—the rectangle that is equal to the square on the ordinate, while it does have as one of its sides the abscissa, has as its other side not the upright side but rather a line that is longer than the upright side. The upright side, although it is not itself the equal rectangle's other side, is, however, the means of getting this other side. To get it, we look at another rectangle. To identify this rectangle, we need to consider the fact that the rectangle which will be equal to the square on the ordinate is not itself equal to—but rather falls short of—the rectangle which is contained by the abscissa and the upright side. That is to say (fig. 8):

> the rectangle which
> is equal to the square on the ordinate and has the abscissa as one of its sides,
>
> is lacking in comparison to
>
> the rectangle which
> also has the abscissa as one of its sides but has as its other side the upright side.

The rectangular lack—that is what we need to attend to.

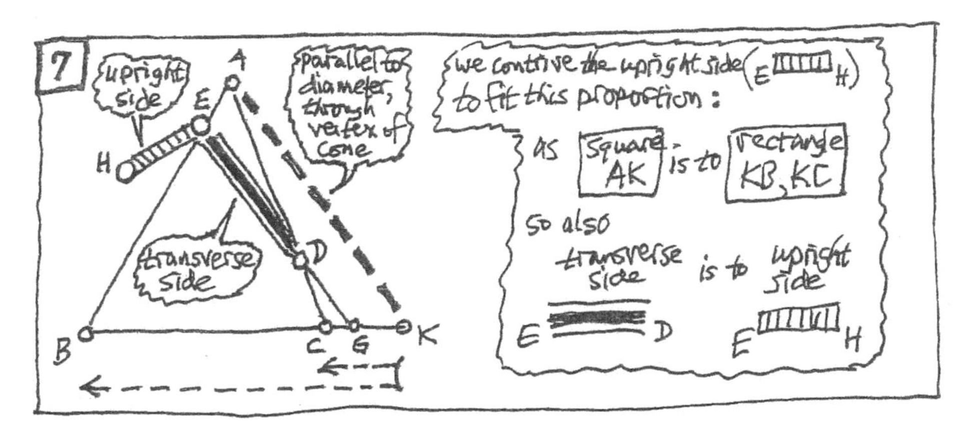

Just how much of a lack is there? Well, the deficiency is rectangular in form (*eidos*), and one of its sides has the same size as the abscissa. Of what size is its other side? Of just the right size to give it a certain shape—namely, this one: the rectangular deficiency, though it may not be the same size as, still looks just like the rectangle contained by the oblique (transverse) side and the upright side. In other words: as the oblique (transverse) side is to the upright side, so is the abscissa (which is the rectangular deficiency's one side) to the rectangular deficiency's other side. The reference rectangle formed by the oblique (transverse) side and the upright side exhibits the form of the rectangular deficiency. (Fig.9.)

Thus, the size of the deficiency for each abscissa will be determined by these two items: the size of the abscissa, and the constant ratio of the constant oblique (transverse) side to the constant upright side.

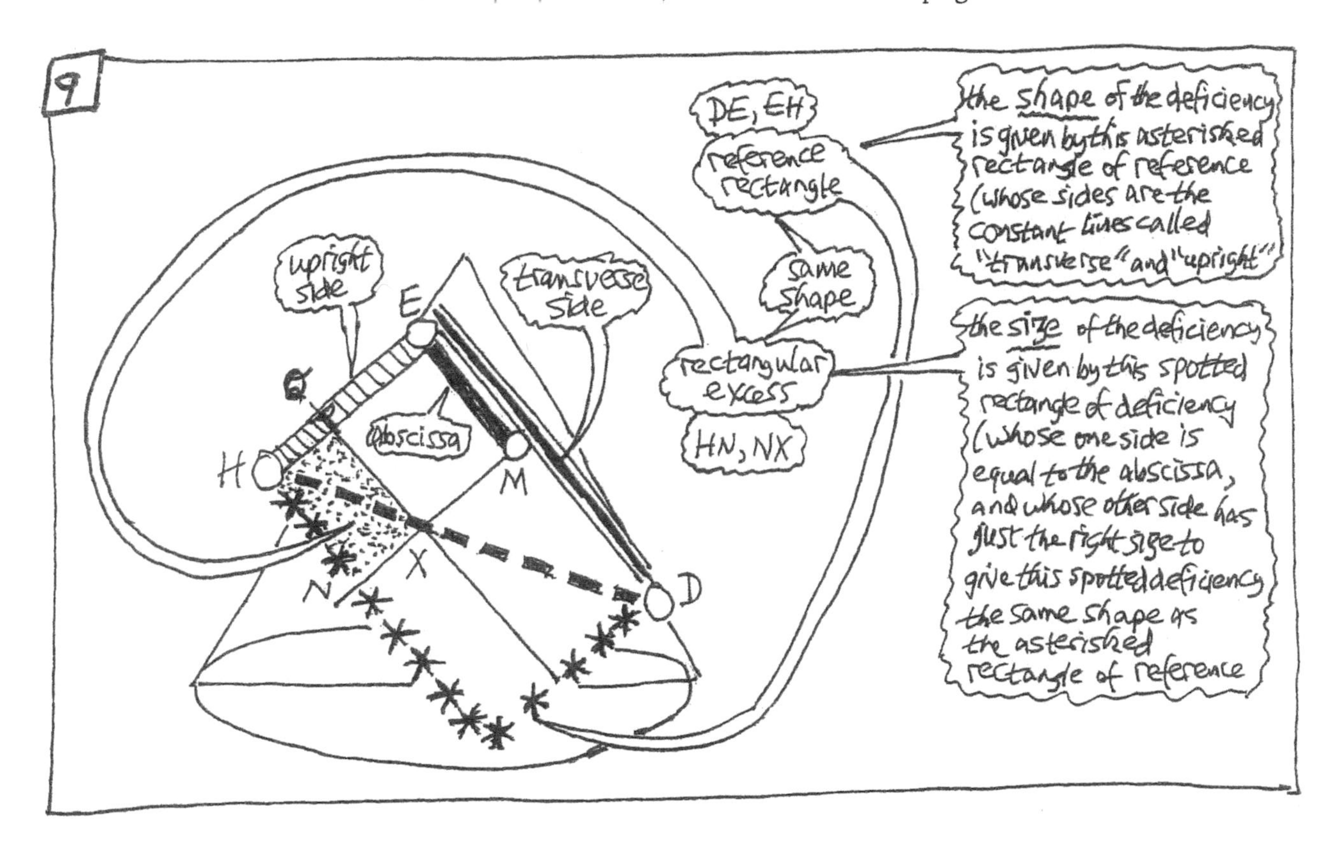

The changing size of the equal rectangle's non-abscissa side (its width) will therefore be determined by the constant upright side. The width, that is to say, will be of just the right size to fit this proportion: the ratio which the oblique (transverse) side has to it (the upright side) will be the same as the ratio which the line that is the difference between the oblique (transverse) side and the abscissa has to the width of the equal rectangle. (What we will get from the oblique [transverse] side, by taking the abscissa away from it, is the line that stretches from the abscissa's non-vertex endpoint to the oblique [transverse] side's non-vertex endpoint.) That is to say (fig. 10), DE — EM (oblique [transverse] side less abscissa) will be to MX (width) as ME (abscissa) is to EH (upright side).

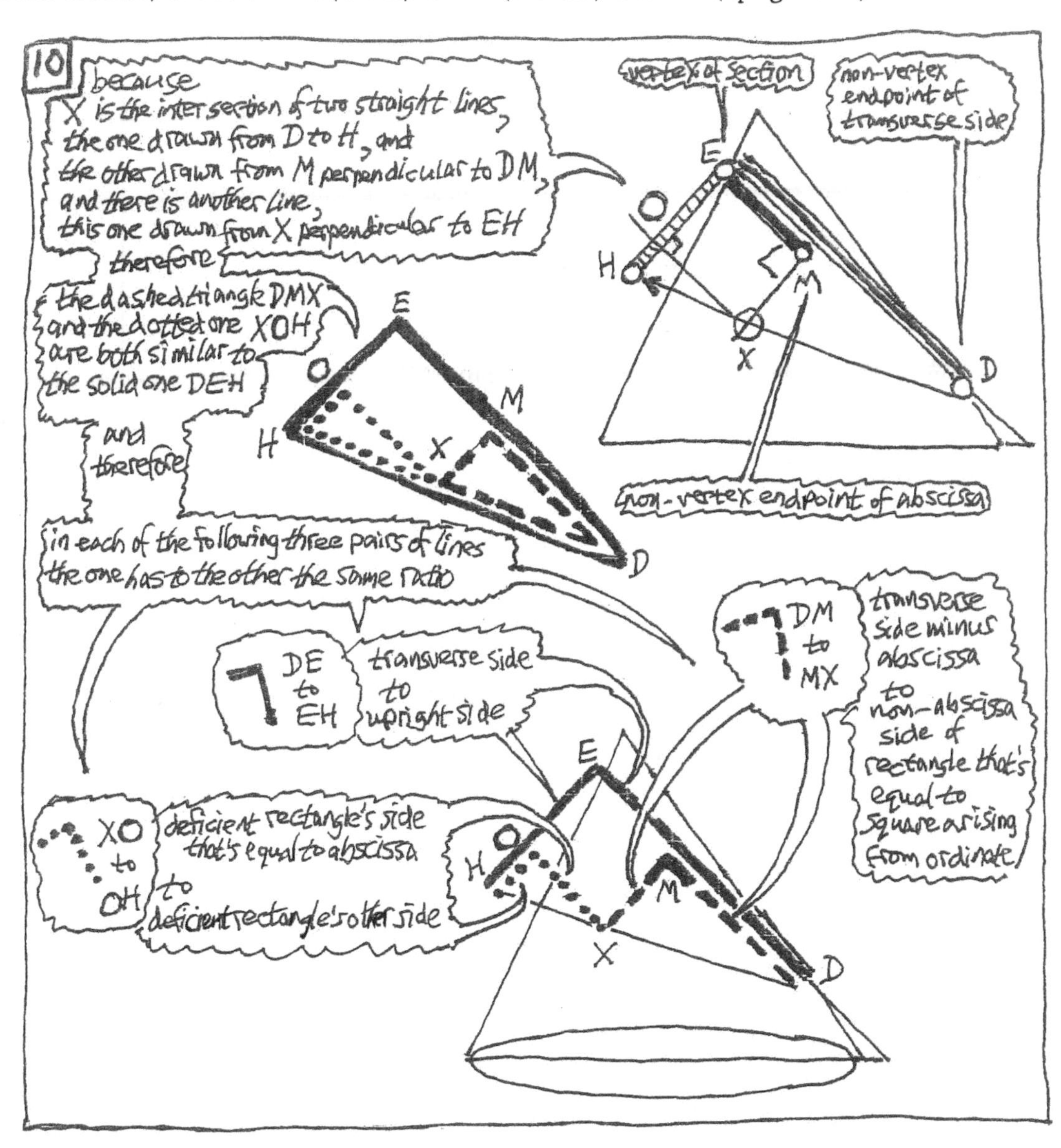

13th proposition: the demonstration

The task of the demonstration is to show that by cutting in the way described, we get a section with the following property (fig. 11):

> *if* straight line EH is contrived so that the ratio which
> the oblique (transverse) side ED has to it
> is the same ratio which
> a square with side AK has to a rectangle with sides KB and KC—
>
> and if EH is used to draw up from the abscissa-point a line MX which
> falls short of the contrived upright side EH just enough so that the ratio which
> the difference between oblique (transverse) side and abscissa will have to it (MX)
> is the same ratio which
> the oblique (transverse) side has to the contrived line (EH),
>
> *then* a square whose side is ordinate LM will be equal to
> a rectangle whose sides are abscissa EM and the line MX.

That way of putting the matter may make things easier to grasp, but its emphasis differs from the emphasis in the text. What is emphasized in the text is not the equality of the square and the rectangle, but rather the equal rectangle's deficiency with respect to another rectangle. The emphasis is two-fold—on size and on shape.

> The equal rectangle is smaller than
> the rectangle whose sides are the abscissa and the upright side.
>
> The deficiency has a certain shape:
> it presents itself in the form of a certain rectangle—
> the rectangle whose sides are the oblique (transverse) side and the upright side.
> In other words,
> the rectangular deficiency's side that is equal to the abscissa
> has a certain ratio to the rectangular deficiency's other side—
> the same ratio which the oblique (transverse) side has to the upright side.

The work of the demonstration, as it was in the previous two propositions, is to establish the equality of a rectangle and a square. The rectangle's sides are EM and MX, and the square's side is the ordinate LM. How does Apollonius show the equality? As in the previous two propositions, square [LM] and rectangle [EM, MX] are shown to be equal to each other by showing that they are both equal to the same thing—namely, rectangle [PM, MR]. The demonstration, again, thus requires the establishment of two equalities (fig. 12).

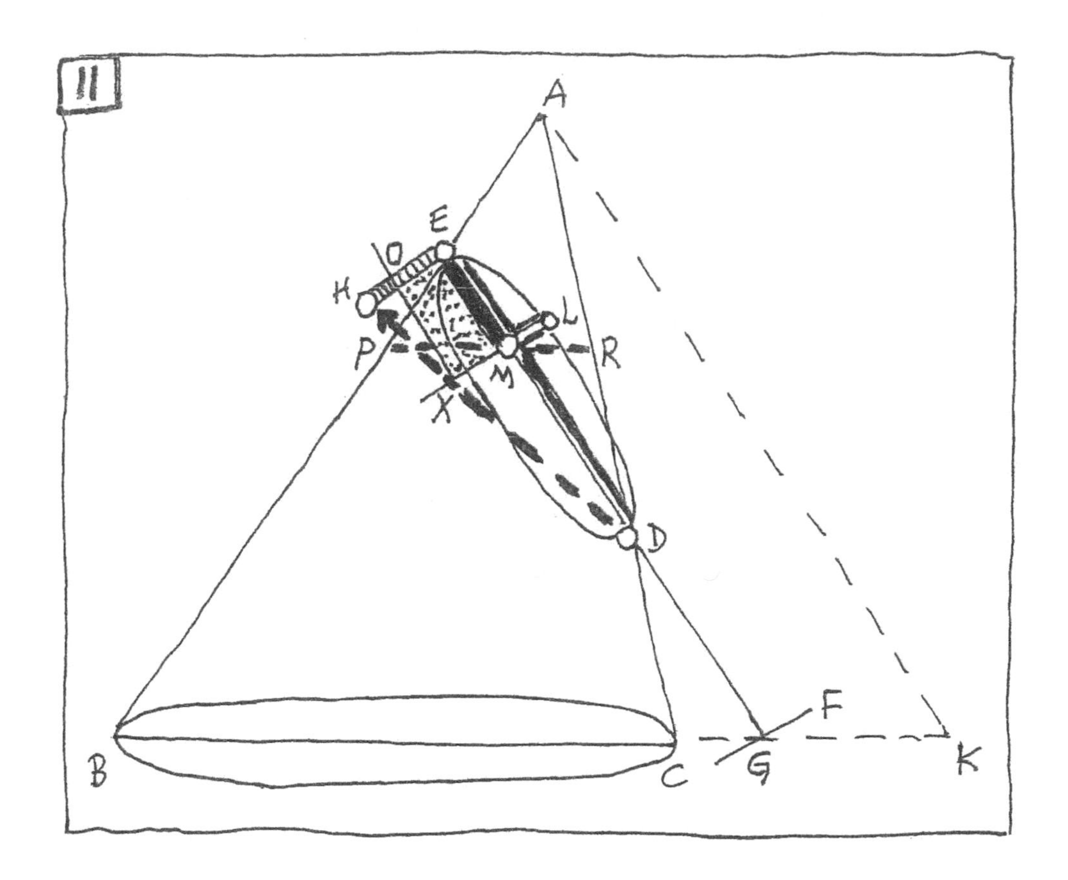
11
A
E
O
H
L
P
R
M
X
D
F
B
C
G
K

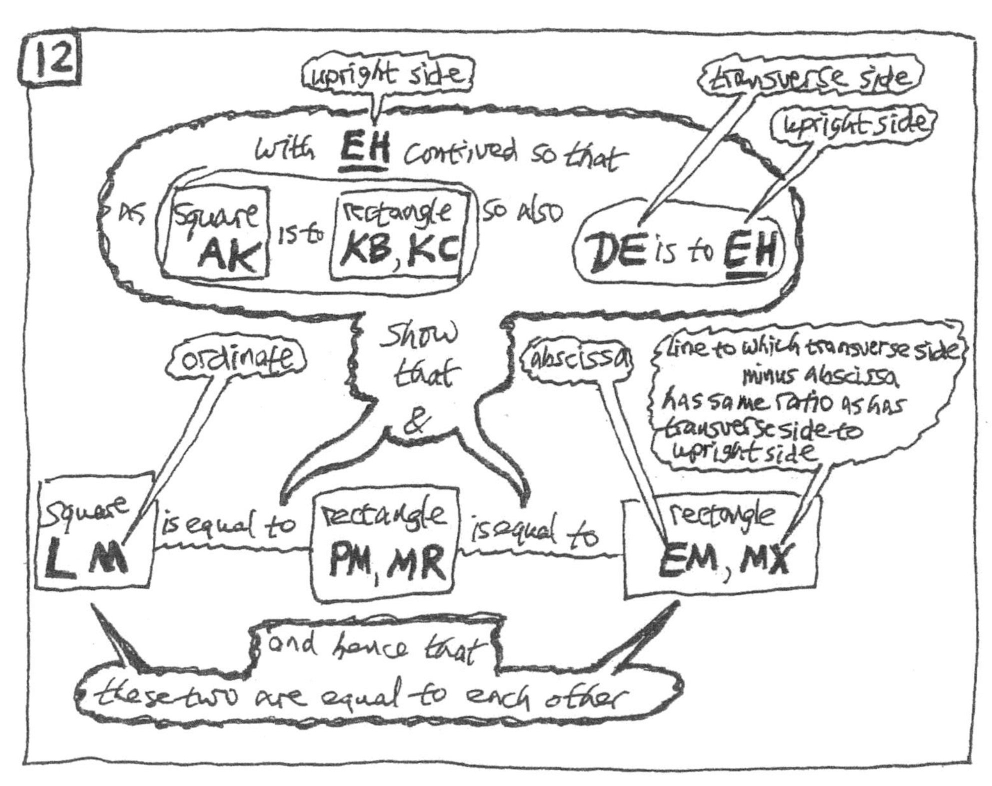
12
upright side
transverse side
upright side
with EH contrived so that
as
square AK
is to
rectangle KB, KC
so also
DE is to EH
show that &
ordinate
abscissa
line to which transverse side minus abscissa has same ratio as has transverse side to upright side
square LM
is equal to
rectangle PM, MR
is equal to
rectangle EM, MX
and hence that
these two are equal to each other

As in the previous two propositions, the establishment of the first equality is the easier part of the demonstration. It does not involve using the line that has been contrived. It merely takes the view from *above* that presents us with a circle.

What we do is the following (fig. 13). Through M, the point where the ordinate meets the diameter, we pass a plane parallel to the base of the cone, thus slicing into the cone a circle. The circle's half-chord is the original section's ordinate LM—which is the side of the square that concerns us. Because this half-chord (LM) is perpendicular to the circle's diameter (PMR), we can say, as before, that the square whose side is LM is equal to the rectangle whose sides are PM and MR.

What remains to be done now is to establish that this rectangle with sides PM and MR, which is equal to the square with side LM, is *also* equal to the rectangle with sides EM and MX.

How do we establish this second equality? As in the previous proposition. Having taken the view from *above* in order to use the square-rectangle property that derives from the proportionality of certain straight lines in a *circle*, we take the view from *beside* in order to use the proportionality of straight lines in similar *triangles*. (Fig. 14.)

The view from above and the view from beside are linked by the straight line PMR; for not only is PMR the diameter of the circle, it is also parallel to the base of the axial triangle—the triangle whose sides when extended are used to determine the size of the line that is contrived to be the upright side.

We need to show for those two rectangles—one whose sides are PM and MR, the other whose sides are EM and MX—that their equality follows from the sameness of the ratios which determine the size of the upright side (EH). The upright side is defined in terms of a relationship among lines that have to do with the axial triangle. We need to show how, from that relationship, another relationship follows. We need to relate the relationship that we have to start with (because we made things that way) to the relationship that we would like to have established (but do not yet have). Since the lines in the relationship that we already have (all of them) are lines that have to do with the axial triangle, we give ourselves a side view of the figure; and, from the axial triangle thus displayed, we take hold of what we need.

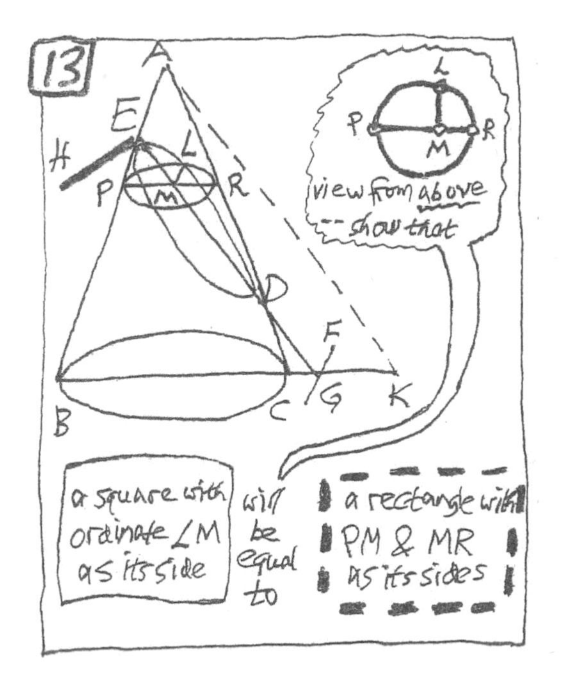
13
A
E
H
L
P
M
R
D
F
B
C
G
K
L
P
M
R
view from above
-- show that
a square with ordinate LM as its side
will be equal to
a rectangle with PM & MR as its sides

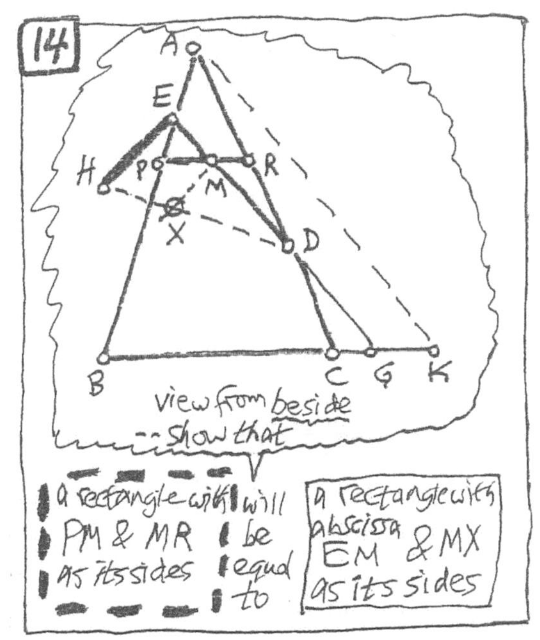
14
A
E
H
P
M
R
X
D
B
C
G
K
view from beside
-- show that
a rectangle with PM & MR as its sides
will be equal to
a rectangle with abscissa EM & MX as its sides

What we have already—by virtue of our slicing the diameter into the axial triangle, and drawing a straight line in the axial triangle through the cone's vertex, parallel to the diameter, as well as our slicing a circle whose diameter is parallel to the triangle's base—is the *similarity of triangles*. From that, we shall get the *equality of rectangles*.

We move from what we already have, to what we do not have but want to have, in the way that is schematized in figure 15. We take the ratio of rectangles used to define the upright side, and we do this:

(**Step I**) The ratio of the square (whose both sides are AK)
to the rectangle (whose sides are KB and KC)
is *taken apart* into its two component ratios, which are
the ratios of their sides (the ratio of AK to KC, and of AK to KB).

Now, taking *the one component* ratio,
we get same ratios by the similarity of triangles:
(**step II**) as AK is to KC, so is DG to GC;
and (**step III**) as DG is to GC, so is DM to MR.
Thus, the one component ratio (the ratio of AK to KC)
is the same as the ratio of DM to MR.

Now, taking *the other component* ratio,
we also get same ratios by the similarity of triangles:
(**step IV**) as AK is to KB, so is EG to GB;
and (**step V**) as EG is to GB, so is EM to MP.
Thus, the other component ratio (the ratio of AK to KB)
is the same as the ratio of EM to MP.

And now (**step VI**), if we *put together* those two resulting ratios of lines
(namely, that of DM to MR and that of EM to MP), we get
the ratio of the rectangles that would be built by using those lines as their sides.
This is the ratio of the rectangle whose sides are EM and MD
to the rectangle whose sides are PM and MR.

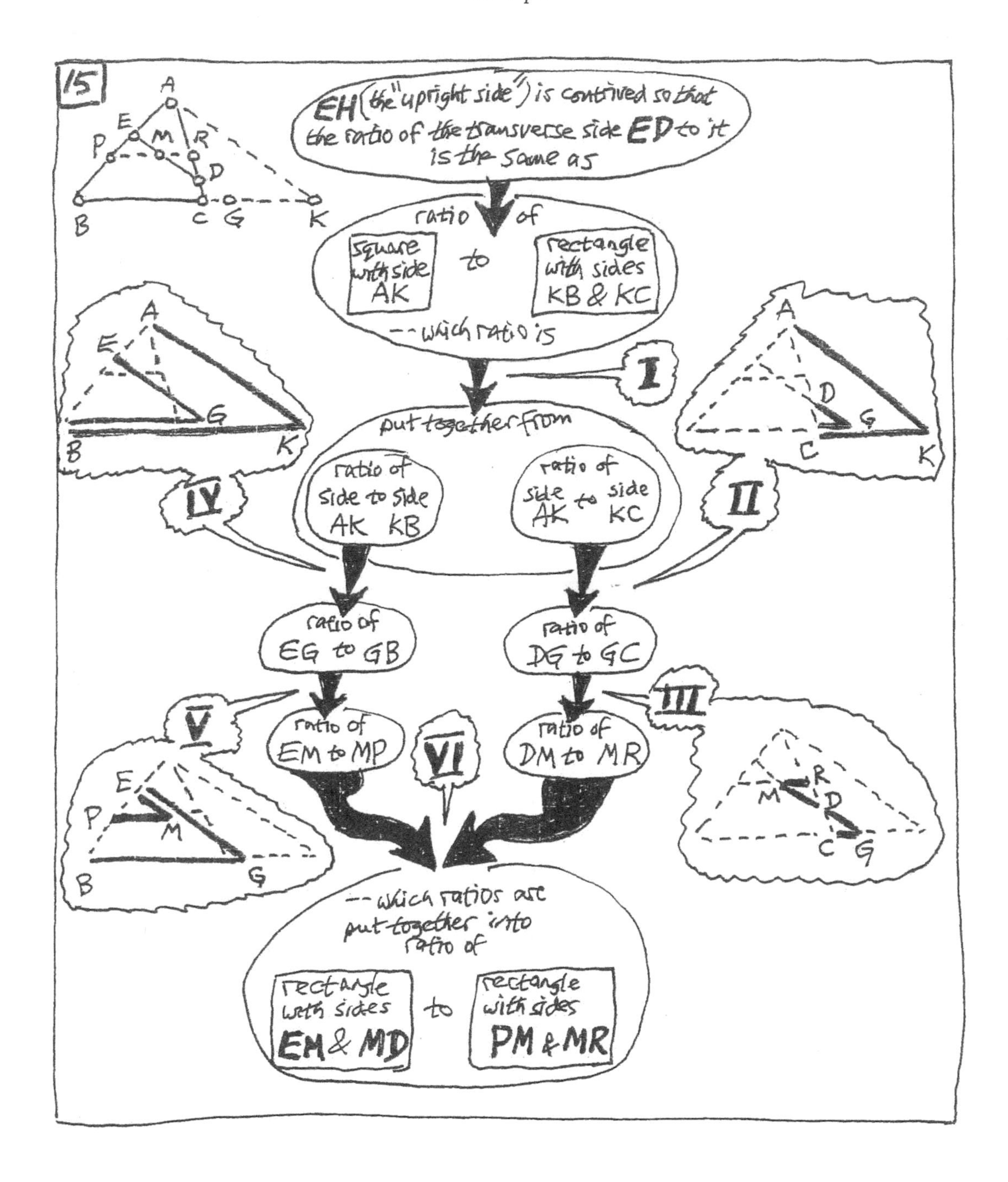
15
EH ("the upright side") is contrived so that the ratio of the transverse side ED to it is the same as
ratio of
square with side AK
to
rectangle with sides KB & KC
-- which ratio is
I
put together from
ratio of side to side AK KB
ratio of side to side AK KC
IV
II
ratio of EG to GB
ratio of DG to GC
III
V
ratio of EM to MP
VI
ratio of DM to MR
-- which ratios are put together into ratio of
rectangle with sides EM & MD
to
rectangle with sides PM & MR

That figure at which we have just looked (figure 15 for the ellipse of Proposition 13) should be compared and contrasted with figure 14 of the previous chapter—which is the corresponding figure for the hyperbola of Proposition 12—and with figure 10 of the chapter before that, which is the corresponding figure for the parabola of Proposition 11.

Now consider figure 16, which presents a map that puts together the several pieces of the ellipse-demonstration of this 13th proposition. It should be compared and contrasted with figure 15 of the previous chapter—which is the corresponding figure for the hyperbola of Proposition 12—and with figure 11 of the chapter before that, which is the corresponding figure for the parabola of Proposition 11.

In the case of the ellipse as in that of the hyperbola, and of the parabola before it, what enables us to articulate the relationship between the ordinate and the abscissa is a third straight line, one which we make just the right size to embody certain ratios. Although we cannot relate the ordinate and abscissa directly, we can do so indirectly—by first contriving a line (the upright side) into which we build the relevant relationships, and by then building boxes out of those several straight lines (namely, out of the line that is the ordinate, the line that is the abscissa, and the line that is contrived to be the upright side).

But the relationship now in the case of the ellipse, as in that of the hyperbola before it, is more indirect than it was in the case of the parabola. Again, there is a more complex relationship built into the articulation of the relationship between the ordinate and abscissa. Again, this complexity is the relationship to the upright side of a side that is not upright but oblique (which name is customarily but less faithfully translated as "transverse").

In the case of the hyperbola, the equal rectangle was larger than that rectangle whose sides are the abscissa and the upright side, and the form of the rectangular excess was given (as the form of any rectangle is given) by the ratio of its sides—which was the ratio of the oblique (transverse) side to the upright side. The case of the ellipse is like that of the hyperbola, but it is not the same. In the case of the ellipse, boxes of the same sort are built, but the equal rectangle is smaller than, instead of being larger than, the rectangle whose sides are the abscissa and the upright side; and the rectangular discrepancy, whose sides have to each other the same ratio as do the oblique (transverse) side and the upright side, is a deficiency rather than an excess.

In an ellipse, as in an hyperbola or a parabola, at different lengths cut off along the diameter, the ordinates themselves have different lengths; the upright side, however, has the same length. That is to say: the parameter which we contrive is the same for every one of that diameter's ordinates in that section. But we could have contrived any number of lines of constant length. Where could we have gotten the notion to contrive *that* one?

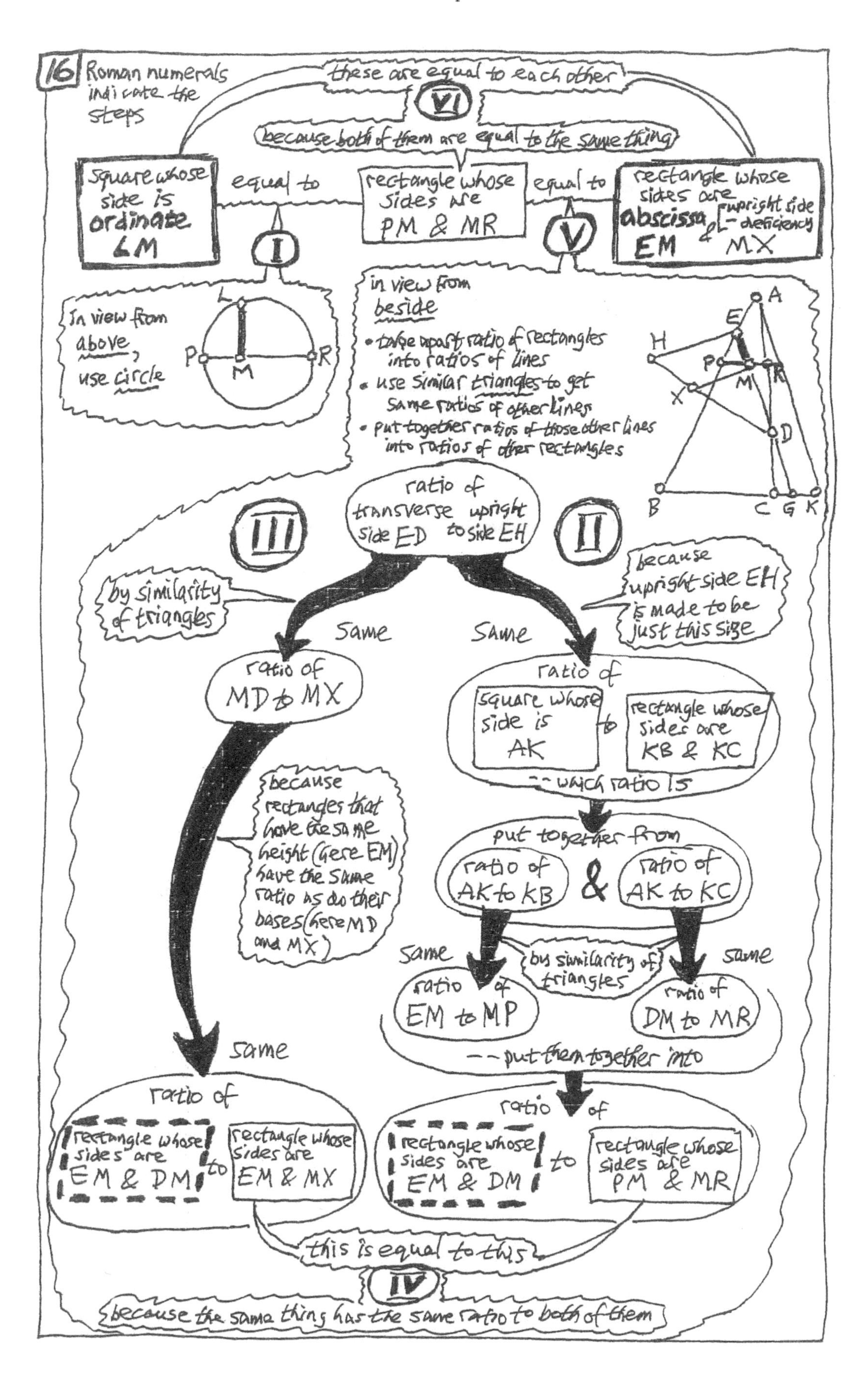
16
Roman numerals indicate the steps
these are equal to each other
VI
because both of them are equal to the same thing
square whose side is ordinate LM
equal to
rectangle whose sides are PM & MR
equal to
rectangle whose sides are abscissa EM & [upright side − deficiency] MX
I
V
In view from above, use circle
L
P
M
R
in view from beside
• take apart ratio of rectangles into ratios of lines
• use similar triangles to get same ratios of other lines
• put together ratios of those other lines into ratios of other rectangles
A
E
H
P
M
R
X
D
B
C
G
K
ratio of transverse side ED to upright side EH
III
II
by similarity of triangles
because upright side EH is made to be just this size
same
same
ratio of MD to MX
ratio of
square whose side is AK
to
rectangle whose sides are KB & KC
-- which ratio is
because rectangles that have the same height (here EM) have the same ratio as do their bases (here MD and MX)
put together from
ratio of AK to KB
&
ratio of AK to KC
same
by similarity of triangles
same
ratio of EM to MP
ratio of DM to MR
-- put them together into
same
ratio of
rectangle whose sides are EM & DM
to
rectangle whose sides are EM & MX
ratio of
rectangle whose sides are EM & DM
to
rectangle whose sides are PM & MR
this is equal to this
IV
because the same thing has the same ratio to both of them

What we are interested in doing is articulating the relationship between the size of the ordinate and the size of the abscissa. In the case of this section, the ellipse, as in that of the hyperbola, the relationship turns out to be quite a mouthful (fig. 17):

> Contrive a line EH of such a size that
> the oblique (transverse) side ED will have to it
> the same ratio to which
> a certain square has to a certain rectangle.
> The square has as its side a line AK drawn
> parallel to the section's diameter,
> from the axial triangle's vertex all the way down to the extension of its base;
> and the rectangle has as its sides
> the lines (KB and KC) that join the bottom-point of that drawn line AK
> and the endpoints of the axial triangle's base
>
> And from the line EH, get a line MX which
> falls short of the contrived upright side EH just far enough so that
> the difference between the oblique (transverse) side and the abscissa
> will have to it (MX)
> the same ratio which
> the oblique (transverse) side has to the upright side.
>
> *If*, now, you take a square whose side is the ordinate LM
> and, equal to that square, you make a rectangle whose one side is the abscissa EM,
> *then* this rectangle's *other* side must be line MX.

In other words:

> a rectangle which has as one of its sides the abscissa
> and which is equal to a square whose side is the ordinate,
> will be smaller than
> a rectangle which also has as one of its sides the abscissa
> but which has as its other side
> (its side standing upright upon the abscissa) the contrived line—
>
> and it will be deficient by just so much, that
> the rectangular deficiency which has as one of its sides
> a line of the same size as the abscissa
> will (except for size) look just like the rectangle whose sides are
> the oblique (transverse) side and the upright side.

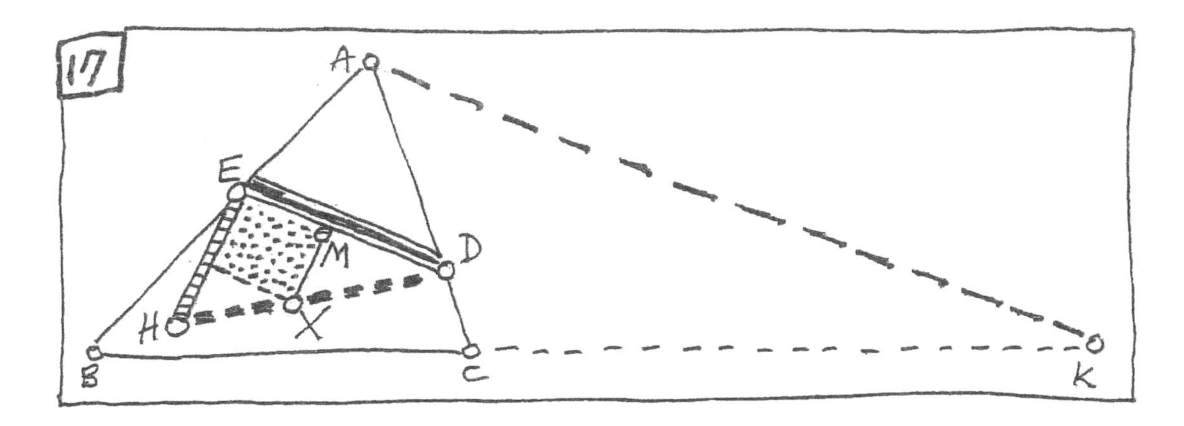

As before with the hyperbola and parabola, so also here with the ellipse, we find a constant relationship between the changing abscissa and the changing ordinate. We relate these changing lines indirectly, by forming boxes out of them together with a constant line (the upright side).

In curves of all three sorts, the relationship is this: the square on the ordinate is equal to a rectangle having the abscissa as one of its sides.

In the parabola, the relationship is not very complicated, since the other side of the equal rectangle is the constant upright side. But in the ellipse, as in the hyperbola, the relationship is more complicated. In these two curves, the equal rectangle that has the abscissa as one of its sides does not have as its other side the upright side; what it has as its other side is a line that is either greater or smaller than the upright side. Apollonius has devised his nomenclature to indicate the three possibilities for a curved non-circular section of a cone: the equal rectangle that has the abscissa as one of its sides is, in one case (that of the *para-bolê*), thrown exactly alongside the upright side; is, in the next case (that of the *hyper-bolê*), thrown over-beyond the upright side; and is, finally, in the third case, (that of the *elleip-sis*), lacking in the length to stretch all the way along the upright side. In the case of the ellipse as in that of the hyperbola, it is because the other side of the equal rectangle is a width which changes, that it cannot be *equal to* any constant upright side. It is, however, *determined by* a constant upright side.

That line, if we do contrive it, will do the work of the demonstration. But what work will show *why* we should have contrived it that way in the first place? We have seen what comes from the contriving of the upright side. We would like to see what the contriving of the upright side itself comes from. That is the work of analysis. We can proceed somewhat as we did with the parabola—but the analysis for the ellipse is longer and more complicated, as it was for the hyperbola. We do for the ellipse what we did for the hyperbola. Instead of moving right on and casting your eye on that, you may want to pause now in your reading, and try your hand at doing on your own an analysis for the ellipse.

analysis to find contrivance for ellipse

As we move down along the conic section's diameter from its vertex E, what stays the same and what becomes different at different points on the diameter? Let us consider things at some point—call it M.

Let us take the *top* view first (fig. 18). It shows us a circle. Even though the circle's diameter PR grows as we move down along the section's diameter, the square on the ordinate LM stays equal to the rectangle contained by PM and MR. The square changes size as we move down along the section's diameter, but its size is always the same as that of the rectangle.

Now let us take the *side* view (fig. 19). As we go down along the conic section's diameter, cutting off a longer and longer abscissa EM, and getting a bigger and bigger diameter PMR for the circle, what happens to that rectangle with sides PM and MR which is equal to the square on ordinate LM? That rectangle changes size, as do both of its sides, PM and MR. Whereas earlier, in the parabola's rectangle, only one of the sides grew and the other stayed the same size, and in the hyperbola's rectangle neither side stayed the same size but both grew, now in the ellipse neither stays the same size but one grows and the other shrinks. PM grows and MR shrinks—and yet something stays the same about each one that changes size. What stays the same about the side PM is its ratio to the growing abscissa (EM): that ratio of PM to EM stays the same because (since PMR stays parallel to the axial triangle's base BC) the little triangle PME (on the left) keeps the same shape while PM descends. And what stays the same about the side MR is its ratio to the shrinking line MD (which is the remainder left by taking the growing abscissa EM away from the constant oblique [transverse] side ED): the ratio of MR to MD stays the same because (again, since PMR stays parallel to the axial triangle's base BC) the little triangle RMD (on the right) keeps the same shape while MR descends.

Thus, the ratio of EM to MP, and the ratio of DM to MR, are both constant. Let us put these two constant ratios together, so that we shall have only one ratio to consider. If these two ratios of straight lines are compounded, they will yield a ratio that is the same as a ratio of rectangles made by taking those straight lines as sides—namely, the ratio of rectangle [EM, MD] to rectangle [PM, MR]. This ratio, since it is compounded out of constant ratios, will itself be constant. Let us see how this can help us.

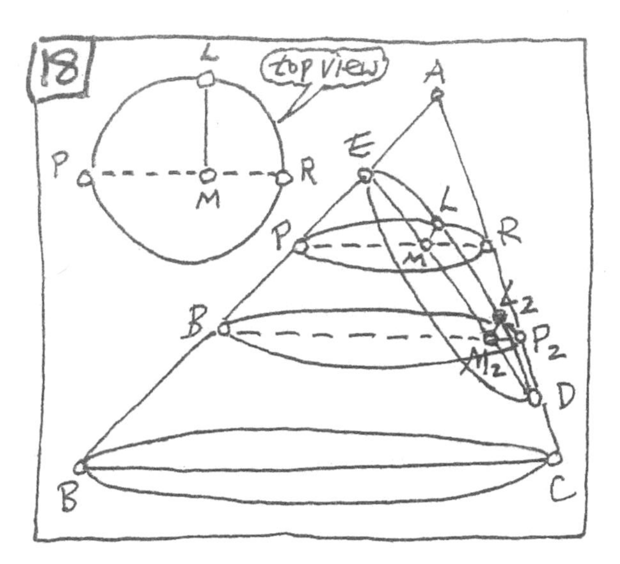

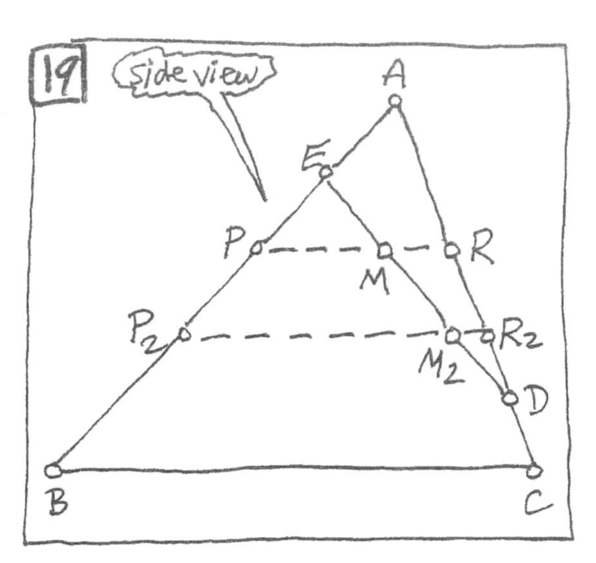

From this constancy of the ratio of rectangle [EM, MD] to rectangle [PM, MR], what we now know is this—that although the ordinate LM changes size as the abscissa EM grows, nonetheless a certain relationship between them stays the same. The relationship is this: the rectangle with sides PM and MR, which is equal to the square whose side is the growing *ordinate*, will have a constant ratio to a rectangle with the following sides—one side (MD) is a shrinking part of a constant whole (this constant whole being the oblique [transverse] side ED), but the abscissa is its growing part, and this growing *abscissa* (EM) is the other side.

So, now (fig. 20) we have the constancy of the ratio of the rectangle with sides MD and EM to the rectangle with the sides PM and MR. The latter term in this ratio is the rectangle contained by the segments (PM and MR) into which the circle's diameter is split by the ordinate. The other term is the rectangle contained by the abscissa (EM) and the difference (MD) between the oblique (transverse) side and the abscissa. That is to say: we have the rectangle which concerns us (the rectangle with sides PM and MR) in a constant ratio to another rectangle, *one* of whose sides is the line which concerns us (abscissa EM), which line is also the non-constant part taken away from a constant whole (DE) to give this rectangle's *other* side (MD).

But this ratio is not what we are after. It is only a step on the way. What we *have* is a rectangle that has a *constant ratio* to the rectangle with sides PM and MR, but what we would *like* to have is a rectangle that is *equal* to the rectangle with sides PM and MR, because then it would be equal to the square that has the ordinate as its side. Moreover, we want that equal rectangle (like the rectangle of constant ratio which we now have) to have the abscissa EM as one of its own sides. If we had the rectangle that we want, then we would have the ordinate and abscissa in relation to each other—not directly, to be sure, but at least through their power to make a square or some other rectangle.

What do we need in order to have what we want? We already have the constancy of the ratio of the rectangle with sides EM and MD to the rectangle with sides PM and MR. We want to have a rectangle which is equal to the second rectangle in the ratio, and which has as *one* of its own sides one of the sides of the first rectangle in the ratio—namely, the abscissa EM. If we had the *other* side, then we would have that rectangle itself—which is what we want. We want a rectangle whose side is equal to the abscissa while the rectangle itself is equal to a square whose side is equal to the ordinate (fig. 21). Our problem, then, is this: to find out what should be its other side.

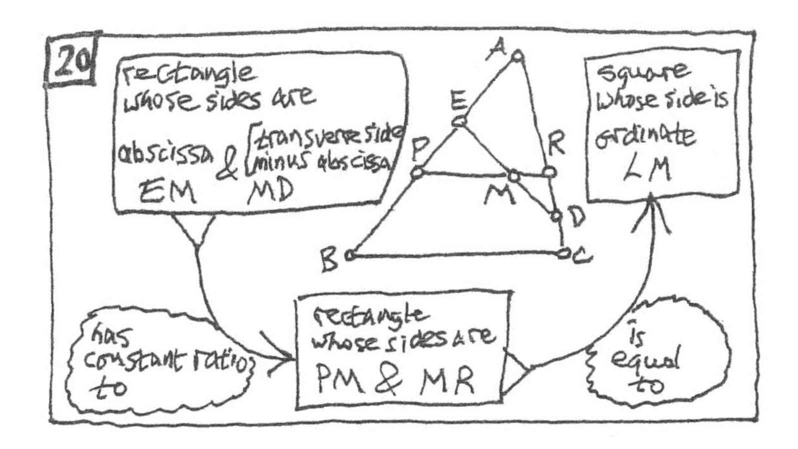
20
rectangle whose sides are
abscissa EM & [transverse side minus abscissa MD
A
E
P
R
M
D
B
C
square whose side is ordinate LM
has constant ratio to
rectangle whose sides are PM & MR
is equal to

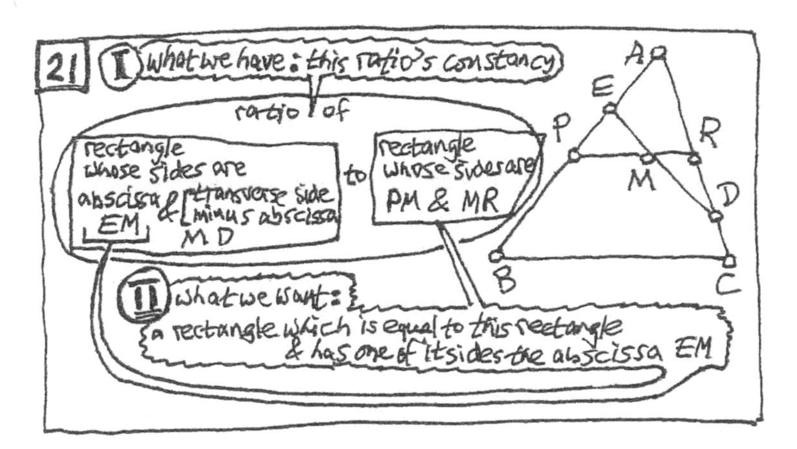
21
I what we have: this ratio's constancy
ratio of
rectangle whose sides are abscissa EM & [transverse side minus abscissa MD
to
rectangle whose sides are PM & MR
A
E
P
R
M
D
B
C
II what we want:
a rectangle which is equal to this rectangle & has one of its sides the abscissa EM

Let us call that unknown other side "width (w)"—and, even though we do not have it, let us *suppose* that we do. That is, let us ask (step I): *what if* we *did* have it? (Fig. 22.)

Now (step II) two magnitudes can be equal to each other only if a third magnitude has the same ratio to each of them that it has to the other. So if we *did* have that unknown side, width (w), then any rectangle (any one at all) would have, to the rectangle that the unknown width (w) makes with the abscissa—to the rectangle, namely, with the sides EM and (w)—the same ratio that it has to its *equal* (to the rectangle, namely, with sides PM and MR). And we already know something about the ratio, to this rectangle with sides PM and MR, of one rectangle in particular; we know the ratio to it of the rectangle, namely, with sides EM and MD. Hence, we know that the rectangle with sides PM and MR can be equal to the rectangle with sides EM and (w) only if the rectangle with sides EM and MD has the same ratio to each of them.

Now (step III), taking the ratio which involves the unknown line, width (w), we see that the two rectangles—one whose sides are EM and MD, and the other whose sides are EM and (w)—both have the same height EM; and so, the ratio of the one rectangle to the other is the same as the ratio of their bases, which is the ratio of line MD to width (w).

Therefore (step IV), although we do not have the unknown line, width (w), we do know its relation to things we do have—we know that *if* we had it, then the ratio to it of MD (which is a line that is part of a constant line) would be the same as the ratio of the rectangle with sides EM and MD to the rectangle with sides PM and MR, which is a ratio that is constant.

But although this ratio is itself constant, it is *not* stated entirely in terms of *lines* that are constant, and so we have not yet determined the unknown line, width (w).

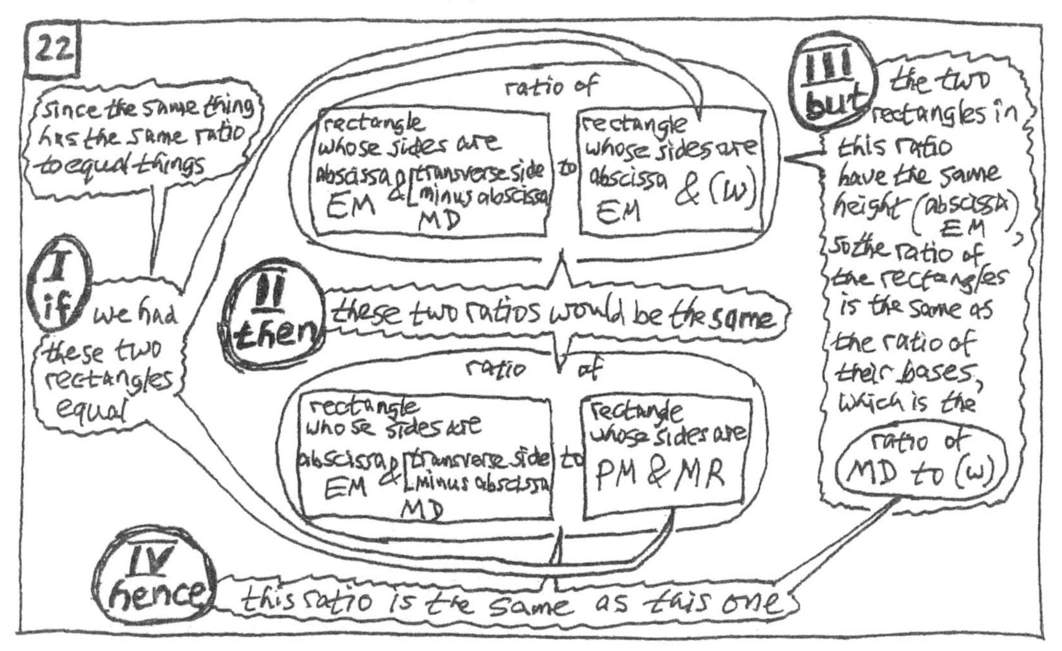

We need to find a ratio which involves only lines that are constant but which is the same as that ratio which we do have. Of the four lines involved in the ratio which we do have (the ratio of the rectangle whose sides are EM and MD to the rectangle whose sides are PM and MR), not one of them is constant.

To find the ratio that we need (fig. 23), we (step I) take apart the ratio that we have. We take apart the single ratio of rectangles into its two component ratios of lines. Since the ratio of the rectangles is the same as the ratio compounded out of the ratios of their sides, the ratio of the rectangles is the same as the ratio compounded out of two ratios—those of EM to MP and of DM to MR, the latter of which ratios, by the similarity of triangles, is the same as the ratio of DG to GB.

Now (step II), we use similar triangles whose sides will be constant lines. And we draw a straight line parallel to the diameter, from the vertex A, down to the extension of the axial triangle's base, so that, we might say, this line AK "externally splits" the base into "segments" KB and KC. And so, by the similarity of triangles, those same ratios that we got in the first step we get now in terms of constant lines, as ratios of AK to KB and of AK to KC.

And now (step III), we put the component ratios back together into one ratio of rectangles, which will be the same *ratio* that we had before, but now with different *terms*, terms now involving none but *constant* lines; this is the ratio that the square whose side is AK has to the rectangle whose sides are KB and KC.

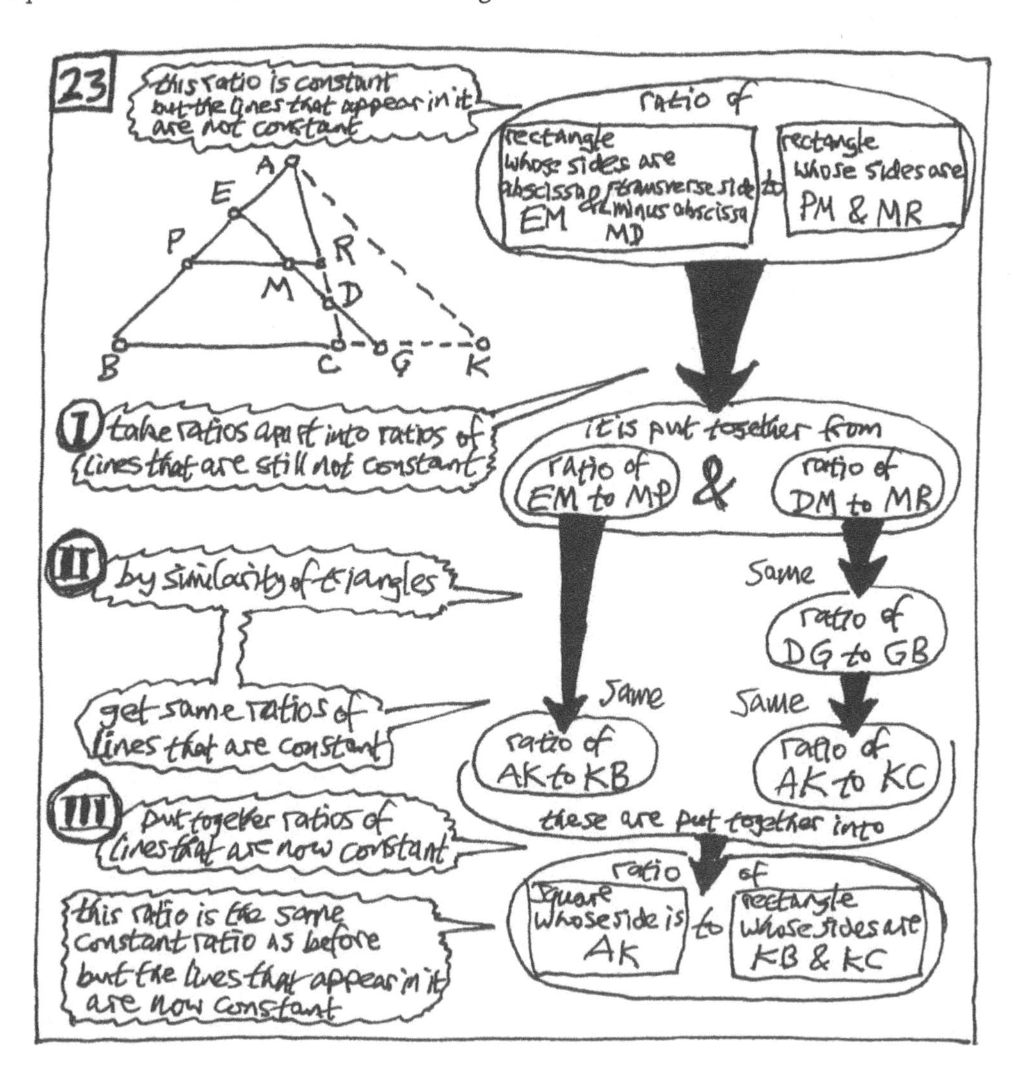

Now we are ready to do the requisite contriving. Figure 24 shows why. We have seen (step I) that if we had the width (w) then the ratio of the line DM to it would be the same as the ratio of the rectangle whose sides are DM and ME to the rectangle whose sides are PM and MR, and (step II) that this ratio of rectangles is the same as the ratio of the square whose side is AK to the rectangle whose sides are KB and KC, and so (step III) that if we had the width (w) then the ratio of line DM to it would be the same as this last ratio.

We therefore now contrive a line (call it MX) having just the right size to be the width (w). It (fig. 25) does so by having just the right size to make two ratios the same. One of them is the ratio, to this contrived line MX, of the line DM (DM being the line that stretches from the abscissa's non-vertex endpoint to the transverse side's non-vertex endpoint—which is to say, DM is the difference between the transverse side and the abscissa). That ratio of DM to MX is (by our contrivance) the same as a ratio set up by our drawing the line AK—namely, the ratio of the square arising from AK to the rectangle contained by the lines (KB and KC) that join the drawn line's bottom-point to the endpoints of the axial triangle's base. This ratio of square to rectangle involves only constant lines, but it is the same ratio as that original constant ratio which involved lines not all of which were constant.

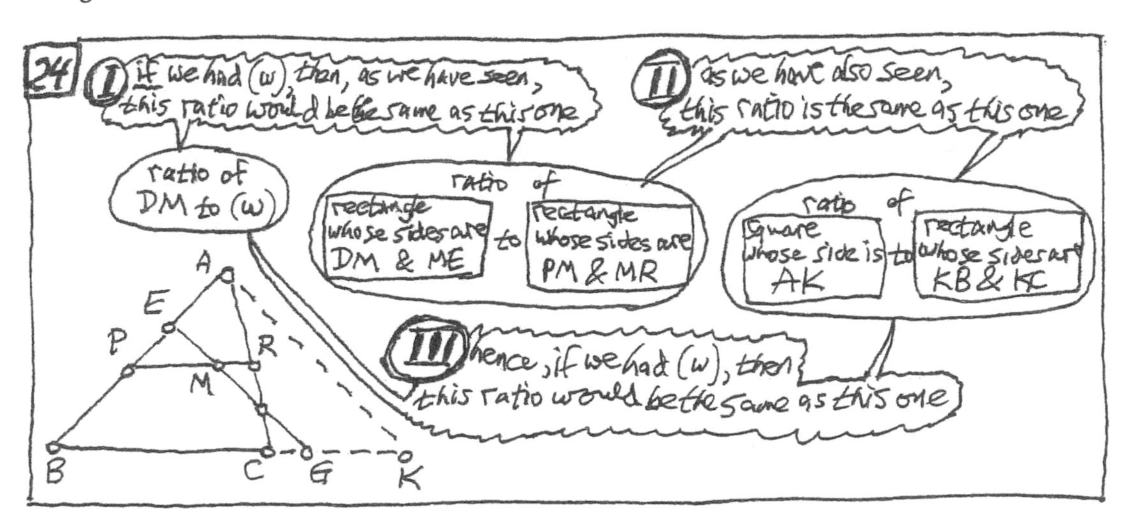

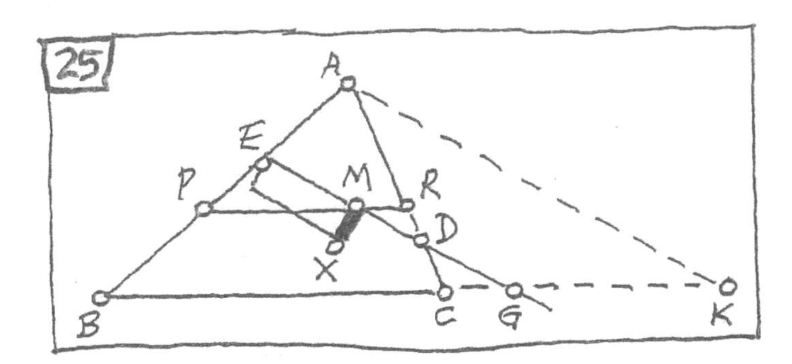

If we do contrive in that way, then we should get equality between the two rectangles—the one having as sides PM and MR, and the other having as sides the abscissa EM and the contrived line MX. This line MX which we have contrived is the width (w) which will make, with the abscissa, a rectangle equal to the square whose side is the ordinate. So, by means of our contrivance (line MX), we have related the abscissa and the ordinate.

Our analytic procedure so far for the ellipse has been the same as it was for the hyperbola, as will be evident from comparing the figures that we have just seen for the analysis of the ellipse with the figures that we saw for the analysis of the hyperbola in the previous chapter. The line MX that we have contrived for the ellipse now does the job that was done earlier by the line that we contrived for the hyperbola.

But this line MX (like its counterpart for the hyperbola)—by contrast with the parabola's upright side—does not have a size that is constant. It cannot be constant because the line DM, which has a ratio to it that is constant, does not itself have a size that is constant.

This shrinking line DM is, however, part of a constant whole, the oblique (transverse) side (ED), of which the abscissa (EM) is its growing part. Can we get some constant line that will, together with the constant oblique (transverse) side (ED), determine the changing width (MX) that is associated with the growing abscissa (EM)? If we can, then we shall have related the changing ordinate and the changing abscissa, indirectly, by means of a line that is constant. Let us therefore continue the analysis.

The equal rectangle's width (MX) is different for every different height (abscissa EM). (Fig. 26.) But while both those lines—height and width—change, a certain ratio stays the same. Consider line MD, which is the difference between the oblique (transverse) side DE and the abscissa EM (the abscissa being the equal rectangle's height). This line MD has a constant ratio to the rectangle's width MX (fig 27). While both these lines (DM and width MX) shrink as height EM grows, the ratio of the one to the other (the ratio of DM to MX) stays the same; this is always the same ratio that the constant oblique (transverse) side ED has to another constant line—call it EH (fig. 28). Thus the constant rectangle which the oblique (transverse) side ED makes with the constant line EH gives the constant shape of the rectangle by which the equal rectangle falls short of the rectangle which the abscissa EM makes with EH (fig. 29). If a rectangle is made with the abscissa EM as height and with this constant line EH as width, then the equal rectangle—whose height is the abscissa EM, and whose width (w) is MX—will fall short of it by a certain amount. This deficiency is equal to a rectangle whose height is also the abscissa EM, but whose own width is just the right size to give it the same shape as a rectangle whose sides are the constant oblique (transverse) side DE and that constant line EH.

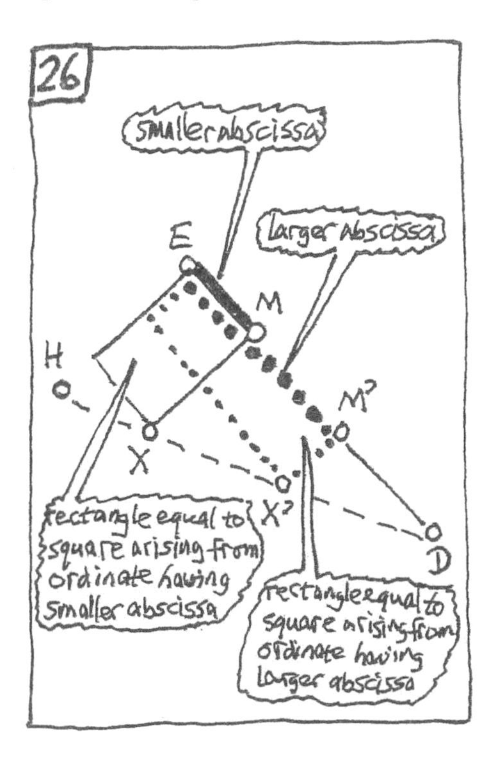

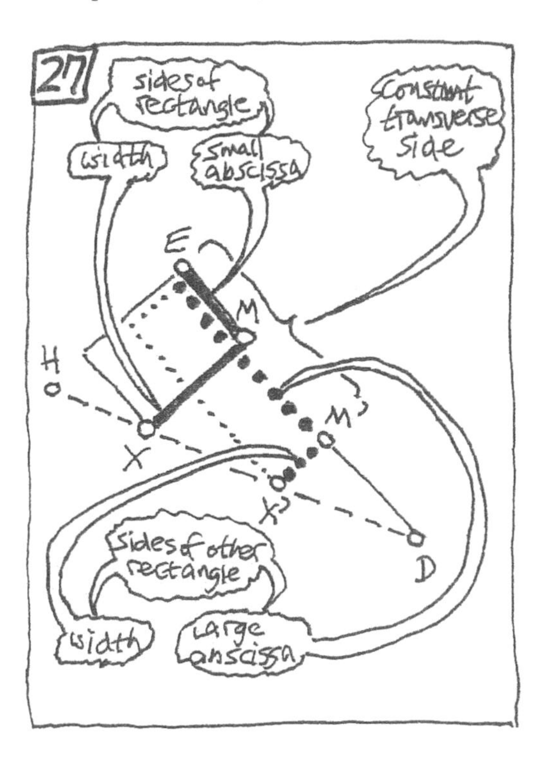

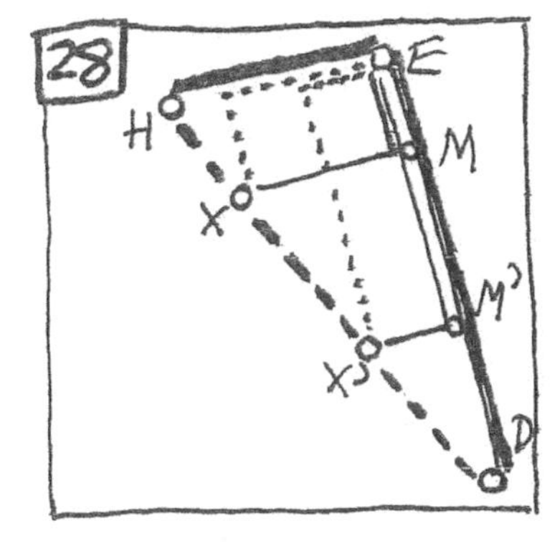

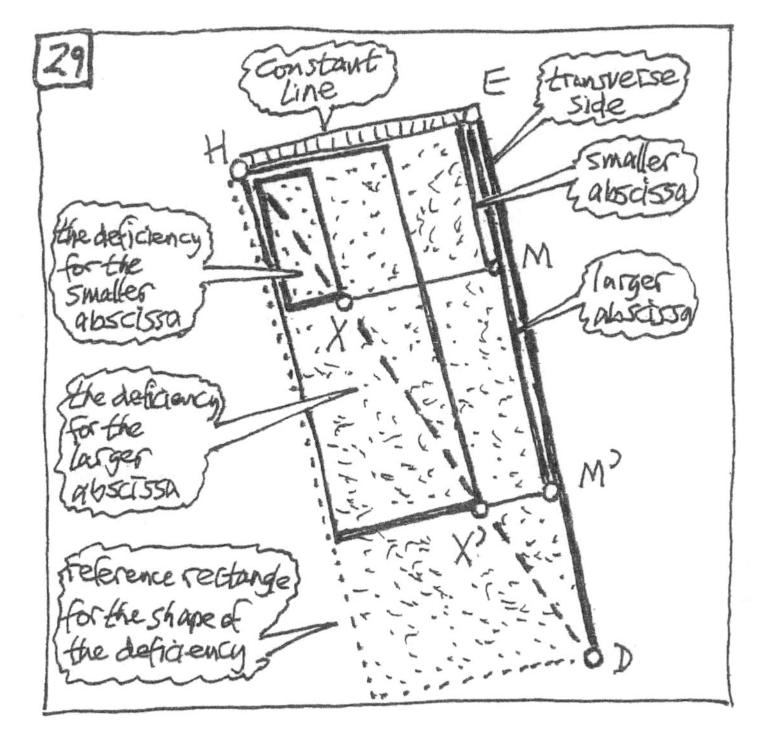

This rectangle to which the rectangular deficiency is always similar has its one side along the diameter; its other side is a line which sticks upright from the diameter. This other side (EH) will be called the *upright side* of this conic section. Earlier, with the parabola, the upright side was a constant line which stuck upright from the diameter and which gave us the width of the equal rectangle. Now, with the ellipse (as with the hyperbola), the upright side is also a constant line which sticks upright from the diameter and which gives us the width of the equal rectangle, but it gives it in a more complicated way than with the parabola. Things are more complicated because the equal rectangle (equal to the square arising from the ellipse's ordinate) now has a width that always, by some definite amount, falls short of the upright side. The ellipse and the hyperbola differ in the one's having a lack where the other has an overshoot, but otherwise the one is like the other with respect to the upright side.

So, this is what we do to get a constant upright side for determining the width MX of the rectangle that is equal to the square on the ordinate for the abscissa EM:

> we contrive the upright side EH so that it will have,
> to the constant oblique (transverse) side ED,
> the following ratio—
>
> the constant ratio which
> the equal rectangle's shrinking width MX
> has to the shrinking line MD
> (MD being the difference between
> the oblique (transverse) side ED and
> the equal rectangle's growing height, abscissa EM).

Let us consider where we are, now that we have a proportion for contriving the upright side—namely, this proportion: oblique (transverse) side ED will have to it (the upright side) the same ratio that square [AK] has to rectangle [KB, KC]. Not only are the ratios identified in this contriving proportion both constant, but the lines mentioned in the ratios are all constant. The ratios are the same as each other because each of them is the same as another ratio—namely, that of DM to MX. Although this ratio of DM to MX was useful for our purposes because it is a constant ratio, it was a mere step on the way—because its terms (DM and MX) are not constant. We have used the ratio of DM to MX as a step on the way to finding a proportion that defines a constant line that will serve as the upright side.

This upright side is *not, itself*, the other side of that rectangle which is equal to the square on the ordinate and has the abscissa as one side. This upright side is constant, whereas that rectangle's other side is not. The upright side is a line that *determines* the rectangle's other side for us. The rectangle's other side is *smaller* than the upright side. The *rectangular deficiency* varies in size, but it has a *constant look* about it. What we can say now is the following.

> *If* there is some width (w) which would make, with the abscissa as height,
> a rectangle equal to the square whose side is the ordinate,
> *then* that line (w) would be such that:
>
> it would be smaller than the constant upright side (u) which is
> determined by the following proportion—
> as square [AK] is to rectangle [KB, KC],
> so also the oblique (transverse) side is to the upright side (u);
>
> and the deficiency of the width (w) with respect to the upright side (u)
> would be determined by the following proportion—namely,
> as the oblique (transverse) side is to the upright side (u),
> so also the difference between the oblique (transverse) side and the abscissa
> is to the width (w). (Fig. 30.)

But can we be sure that a line contrived that way *will in fact* do the work for which we want it? Only if that is *demonstrable*. We must contrive the upright side *first* and then show that we can retrace the steps of the analysis in reverse. That is to say, our analysis cannot substitute for the synthesis in which Apollonius demonstrates the following:

> *if* we cut with a plane properly oriented,
> and place upright a line which we contrive to be of such a size that
> the oblique (transverse) side has the same ratio to it that
> square [AK] has to rectangle [KB, KC],
> *then* from this cutting and contriving, it will follow that
>
> a square whose side is the ordinate
> will be smaller than
> a rectangle whose sides are the abscissa and that line contrived as upright side,
>
> and the deficiency will be equal to
> a rectangle whose sides are the abscissa and a line to which the abscissa has
> the same ratio that the oblique (transverse) side has to the upright side—
> so that this rectangular excess has the same shape as a rectangle whose sides are
> the oblique (transverse) side and the upright side.

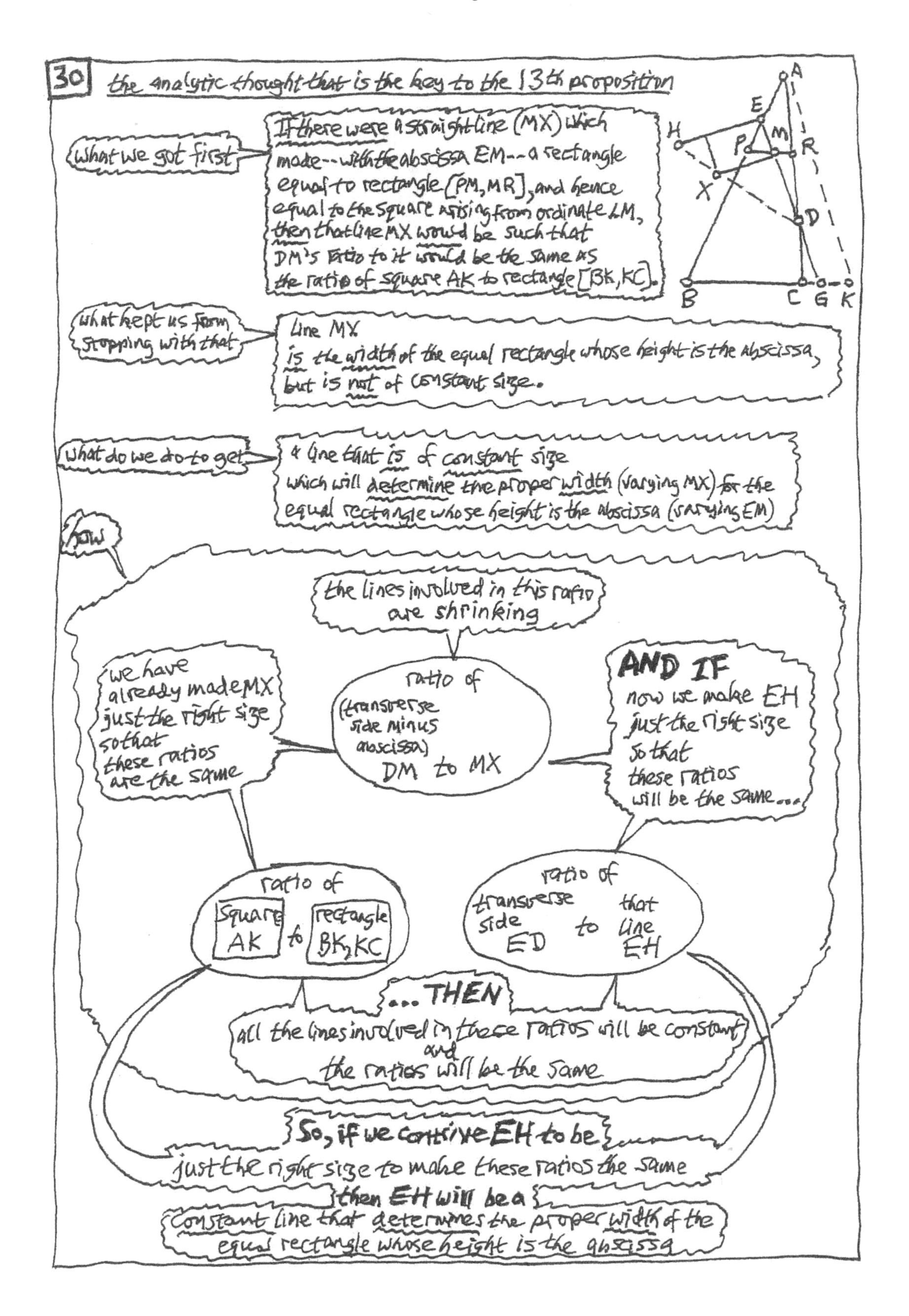
30
the analytic thought that is the key to the 13th proposition
what we got first
If there were a straight line (MX) which made--with the abscissa EM--a rectangle equal to rectangle [PM, MR], and hence equal to the square arising from ordinate LM, then that line MX would be such that DM's ratio to it would be the same as the ratio of square AK to rectangle [BK, KC].
A
E
H
P
M
R
X
D
B
C
G
K
what kept us from stopping with that
line MX is the width of the equal rectangle whose height is the abscissa, but is not of constant size.
what do we do to get
a line that is of constant size which will determine the proper width (varying MX) for the equal rectangle whose height is the abscissa (varying EM)
how
the lines involved in this ratio are shrinking
we have already made MX just the right size so that these ratios are the same
ratio of (transverse side minus abscissa) DM to MX
AND IF now we make EH just the right size so that these ratios will be the same...
ratio of square AK to rectangle BK, KC
ratio of transverse side ED to that line EH
...THEN
all the lines involved in these ratios will be constant and the ratios will be the same
So, if we contrive EH to be
just the right size to make these ratios the same
then EH will be a
constant line that determines the proper width of the equal rectangle whose height is the abscissa

the ellipse and the hyperbola

Neither in this proposition nor in the previous one is it immediately obvious why the straight line AK is drawn. (It is the straight line that is drawn from the axial triangle's vertex, parallel to the diameter, down to the axial triangle's base in the case of the hyperbola, or down to its extension in the case of the ellipse.) In the previous chapter, we saw how to dispense with the line AK in the case of the hyperbola. In the case of the ellipse also, we can dispense with it. The line AK is not needed in order to state the definition of the upright side for either section, nor even in order to state the definition of the upright side for both sections in a single statement. (Fig. 31.)

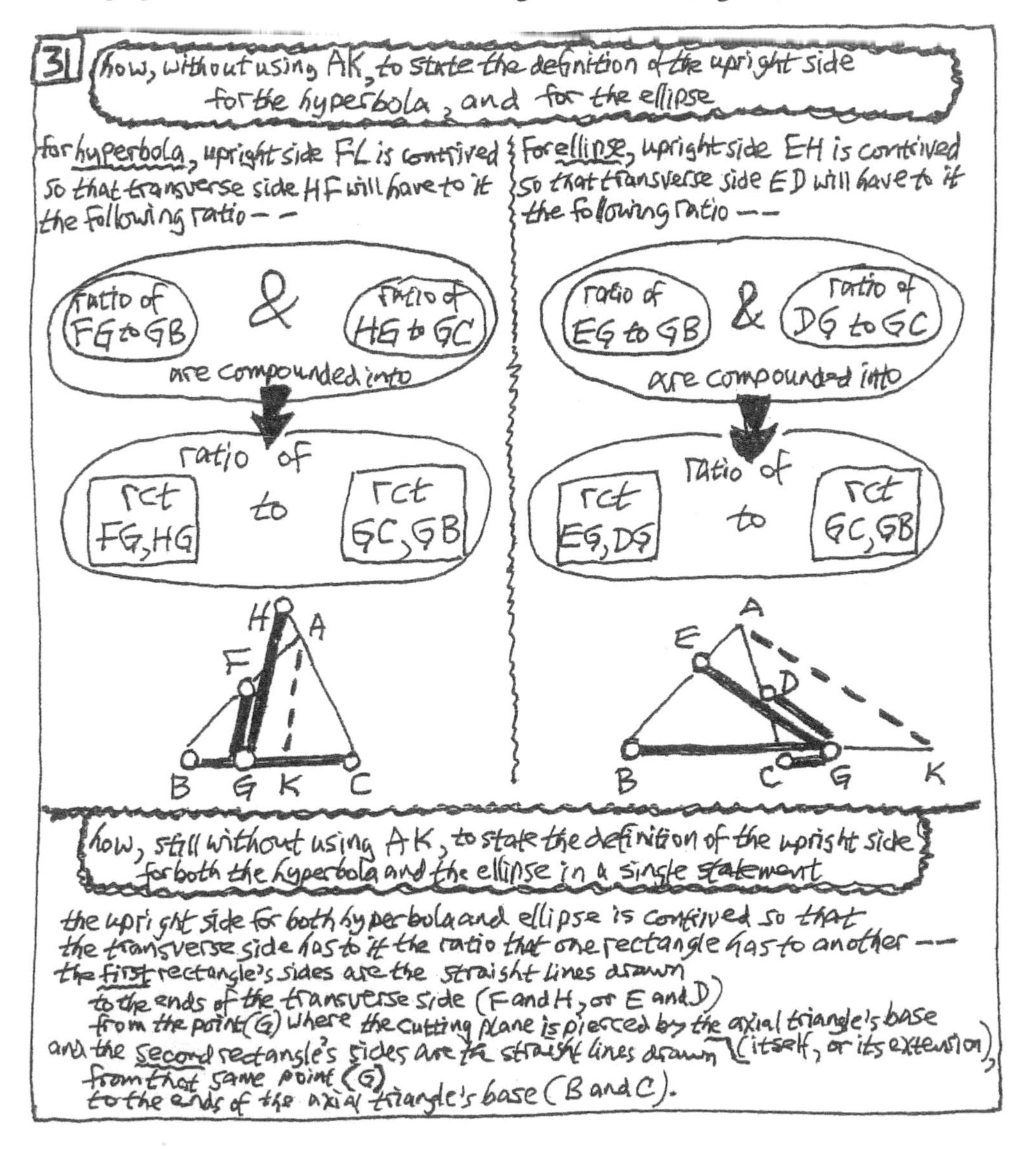

Why, then, is AK drawn in? Well, our use of AK does have the effect of making the ratio for the hyperbola and for the ellipse a ratio not of some *rectangle* to a rectangle but rather of some *square* to a rectangle—*like* the ratio for the *parabola*. (Fig. 32.) A rectangle is determined by *two* lines which contain it, whereas a square is determined by *only one* line from which it arises.

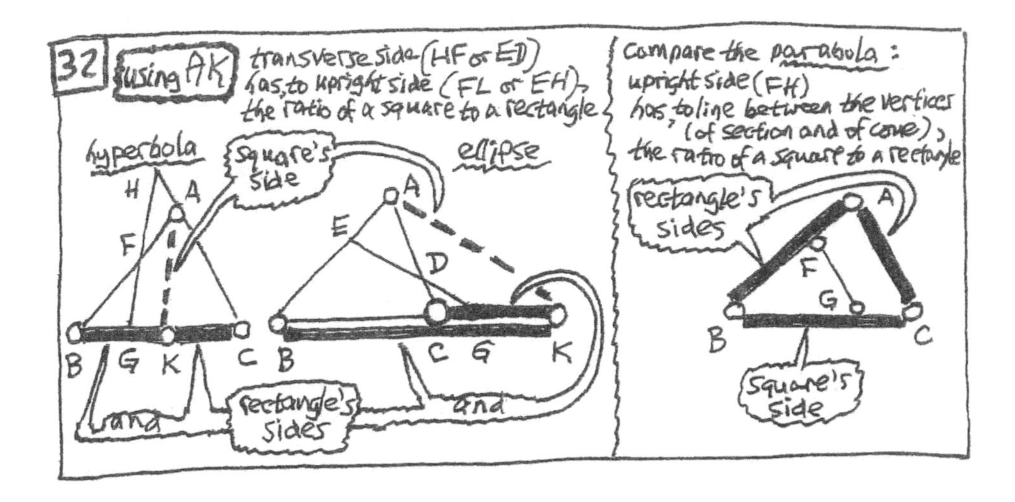

The introduction of AK in the case of the hyperbola and in the case of the ellipse thus has the effect that this one line in each case does the work of two—of FG and HG in the hyperbola, and of EG and DG in the ellipse. The result is that, as in the parabola, the upright side can be defined by a proportion that involves only three straight lines rather than four.

It was disingenuous for our analyses of the hyperbola and of the ellipse to include the drawing in and use of AK instead of FG and HG in the hyperbola, and instead of EG and DG in the ellipse. It would have been more straightforward to do each analysis without line AK. The only reason that was given for drawing in AK in our analyses was the need to get similar triangles involving constant lines—but we could have had similar triangles with the constant lines which were already in the picture. Why, then, in our analysis was the line AK drawn in? Because without the drawing in of AK, the problem-solving analysis would not include, in reverse order, *all* of the steps presented by Apollonius in his demonstrative synthesis. In the propositions of Apollonius, however, line AK was drawn in at the beginning—before the contriving of the upright side—in order to be there for use in contriving the upright side. In the analysis, therefore, the drawing in of line AK should appear at the very end—after the upright side has been found. It should appear as a final touch that simplifies the definition of the upright side by reducing the number of lines that it involves. You might, then, as an exercise, now sketch out for yourself an analysis for the hyperbola—and also one for the ellipse—in which the line AK does not get drawn in until the very end.

The drawing in of line AK is merely one of several features common to the two propositions. The demonstrations for the ellipse in the 13th proposition and for the hyperbola in the 12th proposition are not very different. With some changes in the wording, we could replace these two propositions by one. To speak in a single proposition of excess or deficiency, instead of speaking of excess in one proposition and then speaking of deficiency in another, we would have to change the lettering in the figure specified for the ellipse, so that it would correspond to the lettering in the figure specified for the hyperbola. Apollonius does not label the two figures so as to assign the same letter to corresponding points in the two figures specified by him. He is not concerned to form a pattern in the letters that he uses as signs. Rather, he assigns the letters to the points in the order in which the points come up in the propositions separately. (As was said before, the point X is so called because in the Greek alphabet the letter "Xi" comes after "Nu" and before "Omicron.") As an exercise, you might now try to write a single proposition for the two conic sections. Figures 33 and 34 present a re-labeling that may help you to get started.

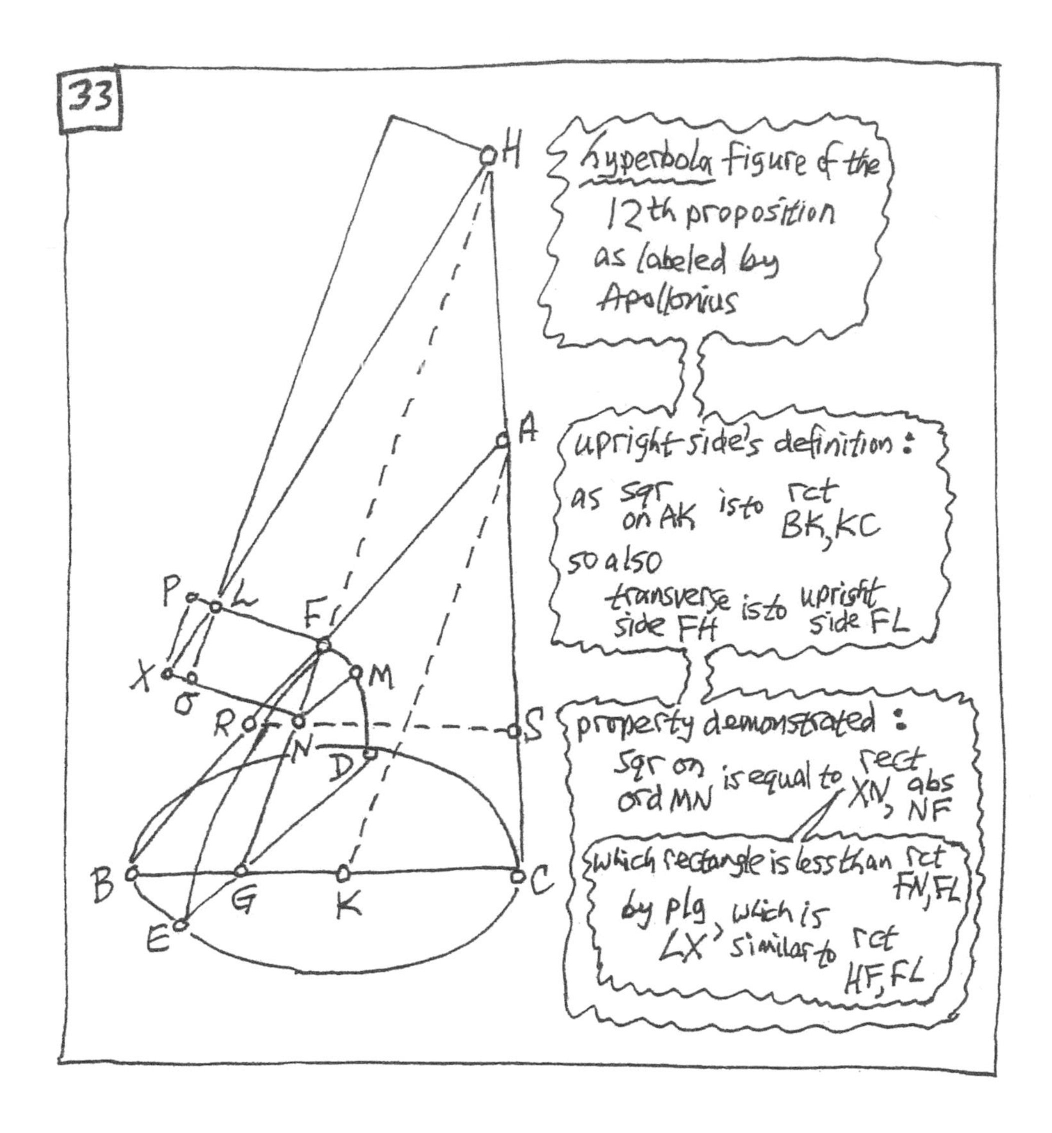
33
hyperbola figure of the 12th proposition as labeled by Apollonius
upright side's definition: as sqr on AK is to rct BK, KC so also transverse side FH is to upright side FL
property demonstrated: sqr on ord MN is equal to rect XN, abs NF
which rectangle is less than rct FN, FL by plg LX, which is similar to rct HF, FL
H
A
P
L
F
X
σ
M
R
N
S
D
B
G
K
C
E

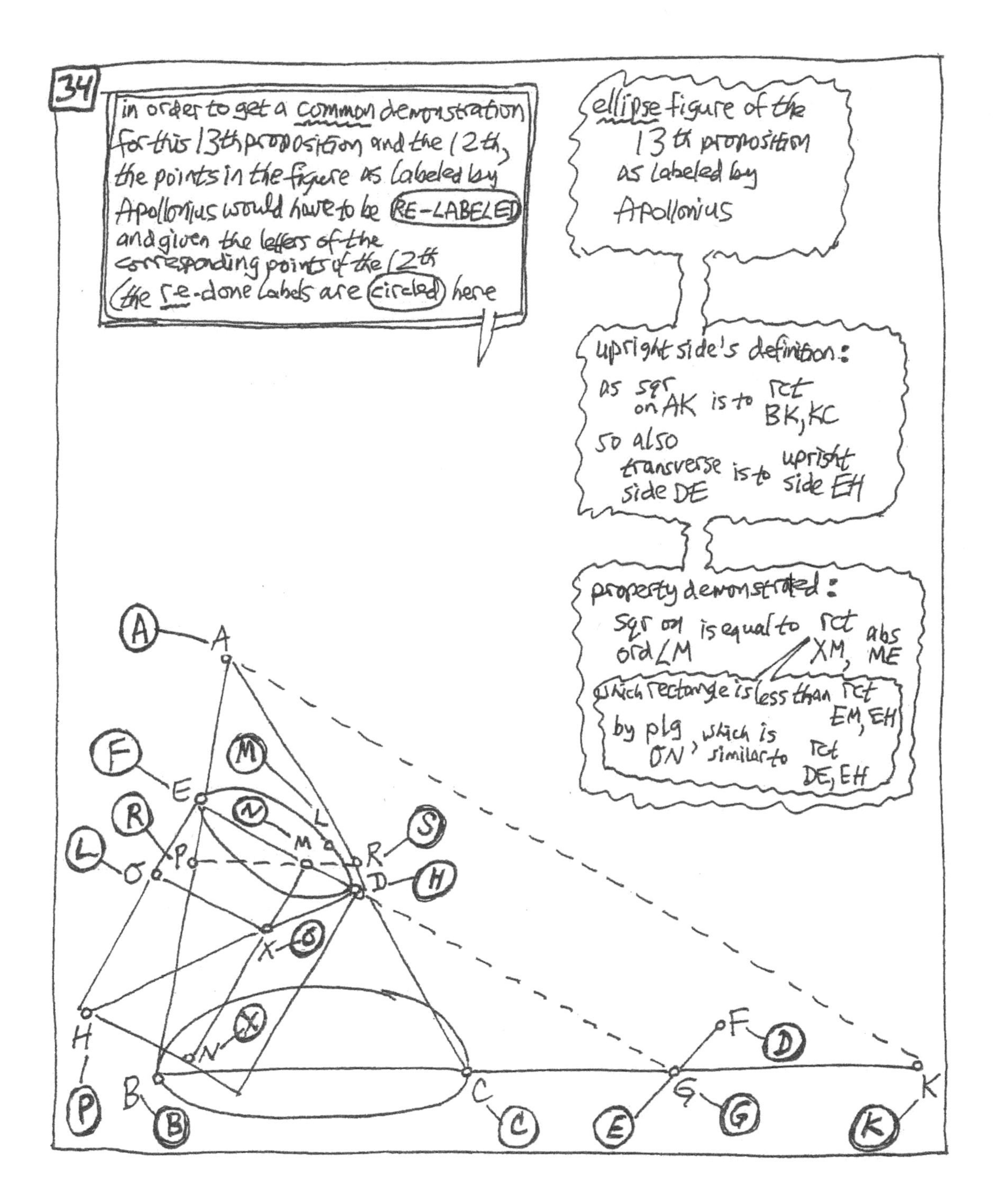
34
in order to get a common demonstration for this 13th proposition and the 12th, the points in the figure as labeled by Apollonius would have to be RE-LABELED and given the letters of the corresponding points of the 12th (the re-done labels are circled here
ellipse figure of the 13th proposition as labeled by Apollonius
upright side's definition: as sqr on AK is to rct BK, KC so also transverse side DE is to upright side EH
property demonstrated: sqr on ord LM is equal to rct abs XM, ME
which rectangle is less than rct EM, EH by plg ON, which is similar to rct DE, EH
A
F
E
M
N
L
R
P
M
L
S
R
O
H
D
X
O
H
X
N
F
D
P
B
B
C
C
E
G
G
K
K

CHAPTER 4. BRINGING THINGS TOGETHER

the upright and oblique sides

In all three sections, as we have seen, Apollonius takes the square arising from the ordinate and makes a rectangle which is equal to it and which has the abscissa as one of its sides. The rectangle, thrown out from the abscissa, stretches along the upright side. In one of the sections, the rectangle's non-abscissa side goes as far as the endpoint of the upright side, and no farther; but in the other sections it either goes beyond the endpoint or does not go as far. The rectangular discrepancy (the excess in the case of the hyperbola, the deficiency in the case of the ellipse) has its shape determined by the following proportion: as the upright side is to the oblique (transverse) side, so the non-abscissa side of the equal rectangle is to the line made by the oblique (transverse) side when the abscissa is added on to it (in the case of the hyperbola) or is taken away from it (in the case of the ellipse).

These two cases for the oblique (transverse) side and the abscissa—that is, the straight lines that are their sum and difference—can both be covered by a single description. They both can be described as a straight line that goes along the diameter between the non-vertex endpoint of the abscissa and the non-vertex endpoint of the oblique (transverse) side.

In both cases, the oblique (transverse) side itself is the line stretching from the section's vertex, in line with the diameter, to the axial triangle's other side or its extension. In the case of the ellipse, therefore, the oblique (transverse) side is the diameter itself.

The 12th proposition (about the hyperbola), before naming the oblique (transverse) side, identifies it as a line subtending an angle. This angle that the oblique (transverse) side goes across is the axial triangle's exterior angle, which is the supplement of the axial triangle's vertex angle (fig. 1). Only at the proposition's end, immediately after the naming of the upright side, is the oblique (transverse) side named. The 13th proposition (about the ellipse) also does not name the oblique (transverse) side until the end of the proposition, immediately after the naming of the upright side. Here, however, a name is less urgent—since the line is early identified by its other name; it is called the diameter (fig. 2).

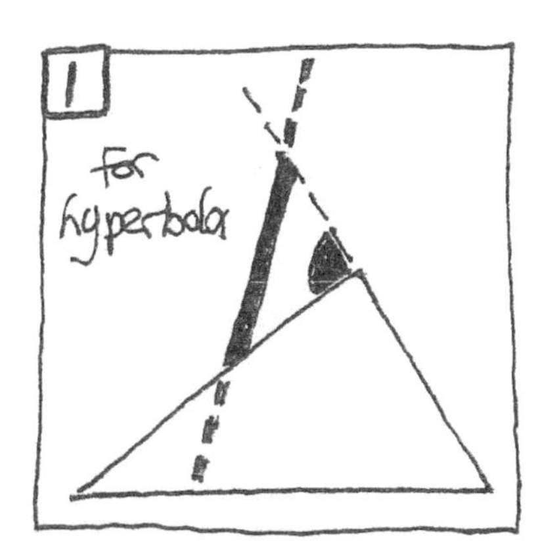

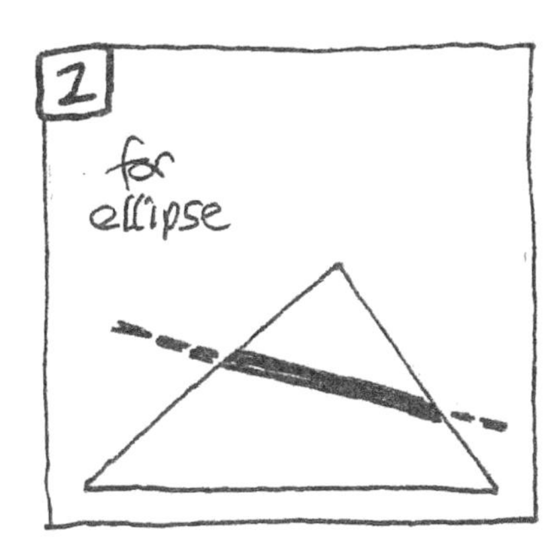

The size of the oblique (transverse) side will depend on the sizes of the angles of the axial triangle, and also on the sizes of the angles that the oblique (transverse) side itself makes with the sides of the axial triangle. It will also depend on the distance of the oblique (transverse) side from the vertex of the axial triangle. On the size of the oblique (transverse) side depends the size of the upright side of the hyperbola or of the ellipse. The parabola, however, has no oblique (transverse) side; the size of its upright side depends directly on the sizes of the angles of the axial triangle and on the diameter's distance from the vertex of the axial triangle (figs. 3 and 4).

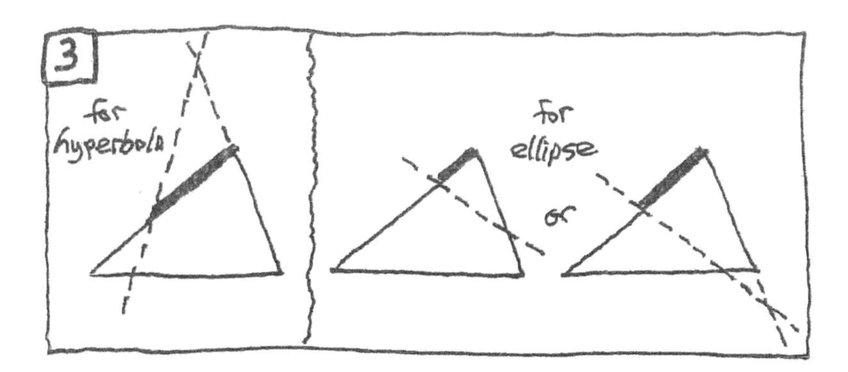

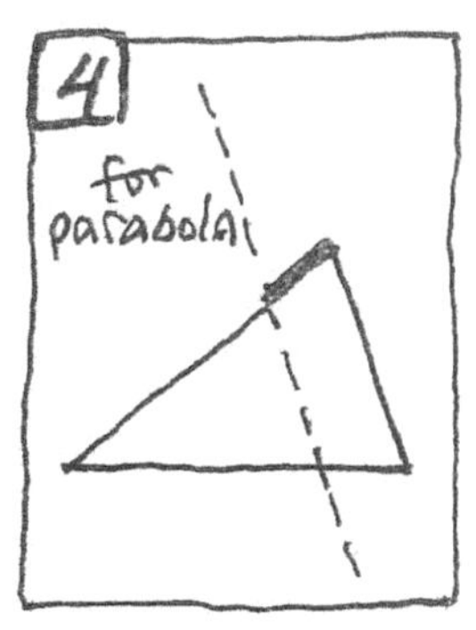

The hyperbola and the ellipse seem, in many respects, to be counterparts of each other, while the parabola stands alone. The hyperbola's abscissa is a varying extension of the constant oblique (transverse) side, whereas the ellipse's abscissa is a varying part of it. In the ratio of contrivance for the upright side, the hyperbola employs the sum of the two lines (the oblique [transverse] side and the abscissa), whereas the ellipse employs their difference—in each case, the result of addition or subtraction is a straight line stretching, in line with the diameter, from the section's vertex to the axial triangle's other side or its extension. But the parabola's diameter nowhere meets the other side of the axial triangle (fig. 5).

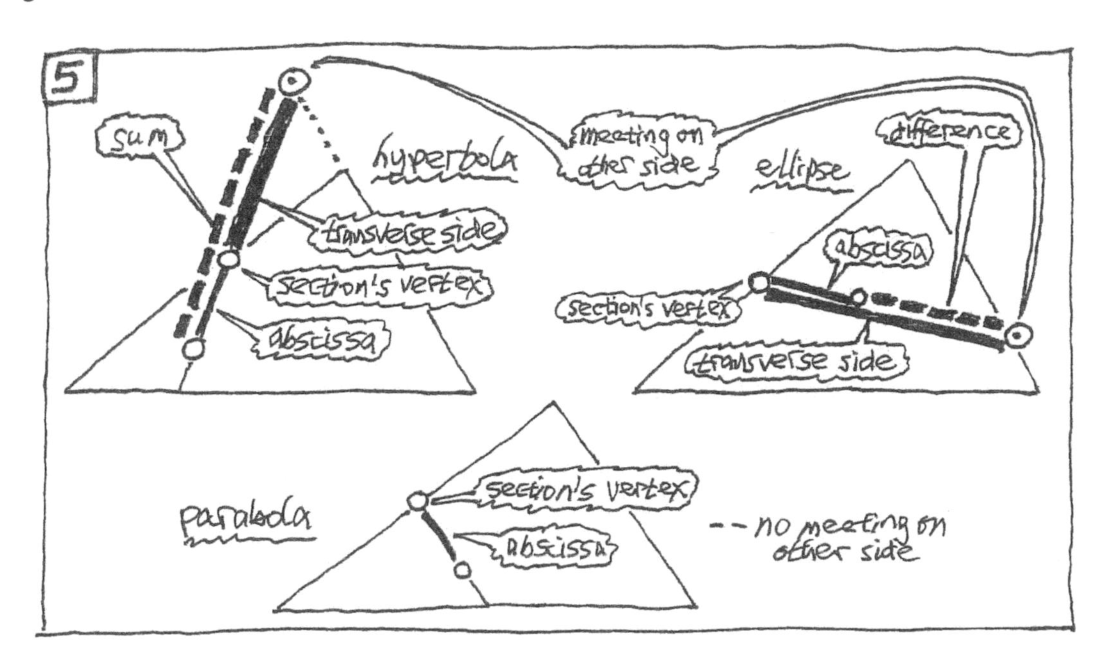

An upright side is used to define each of the three new lines obtained as conic sections. The key to all three sections is the size of the upright side. Though the "whence" of its contriving is obscure, the "whither" is made manifest. The section that is central in the manifest functioning of the upright side is the one presented first. The parabola comes first in making manifest the function of the upright side.

The other two sections are set apart together. They stand apart from the parabola because they both have a line which the parabola does not have—the oblique (transverse) side. This line, unlike the upright side, does not need to be contrived. To see it, all we need to do is look at it—or, at most, to draw out the diameter. The function of this line, however, is less straightforward than is the function of the upright side.

In Greek the "upright" side is called *orthia*. The other side (the one whose name it is customary to translate as "transverse") is called *plagia*. "Upright" is a felicitous rendering of *orthia* (from which we get such English words as "orthodox" and "orthodontist"). It is much more felicitous than the rendering of *plagia* by "transverse." The Latin derivative *trans* calls attention to the line's going "across"; but the intent of Apollonius would be rendered more faithfully by using another Latin derivative: the word *oblique*. The Greek term means that the line is "placed sideways" or is "slanting." The term is used metaphorically in Greek to mean "not straightforward"—or even "crooked and treacherous." In a military context, it refers to flanking attacks against the side, as distinguished from frontal assaults. The root is found in other Greek words, such as these: *plangktor* (a beguiler); *plangktos* (wandering—including wandering in mind, distraught); *eplagza*, the aorist form of *plazo* (to make to wander, to lead astray).

The side that is *orthia* and the side that is *plagia* are named quite carefully. The lines constitute a pair so named in Greek as to suggest a contrast; and when the names are given at the end of the 12th proposition, they are placed by the Greek text thus: "*orthia, plagia*"—in immediate juxtaposition, emphasizing the contrast. The most manifest contrast of the lines has to do with mere position: as we visualize the figures, one side stands straight, and the other lies slantwise. But beside that straightforward contrast which is most directly seen, there is another contrast—one which is presented more obliquely. In order to get straight the full significance of the side that is called "oblique," we shall have to work our way through the rest of Apollonius's book. The book may sometimes seem to be the wanderings of a mind that heedlessly, or even maliciously, renders other minds distraught. To see that it is not, one must look very closely.

There are hints along the way that more is going on than may meet the inattentive eye, but they are merely hints. One such hint is given in the last of the propositions which characterize the conic sections—the 14th proposition, the very next one after the thirteen that we have already studied.

At the end of the 13th proposition, it may have seemed that we finally had all the possible conic sections—but things are not quite so. However, before we go on to the 14th proposition, and see why more is in store for us than we may have thought, we should first see more clearly how these last three propositions (the 11th, 12th, and 13th) constitute a kind of whole.

bringing together the different definitions of the upright side

Although the definition of the *hyperbola*'s upright side and the definition of the *ellipse*'s upright side are very much like each other, they both seem very different from the definition of the *parabola*'s upright side.

The difference is not as great as it appears to be, however. Consider the upright side that we get from the proportion that defines it for the hyperbola (or the ellipse). What will happen to the length of that straight line if we were to keep the section's vertex F fixed at a certain distance from the axial triangle's vertex A, and—using F as a pivot—swing FG closer and closer to being parallel to AC (fig. 6)?

The parallel AK will swing over to the axial triangle's side AC , and so we can show the following—

as this is done:	*what happens is this*:
the cutting-plane that	the upright side FL as defined
makes the hyperbola (or ellipse)	for the hyperbola (or ellipse)
is taken closer and closer	will get closer and closer
in *position* to	in *size* to
the cutting-plane that	the upright side FM as defined
makes the parabola	for the parabola.

Indeed, if we take the one cutting-plane close enough in position to the other, the definitions will make the one upright side as close as we please to being the same size as the other. So, we should not be bothered by the fact that there is a difference between the definition of the upright side for the hyperbola (or ellipse) and its definition for the parabola—though we may want to see more clearly how the difference comes about. Let us now examine the swinging cutting-plane in the case of the hyperbola.

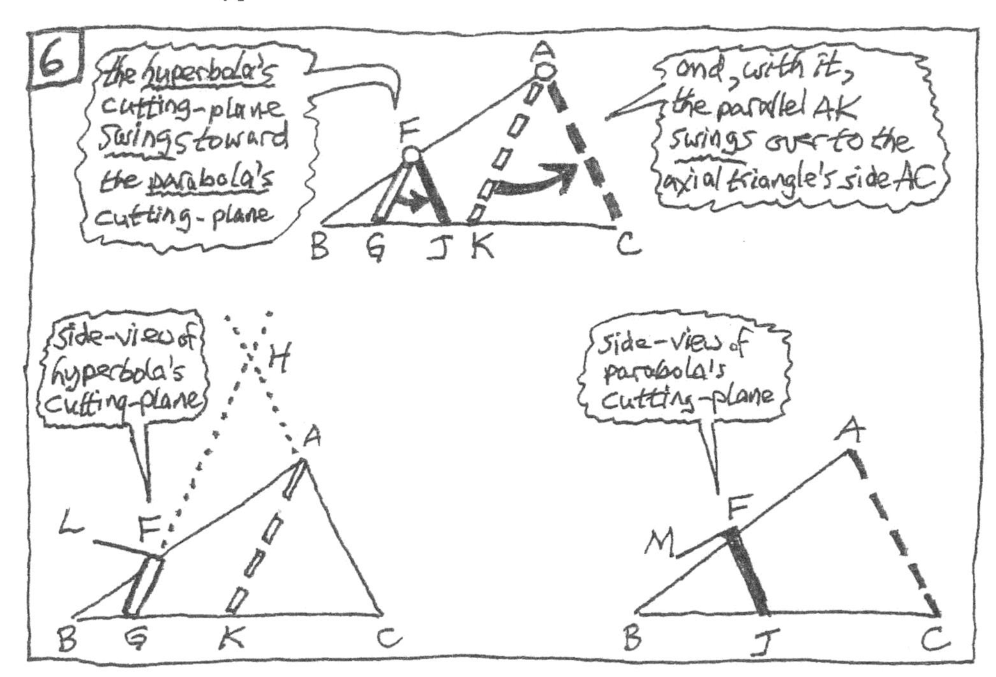

Consider the definitions of the upright sides, which are, for the hyperbola and for the parabola, as follows (see, in figure 7, the labeling in the top right corner for the hyperbola and in the bottom right corner for the parabola):

<table>
<tr><td>transverse HF</td><td>is to</td><td>upright FL of
hyperbola</td><td>as sqr [AK] is to rct [KB,KC];</td></tr>
<tr><td>upright FM of
parabola</td><td>is to</td><td>distance FA
between vertices</td><td>as sqr [BC] is to rct [AB,AC].</td></tr>
</table>

We need to see how to derive, from the proportion that defines the upright side for the *hyperbola*, some other proportion that will look *like* the proportion that defines the upright side for the *parabola*. That is to say, we need to show that the definition of the hyperbola's upright side leads to a proportion of the following sort—namely,

<table>
<tr><td>upright FL of
hyperbola</td><td>is to</td><td>distance FA
between vertices</td><td>as some rct is to some other rct—</td></tr>
</table>

in which, as the hyperbola's diameter FG swings toward the parabola's diameter FJ, the ratio of these rectangles will approach being the following:

<table>
<tr><td></td><td>as sqr [BC] is to rct [AB, AC].</td></tr>
</table>

What rectangles will have a ratio that does that? Well, as the cutting plane approaches parallelism, it approaches making a parabola. What happens to point K? Since AK is parallel to FG, and FG is swinging toward parallelism with AC, it must be the case that point K is approaching point C. What, then, is happening to BK and to AK? It must be the case that BK is approaching BC, and that AK is approaching AC. In other words, the ratio of

<table>
<tr><td></td><td>rct [BC, BK] to rct [AB, AK]</td></tr>
</table>

is the rectangle ratio which, as K approaches C, will approach being the following:

<table>
<tr><td></td><td>as sqr [BC] is to rct [AB, AC].</td></tr>
</table>

So, we need to show that this proportion—namely,

<table>
<tr><td>upright FL of
hyperbola</td><td>is to</td><td>distance FA
between vertices</td><td>as rct [BC,BK] is to rct [AB,AK]—</td></tr>
</table>

follows from the definition of the hyperbola's upright side—which is that

<table>
<tr><td>transverse HF</td><td>is to</td><td>upright FL of
hyperbola</td><td>as sqr [AK] is to rct [KB,KC].</td></tr>
</table>

If only we could show that, then we would be able to see how the definition of the upright side for the parabola fits together with the definition of the upright side for the hyperbola (fig. 7).

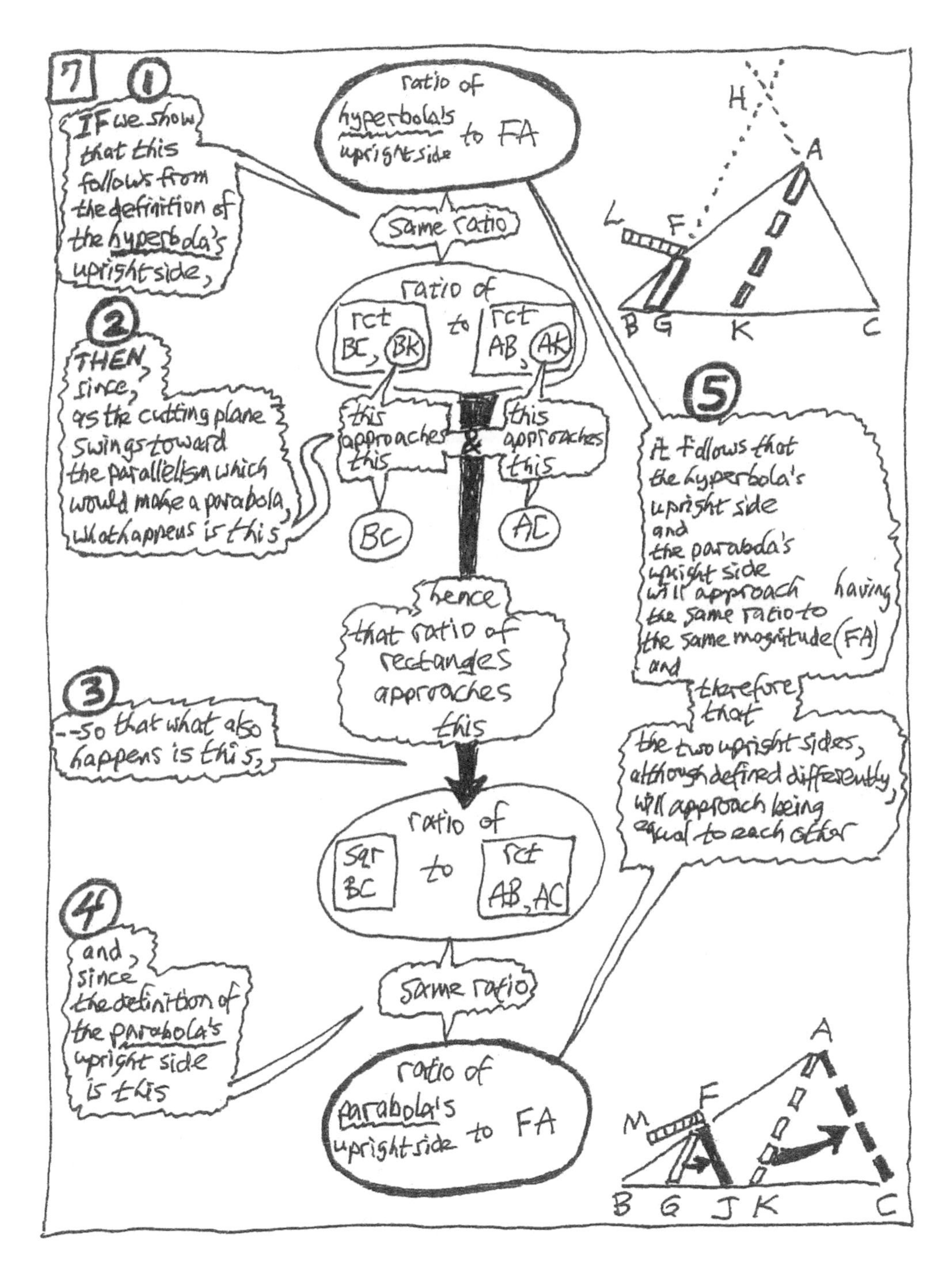

So, what we need to show is this: as the plane which cuts into the cone an hyperbola swings toward the parallelism which would cut into the cone a parabola, the hyperbola's upright side will approach having, to the same magnitude (FA), the same ratio that the parabola's upright side does—and therefore the two upright sides, although differently defined, will approach being equal to each other. How can we show that? As follows.

In order to get a proportion defining the hyperbola's upright side that will have the same form as the proportion defining the parabola's upright side, we need to get a ratio in which the upright side is the antecedent rather than the consequent, and in which the other term in it is FA rather than HF.

So, we begin by turning around the terms in the first ratio of the original proportion defining the hyperbola's upright side; then, in order to keep proportionality, we also turn around the terms in the second ratio of the original proportion. The result is this:

upright FL of *hyperbola*	is to	transverse HF	as rct [BC,BK] is to rct [AB,AK].

Now we will get a left-hand ratio with a different second term—a ratio that will relate FL not to HF but rather to FA. How? Well, if we compound the ratio which we have (the ratio of FL to HF) with a ratio whose antecedent is the same as the consequent of the ratio which we already have, and whose consequent is the same as the consequent of the ratio which we want to get (that is: if we compound the ratio of FL to HF with the ratio of HF to FA), then the ratio which we shall get is the one which we want—the ratio of FL to FA (by the "squooshing" of FL-to-FH and FH-to-FA). But if we compound the first ratio of a proportionality (the ratio on the left), then, in order to keep proportionality, we must also compound with that same ratio the second ratio of the original proportion (the ratio on the right). That is to say: the ratio which will be the same as the new ratio (of FL to FA) is the ratio which is obtained when we take the second ratio of the original proportion (the ratio, that is to say, which rectangle [BK, KC] has to square [AK]) and compound it with the ratio which FH has to FA. (Fig. 8.)

So, we have accomplished, with the definition of the hyperbola's upright side, half of what we wanted to do. We have derived a proportion in which the first ratio is the ratio which the hyperbola's upright side has to the distance FA between the vertices of section and cone. (This corresponds to the first ratio in the original proportion which defines the parabola's upright side—the ratio which the parabola's upright side has to that same distance FA.) But we now have to do something about the second ratio in the proportion which we have derived.

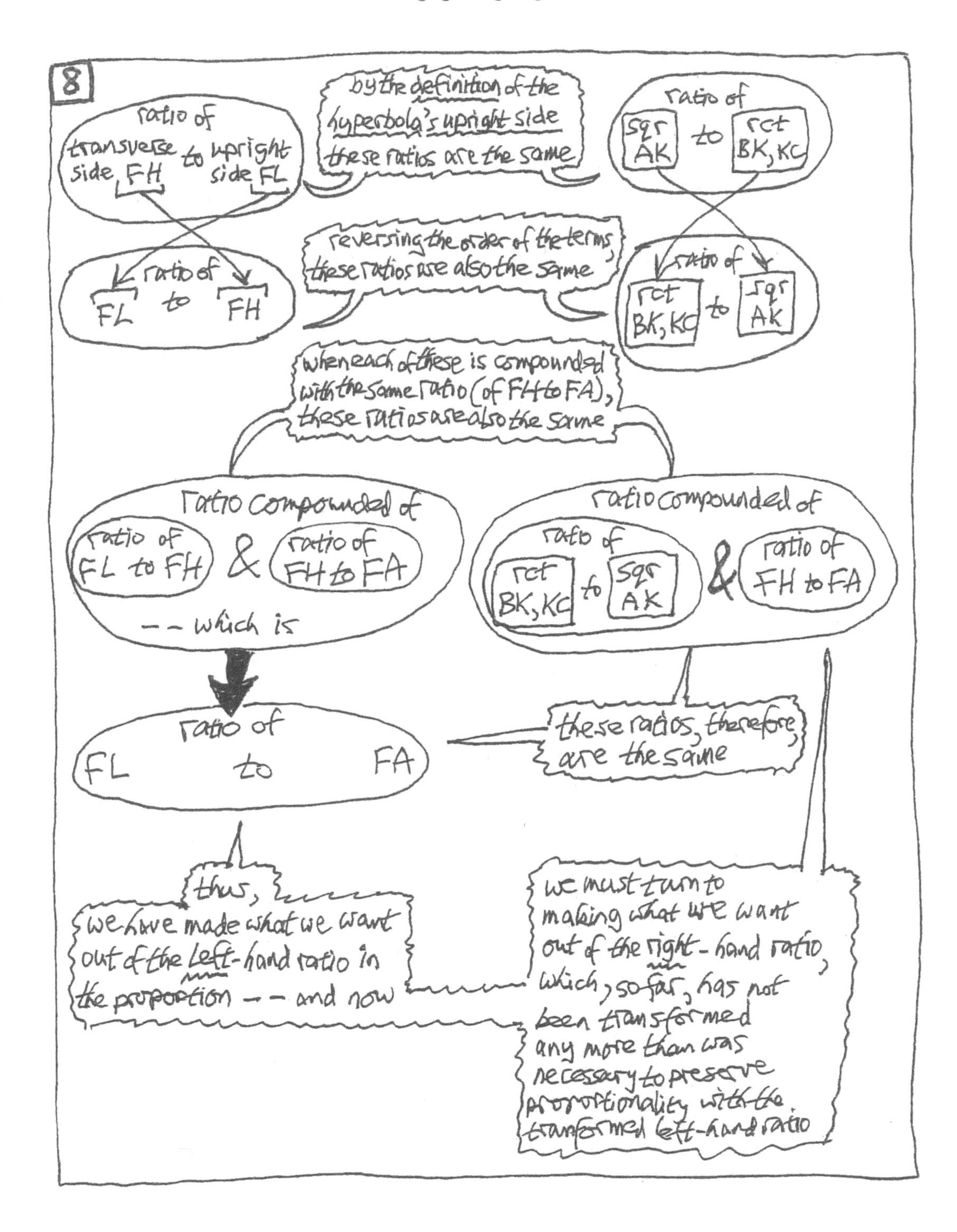
8
ratio of transverse side FH to upright side FL
by the definition of the hyperbola's upright side these ratios are the same
ratio of sqr AK to rct BK, KC
ratio of FL to FH
reversing the order of the terms these ratios are also the same
ratio of rct BK, KC to sqr AK
when each of these is compounded with the same ratio (of FH to FA), these ratios are also the same
ratio compounded of ratio of FL to FH & ratio of FH to FA -- which is
ratio compounded of ratio of rct BK, KC to sqr AK & ratio of FH to FA
ratio of FL to FA
these ratios, therefore, are the same
thus, we have made what we want out of the left-hand ratio in the proportion -- and now
we must turn to making what we want out of the right-hand ratio, which, so far, has not been transformed any more than was necessary to preserve proportionality with the transformed left-hand ratio

We now have to show that the second ratio in the proportion which we have derived (the ratio, that is to say, which is compounded out of the ratio which rectangle [BK, KC] has to square [AK], on the one hand, and, on the other, the ratio which oblique (transverse) side FH has to distance FA) is the same as a ratio of rectangles—rectangles whose sides are lines that, as the hyperbola's diameter swings toward parallelism, approach the lines that form the axial triangle. Before trying to compound the ratios, let us take that ratio which we introduced (the ratio of the oblique [transverse] side FH to the distance FA) and let us find a ratio which is the same as it but which contains the line AK that is used in the consequent of the ratio with which we need to compound it. Through point K, let us (fig. 9) draw a line (call it KQ) parallel to FA, and so (by similar triangles) AK will have to KQ the same ratio which HF has to FA.

We still cannot "squoosh" the ratios, since the consequent of the first ratio is the square with side AK but the antecedent of the second ratio is AK alone. The second ratio needs to have as its antecedent not AK but the square with side AK. So, let us take AK as a common side. Then the ratio of AK to KQ is the same as the ratio of square [AK] to rectangle [AK, KQ]. This we can now compound with the other ratio:

> compounding the ratios
> of (rct [BK, KC] to *sqr [AK]*) & of (*sqr [AK]* to rct [AK, KQ]),
>
> we get the ratio
> of (rct [BK, KC] .... to rct [AK, KQ]).

But we are not yet done. This last ratio involves the line KQ—a line which we introduced to help us move on, but which is not one of the lines with which we want to be left at the end. KQ must go. But how can we get rid of it? Let us de-compound the ratio of rectangles which we now have, into ratios of lines—and then let us find some ratio of lines without KQ that is the same as this ratio with KQ. First, we de-compound. The ratio of rectangle [BK, KC] to rectangle [AK, KQ] is compounded out of these ratios: of BK to AK, and of KC to KQ. This latter ratio (of KC to KQ) is (fig. 10), by similar triangles, the same as the ratio of BC to BA. So, now we take this ratio (of BC to BA), and we compound it with the other ratio (of BK to AK). This compound is the same as the ratio of rectangle [BK, BC] to rectangle [AK, BA]. (Fig 11.)

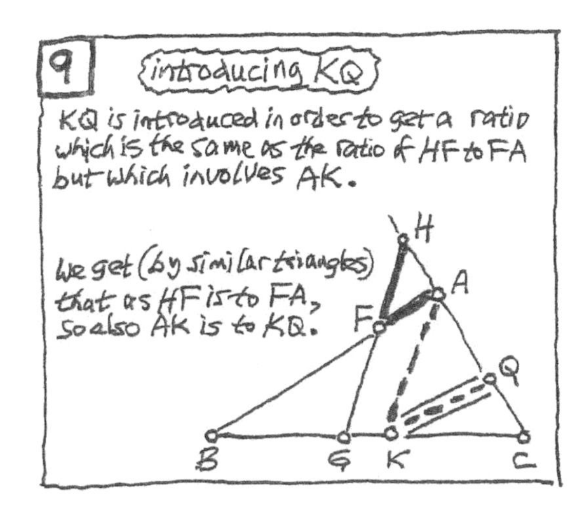

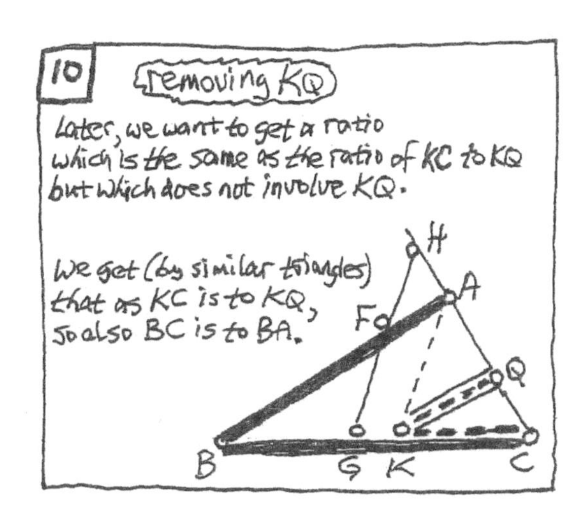

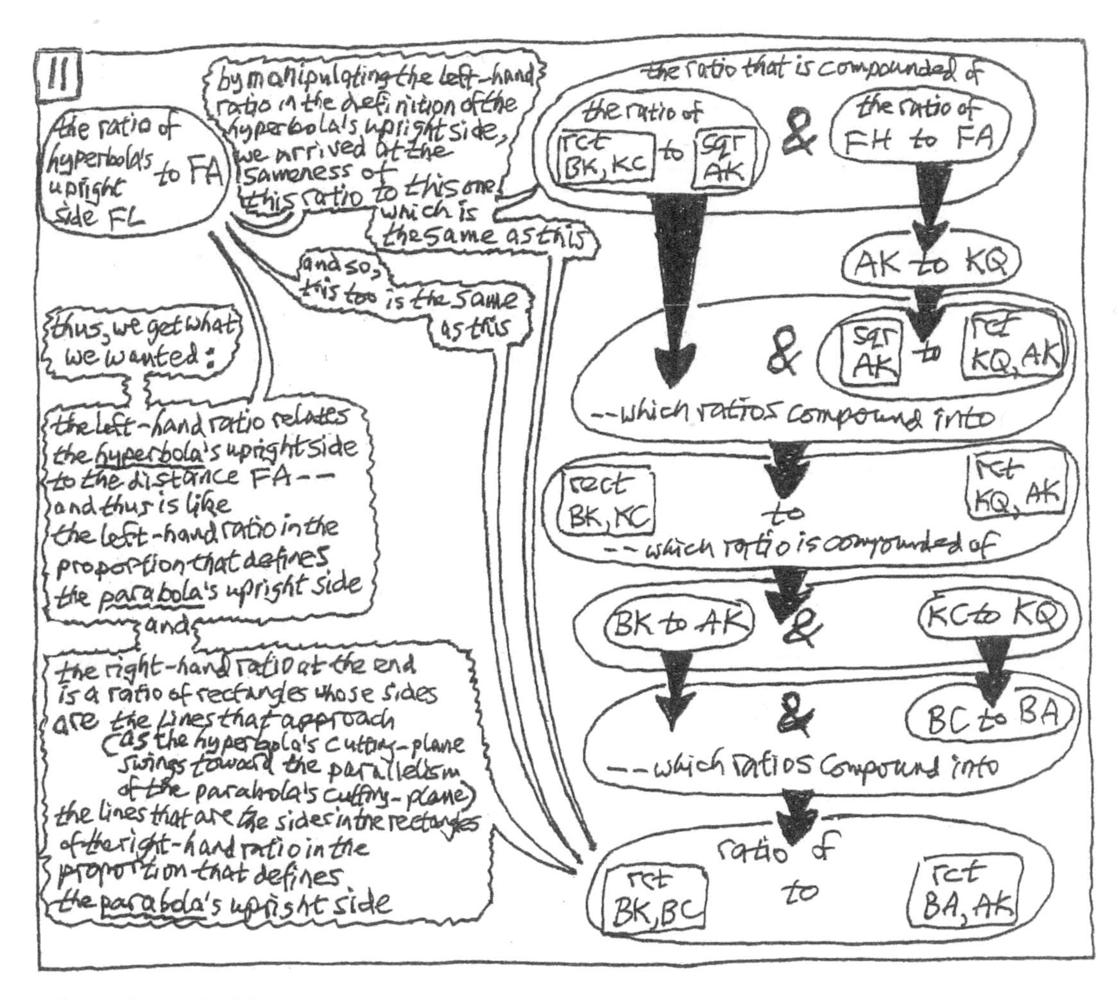

What we have shown is this:

> The ratio (i) of the *hyperbola's* upright side to FA
> is the same as a ratio which approaches a ratio which is the same as
> the ratio (ii) of the *parabola's* upright side to FA.
>
> Thus, since the one ratio approaches the other and has the same consequent,
> the one ratio's antecedent (the hyperbola's upright side) must approach
> the other's antecedent (the parabola's upright side).

And that is to say, about the relation of the *orientation* of the cutting-plane and the *size* of the upright side, the following:

as	so also
the cutting-plane that makes an *hyperbola*	the *hyperbola*'s upright side
approaches in its orientation	will approach in its size
the cutting plane that makes a *parabola*,	the *parabola*'s upright side.

Now that we have examined the swinging cutting-plane in the case of the hyperbola, you may want to work out for yourself, as an exercise, the case of the ellipse.

bringing together the different analyses

Let us label the figure with the same letters for different cuttings of the cone (fig. 12).

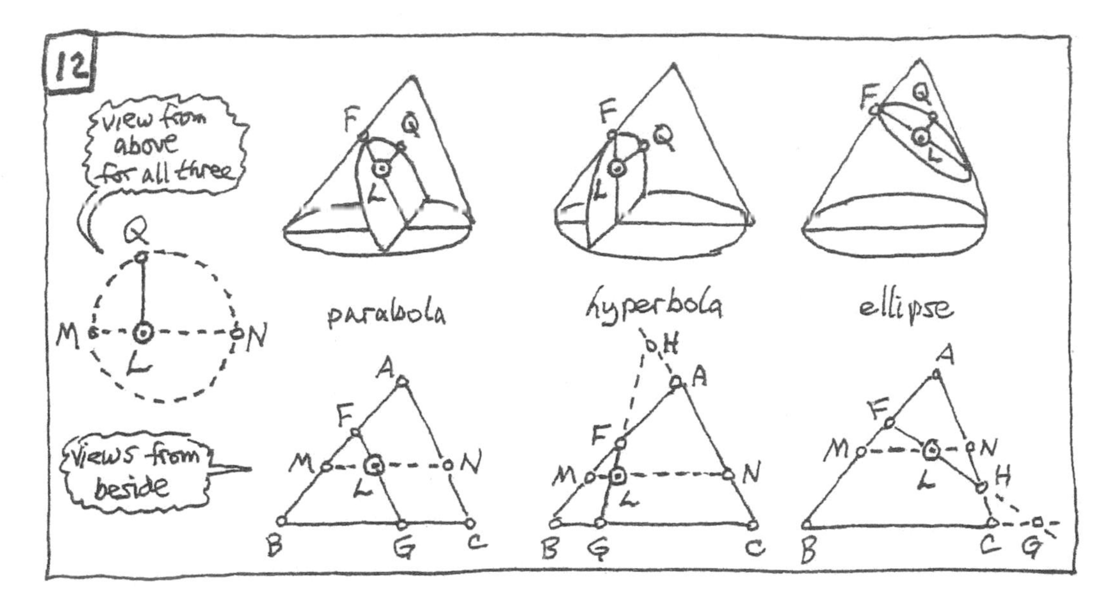

When *viewed from above*, the *circle* gave us equality between the square arising from ordinate QL and the rectangle contained by segments LM and LN. As we moved point L up or down on diameter FG, this rectangle varied because its sides (the segments) varied. What we said about the size of the ordinate was therefore a matter of what stayed the same about segments LM and LN as we moved point L up or down on the diameter; and this was a matter of the relations given to us by the *triangle* when *viewed from beside*.

We took what stays the same about segments LM and LN, and represented it in a rectangle that was equal to the rectangle [LM, LN]—and therefore was also equal to the square arising from ordinate QL. Not only was this rectangle equal to a square whose side is ordinate QL; it also had to have one of its own sides be equal to abscissa FL. By virtue of finding this rectangle's *non-abscissa side*, we figured out a relationship between any *ordinate* in the section and its *abscissa*.

The *simplest* case was the case that we had when only one of the sides of the rectangle [LM, LN] changed as we moved point L up or down. That was the case when L moved on a straight line parallel to one of the sides of the axial triangle—which is to say, when the diameter FG was parallel to the axial triangle's side.

When we begin cutting a cone so as to make neither straight lines nor circles, there are many ways that we might cut. We had to ask ourselves which way to cut first. The parallel cut does stand out as unique—and it is also the one that has the easiest side-view to handle. Thus we had a two-fold reason to begin with the section called the parabola: the orientation of its cutting plane stands out, and the analysis of its property is simplest.

The rightness of the choice seemed to be confirmed by the outcome of analysis. The outcome of the analysis of this conic section, when it is compared with the outcomes of the analyses of the other two, shows this one to have a kind of primacy, since it turns out to be a kind of mean between the remaining two—one of which involves an excess, the other a deficiency. Classical Greek musical scales are arranged similarly: the string length that is presented first is the one that produces the tone in the middle; after it are presented the longer strings, which produce lower tones, and then the shorter strings, which produce higher ones.

We seek new lines, lines that are non-circular curves. We leave for last the one that is most like a circle—the one that is closed. In the "prologue" of the First Book of the *Conics* (that is, in the book's first ten propositions—the ones that merely led up to the characterization of the various sorts of conic sections), the non-circular curves presented first were the ones that are least like a circle. Of these, some are like the closed curve in allowing many different orientations of the cutting plane, and they come after the curve with the one orientation that stands out and is the simplest—the one where the cutting plane is parallel to a side of the axial triangle. To take a plane this way—oriented parallel to a side not the base—is, moreover, the natural next step after taking a plane oriented parallel to the base, which gives a circle. (At least it is the next step after a follow-up that takes the other way to get a circle—namely, by cutting with a plane that makes, in reverse, the same angles which are made by the cutting-plane that is parallel.)

So, we cut parallel to a side and we examine the shape of the section that we get. We already know that it is indefinitely extendable, and also that when (by prolonging the diameter) it is extended, its shape stays the same in the following respect: chords that are drawn through the diameter at the same angle continue to make ordinates.

We examine what stays the same about another aspect of the shape: we ask how the size of the ordinate compares with the size of the line that stretches along the diameter from the vertex down to where it is cut off by the ordinate. We ask how we can relate the sizes of these two lines—the ordinate and the abscissa.

We cut parallel to a side to get the parabolic conic section. If we slice again, in such a way that ordinate QL will now also be a chord in a circle where it is perpendicular to the diameter, then we shall have related ordinate QL to something in the axial triangle—something that in its turn can be related to abscissa FL.

How is ordinate QL related to something in the axial triangle? Here is how: in the circle, a square whose side is ordinate QL will be equal to a rectangle whose sides are the segments ML and LN (fig. 13).

Now we seek a rectangle that, first, has the varying abscissa as one of its sides and, second, also has a constant relationship with the varying rectangle [ML, LN]—and hence with the square that has the varying ordinate as its side. In other words, we seek a width (w) such that:

rectangle [*abs* FL, width(w)]	is equal to	rectangle [ML, LN],	which is equal to	square [*ord* QL].

How do we find such a width (w) that will make such a rectangle?

Well, rectangle [ML, LN] will have a constant ratio to any rectangle to whose sides its own sides have constant ratios. Consider first its *varying* side, ML. Take the ratio of ML to the abscissa (the line we want to involve). This ratio of the rectangle's side ML to the abscissa LF is the same constant ratio which one constant line of the axial triangle has to another—that is to say:

ML has to abscissa LF	the same ratio which	BC has to AC.

As for the rectangle's *other* side LN, it is constant; and therefore it will be in a constant ratio with any constant line. What constant line should we use to get a constant ratio with LN? Why not use the only other constant line which is what LN itself is? And what is line LN? It is the extension of a non-abscissa side of that little triangle MLF whose sides we have already used as terms to get our first constant ratio. That triangle's only other non-abscissa side (LN) has as its extension the line FA, the constant line between the vertices of section and of cone. The rectangle's side LN has, to this line FA, the same constant ratio which one constant line of the axial triangle has to another—that is to say:

LN has to line FA	the same ratio which	BC has to BA.

Hence, if we compound the ratios, the result will be the following proportion:

rct ML, LN	has to	rct abs LF,FA	the same ratio which	sqr BC	has to	rct AC, BA.

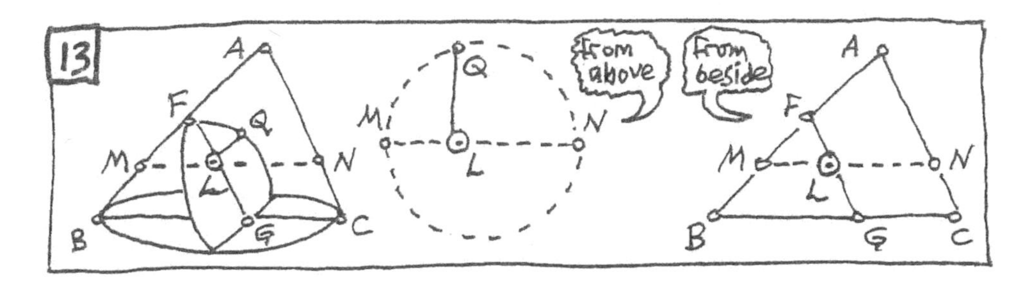

In the proportion that we have just obtained, consider the ratio on the left. Its first term is the rectangle [ML, LN]; its second term is a rectangle that has abscissa LF as its height, and has FA as its width. Now suppose that the second rectangle had as its width, not FA, but a larger line; this would make the rectangle larger. To such a larger second rectangle, the first rectangle would have a smaller ratio. On the other hand, suppose that the second rectangle had as its width a line smaller than FA; this would make the rectangle smaller. To such a smaller second rectangle, the first rectangle would have a larger ratio. So, if we took some line of just the right size, and made *it* the width of a rectangle whose height is the abscissa LF, then rectangle [ML, LN] would have, to that rectangle, a ratio of equality. Now what size line would be just the right size to be that width?

The required width (w) will be obtained by contriving a line that has the same ratio to line FA that square [BC] has to rectangle [BA, AC]. If a line contrived that way is made the width of a rectangle whose height is the abscissa FL, then the rectangle will be equal to the varying rectangle [ML, LN]. Why that is so, is shown in figure 14.

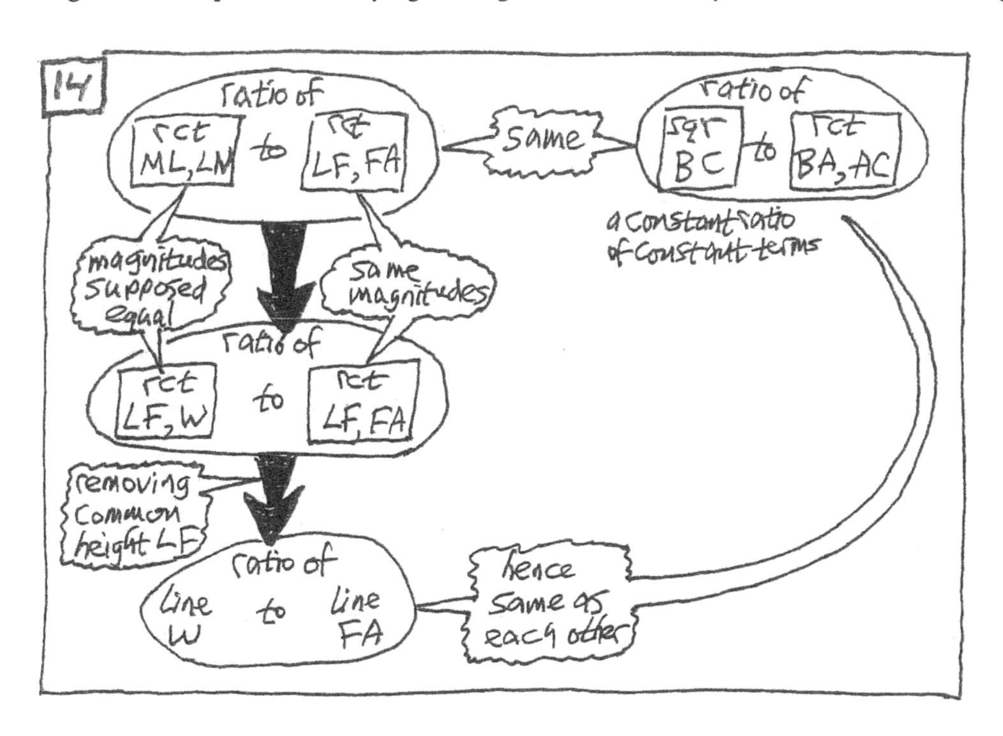

Before we now move on from the parabola to the other sections, it should be noted that if (in order to make a proportion to define the width of the equal rectangle) we had chosen to put LN in ratio, not with FA, but instead with any other constant line, then (since that would have caused the constant ratio in the proportion of rectangles to be some other constant ratio) we would have obtained another definition of the same upright side. That is to say: instead of doing things in the way that we did, we might plausibly have done them in the following way (fig. 15). After using, for the ratio of *ML*, the line FL, we might then *not* have gone right on to decide what to use for the ratio of *LN*. Instead, we might have stayed a while longer with the ratio of *ML*. We might first have noted that the constant ratio of ML to FL is the same as a ratio of constant lines—namely, the ratio of BG to GF. And then, since constant line BG in this last ratio is the counterpart of ML, we might have said; why not use GF, the *other* constant line in that ratio, as the constant other line that we shall use for the ratio of *LN*? And this ratio of LN to GF would be the same as another ratio of constant lines. It would (since the diameter's situation of parallelism makes LN equal to GC) be the same as the ratio of GC to GF. So, when the two ratios (of ML to FL, and of LN to GF) were compounded, we would get the same ratio as the ratio of rectangle [BG, GC] to square [GF]—and this is the ratio that we would use to define the upright side. Thus the upright side would be the line contrived to have the same ratio to *GF* that rectangle [*BG*, *GC*] has to square [*GF*]—instead of being (as before) the line contrived to have the same ratio to *FA* that square [*BC*] has to rectangle [*BA*, *AC*].These two different ways of contriving the upright side (by using FA, and by using FG) will give a line of the same size. In our analysis, we, like Apollonius in his synthetic demonstration, have used FA, for the reason suggested a few moments ago, but we did not have to.

We turn now to the sections cut by a plane that is not parallel to a side—the hyperbola and the ellipse (fig. 16). (For convenience, line AK is drawn into the figure now. As we have seen, however, the work for which this line is needed could be begun with two lines that are already in the picture—the lines FG and HG.)

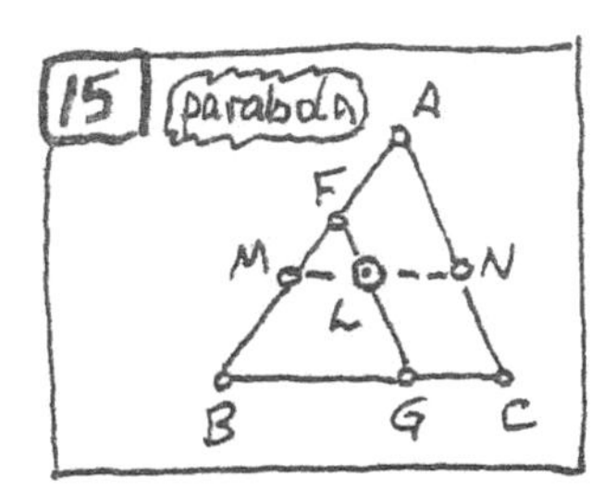

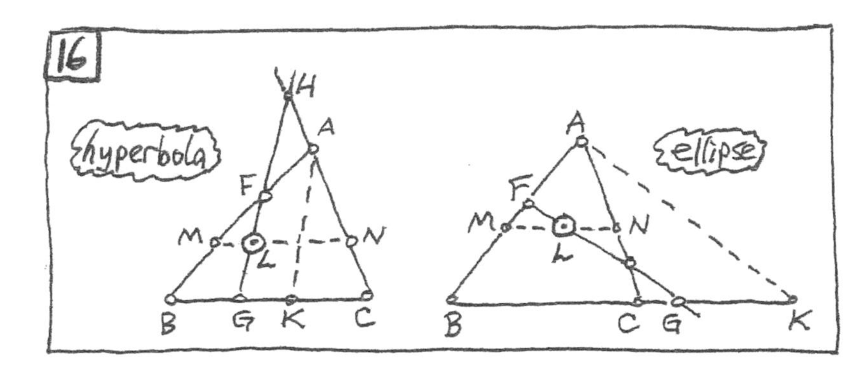

Our first question is this: how shall we relate the abscissa and ordinate? As in the parabola, the square arising from ordinate QL is equal to rectangle [LM, LN]. Hence, any rectangle which is equal to rectangle [LM, LN] will also be equal to the square arising from ordinate QL. We therefore seek such a rectangle—one which has abscissa FL as its height. In order to get the rectangle which we seek, we need only to find its width—call it line (w).

So, our second question is this: how do we find the width (w)? Consider LM and LN, the sides of that rectangle which is equal to the square on the ordinate:

LM varies, but the ratio of FL to it	LN varies, but the ratio of HL to it
is constant, and	is constant, and
is the same as the constant ratio of	is the same as the constant ratio of
constant line AK to constant line KB	constant line AK to constant line KC.

Now, putting things together by compounding ratios, we get that:

> although, like both its sides, the rectangle [LM, LN] varies,
> nonetheless, the rectangle [FL, HL] has to it a ratio that does not vary,
> and this constant ratio is the same ratio which
> the constant square [AK] has to the constant rectangle [KB, KC].

The rectangle [abscissa FL, width (w)] which we are seeking is supposed to be equal to that rectangle [LM, LN]. Hence, any third rectangle—including [FL, HL]—would have the same ratio to either of them. From this, it follows (fig. 17) that:

> the ratio which square [AK] has to rectangle [KB, KC]
> must be the ratio which HL has to width (w).

Now, although this ratio of HL to width (w) stays constant (since it is the same as a ratio of constant terms), its first term HL is still a line that varies—like its other term, the width (w). But we want everything in our determination of the width to be constant. Our final question, then, is this: what to do about that?

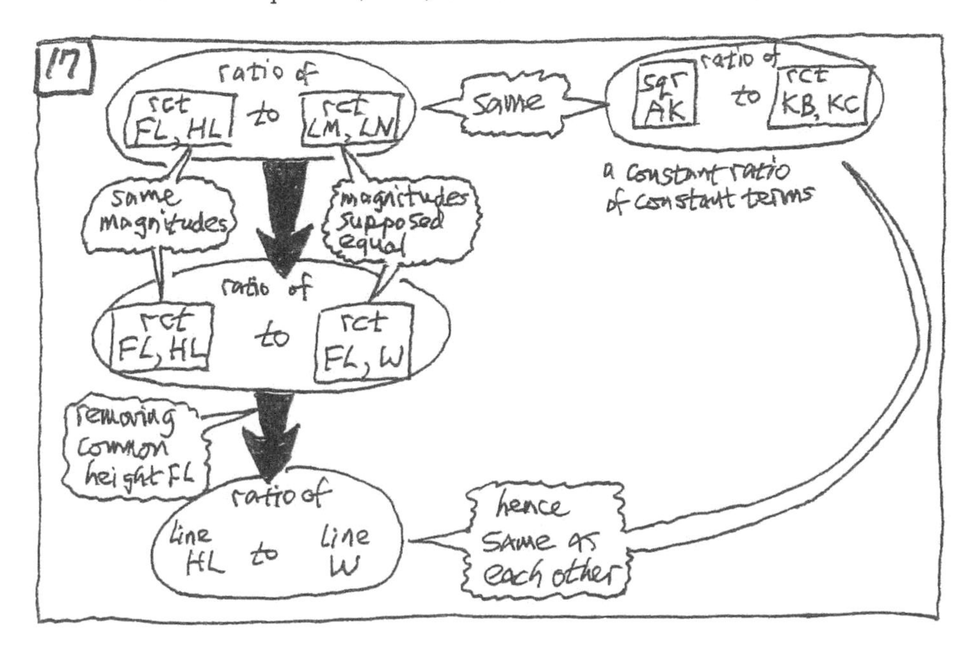

Well, what is HL? In both the hyperbola and the ellipse (fig. 18), it is the line that runs from the *oblique (transverse) side*'s non-vertex endpoint (H) to the *abscissa*'s non-vertex endpoint (L).

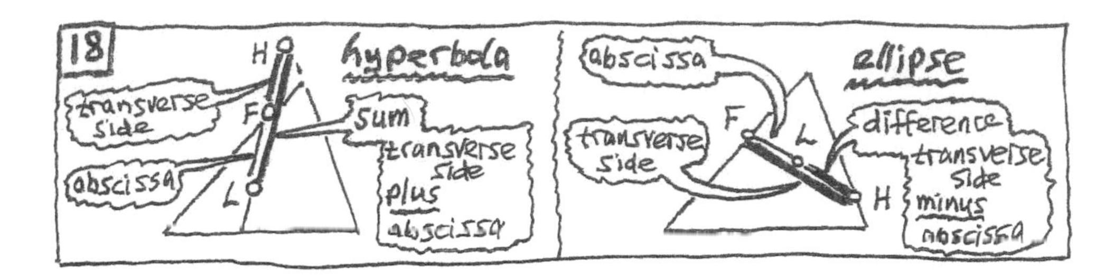

Line HL is thus the line that is given by *the constant oblique (transverse) side* when the varying abscissa is added on to it (in the case of the hyperbola) or is taken away from it (in the case of the ellipse). Line HL, that is to say, varies; but:

in the hyperbola,	while in the ellipse,
it is a varying whole	it is a varying part
having a part that is constant	contained in a whole that is constant.

So, we take the constant ratio of square [AK] to rectangle [KB, KC]. It is the same ratio that line HL (the oblique [transverse] side with the abscissa added on to it or taken away from it) has to the width (w). *This constant ratio* is the ratio that HF (the *constant oblique [transverse] side by itself*) will have to *some* line that is constant. Let us contrive *this constant line*; it is this line that will then be our constant upright side (fig. 19).

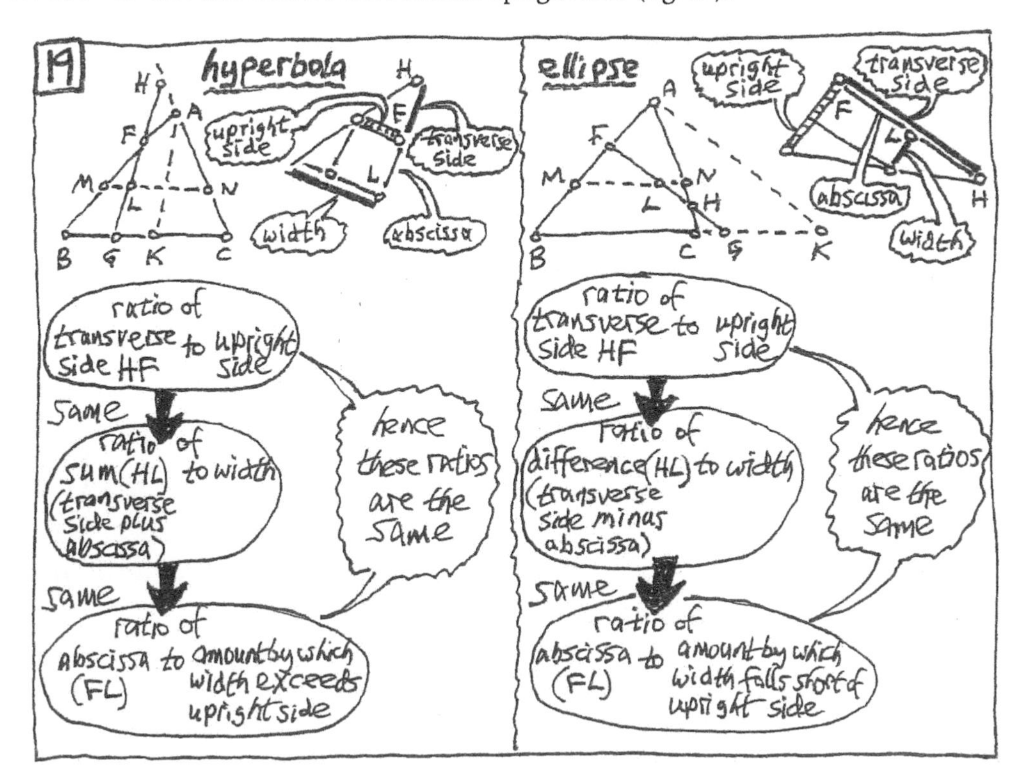

What, then, for the hyperbola and ellipse, is constant about that varying width (w) which, together with the abscissa as height, will make a rectangle that is equal to rectangle [ML, LN]—and is therefore also equal to the square arising from the ordinate? Well, the varying width (w) will be determined by the constant oblique (transverse) side together with the constant upright side (this upright side having been contrived so that the oblique [transverse] side HF will have the same ratio to it that square [AK] has to rectangle [KB, KC])—in the following way:

> the width (w) is unequal to the upright side,
> being greater than it in the hyperbola, smaller in the ellipse;
>
> and the amount by which the width (w) is greater or less than the upright side
> is such that the abscissa has to that amount
> the same ratio which the oblique (transverse) side has to the upright side.

Thus, what is constant about that varying width (w) is shown through its comparison with the constant upright side—and the comparison answers a double question: which of the lines has the greater size, and what is the relative size of the discrepancy? In the parabola, of course, the width of the non-abscissa side of the equal rectangle is the upright side itself.

going in reverse

Until now, our analyses have considered, in reverse, things *within* each of the three propositions. But what about the order in which we have moved *from proposition to proposition*? Even if the parabola should precede the other two, that primary precedence would not necessarily decide which of the other two should precede which. And even if the hyperbola should precede the ellipse, that would not necessarily decide whether both of them should precede the parabola.

But why not begin with the ellipse—and then do the hyperbola—and then last of all do the parabola? If we did begin with the *ellipse*, we would be beginning with that curve which, of all these non-circular curves, is the one that is most like a circle. We would thus be beginning with that curve which is the least new of the new curves, and which is most like the figure of the top-view. Of the trio of conic sections that Apollonius has presented, the ellipse was the first to be encountered in a proposition (the 9th) which deals with the top-view circle property that gets the square on the ordinate equal to a rectangle whose sides are the segments into which the ordinate splits a line parallel to the base in the axial triangle. If we then moved on from the ellipse to the *hyperbola*, we would have another situation in which the cutting plane hits the other side—now hitting not the cone itself, but rather the extension of its surface. The section is a curve of an even newer kind, being indefinitely extendable, like a straight line. By moving on finally to the *parabola*, we would be getting another indefinitely extendable curve, since the cutting plane does not hit the other side at all.

Now consider the properties to be demonstrated for the sections. By beginning with the ellipse, we would be beginning from deficiency. By moving on from the ellipse to the hyperbola, we would be moving from deficiency to excess. Finally, with the parabola, we would be concluding where there is neither excess nor deficiency; instead, the situation alongside the upright side would be just right.

Now let us consider what sort of analysis would be appropriate to treating the sections taken in reverse. Let us not bother with viewing the cone from above—that part of the analysis is simple and is the same for all the sections. Let us proceed directly to the view from beside the cone (fig. 20).

As the abscissa FL lengthens, its ordinate LQ changes. The square on the ordinate therefore also changes, and then so does any rectangle which stays equal to that square. Rectangle [ML, LN] therefore changes. What about the lines which contain that changing rectangle (the lines ML and LN)—what about them stays the same or changes?

One of them (ML) changes, but it has a constant ratio to the abscissa FL. Suppose that the other one (LN) also changes. That is to say: suppose that the diameter, which hits one side of the axial triangle at F, is not parallel to the other side, but hits it or its extension, at some point H. Nonetheless, LN will have a constant ratio to the line that runs along the diameter from L to that point H.

So, as we go down along the diameter, the following two ratios are constant for the *ellipse:*

the ratio of ML to abs FL	and	the ratio of LN to LH.

Those constant ratios of varying lines can be expressed in terms of constant lines:

the ratio of BG to FG	and	the ratio of GC to GH.

Compounding the ratios:

we get that the ratio of	varying rct [ML, LN]	to	varying rct [abs FL, LH]
is the same as			
the constant ratio of	constant rct [BG, GC]	to	constant rct [FG, GH].

We do the same for the *hyperbola*, getting the same constant ratio of rectangles, except that the changing line LH will now be obtained from the constant oblique (transverse) side HF by adding abscissa FL on to it (rather than by taking it away from it, as in the case of the ellipse).

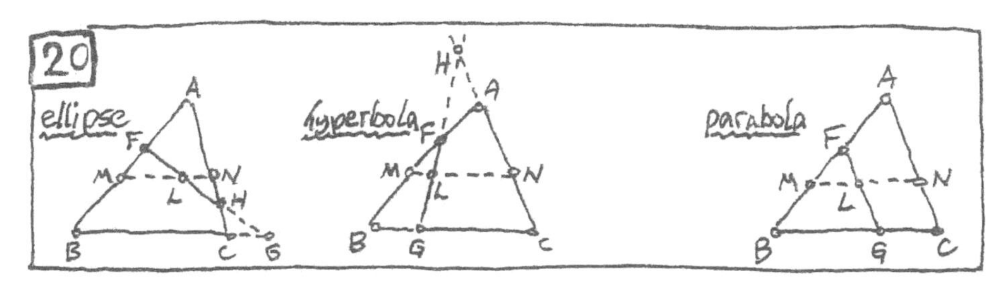

Finally, we do the *parabola*. Now there is no oblique (transverse) side HF, since there is no point of intersection H. And there is only one constant ratio involving varying lines—the ratio of the varying line ML to the varying abscissa FL. The other line (ML's partner LN)—which in the ellipse and in the hyperbola was varying, and had a constant ratio to another varying line—is now (in the parabola) itself constant. But we still need to get a rectangle to which rectangle [ML, LN] has a ratio, and so we need another ratio to compound with the ratio that ML has to FL. We need a constant ratio whose antecedent term is the line LN. Now, since LN is constant, it will have a constant ratio to *any* other constant line. Which one shall we pick then?

Now in the case of this section (the parabola), the constant ratio of ML to FL is again (as in the case of the other sections—the hyperbola and the ellipse) put into constant terms as the ratio of BG to FG. But whereas in the case of the other sections the constant ratio of LN to LH was put into constant terms as the ratio of GC to HG, now the line LN is equal to the constant line GC that was its counterpart in the ratio of constant lines, and there is no line LH or HG. (There is no line LH or HG because there is no point H at which FG intersects AC, no matter how far they may be extended—since diameter FG is parallel to the axial triangle's side AC.) So, for the other line in the constant ratio involving *LN* let us use the same constant line (FG) that is the counterpart, in the ratio of constant lines, of the other line in the constant ratio involving *ML*.

So, as we go down along the diameter, the following two ratios are constant for the parabola:

the ratio of	and	the ratio of
ML to abs FL		LN to the constant line we pick (FG).

Those constant ratios of varying lines can be expressed in terms of constant lines:

the ratio of	and	the ratio of
BG to FG		GC to FG.

Compounding the ratios,

we get that the ratio of	varying rct [ML, LN]	to	varying rct [abs FL, FG]
is the same as			
the constant ratio of	constant rct [BG, GC]	to	constant sqr [FG].

We have now obtained a definition of the upright side for each of the sections by taking them in an order that is the reverse of Apollonius's. How do the definitions obtained in this alternative compare with those that Apollonius came up with? That is shown by figures 21 and 22.

Before turning to figure 22, however, note that *ML* is set in ratio with abscissa FL for *all* the sections, and while *LN* is set in ratio with HL for both *hyperbola and ellipse*, it is not so clear what to set LN in ratio with for the *parabola*. Apollonius uses the constant line FA.

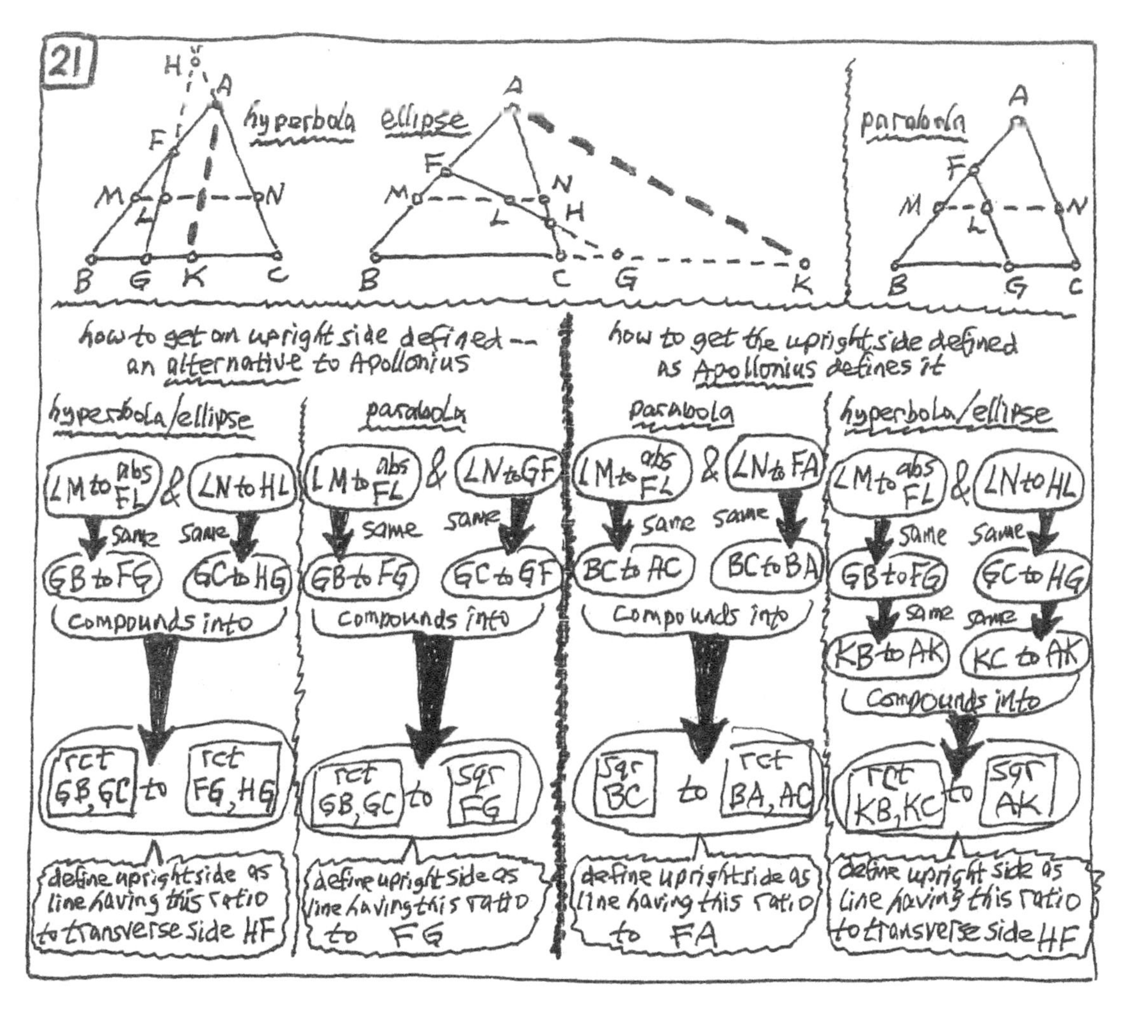

To make manifest what goes into defining the upright side for the several sections, it does not suffice merely to present several such definitions as starting points for demonstrations. If Apollonius had presented an analysis for the parabola, and had done so after presenting one for the other two sections, would that have been a better way to proceed?

Even if it is not clear what is the best way to proceed, it should be clear that we should not take for granted what Apollonius tells us. He does not tell us enough for us to understand what he tells us—unless we on our part ask *why* he is telling it to us, and why he is telling it to us in the *way* that he tells it.

22 the upright side -- comparison of the definitions in Apollonius and the alternative

	in parabola	in hyperbola/ellipse
difference in lines used	Apollonius: FA, BC, BA, AC alternative: FG, GB, GC	Apollonius: HF, KB, KC, AK alternative: HF, GB, GC, FG, HG
results-- advantages & disadvantages	Apollonius: disadvantage that lines used for parabola do not correspond to those used for hyperbola/ellipse alternative: advantage that they do correspond	Apollonius: advantage that fewer lines are used alternative: disadvantage that more lines are used
procedures-- advantages & disadvantages	Apollonius: advantage that similar triangles are used not only with ratio for ML but also with ratio for ∠N alternative: disadvantage that the two ratios are treated differently	Apollonius: disadvantage that extra line AK has to be drawn alternative: advantage that all lines used are already in the picture

Let us look again at the side-view of the sections (fig. 23). Even if we had begun as we have done in this alternative—with the ellipse and hyperbola, where triangles clearly present themselves as giving constant ratios for both segments (for LN as well as ML)—still, we might have wanted to go on to make some modifications that would give the outcome that Apollonius obtained, with its particular combination of advantages and disadvantages.

For the ellipse and the hyperbola, perhaps we would have preferred to have three instead of four constant lines in our ratio. If so, then we could have drawn that line AK, and then, instead of stopping with the ratios

of BG to FG	and	of GC to GH,

we could have gone on to get the ratios

of KB to KA	and	of KC to KA,

which, when compounded, would give us the ratio

of constant rct [KB, KC]	to constant sqr]. [KA]

Perhaps in getting the ratio involving segment LN for the parabola, we would *not* have had a mind to do what would correspond to the ratios which we got for the *other* sections—the ratios, that is to say, which we got from both our first ratios, involving *both* segments ML and LN, *for the ellipse and the hyperbola*. Perhaps in getting the ratio involving segment LN for the parabola, we would be inclined to do what corresponds to the ratio which we first got involving the *other* segment (ML) *for the parabola*.

In other words, suppose that in doing the parabola we had said the following. We already have the ratio which ML has to FL, these lines being sides of a small triangle (that is, being sides parallel to the sides of the bigger axial triangle), and therefore the ratio of the changing sides of the small triangle can be expressed in constant terms which are sides of the axial triangle. That is to say: as ML is to FL, so also BC is to AC. And LN is the extension (up to the axial triangle) of one side of the small triangle; the extension (up to the axial triangle) of the other side of the small triangle is (like LN) a constant line—FA. The constant ratio which constant line LN has to constant line FA can (like the first ratio) also be expressed in terms which are the base and a side of the axial triangle. That is to say: as LN is to FA, so also BC is to BA. If we had been so minded, then we would have had the ratios

of ML to FL	and	of LN to FA,

which are the same as the ratios

of BC to AC	and	of BC to BA,

which, when compounded, would

give us that the ratio	of varying rct [ML, LN]	to varying rct [abs (FL), FA]
is the same as		
the constant ratio	of constant sqr [BC]	to constant rct [AB, AC].

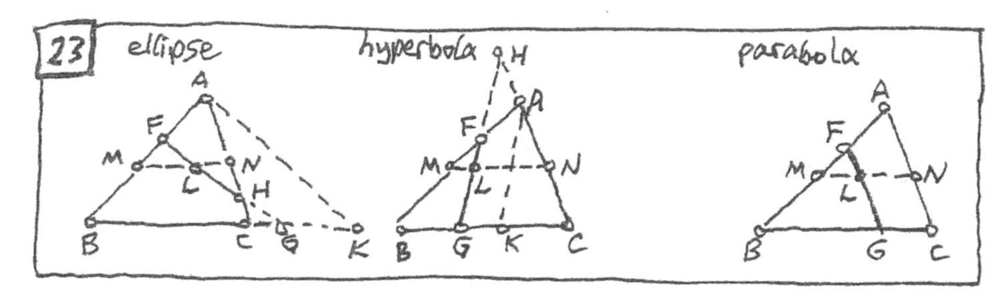

Then we would have, for each of the sections, a constant ratio which rectangle [ML, LN] has to a rectangle whose height is the abscissa (fig. 24).

24

in the section called	when the rectangle's width is equal to	then the ratio is that of
ellipse	the transverse side with abscissa taken away	sqr AK to rct KB, KC
hyperbola	the transverse side with abscissa added on	sqr AK to rct KB, KC
parabola	constant line FA	sqr BC to rct AB, AC

Things would be carried through to completion by way of the following consideration of the width of the rectangle whose height is the abscissa. If we increased the width of that rectangle, then rectangle [ML, LN] would have a smaller ratio to the wider rectangle; and if we decreased the width, then rectangle [ML, LN] would have a greater ratio to the narrower rectangle. So, if we took a line of just the right size and made it the width of a rectangle whose height is the abscissa, then rectangle [ML, LN] would be neither greater than nor smaller than it, but would be just equal to it. What size must that width be to be just right? As we have seen, it must be such as is shown in figure 25.

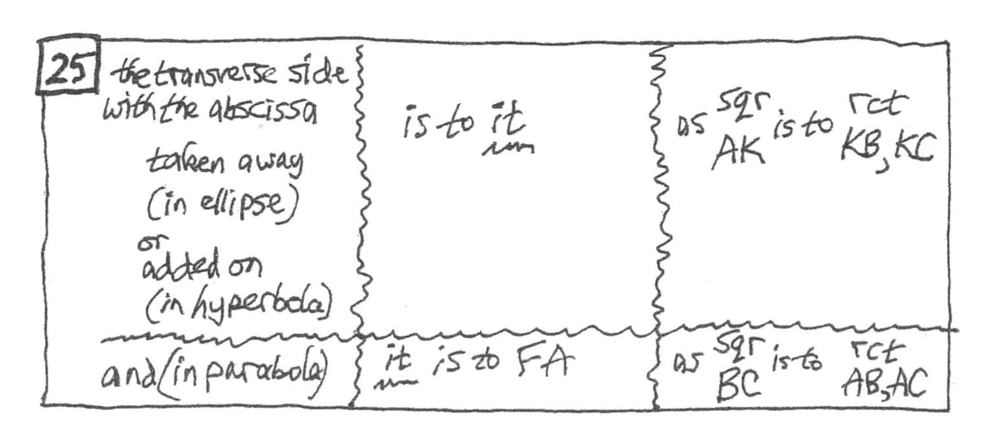

25

the transverse side with the abscissa taken away (in ellipse) or added on (in hyperbola)	is to it	as sqr AK is to rct KB, KC
and (in parabola)	it is to FA	as sqr BC is to rct AB, AC

Whether or not we choose to end in each case as Apollonius did, and whichever conic section we may start with, we get hold of its shape by viewing the cone from above as a stack of larger and larger circles, and from beside (in cross-section through the axis) as a ladder of larger and larger similar triangles (fig. 26).

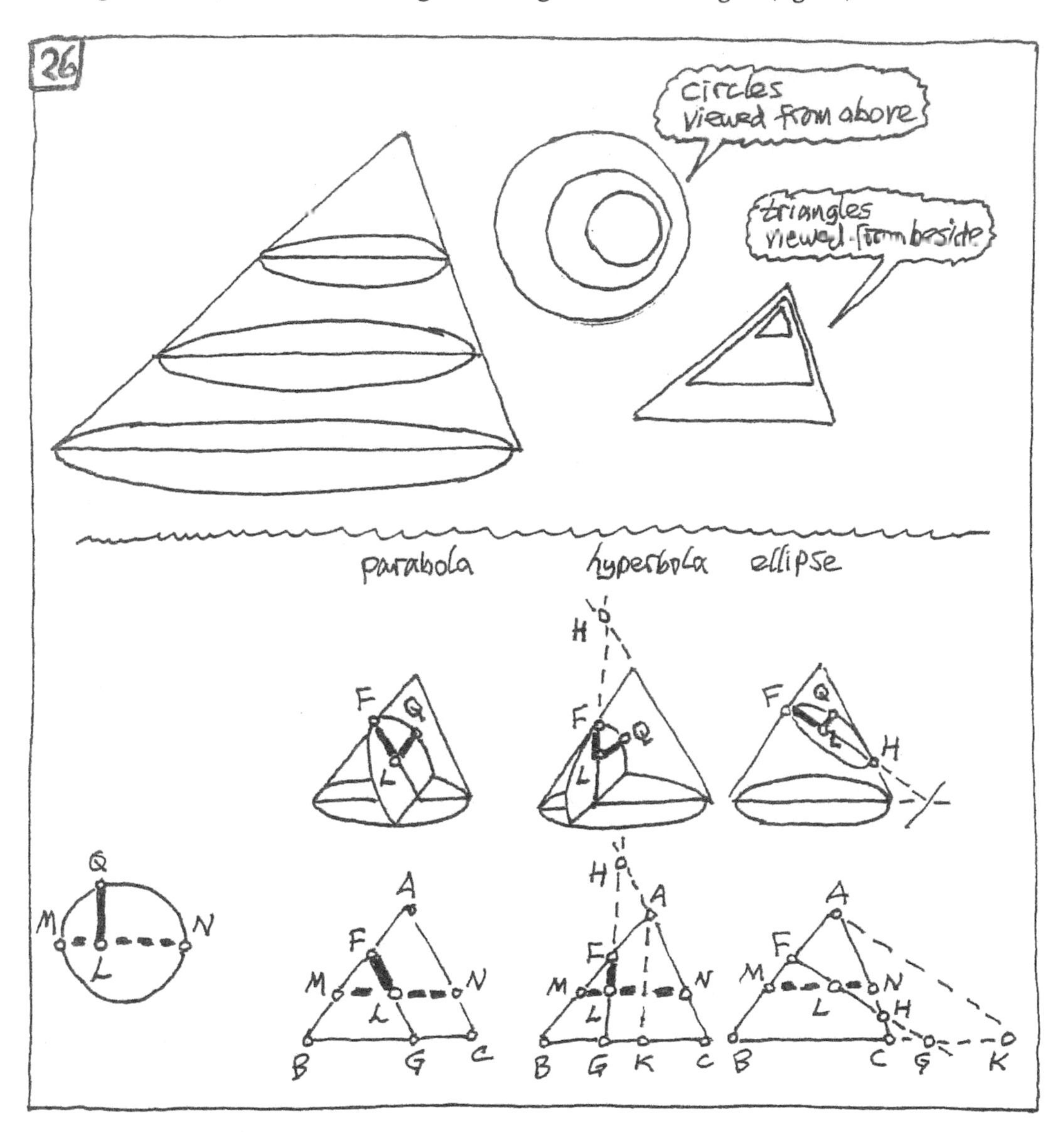

All of our analyses have taken as their point of departure something we have seen by viewing the cone from above:

> equal to the square arising from the *ordinate* QL,
> is the rectangle [ML, LN];

and in each analysis, we have gone on to find something by viewing the cone from beside:

> equal to that rectangle [ML, LN],
> is another rectangle—one whose height is equal to the *abscissa* FL.

The great labor of the three analyses has been to determine what must be the size of the rectangle's non-abscissa side—its width. The three analyses have led to three different such rectangles.

What gives shape to each conic section is constancy in the way it widens as it lengthens—constancy in the relation of its ordinate to its abscissa. This is determined by the size of the line that is the width of a certain rectangle—the rectangle that is equal to the square on the ordinate while having a height of the same size as the abscissa. The width can be determined by a constant line. The size of this constant line which determines the width is itself determined by what happens to segments ML and LN, as abscissa FL lengthens down along the diameter. Though ML and LN may change as FL lengthens, the ratio of ML to LF and the ratio of LN to a certain other line will both stay constant. The constancy of these ratios is what gives constancy to the conic section's shape.

When we did our analysis according to the order of the sections in the propositions of Apollonius, it was for the parabola that we first got an equality of rectangles. This involved a rectangle that had the varying abscissa as its one side, and—as its other side—a *constant* line. So, we were led to find a way, for the hyperbola and the ellipse afterward, to express the constancy of the ratio of rectangles as an equality of rectangles. This involved a rectangle which had the varying abscissa as its one side, and—as its other side—another *varying* line. However, there was a constant ratio that shaped the discrepancy between this rectangle and another rectangle which also had the varying abscissa as its one side, but had a *constant* line as its other side.

In whichever order we may go in our inquiry—whether from parabola to hyperbola to ellipse, or from ellipse to hyperbola to parabola—*we relate the ordinate to the abscissa by means of boxes built from straight lines that embody certain ratios*.

It has all seemed very complicated, however. Is there perhaps some easier way to get ratios that will enable us to establish the relationship?

CHAPTER 5. A ROAD NOT TAKEN

an alternative for the upright side

There is indeed an easier way to establish, by means of ratios, the relationship between ordinate and abscissa. It involves drawing a line that Apollonius himself will not introduce until very late in the First Book. The rest of the lines involved in this alternative, while already in the picture, are not the ones that Apollonius uses in these early propositions that characterize the conic sections. Apollonius could have characterized the sections more easily if that had been all that he had wanted to do.

Between these propositions in which he gets the various sections by cutting cones, however, and those propositions at the end of the First Book to which these propositions lead, Apollonius does things for which he needs to characterize the sections in the very way that he does.

The next section of this chapter will develop an alternative determination of the upright side. It is a single determination for all three sections—one, moreover, that is simpler than any of the several determinations presented by Apollonius. The next section will also show how this alternative determination makes it possible to give a single demonstration for all three sections—one, moreover, that is simpler than any of the several demonstrations presented by Apollonius. But before we move on to the work of the next section, let us consider what the existence of the simpler alternative suggests.

Earlier, we briefly considered, for the hyperbola and the ellipse, an alternative to drawing AK (a line parallel to the diameter, from the cone's vertex down to the axial triangle's base or its extension). We saw that we could avoid drawing that line for the hyperbola and the ellipse, and could also use corresponding lines for both of those sections and also for the parabola, if we used for our ratios—and consequently for our determination of the upright side—the constant lines GB, GC, and either FG twice (for the parabola) or both FG and HG (for the hyperbola and ellipse). This would slightly simplify the apparatus for the demonstrations (one line would not have to be drawn in), and it would bring the three propositions slightly closer together (since corresponding lines would appear more like each other). But the alternative that is about to be presented will be a much greater departure from the way taken by Apollonius than was that earlier alternative.

Figure 1 shows how the upright side is determined in the alternative that will be presented in the next section. This alternative determination will give, for each section, a line of exactly the same size as will be given by the appropriate determination from Apollonius. Figure 2 provides a reminder of how Apollonius determined the upright side.

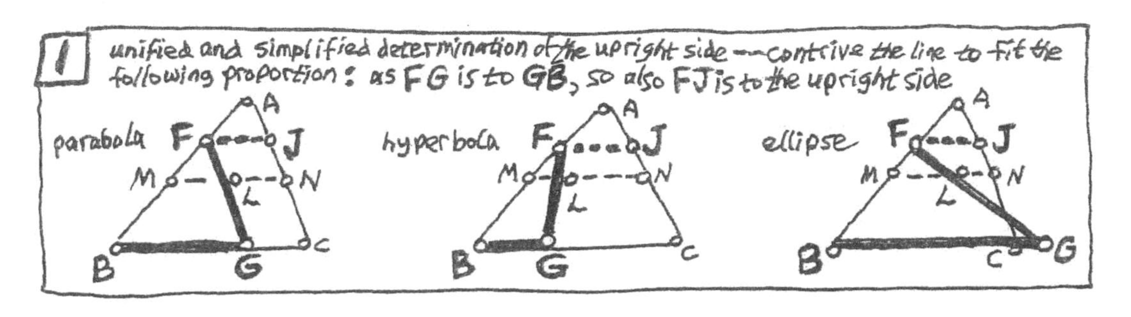

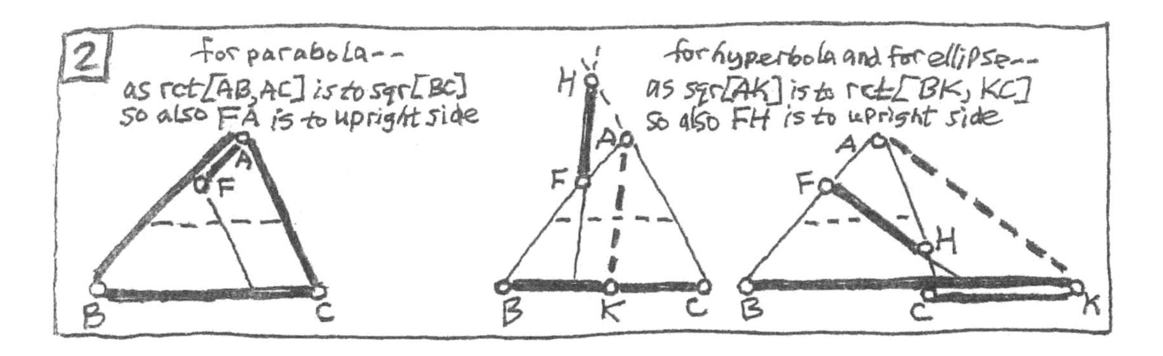

We shall soon see how that alternative determination of the upright side will allow for the sections' demonstrations to be treated not only all together but even as one—and will allow for this one to be treated more simply than any of those given by Apollonius. Why, then, does Apollonius do it his way—why should he be willing to multiply the labor by having three propositions, each one very complex? It is still too soon to say. Later on, we shall be in a better position to consider what it is that we get from doing things in the way that Apollonius does them, and what we would have to give up in return for the unity and simplicity of doing things in our alternative way.

One change made by the unified and simplified alternative is this: the upright side is not related to lines whose endpoints are the corners of the axial triangle. That is to say: the constant ratio that is used now involves lines FG and GB, as well as FJ—rather than lines BC, BA, AC (for the parabola) or lines AK, KB, KC (for the hyperbola and ellipse). By the end of the First Book of the *Conics* we shall see why this matters. Apollonius himself will, eventually, draw the line here called "FJ"—but he will not do so until the 52nd proposition.

Another change made by the alternative is this: the upright side is not related either to the line between the section's vertex and the axial triangle's vertex (in the parabola), or to the oblique (transverse) side (in the hyperbola and ellipse). The line used now is the line drawn from F to J, rather than the line from F to A (parabola) or the line from F to H (hyperbola and ellipse). Our alternative determination of the upright side thus foregoes using the corners of the axial triangle, and (in the hyperbola and the ellipse) it foregoes using the oblique (transverse) side as well. The axial triangle and the oblique (transverse) side—why should Apollonius care very much about keeping them in the story?

As for the axial triangle ABC, it was needed by him in order to get the diameter, and he would need it now to get FG, GB, and FJ—but why not drop it once it has done the job? For this matter of the axial triangle, we must wait until we have read almost to the end of the First Book of the *Conics*.

As for the oblique (transverse) side FH, it was needed by him in order to have something constant with which to shape a rectangular *eidos* of excess or deficiency. But that is not all. The 14th proposition—the proposition with which Apollonius will complete the characterization of the sections—will alert us to the central importance of the oblique (transverse) side in the case of what will be called "opposite sections." As we read the next proposition, we might ask: is the case of the opposite sections an isolated curiosity show, or is it the shape of things to come, revealing a kinship among the sections that is deeper than what meets the eye? We shall learn something about this in the next part of our study, as we proceed through the rest of the First Book of the *Conics*, and especially when we reach the end—the very spectacular end—of the First Book. The very end of the First Book will also be a beginning. It will turn out to be a curtain-raiser—or at least a preview of coming attractions.

a way to simplify and unify

Let us consider just how to develop an alternative to Apollonius's way of determining the upright side—an alternative way that, unlike his way, will be the same for all three conic sections and will be simpler for each one of them.

The property of the parabola is that the upright side makes, with the abscissa, a rectangle equal to the square arising from the ordinate. The property can be demonstrated without doing all that work which we did with Apollonius. The analysis through which we went was somewhat disingenuous: it was more cumbersome than it had to be. Certain complications arose in it because we wanted it to lead to that determination of the upright side from which Apollonius demonstrated the property of the parabola. But we could have taken a more direct way to get a determination of the upright side from which to demonstrate the property—if we had been willing to get a determination different from the one that Apollonius used. Here is how we could have proceeded more directly.

Let us consider the view from beside the equal triangle, since it is there that the way in which we shall now do the job differs from the way of Apollonius.

We wish to relate the ordinate and the abscissa. We have seen, in the view from above, that the ordinate is the side of a square that is equal to the rectangle whose sides are the lines LM and LN. In the view from beside, we have not only those lines (LM and LN) but also the line that is the abscissa (LF). Now, some rectangle with the abscissa LF as one of its sides will have to be equal to the square on the ordinate—but only if that rectangle is also equal to the rectangle whose sides are LM and LN (since things equal to the same thing are equal to each other). Let us call the rectangle's abscissa-side its height. What, then, must be the size of its other side—its width?

Well, no two rectangles can be equal to each other unless their sides are proportional; and if four straight lines are proportional, then the rectangle contained by the terms at the extremes of the proportion will be equal to the rectangle contained by the terms in the middle. So, in the present case, the equality cannot hold between the two rectangles [LM, LN] and [LF, (width)] unless the following proportion holds among their sides:

as LM is to abscissa LF (the height), so also the other side (the width) is to LN.

Even though neither LM nor LF is constant, the ratio of the one to the other is constant—and therefore also constant is the ratio of the width to LN. And so, since LN is constant, the width too is a constant line. So, if we merely contrive a line of such a size that the ratio which it has to constant line LN (the one side of the equal rectangle) is the same as the constant ratio which the other side of the equal rectangle (LM) has to the abscissa FL, then that contrived line will make, with the abscissa, a rectangle which is equal to rectangle [ML, LN], and which is therefore equal to the square on the ordinate. That might also be said in the following way.

Rectangle [ML, LN] is different for each different abscissa FL. That is because at different positions of L on the diameter, although one side of the rectangle (line LN) stays the same, the other side (ML) is different. But it is not only the rectangle's one side LN that stays the same—something else does too. What also stays the same is the ratio which the varying other side ML has to the varying abscissa FL. Hence, there will be some constant line which has, to that constant side LN, that same constant ratio which the varying other side ML has to the varying abscissa FL. This fourth proportional line FV—let us call it the upright side.

And if we take that way to get *to* a determination of the upright side, then we get *from* it the quick and easy demonstration of the property of the parabola that is depicted in figure 3.

The property so quickly and easily demonstrated in the figure is the same property that was demonstrated by Apollonius in the 11th proposition—but the different demonstrations use different determinations of the upright side FV:

In this quick and easy demonstration, the upright side FV is determined as the straight line whose ratio to LN is the same as the ratio which LM has to LF.	In Apollonius's demonstration the upright side FV is determined as the straight line whose ratio to FA is the same as the ratio which sqr [BC] has to rct [BA, AC].

That is to say: the quick and easy demonstration determines the parabola's upright in terms of the following lines—

> the segments (LM and LN) of the circle's diameter, which contain the rectangle equal to the square arising from the ordinate, and
>
> the abscissa (LF);

whereas Apollonius determined it in terms of the following lines—

> the base and sides (BC, and BA and AC) of the axial triangle, and
>
> the distance (FA) from the section's vertex to the axial triangle's vertex.

Now, it is true that Apollonius in his determination of the upright side uses no lines but ones that are constant, whereas two of the lines used in the determination for the quick and easy demonstration are not constant—namely, the two lines LM and LF. But if that is something which matters, then the quick and easy demonstration can be modified, making it only slightly less quick and easy: the constant ratio of LM to LF can be named by speaking instead of the same constant ratio of constant line GB to constant line FG.

It was because of Apollonius's choice of lines to use as terms in determining the parabola's upright side, that he had to do more work to prove the parabola's ordinate-abscissa property. The extra work made no difference at all in the size of the line determined. The line that we got from the determination used by Apollonius has exactly the same size as the line that we get from the determination used in the quick and easy demonstration.

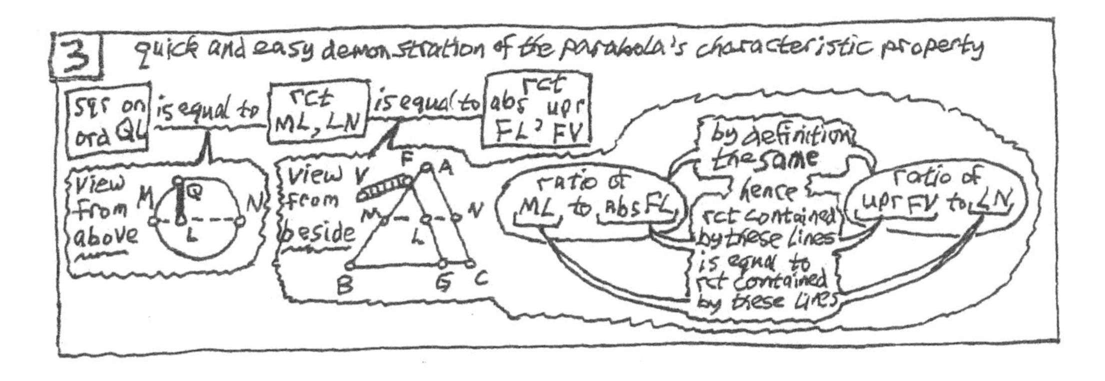

Let us show that the two determinations of the parabola's upright side are equivalent, giving two lines of the same size. From the sameness of ratios which was used *just now by us* to determine the parabola's upright side, we must derive the sameness of ratios which is used *in the 11th proposition by Apollonius* to determine the parabola's upright side. The way that we shall do it is displayed in figure 4.

We wish to take our determination and make a two-fold transformation that will get from it the determination given by Apollonius. (1) We wish to get the ratio which the upright side has to the distance FA between the vertices, rather than to the rectangle's constant side LN. And (2), instead of the ratio which the rectangle's varying side LM has to the abscissa FL, we wish to get the ratio which square [BC] has to rectangle [BA, AC].

This last ratio is a ratio which is compounded of ratios of the lines that constitute the axial triangle. These two component ratios of lines are: the ratio which BC has to BA, and the ratio which BC has to AC. The second of these component ratios is (by similar triangles) the same as the ratio used just now by us to define the parabola's upright side: it is the ratio which the rectangle's varying side ML has to the abscissa FL. So, in order to get the ratio of rectangles used by Apollonius to define the parabola's upright side in his 11th proposition, we need only to take the ratio which we have already, and compound it with the ratio which BC has to BA. But if the *first* ratio of a proportion is compounded with another ratio, then, in order to preserve the proportionality, the *second* ratio of the proportion must also be compounded with that same other ratio. That is to say: the ratio that will be the same as the new ratio will be the ratio which is compounded out of the other ratio in our determining proportion (namely, the ratio which upright side FV has to the rectangle's constant side LN), on the one hand, and, on the other, the ratio which BC has to BA.

So, now we have accomplished half of what we wanted to do. From the proportion by which *we* determined the parabola's upright side, we have derived a proportion in which *one* of the ratios used by us is one of the ratios used in the proportion by which *Apollonius* determined the parabola's upright side. Now we need to go on to derive, from this derived proportion, a proportion in which also the *other* ratio is a ratio used by Apollonius in the determination. In other words: now we know that the ratio which square [BC] has to rectangle [BA, AC] is the same as the ratio compounded by taking *our* determining ratio (namely, the ratio which upright side FV has to the rectangle's constant side LN) and putting it together with the ratio which BC has to BA; and we must go on to show that this compounded ratio in its turn is the same as Apollonius's determining ratio (namely, the ratio which upright side FV has to the distance FA between the vertices). How can we do this?

Well, if we just went ahead and compounded the two ratios of lines that we have got, then we would get a ratio of rectangles—which is not what we want to get. What we want to get from the compounding is rather a ratio of *lines*. So, before we compound, we must first get two ratios of lines like this: each of the ratios must be the same as one of the two ratios of lines that we have got (of FV to LN, and of BC to BA), but the *consequent* term of the *first* ratio must be the same as the *antecedent* term of the *second ratio*. Then, when we compound, we will "squoosh"—and get a ratio of *lines*.

And since the lines which we want to get in the compounded ratio are FV and FA, we need a line—call it (j)—such that: FV has to line (j) the same ratio which FV has to LN, and line (j) has to FA the same ratio which BC has to BA. How can we get such a line? By getting a line which is equal to LN and which also makes with FA a triangle similar to the triangle whose sides are the axial triangle's base BC and its side BA. That is to say: we need only draw a line from F, parallel to the axial triangle's base (BC), across to the point (call it J) where it meets the axial triangle's other side (AC).

Thus we have shown the equivalence of the two determinations of the parabola's upright side—that of our alternative and that of Apollonius.

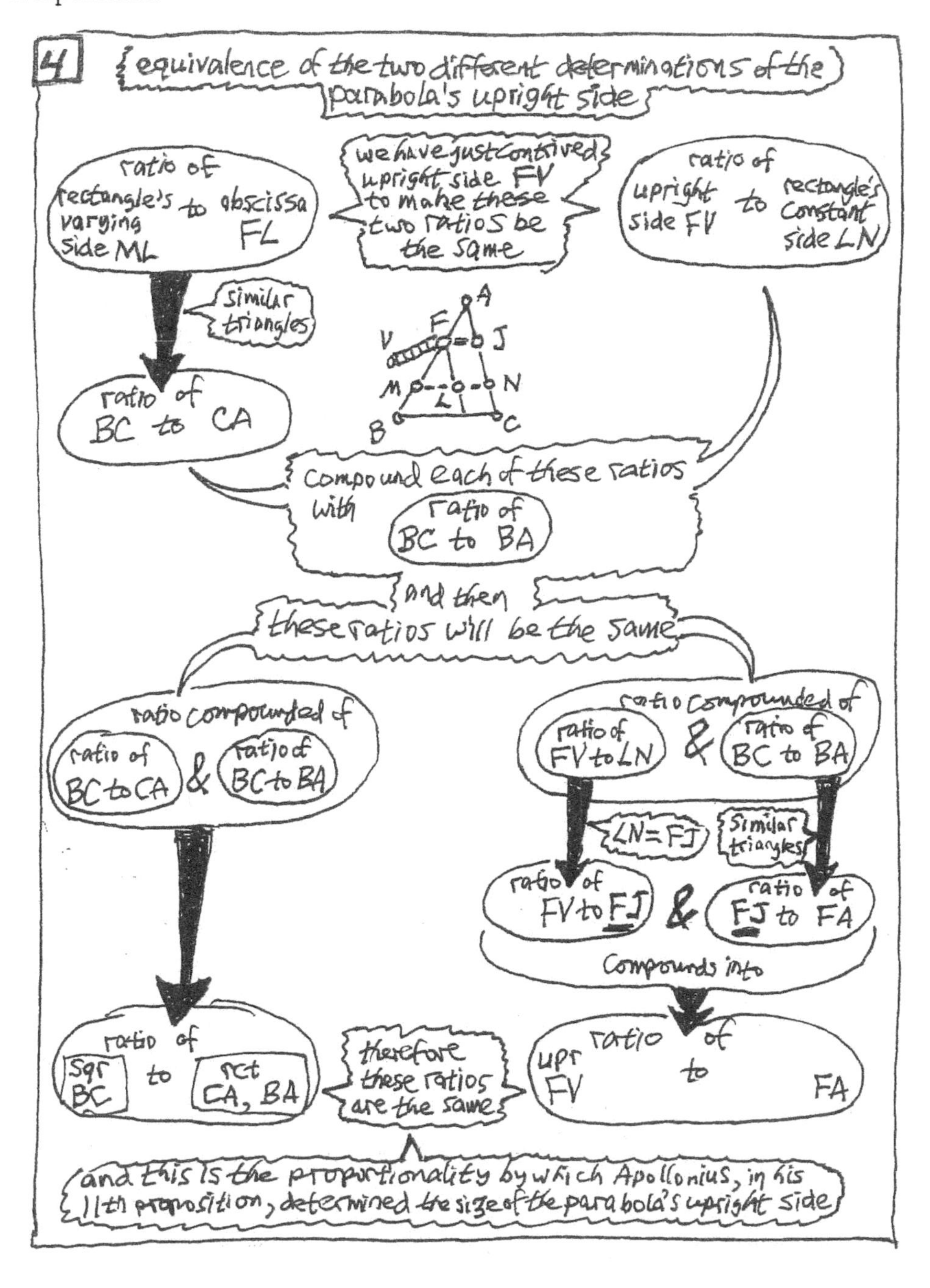

Figure 5 compares the labor of demonstrating the parabola's property by using the two different determinations of the parabola's upright side—the labor-intensive device presented by Apollonius, and the labor-saving alternative device that we presented.

Since the determinations are equivalent, why not determine the upright side by using whatever terms will be easiest to manipulate? After all, the upright side is of interest not because of its position, but rather because of its size. It characterizes a section by relating the sizes of its ordinates and abscissas. When we speak of the "upright" side we refer to *where* it is relative to other lines, but when we determine the line so named we refer to *how large* it is relative to other lines. The upright side is an embodiment of ratios. Why not state the ratios in the terms that are handiest? Let us, then, go further—let us go beyond the case of the parabola—and perform an even greater simplification.

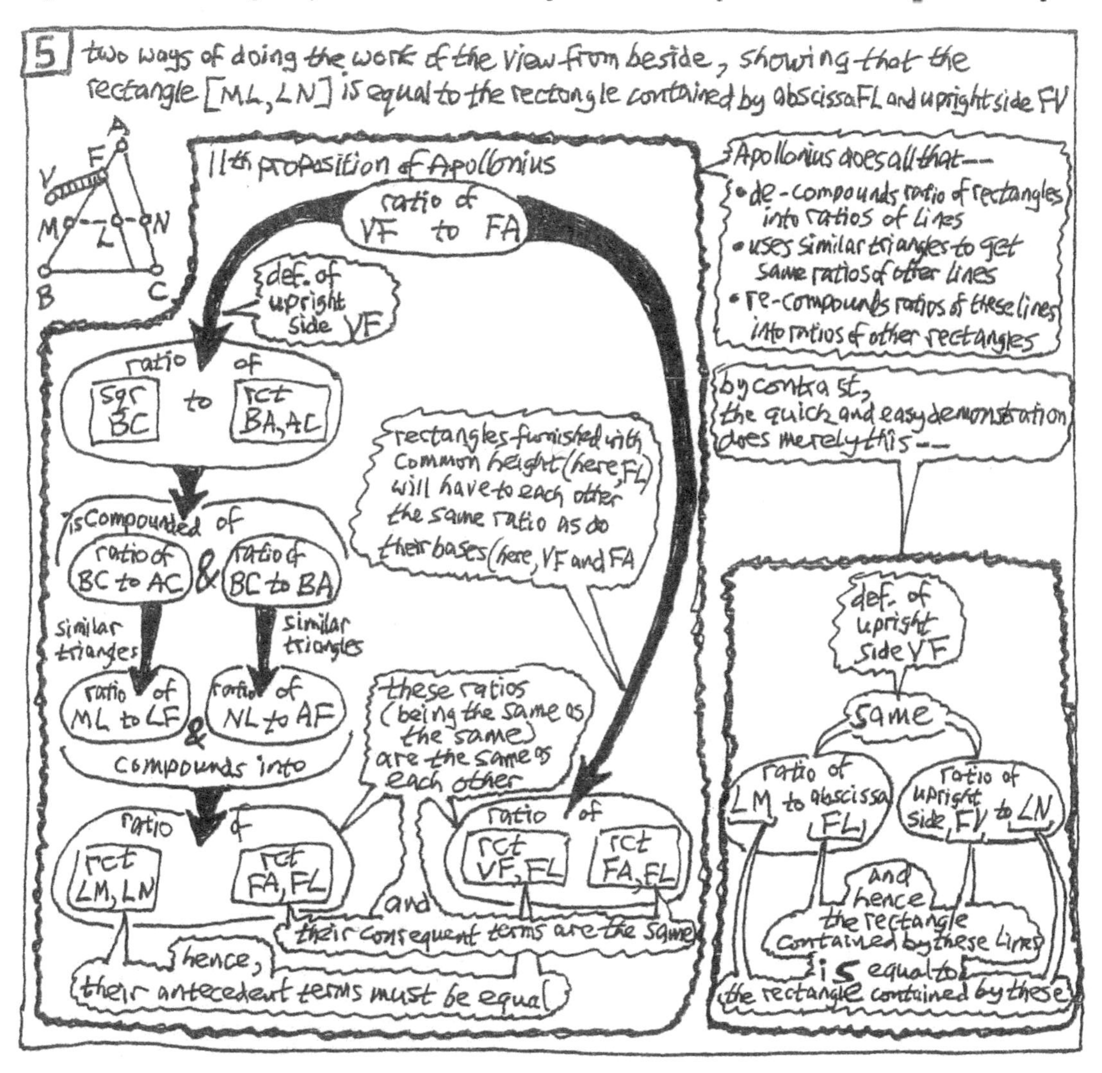

Again and again, we have used the fact (from the top-view circle) that the square arising from the ordinate QL is equal to rectangle [LM, LN]. The only difficulty was to find the width of the rectangle which, having the abscissa as its height, is equal to the rectangle [LM, LN]. What we have learned—after putting a lot of labor into contriving and employing an upright side (FV)—is the following (for which, see figure 6).

The rectangle whose height is equal to the abscissa FL will be equal to rectangle [LM, LN] when

its width FP is,
in the parabola, equal to the upright side FV,
but, in the hyperbola, is larger than it,
and, in the ellipse, is smaller than it;

and the discrepancy between the width FP and the upright side FV (that is to say: the amount OX by which the width is larger in the hyperbola or is smaller in the ellipse) is such as to make, with the abscissa FL, a rectangular form similar to the rectangle contained by the oblique (transverse) side HF and the upright side FV—which means that

the discrepancy OX between width FP and the upright side FV is of such a size that
the discrepancy OX has to the abscissa FL
the same constant ratio that
the upright side FV has to the oblique (transverse) side FH.

Now, the relation between the ordinate and the abscissa, demonstrated by Apollonius as a square-rectangle property in three propositions (the 11th, 12th, and 13th), can be presented in a single demonstration that is almost the same for all three sections—if the determination we use for the upright side is the very same for all three sections, and that demonstration is also simpler than the one Apollonius uses. How? As follows.

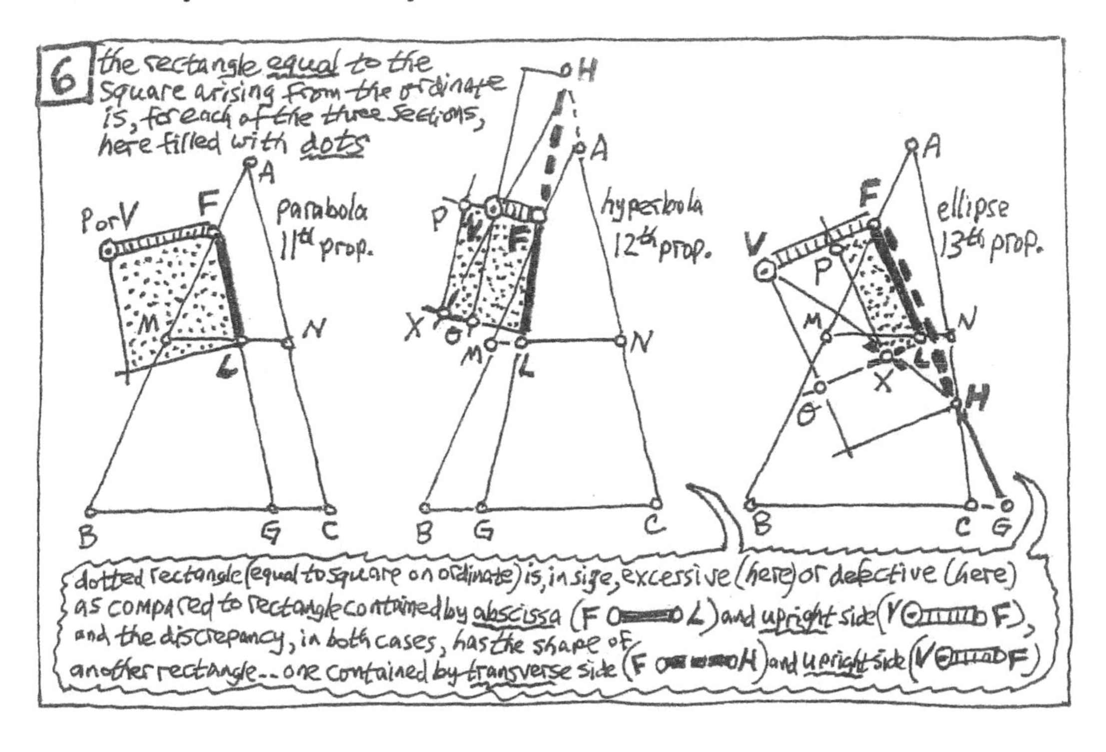

Let us consider the segments (LM and LN) that contain the equal rectangle. What happens to them as abscissa FL lengthens down along the diameter—or shortens, going up along it?

What happens to the one segment LM as its endpoint L ascends along the diameter to F? The size of LM changes, shrinking to nothingness. And what happens to LM as L descends from F? LM, like its abscissa FL, grows, but there is no change in their ratio. The abscissa FL has a constant ratio to the one segment (LM).

And what happens to the other segment (LN) as L ascends to F? The size of LN changes (except in the case of the parabola), but there is a limit on the change in size—as we can see more easily if we draw a line from the section's vertex F, parallel to the base, across to the other side of the axial triangle. Call that line FJ. Now it is easy to see that LN will get as nearly equal to FJ as we please, if L is taken near enough to F. And what happens to LN as L descends from F? LN changes size—growing in the hyperbola, and shrinking in the ellipse. And exactly what size does it assume for a given L? This will depend on the constant ratio of abscissa FL to segment LM. That is to say: the size of LN for a given L will depend on how fast LN changes as L descends. But it will depend also on what size LN is to start with—that is, on the size of the drawn-in constant line FJ.

Let us contrive a line into which we somehow build both of the things which, for any given size of the varying abscissa FL, jointly determine the size of LN—namely, the constant ratio (of FL to LM) and the size of the constant line (FJ). That is, let us take a constant fourth proportional determined in the following way:

constant line FJ will have to this constant fourth proportional line (call it FV)
the same constant ratio which abscissa FL has to segment LM;
however, let us name this constant ratio of varying line FL to varying line LM
in such a way as to use as terms only lines that are constant;
we shall determine FJ by using another ratio than that of FL to LM—namely,
by using the same constant ratio which constant line FG has to constant line GB.

Why have we not just used, for all the sections, the same simplified definition of the upright side which we found for the quick and easy demonstration of the property of the parabola? Why have we just now contrived anew this constant line FV? The reason is this: LN is constant in the case of the parabola, but in the case of the other two sections—the hyperbola and the ellipse—LN is no more constant than LM is. We need to handle the varying of *both* sides of the rectangle—not only LM, but also LN.

In the case of the parabola, where LN is constant, we contrived a constant line FV to embody the changing of LM, in order to relate the abscissa FL to the ordinate. This upright side FV for the parabola was determined as a fourth proportional: constant segment LN has to it the same ratio which abscissa FL has to varying segment ML. But now we want to cover not only the case of the parabola (where LN is constant), but also the cases of the hyperbola and the ellipse (where LN varies). We therefore must determine the constant line FV, not in terms of a line of varying size (LN), but rather in terms of a line of constant size—a constant line to which LN has some ratio which we can characterize. FJ is such a line. FJ is a constant line; and (because it is parallel to MLN) it has, in the parabola, the *same size as the constant line* LN, but, in the hyperbola and the ellipse, it has, *to the varying line* LN, the *same ratio* which the constant line FH has to the varying line LH—FH and LH being the two lines that run from the abscissa's endpoints (F and L) to the point H (where the cutting plane meets the axial triangle's other side).

What we have so far, is depicted in figure 7.

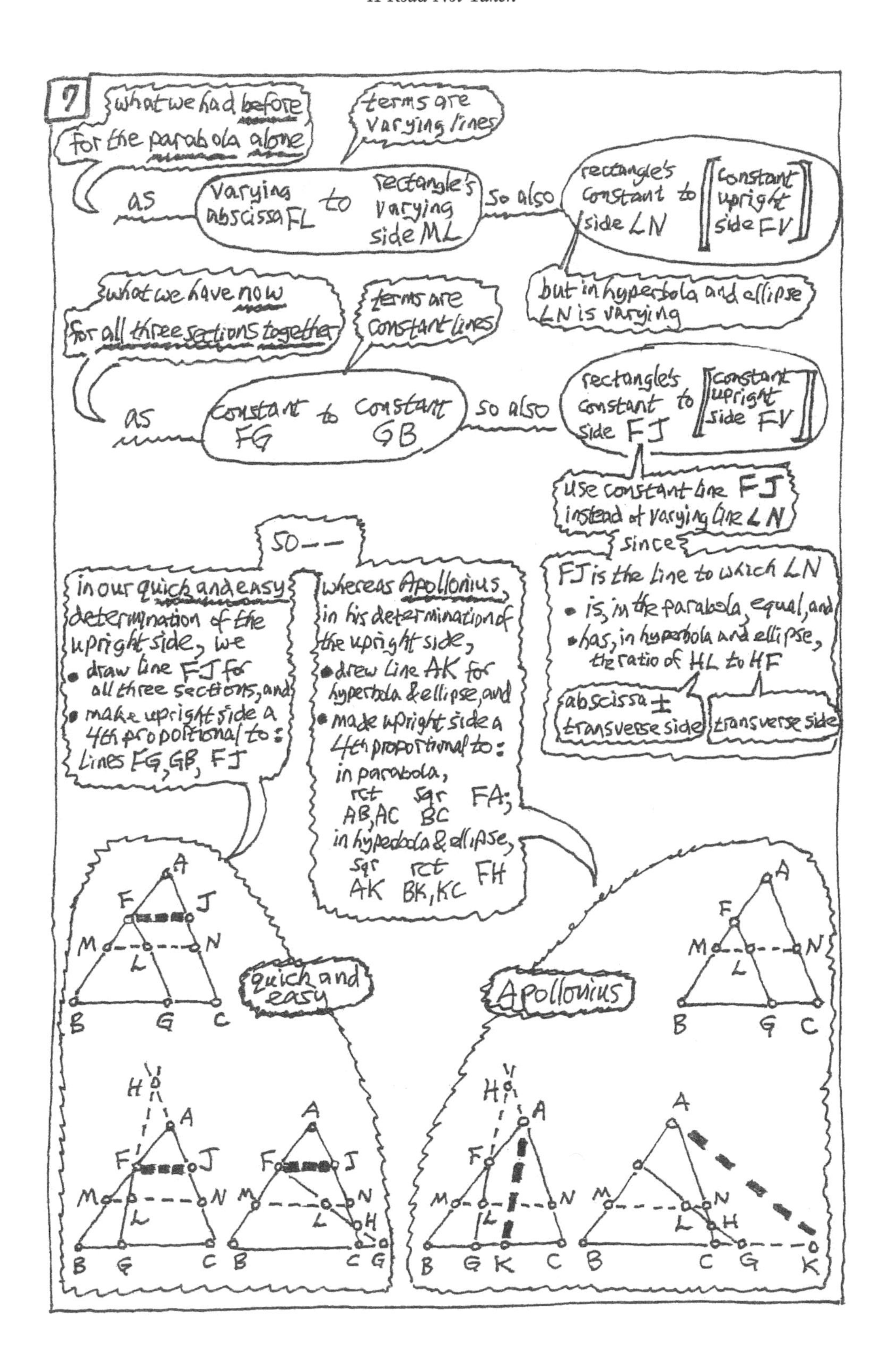
7
what we had before for the parabola alone
terms are varying lines
as
varying abscissa FL to rectangle's varying side ML
so also
rectangle's constant side LN to [[constant upright side FV]]
but in hyperbola and ellipse LN is varying
what we have now for all three sections together
terms are constant lines
as
constant FG to constant GB
so also
rectangle's constant side FJ to [[constant upright side FV]]
use constant line FJ instead of varying line LN
since
FJ is the line to which LN
• is, in the parabola, equal, and
• has, in hyperbola and ellipse, the ratio of HL to HF
abscissa ± transverse side
transverse side
SO--
in our quick and easy determination of the upright side, we
• draw line FJ for all three sections, and
• make upright side a 4th proportional to: lines FG, GB, FJ
whereas Apollonius, in his determination of the upright side,
• drew line AK for hyperbola & ellipse, and
• made upright side a 4th proportional to: in parabola, rct AB, AC sqr BC FA; in hyperbola & ellipse, sqr AK rct BK, KC FH
quick and easy
Apollonius

Now let us begin this again and follow through by operating more systematically and methodically. We want to relate the abscissa and the ordinate. To do this, we find a rectangle which has the abscissa as its side, and which is equal to the square arising from the ordinate. This square arising from the ordinate is equal to the rectangle [LM, LN]. (Fig. 8).

The rectangle which we want is equal to this rectangle, and has the abscissa (FL) as its height.

What then must be its width (w)? Well, what if we suppose that rectangle [FL, (w)] is equal to rectangle [LM, LN]—what then must follow? We know that no two rectangles can be equal to each other unless their sides are proportional; and we know that if four straight lines are proportional, then the rectangle contained by the terms at the extremes of the proportion will be equal to the rectangle contained by the terms in the middle. In the present case, there cannot be the equality of rectangles unless there is the following proportionality of lines:

FL has to LM	the same ratio which	LN has to (w).

First let us look at the ratio on the left. The abscissa FL is not constant, and neither is the segment LM—but the ratio of the one line to the other is constant. It is the ratio of the constant line FG to the constant line GB.

We have looked at the ratio on the left, the one involving the varying segment LM. Now let us look at the ratio on the right, the one involving the other segment LN. This ratio, being the same as the ratio on the left, must be constant. But whereas segment LN is constant in the parabola, it varies in the other sections. The other term in this constant ratio, width (w), therefore must (like the term LN) also be constant in the parabola but vary in the other sections. So, let us state the ratio on the right in other terms: let us state the ratio in terms which—like the ratio itself—are constant. What lines shall we use as terms?

But why ask? After all, we already know that LN has to (w) the same ratio which constant line FG has to constant line GB—is that not enough? No—because this ratio is not the only constant thing that determines the size of LN. (The varying size of LN depends, of course, not only on the sizes of the relevant constant lines, but also on the varying size of the abscissa—but now we are considering only the determinants that are constant. That is to say: we are considering what determines LN for any one size of the abscissa FL.) As L descends from F, line LN's rate of growth (in the hyperbola) or of shrinkage (in the ellipse) depends on how large is the constant ratio which FG has to GB. But LN's size depends not only on the rate at which LN changes as L descends. It depends also on how large LN is at the start—that is, on LN's size just when L in an ascent would reach F (where the shrinking LM vanishes into nothingness). We would like to relate LN to *all* the constant lines which determine it, and not just to FG and GB. How do we do it?

Well, varying line LN runs from the lower end of the abscissa, parallel to the base, across to the axial triangle's other side (the side across from the section's vertex). If we draw a line from the other end of the abscissa (from the upper end, which is the section's vertex)—if, that is to say, we draw a line that (like LN) runs from an end of the abscissa, parallel to the base, across to the other side of the axial triangle—then this parallel line (call it FJ) will be constant. And since it is constant, it will have, to some as yet unknown constant line—call it (u)—the same ratio which LN has to (w), which ratio is the constant ratio of the constant line FG to the constant line GB. (Fig. 9.)

And now we can relate LN to that constant line FJ by which (together with the constant ratio) it is determined.

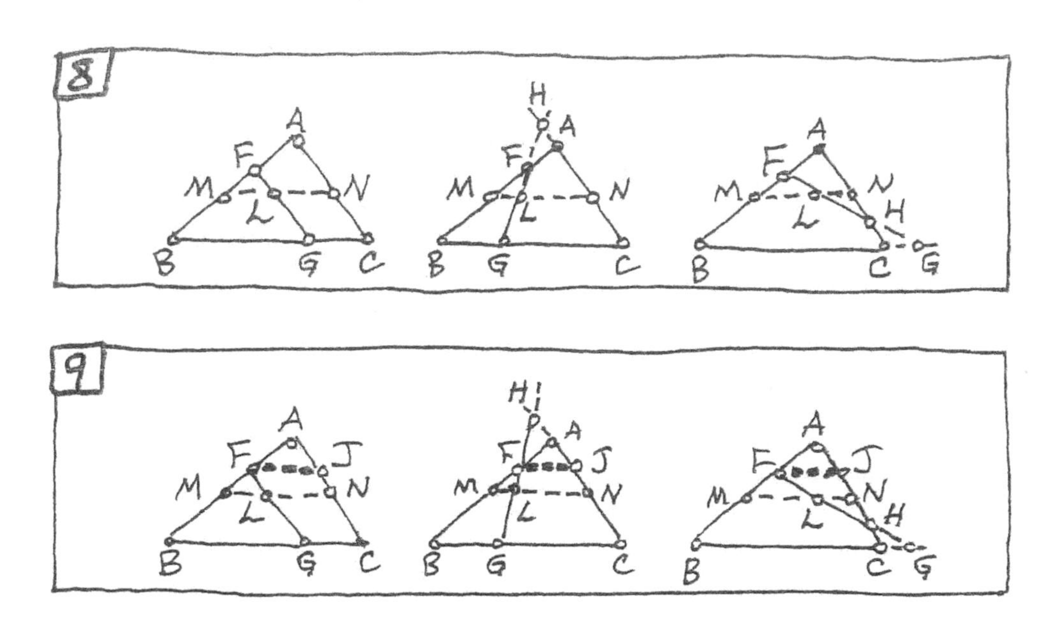

If we take the constant fourth proportional (u), determined entirely in terms of the constant lines which determine the size of the varying line LN, as follows—

as FG is to GB so also FJ is to (u)

—then we can get the ratio that we want, the ratio of LN to (w), in terms of *all* the *constant* lines which determine LN. We can say (fig. 10) that

since FJ is to (u) as FG is to GB as FL is to LM as LN is to (w),

hence as FJ is to (u) .... so also LN is to (w).

And now we can relate LN to FJ by relating line (w) to line (u).

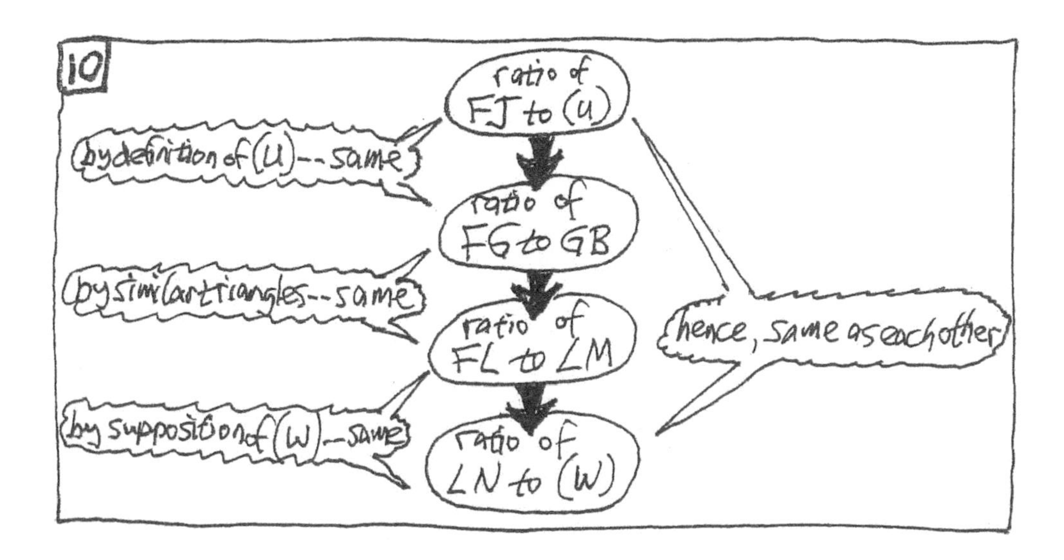

If we take antecedent to antecedent, and consequent to consequent, then from the last proportion, which said that

as FJ is to (u) so also LN is to (w),

we get the following proportion:

as FJ is to LN so also (u) is to (w).

In the case of the parabola, line LN is constant—it is always equal to FJ—and so the width (w) of the equal rectangle is, in the parabola, equal to the constant contrived line (u). We call it the upright side.

In the case of the other two sections, however, LN is not constant. But since FJ is constant, the ratio of FJ to LN varies. Is there any constant line that might serve in some ratio that could be related to this varying ratio of FJ to LN? Well, FJ and LN, being parallel, are corresponding sides in similar triangles—and, in those triangles (which have H as their common vertex), another pair of corresponding sides are FH and HL. Line FH is the oblique (transverse) side; and, as for HL, either (in the case of the hyperbola) the oblique (transverse) side is part of it, or (in the case of the ellipse) it is part of the oblique (transverse) side. (The other part of the whole, in both cases, is the abscissa.) So now, because the triangles are similar, we have that

as FJ is to LN so also FH is to LH.

Now, we have just seen that the ratio of FH to LH is the same as the ratio of FJ to LN, and we saw (just a little while before that) that the ratio of (u) to (w) is also the same as that same ratio of FJ to LN. It follows that the ratios of FH to LH and of (u) to (w)—being the same as the same ratio—are the same as each other:

as FH is to LH so also (u) is to (w).

And from this last proportion (if we take antecedent to antecedent, and consequent to consequent) we get this next—that

as FH is to (u) so also LH is to (w).

That is to say, in the hyperbola and the ellipse,

as the constant oblique (transverse) side (FH) is to the constant upright side (u),
so also the varying line LH will be to the varying width (w) of
the equal rectangle whose height is the abscissa—

the line LH (the line between
the non-vertex end of the oblique [transverse] side and
the non-vertex end of the abscissa)
being the result of the abscissa's being
added on to the oblique (transverse) side (in the hyperbola),
taken away from the oblique (transverse) side (in the ellipse). (Fig. 11.)

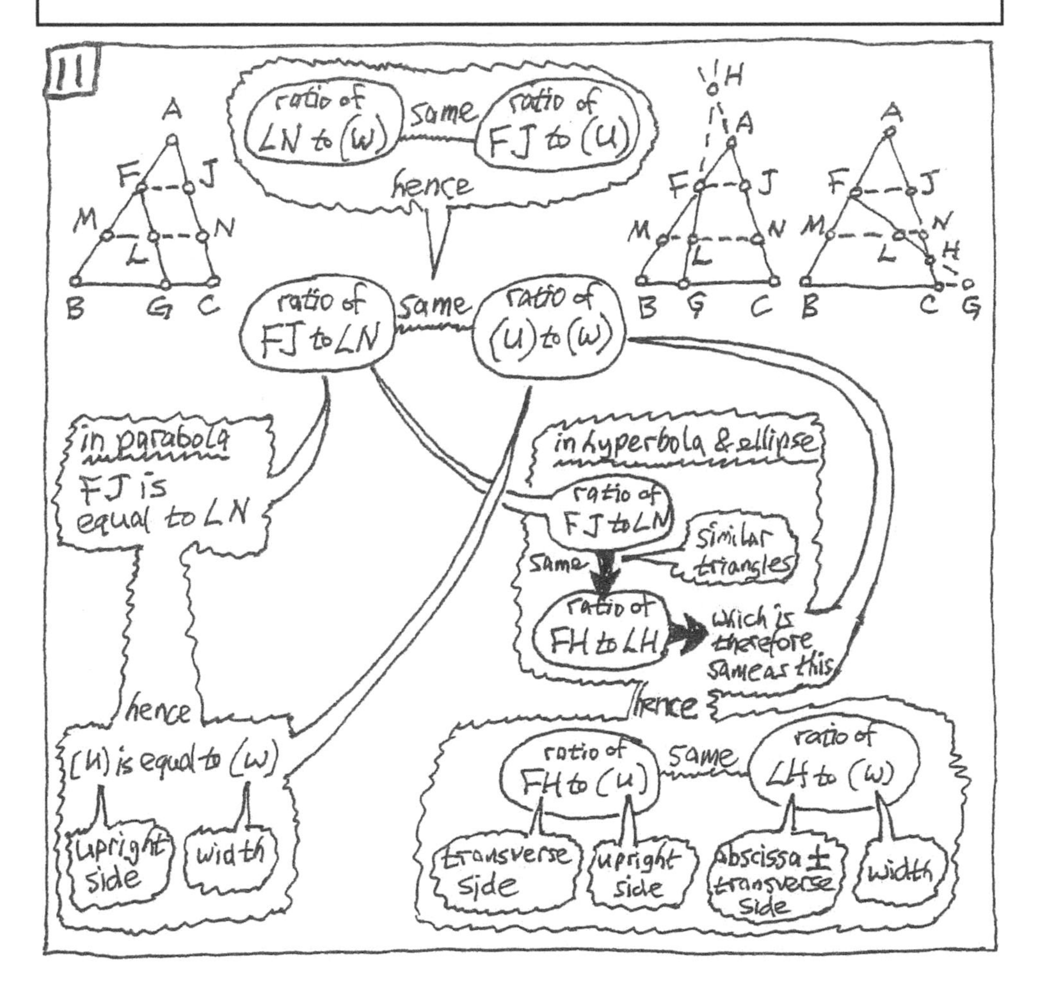

So, we have loosened up a complicated picture into simpler constituent relationships, and have contrived something that allows us to look at fewer items at a time. From such an analysis as the foregoing, having determined the constant line FV so as to build into it the relevant ratios of constant and varying lines, we can obtain the synthesis which is sketched in figure 12. What is shown there, is *all* that would be needed to replace the demonstrations of the 11th *and* the 12th *and* the 13th propositions! It has unity (one demonstration now gives us what formerly required three demonstrations) and it has simplicity (the work that formerly required three demonstrations is now done with less labor than any single one of them took before).

Our alternative shows us that the *parabola* property can be proved with much less labor if we determine the upright side as a fourth proportional in terms of these things:

abscissa FL	rectangle side LM	rectangle side LN,

rather than (like Apollonius) determining it as a fourth proportional in terms of these things:

rectangle on axial triangle's sides AB, AC	square on axial triangle's base BC	distance FA between section's vertex axial triangle's vertex.

If, way back when we first saw Apollonius cut a cone to name this section, we had asked ourselves why Apollonius is willing to labor more than is necessary in order to get the parabola property—why he prefers to use those lines which he uses to define the upright side—we might have thought that he does it to have *constant* lines rather than lines that are varying. Or that he does it because the constant lines he uses are better adapted for bringing the various sections *together*.

But we have seen that neither of these can be his reason. When, in our alternative to the way that was taken by Apollonius, we moved on to dealing with the *other* sections, the upright side was then determined, for all of the sections in the trio, as a fourth proportional in terms of these lines:

constant FG	constant GB	constant FJ.

Those lines are used for *all* of the sections in the trio, and they are all just as *constant* as those which Apollonius used in determining the upright side as a fourth proportional in terms of these things (for the parabola):

rectangle on axial triangle's sides AB, AC	square on axial triangle's base BC	distance FA between section's vertex axial triangle's vertex,

or in terms of these (for the hyperbola and the ellipse):

square on AK	rectangle on BK, KC	oblique (transverse) side. (here called FH).

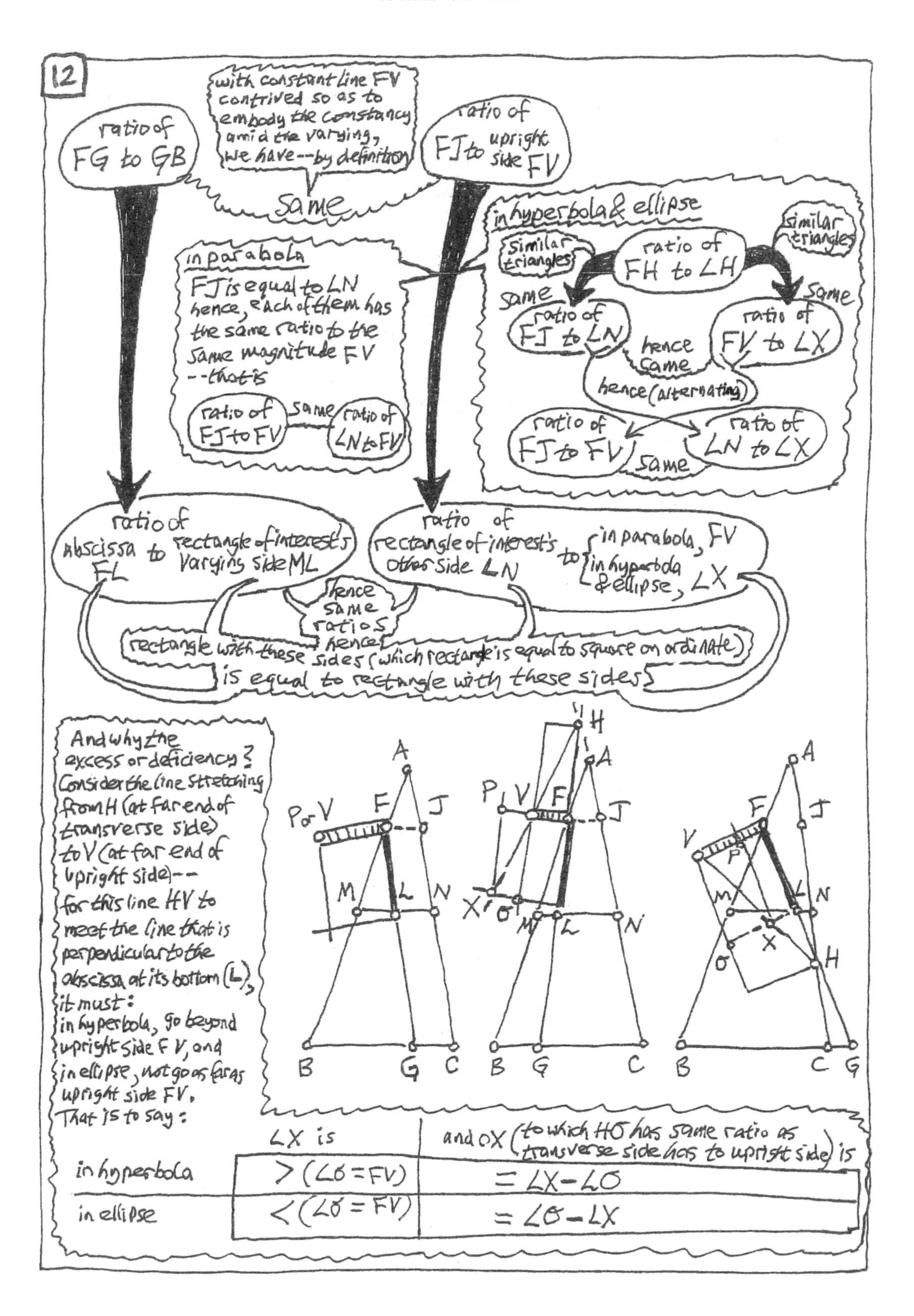
12
ratio of FG to GB
with constant line FV contrived so as to embody the constancy amid the varying, we have--by definition
same
ratio of FJ to upright side FV
in parabola
FJ is equal to LN hence, each of them has the same ratio to the same magnitude FV --that is
ratio of FJ to FV
same
ratio of LN to FV
in hyperbola & ellipse
similar triangles
ratio of FH to LH
similar triangles
same
ratio of FJ to LN
hence same
same
ratio of FV to LX
hence (alternating)
ratio of FJ to FV
same
ratio of LN to LX
ratio of abscissa FL to rectangle of interest's varying side ML
ratio of rectangle of interest's other side LN to {in parabola, FV / in hyperbola & ellipse, LX
hence same ratios
hence
{rectangle with these sides (which rectangle is equal to square on ordinate)} is equal to rectangle with these sides}
And why the excess or deficiency? Consider the line stretching from H (at far end of transverse side) to V (at far end of upright side)-- for this line HV to meet the line that is perpendicular to the abscissa at its bottom (L), it must: in hyperbola, go beyond upright side FV, and in ellipse, not go as far as upright side FV. That is to say:
A
P or V
F
J
M
L
N
B
G
C
H
A
P
V
F
J
X
O
M
L
N
B
G
C
A
F
J
V
P
M
L
N
X
O
H
B
C
G
LX is
and OX (to which HO has same ratio as transverse side has to upright side) is
in hyperbola
> (LO = FV)
= LX − LO
in ellipse
< (LO = FV)
= LO − LX

CHAPTER 6. LOOKING BACK AND LOOKING DEEPER

When we brought the sections together in Chapter 4's treatment of their upright sides and of their analyses, we examined a certain line called (w) for each section. That line was the width of a rectangle whose other side was the abscissa, with the rectangle itself being equal to a square whose side was the ordinate.

In the *parabola*, as we go up and down the diameter, changing the size of the abscissa, the width (w) does *not* change at all. This line (w) stays constant because, in the proportion which determines it, all the other lines are constant (fig. 1). We shall call its constant length the "parameter" of the parabola with its diameter. (The reason for this traditional piece of terminology will be indicated soon.)

In the *hyperbola* and the *ellipse*, however, as we go up or down these sections, line (w) *does* change. It changes because, in the proportion which defines it, not all the other lines are constant (fig. 2).

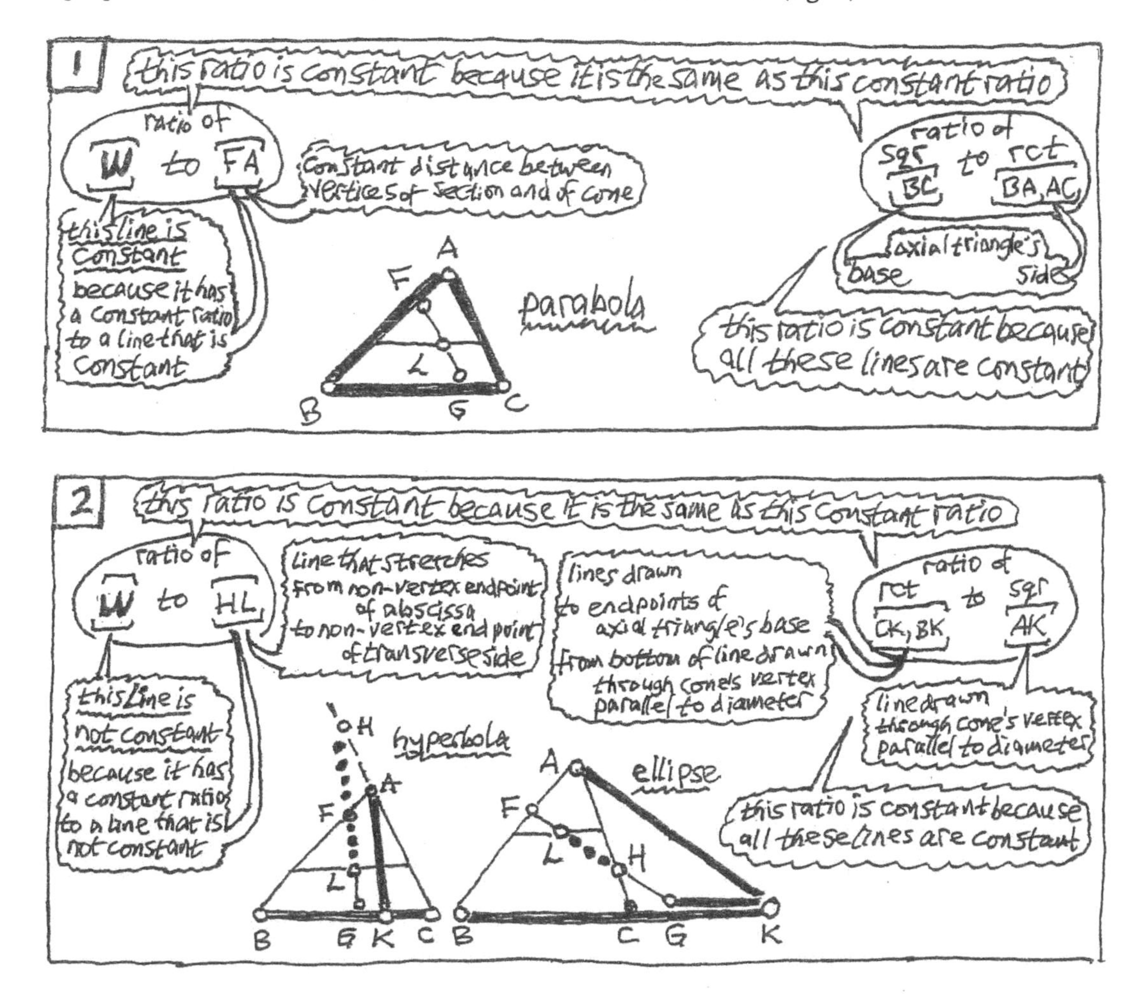

What *is* constant in the hyperbola and the ellipse is the ratio, to the changing width, of the changing line which stretches from the abscissa's non-vertex endpoint L to the oblique (transverse) side's non-vertex endpoint H. This line LH has the following character:

in the hyperbola,	in the ellipse,
while LH changes,	while LH changes,
part of it remains constant;	it is part of a whole that remains constant;
the constant part	the constant whole
is oblique (transverse) side FH	is oblique (transverse) side FH
(the abscissa FL being	(the abscissa FL being
the other part, which changes);	a changing part).

For both of these sections, we contrived a constant line (the "upright side") that will give us the variable width (w) of that equal rectangle (equal, that is, to the square on the ordinate) which has the abscissa as its height. The constant length of the upright side, we shall call the "parameter" of the hyperbola or ellipse with its diameter. The parameter will have, to the constant oblique (transverse) side, the same constant ratio that line (w) has to line LH. Since the ratio remains constant, and its consequent (the oblique [transverse] side) remains constant as well, therefore the antecedent (the parameter) also remains constant (fig. 3).

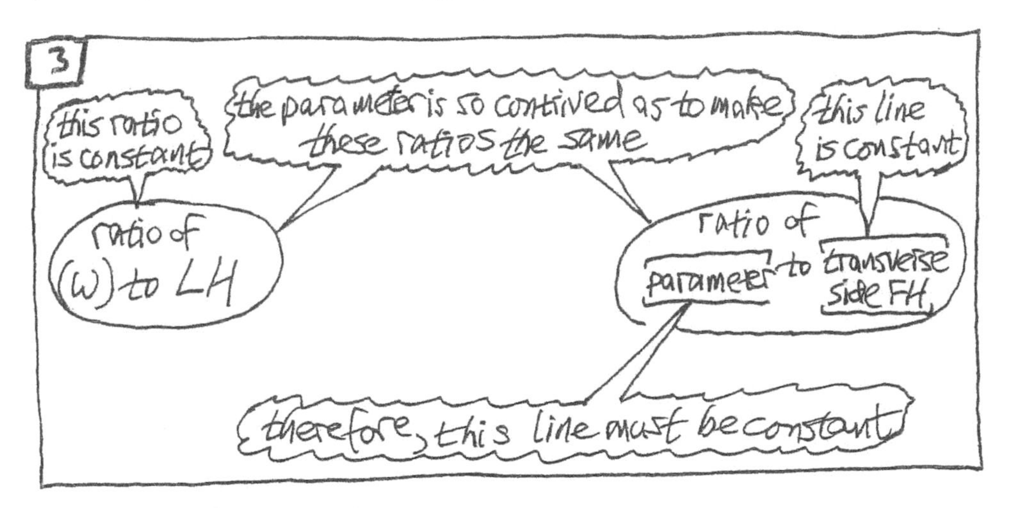

However, while both for the hyperbola and for the ellipse (as well as for the parabola) the parameter is constant, whatever be the length of the abscissa that is taken along the diameter, nonetheless neither for the hyperbola nor for the ellipse will the parameter be the width of that equal rectangle whose height is the abscissa FL. Instead, the rectangle which is contained by the abscissa and parameter will (in the case of the hyperbola) exceed the square arising from the ordinate, or will (in the case of the ellipse) fall short of being equal to it (fig. 4).

The parameter is a certain length of line for that section which has been cut in a certain cone in a certain way—has been cut, that is, so as to give it a certain diameter, which is the line of intersection for the cutting plane and a certain axial triangle. The ordinate and the abscissa are of varying size, but the parameter is of constant size for a given line of section and its diameter. If we were to take another diameter of the same line of section, then, as we shall see, we would have to get another parameter. The parameter is constant for a given line of section *together with a diameter of that curve*. Is it *simply* constant for the *curve*? No. It is, as we say (although Apollonius himself does not use the term), "parametric."

We define the parameter by a proportion among straight lines connected with the axial triangle, and then we lay it as a measure alongside another straight line that relates the straight lines in a conic section. The line laid alongside (*para*) as a measure (*metron*) determines the character of the section.

In the case of the parabola, the parameter will be of the same size as the width of that equal rectangle whose height is equal to the abscissa; but in the other cases, there is a discrepancy in size. The non-parabolic sections are each associated with a figure or shape (in Greek: an *eidos*). Behind the figure, giving to all similar figures the same look or form (also, *eidos*), lurks a ratio (in Greek: a *logos*). The ratio of the oblique (transverse) side to the upright side is the *logos* of the *eidos* (fig. 5).

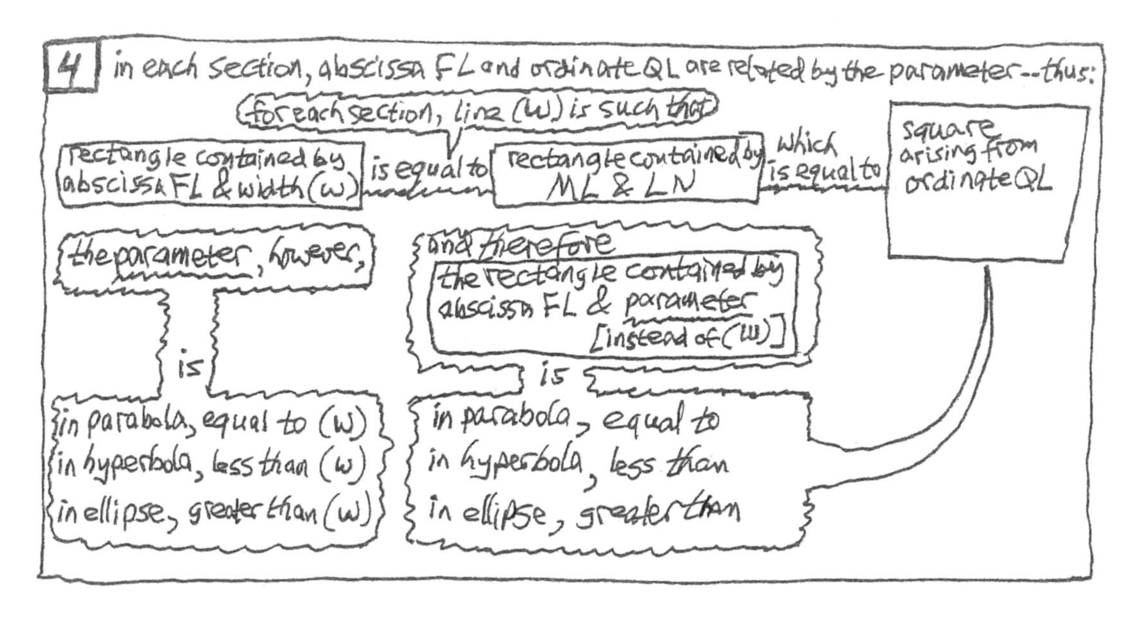

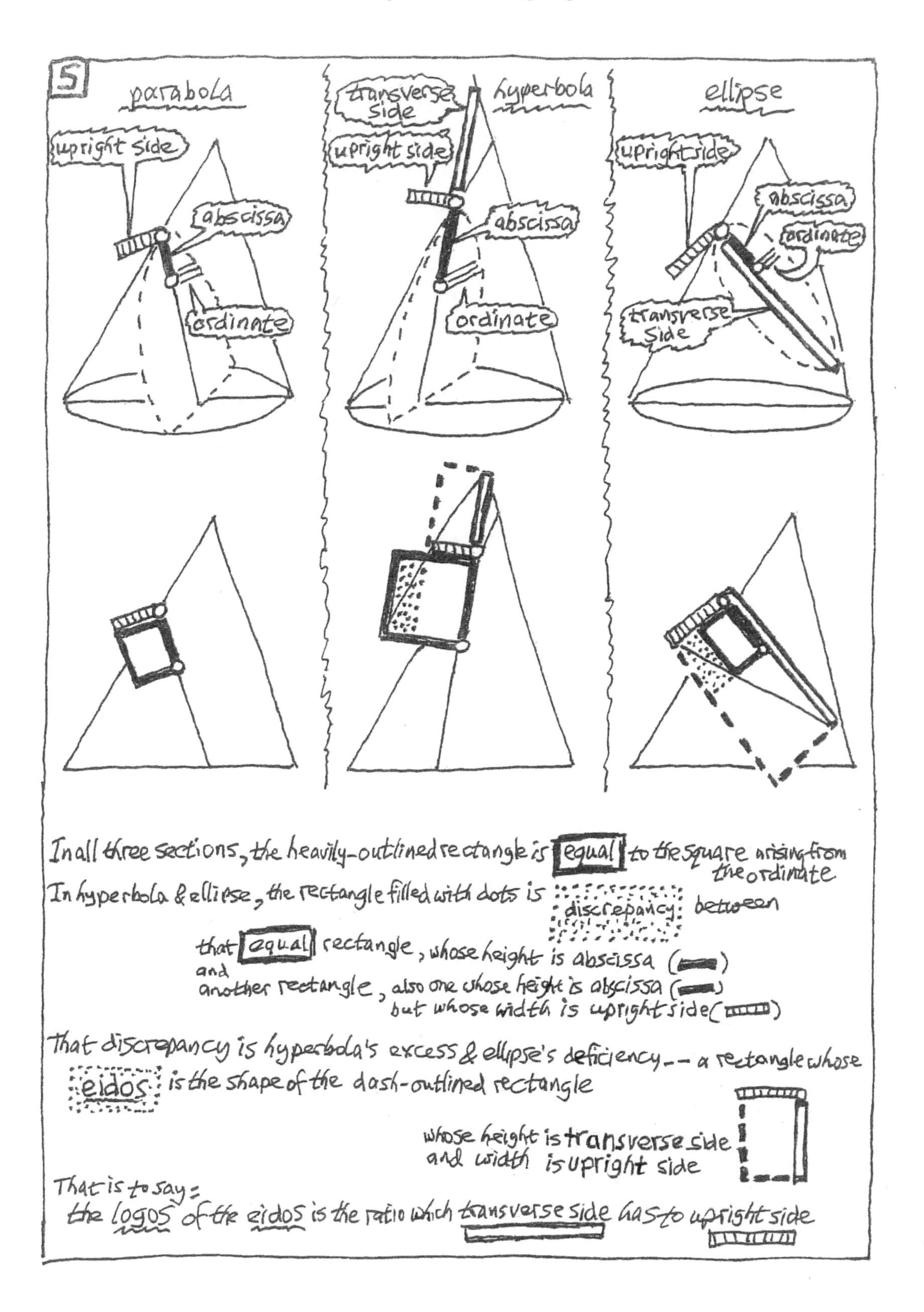
5
parabola
upright side
abscissa
ordinate
transverse side
hyperbola
upright side
abscissa
ordinate
ellipse
upright side
abscissa
ordinate
transverse side
In all three sections, the heavily-outlined rectangle is equal to the square arising from the ordinate
In hyperbola & ellipse, the rectangle filled with dots is discrepancy between
that equal rectangle, whose height is abscissa
and another rectangle, also one whose height is abscissa
but whose width is upright side
That discrepancy is hyperbola's excess & ellipse's deficiency -- a rectangle whose
eidos is the shape of the dash-outlined rectangle
whose height is transverse side and width is upright side
That is to say:
the logos of the eidos is the ratio which transverse side has to upright side

In the circle (where abscissa FL has become ML, and oblique [transverse] side FH has become diameter MLN) the *eidos* is a square; and the logos which the parameter must have to the oblique (transverse) side is one of equality (fig. 6). But should we treat the circle as if it were a special case of the ellipse? If it were to be considered a special case, it could *not* be *merely* a special case of the ellipse. The circle is a really special conic section. It is the foundation of the conic surface, and it is the beginning that rules over our study of the conic sections. In Greek, the *archê* is what initiates, what begins things (giving us our word "archaeology") or what rules over things (giving us our word "architect"). The circle is the *archê* of the curves called conic sections.

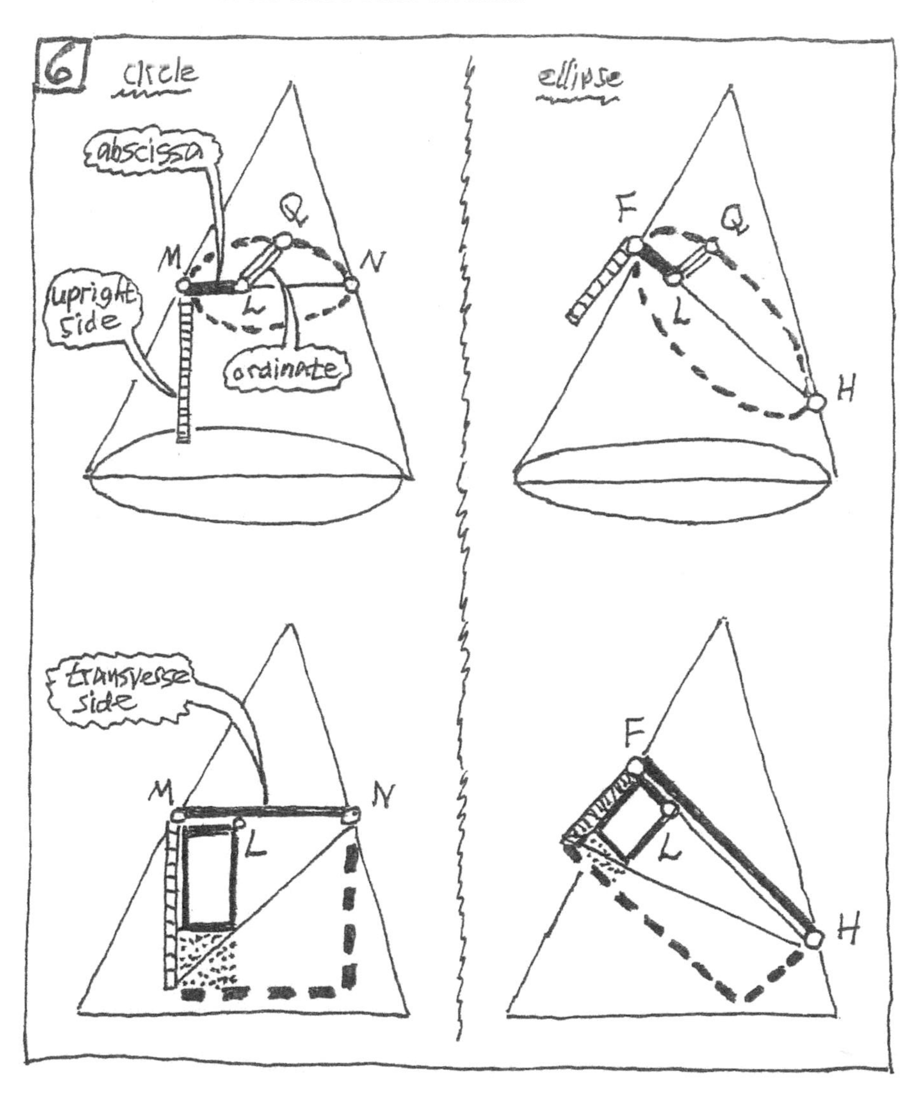

Our study of the conic sections has not yet quite left its beginning behind, although we have seen the sections given definition in these three propositions right after the ten propositions of the prologue. A last definitional proposition, the 14th, will soon give us a glimpse of something suggestive of what lies ahead. After that, Apollonius will give us forty-six more propositions in this first book on conics.

We have been able to get hold of these curved shapes by articulating the relationship between the sizes of certain straight lines. By complicated manipulations we have achieved insights into how the relationship of magnitudes informs the look of figures. By reading further in Apollonius, and reading also some writings by those who came after him, we may transform our view of what he and we have accomplished so far.

Apollonius seems to have been the one who named these curved lines according to whether the rectangles thrown out from their abscissas fall alongside, or fall beyond, or fall short of the straight line that he contrives and calls the "upright" side. Apollonius is therefore also likely to have been the first to contrive this line for all three sections. What led him to do it, he does not say.

But we have considered a way to get *to* the place *from* which Apollonius begins the trip on which he takes his reader. We have considered a way of discovering the device that Apollonius himself merely presents to the reader and shows how to use. Our own thinking took the following route.

The circle comes before the cone: it is from the circle as a base that we get the cone. From the cone, we then can get the circle as a section, to guide our further study of the conic sections. So, we begin by trying to refer the other sections to the circle if we can. The circle has this elementary property: if we raise up a square from a circle's half-chord that is perpendicular to a diameter, and if we build a rectangle from the segments into which the half-chord splits the diameter, then the square will be equal to the rectangle. Now, letting one of those segments of the diameter be taken as the circle's abscissa, and letting the half-chord be taken as the circle's ordinate, we let the sections be cut so as to make each one's ordinate be the same as the circle's. Then we get several proportions by which each section's abscissa (and, in the case of the sections other than the parabola, some part or extension of their diameters) is related both to the cone's axial triangle and also to the rectangle contained by those very segments of the circle's diameter. Thus, that rectangle which has as one side the abscissa, and has as its other side (its upright side) a line of a certain size (the parameter, derived from the axial triangle through the segments of the circle's diameter)—that rectangle either will have the same size as the square arising from the ordinate, or, if it is not equal by itself to the square, then it will be equal when a certain area is added on to the rectangle, or is taken away from it. And this area, when added on, or taken away, will have the same shape as a rectangle contained by the oblique (transverse) side and the upright side.

The lines contrived by Apollonius, and the *eidê* that they constitute for us to look at, are displays of *logoi*. This insight does not appear upon the surface in Apollonius's presentation of the propositions that characterize the three conic sections, but it nonetheless informs them all. What informs the looks of the figures—what he manipulates in order to show them—is the relationship in size of certain straight lines that are attached to the lines of conic section. He does his work with ratios.

That work with ratios has had two parts. (Fig. 7.) In the *first* part, we consider what takes place in a *circle*. The ordinate is the mean proportional between the segments into which the line parallel to the axial triangle's base is split by the point where the ordinate meets the abscissa. In other words, the ratio of the one segment to the ordinate is the same as the ratio of the ordinate to the other segment. In the *second* part, we consider what takes place in that *straight-line* figure, the axial triangle. The one segment has a constant ratio to the abscissa; as for the other segment, either it itself is constant (in the case of the parabola), or else, while not itself constant, it has a constant ratio—

either to a whole line one part of which is the constant oblique (transverse) side and the other part of which is the abscissa (in the case of the hyperbola),	or to a part of a line the whole of which is the constant oblique (transverse) side and the other part of which is the abscissa (in the case of the ellipse).

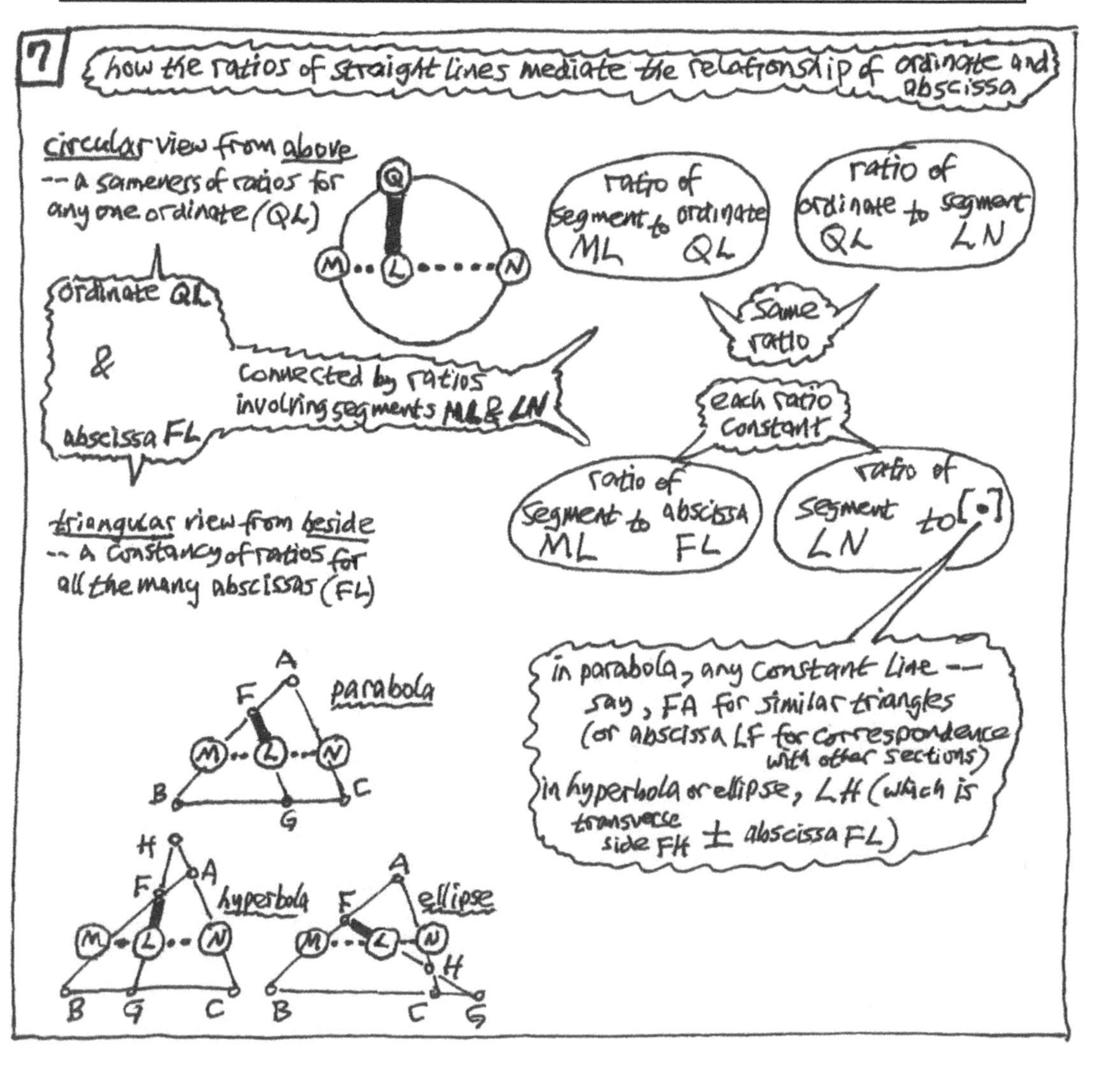

That is to say, the ratios do the following work.

In the circle, the ordinate splits the diameter into segments, with the result that two ratios are the same. Both of the ratios involve the ordinate QL, and each of them involves also one of the segments (ML and LN). Because of the sameness of the ratios (of ML to QL, and of QL to LN), the square on the ordinate QL is the same size as the rectangle contained by the segments (ML and LN).

In the parabola and the hyperbola and the ellipse, we get a line that, with the abscissa FL, will contain a rectangle of the same size as rectangle [ML, LN]. We get such a line in the following way.

Since the ratio of any rectangle to another is the same as the ratio which is compounded out of the ratios of their sides, we need two ratios of lines to compound into a ratio of rectangles. Both of the ratios must be constant; and each of them must involve one of the segments (ML or LN), while one of them must involve also the abscissa FL.

In the axial triangle, by similar triangles, we have the ratios that we need. As point L moves down from A, lengthening the abscissa FL, we have this:

segment ML has a constant ratio to abscissa FL; and the other segment LN either has a constant ratio to any constant line at all (in the parabola), or has a constant ratio to the line LH (in the hyperbola or the ellipse).

In the parabola (since FG is parallel to the axial triangle's side) the segment LN is constant—and that is why it has a constant ratio to any line at all. In the other two sections, the line to which segment ML has a constant ratio is the line that stretches from the abscissa's non-vertex endpoint (L) to the oblique (transverse) side's non-vertex endpoint (H). Line LH is thus the line obtained from the constant oblique (transverse) side HF by adding on to it, or taking away from it, the abscissa FL. That is to say: LH is, in the hyperbola, the sum of HF and FL; and is, in the ellipse, their difference.

The heart of the matter, then, in characterizing the conic sections, is this: the ordinate is the mean proportional between two lines—

one of which	*the other of which either is constant itself,*
is in constant ratio with	*or else is in constant ratio with*
the abscissa,	*the line resulting when the abscissa is*
	added on to or taken away from
	the constant oblique (transverse) side.

What enables us to see this is our combining two points of view. The view from the *top* is what showed us the *ordinate* as having *mean proportionality;* and the view from the *side* is what showed us the *abscissa* as *involved in constancy of relationships*.

Look back now, in this third part of the guidebook, at the figures (at fig. 11 in Chapter 1, at fig. 15 in Chapter 2, and at fig. 16 in Chapter 3) that map the demonstrations of the 11th, 12th, and 13th propositions—the propositions that characterize the parabola, the hyperbola, and the ellipse. Now it will be easier to see what was not easy to see when we first encountered them.

Even way back then, it might have been easy to see what it is that *in all three* of the propositions is *central* to the *top*-view *equality of square and rectangle*. The centrality of the *sameness of two ratios* was not, however, sufficiently emphasized back then, even though we did speak of the ordinate's being a mean proportional.

But what could not have been at all easy to see, back then, was what it is that in all three of the propositions is *central* to the *side*-view *equality of rectangles*. At the heart of the matter is the *constancy of two ratios*. Each involves one of the segments into which the ordinate splits the line that runs across the axial triangle parallel to the base. *The central ratios are buried deep within the body of the demonstration*, where they mediate between the upright side's determining ratio (which is what we start with) and the ratio that contains as one of its terms the mediating rectangle that is equal to the abscissa-sided rectangle (and is therefore equal, in the end, to the ordinate-sided square). The central ratios are shown now in figure 8 surrounded by a heavy-lined oval in the part of the figure labeled by the Roman numeral "I." (The letters in figure 8 refer to the points as they were labeled by Apollonius in characterizing the sections.)

There is something else that is not put in a prominent place but is *central* to the *side*-view *equality*. (For this too, see figure 8.) At the heart of the matter in *distinguishing* the sections is the *constancy of another ratio*. The parabola's side-view map shows its middle part (labeled "II") flowing without any intermediate step, directly from the upright side's determining ratio to the ratio of rectangles that is required for the concluding part (labeled "III"). But in the side-view map for the hyperbola and the ellipse, there is *an intermediate step* in the flow: another constant ratio mediates the movement from the upright side's determining ratio to the ratio of rectangles.

These things should stand out now in retrospect.

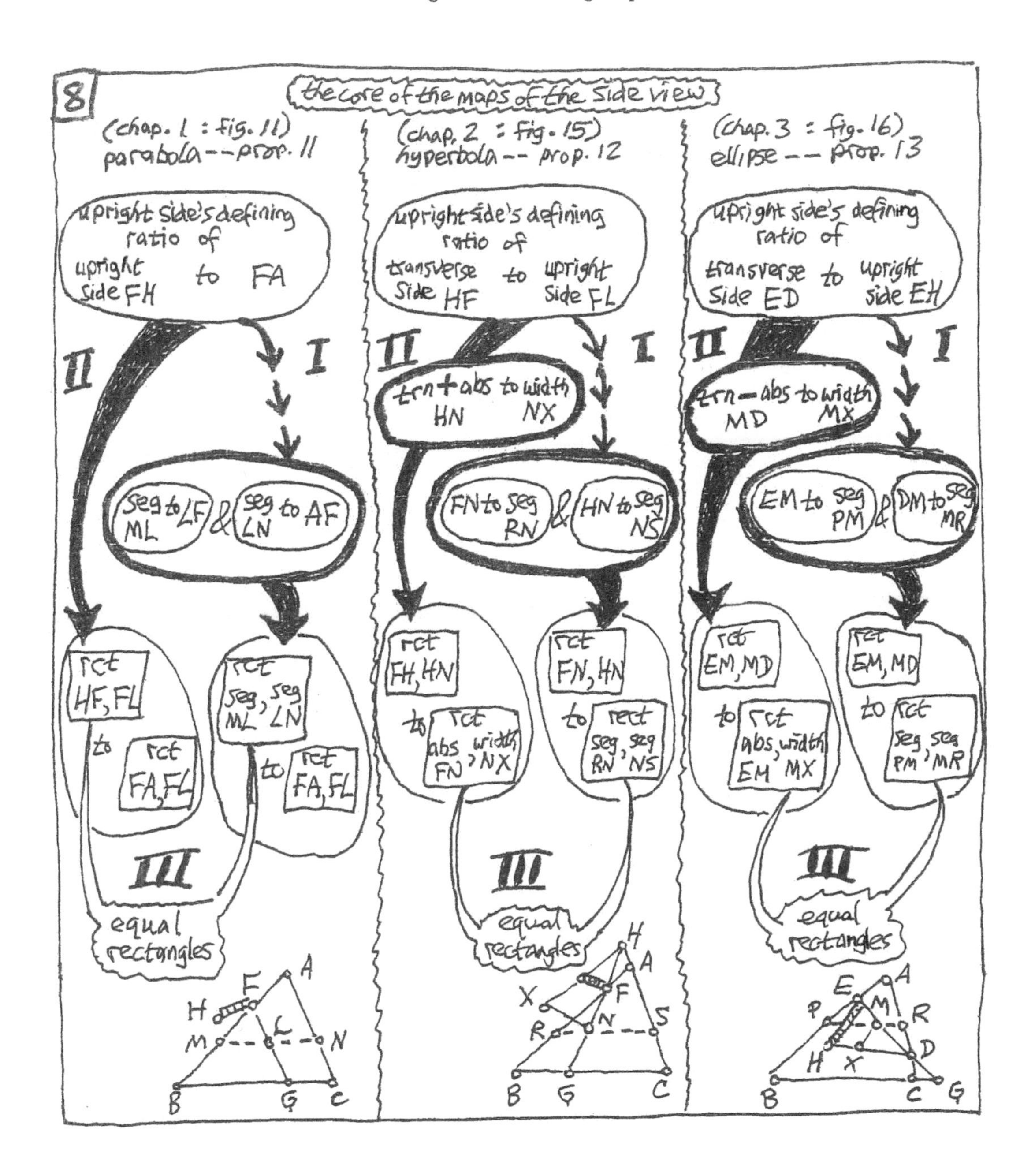

We have had to labor much in order to catch sight of what lies behind the complicated presentation in these last few propositions where Apollonius characterizes the conic sections. At the center of it all, as we have seen, is the sameness of certain ratios of straight lines, and the constancy of certain other ratios of straight lines. The straight lines are not of constant size, but their ratios are expressed in terms of certain constant other lines which are brought in for some purpose that is not made clear in the definitional propositions. That is what is responsible for much of the complexity which makes it hard to see what Apollonius is doing in these propositions. To find out whether Apollonius makes clear just how this initial obscurity serves his purpose as a teacher, the reader must read on. For now, it will suffice to keep in mind the following statement of the chief reason why we need the rest of Book One.

The ordinates of the sections are related to their abscissas by the equality and inequality of certain rectangles which have them as sides (and which involve certain features of other lines as well—their sizes, their ratios, and the shapes of certain rectangles that they make). It can be shown (and later, in the next part of this guidebook, it will be shown) that these relations of ordinate and abscissa are peculiar to the sections. These properties, that is to say, are *distinctive characteristics* of the curves—characteristics which they do *not* share with *other* lines.

But something can have several distinctive characteristics. In the circle, for example, *both* of the following are characteristic properties: (1) the equality of the distances from a given point (called the circle's center) to every one of the points of the containing line; *and* (2) the equality of the following two figures which can be made from straight lines in the circle: the square arising from a half-chord perpendicular to the circle's diameter, and the rectangle contained by the segments into which the half-chord splits the diameter.

As a definition of the circle, the first property is more like the inarticulate primary apprehension of perfect roundness—but it is a step toward the second property. What if we started with the second property? We might define the circle as a closed line, two of whose points are the endpoints of a straight line, such that:

> the perpendicular from any other point of the closed line
> to the straight line
> gives rise to a square which is equal to the rectangle contained by
> the segments into which the perpendicular splits the given straight line. (Fig. 9.)

From that definition, we could demonstrate as a theorem that

> the distance from any point on the curved line
> to the midpoint of the straight line
> is equal to the distance from any other point on the curved line
> to that same midpoint.

We could proceed in this way with the circle, but there does not seem to be any good reason to do so. What about the other conics?

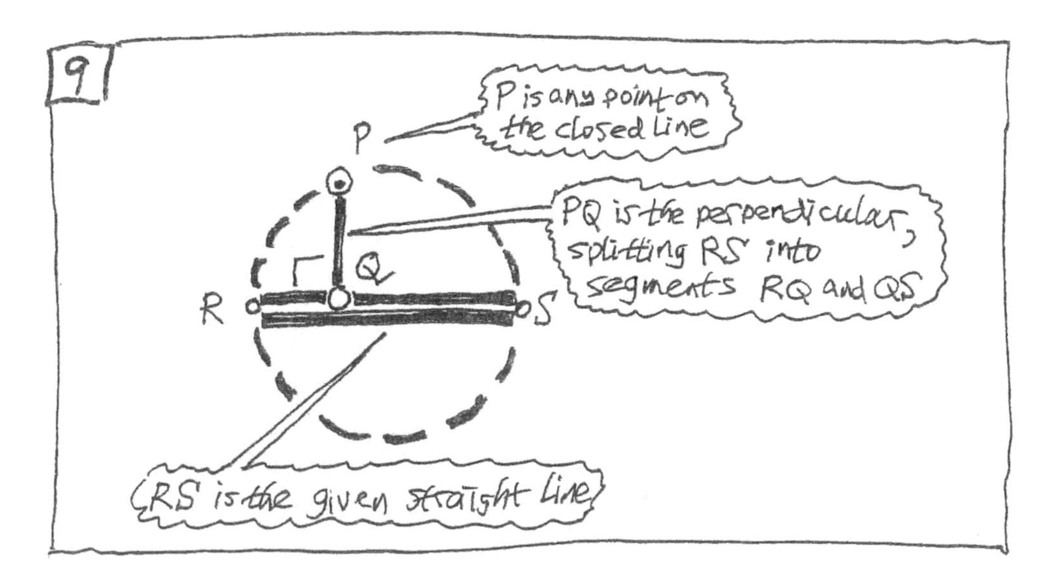

In the conics, which of the following is essential—which of them says what the conic *is*: the particular way that the plane cuts the cone, or the particular square-rectangle property that relates the ordinate and abscissa?

Perhaps we should say that the cutting merely gives us the line; and that once the line has already been given, then the square-rectangle property will characterize it. We can determine whether a given line is a conic section of a determinate sort, by testing whether the square-rectangle property holds for it; or, if we are given a conic section of a determinate sort, we can infer as a consequence that the square-rectangle property holds for it. This would be like the situation in the *Elements*: Euclid's circle is given by the postulate; and his definition is applied as a test or consequence to a line that has already been given.

But the names of the sections (of the new lines, that is to say—the sections other than the ones that are circular or are straight-lined)—the very names of the sections refer, not to the looks of the curved line itself, but rather to the *eidos* formed by straight lines which are *contrived* in it. Although we have the "wheel" (which in Greek is *kyklos*, usually translated "circle"), we do not have the "egg"; we have, rather, the "ellipse." ("Oval," which is Latin for "egg-ish," will hardly supply the precision we want.) We want to name the curves, not by how we generated them, but somehow by a look they have—that is to say, by an *eidos* of *some* kind—and the "ellipse," together with the "parabola" and the "hyperbola," might be even harder to name without the square-rectangle property. Apollonius names the sections neither by saying how they are generated, nor by saying what they themselves look like. He names them, rather, by the looks of figures that he erects alongside them.

Apollonius shows us the curves that we get by making conic sections. But do the looks that belong to those conic sections belong *only* to conic sections? We have not been told. *For all we know, such looks may belong also to some curves which are not conic sections*. Apollonius has shown us this: that if a curve is obtainable by cutting some cone with a plane, then the curve will be a circle, or a parabola, or an hyperbola, or an ellipse. He has *not*, however, shown us the *converse*, which is this: that if a curve is a circle, or a parabola, or an hyperbola, or an ellipse, then it is obtainable by cutting some cone with a plane. That is to say, he has *not* shown us this—that *only* if a curve is obtainable by cutting some cone with a plane, will the curve be a circle, or a parabola, or an hyperbola, or an ellipse.

Eventually, Apollonius *will* show this converse. His *preparations* for showing it will take up most of the rest of the First Book. What he has exhibited in these propositions that characterize the conic sections—namely, what he has displayed in the figures that he has erected alongside the curves that he has cut into cones—these, it will turn out, provide perhaps the best point of departure for showing the converse of the propositions that characterize the conic sections. We have seen, it is true, that there is in fact a *less difficult* way to show the "if" *by itself*; but, as we shall see, Apollonius's way of showing the "if" is a *most elegant* way to show the "*only* if" *besides*. Nothing better indicates this elegance of Apollonius than the set-up he prepares in the next (and last) of the defining propositions, the one that presents the "opposite sections"—the 14th proposition, to which we shall turn next.

CHAPTER 7. THE OPPOSITE SECTIONS, AND LOOKING FORWARD

14th proposition

#14 If the vertically opposite surfaces be cut by a plane not through the vertex, there will, on each of the surfaces, be as section the one called an "hyperbola"; and the two sections' diameter will be the same [line], and also the [lines], [the two of them], along (*para*) which the [lines] drawn down to the diameter parallel to the straight [line] in the cone's base have the power (*dunantai*), [that is, the two lines that are the upright sides, one for each of the two hyperbolic sections,] will be equal; and also the figure's (*eidous*) oblique (*plagia*)] side will be common, being the straight line between the vertices of the sections. And let such sections be called "opposite" (*antikeimenai*). [By now it should be quite apparent why tradition has been inclined to render *plagia* as "transverse."]

—Let there be the vertically opposite surfaces whose vertex the point A is, and let them be cut by a plane not through the vertex; and let it make on the surface as sections the [lines] DEF, GHK. I say that each of the sections DEF, GHK is the one called an "hyperbola."

—For let there be, as the circle along which is carried the straight [line] that carves out the surface, the [circle] BDCF; and on the vertically opposite surface let there be drawn parallel to it, [that is, parallel to that circle,] the plane XGOK—and common sections of the sections GHK, FED and the circles are what the [lines] FD, GK are. [These lines FD, GK] will, then, be parallel. And as axis of the conic surface, let there be the straight [line] LAU; and as centers of the circles, the [points] L, U. And, moreover, from the [point] L, let a [straight line] that is drawn perpendicular to the [line] FD be extended to the points B, C; and through both the [line] BC and the axis let a plane be extended. Then it will make as sections in the circles (on the one hand) the parallel straight [lines] XO, BC, and on the surface (on the other hand) the [lines] BAO, CAX. Then the [line] XO also will be at right [angles] to the [line] GK—since also the [line] BC is at right [angles] to the [line] FD, and each also is parallel to the other. Also—since the plane through the axis falls together with the sections at the points M, N inside the [curved] lines—it is clear that plane also cuts those [curved] lines. Let it cut them at the [points] H, E; therefore the points M, E, H, N are in the plane through the axis and also in the plane in which the [curved] lines are. Straight, therefore, is the line MEHN. And also it is evident that the [points] X, H, A, C are in a straight [line] and also that the [points] B, E, A, O are—for they are on the conic surface and are also in the plane through the axis.

—Let there, then, from the [points] H, E be drawn at right [angles] to the [line] HE the [lines] HR, EP, on the one hand, and, on the other hand, through the [point] A let there be drawn parallel to the [line] MEHN the [line] SAT; also let it be made that [each of those lines EP and HR is a fourth proportional such that] (on the one hand) as is the [square] from the [line] AS to the [rectangle] by BSC so is the [line] HE to EP, and (on the other hand) as is the [square] from the [line] AT to the [rectangle] by OTX so is the [line] EH to HR.

—Since, then, a cone whose vertex the point A is, and whose base the circle BC is, has been cut by a plane through its axis; and [that plane] has made as section the triangle ABC; and [the cone] has also been cut by another plane, one which cuts the base of the cone along a straight [line], namely the [line] DMF that is at right [angles] to the [line] BC; and [this plane] has also made, as section on the surface, the [line] DEF; and the diameter ME, being extended, has, out beyond the vertex of the cone, fallen together with one side of the triangle through the axis; and also, through the point A there has been drawn parallel to the section's diameter, [that is, parallel to] the [line] EM, the [line] AS, and also, from the [point] E there has been drawn at right [angles] to the [line] EM the [line] EP; and also, as is the [square] from AS to the [rectangle] by BSC so is the [line] EH to EP—therefore: the section DEF is an hyperbola; and the [line] EP is the [line] along (*para*) which the lines that are drawn down ordinatewise to the [line] EM have the power (*dunantai*), and the figure's (*eidous*) oblique (*plagia*) side is what the [line] HE is. And likewise also the [section] GHK is an hyperbola—whose diameter the [line] HN is, and the [line] HR is the [line] along which the [lines] that are drawn down ordinatewise have the power, and the figure's oblique side is what the line HE is. I say that equal is the [line] HR to the [line] EP.
—For—since parallel is the [line] BC to the [line] XO—as is the [line] AS to SC so is the [line] AT to TX, and also as is the [line] AS to SB so is the [line] AT to TO. But the ratio of the [line] AS to SC along with (*meta*) the [ratio] of the [line] AS to SB is [this:] the ratio of the [square] from AS to the [rectangle] by BSC—and the [ratio] of the [line] AT to TX along with the [ratio] of the [line] AT to TO is [this:] the ratio of the [square] from AT to the [rectangle] by XTO. Therefore as is the [square] from AS to the [rectangle] by BSC so is the [square] from AT to the [rectangle] by XTO. And also (on the one hand) as is the [square] from AS to the [rectangle] by BSC [so] is the [line] HE to EP; and (on the other hand) as is the [square] from AT to the [rectangle] by XTO [so] is the [line] HE to HR. And also, therefore, as is the [line] HE to EP [so] is the [line] EH to HR. Equal therefore is the [line] EP to the [line] HR.

[One ratio "along with" (*meta*) another—that means: the ratio compounded from the two ratios.]

By the end of the previous proposition, the 13th, Apollonius had exhausted the possibilities for a plane's cutting of lines into a cone's non-planar surface. We can see this by using dichotomy successively, as in figure 1 (on the next page).

Why, then, has he continued cutting? Well, into what does he cut now in the 14th proposition? Not into a cone. Instead, he cuts into a conic surface—or, rather, into vertically opposite conic surfaces, since the generation of a conic surface generates a surface with two parts, these two parts being two surfaces with the same shape, sharing a point.

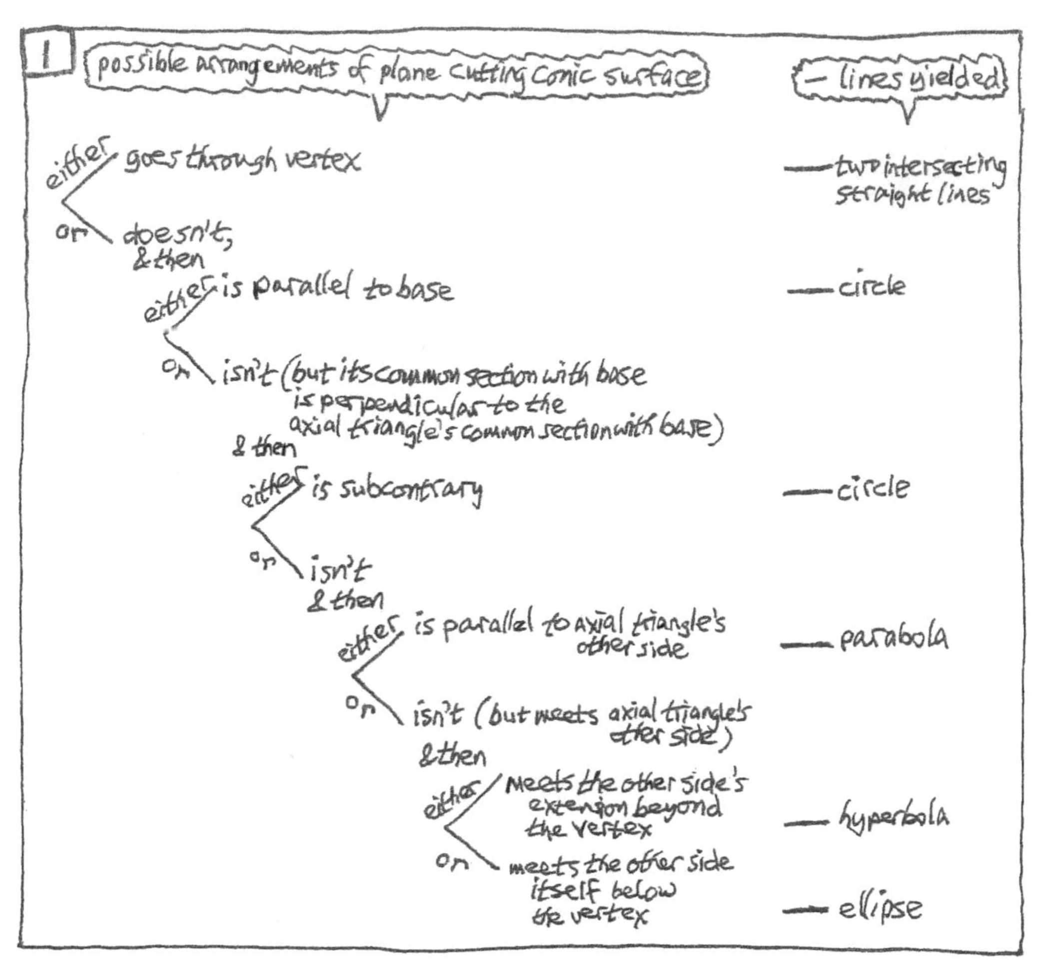

Only once before did Apollonius cut into a conic surface rather than a cone. That was in the 4th proposition, when he cut into a conic surface parallel to its generating circle, and the new circle which he thus obtained became the base of a cone which the newly-cut circle made with the part of the conic surface above the cut. Why does he now again cut into a conic surface and not into a cone—or, rather, why does he cut into vertically opposite conic surfaces, rather than cutting into two cones put together at their vertices?

Suppose that he put two cones together at their vertices. Will he then, by cutting into them, get two sections that have the same size and shape, or are even of the same kind? Not necessarily. When is it that he can be sure that the two sections will be the same? When the surfaces of the two cones are vertically opposite surfaces (fig. 2).

When the surfaces are vertically opposite surfaces, then the axial triangles will be similar (since their bases will be parallel, and the angles at their vertices will be vertical angles). For this reason, the one plane in this case can cut into both surfaces only by cutting an *hyperbola* into both of them.

Moreover, in this case not only must the two sections be hyperbolas—they must also have their *upright sides equal*. Why? Because the single plane which cuts an hyperbola into each of the two surfaces will give them an oblique (transverse) side which is common, as well as giving them axial triangles which are similar—and so, the same magnitude (the common oblique [transverse] side of both sections) has the same ratio to the upright side of each section; and hence the upright sides are equal.

And why is that ratio (of oblique [transverse] side to upright side) the same for both sections? Because in each section, the ratio of the oblique (transverse) side to the upright side is compounded of ratios of lines in that section's own axial triangle; but the axial triangles here are similar, and in similar triangles corresponding lines have to each other the same ratio.

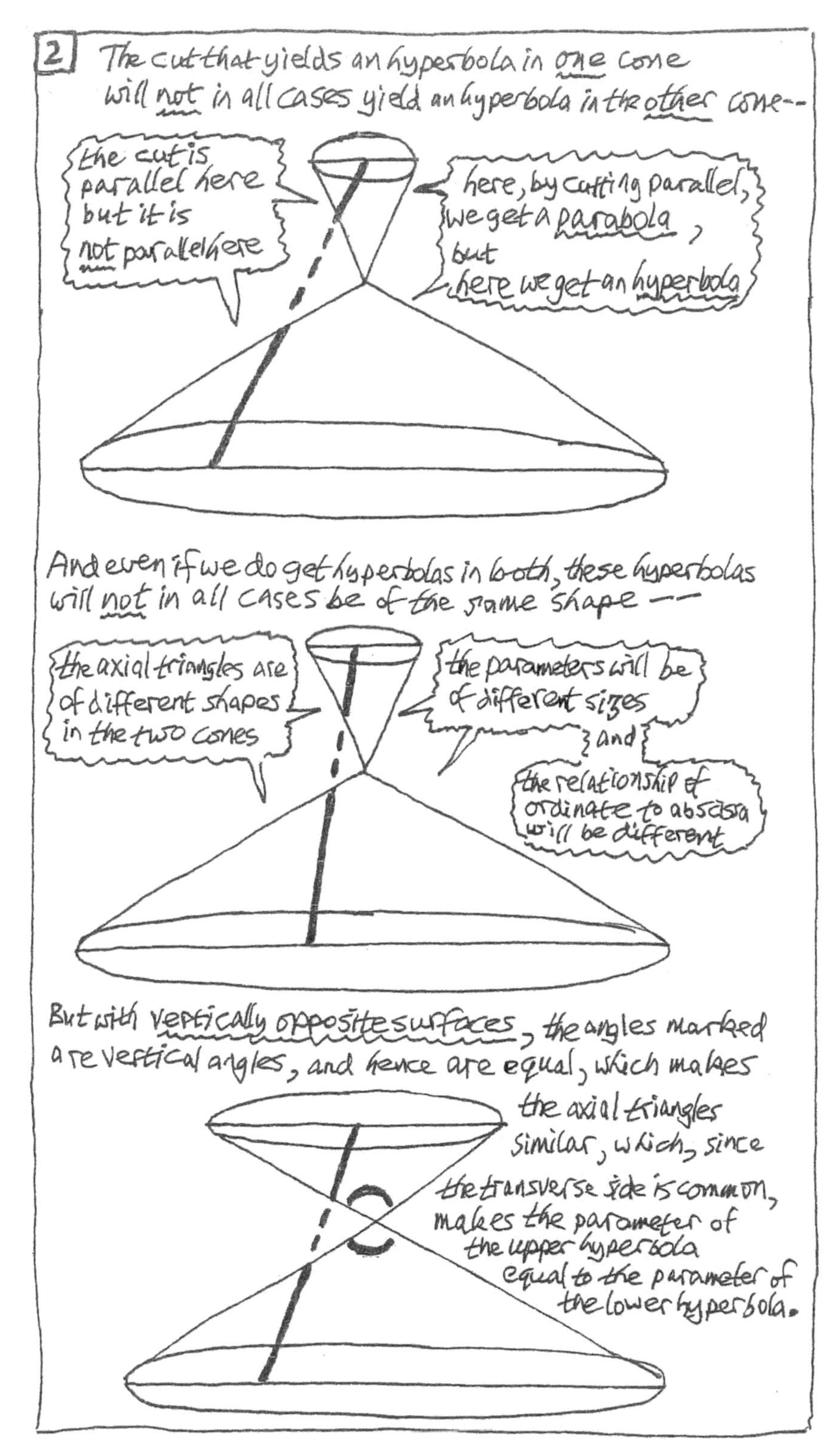

But why bother to do what is done in this 14th proposition? After all, the two lines of section are both of the same kind, and this is a kind of section which we have seen before; moreover, each of the curves has exactly the same size and shape as the other curve (the two of them being hyperbolas with equal upright sides), and their diameters are located in a straight line with each other. Is this case a perfect bore—or are we missing something that might make it interesting?

What we have here is not just a situation in which there are two hyperbolas of the same size and shape. Two hyperbolas of the same size and shape are not as such "opposite sections." Their position matters. (The "posit" in the Latinized "op+posite" corresponds to *keimenai* in the Greek *anti+keimenai*.) But even two hyperbolas of the same size and shape which have their diameters in a straight line are not as such "opposite sections." (Fig. 3.)

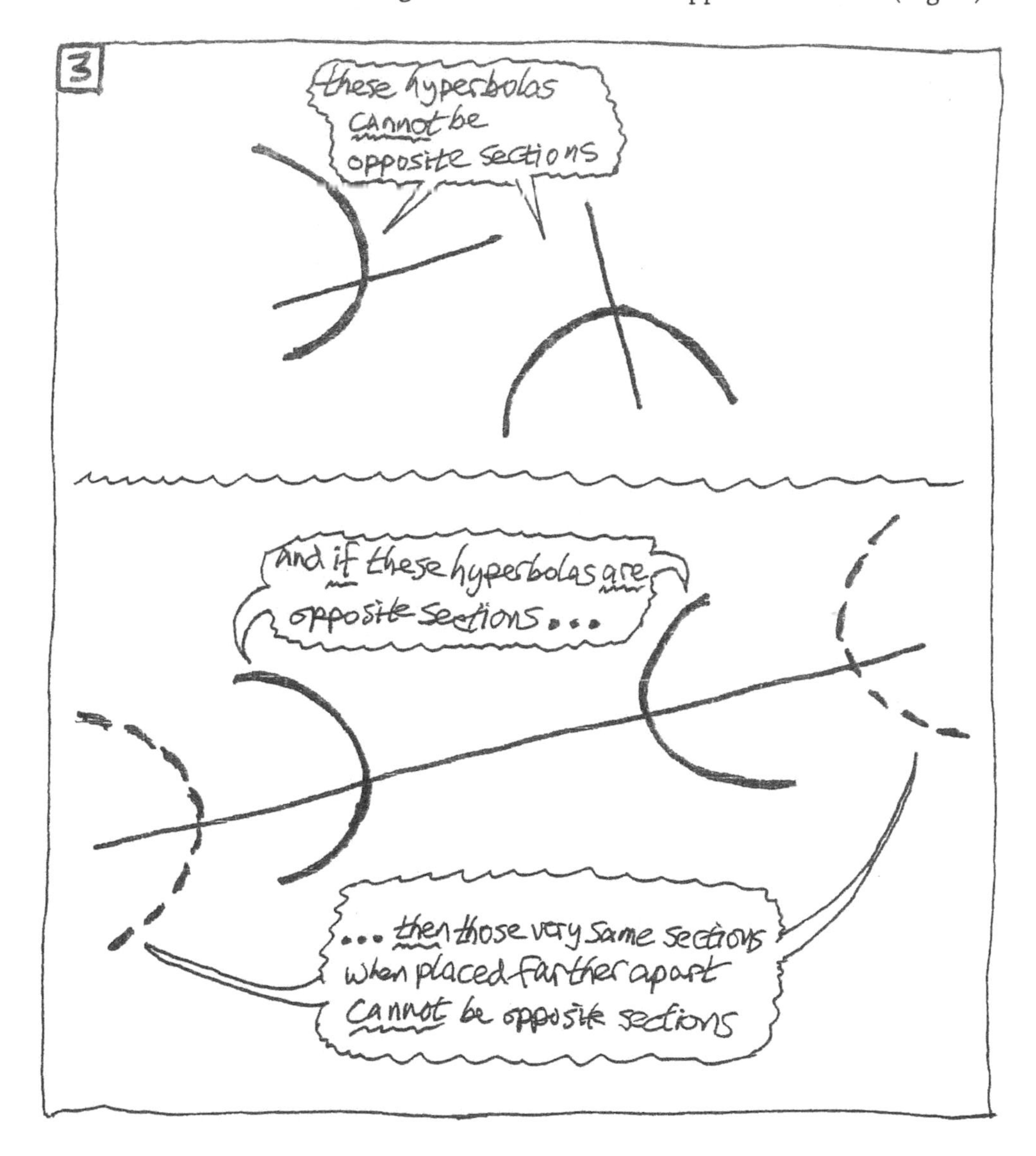

For two hyperbolas to be "opposite sections," not only must they have equal upright sides; the hyperbolas also must have *the very same oblique (transverse) side*. The slicing of a single cut into vertically opposite conic surfaces guarantees that there will be vertically opposite sections—hyperbolas with equal upright sides and the very same oblique (transverse) side. With the opposite sections we do not have a mere throwing together of two figures. We have, rather, a single definite configuration.

The way in which this proposition begins should have put us on the alert. The order of presentation differs from the order in those propositions where we got sections previously. Here, unlike before, Apollonius does the section-cutting prior to getting the axial triangle; indeed, he even obtains ordinates prior to getting the axial triangle. Instead of cutting so that FD will be perpendicular to BC, he draws BC perpendicular to FD (fig. 4). That is to say: here he proceeds so as to cut both surfaces with one slice without regard to what axial triangle, or what set of ordinates, he will have. In this proposition, what is interesting is not the straight line which is the diameter of each section. What is interesting, rather, is the straight line of which the diameters of both curves are prolongations—namely, the oblique (transverse) side.

The opposite sections together have something that keeps them apart: their common oblique (transverse) side. Not only does it keep them apart, however; it also holds them together. They are made one by the same thing that makes them two. Their common oblique (transverse) side makes them *one* pair and makes them one *pair*. Linked and separated by their common oblique (transverse) side, the opposite sections are of interest not for each one's shape by itself, but rather for the figure which they both make together.

What has this to do with the three preceding conic sections, each of which is one because each of them is a single line? Even if we have seen compelling reasons to place the parabola's defining proposition before the defining propositions for the other sections, we have seen no compelling reasons for the hyperbola's defining proposition to be placed before the ellipse's defining proposition. Would it not have been better to present the ellipse before the hyperbola—interchanging the content of the 12th and the 13th propositions—so that the hyperbola would be placed immediately before the special pair of hyperbolas of the 14th proposition? Or is there some way in which this hyperbolic pair is related to the ellipse—some way in which it relates the ellipse to the hyperbola? Does this have anything to do with the oblique (transverse) side?

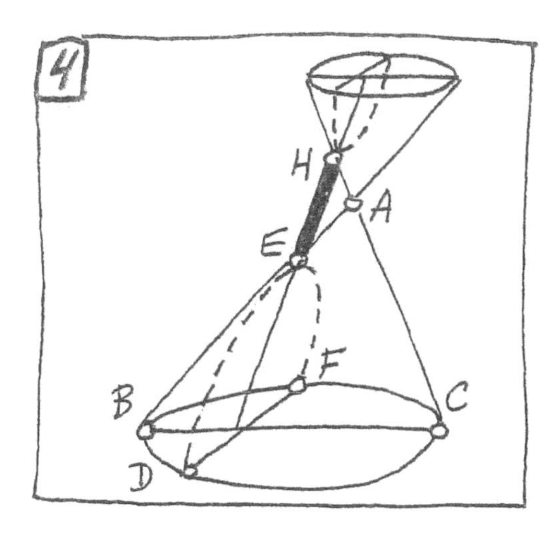

We shall be seeing more of this in what follows. Indeed, even when we do not at first see it in what follows, it will be lurking there.

"Transverse" side, as was mentioned earlier, is the usual translation for what Apollonius names by a Greek word that can be more literally translated "oblique" side. The Greek terminology of Apollonius suggestively indicates a contrast between what is "upright" and what is "oblique." His Greek words have connotations similar to those of the English.

At the end of the 14th proposition, we have finished with the propositions that define the various conic sections. There were definitions long before these defining propositions, however. In his First Definitions, at the beginning of the book, Apollonius gave, after the definition of a diameter, definitions for which he still has not shown the reason. He arranges for us to see it in what follows. Right after the next pair of propositions, we shall see a set of Second Definitions, and, with that preparation, we shall then go on to see much more.

Apollonius tells his tale obliquely. That is clear. But what is not so clear is whether his obliqueness is informed by wisdom and benevolence.

Descartes, for one, did not think so. In his own *Geometry*, and in his sketch of rules for giving direction to the native wit (especially in the 4th rule), Descartes found fault with the ancient mathematicians.

Descartes severely criticized them—for being show-offs. He says that they made analyses in the course of figuring things out, but then, instead of being helpful teachers who show their students how to do what they themselves have done, they behaved like builders who get rid of the scaffolding that has made construction possible. Thus they sought to be admired for conjuring up one spectacular thing to look at after another, without a sign of how they might have found and put together what they present.

Descartes also suggests, however, that they did not fully know what they were doing. They did not see that what they had could be a universal method. They operated differently for different sorts of materials because they treated materials for operation as simply objects to be viewed. Hence they learned haphazardly, rather than methodically, and therefore they did not learn much. Mathematics for them was essentially a matter of wonderful spectacle rather than being material for methodical operation. The characteristic activity of the ancient mathematician was the presentation of theorems, not the transmission and application of the ability to solve problems. They had not discovered the first and most important thing to be discovered—namely, the significance of discovery. They had not discovered the power that leads to discovery and the power that comes from discovery. They were not aware that the first tools to build are tools for making tools. They were too clever to be properly simple, and too simple to be truly clever. They were blinded by a petty ambition. Too overcome by their ambition, they could not be ambitious on the grandest scale.

Was Apollonius as a teacher liable to that criticism by Descartes? By the time that we have finished with our study, will we find Apollonius guilty of that obtuseness of which Descartes, at the beginning of his *Geometry*, accused the classical mathematicians—guilty of the desultory fooling-around and disingenuous showing-off of which Descartes had accused them earlier in his *Rules*? It is too soon for us to form a judgment now upon mathematics presented, in the mode of Euclid and of Apollonius, as synthesis.

In contrast to the constructive work of synthesis, analysis is a loosening-up of things. Is unloosening a work of violence—a kind of chopping-up of shapes and a dragging of lines together? Have we been whirling lines to get a conic surface, into which we then hack sections, which then we hack apart—all of this in order to acquire the skills and the materials to set up shop in the most fundamental kind of construction work?

Or would it be better to say this: that in order to be able to *say* very much, in order to "see" most deeply into things, or to know them (in contrast to merely looking on), we *must* engage in *operation*? How much can we know unless we do perform operations that enable us at the end to view things that we have constructed?

Perhaps the taking apart of a figure is not so much an act of violence as it is an act of abstraction, in which we ignore some parts of the figure in order to concentrate attention on others—as if we kept the figure whole, but colored the lines of interest, so as to have a sequence of panels, each containing the whole drawing but with different panels showing different lines in red. That is to say: perhaps our operation or mental manipulation is only a sequential, differentially attentive seeing—mediation being only a series of *im*-mediacies that are differentiated.

Even if what is most important is not the exercise of manipulative power, but is rather the activity of seeing the truth that abides, we shall be limited to seeing what has already been seen unless we look for what is not manifest. Even the seeing that we get from a demonstration is mediated: "if A, then B" the enunciation tells us in effect, and then the demonstration gives a long mediation to take us from A to B—that is, to see that B is implicit in A. The demonstration leads us from A to B, but we do *not* see *immediately* that A leads to B—we need to look for what is central in the demonstration.

What is it that best helps us to learn: is it learning what has already been learned—learning it as something known—or is it learning how to learn, so that we can get to know not only what has been already learned, but also what has not been learned by any teacher yet?

Even if there is no true learning but the learning of how to learn, Apollonius may be a better teacher for his not giving us questions, but rather only answers that are too hard to sort out and remember *unless* we ourselves figure out what questions to ask. Is it better to provoke questions as he does—or would it be better to present the questions explicitly?

We shall return to this later, in the second volume of this guidebook, when we shall encounter some analysis by Apollonius, and when, after finishing with Apollonius and some transitional authors, we shall consider Descartes' critique of classical mathematics—and his transformation of mathematics, and of the world of learning generally, and therewith of the whole wide world.

PART IV
MEDIATION

Apollonius's *Conics*, Book I: proposition 15 through 51

CHAPTER 1. ANOTHER DIAMETER

transition

Apollonius has shown us the several sorts of conic sections. In some respects, each of them is like all the rest of them; in other respects, some of them are like each other in being different from all the rest of them. How so? That is a question to keep in mind as we proceed.

We have already seen some answers. We have seen (with the 8th and 9th propositions) that both the hyperbola and the parabola are open, eventually reaching a width greater than any given width, whereas the ellipse is a closed curve, having a maximum width. We have also seen (with the 11th, 12th, and 13th propositions) that the parabola has the square on its ordinate equal to the rectangle contained by its abscissa and parameter, whereas both the hyperbola and the ellipse have that square unequal to that rectangle.

However, reading the rest of Book One should not be an exercise in sorting out the properties of parabolas, hyperbolas, and ellipses. If what we wish to do is merely to examine the properties of parabolas, hyperbolas, and ellipses, there are easier ways to do it than by reading Apollonius. The first book of his *Conics* is not a comprehensive treatise on the properties of the various sorts of conic sections. It is more like a story. Its plot has a beginning and a middle and an end.

Having finished the propositions that characterize the kinds of conic sections, we are at the end of something. Or it might be better to say that we are at the end of the *beginning* of something, since Book One of the *Conics* continues. What would be the proper ending for what we have only just begun?

We have seen how a cone can be cut by a plane to yield sections of several sorts that have been named with reference to the square-rectangle property of ordinate and abscissa. That is only the beginning of the story. The ending will show how to find a cone into which a plane can cut a specified curve of any such sort. Why does *that* turn out to be the ending which is required by the beginning—why is the beginning not sufficient as a story by itself?

For the reason indicated previously: once we have learned that lines obtained by cutting a cone in various ways will have certain properties, then what we should like to know is whether the *only* lines that will have those properties are the lines that we can obtain by cutting a cone in one of those ways. If we could show that such lines are indeed the only ones that have those properties, then we would have a proper ending to what we have only just begun. From the truth of a statement of the form "if P, then Q," it does not follow that the statement "if Q, then P" is true. It *may* be true, but it may *not* be. (It is true, for example, that if a line is a circle, then it is closed; but it is not true that if a line is closed, then it is a circle.) The beginning of the book has shown this: if a line is obtainable by cutting a cone in one of various ways, then it has one of various properties. What has not yet been shown is this: if a line has one of those properties, then it must be obtainable by cutting a cone in one of those ways.

Having reached the end of the beginning, we might expect Apollonius to proceed directly to the beginning of the end. He does not do that, however. The end requires preparation. The preparation has a certain indirectness, and, as we shall see, it even takes on a life of its own, so that the end to which we now look forward will *not* turn out to be the *very* end.

The beginning has culminated in the cutting of cones to get the various conic sections. Now, from the 15th proposition through the 51st, the sections will be examined apart from cones. Book One will come to an end by solving the problems of finding the various curves in cones. Given the specification of a conic section of some sort, a cone will be put together, so that the plane of the curve will have the curve as its intersection with the cone. The cone must be found so that the given curve can be found *as* a conic section—that is to say, as a line cut into a cone in a certain way.

In the dozens of propositions that constitute the middle of Book One, we shall be examining curved lines of the various sorts that can be generated by cutting cones or conic surfaces, but we shall not look at any cones or conic surfaces. We shall examine the curves along with their diameters and their upright sides, and also along with various other auxiliary straight lines. There will be no axial triangles, however. We shall *not* be considering the curves *as* cuts: we shall consider their looks apart from the generating of them. Our figures will all be completely planar.

The beginning of the book has shown that certain properties belong to lines that have been cut into a cone by a plane that has been oriented in various ways. The ending is required in order to show that those properties belong *only* to such sections. To get from the beginning to the ending—that is, in order to be able to show that those properties do belong only to such sections—we first must show that because those sections have *those* properties, they have certain *other* properties *also*. These other properties are what we turn to now.

In traversing the book's middle (the bridge that stretches for thirty-seven propositions between the beginning and the end), what we shall consider are the section's diameters and tangents. This will enable us to refer any conic section to any of its diameters, with the tangent at that diameter's vertex as the reference line for its ordinates. This in turn will enable us to solve the problems of finding cones that will show us any specified parabolas, hyperbolas, ellipses, and circles (and also opposite sections) *as* lines cut into the cones (or into the opposite conic surfaces).

To reach the problems that follow the 51st proposition—and to enjoy the surprise ending that in turn will follow them, and will be a new beginning—we must first go through the thirty-seven propositions that lie in-between, figuring out the demonstrations and seeing how each enunciation plays a part in the story. These theorems fall into two stretches. The second stretch begins with the 37th proposition of the book. We now come to the first stretch—the one that goes from the 15th through the 36th proposition. This stretch will occupy us for three chapters of this guidebook.

Now *which* section do we examine *first*? It might seem natural to examine first the first of those new lines which Apollonius obtained by cutting a cone. That was the *parabola*. Nonetheless, Apollonius begins his examination with the *ellipse*. Why? We must wait a while to see.

15th proposition

#15 If in an ellipse, from the midpoint (*dichotomias*) of the diameter a straight line drawn ordinatewise be extended both ways as far as the section, and also there be made a certain [straight line], [that is, there be made a certain straight line which is a third proportional] such that as is the extended [straight line] to the diameter so is the diameter to that certain straight [line]—then whatever straight line, from the section to the extended [straight line], may be drawn parallel to the diameter, will have the power (*dunêsetai*) [to give rise to a square equal to] the [rectangular area] that is laid along (*parakeimenon para*) the third proportional, [with that equal area] having as breadth the [line] taken off by it [taken off, namely, by that straight line parallel to the diameter,] [taken off, that is, from that extended straight line,] on the side toward the section, [and also with that equal area] being lacking (*elleipon*) by a figure (*eidei*) that is similar to the [rectangle] contained by [the following two lines: first,] the [line taken off, on the side toward the section, by a parallel to the diameter, from that extended straight line] to which are drawn the [straight lines parallel to the diameter], and also, [second,] the [third-proportional straight line] along (*para*) which [those drawn straight lines] have the power (*dunantai*); and also, [as for that straight line from a point on the section, drawn parallel to the diameter, down to that ordinatewise straight line extended through the diameter's midpoint], it will, being extended forth as far as the other side of the section, be cut in two [halves] by the [straight line] to which it has been drawn down.

—Let there be an ellipse whose diameter the [line] AB is; and let the [line] AB have been cut in two [halves] at the point C; and, through the [point] C, let there be drawn ordinatewise the [line] DCE, and also let it have been extended both ways as far as the section; and from the point D, let there be drawn, at right [angles] to the [line] DE, the [line] DF; and let [this line DF] be so made that as is the [line] DE to AB so is the [line] AB to the [line] DF; and let there be taken some point on the section—the [point] G, say—and through the [point] G let there be drawn parallel to the [line] AB the [line] GH; and let there be joined the [line] EF; and through the [point] H let there be drawn parallel to the [line] DF the [line] HL, and through the [points] F, L let there be drawn parallel to the [line] HD the [lines] FK, LM.

—I say that the [line] GH has the power (*dunantai*) [to give rise to a square equal to] the [area] DL that is laid along (*parakeitai para*) the [line] DF, while having as breadth the [line] DH, and being lacking (*elleipon*) by a figure (*eidei*), namely the [area] LF, that is similar to the [rectangle] by EDF.

—For let there be—as that [line] along which [is laid that rectangle compared to which] the lines drawn down ordinatewise to the [line] AB have the power [to give rise to squares that are equal to rectangles that are lacking]—the [line] AN, [which is to say that, for the lines drawn down ordinatewise to the ellipse's diameter AB, the line AN is the upright side]; and also let there be joined the [line] BN; and also, through the [point] G let there be drawn parallel to the [line] DE the [line] GX, and through the [points] X, C let there be drawn parallel to the [line] AN the [lines] XO, CP, and through the [points] N, O, P let there be drawn parallel to the [line] AB the [lines] NU, OS, TP.

—Equal therefore is the [square] from the [line] DC to the [area] AP; and [also equal is] the [square] from the [line] GX to the [area] AO. And—since as is the [line] BA to AN so is the [line] BC to CP, and the [line] PT to TN; but equal is the [line] BC to the [line] CA, that is, to the [line] TP; and also, the [line] CP is equal to TA—therefore equal is the [area] AP to the [area] TR, and the [area] XT to the [area] TU. And also—since the [area] OT, to the [area] OR, is equal, but common is the [area] NO—therefore the [area] TU is equal to the area NS. But the [area] TU, to the [area] TX, is equal, and common is the [area] TS—therefore the whole [area] NP, that is, the [area] PA, is equal to the [area] AO [taken] along with the [area] PO; and so the [area] PA exceeds (*huperekhei*) the [area] AO by the [area] OP. And also the [area] AP is equal to the [square] from the [line] CD, and the [area] AO is equal to the [square] from the [line] XG, and the [area] OP is equal to the [rectangle] by OSP—therefore the [square] from the [line] CD exceeds the [square] from the [line] GX by the [rectangle contained] by the [lines] OSP. Also, since the [line] DE has been cut into equal [segments] at the [point] C and into unequal [segments] at the [point] H, therefore the [rectangle] by the [lines] EHD, [taken] along with the [square] from the [line] CH, that is, [taken] along with the [square] from the [line] XG, is equal to the [square] from the [line] CD. Therefore the square from the [line] CD exceeds the [square] from the [line] XG by the [rectangle contained] by the [lines] EHD; and the [square] from the [line] CD exceeds the [square] from the [line] GX by the [rectangle contained] by the [lines] OSP. Therefore the [rectangle] by the [lines] EHD is equal to the [rectangle] by the [lines] OSP. And also, since as is the [line] DE to AB so is the [line] AB to the [line] DF, therefore also as is the [line] DE to the [line] DF so is the [square] from the [line] DE to the [square] from the [line] AB, that is, [so is] the [square] from CD to the [square] from CB. And also, equal to the [square] from CD is the [rectangle] by PCA, that is, the [rectangle] by PCB. Also, therefore, since as is the [line] ED to DF, that is, the [line] EH to HL, that is, the [rectangle] by the [lines] EHD to the [rectangle] by the [lines] DHL, so is the [rectangle] by the [lines] PCB to the [square] from CB, that is, [so is] the [rectangle] by PSO to the [square] from OS. Also, equal is the [rectangle] by EHD to the [rectangle] by PSO; therefore equal is the [rectangle] by DHL to the [square] from the [line] OS, that is, to the [square] from the [line] GH. The [line] GH, therefore, has the power [to give rise to a square equal to] the [area] DL which is laid along the [line] DF and is lacking by a figure (*eidei*), namely the [area] FL, that is similar to the [rectangle] by the lines EDF.

—I say then also, that the [line] GH, when extended as far as the other side of the section, will be cut in two [halves] by the [line] DE.

—For let [line] GH be extended and let it fall together with the section at the [point] W; and also, through the [point] W let there be drawn parallel to the [line] GX the [line] WY, and, through the [point] Y let there be drawn parallel to the [line] AN the [line] YZ. And, since equal is the [line] GX to the [line] WY, therefore equal also is the [square] from the [line] GX to the [square] from the [line] WY. But the [square] from the [line] GX is equal to the [rectangle] by the [lines] AXO, and the [square] from the [line] WY is equal to the [rectangle] by the [lines] AYZ. Proportionally, therefore, as is the line OX to the line ZY so is the line YA to AX, and also as is the [line] OX to the [line] ZY so is the [line] XB to BY, and therefore also as is the [line] YA to AX so is the [line] XB to BY. And also—by taking apart (*dielonti*)—as is the [line] YX to XA so is the [line] YX to YB. [That is to say: as was the whole line (YA) to this whole line's one part (AX), so was the whole line (XB) to this whole line's one part (BY); and therefore—"by taking apart" (traditionally rendered by the Latin *separando*)—as is one whole's other part (YX) to the one part (AX), so is the other whole's other part (YX) to the one part (YB).] Equal, therefore, is the [line] AX to the [line] YB. And also the [line] AC is, to the [line] CB, equal. And also the [line] XC as remainder, to the [line] CY as remainder, is equal; and so, [equal] also is the [line] GH to the [line] HW. The [line] HG, therefore, when extended as far as the other side of the section, is cut in two [halves] by the [line] DH.

15th proposition: enunciation

Sliding the sections out of the cone, the first one that we look at is an ellipse—an ellipse, that is, with its diameter. Its diameter, of course, is the bisector of certain parallel chords. Apollonius now shows that the ordinatewise line that bisects this bisecting diameter will itself work like *another* diameter: it will bisect all chords that are parallel to the original diameter. Chords *parallel* to the *original* diameter are thus *ordinatewise* to the *other* diameter. (Fig. 1.)

Apollonius himself does not here call the ordinatewise bisector of the original diameter another diameter. (And, by the way, what Apollonius calls "the upright side" we shall often call the "parameter" to emphasize its size.) What Apollonius does do in this proposition is demonstrate the following two things: *First*, he contrives a line and shows that it will work like another parameter for the ellipse. The contrived line, that is to say, will give the old square-rectangle property for new straight lines. These new straight lines are: a segment of a chord that is parallel to the original diameter, and a segment of the line that we are calling the *other* diameter. And, *second*, he shows that this other diameter bisects the chords that are parallel to the original diameter.

In effect, what Apollonius shows is this: (I) those chords parallel to the original diameter will cut off on the other diameter (that is, on the ordinatewise bisector of the original diameter) lines of just the right size to be *new abscissas*; and (II) those parallel chords will be situated just right to be bisected by the other diameter—situated just right, that is, for their halves to be *new ordinates*. (Fig. 2.)

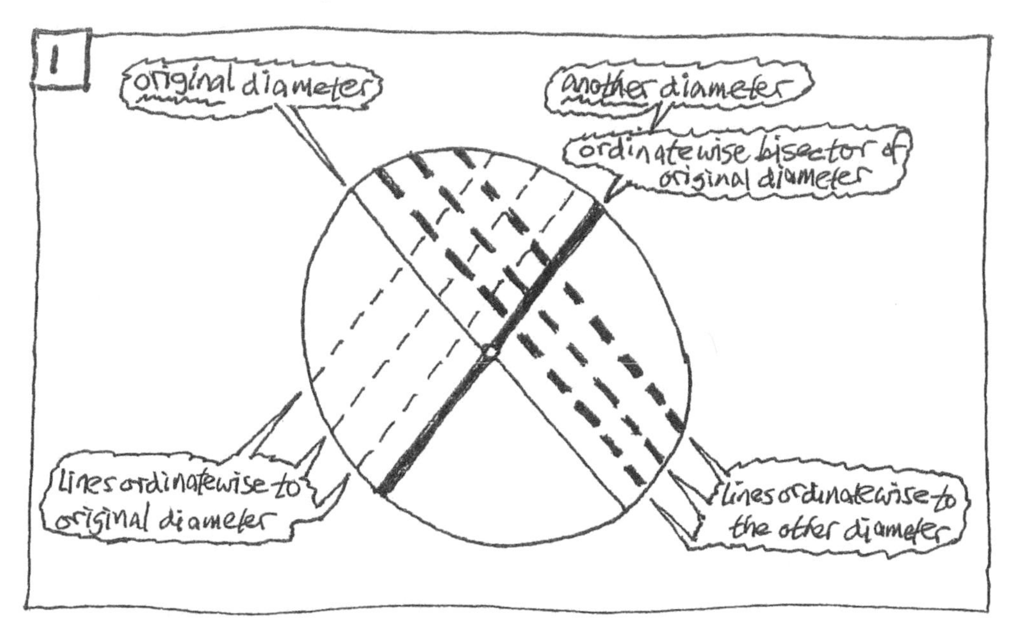

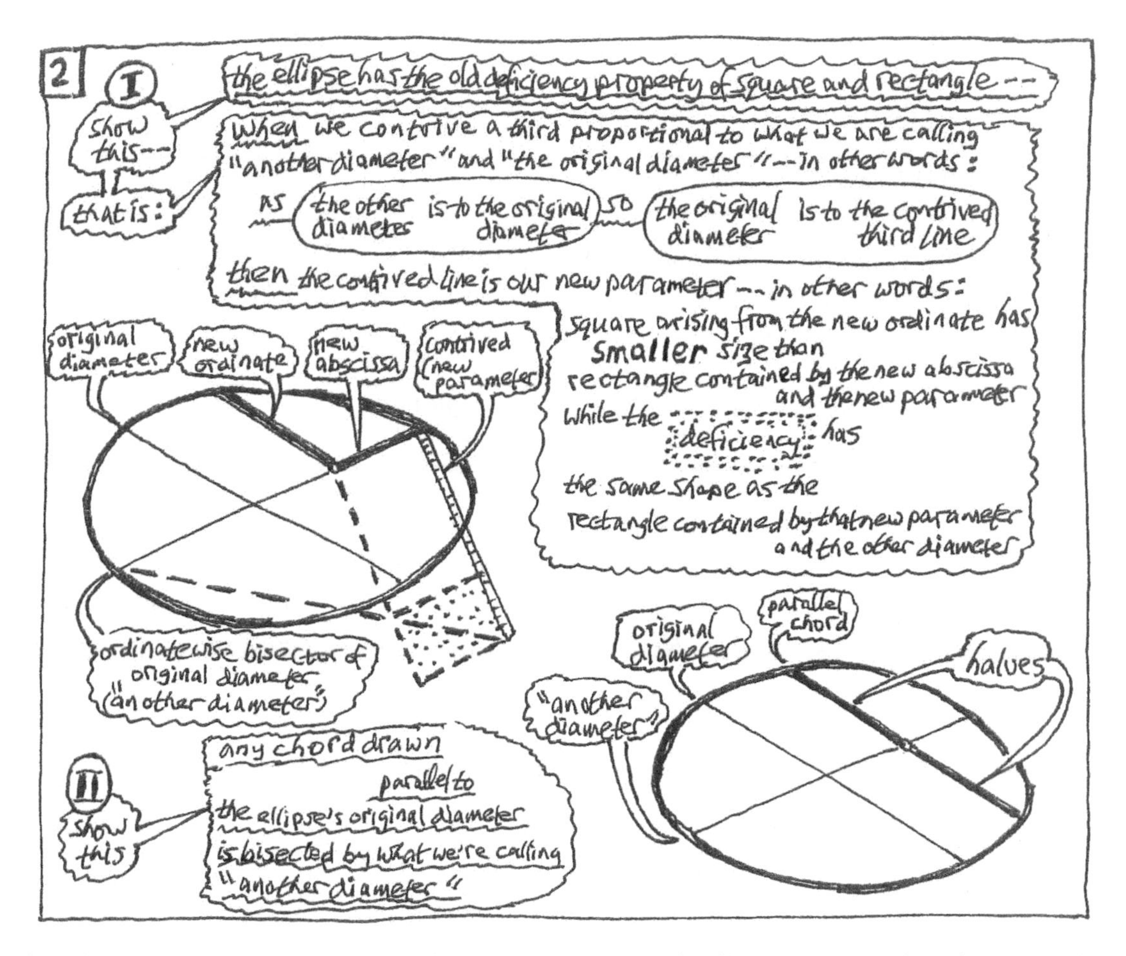

Thus, by taking the ordinate that bisects the original diameter, Apollonius gets another diameter whose own ordinates are parallel to the original diameter. The *original* diameter will be the mean proportional between the *other* diameter and the line contrived by Apollonius to be that *other* diameter's parameter. Now, it would be possible to show that that *other* diameter will also be the mean proportional between the *original* diameter and that *original* diameter's parameter. The other diameter here presented therefore seems to be more than just *an* other diameter; it seems to be *the* other diameter of the original diameter. The original diameter seems to have a special bond yoking it together with this other diameter that is now obtained for the ellipse. The two diameters seem like twins.

This 15th proposition shows us that an ellipse has at least one diameter other than the original diameter given with the ellipse—this other diameter that we now obtain being such that each of the twin diameters has the position that makes it an ordinatewise line for the other, and has the size that makes it a mean proportional between its twin and its twin's own parameter.

Do the twins have other siblings? We do not yet know. Does it help to consider the circle? In a circle, we know that there are as many diameters as we may care to draw (since any other chord through the center of a circle's diameter is another diameter). We also know, however, that in a circle (unlike an ellipse) every diameter's ordinates must meet that diameter at right angles.

15th proposition: the demonstration

Here are the two parts of the demonstration with the points appropriately labeled.

Part I (fig. 3):

given an ellipse with its original diameter AB,
if we take the ordinatewise chord DE that bisects that original diameter,
and if we contrive a line DF of just the right size so that

> the ratio that the ordinatewise bisector DE has
> to the original diameter AB
> is the same as
> the ratio that the original diameter AB has
> to the contrived line DF,

then the ellipse's square-rectangle property will hold when we treat

> DF (the contrived line) as a new parameter,
> GH (drawn parallel to AB) as a new ordinate,
> DH (cut off from DE by GH) as a new abscissa.

Part II (fig. 4):

if we take any chord GW that is parallel to original diameter AB,
then it will be bisected by DE, which is thus
a diameter for chords so drawn ordinatewise.

(Note that the term "parameter" was introduced in the discussion of terminology back in Chapter 1 of Part III, when we were considering the enunciation of the 11th proposition.)

Part I of the demonstration requires some auxiliary lines. The requisite construction is depicted in figure 5.

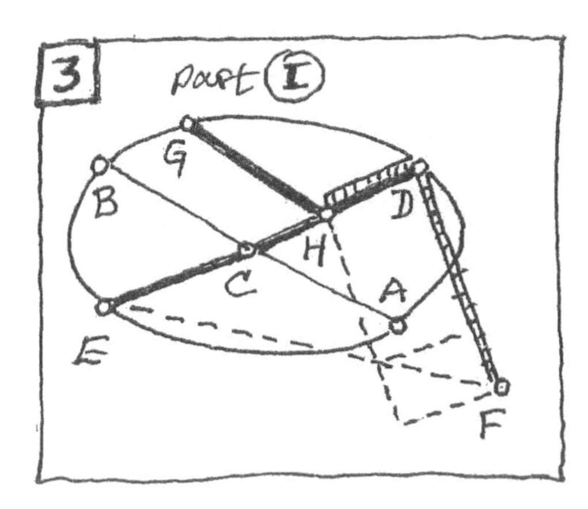

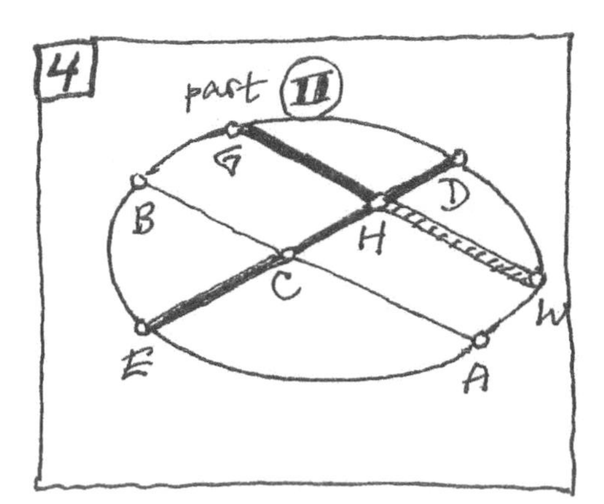

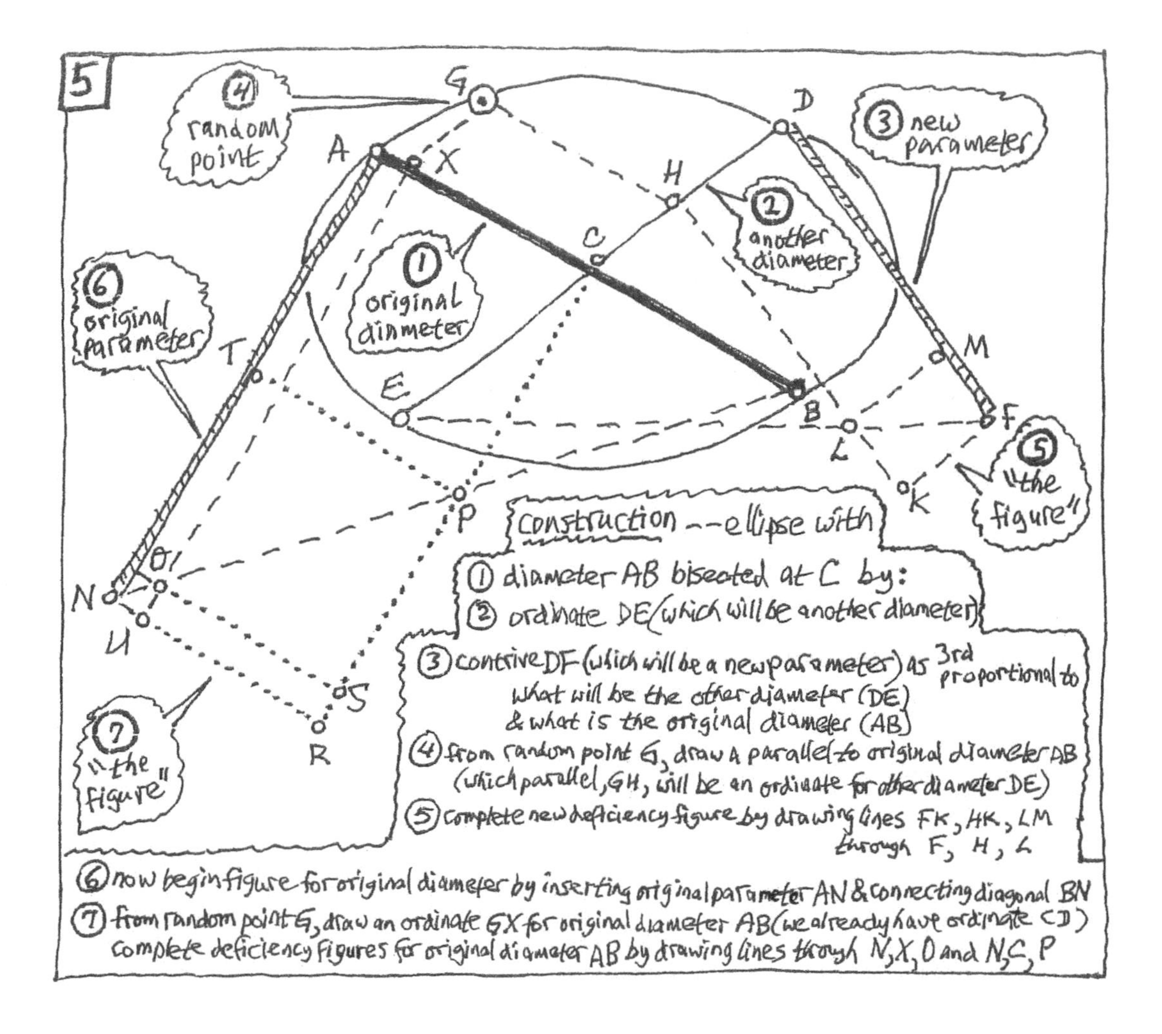

Part I demonstrates that the square on line GH (a new ordinate) is equal to the rectangle contained by the two lines DH (a new abscissa) and HL (a new line that is deficient with respect to new parameter DF). The gist of the reasoning is depicted in figure 6.

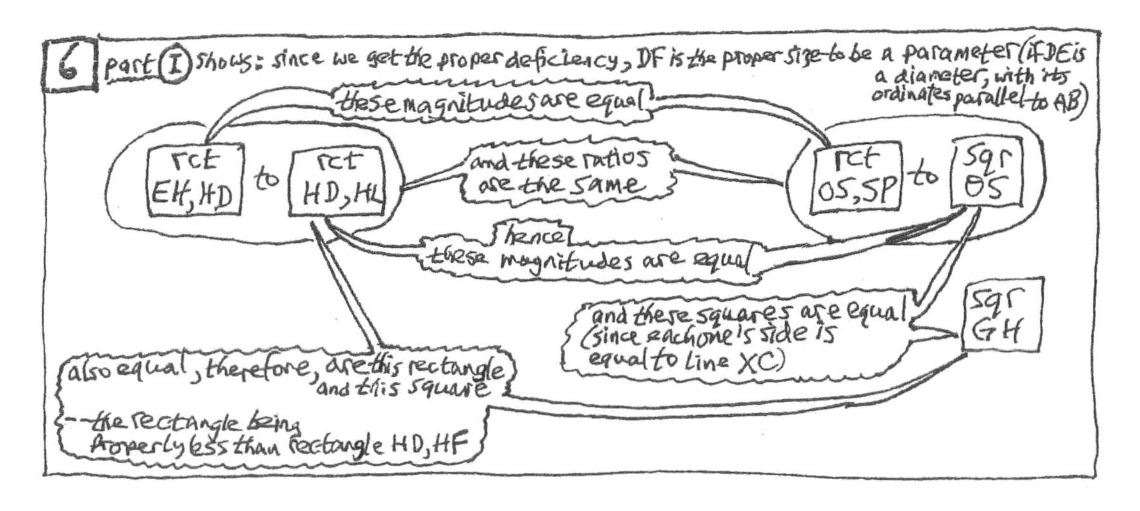

Let us now examine in greater detail Part I of the demonstration.

The first thing shown is the equality of the two rectangles [EH, HD] and [OS, SP]. How?

The point of departure is the ratio of the original diameter or oblique (transverse) side to its own parameter (upright side). This constant ratio, which shapes the deficiency that characterizes the ellipse, is now used as the link among areas related to the line (AC) which is half of the original diameter and which is also a new ordinate for the new abscissa that is half of the new diameter.

From the starting-point in the demonstration (the ratio of the oblique [transverse] side to the upright side of the original diameter), it follows (because there are parallel lines) that the area [AP] is equal to the sum of the two areas [AO] and [PO]. How this equality follows from the starting-point of the proposition, is depicted in figure 7. (The ellipse is not depicted in this figure, but you can find it in the next one, figure 8.)

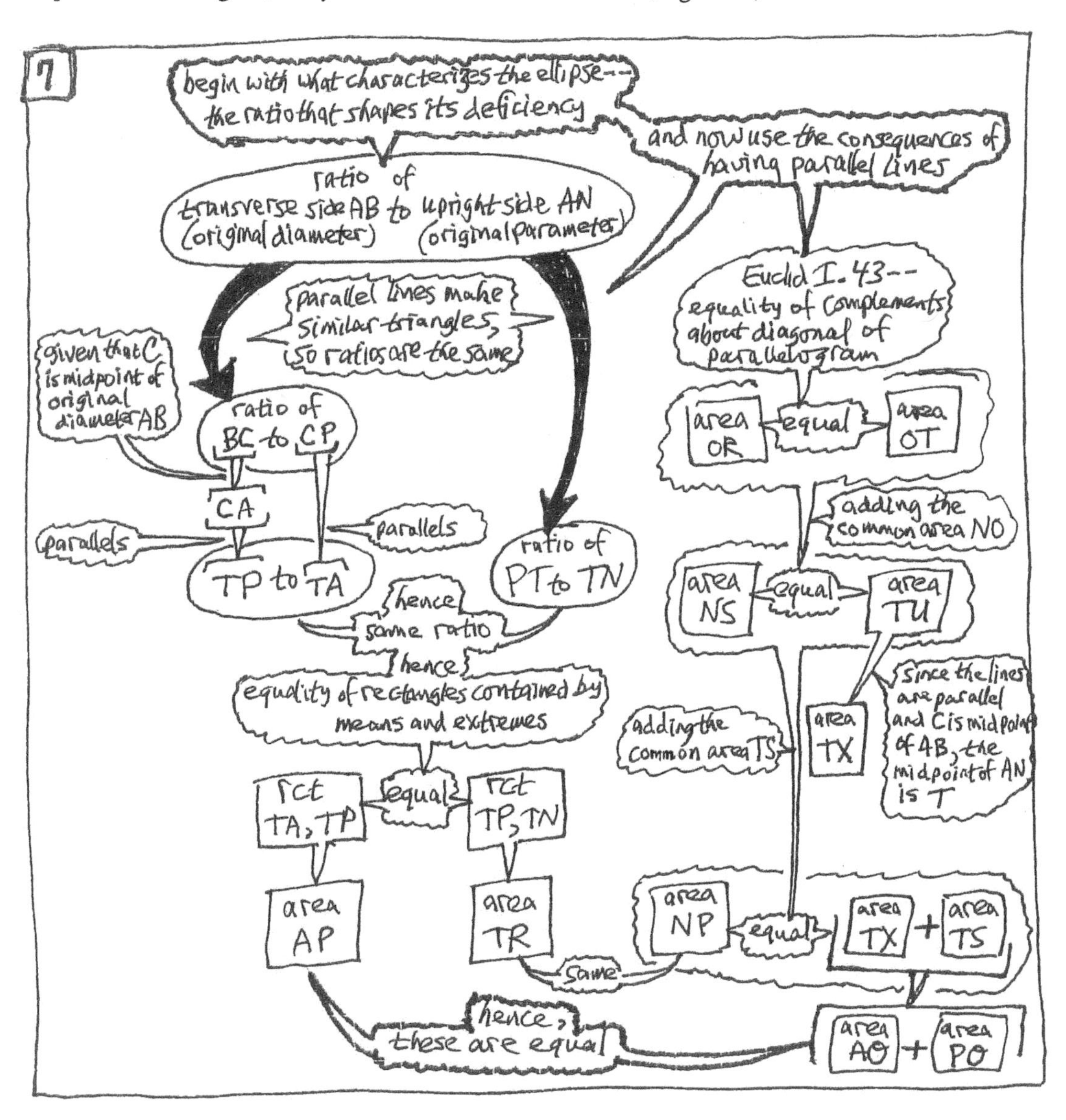

Anything that is equal to area [AP] is equal to anything that is equal to the sum of the two areas [AO] and [PO]. Such things are now obtained by using, on the old diameter, the rectangle-square property of the 13th proposition, since AC and AX are abscissas of the original diameter (and T, X, O are points obtained by using the ratio of the original diameter and its parameter). The use of that rectangle-square property is depicted in figure 8.

Each of those last two squares has as its side an ordinate of the original diameter (CD for the one square, and GX for the other). The connection supplied by those two squares is the central thing in this part of the demonstration, which is depicted in figure 9.

So, that is how Apollonius shows the equality of the rectangles [EH, HD] and [OS, SP].

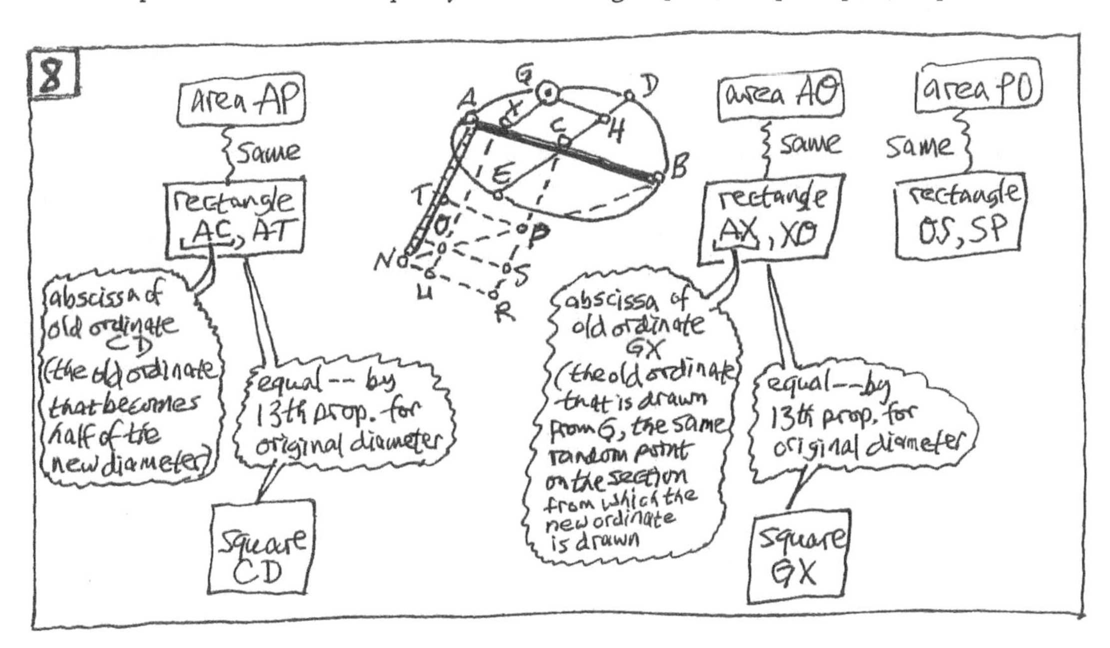

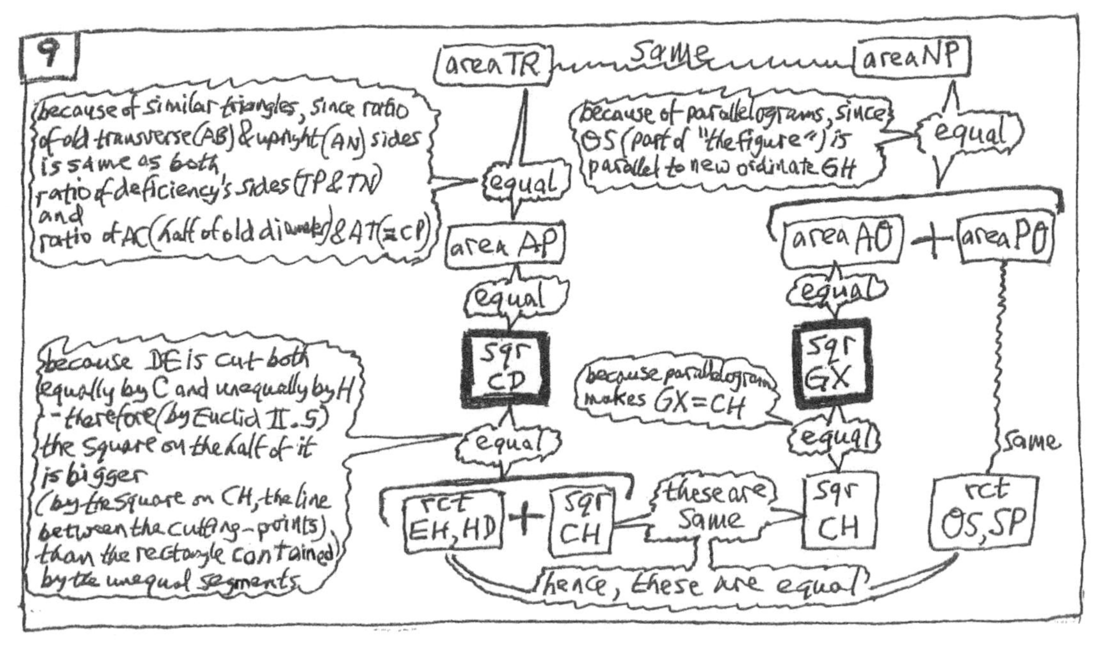

Those rectangles that have been shown to be equal need now to be shown to have the same ratio to two other rectangles. (Once that has been done, we shall be able to conclude that these other two rectangles—namely, rectangle [HD, HL] and square [GH]—are equal to each other.) (Fig. 10.)

Whereas Part I has been based so far upon the ratio of the *original* diameter (AB) to its parameter (AN), the continuation from here will be based upon the ratio of what will be the *new* diameter (DE) to its own parameter (DF). From this ratio of the new diameter to its parameter, it is shown that

> the ratio of the new diameter (DE)
> to the original diameter (AB),
>
> is the same as
>
> the ratio of the line HE
> to the new width (HL) of the equal rectangle—
> that line HE being the difference
> between the new oblique (transverse) side (DE) and the new abscissa (DH).

And then,

> (1) new diameter (DE)
> is connected by Apollonius with
> one of the items in which we are interested—namely, area [OS, SP]—
>
> and (2) the new difference (HE)
> is connected by Apollonius with
> the other item in which we are interested—namely, rectangle [HD, HE].

How are these connections made? (1) The new diameter (DE) is connected with area [OS, SP] in this way: half of that new diameter (DE) is the old ordinate (CD), whose abscissa (AC) is half of the original diameter; this old ordinate (CD) is used in a square-rectangle connection by way of the 13th proposition, which is connected to the area [OS, SP] by way of the similarity of triangles. And (2) the new difference (HE) is connected with rectangle [HD, HE] in this way: HE is used to make a rectangle having HD as its height.

From those connections, what is established (fig. 11) is this: a certain proportion links the rectangles [OS, SP] and [HD, HE].

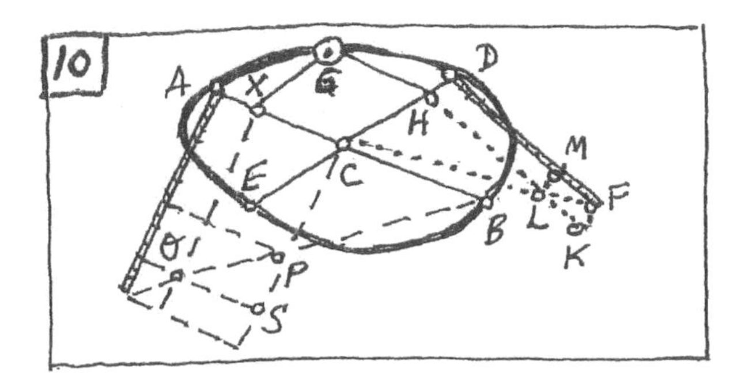

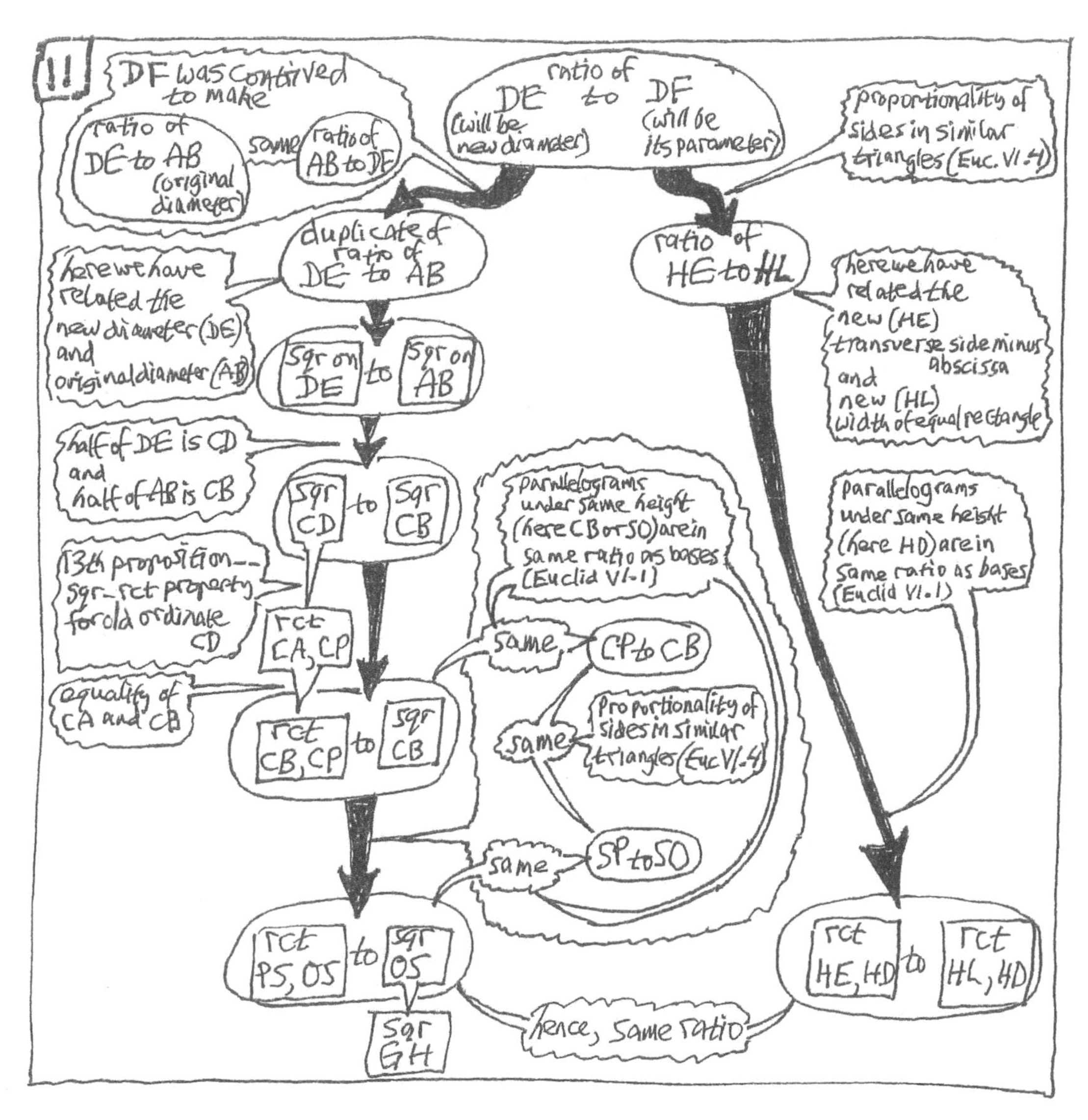

Since we have just seen that the ratios are the same, and we saw earlier that the antecedents of those ratios are equal, it follows that their consequents are also equal: rectangle [HD, HL] is equal to square [GH]. That is to say: take the rectangle whose sides are HD (a new abscissa) and HL (a new line that is deficient with respect to the new upright side DE)—this rectangle is equal to the square whose side is GH (a new ordinate); and it lies along DF (the new upright side) and is deficient by a figure (FL) which is similar, and similarly situated, to the rectangle whose sides are DE (the new oblique [transverse] side) and DF (the new upright side).

Part I of the demonstration has thus shown that a line (HD), which is cut off on the ordinatewise bisector (DE) of the original diameter (AB) by a half-chord (GH) ordered parallel to the original diameter (AB), will exhibit the deficiency-property of the ellipse when we use as parameter a line (DF) contrived as a third proportional to the following two lines: the ordinatewise bisector (DE) of the original diameter, and the original diameter (AB) itself.

Part I of the demonstration has just shown this:

> if, from the ordinatewise line that bisects the ellipse's original diameter,
> a line is cut off by a line drawn parallel to the original diameter,
> then the line cut off is in effect a *new abscissa*.

Part II of the demonstration will now show this:

> a line that is drawn parallel to the original diameter will in effect make
> a *new ordinate* for the ordinatewise bisector of the original diameter;
> that is to say, this ordinatewise bisector of the original diameter will also bisect
> any chord that is drawn parallel to the original diameter.

Part II of the demonstration requires some auxiliary lines, the requisite construction for which is exhibited in figure 12.

Part II begins at the endpoints (G and W) of the line which is drawn from random point G and which will be ordinatewise to the *new* diameter ED. Lines that are ordinatewise to the *original* diameter AB, when they are drawn from those endpoints G and W, will meet this original diameter at points that will be called X and Y.

The line GW is parallel to the original diameter, and the ordinates to the original diameter that are drawn from the endpoints G and W are parallel to each other and to the old ordinatewise line CH. Line GW will therefore be bisected at point H if line XY is bisected at point C. But line XY in fact *is* bisected at point C. Why? Because C cuts AB into two pieces that are equal, and from these equal pieces (AC and CB) the old ordinates (GX and WY) chop off pieces (XA and YB) that are equal.

Part II of the demonstration first shows the equality of the pieces of the original diameter that extend from points X and Y to the respective nearby endpoints (A and B) of the original diameter. How is XA shown to be equal to BY? By using the ellipse property of the 13th proposition with the original diameter, and appealing to the equality of consequent terms when same ratios have the same antecedent terms. That is depicted in figure 13.

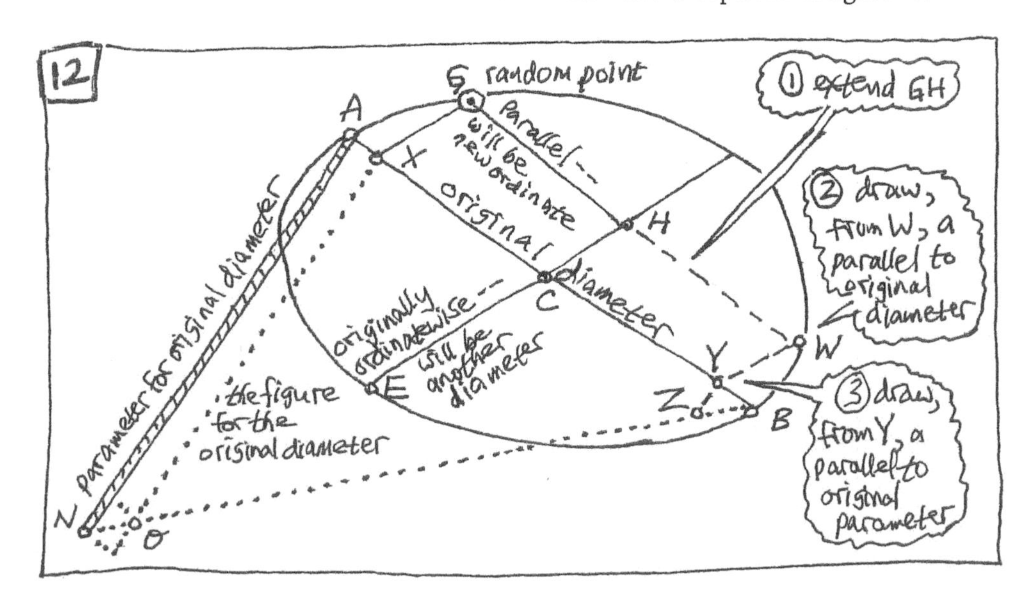

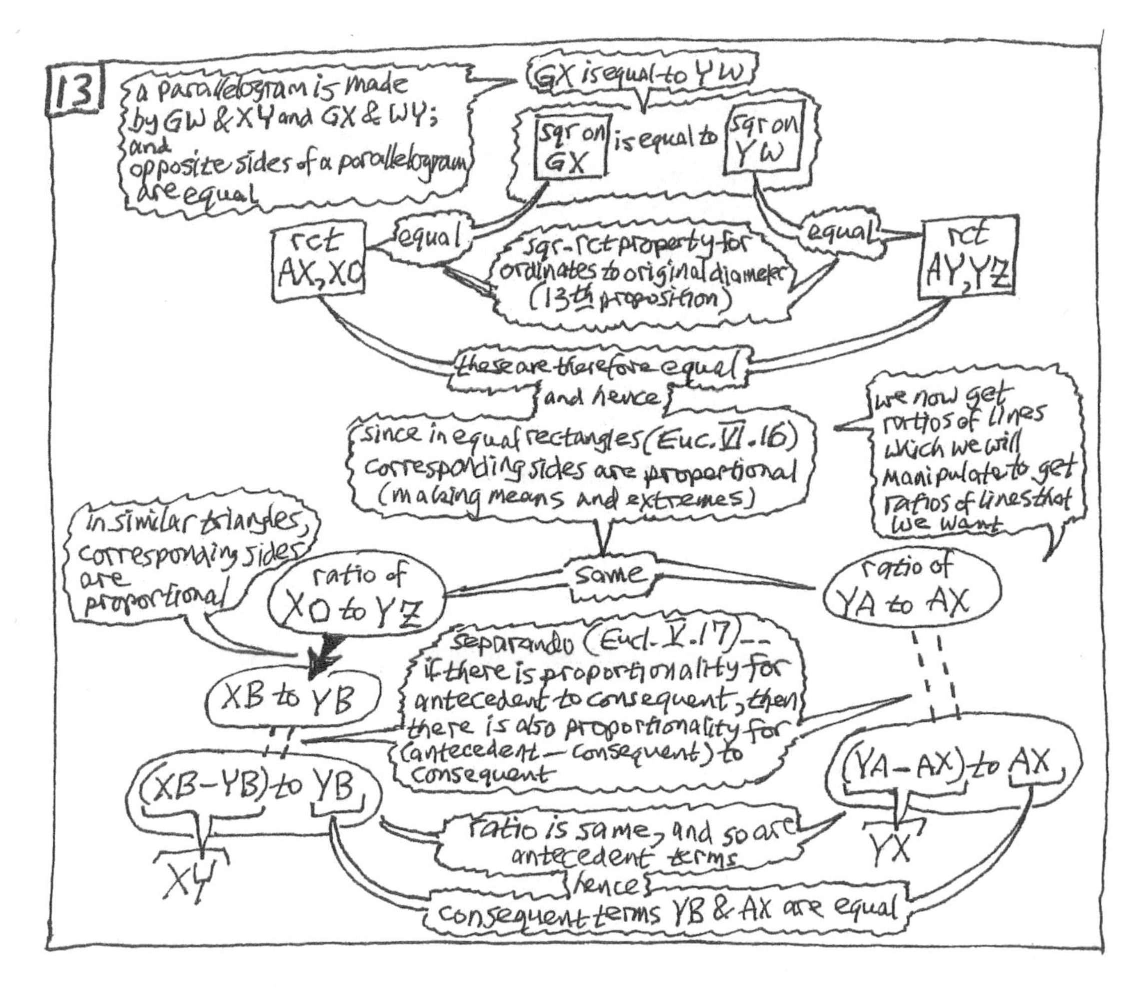

Now the pieces of the original diameter that have been shown to be equal (AX and BY) are taken away from pieces that were given as equal (AC and CB), yielding remainders that are equal (XC and CY); from this equality, it follows that GW is bisected by DE. (Fig. 14.)

It has thus been shown that the original diameter's ordinatewise bisector is another diameter, one whose ordinates are the half-chords drawn parallel to that original diameter.

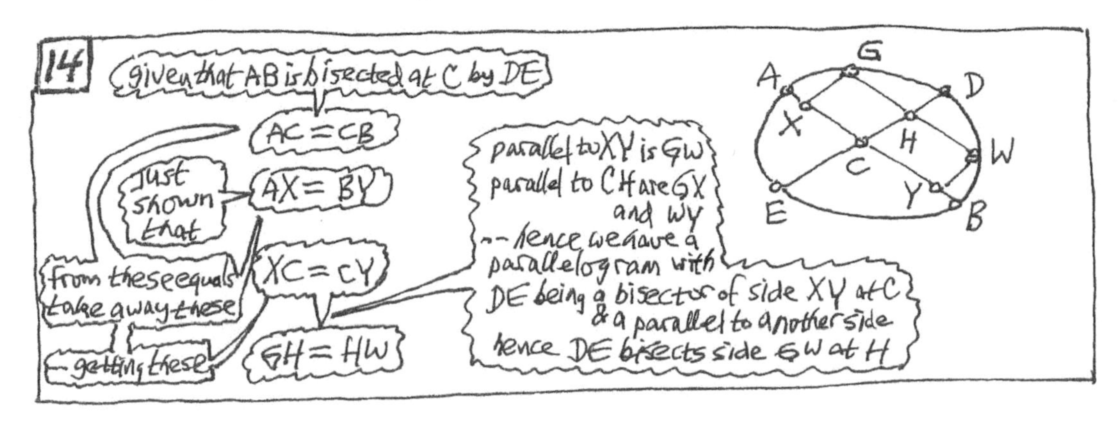

16th proposition

#16 If, through the midpoint (*dichotomias*) of the oblique side (*plagias pleuras*) of the opposite [sections], there be drawn any straight [line] parallel to a [line] drawn down ordinatewise, it will be a diameter of the opposite [sections]—a diameter that is conjugate to the diameter originating previously (*prouparkhousê*).

—Let there be opposite [sections] whose diameter the [line] AB is; and also let there be cut in two [halves] the [line] AB at the [point] C; and also, through the [point] C, let there be drawn parallel to a [line] that has been drawn down ordinatewise the [line] CD. I say that a diameter is what the [line] CD is—a diameter that is conjugate to the [diameter] AB.

—For let there be—[as the lines] along which the lines that are drawn down have the power—the straight [lines] AE, BF [that is to say: each of these straight lines AE and BF is to be an upright side for one of the two hyperbolas lying opposite each other]; and also let the [straight lines] AF, BE being joined be extended; and also let there be taken some chance point on either of the sections, the [point] G, say; and also, through the [point] G let there be drawn parallel to the [line] AB the [line] GH, [with the point H being where this parallel through point G meets the other section], and, from the [points] G, H let there be drawn down ordinatewise the [lines] GK, HL, and, as well, through the [points] K, L let there be drawn parallel to the [lines] AE, BF the [lines] KM, LN. Since, then, equal is the [line] GK to the [line] HL, therefore equal also is the [square] from the [line] GK, [equal, that is,] to the [square] from the [line] HL. But as for the [square] from the [line] GK, it is equal to the [rectangle] by the [lines] AKM; and as for the [square] from the [line] HL, it is equal to the [rectangle] by the [lines] BLN. Therefore the [rectangle] by the [lines] AKM is equal to the [rectangle] by the [lines] BLN. And since equal is the [line] AE to the [line] BF, therefore as is the [line] AE to AB so is the [line] BF to BA. But (on the one hand) as is the [line] AE to AB so is the [line] MK to KB; and (on the other hand) as is the [line] FB to BA so is the [line] NL to LA. Therefore also as is the [line] MK to KB so is the [line] NL to LA. But as is the [line] MK to KB so is—with the [line] KA being taken as common height—the [rectangle] by MKA to the [rectangle] by BKA; and as is the [line] NL to LA so is—with the [line] BL being taken as common height—the [rectangle] by NLB to the [rectangle] by ALB. Therefore also as is the [rectangle] by MKA to the [rectangle] by BKA so is the [rectangle] by NLB to the [rectangle] by ALB. And, alternately (*enallax*), as is the [rectangle] by MKA to the [rectangle] by NLB so is the [rectangle] by BKA to the [rectangle] by ALB. And equal is the [rectangle] by MKA to the [rectangle] by NLB. Equal, therefore, is also the [rectangle] by BKA to the [rectangle] by ALB. Equal, therefore, is the [line] AK to the [line] LB. But also the [line] AC, to the [line] CB, is equal. Therefore also the whole [line] KC, to the whole [line] CL, is equal. And so also is the [line] GX [equal] to the [line] XH. The [line] GH, therefore, has been cut in two [halves] by the [line] XCD; and it is parallel to the [line] AB. A diameter, therefore, is what also the [line] XCD is—a diameter that is conjugate to the [diameter] AB.

16th proposition: enunciation

Sliding the sections out of the cone, the one that we looked at first was the curve called an ellipse, with its diameter. This one that we now turn to look at is the pair of lines called the opposite sections, with its oblique (transverse) side—the oblique (transverse) side being, when extended, the diameter of both of the curves that constitute the opposite sections.

Apollonius shows that with the opposite sections—as with the ellipse—the ordinatewise line that bisects the original diameter will itself work like *another* diameter, and the ordinatewise lines that it will bisect are the chords that are parallel to the original diameter (figs. 15 and 16).

What Apollonius did not do with the ellipse in the previous proposition, he does now with the opposite sections in this proposition: he calls the other diameter by the name "diameter." He says that it is a *diameter* that is *conjugate* to the original diameter.

There is another difference between the two propositions. With the ellipse in the previous proposition, Apollonius first demonstrated that a line contrived to have a certain size works like another parameter—it gives the old square-rectangle property. Also, lines parallel to the original diameter work like new ordinates; and the originally ordinatewise chord that bisects the original diameter works like another diameter, yielding new abscissas for the ellipse. Now with the opposite sections, however, Apollonius does not do what he did previously with the ellipse: he does not contrive a line to work like another parameter. He does not demonstrate that the old square-rectangle property will hold for new ordinates and abscissas. The whole of what he demonstrates now for the opposite sections is only the second part of what he demonstrated previously for the ellipse. What he now demonstrates is only that the ordinatewise bisector of the oblique (transverse) side will bisect lines drawn parallel to the oblique (transverse) side.

Does the difference between what he showed for the ellipse and what he now shows for the opposite sections have anything to do with the fact that the other diameter for the ellipse had a definite size? The other diameter in the ellipse—like the original diameter in the ellipse—was bounded by the curve. By contrast, the other diameter that is obtained in the opposite sections now has no definite size. The chords that it bisects run from one to the other of the hyperbolic curves that constitute the pair of opposite sections—rather than being the chords of one curve (as are the chords that are bisected by the diameter that we see after the oblique [transverse] side is extended in either direction). Note that for a pair of curves, our use of the term "chord" is not being restricted to a straight line that joins two points on a single one of the two constituent curves. What also counts as a chord of the pair of curves, that is to say, is a straight line that joins a point on one curve of the pair and a point on the other.

Let us postpone any further consideration of *what* is demonstrated in this proposition, until after we have considered *how* it is demonstrated and also *what immediately follows* the demonstration.

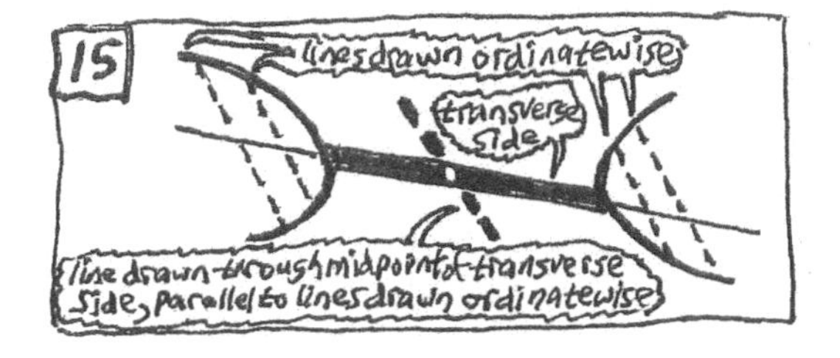

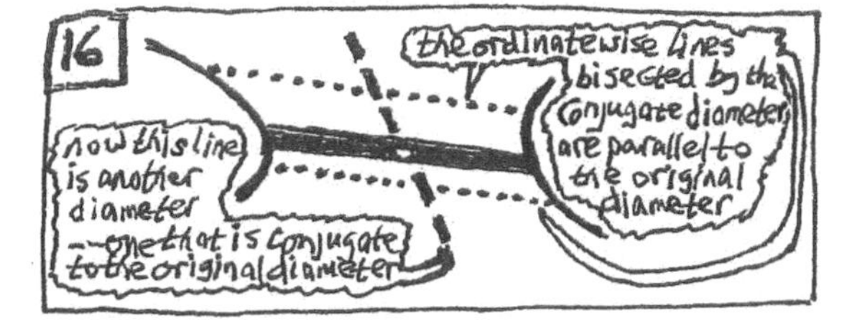

16th proposition: demonstration

The demonstration requires some auxiliary lines, the requisite construction for which is exhibited in figure 17.

Apollonius demonstrates that line GH is bisected by point X—which is to say that line XCD is a diameter, one whose ordinatewise lines stretch from curve to curve, parallel to the original diameter. How does he show that that X bisects GH—in other words, that GX is equal to XH?

He begins, as in the corresponding demonstration for the ellipse, at the endpoints of the line GH that will be ordinatewise to new diameter XD. When, from those endpoints G and H, ordinates are drawn for the original diameter (AB), they will meet this diameter (extended) at points K and L. Line GH is parallel to the original diameter, and when ordinates for the original diameter are drawn from the endpoints G and H, they are parallel to each other and also to the old ordinatewise line XCD. Since opposite sides of parallelograms are equal (Euclid, *Elements*, I.34), line GH will be bisected at point X if line KL is bisected at point C. But line KL in fact *is* bisected at C. Why? Because C cuts AB into two pieces that are equal, and, added to these equal pieces (AC and CB), are other equal pieces (KA and LB), thus making equal sums (KC and CL). (In the *ellipse*, equal pieces were *taken away from* the two halves of the original diameter; now, in the *opposite sections*, the equal pieces are *added on to* the two halves of the original diameter.)

So, the core of the demonstration is to move from the equality of the old ordinates (GK and HL) to the equality of the abscissas (AK and BL) that are cut off by them on the original diameter.

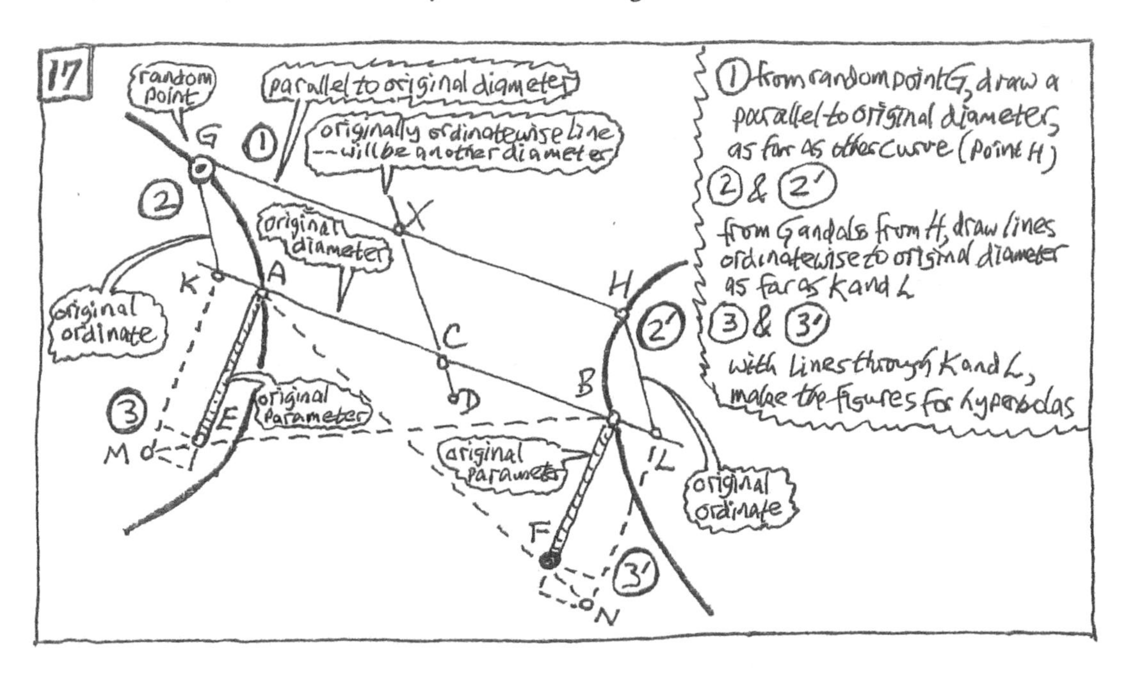

How, then, does the core of the demonstration proceed—how is it shown that if ordinates in opposite sections are equal, then so are their abscissas? The way taken in this proposition for the opposite sections is like the way taken in the previous proposition to demonstrate the equality of the pieces for the ellipse, which were taken away from (rather than added on to) the halves of the original diameter.

Apollonius uses the hyperbola property from the 12th proposition and he uses the equality of the parameters of opposite sections from the 14th proposition; he then appeals to the necessary equality of the terms of a ratio when that ratio is the same as another ratio whose terms are equal. The core of the demonstration is depicted in figure 18. After that, more detail will be depicted in figure 19.

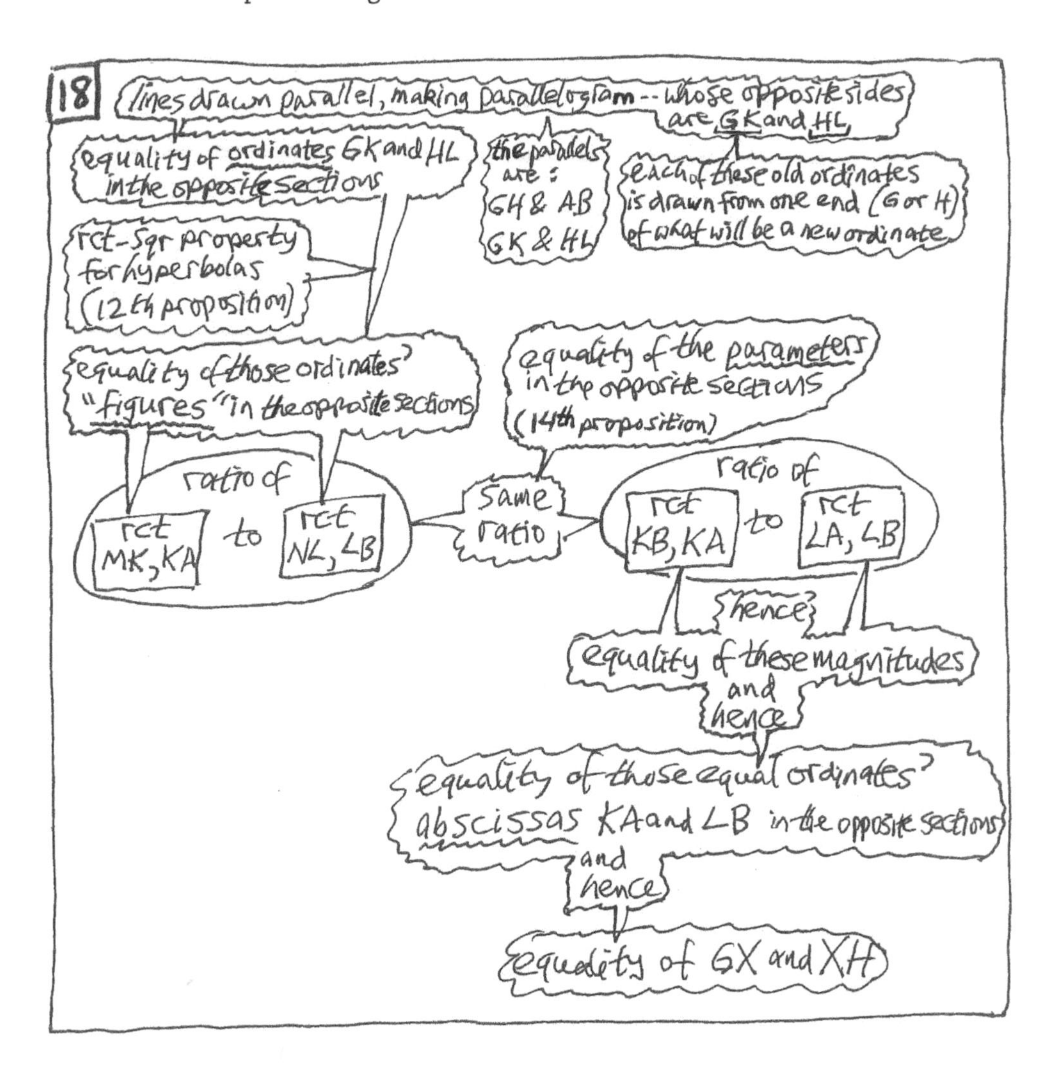

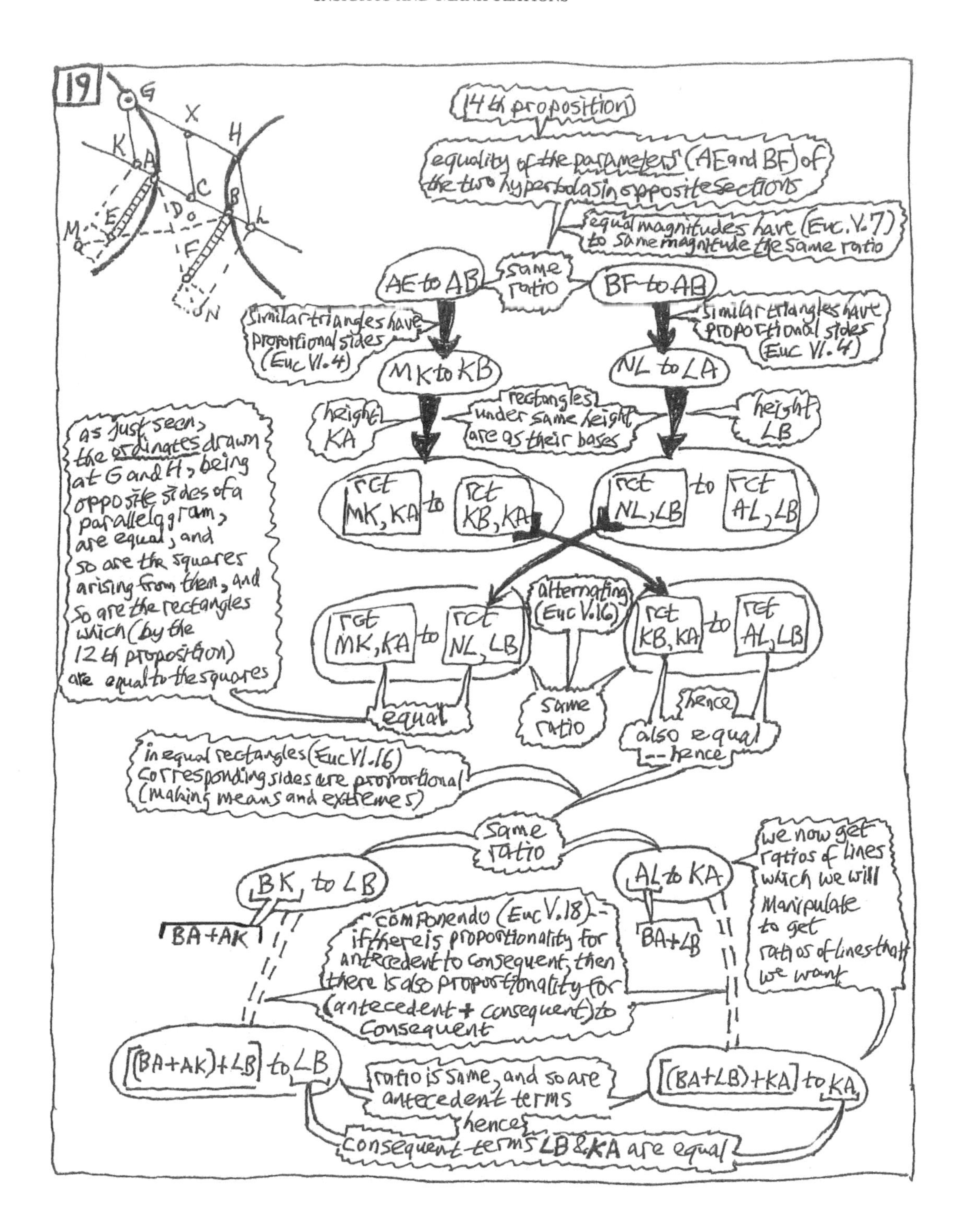

19
G
X
H
K
A
C
D
B
L
M
E
F
N
14th proposition
equality of the parameters (AE and BF) of the two hyperbolas in opposite sections
equal magnitudes have (Euc. V. 7) to same magnitude the same ratio
AE to AB
same ratio
BF to AB
Similar triangles have proportional sides (Euc VI. 4)
similar triangles have proportional sides (Euc VI. 4)
MK to KB
NL to LA
height KA
rectangles under same height are as their bases
height LB
as just seen, the ordinates drawn at G and H, being opposite sides of a parallelogram, are equal, and so are the squares arising from them, and so are the rectangles which (by the 12th proposition) are equal to the squares
rct MK, KA to rct KB, KA
rct NL, LB to rct AL, LB
rct MK, KA to rct NL, LB
alternating (Euc V. 16)
rct KB, KA to rct AL, LB
equal
same ratio
hence
also equal
hence
in equal rectangles (Euc VI. 16) corresponding sides are proportional (making means and extremes)
same ratio
BK, to LB
AL to KA
we now get ratios of lines which we will manipulate to get ratios of lines that we want
BA+AK
BA+LB
componendo (Euc V. 18) - if there is proportionality for antecedent to consequent, then there is also proportionality for (antecedent + consequent) to consequent
[(BA+AK)+LB] to LB
ratio is same, and so are antecedent terms
[(BA+LB)+KA] to KA
hence
consequent terms LB & KA are equal

Now the abscissas of the original diameter that have been shown to be equal (KA and BL) are added on to the pieces of the original diameter that were given as equal (AC and CB), yielding lines that are equal (KC and CL); from this equality it follows that XCD bisects GH. That is depicted in figure 20.

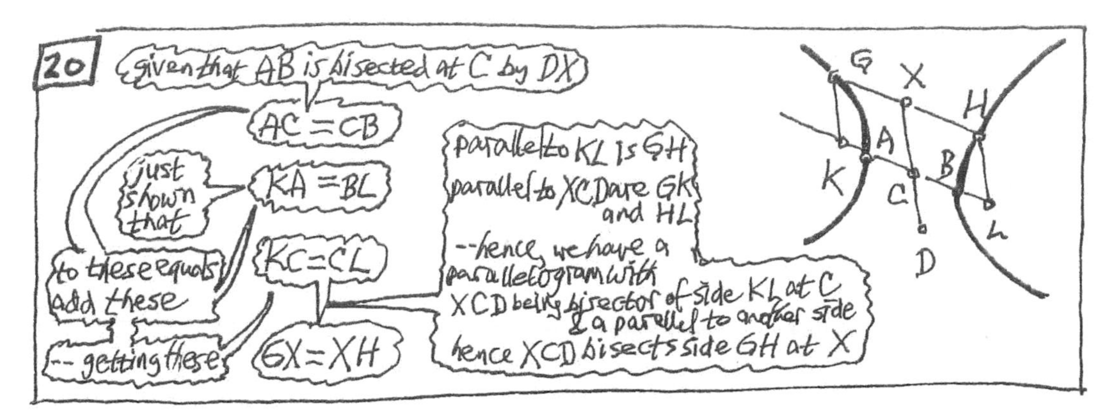

CHAPTER 2. MORE OF WHAT'S WHAT

transition

Immediately after the propositions that characterized the conic sections, there have come, first, a treatment of the ellipse in the 15th proposition, and then, in the 16th, a treatment of the opposite sections. Why turn first to the ellipse; and why not turn, after it, to the single hyperbola or even the parabola? We should keep that question in mind as we come to Apollonius's Second Definitions.

Another thing to keep in mind is that Apollonius uses "hyperbola" as the name of a *single* line. His name for the hyperbolic *pair* of lines is "opposite sections." Only long afterwards did what he calls "opposite sections" get the name "hyperbola," with each hyperbolic line then being called one of its "branches." For Apollonius, an hyperbola is a line produced by the cutting of a *cone* (12th proposition), whereas opposite sections are lines produced by cutting *opposite surfaces* (14th proposition). It may be that for some purposes the terminology of Apollonius is inferior to later terminology; but when we adopt the later terminology because of what it reveals or what it directs our attention to, something may be obscured or we may be distracted from something. Perhaps *each* of the terminologies *both* conceals *and* reveals.

Apollonius did not need modern terminology in order to know that every hyperbolic line can be seen as one constituent of a pair of them—can be seen as being accompanied somehow by an invisible counterpart. The pair of lines that constitute opposite sections is obtained by a single slice through the opposite conic surfaces, but would anybody have looked for the other member of the pair before first having seen the single hyperbolic line by slicing through a cone? And if a complement is demanded for the single hyperbolic line, then why not also demand a complement for the hyperbolic pair? After all, as Apollonius will later show, every pair of lines that constitute opposite sections can be seen as one constituent of a *quartet* of lines—as somehow accompanied by its invisible fellow pair of lines. These two pairs of lines cannot be obtained by a single slice of anything, and yet, as Apollonius will show, they are yoked together in such a way that one pair can be better comprehended by keeping the other pair in mind (fig. 1).

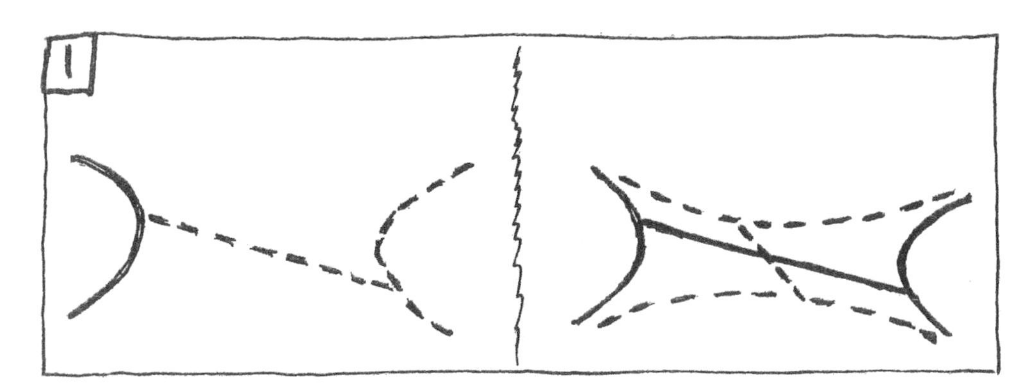

Another thing to keep in mind is that if you take *any* two hyperbolas, they will not necessarily constitute opposite sections; they will only do so if they can be obtained by a single slice of opposite surfaces. But the way that you obtain them is not what makes them interesting. What makes them interesting is what follows from the way that you obtain them. From the fact that the lines of the opposite sections are cut by a single slice, it follows that they have *the same oblique (transverse) side*; and from the fact that they are cut into opposite surfaces, it follows that they have *equal upright sides*.

The opposite sections and their constituent hyperbolas, as well as the ellipse, are shaped by that ratio which has as its terms the oblique (transverse) side and the upright side. Those two straight lines are the sides of the *eidos* of excess or deficiency, and that *eidos* comes to prominence in the Second Definitions that Apollonius places right after these last two propositions.

Second Definitions

Definitions #2 Of the hyperbola and of the ellipse—of each of them—let the midpoint (*dikhotomia*) of the diameter be called a "center of the section," and let the [line] from (*apo*) the center falling to the section be called, [as it will later, in #43, be called] "out from (*ek*) the center of the section" [a term previously used in the case of circles, and traditionally rendered by the Latin *radius*]; and likewise also of the opposite [sections], let the midpoint of the oblique (*plagias*) side be called a "center"; and let the [line], from (*apo*) the center, that, drawn parallel to a [line] which has been drawn down ordinatewise, is the mean proportional between [or, more literally, "having the mean ratio of" (*meson logon ekhousa*)] the sides of the figure (*tou eidou*) and also is cut in two [halves] by the center, [let that line] be called a "second diameter."

Why does Apollonius place these definitions *here*? After all, even though the last few items in the *First* Definitions were not needed at the beginning of the book, he placed them there anyway. Why, then, are these *Second* Definitions not also placed back there at the beginning with the other definitions? And if not placed back there, why are they placed here rather than even later?

Consider what these Second Definitions follow: a proposition about the ellipse and a corresponding proposition about that special pair of hyperbolas which is the opposite sections. That pair of hyperbolic lines, taken together as one thing, is like the ellipse in a respect in which the single hyperbolic line is not. Nor is the parabola.

The parabola, indeed, is very unlike all the other conic sections: it is the only one that does not have a center. The ellipse does have a center, and so does the pair of opposite sections; even the single hyperbola has one. Centers are what the Second Definitions begin with. What we are concerned with here are conic sections that have centers.

The center is defined by Apollonius for the single hyperbola before he defines it for the opposite sections. That seems strange—since, for no apparent reason, an hyperbola's center is located in the region that we might call the "outside" of the hyperbola, whereas there is a manifest reason why the center of the opposite sections must be located outside each of the two hyperbolas. The center must be located outside each of those two constituent curves, in order to be located in the region in-between them.

To justify the location of the center of a single hyperbola, must we think of the curve as being essentially only one branch of an hyperbolic pair? If so, then the modern usage of the term "hyperbola" would seem to be simply an improvement over the classical usage. Or is there some other justification? With regard to that, consider the diameter of the only curve that we know very much about prior to the study of the conic sections—that is, the circle. The circle has more diameters than we can count, the center of any one of them being the same point as the center of any other. Perhaps an hyperbola too can have oblique (transverse) sides that are more than we can count, the midpoint of every one of them being the same point as the center of any other—the very point that these Second Definitions now call the center of the hyperbola.

Earlier (in the 12th proposition) "diameter" was a name used for a straight line on what might be called the "inside" of a single hyperbola; back then, it was not used for the "outside" straight line whose extension it is—namely, the hyperbola's oblique (transverse) side. Now in Second Definitions, however, it is the hyperbola's oblique (transverse) side that is called its diameter.

Likewise for the opposite sections, the oblique (transverse) side is called a diameter (albeit not in these Second Definitions, but just before the Second Definitions, in the 16th proposition). The oblique (transverse) side of the opposite sections is located on the outside of each separate member of the pair of lines in such a way as to be on the inside of the pair taken together as a single thing. Thus, the midpoint of the pair (the center of the opposite sections) is on the inside, not of either line separately, but of the pair of lines taken together: it is, as has been said, in-between the constituent hyperbolas of the pair.

The hyperbola's center (the center of the diameter which is its oblique [transverse] side) is "outside" the section. It may seem like stretching things to call a point the center of something on whose *outside* it is located; but if *some* point is to be designated the center of an hyperbola, then the midpoint of the oblique (transverse) side is the only point that could qualify at all. The hyperbola's oblique (transverse) side has a definite size, and so it has a midpoint. What was previously called the hyperbola's diameter has no midpoint, because it can be prolonged indefinitely.

The parabola's diameter likewise can be prolonged indefinitely; it can, moreover, also be extended indefinitely in the other direction, outside the parabola. In the parabola, there is thus no oblique (transverse) side—and no reason whatsoever to single out any point for designation as the center (fig. 2).

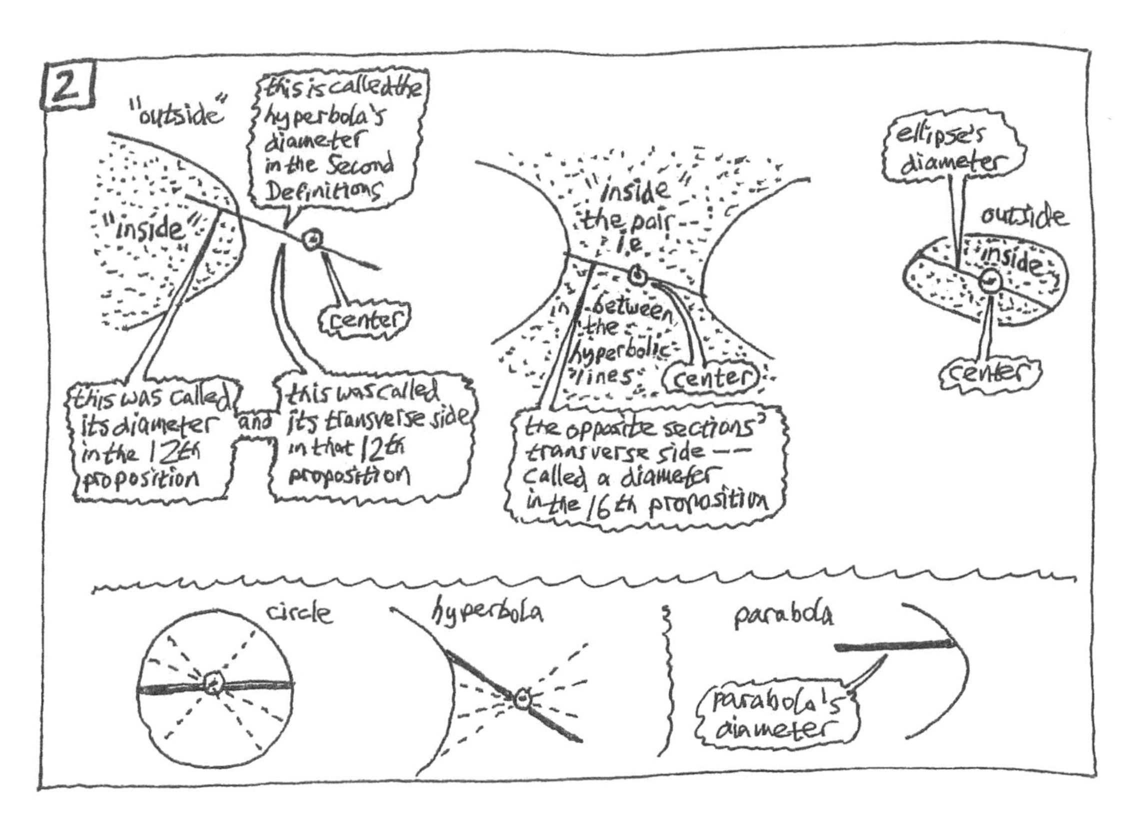

In Second Definitions, as soon as "center" has been defined for hyperbola, ellipse, and opposite sections (that is, for all the non-circular curves except for the parabola), there comes a definition of "second diameter." The definition of centers for all the non-circular curves that can have a center acts as a bridge—a bridge from the 15th and 16th propositions to the definition of second diameters.

Note that in the Greek text the definition does not use the definite article. The definition does not imply that there is a *unique* second diameter; it merely says that the name "second diameter" should be given to a certain line. Translated literally, the words are: "let it be called (a) *second diameter*." A second diameter is such with reference to another diameter that comes first.

What is a second diameter? It is a straight line characterized by the following three things: its position, its size, and its position given its definite size. In particular:

position

it is drawn parallel to the ordinates of an original diameter;

size

the square on it is equal to
the reference rectangle contained by
the oblique (transverse) side and the original upright side—
the figure (*eidos*) that gives the shape of
the area of excess (in the hyperbola) or of deficiency (in the ellipse);

position, given its definite size

it is bisected by the center of the section(s).

Apollonius first spoke about diameters at the beginning of his book, in his First Definitions. After opening with three items concerning the conic surface and the cone, he spoke about diameters, giving a diameter for a single curve; and then—for a pair of curves—giving two kinds of diameters: *oblique (transverse)* and *upright*. (Note that this "upright *diameter*" is not the "upright *side*.") The two kinds of diameters for a pair of curves were distinguished by their difference in position:

an *oblique (transverse)* diameter	an *upright* diameter
(*plagia*: "slanting, sidewise")	(*orthia*)
is located so that its ordinates	is located so that its ordinates
stand within each curve,	run across from curve to curve,
while it itself	while it itself
runs across from curve to curve;	stands in-between the curves (fig. 3).

Next, Apollonius declared that if each of two diameters bisects the chords that are parallel to the other, then they are *conjugate* diameters of the curve or pair of curves whose ordinatewise chords they thus bisect. Being conjugate is a matter of position: when each of two diameters is drawn ordinatewise to the other, then the two diameters are conjugate to each other.

Until the 15th proposition, however, there was no diameter other than the original diameter, the one which was obtained in the course of cutting a cone or conic surfaces. Then, for the ellipse, the 15th proposition gave a diameter other than the original diameter—that other diameter being one which is conjugate to the original diameter (although it was not identified as being conjugate or even as being another diameter). Likewise for the paired hyperbolas that are opposite sections, the 16th proposition gave a diameter other than the original diameter—that other diameter being identified in this case as a diameter which is conjugate to the original diameter.

For the opposite sections, the other diameter was the *upright* diameter which is conjugate to the original *oblique (transverse)* diameter. (The ellipse, unlike the opposite sections, has no diameter that is upright as distinguished from one that is oblique [transverse].)

The upright diameter in First Definitions was given a definite position but not a definite size. The conjugate diameter in First Definitions was likewise given a definite position but not a definite size. The second diameter here in Second Definitions is given not only a definite position but also a definite size (fig. 4).

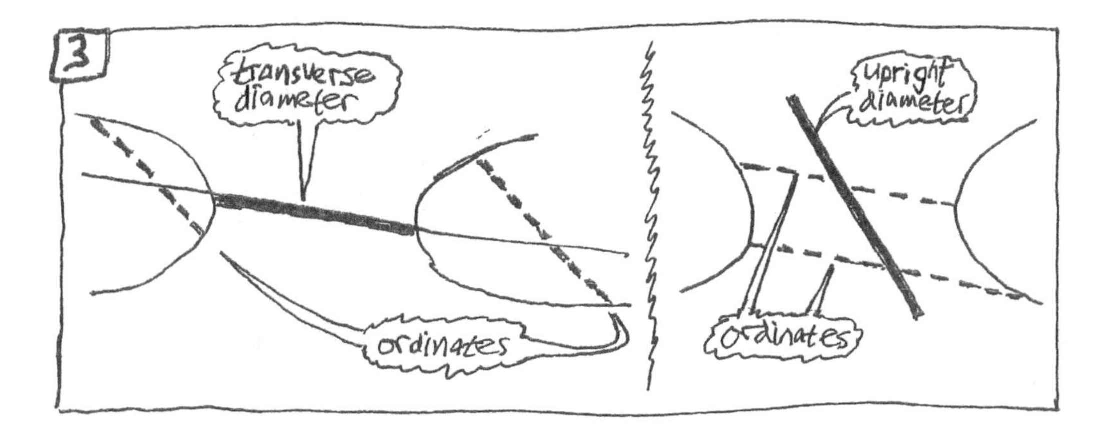

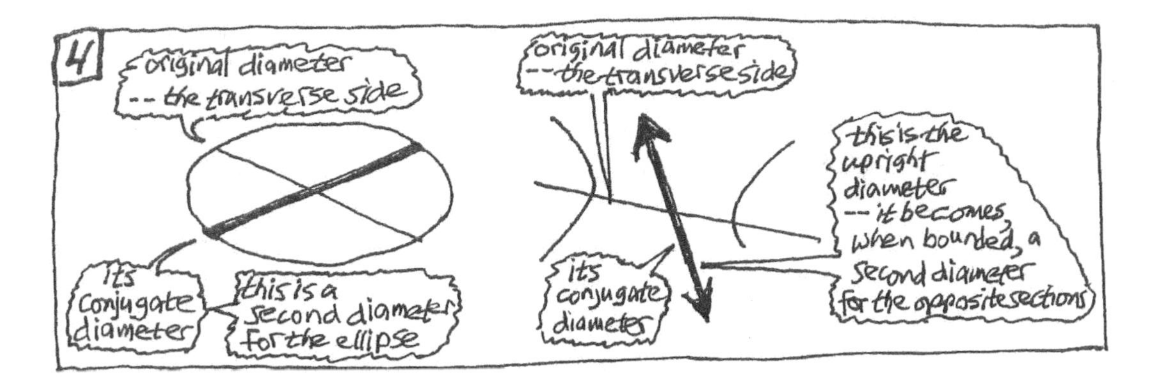

For the ellipse and the opposite sections, it follows from Second Definitions that

> a second diameter is, by its *position*,
> a diameter *conjugate* to the original diameter—
> being a conjugate diameter that
>
> is, by its *size*,
> a *mean proportional between*
> *the oblique (transverse) side and the original upright side*,
> and
> has its *center* on
> *the center of the section(s)*.

(The mean proportionality follows from the definition of the second diameter, since any line is a mean proportional between the sides containing a rectangle that is equal to a square erected on that line.)

For the ellipse, the position and the size of the line called the second diameter are the position and the size of the diameter that is conjugate to the original diameter. Since the ellipse is a single closed curve, the diameter that is conjugate to the original diameter, like the original diameter itself, is bounded by the ellipse. In the pair of open curves that constitutes the opposite sections, however, only the original oblique (transverse) diameter is bounded naturally—is bounded, that is to say, by the pair of curves itself. By contrast, the upright diameter that is conjugate to the opposite sections' original oblique (transverse) diameter cannot be bounded naturally, because the two curves that constitute opposite sections cannot set bounds to a straight line standing upright between those curves.

It is obvious that the ellipse, being a single closed curve, will itself set bounds to any diameter in it. What is not obvious in the ellipse, however—what had to be shown to us—is the relationship in size between its conjugate diameters. That relationship in size is this: an ellipse's original diameter is the mean proportional between its conjugate diameter and this conjugate diameter's parameter. The ellipse's *other* diameter, moreover, is a mean proportional too—between the original diameter to which it is conjugate, and this original diameter's own parameter.

Such a mean proportional, which it is natural to take as giving the size of a second diameter for the ellipse, is taken by Apollonius as giving the size of a second diameter for the opposite sections as well. Apollonius in effect defines a "second diameter" for both the ellipse and the opposite sections as a centered conjugate diameter that is such a mean proportional.

What would seem natural to do for the ellipse alone, Apollonius has done not only for it but also for the opposite sections—without making any reference to either the ellipse or the opposite sections.

We might say that what Apollonius has done is something like the following. First, he places a straight line in the position of that upright diameter which is conjugate to the original oblique (transverse) diameter of the opposite sections. So far, so good: that seems to be the right place to put a second diameter for the opposite sections as well as for the ellipse. Then, since the upright diameter in itself has no definite size, he looks for a way to set bounds to the line. He finds it by taking the mean proportional between the original diameter and its parameter. Now, that may be the right way to determine the size of a second diameter in the ellipse (where the curve itself sets bounds to the conjugate diameter, thus giving it that size), but it does not seem to have anything to do with the opposite sections. Finally, after centering the line which he has laid in place and for which he has chosen a size, he calls it a second diameter without restricting the name to the case of the ellipse. What justifies this, he does not say.

All the sections that have a center (namely, the ellipse, the opposite sections, the hyperbola) have a second diameter, whose size Apollonius determines in the same way for all of them. For the ellipse and also for the opposite sections, the definition of the second diameter in effect sets bounds to the conjugate diameter, but there is no mention of conjugate diameters in the definition. Why not? Because that silence enables the definition to cover the single hyperbola. Since the single hyperbola has no diameter that is conjugate to its original diameter, what is bounded to make *its* second diameter is a line that can only be described as a straight line drawn ordinatewise through the center.

The parabola, however, has no oblique (transverse) side, and hence no center through which to draw a straight line ordinatewise, so there is no way at all to enable the definition to cover the parabola.

With the definition of a second diameter, the pair of hyperbolic lines that constitute the opposite sections has come to resemble the ellipse. That is to say: it, like the ellipse, now has an other diameter which is not merely conjugate—not merely fixed in a certain determinate position—but is also of a certain determinate size. (It is also centered—since centrality is a matter of size and position together.) In the opposite sections, however, unlike in the ellipse, the conjugate diameters need to be distinguished. The oblique (transverse) diameter (the original diameter) is bounded naturally; the upright diameter (the new diameter) is not. The upright diameter is made into a second diameter by a definition that assigns bounds to it. What is a consequence of a natural bounding in the ellipse—namely, the second diameter's mean proportionality between its conjugate diameter and the conjugate diameter's parameter—is here by mere fiat applied to the hyperbola and opposite sections. The definition in size of the second diameter, not being a consequence of the conic's very nature, seems conventional or arbitrary. (Moreover, this arbitrariness with respect to size makes its position also seem arbitrary: by Second Definitions, the second diameter is placed just where the upright diameter would be placed by First Definitions, and is centered at the center of the section, which is located outside the hyperbola, or between the hyperbolas in the pair of opposite sections.)

Apollonius states no reason for his definition of second diameters, but his very silence invites the following question: what does that line's being a special other diameter have to do with that line's being a mean proportional to the sides of the *eidos* (namely, the oblique [transverse] side and the upright side of the original diameter)? Apollonius's very placement of the definition invites us to ask whether this proportionality is not a means to seeing more than meets the eye of a superficial observer.

Our first look at the conic sections apart from cones or conic surfaces has been a look at an ellipse followed immediately by opposite sections. By this juxtaposition of the 15th and the 16th propositions, we see that despite their seeming difference from each other (an ellipse is a single closed line; the opposite sections, an open pair of open lines), and despite their each seeming closer to a parabola than either is to the other (the ellipse has a deficiency, and each of the hyperbolas in the pair of opposite sections has an excess; the parabola—having neither excess nor deficiency—is a mean between the two extremes)—despite those things, the ellipse and the opposite sections have an important resemblance to each other. Both in the ellipse and in the opposite sections, the diameter has a conjugate diameter—another diameter yoked together with the original one in such a way that each of these conjugate diameters is parallel to the other one's ordinates.

But even in the resemblance we see that there is a difference between the ellipse and the opposite sections.

It is true that both in the 15th proposition (for the ellipse), and in the 16th proposition (for the opposite sections), another diameter is provided, one that is conjugate to the original one; but only in the proposition for the opposite sections is what is done identified as the provision of another diameter. There is more need for the identification in the case of the opposite sections: it is stranger for the opposite sections to be said to have a diameter. In opposite sections, the common oblique (transverse) diameter is not placed in the region where the diameter of each hyperbola is placed—it is not placed inside either one of the constituent hyperbolas. Rather, it is placed inside the thing that the two curves taken together constitute—the pair. The diameter that is conjugate to the original oblique (transverse) diameter is (according to First Definitions) an upright diameter—one whose ordinates are not inside either of the constituent hyperbolas, but are inside the pair as a totality.

Although it is position that makes the new diameter conjugate to the original diameter in the ellipse and in the opposite sections alike, the ellipse and the opposite sections also have a difference with respect to the position of the conjugate diameter. Because the ellipse is closed, and can therefore be a setter of bounds, the new diameter's position gives it a definite size, like the original diameter to which it is conjugate. In the opposite sections, however, the new (upright) diameter has no definite size of its own. Because the inside is bounded by the constituent curves on only two sides, the diameter that is conjugate to the original (oblique [transverse]) diameter must be extended indefinitely far to hit the ever-farther-out ordinates drawn from one curve of the pair to the other. That is why the upright diameter, unlike the oblique (transverse) diameter, is indefinitely extendible (fig. 5).

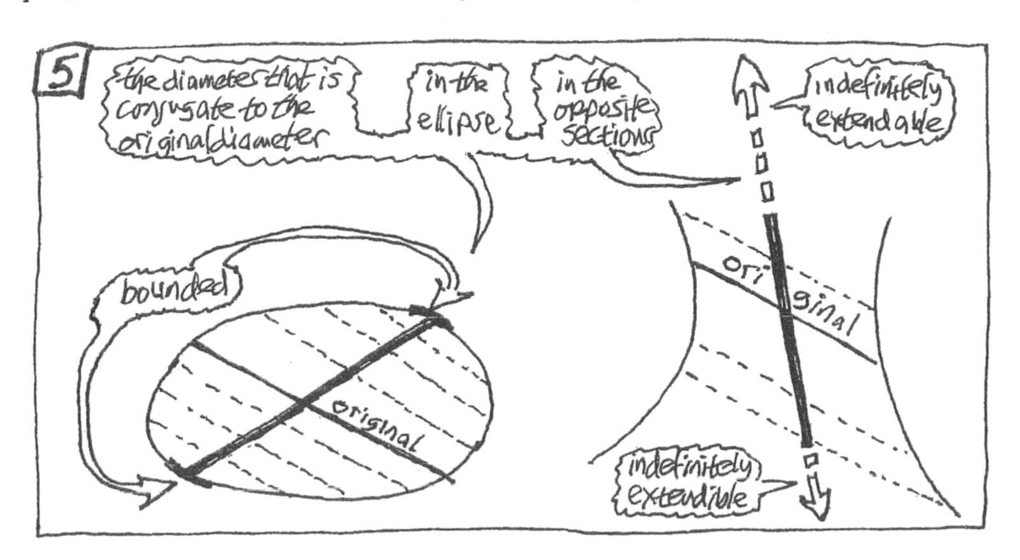

The new diameter that is conjugate to the original diameter has a definite size in the ellipse, though not in the opposite sections; but if the opposite sections' conjugate diameter (the upright diameter) is assigned the definite size that corresponds to that of the ellipse's conjugate diameter, then Apollonius calls it a "second diameter." Apollonius thus invites the question whether what now seems to be an utter difference in the resemblance won't turn out to be less of a difference in the end than it now seems to be.

Just before where we are now, the 16th proposition gave us the conjugate diameter of the opposite sections—that is, it gave us, for the opposite sections, another diameter in *position*. But no *proposition* has yet given us in *size* another diameter for the opposite sections—that is, no *proposition* has yet given us a *second* diameter of the opposite sections. However, the last of the Second Definitions has just defined what it would be if it were presented in a proposition.

Just before the 16th proposition, the 15th proposition gave us (without identification) the conjugate diameter of the ellipse. In the ellipse, each conjugate diameter is the mean proportional between the other conjugate diameter and that other conjugate diameter's parameter. Now, after the Second Definitions, we can therefore say this: in the ellipse, the conjugate diameters are second diameters of each other.

The reason for the size of a second diameter in the case of the ellipse is clear. The ellipse bounds the diameter, which cannot extend beyond the closed curve because there are no ordinates on its outside. In the case of the opposite sections, however, there is no apparent reason for the size of a second diameter; the curves do not bound the conjugate diameter (and so there is also no apparent reason for the centering). There would be a reason for the size of a second diameter of the opposite sections, however, if it turned out that the line will do work, in the opposite sections, that is like the work done by a properly positioned line of that size in the ellipse.

In the case of the opposite sections it is also strange to use the term "diameter" in naming the straight line called the "*oblique (transverse)* diameter." It is strange to call it a diameter, because the ordinates inside the two hyperbolic lines are bisected not by it (the diameter itself), but by its extensions. However, the ordinates that are bisected by the other diameter (the *upright* diameter), while they are outside each of the two hyperbolic lines, *are* in the middle of *the pair taken together*. If it turned out that the line called the "oblique (transverse) diameter" of the opposite sections will do work that is like the work of the line called by that name in the ellipse, then that would also help to justify the terminology of Apollonius.

The new (conjugate) diameter as bounded in the ellipse can be shown to be a mean proportional between the original diameter (the oblique [transverse] side) and that original diameter's parameter. What the 15th proposition showed for the ellipse was not only that the new line was another diameter (that it bisected new ordinates) but also—and even in the first place—that it worked like a diameter in giving a proper deficiency when Apollonius contrived a line to be used as a parameter. (He contrived this line so that the original diameter would be a mean proportional between it and the new line that turned out to be the conjugate diameter.)

What the 16th proposition showed for the opposite sections was that the new line was another diameter—a bisector of new ordinates. What the 16th proposition did *not* show is that it will work like a diameter in giving a proper excess when a line is contrived to be used as a parameter. The definition of a second diameter might lead us to expect that what was *not* shown in the *16th* proposition for the opposite sections *will* be shown *eventually*. And, as a matter of fact, Apollonius will eventually show it. He will show that

> if a line is contrived so that
> the original diameter is a mean proportional between
> the second diameter (as defined) and this contrived line,
> then the contrived line will work like a parameter for the second diameter,
> giving a proper excess.

To learn that fact from him, how long after the 16th proposition must we wait? A long time. We shall have to wait until the final proposition of the book—the 60th proposition. The demonstration of the 60th proposition needs the 59th proposition, which itself needs what stretches from here to it.

Along this stretch will be found some similarities between the ellipse and the pair of hyperbolic lines constituting opposite sections. Beginning with the 38th proposition, we shall see similarities that will give us *some* justification for applying the name "second diameter" to a conjugate diameter of a certain size even when what we have is not an ellipse but opposite sections.

Finally, at the very end of Book One of the *Conics* (with the 60th proposition), we shall see that if we take, for opposite sections, conjugate diameters that are second diameters, then we can get something special. The 60th proposition at last will show something like what might be expected. Something *like* what might be expected—since it will be something that is *also* somewhat *un*expected. As things turn out in the last proposition of the First Book, the second diameter, with the line that is contrived for it, *will* give the proper excess—but *not* for the opposite sections. Rather, it will give it for another pair of hyperbolic lines that makes its initial appearance in the last proposition, which presents a new pair of hyperbolic lines that constitute opposite sections that are *conjugate to* the original opposite sections.

The two pairs of hyperbolic lines (the original opposite sections and the conjugate opposite sections), when taken all together as a single thing, will behave something like an ellipse. The original diameter of the original opposite sections turns out to be the second diameter of the conjugate opposite sections. And *this* diameter of the conjugate opposite sections has as *its* second diameter the original diameter of the original opposite sections.

Introducing the conjugate sections, the spectacular very last proposition of the First Book will bring one story to an end by beginning another. To continue on into the second story, the reader will have to read on into the Second Book of Apollonius's *Conics*.

Now, however, we must see how Apollonius continues with the first story—the one that we are still in the middle of. To understand how Apollonius moves on from where we are now, let us first consider, not how diameters are fixed in size, but rather how the sizes of their ordinates differ.

With the ellipse, the other diameter (the diameter that is conjugate to the original diameter) is inside the ellipse. The ellipse is closed, and so the conjugate diameter has a definite size, since beyond its endpoints (that is, outside the ellipse) there are no ordinates. For each of the conjugate diameters of an ellipse, the lines that are parallel to it and are bounded by the section (that is, the chords which are ordinatewise to the other conjugate diameter) get as small as you please as you take them ever farther from the center; while, as you take them ever closer to the center, you can get them ever larger, but never larger than that conjugate diameter. The ordinates of the other diameter, as well as of the original diameter to which it is conjugate, have a *maximum* size but *no minimum* size (fig. 6).

With the opposite sections, the other diameter (the upright diameter, which is conjugate to the original diameter) is outside each constituent hyperbola. It may be "inside" the pair (even though the pair is not "closed"), but it has ordinates as far out as you may choose to go. These lines (which are ordinatewise to the other conjugate diameter—that is, to the upright diameter—and are parallel to the oblique [transverse] diameter and are bounded by the sections) get as large as you please as you take them ever farther from the center; while, as you take them ever closer to the center, you can get them ever smaller, but never smaller than that (oblique [transverse]) conjugate diameter. The ordinates of the new conjugate diameter have a *minimum* size but *no maximum* size (fig. 7).

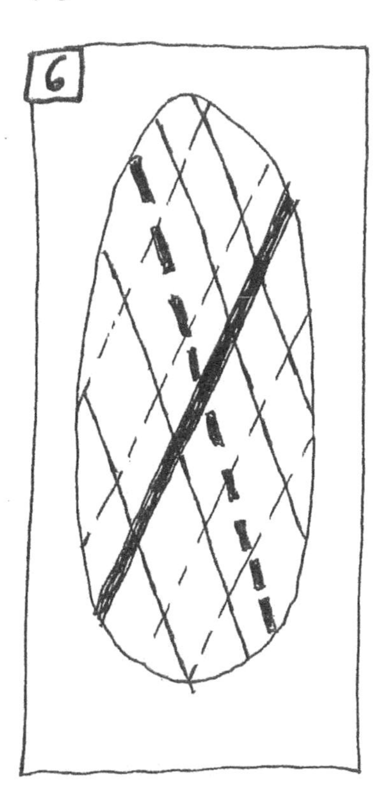

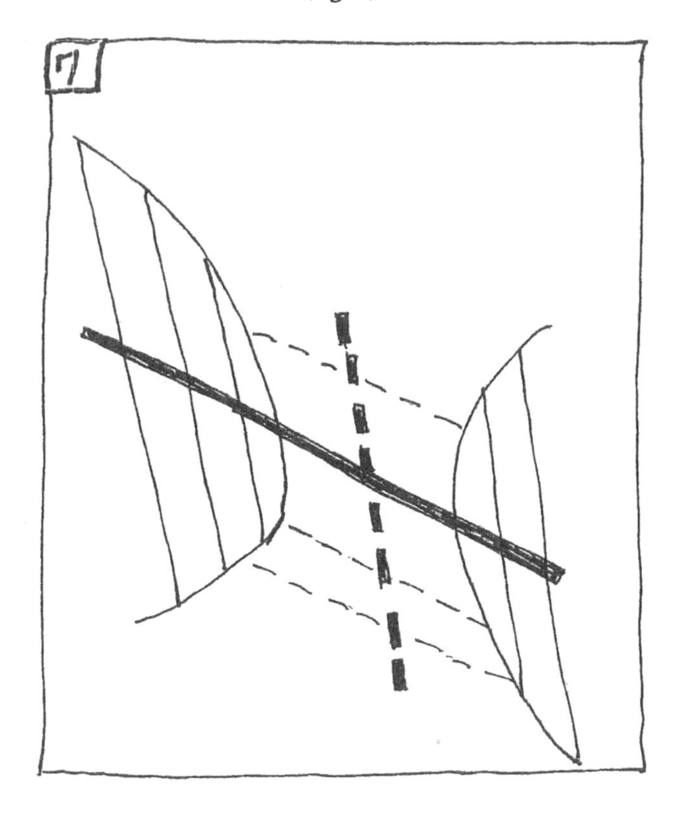

With respect to their ordinates, the pair of opposite sections is like an ellipse in reverse. The other conic sections (the single hyperbola and the parabola) do not have an ordinate either of maximum size or of minimum size. Each conjugate diameter of the ellipse has an ordinate of maximum size, but not of minimum size, while the upright diameter of the opposite sections has an ordinate of minimum size, but not of maximum size. Thus, the new diameter of the opposite sections (the upright diameter, which is conjugate to the original diameter) resembles and is different from the new diameter of the ellipse (the diameter that is conjugate to the original diameter, which is the oblique [transverse] side).

In addition, the opposite sections' two diameters taken together not only resemble but also differ from the two conjugate diameters of the ellipse taken together. Both of the conjugate diameters of the ellipse have an ordinate of maximum size but no minimum size. The conjugate diameters of the opposite sections, by contrast, differ from each other. One of them (the upright diameter) has an ordinate of minimum size but no maximum size; the other one, however, the original (oblique [transverse]) diameter, has, like the diameter of the parabola, no ordinate of maximum size or of minimum size. The upright diameter's ordinates are halves of the lines that are drawn from curve to curve and are parallel to the oblique (transverse) diameter. These get as large as you please as you take them ever farther from the center; as you take them ever closer to the center, they get as small as you please.

By considering how ordinates differ in size at different distances from the center, we can learn something about *tangents*. Let us now move on to see how.

CHAPTER 3. TOUCHINGS AND MEETINGS

17th through 19th propositions

#17 If, in a cone's section, from the vertex of the line there be drawn a straight [line] parallel to a [line] that has been drawn down ordinatewise, it will fall outside the section.
—Let there be a cone's section, whose diameter the [line] AB is. I say that the straight [line] from the vertex—that is, from the point A—which is drawn parallel to a line that has been drawn down ordinatewise will fall outside the section.
—For if possible, let it fall inside—as the [line] AC, say. Since, then, on a cone's section there has been taken a chance point, namely the [point] C, therefore the [line] from the point C, drawn, inside the section, parallel to a line that has been drawn down ordinatewise, will fall together with the diameter AB and will be cut in two [halves] by it. The [line] AC, therefore, being extended, will be cut in two [halves] by the [line] AB—which very thing is absurd. For, being extended, the [line] AC falls outside the section. Not so, therefore, that the straight [line] from the point A that is drawn parallel to a line which has been drawn down ordinatewise will fall inside the line [of section]. Outside, therefore, will it fall—on account of which very thing, it touches, [that is, is tangent to] the section.

#18 If, with a cone's section, a straight [line] that falls together [with it] fall, being extended both ways, outside the section, and there be taken some point inside the section, and through it, [that is, through that point] there be drawn a parallel to that [straight line] which falls together [with the section], then the [parallel line so] drawn, being extended both ways, will fall together with the section.
—Let there be a cone's section and let there be falling together with it the straight [line] AFB; and also let [this straight line], being extended both ways, fall outside the section; and also let there be taken some point inside the section—the [point] C, say; and also, through the [point] C let there be drawn parallel to the [line] AB the [line] CD. I say that the [line] CD, being extended both ways, will fall together with the section.
—For let there be taken some point on the section—the [point] E, say—and also let the [line] EF be joined. And also—since parallel is the [line] AB to the [line] CD, and with [line] AB there falls together a certain straight [line], namely the [line] EF—therefore also the [line] CD, being extended, will fall together with the [line] EF. And if (on the one hand) between the [points] E, F [is where it falls together with the line EF], it is evident (*phaneron*) that also with the section does it fall together; but if (on the other hand) out beyond the point E [is where it falls together with the line EF], it is evident that it will fall together with the section before [it falls together with the line EF]. Therefore the [line] CD—being extended toward those parts [where are points] D, E—falls together with the section. Likewise, then, will we show that [the line CD]—being extended also toward those parts [where are points] F, B—falls together with the section. The [line] CD, therefore, being extended both ways, will fall together with the section.

#19 In every section of a cone, whatever [line] from the diameter may be drawn parallel to a line that has been drawn down ordinatewise will fall together with the section.
—Let there be a cone's section whose diameter the [line] AB is; and also let there be taken some point on the diameter—the [point] B, say; and also, through the [point] B, let there be drawn, parallel to a [line] that has been drawn down ordinatewise, the [line] BC. I say that the [line] BC, being extended, will fall together with the section.

—For let there be taken some point on the section—the [point] D, say. But there is also the [point] A on the section; therefore the straight [line] joined from the [point] A to the [point] D will fall inside the section. And also—since the straight [line] which is drawn from the [point] A, parallel to a [line] that has been drawn down ordinatewise, falls outside the section, and there falls together with it the [straight line] AD, and parallel to the [straight line] that has been drawn down ordinatewise is the [line] BC—therefore also the [line] BC will fall together with the [line] AD. And if (on the one hand) between the points A, D [is where the line BC falls together with the line AD], it is evident that also with the section will the [line BC] fall together; but if (on the other hand) out beyond the [point] D [is where the line BC falls together with the line AD]—as, say, at the [point] E—it is evident that it will fall together with the section before [it falls together with the line AD]. The straight [line] from the point B that is drawn parallel to a [line] that has been drawn down ordinatewise will, therefore, fall together with the section.

The 17th proposition presents a conic section with its diameter (AB), through whose vertex (point A) a straight line is drawn parallel to an ordinate. The demonstration shows that such a straight line must fall outside the section, and thus must be tangent to the curve at the point A.

Note that the ordinates, and hence the tangent, will not always be perpendicular to the diameter of the conic section. Only when the curve is a *circle* do the ordinates of *any* diameter of the section have to be perpendicular to it.

A circle with its diameter, through whose endpoint a straight line is drawn perpendicular to the diameter, is presented in Euclid's *Elements* III.16. Euclid demonstrates by *reductio* that such a straight line must fall outside the circle, and thus be tangent to the circle at the diameter's endpoint. Here, in the demonstration of this 17th proposition, Apollonius too proceeds by *reductio*, as follows:

If the straight line were to fall inside the curve,
and thus were to intersect it at some point (C),
then—since this straight line is
drawn from a point (C) on the conic section, parallel to an ordinate—
it would (by the 7th proposition) meet the diameter (AB) and be bisected by it.

But it is absurd to say that AB would bisect CA prolonged,
since CA prolonged would (by the 10th proposition) fall outside the section.

So, since the straight line that is drawn from point A parallel to an ordinate
cannot fall inside the curve,
it must fall outside—
and thus be tangent to the section.

Let us not dwell upon the demonstrations of the other propositions in this sequence, but concentrate rather upon what they show.

With the 17th proposition, we are no longer considering conic sections of this or that particular sort. We have turned to considering straight lines that touch *any* such curve. But why be interested in tangents—what do they tell us about the curved lines that interest us?

We got hold of the curved lines by talking about the relation between straight lines in them—the relation between diameters and their ordinates. But now, though we are still looking at the curves obtained by cutting a cone, we have left the cutting of the cone behind us, and so we need a new reference line for the orientation of the parallel lines called ordinates. Our question might therefore be restated: what does the *tangent* at the endpoint of a diameter have to do with the relation between that diameter and its *ordinates*?

The 17th proposition tells us this (fig. 1):

if we move an *ordinate* along the diameter toward the vertex, keeping it parallel to its original location, then the ordinate, just as it reaches the vertex-point, will turn into a *tangent*.

Next, the 18th proposition tells us this (fig. 2):

if we move a *tangent* into the section, keeping it parallel to its original location, then the tangent will turn into one *chord* after another.

We know that we shall get parallel chords, but we do not yet know whether these will be ordinates of some diameter—since we do not yet have any assurance that the midpoints of these chords lie on a straight line. So, we can go from ordinate to tangent, but not yet from tangent to ordinate.

But at the end of the 18th proposition we do not even know whether we shall meet the section if we go along a straight line that is drawn ordinatewise from *any* point on the diameter. Earlier, we learned this:

we shall meet the *diameter* if we go along a straight line, parallel to an ordinate, from any point on the *section*.

Now the 19th proposition tells us this:

we shall meet the *section* if we go along a straight line, parallel to an ordinate, from any point on the *diameter*.

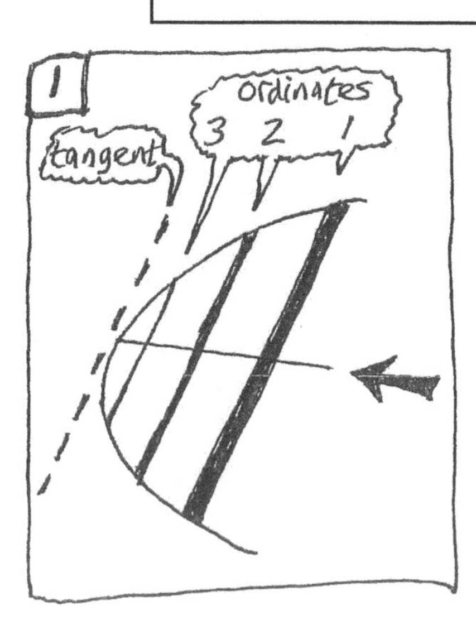

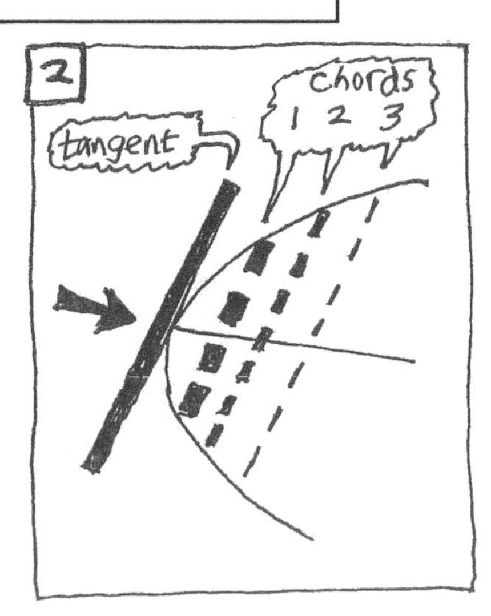

An overview of what these last three propositions have told us, is given in figure 3.

The figures in the explication of these last three propositions have presented matters of general significance without presenting depictions of particulars. For illustrative diagrams that provide help in the reading of Propositions 17 through 19, see the Appendix at the end of this volume.

The 17th through 19th propositions have considered the touchings and meetings of lines. The 22nd through 36th propositions will also do so. The two propositions that intervene between these sequences will now be considered together, as we examine first their enunciations, then their demonstrations, and finally their significance. Here, then, are the 20th and 21st propositions.

3 proposition number	IF parallel to this straight line	& goes through this point	THEN		Later on, compare with proposition number
17th	an ordinate	the endpoint (vertex) of diameter	falls outside both ways --i.e. tangent		
18th	a tangent	any point inside section	meets section both ways --i.e. chord		27th
19th	an ordinate	any point on diameter	meets section		28th

20th and 21st propositions

(The figures in the subsequent explication of these two propositions will present matters of general significance without presenting depictions of particulars. For illustrative diagrams that provide help in the reading of Propositions 21–21, see the Appendix at the end of this volume.)

#20 If, in a parabola, from the section two straight [lines] be drawn down to the diameter ordinatewise, then as the squares from them are to each other, so will be to each other the [lines] cut off by them from the diameter, on the side toward the vertex of the section.
—Let there be a parabola whose diameter the [line] AB is; and also let there be taken some points on it, [that is, on the parabola]—the [points] C, D, say; and also, from the points C, D let there be drawn down ordinatewise, to the [line] AB, the [lines] CE, DF. I say that as is the [square] from DF to the [square] from CE so is the [line] FA to AE.
—For let there be, as the [line] along which the [lines] that are drawn down have the power, the [line] AG. Equal, therefore, is (on the one hand) the [square] from the [line] DF to the [rectangle] by FAG, and (on the other hand) the [square] from the [line] CE to the [rectangle] by the [lines] EAG. Therefore as is the [square] from DF to the [square] from CE so is the [rectangle] by FAH to the [rectangle] by EAG. But as is the [rectangle] by FAG to the [rectangle] by EAG so is the [line] FA to AE. And also, therefore, as is the [square] from DF to the [square] from CE so is the [line] FA to AE.

#21 If, in an hyperbola or ellipse or circle's periphery, straight [lines] be drawn ordinatewise, to the diameter, then the squares from them will be (on the one hand) to the areas (*khôria*) contained by the [lines] taken off by them, [stretching toward the bounds of the oblique side of the figure (*tês plagias pleuras tou eidous*), as is the upright side of the figure (*hê orthia pleura tou eidous*) to its oblique [side] (*tên plagian*); and [those squares] will be (on the other hand) to each other, as are to each other the areas contained by the straight [lines] taken off as has been said. [To clarify: each area is contained by two lines taken off from the diameter by the ordinatewise line, and each of the two lines stretches, along the diameter, from the point where the ordinatewise line meets the diameter to one of the two endpoints of the oblique side.]
—Let there be an hyperbola or ellipse or circle's periphery whose diameter is what the [line] AB is, and whose [line] along which the [lines] that are drawn down have the power is what the [line] AC is; and also let there be drawn down to the diameter ordinatewise the [lines] DE, FG. I say that (on the one hand) as the [square] from the [line] FG is to the [rectangle contained] by the [lines] AGB so the [line] AC is to AB, and (on the other hand) as the [square] from the [line] FG is to the [square] from the [line] DE so the [rectangle] by the [lines] AGB is to the [rectangle] by the [lines] AEB.
—For let there be joined the [line] BC that delimits (*diorizousa*) the figure (*to eidos*); and also, through the [points] E, G let there be drawn parallel to the [line] AC the [lines] EH, GK. Equal, therefore, is (on the one hand) the [square] from the [line] FG to the [rectangle] by KGA, and (on the other hand) the [square] from the [line] DE to the [rectangle] by HEA. And—since as is the [line] KG to GB so is the [line] CA to AB, and as is the [line] KG to GB so is (with the [line] AG being taken as common side) the [rectangle] by KGA to the [rectangle] by BGA—therefore as is the [line] CA to AB so is the [rectangle] by KGA (that is, the [square] from FG) to the [rectangle] by BGA. On account of the same things, then, also as is the [square] from DE to the [rectangle] by BEA so is the line CA to AB. And also, therefore, as is the square from the line FG to the rectangle by BGA so is the square from DE to the rectangle by BEA; alternately, as is the [square] from FG to the [square] from DE so is the [rectangle] by BGA to the [rectangle] by BEA.

[The Greek word *diorizousa* has as its root the word for a boundary (*horos*), which is elsewhere rendered as "definition." If *horoi* were rendered not as "definitions" but rather as "terms," then *diorizousa* might best be rendered as "determines."]

20th and 21st propositions: enunciations

The 20th and 21st propositions tell us something important about the shape of conics. The 20th does so for the parabola; the 21st, for the other ones, which are the conic sections that have a center.

The 20th tells us how a parabola grows. Go along its diameter away from the vertex: when you have gone, say, 9 times as far, then the ordinate will be 3 times as large. To speak more generally: as the abscissas are to each other, so also will be the squares on their ordinates. That is to say, the ratio of the abscissas will be the duplicate of the ratio of the ordinates. In other words, the parabola's shape is such that the ratio of the *ordinates* is the *subduplicate* of the ratios of the *abscissas*.

The 21st proposition shows us the situation in the other sections. It is similar but more complicated. Again, the *ordinates* are in the ratio that is the *subduplicate* of a certain ratio. But, whereas the parabola's ordinates are in the ratio which is the subduplicate *simply* of the ratio of the *abscissas*, now in any of the other sections (hyperbola, ellipse, or circle) the ordinates are in the ratio which is the subduplicate of a ratio that is the *compound* of two ratios. *One* of these two ratios is the ratio of the *abscissas*, but the *other* ratio is the ratio of the lines drawn from the non-vertex endpoint of the abscissas to the non-vertex endpoint of the oblique (transverse) side. These are the lines that result when the abscissa is (in the case of the hyperbola) added on to, or (in the case of the ellipse or the circle) taken away from, the oblique (transverse) side (fig. 4).

We have seen these lines before. When we brought together the different analyses in characterizing the various conic sections (chapter 4 of Part Three of this guidebook), we saw that such a line, which we use when the section is not a parabola (that is, when the equal rectangle's width is not constant) has, to the equal rectangle's width, the same ratio that the oblique (transverse) side has to the upright side (which is a ratio that is constant).

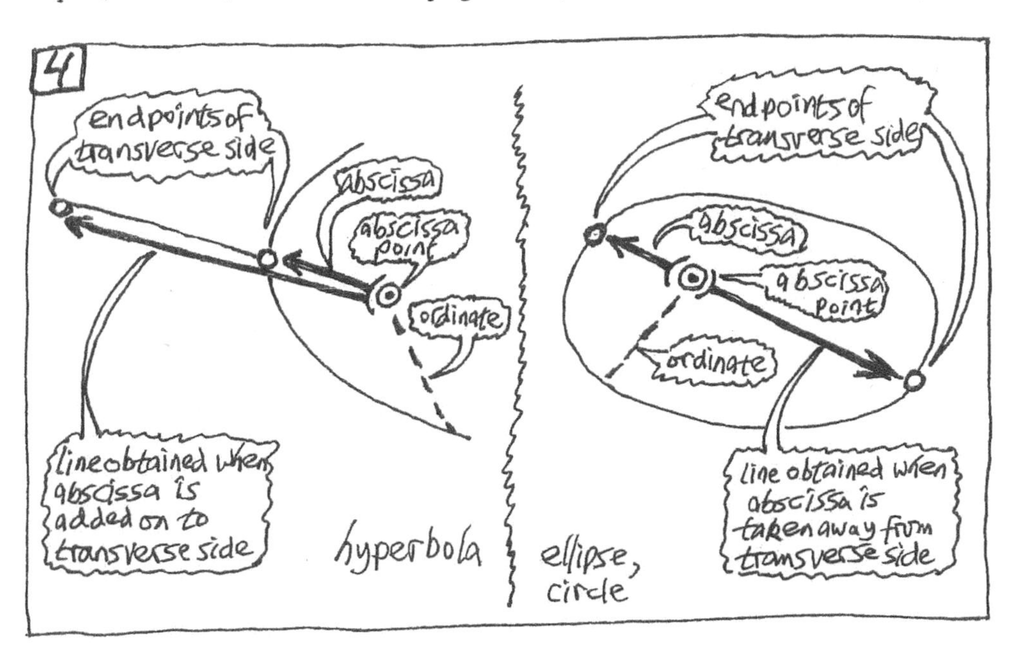

The situation in the parabola is more like that in the other conic sections than might seem from a quick glance at the enunciation. It is true that the ratio in the parabola is stated as a ratio of lines, whereas in the other sections it is stated as a ratio of rectangles. But any ratio of rectangles is the same as the ratio compounded of two ratios of lines—the one ratio being that of the rectangles' heights, and the other ratio being that of their widths. And the ratio of the ordinates in the parabola, like that in the other conic sections, is the subduplicate of a ratio that is compounded of two ratios, one of them being the ratio of the abscissas. The other ratio in the parabola, however, is the ratio of a line to its equal—and such a ratio (the one that any magnitude has to its equal) will, when compounded with any other ratio, give that very same other ratio. In the parabola, therefore, the compounded ratio is the same as the simple ratio of the abscissas because in the parabola, the *non* abscissa side of that rectangle which is equal to the square on the ordinate is the *constant* upright side.

The curves that have a center (and thus have traditionally been called "central" sections) resemble each other in their difference from the parabola: each of them exhibits a version of our fundamental circle property. Our study of the conics has been built upon this property of the circle: a square erected on a half-chord that is perpendicular to a circle's diameter is equal to the rectangle contained by the segments into which the diameter is split by that perpendicular half-chord. In the special case of the circle, a half-chord can be an ordinate only if it is perpendicular to its diameter. In the other sections with a center, however, while the half-chord that is an ordinate can be perpendicular to its diameter, it does not have to be perpendicular. In all the sections with a center, and not only in the circle, the segments into which the ordinate splits the diameter are the lines drawn from the non-vertex endpoint of the abscissa to the endpoints of the oblique (transverse) side. The line to its vertex endpoint is the abscissa, and the line to its non-vertex endpoint is the line that results when the abscissa is added on to, or is taken away from, the oblique (transverse) side. (Fig. 5.)

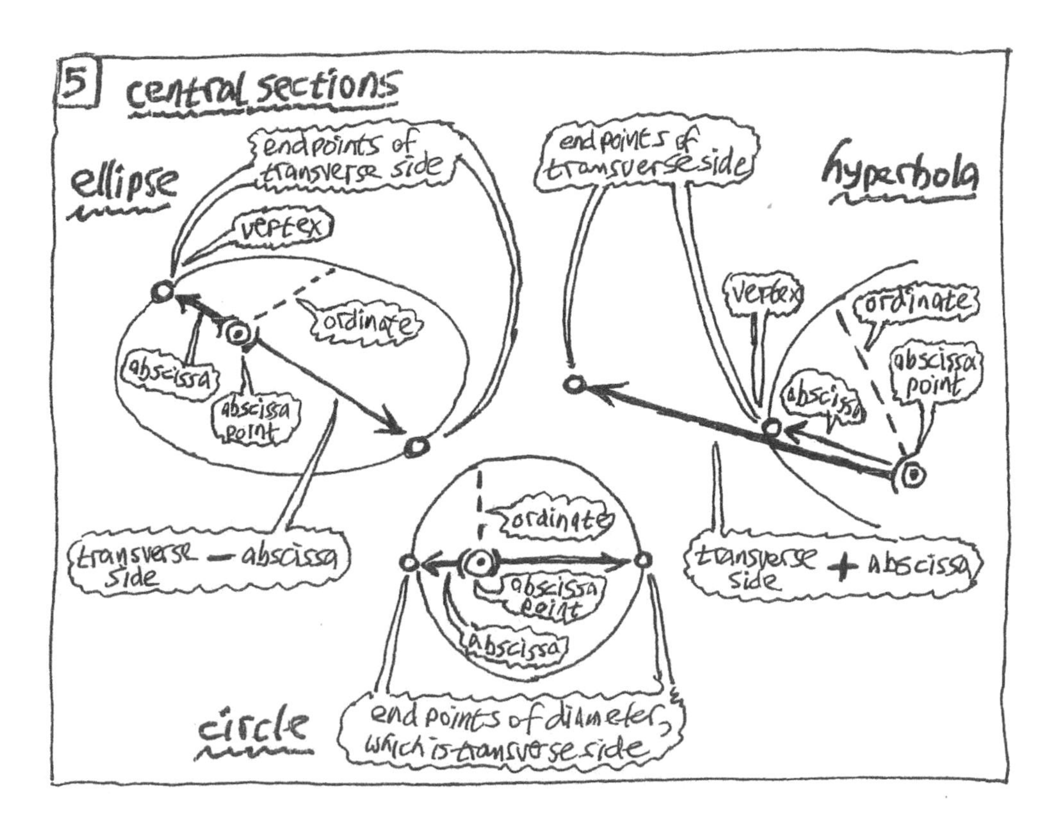

The circle is special in several respects. Not only must its ordinates be perpendicular to its diameter, and any diameter of it be equal to any other diameter of it, but a circle's diameter must also be equal to its upright side. This last is because the proportionality of the squares and rectangles that is found in all the central sections is, in the special case of the circle, an equality. (In all the central sections, including the circle, the square on the ordinate has, to the rectangle contained by the lines from the abscissa-point to the endpoints of the oblique (transverse) side, the same ratio that the upright side has to the oblique (transverse) side; but in the special case of the circle the square and the rectangle are equal, and therefore also equal are the upright side and the oblique (transverse) side.) (Fig. 6.)

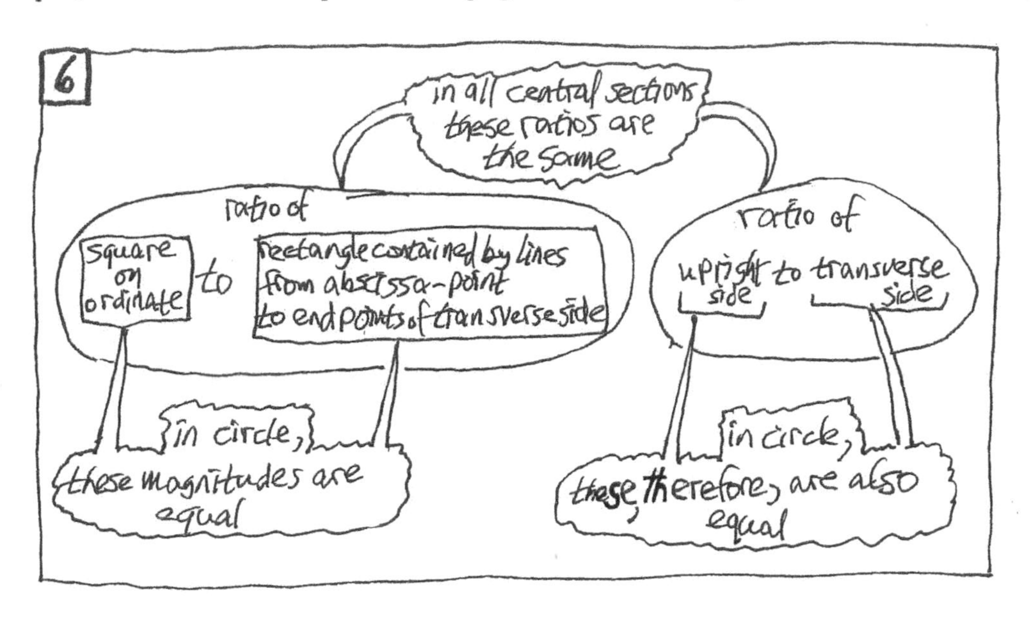

20th and 21st propositions: demonstrations

20th proposition (parabola):

According to the property that defines the parabola in the 11th proposition,
the square on the ordinate is equal to the rectangle that is
contained by the abscissa and the upright side;

hence, the squares on the ordinates are to each other as those rectangles are.

But since all those rectangles have the same width
(namely, the constant upright side),
they are therefore to each other simply as their other sides are—
and these other sides are their abscissas. (See figure 7.)

21st proposition (hyperbola, ellipse, circle):

The enunciation itself says two things, so the demonstration has two parts.

Now, in the sections other than the parabola, we do not have
the equality of the square and the rectangle in question,
the rectangle's width no longer being the constant upright side;
however, what we can do is show the constancy of their ratio.
This was not what was shown in their defining 12th and 13th propositions,
but it does follow from those propositions.
That it does follow, is what needs to be shown first,
in order to be able to show the ratio of the squares to each other.

first part:

Figure 8 depicts how to demonstrate that
the constant ratio which the upright side has to the oblique (transverse) side
is the same ratio which the square on the ordinate has to the rectangle having
as its one side the abscissa, and as its other side the line that stretches
from the non-vertex endpoint of the abscissa
to the non-abscissa end of the oblique (transverse) side
(that is to say: the line that results when the abscissa
is—in the hyperbola—added on to,
or is—in the ellipse—taken away from, the oblique (transverse) side). (See figure 8.)

second part:

It having been shown in the first part that
the ratio of each of those squares to its counterpart rectangle is
the same as the ratio of any other of those squares to its own counterpart rectangle,

it follows that the ratio of the squares to each other is
the same as the ratio of their counterpart rectangles to each other. (See figure 9.)

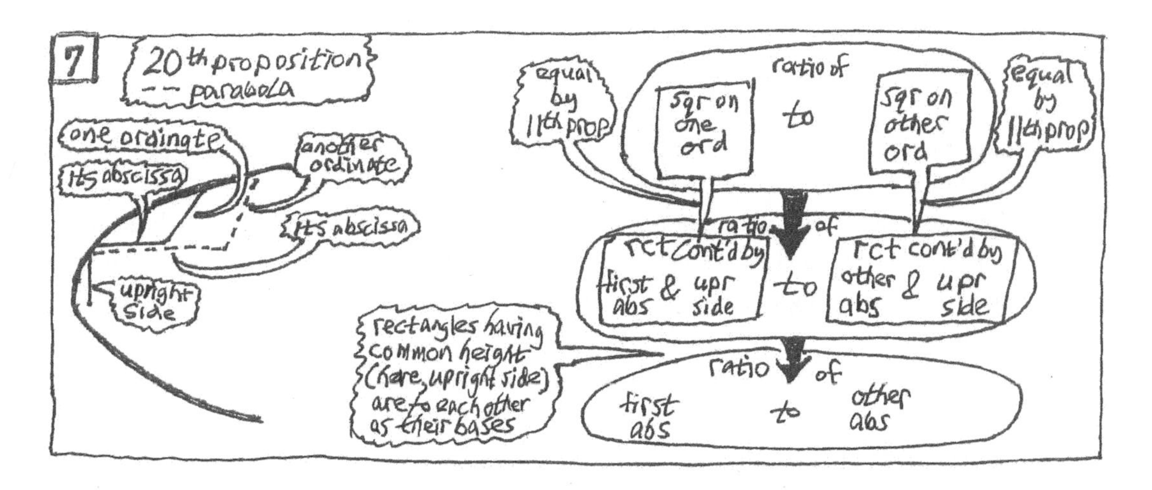
7
20th proposition -- parabola
one ordinate
its abscissa
another ordinate
its abscissa
upright side
equal by 11th prop
ratio of
sqr on one ord
to
sqr on other ord
equal by 11th prop
ratio of
rct cont'd by first & upr side
to
rct cont'd by other & upr side
rectangles having common height (here, upright side) are to each other as their bases
ratio of
first abs
to
other abs

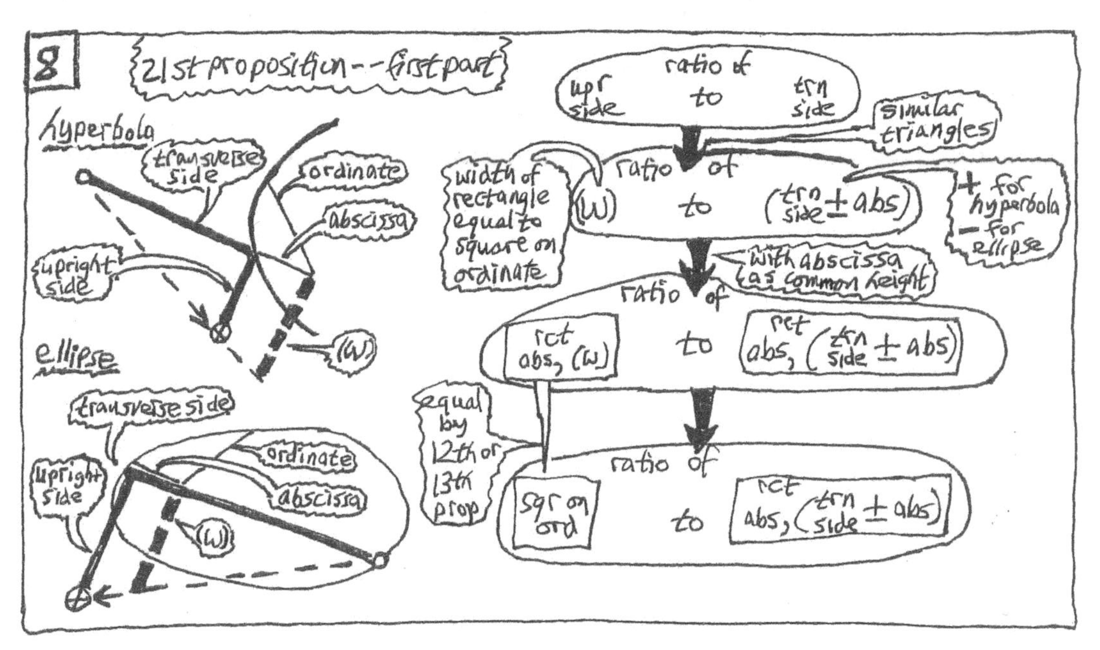
8
21st proposition -- first part
hyperbola
transverse side
ordinate
abscissa
upright side
(W)
ellipse
transverse side
ordinate
abscissa
upright side
(W)
ratio of
upr side
to
trn side
similar triangles
width of rectangle equal to square on ordinate
ratio of
(W)
to
(trn side ± abs)
+ for hyperbola
− for ellipse
with abscissa as common height
ratio of
rct abs, (W)
to
rct abs, (trn side ± abs)
equal by 12th or 13th prop
ratio of
sqr on ord
to
rct abs, (trn side ± abs)

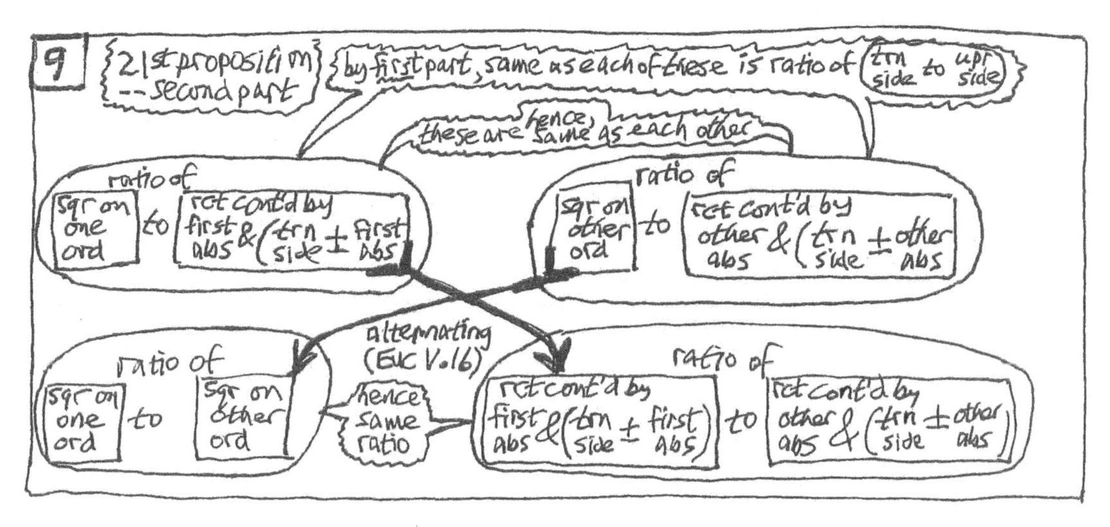
9
21st proposition -- second part
by first part, same as each of these is ratio of (trn side to upr side)
hence, these are same as each other
ratio of
sqr on one ord
to
rct cont'd by first abs & (trn side ± first abs)
ratio of
sqr on other ord
to
rct cont'd by other abs & (trn side ± other abs)
alternating (Euc V.16)
ratio of
sqr on one ord
to
sqr on other ord
hence same ratio
ratio of
rct cont'd by first abs & (trn side ± first abs)
to
rct cont'd by other abs & (trn side ± other abs)

these properties in relation to the defining propositions

The 20th and 21st propositions make it possible to relate the abscissas and the ordinates of each conic section without any mention of an upright side. They are derived by Apollonius from the 11th, 12th, and 13th propositions, which do use an upright side to relate the abscissas and the ordinates of each conic section.

The properties here demonstrated *can* be demonstrated, however, *without* relying on the 11th, 12th, and 13th propositions. In other words: if a cone is cut by a plane that is oriented in any of the various possible ways, then the abscissas and ordinates of the resulting curve can be related *without* introducing *any parameter*.

Figures 10 and 11 depict, first, a demonstration for the case of the parabola (showing the property of the 20th proposition without relying on the 11th proposition), and then a demonstration for the case of the hyperbola, ellipse, or circle (showing the property of the 21st proposition without relying on the 12th proposition or the 13th).

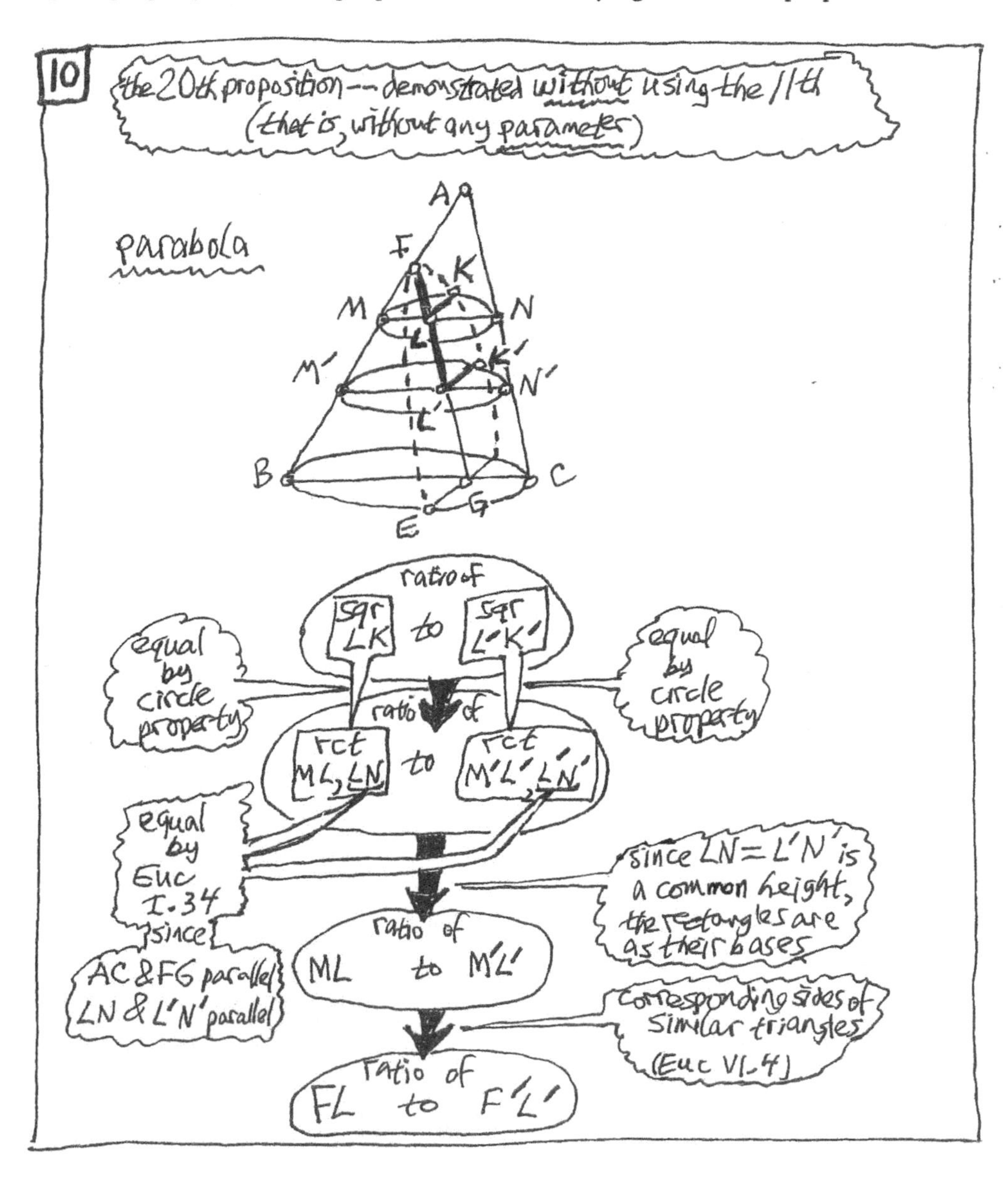

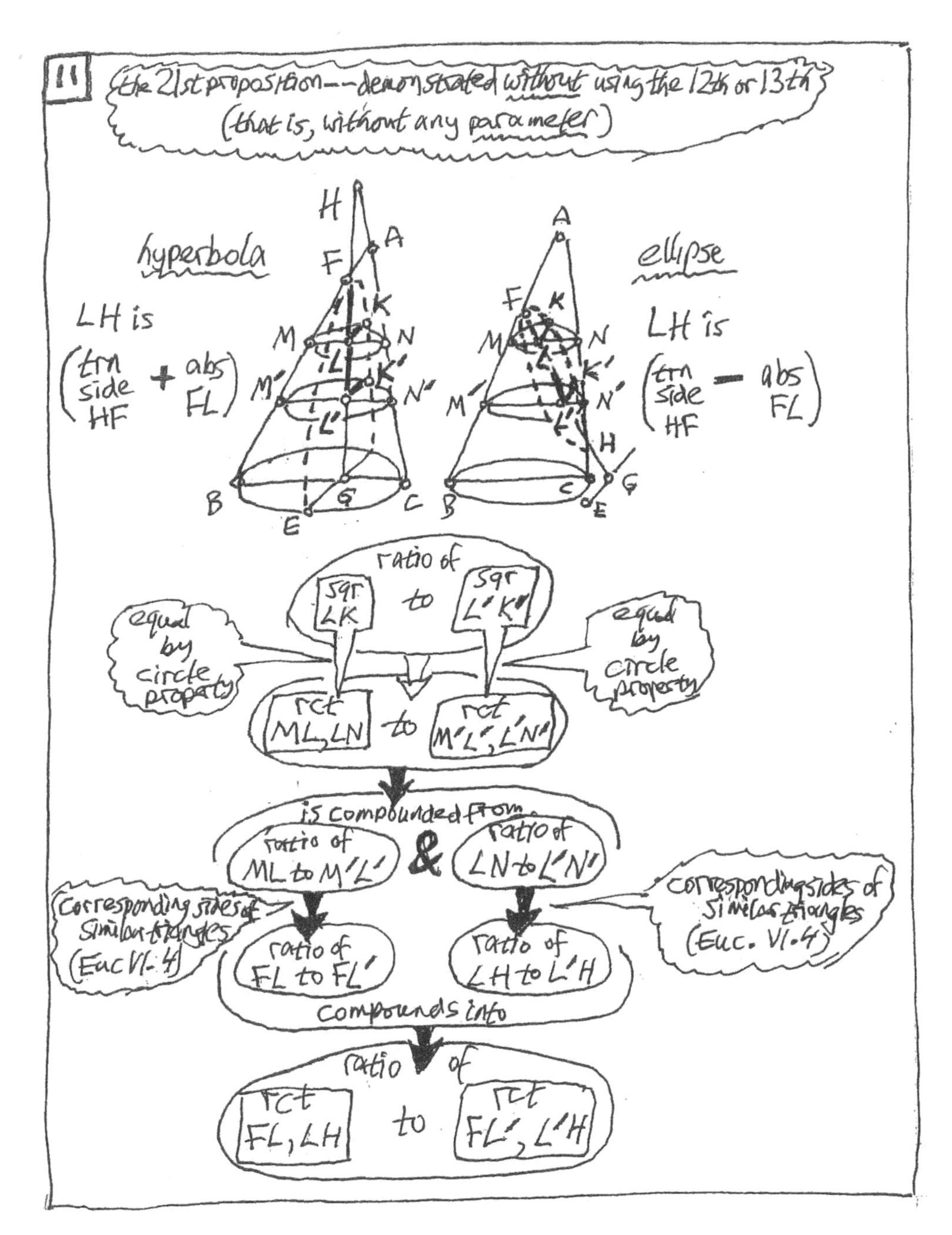

Later on, before the propositions numbered in the fifties (which will return us to the cone), there will be propositions that rely on the early defining propositions (namely, on propositions 11, 12, and 13) to demonstrate properties that could instead be demonstrated by relying only on propositions 20–21 (which themselves, as we have just seen, could be demonstrated without any reliance on the early defining propositions).

Properties of the parabola, for example, which will be demonstrated using the 11th proposition in the 26th proposition, and in the 27th, and in the 32nd, could be demonstrated without any reference to the 11th. Let us skip ahead to see how this could be done for the parabola in the 26th proposition.

In his 26th proposition, Apollonius will show (fig. 12) that

> if a straight line EF is drawn parallel to a parabola's diameter BGA,
> then, when EF is prolonged, it will meet the parabola at some point K.

The gist of the demonstration is this:

> Take some point E on the parallel straight line EF; and
> through it, draw, ordinatewise to the diameter, a line EG.
> Now, there must be some ordinate CH that is
> greater than EG (the line drawn ordinatewise).
> CH, then, being both greater than EG and also parallel to it,
> must be cut by EF prolonged.
> But CH can be cut by EF only if, at some point K, EF cuts the section too.

The demonstration requires showing that that there is some ordinate CH that is greater than line EG. To show it, Apollonius in his 26th proposition uses the parameter provided by his 11th proposition. It can, however, be shown without using the parameter at all, but using instead what Apollonius demonstrates in his 20th proposition. The two ways of showing it, which are presented in the figure are compared in what follows.

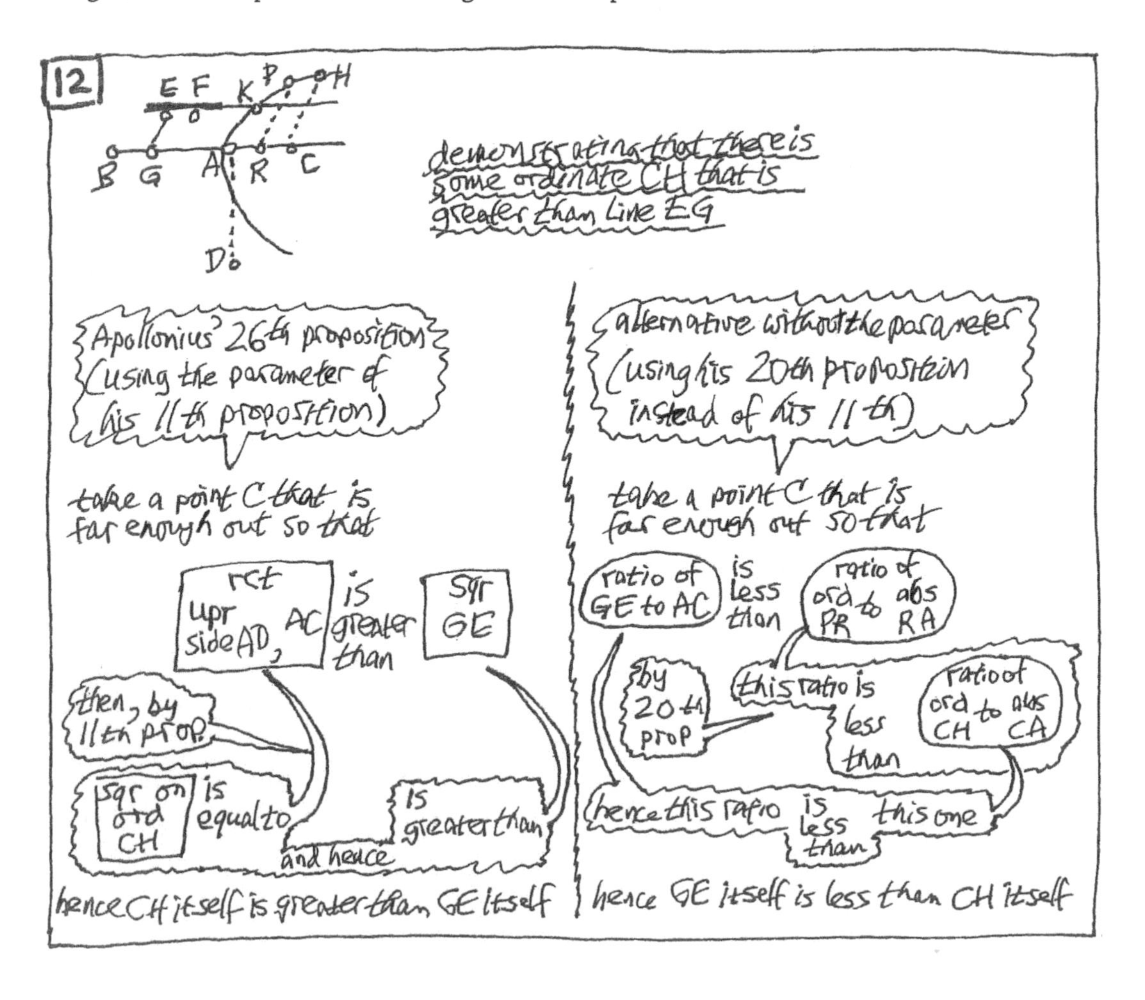

Apollonius— *using proposition 11*	*alternative—* *using proposition 20*
	On the section, take some point P; and from P, draw ordinate PR; and
On the diameter, take some point C that is far enough out so that rct [parameter AD, abscissa AC] is greater than sqr [GE]; and	on the diameter, take some point C that is far enough out so that ratio of GE to abscissa AC is less than ratio of ordinate PR to abscissa RA.
from C, draw ordinate CH.	from C, draw ordinate CH.
Then, *by proposition 11*, sqr [ordinate CH] is equal to rct [parameter AD, abscissa AC]; and hence is greater than sqr [GE].	Then, by *proposition 20*, ratio of ordinate PR to abscissa RA is less than ratio of ordinate CH to abscissa CA; and hence ratio of GE to abscissa AC is less also.
Hence ordinate CH itself is greater than GE itself.	Hence GE is less than ordinate CH.

So, Apollonius could have characterized the various cuttings of the cone by presenting the properties that—for a reason that he does not state—he delays presenting until here in the 20th and 21st propositions. We have seen that propositions 11–12–13 are not necessary for deriving propositions 20–21, and also that 20–21 will suffice for the immediate sequel. Why, then, did Apollonius not just leave out the earlier propositions? Because without 11–12–13, whatever may depend on the parameter would have been deprived of all support.

When Apollonius characterized the various conic sections in 11–12–13, he was doing more than merely relating their abscissas and ordinates. For merely doing that, the parameter or upright side is an unnecessary complication. So, just what was gained by bringing in the upright side to characterize the various conic sections in the 11th, 12th, and 13th propositions?

Consider the case of the parabola. The 20th proposition tells us about the relation between ordinate and abscissa along the diameter of every parabola. From this proposition we learn that the ratio of the ordinates is the subduplicate of the ratio of the abscissas. But this proposition does not tell us what those ratios have to do with the axial triangle and with the line which the cutting plane slices into it. That is to say, the property is presented without reference to what distinguishes one parabola from another. By contrast, the 11th proposition presents a property which enables us, in the very act of saying what makes every parabola a parabola, to speak also of what distinguishes various parabolas from each other. The 11th proposition characterized the parabola by contriving a parametric line into which it built the relation of lines in the particular axial triangle.

The property shown by the 11th proposition is the equality between the square on the ordinate and the rectangle contained by the abscissa and the parameter. The 20th proposition does not determine the particular size of a parameter to which the parabola can be referred. All that the 20th proposition can be used to do is show the equality between the square on the ordinate and a rectangle contained by the abscissa and *some* straight line of *unspecified* size. That can be done in the following way:

> The squares on the ordinates are as the abscissas,
> and if we make rectangles having the abscissas as one side
> and having *any* common other side—call it the width (w)—
> then the abscissas will be as the rectangles;
> and so, the squares will be as the rectangles.
> Then (alternating as in Euclid VI.1)
> any square will be to its rectangle as any other square is to its rectangle.
>
> That constant ratio of square to rectangle will be greater or less
> depending on the size of the line chosen to be the common width (w).
> For *any* parabola, there is *some* size width that will
> make the ratio a ratio of *equality*.

That general truth about parabolas will not, however, specifically determine what will be the size of that line in any particular parabola. That particular size is determined by the particular sides of the axial triangle, together with the particular placement of the line which the cutting plane slices into it. While the 20th proposition gives the relation between ordinate and abscissa for all parabolas generally, the 11th proposition gives it for each parabola in particular.

The 11th proposition thus tells us more than we are told by the 20th. The same is true of what we are told by the 21st proposition in contrast to what we are told by the 12th proposition about the hyperbola and by the 13th about the ellipse.

We shall see, when we reach the propositions numbered in the fifties, that the particularity connected with the parameter has something to do with the movement of reversal mentioned earlier in this guidebook—namely, the movement from showing that

> if a line can be obtained by cutting a cone in a certain way, then it has a certain relation between ordinate and abscissa,

to showing that

> if a line has a certain relation between ordinate and abscissa, then it can be obtained by cutting a cone in a certain way.

In other words, it is the use of the parameter to relate the ordinate and abscissa that makes possible the movement of the book—from showing that the lines which are the various conic sections have certain properties, to showing that the only lines which have those properties are the conic sections.

The propositions that characterized the conic sections were a set-up. For all we knew when we approached those curves with the question "How are ordinates and abscissas related?" a proportion rather than an equation would have sufficed as an answer. Instead of contriving a parameter—that is, instead of getting the line that is the other side of a rectangle which has the abscissa as one side and is equal to the square on the ordinate—we might have been satisfied by the outcome of reasoning like that depicted in figure 13, which brings together what we saw in figures 10 and 11.

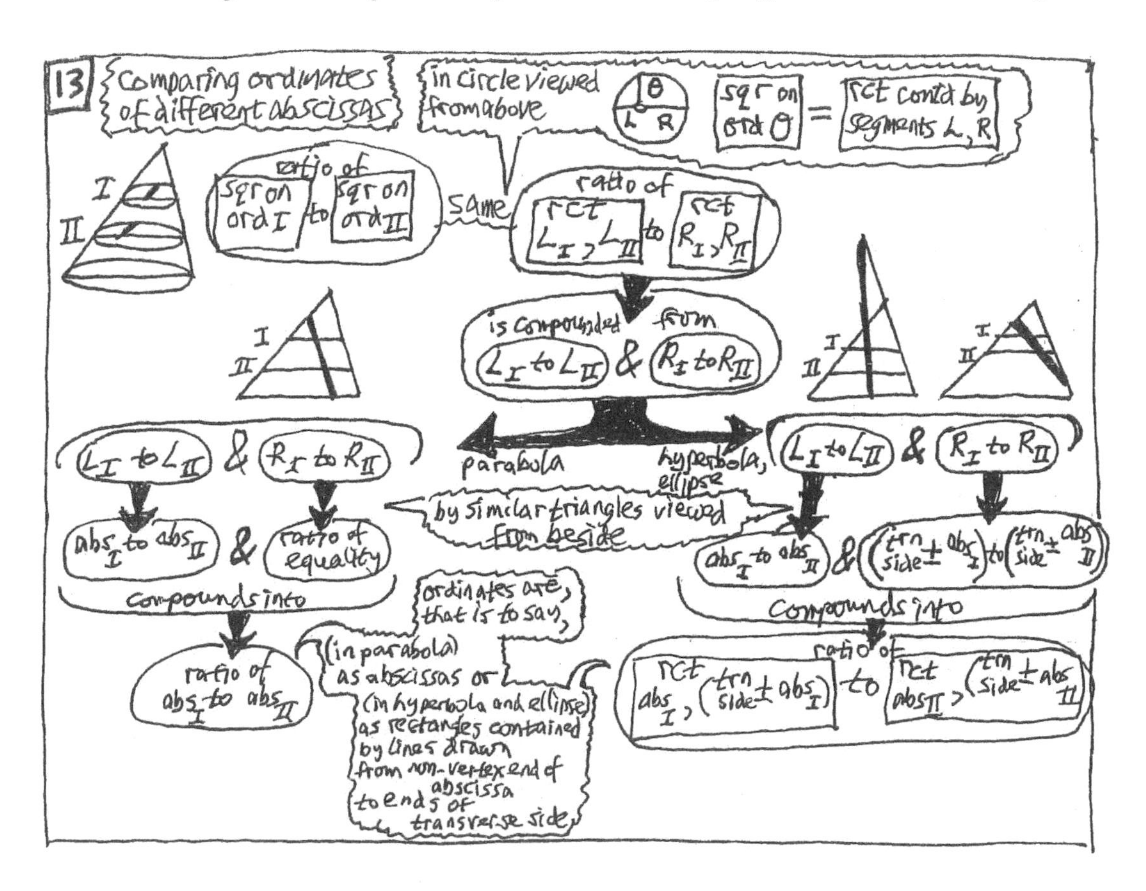

These propositions—the 20th and 21st—show that even without the use of a parameter it is still possible to relate the ordinates and the abscissas of a conic section. Even without the use of a parameter, that is to say, a relationship involving two ordinates can be related to a relationship involving their two abscissas (this latter relationship being one which, in the parabola, involves the abscissas alone; and, in the hyperbola or the ellipse, involves the abscissas taken together with the lines that result when the abscissas are added on to or are taken away from the oblique (transverse) side).

But if we do forego the use of the parameter, then we must forego as well the equality of square and rectangle which gives us that particular relationship between any one ordinate and its abscissa by which any particular parabola or hyperbola or ellipse is distinguished from another curve of the same sort. That is a strong reason for introducing the parameter. It is not, however, a reason for having different definitions of the parameter for conic sections of different sorts.

Very different ratios were used to contrive the parameters of the original diameters for conic sections of different sorts, back when the 11th through 14th propositions characterized the conic sections. But, as we saw back then, it would have been possible for Apollonius to use, for the original diameters of conic sections of the different sorts, a *single* definition for the parameters of sections of *all* the different sorts. That was, however, a road not taken. Apollonius could have avoided using several very different ratios to contrive those parameters, but he used them anyway. Thinking about the 20th and 21st propositions now helps us to see why he did so.

Later, we shall see where Apollonius is going *to*, with the upright side. We shall see how in the end the upright side is needed in order to obtain, as the outcome of a cutting of a cone, one particular conic section as distinguished from any other one of the same sort. Now, however, we receive an indication of where it is that Apollonius is coming *from*, with the upright side—and therefore why it is, that for conic sections of different sorts, he contrives the upright side by using several very different ratios.

If, to determine the size of the upright side, Apollonius had used that single parametric definition which we considered as an alternative to his complex of several definitions, he might have made his presentation simpler. But would he have been doing us a favor? Although the complexity of his presentation made for some initial obscurity, that obscurity was an invitation rather than a barrier. The simpler presentation would have covered up his tracks, keeping us from walking in his footsteps—and thus from discovering for ourselves what lies behind or underneath his presentation. Rather than doing that, Apollonius told us enough for us to figure out what he did not tell us in so many words himself. He was a hard teacher, but perhaps there was some benevolence in his hardness.

We have had to go through seven propositions after completing the propositions that characterized the conic sections, but with the 20th and 21st propositions we are now in possession of a confirming hint about what lay behind the contrivings of upright sides in those propositions where Apollonius characterized the different conic sections. The hint confirms the insight we achieved, in Chapter 6 of Part III of this guidebook, by thinking hard about what Apollonius laid before us in the defining propositions. The simpler presentation, with its single definition for the upright side, might have been more impressive, but it would have been less instructive. What is most instructive in presenting the conic sections, is the inducing of insight into the centrality of certain ratios. Those formative ratios were given an almost literally central place for each of the sections in the complex presentation by Apollonius.

why these two propositions are here

These two propositions are inserted into a discussion of touchings and meetings. Why insert them here between the strings of propositions 17–19 and 22–36, rather than placing them right after the corresponding defining propositions? That is to say, why not put the 20th proposition right after the 11th, and the 21st right after the 12th and 13th? For this reason: while the 20th and 21st propositions tell us interesting things about how the various conic sections grow, that does not seem to be why Apollonius puts them here. This book comprises the propositions that present those properties of the conic sections which contribute to its plot—which moves from cuttings of cones back to cuttings of cones, by way of the second diameter.

Why then not place both of these propositions right after the 14th proposition, where we begin to move away from the cuttings of cones? Partly because that would be a less dramatic breakaway from the cone than we get with the 15th and 16th propositions. Also, since the 20th and the 21st propositions distinguish the parabola from the so-called central sections, it would be better for them not to come before the Second Definitions (which give the definitions of centers and second diameters), and it is good that the Second Definitions come after the display of conjugate diameters in the 15th and 16th propositions.

Why then not place the 20th and 21st propositions right after the sequence 15th proposition-16th proposition-Second Definitions? Because even though the two propositions that we have just been considering might seem to interrupt the discussion here where Apollonius places them, they in fact provide necessary means for continuing this discussion. The 20th and 21st propositions are relied upon, directly or indirectly, by every proposition (except for the 25th and 32nd) in the whole stretch of propositions from the 22nd through the 36th. That reliance is depicted in figure 14.

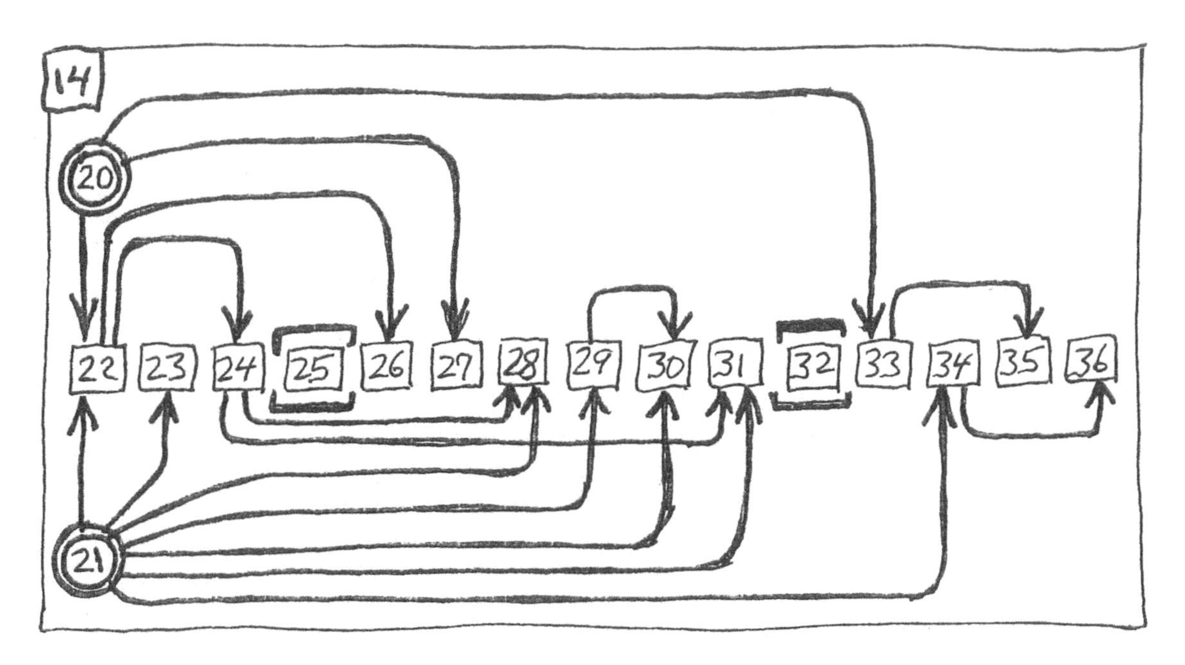

Now before moving on, let us better equip ourselves for the journey by considering the usefulness of that rectangle whose one side is the abscissa and whose other side is the line which results when the abscissa is added on to, or is taken away from, the oblique (transverse) side.

That rectangle is important because of what is depicted in figure 15—namely, the fact that the rectangle has, to the square on the ordinate, a certain constant ratio. This is the same ratio that the sum-or-difference line (oblique [transverse] ± abscissa) has to the width of another rectangle—the rectangle which has the abscissa as a side and is equal to the square on the ordinate.

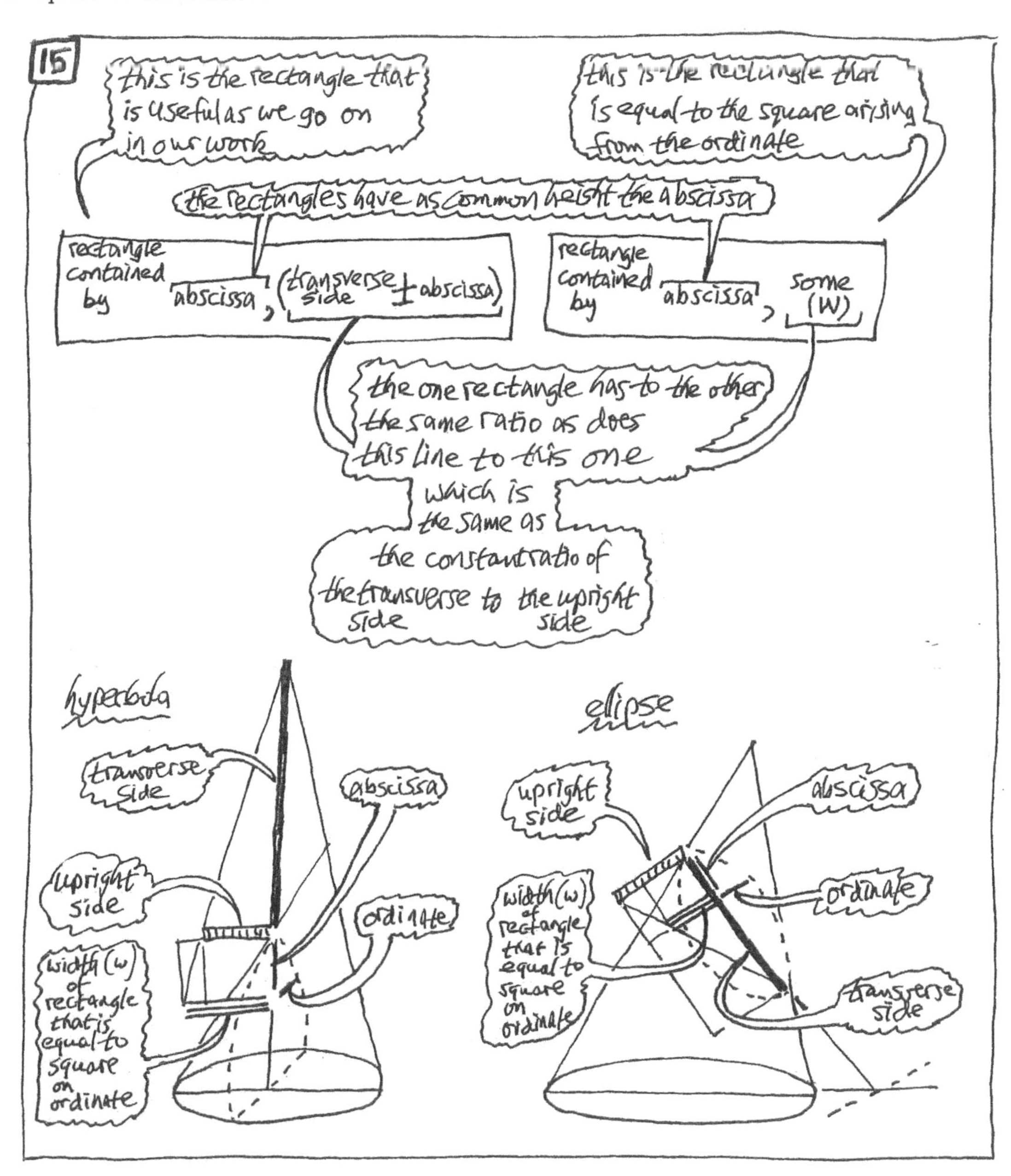

To study the curves after their removal from the cone, we need to get proportions to manipulate. There is a rectangle more useful for that purpose than is the rectangle which is equal to the square on the ordinate. This rectangle which is equal to the square on the ordinate is less useful because while one of its sides is the abscissa, its other side is the width which does *not* appear in the figures once the curves have been removed from the cone. But what rectangle whose one side is the abscissa would in fact be more useful? It is the one whose other side is the sum-or-difference line (oblique [transverse] ± abscissa). Since this line *does* appear in the figures once the curves have been removed from the cone, the rectangle that it makes with the abscissa is the more useful one (fig. 16).

We shall see this in propositions that are coming in the next stretch (such as 23, 28–29, 30), as well as in propositions after that (such as 31 and 34).

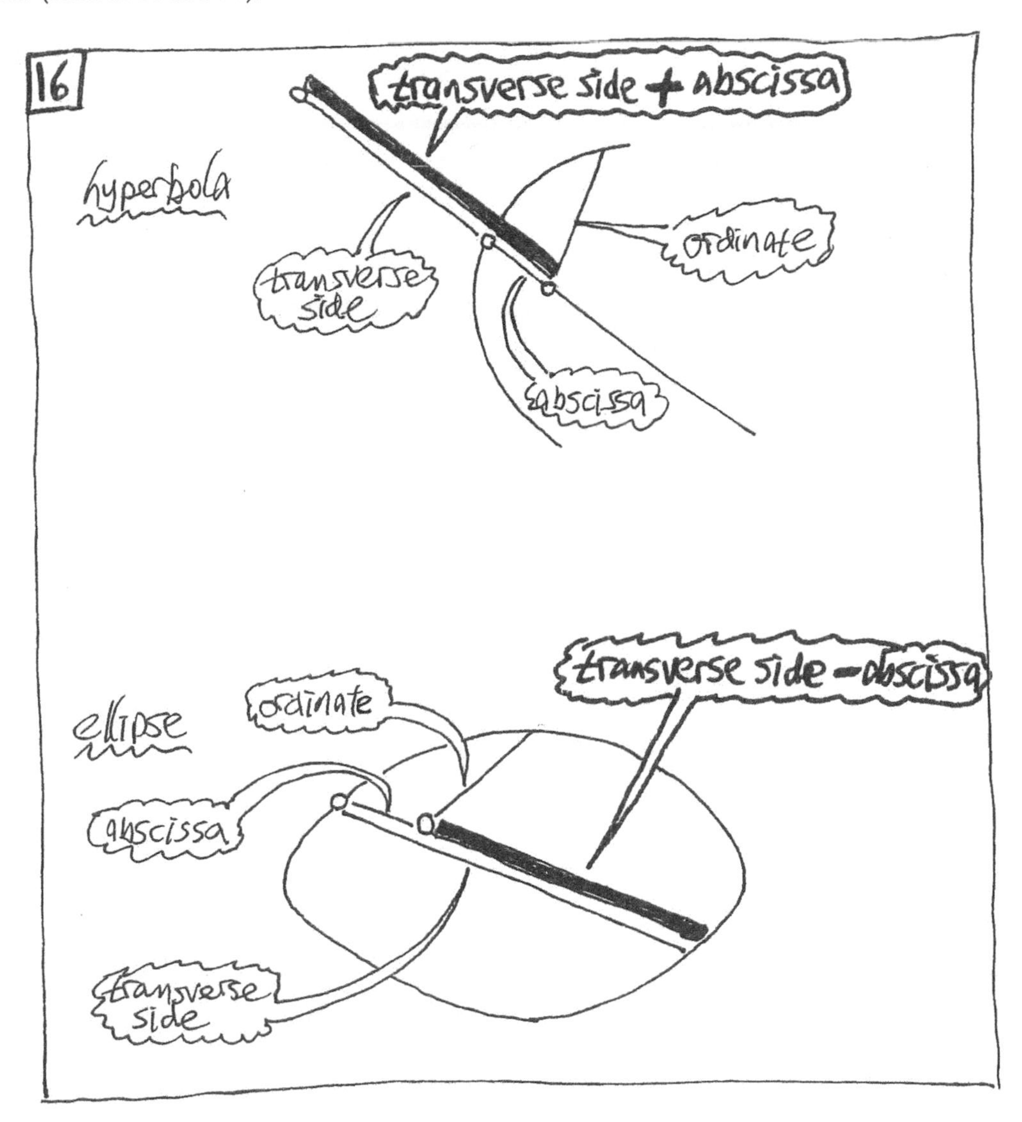

22nd through 30th propositions

This sequence of propositions continues the presentation of straight-line meetings. It deals with:

chords------------straight lines within the curve that meet it at two points;
tangents----------straight lines outside the curve that touch it at one point;
monosecants------straight lines that cut through the curve at only one point.

On a first reading of Apollonius, you may wish to give the demonstrations in this sequence a merely cursory look. The sequence begins with propositions 22–26.

#22 If, with respect to a parabola or hyperbola, a straight [line] cut [the section] at two points—[that straight line] not falling together with the diameter inside—it will, being extended, fall together with the diameter of the section outside the section.
—Let there be a parabola or hyperbola whose diameter the [line] AB is, and let some straight [line] cut the section in two points, namely the [points] C, D. I say that the [line] CD, being extended, will fall together, outside the section, with the [line] AB.
—For let there be drawn down ordinatewise from the [points] C, D the [lines] CE, DB; and let there be, first, the section that is a parabola. Since then, in the parabola, as the [square] from the [line] CE is to the [square] from the [line] DB so the [line] EA is to AB—and greater is the [line] AE than the [line] AB—therefore greater is also the [square] from the [line] CE than the [square] from the [line] DB; and so also the [line] CE is greater than the [line] DB. And also, [these lines CE and DB] are parallels; therefore the [line] CD, being extended, will fall together with the diameter AB outside the section.
—But then let there be an hyperbola. Since then, in the hyperbola, [with the line AF being the oblique side], as the [square] from the [line] CE is to the [square] from the [line] BD so the [rectangle] by FEA is to the [rectangle] by FBA; therefore greater is also the [square] from the [line] CE than the [square] from the [line] DB. And also, [these lines CE and DB] are parallels; therefore the [line] CD, being extended, will fall together with the diameter of the section outside the section.
#23 If, with respect to an ellipse, a straight [line] that lies between the two diameters—[that is, the "one originating previously" (*prouparkhousa*) and another that is "conjugate" (*suzugês*) to it, as in #16]—cut [the section], [that straight line] will, being extended, fall together with each of the diameters outside the section.
—Let there be an ellipse whose diameters the [lines] AB, CD are, and let there cut the section some straight [line]—namely the [line] EF—that lies between the diameters AB, CD. I say that the [line] EF, being extended, will fall together with each of the [lines] AB, CD outside the section.
—For let there be drawn down ordinatewise from the [points] E, F to the [line] AB the [lines] GE, FH (on the one hand), and (on the other hand) to the [line] CD the [lines] EK, FL. Therefore (on the one hand) as is the [square] from the [line] EG to the [square] from the [line] FH so is the [rectangle] by BGA to the [rectangle] by BHA, and (on the other hand) as is the [square] from FL to the [square] from EK so is the [rectangle] by DLC to the [rectangle] by DKC. And (on the one hand) the [rectangle] by BGA is greater than the [rectangle] by BHA—for the [point] G is nearer the midpoint [than the point H is]—and (on the other hand) greater is the [rectangle] by DLC than the [rectangle] by DKC. Greater, therefore, also is (on the one hand) the [square] from the [line] GE than the [square] from FH, and (on the other hand) the [square] from FL than the [square] from EK. Greater, therefore, also is (on the one hand) the [line] GE than the [line] FH, and (on the other hand) the [line] FL than the [line] EK. And also parallel is (on the one hand) the [line] GE to the [line] FH, and (on the other hand) the [line] FL to the [line] EK. Therefore the [line] EF, being extended, will fall together with each of the diameters AB, CD outside the section.

#24 If, with a parabola or hyperbola, a straight [line] that falls together [with the section] at one point, being extended both ways, fall outside the section, then it will fall together with the diameter.
—Let there be a parabola or hyperbola whose diameter the [line] AB is, and let there fall together with it, [that is, with the section], the straight [line] CDE, at the [point] D, and, being extended both ways, let it fall outside the section. I say that it will fall together with the diameter AB.
—For let there be taken some point on the section, the [point] F, say, and let there be joined the [line] DF. The [line] DF, therefore, being extended, will fall together with the diameter of the section. Let it fall together with it at the [point] A. And between the section and the [line] FDA is the [line] CDE. And also, therefore, the [line] CDE, being extended, will fall together with the diameter outside the section.
#25 If, with an ellipse, a straight [line] that falls together [with the section] between the two [conjugate] diameters, being extended both ways, fall outside the section, it will fall together with each of the diameters.
—Let there be an ellipse whose diameters the [lines] AB, CD are; and let there fall together with this [section], between the two diameters, some straight [line] EF, at the [point] G; and let [that line EF], being extended both ways, fall outside the section. I say that the [line] EF will fall together with each of the [lines] AB, CD.
—Let there be drawn down from the [point] G to the lines AB, CD, ordinatewise, the [lines] GH, GK [respectively]. Since parallel is the [line] GK to the [line] AB, but there has fallen together with the [line] GK some [line], namely the [line] GF; therefore also with the [line] AB will [that line GF] fall together. Likewise then, also with the [line] CD will there fall together the [line] EF.

In these first four propositions of the sequence, Apollonius tells us about the orientation of a diameter. In a single curve of section,

if a straight line is a *chord*	if a straight line is a *tangent*
that does not meet the diameter	(hence, does not meet the diameter
inside the section,	inside the section),
then it meets the diameter outside;	then it meets the diameter outside.

What these propositions treat is this:

chord----(#22) of a parabola and of an hyperbola; and #23 of an ellipse with two diameters that are conjugate;
tangent--(#24) of a parabola and of an hyperbola; and (#25) of an ellipse with two diameters that are conjugate.

There is this restriction for the ellipse of the 23rd proposition and of the 25th: the straight line will have the property *if* it meets the curve *between* the endpoints of the two conjugate diameters. In the ellipse, that is to say, the endpoints of the chord, and the point where the tangent touches the curve, are between an endpoint of one of the conjugate diameters and an endpoint of the other one. Why the restriction? Because

any other chord	any other tangent
will cross	will be at an endpoint of
one of the diameters, and	one of the diameters, and
if it happens to be situated	since it must therefore be situated
ordinatewise to this diameter,	ordinatewise to this diameter,
will be parallel to	will be parallel to
that other diameter—	that other diameter—
the one which is conjugate to it;	the one which is conjugate to it.

What we learn from propositions 22–25 is this: a diameter in a single curve of conic section cannot be parallel to a chord or to a tangent (except in the special cases mentioned for the ellipse). The matter is not stated by Apollonius negatively in terms of parallelism, but positively in terms of meetings: chords and tangents, when they are prolonged, will meet the diameter, when it is prolonged.

We have just considered some circumstances when a straight line will meet the section's diameter *outside*—as it does

> if it does not meet the diameter inside
> but does meet the section inside in both directions (that is, is a chord),
>
> or if it is outside the section in both directions away from
> the single point where it meets the section (that is, is a tangent).

Now we move on to consider what happens if a straight line does not meet the diameter anywhere—outside or inside. Having learned that

> if a straight line is parallel to a chord or a tangent,
> then (except for the qualifications in the case of the ellipse)
> it is not parallel to a diameter,

we now learn that

> if a line *is* parallel to a diameter,
> then it meets the section but is *not* a chord or a tangent.

Proposition 26, from which we learn it, puts it this way: in a single curve of section,

> if a straight line is parallel to a diameter,
> then it cuts the curve at exactly one point
> (in other words, it is what's called a "monosecant").

This 26th proposition treats together the parabola and the hyperbola. Right after it (unlike what happens right after propositions 22 and 24) nothing is said about the ellipse. Why not? Because the ellipse is closed. The ellipse can have a chord (23), and it can have a tangent (25); but it cannot have a monosecant.

#26 If, in a parabola or hyperbola, a straight [line] be drawn parallel to the diameter of the section, it will fall together with the section in only one point.

—Let there be, first, a parabola whose diameter the [line] ABC is, and whose upright side the [line] AD is; and parallel to the [line] AB let there be drawn the [line] EF. I say that the [line] EF, being extended, will fall together with the section.

—For let there be taken some point on the [line] EF—the point E, say; and, from the [point] E parallel to a [line] that has been drawn down ordinatewise, let there be drawn the [line] EG; and greater than the [square] from the [line] GE let there be the [rectangle] by DAC; and from the [point] C let there be drawn up ordinatewise the [line] CH. The [square] from HC, therefore, is equal to the [rectangle] by DAC; but greater is the [rectangle] by DAC than the [square] from EG; therefore greater also is the [square] from HC than the [square] from EG; therefore greater also is the [line] HC than the [line] EG. And also [the lines HC and EG] are parallel; therefore the [line] EF, being extended, cuts the [line] HC; and so also with the section will [the line EF extended] fall together. Let it fall together [with the section] at the [point] K. I say, then, that in only one point—the [point] K—will it fall together [with it].

—For if possible, let it fall together [with it] also at the point L. Since then, with respect to a parabola, a straight [line] cuts [the section] at two points, [therefore the straight line], being extended, will fall together with the diameter of the section—which very thing is absurd, for it is supposed to be parallel. The [line] EF, therefore, being extended, will at only one point fall together with the section.

—Let, then, the section be an hyperbola, and the oblique side of the figure (*plagia tou eidous pleura*), let that be what the [line] AB is, and the upright [side] (*orthia*), let that be what the [line] AD is; and let there be joined the [line] DB, and let it be extended. Then—with the same equipment being built (*tôn autôn kataskeuasthentôn*)—let there be drawn, from the [point] C parallel to the [line] AD, the [line] CM. Since then the [rectangle] by MCA is greater than the [rectangle] by DAC; and also, to the [rectangle] by MCA (on the one hand) there is equal the [square] from CH, and the [rectangle] by DAC (on the other hand) is greater than the [square] from GE, therefore greater also is the [square] from CH than the [square] from EG. And so, also the [line] CH is greater than the [line] GE, and the same things as at first—[that is, as in the parabola]—will come together.

We have considered when there are meetings with the diameter outside, and when there is no meeting with it outside or inside. Now we move on (Propositions 27–28) to consider when there is a meeting with the diameter *inside* the section—but only for the parabola and the opposite sections.

For the parabola, proposition 27 tells us this:

> if a straight line meets the diameter inside, then it is a chord.

Proposition 27 (that is to say) now lifts proposition 19's restriction—at least for the parabola; it tells us that *any* line through the diameter, even if it is not drawn ordinatewise, meets the section, and it does so *both* ways).

#27 If, with respect to a parabola's diameter, a straight [line that is inside the section] cut [the diameter], [then that straight line,] extended both ways, will fall together with the section.
—Let there be a parabola whose diameter the [line] AB is; and let there cut this [line] some straight [line] inside the section—namely the [line] CD. I say that the [line] CD, being extended both ways, will fall together with the section.
—For let there be drawn some [line] from the [point] A parallel to a [line] that has been drawn down ordinatewise—the [line] AE, say. The [line] AE, therefore, will fall outside the section. Then the [line] CD, with respect to the [line] AE, either is parallel or is not. If now [that line CD] is parallel to it, [then that line CD] has been drawn down ordinatewise, and so, being extended both ways, will fall together with the section. [That line CD]—not parallel let it now be, [not parallel] to the [line] AE, but, being extended, let it fall together with the [line] AE at the [point] E. Then that it falls together with the section, on the side where the [point] E is, is evident—for if it is thrown together with the [line] AE, much the more does it—before that—cut the section.
—I say that being extended also the other way, [the line CD] falls together with the section. For let there be, as the [line] along which [the lines drawn down ordinatewise] have the power, the [line] MA, and let there be also, [as a line drawn down ordinatewise,] the [line] GF; and let the [square] from AD be equal to the [rectangle] by BAF; and, parallel to ordinatewise, let the [line] BK fall together with the [line] DC at the [point] C. Since equal is the [rectangle] by FAB to the [square] from AD, hence as is the [line] AB to AD so is the [line] DA to AF; and also therefore is the remainder [line] BD to the remainder [line] DF as is the [line] BA to AD. And also therefore as is the [square] from BD to the [square] from FD so is the [square] from BA to the [square] from AD. But since equal is the [square] from AD to the [rectangle] by BAF, hence as is the [line] BA to AF so is the [square] from BA to the [square] from AD, that is, the [square] from BD to the [square] from DF. But as is the [square] from BD to the [square] from DF so is the [square] from BC to the [square] from FG; and as is the [line] AB to AF so is the [rectangle] by BAM to the [rectangle] by FAM. Therefore as is the [square] from BC to the [square] from FG so is the [rectangle] by BAM to the [rectangle] by FAM. And alternately, as is the [square] from BC to the [rectangle] by BAM so is the [square] from FG to the [rectangle] by FAM. But the [square] from FG is equal to the [rectangle] by FAM—on account of the section, [on account, that is, of the section's being a parabola]. Therefore also the [square] from BC is equal to the [rectangle] by BAM. But an upright [side] is what the [line] AM is, and a [line] parallel to ordinatewise is what the [line] BC is. Therefore the section goes through the [point] C; and there falls together with the section the [line] CD at the [point] C.

Next, proposition 28 lifts proposition 18's restriction to a single curve, telling us this:

> if a straight line goes through a point in *one* of the two curves,
> and it is parallel to a tangent in the *other* one,
> then it is a chord.

That is to say, when we draw, through a point in one of the curved lines of the opposite sections, a straight line parallel to a tangent in the other, then, taking the opposite sections *together,* we again get a chord, as in proposition 18—but, still, a chord in *one* of the two curved lines.

#28 If a straight [line] touches one of the opposite [sections], and there be taken some point inside the other section, and through it there be drawn parallel to the touching [line] a straight [line], [this straight line,] being extended both ways, will fall together with the section.

—Let there be opposite [sections] of which the [line] AB is a diameter; and let there touch the section A some straight [line], say the [line] CD; and let there be taken some point inside the other section, say, the [point] E; and, through the [point] E let there be drawn parallel to the [line] CD the [line] EF. I say that the [line] EF, being extended both ways, will fall together with the section.

—Since then it has been shown that the [line] CD, being extended, will fall together with the diameter AB; and there is parallel to it the [line] EF; therefore the [line] EF, being extended, will fall together with the diameter. Let [the line EF] fall together [with the diameter] at the [point] G; and, equal to the [line] GB, let there be put the [line] AH; and through the [point] H let there be drawn parallel to the [line] FE the [line] HK; and let there be drawn down ordinatewise the [line] KL; and equal to the [line] LH, let there be put the [line] GM; and let there be drawn parallel to ordinatewise the [line] MN; and let there be extended straight forth the [line] GN. And—since parallel is the [line] KL to the [line] MN, and the [line] KH to the [line] GN, and a single straight [line] is what the [line] LM is—hence similar is the triangle KHL to the triangle GMN. And equal is the [line] LH to the [line] GM; therefore equal is the [line] KL to the [line] MN; and so also the [square] from KL is equal to the [square] from MN. And—since equal is the [line] LH to the [line] GM, and the [line] AH to the [line] BG, and common is the [line] AB—therefore equal is the [line] BL to the [line] AM. Therefore equal is the [rectangle] by BLA to the [rectangle] by AMB. Therefore as is the [rectangle] by BLA to the [square] from KL so is the [rectangle] by AMB to the [square] from MN; and as is the [rectangle] by BLA to the [square] from LK so is the oblique [side] to the upright [side]; and also, therefore, as is the [rectangle] by AMB to the [square] from MN so is the oblique [side] to the upright [side]. The [point] N, therefore, is on the section. Therefore the [line] EF, being extended, will fall together with the section at the [point] N. Likewise, then, it will be shown that [the line EF], being extended also the other way, will fall together with the section.

We have just brought together the opposite sections with respect to getting a chord in *one* of them. Now we move on to get a chord between the two lines (that is, inside the *pair*)—a chord that goes through the diameter's center. Proposition 29 tells us this about the opposite sections:

> if a straight line goes through a point on *one* of the two curves,
> and also goes through the center—the diameter's center,
> which in Second Definitions is called the center of the *sections*—
> then it is a chord *between* the *two* curves that make up the opposite sections.

When we have obtained such a chord (one that runs between the opposite sections, and also through the center of the sections), we then move on to see that the center of such a *chord* is the very same point as the center of the *sections*. That is to say, Apollonius justifies the beginning of the Second Definitions (defs. 9–10). How? By showing that the diameter's center is the center not only of the diameter, but also of *all* the other lines through the diameter's center that are chords *between* the opposite sections. The point is a point of *symmetry*.

It is a point of symmetry, moreover, not only for the opposite sections, but also for the *ellipse*. Just as, immediately after we obtained the sections, we saw (in the 15th and 16th propositions) a respect in which the opposite sections resemble the ellipse, so we do here too. Here we see how the center that lies between the *opposite sections* resembles the center that lies in the middle of an *ellipse*: all chords through the center, though they are not equal to each other, are nonetheless bisected by the center. From proposition 30, we learn the following about the ellipse and the opposite sections together:

> if a chord meets the diameter at the diameter's midpoint (its center),
> then this meeting-point is also the midpoint of the *chord*.

(Apollonius does not mention the fact that this is also true of the circle. This has happened before. The circle was not mentioned in the 15th proposition either.)

On a first reading, you might, as was suggested, have given the other demonstrations in this sequence a merely cursory look. (The figures in the explication of those Propositions, 22 through 28, have presented matters of general significance without presenting depictions of particulars. For illustrative diagrams that provide help in the reading of those propositions, see the Appendix at the end of this volume.)

But if you do wish more closely to examine the following two concluding propositions, then you will find some help, right after them, in this chapter's final section. Omitting the section, however, will not at all get in the way of your easily moving on into the next chapter of this guidebook.

#29 If in opposite [sections] a straight [line] fall through the center onto either of the sections, [that straight line,] being extended, will cut the other section

—Let there be opposite [sections] whose diameter is what the [line] AB is, and center is what the [point] C is; and let the [line] CD cut the section AD. I say that it will cut also the other section.

—For let there be drawn down ordinatewise the [line] ED; and, equal to the [line] AE, let there be put the [line] BF; and let there be drawn ordinatewise the [line] FG. And—since equal is the [line] EA to the [line] BF, and common is the [line] AB—therefore equal is the [rectangle] by BEA to the [rectangle] by BFA. And—since as is the [rectangle] by BEA to the [square] from DE so is the oblique [side] to the upright [side], but also as is the [rectangle] by AFB to the [square] from FG so is the oblique [side] to the upright [side]—therefore also as is the [rectangle] by BEA to the [square] from DE so is the [rectangle] by AFB to the [square] from FG. But equal is the [rectangle] by BEA to the [rectangle] by AFB; therefore equal is also the [square] from ED to the [square] from FG. Since, then, equal is (on the one hand) the [line] EC to the [line] CF, and (on the other hand) the [line] DE to the [line] FG, and straight is the [line] EF, and parallel is the [line] ED to the [line] FG—also, therefore, the [line] DG is straight. And therefore the [line] CD will also cut the other section.

#30 If in an ellipse or opposite [sections] a straight [line] be drawn both ways from the center—[a straight line, that is,] which falls together with the section—it will be cut in two [halves] at the center.

—Let there be an ellipse or opposite [sections], and let their diameter be what the [line] AB is, and their center be what the [point] C is; and through the [point] C let there be drawn some straight [line] DCE. I say that equal is the [line] CD to the [line] CE.

—For let there be drawn ordinatewise the [lines] DF, EG. And—since as is the [rectangle] by BFA to the [square] from FD so is the oblique [side] to the upright [side], but also as is the [rectangle] by AGB to the [square] from GE so is the oblique [side] to the upright [side]—therefore also as is the [rectangle] by BFA to the [square] from FD so is the [rectangle] by AGB to the [square] from GE; and, alternately, as is the [rectangle] by BFA to the [rectangle] by AGB so is the [square] from DF to the [square] from GE. But as is the [square] from DF to the [square] from GE so is the [square] from FC to the [square] from CG. Alternately, therefore, as is the [rectangle] by BFA to the [square] from FC so is the [rectangle] by AGB to the [square] from CG. Therefore also—for the ellipse, by putting together (*sunthenti*) [traditionally rendered by the Latin *componendo*], and, for the opposite sections, by inverting and converting (*anapalin* and *anastrepsanti*) [traditionally rendered by the Latin *invertendo* and *convertendo*]—as is the [square] from AC to the [square] from CF so is the [square] from BC to the [square] from CG. And alternately. But equal to the [square] from AC is the [square] from CB; therefore equal also to the [square] from FC is the [square] from CG. Therefore equal is the [line] FC to the [line] CG. And parallel are the [lines] DF, GE. Therefore equal also is the [line] DC to the [line] CE.

29th and 30th propositions: demonstrations

29th proposition

This proposition treats the situation, in opposite sections, that is depicted in figure 17.

The center of the opposite sections with diameter AB is point C. From that center C, a straight line is drawn to some point (call it D) on either one of the curves. If that straight line is prolonged toward the other curve, will it intersect it? The demonstration shows that it will. How is that shown? By picking a particular point (call it G) on the other curve, and showing that if the point G that we have picked on the other curve is joined to center C by a straight line, then this line GC will lie in a straight line with DC. (Fig. 18.)

The demonstration answers the following two questions: how are we to pick point G? and how are we to show that GC lies in a straight line with CD?

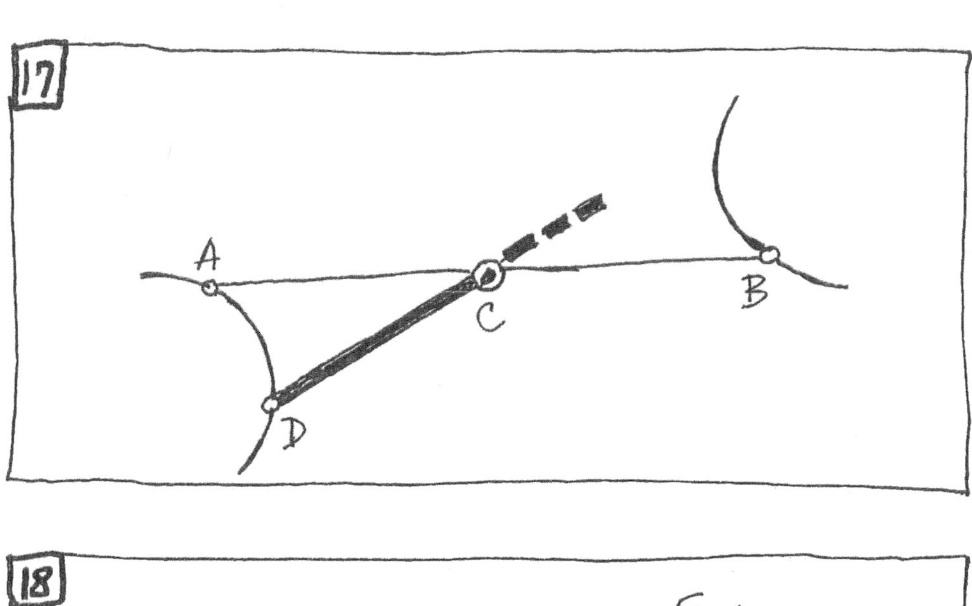

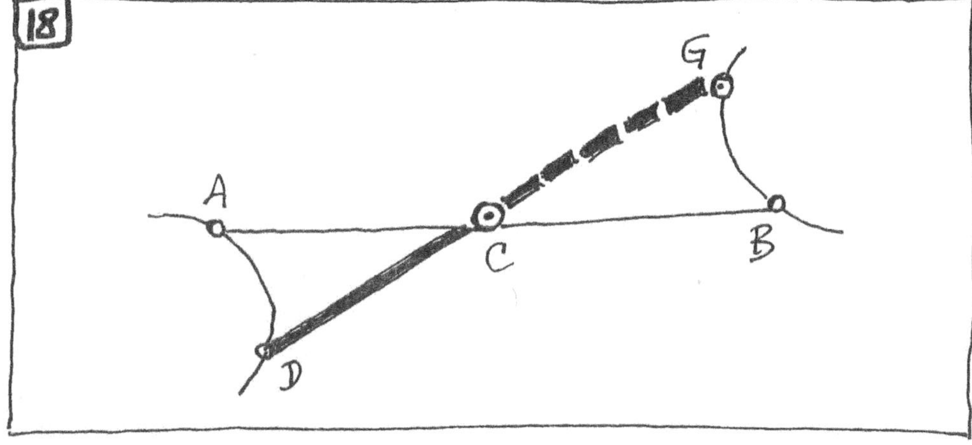

How, then, are we to pick point G? Well, G is to be the point on the other curve that will correspond to point D on the first curve. When we obtained opposite sections, we saw that in them there are similar triangles in reverse; so we would expect to find such triangles here too. (Fig. 19.) The abscissa and ordinate of point G are to correspond to those of point D, so we draw D's ordinate, thus getting its abscissa; and now we can get the abscissa of point G, which will give us its ordinate—and thus (we are assured by the 19th proposition) we shall have point G itself. (Fig. 20.)

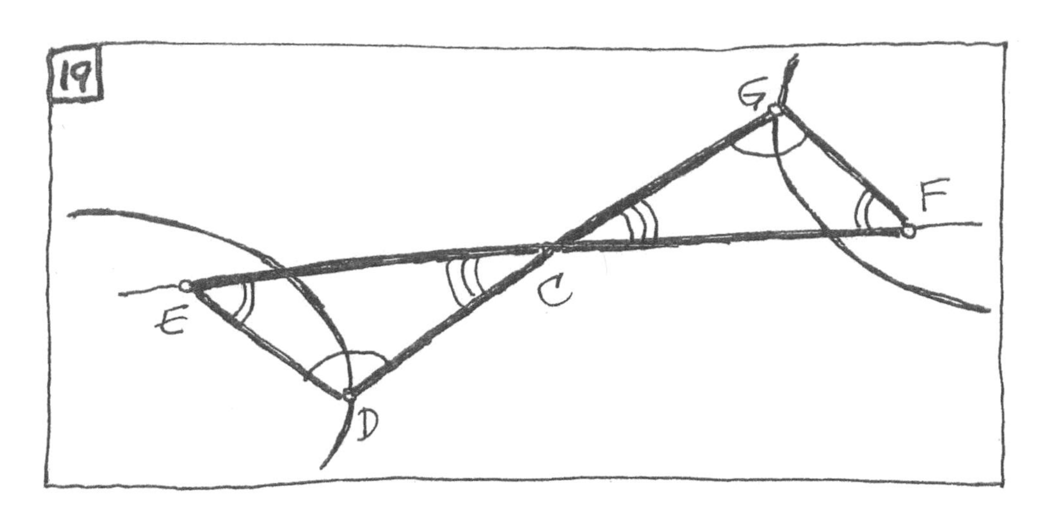

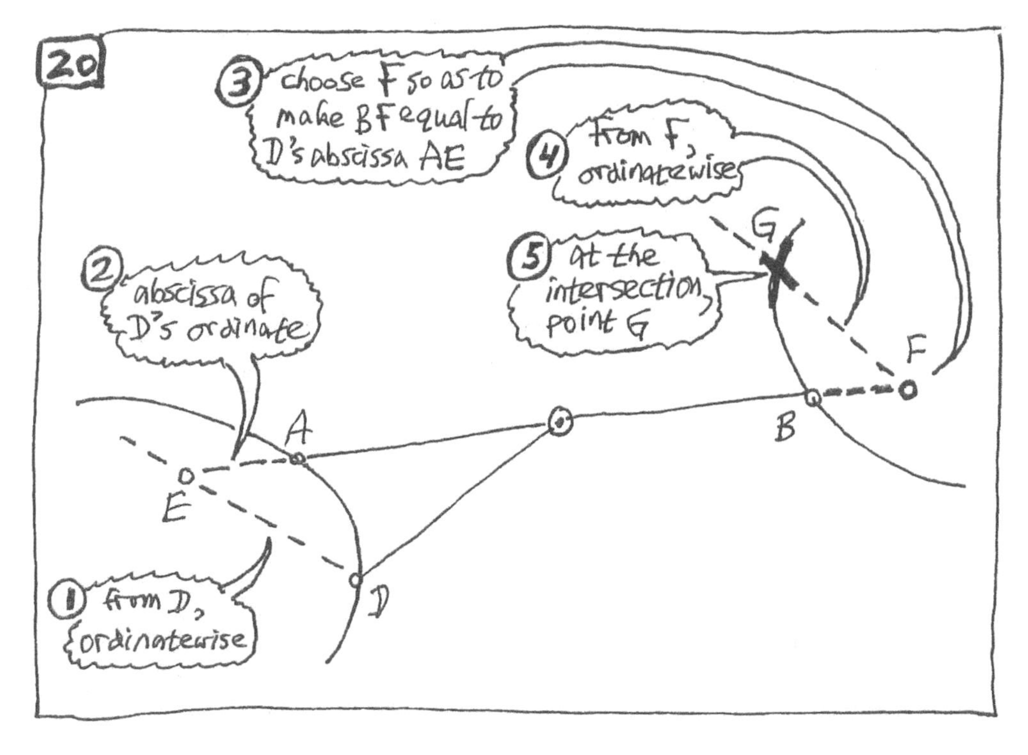

And how are we to show that GC lies in a straight line with CD? Well, Euclid shows (*Elements* I.14) that two straight lines will lie in a straight line if the angles that they make with another straight line are supplementary, as in figure 21. The question, then, is how to show that the two relevant angles are supplementary. Here is how: show that the second angle is supplementary to a third angle that is equal to the first one. (Fig. 22.) It doesn't take much to show why angles (2) and (3) are supplementary. They are supplementary because of how they were constructed—a straight line was drawn from point G to the point C that is on straight line AB. The big thing is to show why angles (1) and (3) are equal. These angles are equal because they are in similar triangles. But why must the triangles be similar? Because two sides and the included angle in one triangle are equal, respectively, to two sides and the included angle in the other triangle. (Fig. 23.)

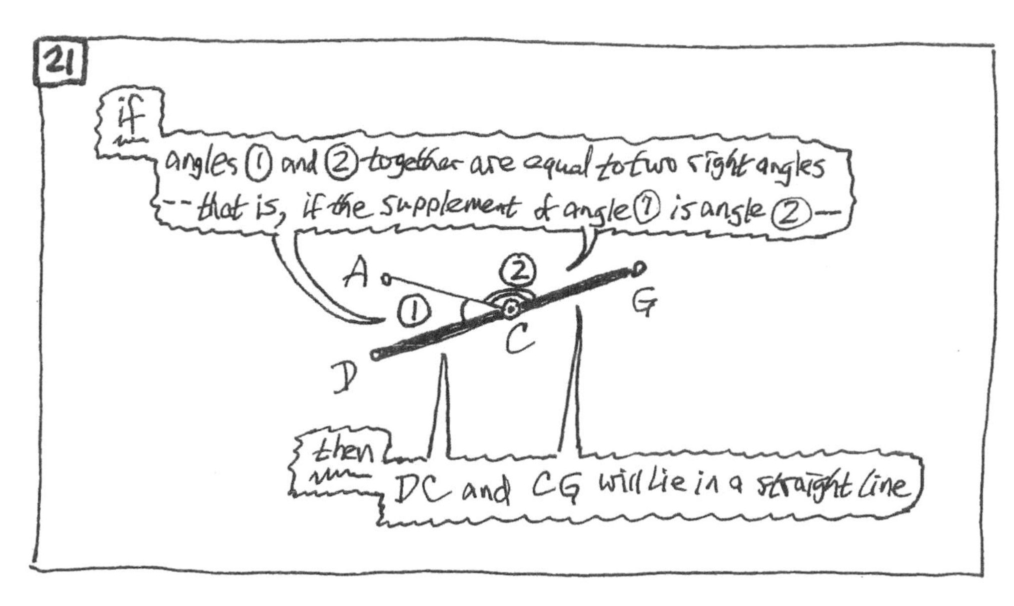

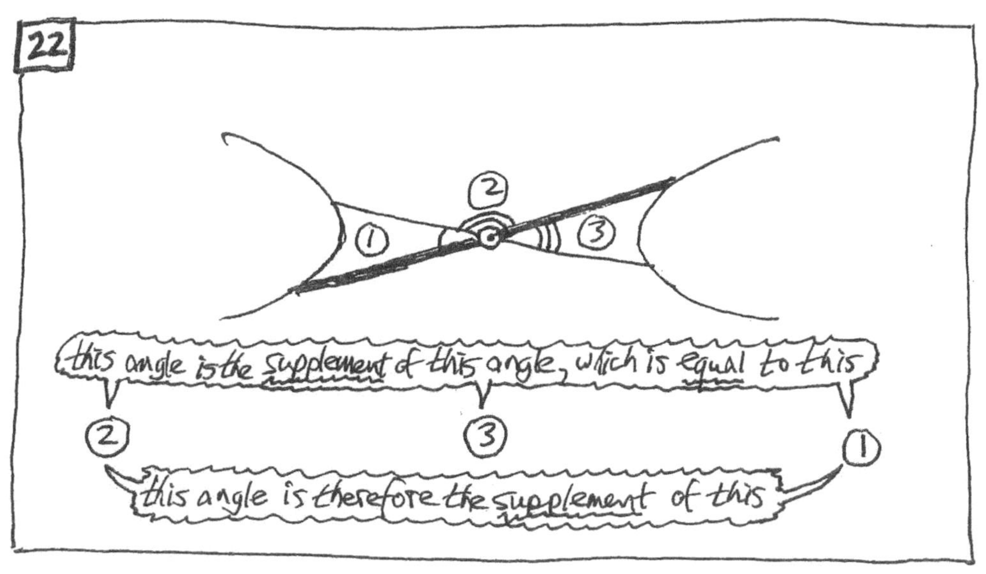

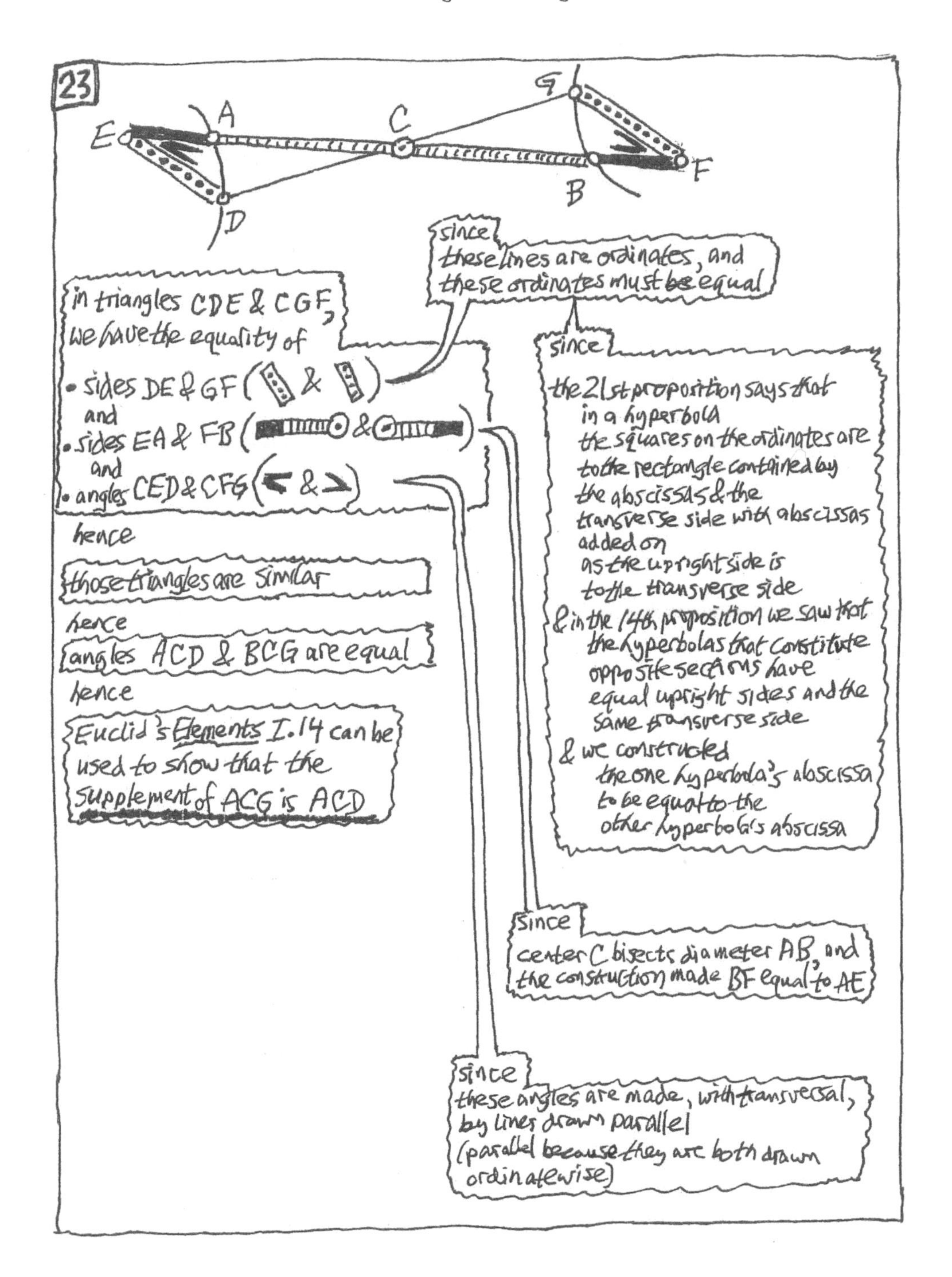

23
E
A
C
G
B
F
D
since
these lines are ordinates, and
these ordinates must be equal
in triangles CDE & CGF,
we have the equality of
• sides DE & GF (&)
and
• sides EA & FB (&)
and
• angles CED & CFG (&)
hence
those triangles are similar
hence
angles ACD & BCG are equal
hence
Euclid's Elements I.14 can be
used to show that the
supplement of ACG is ACD
since
the 21st proposition says that
in a hyperbola
the squares on the ordinates are
to the rectangle contained by
the abscissas & the
transverse side with abscissas
added on
as the upright side is
to the transverse side
& in the 14th proposition we saw that
the hyperbolas that constitute
opposite sections have
equal upright sides and the
same transverse side
& we constructed
the one hyperbola's abscissa
to be equal to the
other hyperbola's abscissa
since
center C bisects diameter AB, and
the construction made BF equal to AE
since
these angles are made, with transversal,
by lines drawn parallel
(parallel because they are both drawn
ordinatewise)

30th proposition

This proposition treats the situation, in an ellipse or in opposite sections, that is depicted in figure 24. The center of diameter AB is point C. Through that center C, a straight line that is drawn in both directions meets the ellipse or the opposite sections at points D and E. Where will this chord DE have its midpoint? The demonstration shows that the midpoint of this chord that passes through the diameter's center will *be* the diameter's center.

But why is this fact not a mere porism that follows immediately from the 29th proposition? After all, the 29th proposition said that if we draw a line from C to a point D on either one of the hyperbolas that constitute opposite sections, then we are sure to get a chord of the opposite sections if we go back and prolong the line in the opposite direction. What guarantees that—does it not guarantee also that C is the midpoint of the line drawn in the 30th proposition? No. If we start from point C and draw a straight line in both directions, there is no guarantee that we shall get a chord of the opposite sections (even though in the closed section, namely the ellipse, we must get a chord). And even if we do get a chord, we cannot be sure either in the ellipse or in the opposite sections that points D and E are counterparts of each other. In the 29th proposition, we began by picking one of the points (G) in such a way as to guarantee that it will be the counterpart of the other point (D). Here in the 30th proposition, however, some labor is necessary in order to demonstrate that the two points that we happen to start with are counterparts of each other.

So, how is it demonstrated that the midpoint of diameter AB (that is, the point C) is also the midpoint of chord DE?

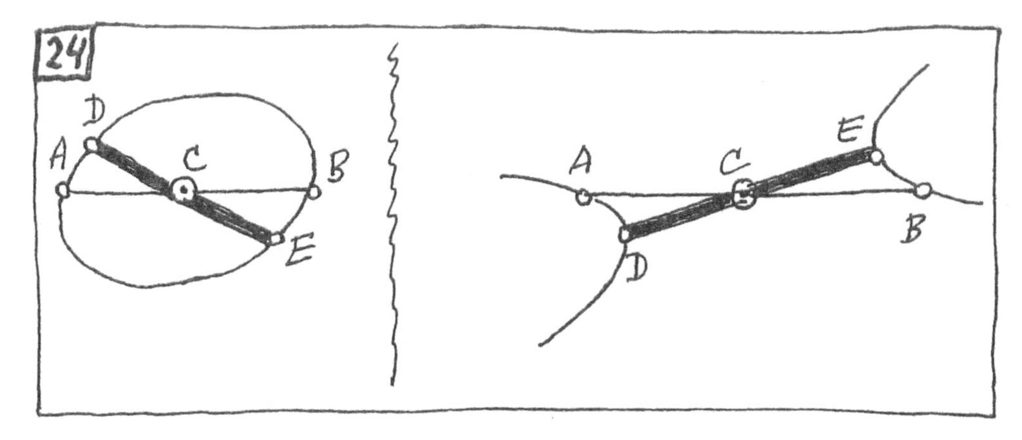

We begin at the endpoints D and E, drawing ordinates (DF and EG)—thus making triangles (CDF and CEG). (Fig. 25.)

Most of the demonstration is devoted to showing that side CF of one triangle is equal to side CG of the other triangle. It then follows, from the consequent congruence of the triangles, that CD is equal to CE (which makes C the midpoint of DE).

The triangles are congruent because side FC in the one triangle is equal to side CG in the other triangle, and the angles at the ends of the side in the one triangle are equal to the angles at the ends of the side in the other triangle. But why are those angles equal? Angles DCF and ECG are equal to each other because they are vertical angles; and angles DFC and CGE are equal to each other because the ordinates DF and EG (which form those angles with the transversal FG) are parallel to each other. But why are the triangles' sides (CF and CG) equal to each other? The demonstration of that equality depends upon the 5th and 6th propositions of the Second Book of Euclid's *Elements*. In effect, what Euclid shows there (as we saw in Part II of this guidebook) is the following:

> When a straight line is cut equally and also unequally—
> whether the unequal cut is internal (II.5) or external (II.6)—
> then the rectangle contained by the unequal segments is (in both cases)
> equal to the difference between two squares,
> namely, the square on the half and the square on the line between the
> two cutting-points
> (one of the points cutting the line equally, and the other cutting it unequally).
>
> And which of the two squares is larger
> will be determined by
> whether the unequal cut is internal (II.5) or external (II.6).

In this 30th proposition of Apollonius, the diameter AB is cut equally at C, and is cut unequally at F and also at G; the unequal cuts are internal in the case of the ellipse, and are external in the case of the opposite sections. The demonstration that CF and CG are equal to each other is depicted in figures 26 and 27.

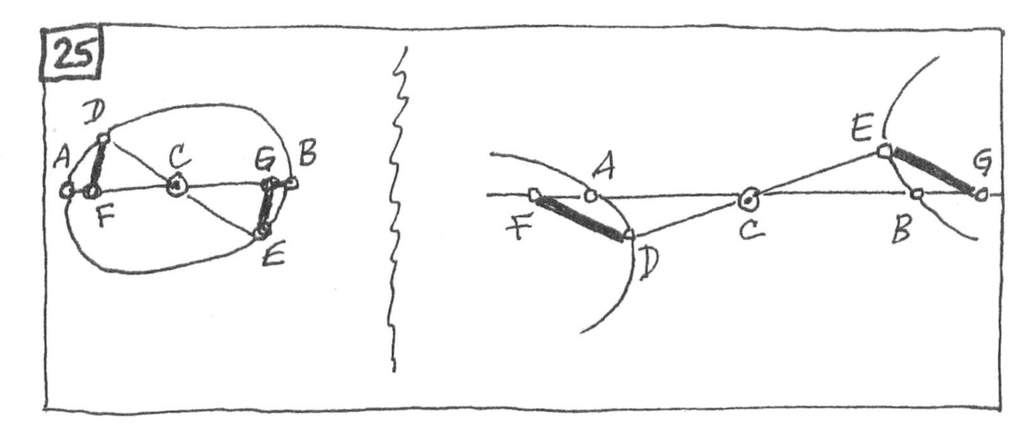

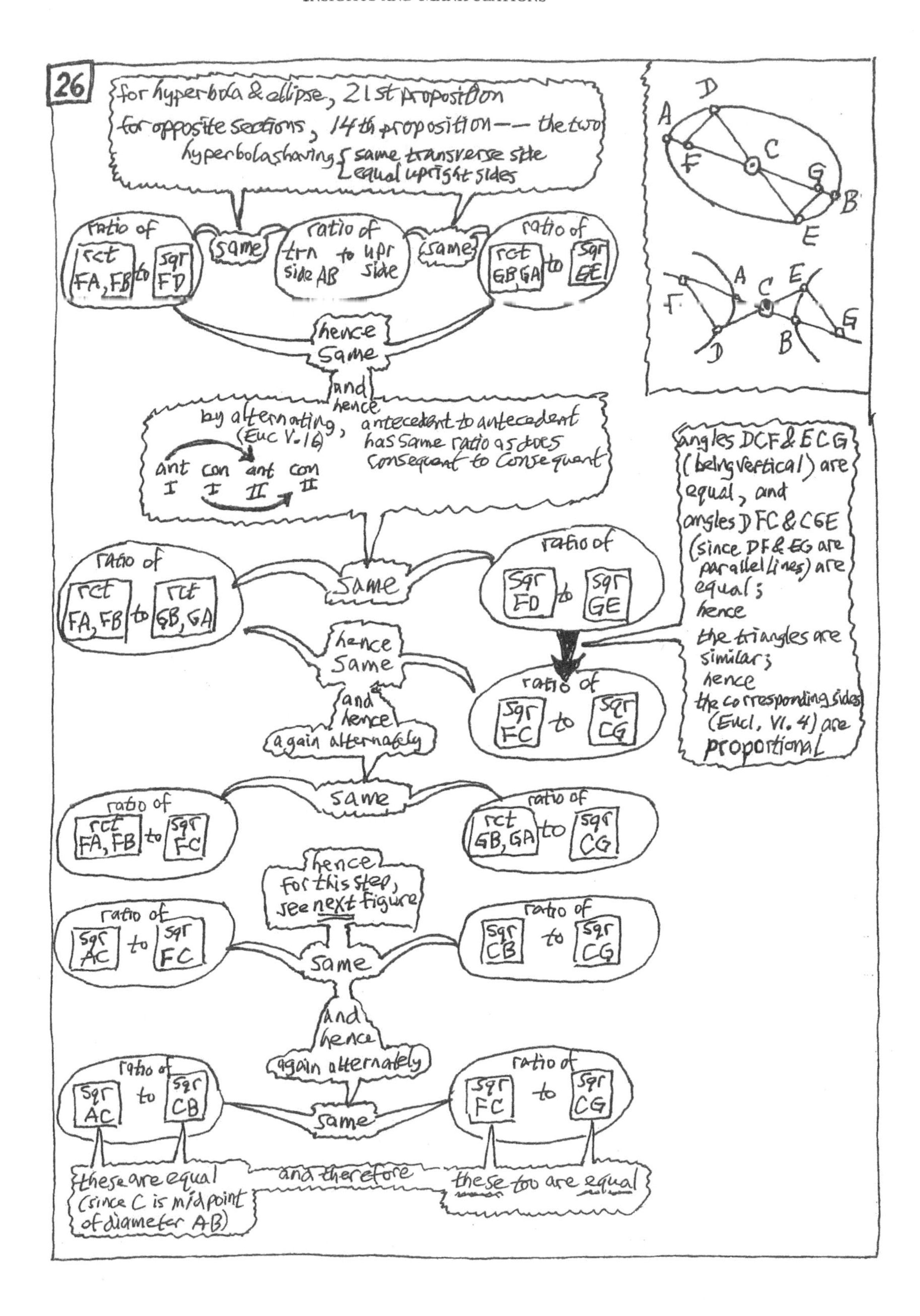
26
for hyperbola & ellipse, 21st proposition
for opposite sections, 14th proposition —— the two hyperbolas having same transverse side equal upright sides
ratio of rct FA, FB to sqr FD
same
ratio of trn side AB to upr side
same
ratio of rct GB, GA to sqr GE
hence same
and hence
by alternating, (Euc V.16) antecedent to antecedent has same ratio as does consequent to consequent
ant I con I ant II con II
ratio of rct FA, FB to rct GB, GA
same
ratio of sqr FD to sqr GE
angles DCF & ECG (being vertical) are equal, and angles DFC & CGE (since DF & EG are parallel lines) are equal; hence the triangles are similar; hence the corresponding sides (Eucl. VI. 4) are proportional
hence same
and hence again alternately
ratio of sqr FC to sqr CG
ratio of rct FA, FB to sqr FC
same
ratio of rct GB, GA to sqr CG
hence For this step, see next figure
ratio of sqr AC to sqr FC
same
ratio of sqr CB to sqr CG
and hence again alternately
ratio of sqr AC to sqr CB
same
ratio of sqr FC to sqr CG
these are equal (since C is midpoint of diameter AB)
and therefore
these too are equal
A
D
C
F
G
B
E

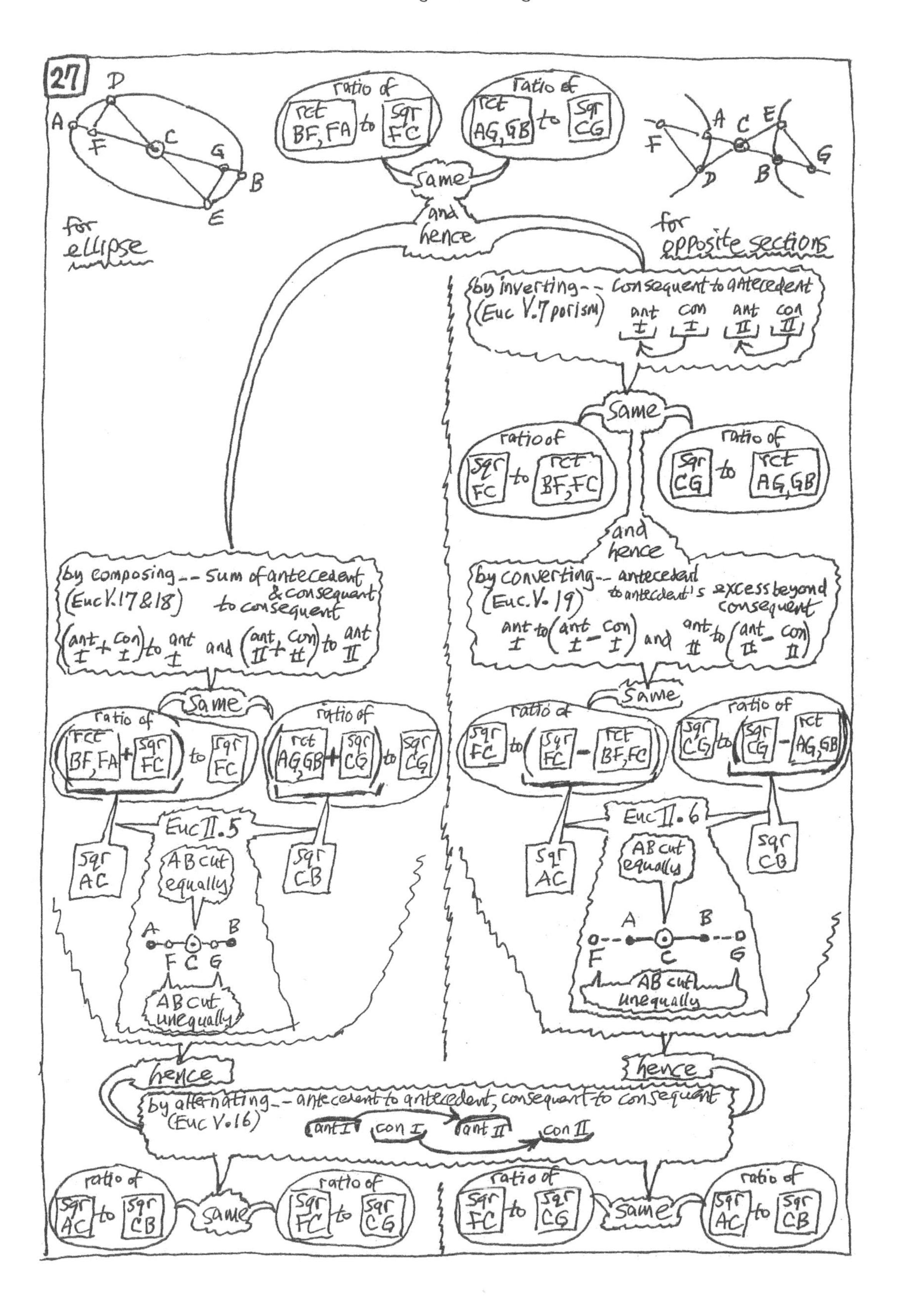

27
for ellipse
for opposite sections
ratio of rct BF, FA to sqr FC
ratio of rct AG, GB to sqr CG
same
and hence
by inverting -- consequent to antecedent (Euc V.7 porism)
ant I con I ant II con II
same
ratio of sqr FC to rct BF, FC
ratio of sqr CG to rct AG, GB
and hence
by composing -- sum of antecedent & consequent to consequent (Euc V.17 & 18)
(ant I + con I) to ant I and (ant II + con II) to ant II
by converting -- antecedent to antecedent's excess beyond consequent (Euc. V.19)
ant I to (ant I − con I) and ant II to (ant II − con II)
same
ratio of (rct BF, FA + sqr FC) to sqr FC
ratio of (rct AG, GB + sqr CG) to sqr CG
ratio of sqr FC to (sqr FC − rct BF, FC)
ratio of sqr CG to (sqr CG − rct AG, GB)
Euc II.5
sqr AC
sqr CB
AB cut equally
AB cut unequally
Euc II.6
AB cut equally
AB cut unequally
hence
by alternating -- antecedent to antecedent, consequent to consequent (Euc V.16)
ant I con I ant II con II
ratio of sqr AC to sqr CB
same
ratio of sqr FC to sqr CG

CHAPTER 4. WHENCE AND WHITHER AT MID-JOURNEY

We have now come half-way through the propositions in the First Book of Apollonius's *Conics:* we have examined 30 of the 60 of them. That is to say, we are exactly in the middle of the work. Mid-way through our journey, where are we? We need to consider both where we are coming *from* and where we are going *to*.

First, let us take an overview of what there is to be seen along the road that has brought us hither, to the mid-course of our journey. After that, we shall take an overview of what there is to see along the road hence.

Right after the First Definitions, which are the very beginning, the first 14 propositions show what sorts of curves can be obtained if cones are cut.

Then, in propositions 15–16 and the Second Definitions, there are presented conjugate diameters and second diameters for those sections which have centers.

Then begins the story of the tangent: we are told (17) that a straight line through a particular point (the diameter's extremity), will, if it is parallel to an ordinate, be a tangent, and (18) that a straight line through any point inside the curve, will, if it is parallel to a tangent, be a chord; next (19), that a straight line through any point that lies on the diameter inside, will, if it is parallel to an ordinate, meet the curve. Next, we are told the relation in size of one ordinate to another—in the parabola first (20) and then in the other curves that are single lines, namely, the ones that have centers (21).

Then come the propositions that we have most recently encountered. Propositions 22–26 deal with the sections that are single curved lines—the parabola, the hyperbola, and the ellipse. With certain qualifications, we learn the following about a straight line in any of these curves: it will not be parallel to the diameter if it is a chord or a tangent; and if it is parallel to the diameter, then it will meet the curve—as a monosecant, however, and not as a chord or a tangent.

The non-parallelism of the chord is shown in propositions 22–23; the non-parallelism of the tangent, in propositions 24–25. The first proposition in each pair treats the parabola together with the hyperbola; the second treats the ellipse. That is to say, in each pair there is one proposition for both of the indefinitely prolongable curves, and one for the closed curve.

The propositions that show the non-parallelism for the ellipse's chord (23) and for the ellipse's tangent (25) contain this qualification: the straight line will have the property *if* the points (in the case of a chord) or point (in the case of a tangent) that it shares with the curve are *between* the endpoints of the two conjugate diameters. (As we have seen, that is because the chord that crosses one of the conjugate diameters can be parallel to the other one, and the tangent at the endpoint of one of the conjugate diameters must be parallel to the other one.)

Another qualification to what is stated above concerns the proposition that shows the parallel's property of being a monosecant (26). The qualification (the restriction of the property to the parabola and the hyperbola) is not stated explicitly. It is suggested by the fact that the proposition for the two indefinitely prolongable curves is not followed by a proposition for the ellipse. (As we have seen, any straight line that meets any closed curve, but is not tangent to it, must meet it again.)

The next two propositions (27–28) are additions to what we were told nine propositions earlier (18–19). What we are told now makes less restrictive what we were told earlier. We are now told that a straight line prolonged both ways will meet the section (that is, will be a chord) if (in a parabola) it merely goes through a point on the diameter, or (in opposite sections) it goes through a point within one of the curved lines and is parallel to a tangent of the other one.

Back in proposition 19, the straight line that met the section was not said to be prolonged both ways, and it was said to be drawn ordinatewise. Chords had not yet become a theme, and proposition 20 (which is used to demonstrate proposition 27) had not yet been demonstrated. Now in proposition 27 the straight line is said to meet the parabola when prolonged both ways, and it is not said to be drawn ordinatewise.

Back in proposition 18, the point within, and the straight line drawn through it to become the chord, both belonged to the same curved line as did the tangent to which the straight line was drawn parallel. Now in proposition 28, the point, and the straight line that is drawn through it to become the chord, belongs to one of the curved lines, while the tangent belongs to the other one.

Proposition 28 brings the opposite sections together with respect to chords—a chord belonging to one section is obtained by drawing a parallel to a tangent belonging to the other opposite section. After this proposition has thus obtained a chord *in* one of the curves that constitute the pair, the next proposition obtains what might be called a chord *between* the curves that constitute the pair.

Proposition 29 shows that such a "chord"—one between the opposite sections—is obtained by prolonging to the other section a straight line drawn through the center from a point on one of the sections. This proposition assures us that for opposite sections, as for the ellipse, if a line through the center of a section is bounded on one end, then it is also bounded by the section on the other end. (But only if it is bounded on one end already, since some straight lines through the center of the opposite sections are bounded by neither of the curves that constitute the pair. We shall later meet the straight lines that make a boundary among the straight lines through the center—a boundary between those that are bounded by the two curved lines and those that are not. These boundary lines, which are the unbounded straight lines through the center that are closest to the curved lines, and to which the tangents get closer and closer as the points of tangency get farther and farther from the center, will be called "asymptotes.")

Proposition 30 shows that in the opposite sections and also in the ellipse, when a chord is drawn through the center, it is bisected by the center. Both in the ellipse and in the opposite sections, the midpoint of the diameter is also the midpoint of every chord through it. The center is even more central than might at first appear. The opposite sections and the ellipse, which look so very different, are very much alike—as was suggested earlier by propositions 15–16 immediately after the various sections were first obtained.

We have just looked back at the road that has brought us from the beginning to here. Now let us take a brief look forward at the road that will take us on from here to the end. We are on the way not only *from* somewhere but also *to* somewhere. The full significance of what will now be pointed out is likely to elude those readers who are beginning this study. This account of what is to come will, no doubt, be more illuminating if re-read later on. At this point, though, it can have the benefit of providing some reason to think this difficult work a tightly constructed integral whole, rather than a tiresome miscellany of one exhausting item after another, and may therefore help to fortify some readers for the rest of the road ahead.

Having just treated tangents and chords and monosecants, Apollonius next (in the 31st proposition) presents a restriction on the hyperbola's tangents—he will show that no tangent will meet the oblique (transverse) side of an hyperbola at a point that is as far away from the section as the center is. He will then show (in the 32nd proposition) that if a straight line at the diameter's extremity has certain properties which, as we saw earlier, will make it a tangent, then there cannot be another such line at that point. Having thus shown the uniqueness of the tangent at a particular point of any conic (namely, at the diameter's extremity), he will then (propositions 33–36) show how to get that unique tangent at any point on any conic. That is to say: he shows, for the parabola (33) and for the central sections (34), that if a straight line at *any* point on the curve has certain properties, then it will be a tangent; and he will also show, for the parabola (35) and for the central sections (36), that if a straight line is a tangent to the curve at any point, then it will have those properties (which is the converse of 33–34), and that there cannot be another such line at that point.

After that treatment of tangents, he will complete the requisite preparations (37–45) that will enable him to get new diameters (46–48) and then their parameters (49–51).

So, being able to refer any conic to any one of its many diameters (using the tangent at that diameter's extremity as the ordinatewise line that orients the ordinates for that diameter), he will be able to return to cones (52–59), showing (what is the converse of 11–14) that if curves are parabolas or hyperbolas or ellipses (or opposite sections), then they can always be obtained by cutting a cone (or conic surfaces).

After that, he will end the First Book of his *Conics* with opposite sections of a sort that he will call "conjugate," making a new beginning (60).

As was pointed out a little while ago, we have worked our way through exactly half the propositions in the First Book of the *Conics*. This midpoint of the First Book's propositions taken numerically is also the book's midpoint considered thematically.

The movement of the First Book is this. The beginning (1–14) tells us that if cones are cut in various ways, then certain curves will be obtained. Propositions 15-51 mediate between the beginning and the ending. The ending (52–59) tells us that if certain curves have been given, then they will be obtained by cutting cones in various ways. A concluding proposition (60) then introduces a new development. The First Book of the *Conics* thus comes full circle: the beginning shows that cuts of cones are lines with certain properties, and the ending shows that lines with those properties are cuts of cones—that is, that the only lines with those properties are cuts of cones. But the middle that helps to carry us around full circle, also helps to carry us straight on forward to the conclusion that introduces the new development to come in the Second Book.

The properties that set us off on our round-trip are straight-line properties that extend a straight-line property of the circle (the square-rectangle property). The properties to which we will be led by the linear development are straight-line properties that are nothing like anything in the circle (they have to do with the so-called "asymptotes" of the hyperbola) but they do come out of the hyperbolic pair's resemblance to the ellipse, and this curve is something of an elongated circle.

What the mediation begins with is this: conjugate diameters as a feature of resemblance between the hyperbolic pair and the ellipse (15–16), followed by the second diameter (Second Definitions), whose definite size seems to suggest a difference between those sections. The suggestion about resemblance (16) is, however, picked up and carried through to a radical conclusion (60). By the very end of the book, that is to say, consideration of conjugate diameters will have led to consideration of conjugate opposite sections: a pair of hyperbolic pairs that is something like an ellipse.

The mediation opens with the 15th proposition and closes with the 51st. Its beginning is 15–16 and the Second Definitions; its middle is 17–45; and its culmination is the six propositions that give new diameters (46–48) and their parameters (49–51).

The key to the mediation is the tangent. The tangent gives a reference line for ordinates without any reference to cones, cutting planes, and axial triangles.

The middle of the mediation is 17–45. It begins with tangency: the 17th proposition gives a tangent for all sections, albeit only at a particular point (the diameter's extremity). The middle proposition of the middle of the mediation is the 31st proposition: it begins a sequence of six propositions on tangency (31–36). The middle pair of these six propositions is 33–34: it gives the tangent for all sections at *any* point at all. (The 34th proposition thus completes, for the central sections, what the 17th begins.)

The 31st proposition also mediates the line of thought that began with the 16th proposition and concludes with the 60th. The tangent property which it presents is this: none of an hyperbola's tangents will intersect with the oblique (transverse) side at a point that is as far away from the curve as the center is. This tangent property points us to the asymptote. A pair of asymptotes is just what unites and separates conjugate opposite sections. The pair of hyperbolic pairs shares a pair of asymptotes—straight lines which the curves nowhere meet but to which they each approach closer than by any given distance. Because they thus approach those straight lines that separate them from each other, they also, though they do not meet each other, do approach each other closer than by any given distance. But the 60th proposition merely points us to the asymptotes. These interesting straight lines are not presented until the next book.

Let us now move on through the remainder of the First Book of the *Conics*. We have come to its 31st proposition. Propositions 31–36 will occupy the next chapter (Chapter Five) in this guidebook *(All the tangents)*. Then, in Chapter Six (*From tangents to new diameters*) will come propositions 37–45, and, in Chapter Seven (*New diameters, new parameters*), will come propositions 46–51. After that, in the chapters of Part Five (*Provisional conclusion*), will come the finding of cones for the curves (propositions 52–59) and the presentation of conjugate opposite sections (proposition 60).

CHAPTER 5. ALL THE TANGENTS

31st proposition

(The figures in the subsequent explication of this proposition will present matters of general significance without presenting depiction of particulars. For an illustrative diagram that provides help in the reading of the proposition, see the Appendix at the end of this volume.)

#31 If on an hyperbola's figure's oblique side (*tês plagias pleuras tou eidous*) there be taken some point that takes off, on the side toward the vertex of the section, not less than half of the oblique side of the figure, and from it [that is, from that point] there fall forth a straight [line] onto the section, then [this straight line], being extended forth, will fall inside the section on the near side of the section.

—Let there be an hyperbola whose diameter the [line] AB is; and let there be taken on it some point—the point C, say—that takes off the [line] CB that is not less than half of the [line] AB; and let there fall forth upon the section some straight [line]—namely the [line] CD. I say that the [line] CD, extended, will fall inside the section.

—For, if possible, outside the section let it fall, as the [line] CDE; and from the chance point E let there be drawn down ordinatewise the [line] EG, and [let there be drawn down ordinatewise] also the [line] DH [but this line be drawn down from the point D].

—And, first, equal let the [line] AC be to the [line] CB. And—since the [square] from EG has, to the [square] from DH, an even greater ratio than does the [square] from FG to the [square] from DH, but (on the one hand) as is the [square] from EG to the [square] from DH so is the [square] from GC to the [square] from CH (on account of the [line] EG being parallel to the [line] DH) and (on the other hand) as is the [square] from FG to the [square] from DH so is the [rectangle] by AGB to the [rectangle] by AHB (on account of the section, [on account, that is, of the section's being an hyperbola])—therefore the [square] from GC has, to the [square] from CH, an even greater ratio than does the [rectangle] by AGB to the [rectangle] by AHB. Alternately, therefore, the [square] from CG has, to the [rectangle] by AGB, an even greater ratio than does the [square] from CH to the [rectangle] by AHB. Therefore, by taking apart (*dielonti*), [which is traditionally rendered by the Latin *separando*,] the [square] from CB has, to the [rectangle] by AGB, an even greater ratio than does the [square] from CB to the [rectangle] by AHB—which very thing is impossible. Not so, therefore, that the [line] CDE will fall outside the section. Inside, therefore, [is where it will fall].

—And also, on account of the same thing, the [straight line] from any of the points on the [line] AC—[that is, the straight line from a point that is closer to point A than is that point C which itself was so taken as to be closer to point A than is the midpoint of the figure's oblique side AB]—will much the more [*a fortiori* is the traditional Latin rendering] fall inside the section, since it will indeed fall inside the [line] CD.

31st proposition: enunciation

If a straight line is drawn to a point on an hyperbola, from a point on the oblique (transverse) side that is no closer to the vertex than the center is, then the straight line will, when prolonged, fall within the section on the near side. The straight line—like a straight line that is drawn parallel to the diameter of a parabola or an hyperbola (26th proposition)—will be a monosecant.

What does that tell us about the hyperbola's tangents? It tells us that at no point of an hyperbola will there be a tangent that intersects the oblique (transverse) side as far away from the curve as the center is. And what does that tell us about the hyperbola's shape? Well, we know that the hyperbola is indefinitely extendible, and we can see that the tangent taken at points farther and farther out will hit the oblique (transverse) side farther and farther away from the vertex in the opposite direction. But not as far as the center—that is what we have just been told. As the hyperbola is extended farther and farther, it gets less and less curvy (without ever going straight). Now, there is a line going up through the center that will not meet the hyperbola anywhere, while the oblique (transverse) side will meet the vertex as it goes through the hyperbola. If the line that stands straight up is rotated (with the center as the pivot) down to meet the oblique (transverse) side, will there be some *last* position before it begins to meet the hyperbola? And if the oblique (transverse) side is rotated up (with the center as the pivot) will there be some *first* position when it no longer meets the hyperbola? Think about that, and you will be on the way to learning about the interesting straight lines that Apollonius will eventually call "asymptotes." See figure 1.

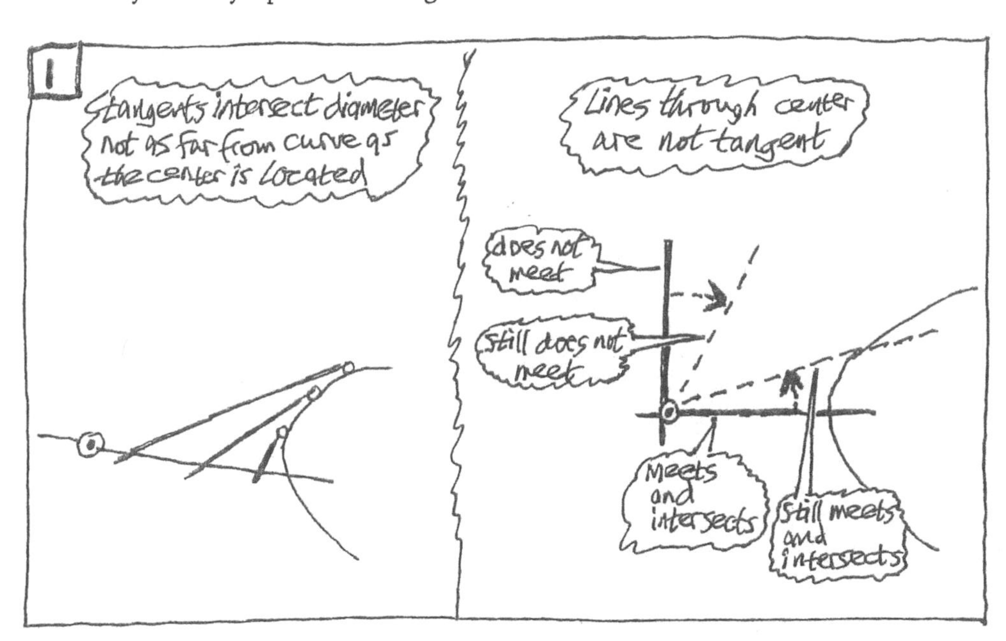

31st proposition: demonstration

Let C be the point on the oblique (transverse) side from which the straight line is drawn to the point D on the hyperbola. Assume that C is the center of the oblique (transverse) side AB. (If C were assumed to be even farther away from B than the center is, then the argument would impel us to the conclusion even more forcefully than it does when C is assumed to be only as far from B as the center is.)

Now what if the straight line CD, when prolonged beyond D, *did* have on it some point (E) that is *outside* the curve? The incompatible consequences that would follow are exhibited in a *reductio ad absurdum*. Here is how.

Draw a straight line ordinatewise from the supposedly outside point E. It will cross the curve at some point (F), and will also cross the diameter at some point (G). EG will be greater than FG; and so, EG's ratio to anything at all will be greater than FG's ratio to that same thing.

As that thing, take the ordinate drawn from the point D where the straight line from C meets the curve. This ordinate will meet the diameter at some point (H) that is closer to B than G is; and so,

GB must be greater than HB, and GA must be greater than HA.

Now a rectangle whose sides are the two greater lines (GB and GA) cannot be smaller than a rectangle whose sides are the two smaller lines (HB and HA). And yet, although the rectangle *cannot* be smaller, it nonetheless *must* be smaller. That is the absurdity.

But why must the one rectangle be smaller than the other one? This follows from EG's being greater than FG, and its consequently having a greater ratio to DH than FG has. The argument is depicted in figure 2.

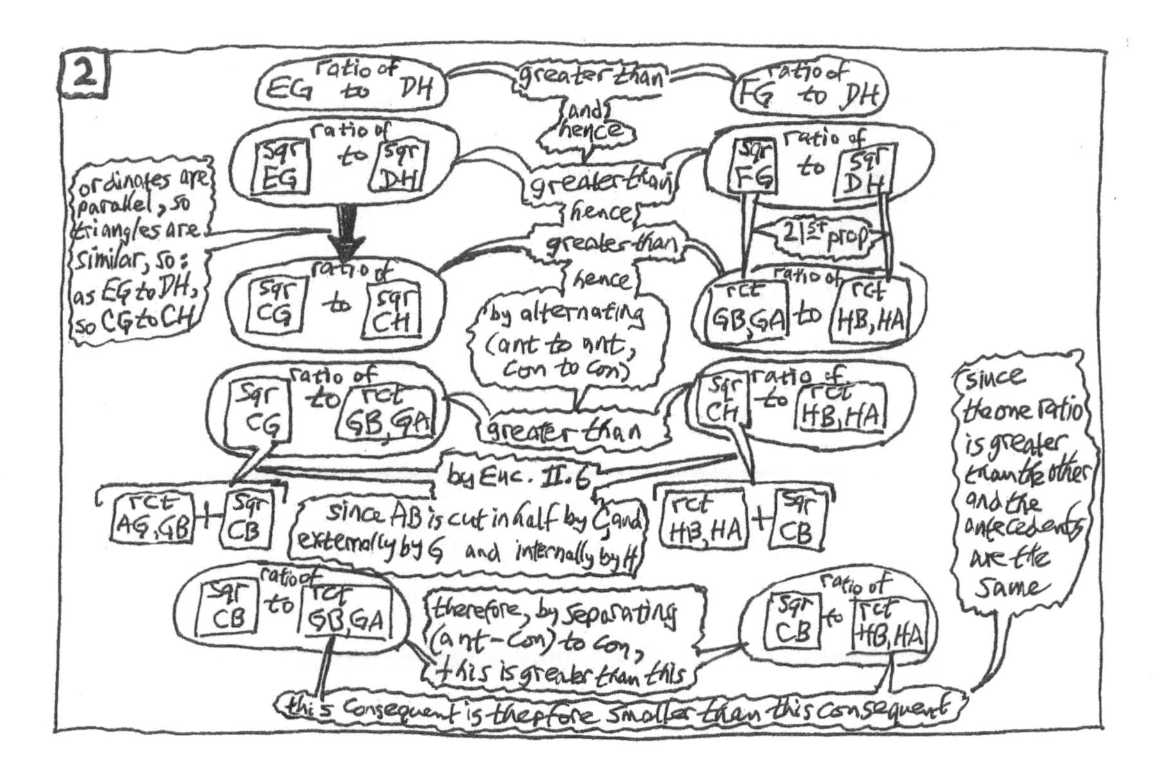
2
ratio of EG to DH
greater than
and hence
ratio of FG to DH
ratio of sqr EG to sqr DH
greater than
hence
ratio of sqr FG to sqr DH
ordinates are parallel, so triangles are similar, so: as EG to DH, so CG to CH
21st prop
greater than
hence
ratio of sqr CG to sqr CH
ratio of rct GB, GA to rct HB, HA
by alternating (ant to ant, con to con)
ratio of sqr CG to rct GB, GA
greater than
ratio of sqr CH to rct HB, HA
since the one ratio is greater than the other and the antecedents are the same
by Euc. II.6
rct AG, GB + sqr CB
since AB is cut in half by C, and externally by G and internally by H
rct HB, HA + sqr CB
ratio of sqr CB to rct GB, GA
therefore, by separating (ant – con) to con, this is greater than this
ratio of sqr CB to rct HB, HA
this consequent is therefore smaller than this consequent

32nd proposition

#32 If, through the vertex of a cone's section, a straight [line] be drawn parallel to a [line] that has been drawn down ordinatewise, then it touches the section; and into the place between the cone's section and the straight [line] another straight [line] will not fall in alongside.

—Let there be a cone's section—first the one called a parabola—whose diameter is what the [line] AB is; and from the [point] A let there be drawn parallel to ordinatewise the [line] AC. Now then, that it falls outside the section has been shown. I say then that also into the place between the straight [line] AC and the section another straight [line] will not fall in alongside.

—For if possible, let one fall in alongside—as the [line] AD, say; and let there be taken some chance point on it—the [point] D, say; and let there be drawn down ordinatewise the [line] DE; and let there be, as the [line] along which the lines drawn down ordinatewise have the power, the line AF. And—since the [square] from DE has, to the [square] from EA, an even greater ratio than does the [square] from GE to the [square] from EA, and the [square] from GE is equal to the [rectangle] by FAE—therefore also the [square] from DE has, to the [square] from EA, an even greater ratio than does the [rectangle] by FAE to the [square] from EA, that is, [it has an even greater ratio] than does the [line] FA to AE. Let there be made, then, [along diameter AB, a line AH just so long that] as is the [square] from DE to the [square] from EA so is the [line] FA to AH; and, through the [just located point] H, let there be drawn parallel to the [line] ED the [line] HLK. Since, then, as is the [square] from DE to the [square] from EA so is the [line] FA to AH, that is, the [rectangle] by FAH to the [square] from AH; and also, as is the [square] from DE to the [square] from EA so is the [square] from KH to the [square] from HA; but equal to the [rectangle] by FAH is the [square] from HL—therefore also as is the [square] from KH to the [square] from HA so is the [square] from LH to the [square] from HA. Equal therefore is the [line] KH to HL—which very thing is absurd. Not so, therefore, that into the place between the straight [line] AC and the section another straight line will fall in alongside.

—Let now the section be an hyperbola or ellipse or circle's periphery, whose diameter is what the [line] AB is, and upright [side] is what the [line] AF is; and let the [line] BF, being joined, be extended; and from the [point] A let there be drawn parallel to ordinatewise the [line] AC. Now then, that it falls outside the section has been shown. I say then that also into the place between the straight [line] AC and the section another straight [line] will not fall in alongside.

—For if possible, let one fall in alongside—as the [line] AD, say; and let there be taken some chance point on it—the [point] D, say; and let there be drawn down ordinatewise from it the [line] DE; and through the [point] E let there be drawn parallel to the [line] AF the [line] EM. And since the [square] from GE is equal to the [rectangle] by AEM, let there be made, [along ordinatewise line EM, a line EN just so long that] equal to the [square] from DE is the [rectangle] by ΛEN; and, thc [linc] AN having been joined, [that is, point A having been joined to the just located point N], let [this line AN] cut the [line] FM at the [point] X; and through the [point] X let there be drawn parallel to the [line] FA the [line] XH, and through the [point] H let there be drawn parallel to the [line] AC the [line] HLK. Since then the [square] from DE is equal to the [rectangle] by AEN, as is the [line] NE to ED so is the [line] DE to EA—and therefore as is the [line] NE to EA so is the [square] from DE to the [square] from EA. But (on the one hand) as is the [line] NE to EA so is the [line] XH to HA, and (on the other hand) as is the [square] from DE to the [square] from EA so is the [square] from KH to the [square] from HA. Therefore as is the [line] XH to HA so is the [square] from KH to the [square] from HA. Therefore a mean proportional is what the [line] KH is of the [lines] XHA. The [square] from HK, therefore, is equal to the [rectangle] by AHX. But also the [square] from LH is, to the [rectangle] by AHX, equal—on account of the section [on account, that is, of the section's being an hyperbola or ellipse or circle's periphery]. The [square] from KH, therefore, is equal to the [square] from HL—which very thing is absurd. Not so, therefore, that into the place between the straight [line] AC and the section another straight [line] will fall in alongside.

32nd proposition: enunciation

If a straight line is drawn ordinatewise through the vertex of any conic section, then it will be a tangent; and in the space between it and the conic section, there cannot be placed another straight line. That is what the enunciation says. But the 17th proposition has already shown that a straight line drawn ordinatewise at the vertex is tangent to the conic section. The news in this 32nd proposition is, therefore, that at the vertex there is only one tangent.

What does that uniqueness of the tangent at the vertex tell us about a conic section's shape? It tells us this: at the vertex, the curve is not pointy. The curve, which is a line that is continuous, also has a curvature that is continuous. In other words, the tangent will swing smoothly (rather than making an abrupt jump) as you pass through the vertex on a trip around the curve. See figure 3.

A straight line that will be tangent to a *particular* curve (the circle) at *any* point on the circumference is exhibited by Euclid in the 16th proposition of the Third Book of his *Elements*. In that same proposition, Euclid shows that that tangent is unique. By contrast, Apollonius here shows the uniqueness of the tangent to *any* conic section at a *particular* point on the curve.

In the proposition that began his story of the tangent (the 17th proposition), Apollonius demonstrated the existence of a straight line that will be tangent to any conic at the vertex. Fourteen propositions have intervened since then, and now the 32nd proposition demonstrates the uniqueness of the tangent belonging to any conic at the vertex. Some of the intervening propositions treated tangent lines. (The ones that did so were the first of them and the last of them, as well as the middle two, and the one mid-way between the middle two and the last one: that is, propositions 18 and 31, as well as 24–25 and 28.) Those intervening propositions allowed the tangent to be at any point, but what they showed was what you will get if you have a tangent already—rather than showing what will give you a tangent at any point. So far, it is only at the vertex (that is, at the endpoint of the diameter) that Apollonius has shown what will give you a tangent. The 32nd proposition therefore confines its showing of uniqueness to the tangent at that particular point.

What is demonstrated in the 32nd proposition could have been demonstrated immediately after the 17th proposition. The intervening propositions do not contain anything that is necessary for the demonstration of the 32nd. By contrast to what he has done here, Apollonius will not delay at all in showing the uniqueness of the conic's tangent in the general case: the two propositions after the 32nd will exhibit straight lines that will be tangent to conic sections at *any* point on them, and uniqueness will this time be shown right away, in the two propositions that come immediately after them. Considerations of logical dependence impose constraints upon the arrangement of the propositions, but such considerations do not by themselves determine the arrangement.

32nd proposition: demonstration

Apollonius shows by *reductio ad absurdum* that between the curve and a straight line that is tangent at the curve's vertex, there cannot be another straight line that goes through the vertex. The demonstration shows that if AD is a straight line drawn non-ordinatewise at the vertex A, then a certain point (K) on this straight line must coincide with a certain point (L) on the curve.

Points K and L are obtained by drawing a straight line ordinatewise through a certain point (H) on the diameter. This straight line meets the supposed tangent at K, and meets the curve at L. Points K and L must coincide because the straight line LH and the supposedly longer straight line HK must be equal to each other. Why must they be equal to each other? Because the squares on them must be equal to each other.

In order to show this, Apollonius draws ordinatewise, from random point D on the supposed tangent, a straight line; it meets the diameter at some point (E). He also draws some other straight lines. Figure 4 shows why the squares on HL and HK are equal to each other.

Why is it true (as figure 4 says) that each of those squares (one of them on HL, and the other on HK) has, in the parabola, a certain ratio to a certain magnitude, and is, in the other curves, equal to a certain magnitude?

For square [HL], both of those things are true because the square on the ordinate of a conic section is equal to a certain rectangle. For the square [HK], both of those things are true because triangles made by parallel straight lines are similar.

The demonstration first treats the parabola all alone, and then it treats the other sections (the "central" sections) all together.

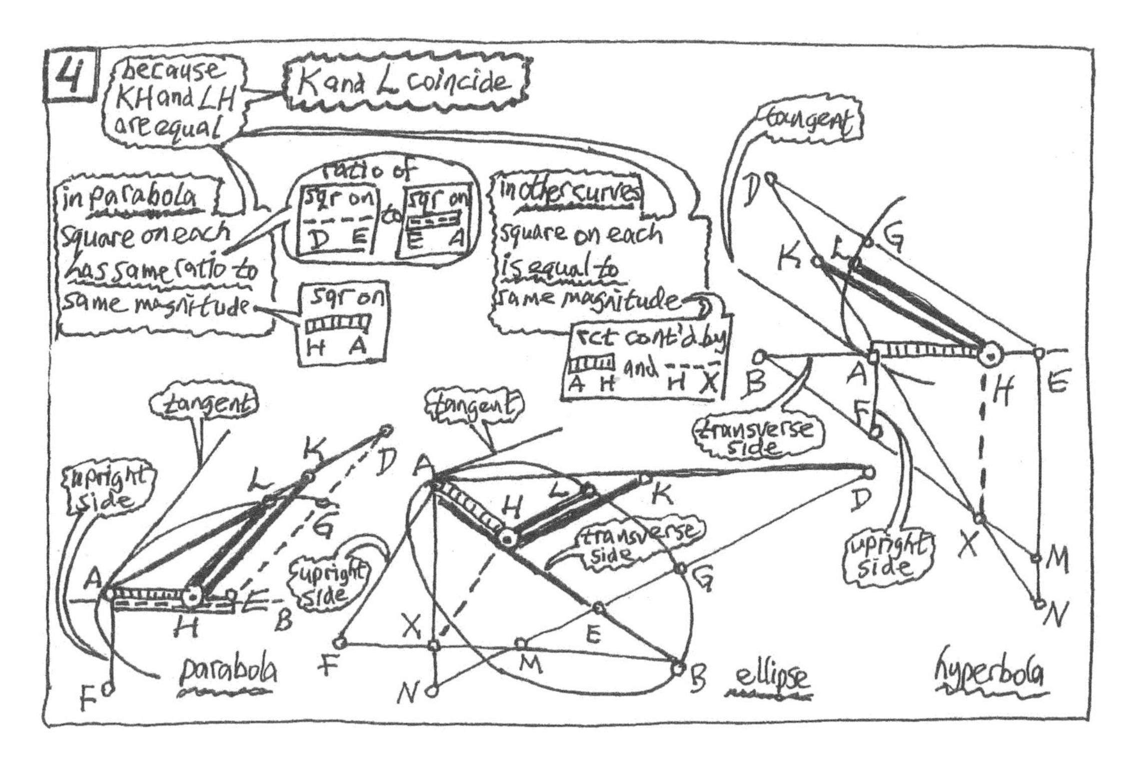

For all the curves, the straight line that supposedly is tangent cannot really be tangent. Why? Because there must be, in addition to the point of tangency, another point that is on the straight line and also coincides with a point on the curve. How is coincidence shown for those points K and L—K (which is on the line that is supposedly tangent) and L (which is on the curve)? It is shown by finding on the diameter some point (H) whose ordinate HL will be equal to a straight line HK that is drawn ordinatewise from H up to the supposed tangent. The point H that will do the trick can be located by determining its distance AH from the vertex. The line AH must be determined from what is known about those lines whose equality needs to be shown (the lines HK and HL). How? Let us first analyze the case of the *parabola*.

Consider what we know about HL. It is an ordinate. If we had AH, then (by the parabola's characteristic property) we know that

the square on ordinate HL would be equal to the rectangle whose sides are abscissa AH and the parameter (AF).

Consider now what we know about HK. The ordinatewise lines through H and D are parallel to each other. So, if we had AH, then (by the property of similar triangles) corresponding sides would be proportional: in other words,

the ratio of HK to HA would be the same as the ratio of DE to EA.

This statement about the line HK—what will it enable us to assert about the square on it? (We wish to speak, not about HK itself, but rather about the square on HK, because what we have said about the other line, HL, is not about HL itself but is rather about the square on it.) Because of the proportionality that involves line HK and the three other lines, we can assert the proportionality of the squares on them—namely:

the ratio of the square on HK to the square on HA would be the same as the ratio of the square on DE to the square on EA.

From that proportion, we know something about what would be necessary in order for square [LH] to be equal to the first term—to be equal, that is, to square [KH]. We know this:

the ratio of it (of square [LH]) to the second term (square [HA]) would have to be the same as the ratio of square [DE] to square [EA].

But in order to find just where H is, we need to determine its location without reference to L. How can we do that?

Well, we know a rectangle that would be equal to square [LH] and would have a common height (HA) with square [HA]. It is the rectangle whose other side (by the parabola's characteristic property) is the parameter AF. Hence,

the ratio of square [LH] to square [HA] would be the same as the ratio of rectangle [HA, AF] to square [HA], which (eliminating the common height) is the same as the ratio of AF to HA.

From this, it follows that the ratio of AF to HA will be the same as the ratio of the square on DE to the square on EA.

So, the equality of KH and LH can be shown if point H is located on the diameter at that place where its distance HA from the vertex fits the following proportion: parameter AF has to that distance HA the same ratio that the square on DE has to the square on EA. Hence, when distance HA is contrived to fit that proportion, then the ordinatewise line HK will be equal to the ordinate HL—and so, point K will be identical with point L. And so, since on the line that is supposedly tangent there is a point that must lie on the curve, the line that is supposedly tangent cannot really be tangent (fig. 5).

But do we not have reason to worry? Consider our contriving of the proportion for distance HA—the contriving, that is to say, which locates a point H that will do the job for the straight line that is *not* a tangent. Might it not do the same for the straight line that *is* a tangent? No, it will not. Why not? Because the true tangent is drawn ordinatewise. Lines drawn ordinatewise from points on the true tangent will therefore lie along the true tangent—and thus will not make triangles with it and the diameter.

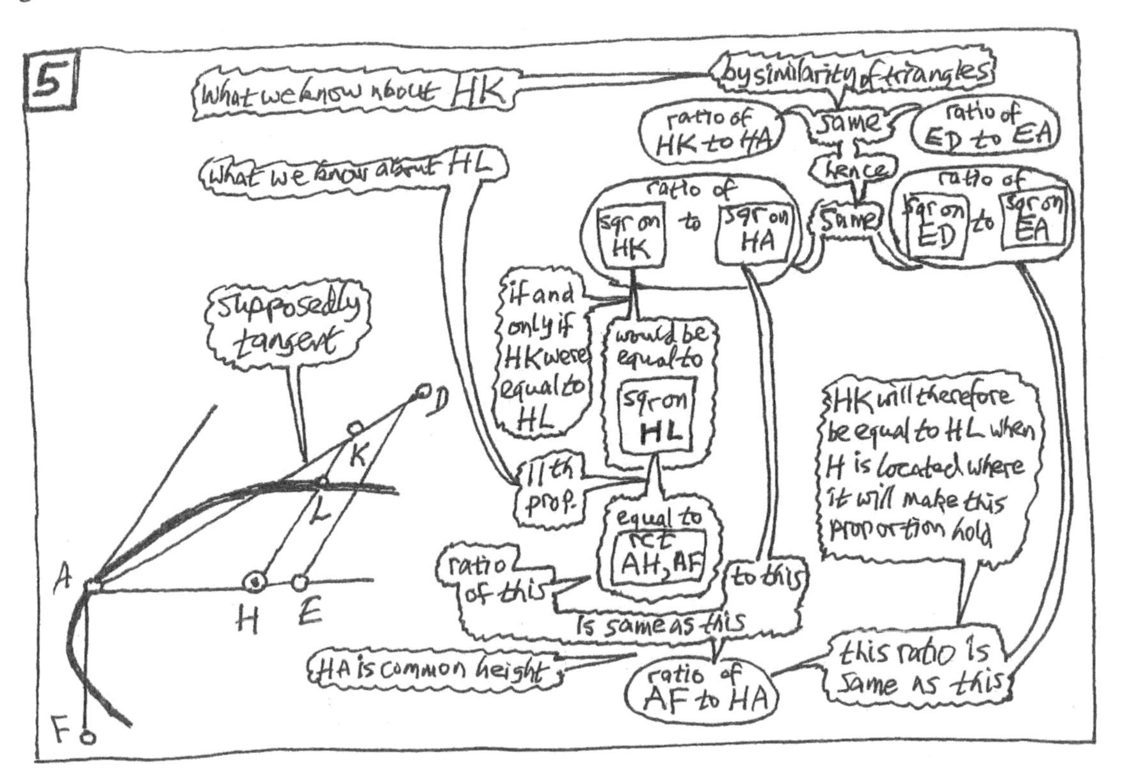

Now, let us turn to an analysis of the situation when the curve is not a parabola, but is rather *one of the other conic sections*.

Our procedure now, with these other curves, is like our previous procedure with the parabola. That is to say: we take what we know about HL (namely, the square-rectangle equality that is the curve's characteristic property) and what we know about HK (namely, a proportion given by the parallelism that makes similar triangles), and we put them together to give us a proportion that will make HK equal to HL.

Here, however, the procedure is a bit more complicated than it was with the parabola. With the parabola, the non-abscissa side of the rectangle that is equal to the square on the ordinate is the upright side (AF) itself; but with the other sections, it is a line determined by a relation among several lines—namely, the oblique (transverse) side, the upright side, and the abscissa.

The procedure, by making a second use of similar triangles, yields a proportion that must hold if HK is equal to HL—a proportion that locates the point H by locating a certain point N.

Point H is where the diameter is intersected by the line that is drawn parallel to the upright side, through point X. This X is the point where BF (the line that joins the oblique [transverse] side's far end and the upright side's far end) is intersected by AN, which is the line that joins the vertex to point N. This N is a certain point that lies on the line that is drawn parallel to the upright side, through point E; on this line, N is the point lying at that distance from E which is contrived to be a third proportional to EA and ED (that is to say: as EA is to ED, so also ED is to distance EN).

As with the parabola, so now with the other sections, in the following manner. If, on the line that is supposedly tangent, you take any point D that is outside the curve, then there is on AD, and closer to A than D is, a point K that is not inside the curve. Then, on a line that is drawn ordinatewise through K, there will be a point H that is also on the diameter, and a point L that is also on the curve. And then either K will lie outside the curve (that is, HL will be smaller than HK), or K will coincide with L (that is, HL will be equal to HK). The coincidence of K and L would mean that the line AD that is supposedly tangent will in fact intersect the curve. So if we can locate on the diameter a point H for which HL must be equal to HK, then we shall be able to conclude that AD cannot be tangent. Where must H be located in order to make HK equal to HL?

With the parabola, H was that point on the diameter whose distance AH from the vertex was the fourth proportional to these magnitudes: square [DE], square [AE], and the upright side AF. In other words, AH was contrived so that as square [DE] is to square [AE], so AF is to it (to AH).

Now, with these other curves, H will be that point on the diameter for which point N will make EN the third proportional to EA and ED. In other words, EN is contrived so that as EA is to ED so ED is to it (to EN). When H is that point, then a point (K) that is on the line AD will coincide with a point (L) that is on the curve; and hence line AD cannot lie entirely outside the curve. The line that supposedly is a tangent cannot really be a tangent. Figure 6 depicts how to find that N which yields H for the hyperbola and ellipse.

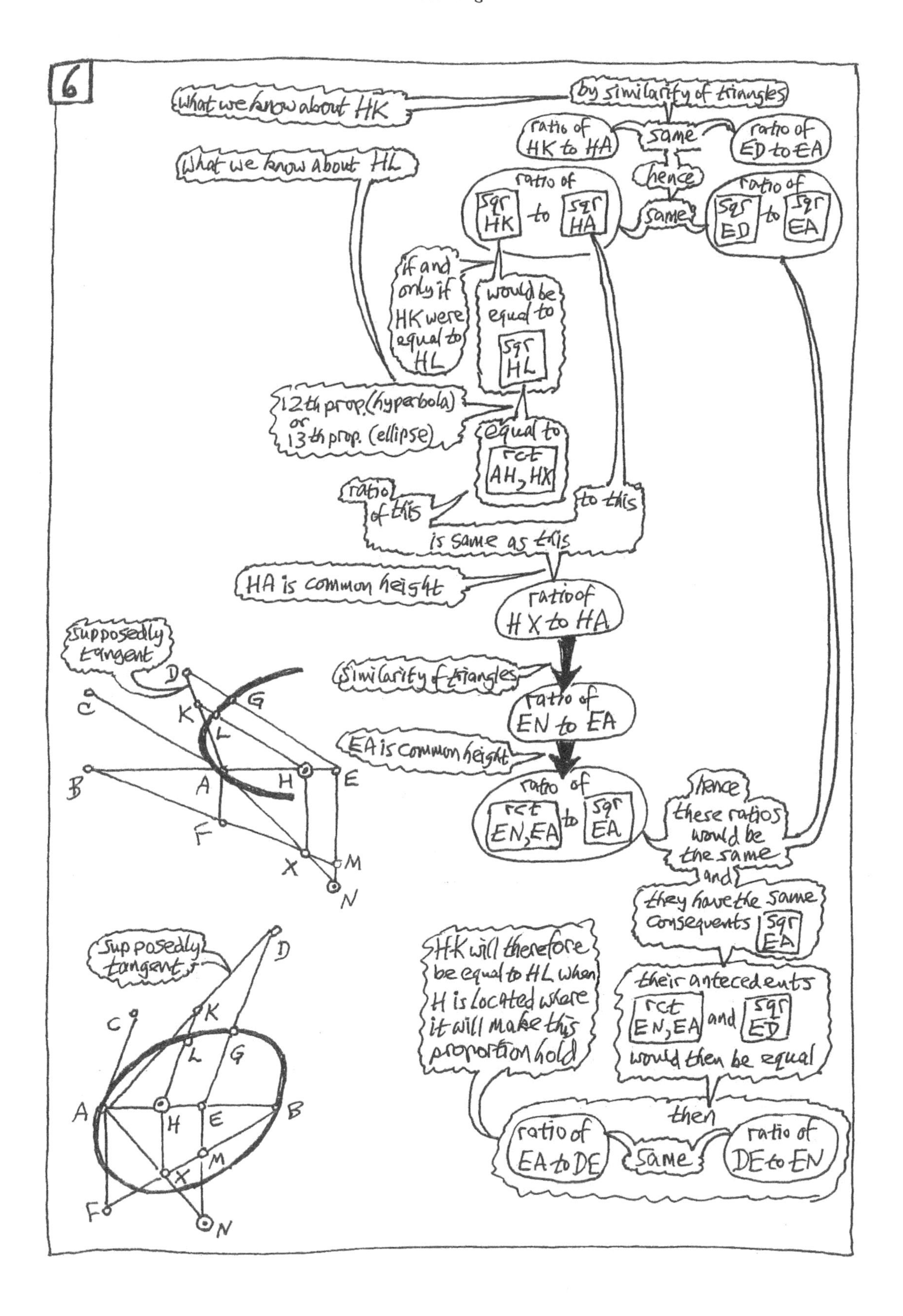
6
What we know about HK
by similarity of triangles
ratio of HK to HA
same
ratio of ED to EA
What we know about HL
hence
ratio of sqr HK to sqr HA
same
ratio of sqr ED to sqr EA
if and only if HK were equal to HL
would be equal to sqr HL
12th prop. (hyperbola) or 13th prop. (ellipse)
equal to rct AH, HX
ratio of this
to this
is same as this
HA is common height
ratio of HX to HA
Similarity of triangles
ratio of EN to EA
EA is common height
ratio of rct EN, EA to sqr EA
hence these ratios would be the same
and
they have the same consequents sqr EA
their antecedents rct EN, EA and sqr ED would then be equal
HK will therefore be equal to HL when H is located where it will make this proportion hold
then
ratio of EA to DE
same
ratio of DE to EN
supposedly tangent
D
C
K
G
L
B
A
H
E
F
X
M
N
Supposedly tangent
D
C
K
L
G
A
H
E
B
M
X
F
N

33rd and 34th propositions: enunciations

The 32nd proposition has just told us that there is only one tangent at the vertex of a conic section. Earlier, the 17th proposition told us what that line is which would be tangent at the vertex: draw a line through the vertex ordinatewise (we were told in effect) and you will obtain a tangent at that particular point. This point, the vertex, is that point on the curve itself which is also on the curve's diameter. So, given the diameter at whose endpoint on the curve we want to put a tangent, the 17th proposition told us how to do it. Apollonius told us how to do it immediately after he had done two things: in propositions 1–14 he had obtained a conic section of every sort, each with its diameter; and then, in propositions 15–16 he had presented conjugate diameters for each of the sections that had one.

Now, in propositions 33–34, Apollonius tells us how to put a tangent to the section at any point on the curve other than the vertex—that is to say, at any point other than the endpoint of that diameter which was obtained in the cutting that yielded the curve. Each of these other points at which he now tells us how to put a tangent will afterward turn out to be the endpoint of another diameter for the curve—a diameter whose ordinates will be parallel to that tangent at its endpoint. By telling us how to obtain a tangent at points on a conic section other than its vertex, Apollonius is preparing us to obtain new diameters for the conic section—diameters other than that one which was obtained in the cutting that yielded the curve, or in that one which is conjugate to the one so obtained.

How, then, is a tangent obtained at a non-vertex point? Not by drawing it in some specified direction, as ordinatewise. It must be done by finding another point—a point that, along with the tangency-point-to-be, determines a straight line that will be tangent.

So far, we have not heard anything about the location of any non-tangency points on a conic's tangent except what we were told by the 31st proposition—namely, that a tangent to an hyperbola cannot intersect the diameter's prolongation in a point that is as far away from the vertex as the center is. But where exactly that intersection happens to be located, we have not yet been told. Apollonius now tells us where it is. Indeed, he tells us where any tangent to *any* conic will intersect the diameter's prolongation. He tells us because this is the point we need in order to draw the tangent at any point other than the vertex.

To get the point that is needed for the drawing of the tangent, Apollonius uses an ordinate of the given diameter. Such an ordinate he used in the 17th proposition to obtain the tangent at the vertex, but he used it then to get the direction in which to draw ordinatewise the straight line that was to be tangent. What he uses now is a particular ordinate, since he needs it not for its direction (which is the same for all of the ordinates of the given diameter), but rather for the size of its abscissa (which is not the same for all those ordinates). The ordinate that he uses is the one that is drawn from the non-vertex point at which a tangent is to be drawn. He needs that ordinate for the distance that it cuts off along the diameter away from the vertex. That distance (the abscissa) will tell us where, on the diameter prolonged outside the curve, to locate the second point that is needed for drawing the tangent (fig. 7).

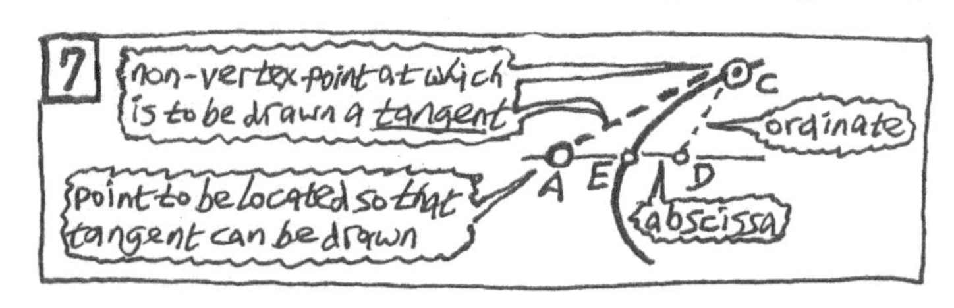

In the parabola, the diameter's prolongation (outside the curve) will be cut by the tangent at a point that is as far from the vertex as the vertex is from the point where the diameter (inside the curve) is cut by the ordinate that is drawn from the tangency-point.

In the hyperbola or the ellipse, the situation is more complicated. The key to the situation is still the abscissa of the tangency-point's ordinate—but now it is not that abscissa taken by itself. Rather, it is the ratio which that abscissa has to the line that results from the oblique (transverse) side when the abscissa is, in the hyperbola, added on to the oblique (transverse) side, or is, in the ellipse, taken away from it (fig. 8).

In other words: To obtain the point where the tangent will meet the diameter's prolongation, first locate the abscissa-point where the diameter is met by the tangency-point's ordinate. And then do the following:

> Take the ratio of the distances from the abscissa-point to
> the endpoints of the oblique (transverse) side,
> and then—on the diameter's prolongation—find the point from which
> the distances to the endpoints of the oblique (transverse) side are
> in that same ratio.
> This will be the very point that is needed,
> for the hyperbola or the ellipse or the circle.
>
> In the parabola, however, merely
> take the distance from the abscissa-point to
> the vertex endpoint of the diameter,
> and then—on the diameter's prolongation—find the point (outside the curve)
> from which the distance to the vertex endpoint of the diameter is
> the same.
> This will be the very point that is needed for the parabola.

Note that all this has been paraphrase, for in none of these propositions does Apollonius pose a problem and give its solution. The problems come much later—near the end of Book One. What we get here, in these propositions in the middle of the book, are theorems. The enunciations, that is to say, are stated more like this:

> A straight line that is drawn from any point on the curve,
> to another point obtained in the way just described,
> will be tangent to the curve.

Apollonius treats the tangent for the parabola in the 33rd proposition; and for the hyperbola, the ellipse, and the circle, in the 34th.

Immediately after that, the uniqueness of the tangent line of the 33rd proposition will be shown in the 35th, which is the converse of that 33rd proposition; the same thing will be done for the 34th proposition in the 36th.

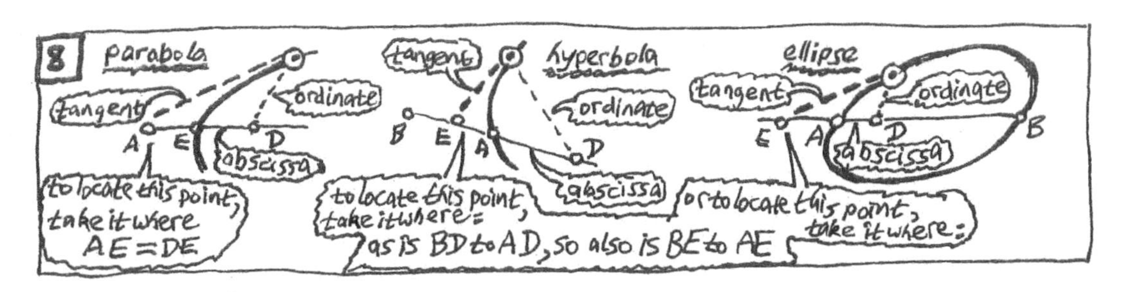

33rd proposition: demonstration

#33 If on a parabola there be taken some point; and from it, [an] ordinatewise [line] to the diameter be drawn down; and there be put, in a straight [line] with the [line] taken off by it, [taken off, that is, by the ordinatewise line,] from the diameter on the side toward the vertex, an equal [line], [with this equal prolongation stretching] from its extremity [that is, from the vertex, which is the extremity of the line taken off]—then the [line] that is joined, from the point generated [by the prolongation], to the point taken [at the outset], will touch the section.
—Let there be a parabola whose diameter the [line] AB is; and let there be drawn down ordinatewise the [line] CD; and, equal to the [line] ED, let there be put the [line] AE; and let there be joined the [line] AC. I say that the [line] AC, being extended, will fall outside the section.
—For if possible, let it fall inside, as the [line] CF; and let there be drawn down ordinatewise the [line] GB. And—since the [square] from BG has, to the [square] from CD, an even greater ratio than does the [square] from FB to the [square] from CD, but (on the one hand) as is the [square] from FB to the [square] from CD so is the [square] from BA to the [square] from AD, and (on the other hand) as is the [square] from GB to the [square] from CD so is the [line] BE to DE—therefore the [line] BE has to ED an even greater ratio than does the [square] from BA to the [square] from AD. But as is the [line] BE to ED so is the [rectangle] by BEA when taken four times, to the [rectangle] by AED when taken four times; and therefore also the [rectangle] by BEA when taken four times has, to the [rectangle] by AED when taken four times, an even greater ratio than does the [square] from BA to the [square] from AD. Alternately, therefore, the [rectangle] by BEA when taken four times has, to the [square] from AB, an even greater ratio than does the [rectangle] from AED when taken four times, to the square arising from AD—which very thing is impossible. For—since equal is the [line] AE to the [line] ED—the rectangle by AED when taken four times is equal to the [square] from AD. But the [rectangle] by BEA when taken four times is less than the [square] from AB—for the point E is not the [line] AB's midpoint (*dichotomia*). Therefore, not inside the section does the [line] AC fall. Therefore it touches [the section].

The demonstration is a *reductio ad absurdum*. It begins by supposing this: when straight line AC is prolonged, it goes inside the curve, thus placing point F inside. That supposition has the following consequence:

the ratio of the quadruple of rectangle [BE, EA] to square [BA]
is greater than
the ratio of the quadruple of rectangle [DE, EA] to square [DA].

Why that follows, is depicted in figure 9.

That consequence of supposing point F to be inside the curve is absurd—as will become clear by putting it together with the fact that point A has been chosen so as to make AE equal to DE. Putting those two things together will have as a consequence something that contradicts another consequence of the fact that point A has been chosen so as to make AE equal to DE. (Fig. 10.)

So, when the second point (A) has been chosen as specified (namely, to be just as far from the vertex as the vertex is from the point where the tangency-point's ordinate meets the diameter), then we shall be led into absurdity if we suppose that when we prolong the straight line that runs from A to C, it will go inside the curve. That is to say: we cannot avoid contradicting ourselves if we deny that AC is tangent.

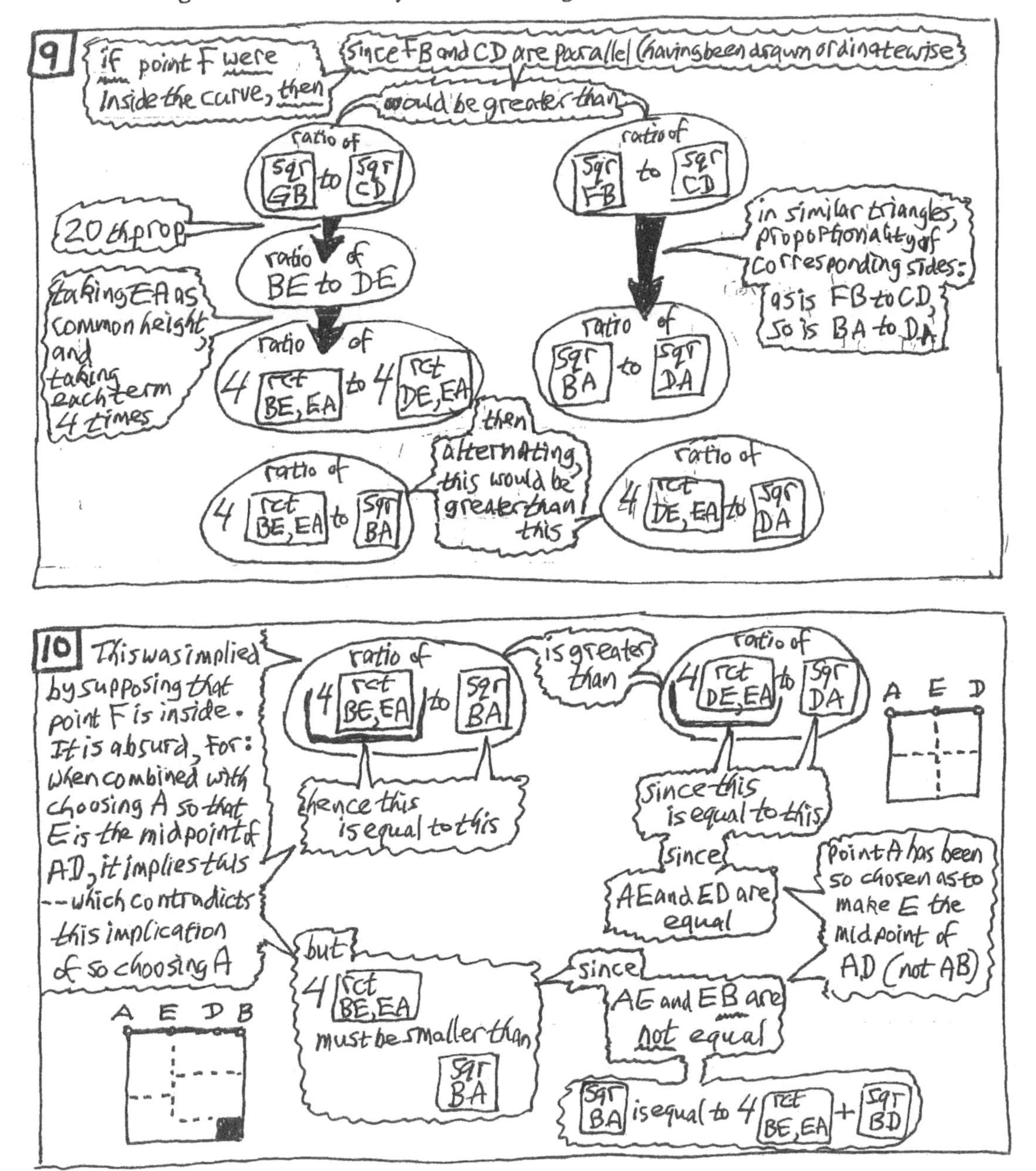

34th proposition: demonstration

#34 If, on an hyperbola or ellipse or circle's periphery, there be taken some point; and from it there be drawn down to the diameter a straight [line] ordinatewise, and the ratio which [the resulting lines] have to each other (the [lines, that is,] which are cut off by the [line] that has been drawn down [and which stretch from the cutting-point] to [each of the two endpoints that are] the bounds of the oblique side of the figure)—if this ratio be the one which the segments of the oblique side have to each other, [these segments of the oblique side being the lines which stretch to the endpoints of the oblique side from some point taken on the oblique side], so that correspondents in the proportion be the segments on the side toward the vertex—then the straight [line] that joins the point taken on the oblique side and the [point] taken on the section will touch the section.

—Let there be an hyperbola or ellipse or circle's periphery whose diameter is what the [line] AB is; and let there be taken some point on the section—the [point] C, say; and from the [point] C let there be drawn ordinatewise the [line] CD; and let there be made [a proper placement for a point E, in line with the diameter, such that] as is the [line] BD to DA so is the [line] BE to EA; and let there be joined the [line] EC. I say that the [line] CE touches the section.

—For if possible, let it cut [the section], as the [line] EDF; and let there be taken some point on it—the [point] F, say; and let there be drawn down ordinatewise the [line] GFH; and let there be drawn through the [points] A, B parallel to the [line] EC the [lines] AL, BK; and let the joined [lines] DC, BC, GC have been extended to the points K, X, M. And—since as is the [line] BD to DA so is the [line] BE to EA, but (on the one hand) as is the [line] BD to DA so is the [line] BK to AN, and (on the other hand) as is the [line] BE to AE so is the [line] BC to CX, that is, the [line] BK to XN—therefore as is the [line] BK to AN so is the [line] BK to NX. Equal, therefore, is the [line] AN to the [line] NX. Therefore the [rectangle] by ANX is greater than the [rectangle] by AOX. Therefore the [line] NX has to XO an even greater ratio than does the [line] OA to AN. But as is the [line] NX to XO so is the [line] KB to BM. Therefore the [line] KB has to BM an even greater ratio than does the [line] OA to AN. Therefore the [rectangle] by KB, AN is greater than the [rectangle] by MB, AO. And so the [rectangle] by KB, AN has, to the [square] from CE, an even greater ratio than does the [rectangle] by MB, AO to the [square] from CE. But (on the one hand) as is the [rectangle] by KB, AN to the [square] from CE so is the [rectangle] by BDA to the [square] from DE—on account of the similarity of the triangles BKD, ECD, NAD—and (on the other hand) as is the [rectangle] by MB, AO to the [square] from CE so is the [rectangle] by BGA to the [square] from GE. Therefore the [rectangle] by BDA has, to the [square] from DE, an even greater ratio than does the [rectangle] by BGA to the [square] from GE. Alternately, therefore, the [rectangle] by BDA has, to the [rectangle] by AG, BG, an even greater ratio than does the [square] from DE to the [square] from EG. But (on the one hand) as is the [rectangle] by BDA to the [rectangle] by AGB so is the [square] from CD to the [square] from GH, and (on the other hand) as is the [square] from DE to the [square] from EG so is the [square] from CD to the [square] from FG; therefore also the [square] from CD has, to the [square] from HG, an even greater ratio than does the [square] from CD to the [square] from FG. Less, therefore, is the [line] HG than the [line] FG—which very thing is impossible. Not so, therefore, that the [line] EC cuts the section. Therefore it touches [the section].

The demonstration of the previous proposition waited awhile to use the consequences of the placement of that second point needed for determining the straight line tangent to the parabola. Here with curves other than the parabola, by contrast, those consequences are the very point of departure for the demonstration.

This second point (here called E) is chosen so as to meet a more complex specification than was met by its counterpart for the parabola. That point for the parabola (called A) was chosen on the diameter's prolongation so as to be as far from the vertex as the vertex itself was from the point where the diameter was met by the tangency-point's ordinate. With the parabola, that is to say, there was an equality of the straight lines that go to the vertex from the points where the diameter is met by these two lines: the line that will be tangent, and the tangency-point's ordinate. Now with these sections other than the parabola, there is (instead of that) a proportion among the straight lines that go to the endpoints of the oblique (transverse) side from the same two points—namely, from point E (where the diameter is met by the line that will be tangent) and from point D (where it will be met by the tangency-point's ordinate).

With the parabola, the simplicity of the specification of the second point made it possible for the demonstration to move (by means of the 20th proposition) to a swift conclusion. With the other curves, the complexity of the specification of the second point requires a long preparation before the demonstration can move (by means of the 21st proposition) to an even swifter conclusion.

The demonstration proper begins with the proportion that specifies the location (E) of the second point on the tangent for these other curves. The proportion is this:

as DB is to DA, so EB is to EA.

The manipulation of that proportion occupies most of the demonstration. What is accomplished by the manipulation is this: the proportion is turned into a statement about a certain ratio of rectangles which was prominent in an earlier proposition about these curves. That ratio (which was prominent in the 21st proposition) is the ratio of rectangle [BD, DA] to rectangle [BG, GA]. Each of the rectangles in the ratio has as its one side the abscissa (DA or GA) and as its other side the line (BD or BG) that results when that abscissa is added on to or is taken away from the oblique (transverse) side (fig. 11).

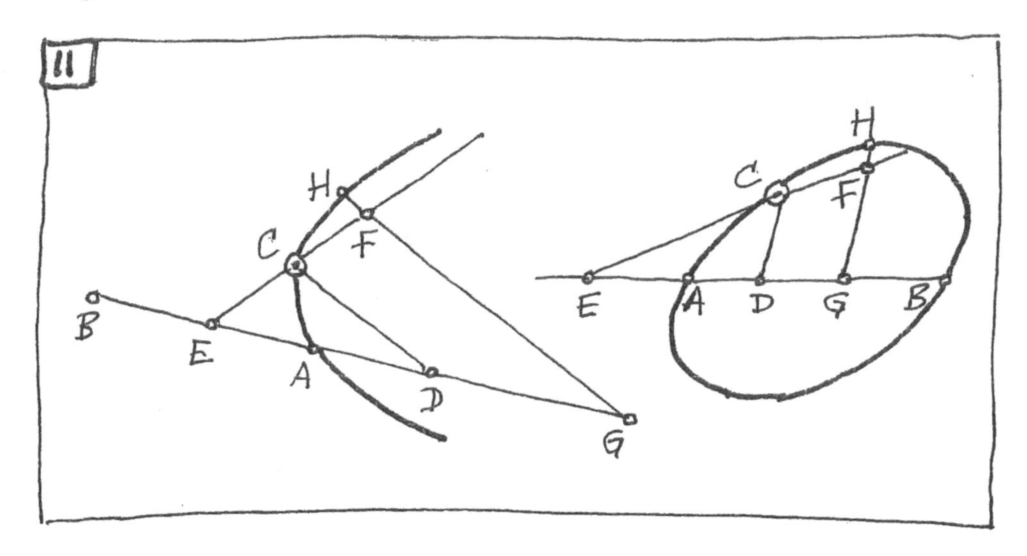

According to the consequence which the manipulation derives from the way that E was chosen (so as to make the proper arrangement with D and with the endpoints of the oblique [transverse] side), that ratio of rectangles ([BD, DA] to [BG, GA]) is greater than the ratio of square [DE] to square [GE]. According to the 21st proposition, that ratio of rectangles is the same as the ratio of the squares on the ordinates, which ordinates here are CD and GH. Putting together this property that was stated in the 21st proposition and that consequence which the manipulation derived from the way that E was chosen, we then use what was used as the point of departure in the demonstration for the parabola—namely, the proportionality of the sides of the similar triangles made by the parallelism of the two ordinatewise lines, one of them drawn through the tangency-point, the other drawn through the point that is supposed to be inside the curve.

Thus, as is depicted in figure 12, it is shown that GH is smaller than FG. So, by supposing F to be *inside* the curve, we are led to the conclusion that F must be farther from G than H is—which is impossible, for (since H is *on* the curve) F would thus be placed *outside* the curve. In other words, even if we begin by supposing F to be inside the curve, we must conclude that it is outside. We cannot avoid contradicting ourselves when we deny that a straight line is tangent to the curve if it joins any point C that is on the curve to a point that is located where E is.

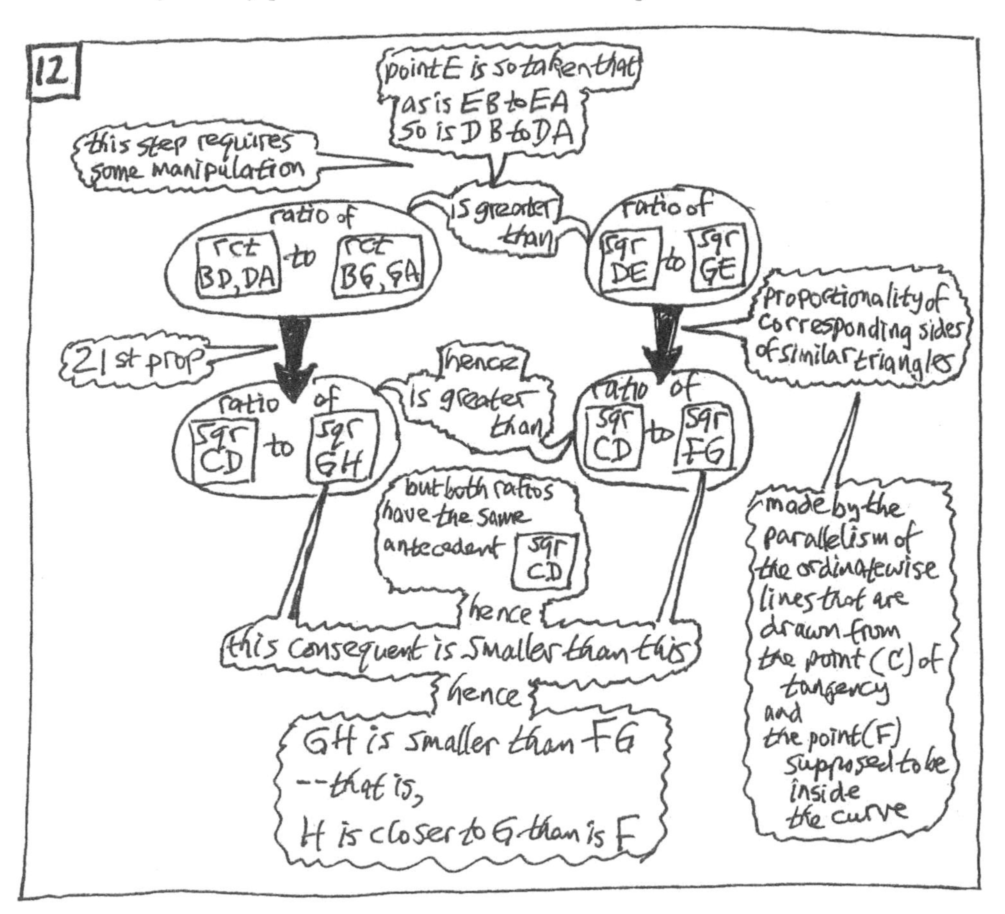

The conclusion follows swiftly from applying the 21st proposition—even more swiftly than the conclusion followed from applying the 20th proposition in the previous proposition. But that application came much sooner in the previous proposition. In the present proposition, the 21st proposition is applied to the statement that the ratio of rectangle [BD, DA] to rectangle [BG, GA] is greater than the ratio of square [DE] to square [GE]. That statement takes a while to derive from the proportion that locates E (the point where the diameter's prolongation is met by the straight line that will be tangent). The derivation takes place in two stages:

> the first goes
> *from* the proportion that determines point E—
> namely: as DB is to DA, so EB is to EA—
> *to* the assertion that
> the ratio of NX to XO is greater than the ratio of OA to AN;
>
> the second moves on from there
> *to* the assertion that
> the ratio of rectangle [BD, DA] to rectangle [BG, GA] is greater than
> the ratio of square [DE] to square [GE].

The lines are exhibited in figure 13, and the derivation is depicted after that, in figure 14. Note that the derivation does not depend on any property of the conic sections. It depends solely on the fact that G is a point on a straight line along which the points A, B, D, and E are arranged according to the proportion.

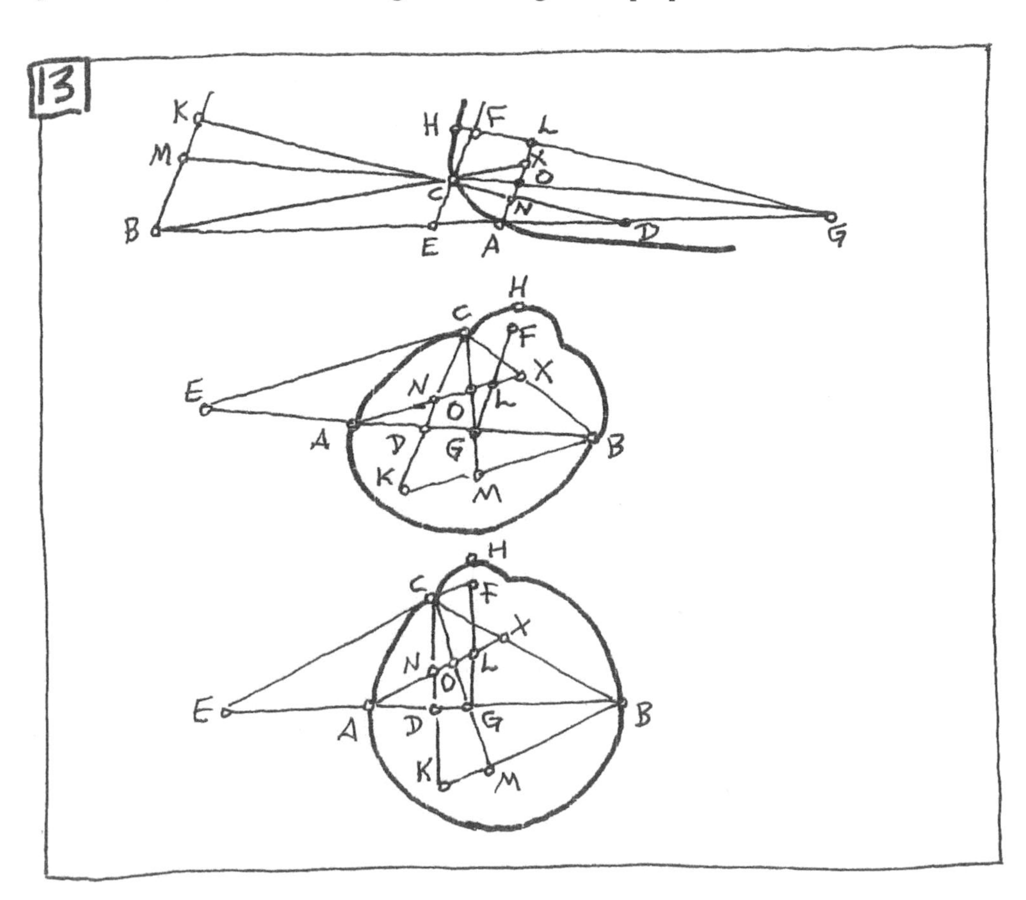

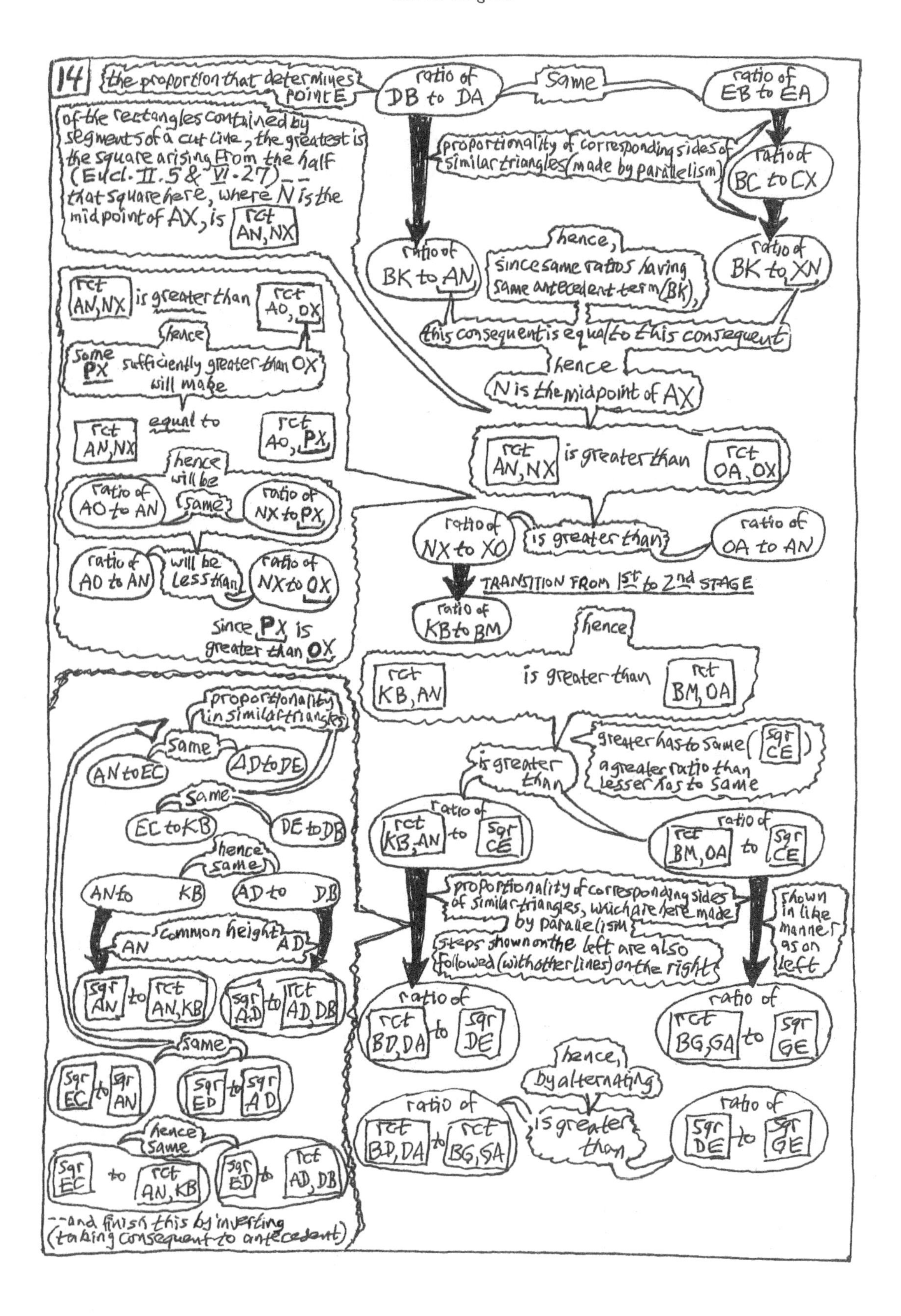
14
the proportion that determines point E
ratio of DB to DA
same
ratio of EB to EA
proportionality of corresponding sides of similar triangles (made by parallelism)
ratio of BC to CX
of the rectangles contained by segments of a cut line, the greatest is the square arising from the half (Eucl. II.5 & VI.27) -- that square here, where N is the midpoint of AX, is rct AN, NX
ratio of BK to AN
hence, since same ratios having same antecedent term (BK),
ratio of BK to XN
this consequent is equal to this consequent
hence
N is the midpoint of AX
rct AN, NX is greater than rct OA, OX
ratio of NX to XO
is greater than
ratio of OA to AN
TRANSITION FROM 1st to 2nd STAGE
ratio of KB to BM
hence
rct KB, AN is greater than rct BM, OA
rct AN, NX is greater than rct AO, OX
hence
some PX sufficiently greater than OX will make
rct AN, NX equal to rct AO, PX
hence will be same
ratio of AO to AN
ratio of NX to PX
ratio of AO to AN
will be less than
ratio of NX to OX
since PX is greater than OX
greater has to same (sqr CE) a greater ratio than lesser has to same
is greater than
ratio of rct KB, AN to sqr CE
ratio of rct BM, OA to sqr CE
proportionality of corresponding sides of similar triangles, which are here made by parallelism
steps shown on the left are also followed (with other lines) on the right
shown in like manner as on left
ratio of rct BD, DA to sqr DE
ratio of rct BG, GA to sqr GE
hence, by alternating
ratio of rct BD, DA to rct BG, GA
is greater than
ratio of sqr DE to sqr GE
proportionality in similar triangles
same
AN to EC
AD to DE
same
EC to KB
DE to DB
hence same
AN to KB
AD to DB
common height
AN
AD
sqr AN to rct AN, KB
sqr AD to rct AD, DB
same
sqr EC to sqr AN
sqr ED to sqr AD
hence same
sqr EC to rct AN, KB
sqr ED to rct AD, DB
--and finish this by inverting (taking consequent to antecedent)

the harmonic mean and the central sections

Let us consider further the arrangement, in the 34th proposition, of those important points that lie in a line with the diameter—namely, E (the point where the line that will be tangent meets the prolongation of the diameter), D (the point where the tangency-point's ordinate meets the diameter), and A and B (the two endpoints of the transverse side). To prepare for considering that arrangement, we need to consider the significance of the "mean" in classical mathematics.

Whenever any magnitude is greater than another, there are many magnitudes whose size is intermediate. Among all those magnitudes that are both greater than the smaller and smaller than the greater, do any of them stand out as being mid-way—as being in some way a "mean" between those two "extremes"? Three such sorts of being mid-way were singled out by the ancient Greek mathematicians. A man named Nicomachus reports (in the twenty-fifth chapter of his *Introduction to Arithmetic*) that they spoke of the *arithmetic* mean, the *geometric* mean, and the *harmonic* mean.

The arithmetic mean	The geometric mean
is a middle in this way:	is a middle in this way:
the *difference* between	the *ratio* of
the greater and the mean	the greater to the mean
is the same as	is the same as
that between	that of
the mean and the smaller.	the mean to the smaller.

The harmonic mean puts those two ways together, in the following way: the *difference* between the greater and the mean has the same *ratio* to the difference between the mean and the smaller, that the greater itself has to the smaller itself. So, for example:

Suppose one line to be 4 meters long and another line to be 10;
then their *arithmetic* mean will be 7 meters long—
since the difference between the greater (10-meter) line and this 7-meter-mean
is the same 3 meters that is
the difference between this 7-meter-mean and the smaller (4-meter) line.

Or suppose one line to be 3 meters long and another line to be 12;
then their *geometric* mean will be 6 meters long—
since the smaller (3-meter) line will be made equal to this 6-meter-mean by
the same doubling that will make
this 6-meter-mean equal to the greater (12-meter) line.

Finally, suppose one line to be 60 meters long and another line to be 20,
then their *harmonic* mean will be 30 meters long—
since the smaller (20-meter) line will be made equal to
the greater (60-meter) line by the same tripling that will make
the 10 meters of difference between
this 30-meter-mean and the smaller (20-meter) line
equal to the 30 meters of difference between
the larger (60-meter) line and this 30-meter-mean. (Fig.15.)

The construction of the arithmetic mean and the geometric mean, when you are given two straight lines as extremes, is no great problem. For the arithmetic mean, take a line obtained by adding on to the smaller line a line that is one-half the difference between the smaller and the greater line. For the geometric mean, use the construction from Euclid (fig. 16) that we considered earlier, in the second part of this guidebook.

But the construction of the harmonic mean is a more complicated business. We shall turn to that in a moment, but first let us consider what is "harmonic" about it. The harmonic mean is harder to grasp than the arithmetic mean and the geometric mean. The harmonic mean originally had another name. It was called the "subcontrary" mean. Its name was changed by a man named Archytas, who called it the "harmonic" mean in his book *On Music*. As has been said before, "harmony" is a Greek term from carpentry, where it referred to the condition in which things fit; the term, imported into music, came to refer to musical fitting. What, then, is the musical significance of the harmonic mean?

Well, take two notes, one with a higher tone (H) and one with a lower tone (L), such that the higher one sounds like the lower one again, only higher. Such notes are said to be an "octave" apart. Now, consider the notes between them (that is to say: those notes that are both lower in tone than the higher, and higher in tone than the lower), and take that one (note M) whose tone stands out as sounding as close as possible to being right in the middle. It will be the one that is called the "fifth." It will not quite seem to be exactly in the middle, for the step up from L to M will seem to be greater than the step up from M to H; but the fifth will seem to be the closest we can get to an harmonious tone that will divide the interval of an octave symmetrically. Our musical scale thus is organized by two intervals—the one called the octave and the one called the fifth.

Now the string-length that will sound a note whose tone is up the scale a fifth turns out to be an harmonic mean (as defined a few paragraphs back). It will be the harmonic mean between the string lengths that will sound notes whose tones are an octave apart.

If, instead of the string-length that is the harmonic mean between two string-lengths that sound the octave, you take the string-length that is the arithmetic mean, then what it will sound is not the fifth, but the "fourth." It will sound less harmonious with them. Among the intermediate string-lengths, that one which sounds most harmonious with the extremes was given the name "harmonic" by the ancient mathematicians. It seemed most fitting.

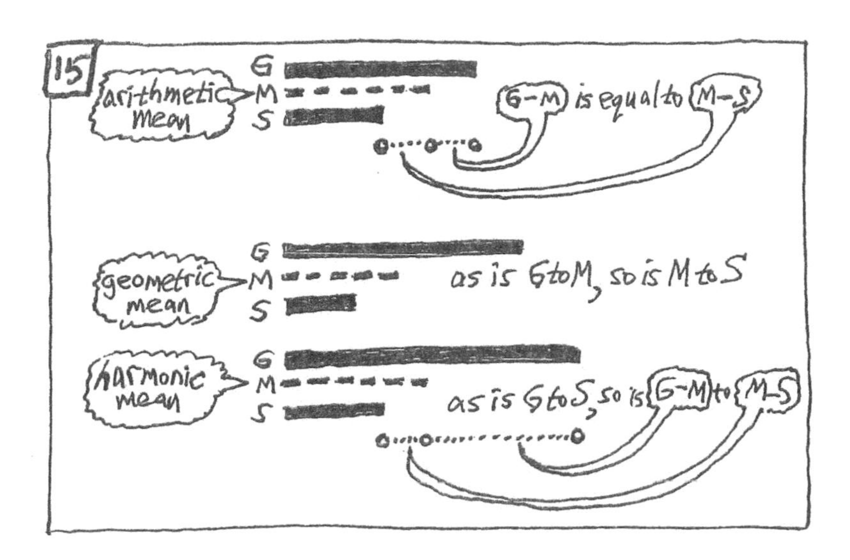

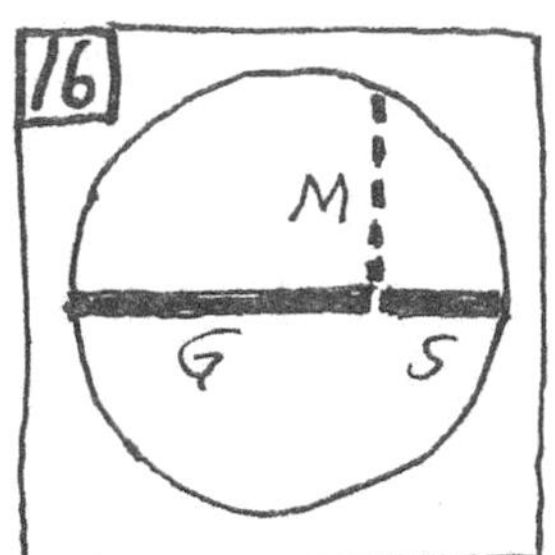

Having been applied when strings had certain lengths, the term "harmonic" then came to be applied to the corresponding proportion in the cutting of a line, and also to the arrangement of the points involved.

Harmonic proportion results when a straight line is cut in the same ratio both internally and externally. The point that cuts it internally is on the line itself; the one that is said to cut it externally is on the line's prolongation.

The lines in the one ratio go from the internal cutting point to the endpoints of the straight line that is cut by it. The lines in the other ratio go from the external cutting point to the same endpoints; and these lines have as their harmonic mean the line between the two cutting points themselves.

The four points that are involved (the one point cutting the straight line internally, the other cutting it externally, and the two endpoints of that straight line which is cut both internally and externally) are then said to be arranged harmonically. Note that by alternating the terms in the proportion, the points are shown to be arranged harmonically in reverse as well (fig. 17).

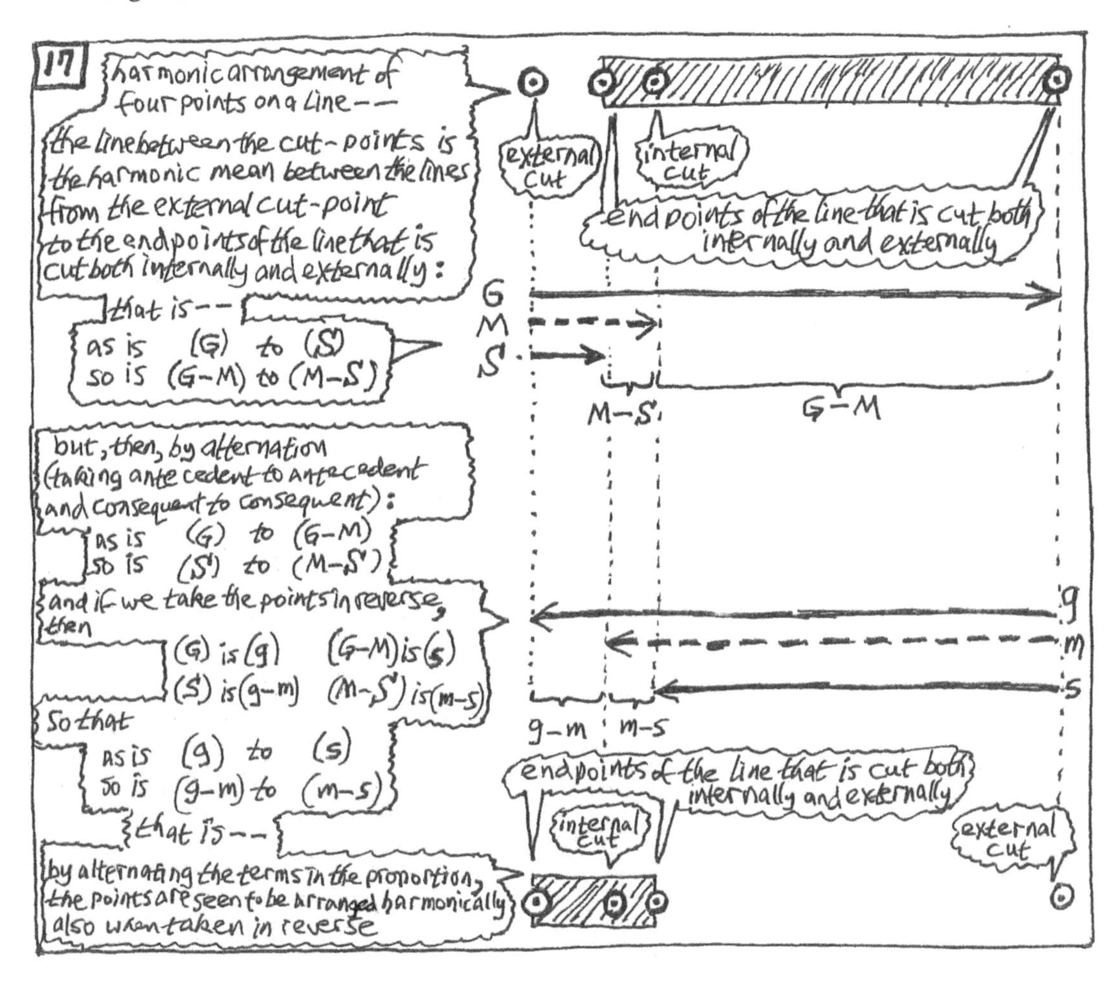

Now the demonstration of the 34th proposition presupposes that certain constructions can be done—namely:

(for the *hyperbola*)	(for the *ellipse*)
given a straight line that is already cut *externally*, to cut it in the same ratio *internally*;	given a straight line that is already cut *internally*, to cut it in the same ratio *externally*.

How this can be done, is depicted in figure 18.

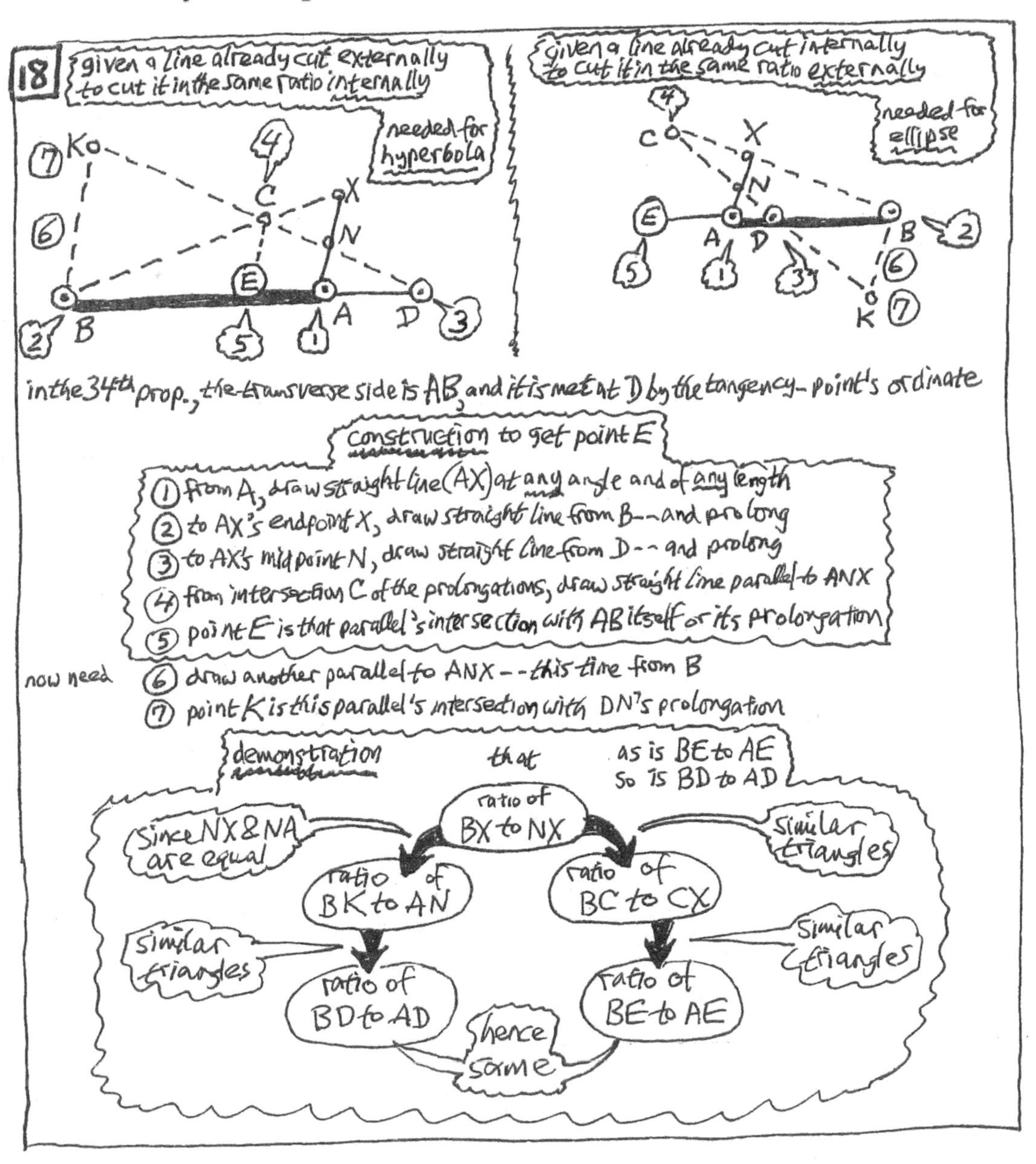

So, the gist of the present proposition might be put as follows. Given the point (C) at which a straight line is to be tangent to a curve, the straight line will be determined by locating another point on the straight line. In the 34th proposition, this other point is E. It is the point where the diameter is met by the tangent. It is located by specifying its place in an arrangement with points D and A and B. (Point D is where the diameter is met by the tangency-point's ordinate, and A and B are the endpoints of the oblique [transverse] side.) Now the hyperbola's oblique (transverse) side is outside the curve, being the diameter's prolongation, whereas the ellipse's oblique (transverse) side is inside the curve, being the diameter itself. The arrangement which locates E, then, is this:

In the hyperbola,	In the ellipse,
the oblique (transverse) side is cut	the oblique (transverse) side is
cut	
internally by	*externally* by
the tangent	the tangent
in the same ratio in which it is	in the same ratio in which it is
cut *externally* by	cut *internally* by
the tangency-point's ordinate;	the tangency-point's ordinate.

That is to say: in both cases, EB will be to EA as DB is to DA. This proportion tells us that points A, B, D, and E are arranged harmonically (fig. 19).

Now what the proposition tells us, in effect, is this. Suppose that a conic section has oblique (transverse) side AB (point A being the vertex), and that from any point C on the curve you draw an ordinate and give the name D to its intersection with the diameter. If, on the oblique (transverse) side, you locate point E where it will cause the oblique (transverse) side (the line that runs from B to A) to be the harmonic mean between the line that runs from B to D and the line that runs from B to E, then the line that runs from E to C will be tangent at C (fig. 20).

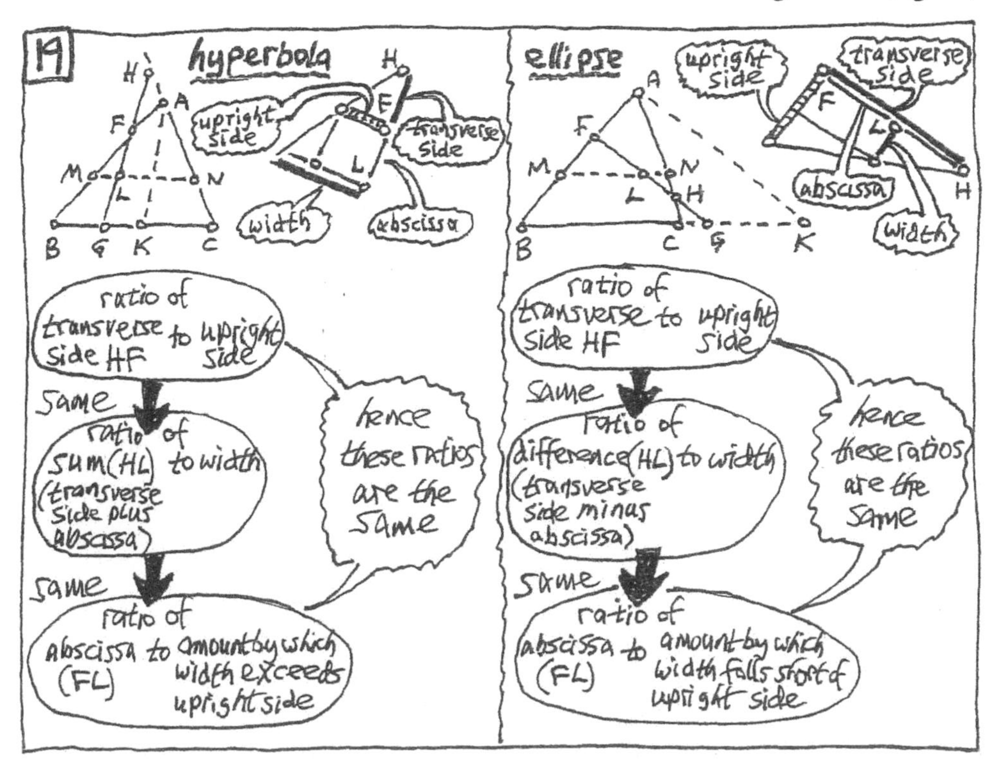

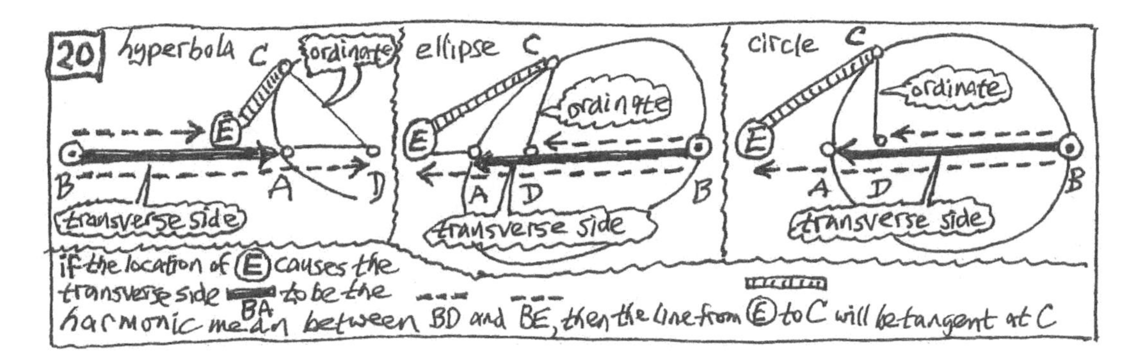

Since the construction for that point E, which brings it about that the ratio of EB to EA is the same as the ratio of DB to DA, thus also brings it about that AB is an harmonic mean between the line from B to D and the line from B to E, we might therefore call it a procedure for constructing a third harmonic proportional.

Now note that the first steps of the 34th proposition are merely—in reverse—the steps for demonstrating that such a third harmonic proportional has been constructed. The only thing done by Apollonius that is special to the conic sections is his arranging for AX to be at a special angle and of a special size. This is arranged by arranging that BX's prolongation and DN's prolongation will have as their point of intersection the tangency-point C. Apollonius could have drawn parallels at B and at E and at A that would have made *any* angle with AB—thus not making the tangency-point C be the point of intersection that Apollonius at the beginning does make it be. But then point C, and the ordinate that is drawn from it, would have had to be introduced at the end, when the 21st proposition is applied, after Apollonius has established that the ratio of rectangle [BD, DA] to rectangle [BG, GA] is greater than the ratio of square [DE] to square [GE]. So, by not using C in the derivation that constitutes most of the demonstration, Apollonius would unnecessarily have introduced an extra point and extra lines.

The point of the proposition is beautifully simple, but the demonstration is a tangle. Up until almost its very end, the demonstration of the 34th proposition is devoted to establishing, not what is true specifically of conic sections, but rather what is true generally of harmonic means—namely, the fact that if points A, B, D, E are arranged harmonically on a straight line, then the ratio of rectangle [DB, DA] to rectangle [GB, GA] is greater than the ratio of square [DE] to square [GE]. After that, it takes very little to obtain an order of size for ratios involving lines in conic sections specifically (fig. 21).

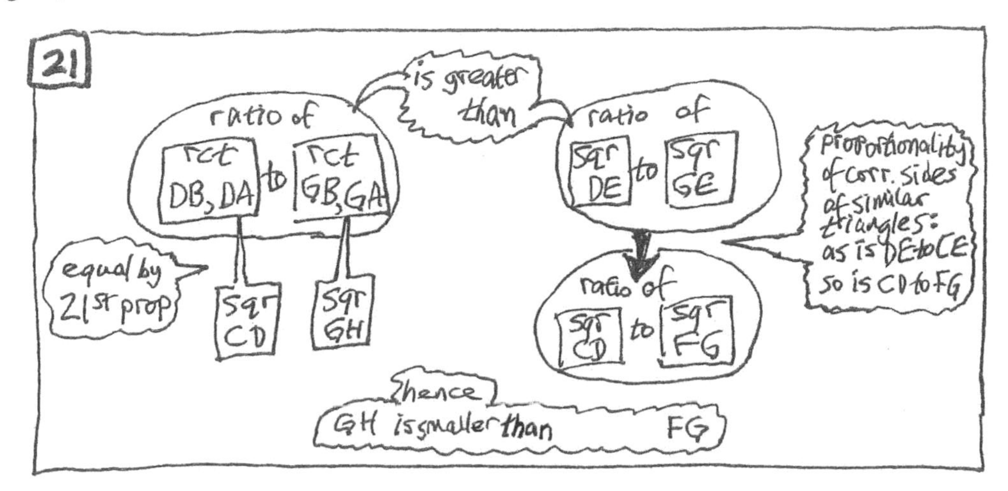

35th and 36th propositions

Once the tangent line has been obtained for the two cases, it is immediately shown to be unique for the two cases. That is to say, the 35th and 36th propositions are the converses of the previous two propositions.

Earlier, in obtaining the line that is tangent at the *vertex* (17th proposition) and in showing it to be unique (32nd proposition), the parabola and the other curves were treated separately but within a single proposition. Now, however, in obtaining the line that is tangent at *any* point (33–34), and in showing it to be unique (35–36), there are separate propositions for the case of the parabola and for the case of the other curves. There is nonetheless some resemblance between the enunciations in the separate propositions.

What is, in the parabola	is, in the other curves
(props. 33 & 35),	(props. 34 & 36),
an *equality* for two lines	a *sameness for the ratio* of two lines
—both of which terminate at	—both of which terminate at
the single endpoint of	one of the two endpoints of
the diameter,	the diameter's oblique (transverse) side.

Figure 22 depicts the enunciations with the points so labeled as to bring out the resemblance.

In the 33rd and 34th propositions, the straight line will be tangent if the points are arranged as stated. In the 35th and 36th propositions, the points will be arranged as stated if the straight line is tangent. Moreover, no other straight line will be tangent at that same contact point.

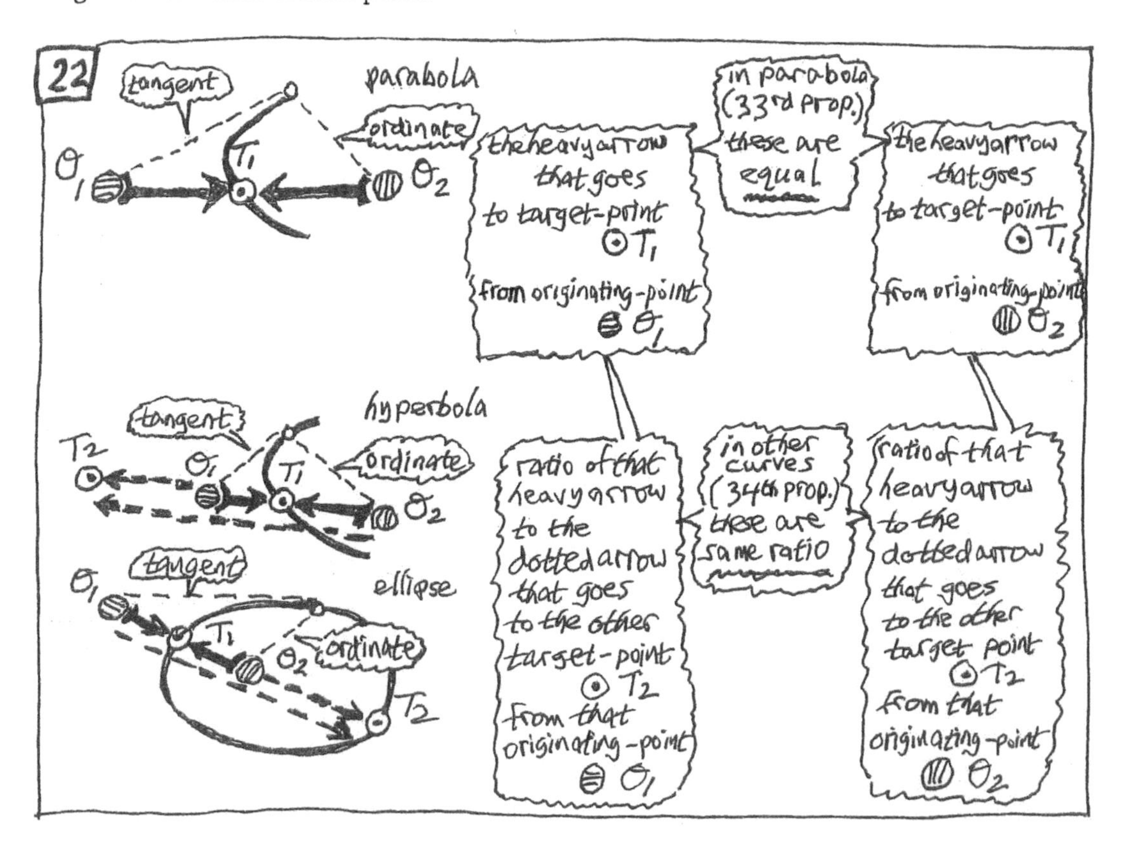

#35 If, with respect to a parabola, a straight [line] touch [the section]—[a straight line, that is,] which falls together with the diameter outside the section—then the straight [line] that is drawn from the touch[point] to the diameter, ordinatewise, will take off an equal [line] from the diameter on the side toward the vertex of the section—[a line equal, that is,] to the line between it [that vertex] and the touching[line]; and into the place between the touching[line] and the section not one straight [line] will fall in alongside.

—Let there be a parabola a diameter of which is what the [line] AB is; and let there be drawn up ordinatewise the [line] BC; and let there be a [line] touching the section, namely the [line] AC.

—I say that the [line] AG is equal to the [line] GB. For, if possible, let it be unequal to it; and, equal to the [line] AG, let there be put the [line] GE; and let there be drawn up ordinatewise the [line] EF; and let there be joined the [line] AF. The [line] AF, therefore, being extended will fall together with the straight [line] AC—which very thing is impossible, for of two straight [lines] there will then be the same bounds, [the endpoints, that is, will be the same]. Not, therefore, unequal is the [line] AG to the [line] GB; therefore it is equal.

—I say next that, into the place between the straight [line] AC and the section, not one straight [line] will fall in alongside. For, if possible, let the [line] CD fall in alongside; and equal to the [line] GD let there be put the [line] GE; and let there be drawn up ordinatewise the [line] EF. Therefore the straight [line] joined from the [point] D to the [point] F touches the section. Being extended, therefore, it will fall outside it. And so it will fall together with the [line] DC, and of two straight [lines] there will be the same bounds—which very thing is absurd. Not so, therefore, that into the place between the section and the straight [line] AC there will fall in, alongside, a straight [line].

#36 With respect to an hyperbola or ellipse or circle's periphery, if there touch [the section] some straight [line] that falls together with the figure's oblique side, and from the touch[point] there be drawn down to the diameter, ordinatewise, a straight [line]—then will it be that as the [line] taken off by the touching[line], on the side toward the bound, [toward, that is, the endpoint] of the oblique side, is to the [line] taken off by the touching[line], on the side toward the other bound of the [oblique] side, so the [line] taken off by the [line] drawn down, on the side toward the bound of the [oblique] side, is to the [line] taken off by the [line] drawn down, on the side toward the other bound of the [oblique] side, so that the correspondents in the proportion are continuous; and into the place between the touching[line] and the cone's section another straight [line] will not fall in alongside.

—Let there be an hyperbola or ellipse or circle's periphery whose diameter is what the [line] AB is; and a touching[line] let the [line] CD be; and ordinatewise let there be drawn down the [line] CE.

—I say that as is the [line] BE to EA so is the [line] BD to DA. For, if not, let it be that, [with there being made a proper placement for a point G, in line with the diameter,] as is the [line] BD to DA so is the [line] BG to GA; and let there be drawn up ordinatewise the [line] GF. Therefore the straight [line] joined from the [point] D to the [point F] will touch the section. Being extended, therefore, it will fall together with the [line] CD. Of two straight lines, therefore, the same points are bounds—which very thing is absurd.

—I say that between the section and the straight [line] CD not one straight [line] will fall in alongside. For if possible, let it fall in alongside, as the [line] CH; and let there be made [a proper placement for a point H, in line with the diameter, such that] as is the [line] BH to HA so is the [line] BG to GA; and let there be drawn up ordinatewise the [line] GF. Therefore the straight [line] joined from the [point] H to the [point] F, being extended, will fall together with the [line] HC. Of two straight lines, therefore, there will be the same bounds—which very thing is impossible. Not so, therefore, that into the place between the section and the straight [line] CD there will fall in, alongside, a straight [line].

(For illustrative diagrams that provide help in the reading of these two propositions, see the Appendix at the end of this volume.)

tangency in the circle and other curves

Apollonius, in the 17th proposition, obtained a tangent at the vertex of the diameter of any conic section; in his 32nd proposition, he showed its uniqueness. Euclid, in the 16th proposition of the Third Book of the *Elements*, obtained a tangent at any point on a circle; in that same proposition, he showed its uniqueness.

Apollonius, in his 33rd proposition, obtained a tangent at any point of a parabola; he showed its uniqueness in his 35th proposition, immediately after obtaining, in his 34th, such a tangent for the other conic sections. It was by determining the point at which the tangent will intersect the diameter, that Apollonius obtained the tangent at any point on any section .

Euclid, in the 17th proposition of the Third Book of his *Elements*, obtained a tangent to a circle from any point outside the circle. From that 17th proposition of Euclid, one can without too much trouble determine the point at which a tangent at any point of a circle will intersect the circle's diameter. The point of intersection will cut the diameter externally in the same ratio in which it is cut internally by the perpendicular that is dropped to the diameter from the tangency-point.

Apollonius, in his 34th proposition, showed not only for the circle, but also for the hyperbola and the ellipse, that the tangent will intersect the diameter at the point that cuts the diameter externally in the same ratio in which it is cut internally by the ordinate that is drawn to the diameter from the tangency-point. To show that fact for the circle alone, Apollonius need not have done very much: that fact can be shown for the first and simplest of the curves merely by going a bit further than Euclid went in the 17th proposition of his elemental book on circles, the Third Book of the *Elements*. If Apollonius had treated the circle in that way, he could then have moved on to treat the ellipse, and then the hyperbola, and then the parabola. But that is not what we have seen him do. Rather, he first presented the tangent's intersection-property for the parabola, in the 33rd proposition, and then he demonstrated the tangent's intersection-property for the central sections all at once, in the same steps for all of them, in the 34th proposition. That is like what he did earlier for the property shown in the 20th and 21st propositions. The 34th proposition is not the only proposition in which Apollonius refrains from beginning with an easy demonstration, for the circle alone, of a property that he prefers to present in a single demonstration that is harder to follow but applies to all the central sections.

Let us see just how Euclid's solution to the problem of drawing a tangent to a circle from an outside point could have been used to derive, for the circle, the property of the tangent that we have seen Apollonius demonstrate in a single demonstration for the circle and the other central sections all together. First, let us look at Euclid's solution. It is depicted in figure 23. After that, in figures 24 and 25, will come a depiction of how Euclid's proposition could have led to the intersection-property of the tangent for the circle.

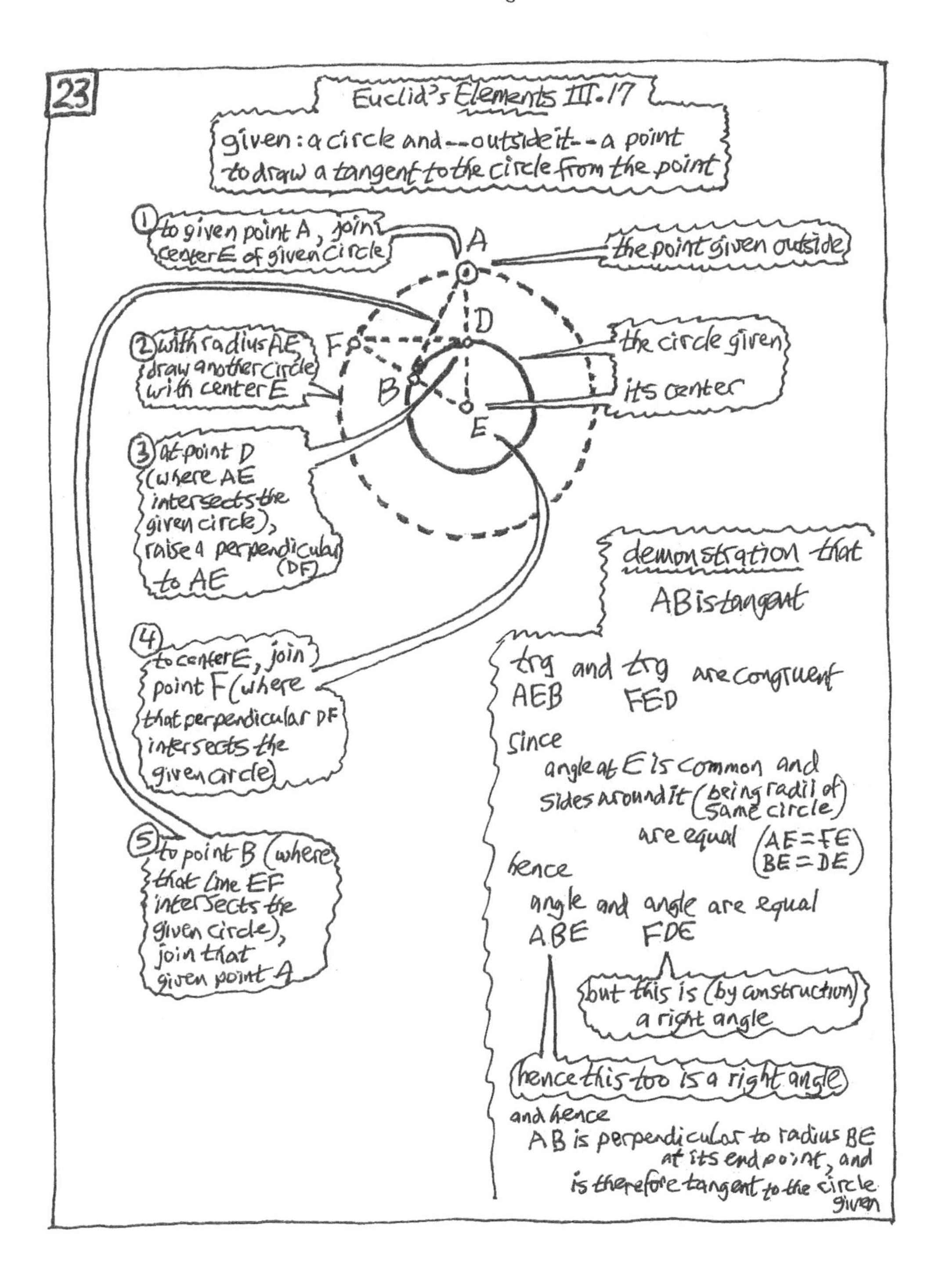
23
Euclid's Elements III.17
given: a circle and -- outside it -- a point
to draw a tangent to the circle from the point
1 to given point A, join center E of given circle
A
the point given outside
D
F
B
E
2 with radius AE, draw another circle with center E
the circle given
its center
3 at point D (where AE intersects the given circle), raise a perpendicular (DF) to AE
4 to center E, join point F (where that perpendicular DF intersects the given circle)
5 to point B (where that line EF intersects the given circle), join that given point A
demonstration that AB is tangent
trg AEB and trg FED are congruent
since
angle at E is common and
sides around it (being radii of same circle) are equal (AE = FE) (BE = DE)
hence
angle ABE and angle FDE are equal
but this is (by construction) a right angle
hence this too is a right angle
and hence
AB is perpendicular to radius BE at its end point, and is therefore tangent to the circle given

Now, how can Euclid's proposition III.17 be used to find that point (E), on the circle's diameter, through which will pass the line that is tangent to the circle at the given point C on the circle? From Euclid's proposition, it follows that the tangent will pass through point E if the placement of point E on the diameter guarantees that—as depicted in figure 24—line GH will pass through the tangency-point C.

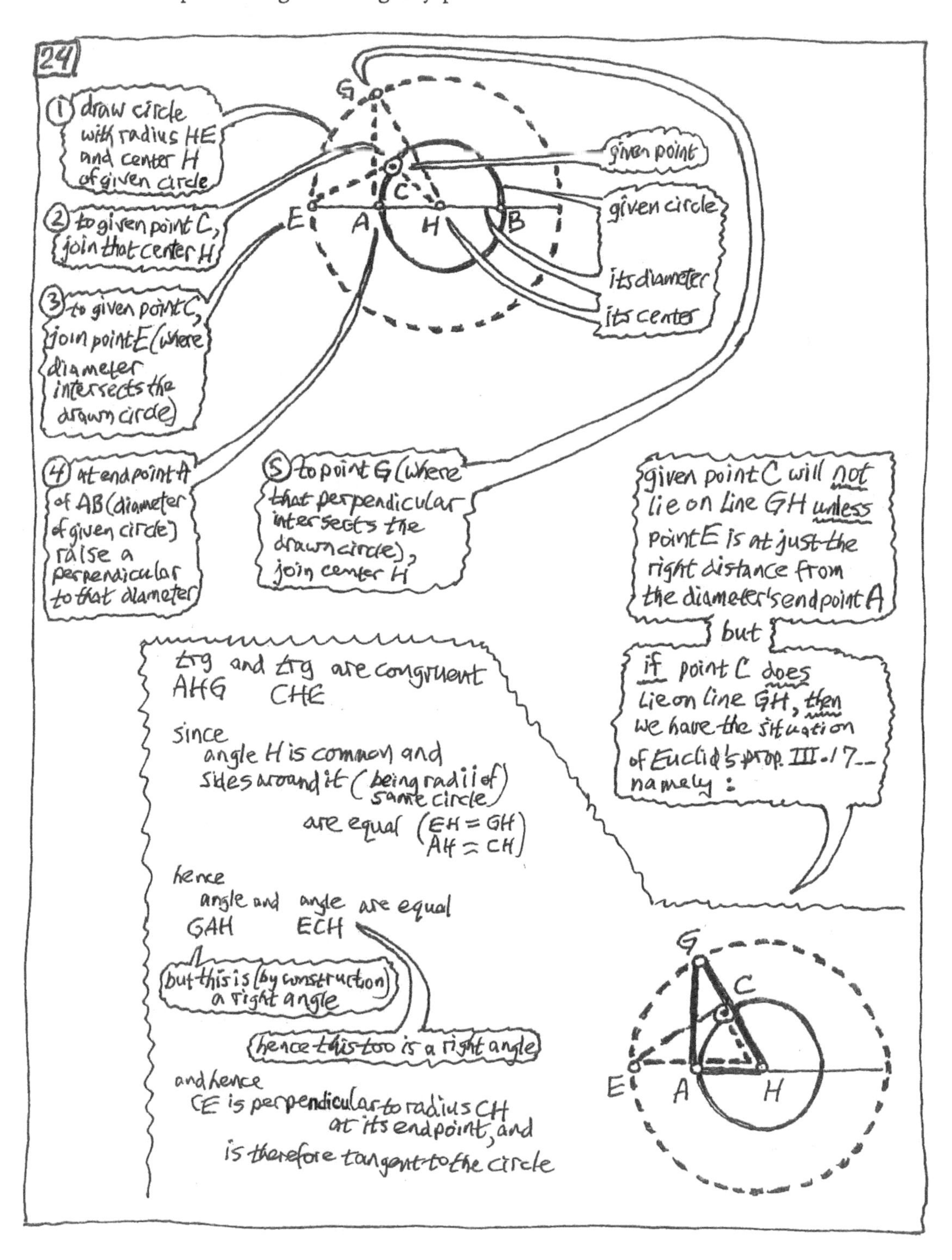

But where on the diameter should point E be placed in order to guarantee that line GH will pass through point C? Point E should be placed so as to cut the diameter externally in the same ratio in which the diameter is cut internally by the perpendicular to it from the point C. Why that will give the requisite guarantee, is depicted in figure 25.

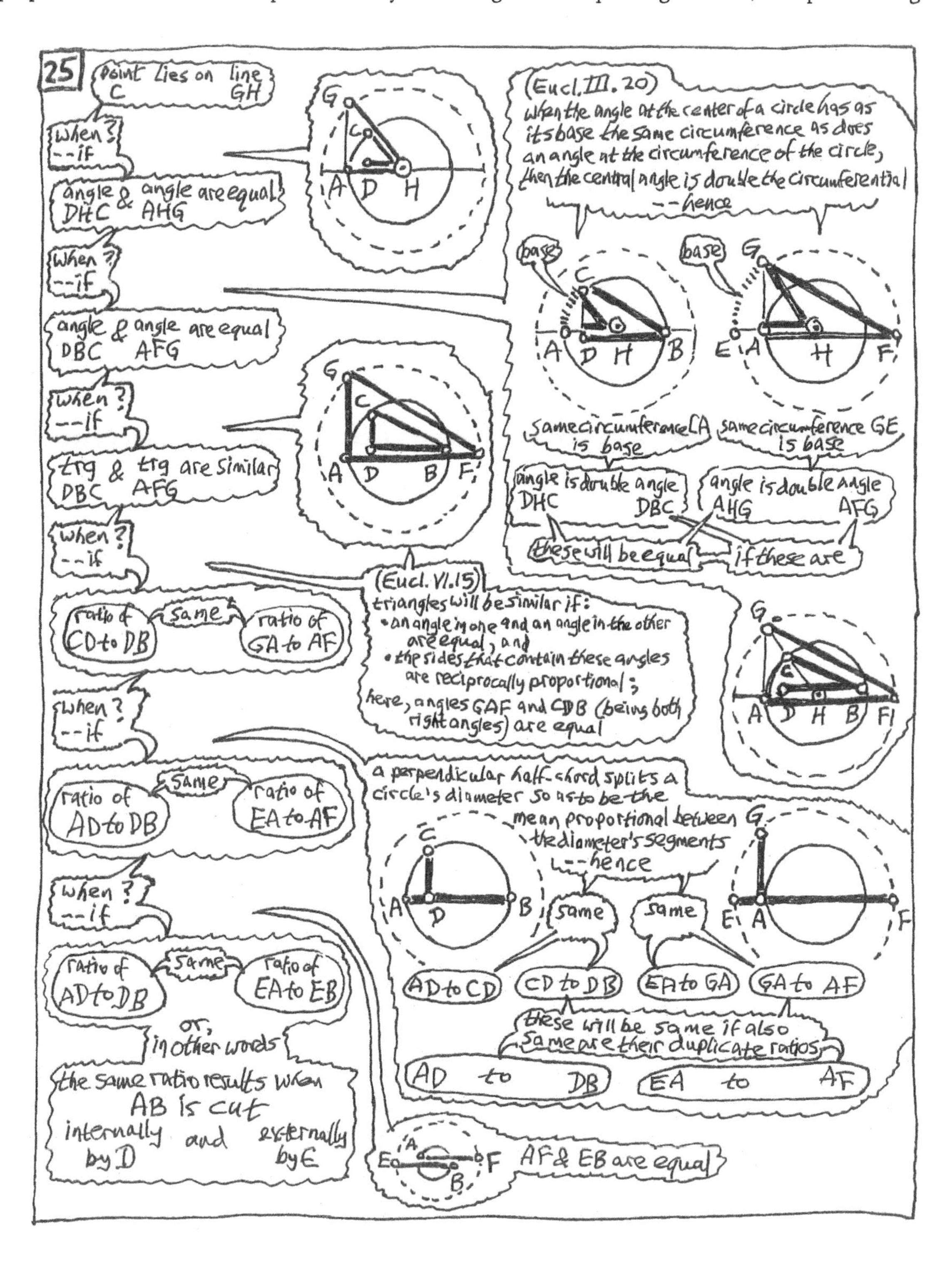

Note that what the previous figure depicted was the steps of a demonstration presented in reverse, indicating how an analysis could have been done. What the next figure depicts (fig. 26) will be reminiscent of what has been said earlier in this guidebook (see Part III, Chapter 1) about analysis and synthesis. So will what the next figure depicts (fig. 27)—namely, what it means to say that presenting the steps of a demonstration in reverse indicates how an analysis could have been done.

But why return to this matter of analysis and synthesis? There is a very good reason. We are now emerging from mid-course in the mediation that takes us from looking at cuts of cones to finding cones that will yield the cuts. Once we have completed the mediation, and then gone all the way to the end of the First Book of the *Conics* of Apollonius, we shall be ready to begin preparing for our study of the transformation of classical geometry. We shall then want to consider some relations between looking at things and finding things, and between finding things and looking for things. For considering those relations, and also for considering radical and comprehensive questions raised by the relationship between looking and doing or making, we shall want to be very familiar with this matter of analysis and synthesis.

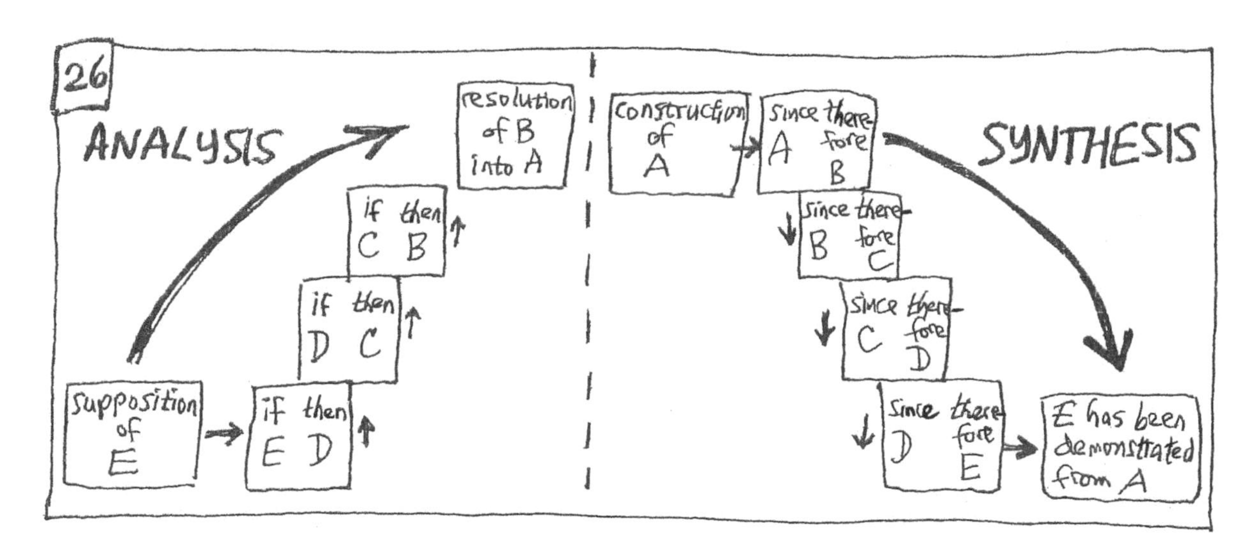

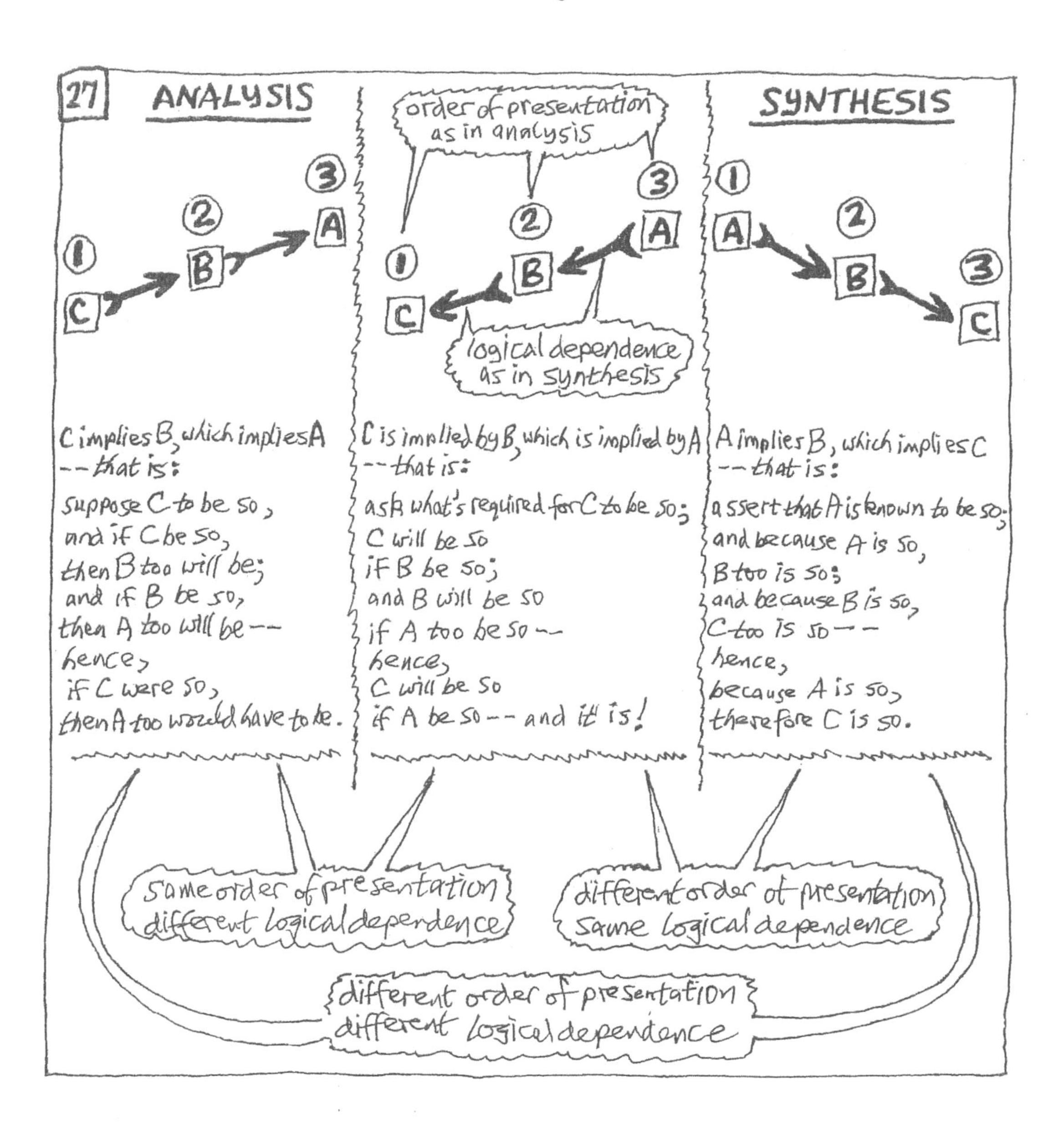
27
ANALYSIS
order of presentation as in analysis
SYNTHESIS
1 2 3
C B A
logical dependence as in synthesis
C implies B, which implies A
-- that is:
suppose C to be so,
and if C be so,
then B too will be;
and if B be so,
then A too will be --
hence,
if C were so,
then A too would have to be.
C is implied by B, which is implied by A
-- that is:
ask what's required for C to be so;
C will be so
if B be so;
and B will be so
if A too be so --
hence,
C will be so
if A be so -- and it is!
A implies B, which implies C
-- that is:
assert that A is known to be so;
and because A is so,
B too is so;
and because B is so,
C too is so --
hence,
because A is so,
therefore C is so.
same order of presentation different logical dependence
different order of presentation same logical dependence
different order of presentation different logical dependence

CHAPTER 6. FROM TANGENTS TO NEW DIAMETERS

our route

From the diameter, there has been obtained the line that is tangent at any point of the conic section, even at a point that is not the endpoint of that diameter. As we have already seen, the line that is tangent at the endpoint of that diameter is ordinatewise to that diameter. As we shall see hereafter, the tangent at a point that is *not* the endpoint of that diameter will *also* be ordinatewise—to a diameter that is *new*, however. Apollonius now goes on to prepare for obtaining such new diameters. That preparation is treated in the present chapter of this guidebook.

What this chapter prepares for will be treated in the following chapter, which will be the last one in the present part of this guidebook. There, those new diameters will finally be obtained by Apollonius, in propositions 46–48, right after which their parameters will be obtained by him, in propositions 49–51, and then, at the end of the 51st proposition, he will do something that is extraordinary for him: he will give a summary of what he has done. He will summarize the six propositions that obtain new diameters and their parameters.

Then, in the next and final part of the first volume of this guidebook, being able to refer any conic to any one of its many diameters (using the tangent at that diameter's extremity as the ordinatewise line which orients the ordinates for that diameter), Apollonius will be able to return to cones, showing, in propositions 52–59, that if curves are parabolas or hyperbolas or ellipses (or opposite sections), then they can always be obtained by cutting a cone (or conic surfaces); and he will, right after that, conclude the First Book of his *Conics* with proposition 60, which, presenting sections that he will call "conjugate," makes a new beginning.

To be able to obtain the new diameters, Apollonius now must first complete the requisite preparations. In the present chapter of this guidebook, treating propositions 37–45, he builds a bridge, on which to move from the tangents treated in the previous chapter, to the new diameters treated in the following chapter. He moves:

from tangents:	parabola (35)	other curves (36)	
by way of a bridge:	parabola (42)	other curves (43)	opposite sections (44)
to new diameters:	parabola (46)	other curves (47)	opposite sections (48).

But where, in all that, are propositions 37–41 and 45? As we shall see, propositions 37–41 constitute an access road that Apollonius will need in order to reach the bridge for the curves other than the parabola. The trip along the access road begins with the next proposition that lies before us—proposition 37. After a brief side-trip onto an overlook in proposition 38, the trip along the access road resumes at proposition 39; and then, after another side-trip onto an overlook in proposition 40, the trip along the access road ends at proposition 41. Proposition 41, besides completing the access road to the bridge for curves other than the parabola, also provides a generalization of the property that characterized such curves back in the 12th and 13th propositions.

The side-trips (propositions 38 and 40) on the access road (propositions 37–41) are not the only side-trips on the way from tangents to new diameters. At the end of the bridge too there is a side-trip—or perhaps it would be better to call it a pulling-over—onto another overlook in proposition 45. These three side-trips provide a sideshow in the spectacle that is laid before us in the First Book of the *Conics*. What the sideshow puts on display is the second diameter. The sideshow contributes to that part of the story which began almost immediately after the propositions that characterized the several conic sections. To be more precise, that part of the story began right after the 16th proposition, with the Second Definitions. It will culminate in the 60th (and last) proposition of the First Book. That part of the story which is about the second diameter displays the kinship of the opposite sections and the ellipse. (Fig. 1.)

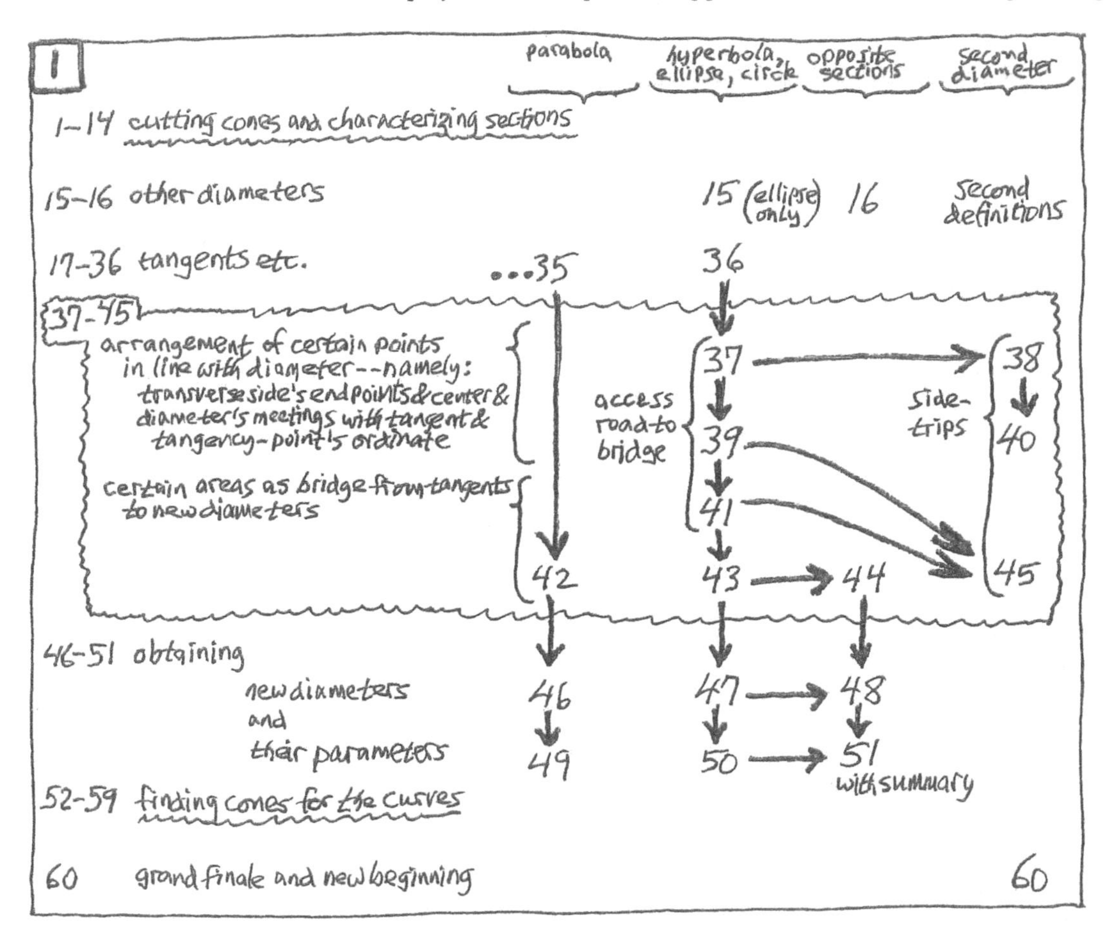

37th and 38th propositions

37th proposition: enunciation

If we were concerned only about the parabola, we could have moved directly from the 35th proposition to the 42nd. The corresponding move for the other curves would be to go from the 36th proposition to the 43rd, but we cannot go directly. We must go first to the 37th, then to the 39th, then to the 41st. Only then can we go to the 43rd. Along the way, we shall see that what is done for the diameter in the 37th will be done for the second diameter in the 38th. (We shall see also that what is done for the diameter in the 39th will be done for the second diameter in the 40th.)

What is the relation of the 37th proposition to the proposition that immediately precedes it? To answer that, we must draw, from any point on the hyperbola or ellipse or circle, a tangent and an ordinate. These two straight lines meet the diameter. For convenience, let us call the points where they meet it "the tangent's *foot*" and "the ordinate's *foot*."

Now, what the 36th proposition told us was this: the feet cut the oblique (transverse) side internally and externally in the same ratio. That is to say:

the line that runs between the two feet is the *harmonic mean* between the lines that run from the ordinate's foot to the oblique (transverse) side's two endpoints.

That harmonic mean will have different sizes for different points on the curve, since for different points on the curve the feet will be located at different places along the diameter and its prolongation. But although the harmonic mean changes as the location of the feet changes, there is another mean that stays the same. That is what we are now told by the 37th proposition. What it tells us is this:

the radius—the line that runs between the oblique (transverse) side's center and its endpoint at the vertex— is the *geometric mean* between the lines that run from the oblique (transverse) side's center to the two feet.

Thus, the 37th proposition, like the 36th, relates the endpoints of the oblique (transverse) side and the points upon it (and upon its prolongation) that are the feet of the tangent and the ordinate. But whereas the 36th proposition did it by using a mean that changes, the 37th does it by using a mean that is constant. The two means for the case of the hyperbola—the harmonic mean and the geometric mean—are shown in figure 2, where the feet of the tangent and the ordinate are the internal and external cut-points of the oblique (transverse) side.

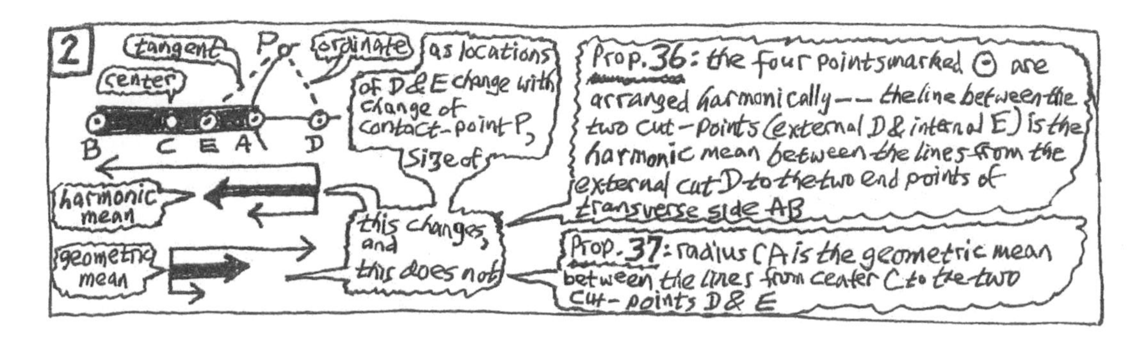

It is true that the enunciation of the 37th proposition does not quite tell us that the radius is the geometric mean between the lines from the center to the feet of the ordinate and tangent. But then the 36th did not explicitly speak of an harmonic mean either. What matters, underneath the surface or behind the superficial looks, is to grasp the *ratios* of the lines between the points. The rest is merely a means of expressing the sameness of the ratios—the proportions in the arrangement (along the prolonged diameter) of the endpoints (or the center and the endpoint) of the oblique (transverse) side and the feet of the tangent and the ordinate.

So, what does the 37th proposition say explicitly? Two things. Both statements take a rectangle

> whose one side is the line from the ordinate's foot to the center,
> and whose other side is a line from the tangent's foot.

When this second line runs from there (the tangent's foot)

to the *center*,	to *the ordinate's foot*,
then the rectangle will be equal to the square on the radius;	*then* the rectangle will have the same ratio to the square on the ordinate that the oblique (transverse) side has to the upright side.

Figure 3 shows the case of the hyperbola.

The first statement is equivalent to speaking of a geometric mean, because when a rectangle is equal to a square, then the rectangle's one side has to the square's side the same ratio that the square's side has to the rectangle's other side. The square's side is thus the mean proportional between the rectangle's sides.

The second statement tells us how the square on the ordinate is related to the *eidos*-ratio of the oblique (transverse) side to the upright side. It is related as follows: the upright side, the oblique (transverse) side, and the square on the ordinate have as fourth proportional the rectangle whose sides are the lines that run from the ordinate's foot to the tangent's foot and to the center.

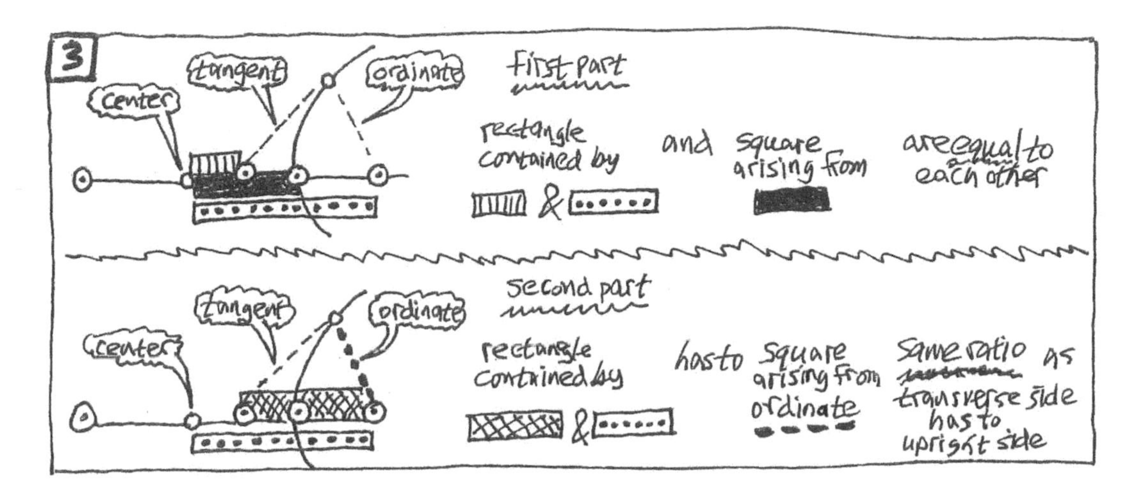

37th proposition: demonstration

#37 If, with respect to an hyperbola or ellipse or circle's periphery, a straight [line] that touches [the section] fall together with the diameter, and from the touch[point] there be drawn down to the diameter a straight [line] ordinatewise—then the straight [line] taken off beside the center of the section by the [line] drawn down will (on the one hand), along with (*meta*) the [line] taken off beside the center of the section, by the touching[line], contain a [rectangle] equal to the [square] from the [line emanating] out-from-the-center, [equal, that is, to the square from the radius] of the section; and will (on the other hand), along with (*meta*) the [line] between the [line] drawn down and the touching[line], contain an area (*khôrion*) that has, to the [square] from the [line] drawn down ordinatewise, that [ratio] which the oblique [side] has to the upright [side].

—Let there be an hyperbola or ellipse or circle's periphery, whose diameter is what the [line] AB is, and touching [the section] let there be drawn the [line] CD; and let there be drawn down ordinatewise the [line] CE, and let [the section's] center be what the [point] F is. I say that equal is the [rectangle] by DFE to the [square] from FB, and also that as is the [rectangle] by DEF to the [square] from EC so is the oblique [side] to the upright [side].

—For—since the [line] CD touches the section, and there has been drawn down ordinatewise the [line] CE—hence as is the [line] AD to DB so is the [line] AE to EB. By putting together *(synthenti)* [traditionally in Latin, *componendo*], therefore, as is the [line] AD, DB (both together) to DB so is the [line] AE, EB (both together) to EB. And, of the [terms] that lead [that is, of the antecedents in the ratios], the halves [are to be considered next].

—On the hyperbola, we shall say this: But of the [line] AE, EB both together (on the one hand) half is the [line] FE, and of the [line] AB (on the other hand) half is the [line] FB; therefore as is the [line] FE to EB so is the [line] FB to BD. By converting (*anastrepsanti*) [traditionally in Latin, *convertendo*], therefore, as is the [line] EF to FB so is the [line] FB to FD; therefore equal is the [rectangle] by DFD to the [square] from FB. And—since as is the [line] FE to EB so is the [line] FB to BD, that is, the [line] AF to DB—therefore, alternately (*enallax*) [*alternando*], as is the [line] AF to FE so is the [line] DB to BE; and, by putting together (*synthenti*) [*componendo*], as is the [line] AE to EF so is the [line] DE to EB; and so, the [rectangle] by AEB is equal to the [rectangle] by FED. But as is the [rectangle] by AEB to the [square] from CE so is the oblique [side] to the upright [side]. Therefore also as is the [rectangle] by FED to the [square] from CE so is the oblique [side] to the upright [side].

—On the ellipse and circle, however, we shall say this: But of the [line] AD, DB both together (on the one hand) half is the [line] DF, and of the [line] AB (on the other hand) half is the [line] FB; therefore as is the [line] FD to DB so is the [line] FB to BE. By converting (*anastrepsanti*) [*convertendo*], therefore, as is the [line] DF to FB so is the [line] BF to FE; therefore equal is the [rectangle] by DFE to the [square] from BF. But (on the one hand) the [rectangle] by DFE is equal to the [rectangle] by DEF and the [square] from FE [taken together], and (on the other hand) the [square] from BF is equal to the [rectangle] by AEB [taken] along with (*meta*) the [square] from FE. Let a common area be taken away, that is, the [square] from EF; therefore the remainder [rectangle] by DEF will, to the remainder [rectangle] by AEB, be equal. Therefore as is the [rectangle] by DEF to the [square] from CE so is the [rectangle] by AEB to the [square] from CE. But as is the [rectangle] by AEB to the [square] from CE so is the oblique [side] to the upright [side]; therefore as is the [rectangle] by DEF to the [square] from EC so is the oblique [side] to the upright [side].

For a map of the demonstration (with illustrative diagrams), see figure 4.

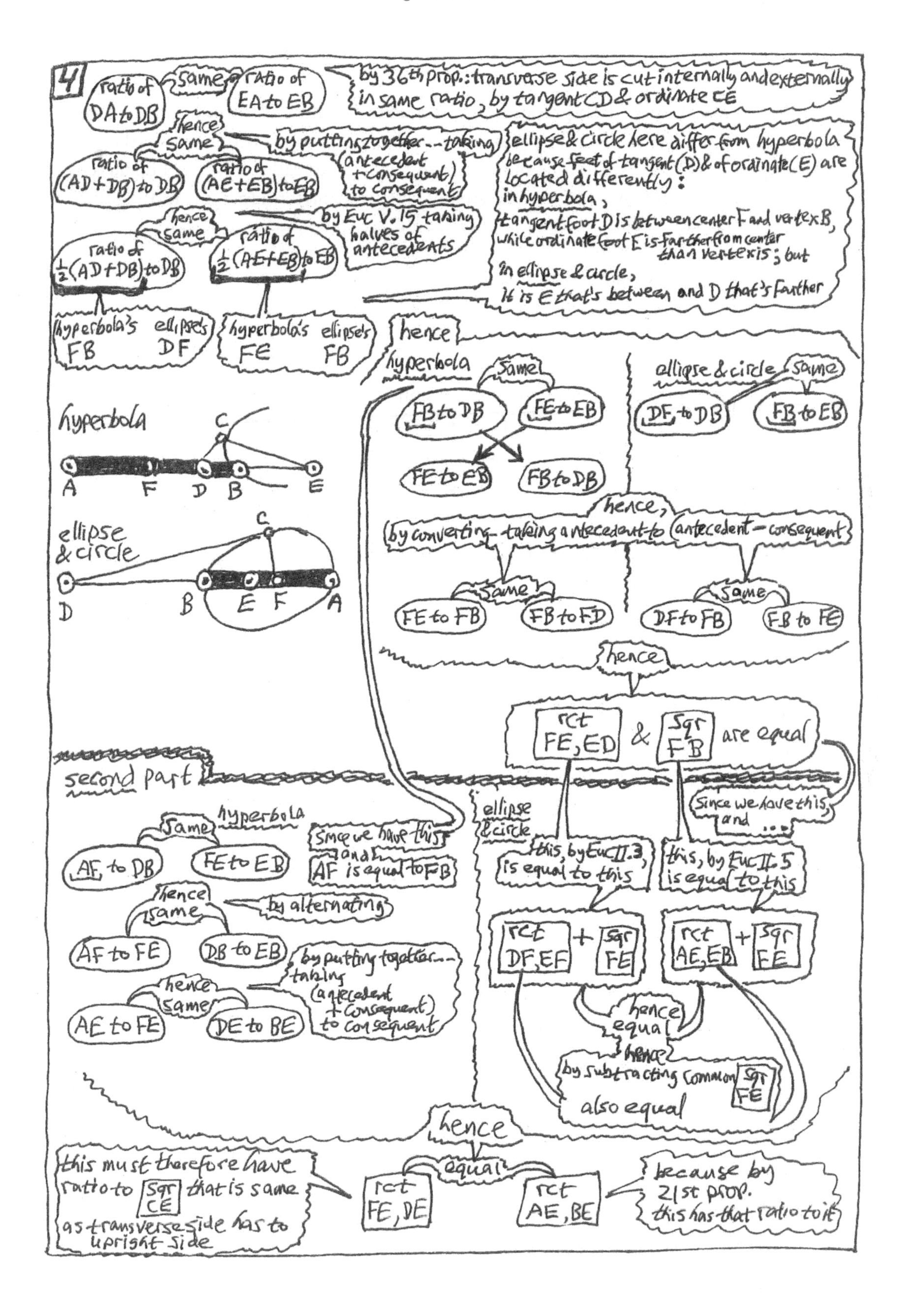
4
ratio of DA to DB
same
ratio of EA to EB
by 36th prop.: transverse side is cut internally and externally in same ratio, by tangent CD & ordinate CE
hence same
by putting together—taking (antecedent + consequent) to consequent
ratio of (AD+DB) to DB
ratio of (AE+EB) to EB
ellipse & circle here differ from hyperbola because feet of tangent (D) & of ordinate (E) are located differently: in hyperbola, tangent foot D is between center F and vertex B, while ordinate foot E is farther from center than vertex is; but in ellipse & circle, it is E that's between and D that's farther
hence same
by Euc V.15 taking halves of antecedents
ratio of ½(AD+DB) to DB
ratio of ½(AE+EB) to EB
hyperbola's FB ellipse's DF
hyperbola's FE ellipse's FB
hence
hyperbola
same
FB to DB
FE to EB
FE to EB
FB to DB
ellipse & circle
same
DF to DB
FB to EB
hyperbola
A F D B E C
ellipse & circle
D B E F A C
hence,
by converting—taking antecedent to (antecedent − consequent)
same
FE to FB
FB to FD
same
DF to FB
FB to FE
hence
rct FE, ED & sqr FB are equal
second part
hyperbola
same
AF to DB
FE to EB
Since we have this and AF is equal to FB
hence same
by alternating
AF to FE
DB to EB
by putting together—taking (antecedent + consequent) to consequent
hence same
AE to FE
DE to BE
ellipse & circle
Since we have this, and ...
this, by Euc II.3, is equal to this
this, by Euc II.5 is equal to this
rct DF, EF + sqr FE
rct AE, EB + sqr FE
hence equal
hence
by subtracting common sqr FE
also equal
hence
equal
this must therefore have ratio to sqr CE that is same as transverse side has to upright side
rct FE, DE
rct AE, BE
because by 21st prop. this has that ratio to it

38th proposition

Earlier in the book, right after the 16th proposition, the Second Definitions introduced a line which was called a "second diameter" of the sections other than the parabola—that is, of the sections with a center (the "central" sections). Now, twenty-two propositions later, Apollonius shows that the line earlier called a "second diameter" resembles the original diameter of the central sections in more than simply being a diameter. How so? In the following way:

As the 37th proposition has, just previously,
related certain points in a certain manner
to the *original diameter* and to the *ordinate to the original diameter*,

so the 38th proposition now
relates corresponding points in a similar manner
to the *second diameter* and to the *parallel to the original diameter*.

In other words:

take what we are told by the 37th proposition;
and where the *original* diameter is mentioned,
now mention instead the *second* diameter,
and where the *ordinate* to the original diameter is mentioned,
mention instead the *parallel* to the original diameter—
the result will be what we are told by the 38th proposition (fig. 5).

(With this small modification, however: in the second part of the enunciation, the terms of the ratio will have to be reversed. In other words, where the ratio of the oblique [transverse] side to the upright side is mentioned, what must be mentioned instead is the ratio of the upright side to the oblique [transverse] side. Why? Because, in the ratio that is the same as this one, the order of size of the rectangle and square is changed in the case of the second diameter.)

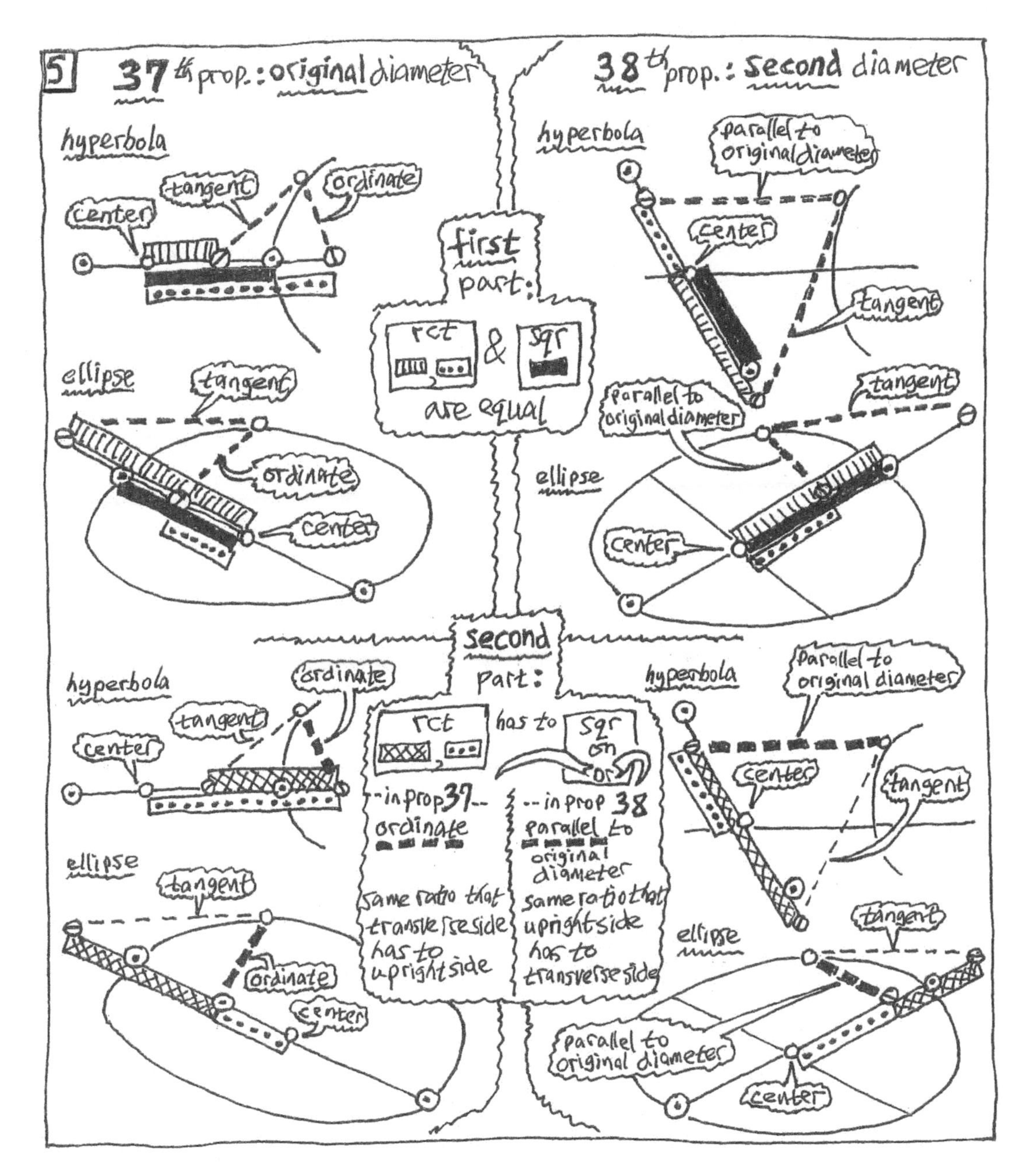

On a first reading of Apollonius, you may wish to give this proposition and its explication a merely cursory look. (Note that, after demonstrating what is stated in the enunciation at the beginning of the proposition, the text of Apollonius was transmitted with an expansion by someone who, while not correct in statement or in demonstration, was trying to insert the following true implication of Apollonius's work: with the same things supposed, the second diameter will be cut internally and externally in the same ratio by the tangent and by the line that is drawn parallel to the original diameter; moreover, a straight line drawn from a point on the curve will be tangent to the curve if it cuts the second diameter in such a way as to satisfy the equality or the proportion mentioned in the enunciation at the beginning.)

#38 If, with respect to an hyperbola or ellipse or circle's periphery, a straight [line] that touches [the section] fall together with the second diameter, and from the touch[point] a straight [line] be drawn down to the same diameter, [to, that is, the second diameter,] parallel to the other diameter—then the straight [line] taken off beside the center of the section by the [line] drawn down, will (on the one hand) along with (*meta*) the [line] taken off beside the center of the section [the line taken off, that is, from the second diameter] by the touching[line], contain an [area] equal to the [square] from the half of the second diameter; and will (on the other hand) along with (*meta*) the [line] [stretching along the second diameter] between the [line] drawn down and the touching[line], contain an area (*khôrion*) that has, as ratio to the [square] from the [line] drawn down, that ratio which the upright side of the figure (*tou eidous*) has to the oblique [side].
—Let there be an hyperbola or ellipse or circle's periphery, whose diameter is what the [line] AGB is, and whose second diameter is what the [line] CGD is, and touching the section let there be the [line] ELF that falls together with the [line] CDE at the [point] F, and parallel to the [line] AB let there be the [line] HE. I say that the [rectangle] by FGH is, to the [square] from GC, equal; and also, that as is the [rectangle] by GHF to the [square] from HE so is the upright [side] to the oblique [side].
—Let there be drawn ordinatewise the [line] ME. Therefore as is the [rectangle] by GML to the [square] from ME so is the oblique [side] to the upright [side]. But as is the oblique [side]—namely the [line] BA—to CD so is the [line] CD to the upright [side]; and therefore as is the oblique [side] to the upright [side] so is the [square] from AB to the [square] from CD; and so also are the quarters, that is, so is the [square] from GA to the [square] from GC; and therefore also as is the [rectangle] by GML to the [square] from ME so is the [square] from GA to the [square] from GC. But the [rectangle] by GML has, to the [square] from ME, the ratio compounded from that which the [line] GM has to ME—that is, to GH—and also from that which the [line] LM has to ME. Inversely (*anapalin*) [*invertendo*], therefore, the ratio of the [square] from CG to the [square] from GA is compounded from that which the [line] EM has to MG—that is, the [line] HG to GM—and also from that which the [line] EM has to ML—that is, the [line] FG to GL. Therefore the [square] from GC has, to the [square] from GA, the ratio compounded from that which the [line] HG has to GM and also that which the [line] FG has to GL—which [compounded ratio] is the same as that which the [rectangle] by FGH has to the [rectangle] by MGL. Therefore as is the [rectangle] by FGH to the [rectangle] by MGL so is the [square] from CG to the [square] from GA. And alternately (*enallax*), therefore, as is the [rectangle] by FGH to the [square] from CG so is the [rectangle] by MGL to the [square] from GA. But equal is the [rectangle] by MGL to the [square] from GA; therefore equal also is the [rectangle] by FGH to the [square] from GC.
—Again—since as is the upright [side] to the oblique [side] so is the [square] from EM to the [rectangle] by GML; and the [square] from EM has, to the [rectangle] by GML, the ratio compounded from the ratio which the [line] EM has to ML, that is, which the [line] HG has to HE, and also the ratio which the [line] EM has to ML, that is, which the [line] FG has to GL, that is, which the [line] FH has to HE, which [compounded ratio] is the same as that which the [rectangle] by FHG has to the [square] from HE—therefore as is the [rectangle] by FHG to the [square] from HE so is the upright [side] to the oblique [side].

For a map of the demonstration (with illustrative diagrams), see figure 6.

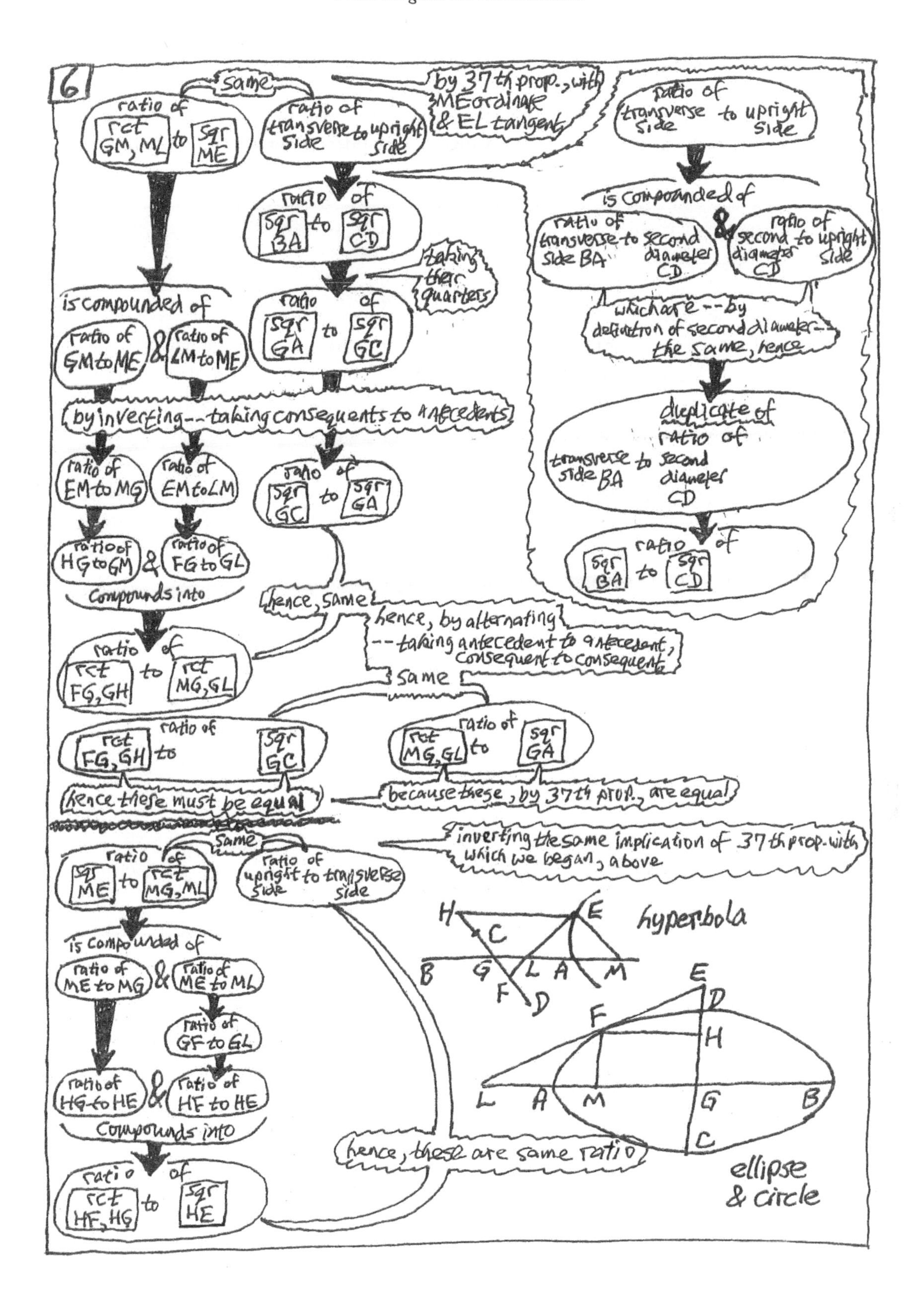
6
same
ratio of rct GM, ML to sqr ME
ratio of transverse to upright side side
by 37th prop., with ME ordinate & EL tangent
ratio of transverse to upright side side
is compounded of
ratio of transverse to second side BA diameter CD
&
ratio of second to upright diameter side CD
which are -- by definition of second diameter -- the same, hence
duplicate of ratio of transverse side BA to second diameter CD
ratio of sqr BA to sqr CD
ratio of sqr BA to sqr CD
taking their quarters
ratio of sqr GA to sqr GC
is compounded of
ratio of GM to ME
&
ratio of LM to ME
by inverting -- taking consequents to antecedents
ratio of EM to MG
ratio of EM to LM
ratio of sqr GC to sqr GA
ratio of HG to GM
&
ratio of FG to GL
compounds into
ratio of rct FG, GH to rct MG, GL
hence, same
hence, by alternating -- taking antecedent to antecedent, consequent to consequent
same
ratio of rct FG, GH to sqr GC
ratio of rct MG, GL to sqr GA
hence these must be equal
because these, by 37th prop., are equal
same
ratio of sqr ME to rct MG, ML
ratio of upright to transverse side side
inverting the same implication of 37th prop. with which we began, above
is compounded of
ratio of ME to MG
&
ratio of ME to ML
ratio of GF to GL
ratio of HG to HE
&
ratio of HF to HE
compounds into
ratio of rct HF, HG to sqr HE
hence, these are same ratio
H
C
E
hyperbola
B
G
L
A
M
F
D
E
D
F
H
L
A
M
G
B
C
ellipse & circle

39th and 40th propositions

On a first reading of Apollonius, you may wish to give the demonstrations of these two propositions a merely cursory look. They are to be found in the next two boxes. After them, this guidebook presents a brief discussion of their significance.

#39 If, with respect to an hyperbola or ellipse or circle's periphery, a straight [line] that touches [the section] fall together with the diameter, and from the touch[point] there be drawn down a straight [line] to the diameter, ordinatewise, then—whichever be taken of the two straight [lines] [that are] (on the one hand) the [line] between the [line] drawn down [at the point where it falls together with the diameter] and the center of the section, and (on the other hand) the [line] between the [line] drawn down [at the point where it falls together with the diameter] and the touching[line] [at the point where it falls together with the diameter]—the [line] drawn down will have to it, [will have, that is, to either one of those two straight lines] the ratio compounded from that [ratio] which the other of the two straight [lines] has to the [line] drawn down and also from that [ratio] which the upright side of the figure has to the oblique [side].

—Let there be an hyperbola or ellipse or circle's periphery, whose diameter is what the [line] AB is, and its center is what the [point] F is; and also, let there be drawn touching the section the [line] CD; and also let there be drawn down ordinatewise the [line] CE. I say that the [line] CE has, to one of the lines FE, ED, the ratio compounded from that which the upright [side] has to the oblique [side] and also from that which the other of the [lines] FE, ED has to the [line] EC.

—For let there be equal the [rectangle] by FED to the [rectangle] by CE, G. [That is to say, let a line G be so made that it will be a fourth proportional to the lines CE, FE, ED.] And—since as is the [rectangle] by FED to the [square] from CE so is the oblique [side] to the upright [side], and equal is the [rectangle] by FED to the [rectangle] by CE, G—therefore as is the [rectangle] by CE, G to the [square] from CE, that is, [as is] the [line] G to EC, so is the oblique [side] to the upright [side]. And since equal is the [rectangle] by FED to the [rectangle] by CE, G, hence as is the [line] EF to EC so is the [line] G to ED. And—since the [line] CE has to ED the ratio compounded from that which the [line] CE has to G and also that which the [line] G has to ED, but (on the one hand) as is the [line] CE to G so is the upright [side] to the oblique [side], and (on the other hand) as is the [line] G to DE so is the [line] FE to EC—therefore the [line] CE has to ED the ratio compounded from that which the upright [side] has to the oblique [side] and also from that which the [line] FE has to EC.

#40 If, with respect to an hyperbola or ellipse or circle's periphery, a straight [line] that touches [the section] falls together with the second diameter, and from the touch[point] there be drawn down a straight [line] to the same diameter, parallel to the other diameter, then—whichever be taken of the two straight [lines] [that are] (on the one hand) the [line] between the [line] drawn down [at the point where it falls together with the second diameter] and the center of the section, and (on the other hand) the [line] between the [line] drawn down [at the point where it falls together with the second diameter] and the touching[line] [at the point where it falls together with the second diameter]—the line drawn down will have to it, [will have, that is, to either one of those two straight lines] the ratio compounded from that which the oblique [side] has to the upright [side] and also from that which the other of the two straight [lines] has to the [line] drawn down.

—Let an hyperbola or ellipse or circle's periphery be what the [section] AB is, and its diameter be what the [line] BFC is, and its second diameter be what the [line] DFE is; and let a touching[line], namely the [line] HLA, be drawn; and also a [line] parallel to the [line] BC, namely the [line] AG, [be drawn]. I say that the [line] AG has, to one of the [lines] HG, FG, the ratio compounded from that which the oblique [side] has to the upright [side] and also from that which the other of the [lines] HG, FG has to the [line] GA.

—Let there be equal to the [rectangle] by HGF the [rectangle] by GA, K. [That is to say, let a line K be so made that it will be a fourth proportional to the lines GA, HG, GF]. And—since as is the upright [side] to the oblique [side] so is the [rectangle] by HGF to the [square] from GA, and equal to the [rectangle] by HGF is the [rectangle] by GA, K—therefore also the [rectangle] by GA, K is to the [square] from GA, that is, the [line] K is to AG, as is the upright [side] to the oblique [side]. And—since the [line] AG has to GF the ratio compounded from that which the [line] AG has to K and also from that which the [line] K has to GF, but (on the one hand) as is the [line] GA to K so is the oblique [side] to the upright [side], and (on the other hand) as is the [line] K to GF so is the [line] HG to GA (on account of the [rectangle] by HGF being equal to the [rectangle] by AG, K)—therefore the [line] AG has to GF the ratio compounded from that which the oblique [side] has to the upright [side] and also from that which the [line] GH has to GA.

By compounding ratios, the 39th proposition follows directly from the 37th, and the 40th from the 38th.

The 39th proposition continues the access road that began with the 37th. The curves for which the access road is needed are the central sections. The access road leads from the tangents of those central sections to the bridge that leads to new diameters whose ordinates will be parallel to those tangents. The 39th proposition tells how the size of the ordinate is related to the arrangement of the following points along the diameter: the center, the ordinate's foot, and the tangent's foot. It says that you will get the following ratio (call it "ratio *alpha*")—

> the ratio that
>   the ordinate
> has to
>   the line running from the ordinate's foot to the center—

if you take the ratio that the upright side has to the oblique (transverse) side, and compound it with the following ratio (call it "ratio *beta*")—

> the ratio that
>   the line running from the ordinate's foot to the tangent's foot
> has to
>   the ordinate.

The proposition says also that if you take that same ratio which the upright side has to the oblique (transverse) side, and you compound it with the *inverse* of ratio *alpha*, then you will get the *inverse* of ratio *beta*. The case of the hyperbola is depicted in figure 7.

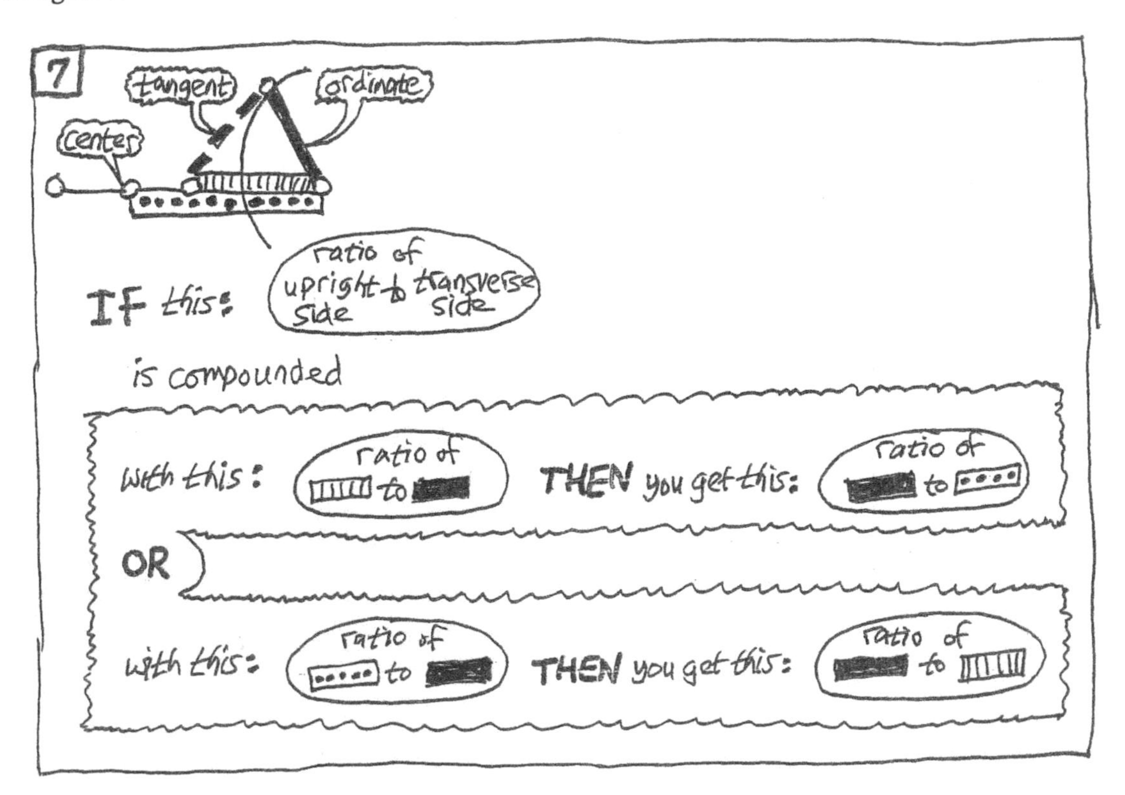

As for the enunciation of the 40th proposition, it will emerge from that of the 39th if we do again what we did in order to make the enunciation of the 38th proposition emerge from that of the 37th—namely, this:

in place of:		
	the original diameter	and the ordinate to the original diameter,
we substitute:		
	the second diameter	and the parallel to the original diameter

(remembering, in doing so, to invert the ratio of the upright side to the oblique [transverse] side, making it the ratio of the oblique [transverse] side to the upright side). The 40th proposition, like the 38th, is a side-trip off the access road.

We see again in the 40th proposition what we saw in the 38th—that as the original diameter works with its ordinate, so also does the second diameter work with the parallel to the original diameter. In other words: for the central sections (that is, the hyperbola, the ellipse, and the circle), certain properties of the ordinates to the section's *oblique (transverse) diameter* are also properties of the ordinates to its *conjugate diameter*.

It therefore no longer seems arbitrary for the upright diameter of the central sections to be defined in *position* as it is—namely, as passing through the center of the oblique (transverse) diameter, which bisects it, and as being parallel to the ordinates of the (oblique) transverse diameter. It does, however, still seem arbitrary for the upright diameter of the central sections to be defined in *size* as it is—namely, as being a mean proportional between the sides of the *eidos*. It is still not clear why, that is to say, when a line is placed where Apollonius says that he will place the second diameter, he will call it a second diameter only when its size makes it a mean proportional between the oblique (transverse) side and the upright side.

41st proposition: enunciation

This proposition is the end of the access road that leads from tangents for the central sections (36th proposition) to the bridge (43rd proposition) that carries us to new diameters for those curves (47th proposition). The 43rd proposition (the bridge for the central sections) will depend upon the 39th proposition and this 41st proposition.

In completing the access road to the bridge for the central sections, the 41st proposition generalizes some features of the property that characterized such curves back in the 12th and 13th propositions.

In order to state the enunciation, given the curve and its diameter, some construction is needed. On an ordinate, draw a parallelogram; and on the radius, draw a second parallelogram. The angles in the one parallelogram are to be equal to those in the other, and their sides are to be related in the following way:

> the ratio of the ordinate side to the remaining side in the one parallelogram
> is compounded of the following two ratios—
> the ratio of the radius side to the remaining side in the other parallelogram,
> and the ratio of the upright side to the oblique (transverse) side.

Now draw a third parallelogram, this one similar to the one on the radius. The radius runs from the center to the vertex-point (which is the point where the diameter meets the curve). The line on which the third parallelogram is to be drawn runs from the center to the abscissa-point (which is the point where the diameter meets the ordinate). Thus, the third parallelogram is drawn on a line that exceeds the radius by the length of the abscissa (in the hyperbola), or falls short of it by that length (in the ellipse).

<table><tr><td>The third parallelogram is the one
on the line running from the center to
the abscissa-point—which is the point
where the diameter meets the ordinate.</td><td>The second parallelogram is the one
on the line running from the center to
the vertex-point—which is the point
where the diameter meets the curve.</td></tr></table>

So much for the construction of the requisite parallelograms. What the proposition does is compare the sizes of the parallelograms, as follows: the first parallelogram (the one on the ordinate) has just that amount of area by which the second parallelogram is larger or smaller than the third. The third parallelogram in the hyperbola exceeds, and in the ellipse falls short of, the second parallelogram.

In figure 8, the parallelogram on the ordinate is outlined with dashes. The parallelogram that is heavily outlined exceeds or falls short of the parallelogram that is outlined with stripes—exceeding (or falling short) when the ordinate meets the diameter farther from (or closer to) the center than the curve does. The area of excess or deficiency—to which the parallelogram outlined with dots is equal—is the gnomon filled with dots.

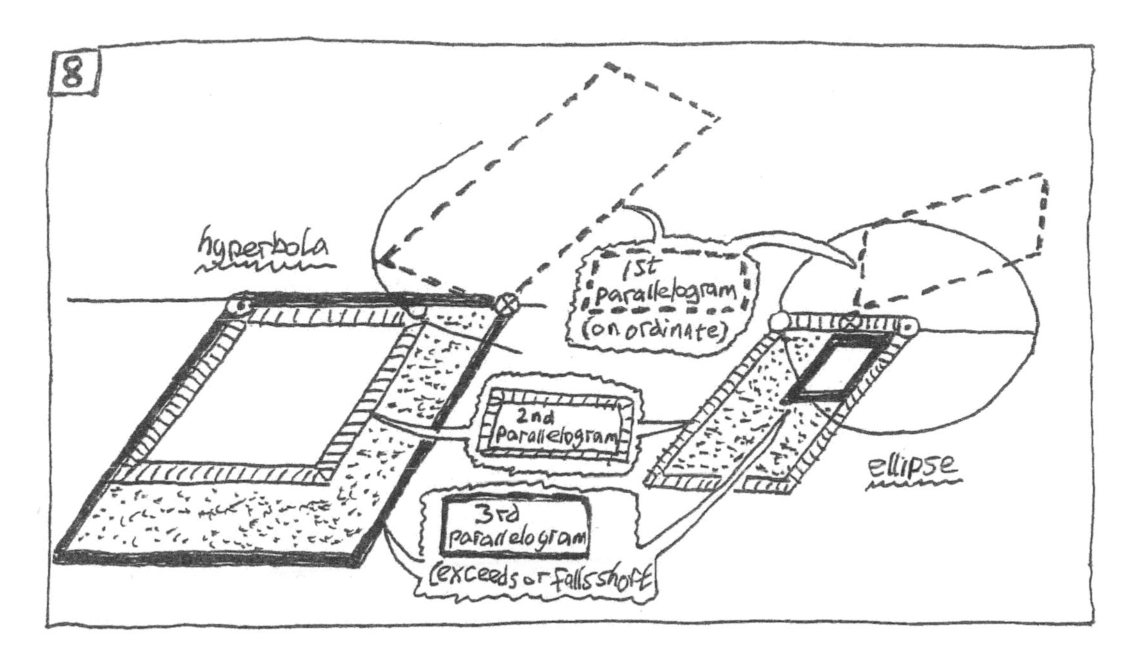

Earlier, the 12th and 13th propositions spoke of an excess or deficiency that involved a parallelogram of a particular kind (the rectangle) and one of the rectangles was a rectangle of a particular kind (the square). The parallelograms in the present proposition are not so restricted.

But if the parallelogram drawn on the ordinate in this proposition is a rectangle—indeed, is a square—then the parallelogram on the radius must also be a rectangle (so that the angles of this parallelogram will be equal to those of the other), and its radius side must have, to its other side, the same ratio that the curve's oblique (transverse) side has to its upright side.

Why this ratio? For the following reason. The ratio of the sides of the parallelogram drawn on the ordinate is compounded of two ratios, and therefore the sides of the parallelogram on the ordinate can be equal to each other only if the first component-ratio is the inverse of the second one—which is to say: those sides can be equal only if the ratio that the radius has to the other side of the parallelogram that is drawn on the radius is the inverse of the ratio that the curve's upright side has to its oblique (transverse) side, this inverse ratio being the ratio of the oblique (transverse) side to the upright side.

Thus, this rectangle on the radius will have the same shape (turned on its side) as that *eidos* of excess and deficiency which characterized the curves in the 12th and 13th propositions. But whereas that *eidos* was constant only in shape, and not in size, this rectangle on the radius is constant in size as well as in shape. It makes a gnomon whose size varies but is always equal to the square on the ordinate. The gnomon is the excess or deficiency (with respect to the rectangle on the radius) of the rectangle on the line from the center to the abscissa-point.

41st proposition: demonstration

The demonstration is depicted in figure 9 (on facing pages—note that the letters that appear in the part of the figure on the left-hand page refer to the diagram in the part of the figure on the right-hand page); and, after that, the text of the demonstration is presented in a two-part box. On a first reading of Apollonius, you may wish to give the demonstration a merely cursory look.

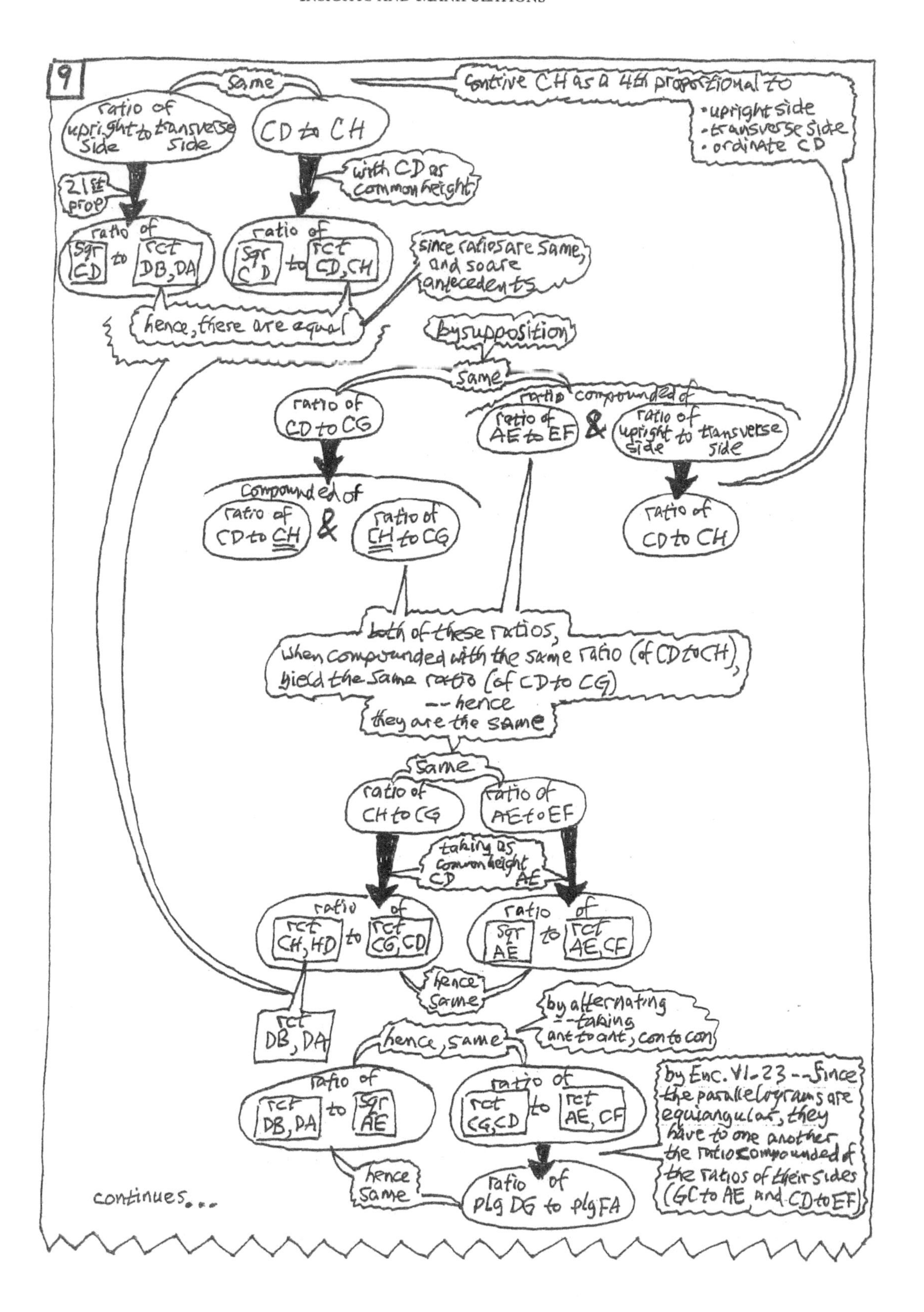
9
same
ratio of upright side to transverse side
CD to CH
contrive CH as a 4th proportional to
•upright side
•transverse side
•ordinate CD
21st prop
with CD as common height
ratio of sqr CD to rct DB, DA
ratio of sqr CD to rct CD, CH
since ratios are same, and so are antecedents
hence, these are equal
by supposition
same
ratio of CD to CG
ratio compounded of
ratio of AE to EF
&
ratio of upright side to transverse side
compounded of
ratio of CD to CH
&
ratio of CH to CG
ratio of CD to CH
both of these ratios, when compounded with the same ratio (of CD to CH), yield the same ratio (of CD to CG) --hence they are the same
same
ratio of CH to CG
ratio of AE to EF
taking as common height CD AE
ratio of rct CH, HD to rct CG, CD
ratio of sqr AE to rct AE, CF
hence same
rct DB, DA
by alternating --taking ant to ant, con to con
hence, same
ratio of rct DB, DA to sqr AE
ratio of rct CG, CD to rct AE, CF
by Euc. VI-23 --Since the parallelograms are equiangular, they have to one another the ratio compounded of the ratios of their sides (GC to AE and CD to EF)
hence same
ratio of plg DG to plg FA
continues...

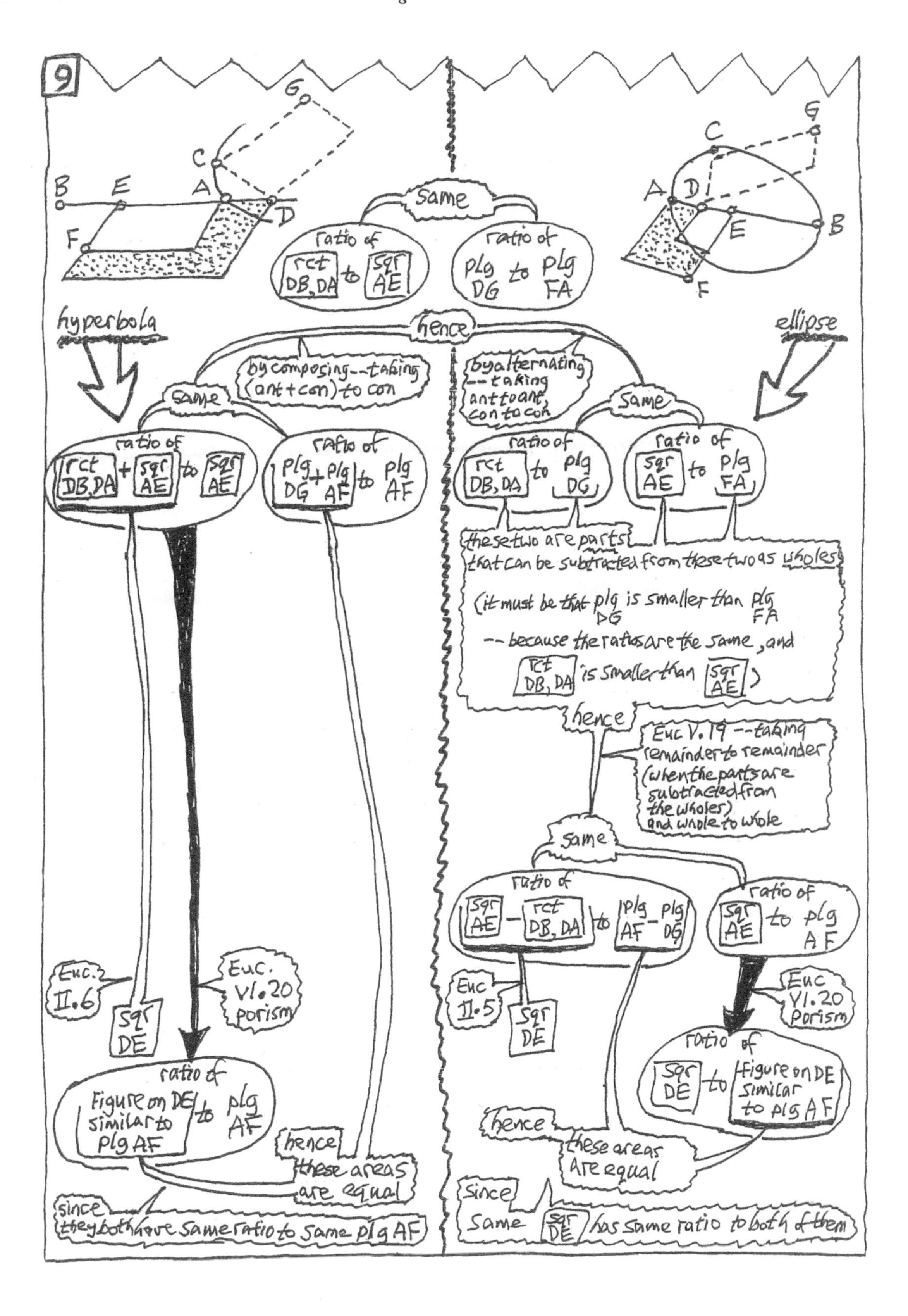
9
G
C
B E A
D
F
same
ratio of
rct DB, DA to sqr AE
ratio of
plg DG to plg FA
hyperbola
hence
ellipse
by composing--taking (ant+con) to con
by alternating --taking ant to ant, con to con
same
same
ratio of
rct DB, DA + sqr AE to sqr AE
ratio of
plg DG + plg AF to plg AF
ratio of
rct DB, DA to plg DG
ratio of
sqr AE to plg FA
these two are parts that can be subtracted from these two as wholes
(it must be that plg DG is smaller than plg FA
-- because the ratios are the same, and rct DB, DA is smaller than sqr AE)
hence
Euc V.19 --taking remainder to remainder (when the parts are subtracted from the wholes) and whole to whole
same
ratio of
sqr AE − rct DB, DA to plg AF − plg DG
ratio of
sqr AE to plg AF
Euc. II.6
sqr DE
Euc. VI.20 porism
Euc II.5
sqr DE
Euc VI.20 Porism
ratio of
Figure on DE similar to plg AF to plg AF
ratio of
sqr DE to Figure on DE similar to plg AF
hence
these areas are equal
hence
these areas are equal
since
they both have same ratio to same plg AF
Since
same sqr DE has same ratio to both of them

#41 If, in an hyperbola or ellipse or circle's periphery, a straight [line] be drawn down ordinatewise onto the diameter; and from the line that has been drawn and also from the [line emanating] out-from-the-center, [that is, from the line traditionally called the "radius"], there be drawn up (*anagraphêi*) equiangular parallelogrammic figures (*eidê*), and the side that is drawn down have, to the figure's (*tou eidous*) remaining side, the ratio compounded from that which the [line emanating] out-from-the-center [the "radius"] has to the figure's (*tou eidous*) remaining side and also from that which the upright side of the figure of the section (*tou eidous tês tomês*) has to the oblique [side]—then the figure (*eidos*) from the [line] [stretching along the diameter] between the center and the [line] drawn down, [a figure, that is to say, which is] similar to the figure (*tôi eidei*) from the [line emanating] out-from-the-center [the "radius"], is, for the hyperbola (on the one hand), greater than the figure (*tou eidous*) from the [line] drawn down, [greater, that is to say] by the figure (*tôi eidei*) from the [line emanating] out-from-the-center [the "radius"], but for the ellipse and the circle's periphery (on the other hand), is, [when taken] along with (*meta*) the figure (*tôi eidei*) from the [line] drawn down, equal to the figure (*tôi eidei*) from the [line emanating] out-from-the-center [the "radius"].

[Note that the Greek verb used for the producing of a straight line is *agô*. Its primary sense is "to drive, lead, convey," and it has been rendered throughout as "to draw." By contrast, the verb used for the producing of a surface (conic or planar) is *graphô*. Its primary sense is "to incise," and it has been rendered as "carved out" (to distinguish it from "draw"). Here, where planar figures have been produced from straight lines, the verb *anagraphô* is used, and this is rendered as "to draw up" (which leaves something to be desired, but the alternatives seem worse).]

—Let there be an hyperbola or ellipse or circle's periphery, whose diameter is what the [line] AB is, and center is what the [point] E is; and, also, let there be drawn down ordinatewise the [line] CD; and, also, from the [lines] EA, CD let equiangular figures (*eidê*) be drawn up (*anagegraphthô*), namely the [figures] AF, DG; and, also, let the [line] CD have to the [line] CG the ratio compounded from that which the [line] AE has to EF and also from that which the upright [side] has to the oblique [side]. I say that (on the one hand) for the hyperbola, the figure (*eidos*) from the [line] ED, that is, the [figure] similar to the [figure] AF, is equal to the [figures] AF,GD [taken together]; and (on the other hand) for the ellipse and the circle, the [figure] from the [line] ED, that is, the [figure] similar to the [figure] AF, is along with (*meta*) the [figure] GD, equal [when they are taken together] to the [figure] AF.

—For let there be made, [as a fourth proportional, the line CH—such that] as is the upright [side] to the oblique [side] so is the [line] DC to CH. And—since as is the [line] DC to CH so is the upright [side] to the oblique [side], but as is the [line] DC to CH so is the [square] from the [line] DC to the [rectangle] by the [lines] DCH, and as is the upright [side] to the oblique [side] so is the [square] from DC to the [rectangle] by BDA—therefore equal is the [rectangle] by BDA to the [rectangle] by DCH. And—since the [line] DC has, to CG, the ratio compounded from that which the [line] AE has to EF and also from that which the upright [side] has to the oblique [side], that is, which the [line] DC has to CH, and further, the [line] DC has to CG, the ratio compounded from that which the [line] DC has to CH and also from that which HC has to CG—therefore the ratio compounded from that which the [line] AE has to EF and also from that which the [line] DC has to CH is the same as the ratio compounded from that which the [line] DC has to CH and also from that which the [line] HC has to CG. Let what is common be taken away, namely, the ratio of the [line] CD to CH; therefore the remaining ratio, of the [line] AE to EF, is the same as the remaining ratio, of the [line] CH to CG. But (on the one hand) as is the [line] HC to CG so is the [rectangle] by HCD to the [rectangle] by GCD, and (on the other hand) as is the [line] AE to EF so is the [square] from AE to the [rectangle] by AEF—therefore as is the [rectangle] by HCD to the [rectangle] by GCD so is the [square] from EA to the [rectangle] by AEF. But the [rectangle] by HCD was shown equal to the [rectangle] by BDA; therefore as is the [rectangle] by BDA to the [rectangle] by GCD so is the [square] from AE to the [rectangle] by AEF. Alternately, as is the [rectangle] by BDA to the [square] from AE so is the [rectangle] by GCD to the [rectangle] by AEF. But as is the [rectangle] by GCD to the [rectangle] by AEF so is the [parallelogram] DG to the [parallelogram] FA, for they are equiangular and have [the one to the other] the ratio compounded from [the ratios of] their sides—[[namely the ratio] of the [line] GC to AE and also [the ratio] of the [line] CD to EF. And therefore as is the [rectangle] by BDA to the [square] from EA so is the [parallelogram] GD to the [parallelogram] AF.

—To be said further, on the hyperbola, [is this]: Therefore, [by putting together (*sunthenti*, or, in the traditional Latin, *componendo*)] as the [rectangle] by BDA [taken] along with (*meta*) the [square] from AE is to the [square] from AE, that is, as is the [square] from DE to the [square] from EA, so the [parallelograms] GD, AF [taken together] are to the [parallelogram] AF. But as the [square] from ED is to the [square] from EA so the figure (*eidos*) from ED, the one that has been drawn up (*anagegrammenon*) similar and [situated] similarly to the [parallelogram] AF, is to the [parallelogram] AF; therefore as the [parallelograms] GD, AF [taken together] are to the [parallelogram] AF, so the figure (*eidos*) from ED, the one similar to the [parallelogram] AF, is to the [parallelogram] AF. Therefore the figure (*eidos*) from ED, the one similar to the [parallelogram] AF, is equal to the [parallelograms] GD, AF [taken together].

—But on the ellipse and the circle's periphery we shall say [this]: Since then as is the whole [square] from AE to the whole [parallelogram] AF so is the taken-away [rectangle] by ADB to the taken-away [parallelogram] DG, and also remainder is to remainder as whole is to whole. And if, from the [square] from AE, there be taken away the [rectangle] by BDA, what remains is the [square] from DE; therefore as is the [square] from DE to the excess by which the [parallelogram] AF exceeds (*huperekhei*) the [parallelogram] DG so is the [square] from AE to the [parallelogram] AF. But as is the [square] from AE to the [parallelogram] AF so is the [square] from DE to the figure (*eidos*) from DE that is similar to the [parallelogram] AF. Therefore as is the [square] from DE to the excess by which the [parallelogram] AF exceeds the [parallelogram] DG so is the [square] from DE to the figure (*eidos*) from DE that is similar to the [parallelogram] AF. Therefore the figure (*eidos*) from the [line] DE that is similar to the [parallelogram] AF is equal to the excess by which the [parallelogram] AF exceeds the [parallelogram] DG. [Taken] along with (*meta*) the [parallelogram] DG, therefore, [that figure from DE] is equal to the [parallelogram] AF.

42nd through 45th propositions

Now we are prepared for the propositions that bridge the way from the propositions which obtain tangents, to the propositions which obtain new diameters.

The bridge for the parabola (whose tangents were obtained in the 35th proposition, and whose new diameters will be obtained in the 48th) is the 42nd proposition.

Next comes the 43rd proposition, the bridge for the curves other than the parabola—the hyperbola, the ellipse, and the circle (whose tangents were obtained in the 36th proposition, and whose new diameters will be obtained in the 47th).

Whereas the bridge for the parabola did not need any preparation, the bridge for the other curves required an access road (propositions 37, 39, 41). The 43rd proposition, the bridge for the curves other than the parabola, is itself the access road to the bridge for the opposite sections, which is the 44th proposition. (There was no proposition that obtained tangents specifically for the opposite sections.)

Finally, just as the access-road propositions (37 and 39) had side-trip propositions (38 and 40), so now also the bridge propositions will have a side-trip proposition—namely, 45. Like the earlier side trips, this one too shows a feature in which the second diameter resembles the original diameter.

After that, we shall be ready to obtain, in the next chapter of this guidebook, new diameters for the curves (46th through 48th propositions), and then their parameters (49th through 51st propositions). Then we shall be ready to return, in the chapters of the final part of the first volume of this guidebook, to the cone.

42nd and 43rd propositions: enunciations

The 42nd proposition treats the parabola; the 43rd treats the other curves. Both propositions show that a certain triangle is equal to a certain quadrilateral. Although the quadrilateral for the parabola is not the same as the quadrilateral for the other curves, the triangle is the same.

Given the curve and its diameter and one of its tangents, we get the triangle as follows. Take any point on the curve—this will be the triangle's apex. Its base will lie along the diameter. Its sides come from the random point: one of them is drawn ordinatewise to the given diameter, the other is drawn parallel to the given tangent. The triangle is heavily outlined in figure 10.

Now, equal to that triangle there is a certain quadrilateral. Its base is that part of the diameter which is the abscissa. One of its sides lies along the line already drawn ordinatewise through the random point, and its other side lies along a line that now needs to be drawn ordinatewise through the vertex (which makes this line another tangent). Finally, another line needs to be drawn to give the quadrilateral's top. This line is drawn through the tangency point given at the beginning. In the parabola, this line is drawn parallel to the diameter. That is to say: in the parabola, it meets the diameter nowhere. In the other curves, however, it meets the diameter at the center. The quadrilateral equal to the heavily outlined triangle is filled with dots in figure 11.

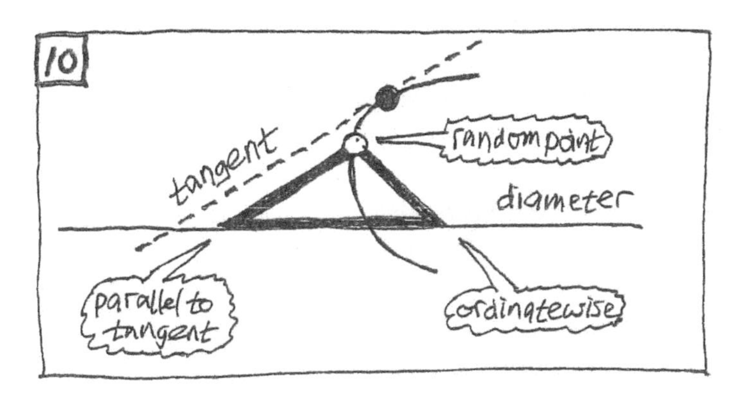

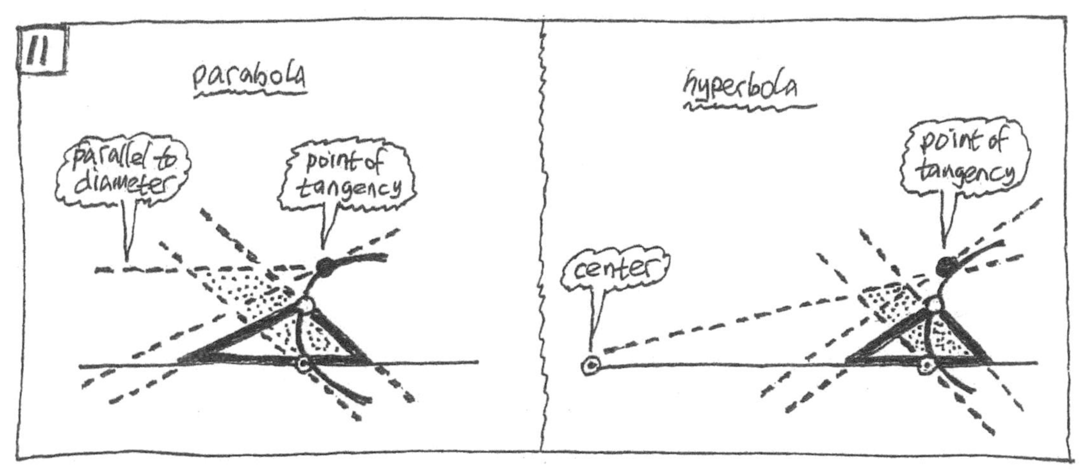

42nd and 43rd propositions: demonstrations

Once the requisite lines have been drawn, the demonstrations proper are not complicated (see the two boxed propositions following). In the accompanying figures 12 and 13, the lines are numbered according to the order of their drawing, and the demonstrations proper are depicted.

#42 If, with respect to a parabola, a straight [line] that touches [the section] fall together with the diameter; and also, from the touch[point] a straight [line] be drawn down onto the diameter ordinatewise, and, some point being taken on the section, there be drawn [from that point] down onto the diameter two straight [lines], one of them parallel to the touching[line], and the other parallel to the [line] that was drawn down from the touch[point]—then the triangle generated by them, [that is, the triangle generated on a segment of the diameter by the two straight lines drawn down onto it from the point taken,] is equal to the parallelogram contained by the [line] drawn down from the touch[point] and by the [line] taken off by the parallel [line], on the side toward the vertex of the section.
—Let there be a parabola, whose diameter is what the [line] AB is; and let there be drawn a [line] touching the section, namely the line AC; and let there be drawn down ordinatewise the [line] CH; and from some chance point let there be drawn down the [line] DF; and (on the one hand) through the [point] D parallel to the [line] AC let there be drawn the [line] DE, and (on the other hand) through the [point] C parallel to the [line] BF [let there be drawn] the [line] CG, and (yet again) through the [point] B parallel to the [line] HC [let there be drawn] the [line] BG. I say that the triangle DEF is equal to the parallelogram GF.
—For—since, with respect to the section, there touches [it] the [line] AC; and there has been drawn down ordinatewise the [line] CH—hence equal is the [line] AB to the [line] BH; therefore double is the [line] AH of the [line] HB. Therefore the triangle AHC, to the parallelogram BC, is equal. And—since as is the [square] from CH to the [square] from DF so is the [line] HB to BF, on account of the section, [on account, that is, of the section's being a parabola], but (on the one hand) as is the [square] from CH to the [square] from DF so is the triangle ACH to the triangle EDF, and (on the other hand) as is the [line] HB to BF so is the parallelogram GH to the parallelogram GF—therefore as is the triangle ACH to the triangle EDF so is the parallelogram HG to the parallelogram FG. Alternately, therefore, as is the triangle AHC to the parallelogram BC so is the triangle EDF to the parallelogram GF. Equal, however, is the triangle ACH to the parallelogram GH; equal, therefore, is the triangle EDF to the parallelogram GF.

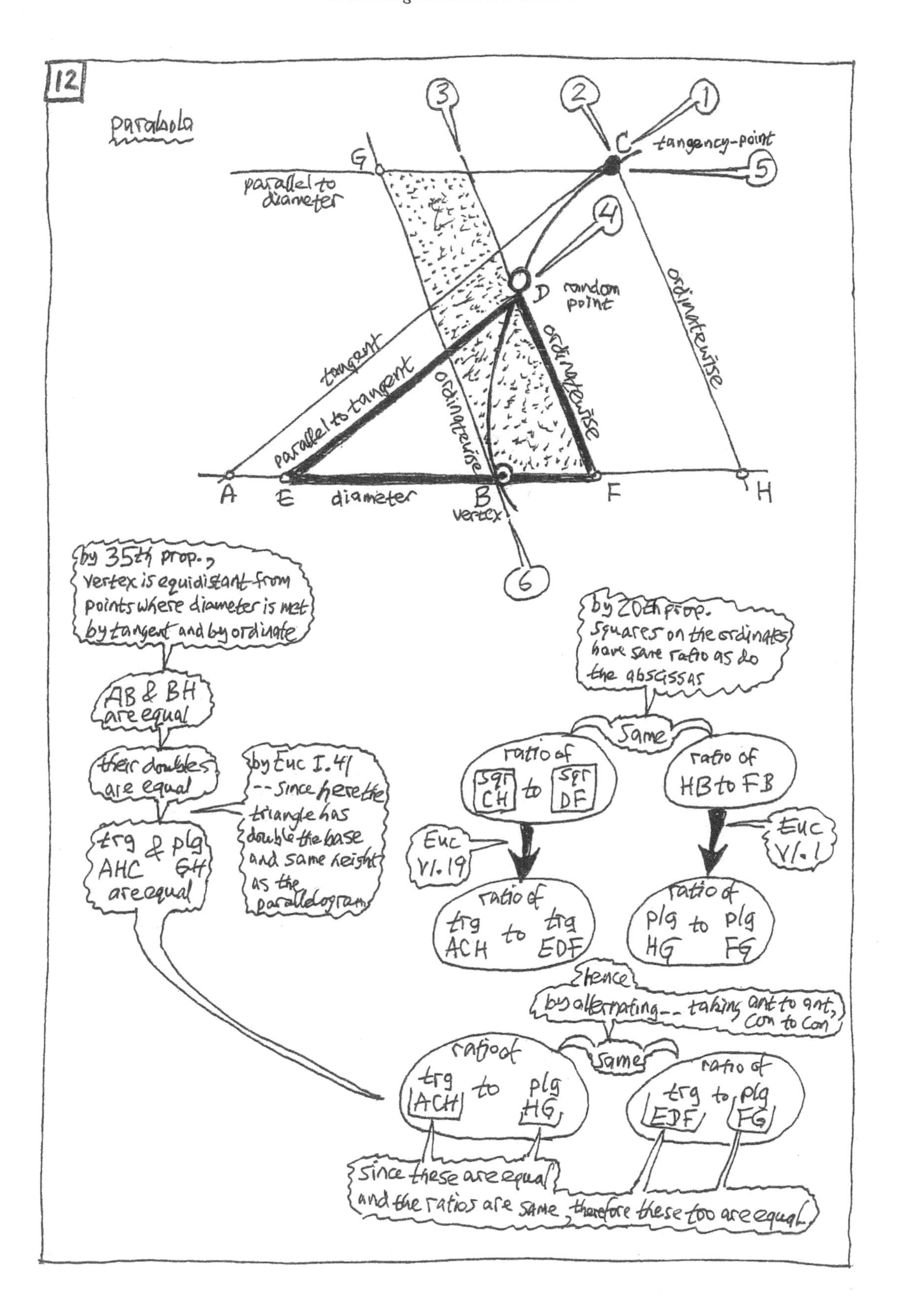
12
parabola
3
2
1
C
tangency-point
G
5
parallel to diameter
4
D
random point
ordinatewise
tangent
parallel to tangent
ordinatewise
ordinatewise
A
E
diameter
B
vertex
F
H
6
by 35th prop., vertex is equidistant from points where diameter is met by tangent and by ordinate
AB & BH are equal
their doubles are equal
by Euc I.41 -- since here the triangle has double the base and same height as the parallelogram
trg & plg AHC GH are equal
by 20th prop. squares on the ordinates have same ratio as do the abscissas
same
ratio of sqr CH to sqr DF
ratio of HB to FB
Euc VI.19
Euc VI.1
ratio of trg ACH to trg EDF
ratio of plg HG to plg FG
hence
by alternating -- taking ant to ant, con to con
same
ratio of trg ACH to plg HG
ratio of trg EDF to plg FG
since these are equal and the ratios are same
therefore these too are equal

#43 If, with respect to an hyperbola or ellipse or circle's periphery, a straight [line] that touches [the section] fall together with the diameter; and also, from the touch[point] there be drawn down a straight [line] ordinatewise onto the diameter; and also, parallel to this [line drawn down ordinatewise] there be drawn through the vertex a [line] that falls together with the straight [line] [or its extension] that has been drawn through the touch[point] and the center, and, some point being taken on the section, there be drawn two straight [lines] onto the diameter, of which [two lines] the one be parallel to the touching[line], and the other be parallel to the [line] drawn down [ordinatewise] from the touch[point]—then the triangle that is generated by them [generated, that is, on a segment of the diameter (or diameter extended) by the two straight lines drawn onto it from the point taken] will be, on the hyperbola, less than the triangle which the line through the center and the touch[point] cuts off, [cuts off, that is, with the help of the line drawn ordinatewise through the point taken], [and will be less, that is,] by the triangle [which arises] from the [line emanating] out-from-the-center [the "radius"] [and which is] similar to the [triangle] cut off. But, on the ellipse and also the circle's periphery, [that triangle which is generated by the two straight lines drawn, through the point taken, onto the diameter] will—[when taken] along with (*meta*) the triangle which is cut off by the [line] to the center [from the touchpoint], [the triangle which is cut off, that is, with the help of the line drawn ordinatewise through the point taken,]—be equal to the triangle [which arises] from the [line emanating] out-from-the-center [the "radius"] [and which is] similar to the [triangle] cut off.

—Let there be an hyperbola or ellipse or circle's periphery whose diameter is what the [line] AB is, and center is what the [point] C is; and let there be drawn touching the section the [line] DE; and let there be joined the [line] CE, and let there be drawn down ordinatewise the [line] EF; and let there be taken some point on the section, namely the [point] G; and parallel to the touching[line] let there be drawn the [line] GH; and let there be drawn down ordinatewise the [line] GK [extended to meet line CE, or its extension, at point M], and through the [point] B let there be drawn up ordinatewise the [line] BL. I say that the triangle KMC differs from the triangle CLB by the triangle GKH.

—For—since (on the one hand) the [line] ED touches [the section], and (on the other hand) the [line] EF is drawn down [ordinatewise]—hence the [line] EF has to FD the ratio compounded from that of the [line] CF to FE and also [that] of the upright [side] to the oblique [side]. But (on the one hand) as is the [line] EF to FD so is the [line] GK to KH; and (on the other hand) as is the [line] CF to FE so is the [line] CB to BL. Therefore the [line] GK will have to KH the ratio compounded from that of the [line] BC to BL and also [that] of the upright [side] to the oblique [side]. And, on account of the things shown in the forty-first theorem, the triangle CKM differs from the triangle BCL by the triangle GHK—for indeed in their doubles, the parallelograms, the same things have been shown.

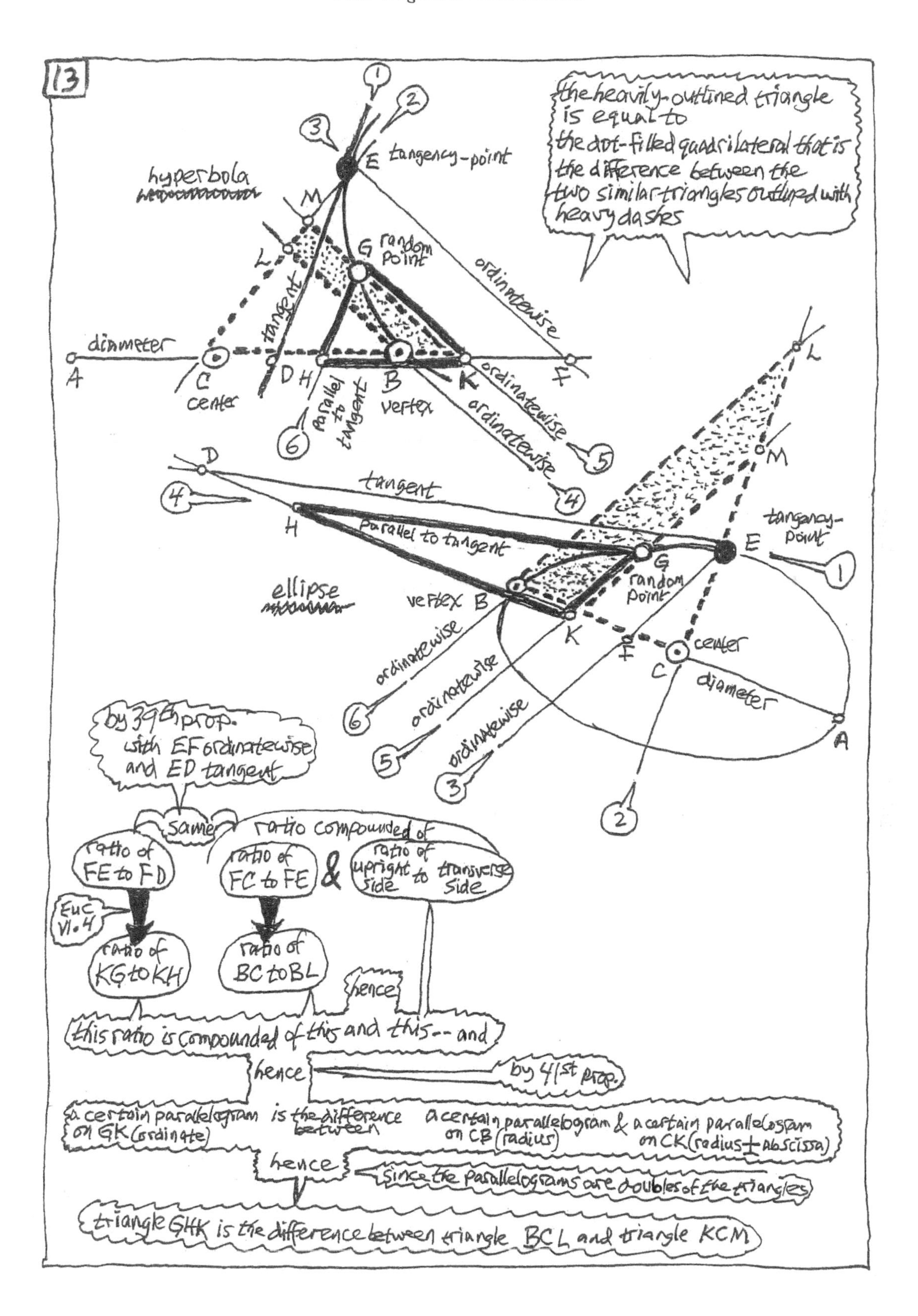
13
the heavily-outlined triangle is equal to the dot-filled quadrilateral that is the difference between the two similar triangles outlined with heavy dashes
hyperbola
tangency-point
random point
tangent
ordinatewise
diameter
center
parallel to tangent
vertex
ellipse
by 39th prop. with EF ordinatewise and ED tangent
same
ratio of FE to FD
ratio compounded of
ratio of FC to FE
&
ratio of upright side to transverse side
Euc VI.4
ratio of KG to KH
ratio of BC to BL
hence
this ratio is compounded of this and this -- and
hence
by 41st prop.
a certain parallelogram on GK (ordinate)
is the difference between
a certain parallelogram on CB (radius)
&
a certain parallelogram on CK (radius ± abscissa)
hence
since the parallelograms are doubles of the triangles
triangle GHK is the difference between triangle BCL and triangle KCM

44th proposition

What the 43rd proposition has just shown about the hyperbola (and also about the ellipse and the circle), the 44th proposition will now show about the opposite sections. That is to say: what was true about the triangles made by the lines drawn with respect to a tangency point, and with respect to a random point on a single hyperbolic curve, is true also about the corresponding triangles made in the pair of hyperbolic curves constituting opposite sections—when the tangency point lies on one of the two curves, and the random point is taken on the other one.

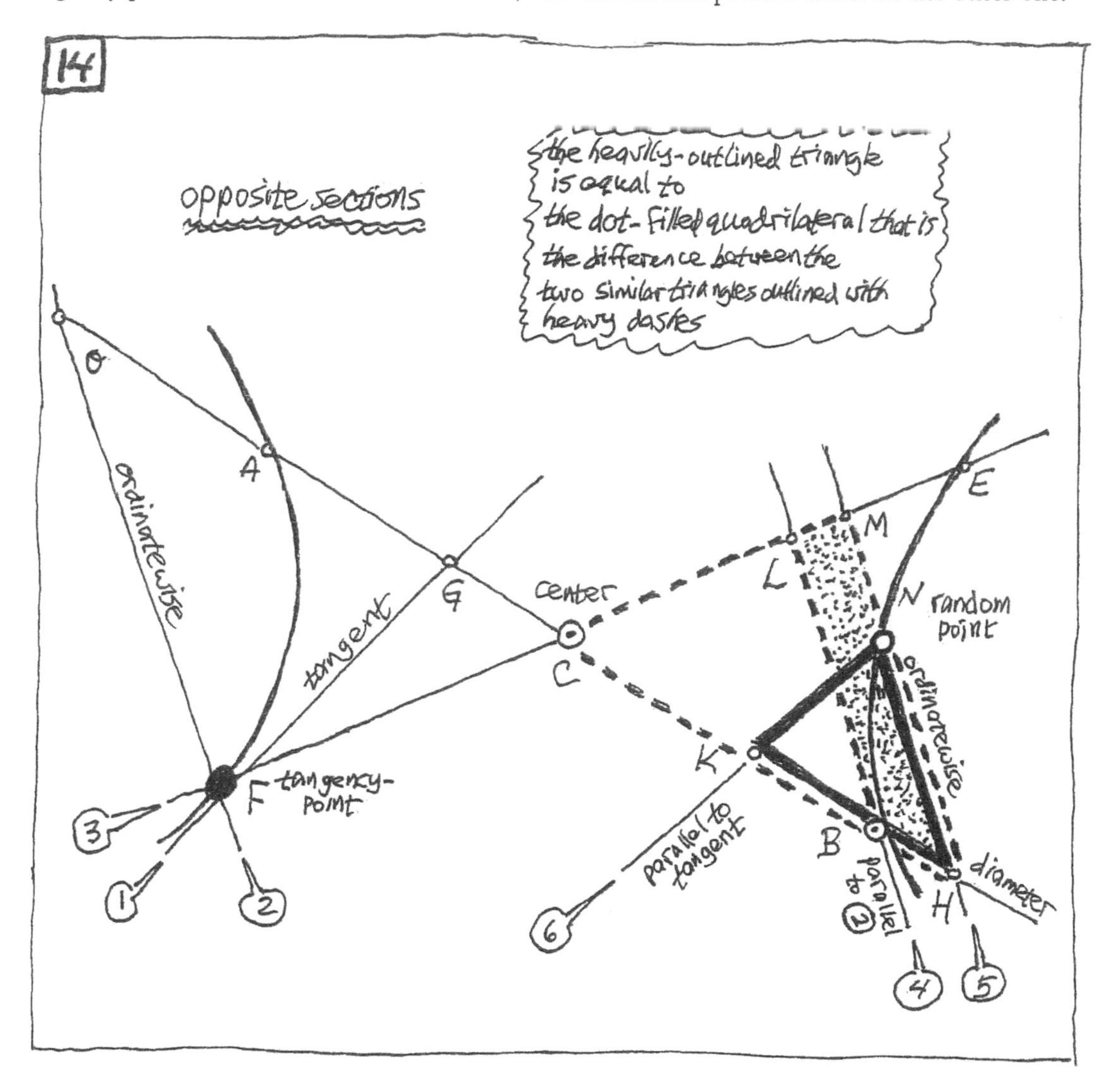

The demonstration of this 44th proposition (for the opposite sections) relies on the 43rd proposition (for the hyperbola). On a first reading of Apollonius, you may wish to give the demonstration a merely cursory look.

#44 If, with respect to one of the opposite [sections], a straight [line] that touches [the section] fall together with the diameter; and also, from the touch[point] there be drawn down some straight [line] ordinatewise onto the diameter; and also parallel to this [line drawn down ordinatewise] there be drawn through the vertex of the other section a [line] that falls together with the straight [line] that has been drawn through the touch[point] and the center, and, there being taken on the section a chance point, there be straight [lines] drawn down onto the diameter, of which [lines] the one be drawn parallel to the touching[line], and the other be drawn parallel to the [line] drawn down ordinatewise from the touch[point]—then the triangle which is generated by them, [which is generated, that is, on a segment of the extended diameter, by the two straight lines drawn onto it from the chance point] will be less than the triangle which the line that has been drawn down cuts off toward the center of the section, [cuts off, that is, with the help of the line through the touchpoint and the center], [and will be less] by the triangle [which arises] from the [line emanating] out-from-the-center [the "radius"] [and which is] similar to the [triangle] cut off.
—Let opposite [sections] be what the [lines] AF, BE are, and their diameter be what the [line] AB is, and center be what the [point] C is; and also, from some point of those on the section FA, say the [point] F, let there be drawn a [line] touching the section, say the line FG, and let there be drawn ordinatewise the [line] FO; and also, let there be joined the [line] CF and also let it be extended, as the [line] CE; and also, through the [point] B let a parallel to the [line] FO what the [line] BL is; and also, let some point on the section BE be what the [point] N is; and also, from the [point] N let there be drawn down ordinatewise the [line] NH, and parallel to the [line] FG let there be drawn the [line] NK. I say that the triangle HKN is, in comparison with the triangle CMH, less by the triangle CBL.
—For, through the [point] E, let there be drawn a [line] touching the section BE, namely the [line] ED, but ordinatewise let the [line] EX be. Since, then, opposite [sections] are what the [lines] FA, BE are, whose diameter is what the line AB is, and the [line] through the center is what the [line] FCE is, and [lines] touching the sections are what the [lines] FG, ED are—hence a parallel to the [line] FG is what the [line] DE is. But the [line] NK is a parallel to the [line] FG; therefore also a parallel to the [line] ED is what the [line] NK is, and the [line] MH [is a parallel] to the [line] BL. Since, then, an hyperbola is what the [line] BE is, whose diameter is what the [line] AB is, and center is what the [point] C is, and a [line] touching the section is what the [line] DE is, and an ordinatewise [line] is what the [line] EX is, and also, a [line] parallel to the [line] EX is what the [line] BL is, and also, there has been taken as a point on the section the [point] N, from which (on the one hand) there has been drawn down ordinatewise the [line] NH and (on the other hand) there has been drawn parallel to the [line] DE the [line] KN; therefore the triangle NHK, in comparison with the triangle HMC, is less by the triangle BCA—for this has, in the 43rd theorem, been shown.

45th proposition

In the 45th proposition we see yet again what we saw in the 38th and also in the 40th—that the second diameter works with the parallel to the original diameter in a way that resembles the way in which the original diameter works with its ordinate.

As the 38th proposition corresponded to the 37th, and the 40th to the 39th, so now the 45th corresponds to the 43rd. One of the cases for the 45th proposition is depicted in figure 15. In it, the heavily outlined triangle is equal to the difference between two triangles—the larger triangle outlined with dashes and the smaller triangle filled with dots. (Because of the arrangement of the lines in this proposition, the figure for this one, unlike the one for the 43rd, contains no quadrilateral equal to the difference between the two triangles. Since the difference cannot be displayed by filling some area with dots, the dots are here used to mark the smaller triangle by filling it.)

Things are not so simple here as they were earlier with the original diameter. It is true that each of the three triangles does have for one side the second diameter (as each of the three triangles earlier had for one side the original diameter), and it is also true that two of the triangles do have for another side the line that passes through the random point and is parallel to the original diameter (as the two triangles earlier had for another side the line that passes through the random point and is ordinatewise). For the third side, however, the situation is now as follows. In the triangle that is the difference between the other two, the third side is the line that runs between the tangency point and the center (rather than being the line that passes through the random point and is parallel to the tangent); and in the triangle that is biggest, the third side is the line that passes through the random point and is parallel to the tangent (rather than being the line that runs between the tangency point and the center). In the remaining triangle, the third side is now the given tangent (rather than being the line that is made tangent by being drawn ordinatewise through the vertex).

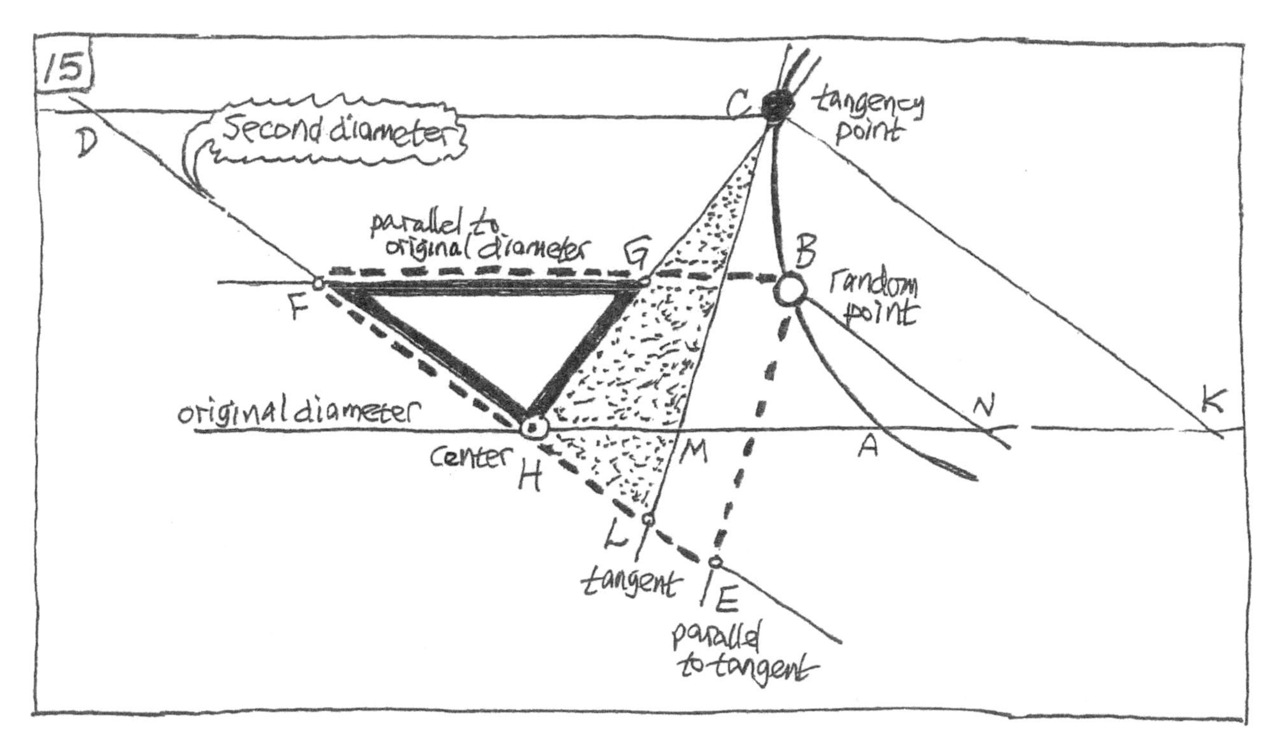

The 38th proposition, which showed for the second diameter what the 37th showed for the original diameter, relied for its demonstration upon the 37th proposition. The 40th proposition, which showed for the second diameter what the 39th showed for the original diameter, relied for its demonstration not upon the 39th but upon the 38th. Now the 45th proposition, which shows for the second diameter what the 43rd showed for the original diameter, relies for its demonstration not upon the 40th nor upon the 43rd, but upon what the 40th relied upon—namely, the 41st and the 39th (which in turn relied upon the 37th). The logical relationship of the these propositions is depicted in figure 16. (We see it here as part of the figure that we examined at the beginning of this chapter.)

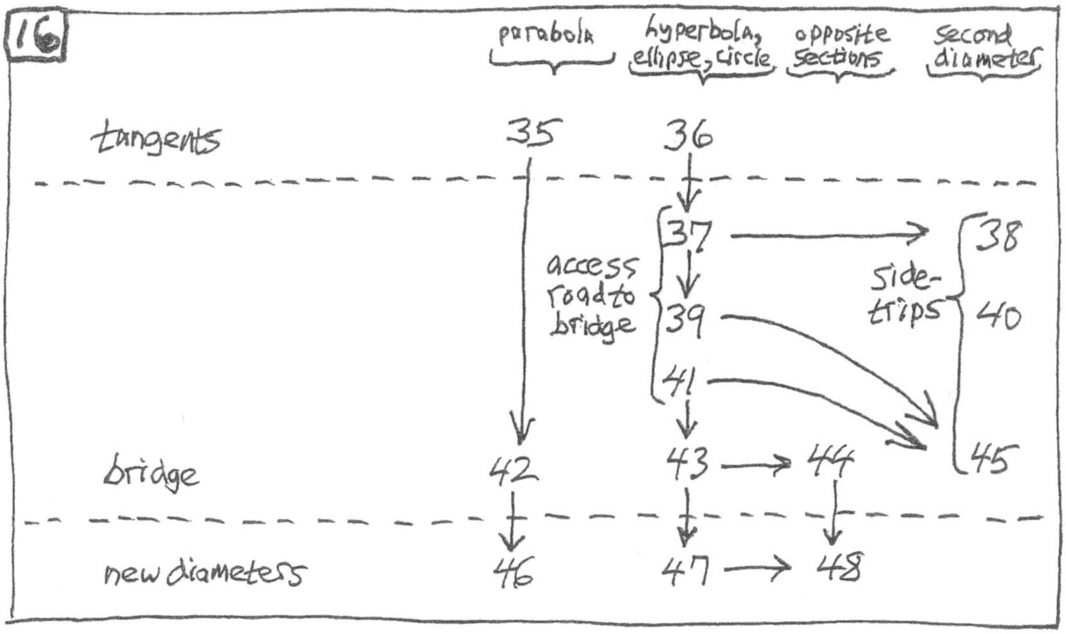

On a first reading of Apollonius, you may wish to give the demonstration in the following box a merely cursory look.

#45 If, with respect to an hyperbola or ellipse or circle's periphery, a straight [line] that touches [the section] fall together with the second diameter; and also, there be drawn down from the touch[point] onto that same diameter some straight [line] parallel to the other diameter; and also, through the touch[point] and the center a straight [line] be extended, and, there being taken on the section a chance point, from which there be drawn onto the second diameter two straight [lines], one of which be parallel to the touching[line] and the other be parallel to the [line] drawn down—then the triangle which is generated by them, [which is generated, that is, on a segment of the second diameter (or of the second diameter extended) by the two straight lines drawn onto it from the chance point], in comparison with the triangle that the [line] drawn down cuts off on the side toward the center [cuts off, that is, with the help of the line that is extended through the touchpoint and the center], will, on the hyperbola (on the one hand), be greater, by the triangle whose base is the touching[line] and whose vertex is the center of the section; and will, on the ellipse and the circle (on the other hand), be, [when taken] along with (*meta*) the triangle that is cut off, equal to the triangle whose base is what the touching[line] is and whose vertex is what the center of the section is.

—Let an hyperbola or ellipse or circle's periphery be what the [line] ABC is, whose diameter is what the [line] AH is, and second [diameter] what the [line] HD is, and center what the [point] H is; and also, let the [line] CML touch [the section] at the [point] C, and the [line] CD be drawn parallel to the [line] AH; and also, let the joined [line] HC be extended; and also let there be taken on the section a chance point, namely the [point] B; and also, from the [point] B let there be drawn the [lines] BE, BF parallel to the [lines] LC, CD. I say that the triangle BEF, on the hyperbola (on the one hand), is, in comparison with the [triangle] GHF, greater by the [triangle] LCH; and, on the ellipse and the circle (on the other hand), is, [when taken] along with (*meta*) the [triangle] FGH, equal to the [triangle] CLH.

—For let the [lines] CK, BN be drawn parallel to the [line] DH. Since, then, there touches the section the [line] CM, and drawn down has been the [line] CK—hence the [line] CK has, to KH, the ratio compounded from that which the [line] MK has to KC and also that which the figure's (*tou eidous*) upright side has to the oblique [side]; but as is the [line] MK to KC so is the [line] CD to DL; therefore the [line] CK has, to KH, the ratio compounded from that of the [line] CD to DL and from that of the upright [side] to the oblique [side]. And the triangle CDL is the figure (*tou eidos*) [arising] from the line KH; and the [triangle] CKH, that is, the [triangle] CDH, is the [triangle arising] from the [line] CK, that is, from the [line] DH. Therefore the triangle CDL—in comparison with the [triangle] CKH—is, on the hyperbola (on the one hand), greater, by the triangle [which arises] from the [line] AH [and is] similar to the [triangle] CDL; and, on the ellipse and the circle (on the other hand), the [triangle] CDH, [when taken] along with (*meta*) the [triangle] CDL—is equal to the same [triangle which arises from the line AH and is similar to the triangle CDL]; for, indeed, in the doubles of them this was shown in the forty-first theorem. Since, then, the [triangle] CDL does, in comparison with the triangle CKH, or in comparison with the [triangle] CDH, differ by the triangle [which arises] from the [line] AH [and is] similar to the triangle CDL, but differs also by the triangle CHL—therefore equal is the triangle CHL to the triangle [which arises] from the line AH [and is] similar to the triangle CDL. Since, then, (on the one hand) the triangle BFE is similar to the [triangle] CDL and (on the other hand) the [triangle] GFH [is similar] to the [triangle] CDH, therefore they have the same ratio. [That is: the ratio of CD to DL is the same as that of BF to FE, and the ratio of CD to DH is the same as that of GF to FH. But there is a link between the ratio of CD to DL and the ratio of CD to DH. Since the ratio compounded from that of CD to DL and that of the upright to the oblique side was shown to be the same as that of CK to KH; but this ratio of CK to KH is also the same as that of CD to DH, it follows that the ratio compounded from that of BF to FE and that of the upright to the oblique side is the same as that of GF to FH. And so, the conclusion here will follow, as it did in #43, from #41.] And (on the one hand) the [triangle] BFE is the [triangle arising] from the [line] NH [that stretches] between the line drawn down and the center, and (on the other hand) the [triangle] GFH is the [triangle] from the [line] BN drawn down, that is, from the [line] FH. And on account of the things shown before, the [triangle] BFE, in comparison with the [triangle] GHF differs by the [triangle which arises] from the [line] AH [and is] similar to the [triangle] CDL—and so differs also by the [triangle] CLH.

CHAPTER 7. NEW DIAMETERS, NEW PARAMETERS

46th through 48th propositions: enunciations

Apollonius now shows how to obtain diameters other than the original one. Although he does not at first call the straight lines obtained by him diameters, nonetheless, that is what they are, for they bisect all the straight lines that are drawn across the curve parallel to a given straight line.

In effect, what Apollonius now shows is this: you will get a *new diameter* if, from any point on the curve, you draw a straight line that,

in the parabola (46),	or, in the hyperbola, ellipse, and circle (47),
is parallel to	meets, at the center,
the original diameter,	the original diameter.

Thus here again, a straight line that is drawn from any of the central sections so as to meet the diameter at the center corresponds to one that is drawn from a parabola so as to meet the diameter nowhere, however far extended.

In all cases, the *ordinates* of the new diameter are parallel to the line that is tangent at the point where the new diameter meets the section. In the opposite sections (48), the ordinates of the new diameter are not located in that hyperbola where the tangent is, but are located rather in the other hyperbola of the pair (fig.1).

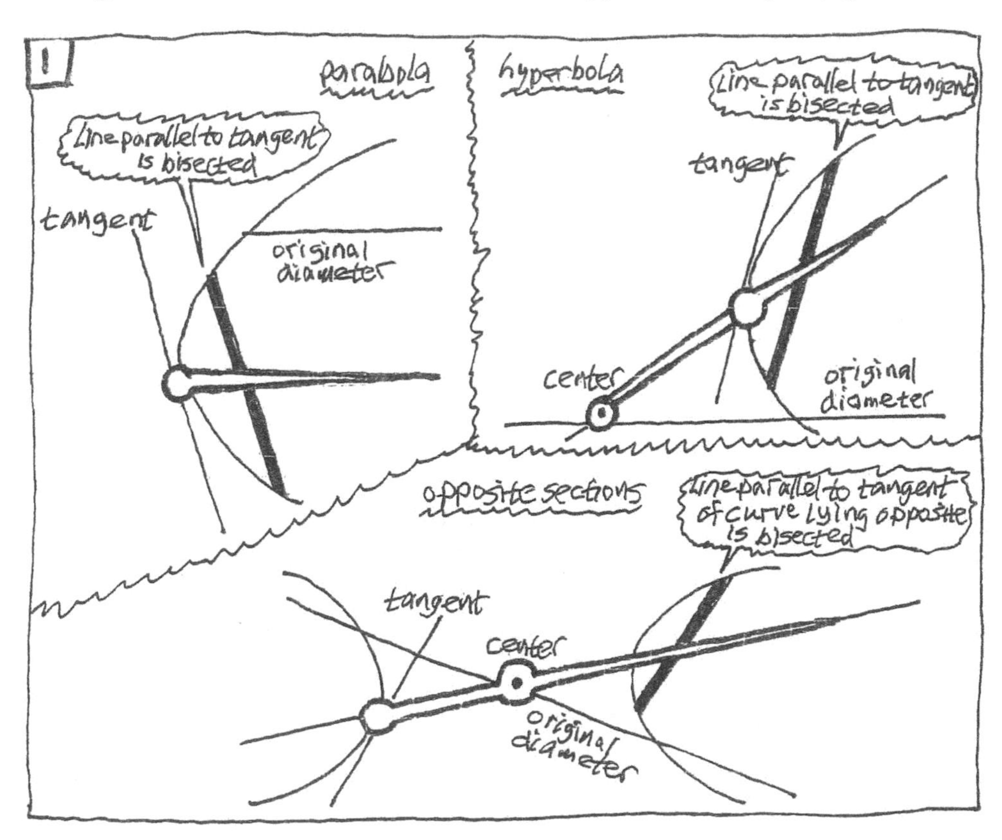

46th through 48th propositions: demonstrations

Each of these propositions in which new diameters are obtained depends on one of the preceding bridge propositions about the equality of certain areas (fig. 2).

The box below contains the proposition for the parabola; the box opposite, the propositions for the hyperbola/ellipse/circle and for the opposite sections. After them, come figures 3 and 4, which depict the drawing of the lines and the gist of the demonstrations. (Or, rather, there are depictions for the parabola and the hyperbola; they should prepare you for the exercise of producing your own depictions for the other cases—the ellipse, the circle, and the opposite sections.)

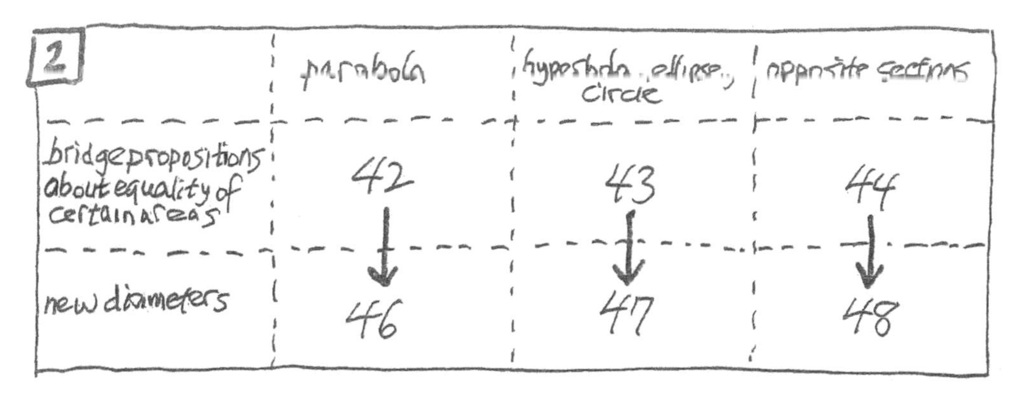

#46 If, with respect to a parabola, a straight [line] that touches [the section] fall together with the diameter, [then] the [line] through the touch[point] that is drawn parallel to the diameter, in the same direction as the section, cuts in two [halves] the [lines] that are drawn in the section parallel to the touching[line].
—Let there be a parabola whose diameter is what the [line] ABD is; and also, let there touch the section the [line] AC, and through the [point] C let there be drawn parallel to the [line] AD the [line] HCM; and also, let there be taken on the section a chance point, the [point] L; and also, let there be drawn parallel to the [line] AC the [line] LNFE. I say that equal is the [line] LN to the [line] NF.
—Let there be drawn ordinatewise the [lines] BH, KFG, LMD. Since, then, on account of the things shown in the forty-second theorem, equal is the triangle ELD to the parallelogram BM, and [so is] the [triangle] ELD to the [parallelogram] BK—therefore the remainder parallelogram GM, to the remainder quadrilateral LFGD, is equal. A common area let there be taken off, namely the pentagon MDGFN; therefore the remainder triangle KFN, to the [triangle] LMN, is equal. And parallel is the [line] KF to the [line] LM; therefore equal is the [line] FN to the [line] LN.

#47 If, with respect to an hyperbola or ellipse or circle's periphery, a straight [line] that touches [the section] fall together with the diameter; and also, through both the touch[point] and the center a straight [line] be drawn in the same direction as the section, [then] it will cut in two [halves] the [lines] that are drawn in the section parallel to the touching[line].

—Let there be an hyperbola or ellipse or circle's periphery whose diameter is what the [line] AB is, and center is what the [point] C is; and also, touching the section, let there be drawn the [line] DE; and also, let there be joined the [line] CE, and also let it be extended; and also let there be taken on the section a chance point, the [point] N, and through the [point] N let there be drawn parallel [to the line DE] the [line] HNOG. I say that equal is the [line] NO to the [line] OG.

—For let there be drawn down ordinatewise the [lines] XNF, BL, GMK. Therefore, on account of the things shown in the 43rd theorem, equal is (on the one hand) the triangle HNF to the [quadrilateral] LBFX and (on the other hand) the triangle GHK to the quadrilateral LBKM; and also therefore the remainder [quadrilateral] NGKF, to the remainder [quadrilateral] MKFX, is equal. A common [area] let there be taken off, namely the pentagon ONFKM; therefore the remainder triangle OMG, to the remainder [triangle] NXO, is equal. And also, parallel is the [line] MG to the [line] NX; therefore equal is the [line] NO to the [line] OG.

#48 If, with respect to one of the opposite [sections], a straight [line] that touches [the section] falls together with the diameter; and also, through both the touch[point] and the center an extended straight [line] cut the other section, then whatever [line] may be drawn in the other section parallel to the touching[line] will be cut in two [halves] by the extended line.

—Let there be opposite [sections] whose diameter is what the [line] AB is, and center is what the [point] C is; and also, let there touch the section A the [line] KL; and also, let there be joined the [line] LC and also let it be extended; and also, let there be taken some point on the section B, the [point] N, say; and also, through the [point] N parallel to the [line] LK let there be drawn the [line] NG. I say that the [line] NO, to the [line] OH, is equal.

—For let there be drawn through the [point] E touching the section the [line] ED; therefore the [line] ED is parallel to the [line] LK; and [is parallel] also to the [line] NG. Since then an hyperbola is what the [line] BNG is, whose center is what the [point] C is; and also, a touching[line] is what the [line] DE is; and also, there has been joined the [line] CE; and also, there has been taken on the section a point, namely the [point] N; and also, through it parallel to the [line] DE has been drawn the [line] NG—hence, on account of what has been previously shown on the hyperbola, equal is the [line] NO to the [line] OG.

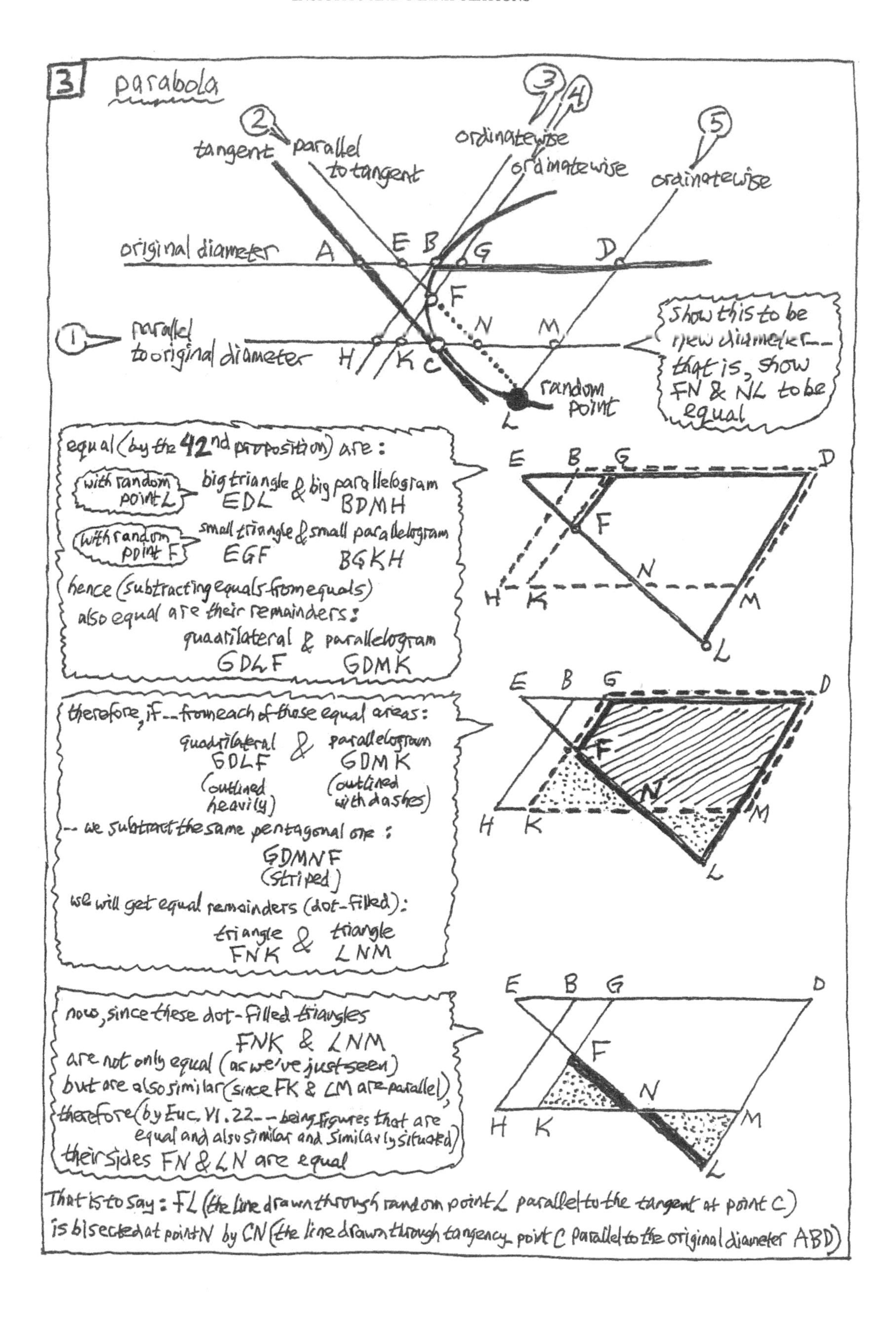
3
parabola
tangent
parallel to tangent
ordinatewise
ordinatewise
ordinatewise
original diameter
A E B G D
F
N M
parallel to original diameter
H K C
random point
L
Show this to be new diameter -- that is, show FN & NL to be equal
equal (by the 42nd proposition) are:
with random point L
big triangle EDL & big parallelogram BDMH
with random point F
small triangle EGF & small parallelogram BGKH
hence (subtracting equals from equals) also equal are their remainders:
quadrilateral GDLF & parallelogram GDMK
therefore, if -- from each of those equal areas:
quadrilateral GDLF (outlined heavily) & parallelogram GDMK (outlined with dashes)
-- we subtract the same pentagonal one: GDMNF (striped)
we will get equal remainders (dot-filled): triangle FNK & triangle LNM
now, since these dot-filled triangles FNK & LNM are not only equal (as we've just seen) but are also similar (since FK & LM are parallel), therefore (by Euc. VI.22 -- being figures that are equal and also similar and similarly situated) their sides FN & LN are equal
That is to say: FL (the line drawn through random point L parallel to the tangent at point C) is bisected at point N by CN (the line drawn through tangency point C parallel to the original diameter ABD)

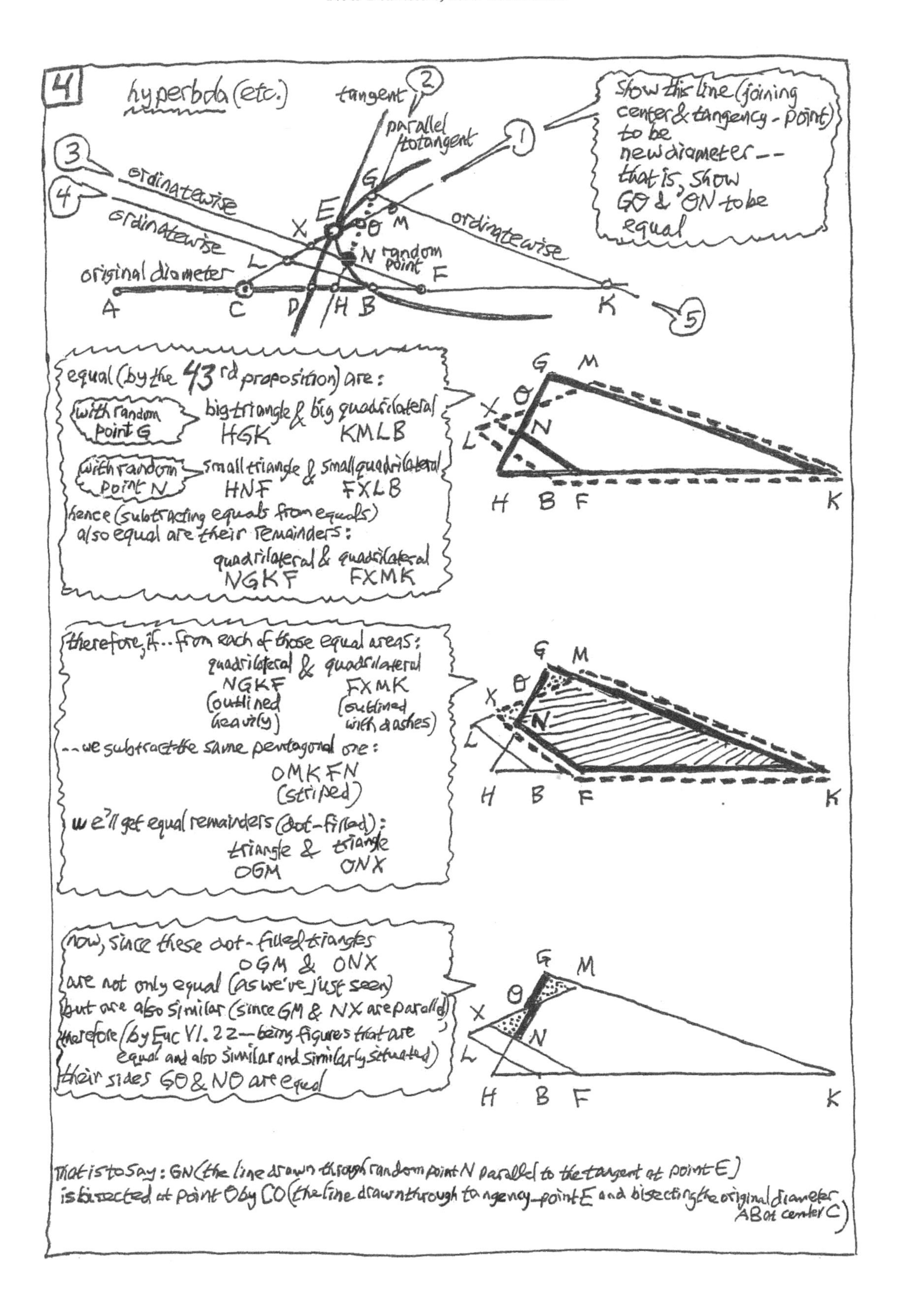
4
hyperbola (etc.)
tangent
2
parallel to tangent
1
Show this line (joining center & tangency-point) to be new diameter -- that is, show GO & ON to be equal
3
4
ordinatewise
ordinatewise
ordinatewise
original diameter
A
C
D
H
B
E
G
X
O
M
N random point
L
F
K
5
equal (by the 43rd proposition) are:
with random point G
big triangle & big quadrilateral
HGK KMLB
with random point N
small triangle & small quadrilateral
HNF FXLB
hence (subtracting equals from equals) also equal are their remainders:
quadrilateral & quadrilateral
NGKF FXMK
G M O X N L H B F K
therefore, if.. from each of those equal areas:
quadrilateral & quadrilateral
NGKF FXMK
(outlined heavily) (outlined with dashes)
--we subtract the same pentagonal one:
OMKFN
(striped)
we'll get equal remainders (dot-filled):
triangle & triangle
OGM ONX
G M O X N L H B F K
now, since these dot-filled triangles
OGM & ONX
are not only equal (as we've just seen)
but are also similar (since GM & NX are parallel)
therefore (by Euc VI.22—being figures that are equal and also similar and similarly situated)
their sides GO & NO are equal
G M O X N L H B F K
That is to say: GN (the line drawn through random point N parallel to the tangent at point E)
is bisected at point O by CO (the line drawn through tangency-point E and bisecting the original diameter AB at center C)

49th through 51st propositions: enunciations

Having just obtained, in propositions 46–48, new diameters with their ordinates, Apollonius will now obtain, in propositions 49–51, parameters for the new diameters. He waits, however, to call a spade a spade. Not until after the new parameters have been obtained—that is, not until the end of the 51st proposition—are lines other than the *original* diameters called by the name "diameter." ("Parameter," however, as was said back in the chapter when we first encountered "the upright side," is not a term employed by Apollonius: its usefulness is its emphasis upon the size that enables a line to serve as an "upright side.")

It was for the original diameter—the diameter that was obtained from the axial triangle in the cone—that the 11th through 14th propositions characterized the several sorts of conic sections by showing the rectangle-square property which relates the sizes of the abscissa and the ordinate. That relation between the original diameter's ordinate and its abscissa was stated in terms of a parametric line contrived by using the cone. Now the rectangle-square property for the abscissa and ordinate is shown for a new diameter—a diameter that is obtained, not from the axial triangle in the cone, but from some other diameter that is already given. The relation between this new diameter's abscissa and its ordinate is stated in terms of a parametric line which is contrived, not by using the cone, but rather by using the two diameters (the original and the new one) and the tangents at their vertices.

This is done for the parabola in the 49th proposition, and then it is done for the central sections. For the hyperbola, ellipse, and circle it is done in the 50th proposition. For the opposite sections it is done in a separate 51st proposition, to show the property when the ordinate of the new diameter is located not in the hyperbola that is used to contrive the parameter, but is located rather in the other hyperbola of the pair.

How to contrive the parameter for the new diameter is depicted in figure 5. This parameter is defined in terms of three lines that stretch from the vertex of the new diameter:

one of them is	*the other two are*	
the stretch (striped in fig.5) that is cut off on the tangent from the new diam's vertex by the orig diam's extension from the orig diam's vertex ;	the stretch (dotted in fig. 5) that is cut off on the tangent from the new diam's vertex by the tangent from the orig diam's vertex,	and the stretch (solid in fig. 5) that is cut off on the new diam's extension from the new diam's vertex by the tangent from the orig diam's vertex.

(The tangent from the vertex of a diameter, whether original or new, is the reference line for its ordinates.) The parameter is so contrived that *it* has, to *twice the striped line*, the same ratio that *the dotted line* has to *the solid line*.

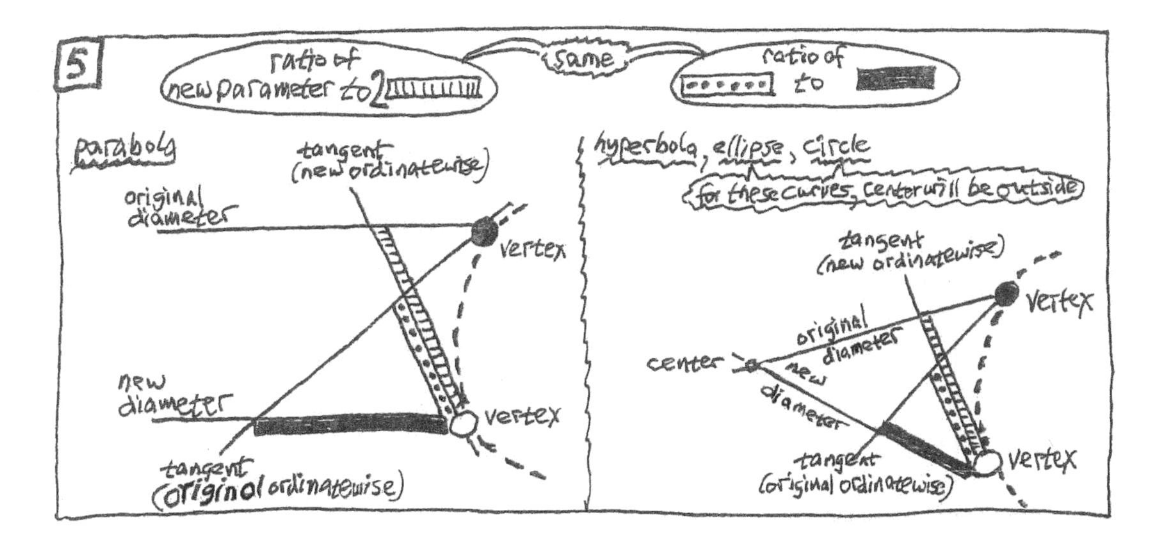

Even without the use of a parameter (as we saw when we considered the 20th and 21st propositions), it is still possible to relate the ordinates and the abscissas of a conic section. Even without the use of a parameter, that is to say, a relationship involving two ordinates can be related to a relationship involving their two abscissas (this latter relationship being one which, in the parabola, involves the abscissas alone; and, in the hyperbola or the ellipse, involves the abscissas together with the lines obtained when the abscissas are added on to or taken away from the oblique [transverse] side). But if we do forego the use of a parameter, then we must forego as well the equality of square and rectangle that gives us that particular relationship between any one ordinate and its abscissa by which any particular parabola or hyperbola or ellipse is distinguished from another curve of the same sort. That is a reason for introducing the parameter. It is not, however, a reason for having different definitions of the parameter for conic sections of different sorts.

For conic sections of different sorts, very different ratios were used by Apollonius to contrive the parameters of the original diameters, back when the 11th through 14th propositions characterized the conic sections—even though he could have used, for the original diameters of conic sections of *all* the different sorts, a *single* definition for their parameters. He did not take that simpler road. Back then, Apollonius used very different ratios to contrive parameters for the original diameters, although he could have avoided doing so; but now, in contriving parameters for the new diameters, he gives a more homogeneous presentation. We have not looked at cones ever since Apollonius slid the curves out of them for our consideration after the 14th proposition. Without any solid figures to look at, we have not had any axial triangles to look at either, and so now the presentation of the parameters for the new diameters can be more homogeneous than was the presentation of the parameters for the original diameters, which were obtained in the axial triangles of cones.

But while the new parameter is not contrived by using different ratios for the sections of different sorts, it is still true that in the central sections the rectangle which is contained by the new parameter and the new abscissa is not equal to the square which arises from the ordinate, but rather overshoots or falls short of it. The central sections are therefore not treated in the same proposition with the parabola.

Note that where the demonstration of the 49th proposition (for the parabola) depends upon the 42nd, the 50th (for the hyperbola, ellipse, and circle) depends upon the 43rd, and the 51st (for the opposite sections) upon the 44th.

49th proposition: demonstration for parabola

The demonstration for the parabola requires the construction of the apparatus of lines shown in figure 6.

What the demonstration shows is that the contrived line acts as a parameter for the new diameter. That is to say, a square on the new ordinate (KL) is equal to a rectangle contained by the new abscissa (DL) and the contrived line (G), because:

> the line has been so contrived that, as it turns out,
> that square will have to that rectangle the same ratio that
> rectangle [KL, LN] has to the double of rectangle [LD, DC] ;
>
> and this latter ratio will be a ratio of equality,
> because the 42nd proposition showed that, in the parabola,
> parallelogram FM is equal to triangle KPM.

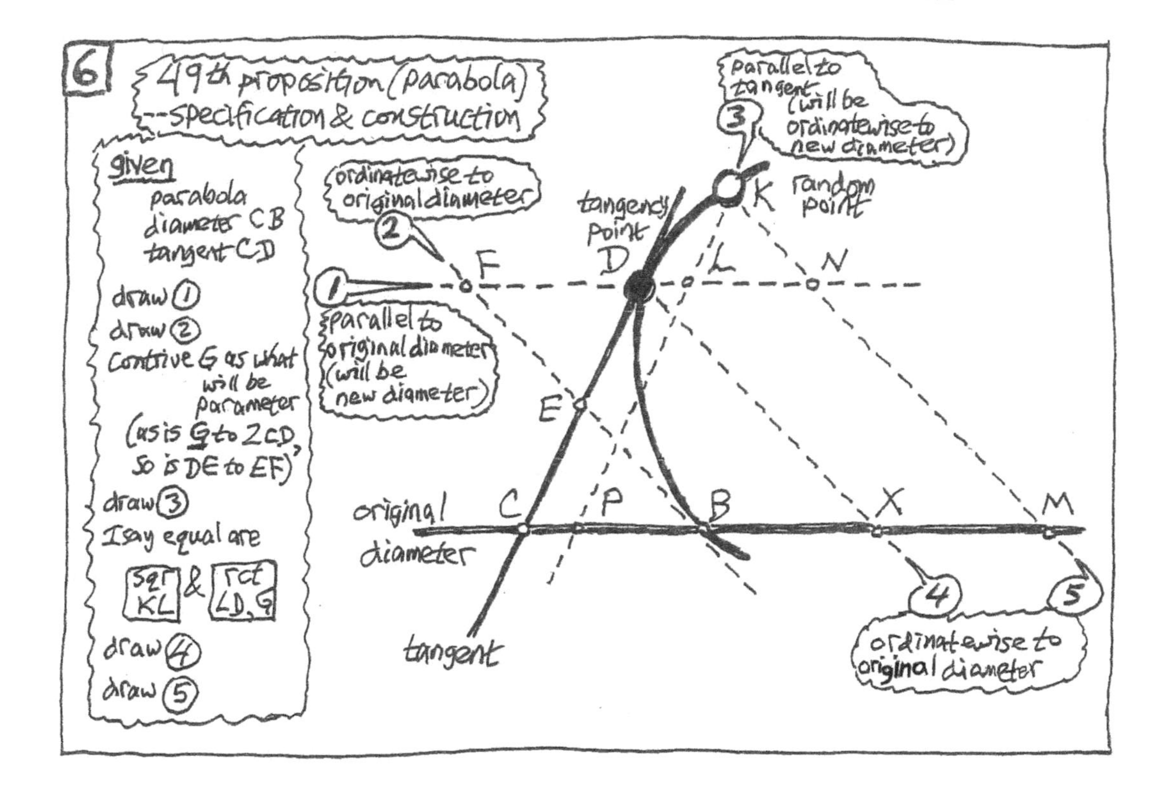

In order to understand the demonstration, the supplement in the following figure will be useful (fig. 7). After it, the demonstration is depicted in figure 8.

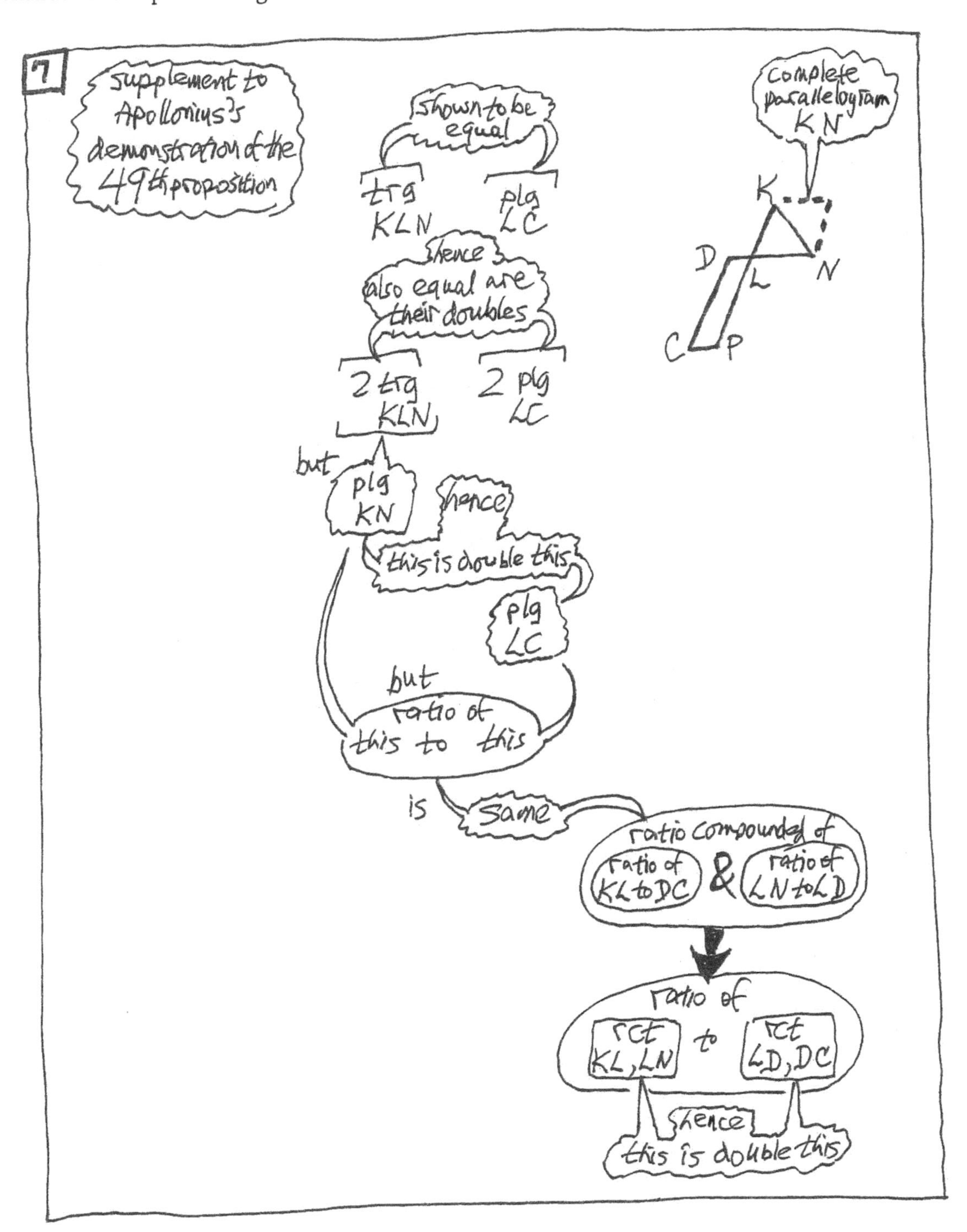

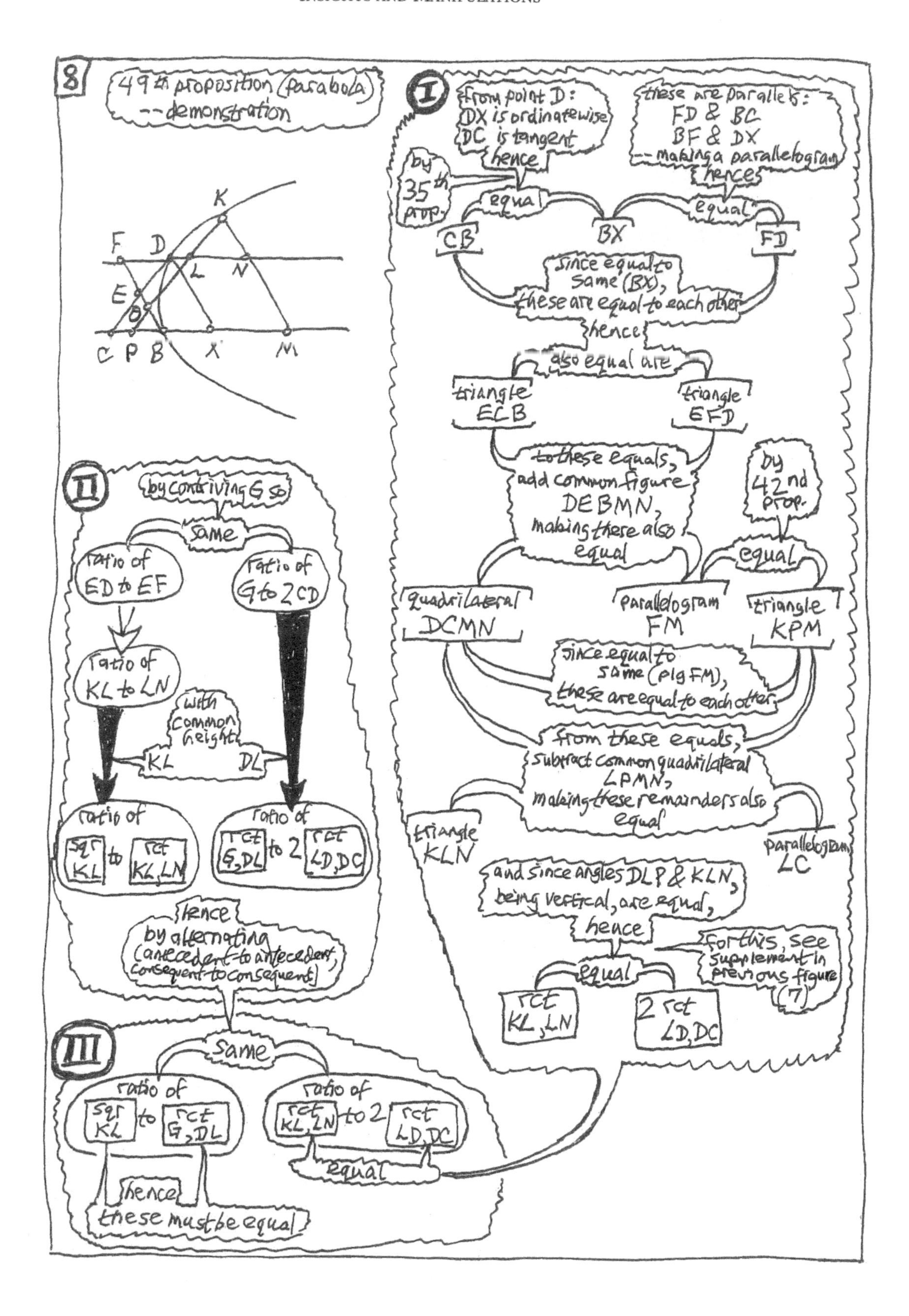
8
49th proposition (parabola) --demonstration
K
F D
L N
E
O
C P B X M
I
from point D: DX is ordinatewise DC is tangent hence
these are parallels: FD & BC BF & DX --making a parallelogram hence
by 35th prop.
equal
equal
CB
BX
FD
since equal to same (BX), these are equal to each other
hence
also equal are
triangle ECB
triangle EFD
to these equals, add common figure DEBMN, making these also equal
by 42nd prop.
equal
quadrilateral DCMN
parallelogram FM
triangle KPM
since equal to same (plg FM), these are equal to each other
from these equals, subtract common quadrilateral LPMN, making these remainders also equal
triangle KLN
parallelogram LC
and since angles DLP & KLN, being vertical, are equal,
hence
for this, see supplement in previous figure (7)
equal
rct KL, LN
2 rct LD, DC
II
by contriving G so
same
ratio of ED to EF
ratio of G to 2CD
ratio of KL to LN
with common height
KL
DL
ratio of sqr KL to rct KL, LN
ratio of rct G, DL to 2 rct LD, DC
hence by alternating (antecedent to antecedent, consequent to consequent)
III
same
ratio of sqr KL to rct G, DL
ratio of rct KL, LN to 2 rct LD, DC
equal
hence
these must be equal

Now here is Apollonius's own presentation for the parabola.

#49 If, with respect to a parabola, a straight [line] that touches [the section] fall together with the diameter; and also, through the touch[point] there be drawn a [line] parallel to the diameter, and from the vertex there be drawn a [line] parallel to a [line] drawn down ordinatewise; and also, there be made [a certain straight line such that] as the segment (*tmêma*) of the touching[line], [that is, that segment of the touching[line] which stretches] between the [line] drawn up ordinatewise and the touch[point], is to the segment (*tmêma*) of the parallel [line], [that is, to that segment of the parallel line which stretches] between the touch[point] and the [line] drawn up [ordinatewise], so [that] certain straight [line] is to the double of the touching[line]—then whatever [line] from the section be drawn [parallel to the touching[line]], onto the straight [line] that is drawn through the touch[point] parallel to the diameter, will have the power (*dunêsetai*) [to give rise to a square equal to] the rectangle contained by [the following two lines]: the straight [line] that has been provided (*peporismenês*) [by the proportion just mentioned], and the [line] taken off by it, [taken off, that is, by the line parallel to the touching[line]], on the side toward the touch[point].

—Let there be a parabola whose diameter is what the [line] MBC is, and a touching[line] [of which] is what the [line] CD is; and also, through the [point] D parallel to the [line] BC let there be drawn the [line] FDN, and ordinatewise let there be drawn up the [line] FB; and also, let there be made [a certain straight line such that] as is the [line] ED to DF, so is [the] certain straight [line], namely the [line] G, to the double of the [line] CD; and also, let there be taken some point on the section, the [point] K, say; and also, let there be drawn through the [point] K parallel to the [line] CD the [line] KLP. I say that the [square] from the [line] KL is equal to the [rectangle] by both the [line] G and the [line] DL—that is, [I say] that, with a diameter being what the [line] DL is, an upright (*orthia*) [side] is what the [line] G is.

—For let there be drawn down ordinatewise the [lines] DX, KNM. And—since the [line] CD touches the section, but there has been drawn down ordinatewise the [line] DX—equal is the [line] CB to the [line] BX. But the [line] BX, to the [line] FD, is equal. Therefore also the [line] CB, to the [line] FD, is equal. And so, [equal] also [is] the triangle ECB to the triangle EFD. A common [area] let there be added on—namely the figure (*skhêma*) DEBMN; therefore the quadrilateral DCMN, to the parallelogram FM, is equal—that is, [is equal] to the triangle KPM. A common [area] let there be taken away, namely the quadrilateral LPMN; therefore remainder the triangle KLN, to [remainder] the parallelogram LC, is equal. And also is equal the angle by DLP to the [angle] by KLN; therefore double is the rectangle by KLN—[double, that is,] of the [rectangle] by LDC. And—since as is the [line] ED to DF so is the [line] G to the double of the [line] CD, and also as is the [line] ED to DF so is the [line] KL to LN—therefore also as is the [line] G to the double of the [line] CD so is the [line] KL to LN. But (on the one hand) as is the [line] KL to LN so is the [square] from KL to the [rectangle] by KLN, and (on the other hand) as is the [line] G to the double of the [line] DC so is the [rectangle] by G, DL to twice the [rectangle] by CDL. And alternately. But equal is the [rectangle] by KLN to twice the [rectangle] by CDL; therefore equal is also the [square] from KL to the [rectangle] by G, DL.

50th proposition: demonstration for hyperbola/ellipse/circle

In studying the 50th proposition, you will probably find it easier to go through the demonstration first with your eye on the hyperbola alone, and then again with your eye on the ellipse and the circle.

The demonstration for the hyperbola requires the construction of the apparatus of lines shown in figure 9. The contriving that is part of the constructing is shown after that, in figure 10.

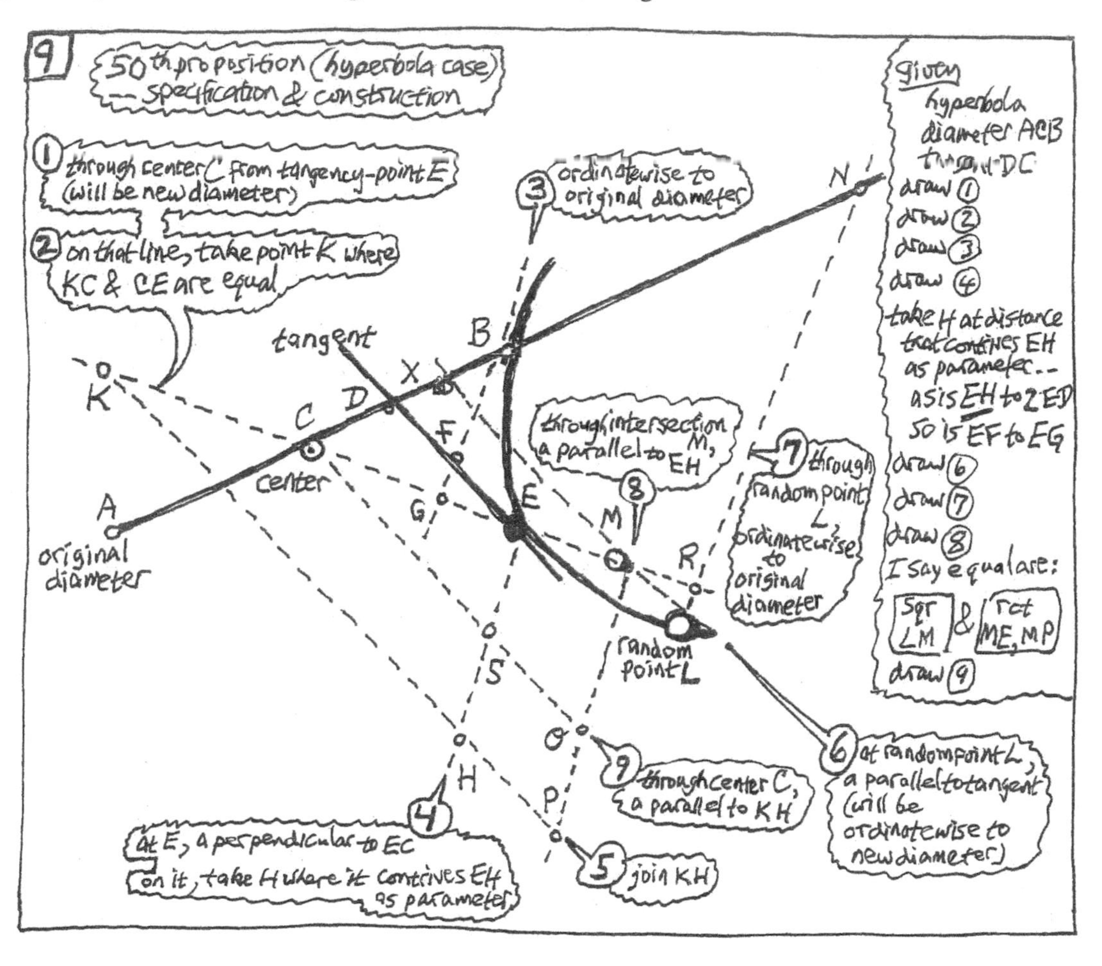

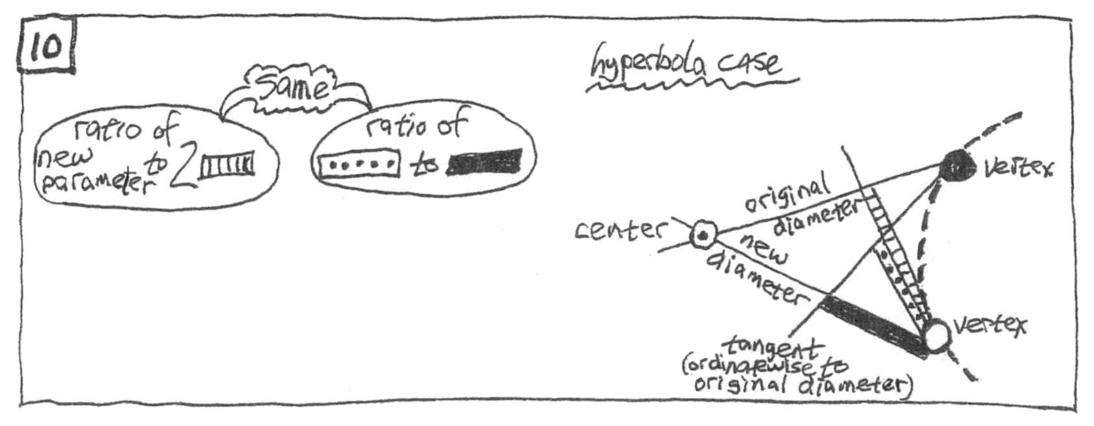

Note that the demonstration depends upon our taking the random point L to be on the *same* side of the original diameter (whose vertex is B) that the tangency point E is on. If, however, we took the random point L to be on the *other* side from the tangency point E, that would not affect the truth of the enunciation; but it would affect the procedure required for demonstrating it. It would require a small change for the hyperbola and a bigger change for the ellipse and circle. (For readers who might be interested, the requisite changes will be laid out right after the text of Apollonius's 50th proposition.)

What the demonstration shows is that the contrived line acts as a parameter for the new diameter. That is to say: the square on the ordinate (LM) is equal to the rectangle that is contained by the new abscissa (EM) and the line MP, and this rectangle [EM, MP] is related to the contrived line (HE) as follows:

> the rectangle [EM,MP] that is equal to the square on the ordinate
> *exceeds* (in the hyperbola) or *falls short of* (in the ellipse and circle)
> the rectangle that is contained by
> the new abscissa EM and the contrived line EH,
>
> and the rectangle that shows the excess or deficiency
> is similar to the rectangle that is contained by
> (as one side) a line EK that is double the radius CE
> and (as the other side) the line HE that is contrived.

That is to say: the 50th proposition shows that

> the *new* diameter
> with the *new* abscissa added on to it (hyperbola)
> or taken away from it (ellipse and circle)
> will have, to the non-abscissa side
> of the rectangle that is equal to the square on the *new* ordinate,
> a ratio that is the same as the ratio that
> the *new* diameter EK has to the *new* parameter (EH)—

which is just what the 12th and 13th propositions showed for the *original* diameter of the hyperbola, ellipse and circle. The original diameter's parameter related its abscissa and its ordinate in just that way.

The core of the demonstration of this 50th proposition is like the core of the 49th. The square on the ordinate LM is equal to the rectangle [(MO+OP), EM] because:

> the line HE has been so contrived that, as it turns out,
> the square will have to the rectangle the same ratio that
> rectangle [LM, MR] has to rectangle [(MX+ED), EM] ;
>
> and this latter ratio will be a ratio of equality,
> because the 43rd proposition showed that
> (in the hyperbola) the difference between triangle RNC and triangle LNX,
> or (in the ellipse and circle) their sum,
> is equal to triangle GBC.

The demonstration is depicted in figures 11, 12, and 13. Note, by the way, that by the middle of the demonstration it has been shown that what the 43rd proposition shows about the old diameter is true also of the new one—namely, that triangle MR is equal to the quadrilateral MEDX.

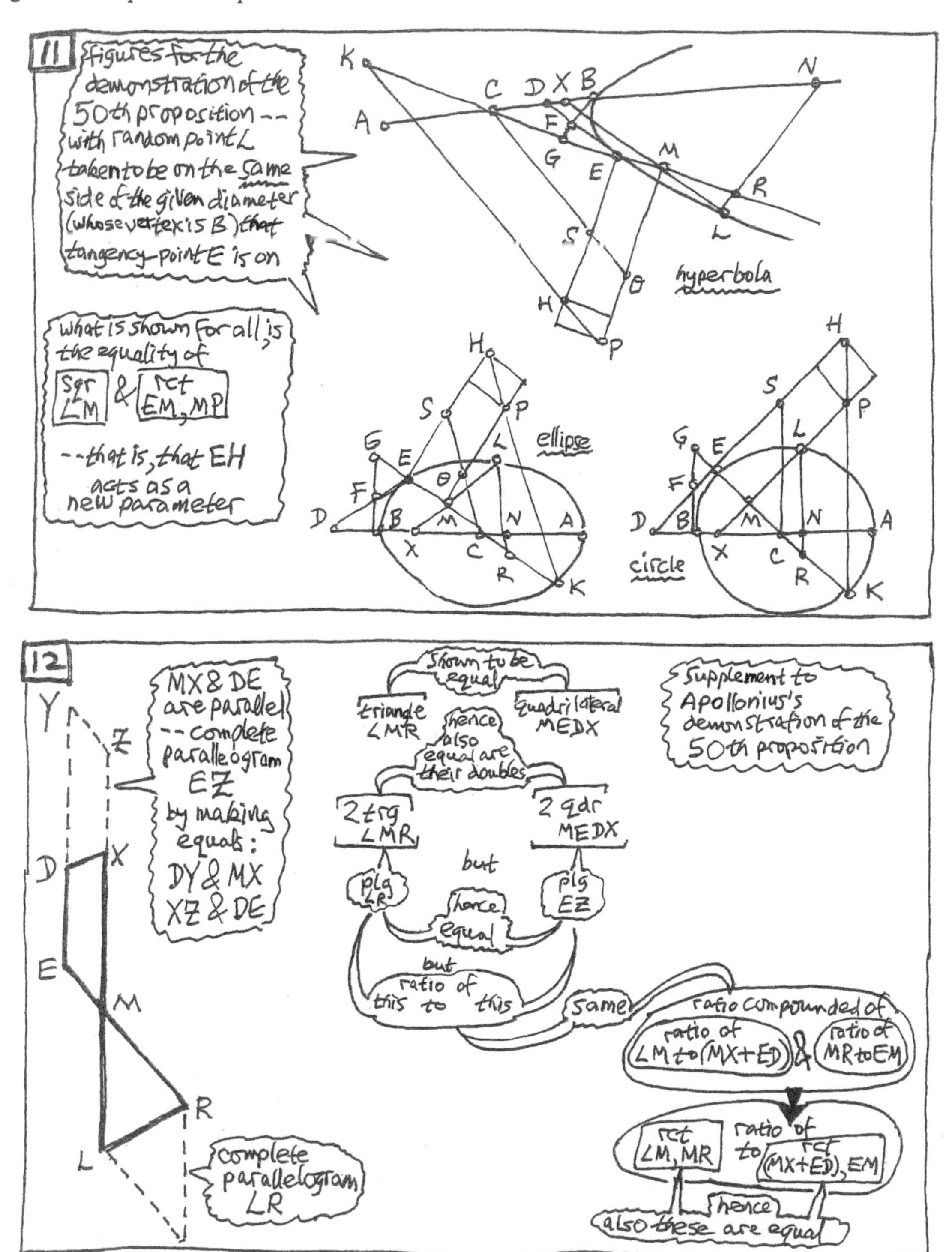

11
figures for the demonstration of the 50th proposition -- with random point L taken to be on the same side of the given diameter (whose vertex is B) that tangency-point E is on
hyperbola
what is shown for all, is the equality of
Sqr LM
&
rct EM, MP
--that is, that EH acts as a new parameter
ellipse
circle
12
MX & DE are parallel -- complete parallelogram EZ by making equals: DY & MX XZ & DE
complete parallelogram LR
Shown to be equal
triangle LMR
quadrilateral MEDX
hence also equal are their doubles
2 trg LMR
2 qdr MEDX
but
plg LR
plg EZ
hence equal
but ratio of this to this
same
ratio compounded of
ratio of LM to (MX+ED)
&
ratio of MR to EM
ratio of
rct LM, MR
to
rct (MX+ED), EM
hence
also these are equal
Supplement to Apollonius's demonstration of the 50th proposition

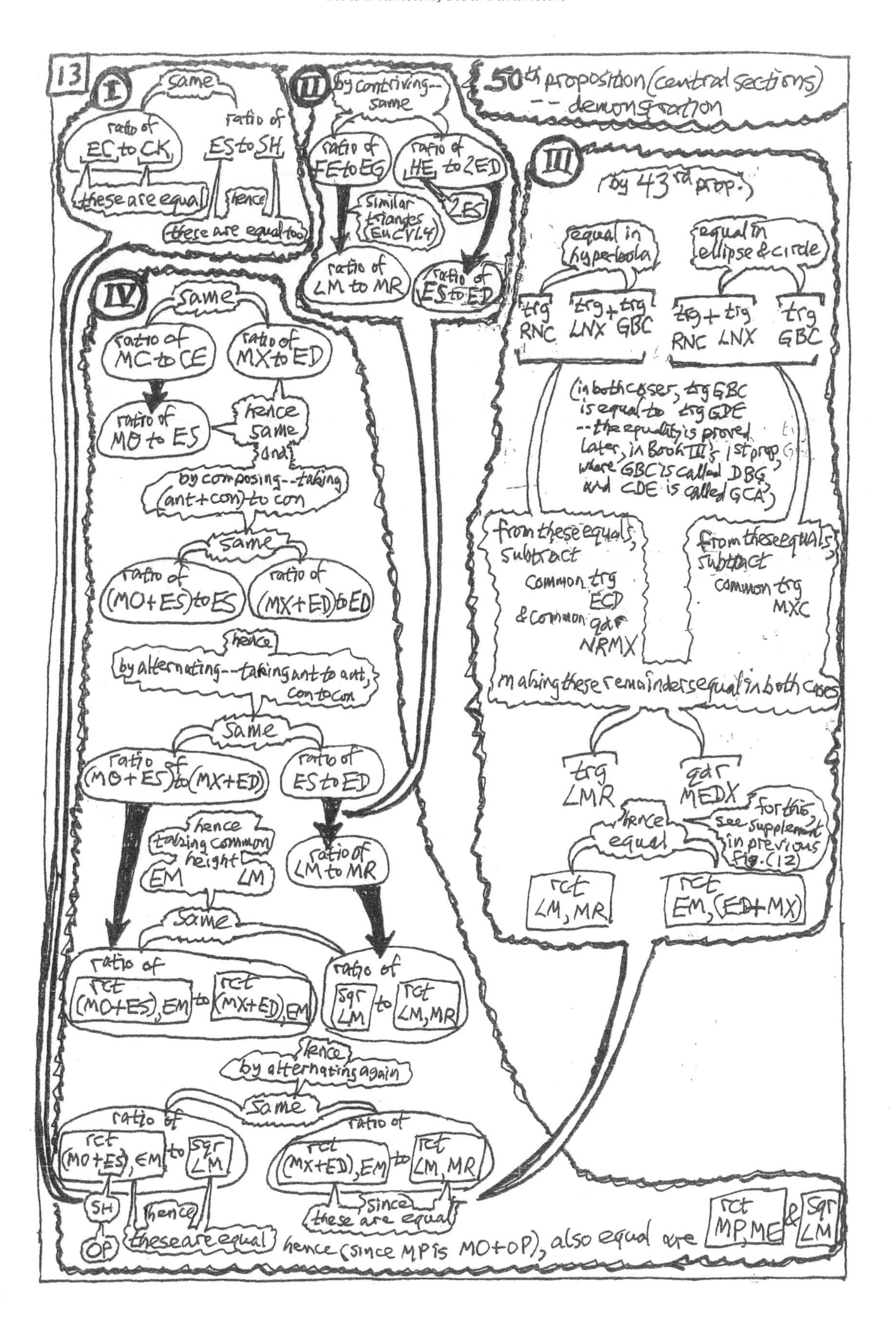
13
50th proposition (central sections) -- demonstration
I
same
ratio of EC to CK
ratio of ES to SH
these are equal
hence
these are equal too
II
by contriving -- same
ratio of FE to EG
ratio of HE to 2ED
2ES
similar triangles (Euc VI.4)
ratio of LM to MR
ratio of ES to ED
III
by 43rd prop.
equal in hyperbola
equal in ellipse & circle
trg RNC
trg LNX + trg GBC
trg RNC + trg LNX
trg GBC
(in both cases, trg GBC is equal to trg GDE -- the equality is proved later, in Book III, 1st prop, where GBC is called DBG, and CDE is called GCA)
from these equals, subtract common trg ECD & common qdr NRMX
from these equals, subtract common trg MXC
making these remainders equal in both cases
trg LMR
qdr MEDX
hence equal
for this, see supplement in previous Fig. (12)
rct LM, MR
rct EM, (ED+MX)
IV
same
ratio of MC to CE
ratio of MX to ED
ratio of MO to ES
hence same and
by composing -- taking (ant+con) to con
same
ratio of (MO+ES) to ES
ratio of (MX+ED) to ED
hence
by alternating -- taking ant to ant, con to con
same
ratio of (MO+ES) to (MX+ED)
ratio of ES to ED
hence taking common height EM LM
ratio of LM to MR
same
ratio of rct (MO+ES), EM to rct (MX+ED), EM
ratio of sqr LM to rct LM, MR
hence by alternating again
same
ratio of rct (MO+ES), EM to sqr LM
ratio of rct (MX+ED), EM to rct LM, MR
since these are equal
SH
OP
hence
these are equal
hence (since MP is MO+OP), also equal are rct MP, ME & sqr LM

Now here is Apollonius's own presentation for the hyperbola, ellipse, and circle. (Note, by the way, that while his demonstration makes use of the equality of the triangles ECD and GBC, he will not prove this equality until the first proposition of Book III. He will do so, however, without relying on the present proposition or on any subsequent one.)

#50 If, with respect to an hyperbola or ellipse or circle's periphery, a straight [line] that touches [the section] fall together with the diameter; and also, through both the touch[point] and the center a straight [line] be extended, and from the vertex a straight [line] that is drawn up parallel to a [line] drawn down ordinatewise fall together with the straight [line] that is drawn through the touch[point] and the center; and also, there be made [a certain straight line such that] as the segment (*tmêma*) of the touching[line], [that segment of it, namely, which is] between the touch[point] and the [line] drawn up [ordinatewise from the vertex], is to the segment (*tmêma*) of the [line] drawn through the touch[point] and the center, [is to that segment of it, namely, which is] between the touch[point] and the [line] drawn up [ordinatewise from the vertex], so [the] certain straight [line] is to the double of the touching[line]—then whatever [line] from the section be drawn onto the [line] drawn through the touch[point] and the center, [that is, be drawn through them] parallel to the touching[line], will have the power [to give rise to a square that is] equal to a certain rectangular area (*khôrion*) which is laid along (*parakeimenon para*) ["is applied to"] the [line] provided (*poristheisan*) [by the proportion just mentioned, making this certain straight line an upright side,] while having as breadth the [line] taken off by it on the side toward the touch[point], [in other words, having as breadth the line taken off *from* the extended line that, going through the center and the touchpoint, will act as a new diameter, and taken off *by* the straight line that, being parallel to the touching line, will, with respect to that new diameter, be drawn ordinatewise], and which [rectangular area], on the hyperbola (on the one hand), overshoots (*huperballon*) by a figure (*eidei*) that is similar to the [rectangle] contained by the double of the [line] between the center and the touch[point] and by the [line] provided (*poristheisês*), but which, on the ellipse and the circle (on the other hand), is lacking (*elleipon*) by such a figure.

—Let there be an hyperbola or ellipse or circle's periphery whose diameter is what the [line] AB is, and center is what the [point] C is, and a touching[line] [of which] is what the [line] DE is; and also, let the joined [line] CE have been extended both ways; and also, let there be put equal to the [line] EC the [line] CK; and also, through the [point] B let there be drawn up ordinatewise the line BFG, and through the [point] E let there be drawn at right angles to the [line] EC the [line] EH; and also, let [this line EH] be so generated that as is the [line] FE to EG so is the [line] EH to the double of the [line] ED; and also, let the joined [line] HK have been extended; and also, let there be taken some point on the section, the [point] L, say; and also, through it let there be drawn parallel to the [line] ED the [line] LMX, and [parallel] to the [line] BG the [line] LRN, and [parallel] to the [line] EH the [line] MP. I say that the [square] from LM is equal to the [rectangle] by EMP.
—For let there be drawn through the [point] C parallel to the [line] KP the [line] CSO. And—since equal is the [line] EC to the [line] CK, but as is the [line] EC to CK so is the [line] ES to SH—therefore equal is also the [line] ES to the [line] SH. And—since as is the [line] FE to EG so is the [line] HE to the double of the [line] ED, and, of the [line] EH, the [line] ES is half—therefore as is the [line] FE to EG so is the [line] SE to ED. As, however, is the [line] FE to EG so is the [line] LM to MR; therefore as is the [line] LM to MR so is the [line] SE to ED. And—since the triangle RNC, in comparison with the triangle GBC, that is, [in comparison with] the triangle CDE, was, on the hyperbola (on the one hand), shown to be greater, and, on the ellipse and the circle (on the other hand) shown to be less—[greater or less, that is] by the triangle LNX—therefore, the common [areas] being taken away, that is, on the hyperbola (on the one hand) the [common] triangle ECD and the [common] quadrilateral NRMX being taken away, but on the ellipse and the circle (on the other hand) the [common] triangle MXC being taken away, [it follows that] the triangle LMR is equal to the quadrilateral MEDX. And parallel is the [line] MX to the [line] DE, but the angle by LMR is, to the angle by EMX, equal; therefore equal is the [rectangle] by LMR to the [rectangle] by the [line] EM and by the [line] ED, MX both together. And—since as is the [line] MC to CE so is both the [line] MX to ED and also the [line] MO to ES—therefore as is the [line] MO to ES so is the [line] MX to DE. And, by putting together (*synthenti*) [traditionally, in Latin, *componendo*], as is the [line] MO, SE both together to ES so is the [line] MX, ED both together to ED; alternately (*enallax*), as is the line MO, SE both together to the [line] XM, ED both together so is the [line] SE to ED. But (on the one hand) as the [line] MO, SE both together is to the line MX, ED both together so the [rectangle contained] by the line MO, ES, both together and by the line EM is to the [rectangle contained] by the [line] MX, ED both together and by the line EM; and (on the other hand) as is the [line] SE to ED so is the [line] FE to EG, that is, the [line] LM to MR, that is, the [square] from LM to the [rectangle] by LMR. And alternately, as the [rectangle contained] by the [line] MO, ES both together and by the [line] ME is to the [square] from ML so the [rectangle contained] by the [line] MX, ED both together and by the [line] ME is to the [rectangle] by LMR. Equal, however, is the [rectangle] by LMR to the [rectangle contained] by the line ME and by the line MX, ED both together; therefore equal also is the [square] from LM to the [rectangle contained] by EM and by the [line] MO, ES both together. And the [line] SE is equal to the [line] SH, and the [line] SH to the [line] OP; therefore equal is the [square] from LM to the [rectangle] by EMP.

50th proposition: demonstration for a differently located random point

As has been already pointed out, Apollonius's own demonstration of the 50th proposition takes the random point L to be on the *same* side of diameter AB that the tangency point E is on, but if we take it on the *other* side, then the demonstration for the hyperbola needs one small change, and the demonstration for the ellipse and circle needs several bigger changes.

Here are the requisite changes, for readers who might be interested. Others will miss nothing essential by skipping over them and proceeding directly to the 51st proposition.

First, the *modification required for the hyperbola*. That is shown in figure 14. After that, you will again see (in figure 15) the depiction of Apollonius's demonstration, but now it is marked to show where the modification is needed.

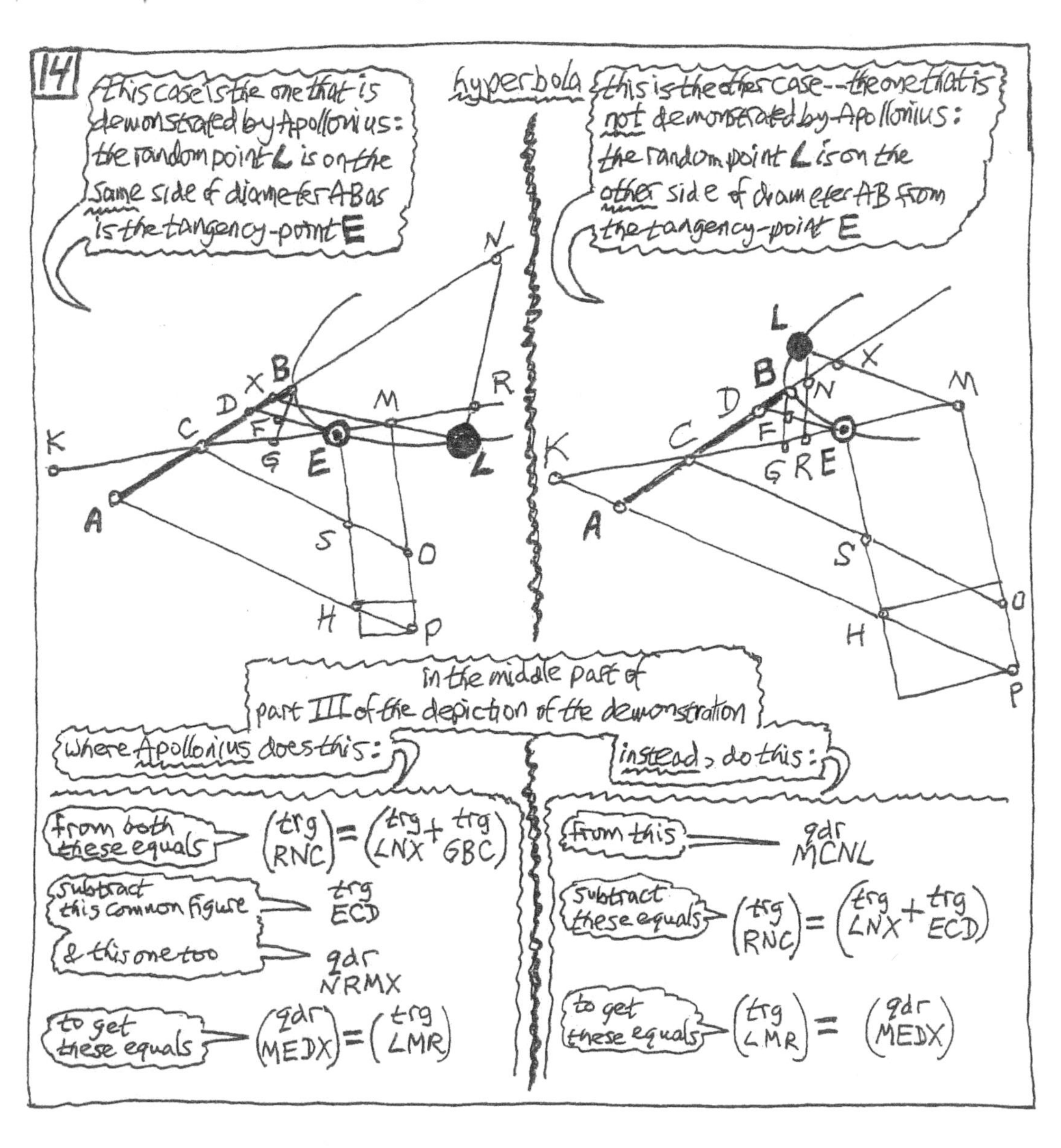

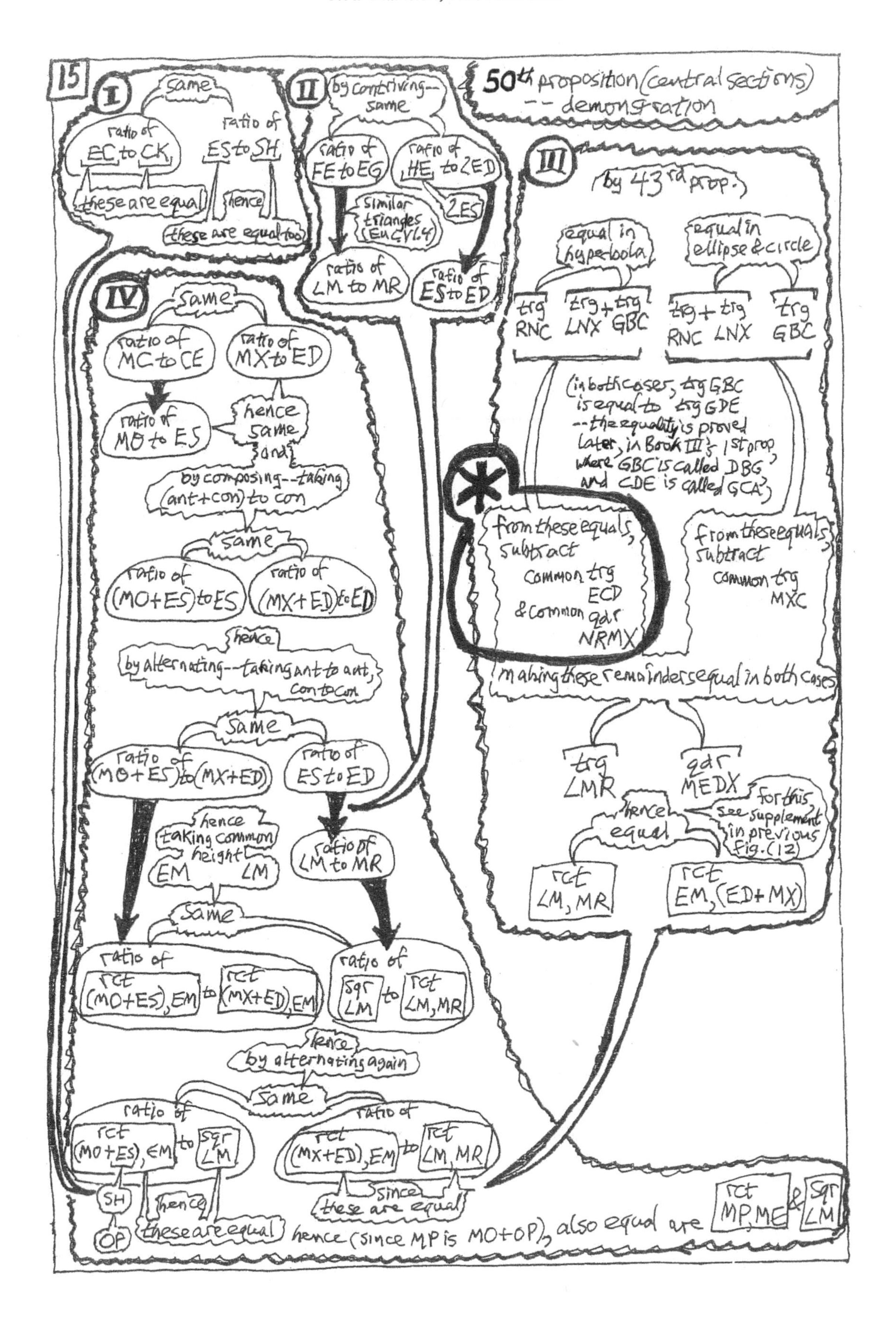
15
50th proposition (central sections) -- demonstration
I
same
ratio of EC to CK
ratio of ES to SH
these are equal
hence
these are equal too
II
by contriving-- same
ratio of FE to EG
ratio of HE, to 2ED
similar triangles (Eucl VI.4)
2ES
ratio of LM to MR
ratio of ES to ED
III
(by 43rd prop.)
equal in hyperbola
equal in ellipse & circle
trg RNC
trg + trg LNX GBC
trg + trg RNC LNX
trg GBC
(in both cases, trg GBC is equal to trg GDE --the equality is proved later, in Book III's 1st prop, where GBC is called DBG and GDE is called GCA)
*
from these equals, subtract common trg ECD & common qdr NRMX
from these equals, subtract common trg MXC
making these remainders equal in both cases
trg LMR
qdr MEDX
hence equal
for this, see supplement in previous Fig. (12)
rct LM, MR
rct EM, (ED+MX)
IV
same
ratio of MC to CE
ratio of MX to ED
ratio of MO to ES
hence same and
by composing--taking (ant+con) to con
same
ratio of (MO+ES) to ES
ratio of (MX+ED) to ED
hence
by alternating--taking ant to ant, con to con
same
ratio of (MO+ES) to (MX+ED)
ratio of ES to ED
hence taking common height EM LM
ratio of LM to MR
same
ratio of rct (MO+ES), EM to rct (MX+ED), EM
ratio of sqr LM to rct LM, MR
hence
by alternating again
same
ratio of rct (MO+ES), EM to sqr LM
ratio of rct (MX+ED), EM to rct LM, MR
SH
OP
hence
these are equal
since these are equal
hence (since MP is MO+OP), also equal are rct MP, ME & sqr LM

Now, the *modification required for the ellipse and circle*. We begin at the same place for these curves as for the hyperbola—namely, at the part marked "III" in the figure depicting Apollonius's demonstration. In part III, on whichever side of the diameter we take random point L for the ellipse and circle,

trg + trg RNC LNX	is shown to be equal to	trg CDE,
	and from each of the equals is subtracted common trg MXC,	

resulting in another equality. But just what are the resulting equals? That will differ in the cases for which the random point L is differently placed. That is shown in figure 16. After that, you will again see in figure 17 the depiction of Apollonius's demonstration, but now it is marked to show where the modification is needed in this case.

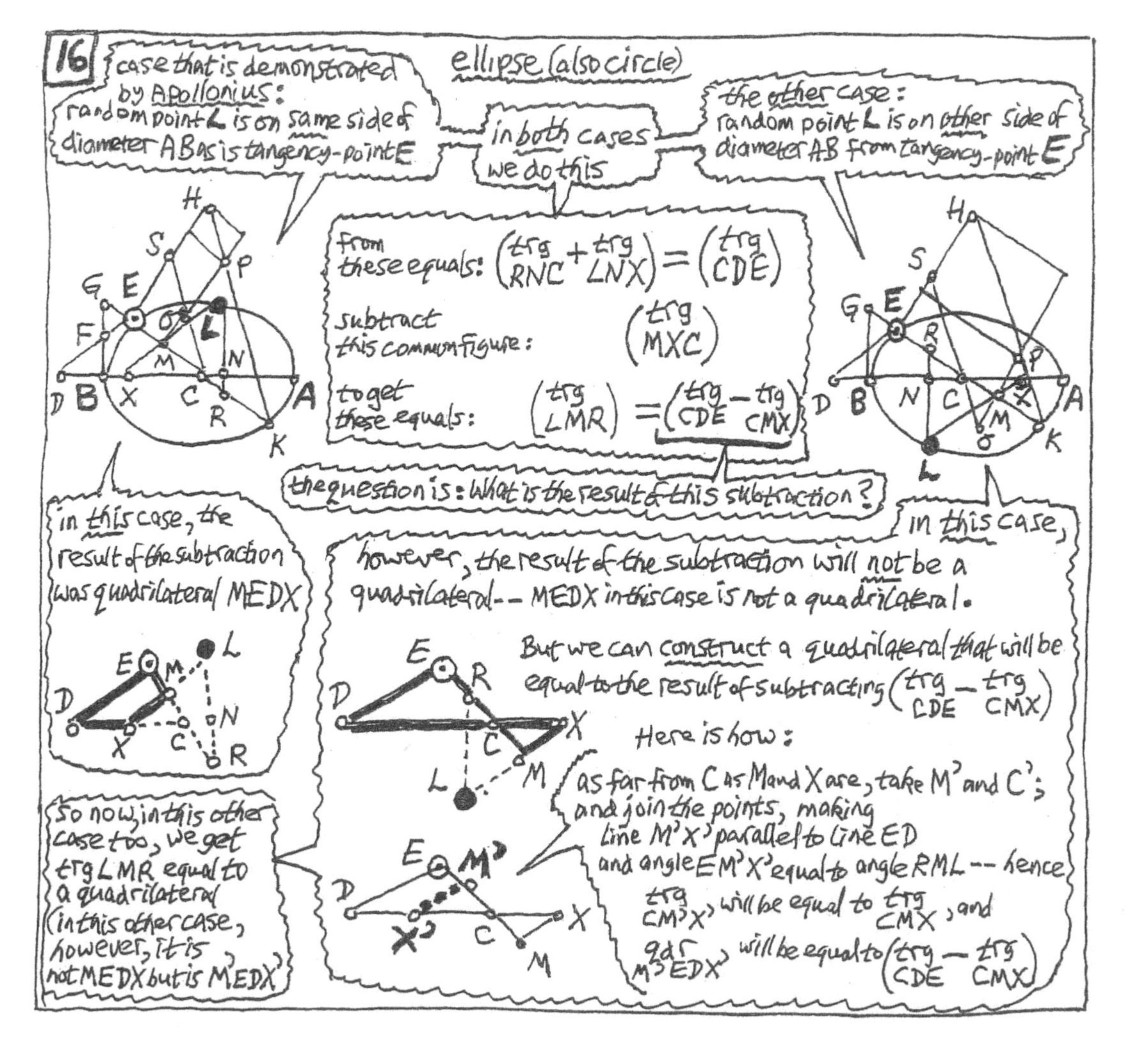

16
case that is demonstrated by Apollonius: random point L is on same side of diameter AB as is tangency-point E
ellipse (also circle)
in both cases we do this
the other case: random point L is on other side of diameter AB from tangency-point E
From these equals: (trg RNC + trg LNX) = (trg CDE)
subtract this common figure: (trg MXC)
to get these equals: (trg LMR) = (trg CDE − trg CMX)
the question is: what is the result of this subtraction?
in this case, the result of the subtraction was quadrilateral MEDX
in this case, however, the result of the subtraction will not be a quadrilateral-- MEDX in this case is not a quadrilateral.
But we can construct a quadrilateral that will be equal to the result of subtracting (trg CDE − trg CMX)
Here is how:
as far from C as M and X are, take M' and C'; and join the points, making line M'X' parallel to line ED and angle EM'X' equal to angle RML -- hence trg CM'X' will be equal to trg CMX, and qdr M'EDX' will be equal to (trg CDE − trg CMX)
So now, in this other case too, we get trg LMR equal to a quadrilateral (in this other case, however, it is not MEDX but is M'EDX')

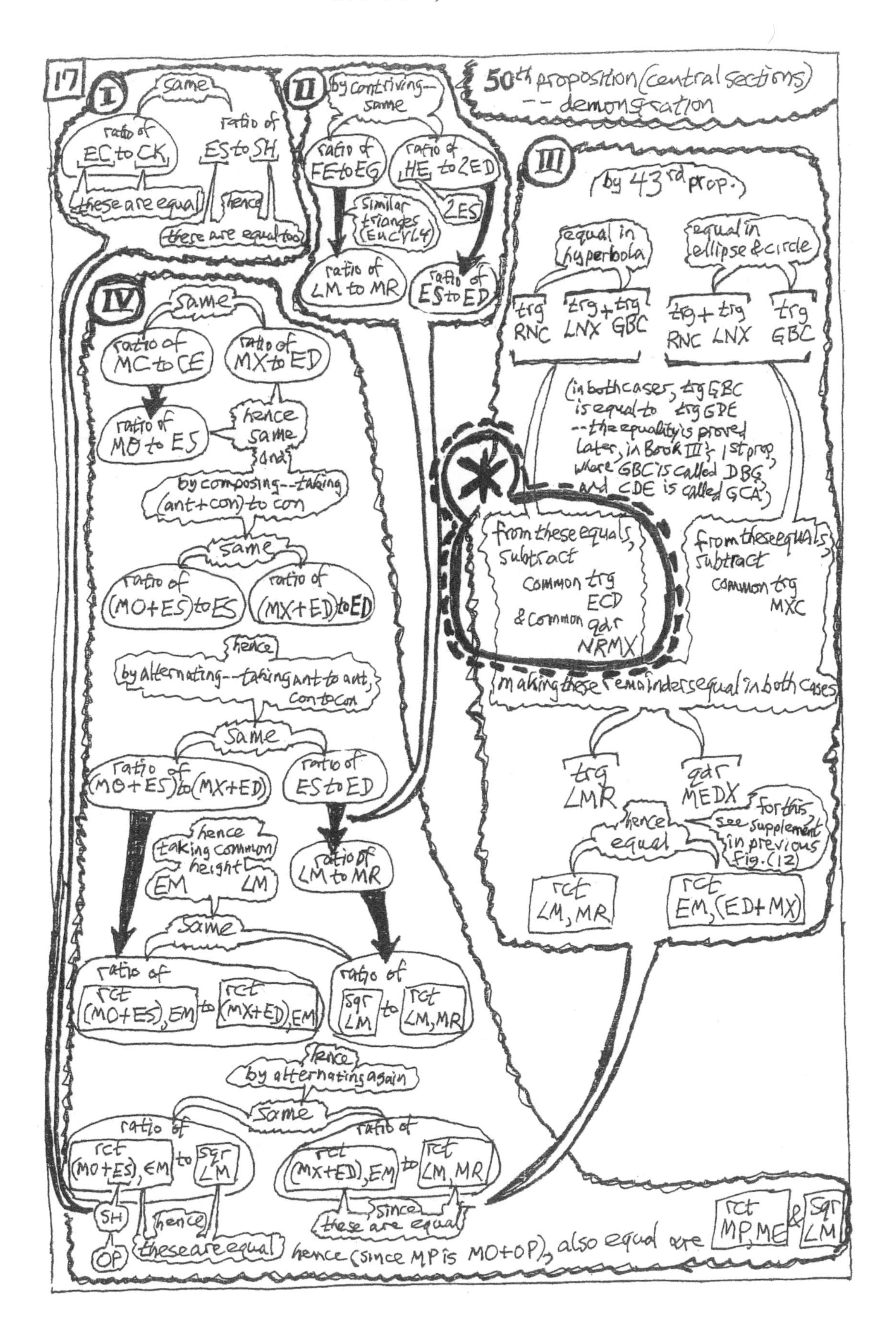
17
50th proposition (central sections) -- demonstration
I
same
ratio of EC to CK,
ratio of ES to SH,
these are equal
hence
these are equal too
II
by contriving-- same
ratio of FE to EG
ratio of HE, to 2ED
similar triangles (Eucl VI.4)
2ES
ratio of LM to MR
ratio of ES to ED
III
(by 43rd prop.)
equal in hyperbola
equal in ellipse & circle
trg RNC
trg LNX + trg GBC
trg RNC + trg LNX
trg GBC
(in both cases, trg GBC is equal to trg GDE --the equality is proved later, in Book III's 1st prop, where GBC is called DBG and CDE is called GCA)
from these equals, subtract common trg ECD & common qdr NRMX
from these equals, subtract common trg MXC
making these remainders equal in both cases
trg LMR
qdr MEDX
hence equal
for this, see supplement in previous Fig. (12)
rct LM, MR
rct EM, (ED+MX)
IV
same
ratio of MC to CE
ratio of MX to ED
ratio of MO to ES
hence same and
by composing--taking (ant+con) to con
same
ratio of (MO+ES) to ES
ratio of (MX+ED) to ED
hence
by alternating--taking ant to ant, con to con
same
ratio of (MO+ES) to (MX+ED)
ratio of ES to ED
hence taking common height EM LM
ratio of LM to MR
same
ratio of rct (MO+ES), EM to rct (MX+ED), EM
ratio of sqr LM to rct LM, MR
hence by alternating again
same
ratio of rct (MO+ES), EM to sqr LM
ratio of rct (MX+ED), EM to rct LM, MR
SH
OP
hence
these are equal
since these are equal
hence (since MP is MO+OP), also equal are rct MP, ME & sqr LM

Now we can use our supplement to Apollonius's demonstration. If the supplement's points M and X are renamed M' and X', then, since M'X' is parallel to ED, and angle EM'X' is equal to angle RML, it will follow that rectangle [LM, MR] is equal to rectangle [EM', (ED + MX)]. (Fig. 18.)

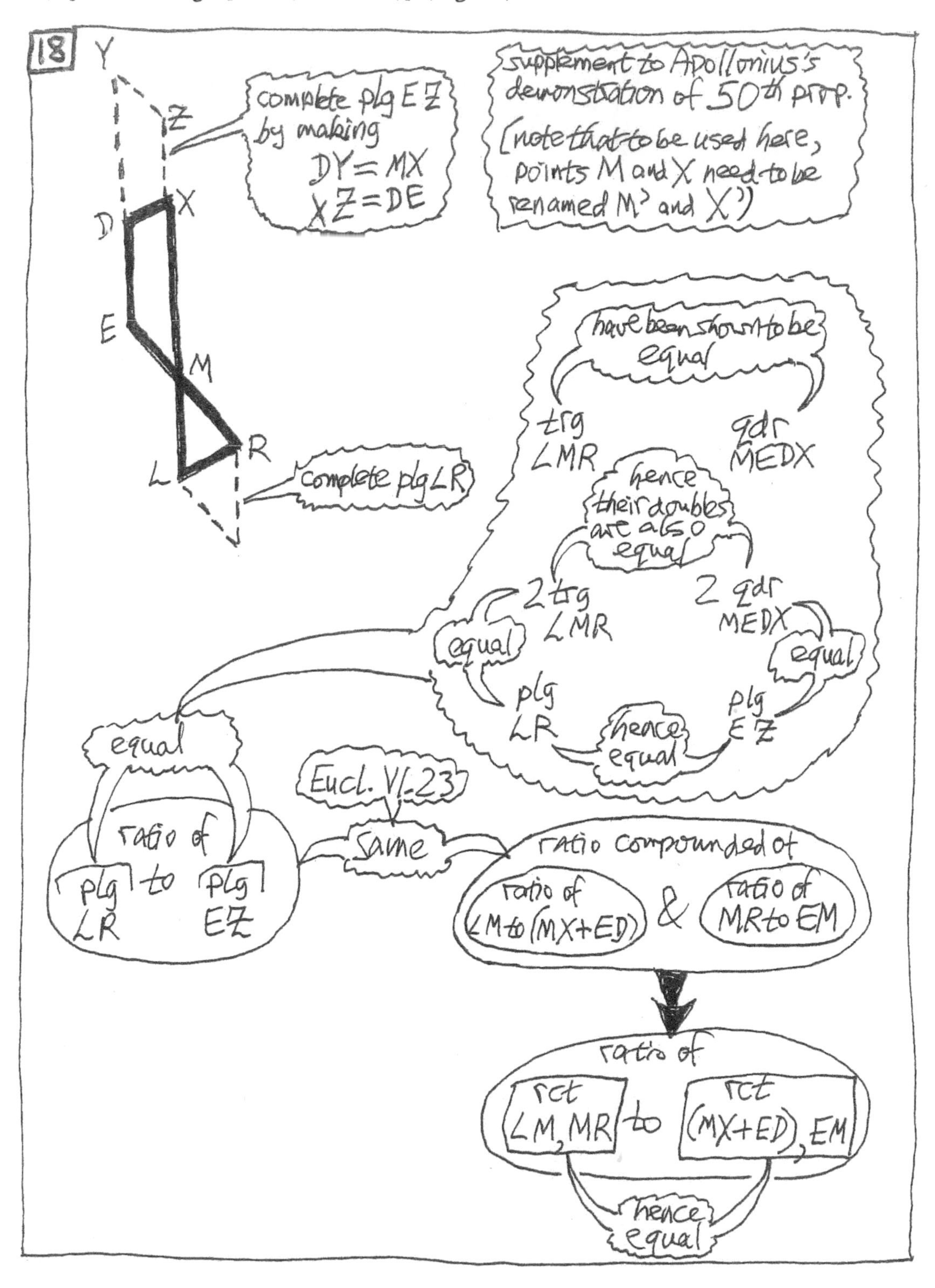

In Apollonius's demonstration, random point L was on the same side as E was, and the line parallel to ED was MX—and so, the rectangle which was equal to rectangle [LM, MR] had as its one side EM. But now, with random point L being on the other side from E, and with the line parallel to ED being M'X', the rectangle which is equal to rectangle [LM, MR] has as its one side EM'. But we want to speak *not* in terms of the extra lines that come when the *primed points are introduced*, but *rather* in terms of the lines determined by the *points that are used in the rest of the demonstration*. In other words, we wish to proceed with the demonstration by obtaining a rectangle which will be equal to rectangle [EM', (ED + MX)] but will have as its side EM instead of EM'. How to proceed is shown in figure 19.

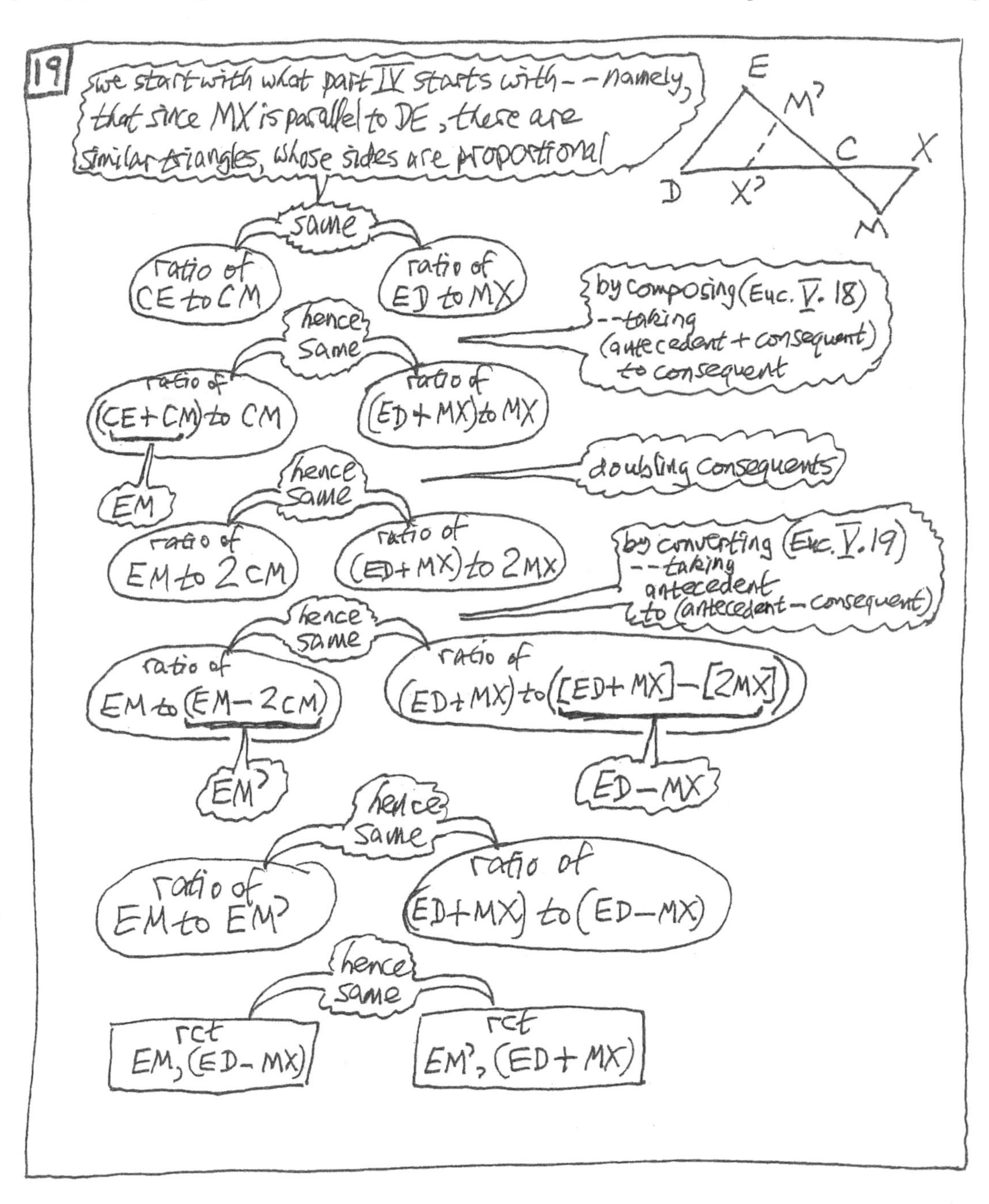

So, what we have done is depicted in figure 20. After that, you will again see (fig. 21) the depiction of Apollonius's demonstration, but now it is marked to show the place where this modification needs to be made.

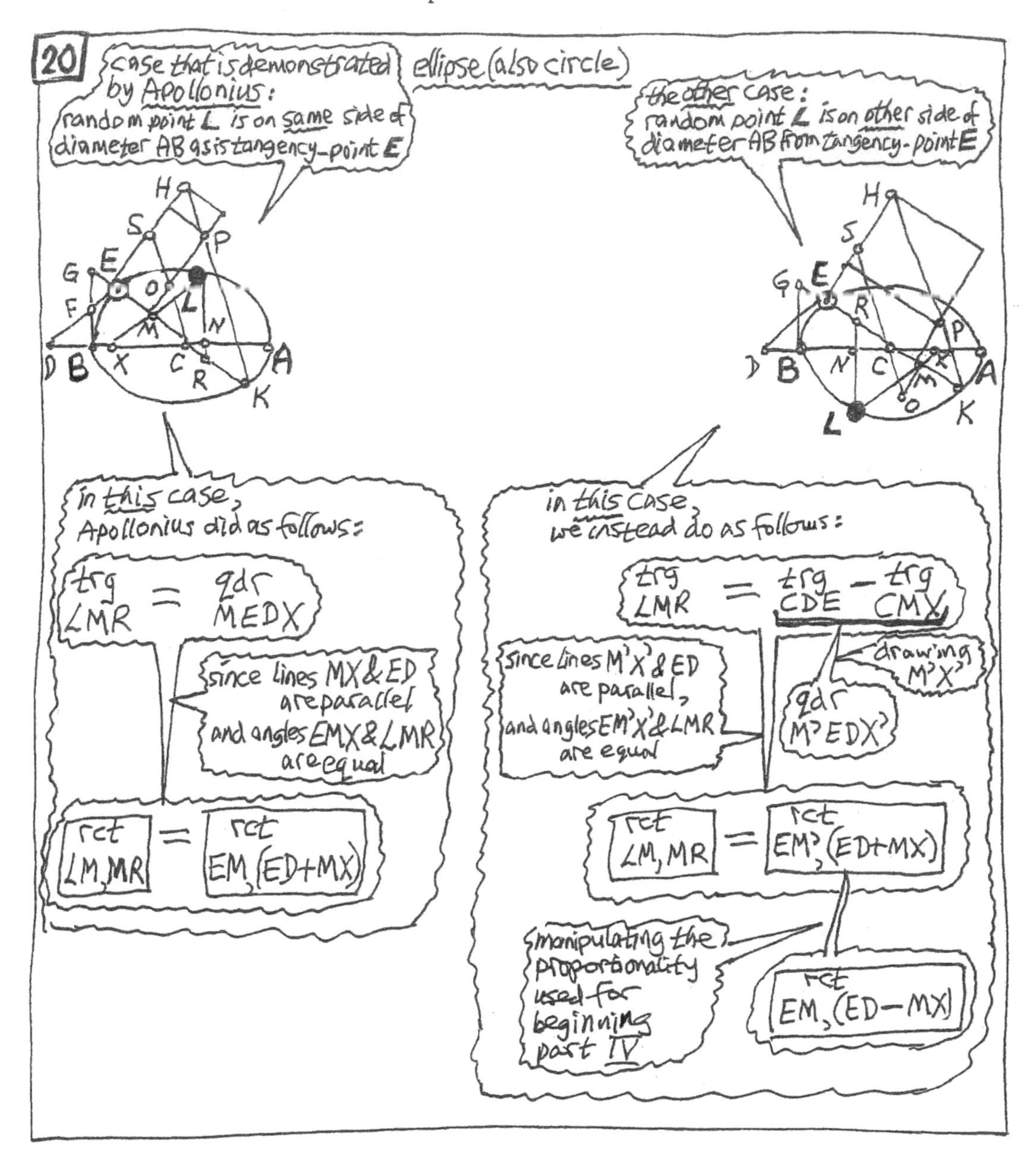

20
case that is demonstrated by Apollonius: random point L is on same side of diameter AB as is tangency-point E
ellipse (also circle)
the other case: random point L is on other side of diameter AB from tangency-point E
in this case, Apollonius did as follows:
trg LMR = qdr MEDX
since lines MX & ED are parallel and angles EMX & LMR are equal
rct LM, MR = rct EM, (ED+MX)
in this case, we instead do as follows:
trg LMR = trg CDE − trg CMX
drawing M'X'
qdr M'EDX'
since lines M'X' & ED are parallel, and angles EM'X' & LMR are equal
rct LM, MR = rct EM', (ED+MX)
manipulating the proportionality used for beginning part IV
rct EM, (ED−MX)

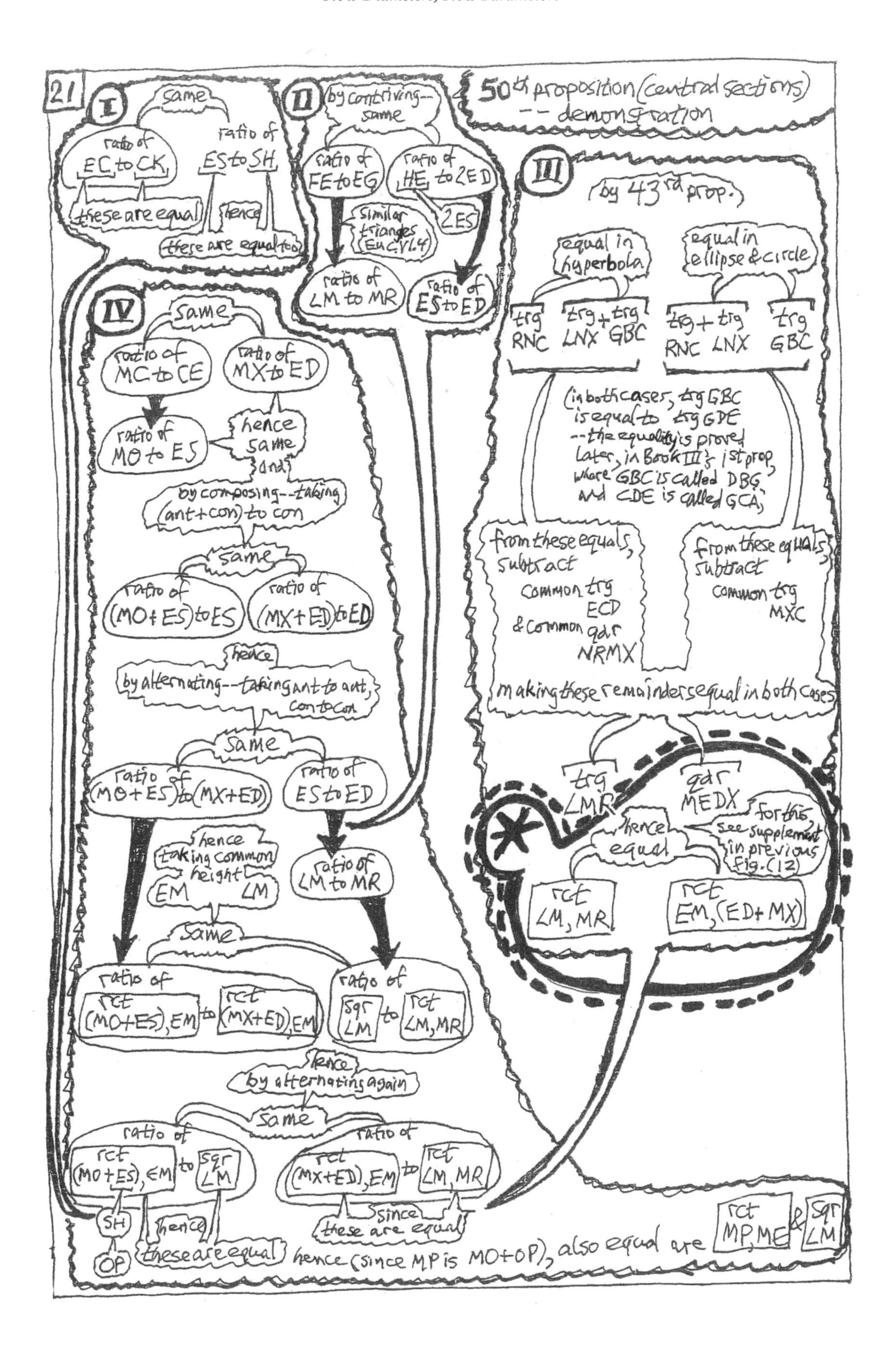
21
50th proposition (central sections)
-- demonstration
I
same
ratio of EC, to CK,
ratio of ES to SH,
these are equal
hence
these are equal too
II
by contriving-- same
ratio of FE to EG
ratio of HE, to 2ED
similar triangles (Euc.VI.4)
2ES
ratio of LM to MR
ratio of ES to ED
III
(by 43rd prop.)
equal in hyperbola
equal in ellipse & circle
trg RNC
trg LNX + trg GBC
trg RNC + trg LNX
trg GBC
(in both cases, trg GBC is equal to trg GDE --the equality is proved later, in Book III's 1st prop, where GBC is called DBG, and CDE is called GCA)
from these equals, subtract common trg ECD & common qdr NRMX
from these equals, subtract common trg MXC
making these remainders equal in both cases
trg LMR
qdr MEDX
for this, see supplement in previous Fig. (12)
hence equal
rct LM, MR
rct EM, (ED+MX)
IV
same
ratio of MC to CE
ratio of MX to ED
ratio of MO to ES
hence same and
by composing--taking (ant+con) to con
same
ratio of (MO+ES) to ES
ratio of (MX+ED) to ED
hence
by alternating--taking ant to ant, con to con
same
ratio of (MO+ES) to (MX+ED)
ratio of ES to ED
hence taking common height EM LM
ratio of LM to MR
same
ratio of rct (MO+ES), EM to rct (MX+ED), EM
ratio of sqr LM to rct LM, MR
hence
by alternating again
same
ratio of rct (MO+ES), EM to sqr LM
ratio of rct (MX+ED), EM to rct LM, MR
SH
OP
hence
these are equal
since
these are equal
hence (since MP is MO+OP), also equal are rct MP, ME & sqr LM

Thus, when the random point L is taken on the other side of diameter AB from the tangency point C, part III of the demonstration will still lead to the conclusion that rectangle [LM, MR] is equal to a rectangle with EM as one side. The other side of this rectangle will be different, however. In Apollonius's demonstration for the ellipse and circle, the result of taking random point L on the *same* side of diameter AB as tangency point E was that the equal rectangle's other side was equal to a line resulting from *adding* MX to ED; but now the result of taking the random point on the *other* side will be that the equal rectangle's other side will be equal to a line resulting from *subtracting* MX from ED. That change in the outcome of part III will require a change in the outcome of part IV (where the outcome of part III is used). To get to that changed outcome, there will have to be a change also near the beginning of part IV. Opposite, in figure 22, you will again see the depiction of Apollonius's demonstration, but now it is marked near the beginning and at the end of part IV.

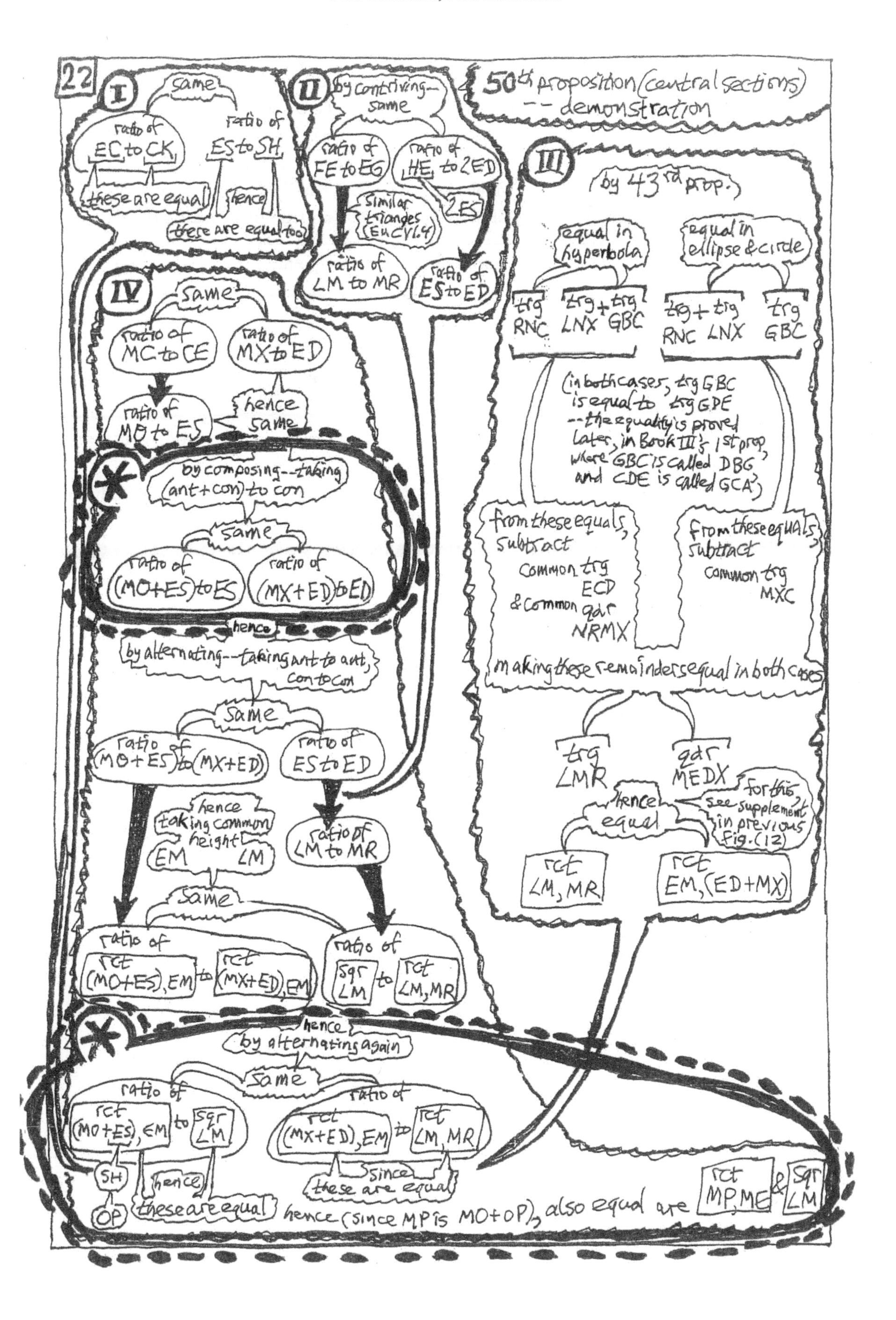
22
I
same
ratio of EC to CK,
ratio of ES to SH,
these are equal
hence
these are equal too
II
by contriving—same
ratio of FE to EG
ratio of HE, to 2ED
2ES
similar triangles (Eucl VI.4)
ratio of LM to MR
ratio of ES to ED
50th proposition (central sections) — demonstration
III
(by 43rd prop.)
equal in hyperbola
equal in ellipse & circle
trg RNC
trg LNX + trg GBC
trg RNC + trg LNX
trg GBC
(in both cases, trg GBC is equal to trg GDE — the equality is proved later, in Book III's 1st prop, where GBC is called DBG and CDE is called GCA)
from these equals, subtract common trg ECD & common qdr NRMX
from these equals, subtract common trg MXC
making these remainders equal in both cases
trg LMR
qdr MEDX
hence equal
for this, see supplement in previous Fig. (12)
rct LM, MR
rct EM, (ED+MX)
IV
same
ratio of MC to CE
ratio of MX to ED
ratio of MO to ES
hence same
by composing—taking (ant+con) to con
same
ratio of (MO+ES) to ES
ratio of (MX+ED) to ED
hence
by alternating—taking ant to ant, con to con
same
ratio of (MO+ES) to (MX+ED)
ratio of ES to ED
hence taking common height EM LM
ratio of LM to MR
same
ratio of rct (MO+ES), EM to rct (MX+ED), EM
ratio of sqr LM to rct LM, MR
hence by alternating again
same
ratio of rct (MO+ES), EM to sqr LM
ratio of rct (MX+ED), EM to rct LM, MR
SH
OP
hence
these are equal
since these are equal
hence (since MP is MO+OP), also equal are rct MP, ME & sqr LM

What, then, is the modification of part IV that is needed in order to accommodate the modification of part III? When the random point L is taken on the other side of the diameter from the tangency point E, we have seen that part III must conclude with rectangle [LM, MR] being equal to rectangle [EM, (ED - MX)] . That is to say: in order to give us the side of the equal rectangle, the line MX must be subtracted from ED—rather than being added to ED (as it is when L and E are on the same side). In part IV, therefore, we must move to a proportion that incorporates the line which is equal to (ED - MX) rather than the line which is equal to (ED + MX). In other words, the way to make use of the proportion that appears near the beginning of part IV is to move on from it by taking the ratio of

(consequent - antecedent) to consequent,

rather than the ratio of

(consequent + antecedent) to consequent.

Only thus will the proportion near the beginning of part IV give us, in the end, a proportion containing a ratio that part III has shown to be a ratio of equality.

This change will still lead to the conclusion that square [LM] is equal to rectangle [EM, MP]; however, the line MP will be different. In Apollonius's demonstration for the ellipse and circle, the result of taking random point L on the *same* side of diameter AB as tangency point E was that the rectangle's side MP was equal to the line resulting from *adding* OM to ES. But the result of taking the random point on the *other* side now will be that the rectangle's side MP is equal to a line resulting from *subtracting* OM from ES. (Fig. 23.)

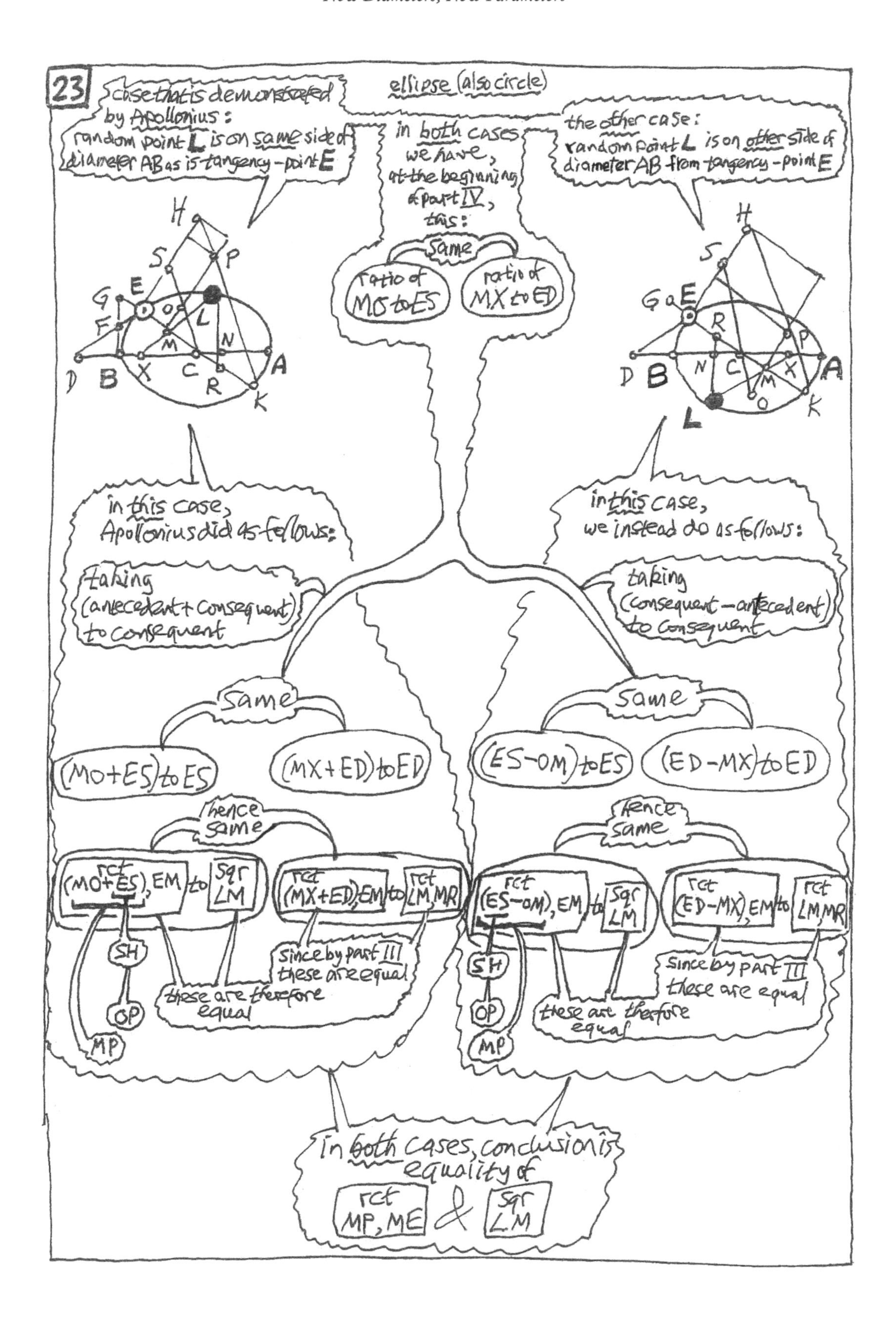
23
ellipse (also circle)
case that is demonstrated by Apollonius: random point L is on same side of diameter AB as is tangency-point E
in both cases we have, at the beginning of part IV, this:
the other case: random point L is on other side of diameter AB from tangency-point E
same
ratio of MO to ES
ratio of MX to ED
H S P G E O L F M N D B X C R A K
H S G E R P D B N C M X A O K L
in this case, Apollonius did as follows:
in this case, we instead do as follows:
taking (antecedent + consequent) to consequent
taking (consequent − antecedent) to consequent
same
(MO+ES) to ES
(MX+ED) to ED
same
(ES−OM) to ES
(ED−MX) to ED
hence same
rct (MO+ES), EM to sqr LM
rct (MX+ED), EM to rct LM, MR
hence same
rct (ES−OM), EM to sqr LM
rct (ED−MX), EM to rct LM, MR
SH
OP
MP
since by part III these are equal
these are therefore equal
SH
OP
MP
since by part III these are equal
these are therefore equal
In both cases, conclusion is equality of
rct MP, ME & sqr LM

51st proposition: demonstration for opposite sections

The demonstration for the opposite sections requires the construction of the apparatus of lines which is shown in figure 24.

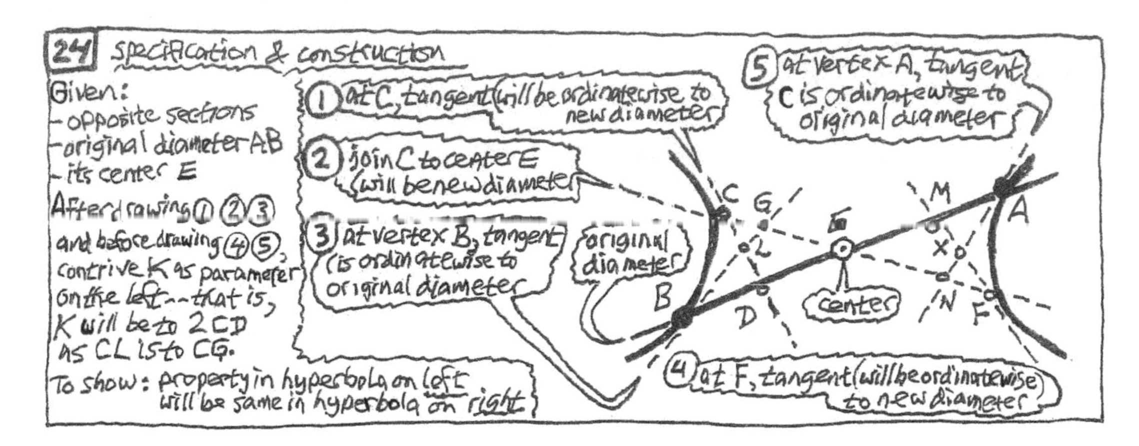

What the demonstration shows is that the line contrived to act as a parameter for one of the hyperbolas in the pair will act as a parameter for the other hyperbola too (figs. 25 and 26).

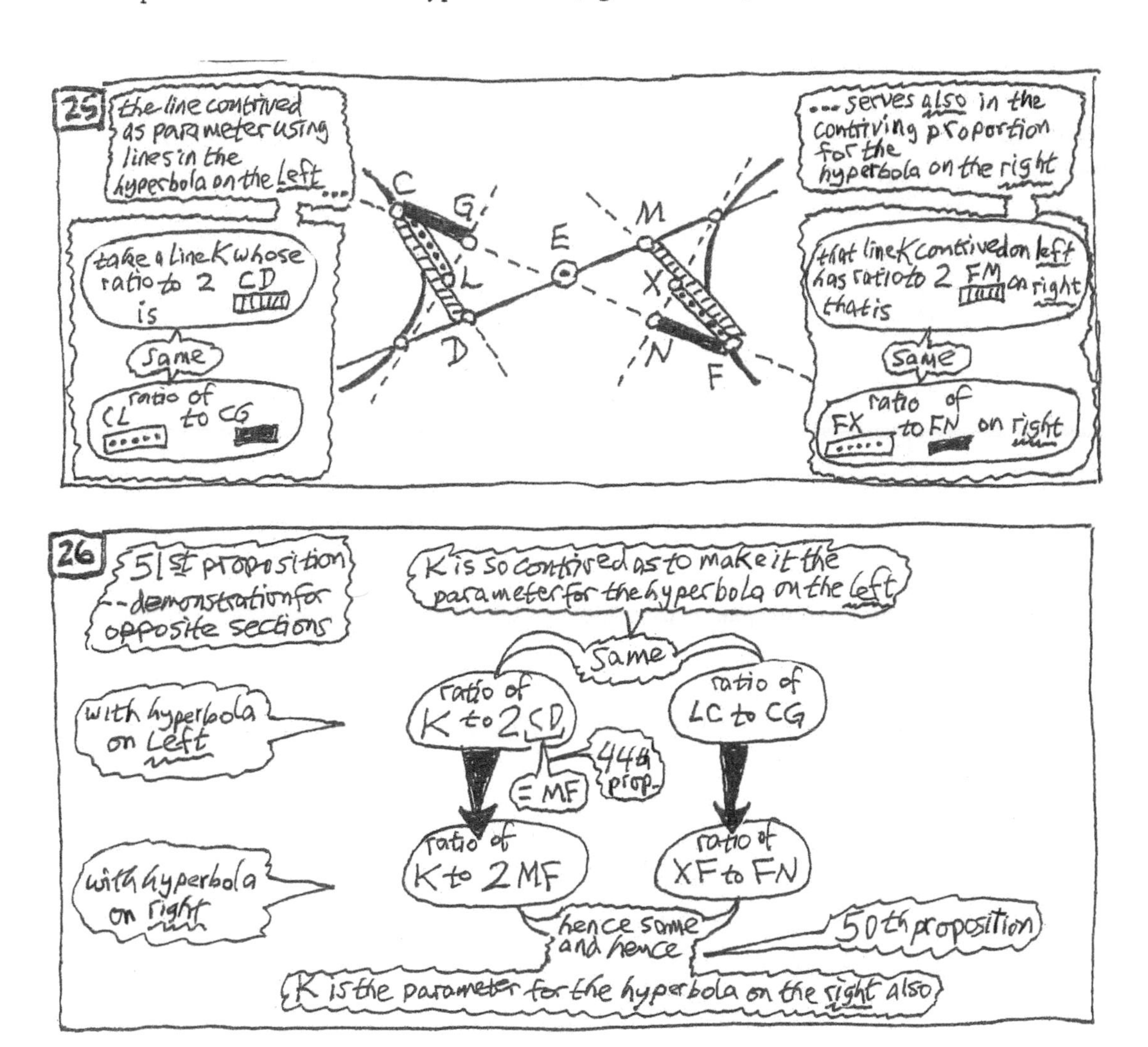

Now here is Apollonius's own presentation for the opposite sections.

#51 If, with respect to either of the opposite [sections], a straight [line] that touches it fall together with the diameter; and also, through both the touch[point] and the center a certain straight [line] be extended as far as the other section, and from the vertex a straight [line] be drawn up parallel to a [line] drawn ordinatewise and fall together with the straight [line] drawn through the touch[point] and the center; and also, there be generated (*genêthê*) [a certain straight line such that] as the segment of the touching[line], [namely that segment of it which stretches] between the [line] drawn up and the touch[point], is to the segment of the [line] drawn through the touch[point] and the center, [namely to that segment of it which stretches] between the touch[point] and the [line] drawn up, so [the] certain straight [line] is to the double of the touching[line]—then whatever straight [line] in the other of the sections be drawn, onto the straight [line] which is drawn through the touch[point] and the center, [whatever such straight line in the other section be drawn, that is,] parallel to the touching[line] will have the power [to give rise to a square that is equal to] a rectangular [area] which is laid along (*parakeimenon para*) ["is applied to"] the [line] previously provided (*prosporistheisan*), [that is, the area is laid along that certain straight line provided by the proportion previously mentioned, which will make that line act as does an upright side], while having as breadth the [line] taken off by it [taken off, namely, by that parallel to the touching[line]] on the side toward the touch[point], [that is, the breadth—being the line taken off from the extension of the line that, going through the center and the touchpoint, will act as a new diameter—will be an abscissa, and the straight line that takes it off, being parallel to the touching line, will, with respect to that new diameter, be drawn ordinatewise], and which [rectangular area] overshoots (*huperballon*) by a figure (*eidei*) that is similar to the [rectangle] contained by both the [straight line that stretches] between the opposite [sections] and the previously provided (*prosoristheisês*) straight [line].

—Let there be opposite [sections] whose diameter is what the [line] AB is, and center is what the [point] E is; and also, let there be drawn touching the section B the [line] CD; and also, let there be joined the [line] CE and let it be extended; and also, let there be drawn ordinatewise the [line] BLG; and also, let there be made [a certain straight line such that] as is the [line] LC to CG so is [that] certain line, namely the [line] K, to the double of the [line] CD. So then: that the [lines] in the section BC which are parallel to the [line] CD and are drawn onto the straight-prolonged [line] EC have the power [to give rise to a square equal to] the areas which are laid along the [line] K, while having as breadths the [line] taken off by them on the side toward the touch[point], and which overshoot by a figure that is similar to the [rectangle] by CF, K—that is manifest; for double is the [line] FC in comparison with the [line] CE. I say, then, that also in the section FA the same thing will come about (*sumbêsetai*). [In other words, the line K, which was made to have just the right size to do something with respect to section BC, has just the right size to do the same thing for section BC's opposite section FA.]

—For let there be drawn through the [point] F, touching the section AF, the [line] MF; and also, let there be drawn up ordinatewise the line AXN. And—since opposite [sections] are what the [lines] BC, AF are, but touching[lines] of them are what the [lines] CD, MF are—therefore equal and parallel is the [line] CD to the [line] MF. Equal, however, is also the [line] CE to the [line] EF; therefore also the [line] E, to the [line] EM, is equal. And—since as is the [line] LC to CG so is the [line] K to the double of the [line] CD, that is, of the [line] MF—therefore also, as is the [line] XF to FN so is the [line] K to the double of the [line] MF. Since, then, an hyperbola is what the [line] AF is, whose diameter is what the [line] AB is, and a touching[line] of which [hyperbola] is what the [line] MF is; and ordinatewise there has been drawn the [line] AN; and as is the [line] XF to FN so is the [line] K to the double of the [line] FM—hence as many [lines] from the section as may be drawn parallel to the [line] FM, onto the straight-prolonged [line] EF, [so many] will have the power [to give rise to a square equal to] the rectangular [area] which is contained by both the straight [line] K and the line taken off by them on the side toward the [point] F, and which overshoots by a figure that is similar to the [rectangle] by CF, K.

—But these things having been shown, it is evident together with them that, in the parabola on the one hand, each of the straight [lines] parallel to that diameter [which originates] out of the [section's] generation (*ek tês geneseôs*) is a diameter, but, in the hyperbola and the ellipse and the opposite [sections] on the other hand, [what is a diameter] is each of the straight [lines] drawn through the center. And [it is evident] also that, in the parabola on the one hand, the lines drawn down to each of the diameters parallel to the touching[lines] will have the power [to give rise to a square equal to] the rectangular [areas] which are laid along the same [diameter], but, in the hyperbola and the opposite [sections] on the other hand, they will have the power [to give rise to a square equal to] the areas which are laid along the same [diameter] and which overshoot by the same figure (*eidei*), but in the ellipse, [they will have the power to give rise to a square equal to] the [areas] which are laid along the same [diameter] and which are lacking by the same figure. And [it is evident] also that all those things about the sections that have been shown before as coming about (*sumbainonta*) when what are thrown alongside together (*sumparaballomenôn*) are the primal diameters (*tôn arkhikôn diametrôn*), those very same things will come about when what are taken alongside together are the other diameters.

[The root *arkh-* has the sense of beginning—of being first, or leading, or governing. It shows up when the student of those human societies which come first in time is called an "archaeologist," or when the one who takes the lead in building is called an "architect," or when the absence of government is called "anarchy." It shows up also in the adjective *arkhikos* here applied to certain diameters. In this translation, "primal diameters" renders what has usually been rendered as "principal diameters." The rendering adopted here directs the attention to primacy in time, without excluding the possibility of primacy in importance too, whereas the usual rendering directs the attention to primacy in importance. The diameters that first appeared in this book were obtained when the conic sections were obtained. Those diameters appeared in the 7th proposition, which was a bridge between the generating of sections that were old lines and the generating of those that were new. Since then, we have seen how to obtain "other" diameters from those diameters—the ones now called "primal" diameters—which we have been obtaining along with the obtaining of the conic sections themselves.]

[*Sumbainonta* (literally: "things that come together"), here rendered as "things that come about," is usually rendered, less literally, as "properties"—which is also the usual rendering of *sumptômata* (literally: "things that fall together"). Apollonius, in his letter introducing the *Conics*, characterizes this first book as containing "the generation of the three sections and of the opposite (sections) and the *arkhika sumptômata* in them worked out more fully and more generally than in writings by others." The usual rendering of this is "the principal properties," but "primal" would not disregard the suggestion of temporal order, which "principal" does disregard. Also, rendering both *sumbainonta* and *sumptômata* as "properties" may obscure something. Although the introductory letter speaks of *arkhika sumptômata* in characterizing the contents of the first book, it speaks merely of *sumbainonta* in characterizing the contents of the second.]

51st proposition: summary, at end, of propositions 46-51

With these things shown, says Apollonius, it is evident that we have *new diameters* (in propositions 46–48): these are the straight lines, drawn from the curve, that run

in the parabola,	parallel to the original diameter;
in the hyperbola, ellipse, circle, and opposite sections,	through the center.

And also is it evident, in effect, that we have *new parameters* (in propositions 49-51)—since the straight lines dropped to each of the diameters, parallel to the tangents, have the power to make squares equal to:

in the parabola,	the rectangles applied;
in the hyperbola and opposite sections,	the areas applied and exceeding by some figure;
in the ellipse,	the areas applied and defective by the same figure.

And evident that all the things already demonstrated about the sections as following when the *original* diameters are used, those very same things follow also when the *other* diameters are taken. The original diameter (the *diametros arkhikê*—the diameter that "initiates" things, the "leading" or "chief" or "principal" diameter) is the diameter established way back in the porism to the 7th proposition; it is the diameter that was obtained along with the section.

Thus at the end of this stretch of seven propositions—three of them giving new diameters and four of them giving parameters for these new diameters—Apollonius gives a summary of what he has done. This is an unusual thing for him to do: Apollonius is sparing with summary. This summary at the end of the 51st proposition ends with a reference to the diameter with which he began, and which was first characterized as such in the porism forty-four propositions back. It was then that, immediately after, new curves were introduced as conic sections. After three propositions which characterized the curves in a general way and four propositions which introduced the particular sections of various sorts, there have come thirty-seven propositions which have led from that showing (how cutting a cone yields the various curves, each one with its original diameter) to what will now be shown—namely, that a curve of a specified sort, with its original diameter and parameter, can always be found in some cone.

Apollonius's summary treats seven propositions that are the outcome of the thirty-seven propositions that followed his introduction of the various particular sections of cones. The summary is not merely a retrospective overview informing us that those propositions have provided new diameters and their parameters. It is also prospective: it tacitly sums up the preparation required for returning to cones. We shall return to cones—we shall be brought full circle—with the propositions that will immediately follow the summary. As we return to cones, it will be useful to remember the following: the original diameter was an axis only if the axial triangle was perpendicular to the cone's base. An axial triangle *may or may not* be perpendicular to the cone's base—in an *oblique* cone, that is. In a *right* cone, however, an axial triangle *must* be perpendicular to the cone's base, and then the original diameter *must* be an axis.

transition

In the first book of his *Conics*, Apollonius tells a story. It has a beginning, and a middle, and an end. It begins by moving from cones to curves, and it ends by moving back from curves to cones.

Way back at the end of the beginning, when we finished the 14th proposition, thus finishing the propositions that characterize the kinds of conic sections, we had seen how a cone that is cut by a plane will give us sections of several sorts named with reference to the rectangle-square property of abscissa and ordinate. The beginning of the story thus showed us this: if a line is obtainable by cutting a cone in one of various ways, then it has one of various properties. What has not yet been shown is this: if a line has one of those properties, then it must be obtainable by cutting a cone in one of those ways. But if that can be shown, then Apollonius should show it. Once we have learned that lines which we can obtain by cutting a cone in various ways will have certain properties, then what we should like to know is whether the *only* lines that will have those properties are the lines that we can obtain by cutting a cone in one of those ways. If such lines are indeed the only lines that have those properties, then by showing it Apollonius would give a proper ending to the story.

When we reached the end of the beginning of the story, we might have expected Apollonius to proceed directly to the beginning of the end. He did not do that, however. The end required preparation. The preparation has had a certain indirectness, and, as we shall see, it has even taken on a life of its own, so that the end to which we now turn will *not* turn out to be the *very* end.

The beginning culminated in cuttings of cones to get the various conic sections. Then, from the 15th proposition through the 51st, the sections were examined apart from cones. Book One will come to an end by solving the problems of finding each of the various curves in some cone. Given a specification for a conic section of some sort, a cone will be put together so that the plane of the curve will have the curve as its intersection with the cone. The cone must be found so that the specified curve can be found *as* a conic section—that is to say, as a line that a plane cuts into a cone in a certain way.

In the dozens of propositions that have constituted the middle of Book One, we have been examining curved lines of the various sorts that can be generated by planes cutting cones or conic surfaces, but we have not looked at any cones or conic surfaces. We have examined the curves along with their diameters, their parameters, and various other auxiliary straight lines. There were no axial triangles, though. We have *not* been considering the lines *as* cuts: we have been considering their looks apart from the generating of them. Our figures have all been completely planar.

The beginning of the book showed that certain properties belong to lines that have been cut into a cone by a plane that has been oriented in various ways, and the ending is required in order to show that those properties belong *only* to such sections. To get from the beginning to the ending—that is, in order to be able to show that those properties do belong only to such sections—we first had to show that *because* those sections have *those* properties, they must have certain *other* properties *also*. These other properties have occupied the middle of the book. In traversing the book's middle (the bridge that stretches for thirty-seven propositions between the beginning and the end), what we have considered are the sections' diameters and tangents. This will enable us to refer any conic section to any of its diameters, with the tangent at that diameter's vertex as the reference line for its ordinates. This in turn will enable us to solve the problems of building cones that will yield, as lines cut into a cone, any curves specified by the ordinate-abscissa properties that gave their names to the conic sections.

Although the solutions to those problems are presented toward the end of Book One (in the 52nd through 59th propositions), the *very* end of the book (the 60th proposition) is *not* another one of those problems. It is a planar problem. The long sequence of theorems that has prepared us for the problems that precede the very end has also set us up for the very end itself. Following the preparatory theorems that have come in the middle, and the problems that will now come toward the end, the very last proposition will come as a grand finale to Book One. Or perhaps it would be better to call it a preview of coming attractions, or a curtain-raiser for the sights to be seen in the Second Book of Apollonius's *Conics*. This final proposition of Book One was so characterized earlier; now that we've reached it, we shall see why.

The surprising ending of the first book will constitute a new beginning that propels us into the second.

Right before the problems that return to the cone, at the end of the 51st proposition, when Apollonius says that lines parallel to the diameter of a parabola are also diameters, as are lines through the center of an hyperbola or ellipse—that is, when he says that they are diameters having, with the lines dropped to them that are parallel to the tangents, the rectangle-square relation between abscissa and ordinate that is mediated by the parameter—he then also says this: all the things demonstrated about the sections as following when the original diameters are used (the original diameters being those obtained originally, in the porism to the 7th proposition, from the axial triangle of the cone into which the sections are cut)—those very same things follow also when the other diameters are used. Just why do all those things follow? They certainly follow insofar as they depend upon the parameter-characteristic, since a curve cannot have the parameter-characteristic of more than one kind of conic section. That can be demonstrated by *reductio ad absurdum*, as in figure 27.

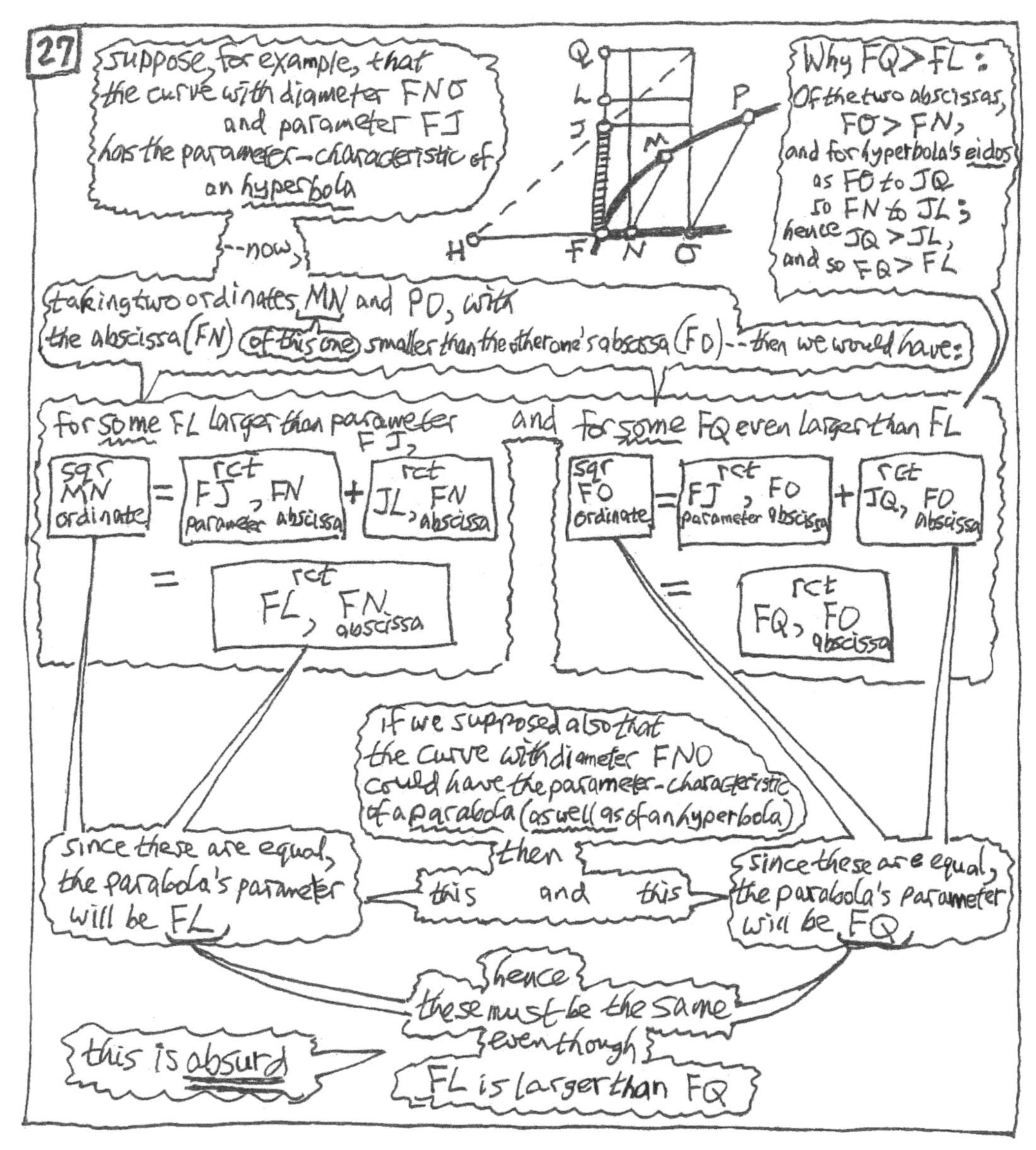

The exhaustiveness of the cone-cutting in the beginning of the book guaranteed that there are no *conic sections* other than those obtained there, with their parameters. All the different ways of cutting a cone yielded all the different conic sections with their original diameters and their parameters. If a curve is a conic section, then it must be either a parabola, or an hyperbola, or an ellipse (or circle)—and no one of them with its original diameter can have the parameter-characteristic of another. Now we have just seen why no curve with *any* of its diameters that has the parameter-characteristic of a parabola or an hyperbola or an ellipse (or circle) can have the parameter-characteristic of another one of them. So, the parameter-characteristic that was shown for any curve that is a conic section in the 11th through 14th propositions seems to define the curve. The 52nd through 59th propositions will show that if a curve with any of its diameters has the parameter-characteristic, then it can be obtained as a conic section. At the end of Book One we shall therefore know that a curve with its diameter has the parameter-characteristic if *and only if* it is a conic section.

PART V
PROVISIONAL CONCLUSION

Apollonius's *Conics*, Book I: proposition 52 to end

CHAPTER 1. INTRODUCTION

Early on, it was shown that all conic sections are lines with certain properties; now, because of the propositions that have followed from those propositions, it can be shown that all lines with those properties are conic sections.

Apollonius, that is to say, will now show that if one of those curves is specified, then it must be obtainable by some cutting of a cone by a plane; and he will also show that opposite sections, by a double operation, are similarly obtainable.

The curve that is sought need only be found in *some* cone, whether or not the specified curve came from that cone originally. It will therefore suffice to show that all lines with those properties are sections of some *right* cone, even though they can be found as sections of oblique cones as well as of right cones. Now since it is enough to show that there is *some* cone that will do the job, and since (as we shall see) some *right* cone will always do it, and since right cones are handy because each axial triangle in a right cone is the same as any other one in it, Apollonius will therefore find the sections in right cones only.

He will do it first for the parabola, then for the hyperbola, and then for the ellipse; and finally, he will do it for the opposite sections. The job will occupy the 52nd through 59th propositions. The 52nd through 59th propositions have the following form: given such-and-such, to do thus-and-so. That is to say, they are not theorems but *problems* (though Apollonius himself does not label them as such).

One of Euclid's postulates enables us to demand that we be allowed to draw a circle when we have been given its center and its radius—that is to say, when we have been given its position and the size of the line that determines its curvature. The propositions that Apollonius has already proved will now enable him to find, as a section of a cone, a curve for which he has been given the following specifications:

—the name of the sort of curve that it is

—the position of one of its diameters and of this diameter's vertex

—the size of the angle at which this diameter is met by its ordinates

—the size of its upright side (that is to say, the parameter which relates the size of its ordinates & the size of the lines cut off by them on their diameter).

The demonstrations for the various conic sections proceed as follows:

Apollonius first constructs a cone,
and cuts it with a plane in a way that yields a section of the sort designated;

then he shows that

not only do
the section's diameter and vertex occupy the designated position,
and its ordinates meet that diameter at the designated angle,

but also the upright side
of the section of that sort, with that diameter and those ordinates,
has the designated size.

The difficulty is to put together a cone which, when cut in the appropriate way, will yield the specified curve. Each of the cones will be a right cone. For each of them, the page will be used as the plane of the axial triangle from which the cone is obtained; and the plane that is perpendicular to the page will be used as the cutting plane (that is, as the plane in which the section lies). (Fig. 1.)

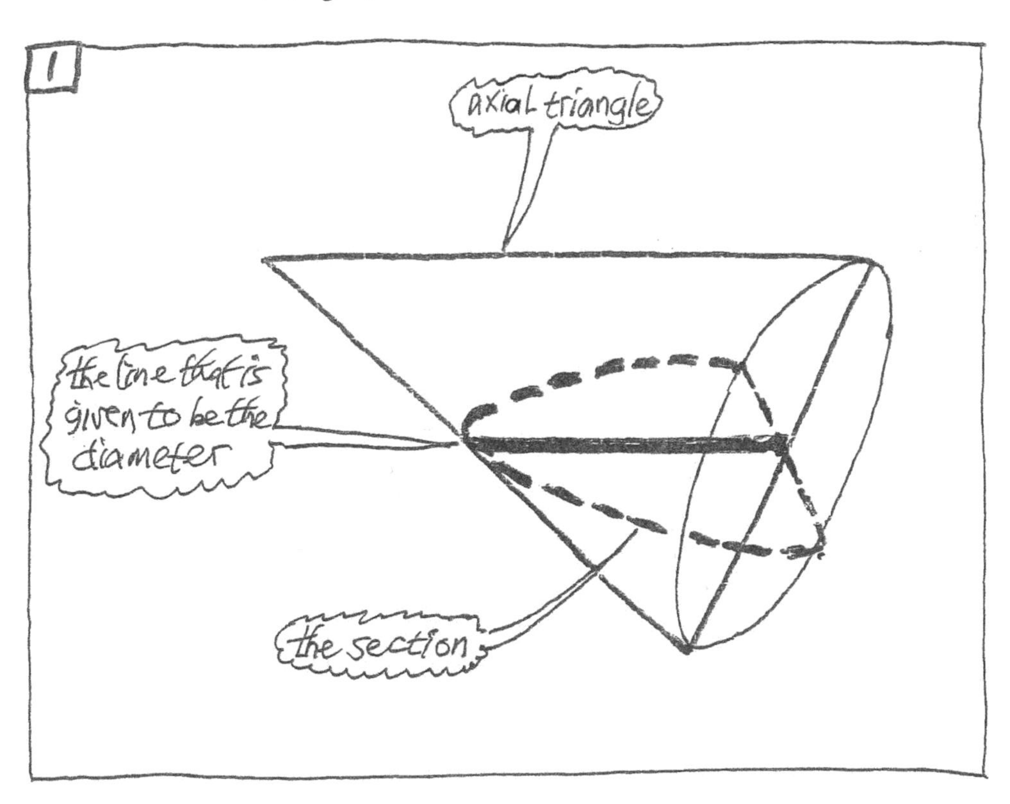

52nd through 59th propositions: the sequence of propositions

This job is distributed over more than four propositions—why? Partly to treat separately the cases when the diameter is and is not to be an axis. Why treat them separately? Because of the following circumstance.

Earlier, we had cones that we cut in various ways to get sections of various sorts with upright sides of various sizes and with ordinates meeting the diameter at some one of various angles. By taking the cutting plane so as to slice the base in a straight line which is perpendicular to an axial triangle's base, we guaranteed that the lines drawn ordinatewise (that is, drawn parallel to that straight line which the cutting plane sliced into the base) would be bisected by the straight line that the cutting plane sliced into the axial triangle. This bisecting straight line was the diameter with which we began—the "original" (or "principal") diameter. Whenever the axial triangle happens to be perpendicular to the base, then the ordinates must be perpendicular to the original diameter. The original diameter, in other words, must be an axis if the axial triangle is perpendicular to the base. Moreover, the original diameter cannot be an axis unless the axial triangle is perpendicular to the base. Now consider what that means for the difference between an oblique cone and a right cone. In an oblique cone, only one of the axial triangles is perpendicular to the base; but in a right cone, every axial triangle is. In a *right* cone, therefore, the original diameter of a section is *always* an *axis*—a diameter which is met by its ordinates at a right angle.

Now we have a difficulty. If we cut a right cone, the axial triangle will of necessity be perpendicular to the base, and hence the section's original diameter will be an axis. What, then, is to be done in the case when we seek a section in a right cone but the *given* line is to be a diameter that will *not* be an axis? In this case, we must first use the given line to find *another* line that *will* be an axis, and then we must show that the section obtained by using this *other* line (this line that will be a diameter that *is* an axis) also has the *given* line as a diameter that is *not* an axis.

That is what Apollonius does. For each sort of section, he first obtains the section in the case when the line that is supposed to be the diameter will meet its ordinates at a *right* angle, and afterward he reduces to that case the other case—the case when the line that is supposed to be the diameter will meet its ordinates at some angle that is *not* a right angle. The two pieces of work appear in separately numbered propositions.

Only the first of the numbered propositions for each sort of section has an enunciation. One sort of section, the ellipse, has an additional complication: for the first case (the case where the given angle at which the ordinates are to meet the diameter is a right angle), there are two sub-cases (propositions 56–57). Why does the ellipse need these two sub-cases? Because it has two axes—one of them (the "major" axis) larger than the other one (the "minor" axis). The one axis is larger than its parameter, and the other axis is smaller than its parameter. Cases need not be distinguished at all for the opposite sections, since cases have already been distinguished for the hyperbola. The layout of the propositions in this sequence is presented in figure 2.

2 given	Find		when the angle made by the ordinates is: right (diameter is an axis)	not right (diameter is not an axis)
• position of a diameter • size of its upright side (parameter)	as sections of a right cone, these curves:	parabola	52	53
		hyperbola	54	55
		ellipse	depending on whether diameter is > or < than parameter: 56(>); 57(<)	58
		opposite sections	59	

CHAPTER 2. FINDING CONES FOR THE CURVES: PARABOLA

the parabola problem solved *(52nd-53rd propositions)*

The problem is to find a parabolic conic section, given the position of one of its diameters (including the position of the diameter's vertex) and given the size of that diameter's upright side (its parameter) for ordinates that meet that diameter at some given angle.

The sought-for parabolic curve is found by Apollonius as a section of a *right* cone—first in the case when it is a right angle at which the ordinates are to meet the diameter, and then in the case when it is some other angle at which the ordinates are to meet the diameter. In the first case (52nd proposition), the diameter is an axis; in the second case (53rd proposition), it is not. In the second case, the line that is given to be a diameter which is *not* an axis is used to find a line to be a diameter which *is* an axis. Finding this other line (the line which will be not only a diameter but also an axis) transforms the situation in the second case into an instance of the first case; and then the only question is whether the solution to the problem as transformed is also a solution to the problem as posed before the transformation.

Apollonius's solution to the problem of finding a specified parabola as a cut in a cone is presented in the following boxes. In the event that the figures in the subsequent explication of these propositions do not provide as much help as you would like in drawing illustrative diagrams for the reading of the propositions, see the Appendix at the end of this volume for such diagrams drawn complete.

#52 A straight [line] being given in a plane, [a line, that is,] bounded at one point—to find in the plane the cone's section called a parabola, whose diameter is what the given straight [line] is, and vertex is what the bound of the straight [line] is, and in which whatever straight [line] may be drawn down from the section onto the diameter at a given angle will have the power [to give rise to a square equal to] the rectangle contained both by the [line] taken off by it on the side toward the section and by some other given straight line. [That is to say: One of the lines containing the rectangle is a line taken off from the given straight line which is to be the diameter; it is bounded on one end by the line which is drawn down onto it at the given angle, and on its other end it is bounded by the point which is to be the vertex of the section. The other containing line is another given straight line which is to be the upright side of this diameter.]
—Let there be given in position the straight [line] AB, a [line] that has been bounded at the point A, and let another [straight line], namely the [line] CD, [be given] in its magnitude.
—And let the given angle first be [an angle that is] right. Required, then, to find, in the plane supposed, a parabola whose diameter is what the [line] AB is, and vertex is what the [point] A is, and upright [side] is what the [line] CD is, and in which the [lines] drawn down ordinatewise will be drawn down at a right angle—drawn down, that is, so that an axis be what the [line] AB is.

—Let there be extended the [line] AB to the [point] E; and also, let there be taken, as the [line] CD's fourth part, the [line] CG, and greater than the [line] CG let the [line] EA be; and also, as a mean proportional of the [lines] CD, EA, let the [line] H be taken [so that, in other words, as is CD to H so is H to EA]. Therefore as is the line CD to EA so is the [square] from H to the [square] from EA. And the [line] CD is, in comparison with the [line] EA, less than its quadruple. And therefore also the [square] from H is, in comparison with the [square] from EA, less than its quadruple. Therefore the [line] H is, in comparison with the [line] EA, less than its double; so that two [lines] EA are greater than the [line] H. It is therefore possible (*dunaton*), from the [line] H and two of the [lines] EA, to construct (*sunstêsasthai*) [more literally: "to constitute"] a triangle. Let there be constructed, then, on the [line] EA, the triangle EAF upright (*orthon*) to the plane supposed, so that equal is (on the one hand) the [line] EA to the [line] AF and (on the other hand) the [line] H to the [line] FE; and also, let there be drawn (on the one hand) parallel to the [line] FE the [line] AK, and (on the other hand) parallel to the [line] EA the [line] FK; and also, let there be envisioned (*noeisthô*) [traditionally rendered as "be conceived"] a cone whose vertex is what the point F is, and whose base is what the circle is, [the circle, namely,] that is about the diameter KA and is upright to the plane through the [lines] AFK. Then upright (*orthos*)] [traditionally rendered as "right"] is what the cone will be; for equal is the [line] AF to the [line] FK.

[The traditional rendering of *noeisthô* as "let there be conceived" has connotations of seizing (Latin *capio*), whereas the Greek term refers to an action of awareness that is more like seeing than it is like seizing—seeing, as we might say, "with the mind's eye."]

—And let the cone have been cut by a plane parallel to the circle KA; and also, let it make as section the circle MNX manifestly upright to the plane [that goes] through the [lines] MFN; and also, let a common section of the circle MNX and of the triangle MFN be what the [line] MN is; a diameter it therefore is, of the circle. And let a common section of the plane supposed and of the circle be what the [line] XL is. Since, then, the circle MNX is upright to the plane supposed, but is upright also to the triangle MFN, therefore their common section the [line] XL is upright to the triangle MFN, that is, to the triangle KFA; and therefore also, to all the straight [lines] that touch it and that are in the triangle, is it upright; and so is it also to both of the [lines] MN, AB.

—Again: since a cone whose base is what the circle MNX is, and vertex is what the point F is, has been cut by a plane upright to the triangle MFN, and [this plane] makes as section the circle MNX, but [the cone] has been cut also by another plane, the plane supposed, which cuts the base of the cone in the straight [line] XL that is at right angles to the [line] MN, which is a common section of the circle MNX and of the triangle MFN, but the common section of the plane supposed and of the triangle MFN, namely the line AB, is parallel to the cone's side FKM—therefore the cone's section that is generated in the plane supposed is a parabola, and its diameter is what the [line] AB is, and the lines drawn down from the section ordinatewise onto the [line] AB will be drawn down at a right angle [to the line AB], for parallel are they to the [line] XL that is at right angles to the [line] AB.

—And—since the three proportionals are what the [lines] CD, H, EA are, [H having been taken as a mean propositional between the other two,] but equal is (on the one hand) the [line] EA to the [line] AF and to the [line] FK, and (on the other hand) the [line] H to the [line] EF and to the [line] AK—therefore as is the [line] CD to AK so is the [line] AK to AF. And also, therefore, as is the [line] CD to AF so is the [square] from AK to the [square] from AF, that is, to the [rectangle] by AFK. Therefore an upright (*orthia*) [side] is what the [line] CD is of the section, for this has been shown in the 11th theorem.

#53 The same things being supposed, let not the given angle be right; and also, let there be put equal to it the angle by HAE; and also, of the [line] CD, half let the [line] AH be; and also, from the [point] H onto the [line] AE let a perpendicular (*kathetos*) be drawn, namely the [line] HE; and also, through the [point] E [let there be drawn] a parallel to the [line] BH, namely the [line] EL; and also, from the [point] A let a perpendicular (*kathetos*) to the [line] EL be drawn, namely the [line] AL; and also, let the [line] EL have been cut in two [halves] at the [point] K; and also, from the [point] K at right angles to the [line] EL let there be drawn the [line] KM, and also let it be extended to the [points] F, G; and also, let there be equal to the [square] from the [line] AL the [rectangle] by LKM, [the point M having been located properly for this]; and also, there being given the two straight [lines] LK, KM, with (on the one hand) the [line] KL [being given] in position and having been bounded by the [point] K, and (on the other hand) the [line] KM [being given] in magnitude; and also, with an angle that is a right [angle], let there be carved out (*gegraphthô*) a parabola whose diameter is what the [line] KL is, and vertex is what the [point] K is, and upright [side] is what the [line] KM is, as has been shown before. But [the parabola] will go through the [point] A, on account of there being equal the [square] from AL to the [rectangle] by LKM; and also, there will touch the section the [line] EA, on account of there being equal to the [line] KL the [line] EK. And the [line] HA, to the [line] EKL, is parallel. Therefore the [line] HAB is a diameter of the section, and the [lines] drawn down onto it from the section that are parallel to the [line] AE will be cut in two [halves] by the [line] AB. But they will be drawn down at the angle by HAE. And also—since equal is the angle by AEH to the angle by AGF, but common is the angle at the [point] A—therefore similar is the triangle AHE to the [triangle] AGF. Therefore as is the [line] HA to EA so is the [line] FA to AG; therefore as is the double of the [line] AH to the double of the [line] AE so is the [line] FA to AG. But the [line] CD is double of the [line] HA; therefore as is the [line] FA to AG so is the [line] CD to the double of the [line] AE. On account of the things shown in the 49th theorem, then, an upright (*orthia*) [side] is what the [line] CD is.

analysis of the parabola problem–first case (52nd proposition)

You may have trouble seeing clearly just what is going on in Apollonius's presentation. That is not surprising. The whole proposition is a putting-together in which the beginning is governed by the end. Some of the items that are put together are given at the outset, and some are items now available for use because previously synthesized or given, but the use does not appear until *after* those available items have been used; so, the proposition appears to be just one item after another until you reach the end and then look back. Looking at it backwards is what makes the proposition intelligible. The synthesis is illuminating only after an analysis.

So, here is a sketch of a line of thought that will help you to understand the solution of the problem in this case, the case when the ordinates are to meet the diameter at right angles.

To solve the problem, it seems best to take the easiest situation. We therefore choose to find the curve as a section of a right cone, rather than of a cone that is oblique. In a right cone, any axial triangle is the same as any other one: we may therefore speak of *the* axial triangle of a right cone. If we can first obtain the axial triangle of a right cone which is proper for our purpose, then we can easily obtain that right cone itself. In such a cone, the diameter sliced by the plane will be an axis of the section—will be, that is, a diameter whose ordinates will make right angles with it.

Let the page serve as the plane of the requisite axial triangle. Now, what kind of a triangle must we get if it is to serve as the axial triangle of a right cone? Well, it will not be the axial triangle of a right cone unless it is an isosceles triangle. The triangle that we need will therefore have to be isosceles.

Where must that triangle be placed if the diameter is to be in the given position? The section cannot be a parabola unless the diameter sliced into the axial triangle by the cutting plane is parallel to one side of the axial triangle. One side of the requisite isosceles triangle will therefore have to be parallel to the line given in position to be the diameter.

So, in order to get a cone to cut, what we need is something like the dashed-line triangle AFK in figure 1. That triangle is not the whole axial triangle NMF (fig. 2), but only its tip. The whole axial triangle is larger, but it has the same shape as its tip and it has its own position fixed by the position of its tip. For that reason, it is sufficient to consider the tip. It is better to proceed by considering the tip rather than the whole axial triangle, because all we need to know are the ratios that its sides have to each other; and if we confine our attention to the tip, that is all that we shall have to worry about.

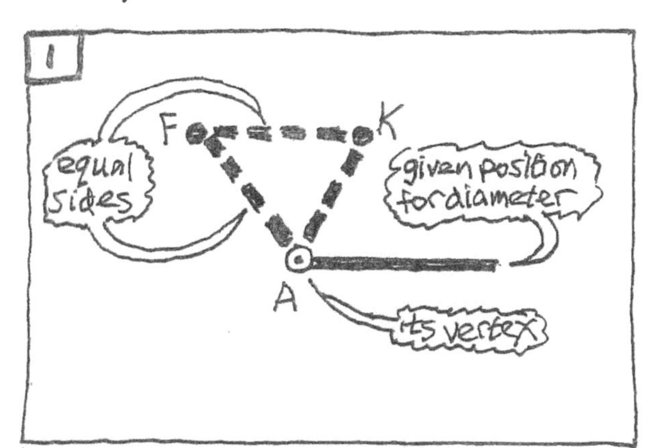

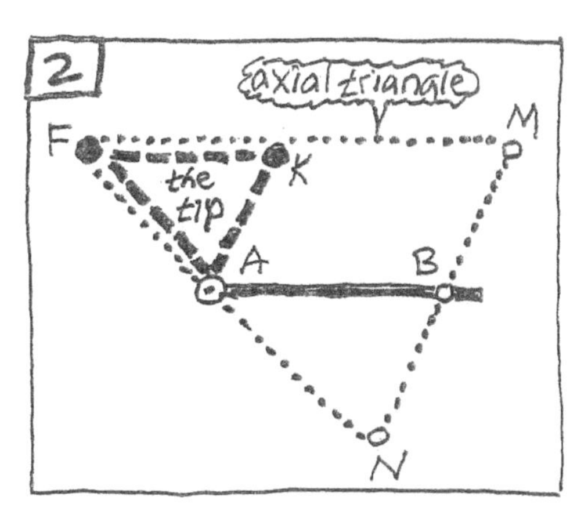

How can we get that triangle AFK which will be the tip of the axial triangle? Let us proceed from something that we already have. Let us prolong the line that gives us the position of what is to be the diameter. Let us prolong it through the point where the section's vertex is to be, and then, upon it, let us construct a triangle AEF that we shall *suppose* to be of the proper shape. And now, if we complete the parallelogram, we shall have a triangle of the same shape, but it will be in the position where we need it. This triangle AFK will be what we need—it will be the tip of the axial triangle (fig. 3).

Now let us draw a circle, perpendicular to the page, having AK as its diameter; and let us use the circle to conceive a cone with F as vertex, letting the generating straight line extend from F through A and far beyond it—to some point N, say (fig. 4).

Now let us pass a plane through the diameter, setting that plane perpendicular to the page. That plane will cut the cone in a parabola (since the diameter AB is parallel to a side FM of the axial triangle NFM). The diameter is in the designated position and the ordinates are at right angles to it, as required (fig. 5).

We have now done everything that was required—except for the one thing that is really tricky. We have taken into account the specifications that deal with position, but not the one that deals with size. That is to say, we have not yet taken into account the specification of the diameter's parameter. So that is what we must do now.

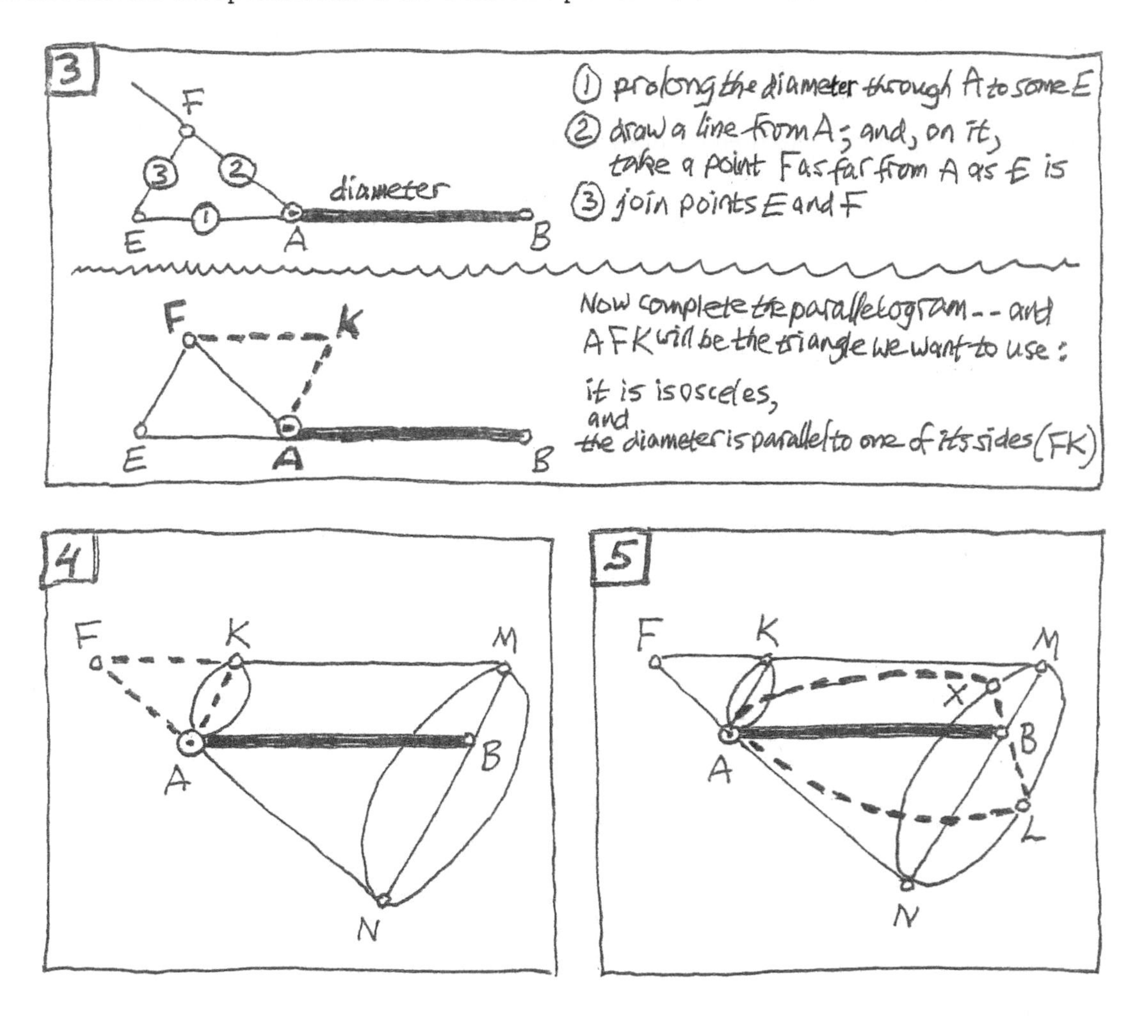

Just how will it fit into what we have done, if we now take into account the size that is specified for the parameter? What will be affected is this. The triangle AEF that we constructed in order to get the tip of the axial triangle cannot be just any isosceles triangle whatever. Taking point E on the diameter's extension at *any* distance from vertex A, and taking line AF so that point F is at *any* distance from that point E, may enable us to cut a cone that yields a parabola whose diameter, with its vertex, is in the specified position, with ordinates meeting that diameter at the specified (right) angle—but that diameter's parameter may *or may not be* of the specified size. It will be of the specified size only if the line of specified size, and the lines that make up the tip of the axial triangle, have suitable relationships of size.

After all, when the parabola's parameter was defined, way back in the 11th proposition, the two ratios whose sameness constituted the defining proportion were in effect the following ratios (if we speak in terms of the lines that make up the triangular tip of the isosceles axial triangle that we are now considering):

> the ratio that the parameter has
> to either of the tip's equal sides,
>
> and the ratio that the square on the tip's bottom has
> to the rectangle (which is a square) contained by the tip's two equal sides.

Let us put that in terms of the lines used to construct triangle AEF—the isosceles triangle from which (by completing the parallelogram) we get the triangular tip AKF (which has a different position but the same size and shape as triangle AEF). CD can be the parameter only if the following is a true statement about CD and the lines used to construct triangle AEF:

> CD is to EA as the square on EF is to the square on AF. (Fig. 6.)

That is not enough to determine the lines used to construct AEF, but it is enough to show that the lines used to construct the tip of the axial triangle must be chosen with a view to the size of the line CD that is specified to be the parameter.

How, then, must we pick points E and F in order for CD to become the parameter? If we think about it, we shall see that the answer is this:

> EA must be greater than one-fourth of CD, and
>
> FE must be the mean proportional between those two lines, EA and CD.

Now let us see why that must be so.

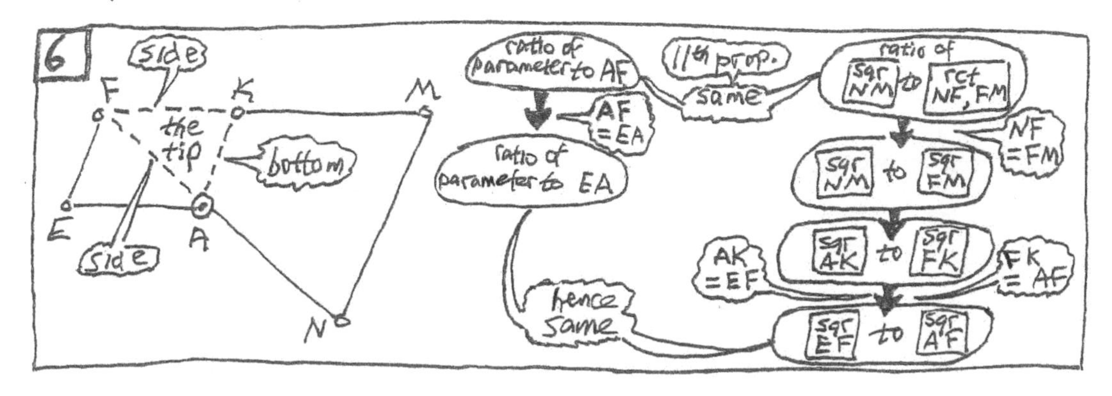

Suppose that triangle AEF were the isosceles triangle that will yield the parabola with diameter AB whose parameter is CD. What would then have to be the sizes of EF and EA?

Well, just because AEF is an isosceles triangle EF would have to be less than twice AF.

Figure 7 shows why that is so.

And because CD is to be the parabola's parameter, the ratio of EF to AE would have to be the same as the ratio of CD to EF— in other words, EF would have to be the mean proportional between CD and AE.

Figure 8 shows why that is so.

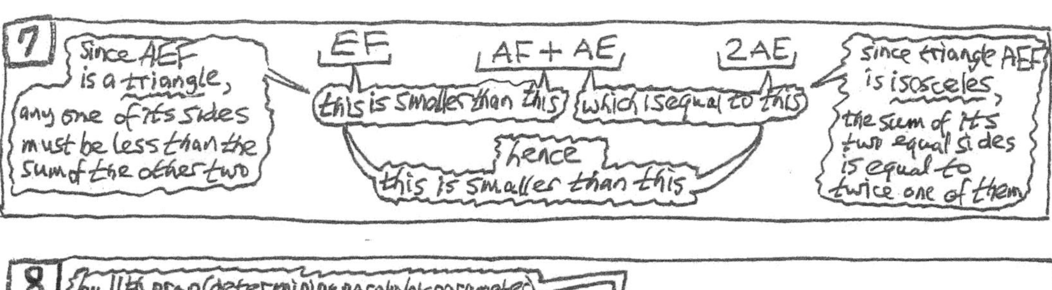

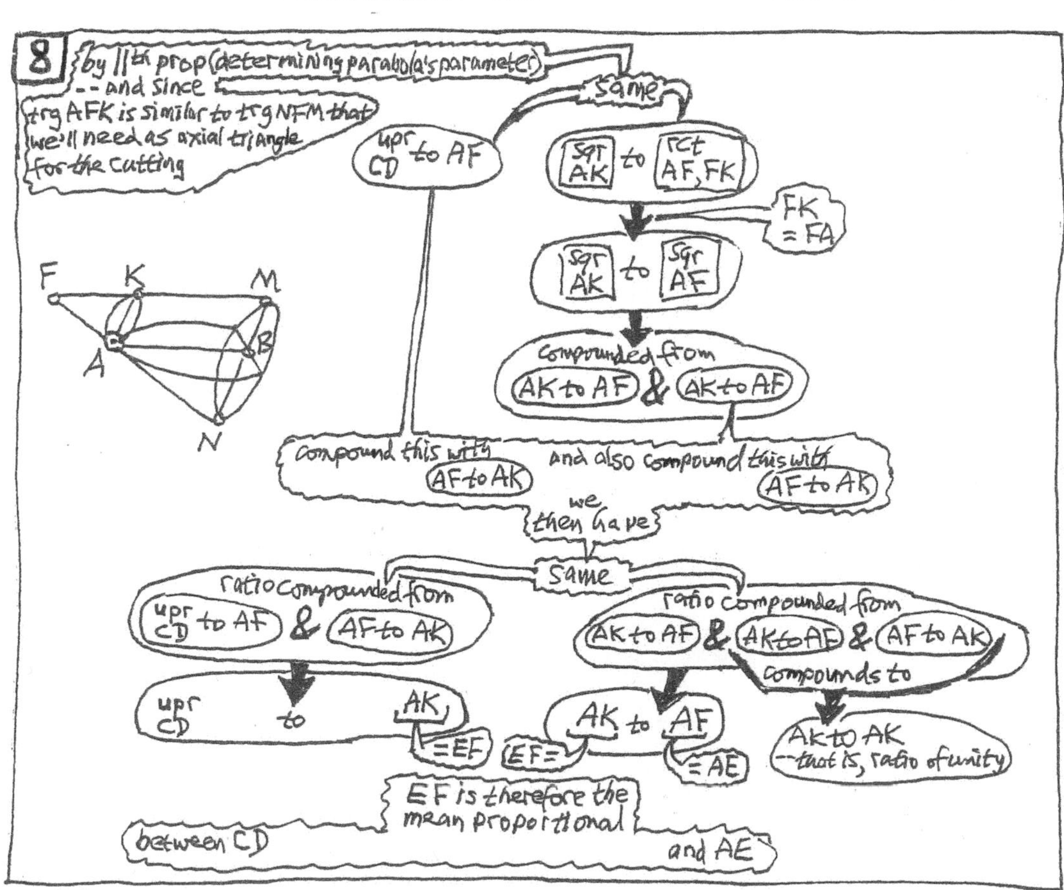

So, now we can be sure of this: to obtain the parabola with axis AB that we are seeking (namely, the parabola whose parameter is CD), we can use isosceles triangle AEF only if the following is true—

as CD is to EF, so also EF is to AE; and the mean proportional EF is less than twice AE.

From those two statements, it follows that AE is less than one-fourth CD. Why that follows, is shown in figure 9.

So, when we construct triangle AEF, we must do it in the manner depicted in figure 10, which takes into account the requirement that the parabola with axis AB must have the specified line CD as its parameter.

Such a procedure determines the distance of F not only from A but also from E. It not only guarantees that *triangle AEF will be isosceles* (since FA is equal to AE); it also guarantees (since FE is the mean proportional between AE and CD) that *CD will be the parameter*.

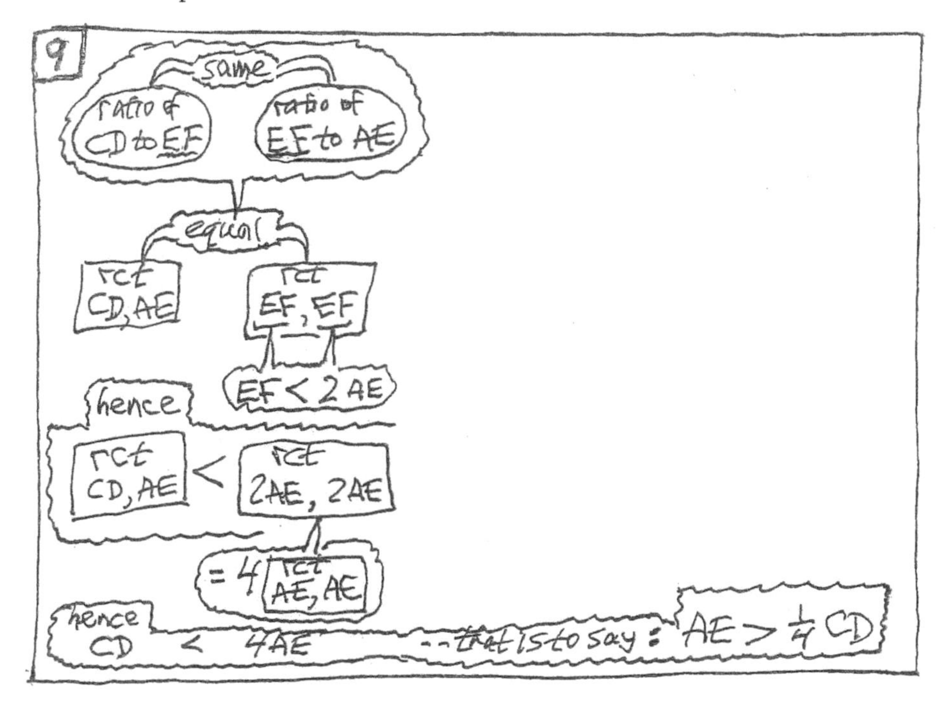

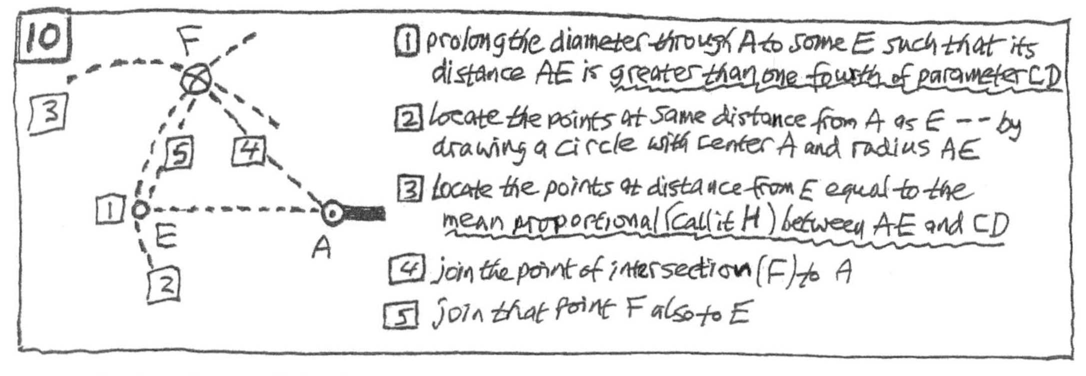

If you now go back and re-read Apollonius's own presentation for the case when the ordinates are to be at right angles to the diameter, the order in which he presents things will no longer leave you unclear about what he is doing.

The order in which Apollonius proceeds in the 52nd proposition is this:

> (I) He begins by letting AE and EF be of the following sizes:
>
> take CG equal to one-fourth of upright side CD,
> and let AE be greater than that CG;
>
> take a line (H) that is the mean proportional between CD and AE,
> and let EF be equal to that H.
>
> He then uses these lines to construct triangle AEF,
> which he then uses to obtain a cone,
> which he shows to be a right cone.
>
> (II) Then, having obtained a right cone in the first part,
> he cuts the cone in the proper way,
> and he shows that
> the section is a parabola with vertex A and specified diameter AB,
> whose ordinates meet it at the specified right angle.
>
> (III) Finally, he shows that
> since H was taken as the mean proportional between CD and AE,
> it follows that CD has to AF the same ratio that
> square [AK] has to rectangle [AF, FK]—
> which is to say that (by the 11th proposition)
> CD is, as specified, the parabola's upright side.

Notice the movement of the synthesis that Apollonius presents. He begins by telling us *that* we are to let AE and EF be of certain sizes, but not *why* we are to make them just those sizes. He concludes by telling us this:

> since we made EF a certain size,
> therefore CD must be the proper size to be the upright side.

Consider, by contrast, our analysis. The question that we asked was this:

> if CD were to be the proper size to be the upright side,
> then what size would we have to make EF?

By answering that question, we know only that

> if CD is the proper size,
> then EF is such-and-such a size.

That is why, even if we showed that last implication, we would still have to show the reverse, which is what Apollonius shows in his synthesis—namely, that

> if EF is such-and-such a size,
> then CD is the proper size.

analysis of the parabola problem—the other case (53rd proposition)

We are now mid-way through the problem of finding a parabolic conic section, when what have been given are: the position of one of its diameters (and of the diameter's vertex), and the size of that diameter's parameter for ordinates that meet that diameter at some given angle. The sought-for parabolic curve is found by Apollonius as a section of a *right* cone. We have already examined how he solves the problem in the case when the ordinates are to meet the diameter at a *right* angle. We now examine how he solves it in the case when the ordinates are to meet the diameter at some angle *other* than a right angle.

Since an axis is a diameter whose ordinates meet it at right angles, the case already examined (the 52nd proposition) was one where the diameter is an axis, and the case to be examined now (the 53rd proposition) is one where the diameter is not an axis. In this latter case, what Apollonius does is this:

he uses the *given* line, which is to be a diameter that is *not* an axis,
to obtain *another* line, which is to be a diameter that *is* an axis;

and once he has obtained this new line, the situation has been turned into an instance of the first case.

Now we shall consider the order in which Apollonius proceeds in the 53rd proposition. We shall start by taking an overview of the main parts, and after that, we shall analyze the particulars within the parts. Here is an overview:

(I) Take the given line AB which is to be that diameter of a parabola
whose ordinates meet it at a certain angle HAE that is not a right angle,
and whose parameter is the given line CD,
and
use AB to obtain KL (a line that is to be
another one of the parabola's diameters—one
whose ordinates meet it at a right angle).
Also obtain KM (the line that will be
the parameter of the new diameter KL that *is* an axis).

(II) Now that we have
the position of KL with vertex K, and the size of KM,
use the previous proposition (the 52nd) to obtain a parabola
whose diameter is KL, with vertex K,
and whose ordinates meet KL at a right angle,
and whose parameter for that diameter KL, with those ordinates, is KM.

(III) Show that this parabola is the very parabola sought for—
namely, the one whose diameter is AB, with vertex A,
and whose ordinates meet AB at non-right angle HAE,
and whose parameter for that diameter AB, with those ordinates, is CD. (Fig. 11.)

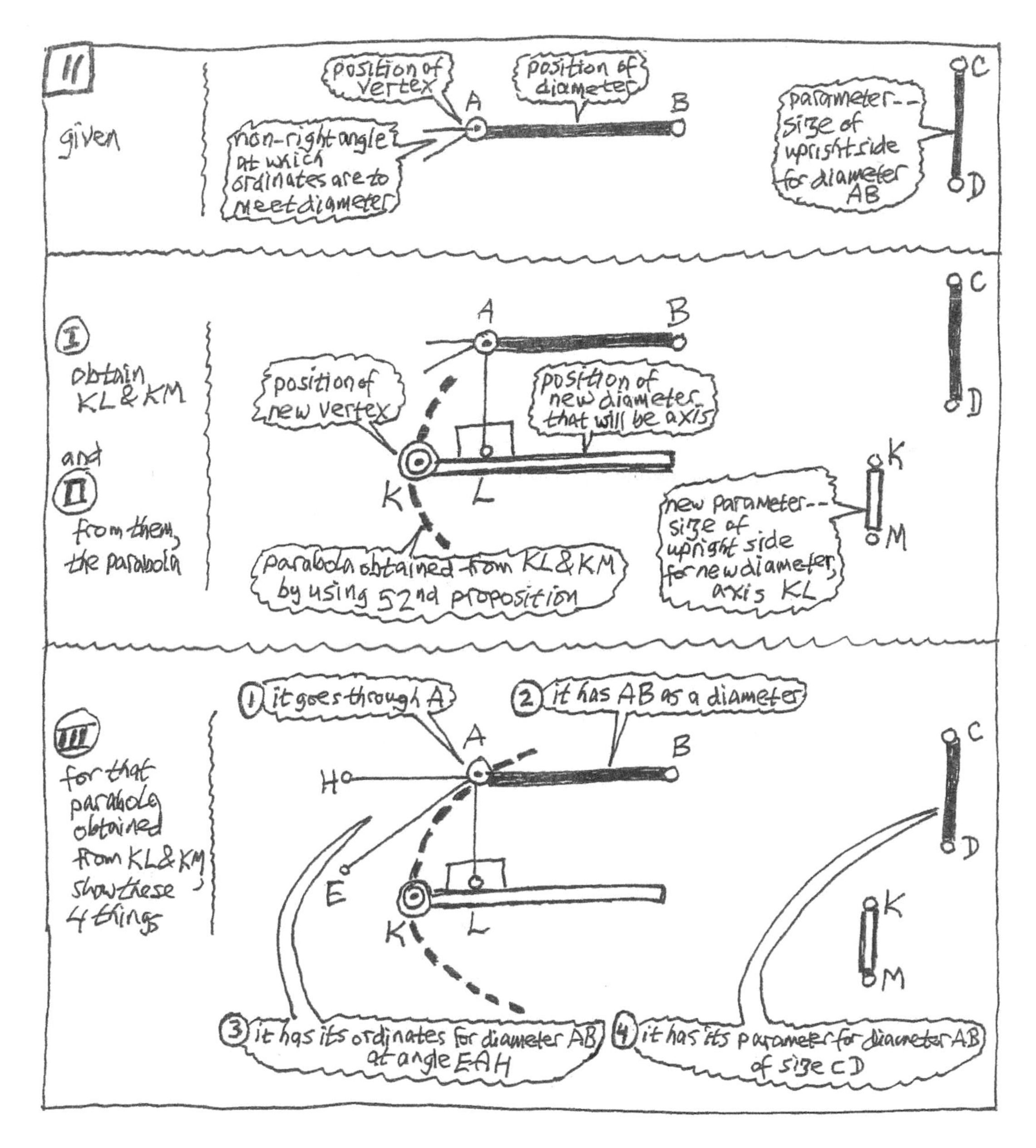

The middle part of Apollonius's presentation is merely an application of the previous proposition, for the 52nd proposition has already solved the problem of finding a parabolic conic section that satisfies the specifications of the position of its *axis* and the size of this axis's parameter. The difficult parts of the presentation are the beginning and the conclusion. The *beginning* uses the items that were first specified for a diameter that is *not* an axis to obtain *new* specifications for a diameter that *is* one. The *conclusion* shows that the parabola that was obtained in the middle part, to satisfy the new specifications, will satisfy the first specifications too.

Again the hardest thing to understand, as you move through Apollonius's presentation, is a beginning in which lines are drawn for no apparent purpose. The purpose only becomes apparent after you have moved through the whole proposition and then look back.

Here is a sketch of an analytic line of thought to help you understand the construction which constitutes the beginning of Apollonius's solution of the parabola problem in the case when the ordinates are to meet the diameter at some angle that is not a right angle.

Suppose that we *had* the axis and vertex of the sought-for parabola. Then, having the axis, we would be able to obtain the parameter for the axis—and then we would be able to obtain the parabola by using the solution from the previous proposition, the 52nd.

How could we obtain the axis's parameter? As follows. Point A is specified to be on the sought-for parabola, and we are supposing that we have the axis with its vertex (call it point K). From A, drop a perpendicular to the axis; and the point where the perpendicular meets the axis, call L. Then AL will be an ordinate and KL will be its abscissa. The square on ordinate AL will (by the 11th proposition) be equal to a rectangle whose one side is the abscissa KL, and whose other side (the "upright" side) is a line of definite size easily obtained from the proportion embodied in the equality of square and rectangle. This upright side, call KM; and then (since square [AL] is equal to rectangle [KL, KM]) the proportion is this: as KL is to AL, so AL is to KM. This line KM, which can easily be found as the third proportional of abscissa KL and ordinate AL, is the parameter of the axis KL. (Fig. 12.)

Now, by merely supposing that we have the axis (and, therewith, its parameter and the sought-for parabola), we wish to arrive at something that we do indeed have that we could use to obtain a line that is in a position to be the axis.

If we had axis KL, then, by the 46th proposition, it would be parallel to AB; so we can get KL by drawing a parallel to AB. But which parallel? And where, along that parallel, would be the position of vertex K? Well, by the 33rd proposition, vertex K would be mid-way between the diameter's intersections with two lines—with the ordinate from any point A on the parabola, and with the tangent from that same point A. But what would be the parabola's tangent from A? It would, by the 17th proposition, be the line through A drawn ordinatewise for diameter AB—and so, we have the tangent from A. Therefore, if we had axis KL, then we would have the intersection of the axis with the tangent from A. Call this intersection the point E. (Fig. 13.)

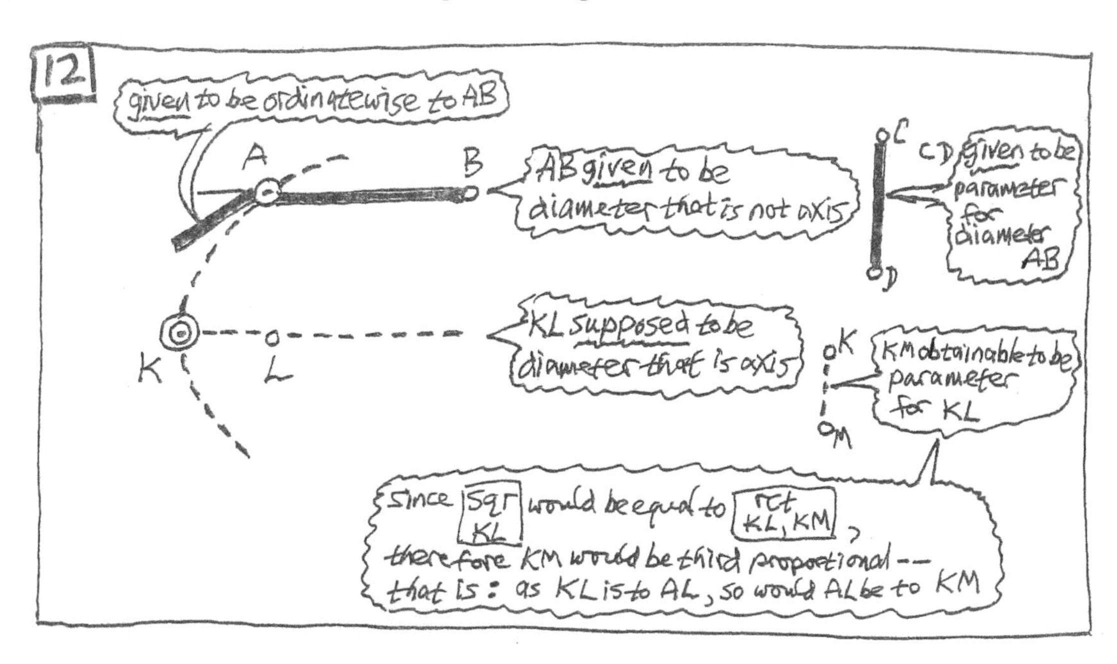

So, if only we had point E we could use it to locate the axis (by just drawing, through E, a parallel to AB) and then use it to locate the axis's vertex K (by just dropping, from A, a perpendicular to the axis—and taking the midpoint between the point L where this perpendicular meets the axis, and the point E). But we do not yet have E unless we suppose that we already have the axis at K, so let us go on supposing it until we figure out some way to locate point E—the point where the axis meets the tangent from the vertex A of the non-axis diameter AB.

Well, what about the tangent at the vertex K of the axis? It would, by the 17th proposition, be drawn ordinatewise for axis KL. That is to say: if at K we raise a perpendicular from the axis KL, it will be the tangent at K. But a line perpendicular to axis KL will also be perpendicular to the extension of diameter AB (since KL is parallel to AB). So, the tangent at vertex K will be perpendicular to the extension of diameter AB. Now for convenience, let us name the point (let's call it F) where that tangent at K meets the other tangent (the tangent from A); let us also name the point (let's call it G) where the tangent at K meets the diameter's extension from A. (Fig. 14.)

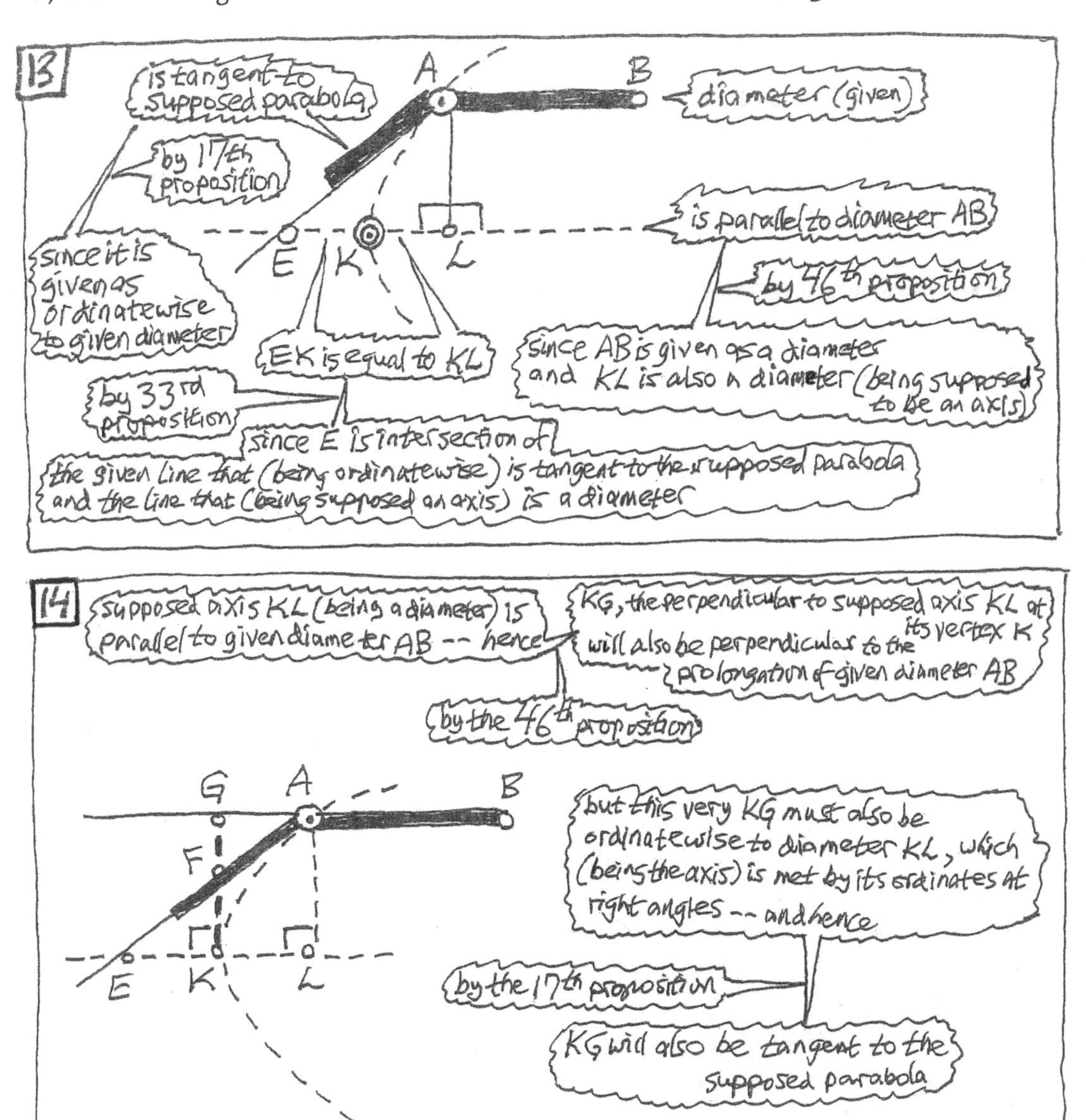

Now, treating KL as an old diameter and AB as a new one, it follows, from the 49th proposition, that the line CD can be the parameter of diameter AB only if

> the ratio of CD to twice AE is the same as the ratio of AF to AG (fig. 15).

That tells us something about point E—especially if we consider that those two lines in the second ratio (AF and AG) are two sides of a triangle. In this triangle, one angle is the given ordinate-angle at A, and another angle of the triangle is the right angle AGF. One of two lines in the first ratio is given (namely, CD—which is the parameter of AB); the other, AE, can be used to make a triangle similar to triangle AFG, and thus we can get it into a proportion. Here is how.

At E, raise a perpendicular from AE. It will intersect with the extension of AB at some point (call it H), making a triangle AHE. Now this triangle AHE shares with triangle AFG the ordinate-angle at A, and it also has, like the other triangle, a right angle; therefore the two triangles are similar—and hence their corresponding sides are proportional. That is to say: KL can be the axis only if

> the ratio of AH to AE is the same as the ratio of AF to AG (fig. 16).

So, if it were true that AB is a non-axis diameter, with vertex A and parameter CD, and KL is the axis, with vertex K, then it would be true that each of two ratios would be the same as the ratio of AF to AG: namely, the ratio of CD to twice AE would be the same as that ratio, and also the ratio of AH to AE would be the same as it. Therefore (since two ratios which are the same as the same ratio must also be the same as each other) CD would have the same ratio to AE's double that AH has to AE itself. From this it follows that CD would have to be double the size of AH.

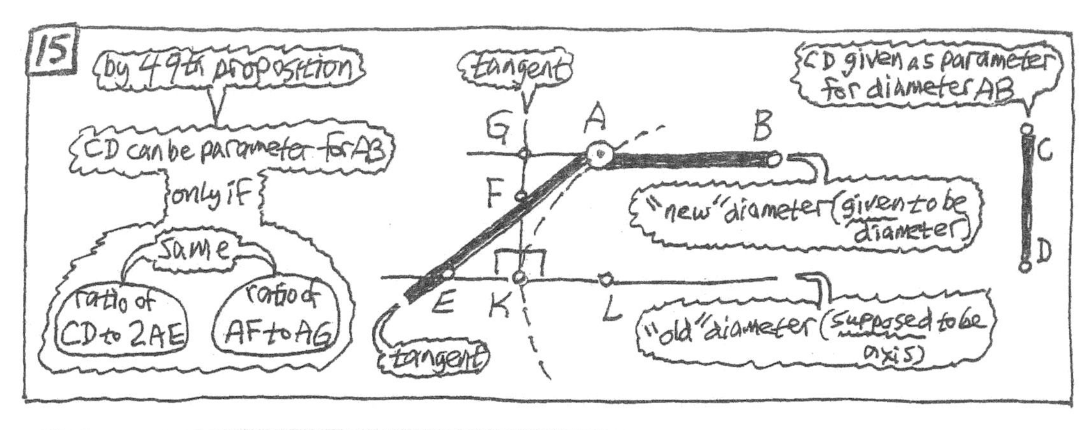

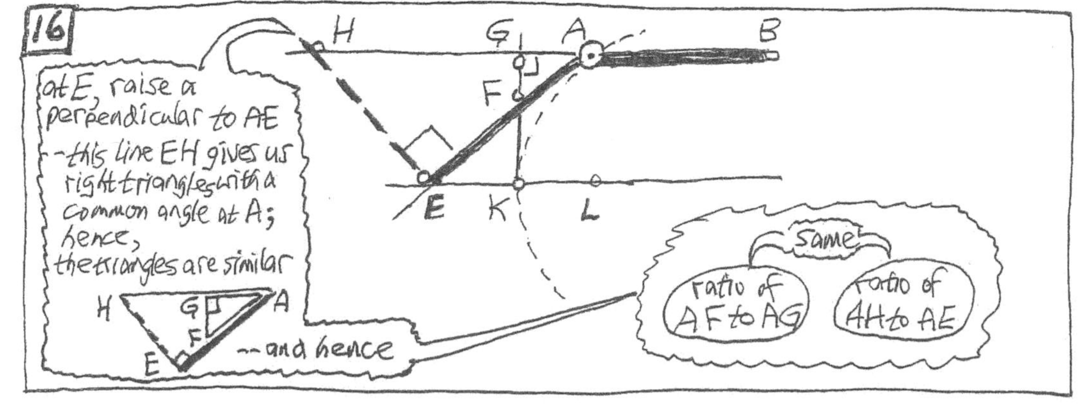

Our conclusion in analyzing the problem is therefore this:

if the axis were KL,
then CD could be the parameter for the non-axis diameter AB
only if AH were equal to half of CD (fig. 17).

What we need to do is therefore this:

take point H on the extension of AB,
at a distance from A equal to half of what is to be AB's parameter CD,

and then perform the steps of the construction in the order that will yield axis KL, with vertex K and parameter KM, for a parabola that will have AB as another of its diameters—one whose vertex will be A, and whose parameter, for ordinates that meet it at a non-right angle equal to the angle at A, will be CD.

In accordance with this, Apollonius begins his presentation by placing the ordinate-angle upside-down on the other side of the vertex from the line that is to be a non-axis diameter (with one of the angle's sides being an extension of the diameter), and then he takes point H on the side that is the extension of the diameter at a distance from the vertex equal to one-half of that diameter's parameter. He does not tell us why he does that, but instead he moves on with the construction that prepares for the cone and its cutting. That *first* part of Apollonius's presentation, which is a construction to obtain lines KL and KM, is depicted in figure 18.

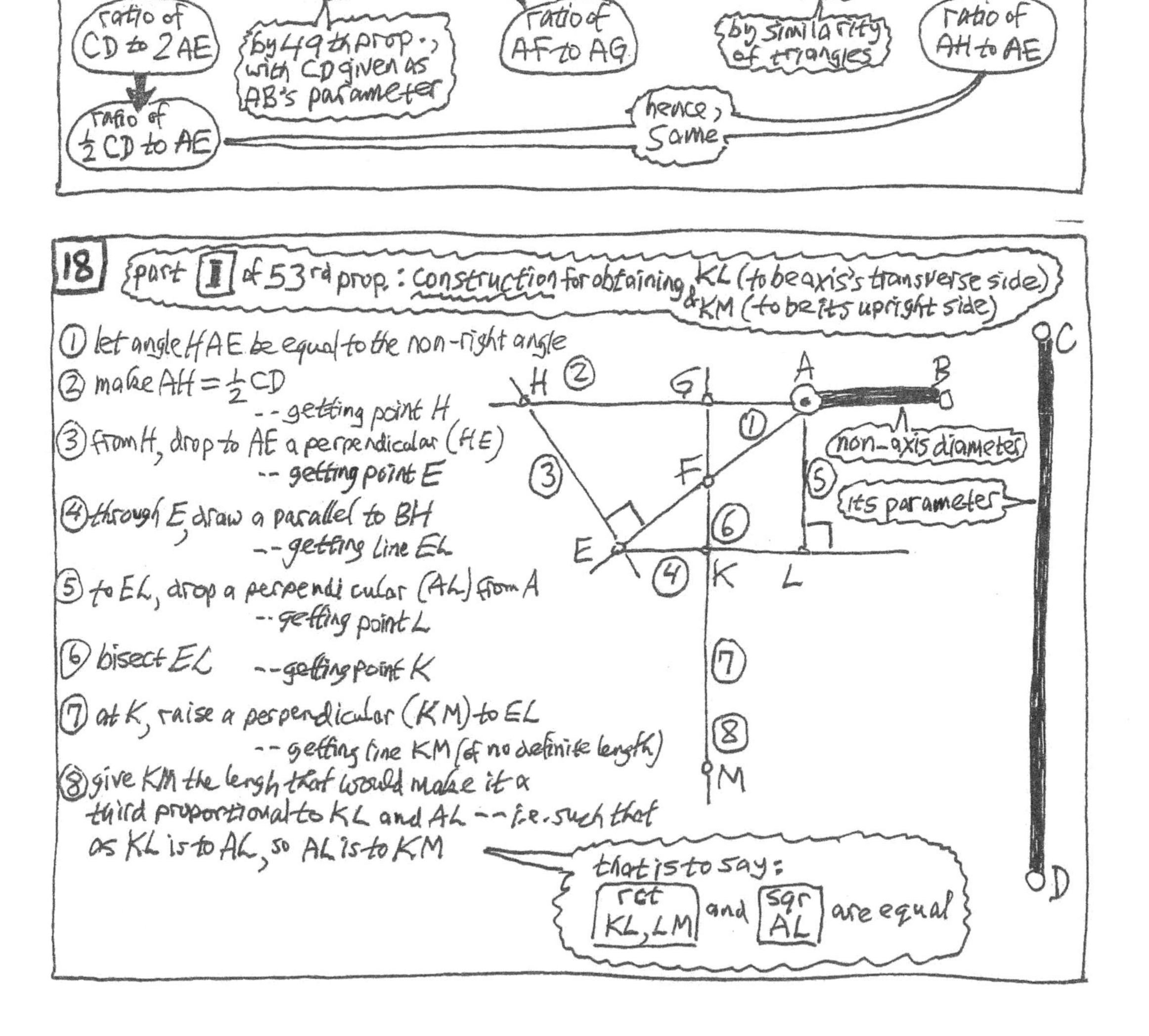

In the *second* part, the lines KL and KM that have been obtained in the first part are used to obtain a parabola—a parabola that has KL as its diameter, with vertex K, and has KM as parameter for that diameter KL, whose ordinates are at right angles, making it an axis. The *third* part shows that the parabola so obtained is indeed the parabola that was sought for—namely, a parabola that has AB as its diameter, with vertex A, and has CD as parameter for ordinates that meet diameter AB at a certain angle that is not a right angle. This last part is depicted in figure 19.

Again in the case of this 53rd proposition, we see Apollonius as synthesist begin by telling us to place lines of certain sizes in certain positions, without telling us *why* to do that just so. He concludes by telling us that since we did that just so, what we have obtained is therefore just what was required. Our question as analysts, however, was the reverse. It was this: if we were to obtain what is required, then what lines of what sizes would have to be placed where? For certain purposes, it may not be necessary to show that

if such-and-such that is obtained were indeed what is required, then the lines would have to be placed just so;

but it is still necessary to show that

if the lines are placed just so, then such-and-such that is obtained is what is required.

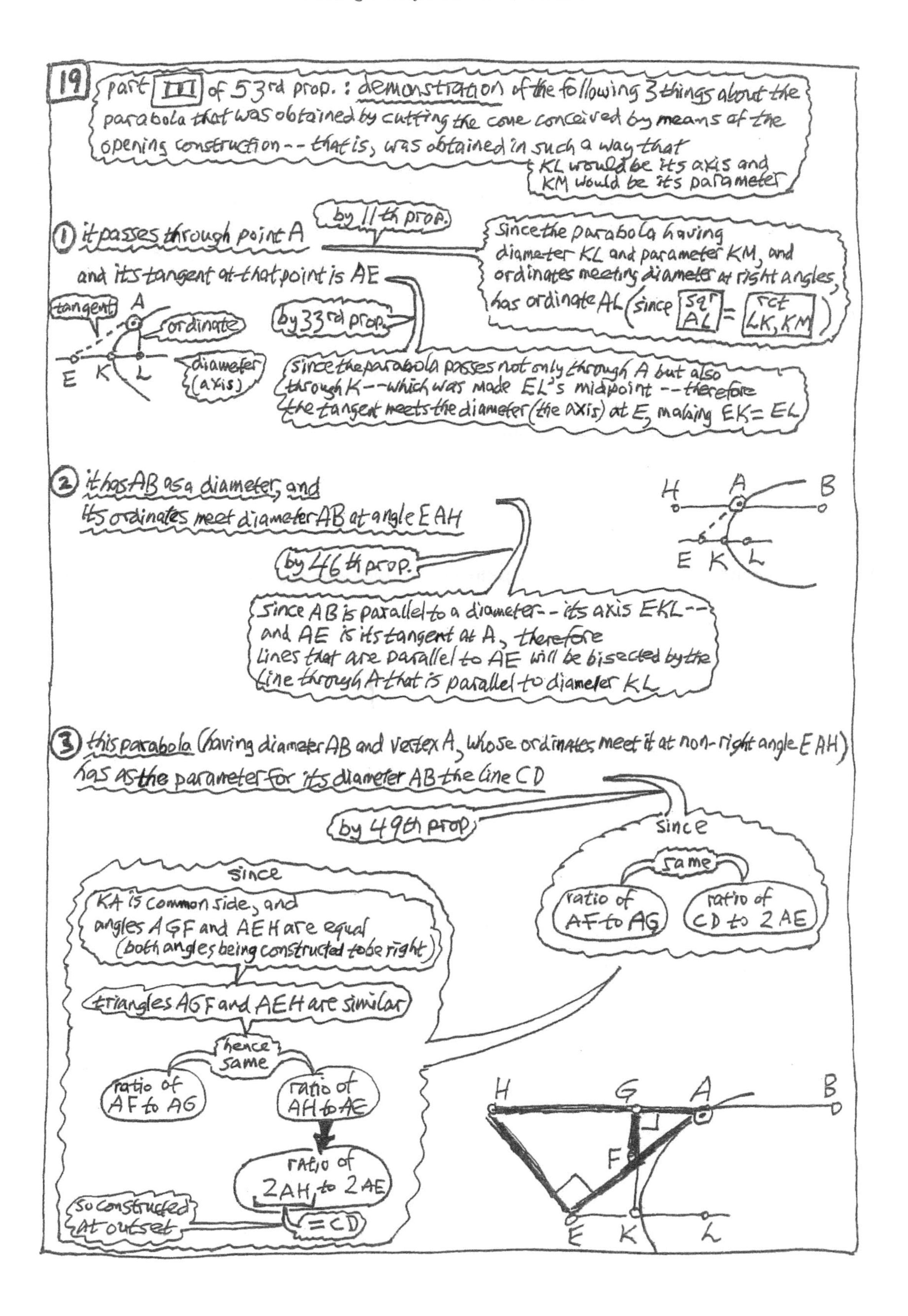
19
part III of 53rd prop. : demonstration of the following 3 things about the parabola that was obtained by cutting the cone conceived by means of the opening construction -- that is, was obtained in such a way that KL would be its axis and KM would be its parameter
1 it passes through point A
by 11th prop.
since the parabola having diameter KL and parameter KM, and ordinates meeting diameter at right angles, has ordinate AL (since sqr AL = rct LK, KM)
and its tangent at that point is AE
by 33rd prop.
tangent
A
ordinate
E K L
diameter (axis)
since the parabola passes not only through A but also through K -- which was made EL's midpoint -- therefore the tangent meets the diameter (the axis) at E, making EK = EL
2 it has AB as a diameter, and its ordinates meet diameter AB at angle EAH
by 46th prop.
since AB is parallel to a diameter -- its axis EKL -- and AE is its tangent at A, therefore lines that are parallel to AE will be bisected by the line through A that is parallel to diameter KL
H A B
E K L
3 this parabola (having diameter AB and vertex A, whose ordinates meet it at non-right angle EAH) has as the parameter for its diameter AB the line CD
by 49th prop.
since
same
ratio of AF to AG
ratio of CD to 2 AE
since
KA is common side, and angles AGF and AEH are equal (both angles being constructed to be right)
triangles AGF and AEH are similar
hence same
ratio of AF to AG
ratio of AH to AE
ratio of 2AH to 2AE
so constructed at outset
=CD
H G A B
F
E K L

finding parabolas to solve the problem of doubling a cube

The solution of the problem that is presented in the 52nd proposition of Apollonius (the problem of finding a certain specified curve as a cut in a cone) makes it possible, by the way, to solve one of the geometrical problems most celebrated among the ancients—namely, the problem of doubling a cube. To solve this problem means to find the line that will be the side of a cube with twice the volume of a cube whose side has been given. Now, in order to comprehend Book One of the *Conics* of Apollonius on its own terms, we do not need to know how its 52nd proposition can be used to solve the problem of doubling a cube; but by considering this now, we shall be better prepared to consider what comes later.

Consider first how plane figures of the same shape are compared in size. If one square, for example, is 25 times as large as another, how do their sides compare? The larger square must have a side 5 times as large as the smaller square's side. That is, the squares are in the *duplicate* ratio of their sides. Now, then, what about similar *solids*? From Euclid's *Elements* (XI.33), we learn that similar parallelepiped solids are to one another in the triplicate ratio of their sides. If, for example, each side of a box (its length, its breadth, and its depth) is 4 times as large as the corresponding side of another box, then the first box will be 64 times as large as the other. The ratio of 64 to 1 is the triplicate of the ratio of 4 to 1. How that ratio of 64 to 1 is the end result when you compound the ratio of 4 to 1 with itself and then compound the result with the ratio of 4 to 1 again, is depicted in figure 20.

So, if a cube is the double of another cube, then the ratio of the cubes (the ratio of 2 to 1) is the triplicate of the ratio of the double-cube's side, which we shall call line (d), to the half-cube's side, which we shall call line (h). (Fig. 21.)

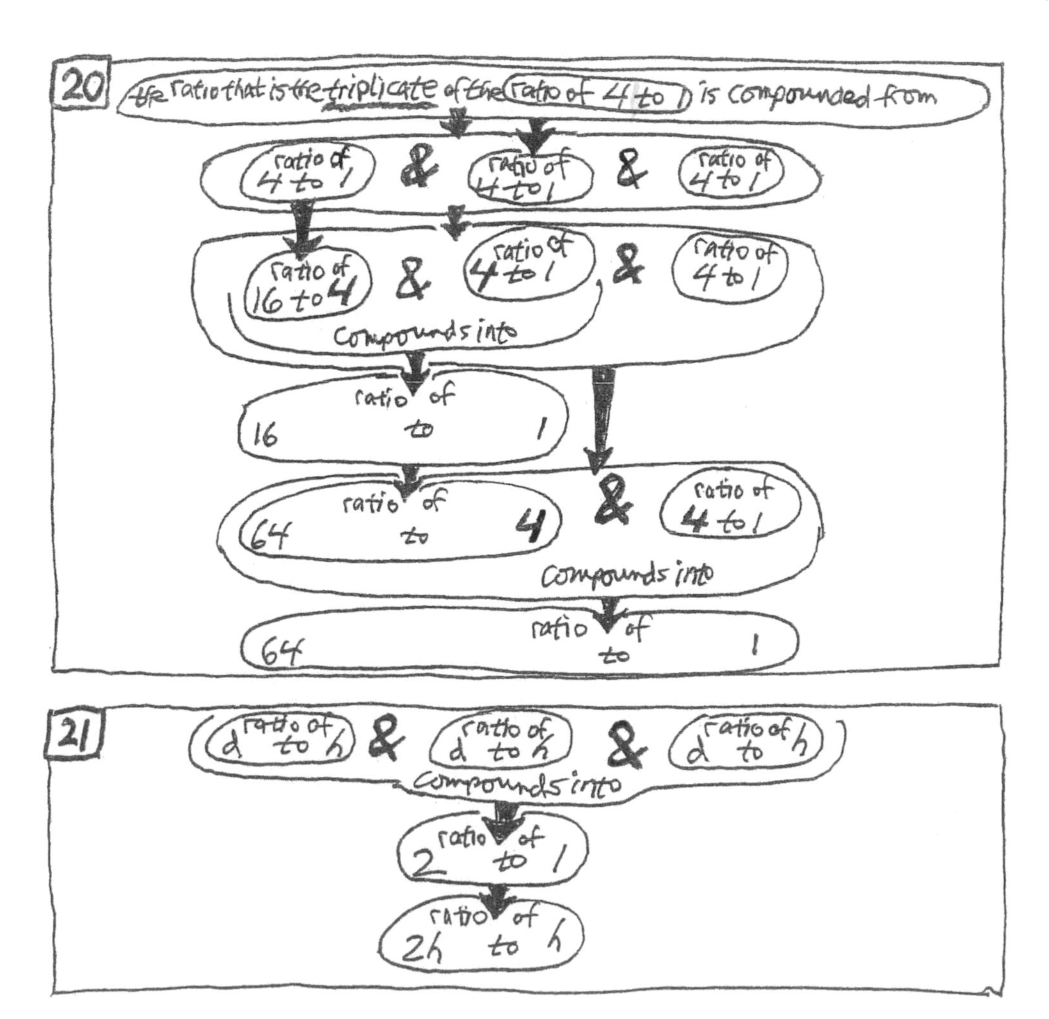

Now, we wish to find the line (d) that has to the given line (h) the ratio that is the subtriplicate of the ratio that line (h)'s double has to line (h) itself. In other words, the ratio of 2(h) to (h) is compounded from three ratios—the first of them is the ratio of (d) to (h), and the other two of them are the same as the ratio of (d) to (h). That is to say, we can obtain line (d) if we can find the line—call it (m)—such that this line (m) and the line (d) are the two mean proportionals between line 2(h) and line (h). (Fig. 22.)

Since the line that we seek—namely, the side (d) of the double-cube—will be the second mean proportional between the lines 2(h) and (h), we need to be able to find two mean proportionals between this pair of lines. But how can that be done?

To find a *single* mean proportional between a pair of lines, we can use Euclid, as in figure 23.

To find *two* mean proportionals between a pair of lines, however, we cannot use Euclid. We can, however, use Apollonius. The 52nd proposition of his *Conics* supplies what we need—as follows.

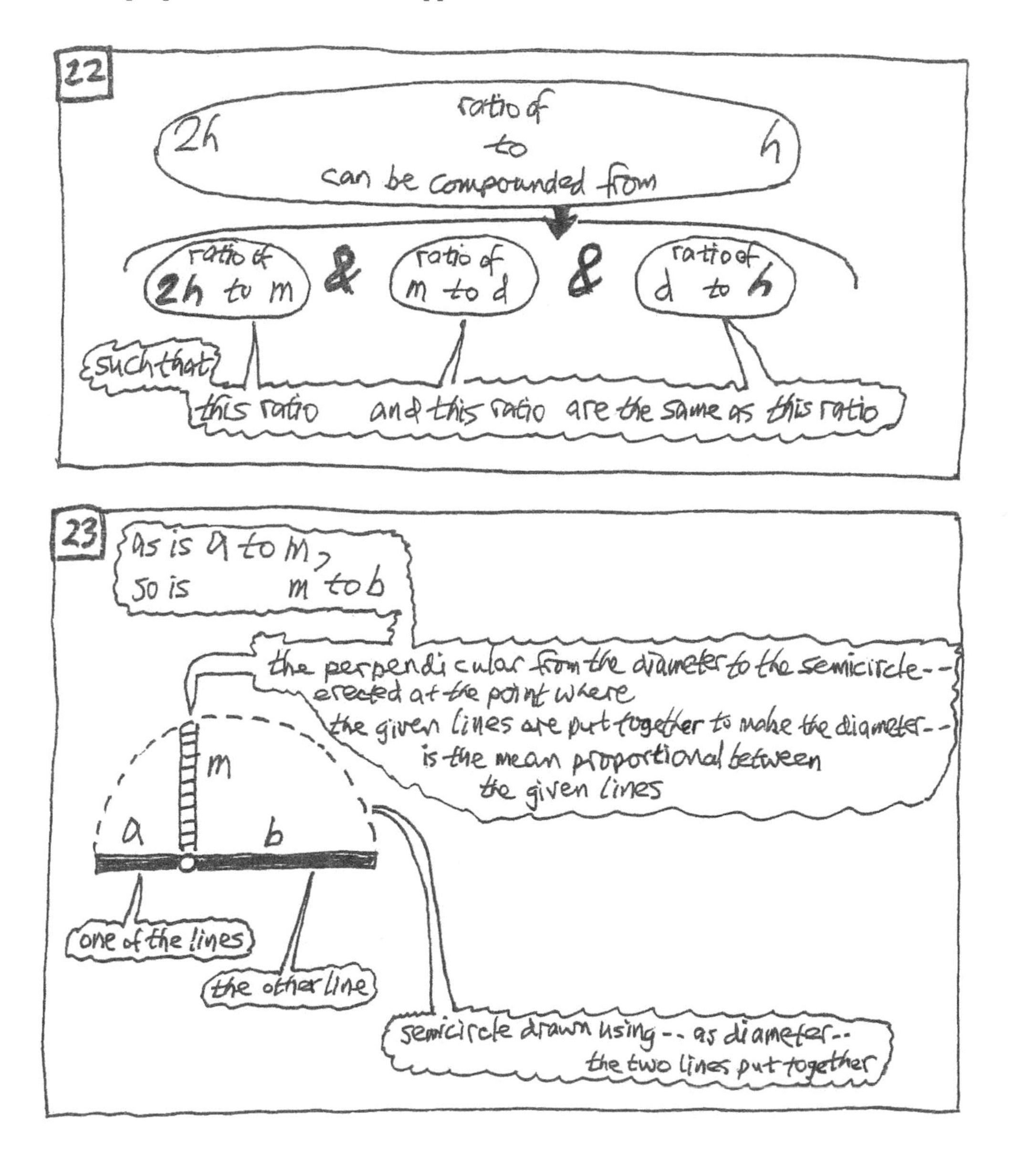

Let two straight lines that are set at right angles to each other be given as axes for two parabolas, with the intersection of the axes being their common vertex; and let line (h) be given as the parameter for the first parabola, and let line 2(h) be given as the parameter for the second one. Now get those parabolas, by using the 52nd proposition of Apollonius. (Fig. 24.)

And now, from the point where the two parabolas intersect each other, drop perpendiculars to the two axes, and call those lines (p) and (q). (Fig. 25.)

Each parabola's ordinate from the intersection-point of the two parabolas will be equal to the other parabola's abscissa for its own ordinate from that intersection-point. The consequence of this fact is depicted in figure 26.

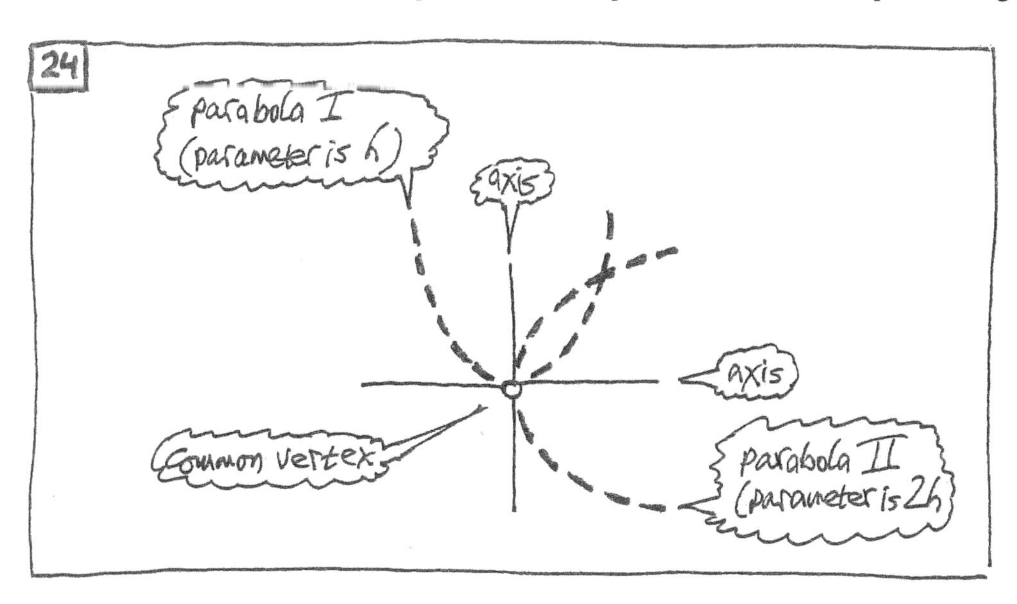

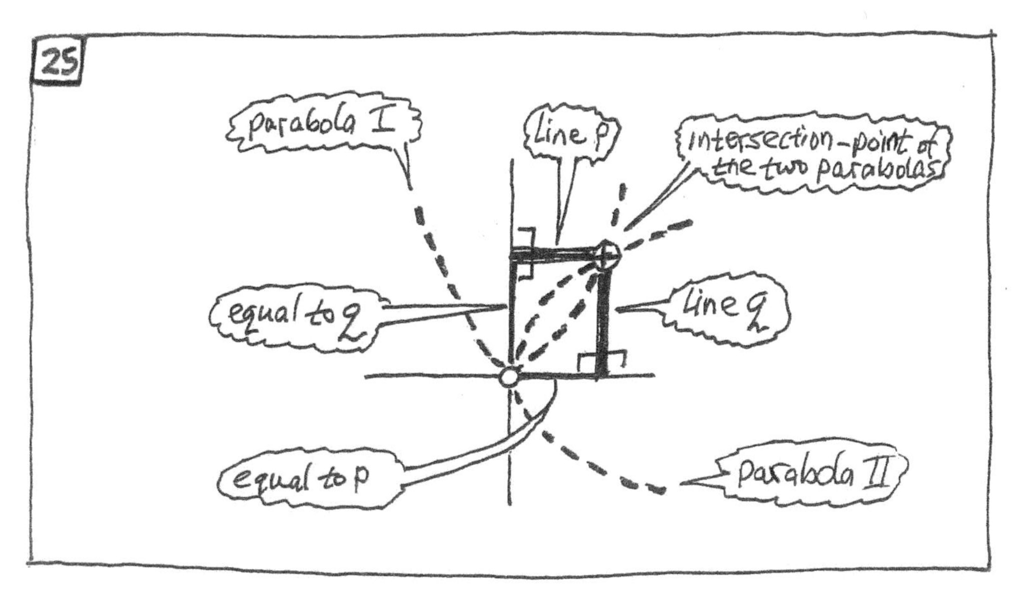

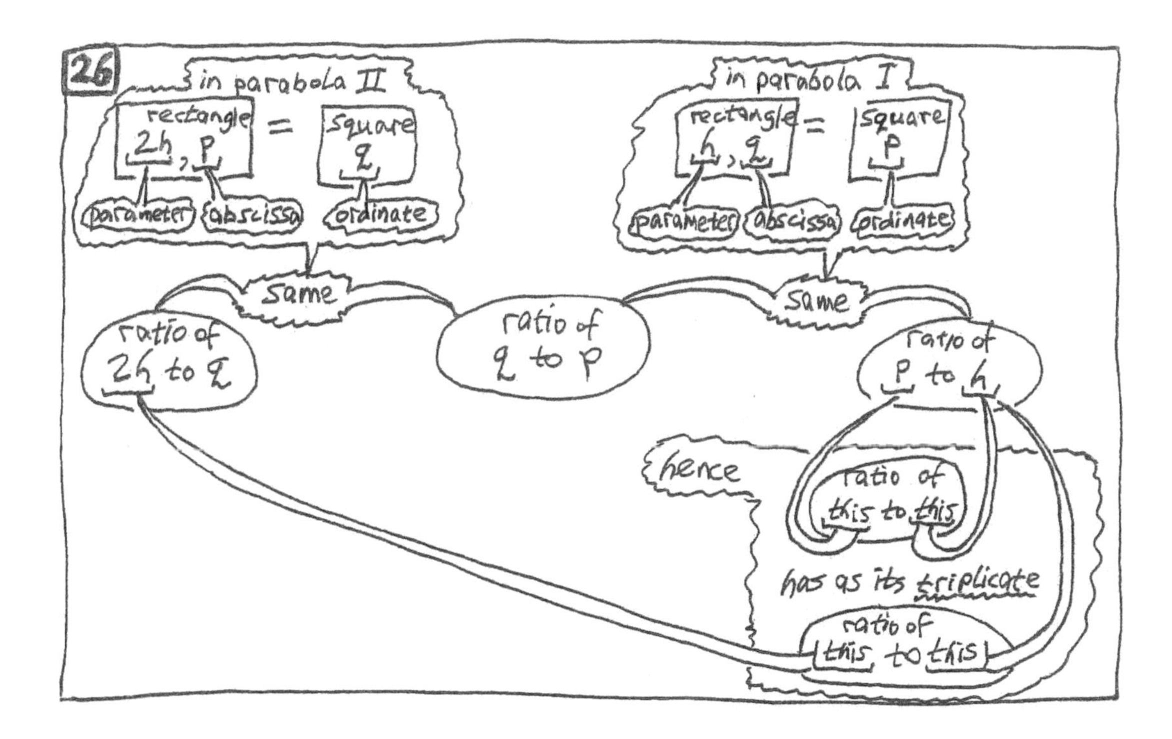

Line (p) and line (q) are two mean proportionals between a pair of lines, one of which is line (h)'s double, and the other of which is line (h) itself. Hence, the ratio of line 2(h) to line (h)—namely, the ratio of 2 to 1—is the triplicate of the ratio of line (p) to line (h). The line (p) will therefore be the side (d) of the cube that is the double of the cube whose side is line (h).

So, our solution to the problem of doubling a cube is this: using the 52nd proposition of Apollonius's *Conics*, we obtain two parabolas whose axes are at right angles to each other, whose vertices coincide, and whose parameters for those axes are (in the first parabola) the half-cube's side, and (in the second parabola) the double of the half-cube's side; and then we shall have the double cube's side if, from the point where the two parabolas intersect, we drop a perpendicular to the first parabola's axis. (Fig. 27.)

The problem of finding any number of mean proportionals between any two given magnitudes will turn out to be more—much more—than a mildly amusing mathematical curiosity. Descartes found it fundamental to his transformation of classical geometry some two millennia after classical geometry was brought to a peak by Apollonius—and it is with Descartes' transformational project that we shall be occupied in the final part of this guidebook's second volume.

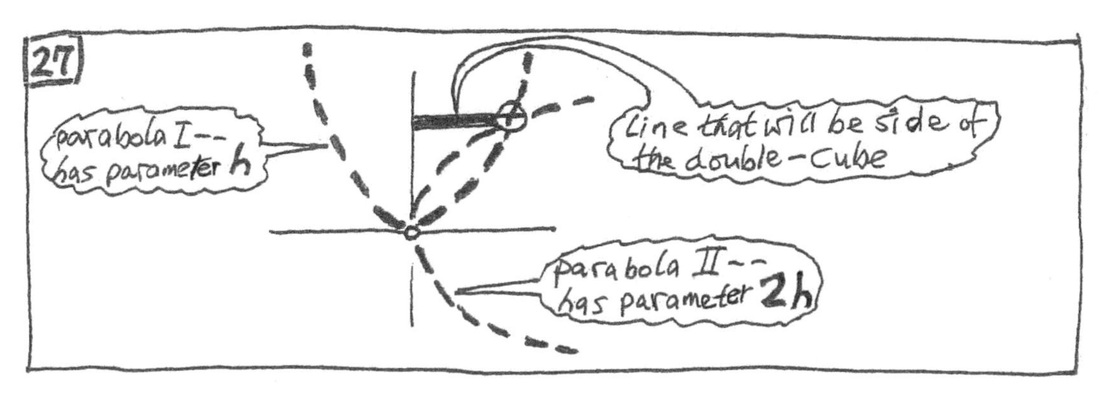

CHAPTER 3. FINDING CONES FOR THE CURVES: HYPERBOLA

the hyperbola problem solved *(54th-55th propositions)*

As in Apollonius's previous problem of finding a conic section that is parabolic, so now in his problem of finding one that is hyperbolic, what is given is this: the position of one of its diameters (including the position of the diameter's vertex), and the size of that diameter's upright side (its parameter) for ordinates that meet that diameter at some given angle. What is given for the hyperbola includes as well the size of the oblique (transverse) side for that diameter.

The sought-for hyperbolic curve, like the previously sought-for parabolic curve, is found by Apollonius as a section of a *right* cone—first, in the case when the ordinates are to meet the diameter at a right angle, and then, in the case when the ordinates are to meet the diameter at some other angle. In the first case (54th proposition), the diameter is an axis; in the second case, it is not. In the second case (55th proposition), the line that is given to be a diameter which is *not* an axis is used to find a line to be a diameter which *is* an axis. Finding this other line (the line that will be not only a diameter but also an axis) transforms the situation of the second case into an instance of the first case; then the only question is whether the solution to the problem as transformed is also a solution to the problem as posed before the transformation.

Apollonius's solution to the problem for the hyperbola is presented in the following boxes. In the event that the figures in the subsequent explication of these propositions do not provide as much help as you would like in drawing illustrative diagrams for the reading of the propositions, see the Appendix at the end of this volume for such diagrams drawn complete.

#54 Two straight [lines] being given, [lines, that is, which are] bounded [and are] at right angles to one another, one of them being extended on the same side as the right angle, [extended, that is, away from the point where the two lines come together]—to find, on the line that has been extended forth, the cone's section called an hyperbola, in the same plane with the straight [lines], so that the [line] extended forth may be a diameter of the section, and a vertex is what the point at the angle may be, and [in which] any [line] that may be drawn down from the section onto the diameter, making an angle equal to the given angle, will have the power [to give rise to a square equal to] a rectangle which is laid along (*parakeimenon para*) ["is applied to"] the other straight [line,] while having as breadth the [line] taken off, by the [line] drawn down, on the side toward the vertex, which [rectangle] overshoots (*huper-ballon*) by a figure (*eidei*) that is similar to, and that is laid similarly to, the [rectangle contained] by the initial straight [lines] (*tôn ex archês eutheiôn*).
—Let there be, as the two given bounded straight [lines] at right angles to one another, the [lines] AB, BC; and also, let there be extended the [line] AB, to the [point] D. Required, then, to find in the plane through the [lines] ABC an hyperbola whose diameter is what the [line] ABD will be, and vertex is what the [point] B will be, and up-right [side] is what the [line] BC will be, and [in which] the [lines] drawn down from the section onto the [line] BD at the given angle will have the power [to give rise to squares equal to] the [rectangles] which are laid along the line BC, while having as breadths the [lines] taken off by them [taken off, that is, by such lines drawn down], on the side toward the [point] B, and which overshoot by a figure that is similar to, and that is laid similarly to, the [rectangle] by the [lines] ABC.

—Let the given angle first be [an angle that is] right; and let there be set up, from the [line] AB, a plane upright (*orthon*) to the plane supposed; and also, in it, about the [line] AB, let there be carved out (*gegraphthô*) a circle, namely the [circle] AEBF, so that the segment (*tmêma*) of the circle's diameter inside the sector (*tmêmati*) AEB has, to the segment of the diameter inside the sector AFB, a ratio not greater than that which the [line] AB has to BC; and also, let there be cut in two [halves] the [line] AEB at the [point] E; and also, let there be drawn from the [point] E onto the line AB a perpendicular, namely the [line] EK, and also let it have been extended to the [point] L; therefore a diameter is what the [line] EL is. If, then, (on the one hand) as is the [line] AB to BC so is the [line] EK to KL, the [point] L is what we make use of; but if (on the other hand) [that proportion does] not [hold], let there be generated [a fourth proportional line KM such that] as the [line] AB is to BC so the [line] EK is to the [line] KM which is less than the [line] KL. And also, through the [point] M let there be drawn parallel to the [line] AB the [line] MF; and also, let there be joined the [lines] AF, EF, FB; and also, through the [point] B [let there be drawn] parallel to the [line] FE the [line] BX. Since, then, equal is the angle by AFE to the angle by EFB—but (on the one hand) the angle by AFE to the [angle] by AXB is equal, and (on the other hand) the angle by EFB to the [angle] by XBF is equal—therefore also the [angle] by XBF to the [angle] by FXB is equal; therefore equal is also the [line] FB to the [line] FX.
—Let there be envisioned (*noeisthô*) a cone whose vertex is what the [point] F is, and base is what the circle about the diameter BX is, [the circle, that is to say,] which is upright (*orthos*) to the triangle BFX. Then the cone will be a right (*orthos*) [cone]—for equal is the [line] FB to the [line] FX.
—Let there have been extended, then, the [lines] BF, FX, MF. And also, let the cone be cut by a plane parallel to the circle BX; then will the section be a circle—let it be the [circle] GPR—and so a diameter of the circle is what the line GH will be. And a common section of the circle GH and of the plane supposed—let the [line] PDR be that. Then the [line] PDR will, to both of the [lines] GH, DB, be upright (*orthos*)—for each circle of the [circles] XB, HG is upright to the triangle FGH, but also the plane supposed is upright to the [triangle] FGH; and therefore also their common section, namely the [line] PDR, is upright to the [triangle] FGH; and therefore does it also with all the straight [lines] that touch it and that are in the same plane make right angles.
—And—since a cone, whose base is what the circle GH is, and vertex is what the [point] F is, has been cut by a plane upright to the triangle FGH, and [the base plane] has been cut also by another plane, namely by the [plane] supposed, along the straight [line] PDR at right angles to the [line] GDH, but the common section of the plane supposed and of the [triangle] GFH, that is, the [line] DB, extended toward the [point] B, falls together with the [line] GF at the [point] A—therefore an hyperbola is what the section will be, on account of the things shown before—[that is,] the [section] PBR will be [an hyperbola] whose vertex is what the point B is and [in which] the lines drawn down onto the [line] BD ordinatewise will be drawn down at a right angle, for parallel are they to the [line] PDR. And—since as is the [line] AB to BC so is the [line] EK to KM, but as is the [line] EK to KM so is the [line] EN to NF, that is, so is the [rectangle] by ENF to the [square] from NF—therefore as is the [line] AB to BC so is the [rectangle] by ENF to the [square] from NF. But equal is the [rectangle] by ENF to the [rectangle] by ANB—therefore as is the [line] AB to CB so is the [rectangle] by ANB to the [square] from NF. And the [rectangle] by ANB has, to the [square] from NF, the ratio compounded from that of the [line] AN to NF and also that of the [line] BN to NF. But (on the one hand) as is the [line] AN to NF so is the [line] AD to DG, and the [line] FO to OG; and (on the other hand) as is the [line] BN to NF so is the [line] FO to OH. Therefore the [line] AB has to BC the ratio compounded from that which the [line] FO has to OG and also the [line] FO has to OH, that is, [the ratio which] the [square] from FO has to the [rectangle] by GOH. Therefore as is the [line] AB to BC so is the [square] from FO to the [rectangle] by GOH. And parallel is the [line] FO to the [line] AD; therefore an oblique side is what the [line] AB is, and an upright [side] what the [line] BC is—for these things have been shown in the 12th theorem.

#55 Let not, then, the given angle be right; and also, let the two given straight [lines] be what the [lines] AB, AC are, and let the given angle be equal to the angle by BAH. Required, then, to carve out (*grapsai*) an hyperbola whose diameter is what the [line] AB will be, and upright [side] is what the [line] AC will be, and [in which] the [lines] drawn down will be drawn down at the angle by HAB.
—Let the [line] AB be cut in [two] halves at the [point] D. And also, upon the [line] AD, let there be carved out (*gegraphthô*) the semicircle AFD. And also, let a certain [line] be drawn to the semicircle, a [line] parallel to the [line] AH, namely the [line] FG which makes the ratio of the [square] from FG, to the [rectangle] by DGA, the same as that of the [line] AC to AB. And also, let there be joined the [line] FHD, and also let it be extended to the [point] D. And also, let the mean proportional of the [lines] FDH be what the [line] DL is. And also, let there be put equal to the [line] DL the [line] DK, and equal to the [square] from AF let there be the [rectangle] by LFM. And also, let there be joined the [line] KM. And also, through the [point] L let there be drawn at right angles to the [line] KF the [line] LN, and also let it be extended to the [point] X. And—there being given two bounded straight [lines] at right angles to one another, namely the [lines] KL, LN—let there be carved out (*gegraphthô*) an hyperbola, whose oblique side is what the [line] KL will be, and upright [side] is what the [line] LN will be, and [in which] the [lines] drawn down onto the diameter from the section will be drawn down at a right angle, while having [the power to give rise to squares equal to rectangles which have] as breadths the [lines] taken away by them on the side toward the [point] L, [that is, rectangles which have as breadths the lines along the diameter from the point L to the points where the diameter is met by the ordinatewise lines,] [which rectangles] overshoot (*huperballonta*) by a figure (*eidei*) similar to the [rectangle] by KLN; and the section will go through the [point] A, for equal is the [square] from AF to the [rectangle] by LFM. And also, there will touch it, [that is, will touch the section,] the [line] AH, for the [rectangle] by FDH is equal to the [square] from DL; and so the [line] AB is a diameter of the section. And also, since as is the [line] CA to the double of the [line] AD, that is, to the [line] AB, so is the [square] from FG to the [rectangle] by DGA—but (on the one hand) the [line] CA has to the double of the [line] AD the ratio compounded from that which the [line] CA has to the double of the [line] AH and also from that which the double of [the] line AH has to the double of the [line] DA, that is, that which the [line] AH has to AD, that is, that which the [line] FG has to GD, therefore the [line] CA has to AB the ratio compounded from that of the [line] CA to the double of the [line] AH and also from that of the line FG to GD, and also (on the other hand) the [square] from FG has to the [rectangle] by DGA the ratio compounded from that which the [line] FG has to GD and also from that which the [line] FG has to GA—therefore the ratio compounded from that of the [line] CA to the double of the [line] AH and also from that of the [line] FG to GD is the same as the [ratio] compounded from that of the [line] FG to GA and also from that of the [line] FG to GD. Let a common ratio be taken away, namely the ratio of the [line] FG to GD; therefore as is the [line] CA to the double of the [line] AH so is the [line] FG to GA; but as is the [line] FG to GA so is the [line] OA to AX; therefore as is the [line] CA to the double of the [line] AH so is the [line] OA to AX. But whenever this be so, then the [line] AC is a [line] along which [are laid overshooting rectangles equal to those squares that arise from the lines which, drawn down from section to diameter at the given angle,] have the power—[or, in other words, line AC does what is done by an upright side]—for this has been shown in the 50th theorem.

analysis of the hyperbola problem–first case (54th proposition)

The core of the solution to the problem now, when the diameter is to be an axis for an hyperbola, is what it was before, when the diameter was to be an axis for a parabola. Now as before, we obtain (in the plane of the page) a triangle that will be the tip of the axial triangle of a certain right cone—the right cone that will yield the specified curve when it is cut by a plane (a plane perpendicular to the page) that passes through the line that is supposed to be the curve's axis.

The tip of the axial triangle must be isosceles (so that the cone will be right, and the diameter will be an axis). Moreover, the diameter must meet one of that triangle's sides in such a way as to diverge from the other side (so that the curve will be an hyperbola), and it must diverge at the appropriate inclination (so that the oblique [transverse] side and the upright side of the hyperbola will each be of the appropriate size).

Once we have the tip of the axial triangle, we can obtain the cone and the conic section (fig. 1).

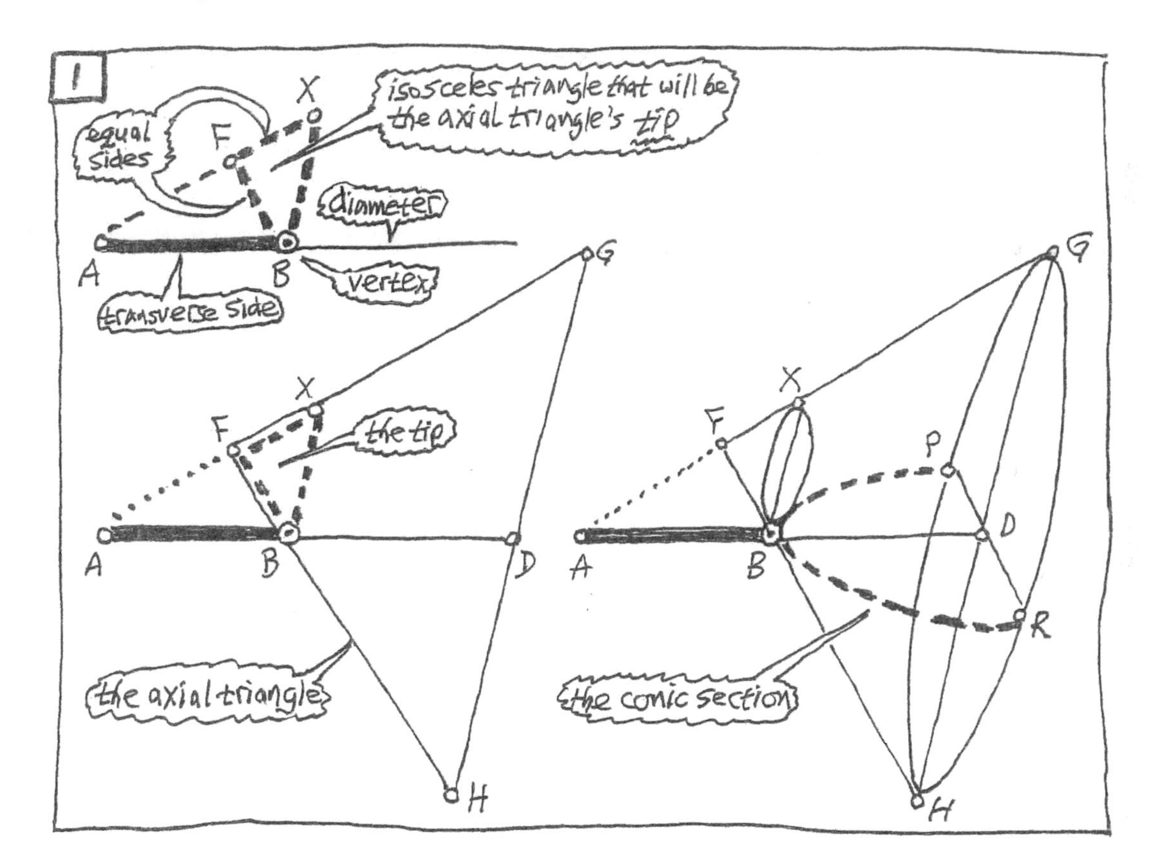

The question, then, is this: How do we obtain the axial triangle's tip? Well, suppose that we had it (call it BFX)—and suppose that, through its apex (F), we draw a straight line parallel to the tip's base (XB); and that we give a name to that straight line's intersection with AB—call it point N. Then (because of how the upright side is defined for the hyperbola in the 12th proposition) the given line BC could only be the upright side of oblique (transverse) side AB if the ratio of rectangle [AN, BN] to square [NF] were the same as the ratio of oblique (transverse) side AB to that line BC. The reason why, is shown in figure 2.

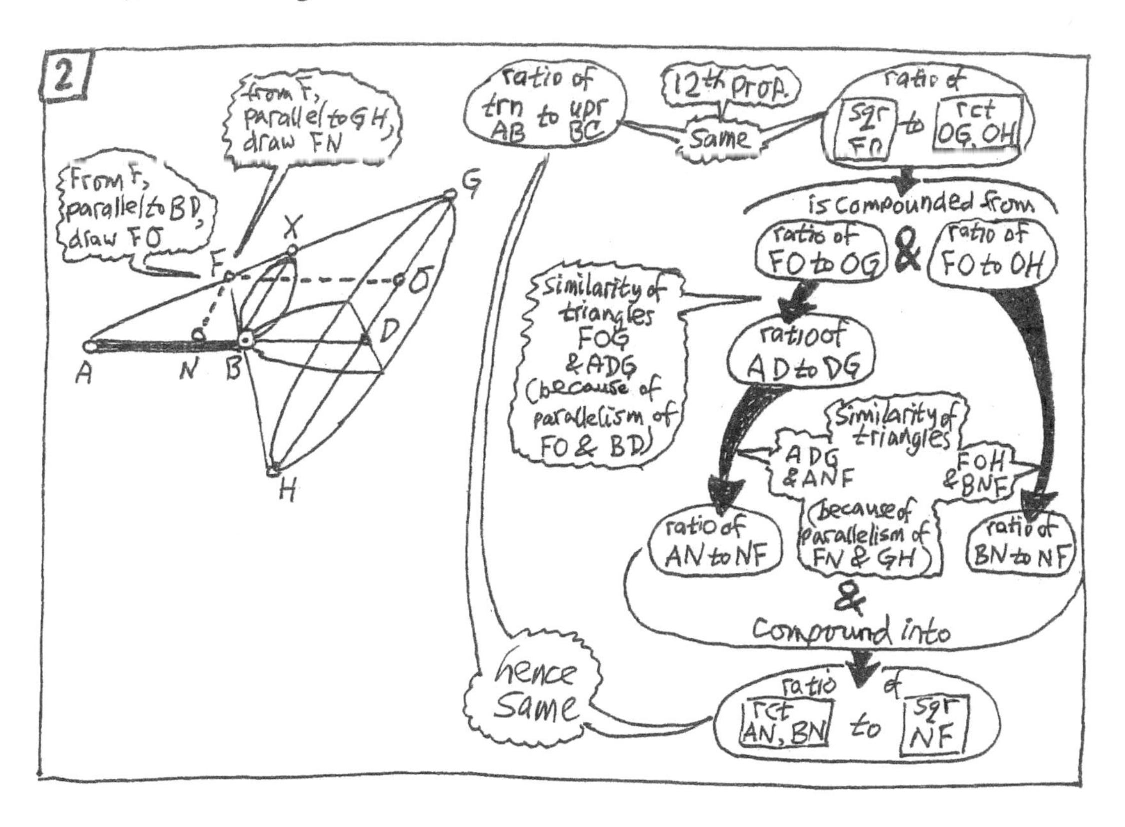

So, the problem reduces itself to this: given AB and BC, find two points—a point N that is on AB, and a point F that is not on AB—such that the following proportion holds:

as AB is to BC, so also rectangle [AN, BN] is to square [NF].

Once we have obtained points N and F, it is easy to obtain the requisite tip of the axial triangle. We just take the following steps: (1) join A to F and prolong the line; then (2) join B to F; then (3) join N to F, and (4) parallel to NF, draw a straight line through B; and now give a name to this straight line's intersection with the prolongation of AF—call it point X. And thus the tip of the axial triangle will be BFX (fig. 3).

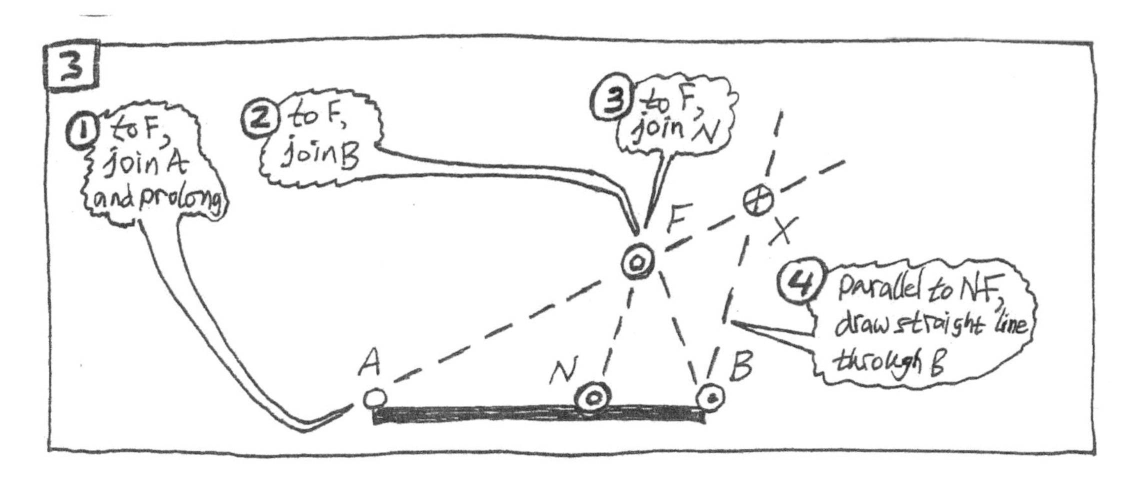

So how can we find point N (on line AB) and point F (not on line AB) such that as AB is to BC, so also rectangle [AN, BN] is to square [NF]? As follows.

First, draw a circle around AB as chord, so that (giving the name EKL to the circle's diameter that is perpendicular to AB) segment EK has to segment KL a ratio that is *not greater* than the ratio that AB has to the parameter BC (fig. 4).

How to draw that circle? As follows. Bisect AB, and at the midpoint (call it K) erect a perpendicular. On that perpendicular bisector, take some point (call it E) such that EK is not less than KA; and now, through this point E and the endpoints of AB, draw a circle. (In the 5th proposition of the Fourth Book of the *Elements*, Euclid shows how to draw a circle through the vertices of any triangle—which is to say, through any three points in a plane.) The circle will meet the perpendicular bisector not only at point E but also at another point (call it L).

Now the ratio of EK to KL may happen to be the same as the ratio of AB to BC. If it is indeed the same ratio, then L will be the sought-for point F, and K will be the sought-for point N. That is to say: L will be the apex F of the axial triangle's tip; and K will be the point N used to obtain the tip's base (BX), in the manner depicted in figure 5.

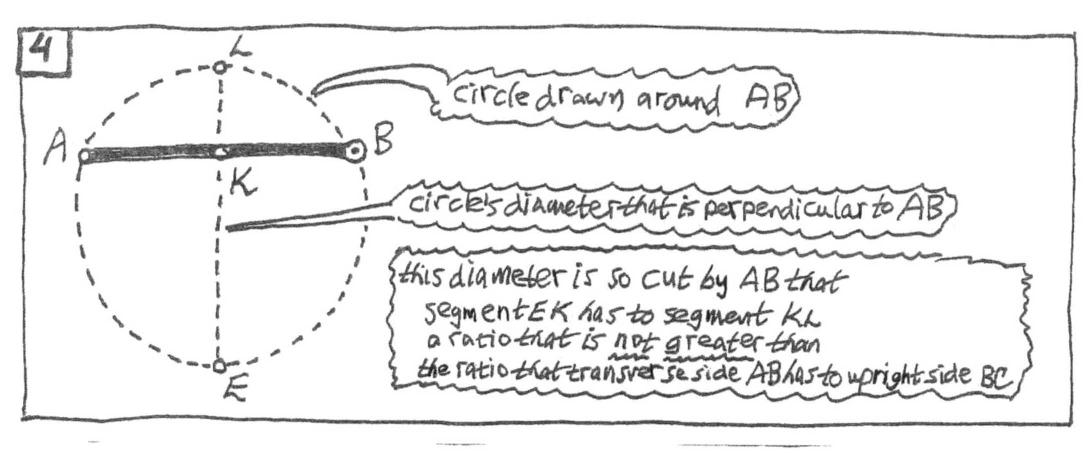

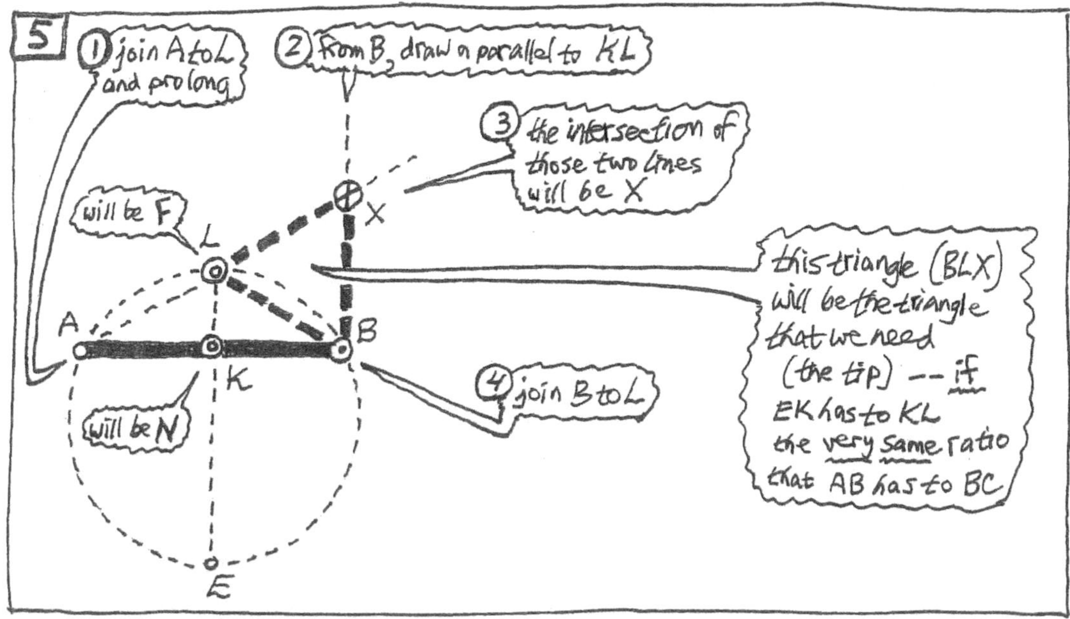

Now why will triangle BLX be the triangle that we need? Because, in the first place, triangle BLX is isosceles. (Why it is isosceles, is shown in figure 6.) And, in the second place, points K and L are located where they make the following proportion hold: as AB is to BC, so also is rectangle [AK, KB] to square [KL]. (Why it holds, is shown in figure 7.)

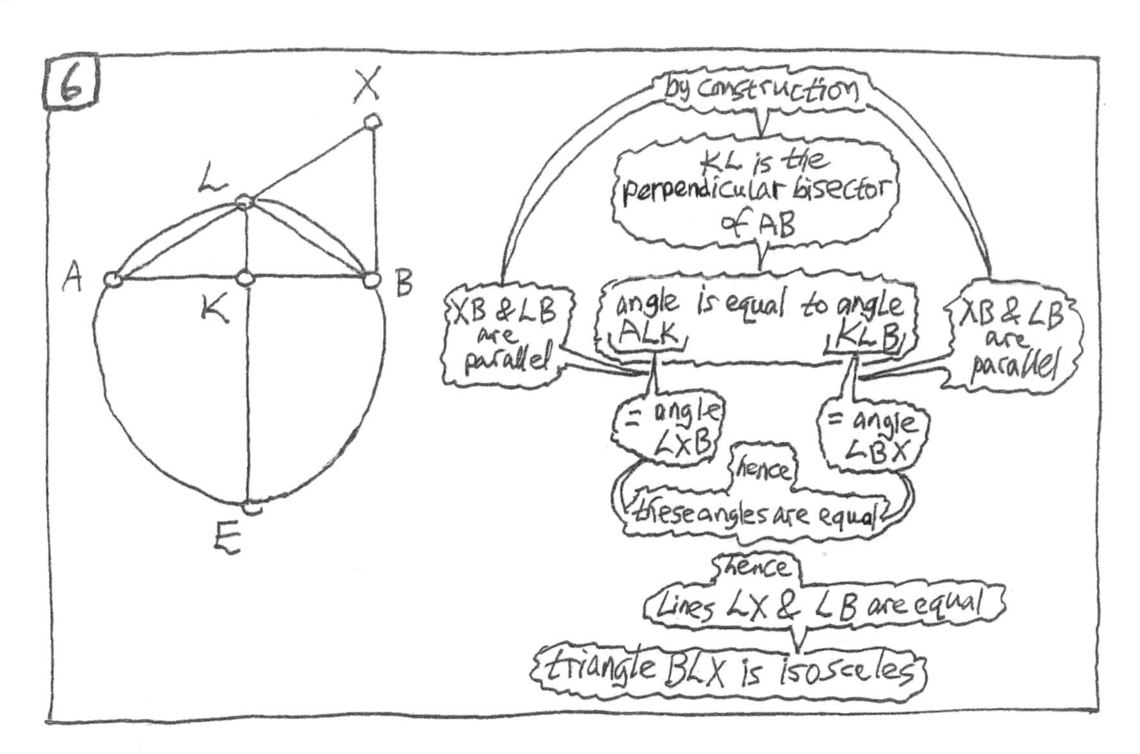

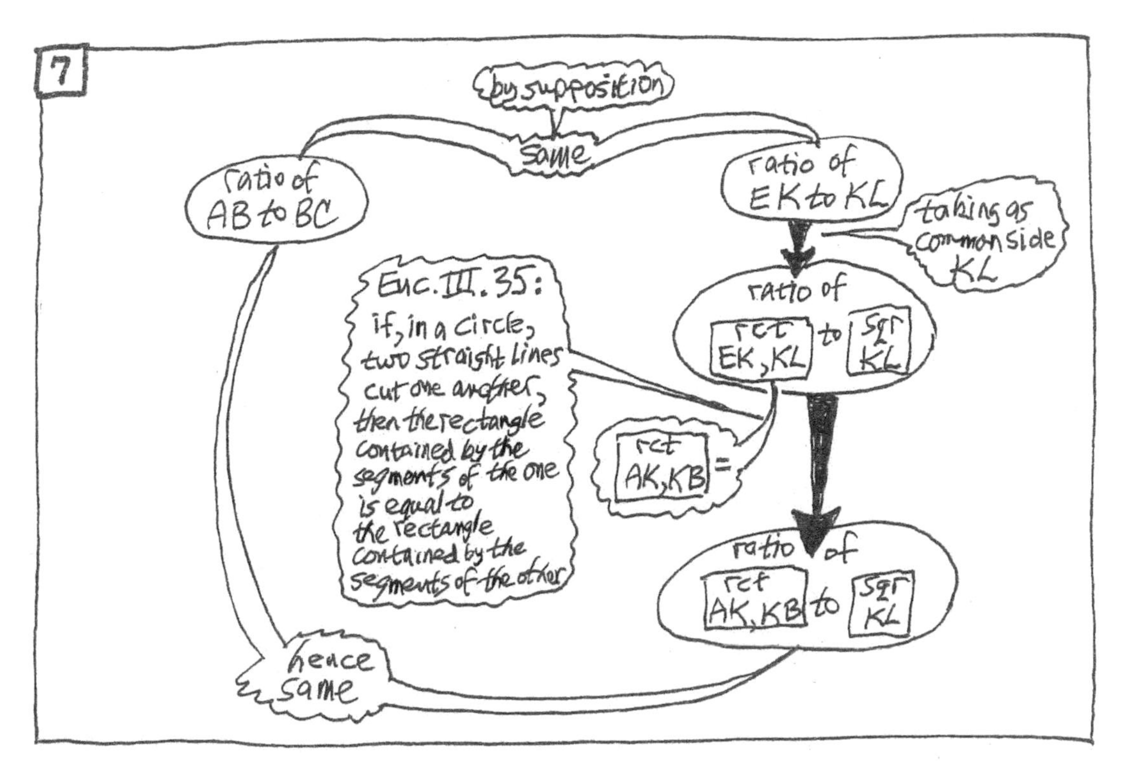

But perhaps the ratio of EK to KL is *not* the same as that of AB to BC. Then the ratio of EK to KL will be *less* than the ratio of AB to BC (since we guaranteed at the outset that it would not be greater).

Now if indeed the ratio of EK to KL is not the same as that of AB to BC, then use a point F that is not L, and use a point N that is not K—obtaining them, with triangle BFX, as follows.

On KL there will be a point—call it M—for which the ratio of EK to KM is the *same* as the ratio of AB to BC. (Why must there be such a point? For the following reason: L is so far from K that EK's ratio to KL is less than AB's ratio to BC, and a point taken close enough to K would make for a ratio as great as you please; so a point M taken at just the right distance from K would make the ratio of EK to KM great enough but not too great. See the 8th proposition of the Fifth Book of Euclid's *Elements*.)

Now do as is indicated in figure 8. BFX will then be the tip of the axial triangle.

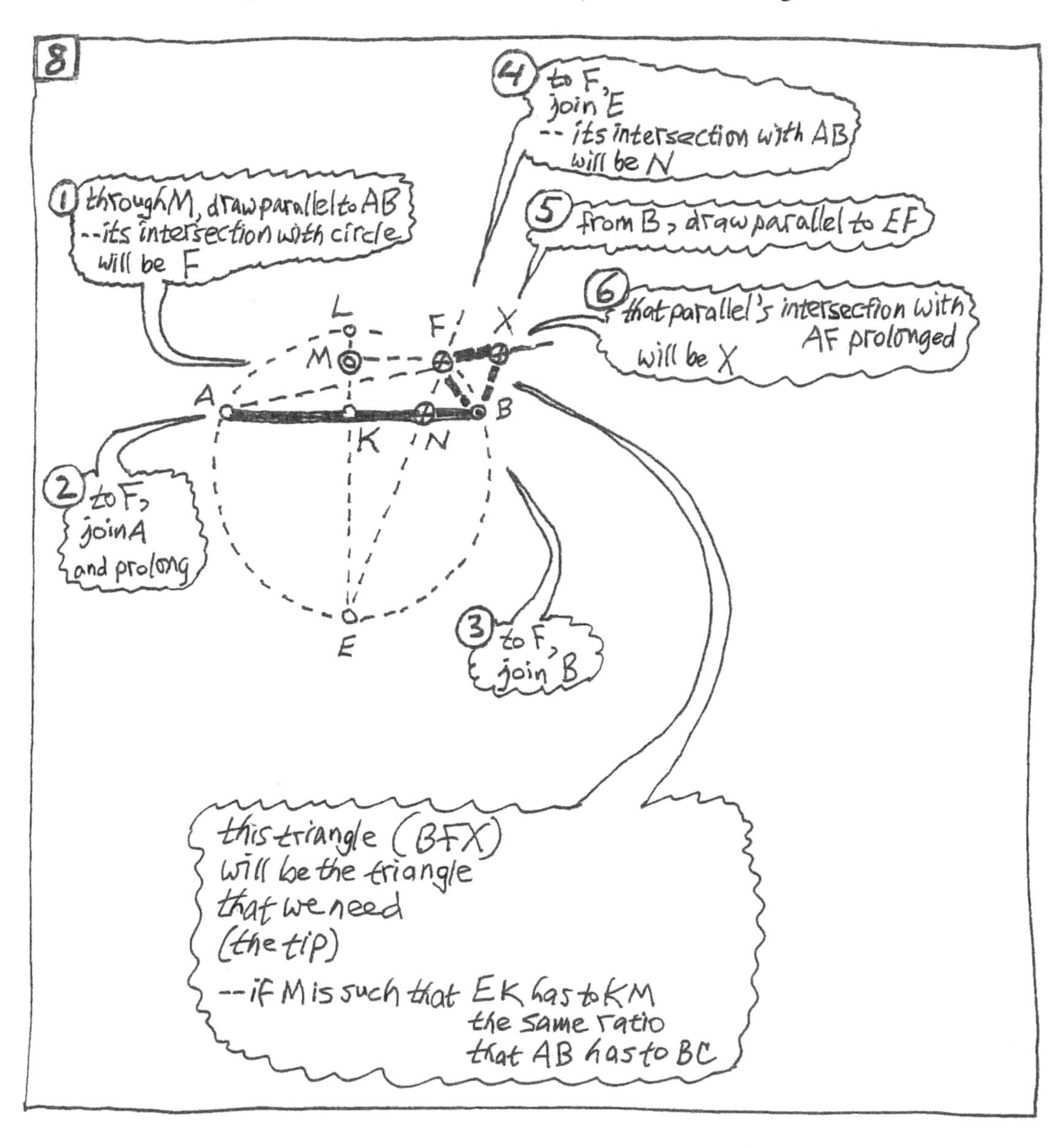

Now, why will triangle BFX be the triangle that we need? Because, in the first place, triangle BFX is isosceles. (Why it is isosceles, is shown in figure 9.) And, in the second place, points F and N are located where they make the following proportion hold: as AB is to BC, so also rectangle [AN, NB] is to square [NF]. (Why it holds, is shown in figure 10.)

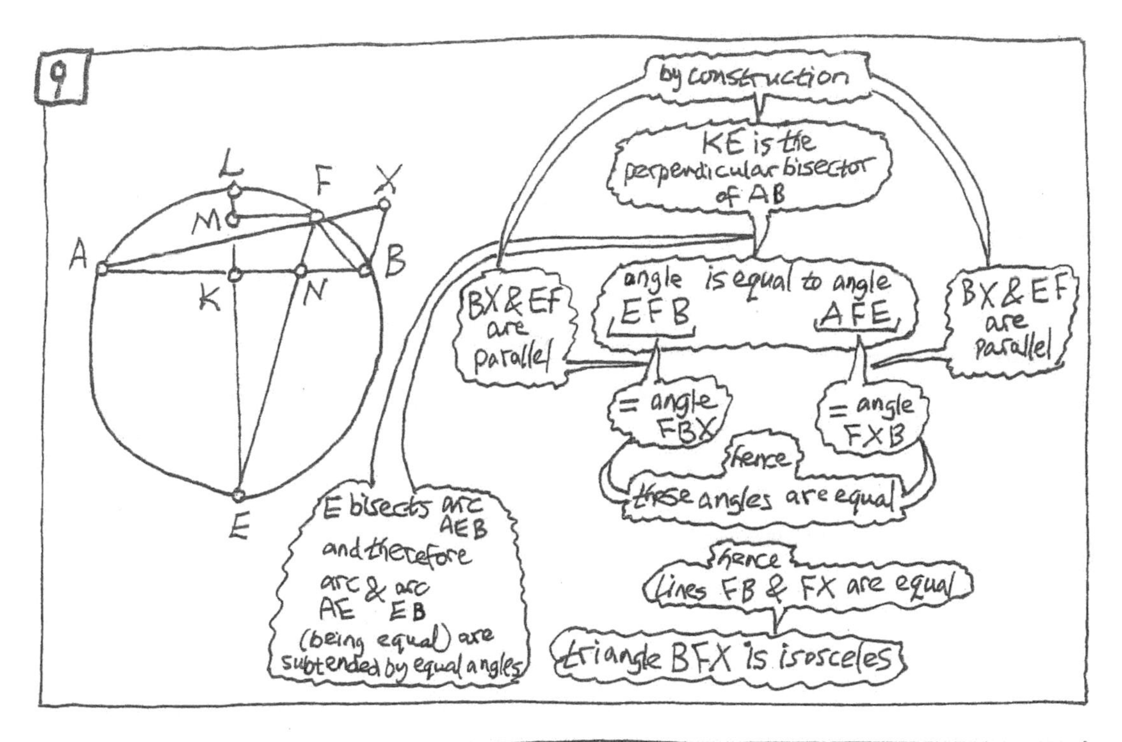

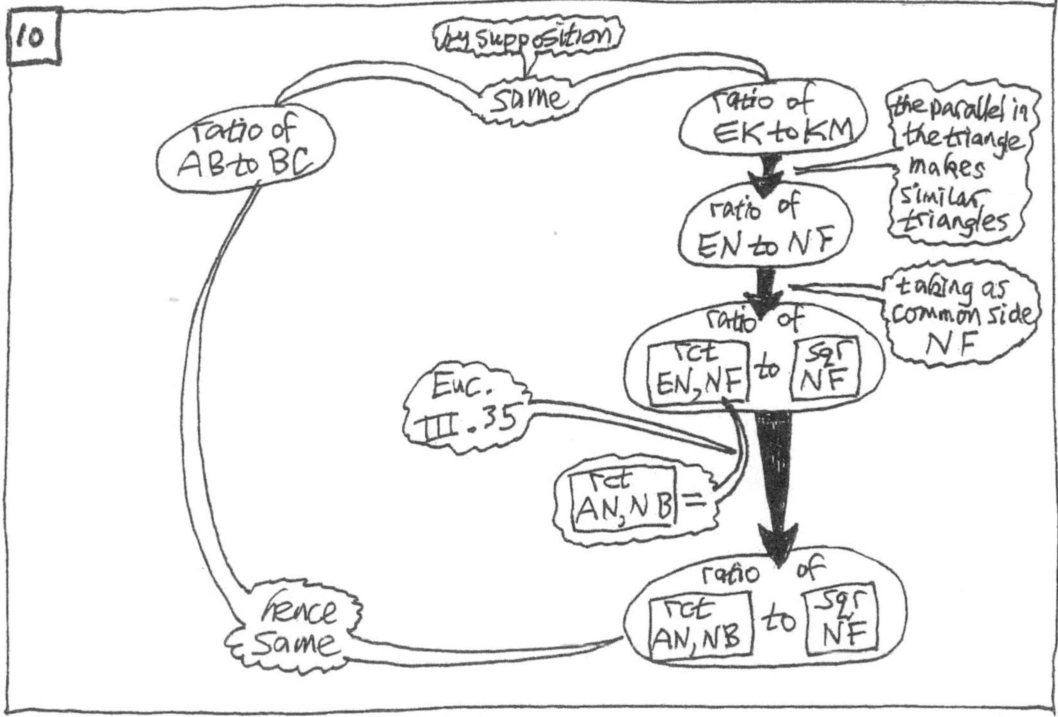

So if we use triangle BFX as the tip of the axial triangle from which we obtain the cone that we cut to obtain the curve, then:

> the tip's being isosceles (and hence the cone's being a right cone)
> guarantees that
> the original diameter will be an axis;
>
> the angle of the cut
> guarantees that
> the curve will be an hyperbola;
>
> the proportion used to construct the tip
> guarantees that
> AB and CD will be that hyperbola's oblique (transverse) and upright sides.

In figuring out what is needed to solve the problem of the hyperbola in the 54th proposition, our order here (as with the parabola before) has been the reverse of the order in the proposition by Apollonius.

> *Our inquiry* first asked:
> what if BC were the upright side of the hyperbolic section—
> what else would then have to be so?
> Then it concluded that a certain key ratio—
> namely, the ratio of rectangle [AN, BN] to square [NF]—
> would have to be the same as the ratio of AB to BC.
>
> In *Apollonius's proposition*, however, first a cone is obtained by
> constructing an axial triangle in such manner that
> the key ratio of rectangle [AN, BN] to square [NF]
> is the same as the ratio of AB to BC.
> Then from this it is shown that another ratio—
> namely, the ratio of square [FO] to rectangle [OG, OH]—
> is also the same as the ratio of AB to BC,
> thus making BC the upright side of the hyperbolic section.

Thus, *our inquiry* moves this way—

> *from*
> BC's being (in supposition) the upright side
> *to*
> the key ratio's being (in consequence) the same as the ratio of AB to BC;

but *Apollonius's proposition* moves this way—

> *from*
> the key ratio's being (by construction) the same as the ratio of AB to BC
> *to*
> BC's being (in consequence) the upright side.

analysis of the hyperbola problem–second case (55th proposition)

In treating the problem of finding the various curves as cuts in a cone, Apollonius does for the hyperbola what he did for the parabola: he presents two cases. His solution presents first the case of the right angle, and then the case of any other angle.

We are now mid-way through the solution to the problem of finding an hyperbolic conic section, when what have been given are: the position and the size of the oblique (transverse) side of one of its diameters (and the position of the diameter's vertex) and the size of the diameter's upright side for ordinates that meet that diameter at some given angle. The sought-for hyperbolic curve is found by Apollonius as a section of a right cone. Having already examined how he solves the problem in the case when the ordinates are to meet the diameter at a *right* angle, we now examine how he solves it in the case when the ordinates are to meet the diameter at some angle *other* than a right angle.

Since a diameter is an axis whose ordinates meet it at right angles, the case already examined (54th proposition) was one where the diameter is an axis, and the case to be examined now (55th proposition) is one where the diameter is not an axis. In this latter case, what Apollonius does is this:

> he uses the *given* line, which is to be a diameter that is *not* an axis, to obtain *another* line, which is to be a diameter that *is* an axis.

Before we move on to see how that is to be done for the hyperbola, take note of this: the lettering that we use in this non-axis case of the hyperbola differs from the lettering that we have used in the hyperbola's axis case. This is to make it easier to follow Apollonius's presentation. With the parabola, Apollonius did not re-letter the figure in the non-axis case; with the hyperbola, however, he does re-letter the figure in the non-axis case. While the oblique (transverse) side is still called AB in this non-axis case, its vertex is now called A. (In the axis case, by contrast, point B was the vertex.) Its upright side is therefore now called AC. (In the axis case, it was called BC.)

The order in which Apollonius proceeds in treating the problem in the non-axis case for the hyperbola is like the order of his treatment of the non-axis case for the parabola. The following is an overview of the main parts in his treatment of the hyperbola problem's non-axis case, here in the 55th proposition:

(I) Take the given line BA (which is to be the oblique [transverse] side of
the diameter of an hyperbola
whose ordinates meet it at a certain angle HAB that is not a right angle,
and whose upright side is AC);
use that line BA to obtain KL (a line that is to be the oblique [transverse] side of
another one of the hyperbola's diameters—one
whose ordinates meet it an angle that is a right angle).
That is to say: use what will be a diameter that is *not* an axis (namely, BA)
in order to obtain what will be a diameter that *is* an axis (namely, KL).
Also obtain LN (the line that will be
the upright side of KL, the new diameter that *is* an axis).

(II) Now that we have the position and the size of KL, with vertex L,
and the size of KN too,
use the previous proposition (the 54th) to obtain an hyperbola
whose oblique (transverse) side is KL, with vertex L,
and whose ordinates meet KL's prolongation at a right angle,
and whose upright side for KL with those ordinates is LN.

(III) Show that this hyperbola is the very hyperbola sought for—
the one whose oblique (transverse) side is BA, with vertex A,
and whose ordinates meet BA's prolongation at non-right angle HAB,
and whose upright side for BA with those ordinates is AC. (Fig. 11.)

With the hyperbola now, as with the parabola earlier, the middle part of Apollonius's presentation is merely an application of the previous proposition, for the 54th proposition has already solved the problem of finding an hyperbolic conic section that satisfies the specifications (the position and the size of the oblique [transverse] side for its *axis*, and the size of this axis's upright side). The difficult parts of the presentation are the beginning and the conclusion. The *beginning* uses the terms that were first specified for a diameter that is *not* an axis to obtain *new* specifications for a diameter that *is* one. The *conclusion* shows that the hyperbola that was obtained in the middle part, to satisfy the new specifications, will satisfy the first specifications too.

Again the hardest thing to understand, as we move through Apollonius's presentation, is a beginning in which lines are drawn for no apparent purpose. The purpose only becomes apparent after we have moved through the whole presentation and then look back.

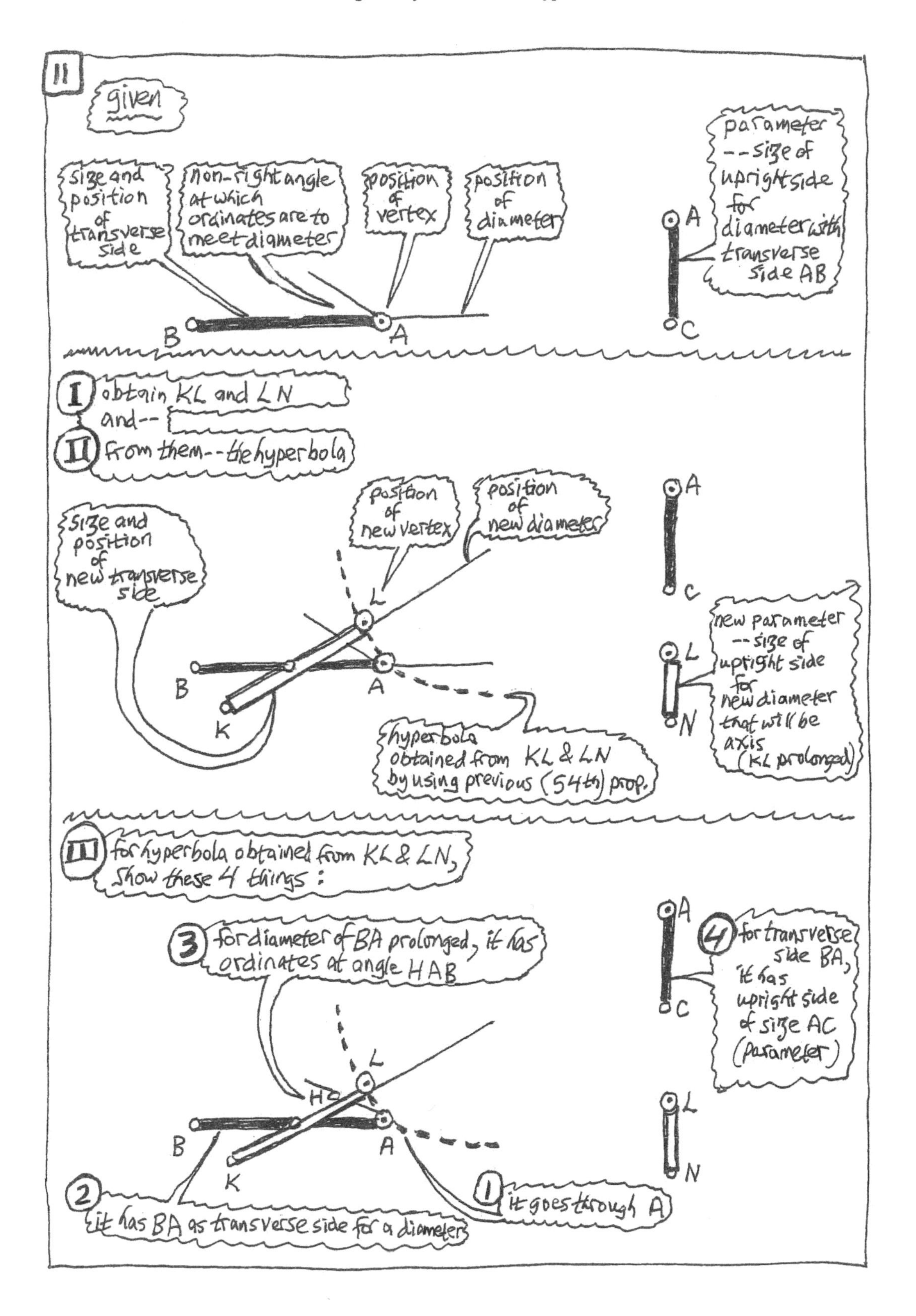
II
given
size and position of transverse side
non-right angle at which ordinates are to meet diameter
position of vertex
position of diameter
parameter --size of upright side for diameter with transverse side AB
B
A
A
C
I obtain KL and LN
and--
II from them--the hyperbola
size and position of new transverse side
position of new vertex
position of new diameter
A
C
L
B
A
K
L
N
new parameter --size of upright side for new diameter that will be axis (KL prolonged)
hyperbola obtained from KL & LN by using previous (54th) prop.
III for hyperbola obtained from KL & LN, show these 4 things:
3 for diameter of BA prolonged, it has ordinates at angle HAB
A
C
4 for transverse side BA, it has upright side of size AC (parameter)
L
H
B
A
K
L
N
2 it has BA as transverse side for a diameter
1 it goes through A

Here is a sketch of an analytic line of thought to help you understand the construction which constitutes the beginning of Apollonius's solution of the problem for the hyperbola in this case when the ordinates are to meet the diameter at some angle that is not a right angle.

Suppose that we *had* the oblique (transverse) side (call it KL) and vertex (L) of the *axis* of the sought-for hyperbola. Then, having the axis's oblique (transverse) side, we would be able to obtain the axis's upright side (call it LN)—and then we would be able to obtain the hyperbola by using the solution from the previous proposition, the 54th.

How could we obtain the axis's upright side if we had its oblique (transverse) side? As follows. Point A is specified to be on the sought-for hyperbola, and we are supposing that we have the oblique (transverse) side KL and its vertex (call it point L). From A, drop a perpendicular to the oblique (transverse) side's prolongation, which is the axis; give a name to the point where the perpendicular meets the axis—call it F. Then AF will be an ordinate, and LF will be its abscissa. By the 12th proposition, since the axis is a diameter of an hyperbola, the square on ordinate AF will be equal to a rectangle whose one side is the abscissa LF, and whose other side (call it FM) is a line of definite size obtainable from the proportion embodied in the equality of square and rectangle. The proportion embodied in that equality is this: as LF is to AF, so AF is to FM. This line FM, which is obtainable as the third proportional to abscissa LF and ordinate AF, also enters into the proportion embodied in the *eidos* property of the 12th proposition. This proportion for the *eidos* of excess is: as the oblique (transverse) side KL is to the upright side LN, so also the oblique (transverse) side taken together with the abscissa (KL+LF = KF) is to the equal rectangle's side FM. From this proportion, the upright side LN can be obtained.

In other words: from abscissa LF and ordinate AF, we get (by the first proportion) side FM; from this side FM and oblique (transverse) side KL and abscissa LF, we get (by the second proportion) upright side LN (fig. 12).

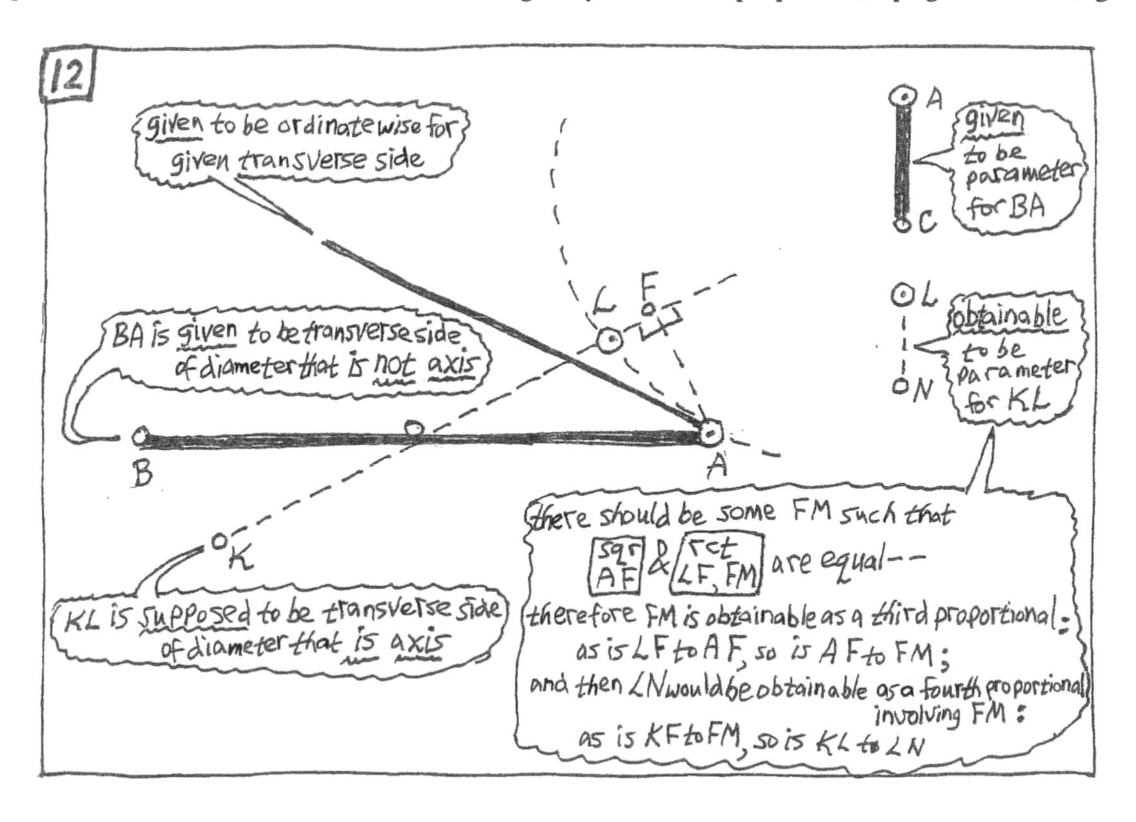

Now let us consider the consequences of supposing that we have the axis's oblique (transverse) side, and therewith its upright side, and hence the sought-for hyperbola. By considering the consequences of supposing that we have the axis's oblique (transverse) side, we wish to arrive at something that we do indeed have that we could use to obtain a line that has the position and the size to be the axis's oblique (transverse) side.

If we did have the oblique (transverse) side KL, then, by the 47th proposition, its midpoint would also be the midpoint (call it D) of BA; so we can get KL by drawing some straight line through BA's center D, and placing its endpoints K and L equidistant from D. But which of the innumerably many such straight lines should we draw, and what should be its size? Where (that is to say) should we place its vertex L? Well, the 37th proposition enables us to say something about the distances between the abscissa-point, the vertex, and the center of an hyperbola's diameter—or, rather, it enables us to do so if we draw a tangent at the point on the hyperbola from which the ordinate is drawn to the abscissa-point, and we take the point (call it H) where that tangent intersects the diameter. What does the proposition enable us to say about D (the center of the axis) and L (the vertex) and F (the abscissa-point for the ordinate from A) and H (the intersection with the line that is tangent at A)? It enables us to say that the square on DL would be equal to the rectangle contained by FD and DH—and, therefore, that DL is the mean proportional between FD and DH. But what would be the hyperbola's tangent at A? It would, by the 17th proposition, be the line through A that is drawn ordinatewise for diameter BA—and so, we have the tangent at A. And, therefore, if we had the axis's oblique (transverse) side KL, we would then have H, the point where the tangent from A intersects the axis's oblique (transverse) side (fig. 13).

So, if only we had point H, we could use it to locate the axis by just joining that point H with the point D (the center of AB). Then we could locate the axis's vertex L by just dropping, from A, a perpendicular to the axis, getting F—and then taking L such that DL is the mean proportional between FD and DH.

But we do not have H unless we suppose that we already have the axis, so let us go on supposing that we have the axis until we arrive at some consequence that will enable us to figure out some way to locate the point (H) where the axis is met by the tangent drawn from the vertex A of the non-axis diameter's oblique (transverse) side BA.

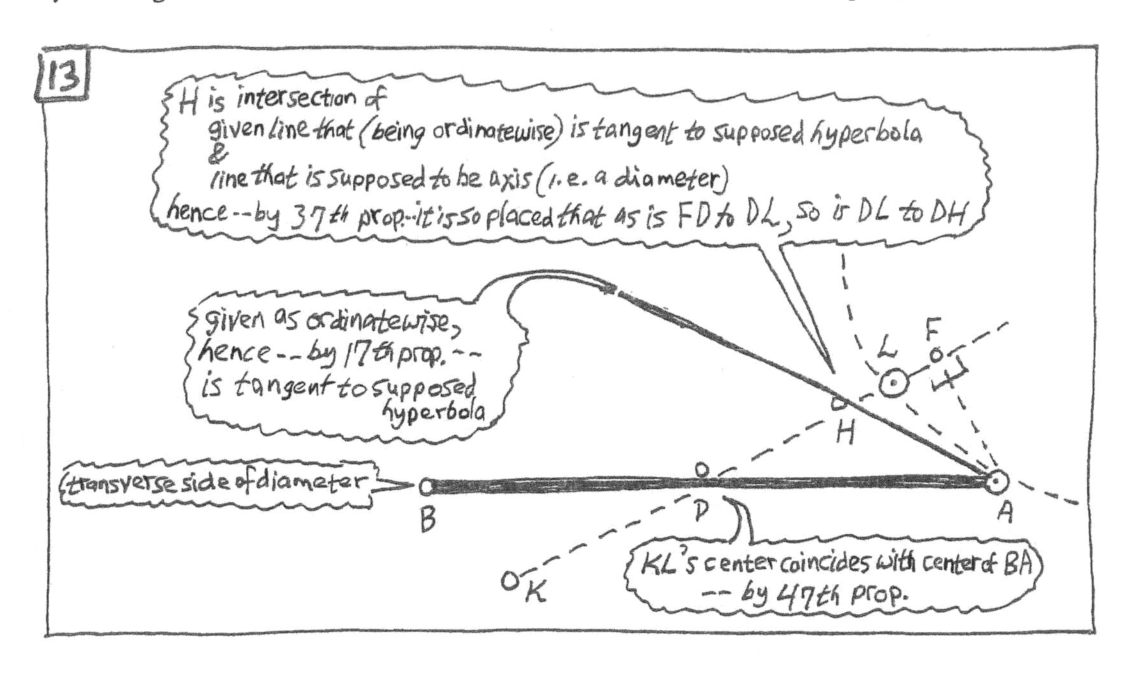

Well, what about the tangent at another point—namely, at the axis's vertex L? That tangent would, by the 17th proposition, be drawn ordinatewise for axis LF. That is to say: if at L we raise a perpendicular from the axis LF, then that perpendicular line—being ordinatewise—will also be tangent at L. But a line perpendicular to axis LF will also be parallel to AF (since the latter was dropped perpendicular to axis LF). So, the tangent at vertex L will be parallel to the line AF. Now, for convenience, let us name the point (call it O) where that tangent from L meets the other tangent, the one from A; let us also name the point (call it X) where the tangent at L meets BA (fig. 14).

Now, treating KL as the oblique (transverse) side of an old diameter, and BA as the oblique (transverse) side of a new diameter, it will follow, by the 50th proposition, that the line AC can be the upright side of BA only if the following proportion holds:

the ratio of AC to twice AH is the same as the ratio of AO to AX (fig. 15).

That tells us something about point H.

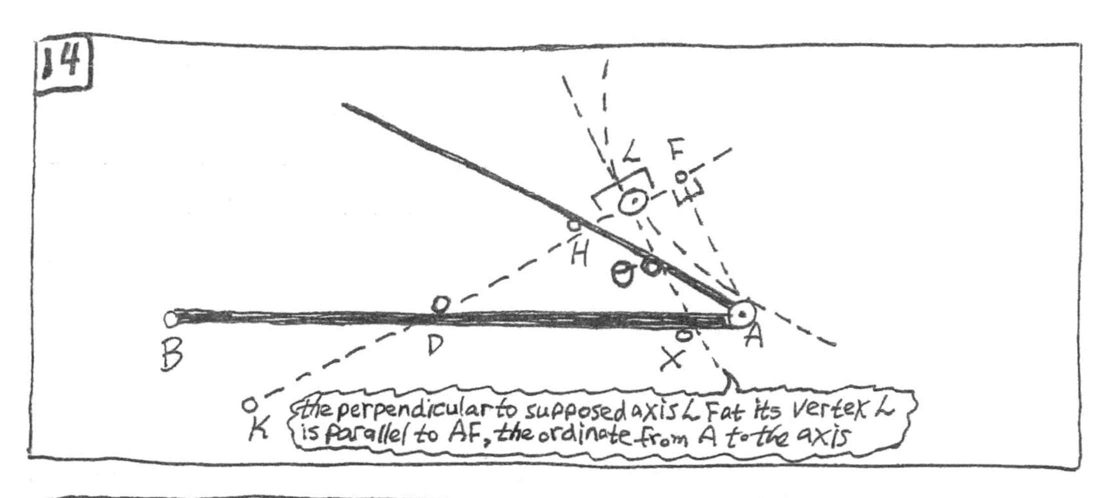

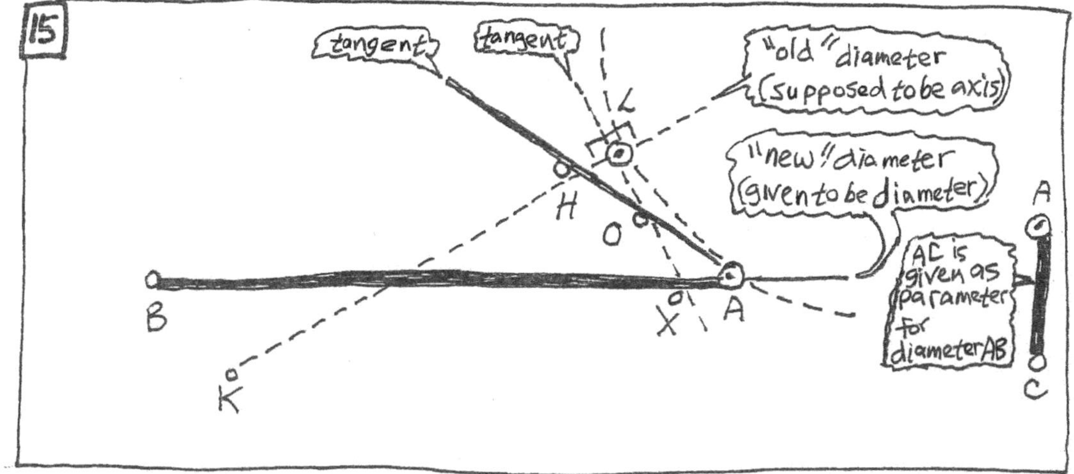

It especially tells us something about point H if we consider that those two lines in the second ratio (AO and AX) are two sides of a triangle. If we make similar triangles, we can involve AC (the upright side of oblique [transverse] side AB) in a proportion with other lines. Let us draw, from F, a line that is parallel to the tangent at A; and let us give a name to the point where that line intersects the diameter for the oblique (transverse) side BA—call the point G. Then (since GA is in line with AX, while AO is parallel to GF, and OX is parallel to FA) the triangles AOX and GFA are similar, and therefore their corresponding sides are proportional:

the ratio of FG to GA is the same as the ratio of OA to AX (fig. 16).

So, if it were true that BA is the oblique (transverse) side of a non-axis diameter, with vertex A and upright side AC, and also that KL is the oblique (transverse) side of the axis, with vertex L, then it would be true that each of the following two ratios would be the same as the ratio of OA to AX: the ratio of AC to twice AH, and the ratio of FG to GA. Therefore AC would have the same ratio to twice AH that FG has to GA. So, what we now know about point H is this:

as AC is to 2AH, so also FG is to GA (fig. 17).

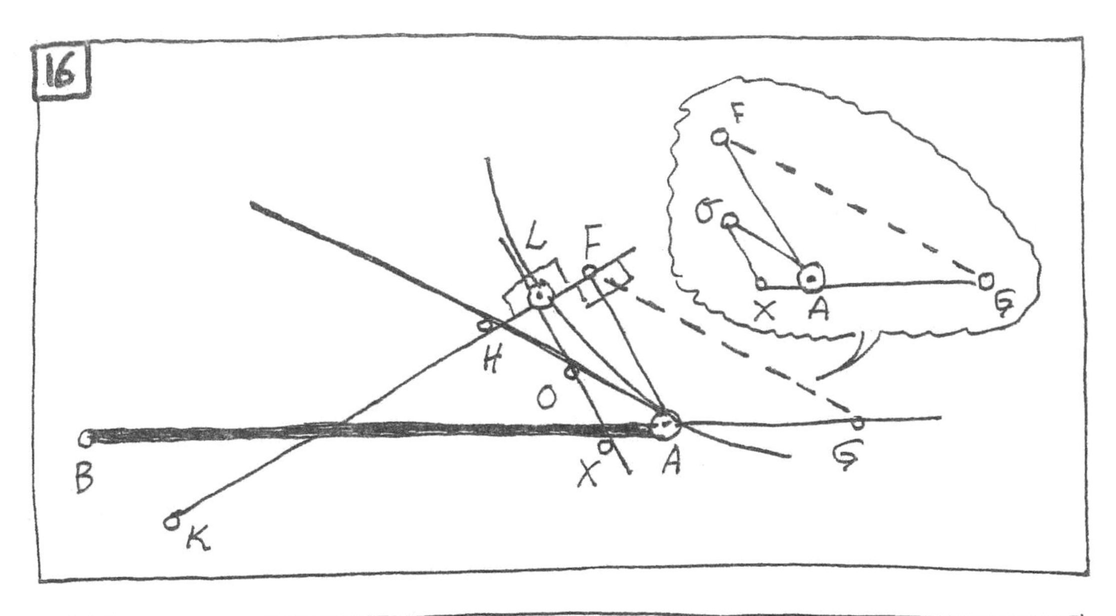

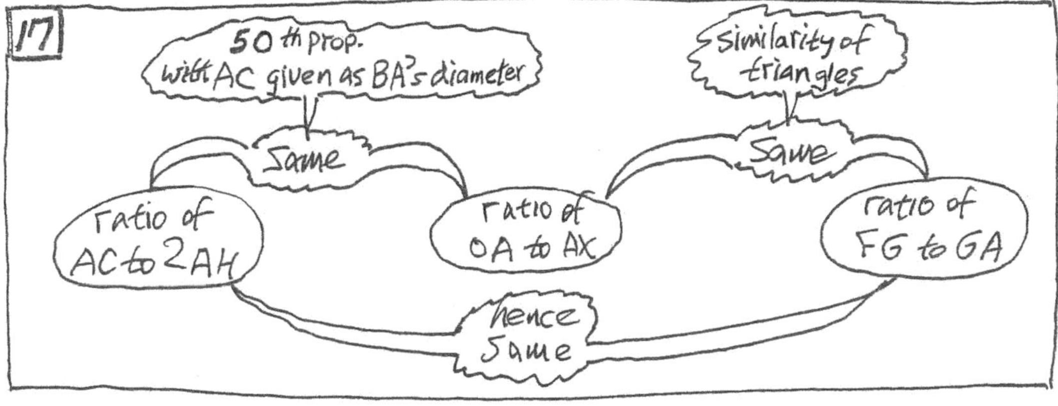

But although we do know something about point H (namely, a proportion that involves AH), we do not yet have that point—because we do not yet have the axis. Does that proportion—from which we know something about that point H which we do not have—tell us anything useful about something else which we do have? Well, we can relate HA to AD—in this way: since HA is parallel to FG, there are similar triangles; and such triangles have their corresponding sides proportional. That is to say:

as HA is to AD, so also FG is to GD (fig. 18).

If we take this last proportion and put it together with the previous one, what will follow is this next one:

as AC is to BA, so also square [FG] is to rectangle [DG, GA].

The reason why, is depicted in figure 19.

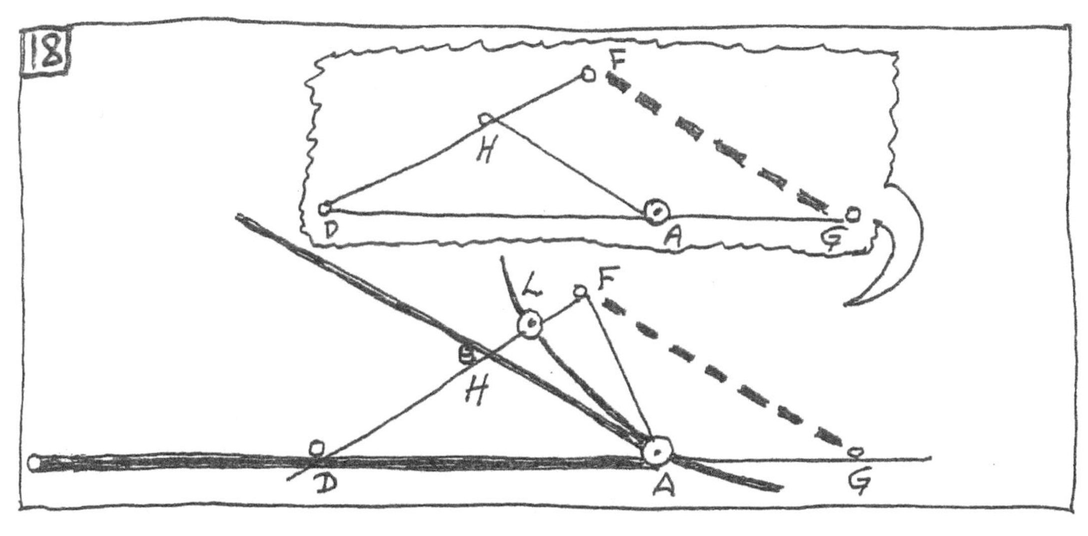

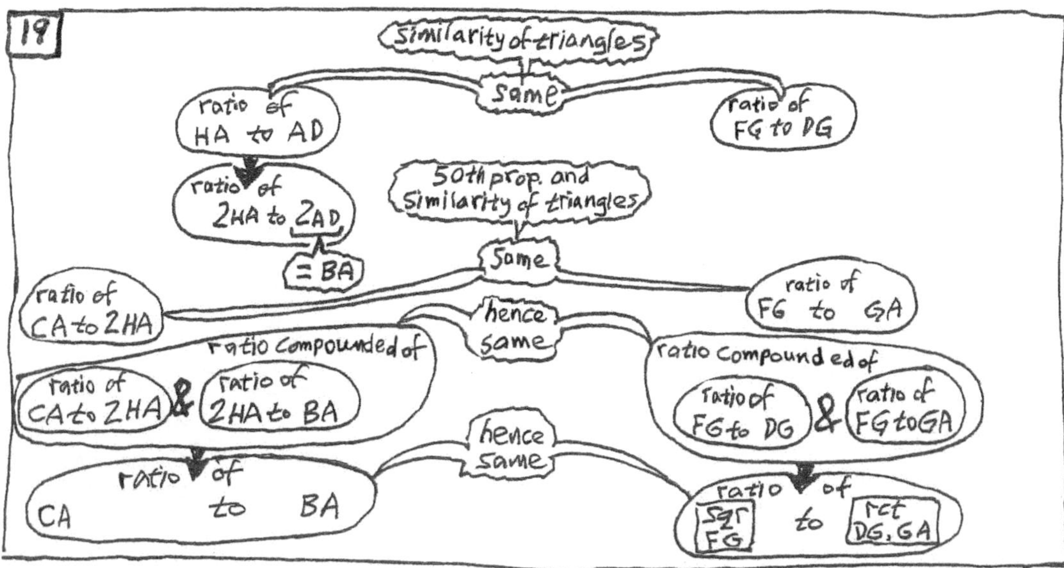

Our conclusion in analyzing the problem is therefore this: we will have the axis if we take the midpoint D of BA and join that point with the point F—the point F being such that

FD and FA meet at a right angle, and square [FG] is to rectangle [DG, GA] as CA is to BA.

We can guarantee that FD and FA will meet at a right angle if we let F be any one of the points on the circumference of a semicircle with DA as diameter (fig. 20).

Assume for a moment that we know how to find out which one of the points on the circumference will be the point that makes the ratio of square [FG] to rectangle [DG, GA] the same as the ratio of CA to BA.

Once we have that point F, what we need to do is to take it and then perform the steps of the construction that will yield the axis's oblique (transverse) side KL, with vertex L and upright side LN, for an hyperbola that will have BA as the oblique (transverse) side of another diameter—a diameter, namely, whose vertex will be A and whose upright side will be AC for ordinates that meet it at a non-right angle equal to the angle at A.

In accordance with this, Apollonius begins his presentation by taking the midpoint D of BA, and then drawing a semicircle around DA, and taking, on the circumference, that point F from which a line parallel to the line that is ordinatewise at A will intersect the prolongation of BA at a point G such that the ratio of square [FG] to rectangle [DG, GA] is the same as the ratio of CA to BA.

Apollonius does not say how to find on the semicircle that point F which will make the proportion hold. He assumes that his readers are skilled enough to supply such a construction for themselves. Since readers, however skilled they are, may by now be somewhat weary, such a construction will now be supplied, along with a sketch of a demonstration that this construction will do the job for which it is needed by Apollonius.

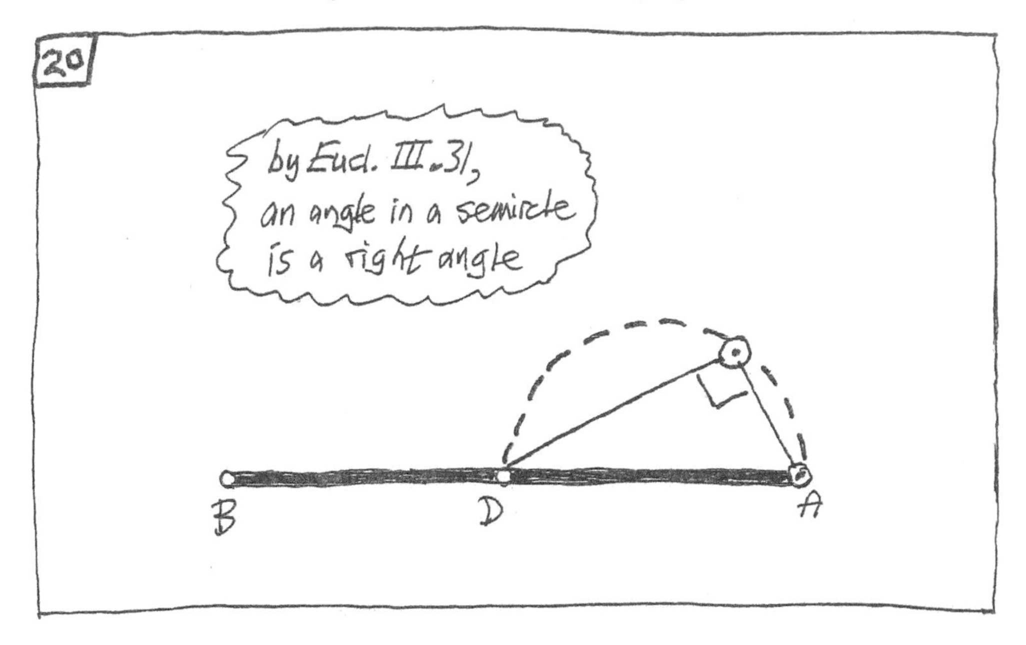

Here is what is given for finding the point F that will fit the ratio specified in the 55th proposition:

—the semicircle with given line DA as diameter; —the non-right angle with DA that is made by some straight line from A; —the ratio of CA to AB.

And what is to be done is this: on the semicircle, find the point F such that—

when a line (FG) is drawn, parallel to the line that is inclined to DA at the given angle, from F as far as the prolongation of DA, then the ratio of square [FG] to rectangle [DG, GA] will be the same as the ratio of CA to AB.

The *construction* is depicted in figure 21. After that, the *demonstration* which does the job is depicted in figure 22.

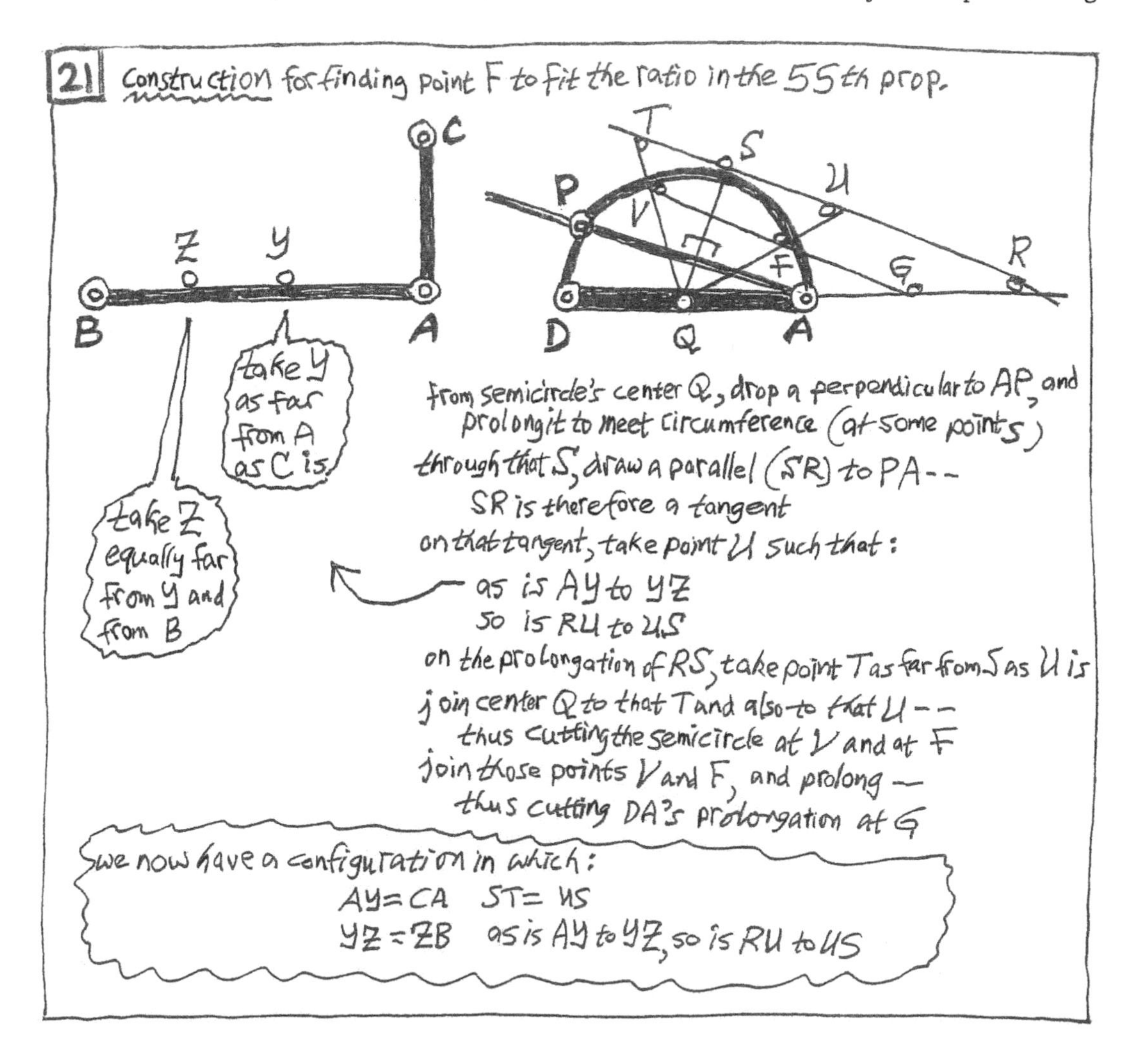

21 Construction for finding point F to fit the ratio in the 55th prop.
C
Z
Y
B
A
take Y as far from A as C is
take Z equally far from Y and from B
T
S
U
P
V
F
G
R
D
Q
A
from semicircle's center Q, drop a perpendicular to AP, and prolong it to meet circumference (at some point S)
through that S, draw a parallel (SR) to PA--
SR is therefore a tangent
on that tangent, take point U such that:
as is AY to YZ
so is RU to US
on the prolongation of RS, take point T as far from S as U is
join center Q to that T and also to that U--
thus cutting the semicircle at V and at F
join those points V and F, and prolong—
thus cutting DA's prolongation at G
we now have a configuration in which:
AY=CA ST=US
YZ=ZB as is AY to YZ, so is RU to US

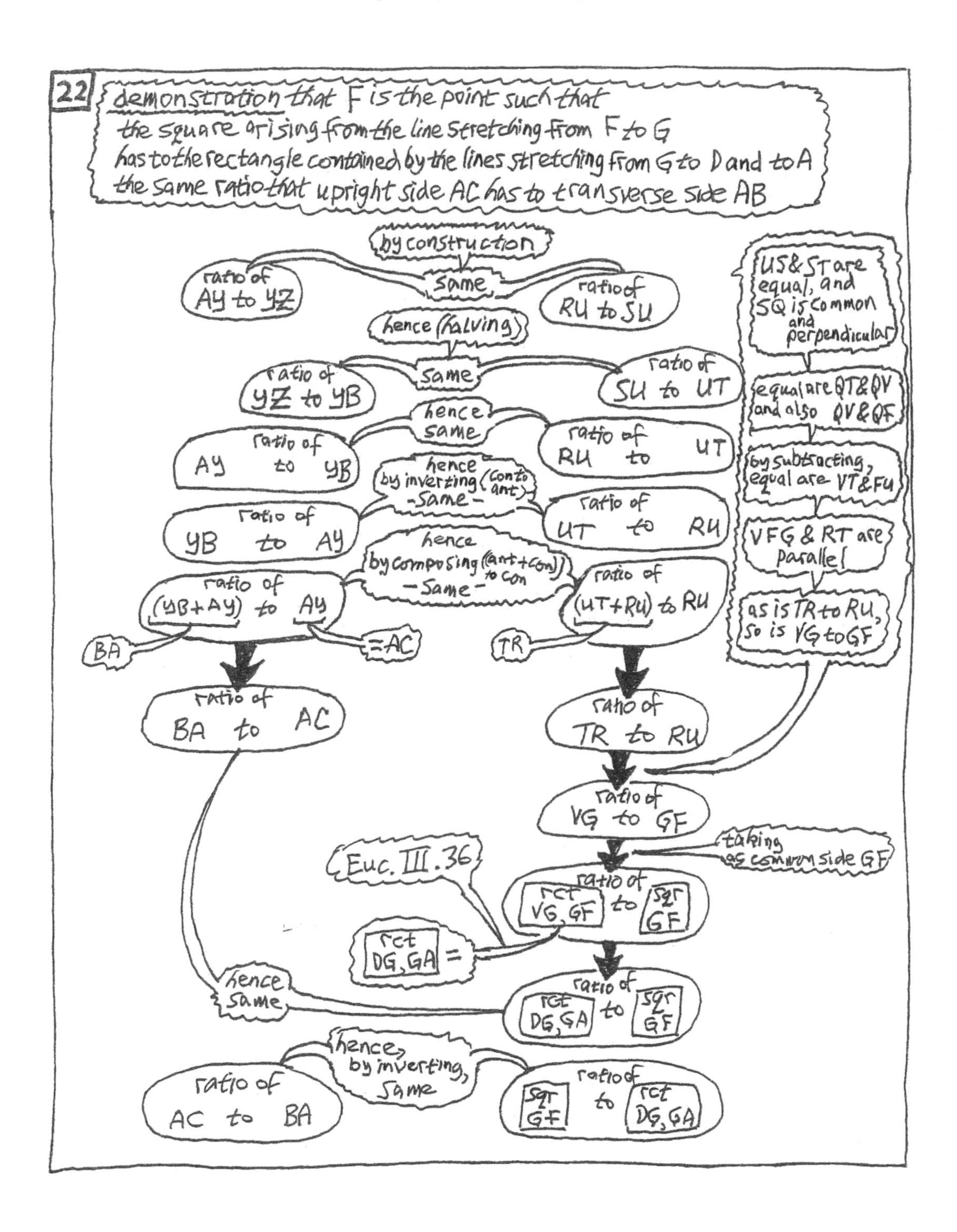
22
demonstration that F is the point such that the square arising from the line stretching from F to G has to the rectangle contained by the lines stretching from G to D and to A the same ratio that upright side AC has to transverse side AB
by construction
ratio of AY to YZ
same
ratio of RU to SU
hence (halving)
ratio of YZ to YB
same
ratio of SU to UT
hence same
ratio of AY to YB
ratio of RU to UT
hence by inverting (con to ant) -same-
ratio of YB to AY
ratio of UT to RU
hence by composing (ant+con) to con -same-
ratio of (YB+AY) to AY
ratio of (UT+RU) to RU
BA
=AC
TR
ratio of BA to AC
ratio of TR to RU
US & ST are equal, and SQ is common and perpendicular
equal are QT & QV and also QV & QF
by subtracting, equal are VT & FU
VFG & RT are parallel
as is TR to RU, so is VG to GF
ratio of VG to GF
taking as common side GF
Euc. III.36
ratio of rct VG, GF to sqr GF
rct DG, GA =
hence same
ratio of rct DG, GA to sqr GF
hence, by inverting, same
ratio of AC to BA
ratio of sqr GF to rct DG, GA

Not only does Apollonius assume that he need not elaborate on how he gets point F, but, as is his way, he also does not say why he takes that point F. Instead, he merely takes it and he then moves on with the construction that prepares for the cone and its cutting.

That preparation, the *first part* of Apollonius's presentation, is a construction to obtain lines KL and LN. It is depicted in figure 23.

The lines KL and LN that have been obtained in the first part are then used in the *second part*, to obtain an hyperbola—an hyperbola that

> has KL as oblique (transverse) side, with vertex L, and
> has LN as upright side for that oblique (transverse) side,
> whose ordinates meet its prolongation at a right angle, making it an axis.

That hyperbola so obtained in the second part is then shown in the *third part* to be the hyperbola that was sought for—namely, an hyperbola that

> has BA as its oblique (transverse) side, with vertex A, and
> has AC as its upright side for
> ordinates that meet its prolongation at a certain angle that is not a right angle.

Following figure 23's depiction of the second part, this last part is depicted in figure 24.

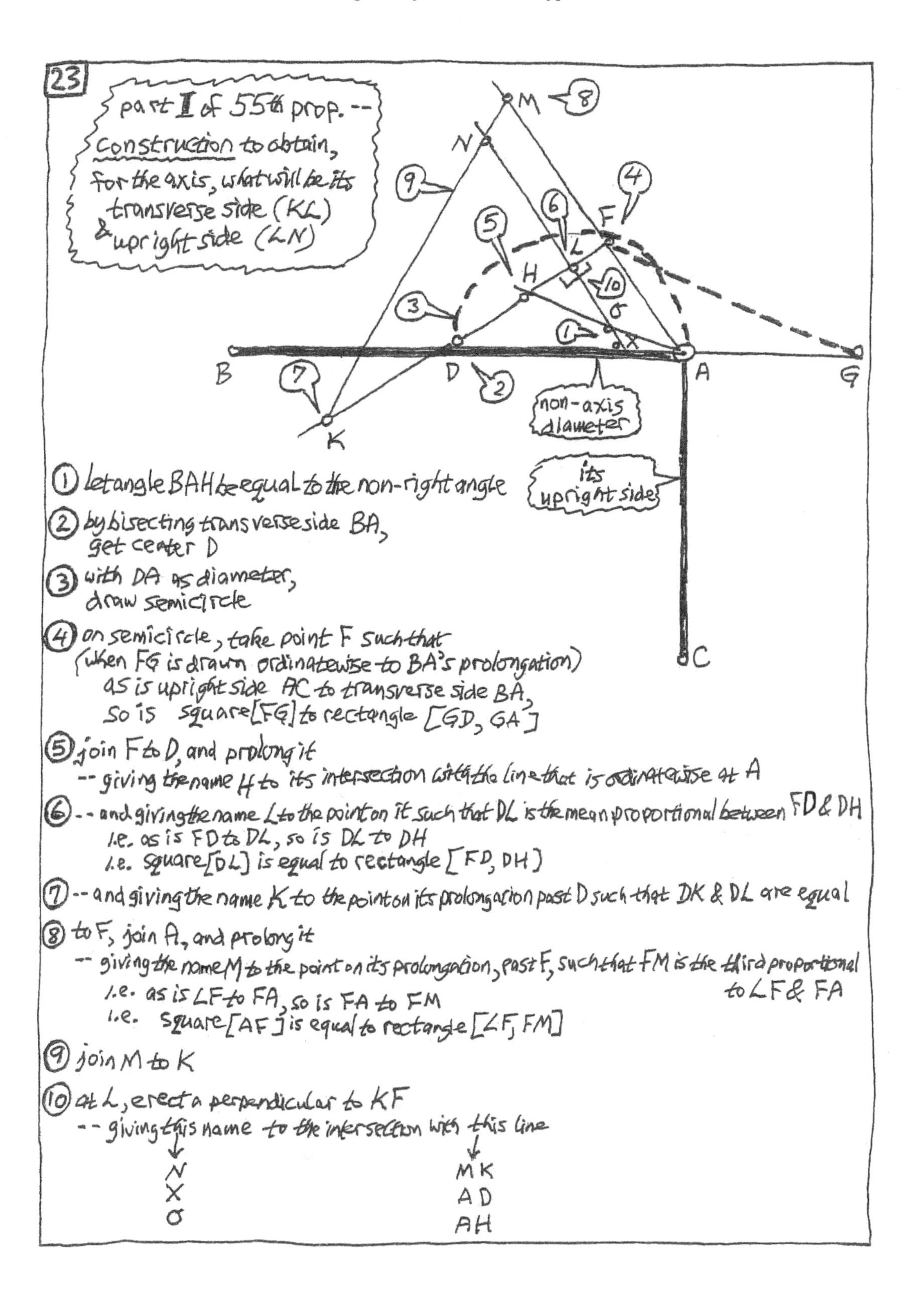
23
part II of 55th prop. --
construction to obtain,
for the axis, what will be its
transverse side (KL)
& upright side (LN)
M
8
N
9
4
6
F
5
L
H
10
σ
3
1
X
B
D
A
G
7
2
non-axis
diameter
K
its
upright side
C
1 let angle BAH be equal to the non-right angle
2 by bisecting transverse side BA,
get center D
3 with DA as diameter,
draw semicircle
4 on semicircle, take point F such that
(when FG is drawn ordinatewise to BA's prolongation)
as is upright side AC to transverse side BA,
so is square[FG] to rectangle [GD, GA]
5 join F to D, and prolong it
-- giving the name H to its intersection with the line that is ordinatewise at A
6 -- and giving the name L to the point on it such that DL is the mean proportional between FD & DH
i.e. as is FD to DL, so is DL to DH
i.e. square[DL] is equal to rectangle [FD, DH]
7 -- and giving the name K to the point on its prolongation past D such that DK & DL are equal
8 to F, join A, and prolong it
-- giving the name M to the point on its prolongation, past F, such that FM is the third proportional
to LF & FA
i.e. as is LF to FA, so is FA to FM
i.e. square[AF] is equal to rectangle [LF, FM]
9 join M to K
10 at L, erect a perpendicular to KF
-- giving this name to the intersection with this line
N
X
σ
MK
AD
AH

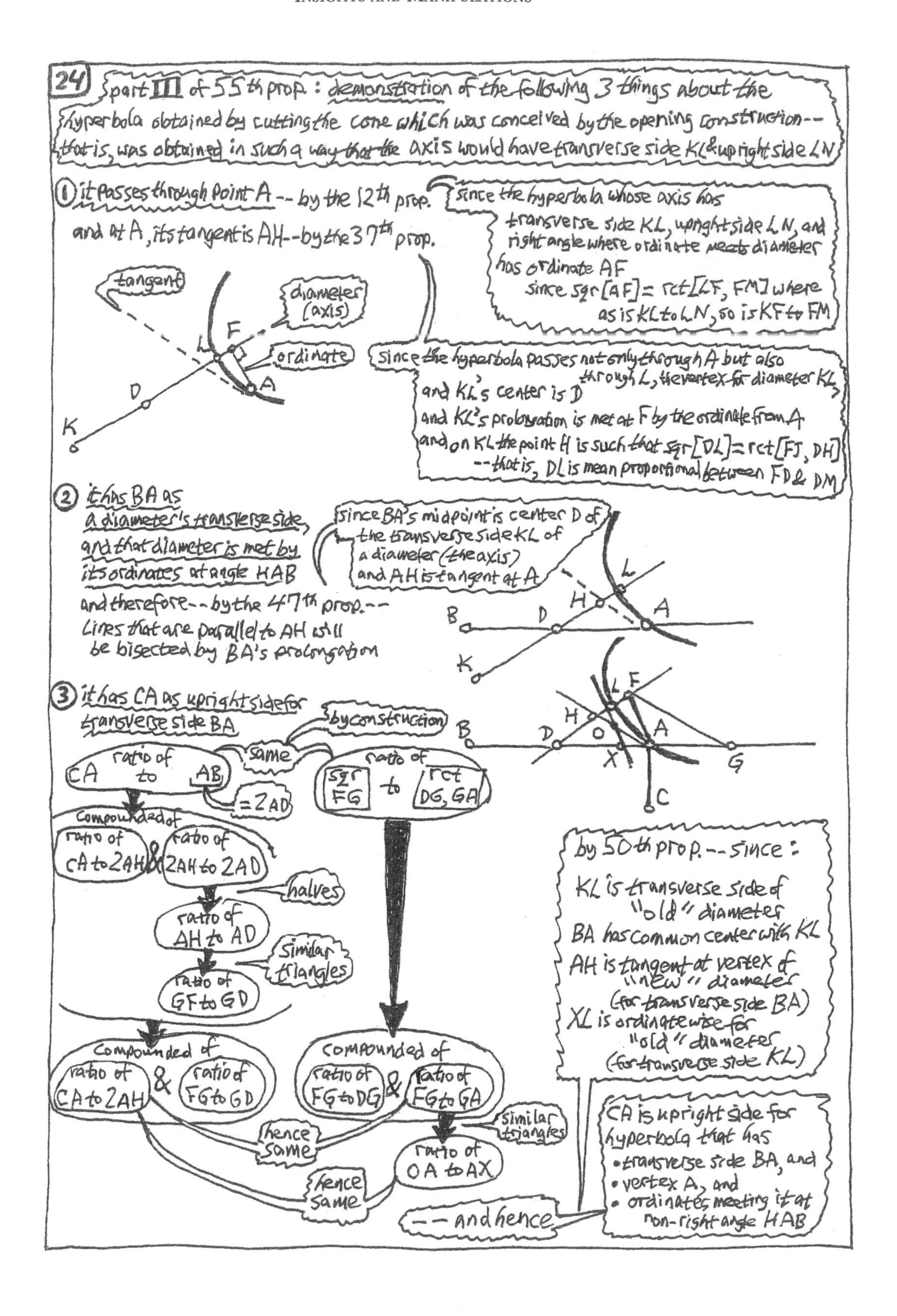

24
part III of 55th prop.: demonstration of the following 3 things about the hyperbola obtained by cutting the cone which was conceived by the opening construction -- that is, was obtained in such a way that the axis would have transverse side KL & upright side LN
① it passes through point A -- by the 12th prop.
and at A, its tangent is AH -- by the 37th prop.
since the hyperbola whose axis has transverse side KL, upright side LN, and right angle where ordinate meets diameter has ordinate AF
since sqr[AF] = rct[LF, FM] where as is KL to LN, so is KF to FM
tangent
diameter (axis)
ordinate
F
L
D
A
K
since the hyperbola passes not only through A but also through L, the vertex for diameter KL, and KL's center is D
and KL's prolongation is met at F by the ordinate from A
and on KL the point H is such that sqr[DL] = rct[FD, DH] -- that is, DL is mean proportional between FD & DH
② it has BA as a diameter's transverse side, and that diameter is met by its ordinates at angle HAB
and therefore -- by the 47th prop. -- lines that are parallel to AH will be bisected by BA's prolongation
since BA's midpoint is center D of the transverse side KL of a diameter (the axis) and AH is tangent at A
L
H
B
D
A
K
③ it has CA as upright side for transverse side BA
by construction
same
ratio of CA to AB
= 2AD
ratio of sqr FG to rct DG, GA
compounded of
ratio of CA to 2AH
&
ratio of 2AH to 2AD
halves
ratio of AH to AD
similar triangles
ratio of GF to GD
compounded of
ratio of CA to 2AH
&
ratio of FG to GD
compounded of
ratio of FG to DG
&
ratio of FG to GA
similar triangles
hence same
ratio of OA to AX
hence same
-- and hence
L F
H
B
D
O
A
X
G
C
by 50th prop. -- since:
KL is transverse side of "old" diameter
BA has common center with KL
AH is tangent at vertex of "new" diameter (for transverse side BA)
XL is ordinatewise for "old" diameter (for transverse side KL)
CA is upright side for hyperbola that has
• transverse side BA, and
• vertex A, and
• ordinates meeting it at non-right angle HAB

CHAPTER 4. FINDING CONES FOR THE CURVES: ELLIPSE

56th-57th-58th propositions

Propositions 56, 57, and 58, which constitute Apollonius's solution to the ellipse problem, are presented in the following boxes.

At this point, you would benefit most from trying to provide for yourself something with respect to the ellipse like what this guidebook has provided for you in your study of Apollonius's solution to the problem of the parabola (propositions 52–53) and of the hyperbola (propositions 54–55). That is to say: you might first do an analysis of the problem, then lay out a synthesis demonstrating that what you have found is indeed a solution, and finally, only after doing that, read the propositions of Apollonius.

If you would like help in drawing illustrative diagrams for the reading of the propositions, see the Appendix at the end of this volume for such diagrams drawn complete.

On a first reading of Apollonius, however, you may wish to pass over the rest of the present chapter entirely, and go directly to the next chapter, which presents his treatment of the opposite sections in the 59th proposition.

#56 Two straight lines being given, [lines, that is, which are] bounded [and are] at right angles to one another—to find, around one of them as diameter, the cone's section called an ellipse, in the same plane with the straight lines, [an ellipse, that is,] whose vertex is what the point at the right angle will be, and [in which] the [lines] drawn down from the section onto the diameter at a given angle will have the power [to give rise to squares equal to] the rectangles which are laid along (*parakeimena para*) the other straight [line], while having as breadth the line taken off by them, [that is, the line taken off from the diameter, by the lines drawn down,] on the side toward the vertex of the section, which [rectangles] are lacking (*elleiponta*) by a figure (*eide*i) that is similar to, and that is laid similarly to, the [rectangle] contained by the given straight [lines].
—Let there be, as the two given straight [lines], the [lines] AB, AC, at right angles to one another, the greater of which [lines] the [line] AB is. Required, then, is, in the plane supposed, to carve out (*grapsai*) an ellipse, whose diameter is what the [line] AB will be, and vertex is what the [point] A will be, and upright [side] is what the [line] AC will be, and [in which] the [lines] drawn down will be drawn down from the section onto the [line] AB at a given angle and will have the power [to give rise to squares equal to] the [rectangles] which are laid along the [line] AC, while having as breadths the [lines] taken off by them [taken off, that is, by the lines drawn down,] on the side toward the [point] A, which [rectangles] are lacking by a figure that is similar to, and that is laid similarly to, the [rectangle] contained by the [lines] BAC.
—And let the given angle first be a right [angle]; and also, let there be set up, from the [line] AB, a plane upright to the [plane] supposed; and also, in it, on the [line] AB, let a sector (*tmêma*) of a circle be carved out (*gegraphthô*), namely the [sector] ADB, whose midpoint (*dichotomia*) let the [point] D be; and also, let there be joined the [lines] DA, DB; and also, let there be put equal to the [line] AC the [line] AX; and also, through the [point] X let there be drawn parallel to the [line] DB the [line] XO, and through the [point] O [let there be drawn] parallel to the [line] AB the [line] OF; and also, let there be joined the [line] DF, and also let it fall together with the [line] AB extended, at the [point] E, say. Then as is the [line] AB to AC so will be the [line] BA to AX, that is, the [line] DA to AO, that is, the [line] DE to EF.

—And also, let there be joined the [lines] AF, FB, and also let them be extended, and, also let there be taken on the [line] FA a chance point, namely the [point] G; and also, through it let there be drawn parallel to the [line] DE the [line] GL; and also let it fall together with the [line] AB extended, at the [point] K, say; then let there be extended the [line] FO and also let it fall together with the [line] GK, at the [point] L, say. Since, then, equal is the arc (*periphereia*) AD to the [arc] DB, equal is the angle by ABD to the [angle] by DFB; and also—since the angle by EFA is equal to the two [angles] [by] FDA, FAD [together], but this angle by FAD (on the one hand) is equal to the [angle] by FBD, and that [angle] by FDA (on the other hand) is equal to the [angle] by FBA—therefore also the [angle] by EFA is equal to the [angle] by DBA, that is to the [angle] by BFD. And also, a parallel is the [line] DE to the [line] LG; therefore the [angle] by EFA, to the [angle] by FGH, is equal, and the [angle] by DFB, to the [angle] by FHG, [is equal]; and so also the [angle] by FGH, to the [angle] by FHG, is equal, and also the [line] FG, to the [line] FH, is equal.

—Let there be carved out (*gegraphthô*), then, about the [line] HG, the circle GHN upright to the triangle HGF; and also, let there be envisioned (*noeisthô*) a cone whose base is what the circle GHN is, and vertex is what the [point] F is; then will the cone be a right [cone], on account of there being equal the [line] GF to the [line] FH.

—And—since the circle GHN is upright to the plane HGF, and also the plane supposed is upright to the plane through the [lines] GH, HF—therefore also their common section, [that is, the section which is common to the plane of the circular base and to the supposed plane], will, to the plane through the [lines] GHF, be upright. Let then their common section be what the [line] KM is; therefore the [line] KM is upright to each of the [lines] AK, KG.

—And—since a cone, whose base is what the circle GHN is, and vertex is what the [point] F is, has been cut by a plane through the axis, and [the cutting plane] makes as section the triangle GHF, and [the cone] has been cut also by another plane, [a plane] through the [lines] AK, KM, which is the [plane] supposed, [which cuts the base plane] in the straight [line] KM which is at right angles to the [line] GK, and the plane [supposed] falls together with the sides FG, FH of the cone—therefore the section generated is an ellipse whose diameter is what the [line] AB is, and whose [lines] drawn down will be drawn down at a right angle, for parallel are they to the [line] KM. And—since as is the [line] DE to EF so is the [rectangle] by DEF, that is, the [rectangle] by BEA, to the [square] from EF, and the [rectangle] by BEA has to the [square] from EF the ratio compounded from that of the [line] BE to EF and also that of the [line] AE to EF, but (on the one hand) as is the [line] BE to EF so is the [line] BK to KH and (on the other hand) as is the [line] AE to EF so is the [line] AK to KG, that is, the [line] FL to LG—therefore the [line] BA has to AC the ratio compounded from that of the [line] FL to LG and also that of the [line] FL to LH, which is the same as that which the [square] from FL has to the [rectangle] by GLH. But whenever this be so, then an upright side of the figure (*eidous*) is what the [line] AC is, as has been shown in the 13th theorem.

#57 The same things being supposed, let the [line] AB be less than the [line] AC; and also, about the diameter AB let it be required to carve out (*grapsai*) an ellipse, so that that an upright [side] is what the [line] AC is. —Let the [line] AB be cut in two [halves] at the [point] D; and also, from the [point] D let there be drawn at right angles to the [line] AB the [line] EDF; and also, let there be equal to the [rectangle] by BAC the [square] from FE, so that equal is the [line] FD to the [line] DE; and also, let there be drawn parallel to the [line] AB the [line] FG; and also, [let the line FG have been made just so long that] as is the [line] AC to AB so is the [line] EF to FG; therefore greater also is the [line] EF than the [line] FG. And—since equal is the [rectangle] by CAB to the [square] from EF—as is the [line] CA to AB so is the [square] from FE to the [square] from AB, and the [square] from DF to the [square] from DA. But as is the [line] CA to AB so is the [line] EF to FG; therefore as is the [line] EF to FG so is the [square] from FD to the [square] from DA. But the [square] from FD is equal to the [rectangle] by FDE; therefore as is the [line] EF to FG so is the [rectangle] by EDF to the [square] from AD. Then, with two bounded straight lines being laid at right angles to one another and the greater being the [line] EF, let there be carved out (*gegraphthô*) an ellipse whose diameter is what the [line] EF is, and upright [side] is what the [line] FG is. Then will the section go through the [point] A, on account of its being [so, that] as is the [rectangle] by FDE to the [square] from DA so is the [line] EF to FG. And equal is the [line] AD to the [line] DB. [The section] will, then, go also through the [point] B. There has been carved out (*gegraphtai*), then, an ellipse about the [line] AB. And—since as is the [line] CA to AB so is the [square] from FD to the [square] from DA, but the [square] from DA is equal to the [rectangle] by ADB—therefore as is the [line] CA to AB so is the [square] from DF to the [rectangle] by ADB. And so an upright [side] is what the [line] AC is.

#58 But let not, then, the given angle be a right [angle]; and also, let there be equal to it the [angle] by BAD; and also, let the [line] AB be cut in two [halves] at the [point] E; and on the [line] AE let there be carved out (*gegraphthô*) the semicircle AFE; and in it let there be drawn parallel to the [line] AD the [line] FG which [by being given the proper size] makes the ratio of the [square] from FG to the [rectangle] by AGE the same as that of the [line] CA to the [line] AB. And let there be joined the lines AF, EF and let them be extended; and let there be taken, as a mean proportional of the [lines] DE, EF the [line] EH; and, equal to the [line] EH, let there be put the [line] EK; and [by properly locating, on the extension of the line AF, the point L] let there be made equal to the [square] from AF the [rectangle] by HFL; and let there be joined the [line] KL; and, from the [point] H let there be drawn at right angles to the [line] HF the [line] HMX, [which is thus] generated parallel to the [line] AFL, for a right [angle] is what the [angle] at the [point] F is. And two given bounded straight [lines] at right angles to one another being what [lines] KH,HM are, let there be carved out (*gegraphthô*) an ellipse, whose oblique diameter is what the [line] KH is, and the figure's (*eidous*) upright side is what the [line] HM is, and the [lines] drawn down will be drawn down onto the [line] HK at a right angle. The section will, then, go through the [point] A, on account of there being equal the [square] from FA to the [rectangle] by HFL. And—since equal is the [line] HE to the [line] EK, and the [line] AE to the line EB—also through the [point] B will go the section. And a center is what the [point] E will be, and a diameter is what the [line] AEB will be. And there will touch the section the [line] DA, on account of there being equal the [rectangle] by DEF to the [square] from EH. And—since as is the [line] CA to AB so is the [square] from FG to the [rectangle] by AGE, but (on the one hand) the [line] CA has to AB the ratio compounded from that of the [line] CA to the double of the [line] DA and also that of the double of the [line] AD to the [line] AB, that is of the [line] DA to AE, and (on the other hand) the [square] from FG has to the [rectangle] by AGE the ratio compounded from that of the [line] FG to GE and also that of the [line] FG to GA—therefore the ratio compounded from that of the line CA to the double of the [line] AD and also that of the line DA to AE is the same as the ratio compounded from that of the [line] FG to GE and also that of the [line] FG to GA. But as is the [line] DA to AE so is the [line] FG to GE; and—this common ratio being taken away [namely, the ratio of the line DA to AE, which is the same as that of the line FG to GE]—then as is the [line] CA to the double of the [line] AD so will be the [line] FG to GA, that is, the [line] XA to AN. But whenever this be so, then an upright side of the figure (*eidous*) is what [line] AC is.

CHAPTER 5. FINDING CONES FOR THE CURVES: OPPOSITE SECTIONS

59th proposition: problem solved for opposite sections

The problem was solved for the single hyperbola in the 55th and 56th propositions. Now, if two hyperbolas have the same oblique (transverse) side, and also have equal upright sides for the ordinates to the prolongations of the common oblique (transverse) side, then they constitute opposite sections. By prolonging the common oblique (transverse) side in the one direction and in the other, you get the diameter of the one hyperbola and the diameter of the other; the oblique (transverse) side's one endpoint and its other endpoint are the vertex of the one hyperbola and the vertex of the other. The ordinates in the one hyperbola meet their diameter at the same angle as do the ordinates in the other hyperbola.

So, find the one hyperbola in a cone, as was done in the 55th and 56th propositions; and also find the other hyperbola in a cone, in the same way—using the same line for the oblique (transverse) side, but using for the vertex its other endpoint, and using an equal line for the upright side, and an equal angle for the ordinatewise line. You will then have obtained two hyperbolas that constitute opposite sections.

That is what Apollonius does in his solution to the problem for the opposite sections, which is presented in the box below (for which, see figure 1). With this treatment of the opposite sections in his 59th proposition, Apollonius completes the solution of the problem of finding the specified curves as conic sections.

#59 Two straight lines being given, [straight lines, that is, which are] bounded [and are] at right angles to one another—to find opposite [sections] whose diameter is what one of the given straight [lines] is and whose vertices are what the bounds of the straight [line] are, and where the [lines] drawn down in each of the sections at the given angle will have the power [to give rise to squares equal to] the [rectangles] which are laid along (*parakeimena para*) the other [one of the straight lines], which [rectangles] overshoot (*huperballonta*) by [a figure that is] similar to the [rectangle] contained by the given straight [lines].
—Let the two given bounded straight [lines] at right angles to one another be what the [lines] BE, BH are, and let the given angle be what the [angle] G is. Required, then, to carve out (*grapsai*) opposite [sections] about one of the [lines] BE, BH so that the [lines] drawn down are drawn down at the angle G.
—And the two given straight [lines] being what the [lines] BE, BH are, let there be carved out (*gegraphthô*) an hyperbola whose oblique diameter is what the [line] BE will be, and whose figure's (*eidous*) upright side is what the [line] HB will be, and [in which] the [lines] drawn down onto the straight prolongation of the [line] BE will be drawn down at the angle G, and let [this hyperbola] be what the [line] ABC is—for how this needs to be generated has been written previously.
—Let there be drawn, then, through the [point] E, at right angles to the [line] BE, the [line] EK which is equal to the [line] BH, and let there be carved out (*gegraphthô*) similarly another hyperbola, the [line] DEF, whose diameter is what the [line] BE is and whose figure's (*eidous*) upright side is what the [line] EK is, and [in which] the [lines] drawn down from the section ordinatewise will be drawn down successively at the angle G. Evident it is, then, that the [lines] B, E are opposite [sections], but their diameter is one [single line], and their upright [sides] are equal.

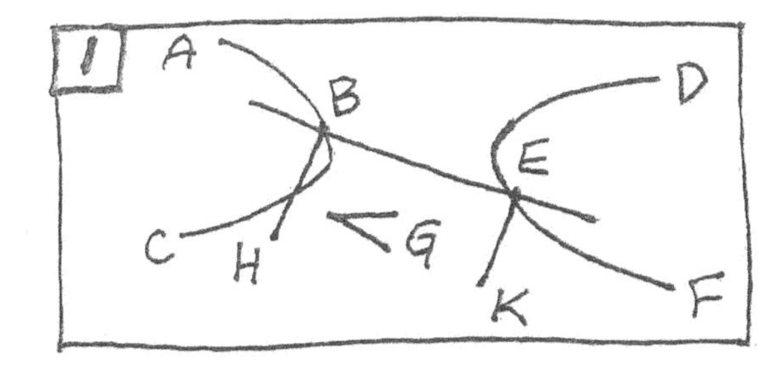

transition from finding cones for the curves

At the beginning of this First Book of his *Conics*, after an introductory sequence of propositions had prepared the way, Apollonius showed (in the 11th through 14th propositions) that all conic sections are lines with certain properties. What followed from that beginning (in the long sequence from the 15th through the 51st proposition) has made it possible for him to show here at the end of the book (in the 52nd through 59th propositions) that all lines with those properties are conic sections. That movement (*from* showing that all conic sections are lines with certain properties, *to* showing that all lines with those properties are conic sections) is the core of the story of the First Book of Apollonius's *Conics*.

The long sequence from the 15th through the 51st propositions was a bridge from the sequence of propositions 11–14 to the sequence of propositions 52–59. Between the cutting of cones (which preceded the 15th proposition) and the finding of cones to cut (which followed the 51st proposition) Apollonius showed things that were necessary in order to be able to find cones whose cuttings will yield specified curves whose specified diameters are not axes. If the conic sections originated *only* as cuts in *right* cones, so that the *only* original diameters were *axes*, then some of the problems solved in propositions 52–59 could have been solved immediately after propositions 11–14, without any need for the intervening propositions. But the existence of non-axial original diameters, for sections originating as cuts in oblique cones, made the intervening propositions necessary. In finding cones from which to obtain specified curves, Apollonius used diameters that were axes; and in finding those axes for curves whose specified diameters were not axes, he made use of what he demonstrated in the course of the long sequence from the 15th proposition through the 51st.

That sequence was not, however, merely a bridge from the sequence 11–14 to the sequence 52–59. It was also a preparation for what will be the glimpsing of wonders that lie beyond the 59th proposition. The bridge propositions were necessary, not in order to treat all kinds of conic sections, but rather to treat all kinds of diameters, not merely axes; and we obtain a deeper insight into the seeming differences of kind in the cuts by considering their diameters of different kinds. As it turns out, the opposite sections, while not being a curve cut into a cone, nonetheless do constitute something like such a curve.

The core of the story of this First Book of Apollonius's *Conics* thus is not the whole of the story of the First Book, just as the First Book is not the whole of Apollonius's *Conics*. After the problems at the end of the First Book, there comes another proposition. This 60th proposition, which (like the proposition just before it) treats opposite sections, does bring the First Book to an end. But it not only provides a spectacular grand finale that brings the First Book to an end, it also begins a line of thought that the Second Book will pursue. This line of thought has something to do with the way that an hyperbola widens. Any parabola or ellipse can be found in any cone (if only you know how to cut it in just the way that you have to)—but not any hyperbola can be found in any cone. That is because a particular hyperbola cannot fit into a particular cone unless the cone widens down from its vertex quickly enough to accommodate the hyperbola.

CHAPTER 6. CONJUGATE OPPOSITE SECTIONS

The rate at which a cone widens has something to do with the rate at which the hyperbola that is cut into it will widen—and the way that any hyperbola widens is peculiar. What determines the widening of an hyperbola is suggested by seeing how opposite sections (which are a cut of opposite conic surfaces, and *not* a cut of a cone) are related in a surprisingly intimate way to a curve that *is* a cut of a cone. The very end of the First Book will now let us in on this surprise, while suggesting a line of thought for continuing the study of things that are conic. So, let us turn now to the final proposition of Apollonius's First Book.

This 60th proposition comes right after Apollonius has shown (in the 59th proposition) how to build cones to obtain the last of the conics—opposite sections, as specified.

Now, before going on, consider what Apollonius did very early in the book, right after he had shown (in the 14th proposition) that by cutting given opposite conic surfaces in a specified way, opposite sections would be obtained. What he showed back then was how to obtain, for the diameter of an ellipse (15th proposition) and for the diameter of opposite sections (16th proposition), another diameter—a diameter conjugate to the original diameter. (In the First Definitions that began the book, he had said that two diameters are conjugate when each is positioned parallel to all those parallel chords which are bisected by the other.) Right after those two propositions showing how there are such conjugate diameters, his Second Definitions identified the center (of the hyperbola, the ellipse, and the opposite sections) and defined the second diameter. A second diameter is a diameter so positioned as to be conjugate to the original diameter, and also so positioned as to have the same center as it does, this latter stipulation being possible because a second diameter is also stipulated to be of a definite size—namely, of such a size as to make it the mean proportional between the oblique (transverse) side and this oblique (transverse) side's upright side. Now that happens to be the very same size that belongs to the conjugate diameter of an ellipse. The conjugate diameter of an hyperbola, however, can be of any size. Even so, the Second Definitions gave no reason at all for the conjugate diameter of that particular size to be called the second diameter of opposite sections. As we observed at the time, what it would seem natural to do for the ellipse alone, was done by the Second Definitions not only for the ellipse but also for the opposite sections, without referring to either the ellipse or the opposite sections. (Indeed, by naming no particular sorts of curves, it does it also for the hyperbola.) Thus, the second diameter was introduced as a strange addendum to the introducing of the conjugate diameter, after the opposite sections had been treated in the last of the introductory propositions involving cones or conic surfaces.

Now in the 60th proposition, after the opposite sections have been treated in the last of the concluding propositions involving cones, Apollonius concludes the First Book with an addendum tying up loose ends that have to do with conjugate diameters of opposite sections. Although in this addendum he doesn't explicitly raise the question of the second diameter, he does supply (to those whom he may have induced to raise the question) what is needed to answer it.

The 60th proposition makes double use of the 59th. The 59th proposition makes it possible to draw opposite sections that have as oblique (transverse) side a line whose position and size are given, with upright side equal to a line whose size is given, and with ordinates that meet its diameter at a given angle (fig. 1).

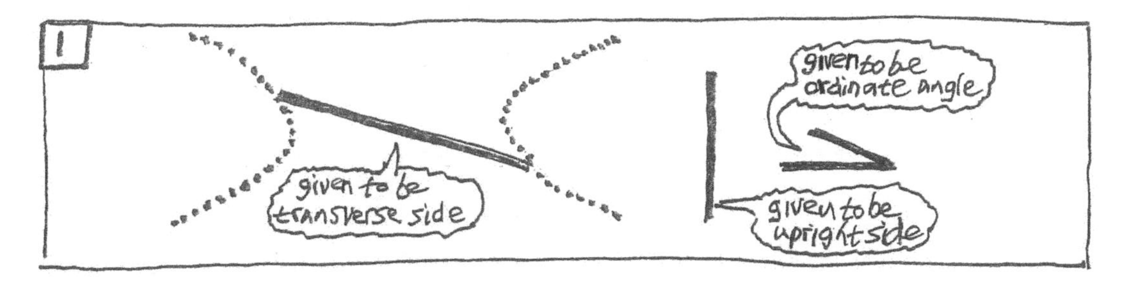

Suppose that we had a given line to be the oblique (transverse) side of opposite sections, and also a given angle to be the angle at which the ordinates meet the diameter. And suppose we were also given the size of the opposite sections' diameter that is conjugate to that oblique (transverse) side (the oblique [transverse] side which has ordinates at that angle). And suppose that, whatever the size of that line which we are given to be that conjugate diameter, we center it on the center of the oblique (transverse) side. Now if we take the conjugate diameter of such a size as would make it the second diameter of the opposite sections that we wish to draw, then we shall have all that we need in order to draw the opposite sections by using the 59th proposition. (That is to say, we shall also have the upright side that we would need, since—by the definition of the second diameter as a mean proportional—the upright side will be that line which, with the oblique (transverse) side, contains a rectangle equal to the square on the second diameter.) And now comes something new.

Having used the two given lines—the one line that was given to be the oblique (transverse) side of a pair of hyperbolas that are opposite sections, and the other line that was given to be their second diameter—suppose now that we draw yet another pair of hyperbolas that are opposite sections. We draw this pair, like the first pair, by using the 59th proposition, with the same two given lines—but this time we interchange what we use the two lines for. That is to say: as oblique (transverse) side for this other pair, we now use the given line that we used as the first pair's second diameter; and as second diameter for this other pair, we now use the given line that was used as the first pair's oblique (transverse) side (fig. 2).

If we do all that, we shall be doing what Apollonius does in the 60th proposition.

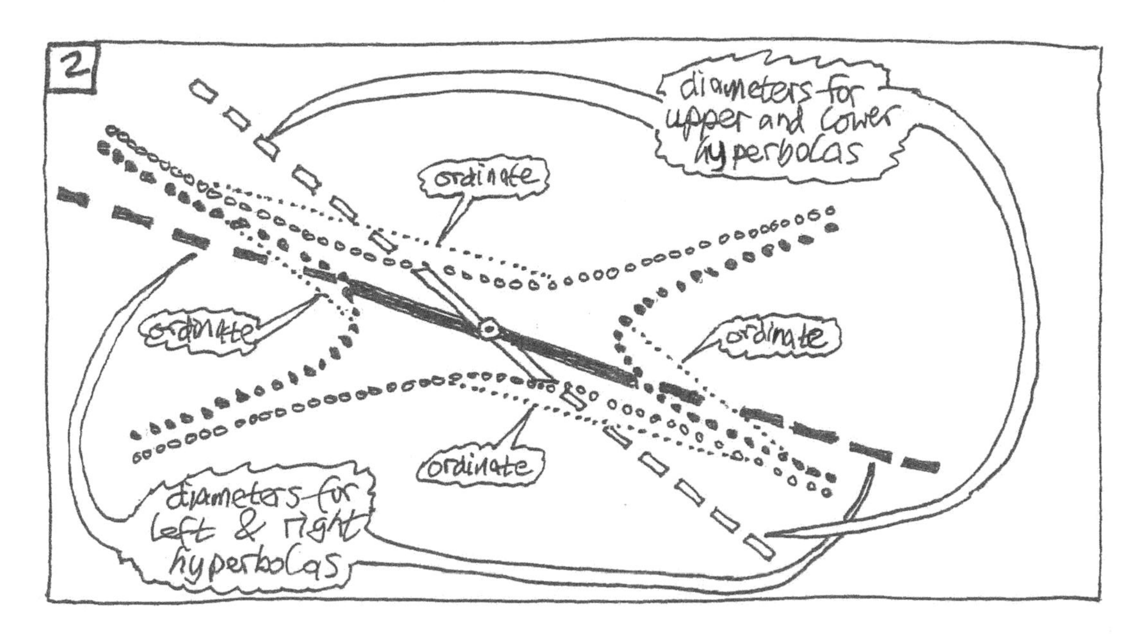

60th proposition

Here, then, is Apollonius's 60th proposition (fig. 3).

#60 Two straight [lines] being given, [straight lines, that is, which] cut one another in two [halves]—to carve out (*grapsai*) about each of them opposite sections, so that their conjugate diameters are what the straight [lines] are, and also, so that the diameter of the one pair of opposite [sections] has the power [to give rise to a square equal to] the figure (*eidos*) of the other pair of opposite [sections], and similarly the diameter of this other pair of opposite [sections] has the power [to give rise to a square equal to] the figure (*eidos*) of that first [pair of opposite sections].
—The two given straight [lines] which cut each other in two [halves]—let the [lines] AC, DE be they. Required, then, to carve out opposite [sections] about each of them as a diameter so that the [lines] AC, DE be conjugate [diameters] in them, and also, so that the [line] DE have the power [to give rise to a square equal to] the figure (*eidos*) of the [lines] about the [line] AC, and the [line] AC have the power [to give rise to a square equal to] the figure (*eidos*) of the [lines] about the [line] DE.
—Let there be equal to the [square] from DE the [rectangle] by ACL. At right angles let there be the [line] LC to the [line] CA. And two straight [lines] being given at right angles to one another, namely the lines AC, CL, let there be carved out (*gegraphthôsan*) opposite [sections], namely the [lines] FAG, HCK—whose oblique diameter is what the [line] CA will be and upright [side] is what the [line] CL will be, and [in which] the [lines] drawn down from the sections will be drawn down onto the [line] CA at the given angle. Then will the [line] DE be a second diameter of the opposite [sections]; for it is the mean proportional of the figure's (*eidous*) sides, and, being parallel to a [line] drawn down ordinatewise, it has been cut in two [halves] at the [point] B.
—Then again, let there be equal to the [square] from AC the [rectangle] by DE, DF, and let there be at right angles the [line] DF to the [line] DE. And two straight [lines] being given laid at right angles to one another, namely the [lines] ED, DF, let there be carved out (*gegraphthôsan*) opposite [sections], namely the [lines] MDN, OEX—whose oblique diameter is what the [line] DE will be and whose figure's (*eidous*) upright [side] is what the [line] DF will be, and [in which] the [lines] drawn down from the sections will be drawn down onto the [line] DE at the given angle. And also, of the [lines] MDN, XEO, [of the lines of section, that is,] a second diameter is what the [line] AC will be, and so the [line] AC (on the one hand) cuts in two [halves] the parallels to the [line] DE between the sections FAG, HCK, and the [line] DE (on the other hand) [does so to] the parallels to the [line] AC.
—Which very thing it was required to do [or "to make"] (*poêsai*). And let these [lines] be called the sections [that are] conjugate.

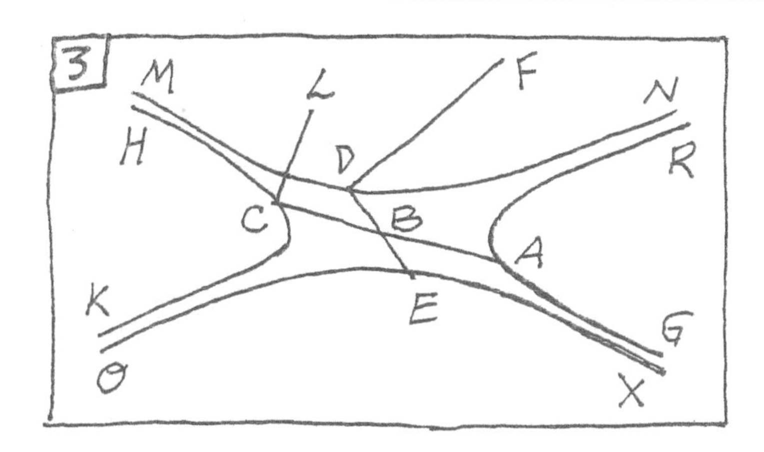

CHAPTER 7. GRAND FINALE AS CURTAIN-RAISER

The preceding chapter in this guidebook dealt with the concern from which arises the concluding proposition of Book One of the *Conics*; that chapter also dealt with how the proposition shows what it shows. Let us now consider the point of what the proposition shows.

In this proposition, Apollonius shows that for any pair of intersecting straight lines that have the same midpoint, there is a pair of opposite sections (in other words, there is a pair of paired hyperbolas) called conjugate opposite sections. They are called conjugate opposite sections because each one of the pair (of paired hyperbolas constituting opposite sections) has the same pair of conjugate diameters as the other. That is to say, four hyperbolas are said to constitute conjugate opposite sections when the following things are true of the oblique (transverse) diameter of each pair of opposite hyperbolas in the quartet: each pair's oblique (transverse) diameter is such that it—

> is parallel to the ordinates of
> the other pair's oblique (transverse) diameter; and
>
> has the same center as does
> the other pair's oblique (transverse) diameter; and
>
> is the mean proportional between
> the other pair's oblique (transverse) diameter
> and the upright side of that other pair's oblique (transverse) diameter.

In other words:

> the *second* diameter of *each* pair of hyperbolas
> is the *oblique (transverse)* diameter of the *other* pair (fig. 1).

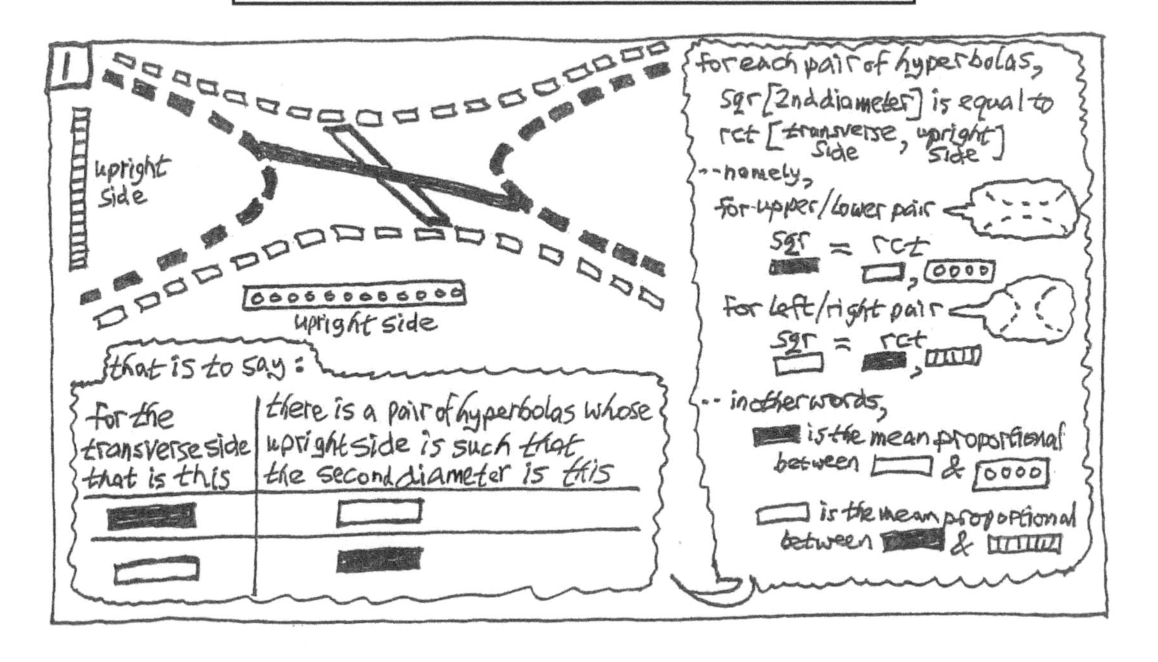

After the 14th proposition had completed the conic cutting that yielded the curves, the First Book became a tool-kit for the cone-construction that is needed at the end of the First Book. This cone-construction is the means of obtaining the curves as specified in the propositions from the 52nd proposition on. Having thus gone full circle from cones to cones, Apollonius will have no more need for cones. The 60th proposition does make use of the solution of the problem of finding a cone that can be cut to find a specified curve, but this final proposition shows what could not be obtained from a conic cutting: the quartet of hyperbolas that constitute conjugate opposite sections. A single act of cutting a cone will yield the parabola, or the hyperbola, or the ellipse, and a single act of cutting opposite conic surfaces will yield the two hyperbolas that constitute the opposite sections; but no single act of cutting a conic solid or opposite conic surfaces will yield four hyperbolas that constitute conjugate opposite sections. The 60th proposition, the last proposition of the First Book, is the curtain-raiser on a spectacular, new, entirely planar show in which conic things appear cut loose from conic solids and conic surfaces.

The curtain-raiser is also the grand finale of the First Book. It is the last problem in the sequence of problems that brings the First Book to completion, and it brings to completion the story of the second diameter. Why is that story complete only now in the 60th proposition? Because the introduction of the second diameter raised a question that could not be answered until the conjugate opposite sections were introduced.

As far apart as you take the endpoints of a conjugate diameter in the opposite sections, there are chords yet farther out that stretch between the paired hyperbolas—chords that would be bisected if you made the conjugate diameter even longer. That is why there has seemed to be no reason for a diameter in that position to be delimited by any definite bounds. Why, then, under the name of second diameter, was that line given, in the Second Definitions, the bounds that seemed appropriate to the ellipse of the 15th proposition but irrelevant to the opposite sections of the 16th? (Fig. 2.)

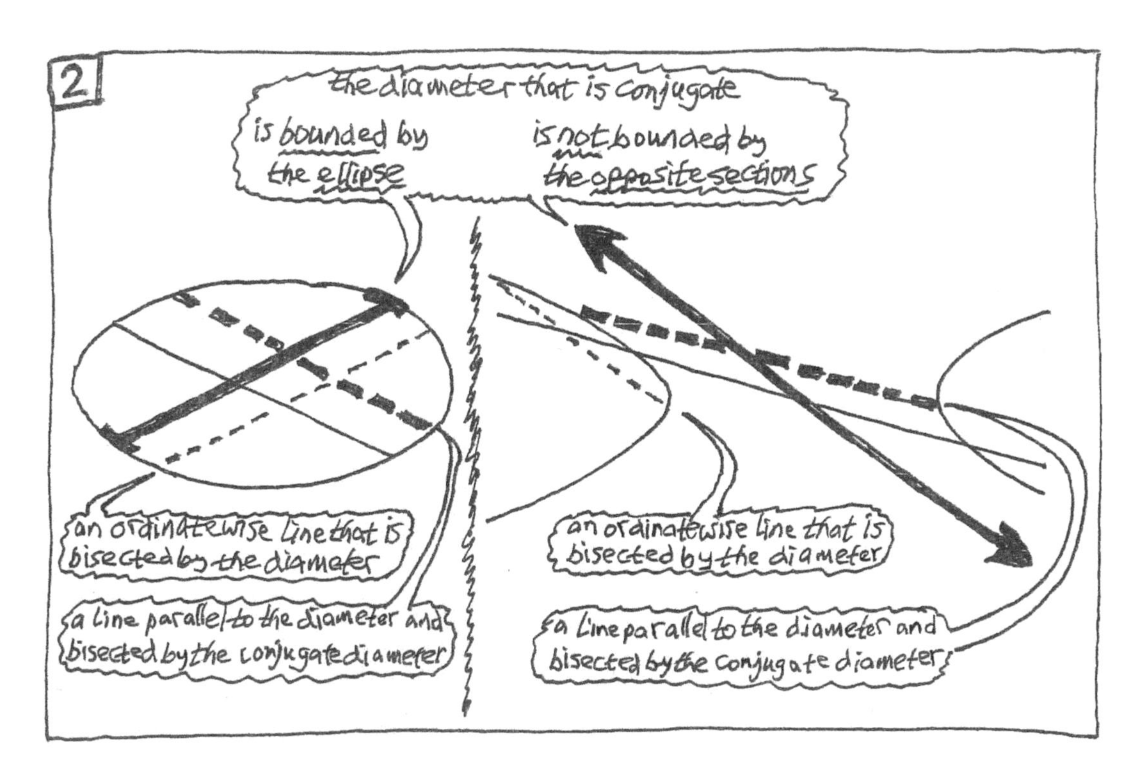

To answer that question, consider what the last proposition has shown us about the conjugate sections.

What makes things "conjugate"? *Con+jug+ate* translates into Latin the Greek *syn+zeug+eis*; it means "yoked together." Yoked together by what?

Well, what yokes together a pair of *diameters*, making them conjugate diameters, is what is essential to their being diameters at all. What makes a straight line a diameter of a conic section is what it does to straight lines parallel to a straight line drawn ordinatewise: a diameter bisects chords positioned ordinatewise. A diameter that is conjugate to a given diameter of a conic section is *the* other diameter of the given diameter; it is that *one* (of all other diameters of the section) which is positioned *ordinatewise* to the given diameter. Moreover, for a diameter to be conjugate to a given diameter, what is required is not only that it be ordinatewise to the given diameter, but also that the given diameter be ordinatewise to it—in other words, the relation is reciprocal: each of the conjugate diameters is ordinatewise to the other. So, two straight lines are conjugate diameters if each one is a diameter and the other one is ordinatewise to it—in other words, if each one is a diameter and the other one is in a position to be a line with reference to which the first one *is* a diameter. *Diameters* that are conjugate are thus as intimately connected as they could be. Note that what connects them is *position*; conjugate diameters are *not* as such of some definite *size*.

Now, what about *opposite sections* that are conjugate: how intimately connected are they—what yokes *them* together? Well, conjugate sections are a pair of opposite sections that share the same pair of conjugate diameters. One of the diameters binds one pair of hyperbolas together as opposite sections—that is to say: it is the oblique (transverse) side of the first pair of hyperbolas. And the other diameter, which is conjugate to that oblique (transverse) side of the first pair of hyperbolas, is also itself the oblique (transverse) side of the other pair of hyperbolas. And this other oblique (transverse) side has as the diameter conjugate to it the oblique (transverse) side of the first pair of hyperbolas. In other words, each pair's oblique (transverse) diameter has as its own conjugate diameter the other pair's oblique (transverse) diameter. Although the oblique (transverse) diameter that makes one pair of hyperbolas opposite sections has a definite size, there is no definite size for the diameter that is conjugate to that oblique (transverse) diameter; but if that conjugate diameter is also to be an oblique (transverse) diameter that will make another pair of hyperbolas opposite sections—ones that will, together with the first ones, constitute conjugate opposite sections—then it too must be bounded. That conjugate diameter must have a certain size, not merely a certain position.

So now we have a clue to an answer to our lingering question about the second diameter. Let us restate the question before we go on to the answer.

A conjugate diameter as such has a certain position. Its position in the ellipse gives it a certain size—but *only* in the ellipse. A conjugate diameter with that size is called a second diameter—but *not* only in the ellipse. The question is: why not only in the ellipse? Or, to sharpen the question: why in the opposite sections?

Now, in the 60th proposition, Apollonius has finally suggested an answer to our question about the second diameter:

> in the conjugate opposite sections,
> we are shown a straight line
> whose position makes it
> a diameter conjugate to the oblique (transverse) diameter of one hyperbolic pair,
> and whose size makes it also
> the oblique (transverse) diameter of the other hyperbolic pair;
>
> and that size which will fit it to be
> a second oblique (transverse) diameter
> in the conjugate opposite sections
> is the very size that fits the definition of
> *the second diameter* for the first of the oblique (transverse) diameters.

But why did we ask only about the opposite sections, and not about the parabola and hyperbola too? For this reason: a parabola does not have a center, and so it does not have a conjugate diameter; and a single hyperbola, while it does have a center and does also have a conjugate diameter, needs to be considered as one component in opposite sections before we can see the significance of calling a point that is in a certain position its "center" and a line that is in a certain position its "conjugate diameter." In like manner now with respect to the significance of giving the name of "the second diameter" to a line that is in a certain position and is of a certain size: we can see that significance only after we have considered the opposite sections as one component in conjugate opposite sections. The 60th proposition links the ellipse and opposite sections by way of the second diameter—and the second diameter only makes sense when we consider the conjugate opposite sections.

And yet, although the 60th proposition links the opposite sections and the ellipse by way of the second diameter, there is still a manifestly great difference between a single curve that is closed (the ellipse) and an open pair of curves each one of which is open (the opposite sections).

But is the difference between the ellipse and the opposite sections as great as it appears to be? Let us look a little closer at this configuration that is brought to our attention in the 60th proposition. Latent in this configuration of the conjugate opposite sections is the hint of a surprising likeness between an ellipse and a certain quartet of hyperbolas. The component hyperbolas are four and yet they are somehow also one, in such a way that what holds them apart is also what holds them together; this same thing will provide a theme for the further study of conics in the Second Book.

Apollonius has already shown in many propositions how the two hyperbolas that constitute opposite sections act in certain respects as much like a single thing as a parabola or hyperbola or ellipse does. Later on, he will show how the four hyperbolas that constitute conjugate opposite sections do so too. (The culmination of this will come in the 15th proposition of the Third Book.)

But there is this difference (mentioned previously) between the opposite sections and the conjugate opposite sections: although we can get the paired hyperbolas by a single conic cutting (we can get two curves by one cutting of opposite conic surfaces), we cannot get the paired pairs of hyperbolas by a single conic cutting. There is a way, however, to get a glimpse of the conjugate opposite sections as somehow resembling a single section.

Sometimes it is easy to relate features of curves in the plane but not so easy to find the corresponding related features in the cones or conic surfaces from which they come. While the cone supports our talk about the features of the curves obtained from it, our talk about the curves can get very complicated if we keep referring to the cone as we investigate configurations that contain those curves. If we had not, as it were, climbed up on the cone to look at those configurations, then we would have plummeted downward when, on looking down, we recognized our unsupported situation. But although we have climbed up on the cone, nonetheless to keep on looking down upon the steep and tangled path by which we have scrambled up could make us dizzy and impede our ascending any higher.

So, let us do this. Take the paired hyperbolas that constitute opposite sections, and, with this pair's oblique (transverse) diameter and its second diameter, get the other paired hyperbolas that constitute opposite sections by virtue of having those same conjugate diameters. Now take these four curves that constitute conjugate opposite sections—and flip them inside out. There will then be gaps between them (fig. 3).

But if, before you flip them, you prolong the curves enough, then you can make the gaps as small as you please, and thus the four-curve configuration will get as close as you please to being a single closed curve (fig. 4).

On the other hand, the shorter the stretch that you take along the four curves before you flip them (if you flip them around their midpoints rather than around the line connecting their endpoints), the closer the four-curve configuration will come to small stretches of a certain small ellipse (fig. 5).

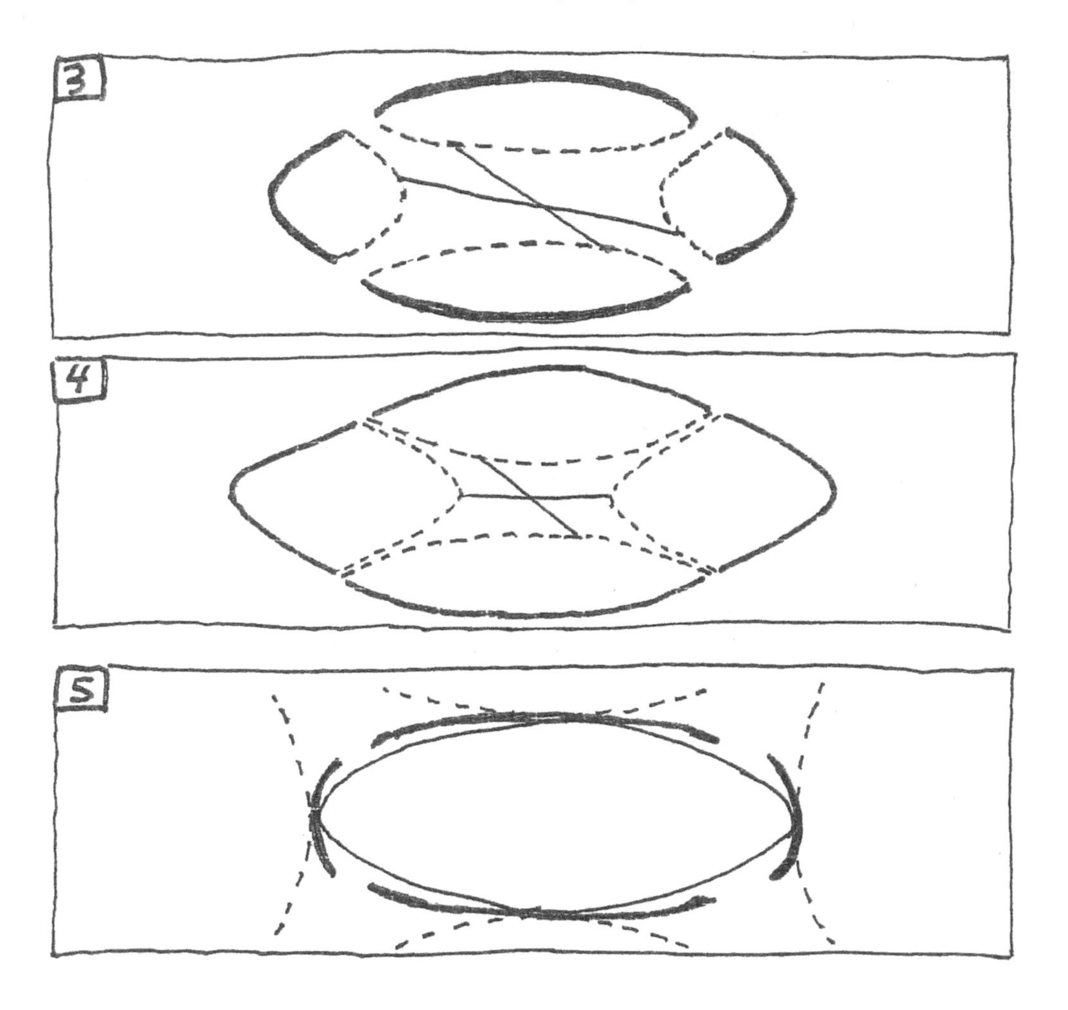

Now consider, not just any conjugate diameters, but, rather, conjugate diameters that are conjugate axes. That is to say: consider conjugate opposite sections whose common conjugate diameters are at right angles to each other. As it turns out, these conjugate diameters that are conjugate axes for the conjugate opposite sections will also be conjugate axes for that small ellipse inside the conjugate opposite sections. The vertices of the quartet of hyperbolas that constitute the conjugate opposite sections will be the vertices of the small ellipse inside (fig. 6).

But why is it that the four hyperbolas, when flipped inside out, will come as close together as you please if you prolong them far enough before you do the flipping? Because the farther out you go, on any one of two adjacent curves of the conjugate opposite sections before you flip them, the closer you will get to the other adjacent curve. The two curves never meet, and yet, far enough out, they do get as close as you please to traveling side-by-side in the same direction—indeed, along the same straight line. Each, that is to say, approaches a straight line between them—a straight line through the center that is parallel to a straight line joining the axial vertex of one of the two adjacent curves to the axial vertex of the curve opposite the other of the two adjacent curves (fig. 7).

In the Second Book of the *Conics*, Apollonius will show that the hyperbolas in conjugate opposite sections are intimately connected because they approach as close as you please (without touching) to the same straight lines—and hence also to each other. The hyperbolas have the utmost intimacy short of touching—that is to say, they are conjugate.

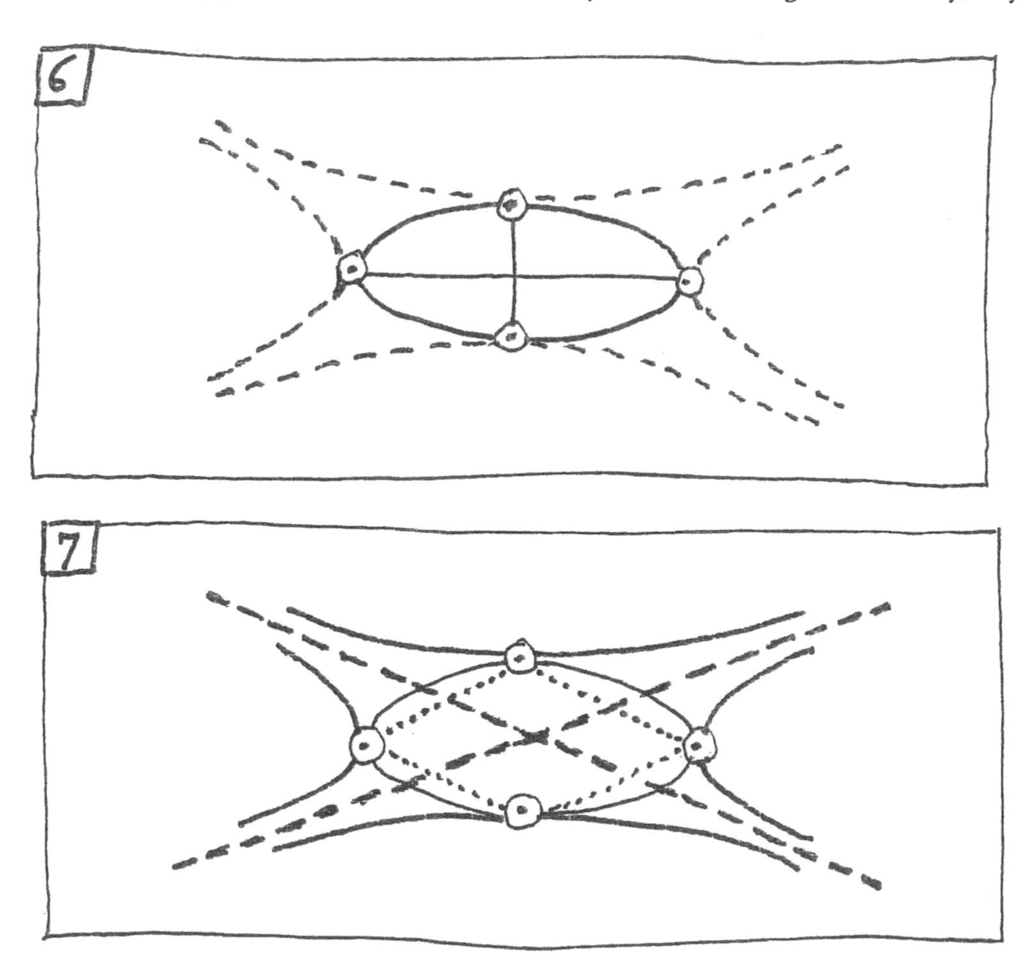

We have seen that a pair of hyperbolas that are opposite sections are held together by a straight line that holds them apart—their common oblique (transverse) side. It holds them a definite distance apart. Now we see another straight line holding hyperbolas together by holding them apart. But this straight line holds apart, not two hyperbolas that are opposite sections, but rather two adjacent hyperbolas in conjugate opposite sections. And it does not bring them together by holding them a definite distance apart (its length) but rather it holds them apart by being a boundary which is reached by neither of them but to which each of them does get closer than any definite distance.

There are thus two ways in which a straight line can be a barrier that binds: it can do so through the *length* of it, and it can do so *along* it (fig. 8).

Straight lines that are nowhere met by an hyperbola that nonetheless gets as close to them as you please are called the hyperbola's *asymptotes*. (*A+sym+ptote* is Greek for "not falling together.") Apollonius will show, early in the Second Book, that hyperbolas have such asymptotes almost touching them; right after showing this, he will show (15th proposition) that opposite sections have common asymptotes and (17th proposition) that conjugate opposite sections also do (fig. 9).

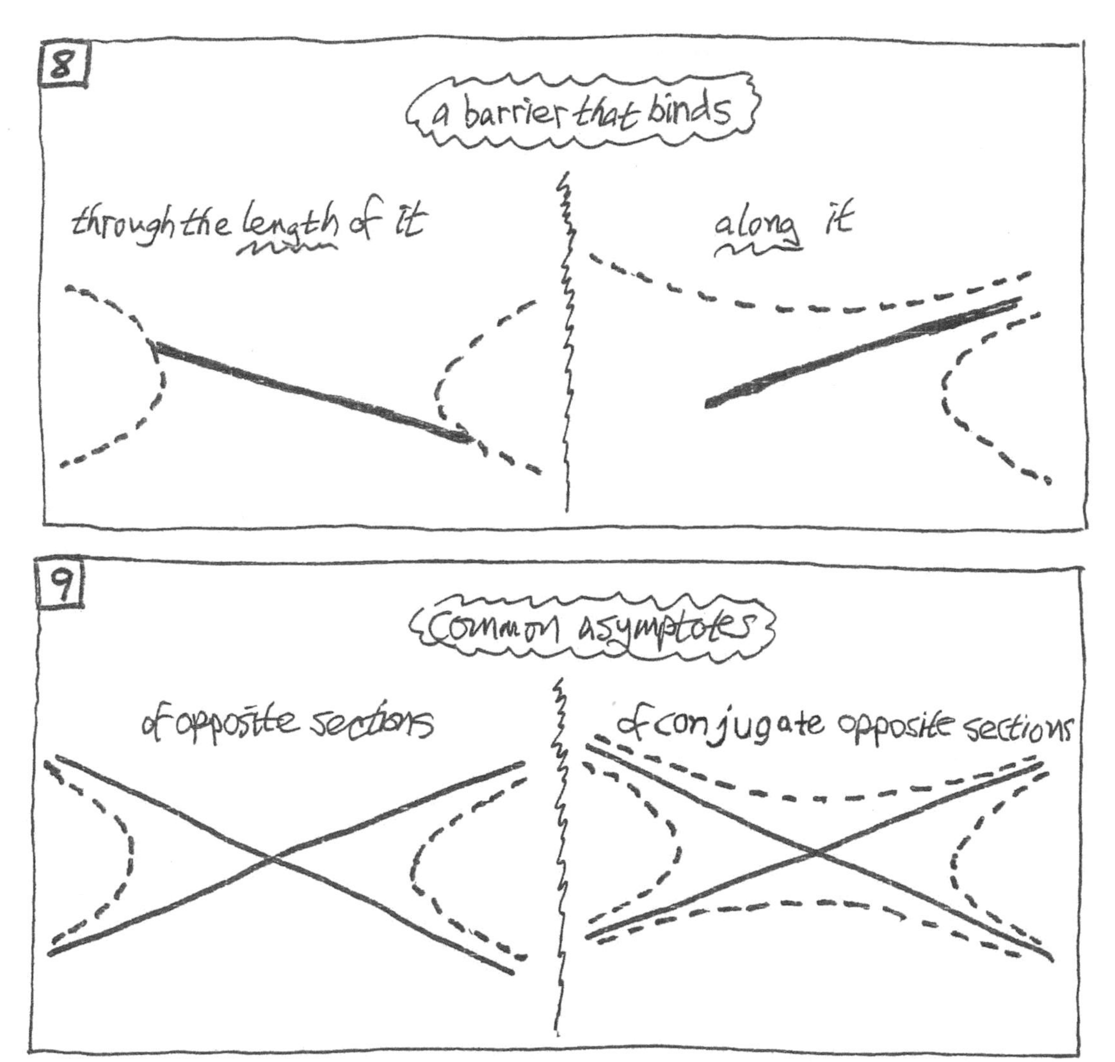

So, what has happened along our way from the last of the conic cuttings (in the 14th proposition) to the last of the conic findings (in the 60th proposition) is the following.

The 15th proposition showed us the conjugate diameter of the original diameter for the ellipse. The conjugate diameter was parallel to the ordinates of the original diameter. Both of the diameters were bounded by the ellipse.

The 16th proposition showed us the conjugate diameter of the original diameter for the opposite sections. The conjugate diameter was parallel to the ordinates of the original diameter. Only the original diameter was bounded by the opposite sections. The conjugate diameter was therefore not bounded at all.

The conjugate diameter of the original diameter was then bounded in the Second Definitions, which called it the second diameter. That bounding was not restricted to the ellipse, but it seemed to make sense only for the ellipse—and not for the opposite sections.

As we proceeded through the First Book, this line's position made more and more sense for all central sections: a line that passes through the center and is parallel to the ordinates behaved in more and more ways like another diameter. But nothing until the 60th proposition made sense of the line's size.

If you flip the conjugate opposite sections, the pair of pairs of curves when flipped will get more and more like a single closed curve as you prolong the hyperbolas before you flip them, since any two adjacent hyperbolas that get flipped inside out will approach as a limit the same straight line, their common asymptote. When the curves are prolonged, the gaps between them after flipping are smaller.

When the curves are shortened, however, although the gaps between them after flipping will get larger, the four curves do get closer to small arcs of a small ellipse inside the space between the conjugate opposite sections—the ellipse whose conjugate axes are the same as those of the conjugate opposite sections.

The 60th proposition returns in a way to the 16th (and, with it, to the 15th). After the First Book is brought to an end by the 60th proposition, the Second Book will not move inward to the space between the quartet of hyperbolas; rather, it will move outward to show that while an hyperbola grows indefinitely wide as it grows indefinitely long, nonetheless its growing wide is limited. It grows within the sides of an angle. That angle can be obtained by slicing the cone with a plane that goes through the vertex parallel to the plane that yields the section. But we need not leave the plane in order to get that angle. We can get the asymptotes in the plane of the section. Apollonius now moves on to doing that, in the next book of his *Conics*.

APPENDIX

Diagrams for Selected Propositions:

17–28
31
35–36
#52–58

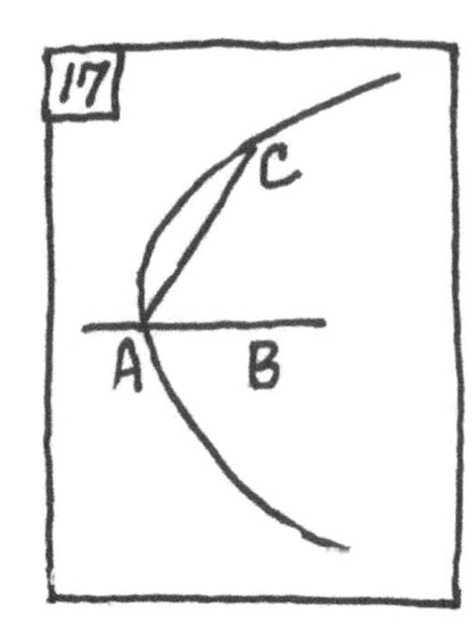
17
C
A
B

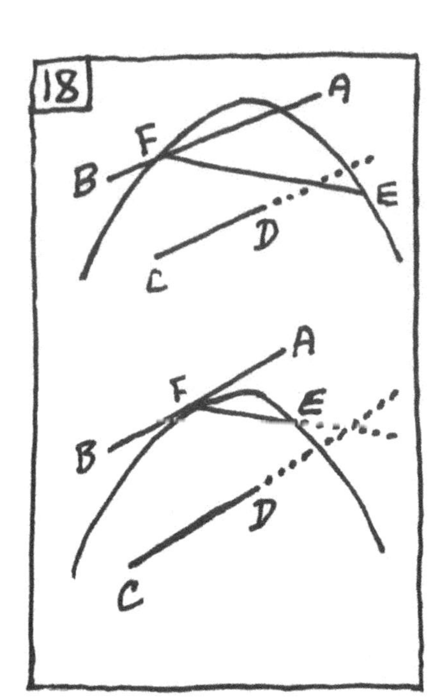
18
A
F
B
E
D
C
A
F
E
B
D
C

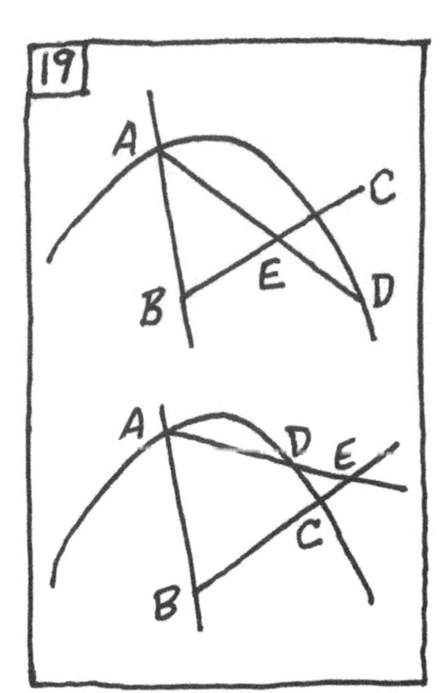
19
A
C
E
B
D
A
D
E
C
B

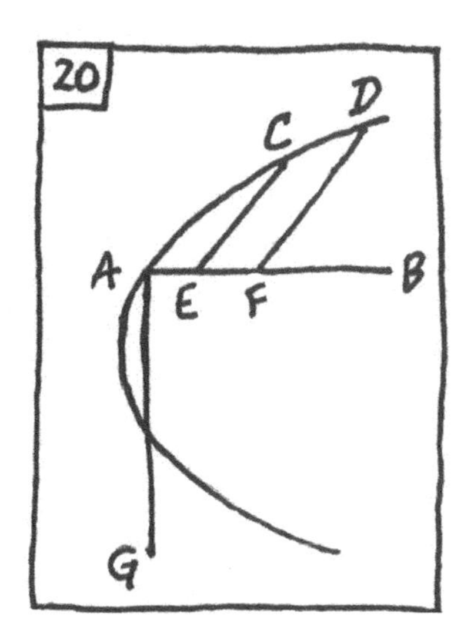
20
C
D
A
E
F
B
G

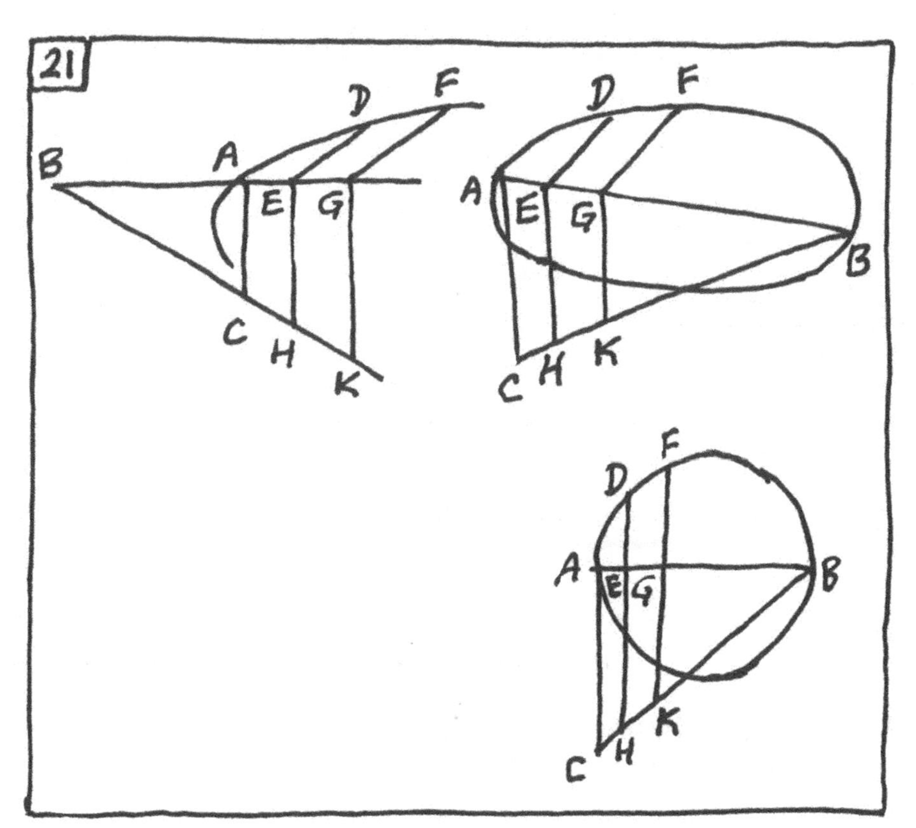
21
B
A
D
F
E
G
C
H
K
D
F
A
E
G
B
C
H
K
D
F
A
E
G
B
C
H
K

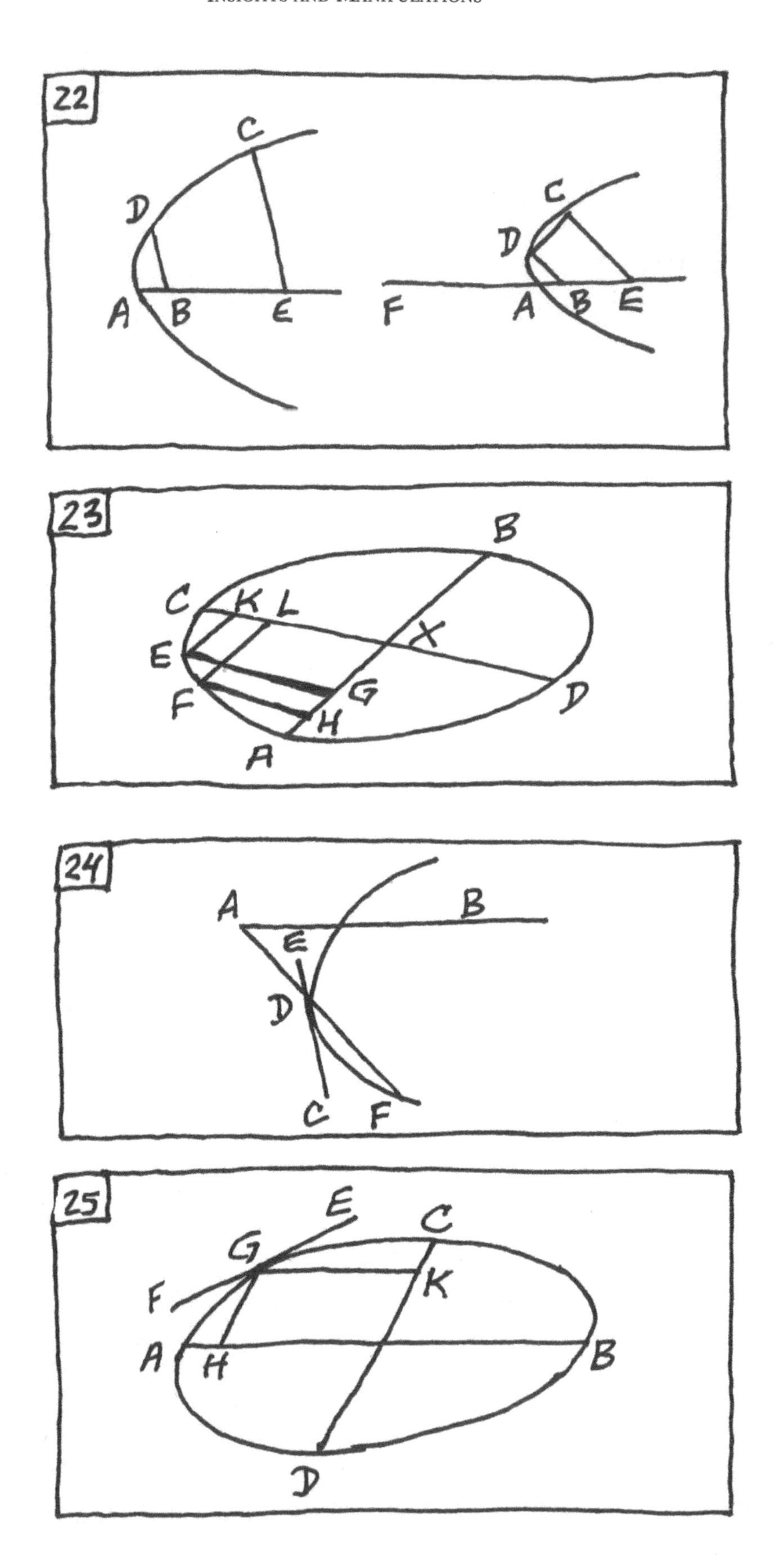
22
C
D
A
B
E
F
A
B
E
C
D
23
B
C
K
L
X
E
G
D
F
H
A
24
A
B
E
D
C
F
25
E
C
G
K
F
A
H
B
D

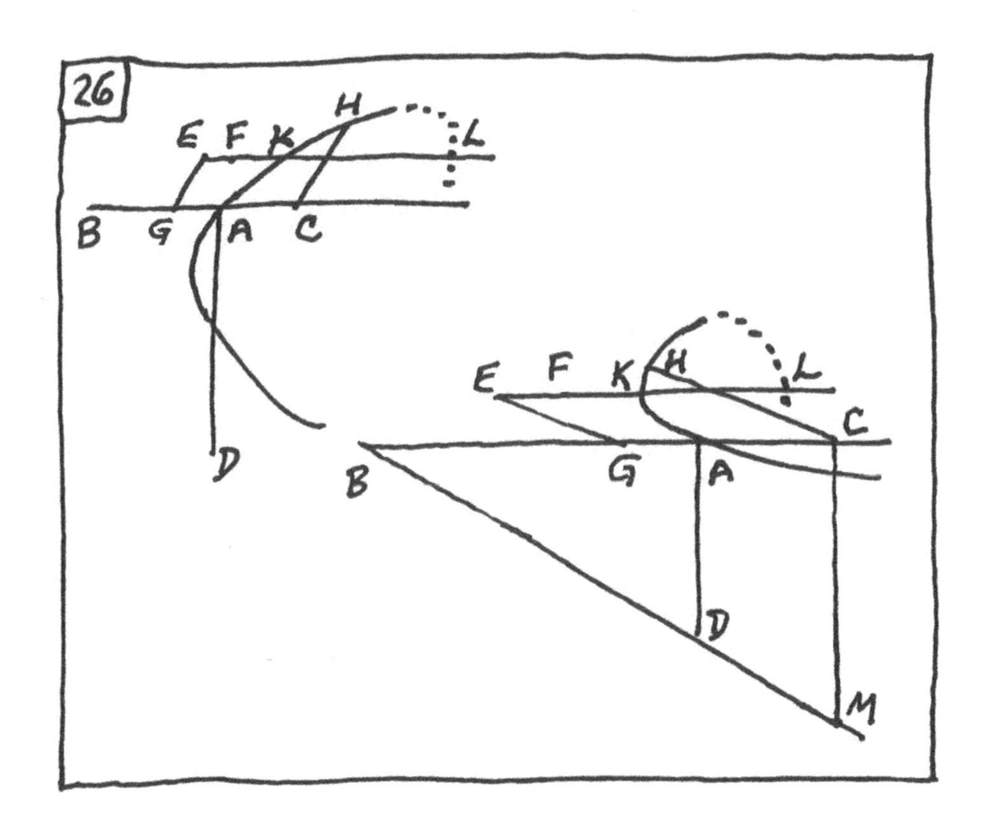
26
E F K H L
B G A C
D
E F K H L
C
B G A
D
M

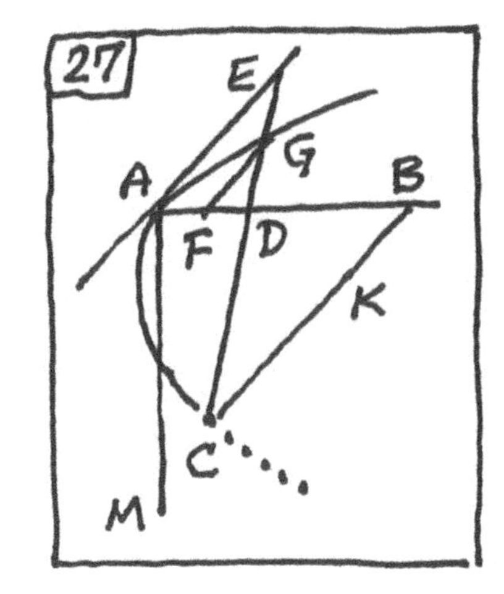
27
E
G
A B
F D
K
C
M

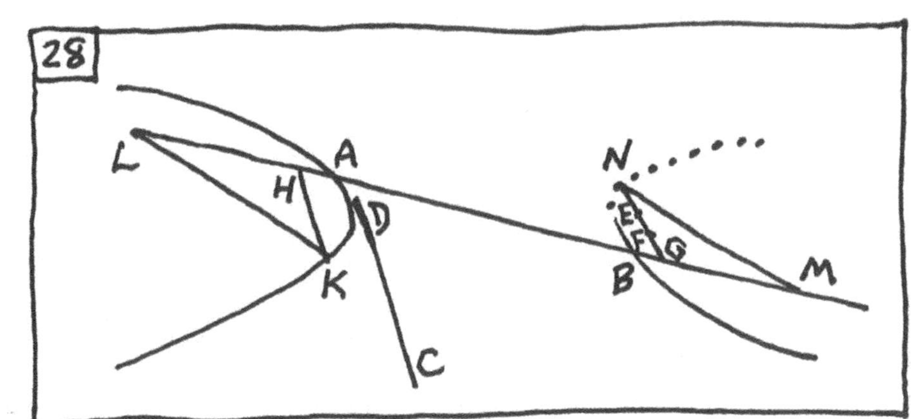
28
L A
H D
K
C
N
E
F G
B M

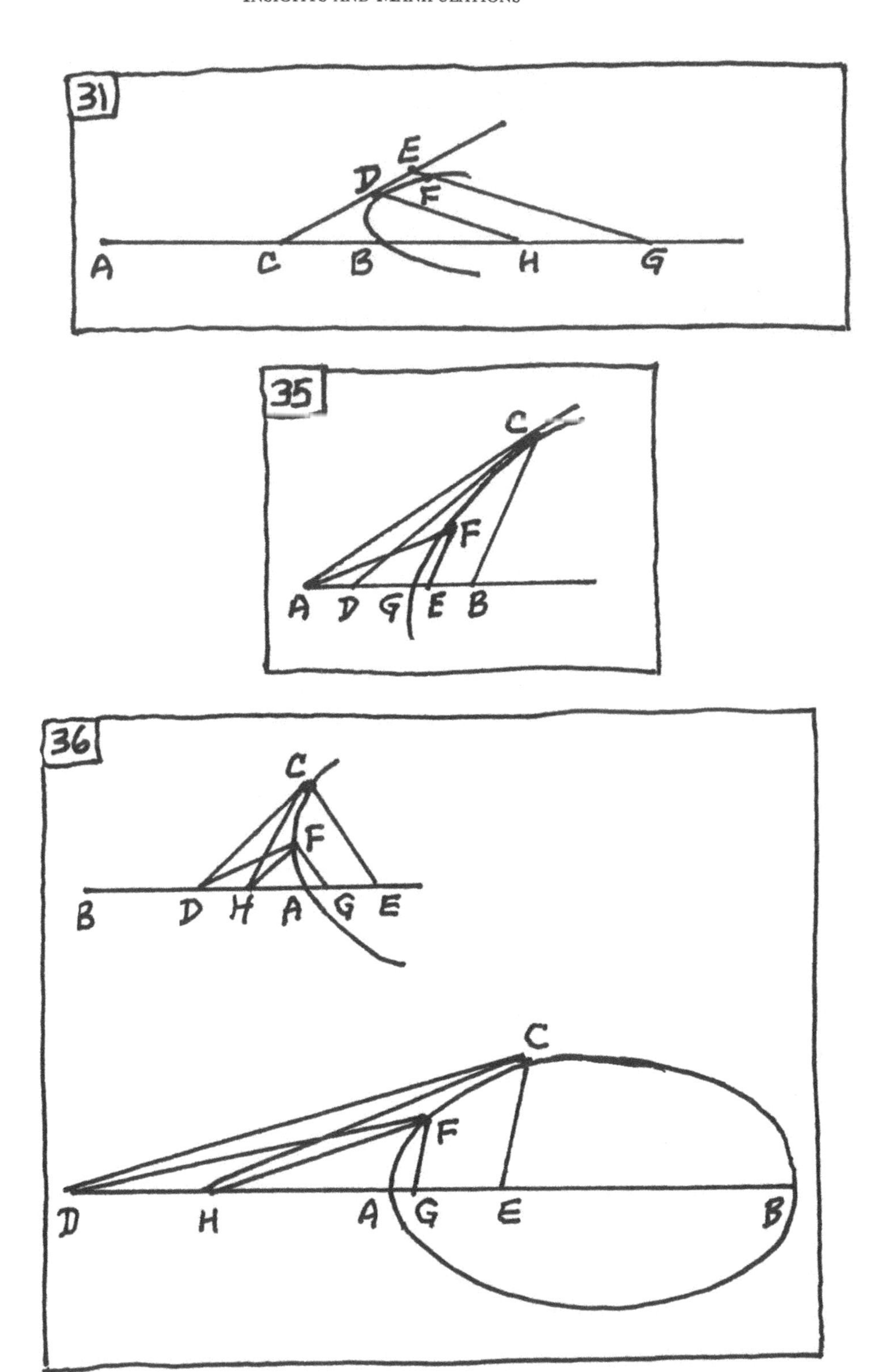
31
E
D
F
A
C
B
H
G
35
C
F
A
D
G
E
B
36
C
F
B
D
H
A
G
E
C
F
D
H
A
G
E
B

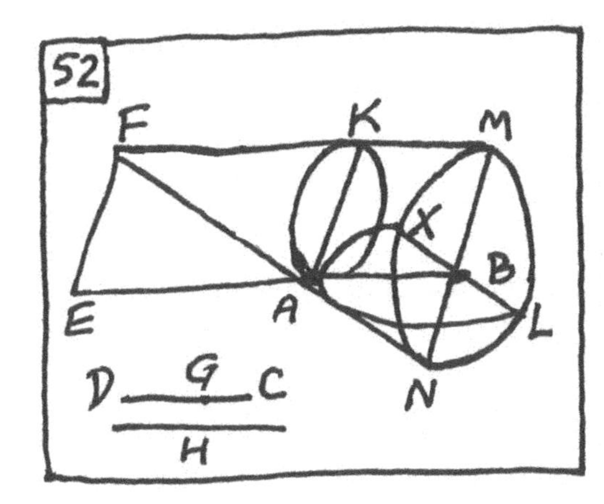
52
F
K
M
X
B
E
A
L
D
G
C
N
H

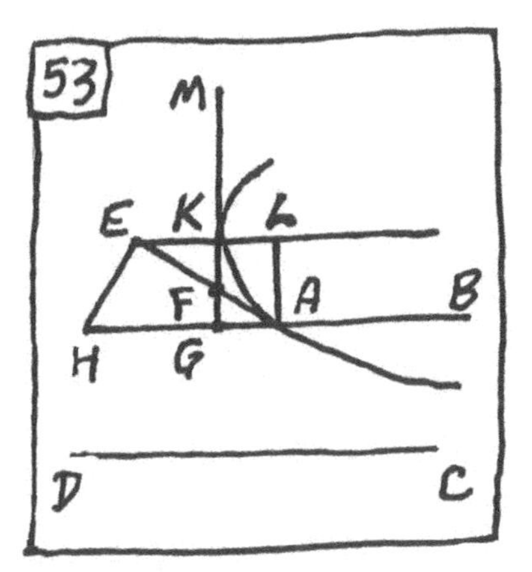
53
M
E
K
L
F
A
B
H
G
D
C

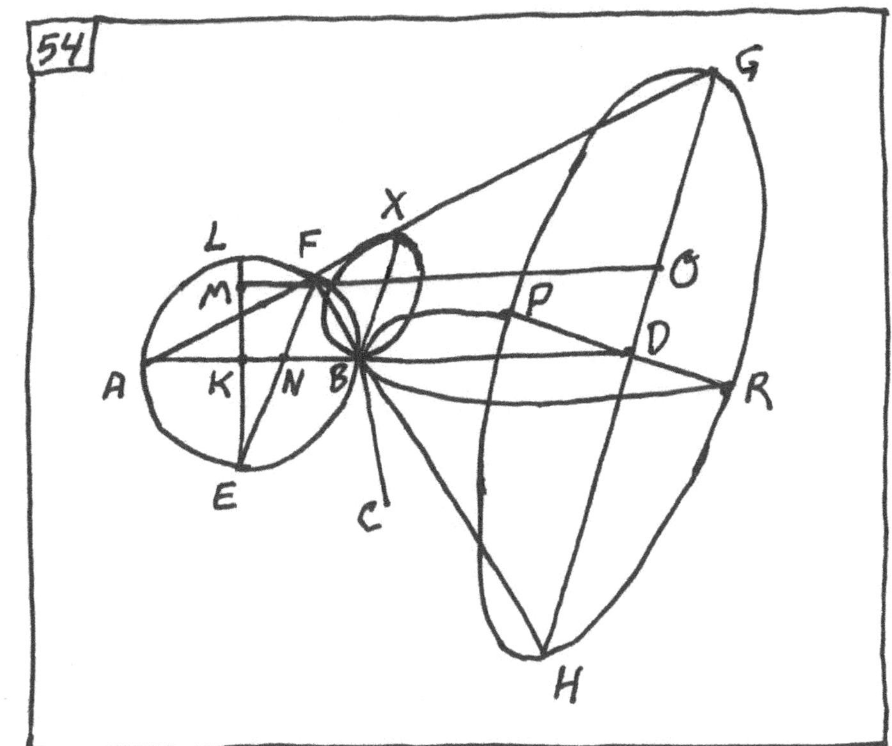
54
G
X
L
F
O
M
P
D
A
K
N
B
R
E
C
H

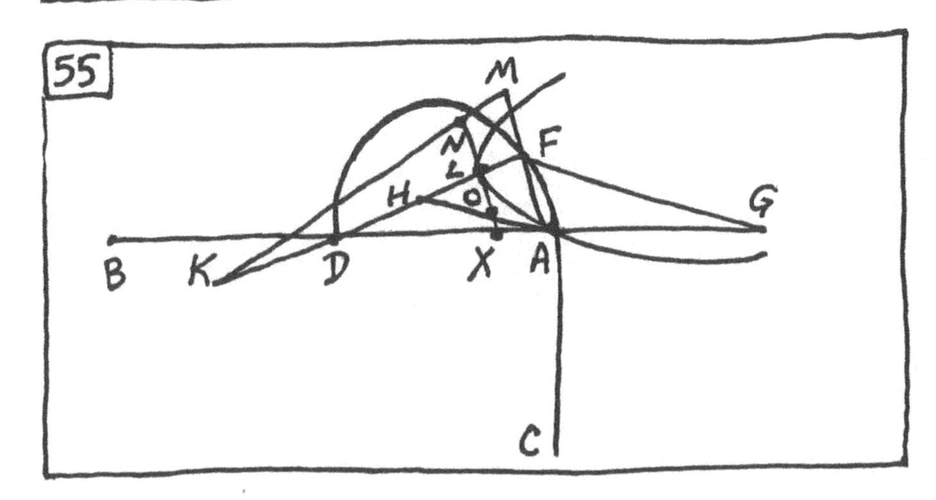
55
M
N
F
L
H
O
G
B
K
D
X
A
C

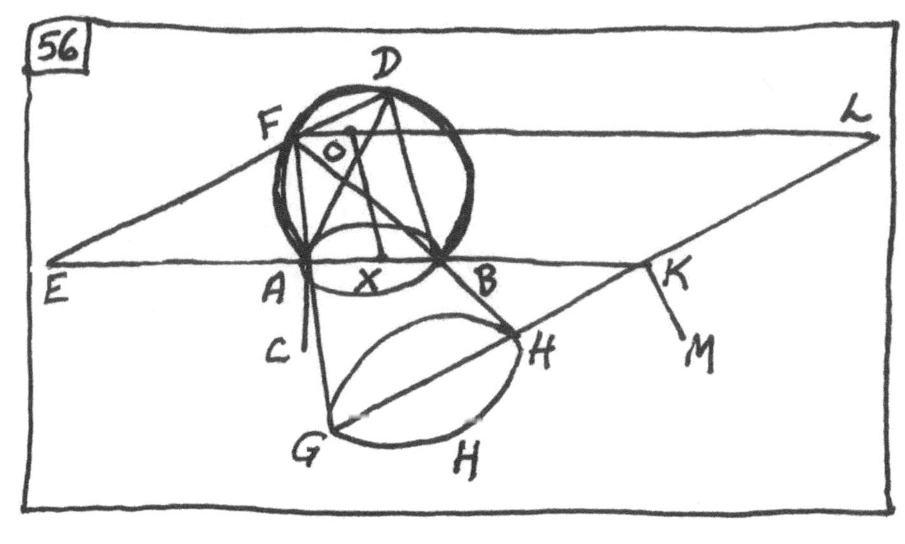
56
D
F
O
L
E
A
X
B
K
C
H
M
G
H

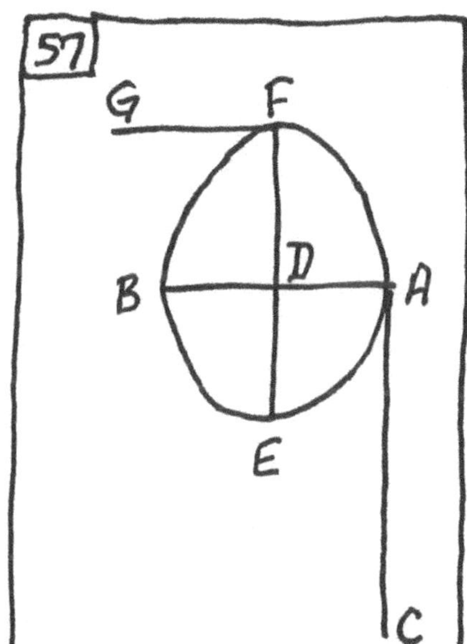
57
G
F
D
B
A
E
C

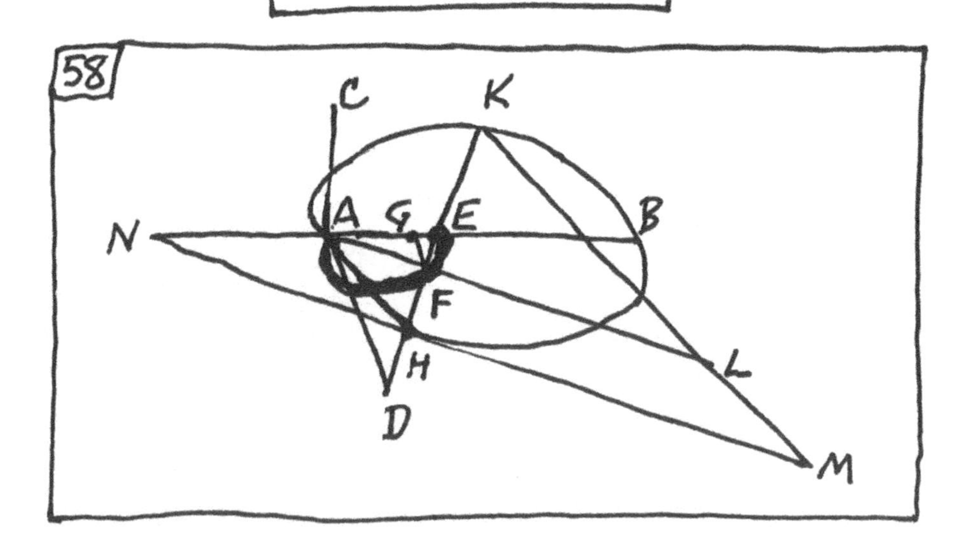
58
C
K
N
A
G
E
B
F
H
L
D
M

VOLUME TWO: THE WAY TO CARTESIAN MASTERY

PART VI

MORE VIEWS OF CONICS,
AND SOME GEOMETRICAL PROBLEMS

Apollonius's *Conics*, Books II–III: selections

CHAPTER 1. ASYMPTOTES AT THE BEGINNING OF BOOK II OF THE CONICS

As we move on from Book One of Apollonius's *Conics*, let us remind ourselves of what it shows us. Its beginning culminates in propositions that characterize the different sorts of curves that are conic sections. From the succeeding propositions in Book One, we learn that conic sections are the only curves that can have the characteristics of those curves which were presented in the beginning of the book. The key to understanding those characteristics is a certain "upright" side, while a certain "slantwise" (or "transverse") side is the key to a deeper examination of the curves. When these two sides interchange their functions in a certain manner, they provide a new spectacle, in which there lurk two lines that constitute as tight a frame as possible for a unitary configuration that is manifestly not a single curve.

Now, at the beginning of Book Two of the *Conics*, the lurking lines manifest themselves as "asymptotes." The 1st proposition shows that certain straight lines are an hyperbola's asymptotes, and the 2nd proposition justifies considering them to be "the" asymptotes of the hyperbola. Later on, the 15th proposition shows that the opposite sections have the same asymptotes; and the 17th, that conjugate opposite sections too have the same asymptotes.

We shall ignore almost everything that lies between the first seventeen propositions and the ten propositions (stretching from the 44th through the 53rd) with which the Second Book concludes. Of these concluding ten propositions, nine are problems, and the one of them that is not a problem (the 52nd) is a theorem that is needed for the last of those problems. (The 48th is also not in itself a problem, but because it shows the uniqueness of the solution to the problem of the proposition that immediately precedes it, it is, in effect, part of the treatment of that problem.)

Almost everything that comes between the beginning of Book Two and its conclusion will be ignored. We shall not ignore everything, however. We shall have to pause after the 17th proposition, in order to look at the proposition that comes right after the central proposition of the second Book, as well as to look at the one that comes immediately after that, and also to look at the converse, which in its turn is next after that.

In Book Two, that is to say, we shall be proceeding more briskly than we did in Book One. Our examination of Book One was sufficient to enable us to appreciate the articulation of an Apollonian book as an integral whole. We need not repeat the experience. Instead, we shall first be taking a fairly quick look at the beginning of Book Two, for which the very last proposition at the end of Book One was a preview. Some additional propositions at the beginning of Book Two will be required for our examination of the conclusion of Book Two, to which we shall then turn. In the conclusion of Book Two, Apollonius analyzes problems. The examination of this analytic work presented by Apollonius will prepare us for the later critique of Apollonius presented by Descartes.

Because we shall be using books Two and Three of the *Conics* as mere quarries for propositions, the explications that accompany the texts will not be as expansive as they were with Book One, and whatever step-by-step depictions are embedded in the explications here may leave readers wanting more complete diagrams as they study the propositions. Therefore, as with some of the propositions in Book One, complete diagrams for those propositions of Book Two which are selected for inclusion in this chapter and the next, and for those propositions of Book Three which are selected for inclusion after that, in Chapters 3 and 6, are included in *the Appendix at the end of this second volume of the guidebook*.

Here, then, is the beginning of Book Two.

the first four propositions of Book Two: asymptotes

The first two of the propositions are presented in the following boxes.

(II) #1 If, with respect to an hyperbola, a straight [line] touch [the section] at its vertex, and from the touching[line], on both sides of the diameter, there be taken off a [line] equal to the [line] having the power [to give rise to a square that is equal to] a fourth of the figure (*eidous*)—then the straight [lines] that are drawn from the center of the section, to the bounds taken on the touching[line], will not fall together with the section.
—Let there be an hyperbola whose diameter is what the [line] AB is, and center is what the [point] C is, and upright [side] is what the [line] BF is; and also let there touch the section at the [point] B the [line] DE; and also, let there be equal to the fourth of the figure (*eidous*) [contained] by the lines ABF the [square] from each of the [lines] BD, BE; and also, let the joined [lines] CD, CE be extended. I say that [these lines CD and CE, extended,] will not fall together with the section.
—For, if possible, let the [line] CD fall together with the section at the [point] G; and also, from the [point] G let there be drawn down ordinatewise the [line] GH; it is therefore parallel to the [line] DB. Since then as is the [line] AB to BF so is the [square] from AB to the [rectangle] by ABF—but of the [square] from AB (on the one hand) a fourth part is the [square] from CB, and of the [rectangle] by ABF (on the other hand) a fourth is the [square] from BD—therefore, as is the [line] AB to BF so is the [square] from CB to the [square] from DB, that is, the [square] from CH to the [square] from HG. And also, as is the [line] AB to BF so is the [rectangle] by AHB to the [square] from HG. Therefore, as is the [square] from CH to the [square] from HG so is the [rectangle] by AHB to the [square] from HG. Therefore equal is the [rectangle] by AHB to the [square] from CH—which very thing is absurd (*atopon*). Not so, therefore, that the [line] CD falls together with the section. Likewise, then, shall we show that neither does the [line] CE [fall together with the section]. Therefore [lines that are] incapable of falling together with (*asumptotai*) the section are what the [lines] CD, DE are.

[These lines CD and CE, here characterized as "asymptotes," are not merely any lines that are "incapable of falling together with" the section. The term is indeed used with such generality by Euclid to refer to any straight lines or planes that do not cut other lines or surfaces. But Apollonius later makes clear (at the end of the demonstration of #14 in this second book of his *Conics*) that lines like these asymptotes are, in a special sense, "*the* asymptotes" to the section. That is to say: among lines that cannot meet the section however far extended, they are those that can get closer to it than by any given distance, however small that distance may be, if they are extended far enough—and any other lines that cannot meet the section however far extended, but that can get closer to it than by any given distance if extended far enough, nonetheless cannot be closer than are these. The term "asymptote" subsequently came to be used only in this restricted sense.]

(II) #2 With the same things being so, what is to be shown is that another "asymptote" is not what a [line] is which cuts the angle contained by the [lines] DCE.

—For, if possible, let [another asymptote] be what the [line] CH is; and also, through the [point] B let there be drawn parallel to the [line] CD the [line] BH; and also, let [the line BH] fall together with the [line] CH at the [point] H; and also, equal to the [line] BH let there be put the [line] DG; and also, let the joined [line] GH be extended to the [points] K, L, M. Since then the [lines] BH, DG are equal and parallel, the [lines] DB, HG also are equal and parallel. And since the [line] AB is cut in two [halves] at the [point] C and there is put onto it a certain [line], namely the [line] BL, [that is, in traditional parlance, BL is "added to" AB], the [rectangle] by ALB is, [when taken together] with (*meta*) the [square] from CB, equal to the [square] from CL. Likewise then, since parallel is the [line] GM to the [line] DE, and equal is the [line] DB to the [line] BE, therefore equal is also the [line] GL to the [line] LM. And since equal is the [line] GH to the [line] DB, therefore greater is the [line] GK than the [line] DB. But also is the [line] KM greater than the [line] BE, since also the [line] LM is greater; therefore the [rectangle] by MKG is greater than the [rectangle] by DBE, that is, greater than the [square] from DB. Since, then, as is the [line] AB to BF so is the [square] from CB to the [square] from BD, but (on the one hand) as is the [line] AB to BF so is the [rectangle] by ALB to the [square] from LK, and (on the other hand) as is the [square] from CB to the [square] from BD so is the [square] from CL to the [square] from LG—therefore also as is the [square] from CL to the [square] from LG so is the [rectangle] by ALB to the [square] from LK. Since, then, as is a whole, namely the [square] from LC, to a whole, namely the [square] from LG, so is what is taken away, namely the [rectangle] by ALB, to what is taken away, namely the [square] from LK—therefore also the remainder, namely the [square] from CB, is to the remainder, namely the [rectangle] MKG, as the [square] from CL is to the [square] from LG, that is, the [square] from CB to the [square] from DB. Therefore equal is the [square] from DB to the [rectangle] by MKG—which very thing is absurd; for greater than it [greater, namely, than that square] has [this rectangle] been shown [to be]. Not so, therefore, that the [line] CH is an asymptote to the section.

The 1st proposition shows that certain lines drawn from an hyperbola's center will not meet the section no matter how far it or they may be extended. The enunciation is depicted in figure 1.

The demonstration shows the absurdity that would follow from supposing that at some point the section will be met by the line CD, drawn in the designated manner. This *reductio* proceeds as follows: if possible, let CD meet the section at some point (call it G); then, from this supposed meeting-point G, draw a line (GH) parallel to the tangent that goes through the vertex (B); and the point where this line parallel to the tangent intersects the diameter, call point H. This is depicted in figure 2.

Now from the existence of that line GH which has just been drawn, two contradictory conclusions follow.

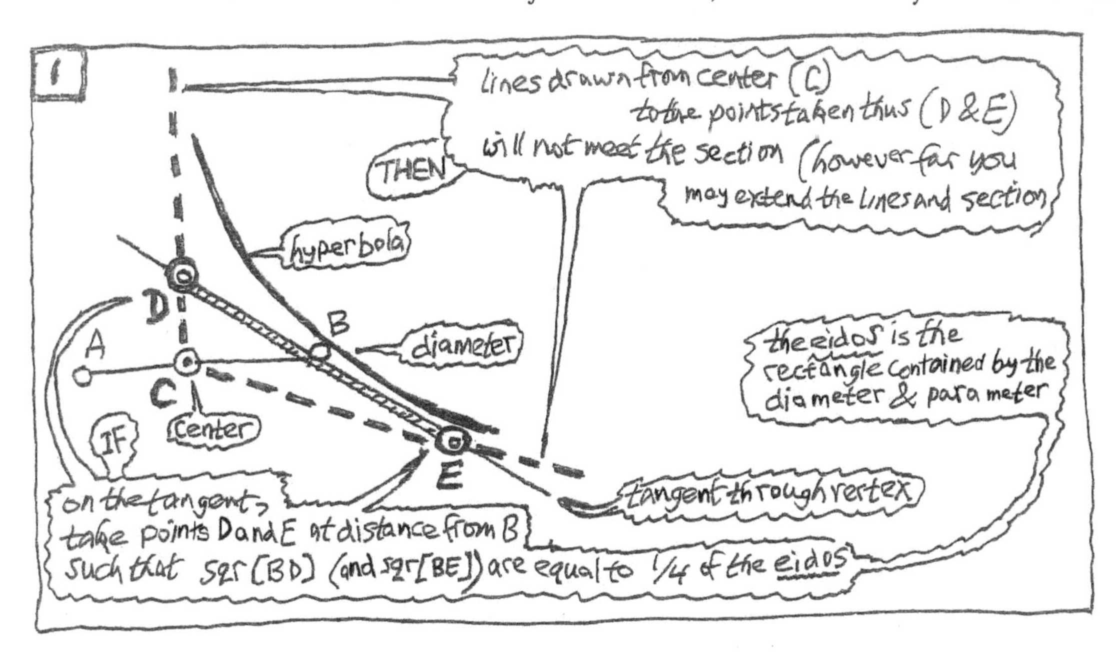

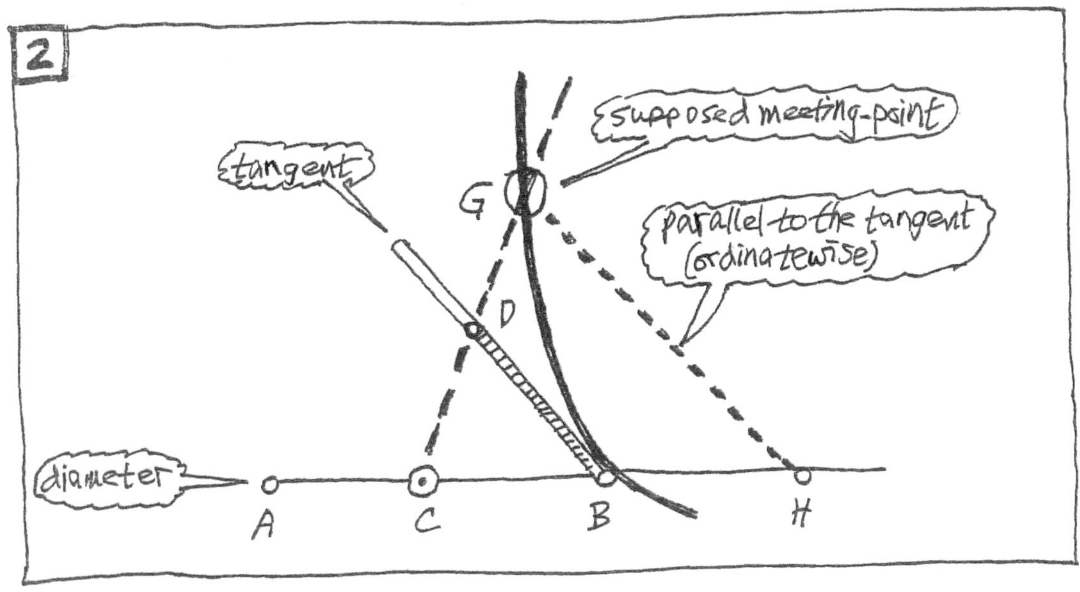

The *first* conclusion is that there is a certain *equality*. Since point G is on the section, and point H is on the diameter, and GH is parallel to the tangent that passes through the vertex, therefore line GH is an ordinate; and hence the rectangle contained by the lines stretching from H to the diameter's endpoints (A and B)—namely, the rectangle [HA, HB]—must be equal to the square [HC] raised on the line stretching from H to the diameter's midpoint C. That conclusion is derived by applying the 21st proposition of Book One of Apollonius's *Conics*.

However, *another* conclusion is derived by applying the 6th proposition of Book Two of Euclid's *Elements*: rectangle [HA, HB] must be *smaller* than that square [HC].

And thus we are forced to *contradictory* conclusions: the rectangle is shown to be *smaller* than the square to which it has been shown to be *equal*. By drawing the line GH from that point G where the section is supposedly met by line CD extended, we have been led into absurdity. That is why there cannot be any such meeting-point (fig. 3).

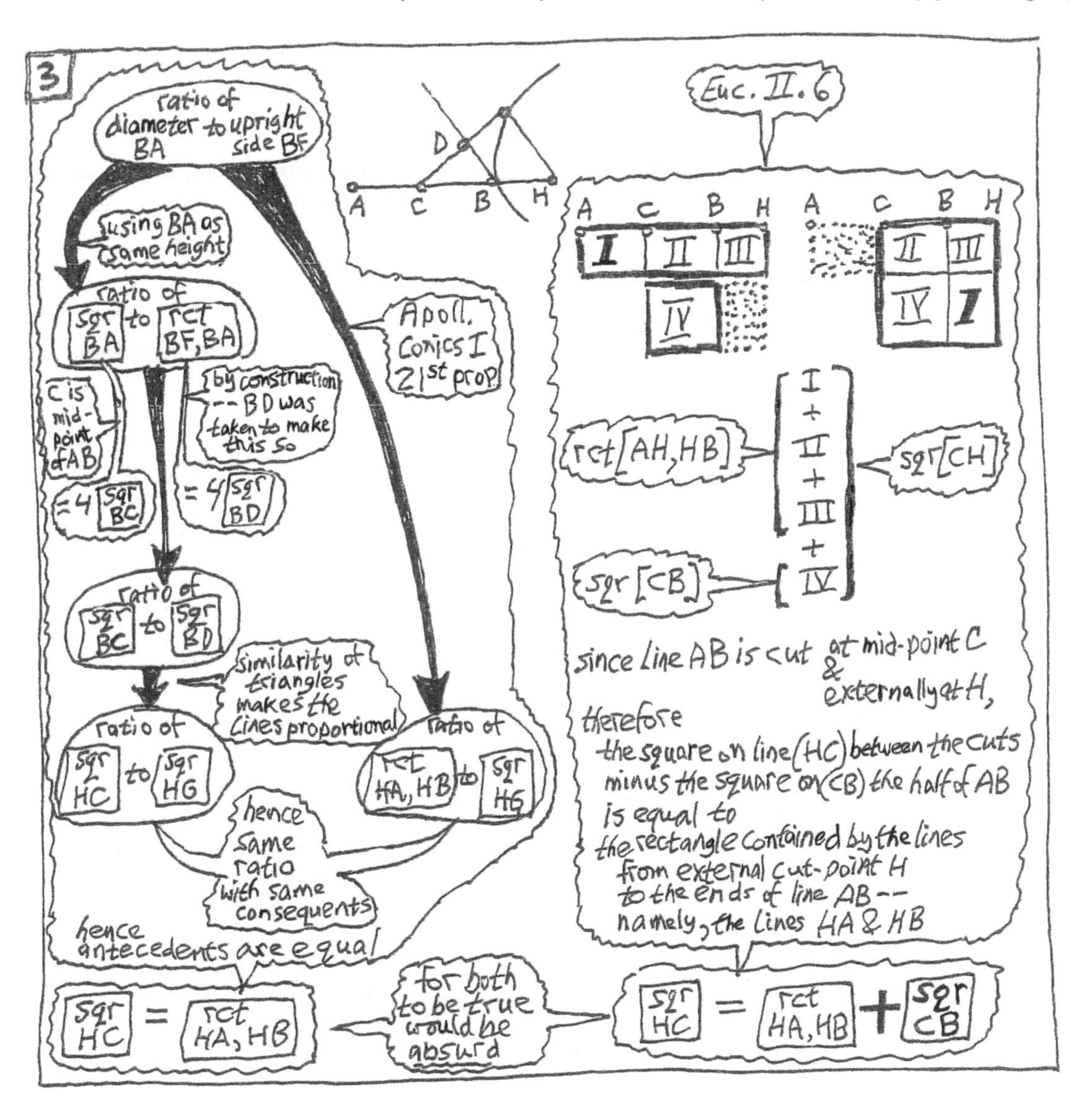

Points D and E on the tangent are taken far enough away from the point of tangency to spread apart the lines CD and CE far enough for them to clear the hyperbola—that is, far enough that neither of them will fall together with the section. That is why Apollonius finishes up the proposition by calling them "asymptotes" to the section. The Greek *a+sym+ptôtoi* corresponds to the English "not+together+able to fall." These lines that are obtained here will have their peculiarly interesting characteristic shown later, in the 14th proposition's porism, which will also explicitly mention the fact that these lines are not the hyperbola's only asymptotes; there are other pairs of lines through the center that also cannot fall together with the section. The other pairs cannot do so if they are spread *farther* apart than the pair in this proposition. But are there other pairs which cannot fall together with the section even though they are *not* spread *as far* apart as the pair in this proposition (fig. 4)?

These lines obtained here—are they distinctive in being asymptotes that are *closest* to the section, and therefore ones that are properly distinguished by being called "*the* asymptotes"? The very next proposition shows that they are. (In modern parlance, it should be noted, it is only the asymptotes here called "*the* asymptotes" that are called "asymptotes" at all.)

Before moving on to consider why these asymptotes that have been obtained here must be *the* asymptotes (that is, the asymptotes that are the *closest* to the section), let us look back and see why one might choose to begin, as Apollonius does, by taking those particular points D and B as the points through which to draw the lines from center C. Why must we have expected these lines to be asymptotes at all?

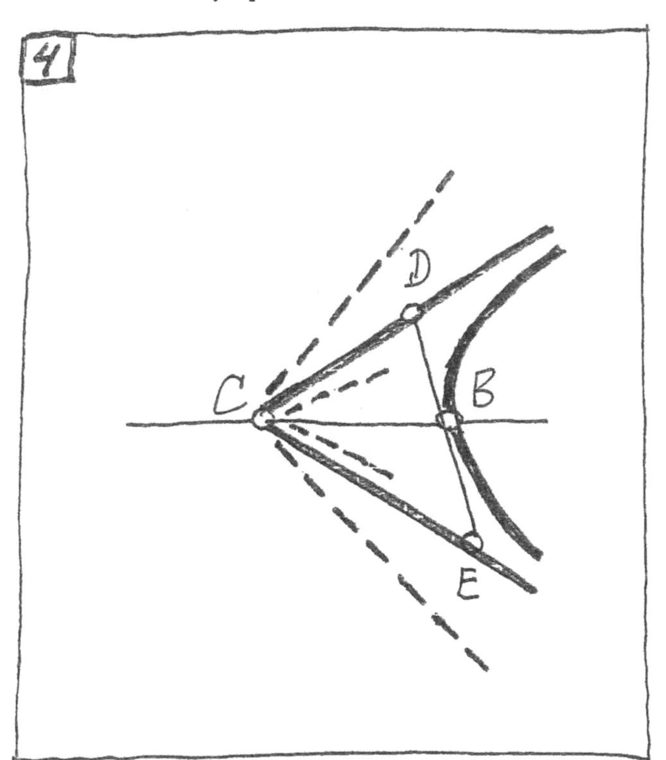

It is clear that a line drawn ordinatewise through the center does not meet the hyperbola, and neither does another line that inclines away from that line merely a little (fig. 5). A line from the center that does indeed meet the section will have to cross a line through the vertex that is parallel to the ordinatewise line—that is, somewhere it will cross that line which is tangent at the vertex. So too, a line from the center that does not meet the section—even though it inclines away from the ordinatewise line through the center—will cross that tangent farther from the vertex than will the line that does meet the section (fig. 6). Now, what can be said about DB if CD does meet the section at some point G? That would provide some insight into the condition for CD *not* to meet the section at any point.

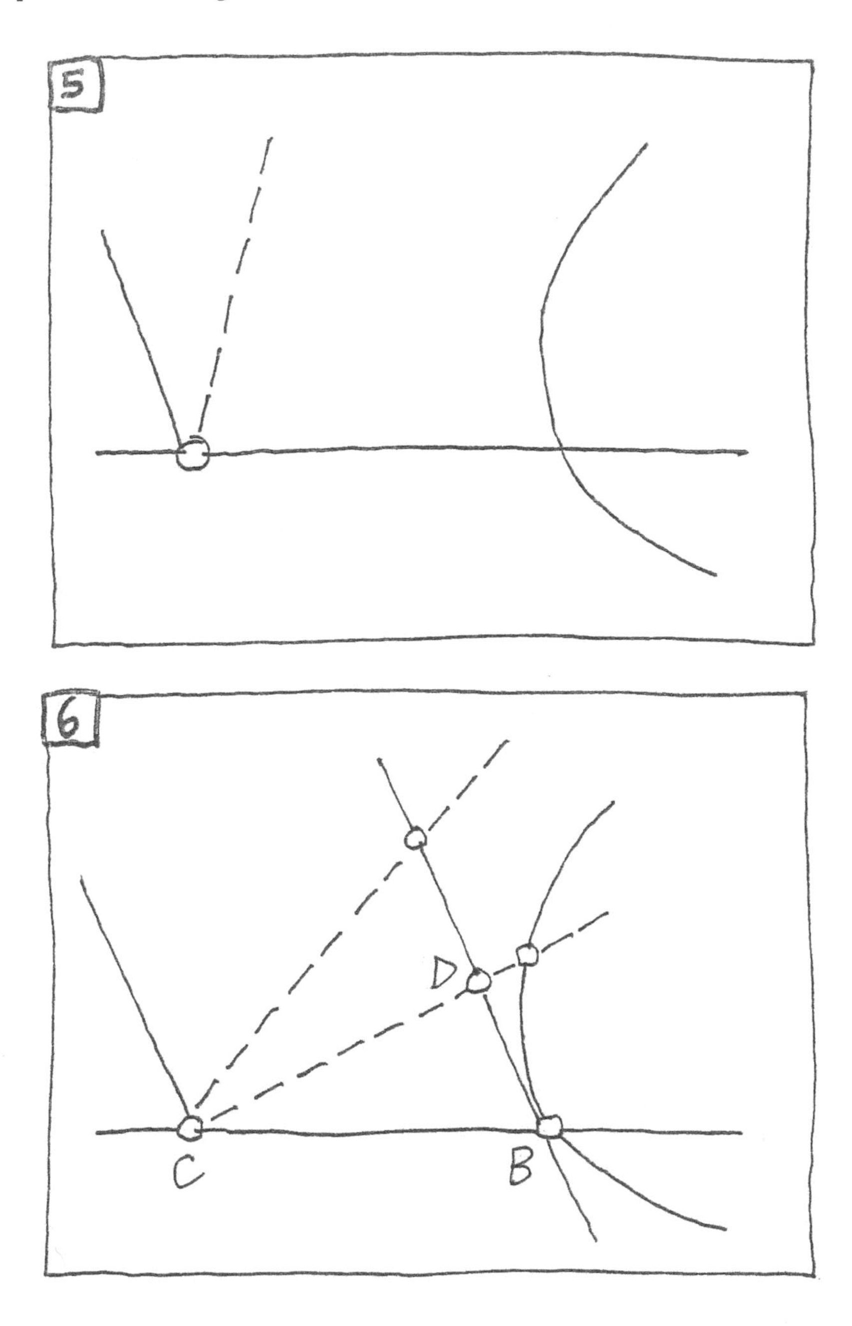

Well, if a straight line emanating from the center does indeed meet the section at some point G, then from that point G (as from any other point on the section) an ordinate can be drawn—that is, there is an ordinatewise line from G that meets the diameter at some point H (fig 7). Now, for that point H (as for any other point on the diameter extended) the 6th proposition of Book Two of Euclid's *Elements* shows this:

square HC	is equal to	rectangle HA, HB	+	square CB.

Now what does that have to do with DB? Consider the situation in the similar triangles:

> the ratio of CH to GH is the same as
> the ratio of BC to BD.

Making squares on these lines,

> the ratio of square [CH] to square [GH] is the same as
> the ratio of square [BC] to square [BD].

Hence, the equality involving square [HC] implies that

> the ratio of (rectangle [HA, HB] + square [CB]) to square [GH] is the same as
> the ratio of square [BC] to square [BD].

But that square [BC] is one-fourth of square [BA] (fig. 8). And BA, which is the diameter, is related to the rectangle [HA, HB] contained by the lines from point H to the endpoints of the diameter. The 21st proposition of the First Book of the *Conics* tells us this about that rectangle's relation to the square on the ordinate (GH): the ratio of the rectangle [HA, HB] to the square [GH] is the same as the ratio of the diameter BA to its parameter BF.

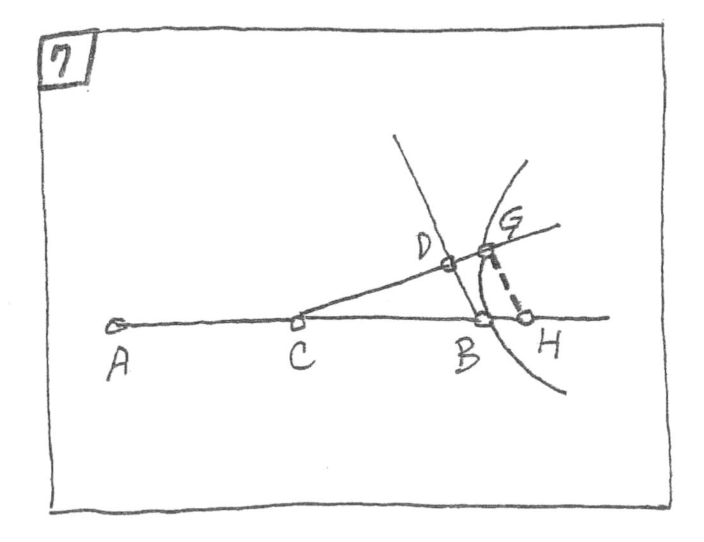

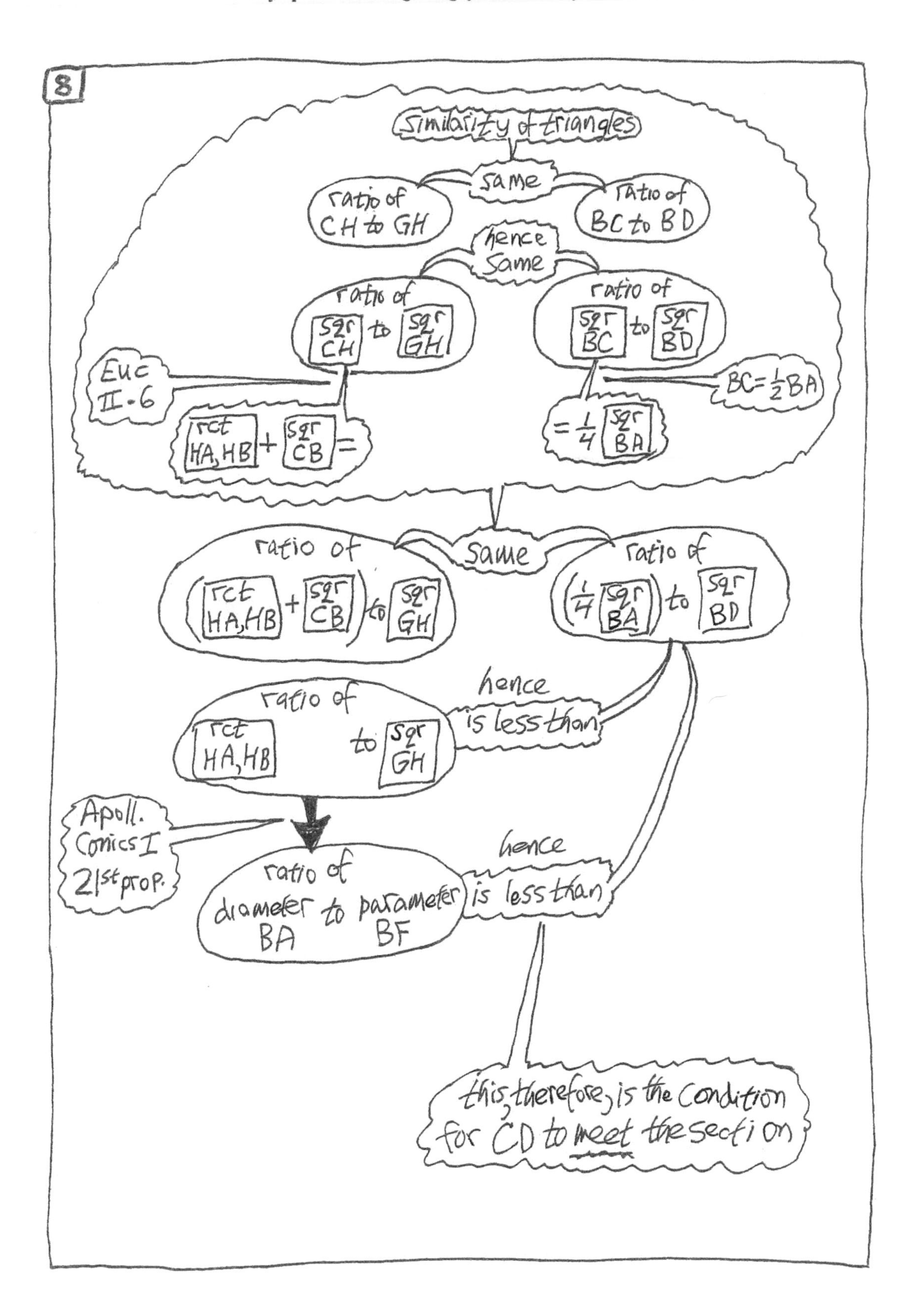
8
Similarity of triangles
same
ratio of CH to GH
ratio of BC to BD
hence same
ratio of sqr CH to sqr GH
ratio of sqr BC to sqr BD
Euc II.6
BC=½BA
rct HA,HB + sqr CB =
= ¼ sqr BA
same
ratio of (rct HA,HB + sqr CB) to sqr GH
ratio of (¼ sqr BA) to sqr BD
hence is less than
ratio of rct HA,HB to sqr GH
Apoll. Conics I 21st prop.
hence is less than
ratio of diameter BA to parameter BF
this, therefore, is the condition for CD to meet the section

So, the consequence of all that is the following condition for CD to meet the section:

> the ratio of diameter BA to parameter BF,
> which is the same as
> the ratio of the rectangle [HA, HB] to square [GH],
> must be less than
> the ratio of (rectangle [HA, HB] + square [CB]) to square [GH],
> which is the same as the ratio of one-fourth square [BA] to square [BD]—

which is to say:

> the ratio of diameter BA to parameter BF
> must be less than
> the ratio of one-fourth square [BA] to square [BD].

Now, since the condition for CD to *meet* the section relates BD, in a certain way, to the unchanging diameter and parameter, we might suspect that CD would *not* meet the section if BD were not related in that way to the diameter and parameter—that is, if:

> the ratio of diameter BA to parameter BF
> were not less than
> the ratio of one-fourth square [BA] to square [BD].

Let us take the simplest case, the one where the ratios are the same, and simplify the relationship. See figure 9.

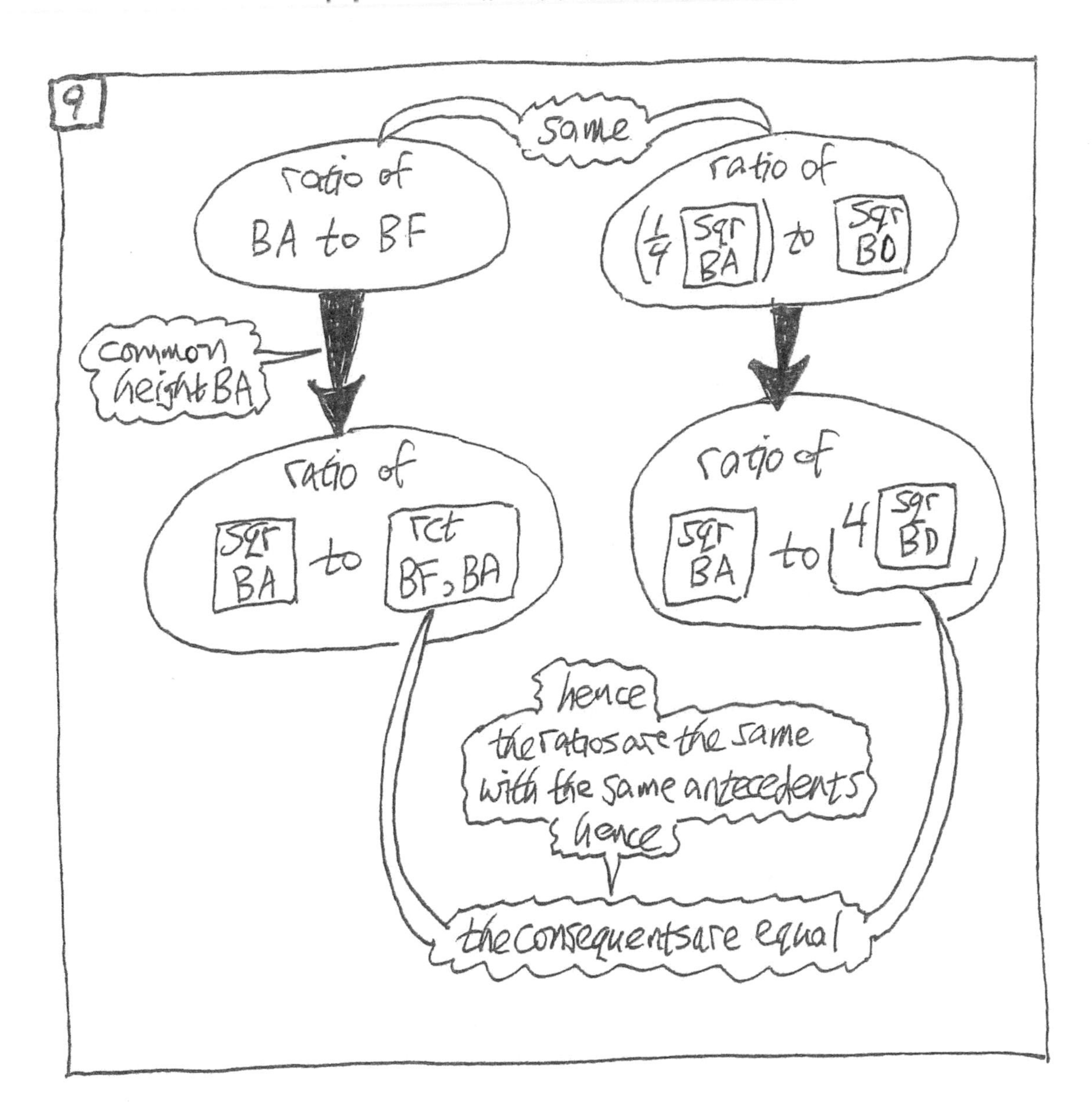
9
same
ratio of
BA to BF
ratio of
($\frac{1}{4}$ sqr BA) to sqr BO
common
height BA
ratio of
sqr BA to rct BF, BA
ratio of
sqr BA to 4 sqr BD
hence
the ratios are the same
with the same antecedents
hence
the consequents are equal

What all of that adds up to, is this (fig. 10):

> Point G cannot be on the section unless
> the line that is drawn ordinatewise from it
> will meet the diameter at some point H,
> and point H can be on the diameter
> only if the ratio of the rectangle [HA, HB] to square [GH],
> which is the ratio of diameter AB to parameter BF,
> is less than the ratio of one-fourth square [AB] to square [BD].

You can therefore be sure that you will not meet the section by drawing a line from C through point D if you take point D as follows:

> locate it on the tangent through B,
> at that distance from B which will make
> square [BD] equal to one-fourth rectangle [BF, BA].

It is clear that BD has been taken to be of such a size as to guarantee that CD when extended cannot meet the section. If CD were to cross the tangent even farther away from B than it does, then CD would be located even farther away from the section than it is. Line BD is therefore large enough to do the job. But is it larger than it needs to be? In other words, is there a line from C that cannot meet the section when extended—and yet lies between CD and CB?

It turns out that although, of all the lines from C that cannot meet the section, there is no farthest line, nonetheless there is a closest one. That closest one is line CD—the asymptote presented in the 1st proposition. That is what is shown in the 2nd proposition. In other words, if a line from C does not meet the section, then it must cross the vertex's tangent at a distance no smaller than the line whose square is equal to one-fourth of the hyperbola's *eidos* (that is to say, one-fourth of the rectangle contained by the diameter and its parameter). Line BD is not only long enough, it is *just* long enough; if it were any shorter, it would not do.

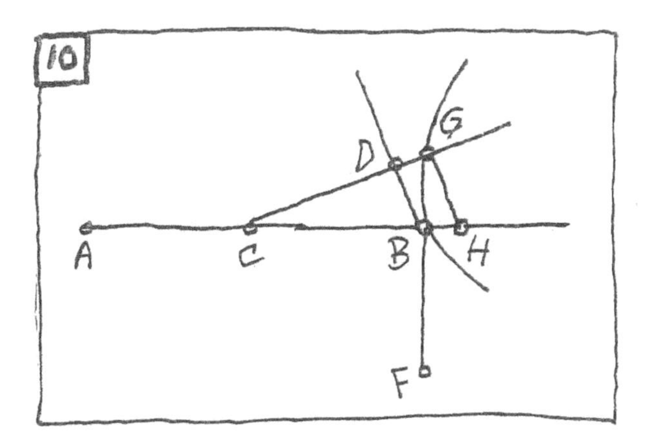

The 2nd proposition shows the absurdity that would follow from supposing that any line that lies between the asymptotes of the 1st proposition could be another asymptote. The *reductio* proceeds as follows.

Starting from C, draw any line that lies between the asymptotes CD and CE, and suppose that it is indeed possible for this line to be another asymptote. Suppose, in other words, that it nowhere meets the section. Now, this line may not meet the section, but it will be met by a line from vertex B that is drawn parallel to line CD (the original asymptote—the one presented in the 1st proposition). The meeting-point (call it H) will, like all the points on the supposed asymptote, lie outside the section. A line that goes through this meeting-point H, if it is parallel to that tangent (DBE) which goes through the vertex B, will be drawn ordinatewise. It will meet the section (at a meeting-point we shall call K), and it will meet the diameter (at a meeting-point we shall call L), and will meet the asymptotes (at meeting-points we shall call M and G). (See fig. 11.) Now, from all that, there follow two conclusions which are mutually contradictory.

On the one hand, since line GH is equal to line DB (because of the parallelism) and line GK is greater than line DB (because point H, like every other point of CH, is supposedly outside the section), it follows that rectangle [KM, KG] is greater than square [DB].

On the other hand, contradicting that conclusion, is a conclusion that follows from the size of the line DB that gave us the asymptotes in the 1st proposition: rectangle [KM, KG] is equal to square [DB].

Thus, we are led into asserting that rectangle [KM, KG] is *greater* than square [DB] *and also* is *equal* to it. But what leads us to that absurd assertion is the supposition that while CH goes inside the angle contained by the original asymptotes (CD and CE), it nonetheless nowhere meets the section. What we supposed, therefore, just cannot be so. That is to say: line CH cannot be an asymptote if line CD is one. (Indeed, neither can any line be an asymptote if it cuts into the angle that is contained by asymptotes CD and CE elsewhere than at its vertex.)

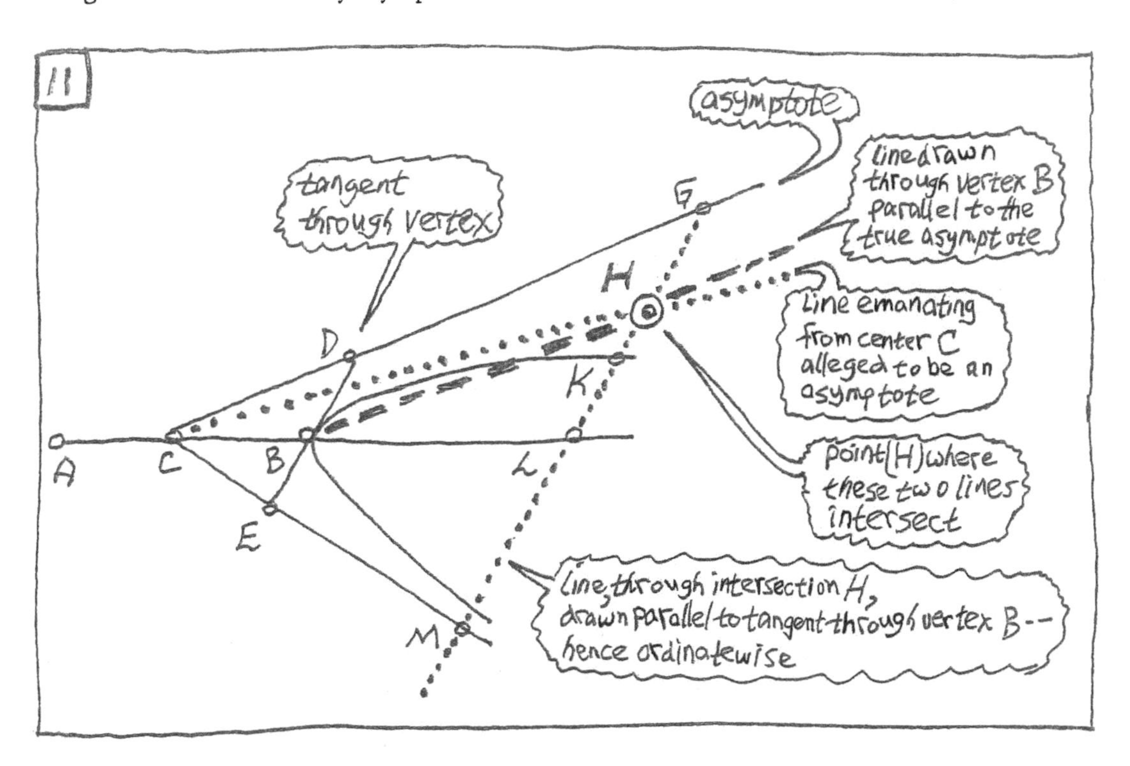

How, then, does the demonstration work? Now, as in the previous proposition, Euclid's II.6 is applied (along the *diameter*) to an *external* cut in a line that is bisected; but now, in addition, Euclid's II.5 is applied (along a line drawn *ordinatewise)* to an *internal* cut in a line that is bisected. Toward the end of the demonstration, the Euclidean maneuver that was previously employed (II.6) enables us to see that when from square [CL] there is taken away the rectangle [AL, LB], then the remainder is equal to the square [CB]; but what we need to see also is that when from square [LG] there is taken away the square [LK], then the remainder is equal to rectangle [KM, KG]—and this requires, not II.6, but II.5. (Fig. 12.)

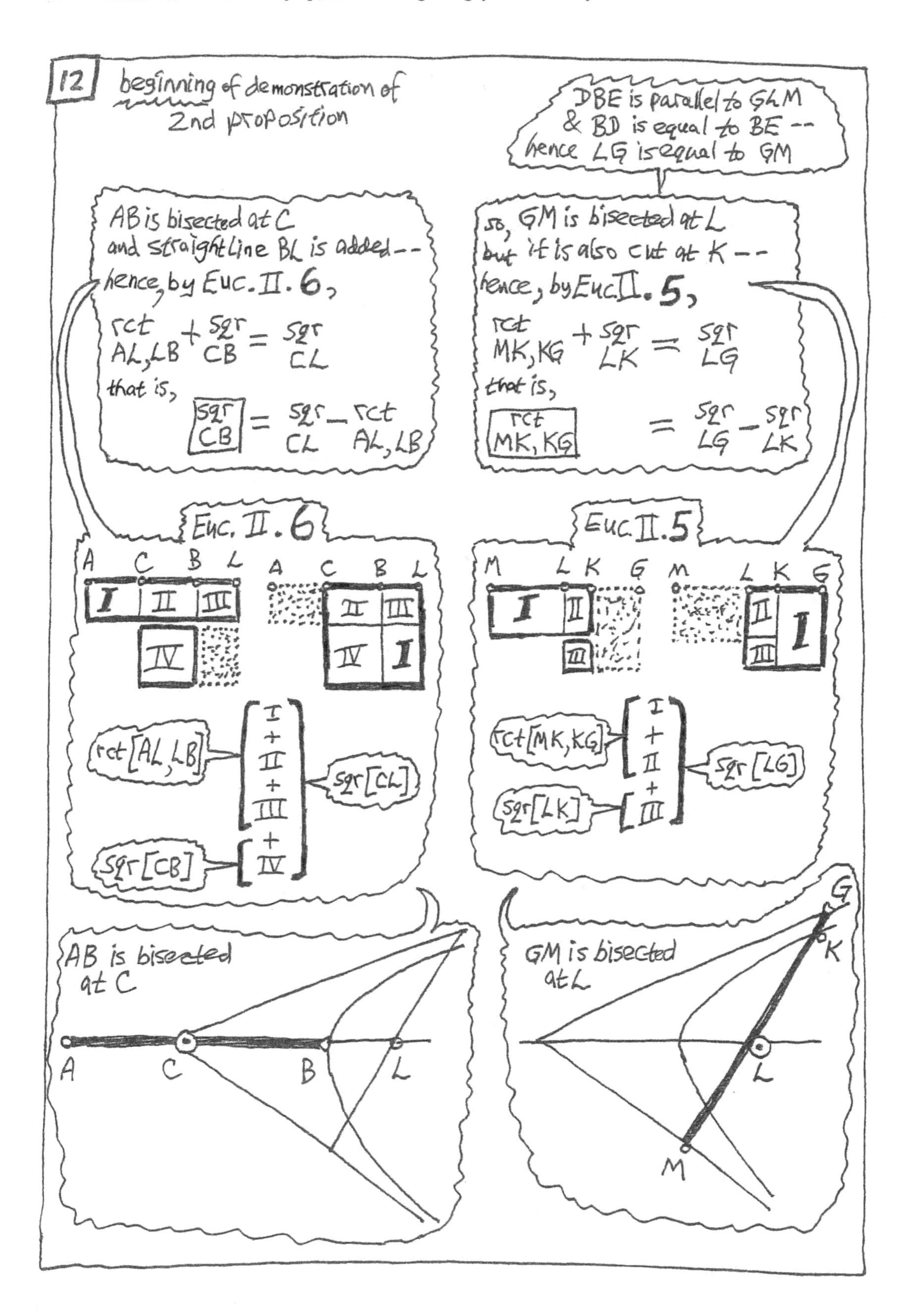
12
beginning of demonstration of 2nd proposition
DBE is parallel to GLM & BD is equal to BE -- hence LG is equal to GM
AB is bisected at C and straight line BL is added -- hence, by Euc. II. 6,
rct AL,LB + sqr CB = sqr CL
that is,
sqr CB = sqr CL – rct AL,LB
so, GM is bisected at L but it is also cut at K -- hence, by Euc II. 5,
rct MK,KG + sqr LK = sqr LG
that is,
rct MK,KG = sqr LG – sqr LK
Euc. II. 6
A C B L A C B L
I II III IV
rct[AL,LB]
sqr[CL]
sqr[CB]
Euc. II. 5
M L K G M L K G
rct[MK,KG]
sqr[LG]
sqr[LK]
AB is bisected at C
A C B L
GM is bisected at L
G K L M

Those applications of two propositions juxtaposed by Euclid are presented together at the beginning of the demonstration, but what we are shown in the opening of the demonstration is not used until its closing. In the demonstration's central part, we consider the supposition that line CH is an asymptote. The line that is an asymptote in supposition cannot be an asymptote in fact unless every one of its points is between the section and that asymptote CD which was presented in the previous proposition. But if every point on the supposed asymptote lies between the section and line CD, then point H (being one of those points) must lie there; then, since GH is equal to DB, it follows that line GK is greater than DB—and therefore that rectangle [KM, KG] is greater than square [BD] (fig. 13).

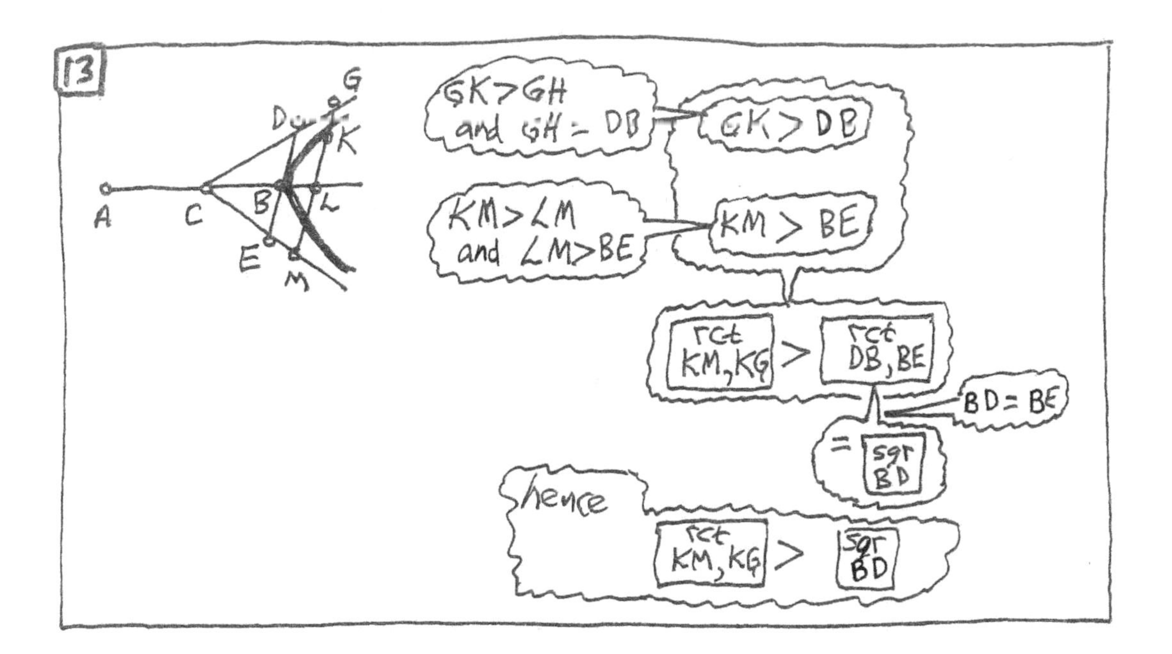

We can now see that if that rectangle [KM, KG] were *not* greater than that square [BD], then point H could not lie between the section and that asymptote (CD) which was presented in the previous proposition. In other words, CH cannot be another asymptote if rectangle [KM, KG] is less than or equal to square [BD].

But in fact, as we are shown in the concluding part of the proposition, that rectangle must be equal to that square. Why? Because the 1st proposition made BD of such a size that the square on it is equal to one-fourth of the rectangle contained by BA and BF. From that fact, the equality of rectangle [KM, KG] and square [BD] follows—when you apply to the diameter Euclid's II.6 and Apollonius's I.21, while applying Euclid's II.5 to the line that is drawn ordinatewise. Applying those propositions gives flesh to the argument whose skeleton is this:

> the ratio of square [BC] to square [BD]
> is the same as its ratio to
> square [BD]'s equal—which is one-fourth of the rectangle [BA, BF];
>
> but the ratio of square [BC] to one-fourth of rectangle [BA, BF]
> is the same as its ratio to
> rectangle [MK, KG]—which is therefore square [BD]'s equal.

See figure 14.

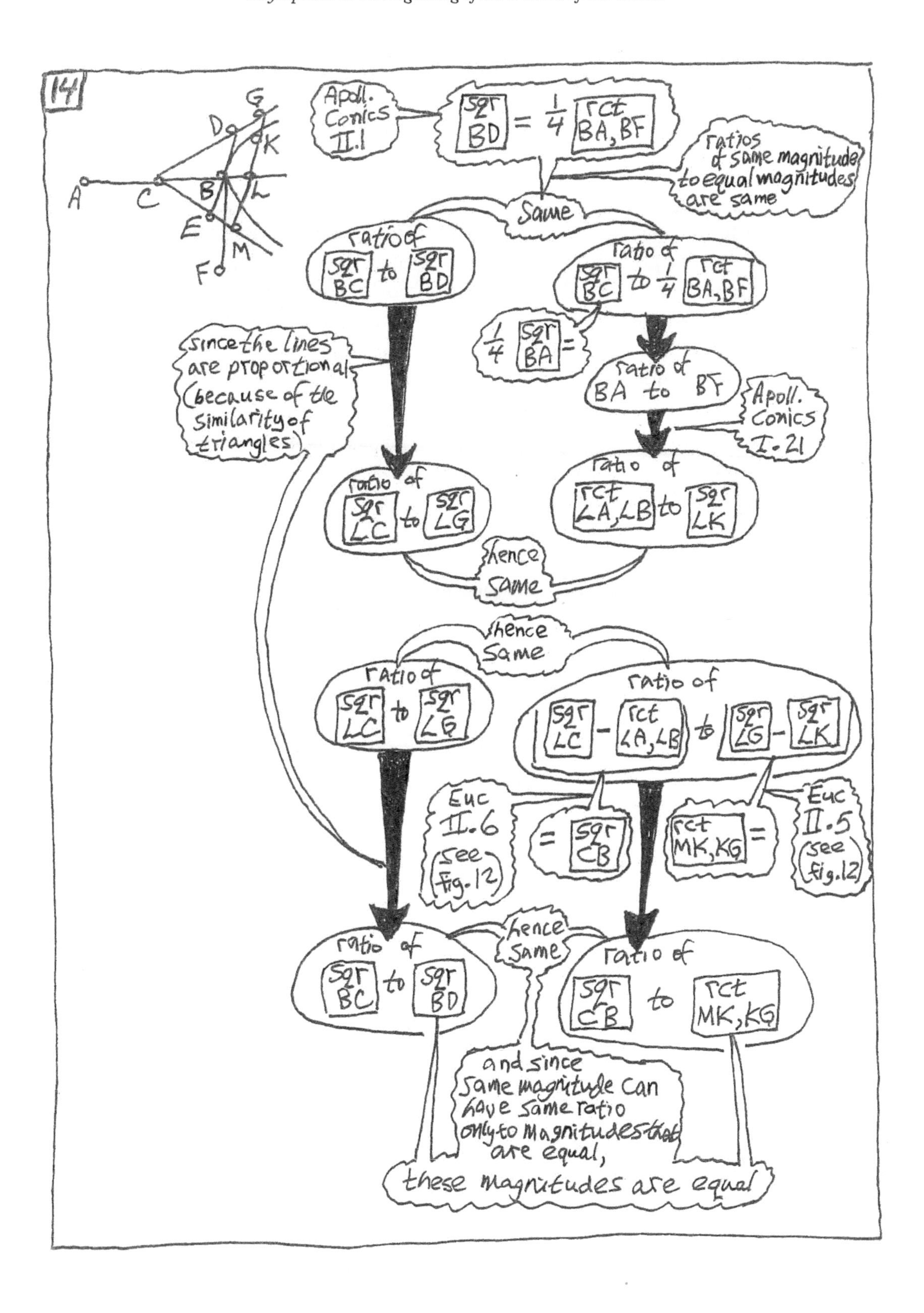

14
A
C
B
L
D
G
K
E
M
F
Apoll. Conics II.1
sqr BD = 1/4 rct BA, BF
ratios of same magnitude to equal magnitudes are same
Same
ratio of sqr BC to sqr BD
ratio of sqr BC to 1/4 rct BA, BF
1/4 sqr BA =
ratio of BA to BF
Apoll. Conics I.21
since the lines are proportional (because of the similarity of triangles)
ratio of sqr LC to sqr LG
ratio of rct LA, LB to sqr LK
hence same
hence same
ratio of sqr LC to sqr LG
ratio of sqr LC - rct LA, LB to sqr LG - sqr LK
Euc II.6 (see fig. 12)
= sqr CB
rct MK, KG =
Euc II.5 (see fig. 12)
ratio of sqr BC to sqr BD
hence same
ratio of sqr CB to rct MK, KG
and since same magnitude can have same ratio only to magnitudes that are equal,
these magnitudes are equal

The lines that were shown in the 1st proposition to be the hyperbola's asymptotes are thus shown in the 2nd proposition to be the asymptotes that are closest to the section. But just how close is that? Apollonius will not tell us until the 14th proposition.

What he does go on to tell us in the immediate sequel (that is, in the 3rd proposition) is the converse of what he has already told us.

(II) #3 If, with respect to an hyperbola, a straight [line] touch [the section], [the straight line] will fall together with both of the asymptotes, and it will be cut in two [halves] at the touch[point], and the square from each of its segments will be equal to the fourth of the figure (*eidous*) for the diameter drawn through the touch[point].
—Let an hyperbola be what the [line] ABC is, and its center be what the [point] E is, and asymptotes be what the [lines] FE, EG are, and also, let there touch it at the [point] B some line, say, the [line] HK. I say that, extended, the [line] HK will fall together with the [lines] FE, EG.
—For, if possible, let it not fall together with them; and also, let the [line] EB, joined, be extended; and also, let there be put equal to the [line] BE the [line] ED; therefore a diameter is what the [line] BD is. Let there be put equal to the fourth of the figure for the [line] BD, then, the [square] from each of [lines] HB, BK; and also, let there be joined the [lines] EH, EK. Asymptotes, therefore, are what [these lines EH, EK] are—which very thing is absurd; for what are supposed to be asymptotes are the [lines] FE, EG. Therefore the [line] KH, extended, will fall together with the asymptotes EF, EG at the points F, G.
—I say then also that the [square] from each of the [lines] BF, BG will be equal to the fourth of the figure for the [line] BD.
—For let it not be, but, if possible, let there be equal to the fourth of the figure the [square] from each of the [lines] BH, BK. Therefore asymptotes are what the [lines] HE, EK are—which very thing is absurd. Therefore the [square] from each of the [lines] FB, BG will be equal to the fourth of the figure for the [line] BD.

Here we again have an hyperbola and its center and one of its tangents. What we showed before, was this:

> if lines are drawn from the hyperbola's center
through two points that are taken on the tangent in a certain way,
then these lines will be the hyperbola's asymptotes;

what we show now, is this:

> if lines from the hyperbola's center are the asymptotes,
then these lines will pass through
two points taken on the tangent in that same way.

The asymptotes meant here are those lines from the center which are the closest ones of all the straight lines which do not meet the hyperbola. Otherwise, these asymptotes could pass through points on the tangent that are farther away from the point of tangency than are the points on the tangent that would yield the asymptotes of the 1st proposition. What is here demonstrated is the converse of the 1st proposition as linked with the 2nd (which restricts the meaning of "the asymptotes"). The converse of the 1st proposition needs to be demonstrated because the asymptotes could have been drawn through points on another one of the hyperbola's tangents, and they thus perhaps would intersect this present tangent at points other than the points on the tangent that would yield the asymptotes of the 1st proposition. The demonstration, consisting of two *reductio*s, is depicted below.

Let there be: hyperbola ABC; center E; asymptotes EF and EG; HK tangent at B.
And join B to E, prolonging it to D, where DE=EB—hence DB is a diameter.

Tangent KH prolonged will meet those asymptotes at some points F and G.	Tangent KH will meet those asymptotes at points F and G, such that each of the squares on BF and BG is equal to 1/4 of fig on diam BD.
For, if possible, let KH *not* meet them;	For, if possible, let each of the squares on BF and BG *not* be equal to 1/4 of fig on diam BD;
take H and K on the tangent where each of the squares on BH and BK is equal to 1/4 of fig on diam BD.	take H and K on the tangent where each of the squares on BH and BK is equal to 1/4 of fig on diam BD

Then in each case, this absurdity follows:
asymptotes will be obtained (by the 1st proposition)
if center E is joined to points H and K,
and will also be obtained (by the 2nd proposition)
if center E is joined to points F and G.

Having presented the hyperbola's asymptotes in the first three propositions, Apollonius in the 4th proposition solves the problem of obtaining an hyperbola when what is given are the following items:

> two straight lines, containing an angle, which are to be the asymptotes, and
>
> a point within the angle, through which the hyperbola is to pass.

(II) #4 Given two straight [lines] containing an angle, and also a point inside the angle, to carve out (*grapsai*) through the point the cone's section called an hyperbola so that asymptotes of it are what the given straight [lines] are.

—Let there be two straight [lines], namely the [lines] AC, AB, containing a chance angle at the [point A]; and also, let there be given some point, say the [point] D; and let there be required, through the [point] D, to carve out (*grapsai*), for the [lines] CAB as asymptotes, an hyperbola.

—Let there be joined the [line] AD; and also, let it be extended to the [point] E; and also, let there be put equal to the [line] DA the [line] AE; and also, through the [point] D let there be drawn parallel to the [line] AB the [line] DF; and also, let there be put equal to the [line] AF the [line] FC; and also, let the joined [line] CD be extended to the [point] B; and also, equal to the [square] from CB, let there be generated (*gegonetô*) the [rectangle] by DE, G [in other words, obtain a line G that, when made a side of a rectangle with side DE, will make this rectangle equal to that square]; and also, with there being extended the [line] AD, let there be carved out (*gegraphthô*) about it, through the [point] D, an hyperbola, so that the [lines] drawn down have the power [to give rise to squares equal to] the [rectangles] which are laid along ["are applied to"] the [line] G, which [rectangles] overshoot by a figure (*eidei*) that is similar to the [rectangle] by DE, G.

—Since, then, parallel is the [line] DF to the [line] BA, and equal is the [line] CF to the [line] FA, therefore equal is also the [line] CD to the [line] DB—and so the [square] from the [line] CB is a quadruple of the [square] from CD. And also the [square] from the [line] CB is equal to the [rectangle] by DE, G. Therefore each of the [squares], [whether] from CD [or from] DB, is a fourth part of the figure (*eidous*) by DE, G. Therefore the [lines] AB, AC are asymptotes of the carved-out (*grapheisês*) hyperbola.

Why, in addition to being given the lines that are to be asymptotes, is it necessary to be given a point within the asymptotic angle? Because as many different hyperbolas as you please can be found with the same given lines as asymptotes. What makes one of them different from another, is how far from the center the vertex lies. Two different hyperbolas with the same asymptotes differ only in location from two hyperbolas obtained from the same cone by cutting it with planes that are parallel to each other. In other words, you can find as many hyperbolas as you please that have a given asymptotic angle of the same size, while having diameters of different sizes but of a particular orientation with respect to the sides of the asymptotic angle (fig. 15).

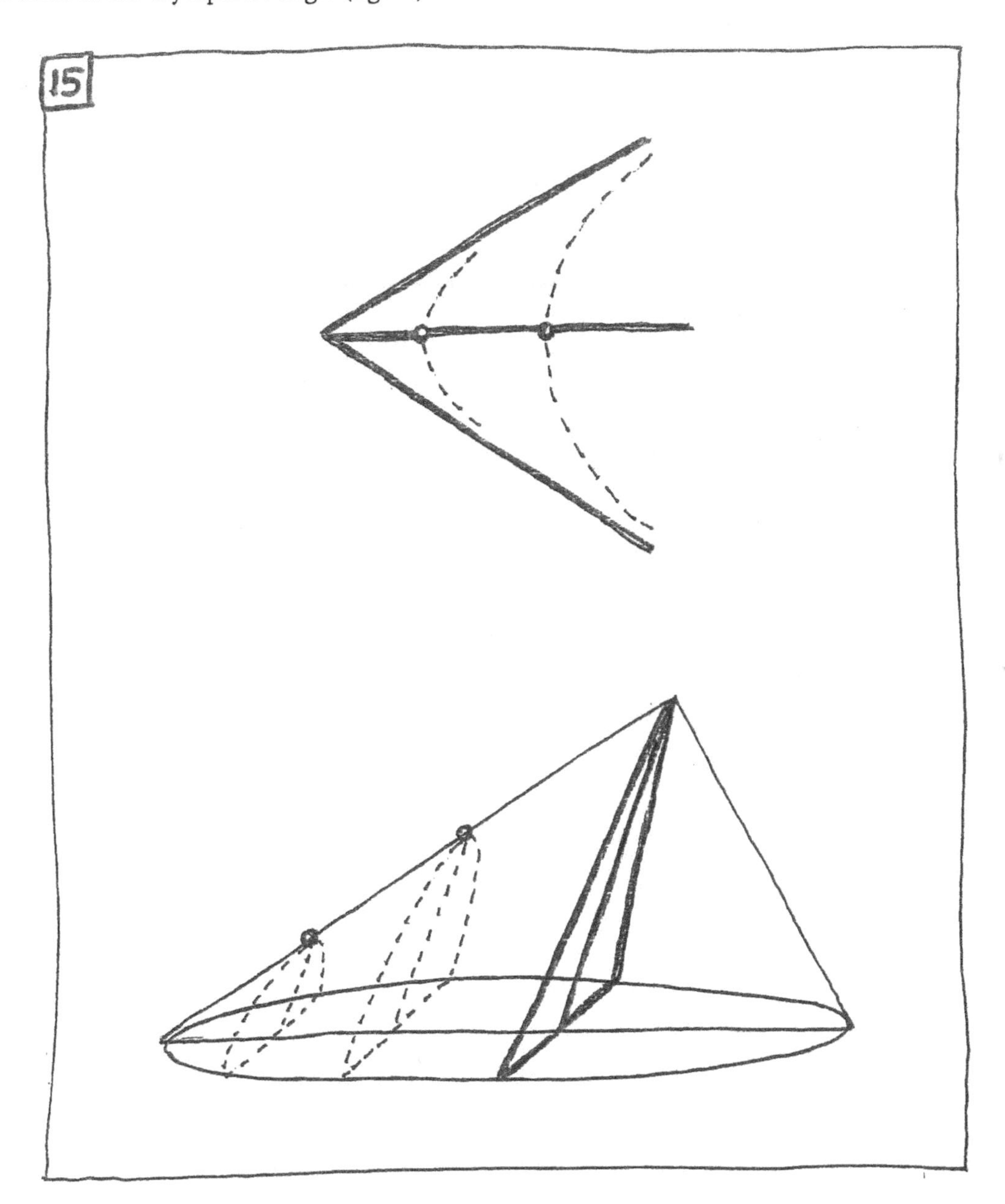

The 4th proposition presents the solution to the problem, but it does not present any analysis that leads to the solution. How would one proceed if one were to incorporate an analysis into the presentation? Somewhat as follows.

We could obtain the hyperbola if we had its diameter (with one end's being designated as the vertex) and if we also had its parameter.

From what we are given in the problem, it would be easy to get a diameter. The point from which the given lines diverge is the center, and a line from there to the given point would be half of that diameter which would have the given point as vertex (fig. 16).

So we have got the requisite diameter. Now we could get the requisite parameter if we had the tangent through the vertex D. How? Well, suppose we had the tangent BC (fig. 17). Then the square on CD would be equal to one-fourth of the rectangle contained by the diameter DE and the parameter (call it G). That is to say: as DE is to twice CD, so also twice CD is to G. So, since twice CD is CB, we'll get G if we merely contrive a third proportional to DE and CB.

So, all that's left to get is the tangent through vertex D. We would have it if we had merely one other point on it; and the only other points on it that we have any reason to try to get, are its intersections (C and B) with the lines that are to be the hyperbola's asymptotes. What we know about C and B is that they are equally distant from D, the point of tangency. From that fact, we can locate C and B on the lines that are to be asymptotes. We need only put them where they will make CD equal to DB. Where will that be? In order to answer, let us consider similar triangles. From D, let us draw a line parallel to one of the given lines. That line which we draw will intersect the other given line at some point (call it F). (See figure 18.) Since the line DF is parallel to line AB, the triangle FDC will be similar to triangle ABC. Hence, CD will be equal to DB if, and only if, CF is equal to FA. The construction that is required, therefore, is the one that is the following (fig. 19).

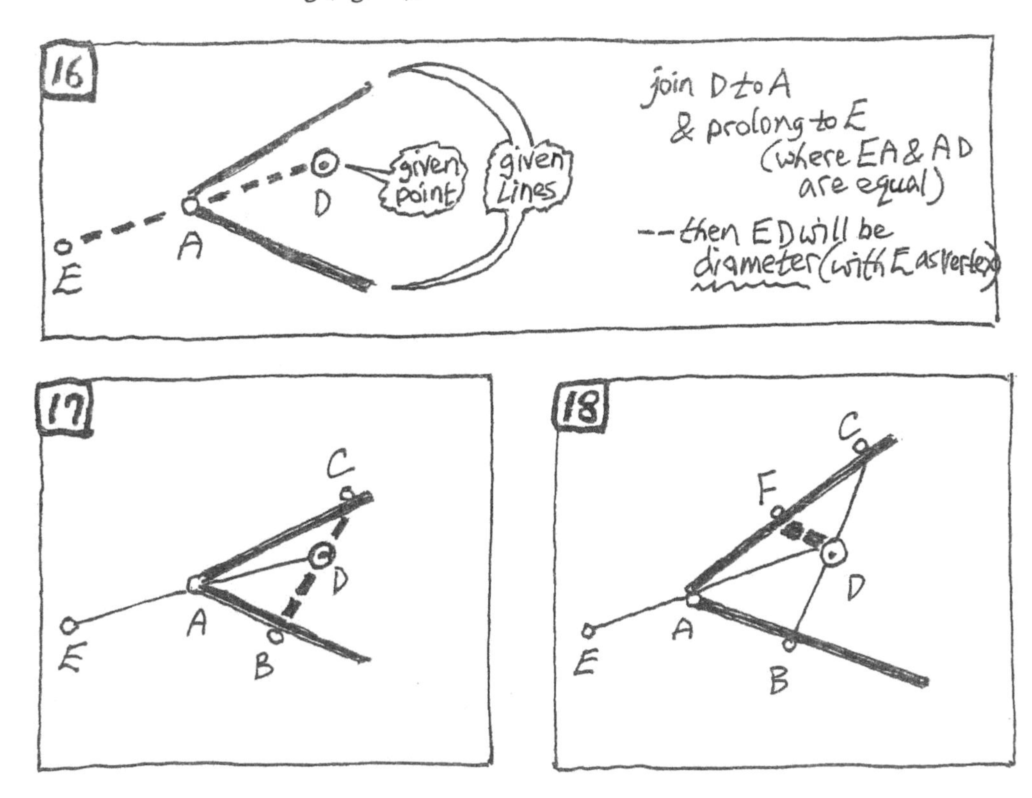

(1) join DA
(2) prolong it to E, where AE=AD
(3) through D, draw a parallel to the lower given line, making intersection F
(4) take point C where FC=FA
(5) join this point C to that point D
(6) prolong this line CD, making intersection B
(7) contrive line G so that rectangle [DE, G] = square [CB]
(8) about AD extended, draw an hyperbola through D (as vertex)

That construction is the first part of Apollonius's proposition. The second part is the demonstration that the hyperbola so obtained is indeed the one sought (fig. 20).

From square [CD]'s equality to one-fourth of rectangle [DE, G], the conclusion of the demonstration follows: lines AC and AB must be the asymptotes of that hyperbola whose diameter is DE and whose parameter is G; and that hyperbola is the one that was obtained by the construction.

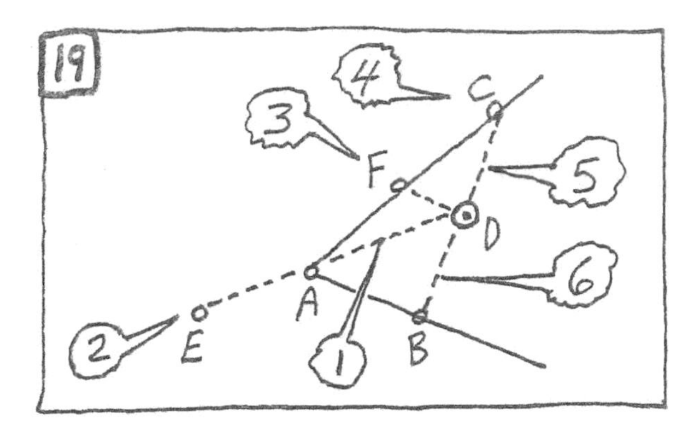

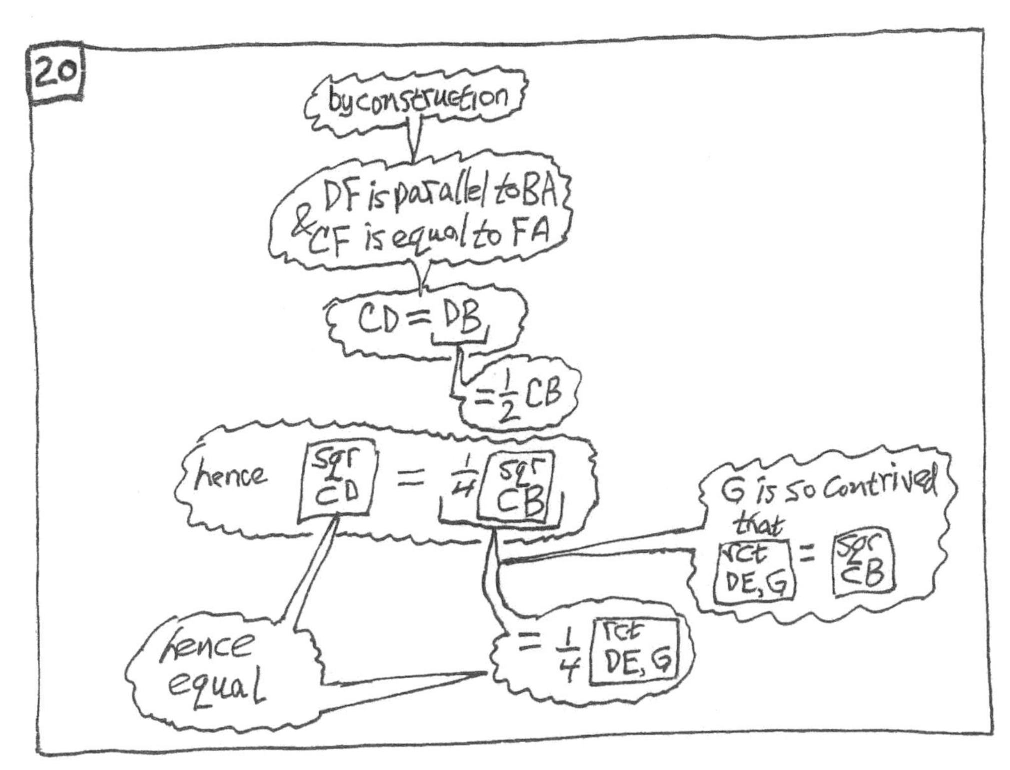

some propositions needed later

Now we consider the 5th through 9th propositions, because some later propositions will depend upon them.

The 5th and 6th together tell us something about every conic curve (the 5th, about the open curves—the parabola and the hyperbola; the 6th, about the closed ones—the ellipse and the circle): if a line is a diameter, then the tangent through the point where the diameter meets the section will be parallel to a chord that is bisected by that line. In other words, the 5th and 6th propositions together tell us this: any chord bisected by a diameter is drawn ordinatewise to that diameter. The 7th is the converse for all of the curves. It tells us this: if a straight line joins a tangency-point to the midpoint of a chord that is parallel to the tangent, then the line is a diameter.

(II) #5 If a parabola's or hyperbola's diameter cut in two [halves] some straight line [inside the section], the line touching the section at the bound of the diameter will be parallel to the straight [line] that is cut in two [halves].
—Let a parabola or hyperbola be what the [line] ABC is, whose diameter is what the [line] DBE is; and also, let there touch the section the [line] FBG, and let there be drawn some straight [line] inside the section, namely the [line] AEC, that makes equal the [line] AE to the [line] EC. I say that parallel is the [line] AC to the [line] FG.
—For if not, let there be drawn, through the [point] C parallel to the [line] FG, the [line] CH; and also, let there be joined the [line] HA. Since, then, a parabola or hyperbola is what the [line] ABC is, whose diameter the [line] DE is and touching[line] the [line] FG is, and also, a parallel to it [a parallel, that is, to the line FG] is what the [line] CH is—hence equal is the [line] CK to the [line] KH. But also [equal] is the [line] CE to the [line] EA. Therefore the [line] AH is parallel to the [line] KE—which very thing is impossible, for [line AH] falls together, when extended, with the [line] BD.

(II) #6 If an ellipse's, or circle's periphery's, diameter cut in two [halves] some straight [line] that is not through the center, the line touching the section at the bound of the diameter will be parallel to the straight [line] that is cut in two [halves].
—Let there be an ellipse or circle's periphery whose diameter is what the [line] AB is; and also, let the [line] AB cut the [line] CD—[which is a line] not through the center—in two [halves] at the [point] E. I say that the [line] touching the section at the [point] A is parallel to the [line] CD.
—For let it not be, but, if possible, let a parallel to the [line] touching at the [point] A be what the [line] DF is; therefore equal is the [line] DG to the [line] FG. But [equal] also is the [line] DE to the [line] EC; therefore parallel is the [line] CF to the [line] GE—which very thing is absurd. For if the point G is the center of the section AB, the [line] CF will fall together with the [line] AB; or, [if the point G is] not [the center], let there be supposed [to be the center] the [point] K—and also, let the joined [line] DK be extended to the [point] H; and also, let there be joined the [line] CH. Since, then, equal is the [line] DK to the [line] KH, but [equal] also is the [line] DE to the [line] EC, therefore parallel is the [line] CH to the [line] AB. But [parallel to the line AB is] also the [line] CF—which very thing is absurd. Therefore the [line] touching at the [point] A is parallel to the [line] CD.

(II) #7 If, with respect to a cone's section or circle's periphery, a straight [line] touch [it], and also, a [line] parallel to this [line] be drawn in the section and also be cut in two [halves], then the straight [line] joined from the touch[point] to the midpoint (*dichotomian*) will be a diameter of the section.
—Let a cone's section or a circle's periphery be what the [line] ABC is, and a [line] touching it be what the [line] FG is; and also, a parallel to the [line] FG be what the [line] AC is; and also, let the [line] AC be cut in two [halves] at the [point] E; and also let there be joined the [line] BE. I say that the [line] BE is a diameter of the section.

—For let it not be, but, if possible, let a diameter of the section be what the [line] BH is. Therefore equal is the [line] AH to the [line] HC—which very thing is absurd; for the [line] AE, to the [line] EC, is equal. Therefore not so, that the [line] BH will be a diameter of the section. Likewise, then, will we show that neither will there be any other [diameter] than the [line] BE.

These propositions are then used to demonstrate the 8th proposition, which tells us that a chord of an hyperbola will, when prolonged in both directions, meet the asymptotes, and that the two prolongations will be equal. This is then used to demonstrate the 9th proposition, which tells us this about the point that is common to an hyperbola and to a straight line that meets the asymptotes: if that common point is equidistant from the meeting-points, then the line will be tangent at the common point. The 9th proposition, in effect, slides the chord along, keeping it, in its later positions, parallel to itself when it was in its earlier positions, until the chord vanishes and its equal prolongations become emanations from a single point. This is something like the situation in the 17th and 18th propositions of Book One.

(II) #8 If, with respect to an hyperbola, a straight [line] fall together with [the section] in two points, [the straight line], extended both ways, will fall together with the asymptotes, and the [lines] taken off from it by the section, on the side toward the asymptotes, will be equal.
—Let there be as an hyperbola the [line] ABC, and as asymptotes the [lines] ED, DF; and also, let there fall together with the [line] ABC some [line], namely the [line] AC. I say that it, extended both ways, will fall together with the asymptotes.
—Let the [line] AC be cut in two [halves] at the [point] G; and also, let there be joined the [line] DG. A diameter of the section, therefore, is what [the line DG] is; therefore, the [line] touching at the [point] B is parallel to the [line] AC. Then let there be as a touching[line] the [line] HBK; then will [this touching[line] HBK] fall together with the [lines] ED, DF. Since, then, parallel is the [line] AC to the [line] KH; and also, the [line] KH falls together with the [lines] DK, DH; therefore also will the [line] AC fall together with the [lines] DE, DF.
—Let it fall together with them at the [points] E, F; and equal is the [line] HB to the [line] BK; therefore equal also is the [line] FG to the [line] GE. And so, [equal] also is the [line] CF to the [line] AE.

(II) #9 If a straight [line] that falls together with the asymptotes be cut in two [halves] by the hyperbola, in only one point does it touch the section.
—For let a straight [line], namely the [line] CD that falls together with the asymptotes CAD, be cut in two [halves] by the hyperbola at the point E. I say that at another point it does not touch the section.
—For if possible, let it touch it at the [point] B. Therefore equal is the [line] CE to the [line] BD—which very thing is absurd, for the [line] CE is supposed equal to the [line] ED. Therefore not at another point does it touch the section.

Our examination of the asymptotes in Book Two is of interest as the completion of what was implicit in the culminating proposition of Book One. It is also of use, along with some other propositions in the middle of Book Two, in preparing the way for the stretch of problems that conclude Book Two, as well as in preparing the way for theorems at the end of the Third Book which in their turn prepare us to consider the unsolved more difficult problem that was Descartes' point of departure in his transformation of geometry. The plan of our route through the Second Book of Apollonius's *Conics* (with arrows that go from one proposition to another proposition that depends upon it) is depicted in figure 21. (The figure includes only those propositions from the end of Book Two whose connection with what comes before needs to be clarified here.)

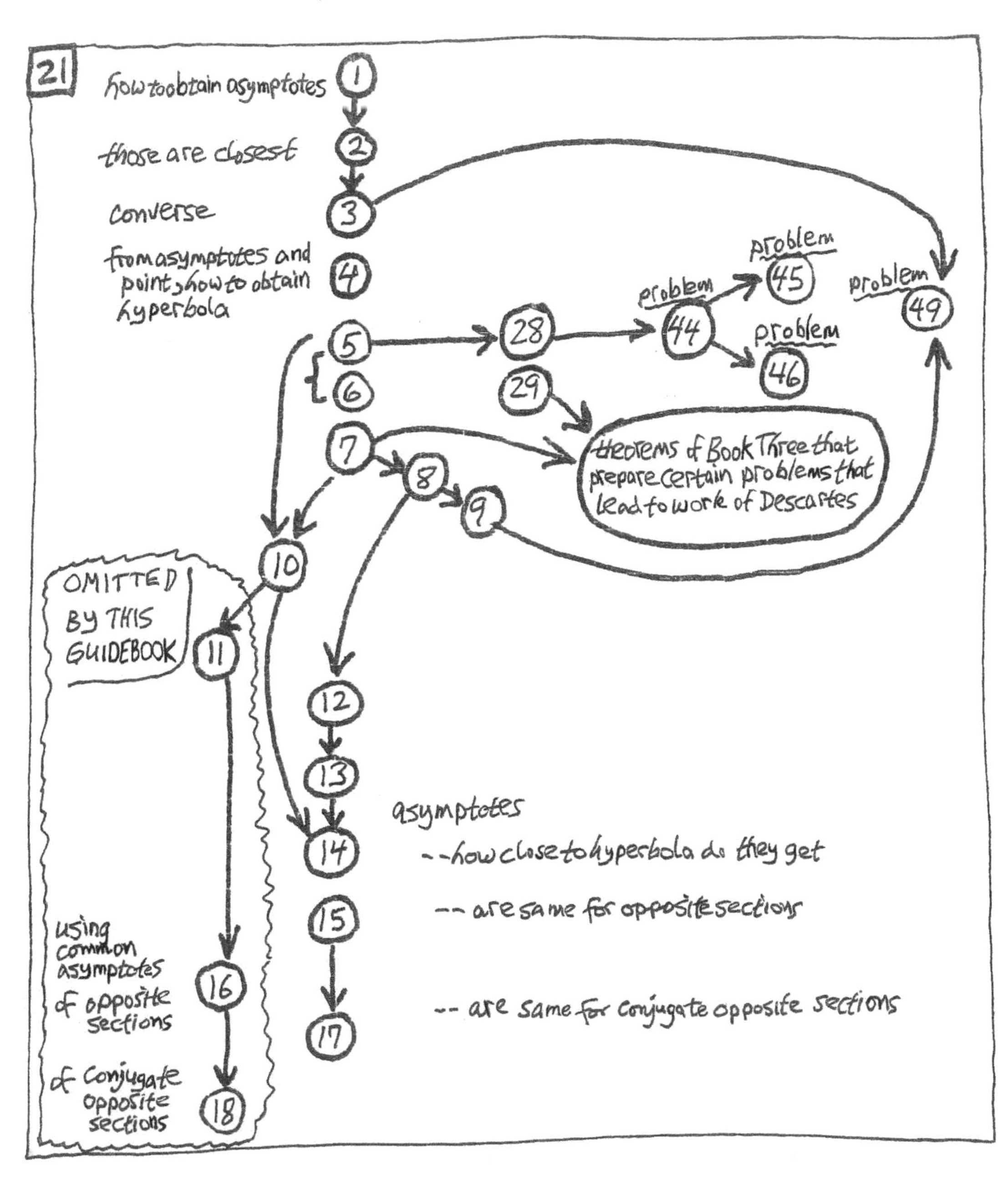

So, let us now look at the 28th and 29th propositions. They tell us something that will come in handy later on—namely, how to obtain with ease a diameter for any conic section (fig. 22).

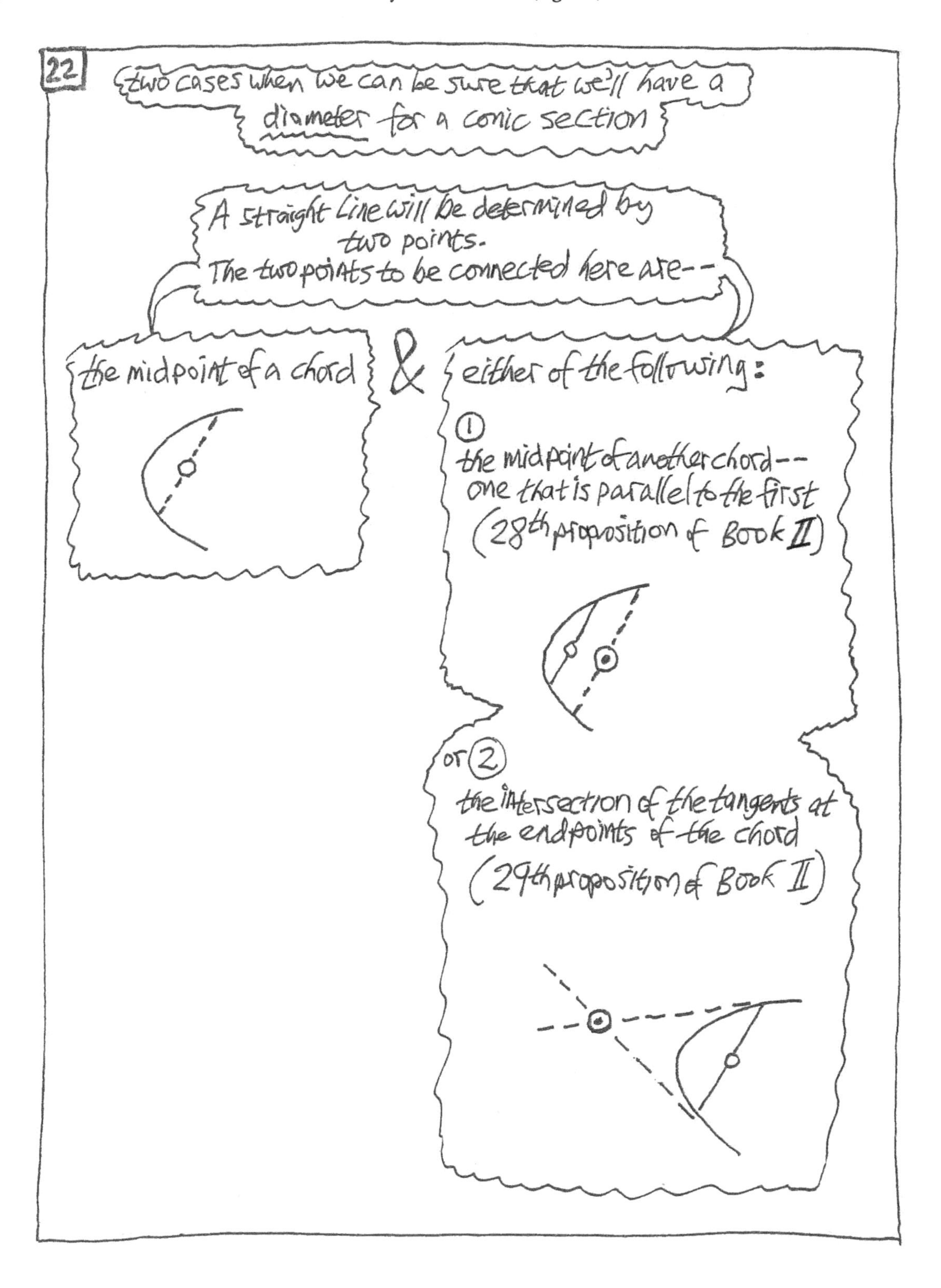

From the 5th proposition, the 28th flows. The 28th proposition shows that you will get a diameter of any conic section if you draw a straight line connecting a certain pair of points. Just draw a straight line connecting the midpoint of any chord of the section to the midpoint of any other chord that is parallel to the first one.

The next proposition, the 29th, shows that you will also get a diameter when the midpoint of any chord of the section is connected, by a straight line, to the intersection of the section's tangents that are tangent at the endpoints of the chord.

(II) **#28** If, in a cone's section or circle's periphery, with respect to two parallel straight [lines], some straight [line] cut [those two parallels] in two [halves], then a diameter of the section is what [this straight line that cuts them] will be.

—For in a cone's section let two parallel straight [lines], namely the [lines] AB, CD, be cut in two [halves] at the [points] E, F; and also, let the joined [line] EF be extended. I say that a diameter of the section is what [this line EF] is.

—For if not, let there be, if possible, as a diameter the [line] GFH. Therefore the [line] touching at the [point] G is parallel to the [line] AB. And so, the same [line] is parallel to the [line] CD. And a diameter is what the [line] GH is; therefore equal is the [line] CH to the [line] HD—which very things is absurd, for the [line] CE is supposed equal to the [line] ED. Therefore not a diameter is the [line] GH. Likewise, then, will we show that neither will there be any other except the [line] EF. Therefore the [line] EF will be a diameter of the section.

(II) #29 If, in a cone's section or circle's periphery, two straight [lines] that touch [it] fall together, then a straight [line] drawn from their [point of] falling together, to the midpoint (*dichotomian*) of the [line] that joins the touch[points], is a diameter of the section.

—Let there be a cone's section or circle's periphery, touching which let straight [lines] be drawn, namely the lines AB, AC that fall together at the [point] A; and also, let the joined [line] BC be cut in two [halves] at the [point] D, and let there be joined the [line] AD. I say that a diameter of the section is what [this line AD] is.

—For, if possible, let a diameter be what the [line] DE is, and let there be joined the [line] EC; [this line EC] will then cut the section. Let it cut at the [point] F, and, through the [point] F parallel to the [line] CDB let there be drawn the [line] FKG. Since, then, equal is the [line] CD to the [line] DB, equal also is the [line] FH to the [line] HG. And since the [line] touching at the [point] L is parallel to the [line] BC, but also is the [line] FG parallel to the [line] BC, therefore also the [line] FG is parallel to the [line] touching at the [point] L. Therefore equal is the [line] FH to the [line] HK—which very thing is impossible. Therefore not a diameter is the [line] DE. Likewise, then, will we show that neither will there be any other [diameter] except the [line] AD.

asymptotes again

Now let us return to the propositions at the beginning of the Second Book, in order to complete our examination of the asymptotes. This completion (starting from the 10th proposition) will have the structure depicted in figure 23.

The 10th proposition continues the consideration of distances between asymptote and section, those distances being taken along the prolongations of chords (fig. 24). Any such pair of lines will contain a rectangle that is equal to one-fourth of the rectangle that is contained by the diameter and the parameter.

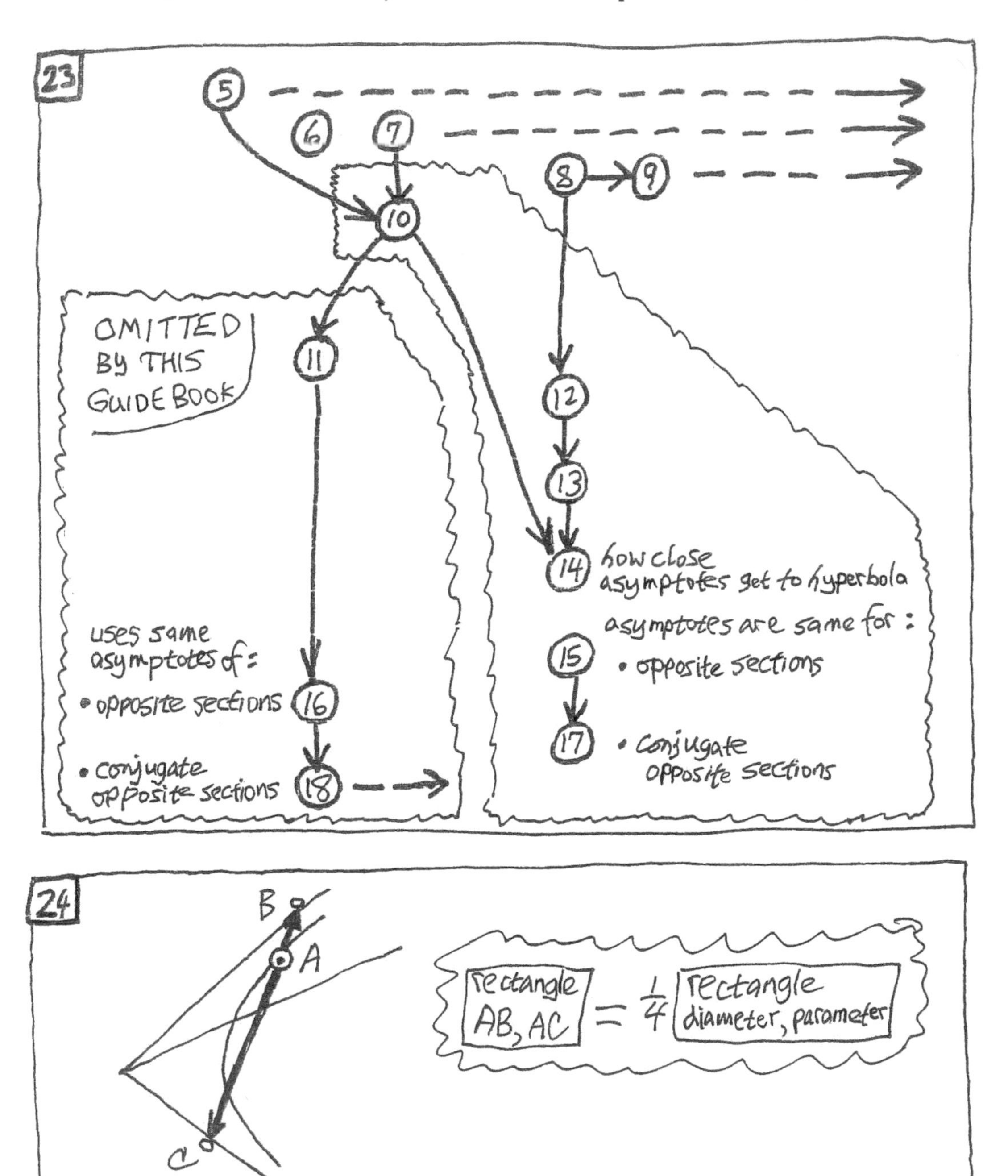

What that means is this. Suppose we move the point along the curve, toward the vertex of that diameter for which the lines in the pair are drawn ordinatewise. Then when the lines become equal to each other, we shall be at the vertex (since the lines will lie along a tangent). That is what we were told by the 9th proposition (fig. 25).

Suppose we now move back again, keeping the lines lying along a line that remains ordinatewise as it moves. Then however unequal the lines become, something will remain the same about their sizes: as the distance of the point from one asymptote grows, the distance of the point from the other asymptote shrinks, so that the size of the rectangle that they make remains the same. That is what we are told now by the 10th proposition (fig. 26).

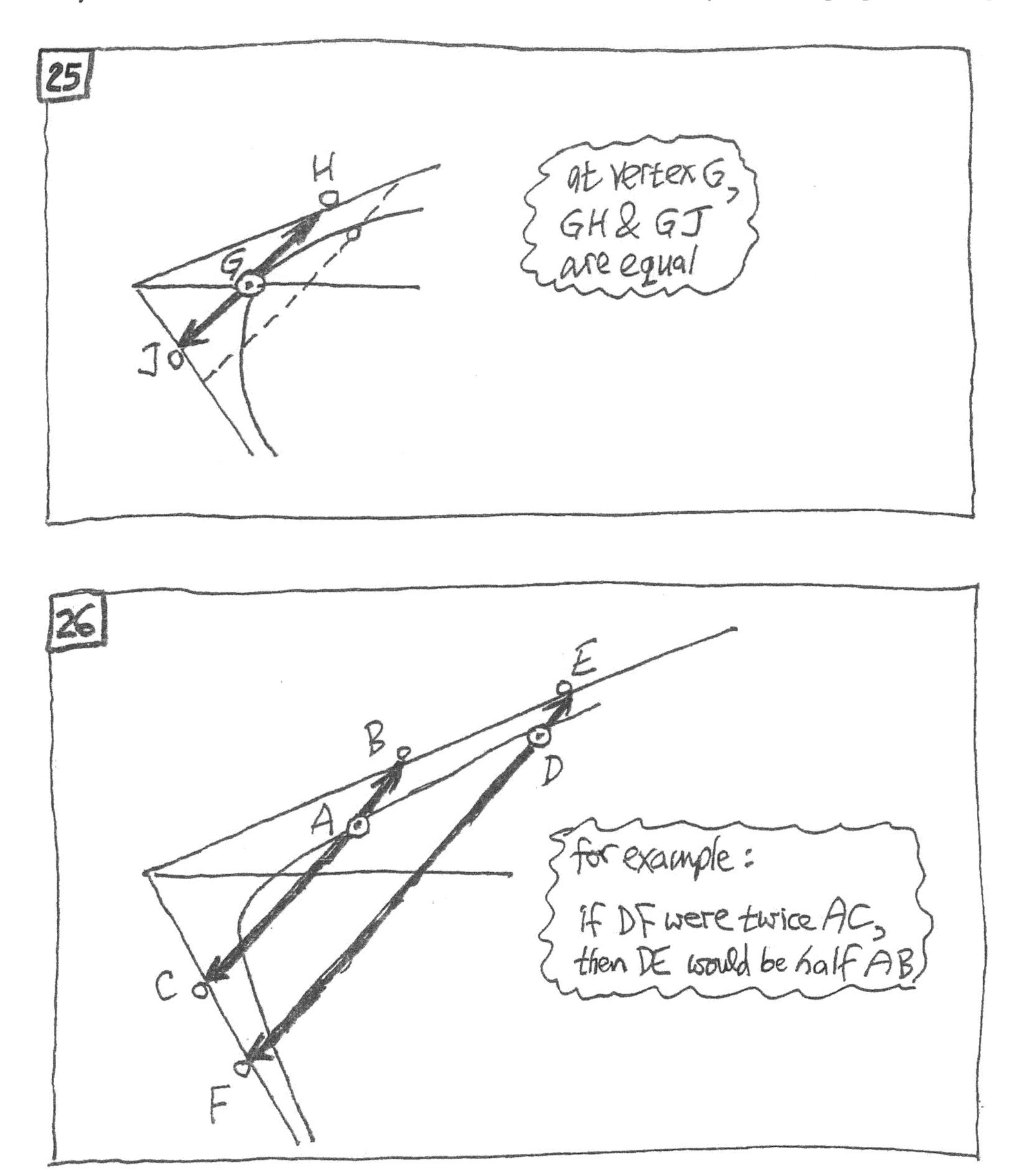

The 10th proposition is presented in the box below.

(II) **#10** If some straight [line] that cuts the section fall together with both of the asymptotes, the rectangle contained by the straight [lines that are] cut off between the asymptotes and the section is equal to the fourth of the figure (*eidous*) generated on the diameter that cuts in two [halves] the [lines] drawn parallel to the drawn straight [line].
—Let there be as an hyperbola the [line] ABC, and as asymptotes of it the [lines] DE, EF; and also, let there be drawn some [line] DF that cuts the section and the asymptotes; and also, let there be cut in two [halves] the [line] AC at the [point] G, and also, let there be joined the [line] GE; and also, let there be put equal to the [line] BE the [line] EH, and let there be drawn from the [point] B at right angles to the [line] HEB the [line] BM; therefore a diameter is what the [line] BH is, and an upright [side] is what the [line] BM is. I say that the rectangle by the [lines] DAF is equal to the fourth of the [rectangle] by the [lines] HBM, and likewise, then, also the [rectangle] by the [lines] DCF [is equal to it].
—For let there be drawn through the [point] B touching the section the [line] KL; therefore parallel it is to the [line] DF. And since it has been shown that as is the [line] HB to BM so is the [square] from EB to the [square] from BK, that is, the [square] from EG to the [square] from GD, and as is the [line] HB to BM so is the [rectangle] by HGB to the [square] from GA—therefore as is the [square] from EG to the [square] from GD so is the [rectangle] by HGB to the [square] from GA. Since, then, as is the whole [square] from EG to the whole [square] from GD, so is the taken-away [rectangle] by HGB to the taken-away [square] from AG; and therefore also what remains, the [square] from EB, is to what remains, the [rectangle] by DAF, as the [square] from EG is to the [square] from GD, that is, the [square] from EB to the [square] from BK. Therefore equal is the [rectangle] by FAD to the [square] from BK. Likewise, then, will it be shown also that equal is the [rectangle] by DCF to the [square] from BL; therefore also equal is the [rectangle] by FAD to the [rectangle] by FCD.

A line is drawn to cut both the hyperbola (at A and C) and its asymptotes (at D and F). By joining the center E to AC's midpoint G, one obtains the diameter EG at whose vertex B the tangent KL will be parallel to AC, according to the 5th proposition (fig. 27).

And, letting BM be the parameter, we have what is depicted in figure 28.

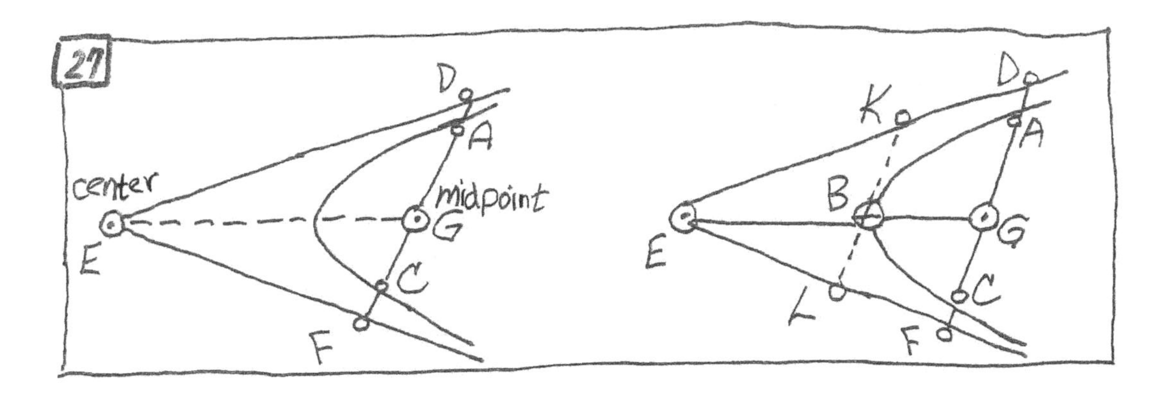
27
D
A
center
midpoint
E
G
C
F
K
B
L

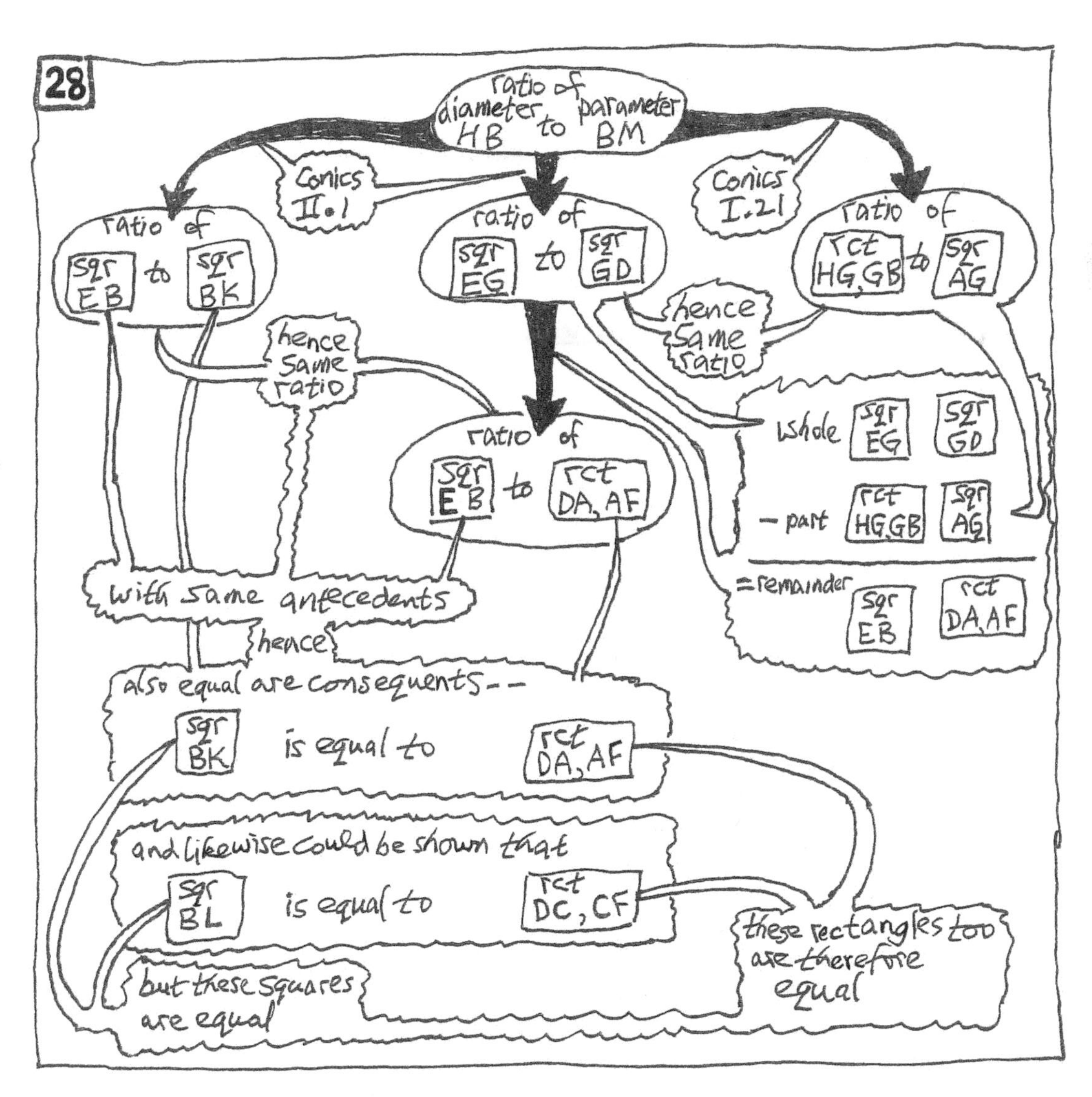
28
ratio of diameter HB to parameter BM
Conics II.1
Conics I.21
ratio of sqr EB to sqr BK
ratio of sqr EG to sqr GD
ratio of rct HG,GB to sqr AG
hence same ratio
hence same ratio
whole sqr EG sqr GD
− part rct HG,GB sqr AG
=remainder sqr EB rct DA,AF
ratio of sqr EB to rct DA,AF
with same antecedents
hence
also equal are consequents--
sqr BK is equal to rct DA,AF
and likewise could be shown that
sqr BL is equal to rct DC,CF
these rectangles too are therefore equal
but these squares are equal

Passing over the 11th proposition, we see in the 12th proposition that the distances to the asymptotes from a point on the hyperbola will make a rectangle that is the same size for every point even if the two distances do not lie along a single straight line that is parallel to the single straight line that is its counterpart at any other point—that is, so long as each of the two distances making an angle at one point is parallel to its counterpart at any other point (fig. 29).

In order to demonstrate the proposition, a line is drawn between the two points (D and G) on the hyperbola, and then prolonged until it meets the asymptotes (at A and C). See figure 30.

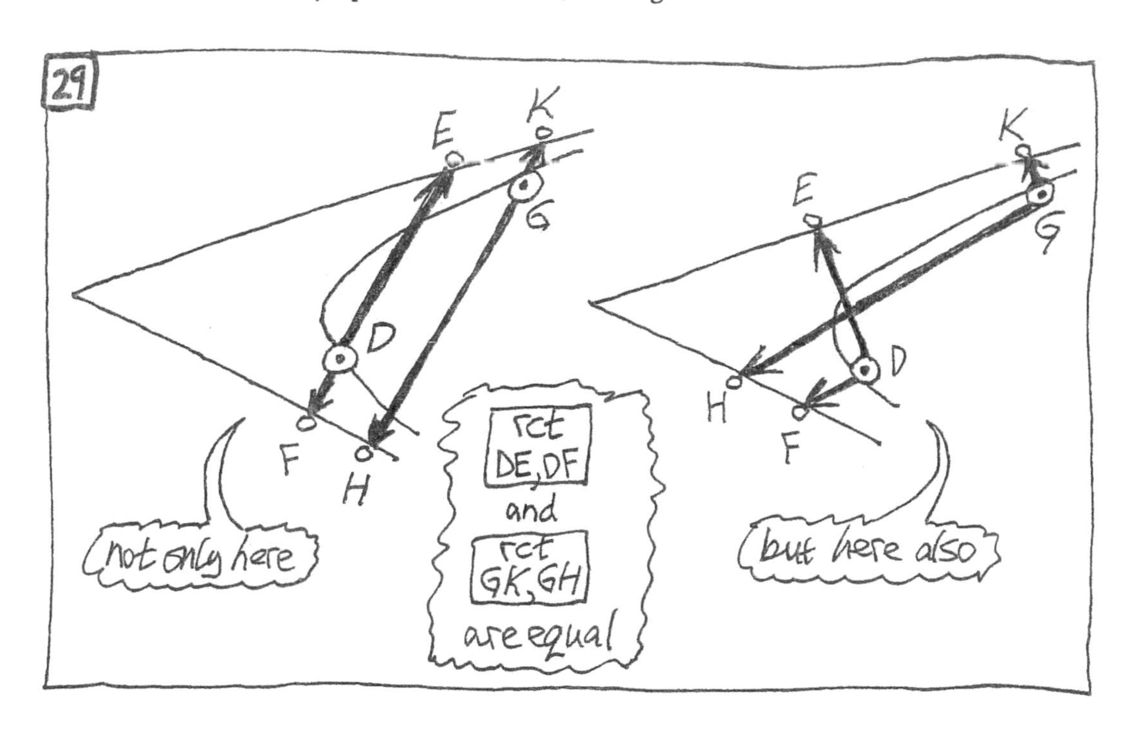

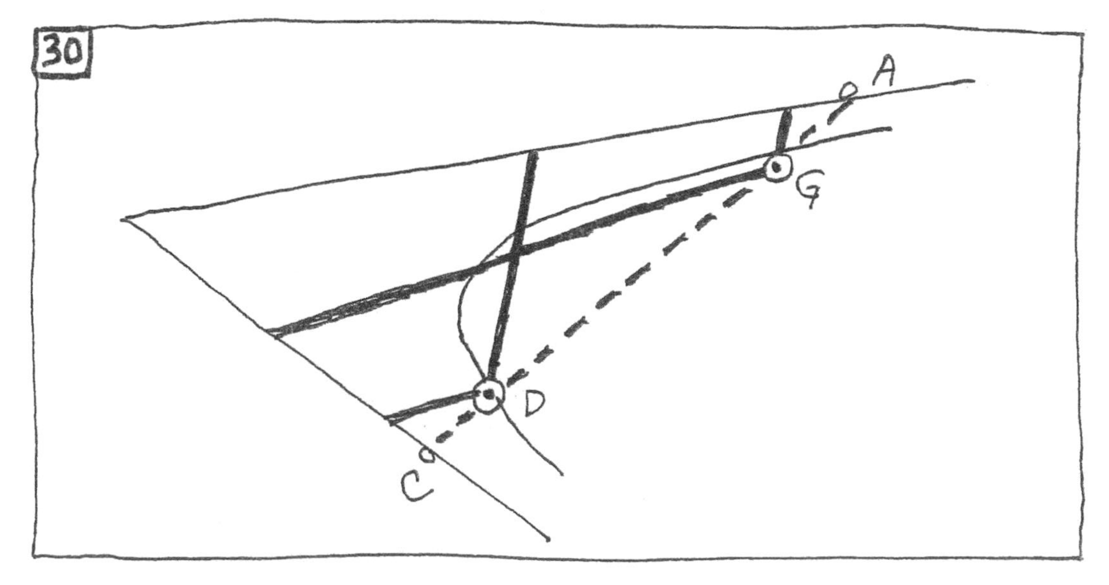

The demonstration of the 12th proposition is depicted in figure 31. The proposition itself is presented in the box following.

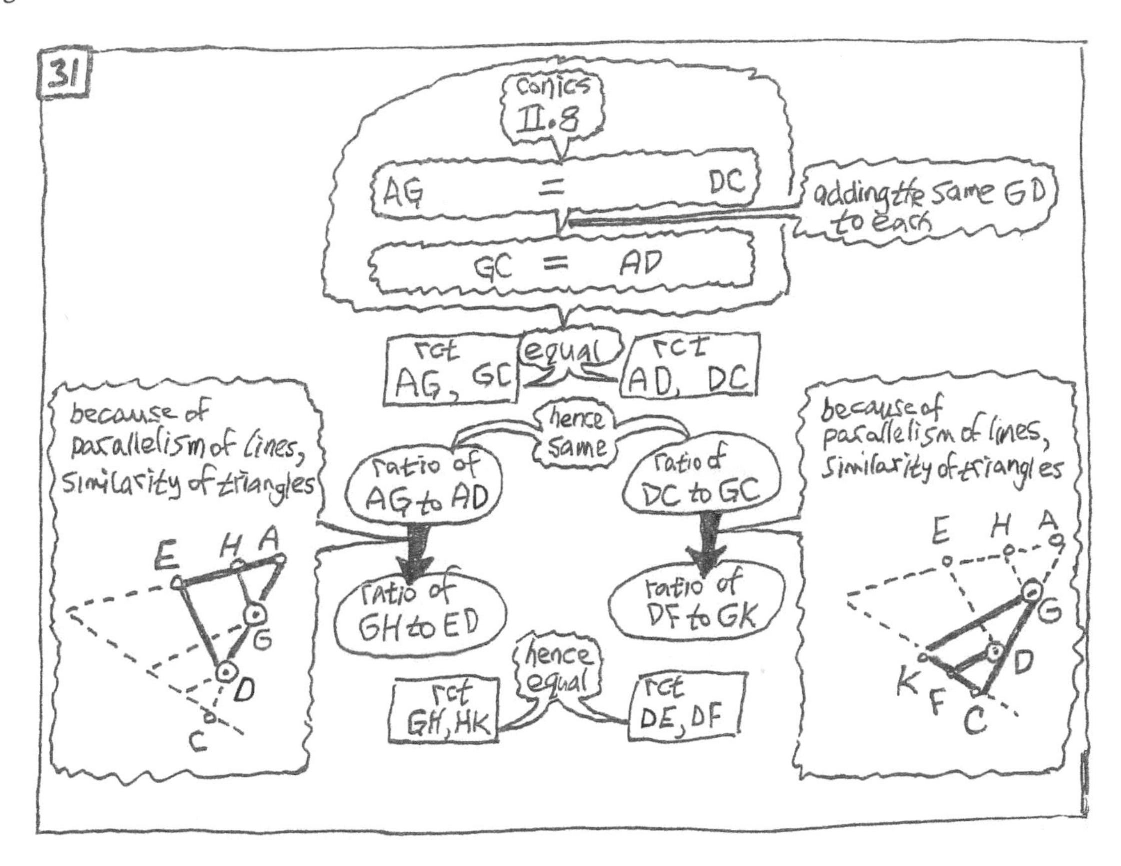

(II) #12 If, to the asymptotes, from some point of those on the section, 2 straight [lines] be drawn at chance angles, and also, to those [two straight lines] parallels be drawn from some point of those on the section—then the rectangle contained by the parallels will be equal to the [rectangles] by those [straight lines] to which they were drawn parallel.
—Let there be an hyperbola whose asymptotes are what the [lines] AB, BC are; and also, let there be taken some point on the section, namely the [point] D; and also, from it to the [lines] AB, BC let there be drawn down [at chance angles] the [lines] DE, DF, and let there be taken some other point on the section, namely the [point] G; and also, through the [point] G parallel to the [lines] ED, DF let there be drawn the [lines] GH, GK. I say that equal is the [rectangle] by EDF to the [rectangle] by HGK.
—For let there be joined the [line] DG, and also let it be extended to the [points] A, C. Since, then, equal is the [rectangle] by ADC to the [rectangle] by AGC, therefore as is the [line] AG to AD so is the [line] DC to CG. But (on the one hand) as is the [line] AG to AD so is the [line] GH to ED, and (on the other hand) as is the [line] DC to CG so is the [line] DF to GK; therefore as is the [line] GH to DE so is the [line] DF to GK. Therefore equal is the [rectangle] by EDF to the [rectangle] by HGK.

That 12th proposition will now be used to demonstrate the 13th—which in turn will be used to demonstrate the 14th.

The 13th proposition is presented in the box below. The demonstration shows by *reductio* that we shall involve ourselves in self-contradiction if we deny that EF meets the section at some point—or if, upon our accepting the existence of some meeting-point M, we deny that M is the only meeting-point (fig. 32).

(II) #13 If, in the place bounded off by the asymptotes and the section, there be drawn some straight [line] parallel to one or the other of the asymptotes, [this parallel] will fall together with the section in one point only.
—Let there be an hyperbola whose asymptotes are what the [lines] CA, AB are; and also, let there be taken some point, namely the point E; and also, through it parallel to the [line] AB let there be drawn the [line] EF. I say that [this line EF] will fall together with the section.
—For, if possible, let it not fall together; and also, let there be taken some point on the section, namely the [point] G; and also, through the [point] G parallel to the [lines] CA, AB let there be drawn the [lines] GC, GH; and also, let the [rectangle] by CGH be equal to the [rectangle] by AEF; and also, let there be joined the [line] AF; and also, let [this line AF] be extended; then it will fall together with the section. Let it fall together with it at the [point] K; and also, through the [point] K parallel to the [lines] CAB let there be drawn the [lines] KL, KD; therefore the [rectangle] by CGH is equal to the [rectangle] by LKD. But [this rectangle by CGH] is supposed equal also to the [rectangle] by AEF; therefore the rectangle contained by DKL, that is the rectangle contained by KLA, is equal to the rectangle contained by AEF—which very thing is impossible, for greater is both the [line] KL than the [line] EF and the [line] LA than the [line] AE. Therefore there will fall together the [line] EF with the section. Let it fall together with it at the [point] M. I say, then, that at another [point] it will not fall together with it.
—For, if possible, let it fall together with it also at the [point] N; and also, through the [points] M, N let there be drawn parallel to the [line] CA the [lines] MX, NB. Therefore the [rectangle] by EMX is equal to the [rectangle] by ENB—which very thing is impossible. Therefore not at another point will [the line EF] fall together with the section.

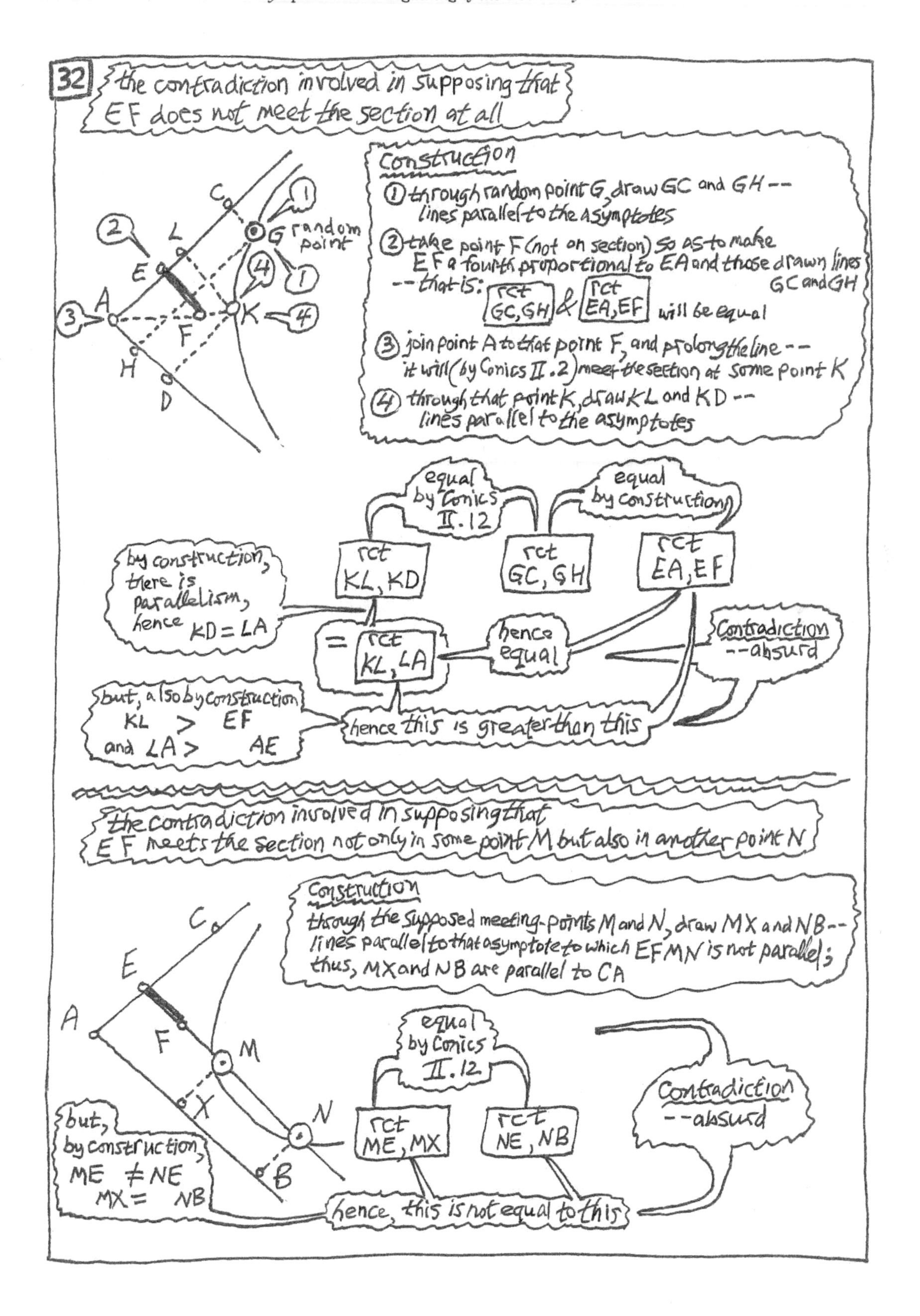
32
the contradiction involved in supposing that EF does not meet the section at all
C
L
G random point
E
A
F
K
H
D
Construction
① through random point G, draw GC and GH -- lines parallel to the asymptotes
② take point F (not on section) so as to make EF a fourth proportional to EA and those drawn lines GC and GH -- that is: rct GC, GH & rct EA, EF will be equal
③ join point A to that point F, and prolong the line -- it will (by Conics II.2) meet the section at some point K
④ through that point K, draw KL and KD -- lines parallel to the asymptotes
equal by Conics II.12
equal by construction
rct KL, KD
rct GC, GH
rct EA, EF
by construction, there is parallelism, hence KD = LA
= rct KL, LA
hence equal
Contradiction --absurd
but, also by construction KL > EF and LA > AE
hence this is greater than this
the contradiction involved in supposing that EF meets the section not only in some point M but also in another point N
Construction
through the supposed meeting-points M and N, draw MX and NB -- lines parallel to that asymptote to which EFMN is not parallel; thus, MX and NB are parallel to CA
C
E
A
F
M
X
N
B
equal by Conics II.12
rct ME, MX
rct NE, NB
Contradiction --absurd
but, by construction, ME ≠ NE MX = NB
hence, this is not equal to this

Now we are ready to answer the following question about the lines that were shown to be the hyperbola's asymptotes in the 1st proposition, and were shown in the 2nd proposition to meet no other lines that are asymptotes. Just how close to the section do these asymptotes get? As we shall soon see, they get ever closer to the section. But things can get ever closer together and yet always be farther apart than some definite distance. On the other hand, things that get ever closer together *may* eventually get closer together than any predetermined distance. The difficulty is that they do not have to. An example showing that they do not have to, is displayed in figure 33.

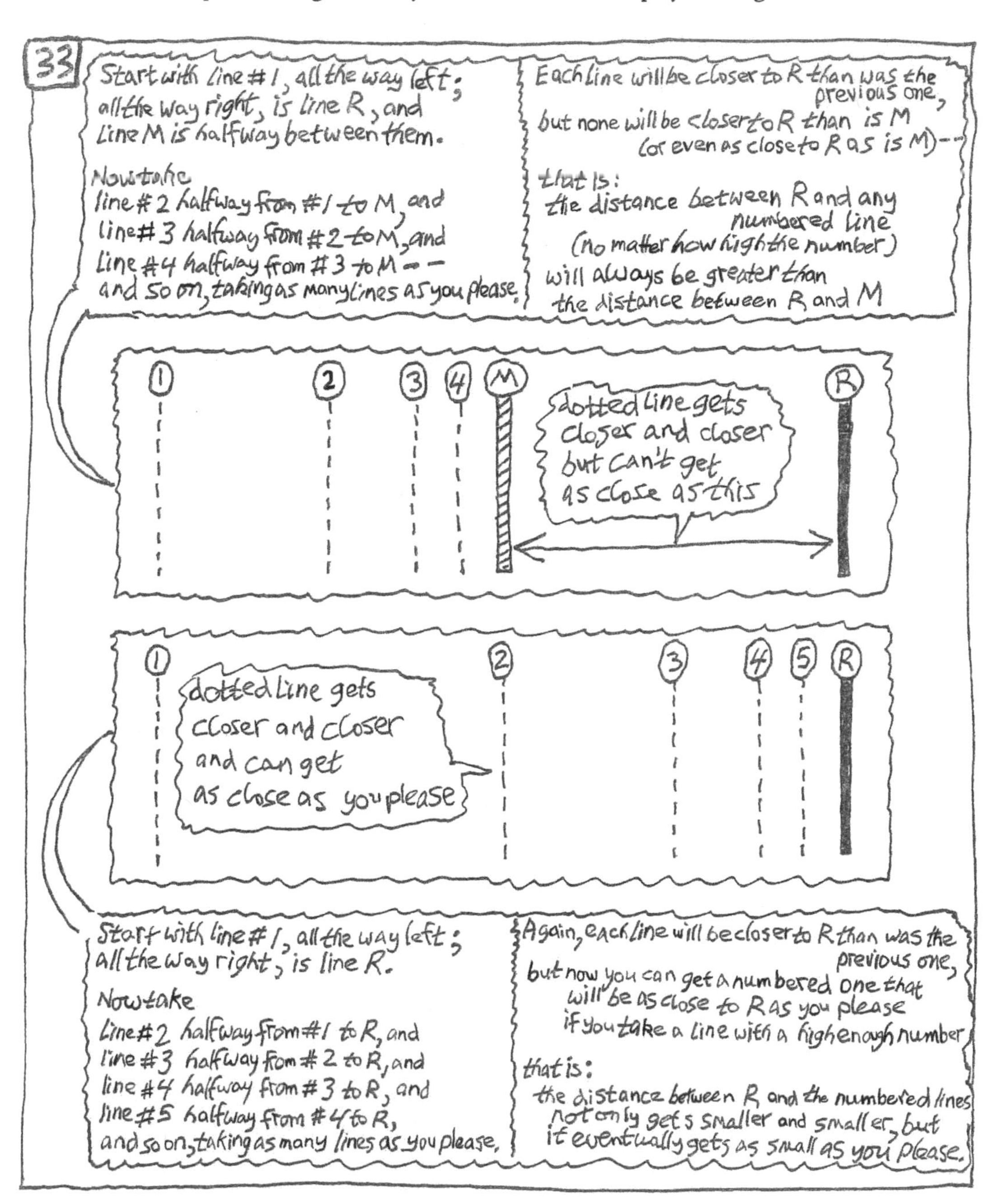

It is because things can get ever closer together without their ever getting as close together as things separated by some predetermined distance, that there are two parts to the 14th proposition (fig. 34). The 14th proposition shows, first, that indefinite prolongation of the asymptotes and the section gets them ever closer together; it then shows also that they eventually get as close together as you please.

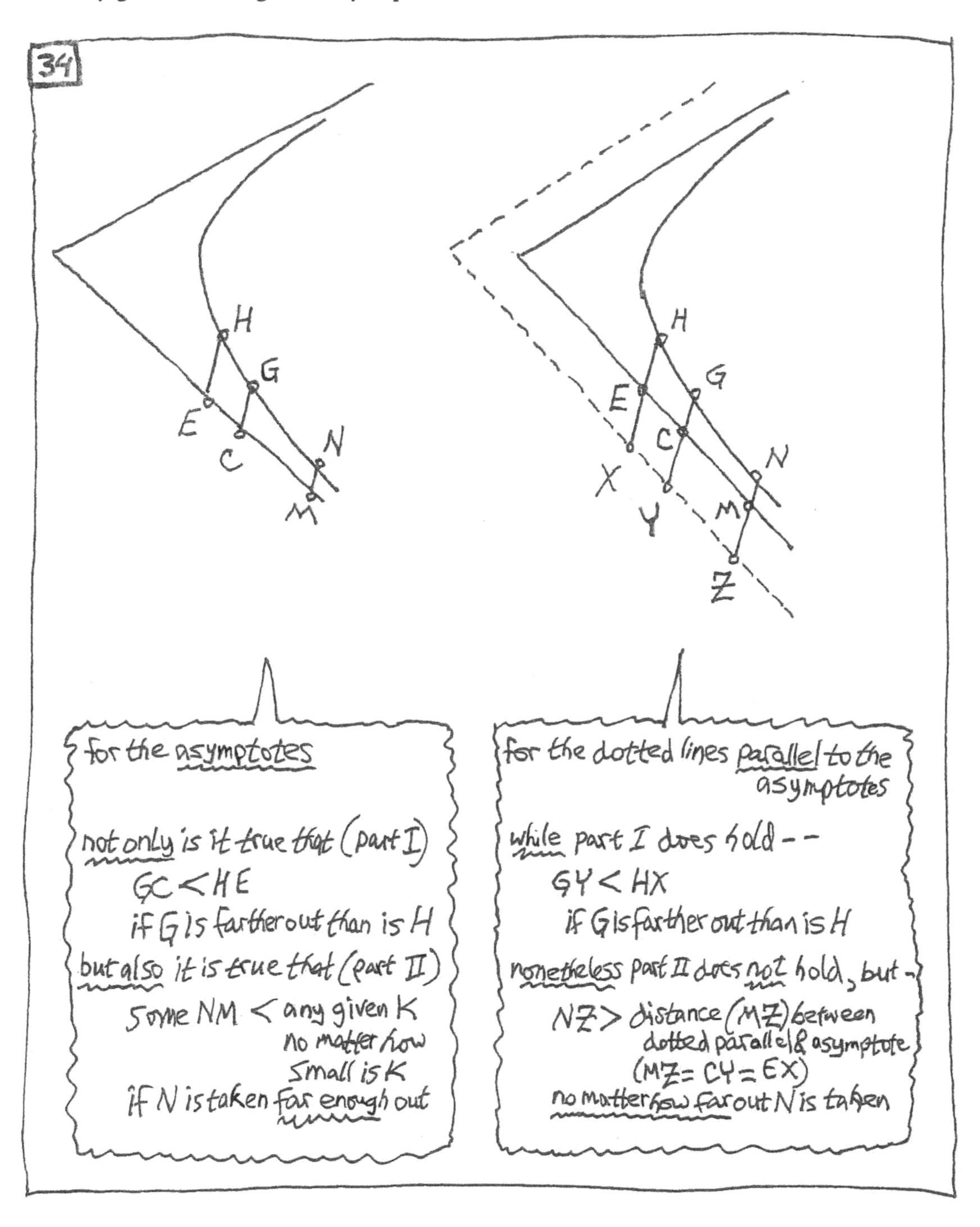

The two parts of the demonstration of the 14th proposition are depicted in figure 35. The proposition itself is presented in the box following.

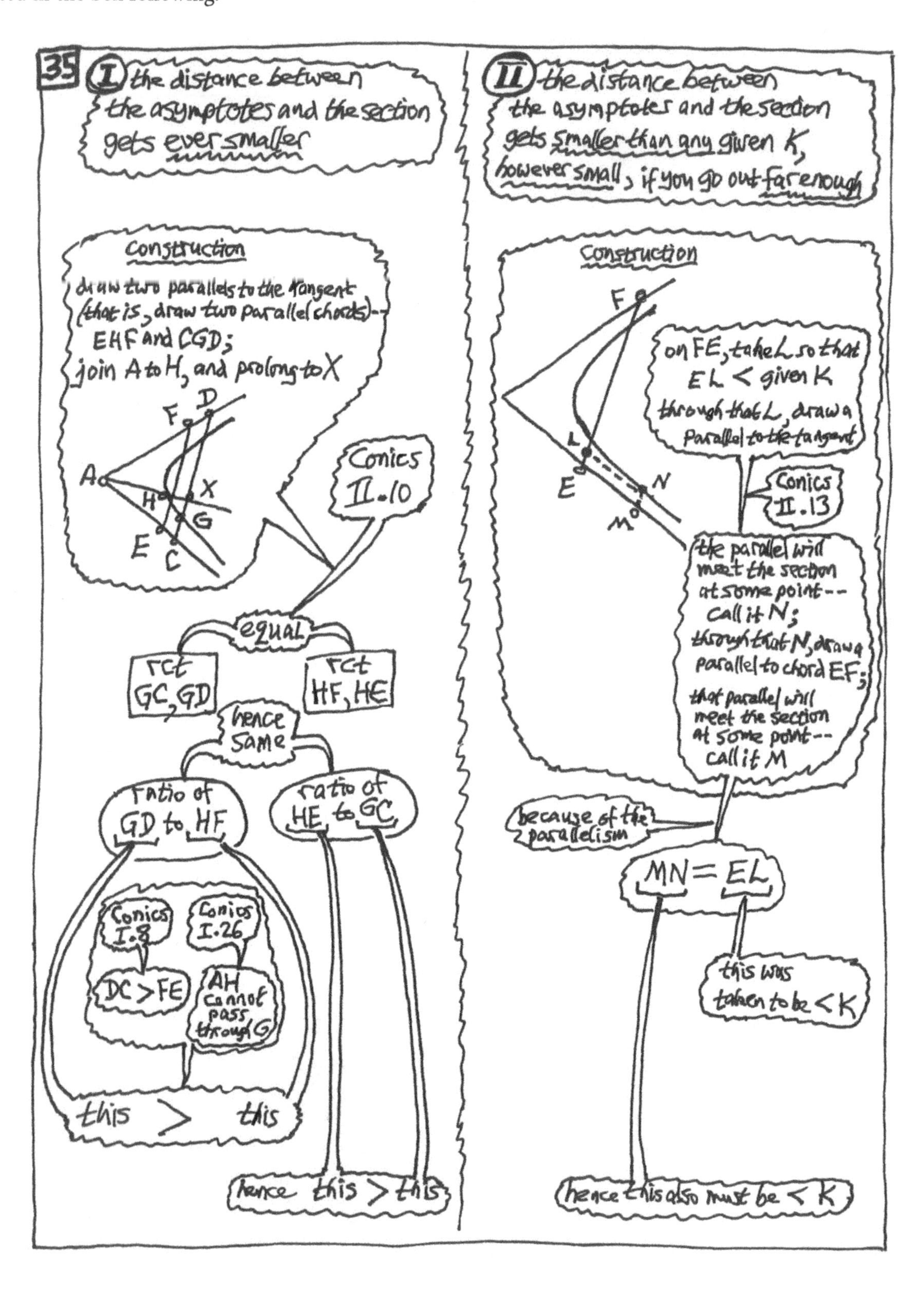

(II) #14 The asymptotes and the section, when boundlessly extended, draw nearer to each other; and a distance less than any given distance do they reach.

—Let there be an hyperbola asymptotes of which are what the [lines] AB, AC are, and a given distance be what the [distance] K is. I say that the [lines] AB, AC and the section, when extended, draw nearer to each other, and will reach a distance less than the [distance] K.

—For let there be drawn parallel to the touching[line] the [lines] EHF, CGD; and also, let there be joined the [line] AH; and also, let [this line AH] be extended to the [point] X. Since, then, the [rectangle] by CGD is equal to the [rectangle] by FHE, therefore as is the [line] DG to FH so is the [line] HE to CG. But greater is the [line] DG than the [line] FH; therefore greater is also the [line] EH than the [line] CG. Likewise, then, will we show that also in succession the lines are less.

—Then let there be taken a distance less than the [distance] K, namely the [distance] EL, and through the [point] L parallel to the [line] AC let there be drawn the [line] LN; therefore it will fall together with the section. Let it fall together with it at the [point] N; and also, through the [point] N parallel to the [line] EF let there be drawn the [line] MNB. Therefore the [line] MN is equal to the [line] EL, and on account of this is less than the [distance] K.

—Porism: From this, then, is it evident that nearer to the section than all the asymptotes are the [lines] AB, AC, and the angle contained by the [lines] BAC is quite clearly less than the angle contained by other asymptotes to the section.

[Of all those straight lines that are "asymptotes" in the sense of "not falling together with" the section, that pair which is nearer to the section than all others we consider to be "*the* asymptotes."]

Our examination of the hyperbola's asymptotes now concludes with propositions showing that the same asymptotes belong to the hyperbolas which constitute opposite sections, and also that the same asymptotes belong to the hyperbolas which constitute conjugate opposite sections. These propositions (the 15th and the 17th) are presented in the following boxes.

(II) **#15** Of opposite sections, common are the asymptotes.
—Let there be opposite sections a diameter of which is what the [line] AB is, and center is what the [point] C is. I say that the asymptotes of the sections A, B are common.
—Let there be drawn, through the points A, B [lines] that touch the sections, namely the [lines] DAE, FBG; therefore parallel are they. Let, then, each of the [lines] DA, AE, FB, BG be taken off, each having the power [to give rise to a square equal to] a [rectangle that is] equal to the fourth of the figure (*eidous*) that is laid along ["is applied to"] the [line] AB; therefore equal are the [lines] DA, AE, FB, BG. Then let there be joined the [lines] CD, CE, CF, CG. Then it is evident that in a straight [line] is the [line] DC with the [line] CG, and the [line] CE with the [line] CF—on account of the parallels. Since, then, it is an hyperbola whose diameter the [line] AB is, and whose touching[line] the [line] DE is, and also, each of the [lines] DA, AE has the power [to give rise to a square equal to] the fourth of the figure (*eidous*) that is laid along ["is applied to"] the [line] AB, therefore asymptotes are what the [lines] DC, CE are. On account of the same things, then, also asymptotes to the section B are what the [lines] FC, CG are. Therefore of opposite [sections] common are the asymptotes.

(II) #17 Of opposite [sections] that are in conjugacy (*kata suzugian*), common are the asymptotes.
—Let there be conjugate opposite [sections] whose conjugate diameters are what the [lines] AB, CD are, and center is what the [point] E is. I say that common are their asymptotes.
—For let there be drawn [lines] touching the sections through the points A, B, C, D—namely the [lines] FAG, GDH, HBK, KCF; therefore a parallelogram is what the [polygon] FGHK is. Then let there be joined the [lines] FEH, KEG; therefore straight are they, and are also diagonals of the parallelogram, and they are all cut in two [halves] at the point E. And since the figure (*eidos*) on the [line] AB is equal to the square from the [line] CD, and equal is the [line] CE to the [line] ED; therefore each of the squares from FA, AG, KB, BH is equal to a fourth of the figure (*eidous*) for the [line] AB. Therefore asymptotes of the sections A, B are what the [lines] FEH, KEG are. Likewise, then, will we show that the same [lines] are also asymptotes of the sections C, D. Of opposite [sections] that are in conjugacy, therefore, common are the asymptotes.

CHAPTER 2. PROBLEMS AT THE END OF BOOK II OF THE CONICS

In the Second Book of his *Conics*, Apollonius explores asymptotes and conjugate opposite sections.

The 43rd proposition is the last theorem in the succession of theorems which stretches from the beginning of the Second Book. This theorem, which is not the last proposition in the Second Book, presents a situation that will yield conjugate diameters for conjugate opposite sections. What it tells us is this:

> if one of the conjugate opposite sections
> is cut in two points by a straight line EF,
> and through the center Y are drawn two straight lines—
> one of them drawn to the midpoint G of the cutting line,
> and the other one drawn parallel to the cutting straight line—
> then these straight lines YG and YC
> will form conjugate diameters of the opposite sections (fig. 1).

That 43rd proposition is followed by the last ten propositions of the Second Book. These propositions (the 44th through 53rd) are not theorems but problems—except for the 48th and the 52nd, which, even though they are theorems, merely complement the work of the problems. What is treated in these ten propositions that conclude the Second Book is depicted on the following page. For the purposes of this guidebook, our treatment of Book II needn't go beyond the 49th proposition (which is the first of the problems of drawing a tangent to a given conic section).

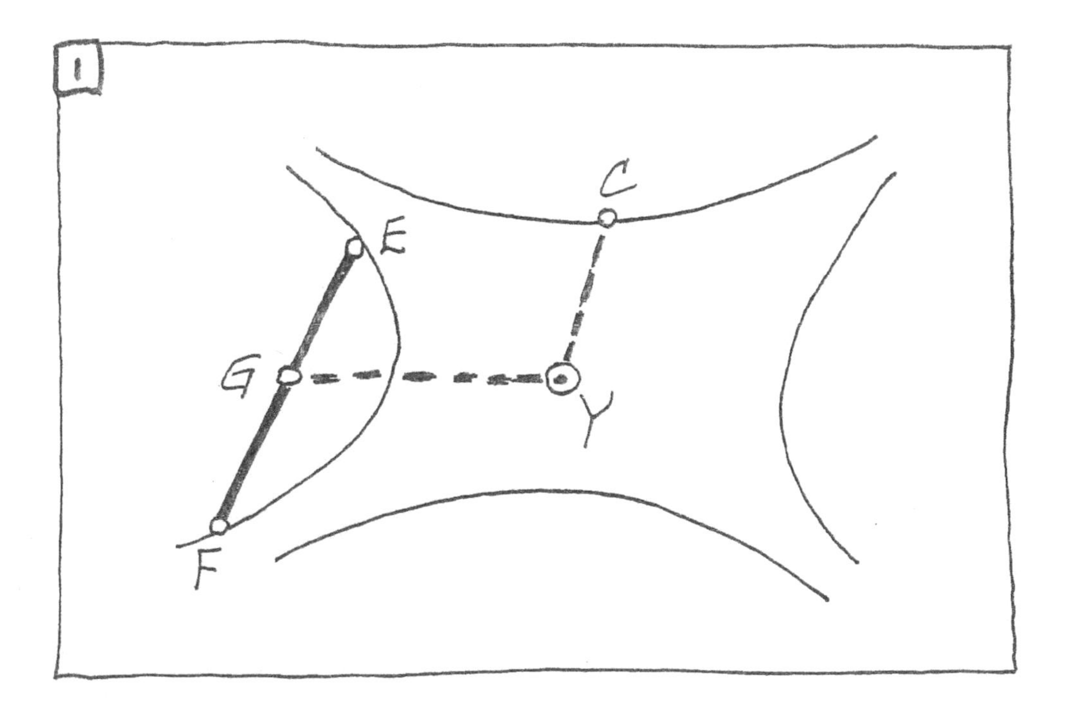

to find		proposition
a diameter	of a given conic section	44
the center	of a given conic section that has a center i.e., of an ellipse or hyperbola	45
the axis	of a given parabola	46
	of a given hyperbola or ellipse	47
	(theorem) the completeness of the previous proposition's solution for hyperbola and ellipse: i.e., there are no other axes	48
that tangent to a given section		
	which is drawn from a given point not within the section	49
	which makes a given acute angle with	
	the axis, on the same side as the section	50
	that diameter—of a parabola or hyperbola— which goes through the tangency-point	51
	(theorem) a restriction on the possible size of the given acute angle in the situation of the next proposition, when the given section is an ellipse: if an ellipse's tangent makes an angle with that diameter which goes through the tangency-point, then this angle is not less than the angle that is adjacent to the one contained by the straight line deflected at the middle of the section	52
	that diameter—of an ellipse— which goes through the tangency-point (the given acute angle being restricted in size as shown in the previous proposition)	53

In solving these problems, Apollonius first presents an analysis ("let it have been done," he says) and then a synthesis (constructing what is sought by proceeding in reverse through the steps already presented in the analysis). Before you go on, you may find it helpful to re-read what Part Three of this guidebook said at the end of Chapter 1 (*The parabola*), under the heading *analysis and synthesis*.

Having earlier considered analysis as a means of seeking theorems and their demonstrations, let us now move on to consider it as a means of providing solutions of problems.

A classic statement on the distinction between analysis and synthesis in relation to the distinction between theorems and problems is presented in the *Collection* of Pappus, which dates from around 300 AD, which is some five centuries after the time of Apollonius. In the preface to the Third Book of that work, Pappus says:

> With some of the ancients saying that
> all the inquiries (*zêtoumena*) in geometry are problems,
> and some saying that all of them are theorems,
> those who wish to discriminate them more technically
> hold it worthwhile (*axiousi*)
> to call that a "problem" for which
> this is thrown forth (*proballetai*):
> to do, indeed to furnish (*poiêsai kai kataskeuasai*) something;
> and to call that a "theorem" in which
> this is gazed at (*theôreitai*):
> what wholly in consequence follows upon
> some things' being supposed (*hypokeimenôn*).
>
> He, then, who proposes (*proteinôn*) the theorem—
> whatever the way (*tropon*) he has seen it—
> is held to be required (*axioi*) to seek (*zêtein*)
> the conclusion inherent in the premise,
> and he would not be proposing it soundly otherwise;
> and he who proposes the problem
> is excusable and not called to account
> if he commands (*prostaxêi*) something
> somehow impossible to furnish,
> for the inquirer's work is to determine just this—
> what is possible and what is impossible,
> and, insofar as it is possible,
> when and how and in how many ways it is possible.

(Note that here, and subsequently, the translations from the Greek of Pappus—as well as, later on, from the Greek of Diophantus, the Latin of Viète, and the French of Descartes—will be so presented in this guidebook that the layout of the text can help to make the structure of the thought easier to see than it would be if the print stretched uniformly across the page from margin to margin.)

Later on, in the preface to the Seventh Book of his *Collection*, Pappus speaks of several books that constitute a library for doing analysis—"special material furnished for those who, going beyond the common elements, wish to obtain capability having to do with finding (*dynamein heuretikên*) in respect to lines of problems set before them." He goes on to say this:

> Now analysis is a road (*hodos*)—
> from the thing that is sought [but taken] as agreed to,
> through the things that follow [from it] in order,
> to something agreed to with respect to synthesis.

For in the analysis—
having supposed (*hypothemenoi*) as having come into being
the thing that is sought—
we look out for that from which this comes about,
and we look out again for the antecedent of that,
until, thus stepping back up, we face something out of the things that
are already known or have the rank of something primary (*taxin arkhês*);
and such an access-road (*ephodos*), we call "analysis"—
as being a "loosening or solving" that "backs up" (*anapalin lysis*).

But in the synthesis,
by a reversal of what is gotten hold of in the analysis—
taking as a support the last thing that has come into being already,
and setting in natural order as consequents
those things that were antecedents,
and putting them together onto one another (*allêlois episynthentes*)—
we finally arrive at the furnishing of the thing sought;
and this, we call "synthesis."

And analysis is of two-fold kind:
the one, having to do with seeking (*zêtetikon*) of the truth—
which is called
"having to do with gazing" (*theôretikon*);
the other, having to do with supplying ways and means (*poristikon*)—
which is called
"having to do with what's thrown forth" (*problêmatikon*).

In the theoretical kind, supposing the thing sought
as being something that is, and as being what is true,
and then—through the things that follow in order
as being true, and as being in accordance with supposition—
advancing to something that is agreed to,
if that thing that is agreed to is true,
then also true will be the thing sought,
and the demonstration will be the reverse of the analysis
(*hê apodeixis antistrophos tê analysei*),
but if we happen upon something that is agreed to be false,
then also false will be the thing sought.

In the problematical kind, supposing the thing commanded
as being apprehended,
and then— through the things that follow in order as being true—
advancing to something that's agreed to,
if the thing that's agreed to is possible and suppliable,
i.e., if it is what the mathematicians call "given,"
then also possible will be the thing commanded,
and again the demonstration will be the reverse of the analysis,
but if we happen upon something that is agreed to be impossible,
then also impossible will be the thing thrown forth, the problem.

Thus, analysis of the kind that is *zetetic* has to do with seeking demonstrations for what stays ever the same to be gazed at; it has to do with being *theoretic*. (In Greek, the verb for "seeking" is *zêteô*; for "gazing," is *theôreô*.) Analysis of the kind that is *poristic* has to do with supplying solutions to challenges thrown forth; it has to do with dealing with what is *problematic*. (In Greek, the verb for "supplying ways and means" is *poridzô*; the word for "something that is thrown forth" is *problêma*.)

"Given A, to get Z"—that is the challenge *thrown* forth. That's your *problem*. How will you solve it? You must first go about *loosening up* the tight tangle, and then you must go about *putting things together*. You must first do *analysis*—find *some* way which will get you from what is *sought*, to what is *given*; and then do *synthesis*—reverse your steps, going back again along that same way (going from what is *given*, to what is *sought*).

According to a nineteenth-century German scholar named Hankel, each of these two parts of the solution may, in turn, be divided into two parts—as follows. (See Heath's note on pp. 138–42 of the first volume of his edition of Euclid's *Elements*.)

PROBLEM: From A, to get Z.

ANALYSIS: Find *some* way *from* Z to A.

TRANSFORMATION:	*If* we *did* have Z, then we *would* have B; and then we would *also* have C/D.
RESOLUTION:	But A would give us M, which would give us C/D.

SYNTHESIS: Go back along that same way,
taking the same steps in reverse—
that is, going from A to Z.

CONSTRUCTION:	From A, get M, thus constructing C/D.
DEMONSTRATION:	Since we now *have* C/D, we therefore have B; hence we have what we sought— namely, Z.

In the analysis, the getting of Z is transformed into the getting of other things, and then the getting of those other things is resolved into their constructibility from A; then, in the synthesis, the construction that has been shown to be needful and possible is actually performed, after which the demonstration shows, from the construction obtained from A, that we get Z.

The transformation begins the analysis by moving from what is sought; then, taking the same steps in reverse order, the demonstration ends the synthesis by moving from what is given. The resolution, which ends the analysis, and the construction, which begins the synthesis, take the same steps in the same order, both moving from what is given. Thus, the solution will have the structure depicted in figure 2.

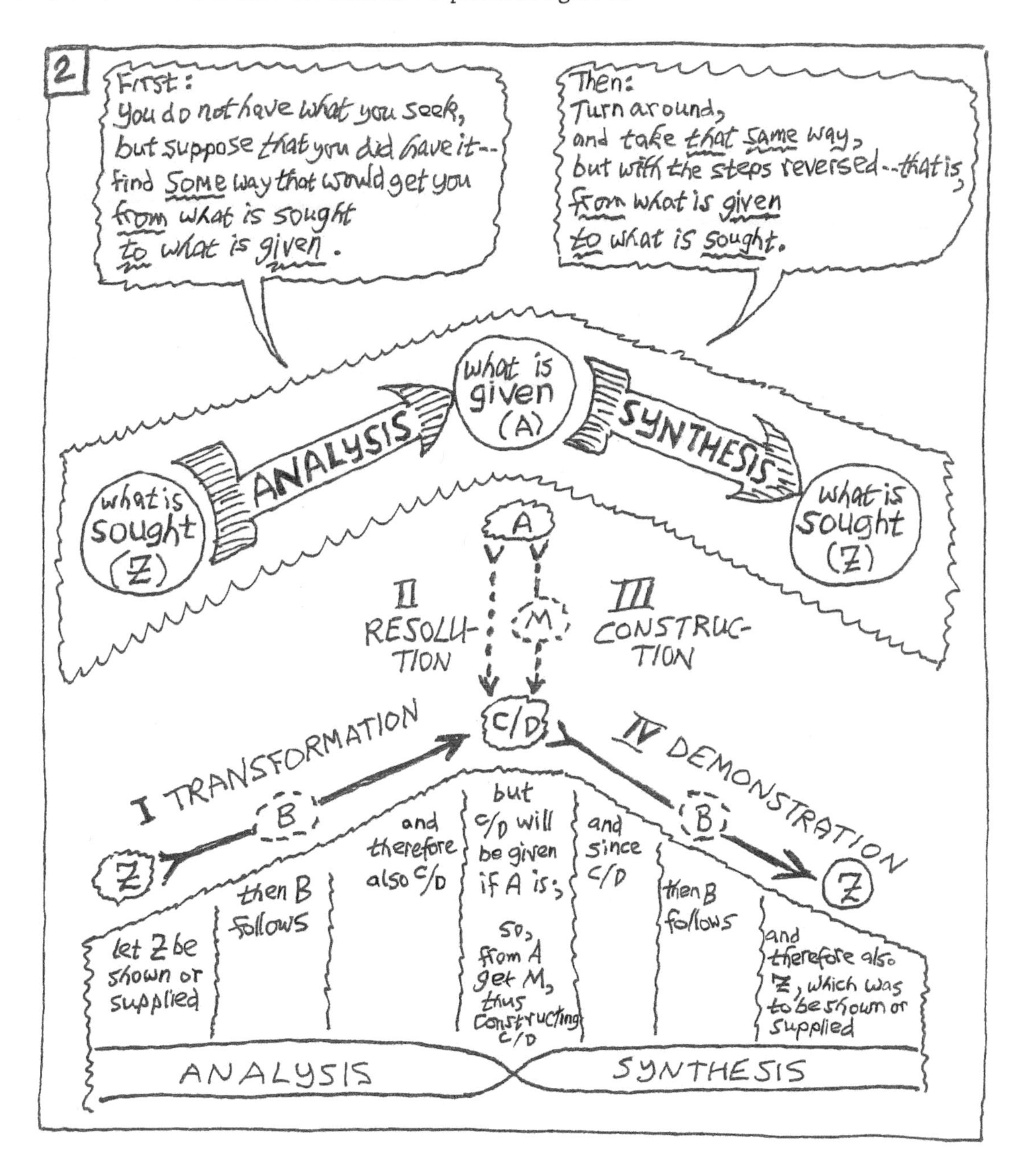

Now let us consider the 44th proposition of the Second Book, and also the 45th. (Remember: we are using books Two and Three of the *Conics* as mere quarries for propositions—and therefore, as with some of the propositions in Book One, complete diagrams for those propositions of Book Two which are selected for inclusion in this chapter and the previous one, and for those propositions of Book Three which are selected for inclusion after this, in Chapters 3 and 6, are included in the Appendix at the end of this second volume of the guidebook.)

It should be noted that Apollonius himself does not identify propositions as "problems"; as is his practice elsewhere, he merely numbers the items one after another, without calling them "theorems" or "problems"—or even "propositions."

(II) #44 Of a given section of a cone, the diameter—to find it.
—Let there be the given section of a cone, on which [section] are the points A, B, C, D, E. Required, then, to find its diameter.
—Let [the diameter] have been generated, and let the [line] CH be it. Then with there being drawn ordinatewise the [lines] DF, EH, and also with them being extended, there will be equal (on the one hand) the [line] DF to the [line] FB and (on the other hand) the [line] EH to the [line] HA. If, then, we set in order the [lines] BD, EA to be in position as parallels, there will be given the points H, F. And so, there will be in position the [line] HFC.
—Constructed will it then be (*suntethêsetai*) thus: let there be the given section of a cone on which [section] are the points A, B, C, D, E, and let there be drawn parallel the [lines] BD, AE, and let them be cut in two [halves] at the [points] F, H. And the [line] FH, joined, will be a diameter of the section. And in the same way, will we indeed boundlessly find diameters.

(II) #45 Of a given ellipse or hyperbola, the center—to find it.
—And this is evident, for if drawn through be two diameters of the section, namely the [lines] AB, CD, then the [point] at which they cut each other will be the section's center, as is supposed.

The layout of the 44th proposition is depicted in the following box. Things are so simple that there is not much of a demonstration in this proposition; and things are even simpler in the 45th proposition.

PROBLEM : Given a section of a cone with its points,
to find a diameter of the section.

ANALYSIS: *Suppose* we do have the thing *sought*; and
find *some* way to get
from this thing that is *sought* (a diameter of the section)
to things that are *given* (points on the section).

TRANSFORMATION :
Let it be that we have the diameter—
that is, ask: what if we had the diameter;
well, *if* we did have the diameter,
then we would have the ordinates,
and then we would have the midpoints of the ordinates.

RESOLUTION:
And we would have midpoints of ordinates
if we took midpoints of parallel chords.

SYNTHESIS: Do the construction indicated by the resolution,
and then take the way taken in the transformation,
but go in *reverse*—that is,
from the things that are *given*
to the thing that is *sought*.

CONSTRUCTION:
Draw two chords that are parallel;
bisect them;
join the midpoints.

DEMONSTRATION:
The line joining the midpoints will (by II.28)
be a diameter of the section (according to First Definitions 4).

There is more, however, to the 46th proposition (and to the 47th and the 48th after it). The 46th is presented below; and the proposition is laid out in the box that follows it.

(II) #46 Of a given section of a cone, the axis—to find it.
—Let the given section of a cone first be a parabola, on which are the [points] F, C, E. Required, then, to find its axis.
—For let there be drawn its diameter, the [line] AB, say. Then if the [line] AB is an axis, there would be generated the thing ordered; but if not, let it have been generated, and let there be as axis the [line] CD; therefore the [line] CD is an axis parallel to the [line] AB—and the [lines] that are drawn perpendicular to it, it cuts in two [halves]. And the perpendiculars to the [line] CD are perpendiculars also to the [line] AB; and so, the [line] CD cuts the perpendiculars to the [line] AB in two [halves]. Then if I will set in order the [line] EF as a perpendicular to the [line] AB, it will be [given] in position, and on account of this there will be equal the [line] ED to the [line] DF; therefore given is the [point] D. Therefore through the given [point] D parallel in position to the [line] AB there has been drawn the [line] CD; therefore [given] in position is the [line] CD.
—Constructed will it then be (*suntethêsetai*) thus: let there be the given parabola on which are the [points] F, E, A; and also, let there be drawn as a diameter of it the [line] AB; and also, let there be drawn perpendicular to it [perpendicular, that is, to diameter AB] the [line] BE; and also, let [line BE] be extended to the [point] F. Then if (on the one hand) equal is the [line] EB to the [line] BF, it is evident that the [line] AB is an axis. But if (on the other hand) not equal, let the [line] EF be cut in two [halves] by the [point] D; and also, let there be drawn parallel to the [line] AB the [line] CD; then it is evident that the [line] CD is an axis of the section—for, being parallel to the diameter, that is, being [itself] a diameter, it cuts the [line] EF in two [halves] and also at right angles. Therefore the given parabola's axis has been found as the [line] CD.
—And it is evident that there is [only] one axis of the parabola. For if there will be another, as the [line] AB, it will be parallel to the [line] CD. And it cuts the [line] EF; and so, also [cuts it] in two [halves]. Therefore equal is the [line] BE to the [line] BF—which very thing is absurd.

PROBLEM: Given a section of a cone that is a parabola,
to find the axis.

ANALYSIS: Suppose we have the thing that is sought (the parabola's axis);
and find some way to get
from it to the thing that is given (the parabola).

TRANSFORMATION:
Let it be that we have the axis;
if we did have it, then we would also have another diameter
(namely: any parallel to the axis that meets the parabola;
since Apollonius demonstrates earlier (I.46) that
any parallel to a diameter of a parabola is also a diameter
—and an axis is a diameter)
and any such other diameter's perpendiculars
would be perpendicular also to the axis
(since a perpendicular to one of two parallels
is perpendicular to the other parallel also)
and would be bisected by the axis
(since the axis is a diameter—
that is, a bisector—
of chords that are perpendicular to it);
that is to say, we would have the ordinates of the axis.

RESOLUTION:

And we would have the ordinates of the axis
if we took the perpendiculars to any diameter.

SYNTHESIS: Do the construction indicated by the resolution,
and then take the way taken in the transformation,
but go in reverse—
that is, go from the things that are given
to the thing that is sought.

CONSTRUCTION:

For the given parabola, draw a diameter
(as II.44 says can be drawn for any conic section—
by connecting the midpoints of two parallel chords);
and, perpendicular to that diameter, draw a chord;
now if the chord is bisected by the diameter
then the diameter is the axis—
but if not, then bisect the chord, and
through the midpoint thus obtained,
draw a line parallel to the diameter.

DEMONSTRATION:

This line is the *axis*, for
it is a diameter (since it is parallel to a diameter of the parabola)
and the chord that it bisects (its ordinate) is perpendicular to it;
and it is *the* axis—there is no other—for
another axis would have to be yet *another* line that
passes through the chord's midpoint
and is parallel to the diameter, which is absurd.

What the 46th proposition did for the parabola, the 47th and 48th do for the hyperbola or ellipse. The 47th finds an axis (and its conjugate) for the hyperbola or ellipse, and the 48th shows that the section has no other axis. The box below presents the 47th, which is depicted in the box that follows it. The 48th will be presented in the box after that, and then depicted in figure 3.

(II) #47 Of the given hyperbola or ellipse, the axis—to find it.
—Let an hyperbola or ellipse be what the [line] ABC is; required, then, to find its axis.
—Let it have been found; and let the [line] KD be it, and let the section's center be what the [point] K is; therefore the [line] KD cuts the lines drawn down ordinatewise in two [halves] and at right angles.
—Let there be drawn perpendicular the [line] CDA, and let there be joined the [lines] KA, KC; then, since equal is the [line] CD to the [line] DA, therefore equal is the [line] CK to the [line] KA. Then if we set in order [that is, fix in location] the given [point] C, there will be given the [line] CK. And so, the circle that is carved out (*graphomenos*) with center the [point] K and with [radial] distance the [line] KC will also go through the [point] A and will be given in position. And there is also the section ABC given in position; therefore given is the [point] A. But there is also the [point] C given; therefore in position the [line] CA is [given]. And also, equal is the [line] CD to the [line] DA. Therefore given is the [point] D. But also the [point] K is given; therefore given in position is the [line] DK.
—Constructed will it then be (*suntethêsetai*) thus: let there be as the given hyperbola or ellipse the [line] ABC, and also, let there be taken as its center the [point] K, and let there be taken on the section as a chance point the [point] C; and also, with center the [point] K and with [radial] distance the [line] KC, let a circle be carved out (*gegraphthô*), namely the [circle] CEA; and also, let there be joined the [line] CA and let it be cut in two [halves] at the [point] D; and also, let there be joined the [lines] KC, KD, KA; and also, let the [line] KD be drawn through to the [point] B.
—Since, then, equal is the [line] AD to the [line] DC, and common is the [line] DK, therefore the two [lines] CDK are equal to the two [lines] ADK, and also base the [line] KA is equal to [base] the [line] KC. Therefore the [line] KBD cuts the [line] ADC in two [halves] and also at right angles. Therefore an axis is what the [line] KD is.
—Let there be drawn through the [point] K parallel to the [line] CA the [line] MKN; therefore the [line] MN is an axis of the section, conjugate to the [line] BK.

PROBLEM: Given an hyperbola or ellipse,
to find the axis.

ANALYSIS: Suppose we have what is sought (the axis); and
find some way to get from it to what is given (the section).

TRANSFORMATION:

Let it be that we have the axis KD—the center being point K;
if we did have it, then it would bisect its ordinates, at right angles;
now, perpendicular to the axis, let a chord (CDA) be drawn;
and let center K be joined to the chords' endpoints,
making KA and KC, which are equal
(since CD and DA are equal).

RESOLUTION:

If given point C is fixed on the section, then CK will be given,
and hence a circle around center K having this line KC as radius
will also pass through point A and be given in position;
hence (since the section ABC is also given in position)
point A is given;
hence (since point C is also given) CA is given in position;
hence (since CD is equal to DA) point D is given;
hence (since point K also is given) DK is given in position.

SYNTHESIS: Do the construction indicated by the resolution,
and then take the way taken in the transformation,
but go in reverse—
that is, go from the things that are given
to the thing that is sought.

CONSTRUCTION:

Take the center K of the given hyperbola or ellipse (ABC),
and take a random point C on the section;
and, with center K and radius KC, carve out the circle CEA;
join CA and bisect it (at D);
join K to C, to D, and to A; and draw KD through to B.

DEMONSTRATION:

AD is equal to DC,
and DK is common,
and KA is equal to KC;
hence KBD bisects ADC at right angles;
hence KD is an axis.

Now drawing, through K, a line (MKN) parallel to CA—
this line MN will be the axis of the section that is conjugate to BK.

(II) **#48** Then with these things having been shown, let the next step be to show that other axes of the same sections [that is, of the hyperbola or of the ellipse] there are not.

—For, if possible, let also another axis be what the [line] KG is. Then according to the same things as before, with the [line] AH being drawn perpendicular, equal will be the [line] AH to the [line] HL; and so, also the [line] AK [will be equal] to the [line] KL; but [the line AK is equal] also to the [line] KC; therefore equal [will be] the [line] KL to the [line] KC—which very thing is absurd.

—That the circle AEC does not collide with the section also in another point between the [points] A, B, C—now that is evident on the hyperbola. But on the ellipse, let there be drawn as perpendiculars the [lines] CR, LS. Then—since equal is the [line] KC to the [line] KL, for they are [lines] out-from-the-center, [that is, are radii]; and also, equal is the [square] from CK to the [square] from KL; but (on the one hand) equal to the [square] from CK are the [squares taken together] from CR, RK and (on the other hand) equal to the [square] from LK are the [squares taken together] from KS, SL—therefore the [squares taken together] from CR, RK are equal to the [squares taken together] from LS, SK. Therefore, that by which the [square] from CR differs from the [square] from SL, by this does the [square] from KS differ from the [square] from RK. Again, since the [rectangle] by MRN along with (*meta*) the [square] from RK is equal to the [square] from KM, and also the [rectangle] by MSN along with (*meta*) the [square] from SK is equal to the [square] from KM—therefore the [rectangle] by MRN along with (*meta*) the [square] from RK is equal to the [rectangle] by MSN along with (*meta*) the [square] from SK. Therefore that by which the [square] from SK differs from the [square] from KR, by this does the [rectangle] by MRN differ from the [rectangle] by MSN. But it was shown that, that by which the [square] from SK differs from the [square] from KR, by this does the [square] from CR differ from the [square] from LS; therefore that by which the [square] from CR differs from the [square] from SL, by this does the [rectangle] by MRN differ from the [rectangle] by MSN. And since [lines] drawn down are what the [lines] CR, LS are, therefore as is the [square] from CR to the [rectangle] by MRN so is the [square] from SL to the [rectangle] by MSN. But there was shown also for both [squares] the same excess (*huperechê*)—[that is, the former square's excess over the former rectangle is equal to the latter square's excess over the latter rectangle]—therefore equal is (on the one hand) the [square] from CR to the [rectangle] by MRN and (on the other hand) the [square] from SL to the [rectangle] by MSN. Therefore a circle is what the [line] LCM is—which very thing is absurd, for it is supposed [to be] an ellipse.

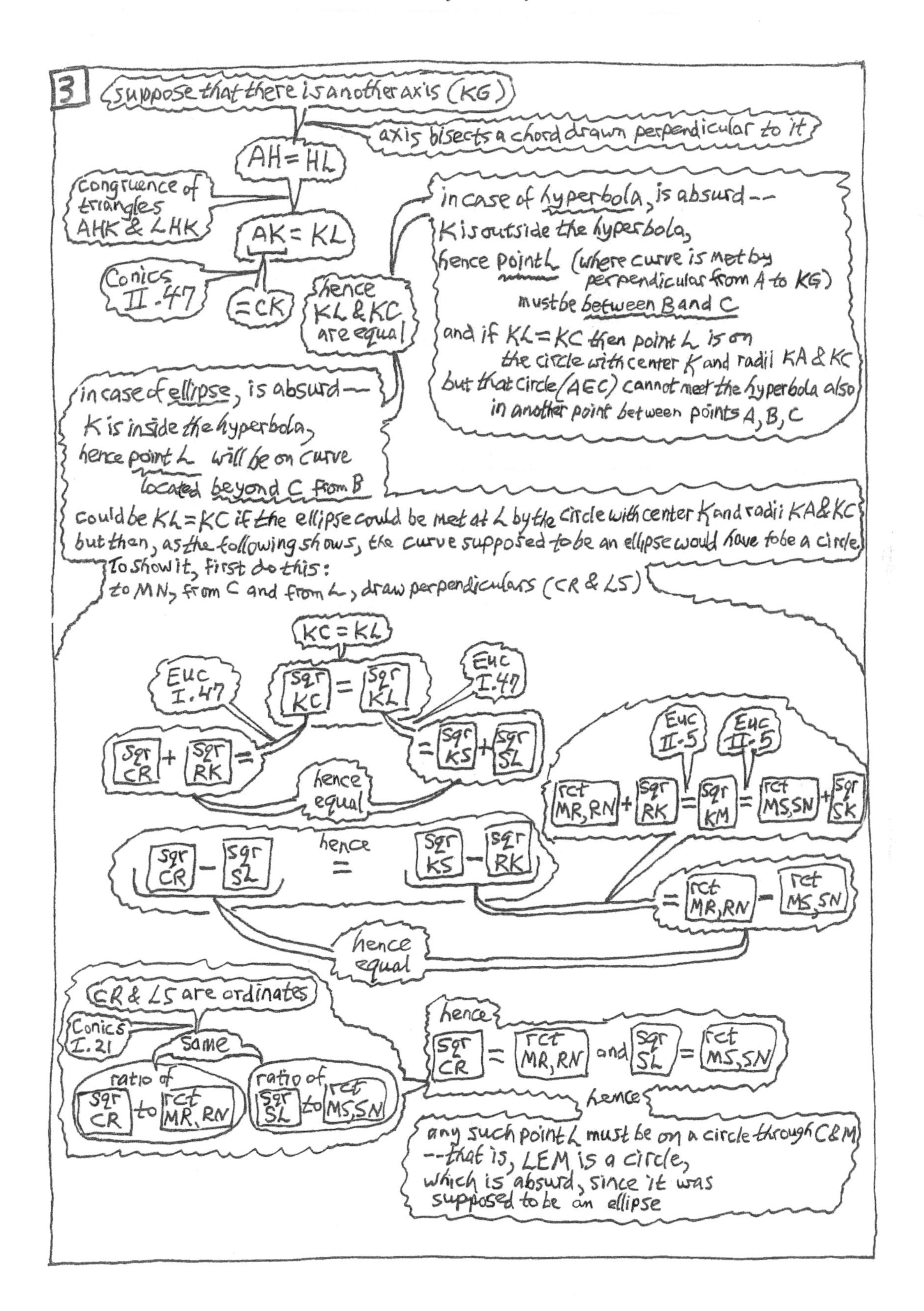
3
suppose that there is another axis (KG)
axis bisects a chord drawn perpendicular to it
AH = HL
congruence of triangles AHK & LHK
AK = KL
Conics II.47
= CK
hence KL & KC are equal
in case of hyperbola, is absurd -- K is outside the hyperbola, hence point L (where curve is met by perpendicular from A to KG) must be between B and C
and if KL = KC then point L is on the circle with center K and radii KA & KC
but that circle (AEC) cannot meet the hyperbola also in another point between points A, B, C
in case of ellipse, is absurd -- K is inside the hyperbola, hence point L will be on curve located beyond C from B
could be KL = KC if the ellipse could be met at L by the circle with center K and radii KA & KC
but then, as the following shows, the curve supposed to be an ellipse would have to be a circle.
To show it, first do this: to MN, from C and from L, draw perpendiculars (CR & LS)
KC = KL
Euc I.47
sqr KC = sqr KL
Euc I.47
sqr CR + sqr RK =
= sqr KS + sqr SL
hence equal
Euc II.5
Euc II.5
rct MR, RN + sqr RK = sqr KM = rct MS, SN + sqr SK
sqr CR − sqr SL
hence =
sqr KS − sqr RK
= rct MR, RN − rct MS, SN
hence equal
CR & LS are ordinates
Conics I.21
same
ratio of sqr CR to rct MR, RN
ratio of sqr SL to rct MS, SN
hence
sqr CR = rct MR, RN and sqr SL = rct MS, SN
hence
any such point L must be on a circle through C & M
-- that is, LEM is a circle,
which is absurd, since it was supposed to be an ellipse

The last of the problems that we shall consider is presented in the 49th proposition. The beginning of the proposition treats the first of the conic sections (the parabola). It is presented in the box below, and then its structure is laid out after that, following the depiction in figure 4. Then the rest of the proposition is presented in the boxes on the following pages.

(II) #49 A cone's section being given, and also a point not inside the section, to draw from the point a straight [line] touching the section at one point.
—Let the given section of a cone first be a parabola whose axis the [line] BD is. Required, then, from the given point, which is not inside the section, to draw a straight [line] as previously set forth (*prokeitai*).
—The given point, then, is: either on the line, [that is, on the curve,] or on the axis, or in the remaining place outside.
—Now let [the given point] be on the line; and also, let the [point] A be it; and also, let [the required line] have been generated; and also, let the [line] AE be it; and also, let there be drawn a perpendicular [to the axis BD], namely the [line] AD; then [this line AD] will be [given] in position. And equal is the [line] BE to the [line] BD; and also, there is given the [line] BD; therefore given is also the [line] BE. And also is the [point] B given; therefore given is also the [point] E. But [given] also is the [point] A; therefore [given] in position is the [line] AE.
—Constructed then will be (*suntethêsetai*) [the touchingline] thus: from the [point] A let there be drawn a perpendicular [to the axis BD], the [line] AD; and also, let there be put equal to the line [BD] the [line] BE; and also, let there be joined the [line] AE. Then it is evident that [the line AE] touches the section.
—Again, let the given point be on the axis—say the [point] E; and also, let [the required line] have been generated, and let there be drawn a touching[line], say the [line] AE; and also, let a perpendicular [to the axis] be drawn, say the [line] AD; therefore equal is the [line] BE to the [line] BD. And given is the [line] BE; therefore given is also the [line] BD. And also is given the [point] B; therefore given is also the [point] D. And there is upright [to the line BD] the [line] DA; therefore [given] in position is the [line] DA; therefore given is the [point] A. But [given] also is the [point] E; therefore [given] in position is the [line] AE.
—Constructed then will be (*suntethêsetai*) [the touchingline] thus: let there be put equal to the [line] BE the [line] BD; and also, from the [point] D let there be put upright to the [line] ED the [line] DA; and also, let there be joined the [line] AE. Then it is evident that the [line] AE touches.
—And it is evident also that, even if the given point is the same as the [point] B, the [line] drawn upright [to the line BD] from the [point] B touches the section.

—Then let the given point be the [point] C [in the remaining place outside]: and also, let [the required line] have been generated; and also, let the [line] CA be it; and also, through the [point] C parallel to the axis, that is, [parallel] to the [line] BD, let there be drawn the [line] CF; therefore [given] in position is the line CF. And also, from the [point] A let there be drawn ordinatewise to the [line] CF the [line] AF; there will then be equal the [line] CG to the [line] FG. And also, there is given the [point] G; therefore given is also the [point] F. And also, the [line] FA has been drawn up ordinatewise, that is, has been drawn parallel to the [line] touching at the [point] G; therefore [given] in position is the [line] FA. Therefore given is also the [point] A; but [given] also is the [point] C. Therefore [given] in position is the [line] CA.

—Constructed will be (*suntethêsetai*) [the touchingline] thus: let there be drawn through the [point] C parallel to the [line] BD the [line] CF; and also, let there be put equal to the [line] CG the [line] FG; and also, parallel to the [line] touching at the [point] G let there be drawn the [line] FA; and also, let there be joined the [line] AC. Then it is evident that [this line AC] will do the problem (*problêma*).

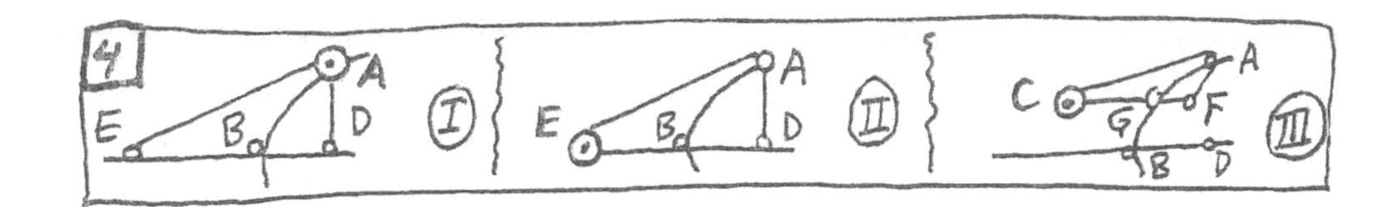

To draw the tangent to a given parabola from a given point not within it—
three cases, depending on the location of the given point:

(I) the given point A *is on the curve itself* *but not on the axis (BD)*	(II) the given point E *is outside the curve* *and on the axis (BD)*	(III) the given point C *is outside the curve* *but not on the axis (BD)*
—ANALYSIS— *TRANSFORMATION:* Let it have been done— let the tangent be AE. From given point A, draw AD perp to axis; then AD is given in position.	*—ANALYSIS—* *TRANSFORMATION:* Let it have been done— let the tangent be EA. From given point A, draw AD perp to axis; then AD is given in position.	*—ANALYSIS—* *TRANSFORMATION:* Let it have been done— let the tangent be CA. From given point C, draw CF parallel to axis; then CGF is given in position; from given point A, draw AF ordinatewise (parallel to tangent at G).
Then since BD is also given, and by (I.35) BE is equal to BD, also given is BE.	Then since BE is also given, and (by I.35) BD is equal to BE, also given is BD.	Then since CG is also given, and (by I.35) GF is equal to GC, also given is GF.
RESOLUTION: Since BE is given, and point B is given, then also given is point E;	*RESOLUTION:* Since BD is given, and point B is given, then also given is point D; hence DA is given in position, and thus point A is also given;	*RESOLUTION:* Since GF is given, and FA is ordinatewise, then also given is point A;
hence AE is given in position. (So if we had AE, then we'd also have BE; and we would have BE if we took E as far from B as D is, along the axis.)	hence EA is given in position. (So, if we had EA, then we'd also have BD; and we would have BD if we took D as far from B as E is, along the axis.)	hence CA is given in position. (So, if we had CA, then we'd also have GF; and we would have GF if we took F as far from G as C is, along the axis's parallel.)

—*Synthesis*—	—*Synthesis*—	—*Synthesis*—
Construction:	*Construction:*	*Construction:*
From A, draw AD perp to axis; make BE equal to BD;	Make BD equal to BE; from D, draw DA perp to axis;	From C, draw CF parallel to axis; make GF equal to GC; at F, draw FA parallel to tangent at G;
Demonstration:	*Demonstration:*	*Demonstration:*
Join given A to that E. (By I.33) AE is tangent.	Join given E to that A. (By I.33) EA is tangent.	Join given C to that A. (By I.33) CA is tangent.

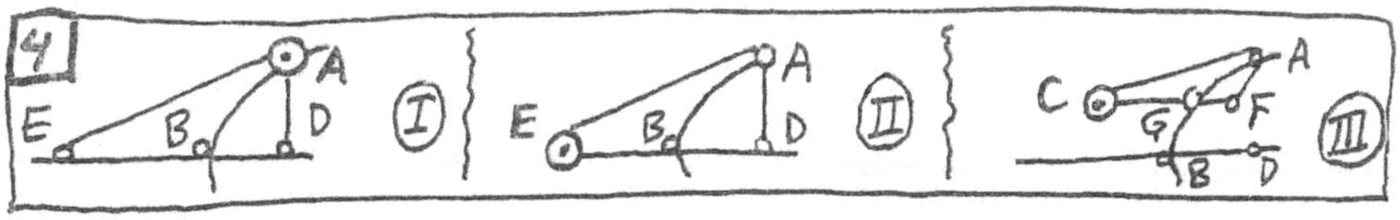

—Again: let it be an hyperbola whose axis the [line] DBC is, and center the [point] H is, and asymptotes the [lines] HE, HF are. Then the given point will be given either on the section, or on the axis, or inside the angle by the lines EHF, or in the adjacent place, or on one of the asymptotes containing the section, or in the [place] between the lines containing the angle vertical to the angle by EHF.
—Let [the given point] be, first, on the section, as the [point] A: and also, let [the required line] have been generated; and also, let there be as touching[line] the [line] AG; and also, let there be drawn perpendicular [to the axis DBC] the [line] AD; and also, as oblique side of the figure (*plagia tou eidous pleura*) let there be the [line] BC; then as is the [line] CD to DB so will be the [line] CG to GB. And the ratio of the [line] CD to DB is given; for given are both [lines]; therefore the ratio also of the [line] CG to GB is given. And also given is the [line] BC; therefore given is the [point] G. But [given] also is the [point] A; therefore [given] in position is the [line] AG.
—Constructed will be (*suntethêsetai*) [the touchingline] thus: from the [point] A let there be drawn perpendicular [to the axis DBC] the [line] AD; and also, let there be the same as the ratio of the [line] CD to DB the ratio of the [line] CG to GB; and also, let there be joined the [line] AG. Then it is evident that the [line] AG touches the section.
—Then again, let the given point be on the axis—say the [point] G; and also, let [the required line] have been generated; and also, let there be drawn as touching[line] the [line] AG; and also, let there be drawn perpendicular [to the axis DBC] the [line] AD. Then, according to the same things, as is the [line] CG to GB so will be the [line] CD to DB. And also, there is given the [line] BC. Therefore given is the [point] D. And also, there is upright [to the axis DBC] the [line] DA; therefore [given] in position is the [line] DA. And [given] in position is also the section; therefore given is the [point] A. But [given] is also the [point] G; therefore [given] in position is the [line] AG.
—Constructed will then be (*suntethêsetai*) [the touching[line]] thus: let the other things be supposed the same; and also, let there be made the same as the ratio of the [line] CG to GB the ratio of the [line] CD to DB; and also, upright [to the axis DBC] let there be drawn the [line] DA; and also, let there be joined the [line] AG. Then it is evident that the [line] AG does the problem—and that from the [point] G there will be drawn another [line] touching the section, [this one] to the other side.

—The same things being supposed, let the given point, namely the [point] K, be in the place inside the angle by the lines EHF: and let it be required to draw from the [point] K a [line] touching the section. Let [the required touching[line]] have been generated; and also, let the [line] KA be it; and also, let the joined [line] KH be extended; and also, let there be put equal to the [line] LH the [line] HN; therefore all are given. Then also will the [line] LN be given. Then let there be drawn the [line] AM ordinatewise to the [line] MN; then also as is the [line] NK to KL so will be the [line] MN to ML. And the ratio of the [line] NK to KL is given; therefore the ratio also of the [line] NM to ML is given. And also there is given the [point] L; therefore given also is the [point] M. And there has been drawn up, parallel to the [line] touching at the [point] L, the [line] MA; therefore [given] in position is the [line] MA. And given in position is also the section ALB; therefore given is the [point] A. But also the [point] K is given; therefore given is the [line] AK.
—Constructed will then be (*suntethêsetai*) [the touching[line]] thus: let the other things be supposed the same; and also, as the given point the [point] K; and also, let the joined [line] KH be extended; and also, equal to the [line] HL let there be put the [line] HN; and also, let [the locating of the point M] be so done that as is the [line] NK to KL so is the [line] NM to ML; and also, parallel to the [line] touching at the [point] L, let there be drawn the [line] MA; and also, let there be joined the [line] KA; therefore the [line] KA touches the section. And it is evident that also another [line] will be drawn from the [point] K touching the section, [this one drawn] to the other side.
—The same things being supposed, let the given point be—on one of the asymptotes containing the section—the [point] F: and let it be required to draw from the [point] F a [line] touching the section. And let [the required touching[line]] have been generated; and also, let it be the [line] FAE; and also, through the [point] A let there be drawn parallel to the [line] EH the [line] AD; then equal will be the [line] DH to the [line] DF, since also the [line] FA is equal to the [line] AE. And also, given is the [line] FH; therefore given is the [point] D. And, through a given [point], namely the [point] D, and parallel in position to the [line] EH, has been drawn the [line] DA; therefore [given] in position is the [line] DA. And [given] in position is also the section; therefore given is also the [point] A. But [given is] also the [point] F. Therefore [given] in position is also the [line] FAE.
—Constructed will then be (*suntethêsetai*) [the touching[line]] thus: let there be as the section the [line] AB, and also the [lines] EH, HF as asymptotes, and also, as the given point on one of the asymptotes that contain the section, the [point] F; and also, let the [line] FH be cut in two [halves] at the [point] D; and also, through the [point] D let there be drawn parallel to the [line] HE the [line] DA; and also, let there be joined the [line] FA. And since equal is the [line] FD to the [line] DH, therefore equal also is the [line] FA to the [line] AE. And so, on account of the things shown before, the [line] FAE touches the section.

—The same things being supposed, let the given point be in the place under the angle adjacent to the lines containing the section, and also, let the [point] K be it: required, then, from the [point] K to draw a [line] touching the section. And let [the required touching[line]] have been generated; and also let the [line] KA be it; and also, let the joined [line] KH be extended; then it will be [given] in position. If, then, on the section there be taken a given point, the [point] C, and through the [point] C there be drawn parallel to the [line] KH the [line] CD, it will be [given] in position. And if the [line] CD be cut in two [halves] at the [point] E, and the [line] HE, joined, be extended, it will be [given] in position as a diameter that is conjugate to the [diameter] KH. Then let there be put equal to the [line] BH the [line] HG; and also, through the [point] A let there be drawn parallel to the [line] BH the [line] AL; then—on account of the [lines] KL, BG being conjugate diameters, and also, a touching[line] being what the [line] AK is, and also, the [line] AL being a [line] drawn parallel to the [line] BG—hence the [rectangle] by the [lines] KHL is equal to the fourth part of the figure (*eidous*) for the [line] BG. Therefore given is the [rectangle] by KHL. And also given is the [line] KH; therefore given is also the [line] HL. But it is [given] also in its position; and also given is the [point] H; therefore given is also the [point] L. And, through the [point] L has been drawn parallel in position to the [line] BG the [line] LA; therefore [given] in position is the [line] LA. And also [given] in position is the section; therefore given is the [point] A. But [given] is also the [point] K; therefore [given] in position is the [line] AK.
—Constructed will then be (*suntethêsetai*) [the touching[line]] thus: let the other things be supposed the same, and let the given point, the [point] K, be in the place chosen before; and also, let the [line] KH, joined, be extended; and also, let there be taken some point, say the [point] C; and also, let there be drawn parallel to the [line] KH the [line] CD; and also, let the [line] CD be cut in two [halves] by the [point] E; and also, let the [line] EH, joined, be extended; and also, equal to the [line] BH let there be put the [line] HG; therefore the [line] GB is an oblique (*plagia*) diameter conjugate to the [diameter] KHL. Then, equal to the fourth of the figure (*eidous*) alongside the [line] BG let there be put the [rectangle] by KHL; and also, through the [point] L let there be drawn parallel to the [line] BG the [line] LA; and also, let there be joined the [line] KA; then it is evident that the [line] KA touches the section, on account of the converse of the theorem (*tên antistrophên tou theôrêmatos*) [the theorem, that is, which is the 38th item of the First Book].
—And if [the given point] be given in the place between the [lines] FHP, impossible will the problem be. For the touching[line] will cut the [line] GH. And so, it will fall together with each of [lines] FHP—which very thing is impossible, on account of the things shown in the 31st [item] of the First Book and in the third [item] of this.

—The same things being supposed, let the section be an ellipse, and the given point be on the section, say the [point] A, and let it be required to draw from the [point] A [a line] touching the section.
—Let [the required line] have been generated; and also, let the [line] AG be it; and also, from the point A let there be drawn ordinatewise to the axis BC the [line] AD; then there will be given the [point] D; and also, as is the [line] CD to DB so will be the [line] CG to GB. And also, the ratio of the [line] CD to DB is given; therefore the ratio also of the [line] CG to GB is given. Therefore given is the [point] G. But [given] also is the [point] A. Therefore given in position is the [line] AG.
—Constructed will then be (*suntethêsetai*) [the touching[line]] thus: let there be drawn perpendicular [to the axis BC] the [line] AD; and also, the same as the ratio of the [line] CD to DB let there be the ratio of the [line] CG to GB; and also, let there be joined the [line] AG. Then it is evident that the [line] AG touches [the section], just as also on the hyperbola.
—Then again, let the given point be the [point] K—[a point, that is to say, not on the section, but not inside either]—and let it be required to draw a touching[line]. Let [this touching[line]] have been generated; and also, let the [line] KA be it; and also, let the [line] KLH, joined to the center H, be extended to the [point] N [on the section]; then it will be [given] in position. And also, if the [line] AM be drawn ordinatewise, then as is the [line] NK to KL so will be the [line] NM to ML; and the ratio of the [line] KN to KL is given; therefore the ratio also of the [line] MN to LM is given. Therefore given is the [point] M. And there has been drawn up [ordinatewise, that is,] the [line] MA, for it is parallel to the line that touches at the [point] L; therefore [given] in position is the [line] MA. Therefore given is the [point] A. But [given] also is the [point] K; therefore given in position is the [line] KA.
—And the construction (*sunthesis*) is the same as the one before it.

It would be a useful exercise if, before going on, you were to try your hand at writing up a layout of these latter parts of the 49th proposition, which treat the hyperbola and the ellipse.

You might also, before going on, consider this: For propositions that are theorems, Apollonius does not include analyses. He includes them only for propositions that are problems. Why might that be appropriate?

It should be noted that "proposition" is not his word but ours, borrowed from Latin. Its counterpart in Greek would be *pro+thesis* or *pro+tasis*. But while this term is used by ancient Greek authors in grammar, it is not used in mathematics. The Greek text gives numbers to the items that we call "propositions," but it gives them no labels—nothing that corresponds to "proposition" or "theorem" or "problem." It was from Pappus that we learned that propositions are of two sorts, theorems and problems. In a theorem, we see something that is shown from something that we have already seen. A problem calls for the supplying of something; the solution finds the thing or makes it. A theorem presents something to look *at*. A problem presents something to look *for*. "Theoretical" analysis is undertaken in order to "seek" what would demonstrate a theorem. "Problematical" analysis is undertaken in order to "supply" what would solve a problem.

Both theorems and problems require a demonstration, and demonstration is synthesis; but while analysis is an extrinsic preliminary to demonstration, it is an intrinsic part of solution. In preparing us to find out what there is to look at and how we can show it, analysis is useful; but in problem-solving it is not so much a preparation as a proper part of the activity itself. The activity at the peak of classical mathematics is vision; the solution of problems is a subordinate activity. Problem-solving is mere provision of what is required for vision. In demonstrating a theorem, we lose ourselves in looking: a sequence of views is passed before us so that at last we can see the spectacle that is stated at the outset in the enunciation. In solving a problem, we do not lose ourselves in looking: *we* ourselves are *part* of the problem. The solution of the problem is the linking of what we do not have yet to what we do have already. The synthesis shows that the problem is indeed solved by the getting that is the analysis; but before we can see that the getting works, we must see what the getting is. It is therefore appropriate to include an analysis in a proposition that presents a solved problem.

Now, it is necessary to have syntheses to demonstrate theorems, and it is appropriate to have analyses to solve problems, but why bother to add syntheses to the solution of problems? After all, once you know *what* the solution will be, there's no difficulty in showing *that* it is the solution. Indeed, Hankel, the nineteenth-century German scholar to whom we referred earlier, is quoted by Jacob Klein as complaining of the "Greek ethnic peculiarity" which loads down every analytical solution in geometry with the "useless ballast" of a "completely worked out synthesis which says everything over again in reverse order." (*Greek Mathematical Thought and the Origins of Algebra, p.163* [Mineola, NY: Dover, 1992; reprint of MIT Press edition pub. 1968, translated by Eva Brann from the German text pub. 1934–1936]). The answer is, that analysis tells us merely how something might be obtained—and what is merely obtainable is not yet obtained. The statement "if Z, then A" can be true while "if A, then Z" is not. If every road we took were paved with nothing but manipulation of equations, then we could take for granted the convertibility of our "if-then" statements, since the manipulation of equations is reversible. But magnitudes of just *any* sort do not seem to be equateable with magnitudes of just any *other* sort; so, until we have actually traveled back again along the road that we took in the analysis, we can't be sure that it is a two-way street.

CHAPTER 3. POINTS OF APPLICATION IN BOOK III OF THE CONICS

Near the end of the Third Book of his *Conics*, Apollonius shows some properties of a certain pair of points along the axis of any conic section that has a center. Each such point cuts the axis, externally (in an hyperbola) or internally (in an ellipse), into two segments that will contain a rectangle equal to one-fourth of that interesting figure which is the rectangle contained by the conic section's oblique (transverse) side and upright side—or, we might say, by a diameter and its parameter. An examination of Apollonius's treatment of these "points of application" will prepare us to consider the modern alternative to the study of Apollonius.

The points of application arise in the 45th proposition (which depends on the 42nd). A little later, in the 51st and 52nd propositions, we learn how the different curves can be neatly characterized by a certain property of the distances to the points of application from any point on one of the central sections. In order to demonstrate those propositions, a few others are needed. One of these, the 48th, is of particular interest because of its implications for the study of physical reflections. How the propositions depend on one another is depicted in figure 1.

For convenience in seeing at a glance what is going on in these propositions that lead up to the 51st and 52nd, we shall look at the case of the ellipse (except in proposition 45 where we shall locate the points of application in hyperbolas too). Keep in mind as we proceed that although our depictions will show only the ellipse, the propositions themselves cover all of the central sections—all the sections, that is to say, except for the parabola. The reader may want to take it as an exercise to produce depictions that show the other cases. (But remember, as was said before, that complete diagrams for those propositions of Book Three which are selected for inclusion here in this chapter, and also later in Chapter 6, are included, along with those selected for inclusion from Book Two, in the Appendix at the end of this second volume of the guidebook.)

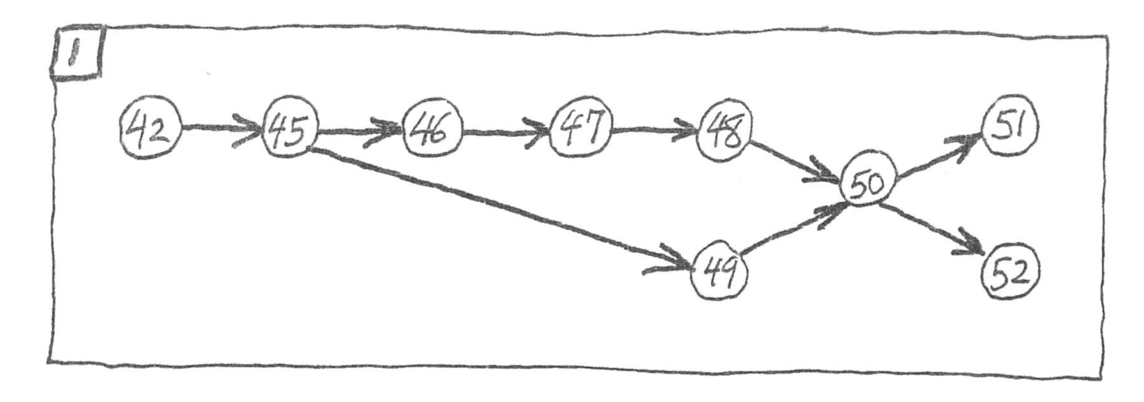

What the 42nd proposition shows is this: Suppose that we have any of the central sections, with a diameter and also lines drawn ordinatewise from that diameter's endpoints; then a tangent will cut off on each of those ordinatewise lines a segment such that the two segments will contain a rectangle equal to one-fourth of "the figure" to that diameter—that is, they will contain a rectangle equal to one-fourth of the rectangle that is contained by that diameter and its parameter (fig. 2).

Using that 42nd proposition, the 45th proposition introduces certain special points on the axis of any of the central sections, and shows the following.

> Suppose that we have the axis
> and also have lines drawn ordinatewise from the axis's endpoints
> (an axis, we must remember, has its ordinates perpendicular to it),
> and suppose we draw a line from either special point
> to each point where the tangent meets one of the two ordinatewise lines.
>
> Then the angle made by the two lines that are drawn
> from the special point to the two meeting-points
> will be a right angle.
>
> Where on the axis are those special points?
> At a place where each of them cuts the axis into two segments that
> will contain a rectangle equal to one-fourth of the figure to the axis.

Again, remember that, for convenience, it is the case of the ellipse at which we shall look in our depiction (fig. 3).

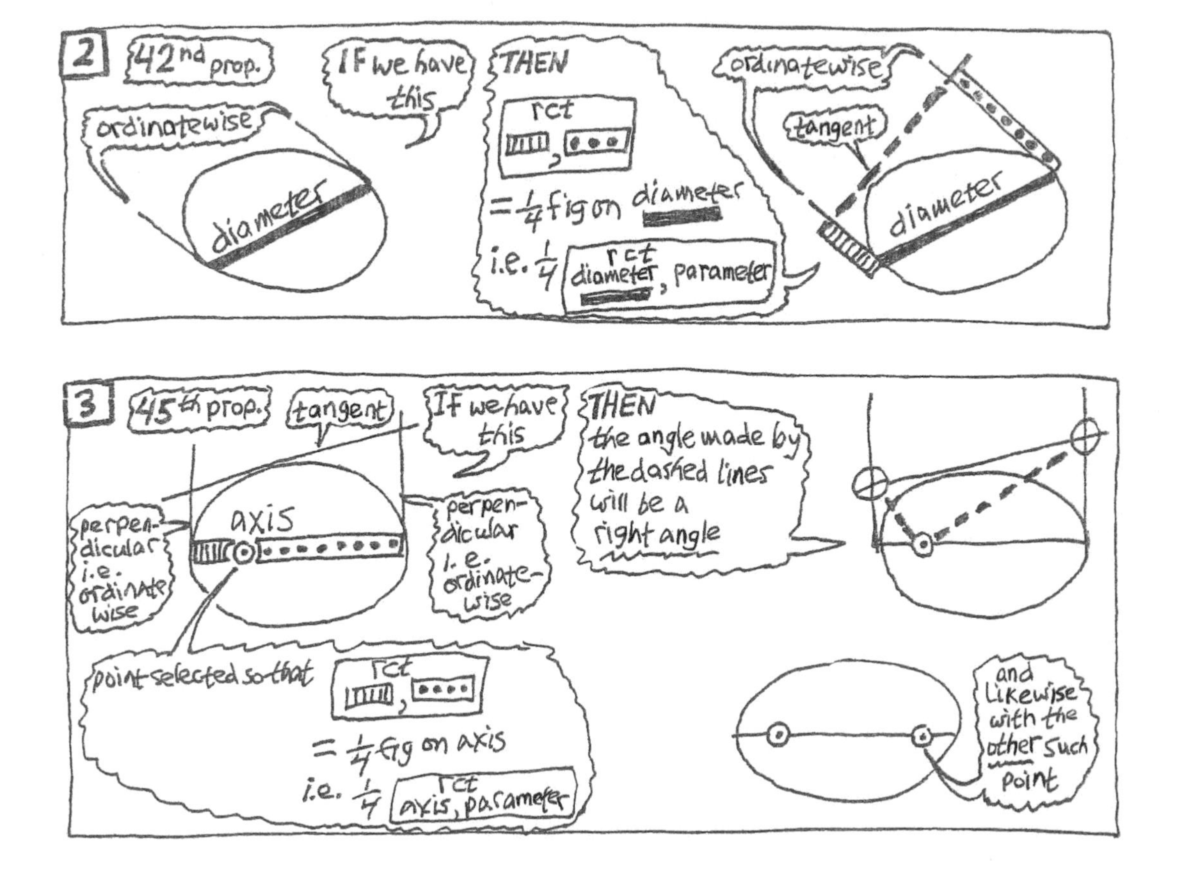

What is shown about the "points of application" introduced in the 45th proposition follows from the fact which is established in the 42nd proposition. Let us consider the demonstration which establishes that fact: see figure 4. After that, we shall consider the 45th proposition and its consequences.

(III) #42 If, in an hyperbola or ellipse or circle's periphery or the opposite [sections], there be drawn, from the extremities of the diameter, [lines] parallel to a [line] drawn down ordinatewise, and some other [line] be drawn—as has chanced—that touches [the section], [this chance touching[line]] will cut off from them [from, that is, the ordinatewise parallel lines] straight [lines] that contain a [rectangle] equal to the fourth part of the figure (*eidous*) for the same diameter.

—For let there be some one of the aforesaid sections, whose diameter is what the [line] AB is, and from the [points] A, B let there be drawn parallel to a [line] drawn down ordinatewise the [lines] AC, DB, and let some other [line] touch [the section] at the [point] E, namely the [line] CED. I say that the rectangle by AC, BD is equal to the fourth part of the figure for the [diameter] AB.

—For let there be as center the [point] F, and through it let there be drawn parallel to the [lines] AC, BD the [line] FGH. Then since the [lines] AC, BD are parallel, and the [line] FG is also parallel, therefore [the line FG] is a diameter conjugate to the [diameter] AB; and so, the [square] from FG is equal to the fourth of the figure (*eidous*) for the [diameter] AB.

—If, then, the [line] FG—on the ellipse and also on the circle—goes through the [point] E, equal will there be generated the [lines] AC, FG, BD, and it is evident from that by itself that the [rectangle] by AC, BD is equal to the [square] from FG, [is equal, that is,] to the fourth of the figure for the [diameter] AB.

—Then let [the line FG] not go [through the point E]; and also, let there fall together the [lines] DC, BA extended, say at the [point] K; and also, through the [point] E let there be drawn parallel to the [line] AC the [line] EL, and parallel to the [line] AB the [line] EM. Then since equal is the [rectangle] by KFL to the [square] from AF, hence as is the [line] KF to FA so is the [line] FA to FL, and the [line] KA is to AL as is the [line] KF to FA, that is, to FB. Inversely (*anapalin* [traditionally, in Latin, *invertendo*]) as is the [line] BF to FK so is the [line] LA to AK. Putting together or separating (*sunthenti* or *dielonti* [traditionally, in Latin, *componendo* or *separando*]) as is the [line] BK to KF so is the [line] LK to KA. Therefore also as is the [line] DB to FH so is the [line] EL to CA. Therefore the [rectangle] by DB, CA is equal to the [rectangle] by FH, EL, [is equal,] that is, to the [rectangle] by HFM. But the [rectangle] by HFM is equal to the [square] from FG, [is equal,] that is, to the fourth of the figure for the [diameter] AB; therefore also the [rectangle] by DB, CA is equal to the fourth of the figure for the [diameter] AB.

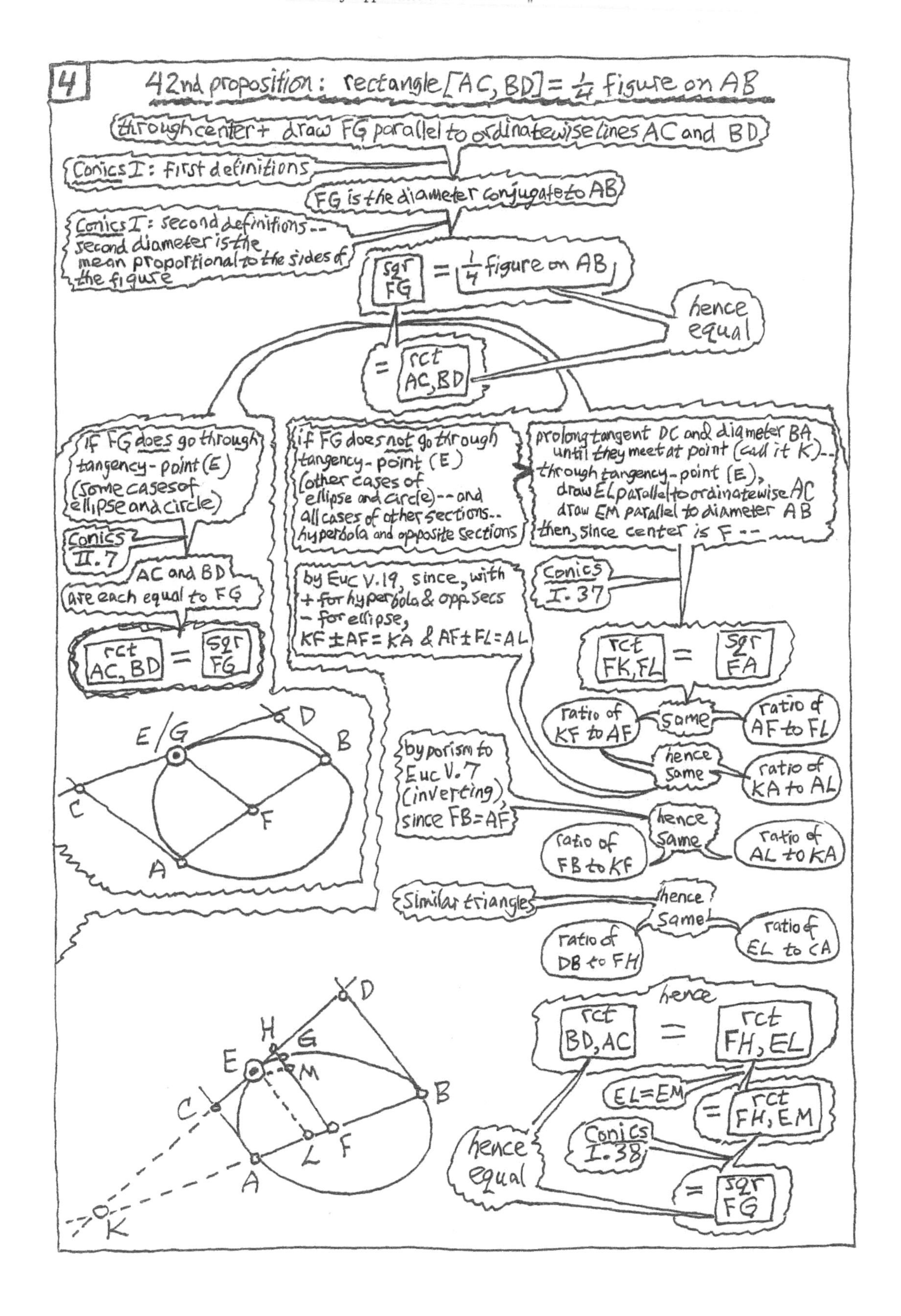

4
42nd proposition: rectangle[AC, BD] = 1/4 figure on AB
through center + draw FG parallel to ordinatewise lines AC and BD
Conics I: first definitions
FG is the diameter conjugate to AB
Conics I: second definitions -- second diameter is the mean proportional to the sides of the figure
sqr FG = 1/4 figure on AB
hence equal
= rct AC, BD
if FG does go through tangency-point (E) (some cases of ellipse and circle)
if FG does not go through tangency-point (E) (other cases of ellipse and circle) -- and all cases of other sections.. hyperbola and opposite sections
prolong tangent DC and diameter BA until they meet at point (call it K) -- through tangency-point (E), draw EL parallel to ordinatewise AC draw EM parallel to diameter AB then, since center is F --
Conics II.7
AC and BD are each equal to FG
rct AC, BD = sqr FG
by Euc V.19, since, with + for hyperbola & opp. secs - for ellipse, KF ± AF = KA & AF ± FL = AL
Conics I.37
rct FK, FL = sqr FA
ratio of KF to AF
same
ratio of AF to FL
hence same
ratio of KA to AL
by porism to Euc V.7 (inverting), since FB = AF
hence same
ratio of FB to KF
ratio of AL to KA
Similar triangles
hence same
ratio of DB to FH
ratio of EL to CA
hence
rct BD, AC = rct FH, EL
EL = EM
= rct FH, EM
Conics I.38
hence equal
= sqr FG
E/G
D
B
C
F
A
D
H
G
E
M
C
B
L
F
A
K

In the 45th proposition, what we want to do is take a point F (and another point G as well) that will be located on the prolongation of axis AB (for the hyperbola) or on the axis itself (for the ellipse) just where its distances to the axis's endpoints (A and B) will make a rectangle of a certain size. If such a rectangle is "thrown alongside" the axis, it will "exceed" (for the hyperbola)—or "be deficient" (for the ellipse)—by a square (fig. 5).

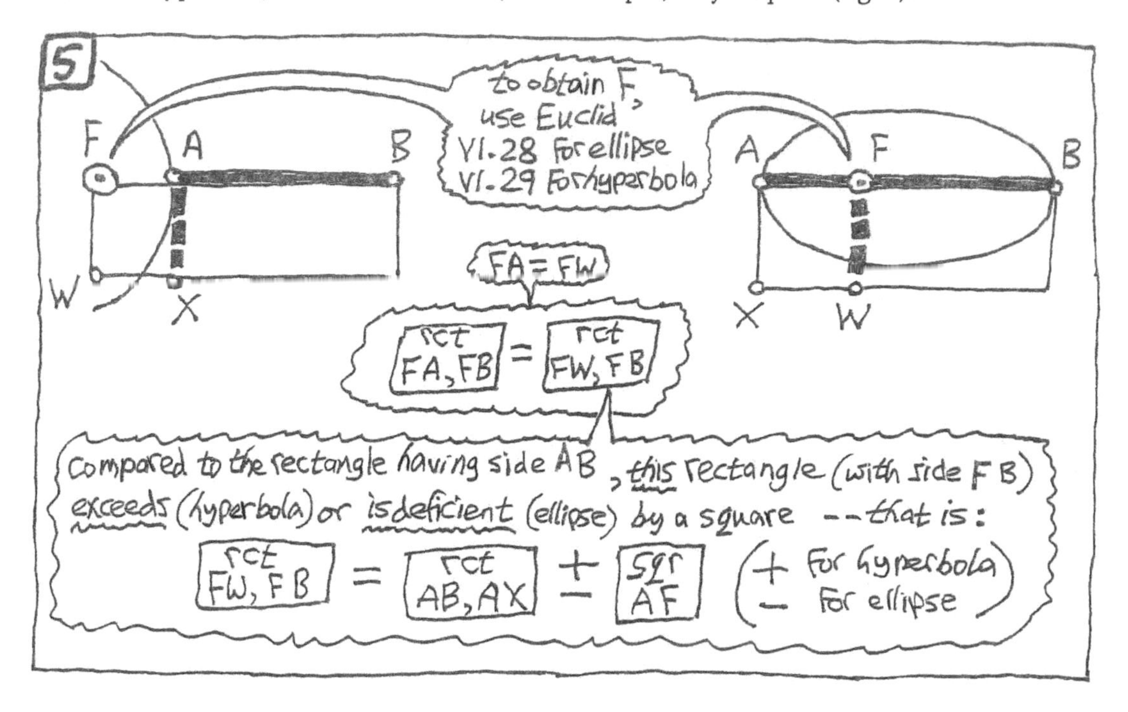

Way back when we were considering terminology in the enunciation of the 11th proposition of the First Book of the *Conics* (when we were first obtaining the conic sections at the beginning of Part Three of this guidebook), we took note of the fact that Euclid shows how figures can be "thrown alongside" a straight line in such a way as to be "excessive" (*hyperballon*, *Elements* VI.29) or "deficient" (*elleipon*, *Elements* VI.28) by a parallelogram whose shape has been given. Here in the 45th proposition of the Third Book, Apollonius now employs just that capability which Euclid provides in those propositions. On each side, both on the left (at F) and on the right (at G), a rectangle that is equal to one-fourth of the figure is "thrown alongside" the axis)—being excessive by a square figure in the hyperbola and in the opposite sections (requiring *Elements* VI.29), but being deficient by a square figure in the ellipse (requiring *Elements* VI.28). These two propositions in the Sixth Book of Euclid's *Elements* carry forward the work, begun in its Second Book, that enables us to relate the sizes of lines indirectly by relating the sizes of the boxes that they can contain.

What Apollonius, following Euclid, calls "throwing alongside" (*parabolê*) is traditionally rendered as "application"—and so "the points of application" is the traditional rendering of what Apollonius calls "the points generated from the throwing alongside" (*ta ek tês parabolês ginomena semeia).*

The 45th proposition is depicted in figure 6, and is presented in the box after it.

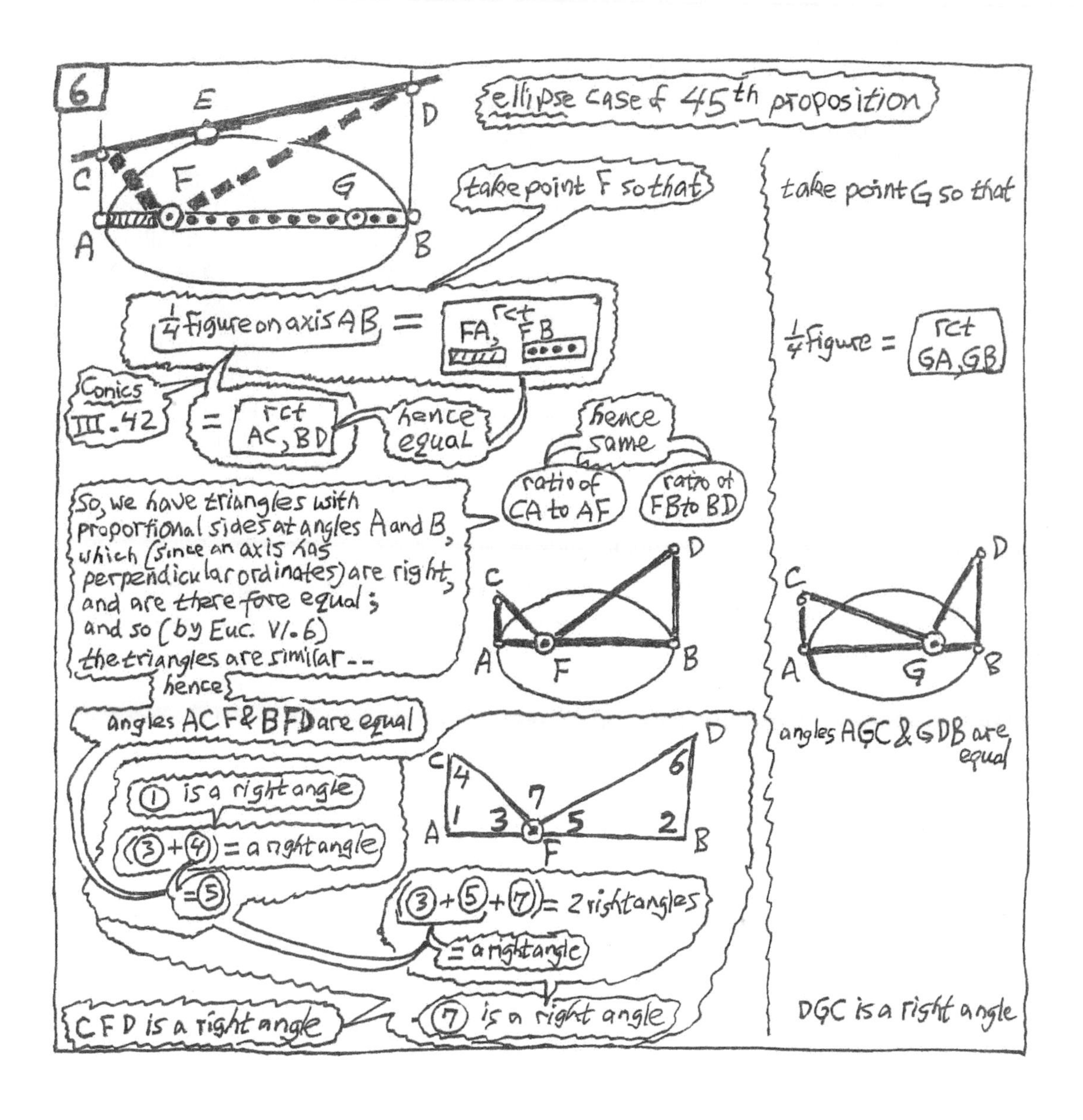
6
ellipse case of 45th proposition
E
C
D
F
G
A
B
take point F so that
1/4 figure on axis AB = rct FA, FB
Conics III.42
= rct AC, BD
hence equal
hence same
ratio of CA to AF
ratio of FB to BD
So, we have triangles with proportional sides at angles A and B, which (since an axis has perpendicular ordinates) are right, and are therefore equal; and so (by Euc. VI.6) the triangles are similar--
hence
angles ACF & BFD are equal
① is a right angle
(③+④) = a right angle
=⑤
(③+⑤+⑦)= 2 right angles
= a right angle
⑦ is a right angle
CFD is a right angle
take point G so that
1/4 figure = rct GA, GB
angles AGC & GDB are equal
DGC is a right angle

(III) #45 If, in an hyperbola or ellipse or circle's periphery or the opposite [sections], there be drawn, from the extremity of the axis, straight [lines] at right angles [to the axis], and a [rectangle which is] equal to the fourth part of the figure (*eidous*) be thrown alongside ["be applied to" the axis] on each side—[a rectangle, that is,] which overshoots (*huperballon*) by a square figure (*eidei*) on the hyperbola and opposite sections, but which is lacking (*elleipon*) on the ellipse—and there be drawn some straight [line] that touches the section and that falls together with the straight [lines] that are at right angles, then the straight [lines] drawn from the points of falling together to the points [of application, that is, to the points] generated from the throwing alongside (*ta ek tês parabolês ginomena semeia*), make right angles at the said points.
—Let there be one of the said sections whose axis is what the [line] AB is, and at right angles let the [lines] AC, BD be, and a touching[line] let the [line] CED be, and a [rectangle] equal to the fourth part of the figure (*eidous*) let there be thrown alongside on each side as has been said—namely the [rectangle] by AFB and the [rectangle] by AGB—and let there be joined the [lines] CF, CG, DF, DG. I say that each angle—both the one by CFD and the one by CGD—is a right [angle].
—For since the [rectangle] by AC, BD has been shown equal to the fourth part of the figure for the [diameter] AB, and also the [rectangle] by AFB is equal to the fourth part of the figure, therefore the [rectangle] by AC, BD is equal to the [rectangle] by AFB. Therefore as is the [line] CA to AF so is the [line] FB to BD. And right [angles] are what the angles at the points A, B are; therefore equal is the [angle] by ACF to the [angle] by BFD, and [equal is] the [angle] by AFC to the [angle] by FDB. And since the [angle] by CAF is right, therefore the [angles] by ACF, AFC are [together] equal to one right [angle]. And also the [angle] by ACF has been shown equal to the [angle] by DFB; therefore the [angles] by CFA, DFB are [together] equal to one right [angle]. Therefore, as a remainder, the [angle] by DFC is right. Likewise, then, will also the [angle] by CGD be shown to be right.

Now let us consider how to employ Euclid's propositions in a construction that yields Apollonius's points of application.

A point of application is a point F on the axis AB (or on its extension) such that lines drawn from F to both endpoints of the axis will make a rectangle [FA, FB] equal to one-fourth of "the figure." ("The figure" when we speak of conic sections refers to the rectangle contained by the oblique [transverse] side and the upright side.) In the work of locating a point of application, Part I obtains something equal to one-fourth of the figure; this something is the square on the line CD, which is that perpendicular from the center whose length is one-half of the conjugate axis. Part II then obtains something equal to that square [CD] and equal also to the rectangle [FA, FB]; and so, this rectangle [FA, FB] will be equal to one-fourth of "the figure." What is the something that does this mediating work in Part II? It is the rectangle with sides AF and AU. We make this rectangle equal to square [CD], and we ensure its equality to rectangle [FA, FB] by ensuring that AU will be equal to FB—that is, we bring it about that rectangle [FB, BV] will be a square. See figure 7.

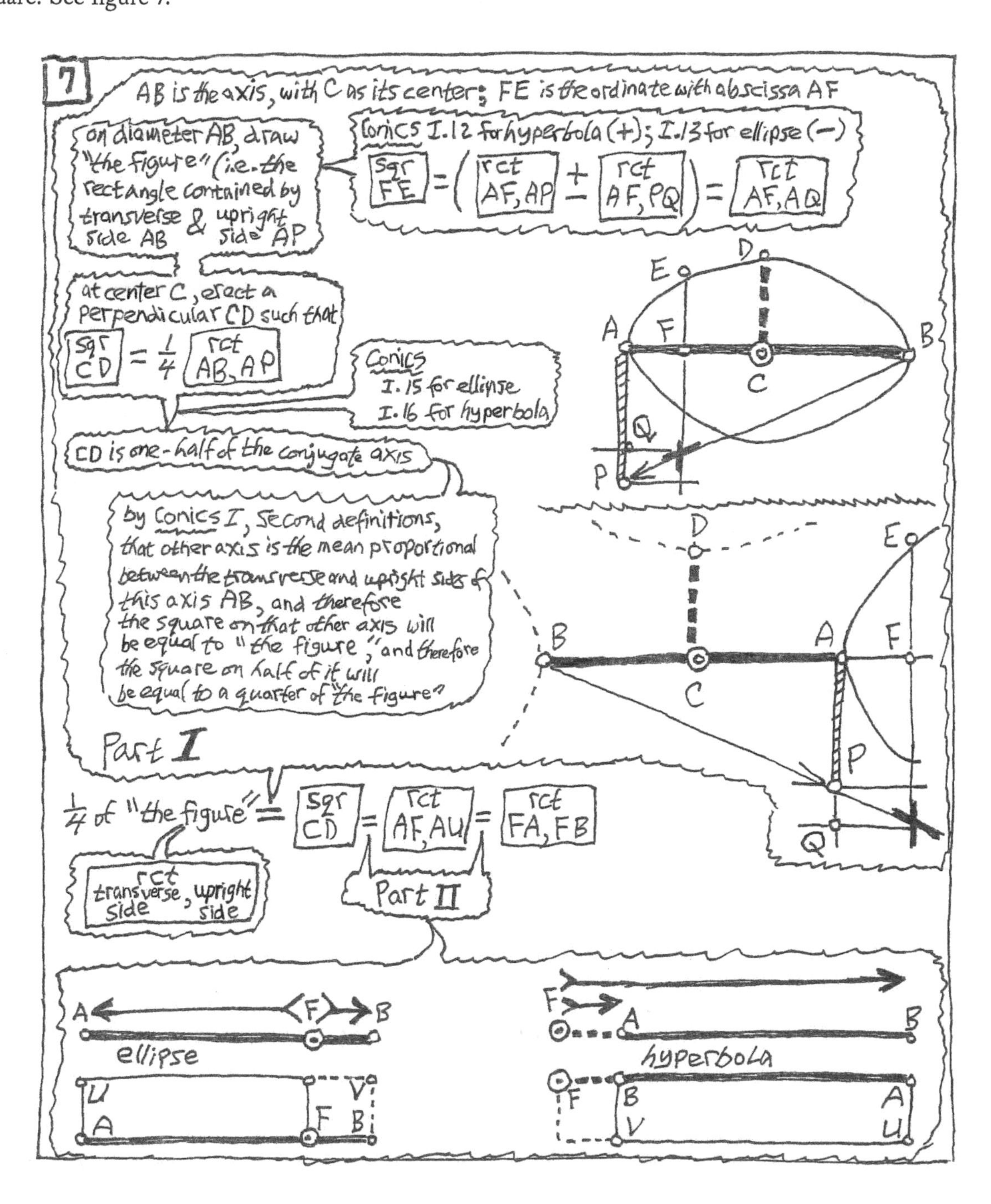

Part II of the work of locating a point of application is depicted in figure 8, which clarifies how the work done in Euclid's VI.28 and 29 enables us to locate such a point. Point F cuts axis AB, internally in the ellipse and externally in the hyperbola; point O is located so that not only is rectangle [AF, AU] equal to square [CD], but also AU (which is equal to VB) is equal to FB (by making rectangle [FB, BV] a square). That is to say: The rectangle [AF, AU] is, with respect to rectangle [AB, AU], excessive or deficient in the following way—the side of [AF, AU] that lies along AB (namely, side AF) exceeds or falls short of AB, so that the rectangle[AF, AU] itself exceeds or falls short of a rectangle whose one side is AB and whose other side is equal to the other side (AU) of rectangle [AF, AU]. In other words, we use Euclid VI.28 or 29, applying to AB a rectangle [AF, AU] which is equal to square [CD] and is deficient or excessive (with respect to rectangle [AB, AU]) by a square, so that AU will be equal to BF. In short: AU is equal to the remainder when we take the smaller of AB and AF away from the larger.

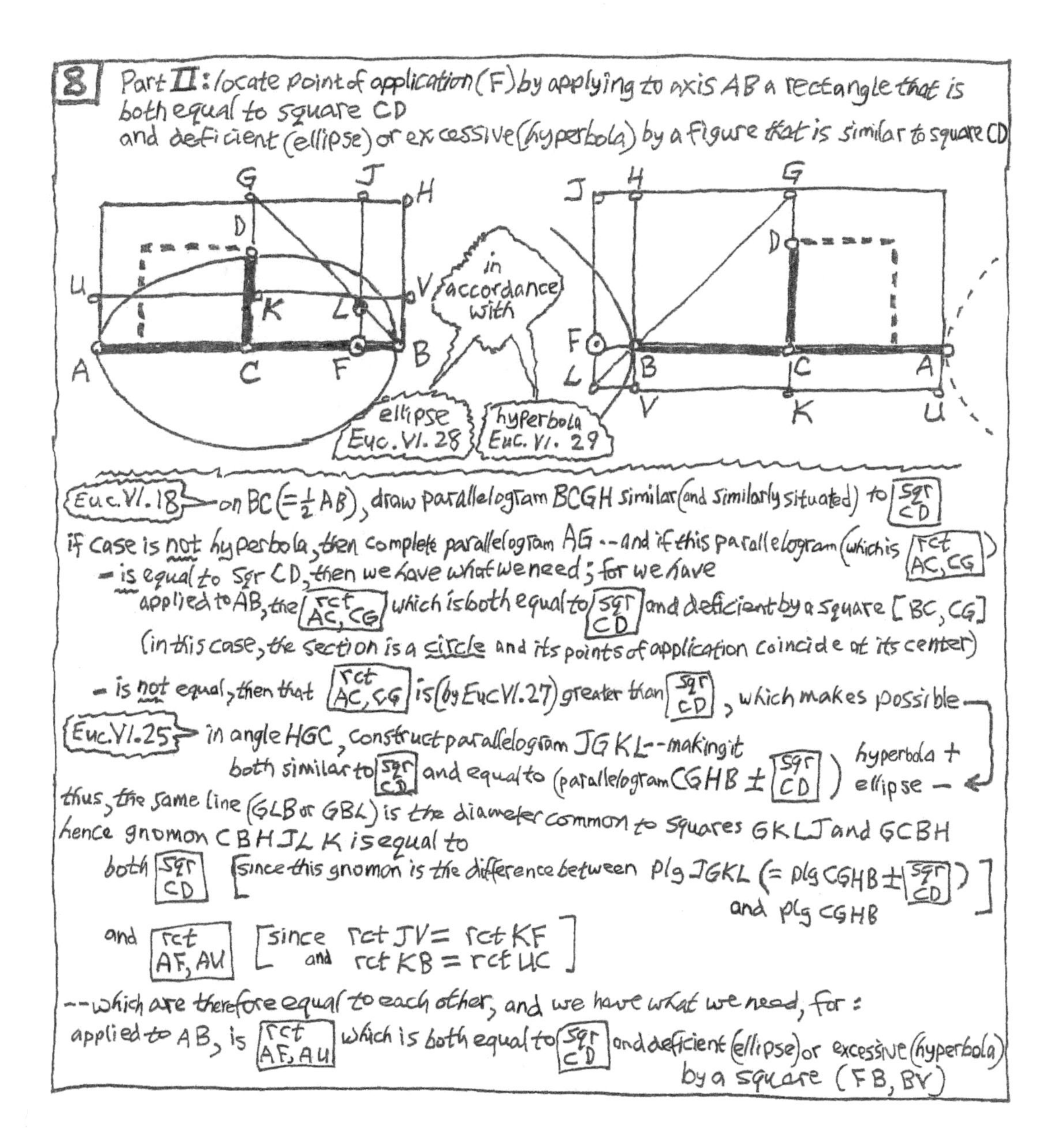
8
Part II: locate point of application (F) by applying to axis AB a rectangle that is both equal to square CD
and deficient (ellipse) or excessive (hyperbola) by a figure that is similar to square CD
G
J
H
D
U
V
K
L
A
C
F
B
in accordance with
ellipse Euc. VI. 28
hyperbola Euc. VI. 29
J
H
G
D
F
L
B
C
A
V
K
U
Euc. VI. 18 — on BC (= ½ AB), draw parallelogram BCGH similar (and similarly situated) to sqr CD
if case is not hyperbola, then complete parallelogram AG -- and if this parallelogram (which is rct AC, CG)
– is equal to sqr CD, then we have what we need; for we have applied to AB, the rct AC, CG which is both equal to sqr CD and deficient by a square [BC, CG]
(in this case, the section is a circle and its points of application coincide at its center)
– is not equal, then that rct AC, CG is (by Euc VI.27) greater than sqr CD, which makes possible
Euc. VI. 25 — in angle HGC, construct parallelogram JGKL -- making it both similar to sqr CD and equal to (parallelogram CGHB ± sqr CD) hyperbola + ellipse –
thus, the same line (GLB or GBL) is the diameter common to squares GKLJ and GCBH
hence gnomon CBHJLK is equal to
both sqr CD [since this gnomon is the difference between plg JGKL (= plg CGHB ± sqr CD) and plg CGHB]
and rct AF, AU [since rct JV = rct KF and rct KB = rct UC]
-- which are therefore equal to each other, and we have what we need, for:
applied to AB, is rct AF, AU which is both equal to sqr CD and deficient (ellipse) or excessive (hyperbola) by a square (FB, BV)

The 46th and 47th propositions are depicted in figures 9 and 10 for the case of the ellipse, and are presented in the box that follows them.

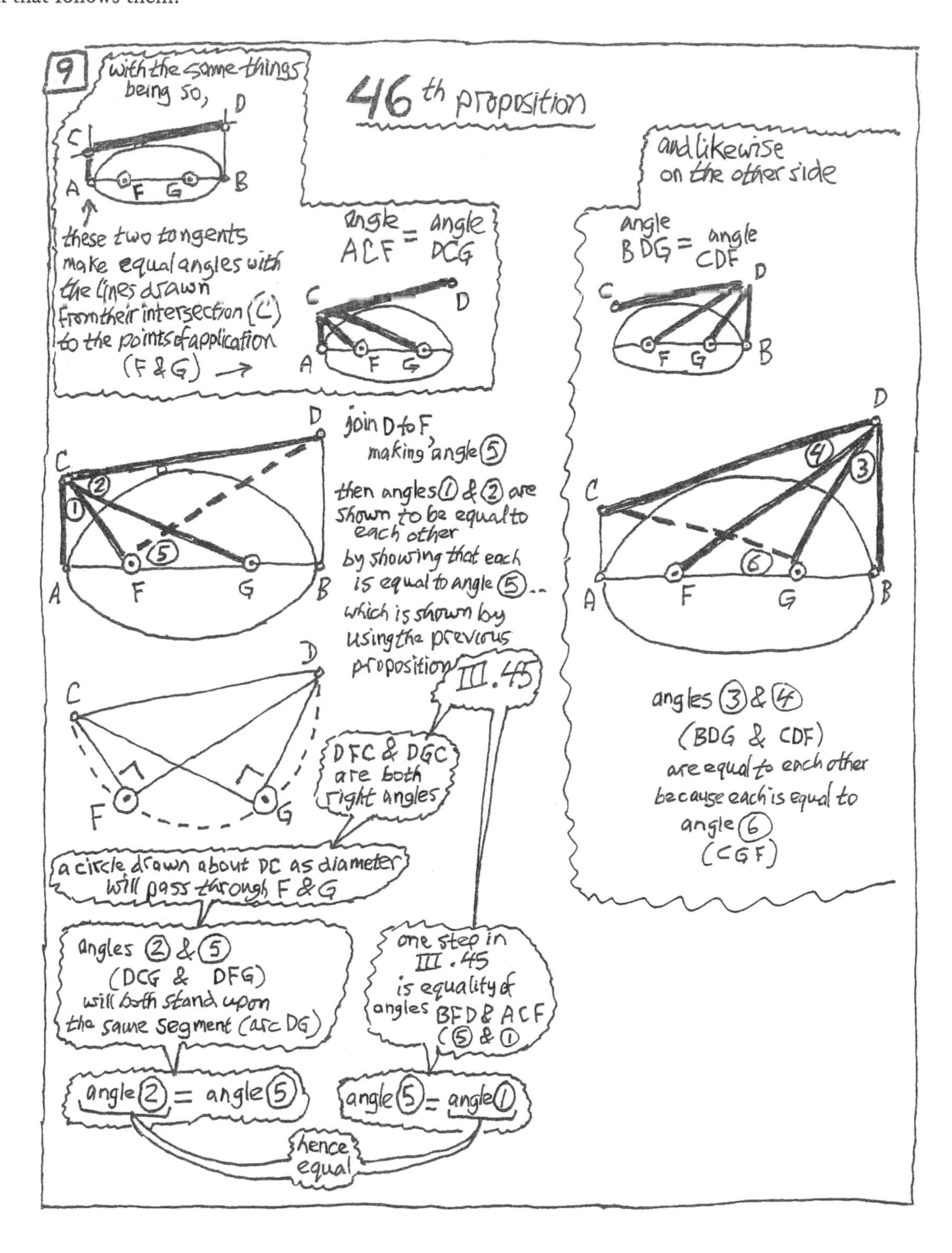

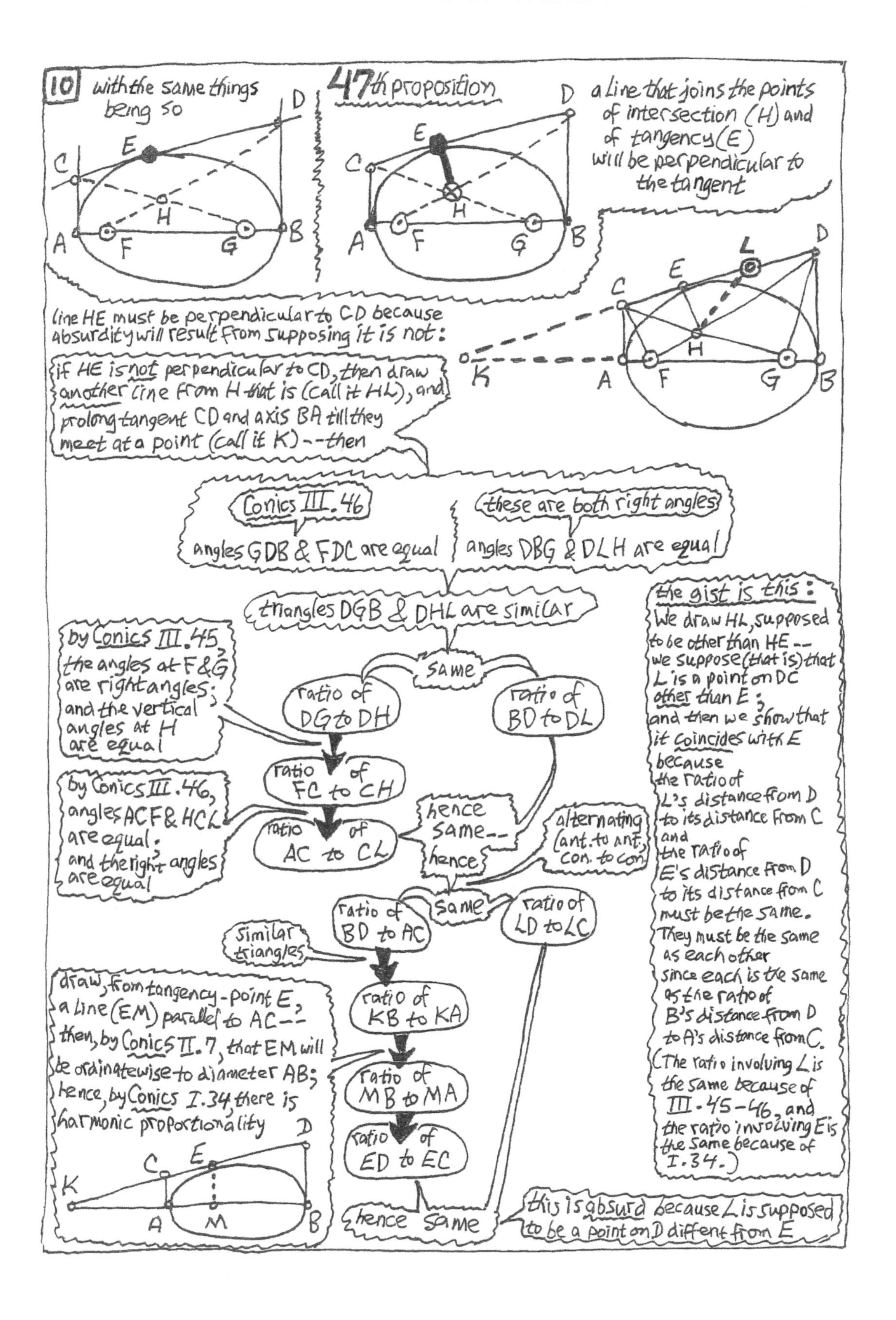

10
with the same things being so
47th proposition
a line that joins the points of intersection (H) and of tangency (E) will be perpendicular to the tangent
line HE must be perpendicular to CD because absurdity will result from supposing it is not:
if HE is not perpendicular to CD, then draw another line from H that is (call it HL), and prolong tangent CD and axis BA till they meet at a point (call it K)--then
Conics III.46
angles GDB & FDC are equal
these are both right angles
angles DBG & DLH are equal
triangles DGB & DHL are similar
same
ratio of DG to DH
ratio of BD to DL
by Conics III.45, the angles at F&G are right angles; and the vertical angles at H are equal
ratio of FC to CH
by Conics III.46, angles ACF & HCL are equal; and the right angles are equal
ratio of AC to CL
hence same-- hence
alternating (ant. to ant., con. to con.)
same
ratio of BD to AC
ratio of LD to LC
similar triangles
ratio of KB to KA
draw, from tangency-point E, a line (EM) parallel to AC-- then, by Conics II.7, that EM will be ordinatewise to diameter AB; hence, by Conics I.34 there is harmonic proportionality
ratio of MB to MA
ratio of ED to EC
hence same
this is absurd because L is supposed to be a point on D different from E
the gist is this:
We draw HL, supposed to be other than HE -- we suppose (that is) that L is a point on DC other than E; and then we show that it coincides with E because the ratio of L's distance from D to its distance from C and the ratio of E's distance from D to its distance from C must be the same. They must be the same as each other since each is the same as the ratio of B's distance from D to A's distance from C. (The ratio involving L is the same because of III.45-46, and the ratio involving E is the same because of I.34.)

(III) #46 The same things being so, the [lines] joined make equal angles with the touching[lines].
—For, the same things being supposed, I say that equal is the [angle] by ACF to the [angle] by DCG, and the [angle] by CDF to the [angle] by BDG.
—For—since there has been shown to be right each of the [angles] by CFD, CGD—the circle carved out (*graphomenos*) about the [line] CD as a diameter will go through the points F, G; therefore equal is the [angle] by DCG to the [angle] by DFG, for in the same segment of the circle are they. But the [angle] by DFG was shown equal to the [angle] by ACF; and so, the [angle] by DCG is equal to the [angle] by ACF. And likewise [equal is] also the [angle] by CDF to the [angle] by BDG.

(III) #47 The same things being so, the [line] that is drawn from the [point of the] falling together of the joined lines to the touch[point] will be at right angles to the touching[line].
—For let there be supposed the same things as those previously [supposed]; and also, let the [lines] CG, FD fall together with one another at the [point] H, and [let there fall together with one another] the [lines] CD, BA, extended, at the [point] K; and also, let there be joined the [line] EH. I say that perpendicular is the [line] EH to the [line] CD.
—For, if not, let there be drawn from the [point] H perpendicular to the [line] CD the [line] HL. Then—since equal is the [angle] by CDF to the [angle] by GDB, and also the right [angle] by DBG is, to the right [angle] by DLH, equal—therefore similar is the triangle DGB to the triangle LHD. Therefore as is the [line] GD to DH so is the [line] BD to DL. But as is the [line] GD to DH so is the [line] FC to CH, on account of the [angles] at the [points] F, G being right and also the [angles] at the [point] H being equal; and as is the [line] FC to CH so is the [line] AC to CL, on account of the similarity of the triangles AFC, LCH; therefore also as is the [line] BD to DL so is the [line] AC to CL. Alternately (*anallax*), as is the [line] DB to CA so is the [line] DL to LC. But as is the [line] DB to CA so is the [line] BK to KA; therefore also as is the [line] DL to CL so is the [line] BK to KA. Let there be drawn from the [point] E parallel to the [line] AC the [line] EM; therefore it will have been drawn down ordinatewise to the [line] AB; and as is the [line] BK to KA so will be the [line] BM to MA. And as is the [line] BM to MA so is the [line] DE to EC; therefore also as is the [line] DL to LC so is the [line] DE to EC—which very thing is absurd. Not so, therefore, that the [line] HL is perpendicular, nor is any [line] other than the [line] HE.

The 48th proposition demonstrates that the straight lines drawn from a tangency-point to the application-points will make equal angles with the tangent. This proposition is depicted in figure 11 for the case of the ellipse, and is presented after that in the box below.

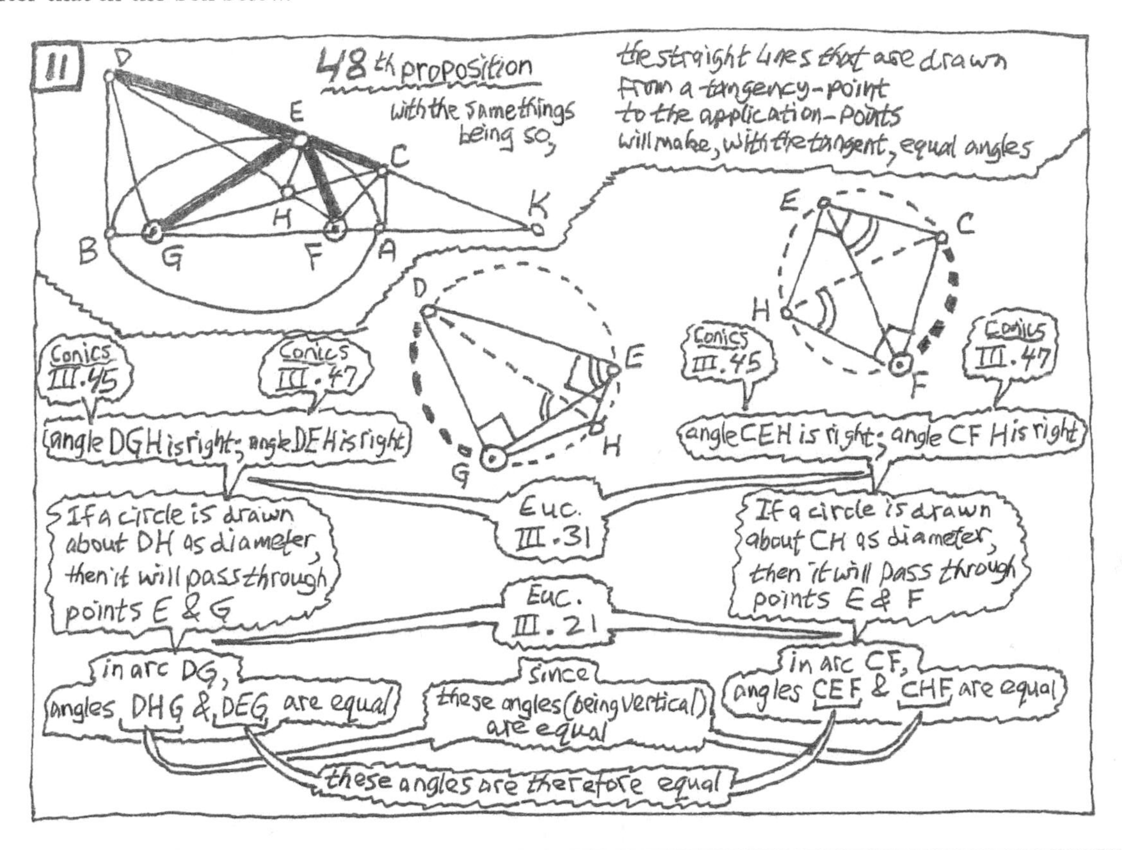

(III) #48 The same things being so, it is to be shown that the [lines] from the touch[point] to the points generated out of the throwing alongside [the points generated, in traditional parlance, "by the application"] (*ek tês parabolês*) make equal angles with the touching[line].

—For let there be supposed the same things, and let there be joined the [lines] EF, EG. I say that equal is the [angle] by CEF to the [angle] by GED.

—For—since right [angles] are what the angles by DGH, DEH are—the circle carved out (*graphomenos*) about the [line] DH as diameter will go through the points E, G; and so, equal will be the [angle] by DHG to the [angle] by DEG, for they are in the same segment. Likewise, then, also the [angle] by CEF is, to the [angle] by CHF, equal. But the [angle] by CHF, to the [angle] by DHG, is equal, for they are vertical [angles]; therefore also the [angle] by CEF is, to the [angle] by DEG, equal.

Apollonius speaks of points of application only for the sections other than the parabola. Why not for the parabola? Because a parabola has no oblique (transverse) side. A parabola's axis (like its other diameters) has no definite size: it has only one determinate endpoint, and therefore it cannot be split by a point whose distances to two determinate endpoints will equal one-fourth of the figure that is contained by the diameter so determined and its parameter.

Nonetheless, there is a certain point on the parabola's axis—a single point—that has some properties resembling the properties of a point of application of any other section. We can obtain that point for the parabola in the following way.

Consider an ellipse. We can for convenience, imagine it as obtained by cutting a right cone, so the diameter AD will be an axis, with a point of application at F. See figure 12.

Now, keeping the ellipse's vertex fixed, let us imagine the plane that cuts it into the cone as swinging down toward parallelism, the position in which the section cut by the plane would be not an ellipse but a parabola. As the cutting plane swings down, the axis lengthens. See figure 13, in which the axis lengthens from AP to AP' as the cutting plane swings down toward the position AQ, where the section cut would be a parabola.

Now, by extending the conic surface indefinitely, we can make the axis of the ellipse as long as we please if we swing the cutting plane close enough to the axis AQ of the parabola whose vertex is A. But if the cutting plane of the ellipse gets close enough to that position where the section it cuts will be a parabola, what will happen to the parameter of the ellipse? It can be shown that the parameter of the ellipse will get as close as you please to being quadruple AF, the distance from the vertex to the point of application. Figure 14 indicates why.

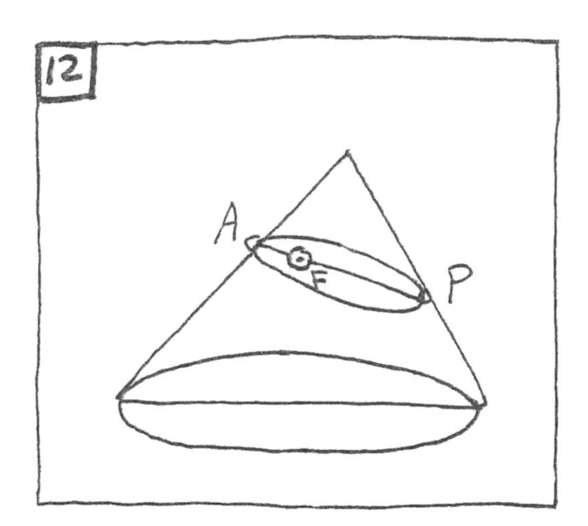

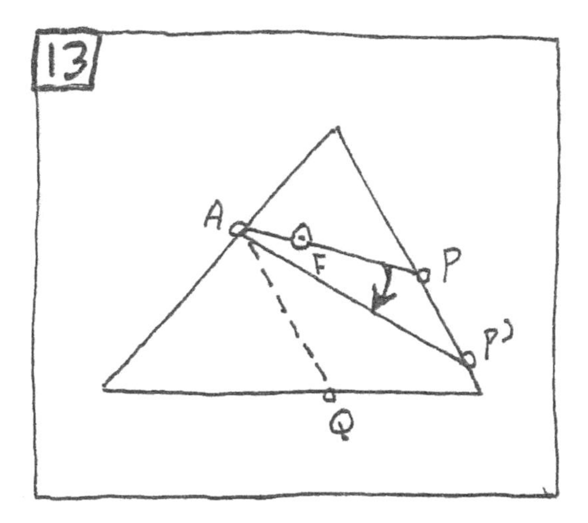

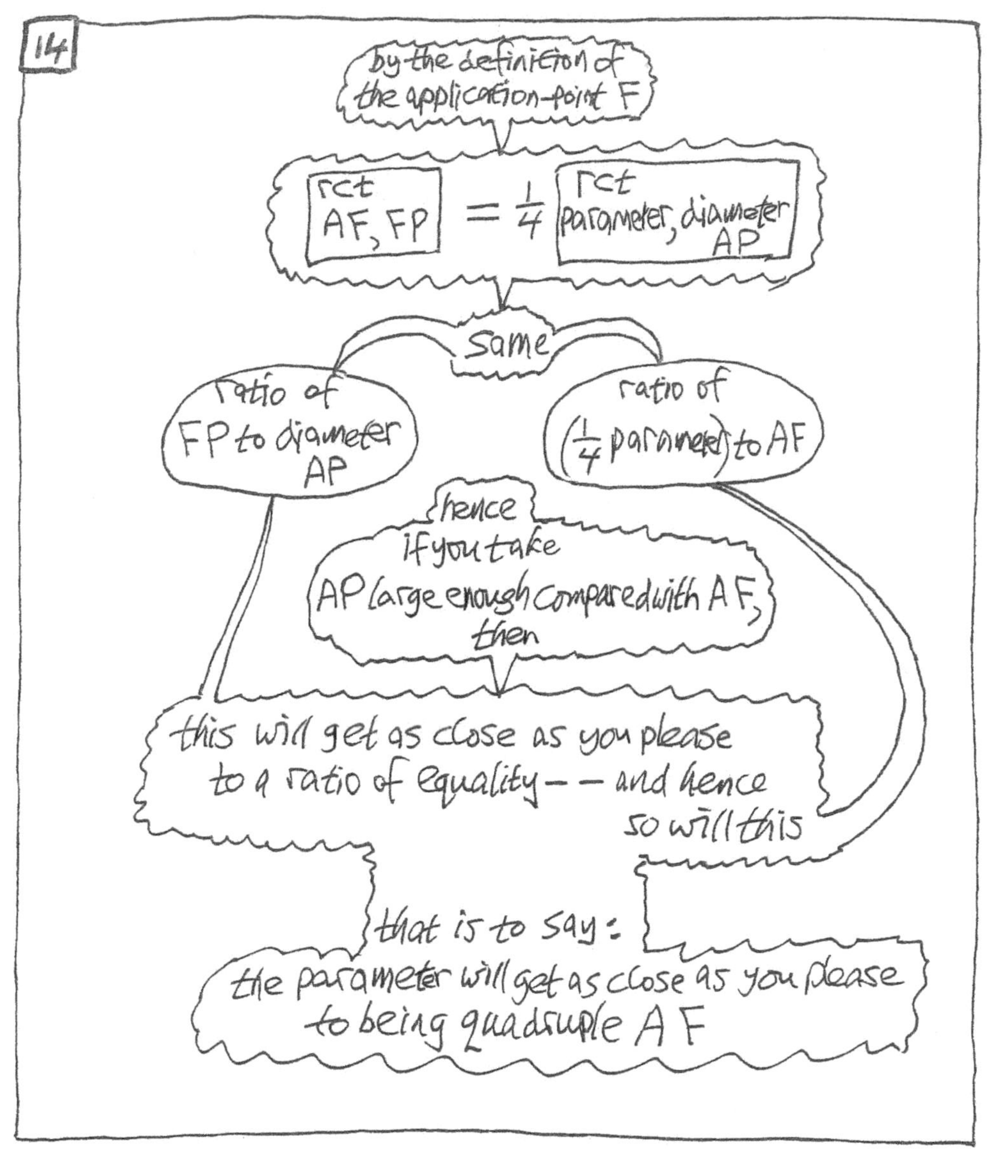

That is to say: as the ellipse gets as close as you please to the parabola, while the ellipse's vertex stands still, the ellipse's parameter gets as close as you please to being four times as large as the distance of its point of application F from its vertex A. How, then, will you obtain the point in the *parabola* that corresponds to the ellipse's point of application? By taking that point on the parabola's axis whose distance from its vertex is exactly one-fourth the size of its parameter.

Now let us see how that point in the parabola works like the point of application of an ellipse.

Certain lines that (in the 45th, 47th, and 48th propositions) were associated with the point of application F in the ellipse have as their counterparts certain lines which can be associated with that point F in the parabola.

First, consider the ellipse in the *45th proposition*. See figure 15. The line CE that is tangent to the ellipse intersects, at points C and D, the perpendicular lines standing at the axis's endpoints (A and B). The 45th proposition says that the line which emanates from F (the ellipse's point of application) and is drawn to point C will be perpendicular to the line which emanates from that same point F and is drawn to point D.

In the parabola (fig. 16), it can be shown that a tangent CE will be perpendicular to the line from F to that point C where the tangent meets the line at A which is perpendicular to the axis; so CF will also be perpendicular to the line from F that goes parallel to the tangent.

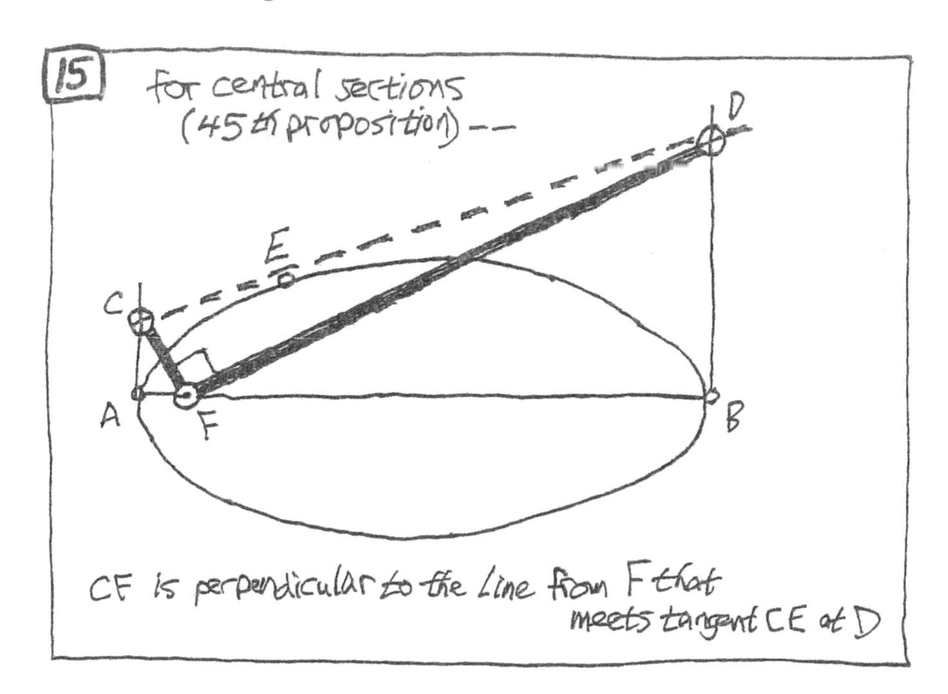

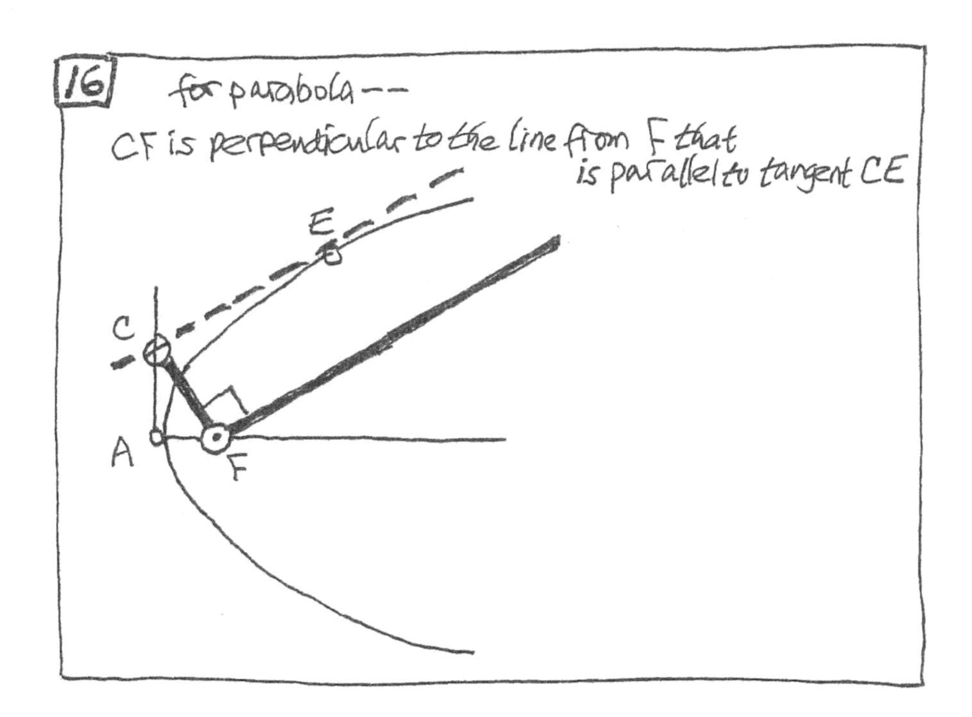

16
for parabola--
CF is perpendicular to the line from F that
is parallel to tangent CE
E
C
A
F

The counterpart of the line from F that, in Proposition 45, went off to intersect the ellipse's tangent, thus, in the parabola, is parallel to the tangent. The same is true, we shall now see, in proposition 47; and again the counterpart of the line from F that, in proposition 48, went off to intersect the ellipse's axis, we shall afterward see, in the parabola, is parallel to the axis. The reason for those intersections in the ellipse is that the ellipse's axis comes to an end, and has two points of application. The parabola's axis, however, can be prolonged indefinitely, and it therefore does not have points of application. Nonetheless, as we have just seen, and shall now see again, point F in the parabola works something like a point of application in the ellipse (or, for that matter, in the hyperbola too).

Next, consider the ellipse in the *47th proposition*. See figure 17. The line from C to G intersects the line from F to D. The 47th proposition says that the line drawn from their intersection-point H to the tangency-point E will make a right angle with the tangent.

In the parabola (figure 18), draw a line from the tangency-point E to the point H where two lines intersect—one of them, from point C (where the tangent meets the line at A that is perpendicular to the axis), drawn parallel to the axis; the other one of them, from F, drawn parallel to the tangent. That line from the tangency-point E to the intersection-point H, it can be shown, will be perpendicular to the line from F that is parallel to the tangent; so EH will also be perpendicular to the tangent.

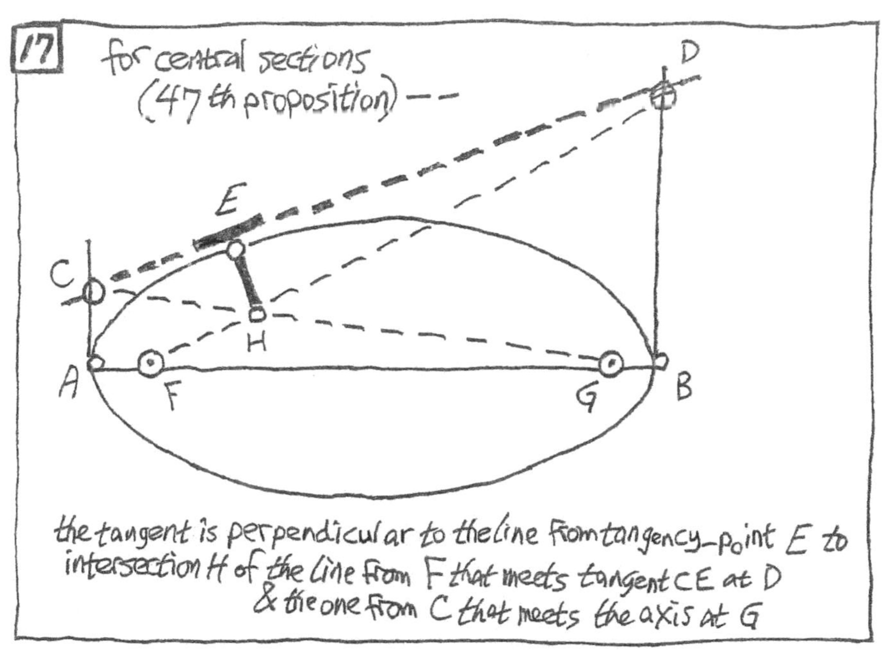

17
for central sections
(47th proposition) --
D
E
C
H
A
F
G
B
the tangent is perpendicular to the line from tangency-point E to intersection H of the line from F that meets tangent CE at D & the one from C that meets the axis at G

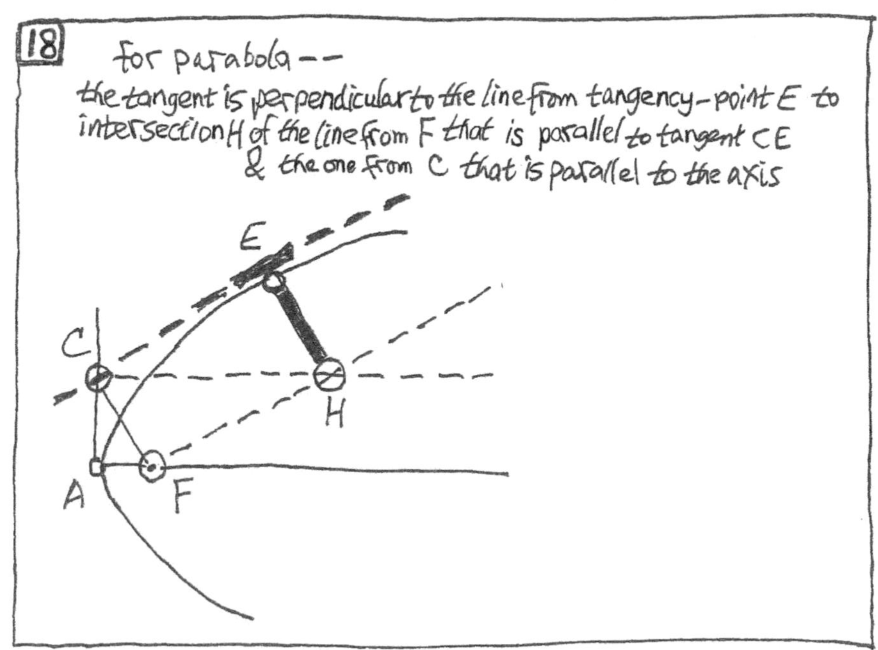

18
for parabola --
the tangent is perpendicular to the line from tangency-point E to intersection H of the line from F that is parallel to tangent CE & the one from C that is parallel to the axis
E
C
H
A
F

Finally, consider the ellipse in the *48th proposition*, and consider the parabola. See figure 19. In the ellipse (and other central sections), the angle which the tangent makes with the line from the tangency-point meeting the axis at F is equal to the angle which that tangent makes with the line from that tangency-point meeting the axis at G. In the parabola, that statement will be true if we replace the last few words—"meeting the axis at G"—by the words "running parallel to the axis."

Again, point F in the parabola works something like the ellipse's point of application.

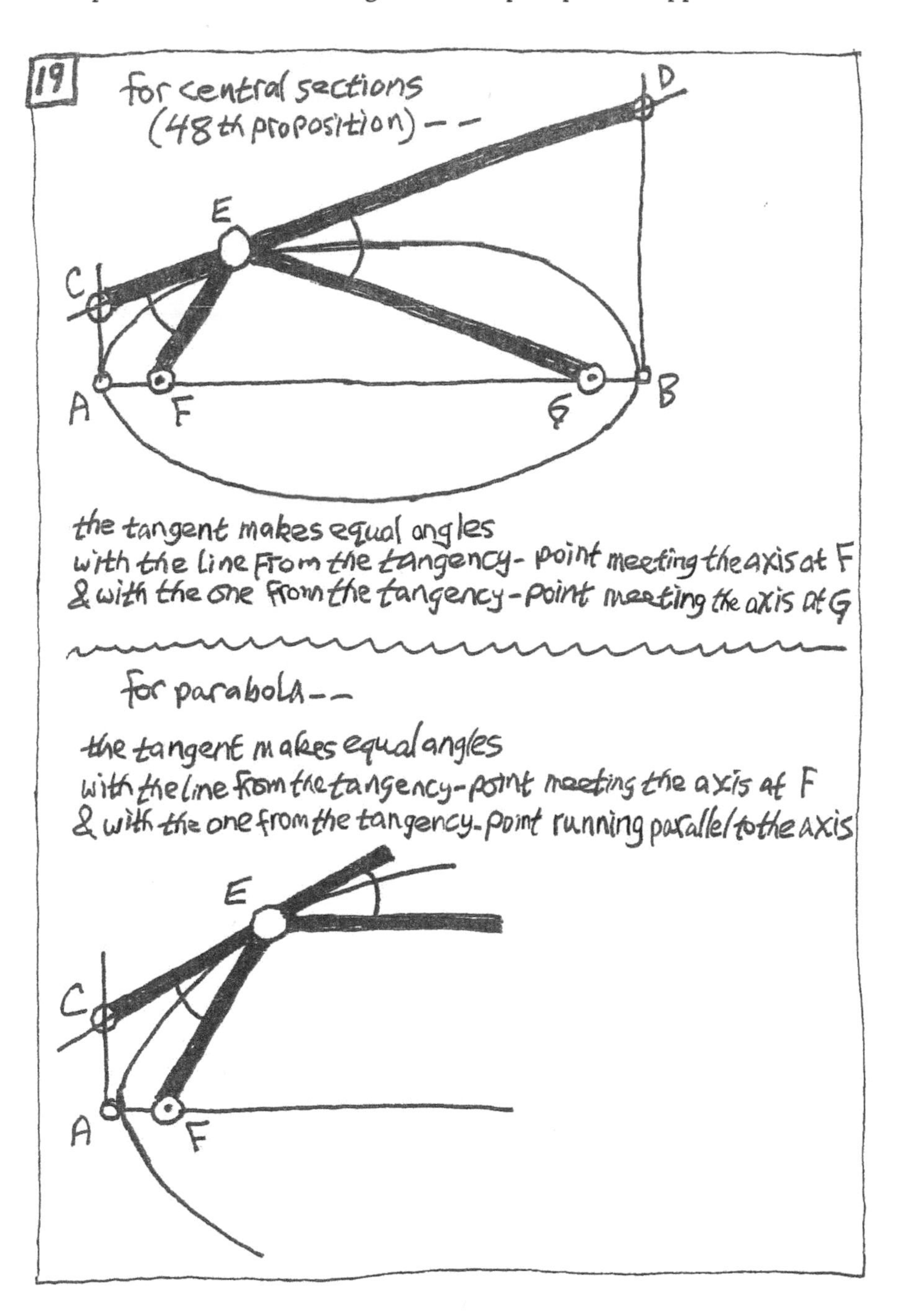

19
for central sections
(48th proposition)--
D
E
C
A
F
G
B
the tangent makes equal angles
with the line from the tangency-point meeting the axis at F
& with the one from the tangency-point meeting the axis at G
for parabola--
the tangent makes equal angles
with the line from the tangency-point meeting the axis at F
& with the one from the tangency-point running parallel to the axis
E
C
A
F

Apollonius never speaks of any point of application in a parabola. A rectangle cannot be "applied" when a line does not have two determinate endpoints; and the only determinate endpoint of a parabola's axis is the vertex. To someone interested in the use of the properties of curves in studying physical situations, however, the point in the parabola that *corresponds* to a point of application in the ellipse is of great interest.

Consider sunbeams, for example. If a beam of light falls perpendicularly upon a flat mirror, then it is reflected directly back upon the line that is perpendicular to the surface. If it is incident along a line that makes an angle with the perpendicular, then it is reflected back along a line that make an equal angle with the perpendicular. If the beam falls upon a surface that is curved, its equal angles of incidence and reflection are made with the perpendicular to the *tangent* to the surface. (The perpendicular to the tangent to the surface is called the "normal" to the surface.) See figure 20.

Since the sun is so immensely far away, beams of sunlight come to us as if they are parallel to each other; so if you take a surface of parabolic shape (a "paraboloid"—you get it by revolving a parabola about its axis, just as you get a sphere by revolving a circle about its diameter), and if you turn the inside toward the sun, with the axis pointing at the sun, then the beams of sunlight after reflection will all meet at the same point—that very point on the parabola's axis which is at a distance from its vertex equal to one-fourth of its parameter. During the day, the sunlight concentrated at that point can get hot enough to burn something flammable that is placed there. Or if at night a lamp is placed there, then the beams that emanate from it will, upon reflection, be all sent out in parallel, thus giving the illumination a single direction, the direction in which the axis is pointed (fig. 21).

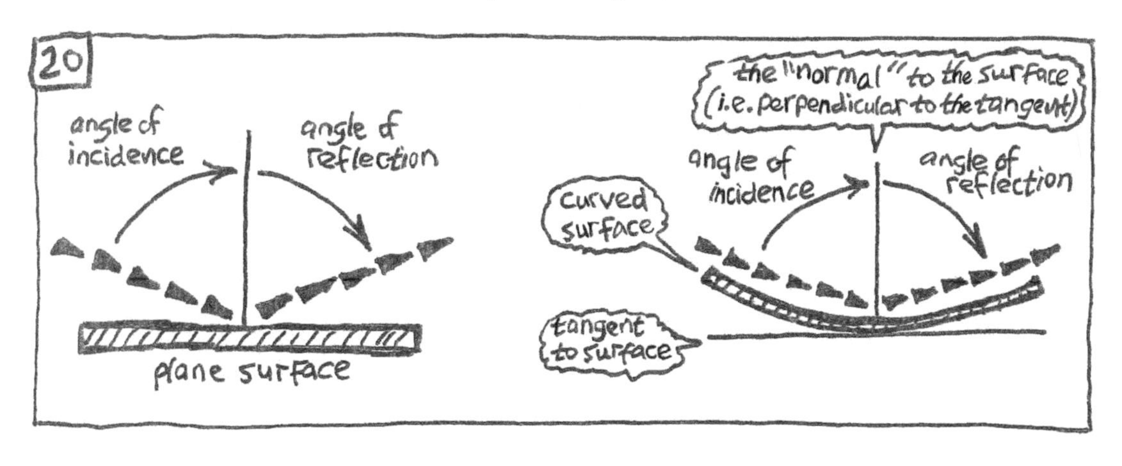

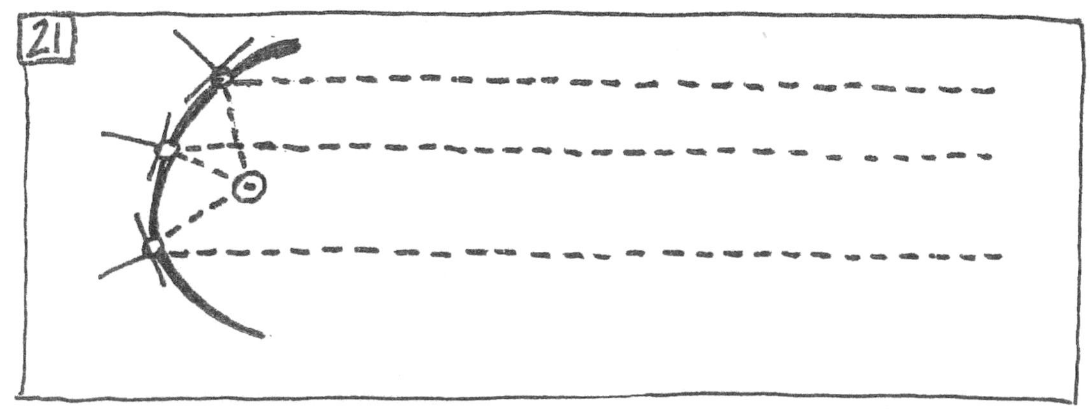

The burning-point of the parabola was known about very long ago. It came to be named, in Latin, the *focus*—that is, the "hearth"—of the parabola.

The term *focus* was not confined to the parabola, naming that single point in it whose properties resemble the properties of a point of application in the hyperbola and ellipse. The term (whose plural is *foci*) was used also to name these points of application in the hyperbola and ellipse—these points too being of interest in using the properties of curves to study physical situations. Sound, like light, will, when reflected from a surface, make the same angle that it made in falling upon the surface; hence it is possible to construct an elliptical "whispering gallery" that embodies what Apollonius shows in his 48th proposition. If a single curved wall encloses a room with an elliptical floor, then someone who is standing at one focus can whisper too softly to be heard by someone standing close by, and yet nonetheless be heard by someone else who is standing far across the room at the other focus. Why? Because the *emitted* sound *scatters* the whisper out in all directions, with only a little of it going in any particular direction, such as the one that will bring it to the person who is standing close by—whereas the *reflected* sound will all be *concentrated* at the other focus far across the room, making it loud enough to be heard by someone who is standing there (fig. 22).

Now even apart from its direct usefulness in studies that are physical, the focus might provide an important means for examining the conics in their resemblances and their differences. Consider the following:

The hyperbola and the ellipse resemble each other in having two foci,
these two foci being equidistant from the center.
The hyperbola and ellipse differ from each other in having
the distance between the foci
greater than the axis (hyperbola) or less than it (ellipse).
The less is the distance that separates the foci of an ellipse,
the more does the ellipse look like a circle.
In the circle, there is no distance at all between foci;
its center is its focus.
Likewise in the conic section that has no center, the parabola,
there is a single focus—
not because what would be two foci in an ellipse
have come to coincide (as in the circle),
but because in the parabola one of those two foci has gone away.

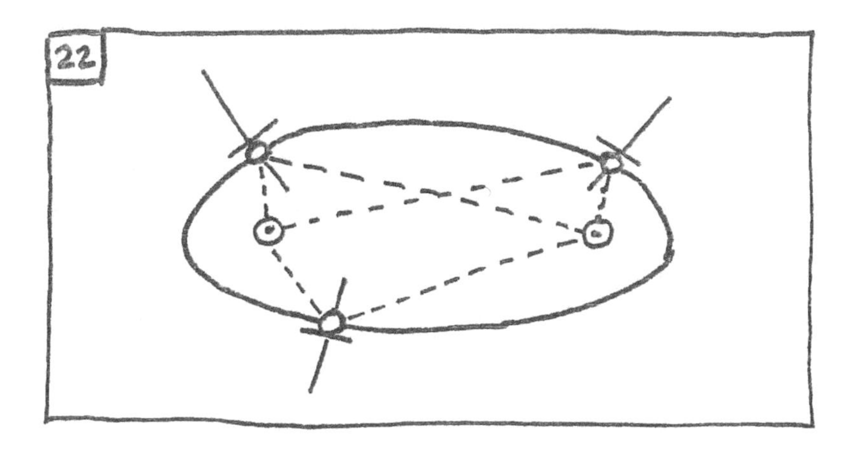

And yet, although the foci can be used to characterize the various kinds of conic sections, and even to distinguish one section of some kind from another one of the same kind, nonetheless a consideration of the foci cannot simply substitute for obtaining the curves by cutting cones. In order to understand why, we should first look at the propositions (the 51st and 52nd) in which Apollonius demonstrates those properties of the points of application which nowadays are often used to define the hyperbola and ellipse. To do those demonstrations, we shall need to demonstrate the 50th proposition. This demonstration will depend upon the last proposition that we examined—the 48th. The 48th proposition will not suffice, however, for demonstrating the 50th. The 50th proposition depends not only on the 48th but also on the 49th. The 49th proposition is depicted in figure 23, and the 50th in figure 24 (in both figures, for the case of the ellipse); then the propositions are presented in the boxes on the following pages.

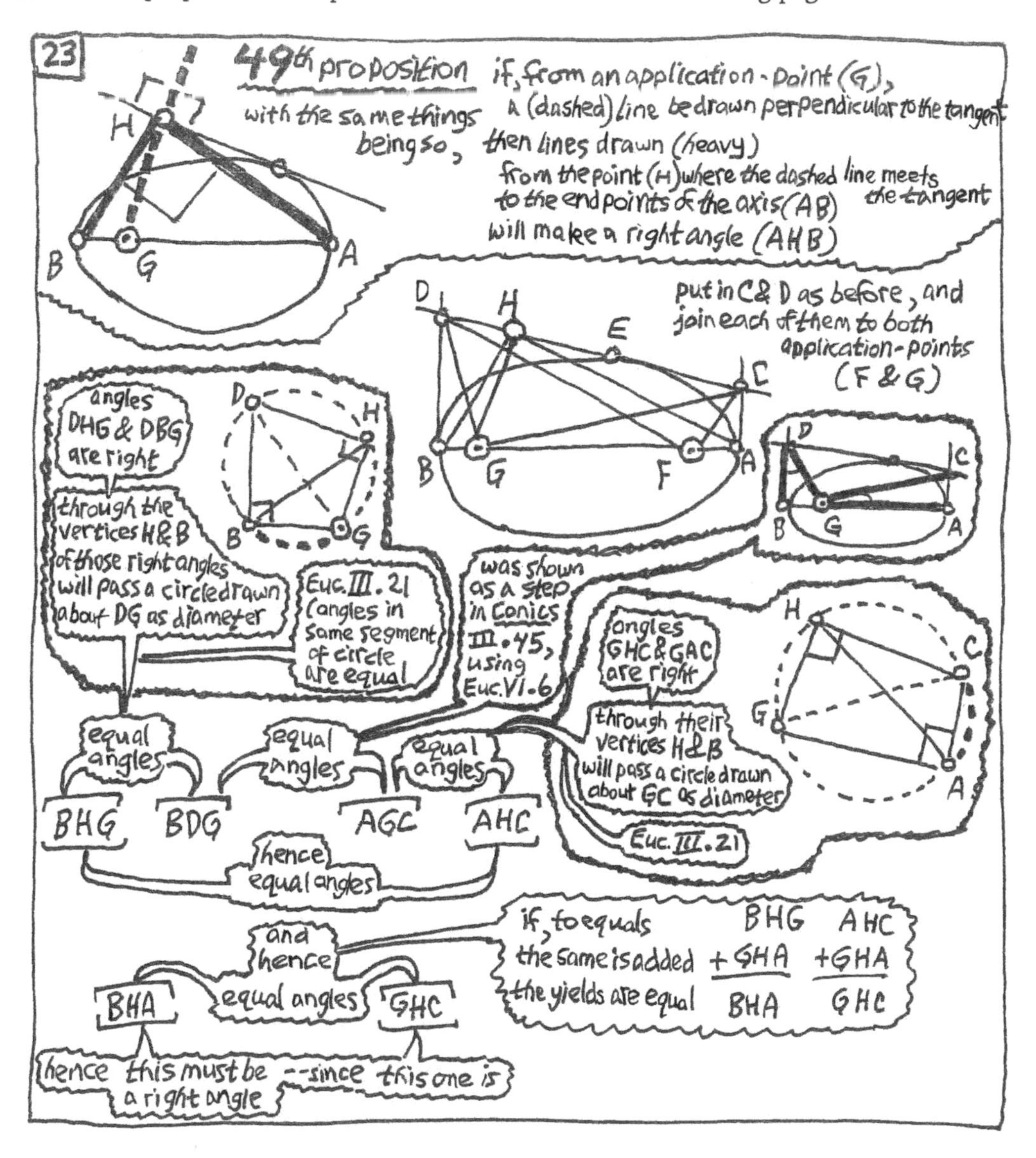

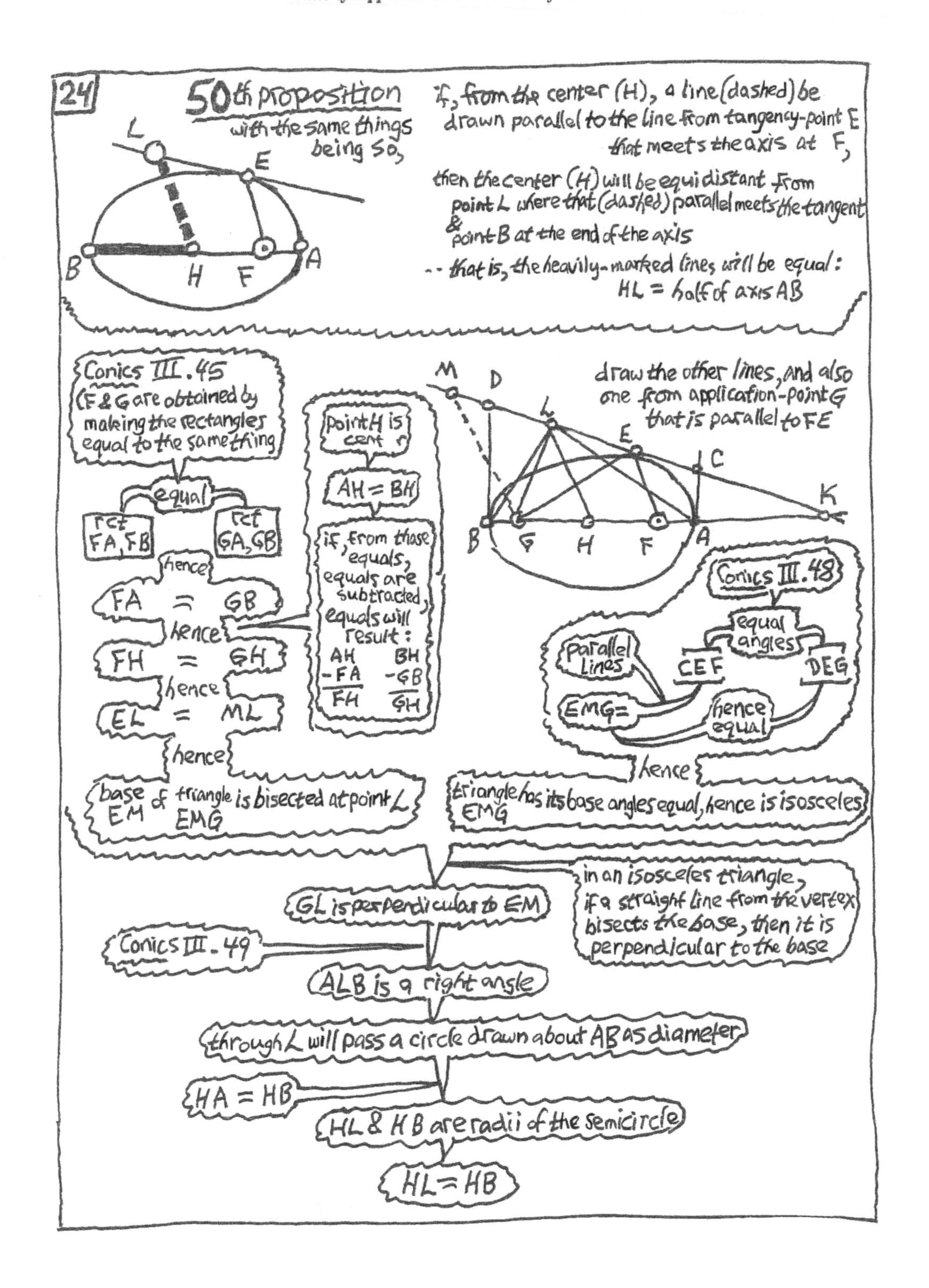
24
50th proposition
with the same things being so,
if, from the center (H), a line (dashed) be drawn parallel to the line from tangency-point E that meets the axis at F,
then the center (H) will be equidistant from point L where that (dashed) parallel meets the tangent & point B at the end of the axis
-- that is, the heavily-marked lines will be equal: HL = half of axis AB
L
E
B
H
F
A
Conics III.45 (F & G are obtained by making the rectangles equal to the same thing
equal
rct FA, FB
rct GA, GB
hence
FA = GB
hence
FH = GH
hence
EL = ML
hence
base EM of triangle EMG is bisected at point L
Point H is cent r
AH = BH
if, from those equals, equals are subtracted, equals will result:
AH − FA FH
BH − GB GH
draw the other lines, and also one from application-point G that is parallel to FE
M
D
L
E
C
K
B
G
H
F
A
Conics III.48
equal angles
CEF
DEG
parallel lines
EMG=
hence equal
hence
triangle EMG has its base angles equal, hence is isosceles
in an isosceles triangle, if a straight line from the vertex bisects the base, then it is perpendicular to the base
GL is perpendicular to EM
Conics III.49
ALB is a right angle
through L will pass a circle drawn about AB as diameter
HA = HB
HL & HB are radii of the semicircle
HL = HB

(III) #49 The same things being so, if—from some one of the points [that are generated out of the throwing alongside (that is, by the "application")]—a perpendicular be drawn to the touching[line], then the [lines] from that generated point to the bounds of the axis make a right angle.
—For let there be supposed the same things; and also, from the [point] G perpendicular to the [line] CD let there be drawn the [line] GH; and also, let there be joined the [lines] AH, BH. I say that the angle by AHB is right.
—For since right is the [angle] by DBG and also the [angle] by DHG, the circle carved out (*graphomenos*) about the [line] DG as diameter will go through the [points] H, B, and equal will be the [angle] by GHB to the [angle] by BDG. But the [angle] by AGC was, to the [angle] by BDG, shown equal; therefore also the [angle] by BHG is, to the [angle] by AGC, that is, to the [angle] by AHC, equal. And so, [equal] also is the [angle] by CHG to the [angle] by AHB. But right is the [angle] by CHG; therefore right is also the [angle] by AHB.

(III) #50 The same things being so, if from the center of the section there falls forth to the touching[line] some parallel to that straight [line] which is drawn through the touch[point] and one of the points [of "application"], then equal will [that parallel] be to the half of the axis.
—For let there be the same things as the things before; and also, as center let there be the [point] H; and also, let there be joined the [line] EF; and also, let the [lines] DC, BA fall together at the [point] K; and also, through the [point] H parallel to the [line] EF let there be drawn the [line] HL. I say that equal is the [line] HL to the [line] HB.
—For let there be joined the [lines] EG, AL, LG, LB; and also, through the [point] G parallel to the [line] EF let there be drawn the [line] GM. Since, then, the [rectangle] by AFB is equal to the [rectangle] by AGB, therefore equal is the [line] AF to the [line] GB. But also is the [line] AH, to the [line] HB, equal; therefore also is the [line] FH, to the [line] HG, equal. And so, also is the [line] EL, to the [line] LM, equal. And—since also the [angle] CEF was shown, to the [angle] DEG, equal, and the [angle] by CEF is equal to the [angle] by EMG—therefore equal also is the [angle] by EMG to the [angle] by MEG. Therefore equal also is the [line] EG to the [line] GM. But also the [line] EL, to the [line] LM, was shown equal; therefore perpendicular is the [line] GL to the [line] EM. And so, on account of what was shown before, a right [angle] is what the [angle] by ALB is, and the circle carved out (*graphomenos*) about the [line] AB as diameter will go through the [point] L. And equal is the [line] HA to the [line] HB; and therefore also the [line] HL, being [a line] out-from-the-center, [being, that is, a radius] of the semicircle, is equal to the [line] HB.

Now we are ready for the culmination of Apollonius's treatment of the points of application in the Third Book of his *Conics*. This is the demonstration of what has come to replace the cutting of cones, or the comparing of squares and rectangles, in the characterization of curves called "conic sections." From the 51st proposition, we learn that the straight lines drawn from any point on an hyperbola (or opposite sections) to the points of application have a constant difference; from the 52nd, that those lines drawn from any point on an ellipse have a constant sum.

The 51st and 52nd propositions are depicted in figures 25 and 26, and are presented after that in the boxes on the following pages.

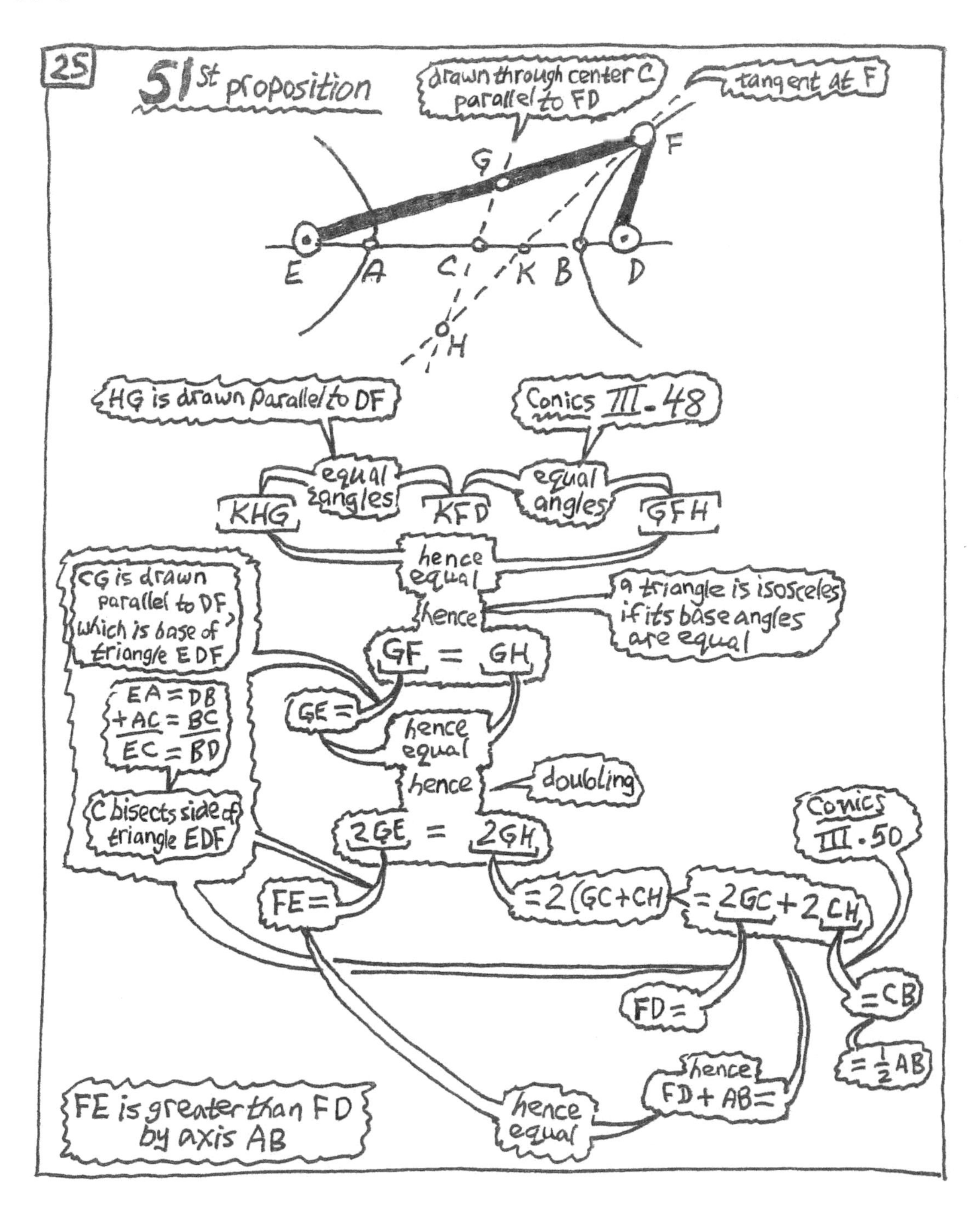

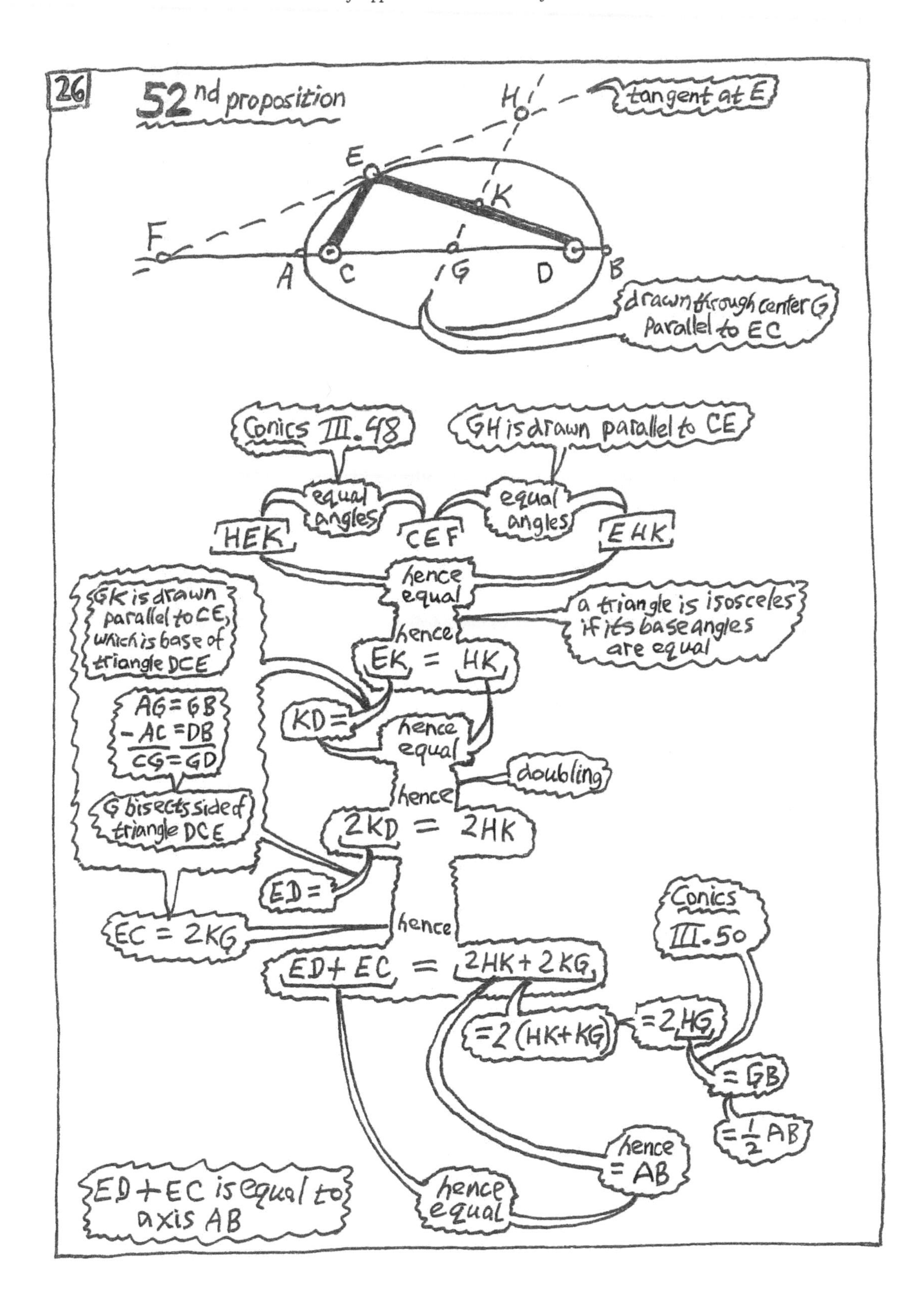
26
52nd proposition
tangent at E
H
E
K
F
A
C
G
D
B
drawn through center G
Parallel to EC
Conics III.48
GH is drawn parallel to CE
equal angles
equal angles
HEK
CEF
EHK
hence equal
a triangle is isosceles if its base angles are equal
GK is drawn parallel to CE, which is base of triangle DCE
hence
EK = HK
KD =
hence equal
AG = GB
− AC = DB
CG = GD
doubling
hence
G bisects side of triangle DCE
2KD = 2HK
ED =
EC = 2KG
hence
ED + EC = 2HK + 2KG
Conics III.50
=2(HK+KG)
=2HG
=GB
=½AB
hence =AB
hence equal
ED + EC is equal to axis AB

(III) **#51** If, with respect to an hyperbola or the opposite [sections], there be thrown alongside (*parablêthêi*) the axis, [that is, there be "applied to" the axis], on both sides, [a rectangle that is] equal to the fourth part of the figure (*eidous*) and that overshoots (*huperballon*) by a square figure (*eidei*), and from the points generated out of the throwing alongside, [from, that is, "the points of application,"] straight [lines] be deflected to either one of the sections—then the greater has its size over (*huperechei*) [the size] of the lesser, [that is, exceeds it], by [just the size of] the axis.
—For let there be an hyperbola or opposite [sections] whose axis is what the [line] AB is, and center is what the [point] C is; and also, equal to the fourth part of the figure (*eidous*) let each of the [rectangles] by ADB, AEB be; and also, from the points E, D let there have been deflected to the line [of section] the [lines] EF, FD. I say that the [line] EF has [its size] over [the size of] the [line] FD, [that is, exceeds it], by [just the size of] the [line] AB.
—Let there be drawn, through the [point] F as a touching[line] the [line] FKH, and through the [point] C parallel to the [line] FD the [line] GCH; therefore equal is the [angle] by KHG to the [angle] by KFD, for they are alternate (*enallax*). And the [angle] by KFD is equal to the [angle] by GFH. Therefore also the [angle] by GFH is equal to the [angle] by GHF. Therefore equal is the [line] GF to the [line] GH. But the [line] FG is, to the [line] GE, equal—since [equal] also is the [line] AE to the [line] BD, and also the [line] AC to the [line] CB, and also the [line] EC to the [line] CD—and therefore also the [line] GH to the [line] EG is equal. And so, the [line] FE, in comparison with the [line] GH, is double. And also, since the [line] CH has been shown equal to the [line] CB, therefore the [line] EF is double of the [lines] GCB both together. But of the [line] GC, double is what the [line] FD is; and of the [line] CB, double is what the [line] AB is. Therefore the [line] EF is equal to the [lines] FD, AB both together. And so, the [line] EF has [its size] over [the size of] the [line] FD, [that is, exceeds it], by [just the size of] the [line] AB.

(III) #52 If, in an ellipse, there be thrown alongside (*parablêthêi*) ["applied to"] the greater of the axes, on both sides, a [rectangle] that is equal to the fourth part of the figure (*eidous*) and that is lacking (*elleipon*) by a square figure (*eidei*), and from the points generated out of the throwing alongside, [from, that is, "the points of application,"] straight [lines] be deflected to the line [of section]—then they will be equal to the axis.
—Let there be an ellipse whose greater axis the [line] AB is; and also, equal to the fourth part of the figure (*eidous*) let each of the [rectangles] by ACB, ADB be; and also, from the [points] C, D let there be deflected to the line [of section] the [lines] CED. I say that the [lines] CED [together] are equal to the [line] AB.
—Let there be drawn as a touching[line] the [line] FEH; and also, center let the [point] G be; and also, through it parallel to the [line] CE [let there be drawn] the [line] GKH. Since, then, equal is the [angle] by CEF to the [angle] by HEK, and the [angle] by FEC to the [angle] by EHK is equal, therefore also the [angle] by EHK to the [angle] by HEK is equal. Therefore equal is the [line] HK to the [line] KE. And—since also the [line] AG to the [line] GB is equal, and also the [line] AC to the [line DB]—therefore also the [line] CG to the [line] GD is equal; and so, also the [line] EK to the [line] KD. And also, on account of this, double is the [line] ED of the [line] HK, and the [line] EC [double] of the line KG, and also the [lines] CED both together are [double] of the [line] GH. But also the [line] AB is double of the [line] GH; therefore equal is the [line] AB to the [lines] CED [both together].

These last two propositions make possible devices for drawing ellipses and hyperbolas.

For an ellipse, fix two pins where the foci are to be; around them, place a loop of string, and pull it taut with a pencil-point (fig. 27).

Now move the pencil, keeping the loop of string taut. The size of the loop will stay the same, and so will the size of that part of the loop which stretches between the two fixed pins. The distance by which the pencil-point gets farther away from one pin will therefore be the same distance by which it gets closer to the other pin. That is to say, every point on the line that is traced out by the pencil will yield the same sum when its distance to one fixed point is added onto its distance to the other fixed point (fig. 28).

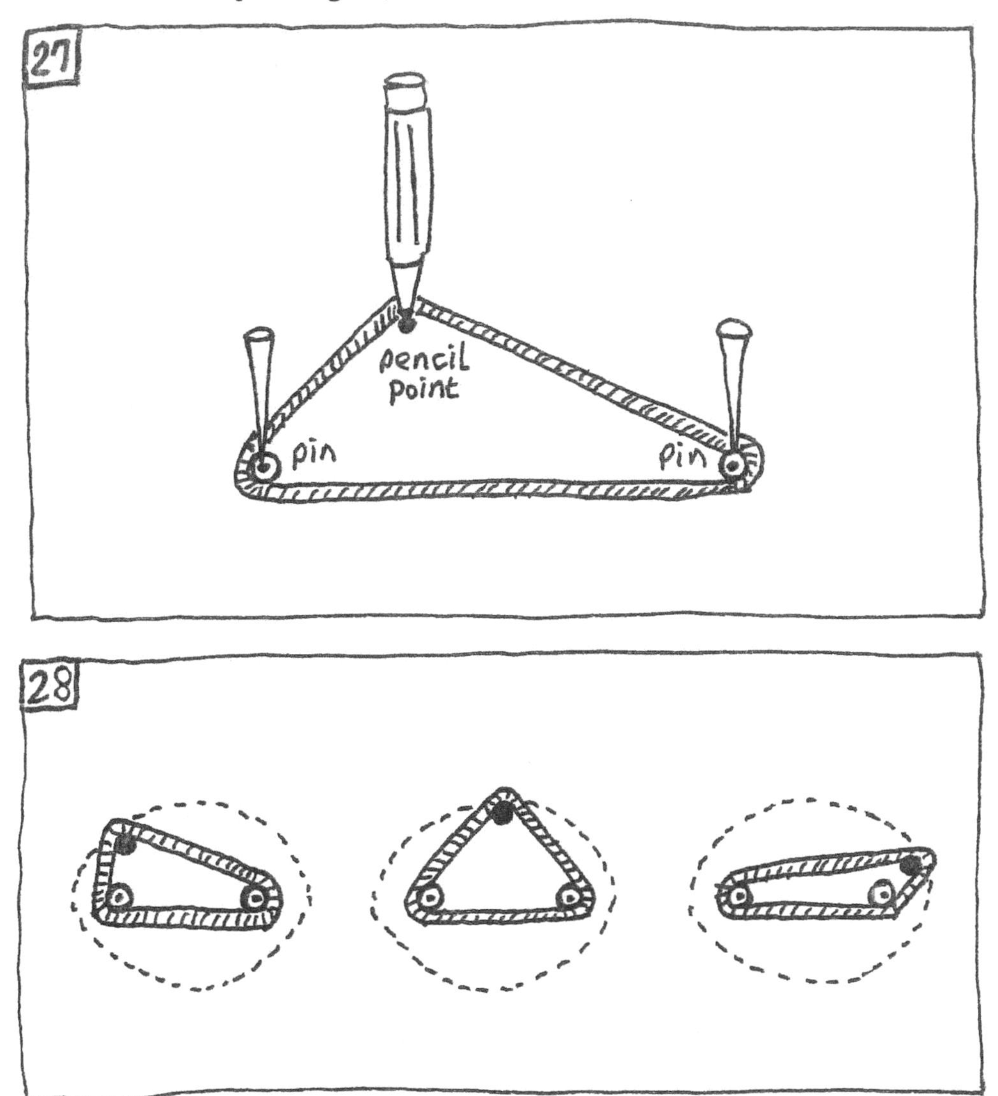

For an hyperbola, we want to devise a way of arranging the loop of string so that the increase in distance by which the pencil-point gets farther away from one pin will be the same as the increase in distance by which it gets farther away from the other pin. That way, the difference between its distance from one pin and its distance from the other one will stay the same as the pencil-point moves. How to arrange that, is shown in figure 29.

The loop is taut at one end around the pin; it is taut as all of it goes together past the other pin, and it is held down taut at the other end (but is allowed to move) by the hand that does not hold the pencil, while the other hand moves the pencil-point to trace out a line. That way, the same additional length of string is being fed to the pencil point from around the left pin as from around the right pin, since it is being supplied equally from the left side and from the right side of the part of the loop that is tightly held straight down by the hand. Three points on this line are shown in figure 30. (Note that the curve thus drawn can be extended only by taking a longer loop of string. This is true also for the curve that we shall look at next, the parabola.)

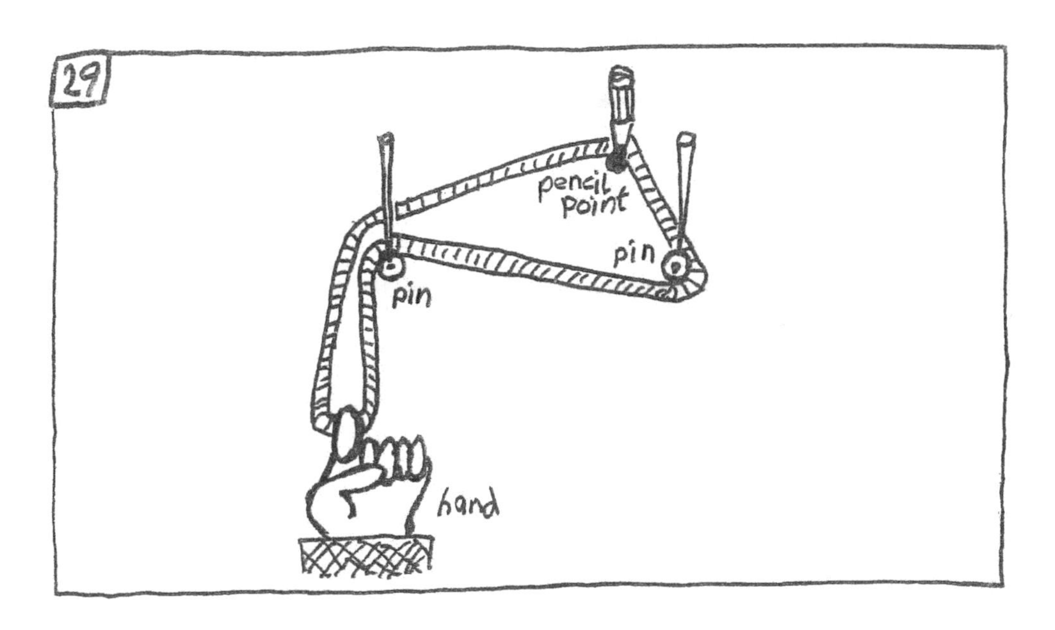

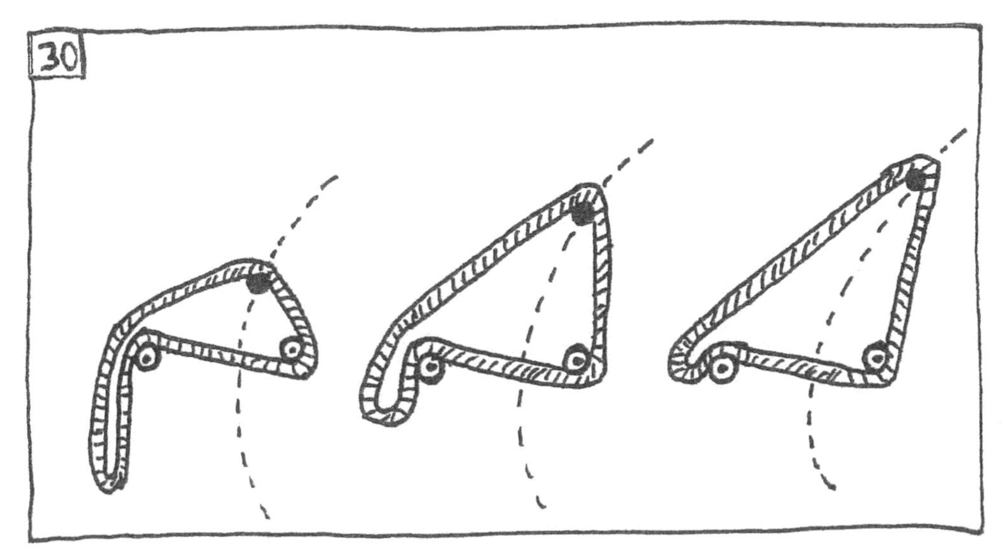

Is there a device for drawing a parabola? We saw that there is a point on a parabola's axis that has properties resembling the properties demonstrated by Apollonius for the points of application; but there is only one such point. The circle too had only one such point, but the circle resembles an ellipse, so we might say that since the circle has two foci that happen to coincide, the addition-property of the 52nd proposition then holds for the circle too. But neither the addition-property of the 52nd proposition nor the subtraction-property of the 51st holds for the single-focus parabola. It can be shown, however, that the distance from any point on a parabola to the fixed point that is called its "focus" will be equal to the distance from that point on the parabola to a certain fixed straight line (called its "directrix"). This property makes possible a device for drawing parabolas. This device, like the term "focus," was the work of Kepler. (The devices for drawing ellipses and hyperbolas can be found in Descartes' optical work, the *Dioptrics*.) Here is Kepler's device for drawing a parabola.

Fix the line that is to be directrix, and let a T-square of length AB slide along it; and at the far end of the T-square (at moving point A) fasten a string that has the same length as the T-square (fig. 31).

Now swing the loose end of the string tautly up until it is perpendicular to the T-square, and tack it down there—calling this point F. It is to be the focus (fig. 32).

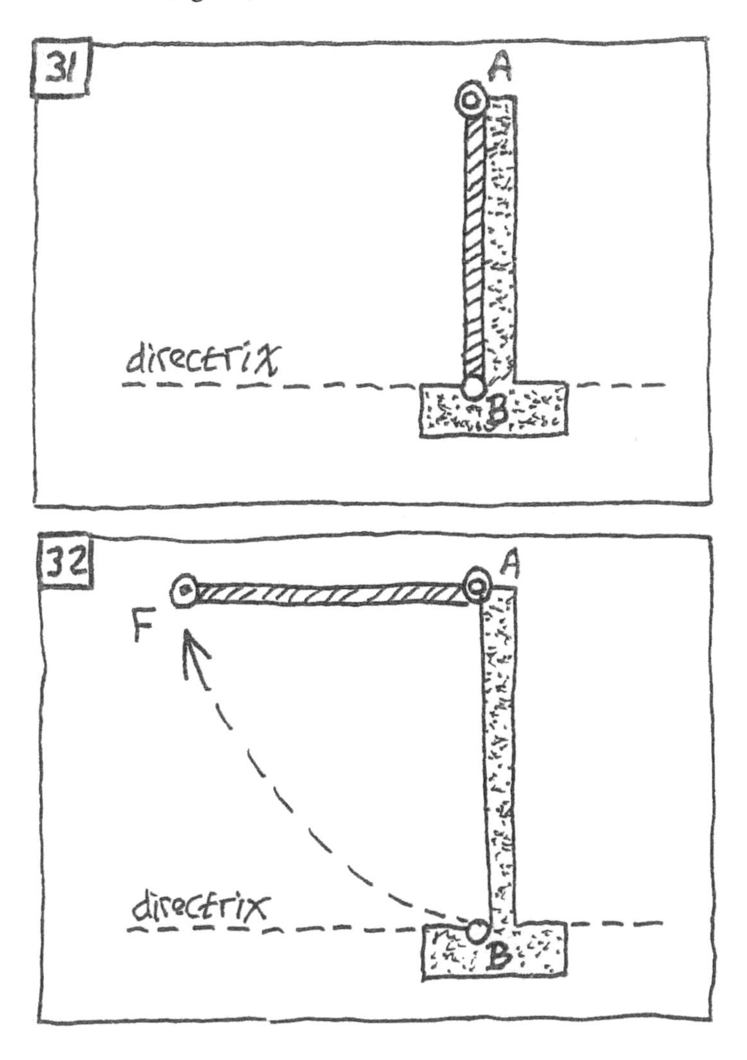

The string is to be kept fastened at its endpoints, F (on the paper) and A (on the sliding T-square), and the pencil-point is to be kept against the string and above it, starting just to the left of point A. Pushing the pencil-point straight down along the T-square to point X will make the T-square move leftward, closer to F. That is because the string is of fixed length and is attached to point F (whose position on the paper is fixed) and to point A (which moves with the T-square). See figure 33.

The pencil-point is at X, and the total length of the string is the constant AB, which is equal to the sum of AX and XF. How far from fixed point F will the pencil-point be? See figure 34.

The pencil-point will be as far from F as it is from B—which is to say that every point on the line traced out by the pencil-point will be equally distant from the fixed point and from the fixed straight line. The fixed point will be the focus and the straight line will be the directrix of a parabola, which is the line that will be drawn by Kepler's device.

So now we have devices that enable us to draw the curves that are conic sections. The power that they give us might tempt us to say what many people nowadays say—that just as a circle is merely the collection of all points that are equidistant from a fixed point, so an hyperbola or an ellipse is merely the collection of all points whose distances from two fixed points have a constant difference or a constant sum, and a parabola is merely the collection of all points that are equidistant from a fixed point and a fixed straight line. We need to ask what would be said about that by Euclid, by Apollonius, and by their followers.

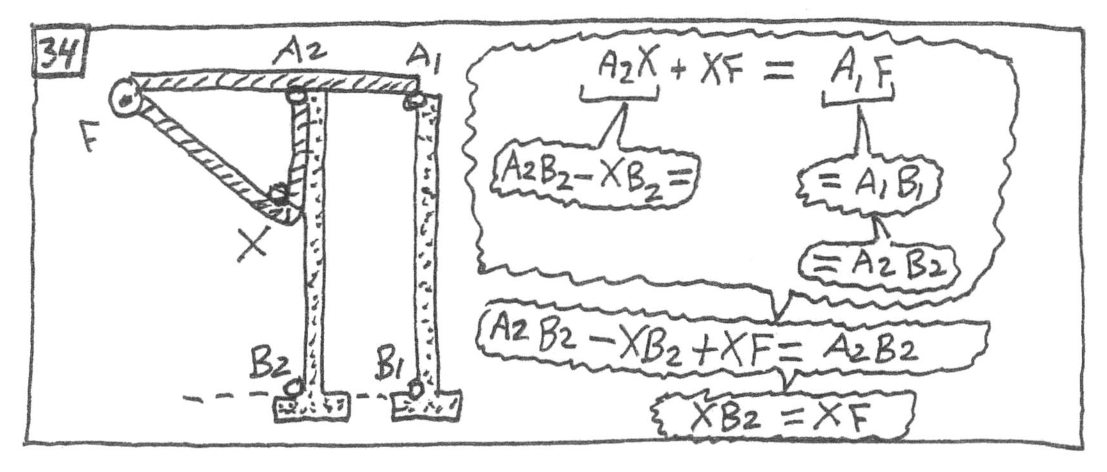

CHAPTER 4. APOLLONIUS'S INTRODUCTORY LETTER, AND LOCI

Apollonius transmitted his *Conics* to a friend along with a letter that gives a sketch of its contents, which you will find in the following box.

> [LETTER PREFATORY TO BOOK ONE]
> ....From the eight books (of the *Conics*) the first four fall [under the heading of] an elementary (*stoikheiôôdê*) introduction. And the first book contains the generatings of the three sections and the opposite [sections], and also the primal properties (*archika sumptômata*) in them, [in, that is, those sections], worked out more fully and universally (*katholou*) than by the writings of others. And the second book [contains] what turns out (*sumbainonta*) about the diameters and axes of the sections, and also the asymptotes and other things that hold general and necessary use for delimitations (*diorismous*) [of possibility]; and what it is that I call diameters, and what axes, you will know from this book. And the third book [contains] many mind-boggling (*paradoxa*) theorems useful for the construction (*suntheseis*) of solid loci and for delimitations [of possibility], most of which, and the most beautiful, are new—having come to know which, we were conscious that not constructed by Euclid was the locus on three and four lines, but only a chance little piece of it, and this not successfully, for it was not possible, without the things additionally found out by us, for this construction to be completed. And the fourth book [contains] in how many ways the sections of the cones come together with each other and also with the periphery of the circle, and other way-out things, none of which has been written up by those before us: a cone's section or circle's periphery—in how many points do they come together. The rest of the books are substantially more abundant—for there is one about maxima and minima treated more fully, and one about equal and similar sections of a cone, and one about delimitation theorems, and one about conic delimitation problems. ....

In this prefatory letter, Apollonius says that the Third Book contains theorems useful for constructing what are called "solid loci." These theorems are for the most part new, he says, and without them Euclid could not have done much on the "three-line and four-line locus," nor done well what he did manage to do. What is Apollonius talking about?

First, what is a "locus"? The Latin word for *place* is "locus"; it gives us English words like "local." It translates the Greek word *topos*, which gives us English words like "topography." In a mathematical context, it refers to a line as a place where every point has a certain property.

For example, a circular line is a locus of points with the property that they are all located equally distant from a single point (the circle's center).

Note that a circular line is *not* the same as a collection of points that are located equally distant from a single point. Such a collection of twelve points is depicted in figure 1, but there is no circular line in the picture.

If you take as many such points around a center as you please, there will still be *more* of them that you could take. If any collection of points equidistant from a single given point were a circular line, then Euclid would not need to list, as a prerequisite for his *Elements*, "to draw a circle." Why does Euclid make this a prerequisite (his third "postulate")? Is it merely because no *finite* collection of points located at a given distance from a given point will constitute a circular line? One might think that since *all* the points located at a given distance from a given point do not constitute a finite collection, perhaps the difference between lines and certain collections of points will be obliterated if only the collections are infinite.

But that is not so. A collection of points that lie on a line will not constitute the line merely by being infinite, not even if the collection is infinite everywhere (that is: not even if between any two of the points there is yet another one of them, *ad infinitum*). An infinite collection of points, even if it is infinite everywhere, is not as such continuous; a line, however, *must* be continuous.

Here is an example showing that continuity is not guaranteed by the infinity of a collection of points, even of one that is infinite everywhere. Take the following collection of points lying on a straight line AB: the points at each end of the line; the point midway between them; the points midway between every adjacent pair of the points named so far; again, the points midway between every adjacent pair of the points named so far; again, etc., etc., etc. There are points which lie on the straight line but which cannot be included in that collection no matter how long we go on. Figure 2 shows how to locate one such point. That point cannot be in the collection because the ratio of the diagonal of a square to its side is not the same as any numerical ratio.

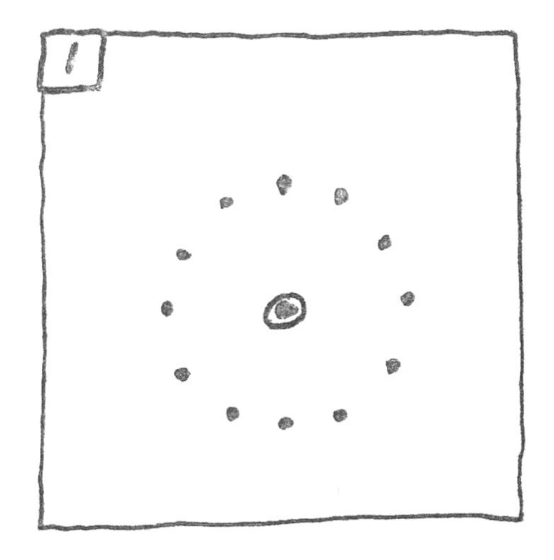

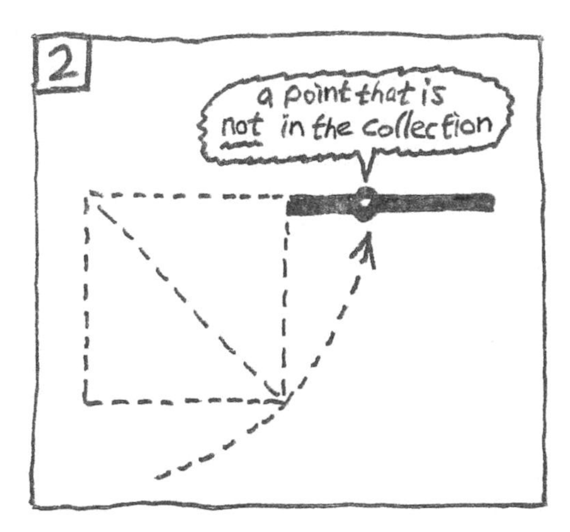

A locus problem requires us to find a line such that every one of its points has a certain property. Since a mere collection of points is not as such a line, the solution of a locus problem in classical geometry therefore involves getting a line and not just points. You must:

<table>
<tr><td>first</td><td>show</td><td>that</td><td>if points have a certain property,
then they are on a certain line;</td></tr>
<tr><td>then</td><td>show</td><td>how to get that line,
and
that</td><td>

if points are on that line,
then they have that property.</td></tr>
</table>

You must, that is to say, do the following three things:

find a way to get points that have the property,

find a way to get a line containing all such points,

determine what kind of a line it is in various cases.

Thus the solution of a locus problem requires an analysis followed by a synthesis:

(*analysis*) find out—

if there were a line on which were located all the points that have the property
then what sort of line would it be;

(*synthesis*) show—

if you do such-and-such to get a line of that sort,
then all of the line's points will have the property.

CHAPTER 5. PAPPUS ON CONICS AS LOCI OF A CERTAIN KIND

Let us now consider some things that are said by Pappus in the Fourth Book of his *Collection*. (Note again that here, and subsequently, translations from the Greek of Pappus—as well as, later on, from the Greek of Diophantus, the Latin of Viète, and the French of Descartes—will be so presented in this guidebook that the layout of the text can help to make the structure of the thought easier to see than it would be if the print stretched uniformly across the page from margin to margin.) Pappus says:

We say that there are three kinds of problems in geometry,
and that some of them are called "plane," some "solid," some "linear."
Those capable of being solved through
a straight line and a circumference of a circle
would properly (*eikotôs*) be called "*plane*"—
for those lines through which
such problems are found out (*heurisketai*)
have their genesis in a plane.
But those problems which are solved when
one or even several of the conic sections
are employed for their finding out—
such are called "*solid*,"
since for the construction it is necessary to make use of
surfaces of solid figures, I mean the conic surfaces.
A third kind of problem remains—the one called "*linear*,"
since for the construction are employed
other lines besides those mentioned,
lines having their genesis
more multifarious (*poikilôter*) and more forced,
being generated from more irregular surfaces and complicated motions;
such are ... spirals and quadratrices and cochloids and cissoids.

We shall not examine plane problems. Circular or straight lines are easy to get: allowing them to be drawn is made a prerequisite in Euclid's *Elements*. What we shall consider are solid problems. *Problems* are called "solid" when their solutions use conic sections, since these lines that they use are not obtained in a plane by appealing to Euclid's postulates, but rather are obtained by a plane's cutting of a solid figure (namely, a cone). Being obtained from a solid figure, these *lines* are called "solid" loci. We have considered them as cuts in cones or in conic surfaces, but so far we have only glanced at them as loci. We shall now consider them more closely as loci.

In the Seventh Book of his *Collection*, Pappus shows that the locus of points every one of which has this property—

its distance to a straight line that is given in position has a given ratio to its distance to a given point that is not on the straight line—

is a conic section. (Later on, the given point came to be called the "focus" and the given straight line the "directrix.") Pappus shows that

if the given ratio is a ratio of equal to equal, then the locus is a parabola; and if it's a ratio of greater to less (or of less to greater), then the locus is an ellipse (or an hyperbola). See figure 1.

For our present purpose, it will suffice to examine only the case of the parabola. For the sake of clarity, we shall consider a modification of Pappus's presentation. We begin with a construction, from which we shall show that if a point satisfies the aforementioned property (that is to say, if the aforementioned ratio is a ratio of equality), then the point will lie on a parabola. This is depicted in figure 2.

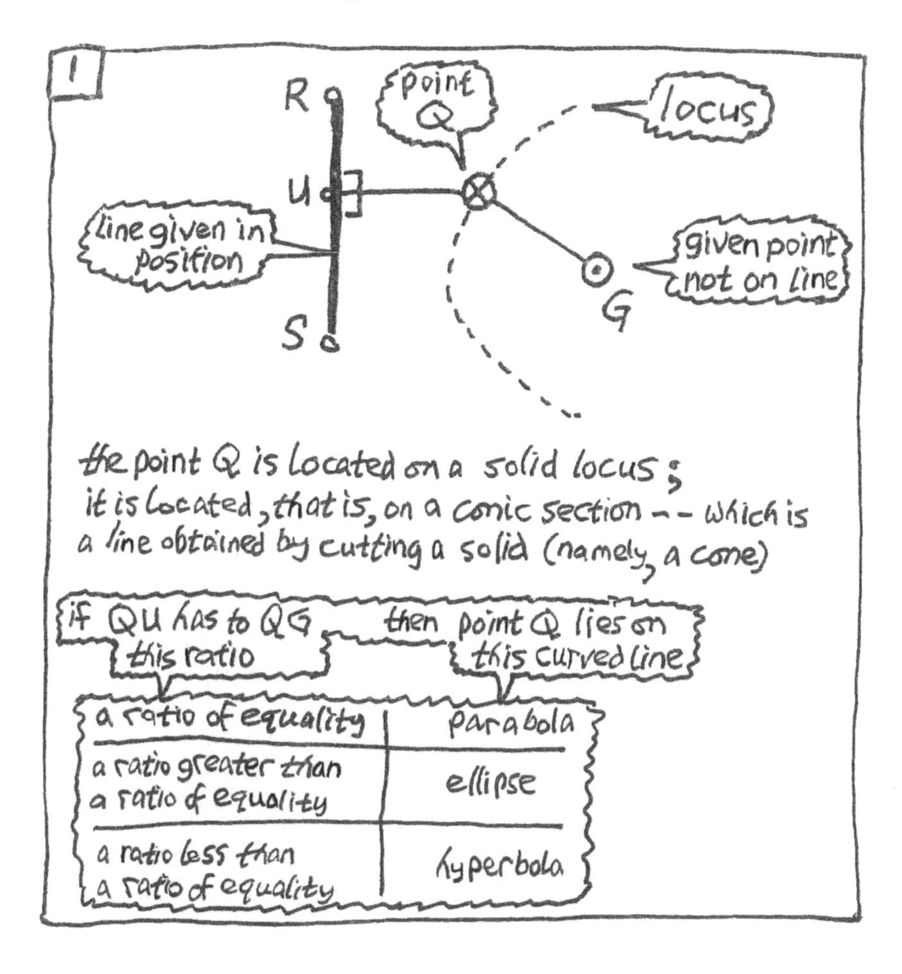

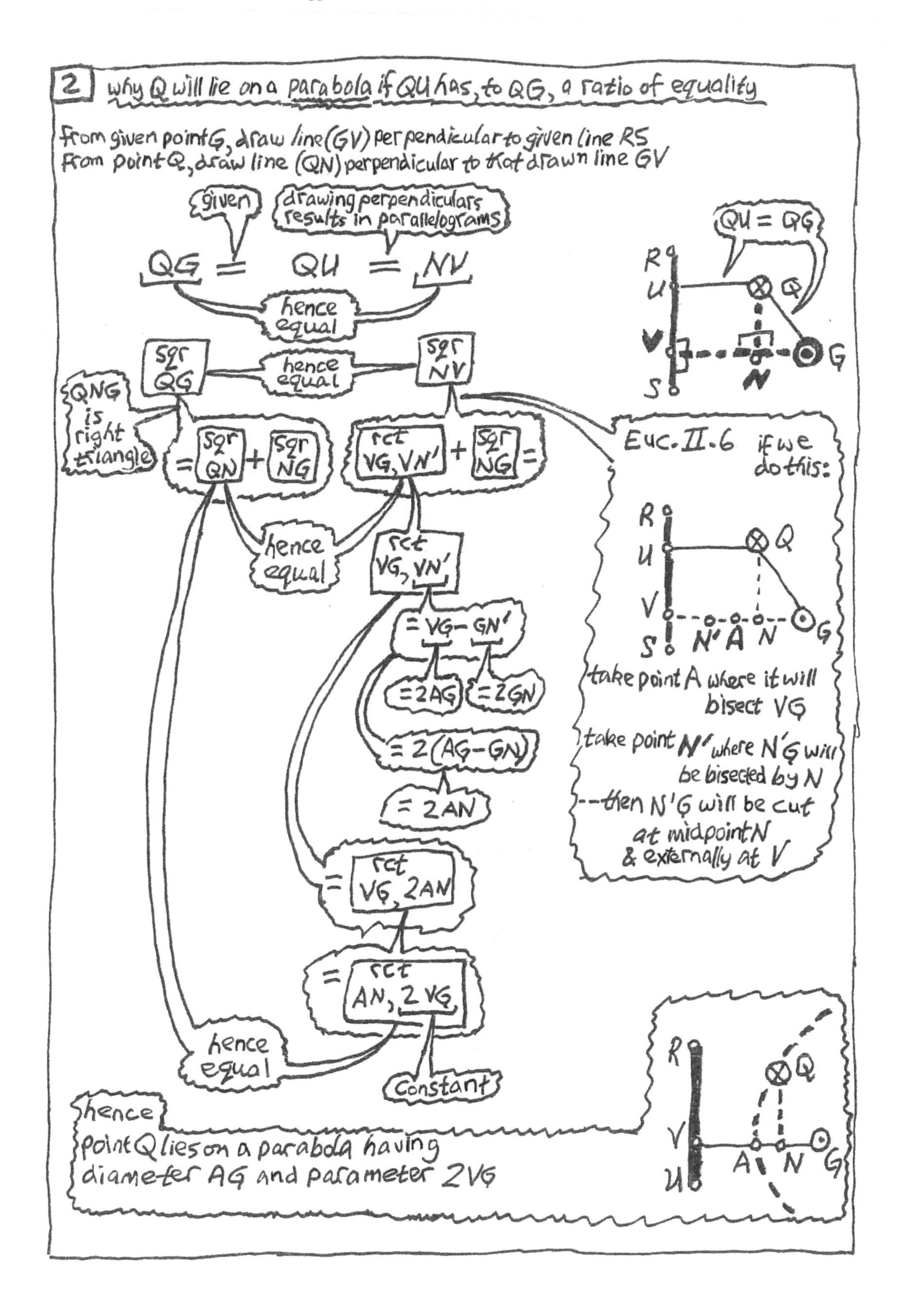
2 why Q will lie on a parabola if QU has, to QG, a ratio of equality
From given point G, draw line (GV) perpendicular to given line RS
From point Q, draw line (QN) perpendicular to that drawn line GV
given
drawing perpendiculars results in parallelograms
QG = QU = NV
hence equal
sqr QG
hence equal
sqr NV
QNG is right triangle
= sqr QN + sqr NG
rct VG, VN' + sqr NG =
hence equal
rct VG, VN'
= VG − GN'
= 2AG
= 2GN
= 2(AG − GN)
= 2AN
= rct VG, 2AN
= rct AN, 2 VG
hence equal
constant
QU = QG
Euc. II.6 if we do this:
take point A where it will bisect VG
take point N' where N'G will be bisected by N
--then N'G will be cut at midpoint N & externally at V
hence
point Q lies on a parabola having diameter AG and parameter 2VG

But what has just been depicted in figure 2 is merely the beginning for Pappus. His ending comes only after he goes on to say "the locus (the place—or, in Greek, *topos*) will be synthesized thus." This might seem to be redundant. Why does he engage in apparently unnecessary repetition? Because with a collection of points, no matter how numerous, you may have merely a stretch of dots—or at best dashes—separated by gaps; whereas if you cut a cone with a plane, then you will get a line. So, after first showing what we have just seen—namely, that

> if the property (QU's equality to QG) holds for a point Q,
> then point Q is located on *some parabola*—

Pappus then goes on; he first *constructs a certain parabola* in a cone, and then he *demonstrates* that

> if a point Q is located on *this* parabola,
> then the property (QU's equality to QG) holds for point Q.

Again, for the sake of clarity, we shall consider a modification of Pappus's presentation. This synthesis for the parabola is depicted in figures 3 and 4. Likewise we could show, as was said above, that when the given ratio is greater (or less) than a ratio of equality, then the locus is an ellipse (or an hyperbola).

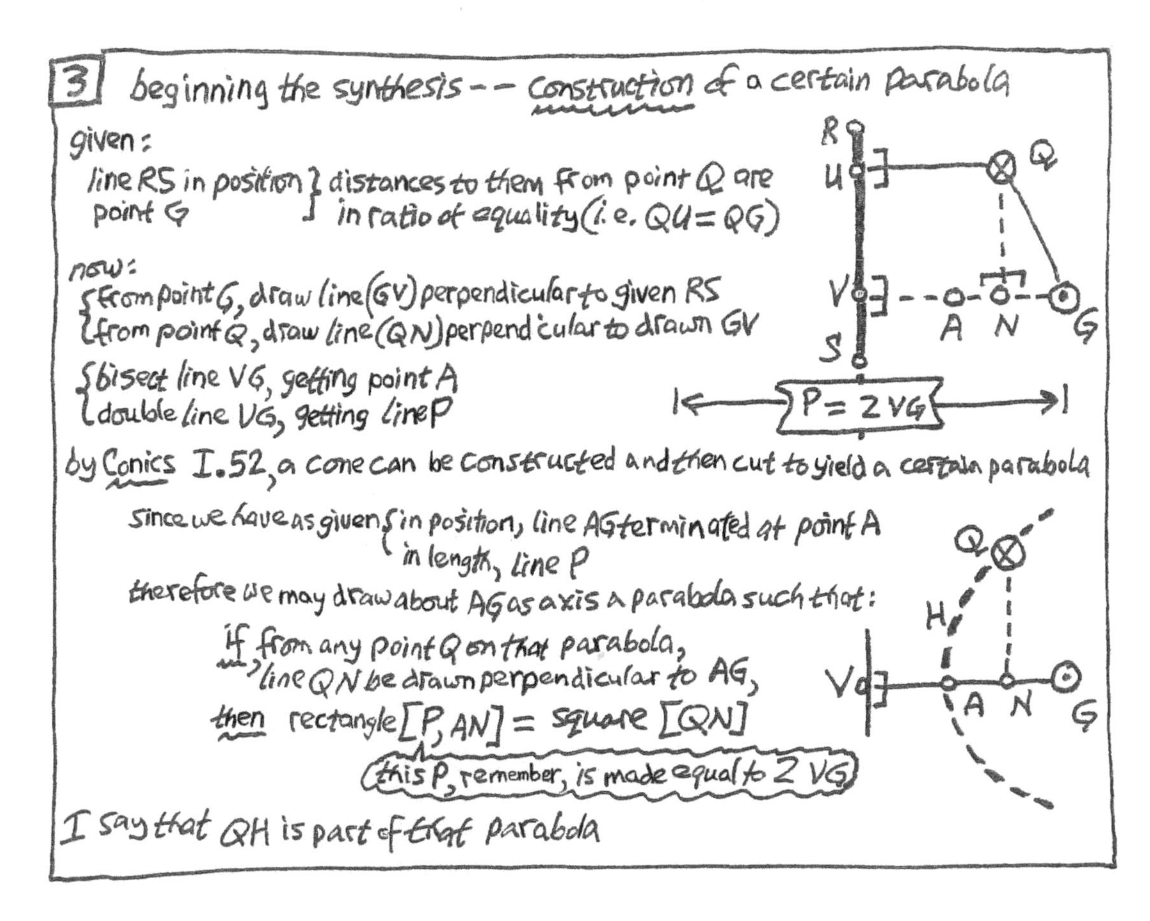

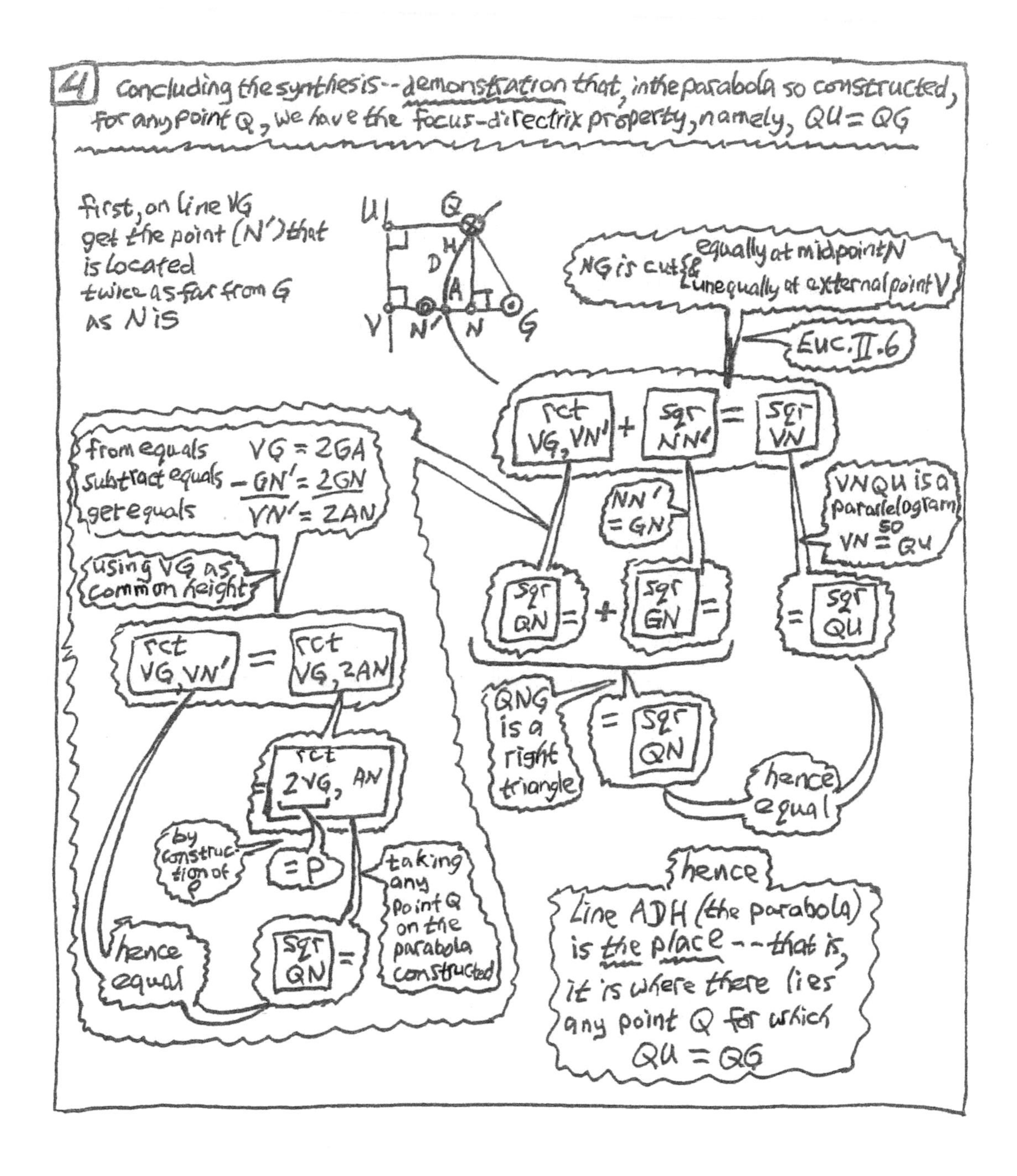
4
Concluding the synthesis--demonstration that, in the parabola so constructed, for any point Q, we have the focus-directrix property, namely, QU = QG
first, on line VG get the point (N′) that is located twice as far from G as N is
U
Q
H
D
A
V
N′
N
G
NG is cut equally at midpoint N & unequally at external point V
Euc. II.6
rct VG, VN′
+
sqr NN′
=
sqr VN
from equals VG = 2GA
subtract equals − GN′ = 2GN
get equals VN′ = 2AN
NN′ = GN
VNQU is a parallelogram so VN = QU
using VG as common height
sqr QN
+
sqr GN
=
sqr QU
rct VG, VN′
=
rct VG, 2AN
QNG is a right triangle
sqr QN
rct 2VG, AN
by construction of P
= P
taking any point Q on the parabola constructed
hence equal
sqr QN
hence equal
hence
Line ADH (the parabola) is the place--that is, it is where there lies any point Q for which QU = QG

CHAPTER 6. THE VERY END OF BOOK III OF THE CONICS, AND CONICS AS LOCI OF ANOTHER KIND

the property at the very end of the Third Book of the Conics

Pappus also discusses loci in relation to other properties. In particular, he speaks of the so-called locus with respect to three and four lines, and the locus with respect to five, or six, or more than six lines.

The problems presented by those loci are the occasion for Descartes' innovative *Geometry*. The power of Descartes' new way is displayed by showing the reader how to solve with relative ease such problems as had confounded the best efforts of mathematicians for a millennium and a half. Before we turn to Descartes' solution, however, we must see the difficulty with which he contended, and also what enabled him to work a transformation that provided powerful new means for solving problems generally.

The point of departure for the locus with respect to three and four lines comes at the end of the Third Book of Apollonius's *Conics*. It follows the propositions previously treated in this guidebook (in chapter 3 of this part) where Apollonius presents properties of the distances from any point on an hyperbola (and opposite sections), or on an ellipse, to the points that we now call the foci. These are the 51st and 52nd propositions of the Third Book. To save time now we shall ignore the last two propositions of that book (II.55–56); they show, for the opposite sections, what the third proposition from the end (III.54) shows for the rest of the conic sections. Let us turn to this 54th proposition. First we must establish its prerequisites. The demonstration of III.54 requires what is demonstrated in III.16, which in turn requires what is demonstrated in III.1 and III.2 (fig.1).

It will be convenient for our purpose, in treating these propositions from Book Three, to give our attention for the most part to the case of the hyperbola. (But remember, as was said before, that complete diagrams for those propositions of Book Three which are selected for inclusion here in this chapter can be found among those in the Appendix at the end of this second volume of the guidebook.)

The first two propositions of the Third Book treat a situation where there is, through each of two points on a conic section, a diameter and a tangent (fig. 2).

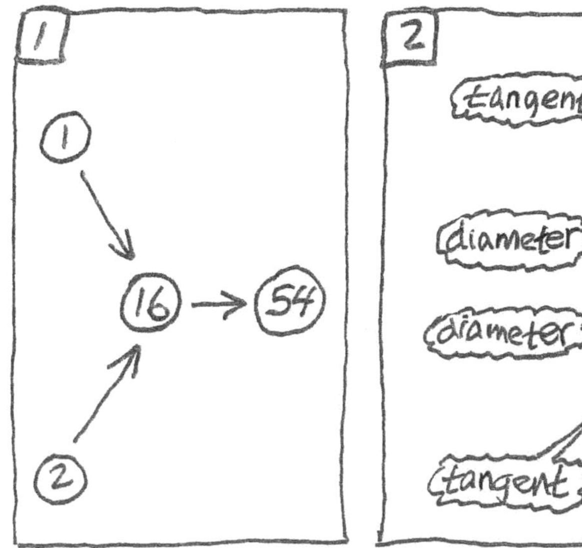

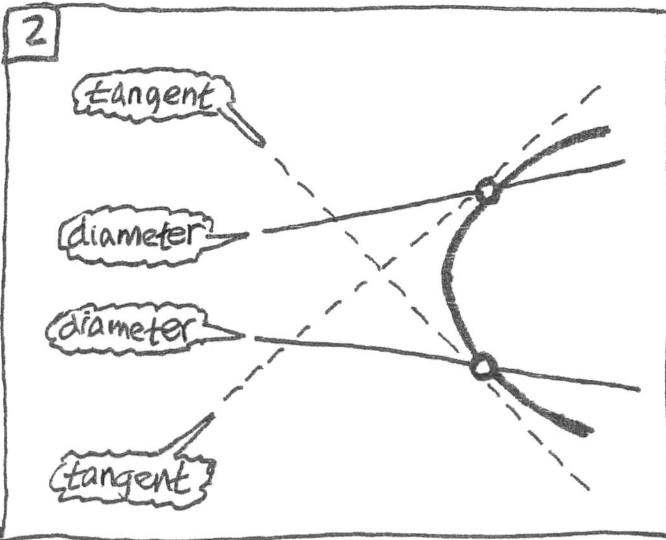

The 1st proposition shows the equality of the triangles (striped in figure 3) whose sides lie along the tangents and whose bases lie along the diameters.

The demonstration is depicted in figure 4; and after that, Apollonius's proposition is presented in the box on the following page.

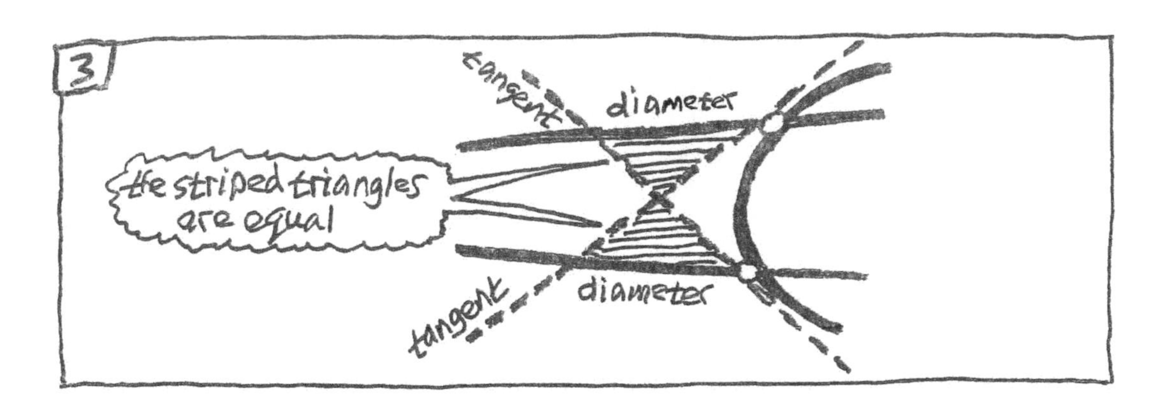

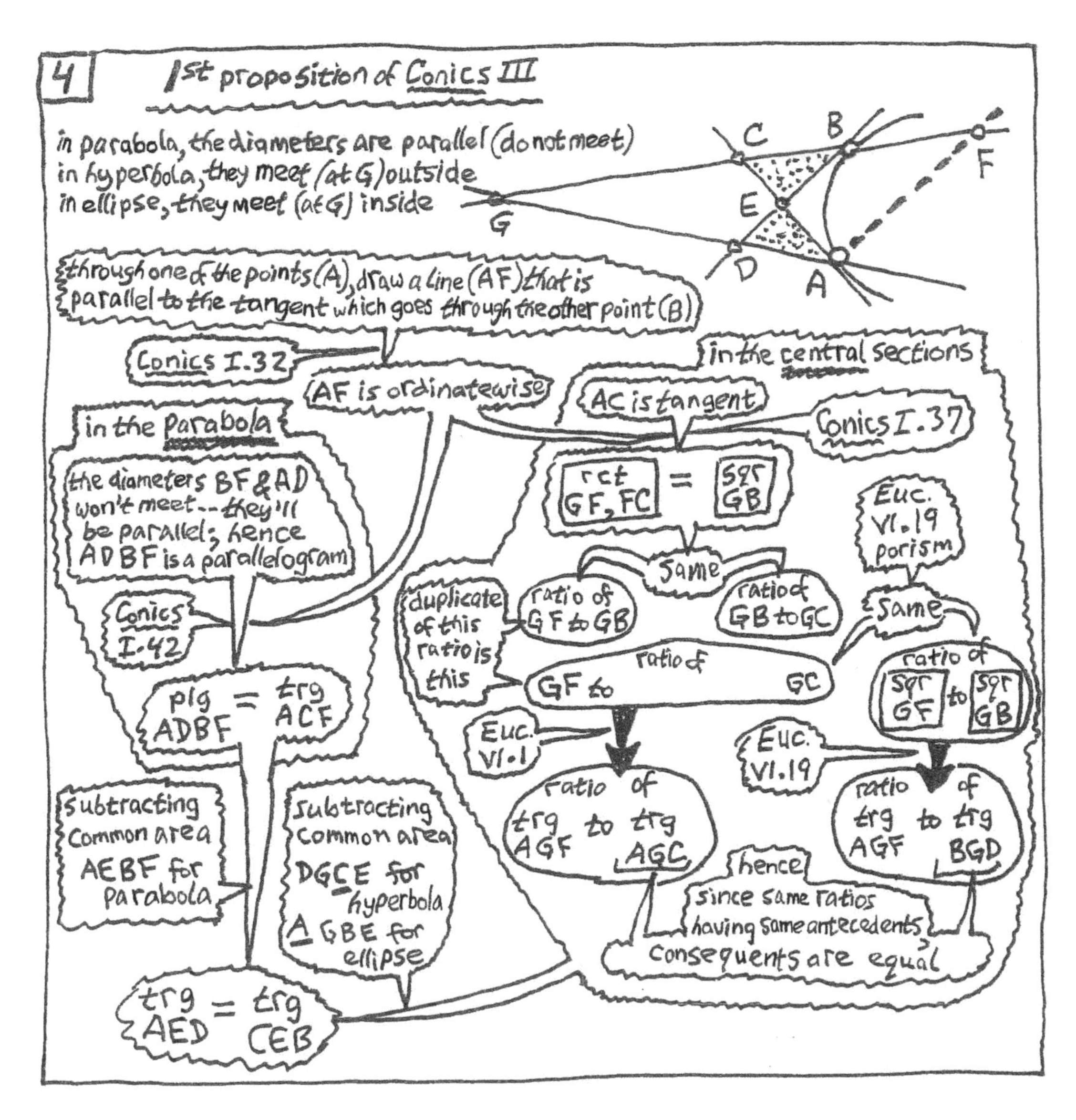

(III) #1 If, with respect to a cone's section or a circle's periphery, straight [lines] that touch [it] fall together, and there be drawn through the touch[points] diameters that fall together with the touching[lines], equal will be the triangles generated vertically.

—Let there be as a cone's section or circle's periphery the [line] AB; and also, let there touch the [line] AB both the [line] AC and the [line] BD, which fall together at the [point] E; and also, through the [points] A, B let there be drawn as diameters of the section the [lines] CB, DA that fall together with the touching[lines] at the [points] C, D. I say that equal is the triangle ADE to the [triangle] EBC.

—For let there be drawn from the [point] A parallel to the [line] BD the [line] AF; therefore ordinatewise has it been drawn down. Then on the parabola, on the one hand, equal will be the parallelogram ADBF to the triangle ACF; and also, with the common [area] being taken away, the remainder triangle ADE is equal to the [remainder] triangle CBE. And on the remaining [curves], on the other hand, let there fall together the diameters at the center G. Since, then, there has been drawn down the [line] AF, and the [line] AC touches, hence the [rectangle] by FGC is equal to the [square] from BG. Therefore as is the [line] FG to GB so is the [line] BG to GC; therefore also as is the [line] FG to GC so is the [square] from FG to the [square] from GB. But as is the [square] from FG to the [square] from GB so is the triangle AGF to the triangle DGB, and as is the [line] FG to GC so is the triangle AGF to the triangle AGC; and therefore also as is the [triangle] AGF to the [triangle] AGC so is the [triangle] AGF to the [triangle] DGB. Equal therefore is the [triangle] AGC to the [triangle] DGB. Let, as a common [area], there be taken away the [area] AGBE; therefore the remainder triangle AED is equal to the [remainder triangle] CEB.

In that same situation (where there are two points on the section, each of them having through it a diameter and a tangent), the 2nd proposition draws, through a random third point on the section, a parallel to each of the two tangents (fig. 5).

This 2nd proposition uses the 1st proposition to show the equality of the heavy-lined quadrilateral and the dotted-line triangle (fig. 6).

The demonstration is depicted in figure 7; Apollonius's proposition is presented in the box on the following page.

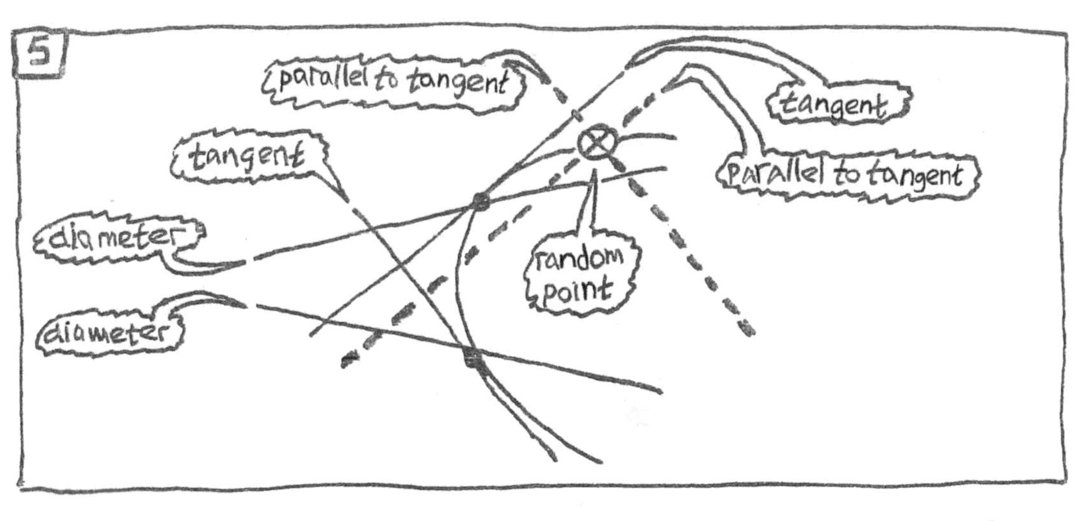

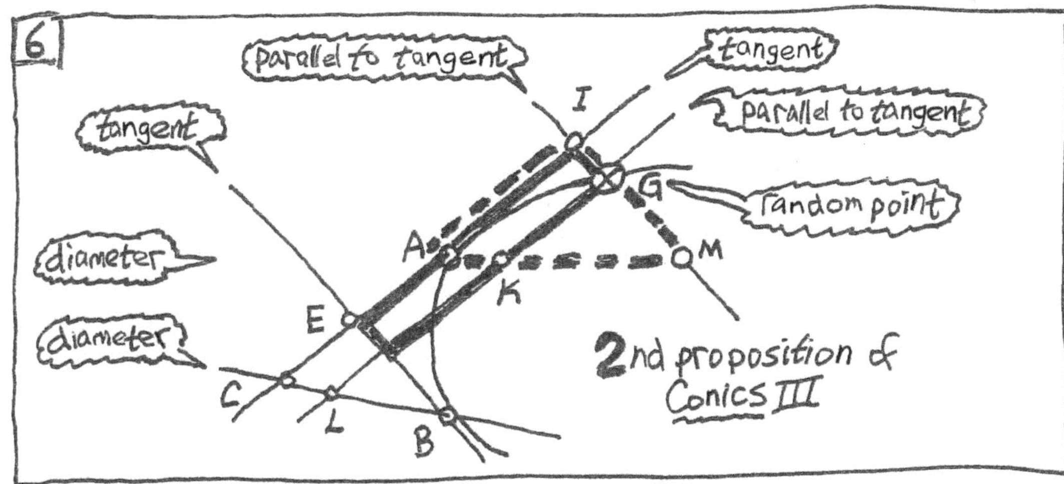

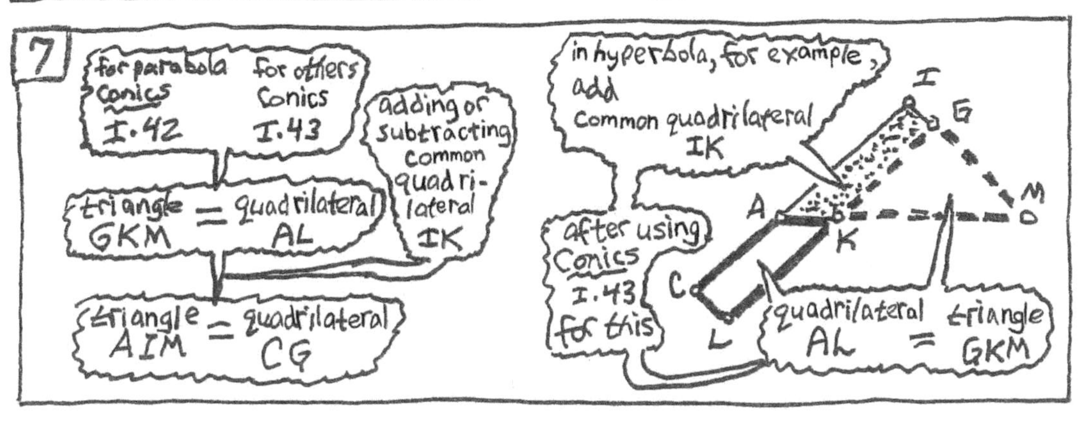

(III) **#2** The same things being supposed, if, on the section [of a cone] or the circle's periphery, there be taken some point; and also, through it, [lines] parallel to the touching[lines] be drawn as far as the diameters, then the quadrilateral that is generated on both one of the touching[lines] and one of the diameters will be equal to the triangle that is generated on both the same touching[line] and the other one of the diameters.
—For let there be as a cone's section or circle's periphery the [line] AB; and also, as touching[lines] the [lines] AEC, BED and as diameters the [lines] AD, BC; and also, let there be taken some point on the section, say the [point] G; and also, let there be drawn parallel to the touching[lines] the lines GKL, GMF. I say that equal is the triangle AIM to the quadrilateral CLGI.
—For, since the triangle GKM has been shown equal to the quadrilateral AL, let the common quadrilateral IK be put together with it or taken away from it, and there will be generated the triangle AIM equal to the quadrilateral CG.

The 16th proposition, which depends on the 1st and 2nd propositions, treats a situation where there is, through each one of two points on a conic section, a tangent, and there is also, through a random third point on the section, a line which runs parallel to one of the tangents and cuts the section and the other tangent (fig. 8).

The lines that go

from point C	(where one tangent intersects the other)
to points A and B	(both of them on the section)

stay the same, while the lines that go

from point E	(where one tangent intersects that line through random point D which is drawn parallel to the other tangent)
to points A and D and F	(all of them on the section)

are lines that do not stay the same size, but are of different sizes for different random points. Nonetheless—and this is what is shown by the 16th proposition—there is something that stays the same about the sizes of those lines (EA, ED, and EF) whose sizes do not stay the same:

> the ratio of the square on EA to the rectangle contained by ED and EF
> stays the same as
> the ratio of the square on CA to the square on CB;
>
> and this latter ratio, since lines CA and CB stay the same,
> stays the same for every different random point D.

Note that the lines are set out in the reverse order in the proposition; what the proposition says is this:

> as square [CB] is to square [CA], so rectangle [EF, ED] is to square [EA].

The demonstration is depicted in figure 9; Apollonius's proposition is presented in the following box.

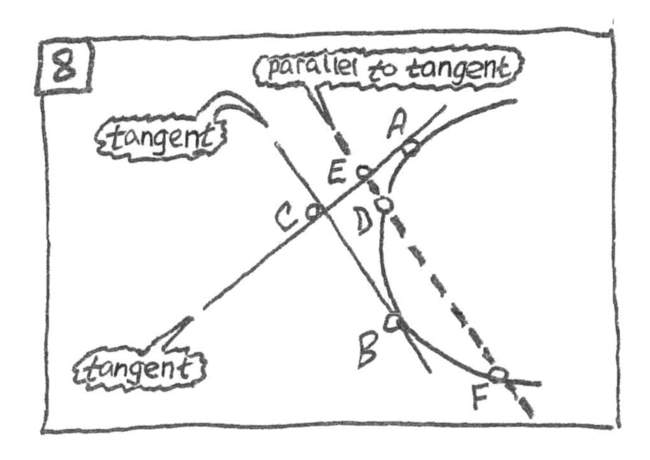

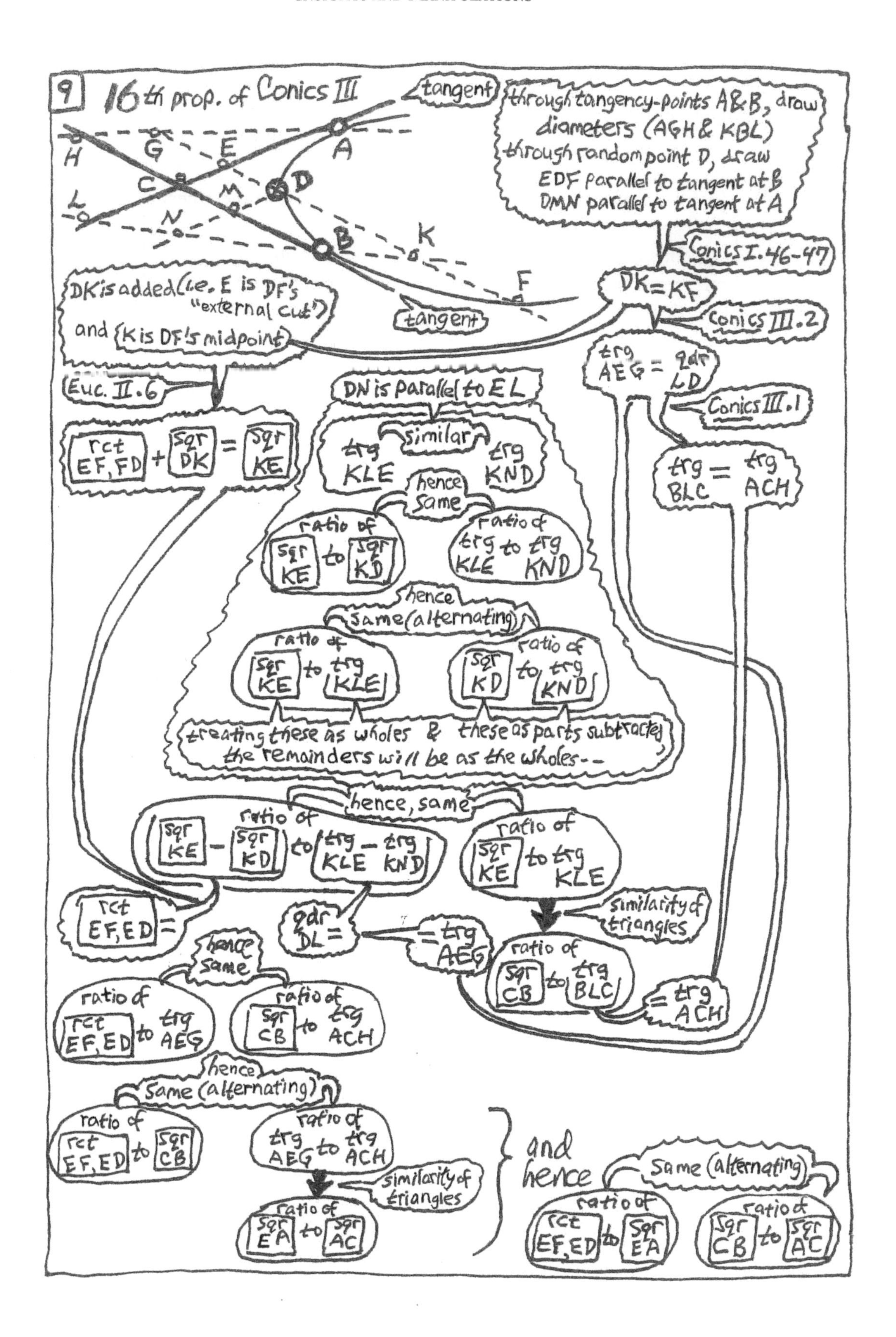
9
16th prop. of Conics III
tangent
through tangency-points A&B, draw diameters (AGH & KBL)
through random point D, draw EDF parallel to tangent at B
DMN parallel to tangent at A
H
G
E
A
C
M
D
L
N
B
K
F
tangent
Conics I.46-47
DK=KF
DK is added (i.e. E is DF's "external cut")
and K is DF's midpoint
Conics III.2
trg AEG = qdr LD
Conics III.1
Euc. II.6
rct EF, FD + sqr DK = sqr KE
DN is parallel to EL
trg KLE similar trg KND
hence same
ratio of sqr KE to sqr KD
ratio of trg KLE to trg KND
trg BLC = trg ACH
hence same (alternating)
ratio of sqr KE to trg KLE
ratio of sqr KD to trg KND
treating these as wholes & these as parts subtracted
the remainders will be as the wholes--
hence, same
ratio of sqr KE – sqr KD to trg KLE – trg KND
ratio of sqr KE to trg KLE
rct EF, ED =
qdr DL =
= trg AEG
similarity of triangles
ratio of sqr CB to trg BLC
= trg ACH
hence same
ratio of rct EF, ED to trg AEG
ratio of sqr CB to trg ACH
hence same (alternating)
ratio of rct EF, ED to sqr CB
ratio of trg AEG to trg ACH
similarity of triangles
ratio of sqr EA to sqr AC
and hence
same (alternating)
ratio of rct EF, ED to sqr EA
ratio of sqr CB to sqr AC

(III) #16 If, with respect to a cone's section or circle's periphery, two straight [lines] that touch [it] fall together, and from some point of those on the section there be drawn a straight [line] that is parallel to some one of the touching[lines] and that cuts the section and also the other of the touching[lines]—then as the squares from the touching[lines] are to each other so the area (*chôrion*) contained by the [lines] between the section and the touching[line] will be to the square from the [line] taken off on the side toward the touch[point].
—Let there be as a cone's section or circle's periphery the [line] AB; and also, [let there be] as [lines] that touch it the [lines] AC, CB that fall together at the [point] C; and also, let there be taken some point on the section AB, say the [point] D; and also, through it let there be drawn parallel to the [line] CB the [line] EDF. I say that as is the [square] from BC to the [square] from AC so is the [rectangle] by FED to the [square] from EA.
—For let there be drawn through the [points] A, B as diameters the [line] AGH and the [line] KBL, and through the [point] D as a parallel to the [line] AL the [line] DMN; it is evident from this by itself that equal is the [line] DK to the [line] KF, and [equal is] also the triangle AEG to the quadrilateral LD, and also the triangle BLC to the triangle ACH. Since, then, the [line] FK is, to the [line] KD, equal, and also there is put together [with the line FKD] the [line] DE, hence the [rectangle] by FED is, along with (*meta*) the [square] from DK, equal to the [square] from KE. And—since similar is the triangle ELK to the [triangle] DNK—as is the [square] from EK to the [square] from KD so is the triangle EKL to the [triangle] DNK. And also, alternately (*anallax*); and also, as is the whole [square] from EK to the whole triangle ELK so is the taken-away [square] from DK to the taken-away triangle DNK. Therefore as is the remainder [rectangle] by FED to the remainder [quadrilateral] DL so is the [square] from EK to the [triangle] ELK. But as is the [square] from EK to the [triangle] ELK so is the [square] from CB to the [triangle] LCB; therefore also as is the [rectangle] by FED to the quadrilateral LD so is the [square] from CB to the triangle LCB. But equal is (on the one hand) the [quadrilateral] DL to the triangle AEG and (on the other hand) the [triangle] LCB to the [triangle] AHC; and therefore also as is the [rectangle] by FED to the triangle AEG so is the [square] from CB to the [triangle] AHC. Alternately, as is the [rectangle] by FED to the [square] from CB so is the triangle AEG to the [triangle] AHC. But as is the [triangle] AGE to the [triangle] AHC so is the [square] from EA to the [square] from AC; and therefore also as is the [rectangle] by FED to the [square] from CB so is the [square] from EA to the [square] from AC. And alternately.

Now we are prepared for the 54th proposition—the point of departure for the locus with respect to three and four lines. The 54th proposition will enable us to infer that a conic section or a circle is a line that has a certain property needed for dealing with the three-line locus. Certain straight lines associated with different points on a conic section have the same relations at all those different points.

In order to state the property, we need first to get certain lines for any conic section or circle, as follows (fig. 10):

> take any two tangents (their tangency-points, we'll call A and C);
>
> connect those tangency-points to each other by a straight line (AC),
> and take its midpoint (E);
>
> connect that midpoint E to the tangents' intersection-point (D);
>
> through each tangency-point,
> draw a parallel to the tangent at the other one;
>
> now take a random point (H),
> and to it connect each tangency-point (A and C)—
> and keep going until the extension of each connecting line cuts off
> that line through the other tangency-point which is
> parallel to the tangent through this tangency-point;
> and thus you will get, as the cut-off points, F and G.

Now we can state the property that is needed for the three-line locus:

> from different random points H,
> we'll get different sizes for both of the lines cut off (AF and CG);
>
> and yet, the rectangles contained by
> all those different pairs of lines AF and CG,
> which we get from different random points H,
> all have the very same size.

How do we know that rectangle [AF, CG] is constant? Because III.54 tells us that its ratio to a constant square is a constant ratio. Let us see why.

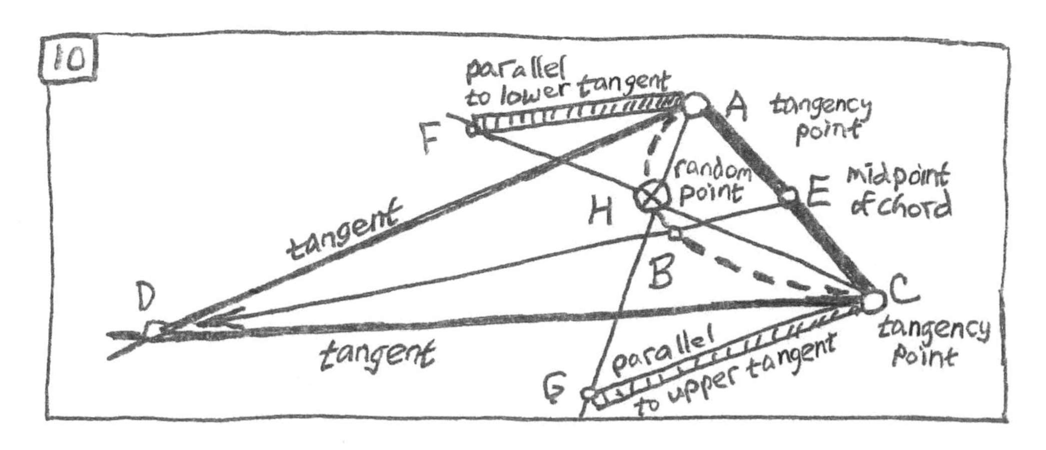

The rectangle in which we are interested is the rectangle contained by the lines AF and CG, each of which lines is the line which,

> running through one of the tangency-points,
> is parallel to the tangent at the other one
> and
> is cut off by the line that runs, from this other tangency-point,
> through the random point.

That rectangle [AF, CG] is the antecedent in a ratio. The consequent in that ratio is the square whose side is the line between the tangency-points. Since this line (chord AC) is constant, so also is the square upon it; therefore the rectangle with varying sides (the rectangle in which we are interested) will be constant in size if its ratio to the constant square is the same as some ratio that is constant. What ratio is this? It is a ratio that is constant because it is compounded of two constant ratios of rectangles—and these two ratios of rectangles are constant because all the sides of the rectangles are constant.

The ratio that we get from compounding is a *constant* ratio (fig. 11). We demonstrate that it is *the same as* the ratio of

> the rectangle [AF, CG] in which we are interested
> to the constant square on AC.

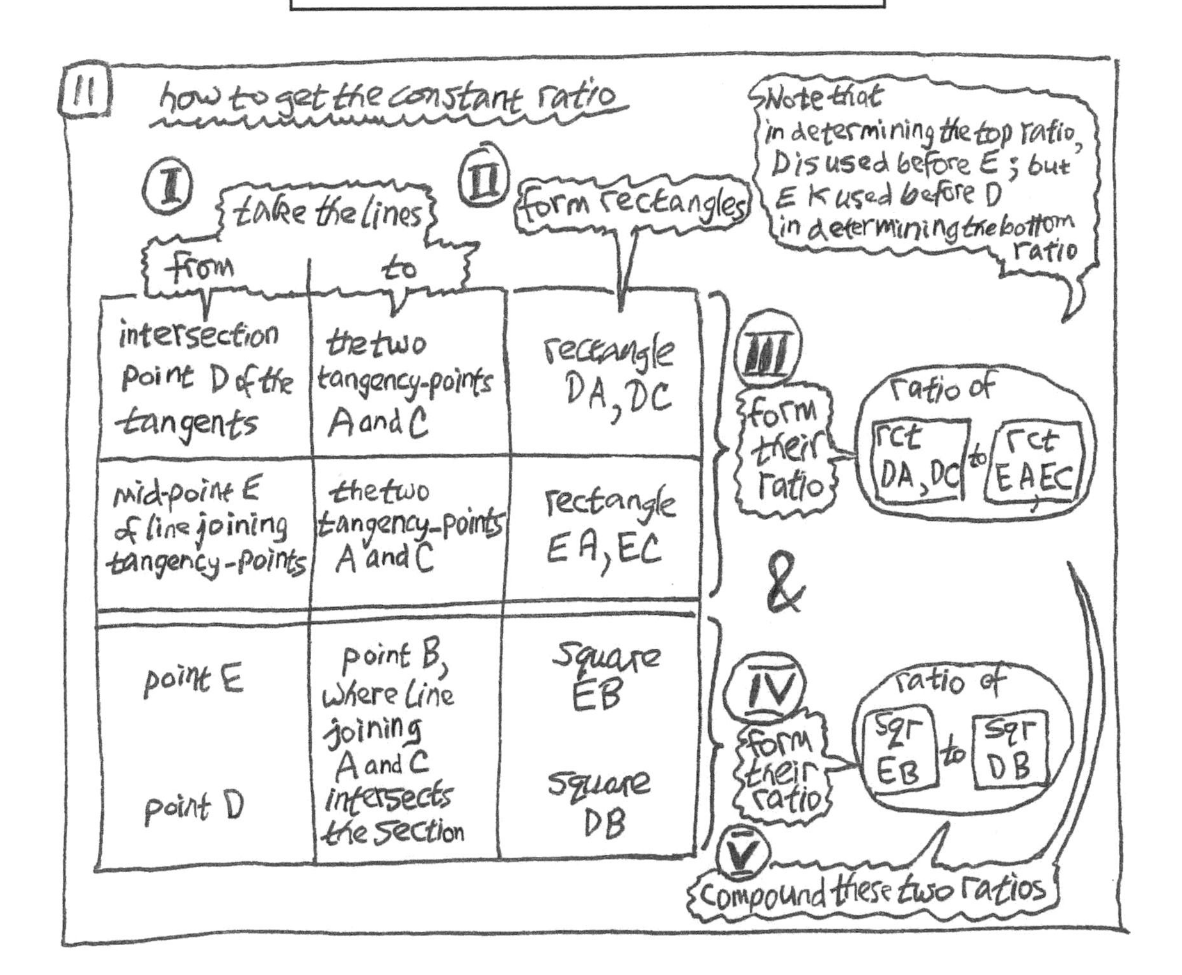

For the demonstration, two more lines need to be drawn in: two lines that are parallel to AC (fig. 12). One of the parallels is drawn through the random point H; the other is drawn through the point B where the conic section is intersected by the line connecting the chord's midpoint E and the tangents' intersection D. These two parallel lines will also be needed later when we seek to use Apollonius's proposition III.54 to state the three-line locus property.

Apollonius's demonstration of III.54 turns upon the following sameness of two ratios:

rectangle	to	rectangle	is the same as	rectangle	to	rectangle
[NC, MA]		[MB, BN]		[LC, KA]		[LH, HK].

The first part of the demonstration establishes that these two ratios are the same as each other. The rest of the demonstration gets other ratios that are the same as these—by decompounding, then using similar triangles, then recompounding. The demonstration is depicted in figure 13; and after that, proposition III.54 is presented in the boxes on the following pages.

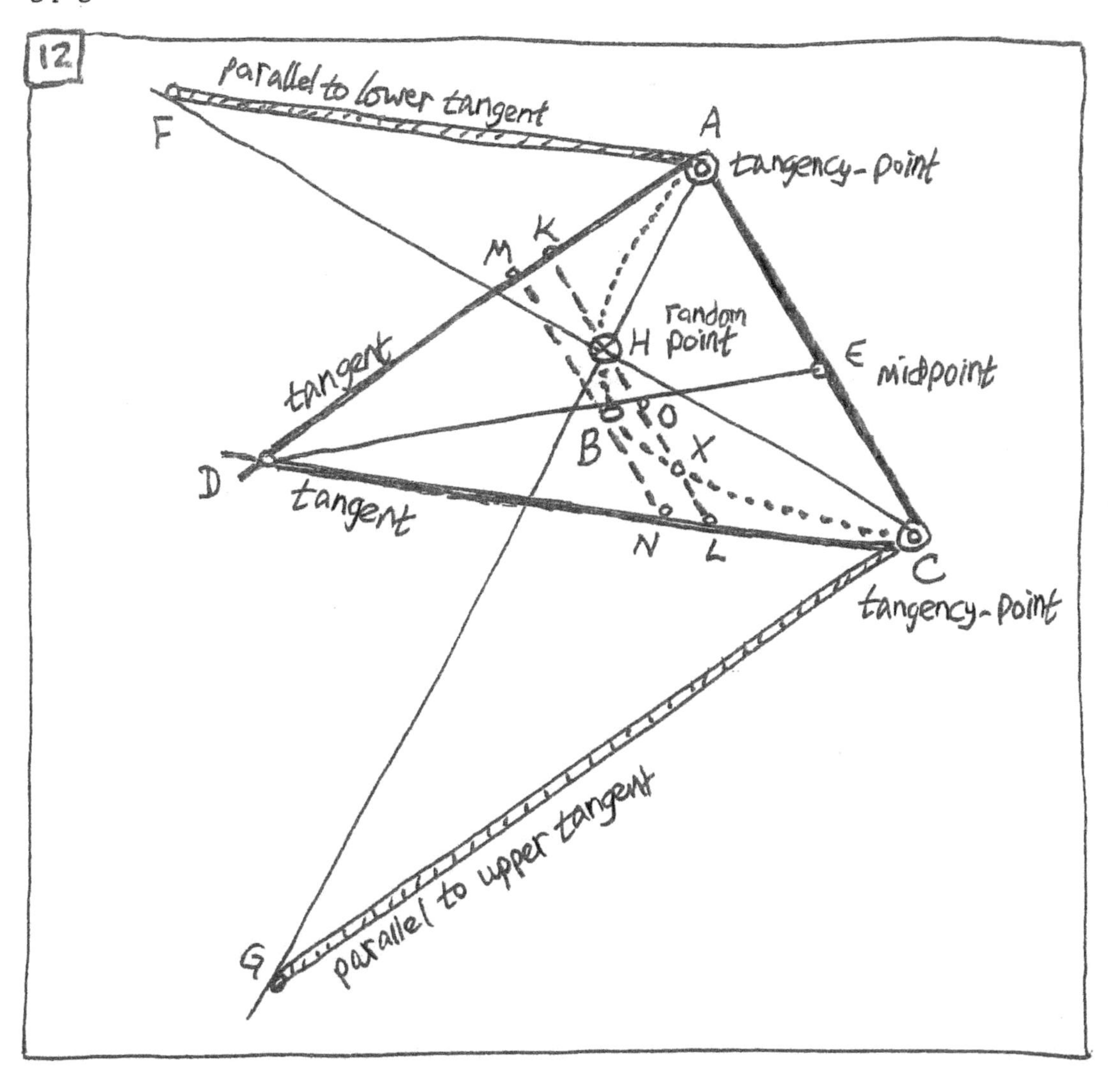

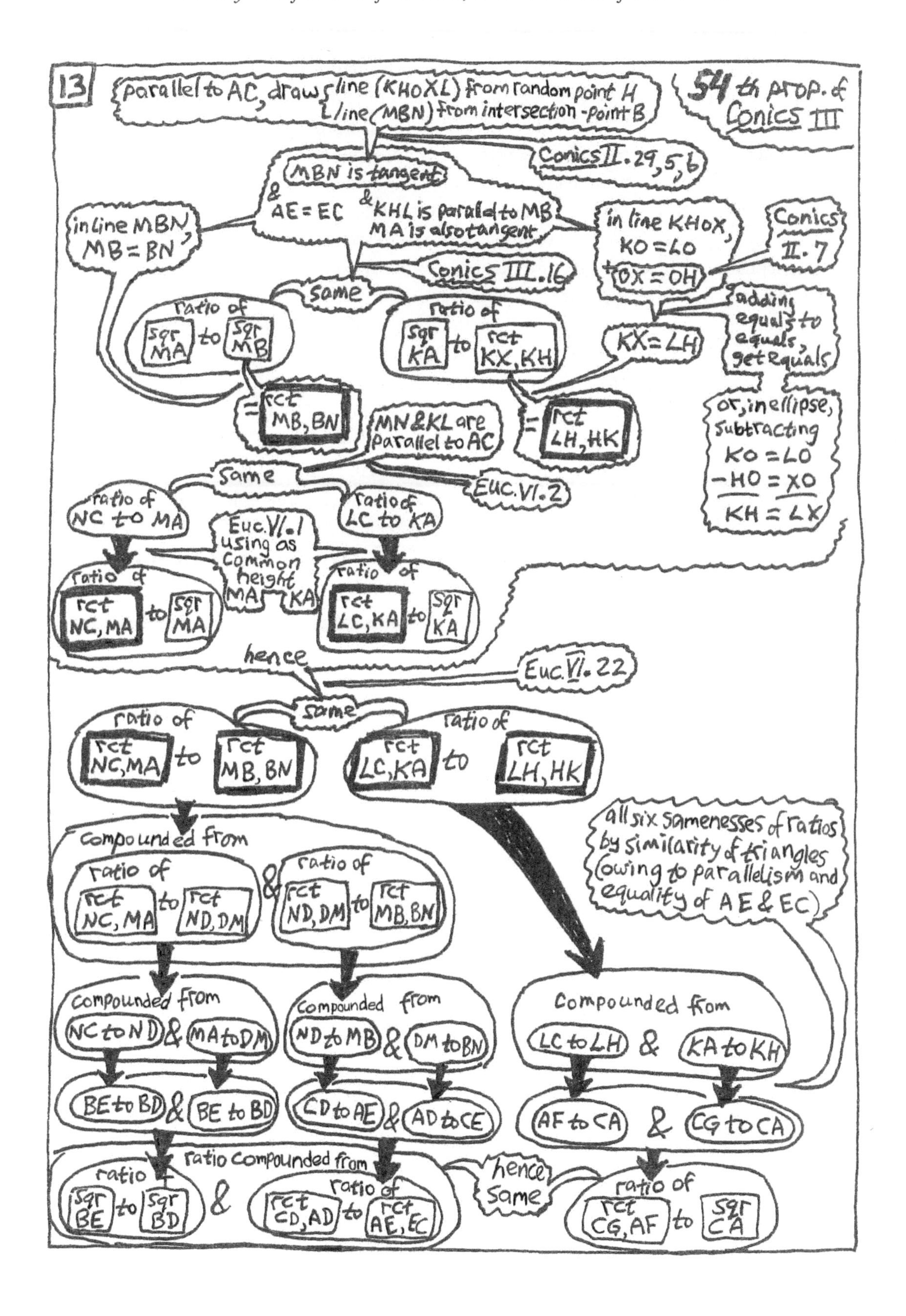
13
parallel to AC, draw line (KHOXL) from random point H
line (MBN) from intersection-point B
54th prop. of Conics III
Conics II.29,5,6
MBN is tangent
& AE = EC
& KHL is parallel to MB
MA is also tangent
in line MBN, MB = BN
in line KHOX, KO = LO
+ OX = OH
Conics II.7
Conics III.16
same
ratio of sqr MA to sqr MB
ratio of sqr KA to rct KX, KH
KX = LH
adding equals to equals, get equals
= rct MB, BN
= rct LH, HK
MN & KL are parallel to AC
or, in ellipse, subtracting
KO = LO
− HO = XO
KH = LX
same
Euc. VI.2
ratio of NC to MA
ratio of LC to KA
Euc. VI.1 using as common height MA KA
ratio of rct NC, MA to sqr MA
ratio of rct LC, KA to sqr KA
hence
Euc. VI.22
same
ratio of rct NC, MA to rct MB, BN
ratio of rct LC, KA to rct LH, HK
compounded from
ratio of rct NC, MA to rct ND, DM
& ratio of rct ND, DM to rct MB, BN
all six samenesses of ratios by similarity of triangles (owing to parallelism and equality of AE & EC)
compounded from NC to ND & MA to DM
compounded from ND to MB & DM to BN
compounded from LC to LH & KA to KH
BE to BD & BE to BD
CD to AE & AD to CE
AF to CA & CG to CA
ratio compounded from
ratio of sqr BE to sqr BD
& ratio of rct CD, AD to rct AE, EC
hence same
ratio of rct CG, AF to sqr CA

(III) **#54** If, with respect to a cone's section or circle's periphery, two straight [lines] that touch [it] fall together, and through the touch[points] there be drawn parallels to the touching[lines]; and also, from the touch[points] across to the same point of the line [of section] straight [lines] be drawn, cutting the parallels—then the rectangle contained by the [lines] cut off has, to the square from the [line] joining the touch[points], the ratio compounded from [the following ratios]: [the one ratio is] that of [one square]—[namely] the [one to which] the inside segment of the [line] that joins the [point] of falling together of the touching[lines] and the midpoint (*dichotomian*) of the [line] that joins the touch[points] has the power [to give rise]—to [another square, namely] the [one to which] the remaining [segment] has the power [to give rise]; and [the other ratio is] that which the [rectangle] contained by the touching[lines] has to the fourth part of the square from the [line] joining the touch[points].
—Let there be as a cone's section or circle's periphery the [line] ABC, and also let there be as touching[lines] the [lines] AD, CD; and also, let there be joined the [line] AC; and also, let it be cut in two [halves] at the [point] E; and also, let there be joined the [line] DBE; and also, let there be drawn from the [point] A parallel to the [line] CD the line AF, and from the [point] C parallel to the [line] AD the [line] CG; and also, let there be taken some point on the line [of section], say the [point] H, and let the joined [lines] AH, CH be extended to the [points] G, F. I say that the [rectangle] by AF, CG has, to the [square] from AC, the ratio compounded from that which the [square] from EB has to the [square] from BD and also that which the [rectangle] by ADC has to the fourth of the [square] from AC, that is, [has] to the rectangle contained by AEC.

—For let there be drawn, from the [point] H parallel to the [line] AC the [line] KHOXL, and from the [point] B [parallel to that same line AC] the [line] MBN; then it is evident that the [line] MN touches [the curve]. Since, then, equal is the [line] AE to the [line] EC, equal is also the [line] MB to the [line] BN, and also the [line] KO to the [line] OL, and also the [line] HO to the [line] OX, and also the [line] KH to the [line] XL. Since, then, the [lines] MB, MA touch [the curve], and parallel to the [line] MB has been drawn the [line] KHL, hence as is the [square] from AM to the [square] from MB, that is, to the [rectangle] by MBN, so is the [square] from AK to the [rectangle] by XKH, that is, to the [rectangle] by LHK. And as is the [rectangle] by NC, MA to the [square] from MA so is the [rectangle] by LC, KA to the [square] from KA; therefore, through what is equal (*di isou*) [traditionally rendered by the Latin *ex aequali*], as is the [rectangle] by NC, MA to the [rectangle] by NBM so is the [rectangle] by LC, KA to the [rectangle] by LHK. But the [rectangle] by LC, KA has to the [rectangle] by LHK the ratio compounded from that of the [line] CL to LH, that is, of the [line] FA to AC, and also that of the [line] AK to KH, that is, of the [line] GC to AC, which is the same as that which the [rectangle] by GC, FA has to the [square] from CA; therefore as is the [rectangle] by NC, MA to the [rectangle] by NBM so is the [rectangle] by GC, FA to the [square] from CA. But the [rectangle] by CN, MA to the [rectangle] by NBM has—with the [rectangle] by NDM being taken as a mean—the ratio compounded from that which the [rectangle] by CN, AM has to the [rectangle] by NDM and also that which the [rectangle] by NDM has to the [rectangle] by NBM; therefore the [rectangle] by GC, FA has to the [square] from CA the ratio compounded from that of the [rectangle] by CN, AM to the [rectangle] by NDM and also that of the [rectangle] by NDM to the [rectangle] by NBM. But (on the one hand) as is the [rectangle] by NC, AM to the [rectangle] by NDM so is the [square] from EB to the [square] from BD, and (on the other hand) as is the [rectangle] by NDM to the [rectangle] by NBM so is the [rectangle] by CDA to the [rectangle] by CEA; therefore the [rectangle] by GC, FA has to the [square] from AC the ratio compounded from that of the [square] from BE to the [square] from BD and also that of the [rectangle] by CDA to the [rectangle] by CEA.

What Apollonius shows in the 54th proposition for all but the opposite sections, he shows for the opposite sections in his next two propositions (which we shall omit). The 55th proposition treats the case of the opposite sections when one of the two tangents is located in each of the curves in the pair; the 56th proposition treats the case of the opposite sections when both of the tangents are located in one curve of the pair and the random point is located in the other curve.) With these propositions, Apollonius concludes the Third Book of his *Conics*.

the conics' locus property with respect to certain lines

Now that with the demonstration of III.54 we have obtained what we need from Apollonius, we can show the conic section's locus property with respect to certain lines.

The property shown by Apollonius for all the single-curve sections in III.54 (and for opposite sections in III.55–56) implies that if a line is a conic section, then it is a locus with respect to three straight lines fixed in position. These fixed lines in the implication are lines that were important in III.54: they are any chord of the section, together with the tangents at the endpoints of that chord.

The locus property of the curve has to do with the distances to the three fixed lines from any point on the curve. For different points on the curve, those distances are different; but if you use any two of the distances as the sides of a rectangle, and you use the remaining third distance as the side of a square, then the ratio of the rectangle to the square will be the same for all the different points on the curve.

The distances need not be taken along a perpendicular—that is, they need not be taken at a right angle—but whatever angle you use for taking the distance from any one point to any one of the three lines, you must use the same angle for taking the distance from any other point to that same line (fig. 14).

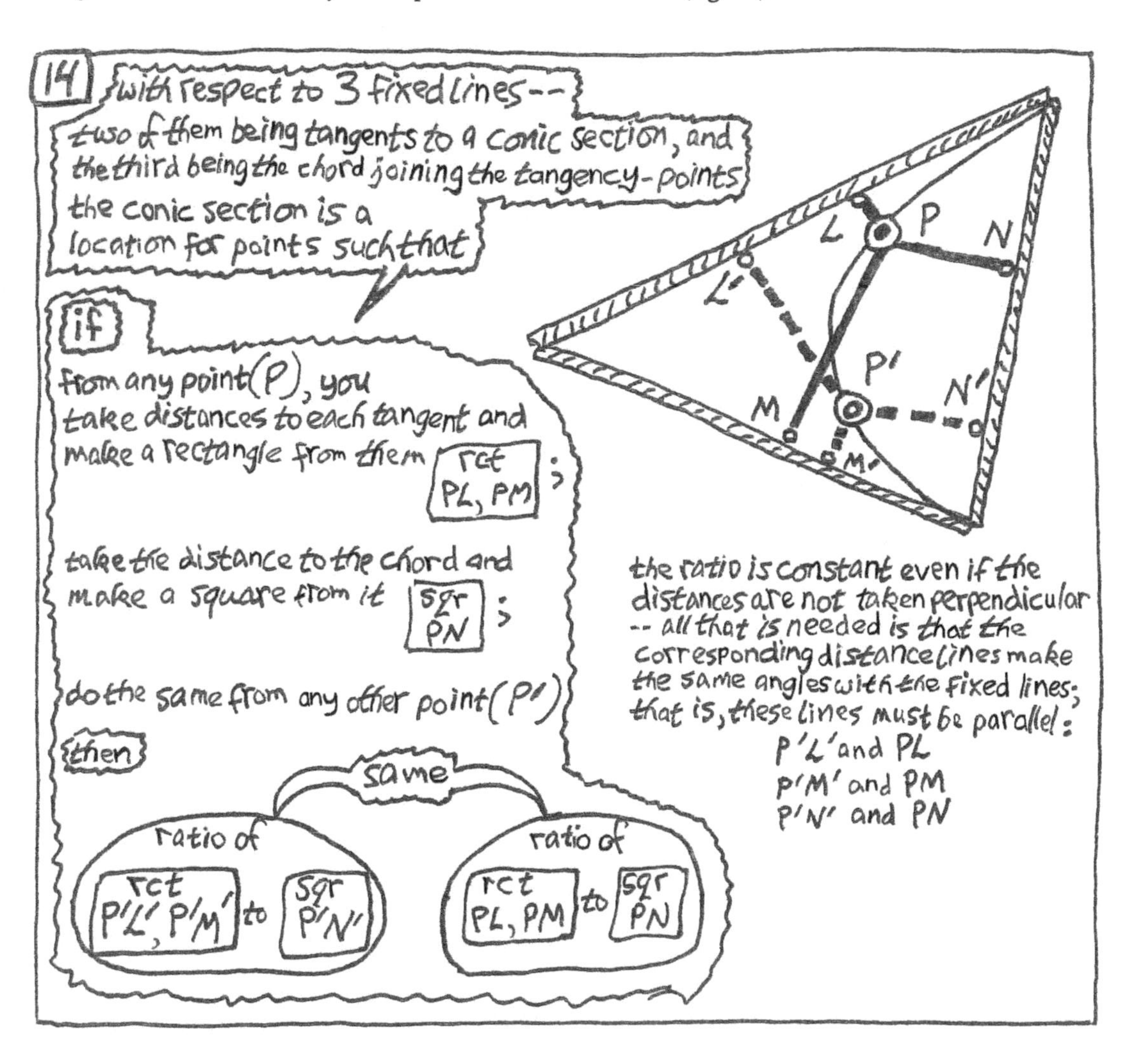

From III.54—using the sameness of ratios that is obtained by taking similar triangles and by compounding ratios of their sides—we shall show that if a line is a conic section, then it is a locus with respect to three lines.

The *converse* (namely, that if a line is a locus with respect to three lines, then it is a conic section) is what is needed for solving the three-line locus *problem*. For now, however, let us consider, not the converse and the problem that it helps to solve, but, rather, let us now consider a demonstration of the relevant theorem itself. That is, let us see how III.54 can show us that if a line is a conic section, then it is a locus with respect to three lines.

Apollonius's III.54 is of use here because of what it implies about the rectangle [AF, CG]. What III.54 said was this: in conic sections and in circles, the ratio which that rectangle [AF, CG] has to square [AC] is the same as the ratio compounded of the following two ratios—namely, (fig. 15)

square [BE]	to	square [BD]	&	rectangle [DA, DC]	to	1/4 square [AC].

Now, consider that tangents AD and AC are fixed and given, and that therefore also given is straight line DE which bisects the line AC joining the tangency-points. That is to say, the following lines are all of them fixed and given: AC, BE, BD, DA, DC. And therefore also fixed and given are the squares that arise from them and the rectangles that they contain. Now then, what does III.54 imply about the rectangle [AF, CG]?

What it implies is this. As H is taken at different points along the curve, the lines through it from the tangency-points will go on to chop off different lengths from the lines that are parallel to the tangents; that is, AF and CG will change size. Even so, however, the rectangle [AF, CG] which they contain will not change size. (That is because its ratio to a square of constant size—namely, square [AC]—is the same as the ratio compounded out of ratios of squares and rectangles which are of constant size.) This constancy of the ratio of rectangle [AF, CG] to square [AC] is what we need for the demonstration of the conic's three-line locus property.

Let us now consider the distances involved in the locus property.

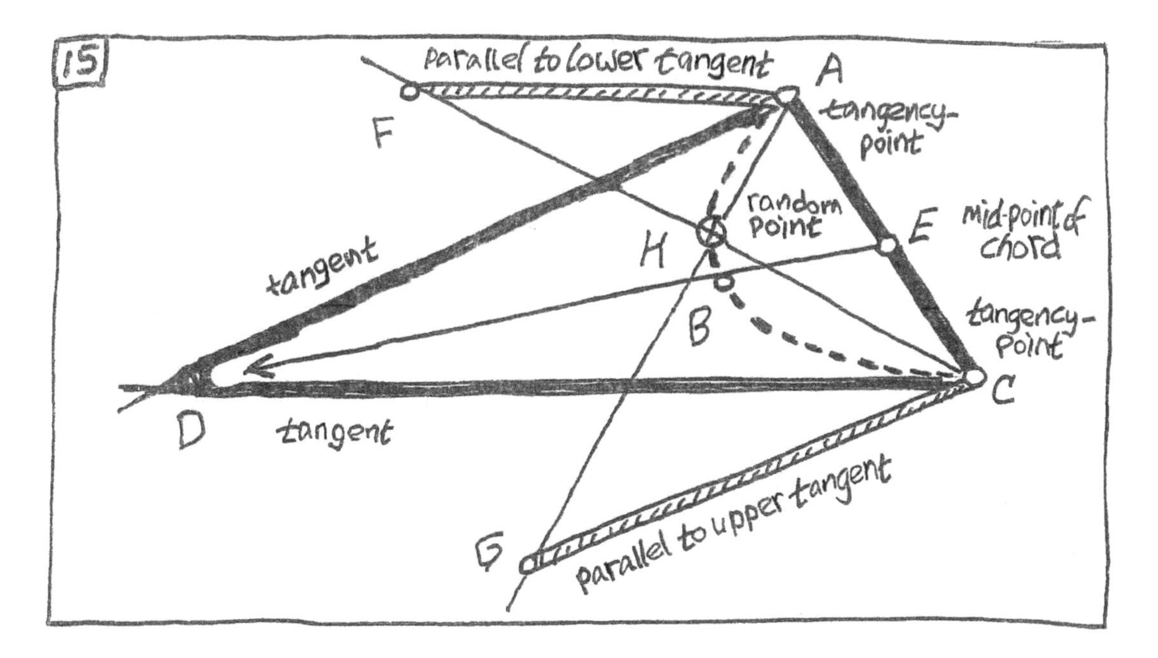

We have a curve ABC that is a conic section. We want to have distances, from a random point H on the curve, to three straight lines (a chord AC and, at its endpoints, the tangents AD and CD). Through random point H, we draw straight lines that give the distances at particular angles that will be convenient for the demonstration—namely (see figure 16) these:

by drawing a parallel to the chord AC,
we get distances HP and HQ; and

by drawing a parallel to the line DE that
joins the midpoint E of the chord and
the intersection point D of the tangents at the chord's endpoints,
we get distance HX.

We want to show a constancy in the relationships among those distances (HP, HQ, HX). This last distance HX is convenient for the demonstration, but the other two (HP and HQ) are not. So what we do is this:

by drawing parallels to the tangents,
we get, along the chord AC that joins the tangency-points,
segments AY (equal to distance HP)
and ZC (equal to distance HQ).

By showing constancy in the relationship among the distances AY, CZ, and HX, we shall show it also in the relationship among the distances (HP, HQ , and HX) from the random point to the three lines.

A demonstration of the constancy for AY, CZ, and HX is depicted in figure 17.

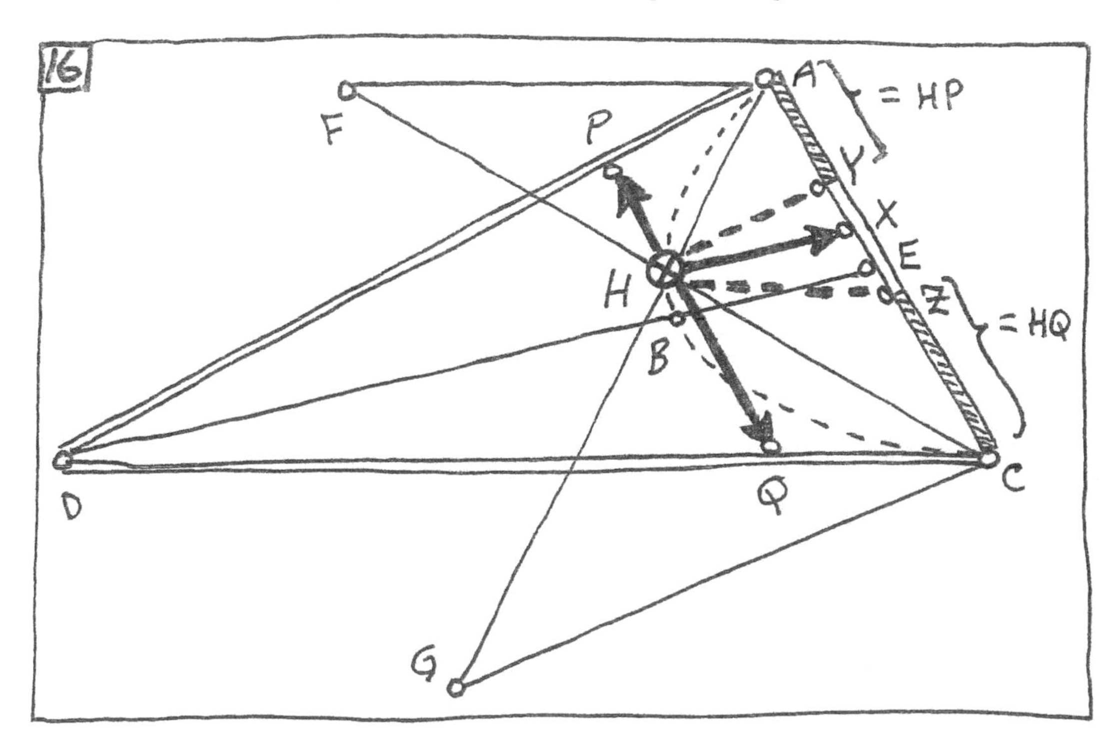

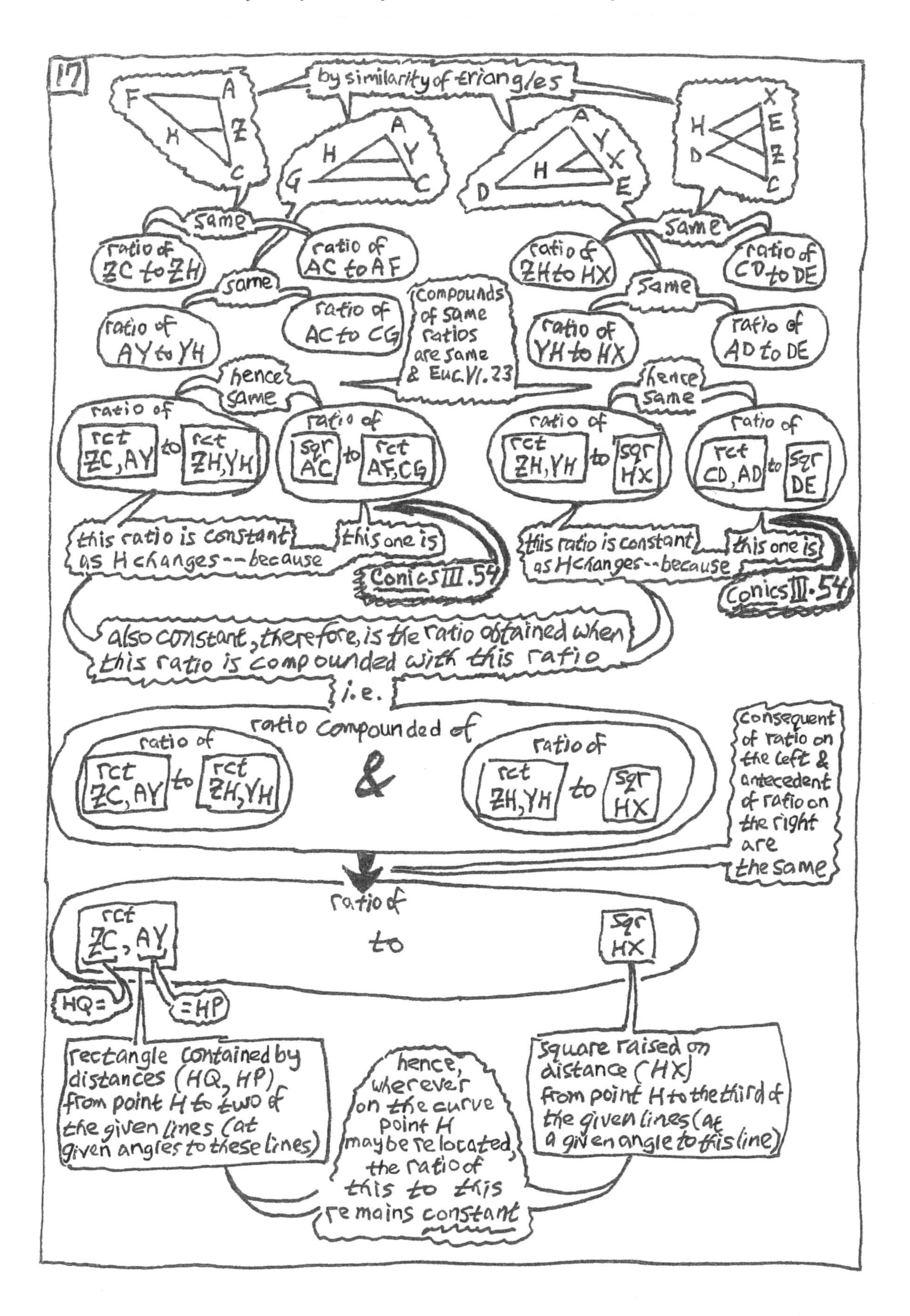
17
by similarity of triangles
F A H Z C
H A Y G C
A Y X H D E
X E H Z D C
same
ratio of ZC to ZH
ratio of AC to AF
same
ratio of ZH to HX
ratio of CD to DE
same
ratio of AY to YH
ratio of AC to CG
Compounds of same ratios are same & Euc.VI.23
same
ratio of YH to HX
ratio of AD to DE
hence same
ratio of rct ZC, AY to rct ZH, YH
ratio of sqr AC to rct AF, CG
hence same
ratio of rct ZH, YH to sqr HX
ratio of rct CD, AD to sqr DE
this ratio is constant as H changes--because
this one is
Conics III.54
this ratio is constant as H changes--because
this one is
Conics III.54
also constant, therefore, is the ratio obtained when this ratio is compounded with this ratio
i.e.
ratio compounded of
ratio of rct ZC, AY to rct ZH, YH
&
ratio of rct ZH, YH to sqr HX
consequent of ratio on the left & antecedent of ratio on the right are the same
ratio of
rct ZC, AY
to
sqr HX
HQ=
=HP
rectangle contained by distances (HQ, HP) from point H to two of the given lines (at given angles to these lines)
hence, wherever on the curve point H may be relocated, the ratio of this to this remains constant
square raised on distance (HX) from point H to the third of the given lines (at a given angle to this line)

Now we need to show that the constancy of the rectangle-square ratio is true not only for distances taken at those particular angles which we have chosen, but for distances taken at *any* angles. We do it by showing that *because* it is true at those particular angles which we *have* chosen, it will be true *also* at any angles which we *might* choose. The ratio for one trio of angles will not be the same as it would be for another trio of angles, but it will be constant for any one trio of angles which we choose. Figure 18 shows why.

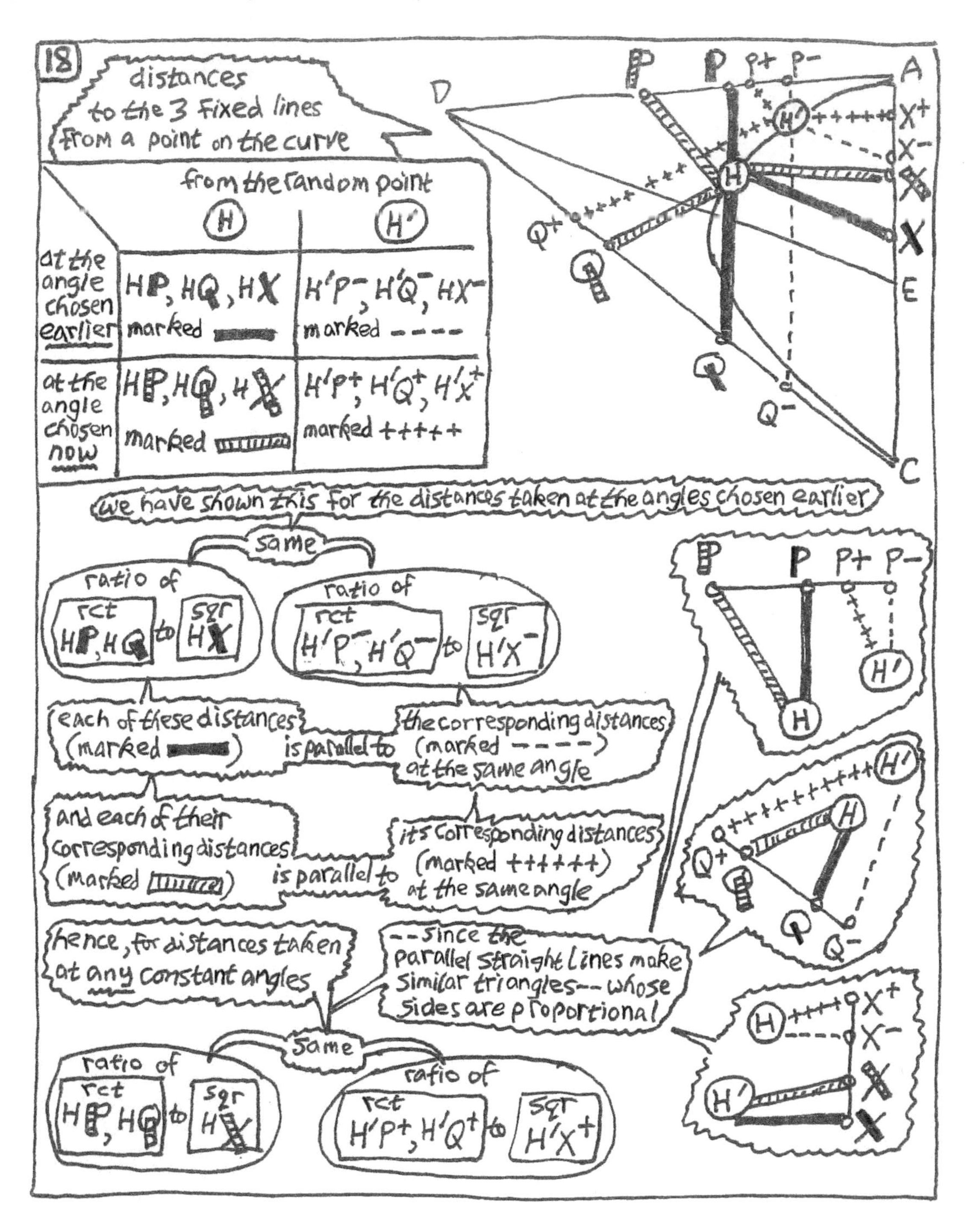

We have established the theorem which states the three-line locus property of conic sections. What the theorem says is this: if a line is a conic section, then it must be a locus with respect to three lines.

But if a line is a locus with respect to three lines, must it then be a conic section? That is the question answered by the *converse* of that theorem which states the three-line locus property of conic sections. What the converse says is this: consider a line whose points have the property that

> *when* we take the distances
> from any one of the points
> to each of three fixed straight lines
> (not necessarily taking distances perpendicular to the three fixed lines,
> or even at the same angle for all three distances),
> and we use two of the distances as sides of a rectangle
> and use the third distance as the side of a square,
> and we take the ratio of the rectangle to the square,
>
> and we do that with any other of the points,
> (making sure, however, that
> the corresponding distances for this point
> are on lines which are parallel to those used for
> the distances for the first point),
>
> *then* that ratio of rectangle to square
> is the same for the different points;

now (says the converse) *if* that property belongs to the points of a line, *then* the line is a conic section.

To establish the converse of the three-line locus property is to solve the three-line locus problem. The solution of the problem requires us to:

> find a way to get
> points that have the relevant property with respect to three fixed lines (fig. 19); and
>
> find a way to trace
> a line that contains all such points; and
>
> determine, in the various possible cases, what kind of line it will be.

A way to find such points, as the first step in solving the three-line locus problem, is the following.

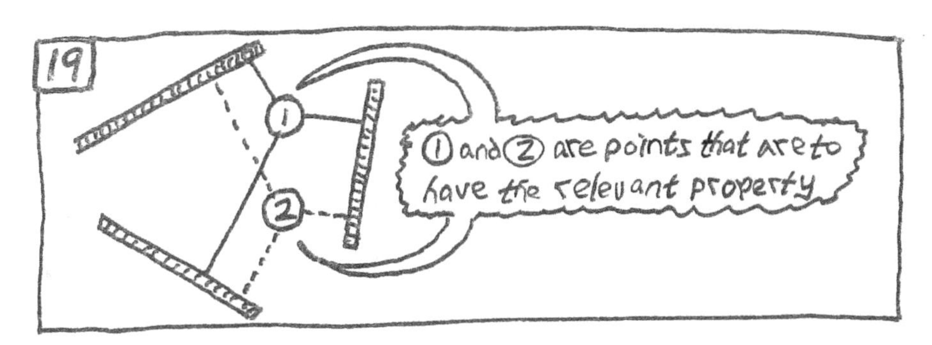

19
1
2
① and ② are points that are to have the relevant property

We are given three straight lines which, when prolonged, make a triangle—say, triangle ADC. Vertex D is joined to the midpoint E of the opposite side AC (let us call AC the base, and let us call the other two sides merely the sides), making AE equal to EC. We are also given the three angles at which the three given lines will be met by the three distance-lines drawn to them from any of the points that interest us: the distance-line to the base will be parallel to the line from vertex D that bisects the base, and the distance-lines to each of the two sides will be parallel to the base. See figure 20—in which:

> the angle at which the distance-line to AC will meet AC
> is equal to that angle (DEA) at which AC is met by
> the line joining vertex D to AC's midpoint E;
>
> the angle at which the distance-line to DA will meet DA
> is equal to that angle (CAD) at which DA is met by AC;
>
> the angle at which the distance-line to DC will meet DC
> is equal to that angle (ACD) at which DC is met by AC.

Finally, we are given some constant ratio—of, say, r to s.

Now what we need is a way to find points, as many as we please, every one of which has the following property: the three distance-lines from the point to the three given lines (the distances being taken along lines that meet the given lines at the given angles) are such that for any point the ratio

of the rectangle contained by the distance-lines from that point to the given triangle's sides	to the square on the distance-line from that point to the given triangle's base

is always the same as the given constant ratio of r to s.

What we shall do is this:

> (I) obtain a certain point H that will turn out to have the property;
>
> (II) show how to use that point H to get other such points H', H", H"', etc.

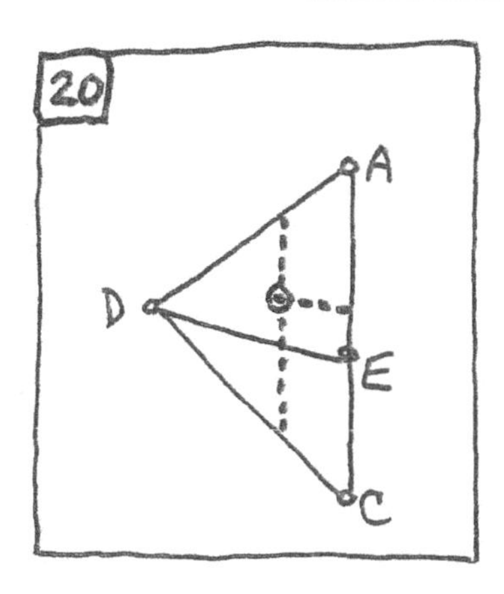

First, point H is obtained, as follows.

> Through vertex D, draw a parallel to base AC.
> On this parallel, take point O such that
>
> the ratio of square [DO] to square [DE]
> is the same as
> the given ratio (of *r* to *s*).
>
> Join OE, intersecting side AD—call the intersection W.
> Through that intersection-point W, draw a parallel to base AC,
> intersecting the other side CD—call the intersection V;
> now also intersected by the parallel (WV) to base AC is
> the line from vertex D that bisects the base (at E)—
> and this intersection (of WV with DE), call point H.
> This point (H) will have the property. (Fig. 21.)

Note that point H corresponds to what we were calling B (while W corresponds to P; and V, to Q).

That point H is now used to get another point H' that will also have the property (namely, the constancy of the ratio involving the distance-lines).

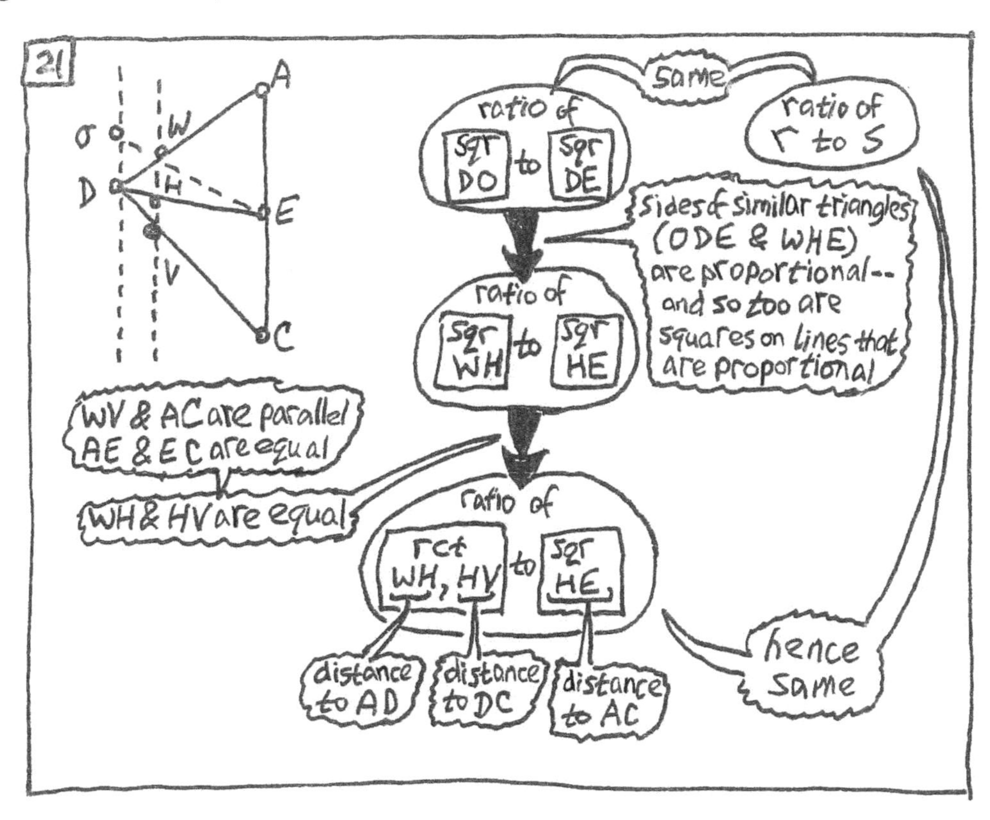

We prepare to get H' by first getting two other points (F and G) from point H. To point H, we join each endpoint of the base—that gives lines AH and CH; and then, we do the following (fig. 22):

Extend through H that line to H from endpoint A; it will then intersect a line which emanates from the other endpoint (C) and which is drawn parallel to the side AD emanating from A. Call the intersection G.	Extend through H that line to H from endpoint C; it will then intersect a line which emanates from the other endpoint (A) and which is drawn parallel to the side CD emanating from C. Call the intersection F.

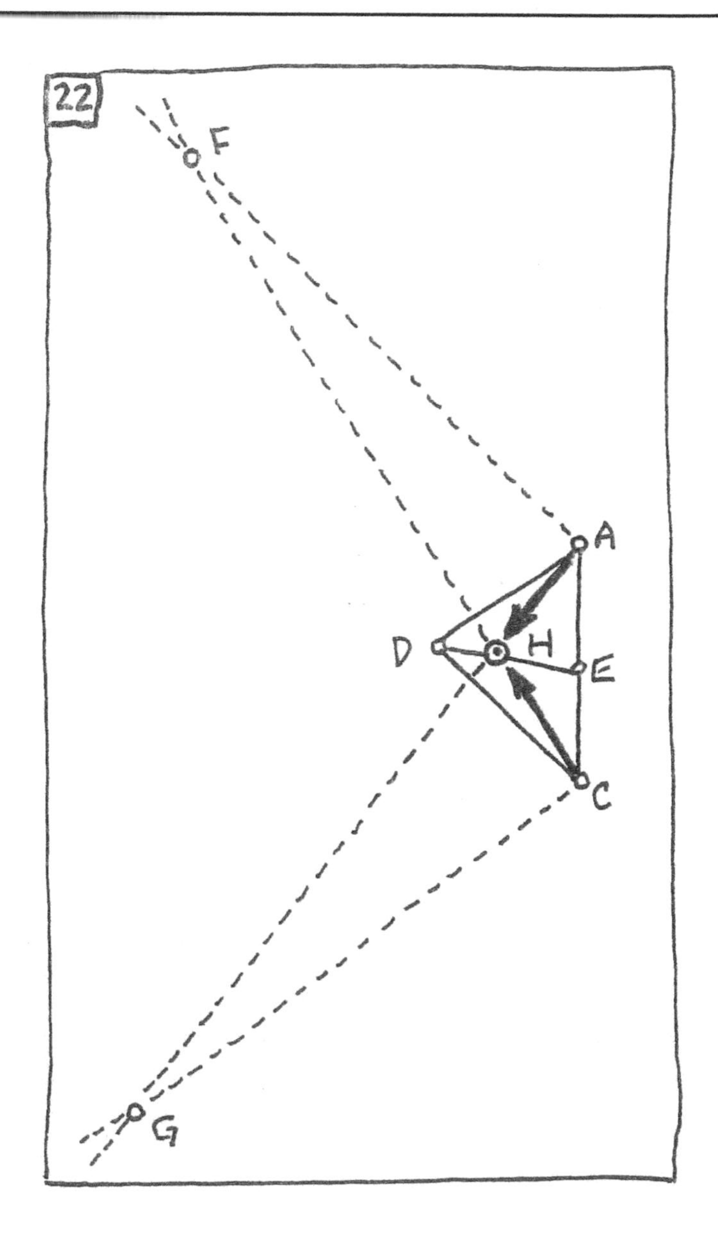

Now that the point H has given us F and G, we are prepared to obtain the point H'. On line AF or its prolongation, we take any point F'; and, on line CG or its prolongation, we take that point G' for which

rectangle AF', CG'	is equal to	rectangle AF, CG.

We must be careful, however: if F' is taken inside angle ADC (or, in other words, is taken on the other side of A from F), then G' should also be taken inside that angle (that is, taken on the other side of C from G), and if F' is taken outside that angle, then G' should too. Now, we join each of those points F' and G' to that endpoint of the base to which it has not yet been joined; and these lines that we make (namely, lines F'C and G'A) will intersect. Their intersection-point will be H'. This point H' will, like H, have the property. Going on to take any other point (F'') on FA or on its prolongation, together with its proper G'' on GC or on its prolongation, will yield yet another point H'' that has the property. Thus we can obtain as many points as we please, each one having the property (fig. 23).

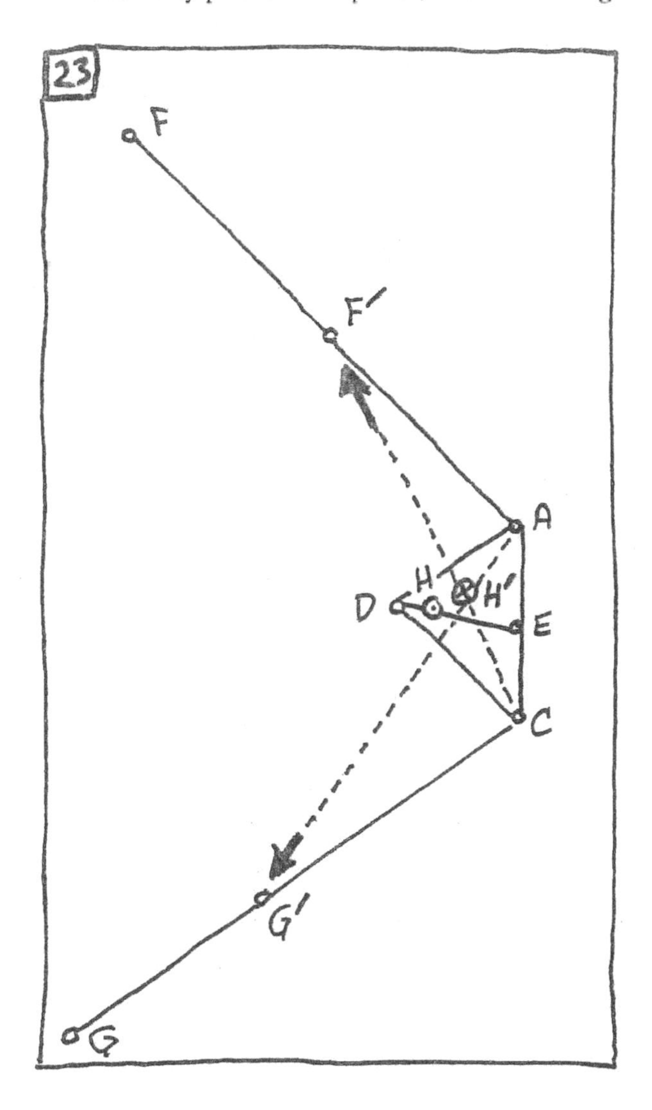

The problem will therefore be solved if we show how to obtain a line that will pass through all those points. Such a line can be obtained if we

take line HE to be given as a diameter, having point H as vertex,
and take line AE to be given as its ordinate,
and use the 52nd and following propositions of the First Book
to obtain a conic section having
that diameter (with that vertex) and that ordinate.

But those propositions can be used only if, besides a diameter and the angle made with it by its ordinate, we have been given also the size of its parameter and the kind of section it is to be. We must therefore figure out

how the size of rectangle [AF, CG],
when it is considered together with the diameter HE and its ordinate EA,
will tell us what size parameter we must use
to obtain what kind of section.

Only after we have figured that out, will we have the solution of the three-line locus problem.

The key to solving the three-line locus problem is the conic sections' property that makes them three-line loci. We considered that property of the conic sections by examining the property of the conic sections that Apollonius demonstrates at the very end of the Third Book of the *Conics* (propositions III.54–56). The three-line locus is a special case of the *four*-line locus. Now let us consider the latter (fig. 24).

We have seen that	Now we shall see that
a conic section is	a conic section is
(with respect to	(with respect to
the *three* lines that are	the *four* lines that are
the sides of a triangle	the sides of a quadrilateral
that is formed by a	that is
a chord and the tangents at	inscribed
the chord's endpoints)	in it)
a locus of points such that:	a locus of points such that:
when the distances	when the distances
from a point	from a point
to those three lines	to those four lines
are used as sides	are used as sides
(the two to the tangents	(any two of them
being used for a rectangle,	being used for one rectangle,
and the third	and the other two
being used for a square)	being used for another rectangle)
then the rectangle has	then the one rectangle has
to the square	to the other rectangle
a ratio that is constant.	a ratio that is constant.

(The ratio is the same for every point, provided that the angle which is used for the distance to the same fixed straight line is the same at any one point as it is at any other.)

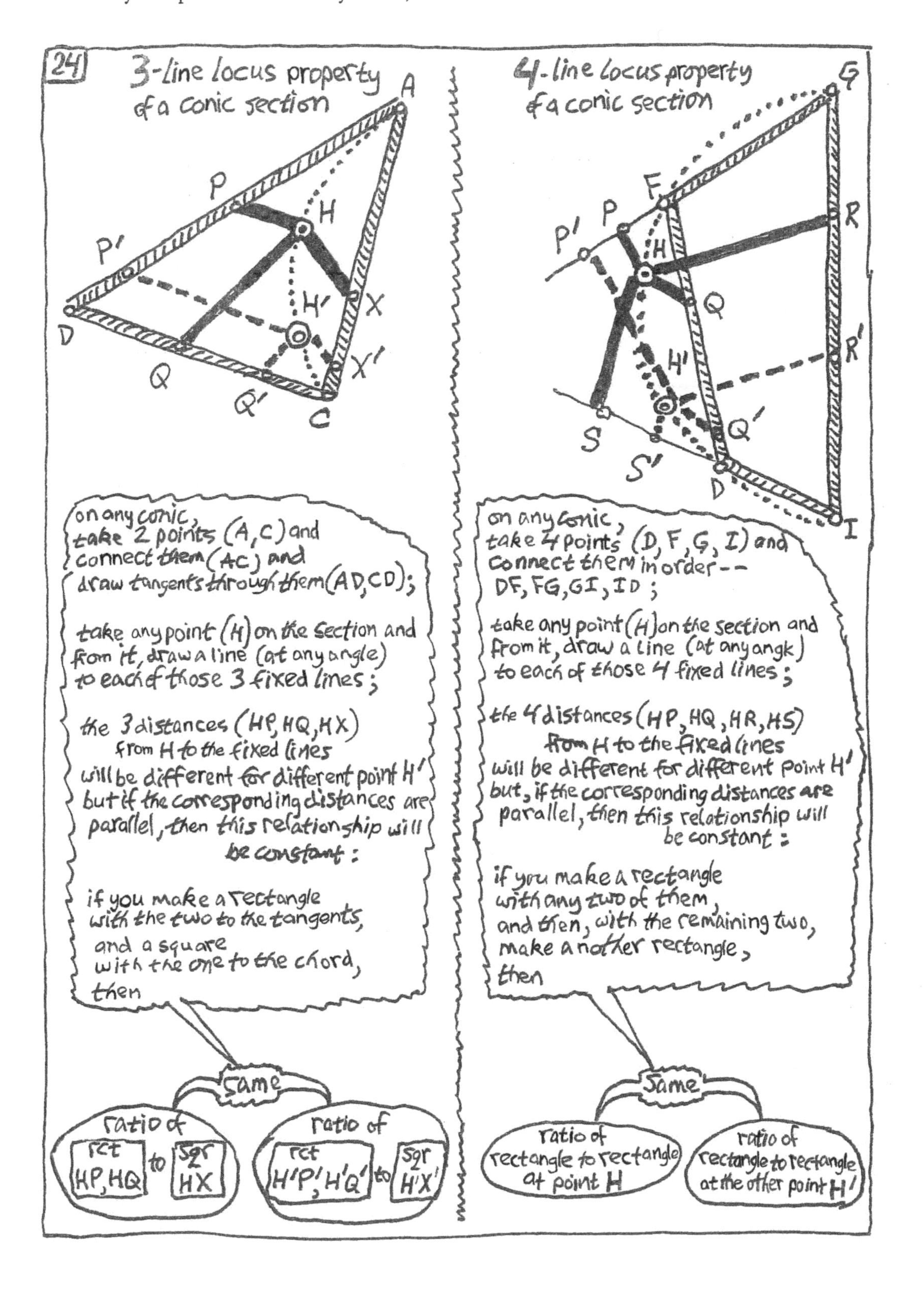

24
3-line locus property of a conic section
A
P
H
P'
X
D
H'
X'
Q
Q'
C
on any conic, take 2 points (A,C) and connect them (AC) and draw tangents through them (AD,CD);
take any point (H) on the section and from it, draw a line (at any angle) to each of those 3 fixed lines;
the 3 distances (HP,HQ,HX) from H to the fixed lines will be different for different point H' but if the corresponding distances are parallel, then this relationship will be constant:
if you make a rectangle with the two to the tangents, and a square with the one to the chord, then
same
ratio of rct HP,HQ to sqr HX
ratio of rct H'P', H'Q' to sqr H'X'
4-line locus property of a conic section
G
F
P
P'
R
H
Q
R'
H'
S
S'
Q'
D
I
on any conic, take 4 points (D, F, G, I) and connect them in order -- DF, FG, GI, ID;
take any point (H) on the section and from it, draw a line (at any angle) to each of those 4 fixed lines;
the 4 distances (HP, HQ, HR, HS) from H to the fixed lines will be different for different point H' but, if the corresponding distances are parallel, then this relationship will be constant:
if you make a rectangle with any two of them, and then, with the remaining two, make another rectangle, then
Same
ratio of rectangle to rectangle at point H
ratio of rectangle to rectangle at the other point H'

Now let us consider how to show that a conic section is a locus such that the distances from any point on it to two of the four fixed lines will contain a rectangle having a constant ratio to a rectangle contained by the distances from that point to the other two of the fixed lines. We could use any angles for the distances to the four fixed lines; but, since right angles are the most convenient, all our distances will be perpendicular. (Again we shall omit the case of the opposite sections.)

We shall establish the property of points on a conic with respect to the *four* fixed lines by taking as an example, in figure 25, the ratio that

rectangle	has to	rectangle
HP, HS		HQ, HR.

The heavy lines in the figure—HP, HQ, HR, HS—are the distances from point H to each of the four fixed lines, which are prolonged where necessary. Those are the distances used as sides to form the rectangles in the ratio. The constancy of that ratio of rectangles is what we want to show.

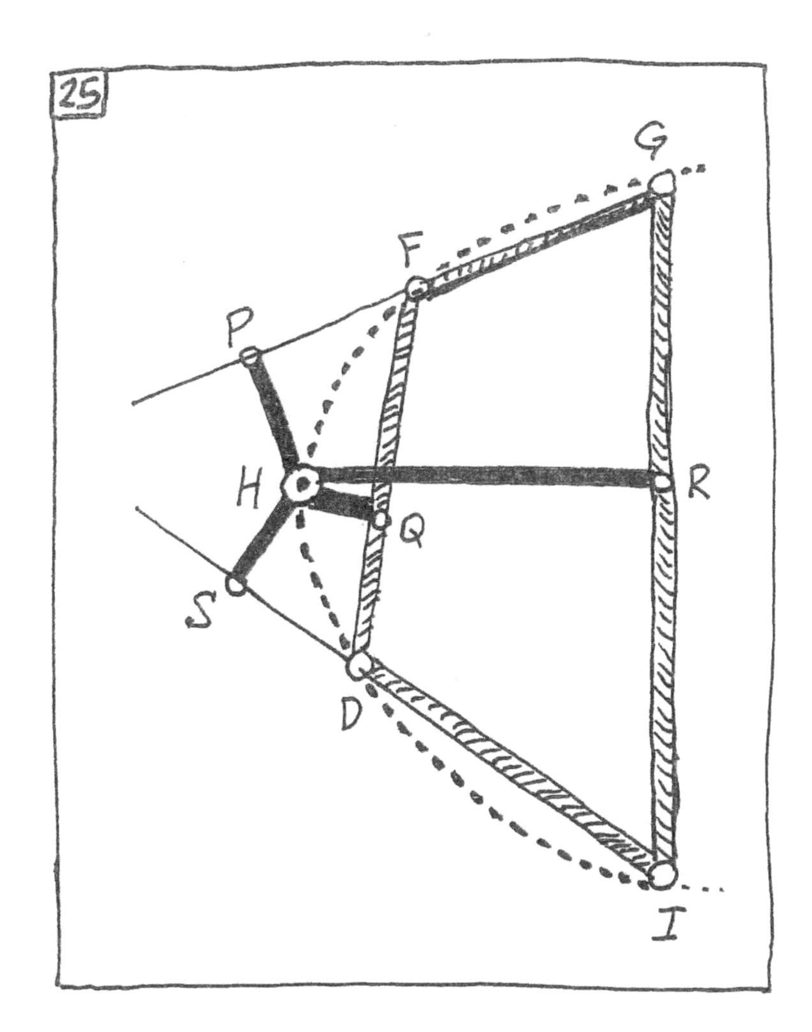

To show this property of points on a conic with respect to *four* fixed lines, we use the property of points on a conic with respect to the *three* fixed lines of Apollonius III.54. *One* of the *three* fixed lines in the *earlier* property was a chord of the section, and *each* of the *four* fixed lines that we have *now* is a chord. We therefore already have (quadruply) one of the three fixed lines of III.54. What we still need is, for each of the chords, two tangents—one of them at each of the endpoints of the chord. Those tangents are the dashed lines in figure 26; and the heavy solid lines (HV, HX, HY, HZ) in the figure are the distances from point H to each of those tangents.

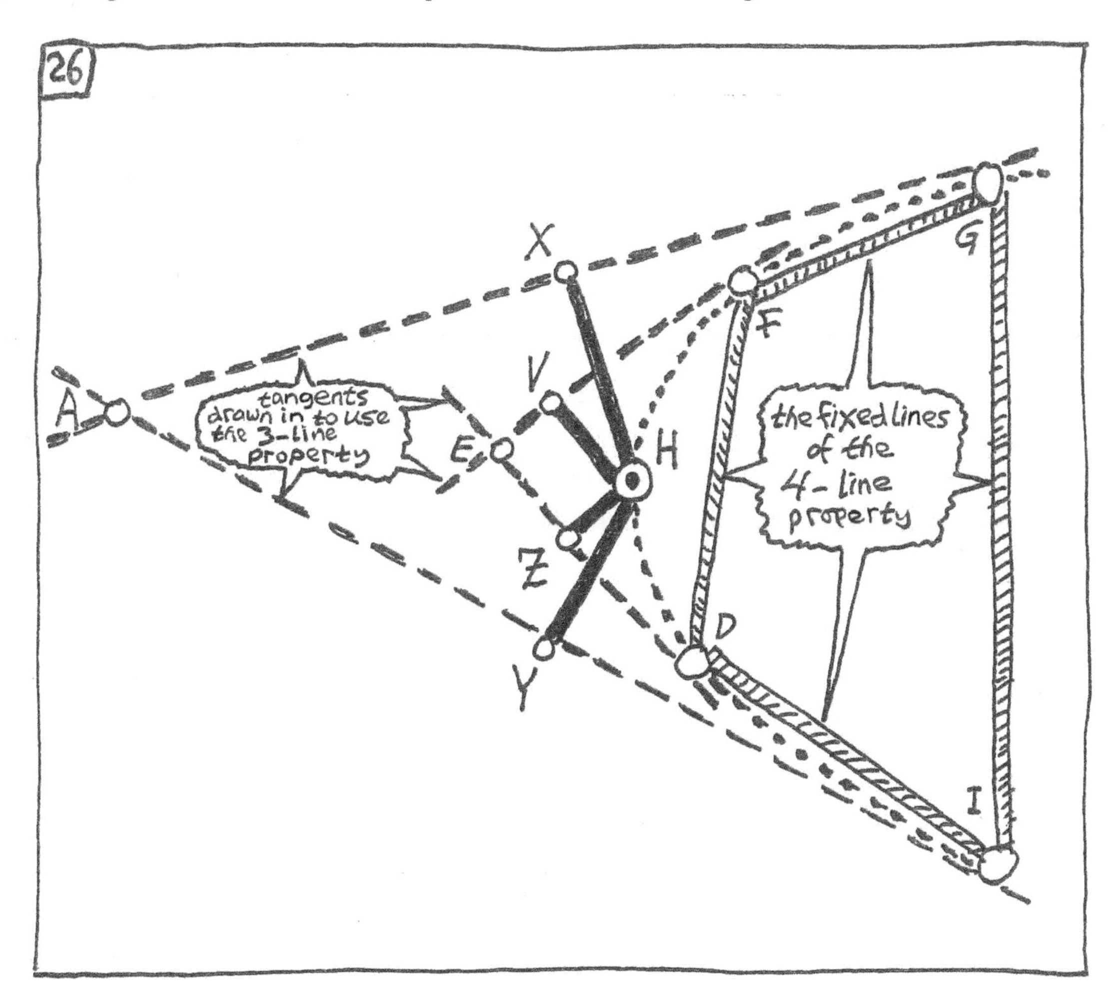

In order for the chord-distances (HP, HQ, HR, HS) to be related to each other, we use III.54, which relates each chord-distance to two tangent-distances (the tangents being those at the chord's endpoints):

for this chord......	GF	FD	GI	ID
this chord-distance...	HP	HQ	HR	HS
and				
these tangent-distances....	HV, HX	HV, HZ	HX, HY	HY, HZ.

For example, take GF, the first of the chords (fig. 27). From III.54, it follows that, for chord GF, the ratio which

rectangle	has to	square
HV, HX		HP

is constant. And similarly for every other chord—so that, for each of the distances to the four fixed lines in which we are interested, the three-line locus property gives us a constant ratio. Thus, we get four constant ratios.

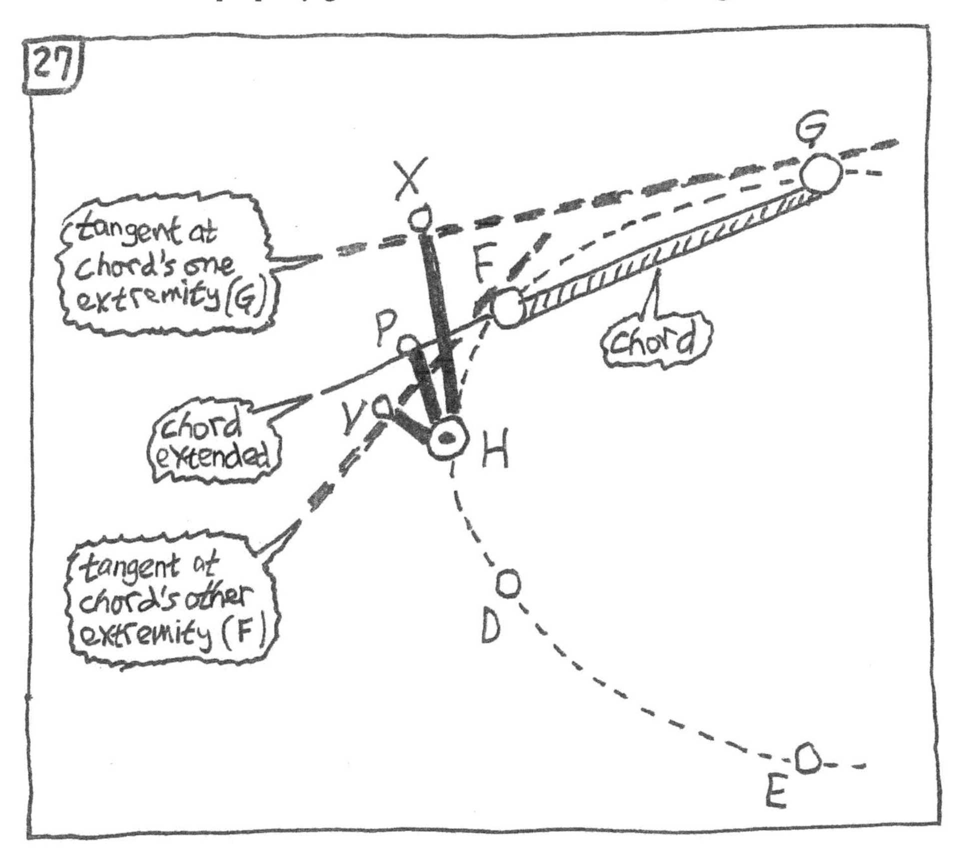

A ratio that is compounded of two constant ratios is constant too. When we compound ratios that are constant because of the *three*-line locus property of a conic section, the ratios that result are constant ratios; when we compound these ratios in turn, the ratios that result are again constant; and the process is repeated until finally we get a single constant ratio that states the *four*-line locus property which we seek to establish. The procedure is sketched in figure 28.

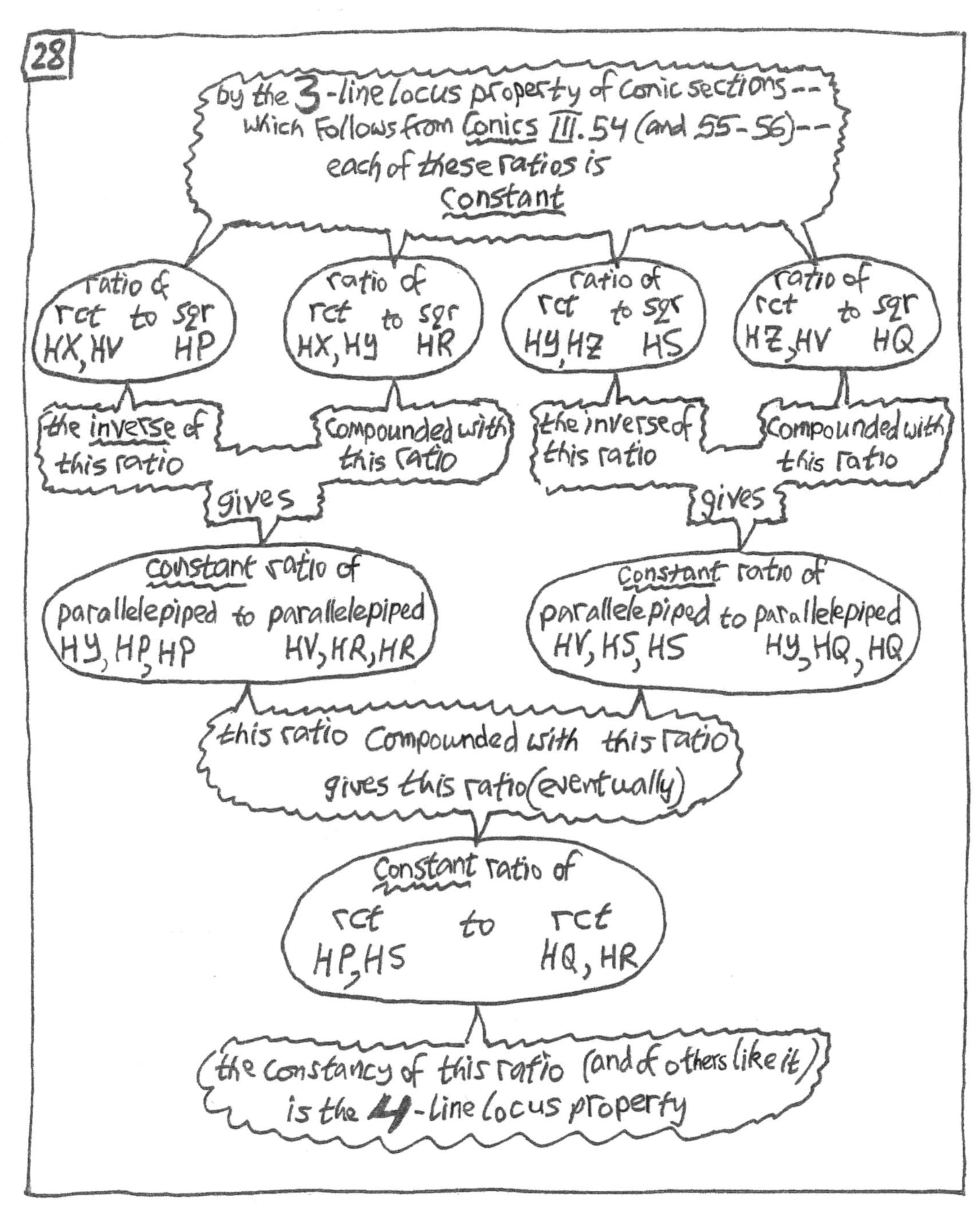

In performing the compounding, there are many complications—so it will not be apparent from the sketch how to get from one step to the next. We must fill in some details.

To simplify the work, let us abbreviate the names that we use for the lines that emanate from point H: let us identify them simply by their other endpoint. Rather than referring to "HX," for example, let us speak simply of line "X." So, the lines from point H to the tangents will be (taking them successively) X, V, Y, Z; and the lines going from point H to the chords will be P, S, Q , R—in the following way:

this line	goes from point H to the chord that connects tangency-points at these distances
P	X and V
S	Y and Z
Q	V and Z
R	X and Y.

A bit clearer view of the movement of thought that establishes the four-line locus property (whence it comes; whither it goes; and the steps between) can now be obtained by looking at figure 29.

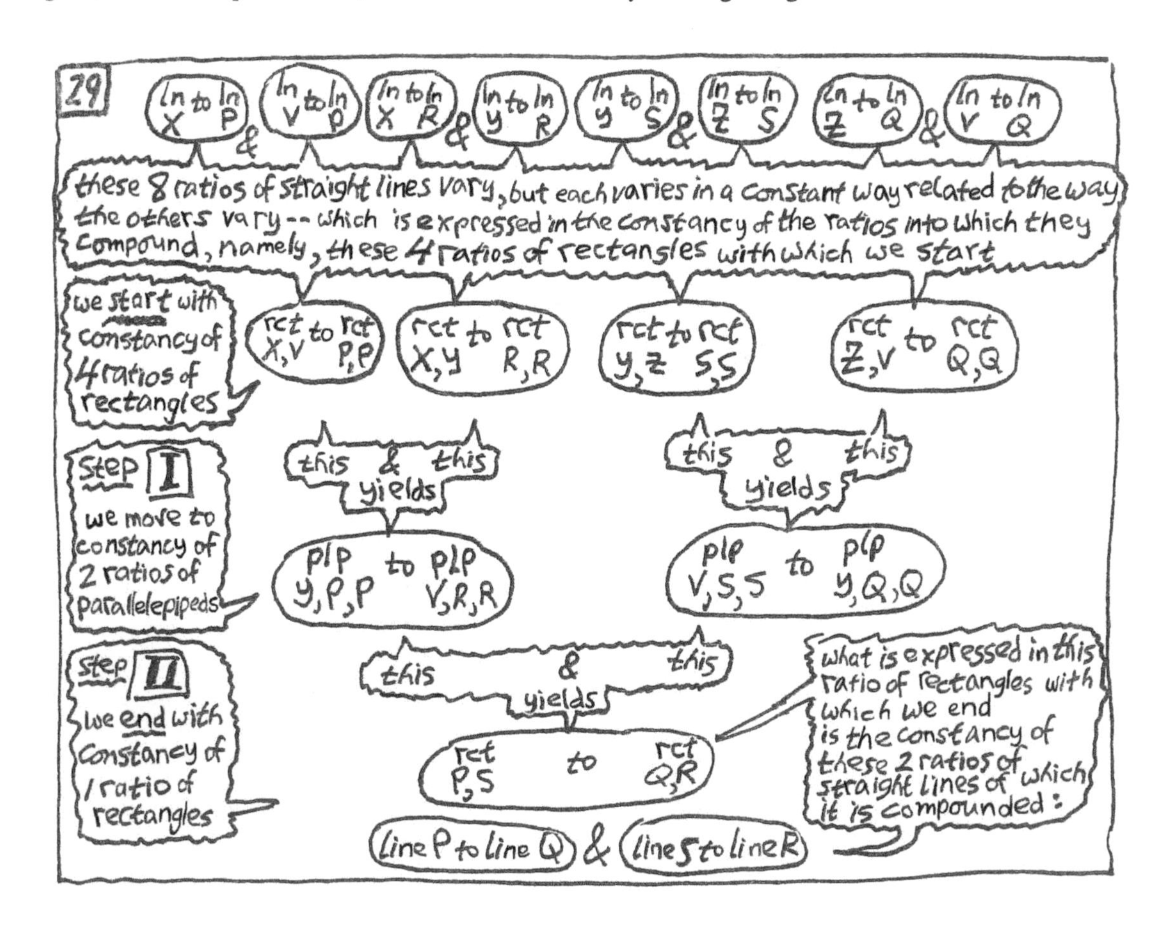

Figures 30 and 31 will show how to manage to "squoosh" the ratios in step-I and step-II. When you turn to these figures, note that in step-I we do *not* end up with the ratio compounded of the ratios that we start with, for we *invert* some of the ratios before compounding. That is, we have some ratio (call it the ratio of *alpha to beta*) and, *instead of* compounding *that* ratio with another ratio, we compound the ratio of *beta to alpha* with the other ratio. But unless the terms of a ratio are equal to each other, the ratio that results from inverting its terms will be greater or less than the original ratio. Why does it not matter here that we change the ratio around before we do the compounding? Because if the ratio of alpha to beta is constant, then so is the ratio of beta to alpha. The *only* thing that we care about in this, is whether the ratios that we get as a result of the manipulation are *constant*.

Those figures 30 and 31 will enable us to end up with a single ratio of one magnitude to another—rather than having to put up with many layers of nested parentheses, as in the following expression:

```
the ratio compounded of
          (((the ratio compounded of
               ((the ratio compounded of
                    (the ratio of P to X)
                    and
                    (the ratio of P to V)))
               and
               ((the ratio compounded of
                    (the ratio of X to R)
                    and
                    (the ratio of Y to R))))))
          and
          (((the ratio compounded of
               ((the ratio compounded of
                    (the ratio of S to Y)
                    and
                    (the ratio of S to Z)))
               and
               ((the ratio compounded of
                    (the ratio of Z to Q)
                    and
                    (the ratio of V to Q)))))).
```

Now let us see how the manipulations are done in step-I and step-II.

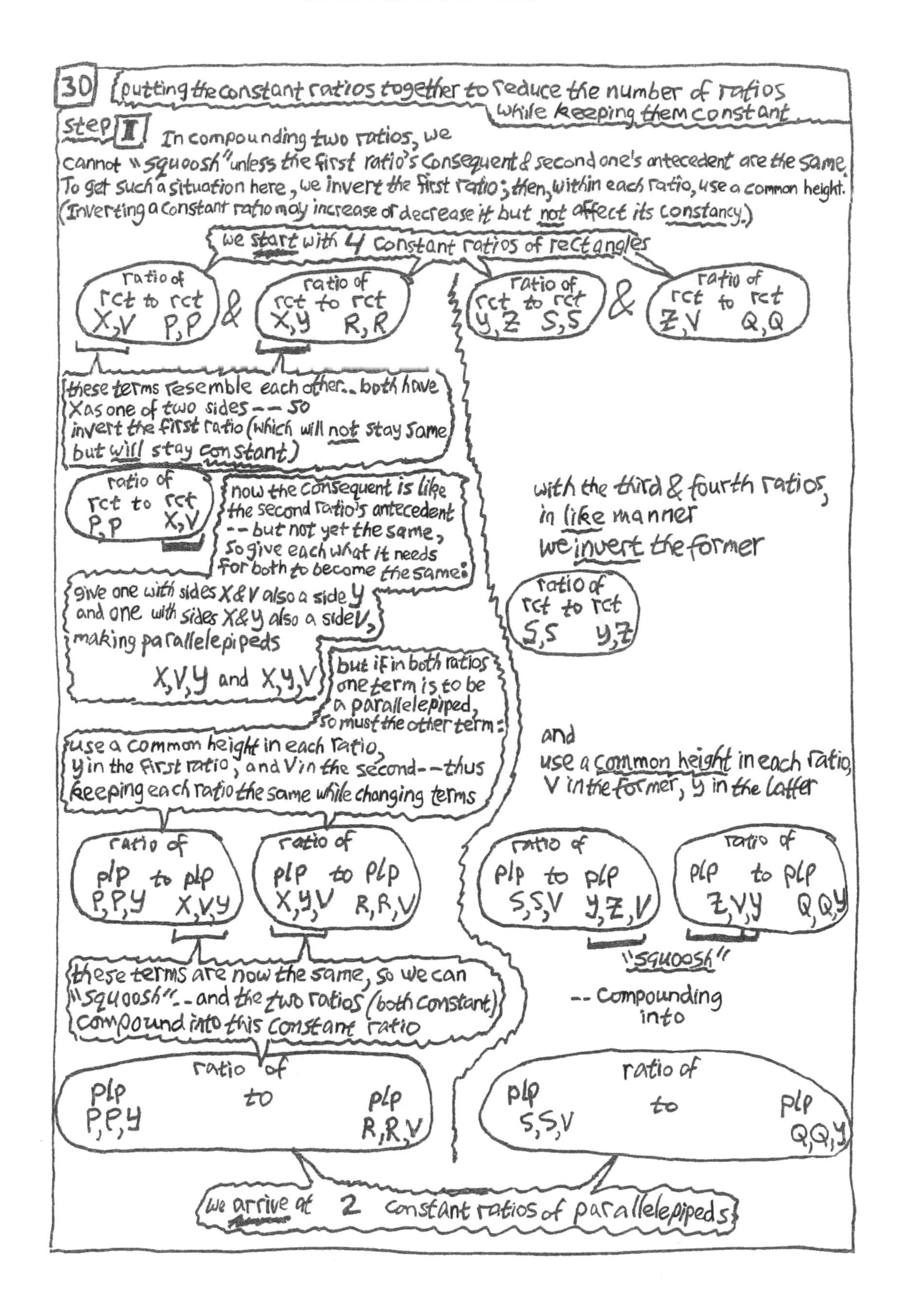
30
putting the constant ratios together to reduce the number of ratios while keeping them constant
step II In compounding two ratios, we cannot "squoosh" unless the first ratio's consequent & second one's antecedent are the same. To get such a situation here, we invert the first ratio; then, within each ratio, use a common height.
(Inverting a constant ratio may increase or decrease it but not affect its constancy.)
we start with 4 constant ratios of rectangles
ratio of rct to rct X,V P,P
&
ratio of rct to rct X,y R,R
ratio of rct to rct y,Z S,S
&
ratio of rct to rct Z,V Q,Q
these terms resemble each other.. both have X as one of two sides -- so invert the first ratio (which will not stay same but will stay constant)
ratio of rct to rct P,P X,V
now the consequent is like the second ratio's antecedent -- but not yet the same, so give each what it needs for both to become the same:
give one with sides X & V also a side y and one with sides X & y also a side V, making parallelepipeds X,V,y and X,y,V
but if in both ratios one term is to be a parallelepiped, so must the other term:
use a common height in each ratio, y in the first ratio, and V in the second--thus keeping each ratio the same while changing terms
with the third & fourth ratios, in like manner we invert the former
ratio of rct to rct S,S y,Z
and use a common height in each ratio, V in the former, y in the latter
ratio of plp to plp P,P,y X,V,y
ratio of plp to plp X,y,V R,R,V
ratio of plp to plp S,S,V y,Z,V
ratio of plp to plp Z,V,y Q,Q,y
these terms are now the same, so we can "squoosh"..and the two ratios (both constant) compound into this constant ratio
"squoosh" -- compounding into
ratio of plp P,P,y to plp R,R,V
ratio of plp S,S,V to plp Q,Q,y
we arrive at 2 constant ratios of parallelepipeds

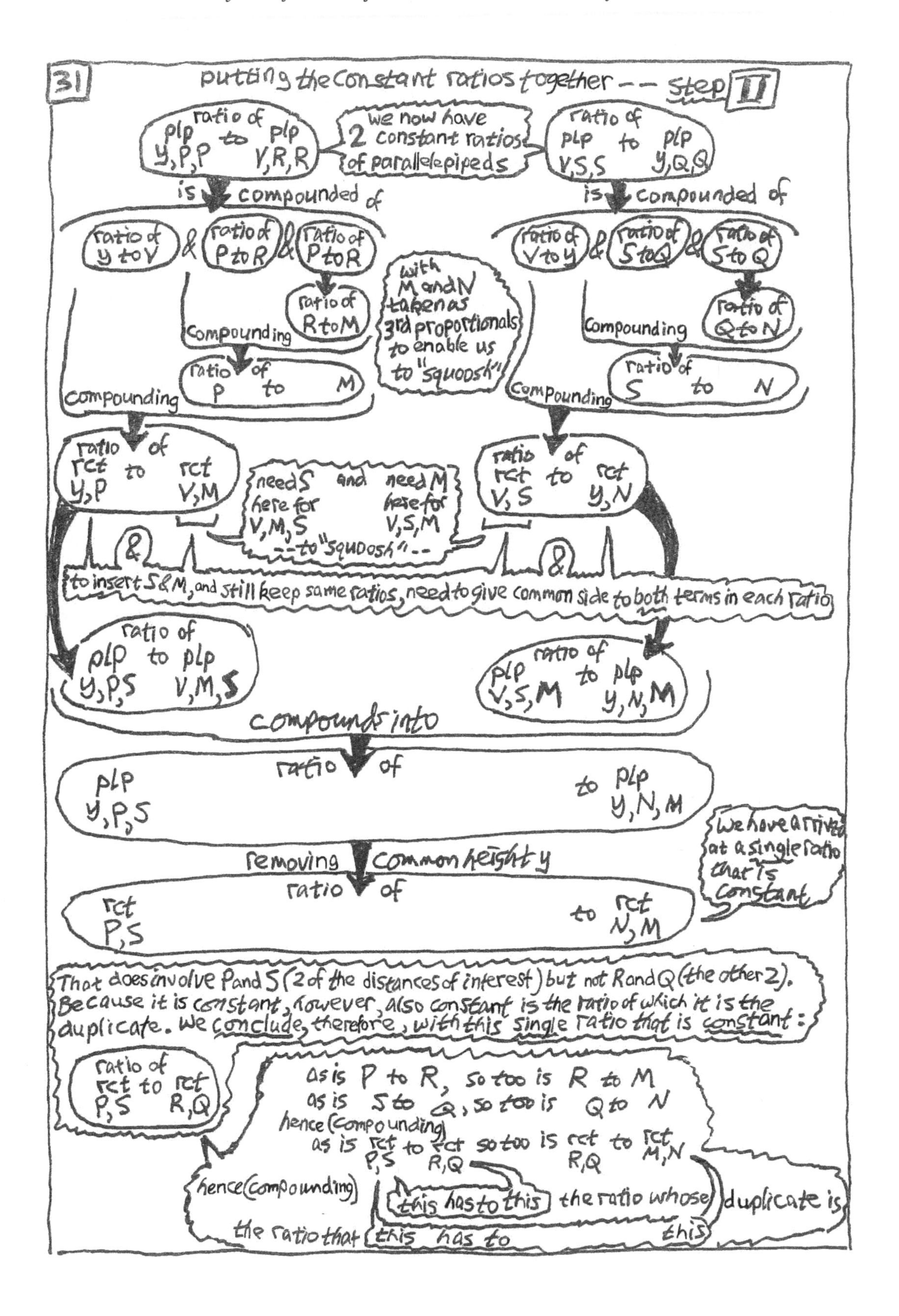
31
putting the constant ratios together -- step II
ratio of plp y,P,P to plp V,R,R
we now have 2 constant ratios of parallelepipeds
ratio of plp V,S,S to plp y,Q,Q
is compounded of
is compounded of
ratio of y to V
ratio of P to R
ratio of P to R
ratio of R to M
ratio of V to y
ratio of S to Q
ratio of S to Q
ratio of Q to N
with M and N taken as 3rd proportionals to enable us to "squoosh"
compounding
ratio of P to M
compounding
ratio of S to N
compounding
compounding
ratio of rct y,P to rct V,M
ratio of rct V,S to rct y,N
needS here for V,M,S and need M here for V,S,M --to "squoosh"--
&
&
to insert S&M, and still keep same ratios, need to give common side to both terms in each ratio
ratio of plp y,P,S to plp V,M,S
ratio of plp V,S,M to plp y,N,M
compounds into
ratio of plp y,P,S to plp y,N,M
removing common height y
ratio of rct P,S to rct N,M
We have arrived at a single ratio that is constant
That does involve P and S (2 of the distances of interest) but not R and Q (the other 2). Because it is constant, however, also constant is the ratio of which it is the duplicate. We conclude, therefore, with this single ratio that is constant:
ratio of rct P,S to rct R,Q
as is P to R, so too is R to M
as is S to Q, so too is Q to N
hence (compounding)
as is rct P,S to rct R,Q so too is rct R,Q to rct M,N
hence (compounding)
this has to this
the ratio whose duplicate is
the ratio that this has to this

CHAPTER 7. UNSOLVED GEOMETRICAL PROBLEMS: PAPPUS'S MANY-LINE LOCUS

By making use of Apollonius's III. 54, we have been able to see that if a line is a conic section, then it is

a three-line locus with respect to
any of its chords taken together with
the tangents at the endpoints of the chord,

and we have seen that it is therefore also

a four-line locus with respect to
any quadrilateral inscribed in it.

(Each of the two lines which together constitute opposite sections is, of course, in itself a three- and a four-line locus by virtue of being an hyperbola. But, besides that, it can be shown that the two lines which together constitute opposite sections are, together, a single three-line locus when one of the tangents is taken in each of them, and they also are, together, a single four-line locus when two of the four chords are taken in each of them; and each of them is also a four-line locus when the quadrilateral is inscribed in the other one. As an exercise, you might want to go back sometime and work out for yourself a demonstration of the case of the opposite sections.)

Now what about the converse: if a line is a locus with respect to three lines that form some triangle, or with respect to four lines that form some quadrilateral, what kind of a line is it? Is that line a conic section? We have considered how we might go about solving the three-line locus problem, though we have not actually set out a solution. As for the four-line locus, merely seeing that the conic sections have that property has been an exhausting enough task, even without our going on to the question whether the conic sections are the only lines that have it.

In the Seventh Book of his *Collection*, Pappus reports on problems in the studies of the ancients. After quoting Apollonius's introduction to the *Conics*, and criticizing Apollonius for boastfulness and for ingratitude toward his predecessors, Pappus gives the following statement of the three-line and four-line locus.

With *three* straight lines being given in position,
if, from some one and the same point,
straight lines be drawn, at given angles, onto the three straight lines,
and there be given the ratio
of the rectangle contained by two of the drawn lines
to the square on the remaining drawn line,
then the point will lie on a solid locus given in position—
that is, on one of the three conic curves.

And *if* straight lines be drawn, at given angles,
onto *four* straight lines given in position,
and there be given the ratio
of the rectangle contained by two of the drawn straight lines
to the rectangle contained by the remaining two drawn lines,
then likewise the point will lie on a conic section given in position.

Pappus then goes on to speak about the status of the inquiry into the loci with other than three or four straight lines.

Now *if* the straight lines be drawn onto *only two* given straight lines,
then the locus has been shown to be plane.
But *if* the straight lines be drawn onto *more than four*,
then the point will lie on loci as yet unknown, called only curves,
of as yet unidentified sorts, without any properties, not one of which—
not the first and seemingly most obvious—
have they synthesized, having shown it to be useful.
The propositions (*protaseis*) of them are these:
If, from some point, straight lines be drawn at given angles
onto *five* straight lines given in position,
and the ratio of the rectangular parallelepiped contained by three of the drawn lines
to the rectangular parallelepiped contained by
the remaining two drawn lines and some given straight line,
then the point will lie on a curve given in position.
And *if* the straight lines be drawn onto *six*,
and the ratio of the aforesaid solid contained by three of the drawn straight lines
to the solid contained by the remaining three,
then again the point will lie on a curve given in position. ...
But *if* upon *more than six*,
then, on the one hand, they're no longer able to say:
"if there be given the ratio of some figure contained by four drawn lines
to the figure contained by the ones remaining"—
since there isn't any figure contained by more than three dimensions (*diastasemôn*)—
but, on the other hand, some recent writers have agreed among themselves to give an
interpretation to such things,
(*sygkekhôrêkasi ... heautois hoi brakhu pro hêmôn hermêneuein ta toiauta*)
though not in any way signifying one thoroughly graspable thing (*dialêpton*)
when saying: "the thing contained by these" with reference to
"the square on this line" or "the rectangle on these lines."
It was, however, possible (*parên*) through compound ratios (*logôn*)
to speak (*legein*) and to show these things generally,
both for the aforesaid propositions (*protaseis*) and for these, in this way:
if, from some point, straight lines be drawn at given angles onto straight lines given in position,
and there be given the ratio compounded
of that ratio which the one drawn line has to the other one,
and of that ratio which another has to another,
and of that ratio which a different one has to yet a different one,
and of that ratio which the remaining one has to a given one,
if there be seven—and, if there be eight, even of that ratio which
another one remaining has to yet another remaining—
then the point will lie on a given curve,
and likewise as many as they may be, whether their number be odd or even.
Of these propositions following upon the locus with respect to four lines,
as was said, not one have they synthesized, so as to know the curve.

To understand the interpretive difficulty to which Pappus refers, let us consider again the compounding of ratios, reminding ourselves of some things said earlier.

Euclid defines duplicate ratio (in the definitions for *Elements* V); and then he also defines triplicate ratio. The primary notion seems to be this: K,L,M all being magnitudes of the same kind,

we call ....the ratio that K........has...to.....................................M

the duplicate of........the ratio that K.........has...to..L.

when ....as K.........is......to..L...so..L.........is...to..M.

For example, the size-relationship of the nontuple to the triple is the same as that of the triple to the unit. That is to say, nontupling is tripling done twice. In other words, nontupling is tripling duplicated.

Tripling duplicated is different from doubling a triple. Doubling a triple is sextupling, not nontupling. Doubling a triple does not involve the same size-relationship (that of the triple) taken twice. Rather, it involves two different size-relationships—that of the triple and that of the double. When the two size-relationships are numerical and simple, it is easy to see how to put them together.

But what if the ratios that we have are the ratios of "a to b" and "c to d"? Then what we do is turn the operation into something like duplication—something that is not duplication, but is like it in that intervals are put together by finding a link. We find another ratio that is the same as the first ratio, and another ratio that is the same as the second ratio—those new ratios being linked by making the consequent in the first ratio the same as the antecedent in the second ratio. That is to say:

taking any Q, we find the P, and also the R, for which

as a is to b, ... so..............P is to Q,
and
as c is to d, ... so..........................Q is to R.

Thus we might say that the interval (not the difference, but the interval) from a to b is the same as the interval from P to Q, and the interval from c to d is the same as the interval from Q to R.

How, then, might we get the interval that results from putting together the intervals from a to b and from c to d? By putting together the intervals from P to Q and from Q to R—which is to say, by taking the interval from P all the way to R.

Thus compounding two ratios, all of whose terms are of the same kind, we get a ratio of terms of that same kind. That involves a process mediated by the finding of two fourth proportionals. (In the example, P is the fourth proportional to the terms b, a, and Q, while R is the fourth proportional to terms c, d, and Q.) Because of the contriving of mediating terms that replace terms with which we started, the origin is not apparent in the outcome. What has been done is not immediately obvious.

We have just seen how to represent as a ratio of two *lines* the ratio compounded of two ratios each of which is a ratio of two lines—the ratio compounded, that is to say, of the ratios of line a to line b and of line c to line d. That compounded ratio will be the ratio of line P to line R—where (taking any line Q at all) line P is the fourth proportional to lines b, a, Q, while line R is the fourth proportional to lines c, d, Q.

But suppose that when we compound ratios, we come up with magnitudes of a *different* kind from the magnitudes in the ratios being compounded. The compounding of ratios can be more immediately grasped if the process that we use is not the finding of fourth proportionals for lines, but is rather the building of rectangles. In Part Two of this guidebook, we saw Euclid's demonstration (VI.23) that rectangles have to each other the ratio compounded of the ratios of their sides. The ratio compounded of the ratios of line a to line b and of line c to line d will be the same as the ratio of two rectangles—namely, the ratio of rectangle [a, c] to rectangle [b, d]. This is not a *definition* of the compounding of two ratios (if it were, then there would be no need for a demonstration), but it is a good *representation* of the compounding. Building the rectangles by putting together the lines corresponds to putting together the ratios of the lines. Why? Because what determines the ratio of the rectangles is not (by itself) the ratio of the lines used as widths, nor (by itself) the ratio of the lines used as heights—but, rather, both ratios *put together*.

This building procedure is also good for compounding *three* ratios. Three-dimensional boxes (rectangular parallelepipeds), like two-dimensional boxes (rectangles), are to each other in the ratio compounded of the ratios of their sides. Just as, if you take a rectangle, and you double its width while you triple its height, then you will get a rectangle that is six times as large—so, in like manner, if you take a parallelepiped, and you double its width while you triple its height and you quadruple its depth, then you will get a parallelepiped that is twenty-four times as large.

If more than three ratios are compounded, however, this building procedure cannot be used as a complete substitute for finding fourth proportionals. The difficulty is this. It is true that any single ratio at all can be represented not only as a ratio of lines, but also as a ratio of rectangles, or as a ratio of parallelepipeds—and so can any ratio that is compounded out of ratios, no matter how many component ratios may be taken together; but a ratio that is compounded out of more than three ratios of lines can be represented as a ratio of lines, or of rectangles, or of parallelepipeds, only if there has been some employment of the process of finding a fourth proportional, to make some "squooshing" possible. We could have an immediate apprehension of the ratio compounded out of more than three ratios of lines only if we could view more than three lines perpendicular to each other, and build a box with that as its corner; but we cannot.

In compounding two ratios of lines, we can get a ratio of lines as our result—but only if we do not stay with the original lines. (Three of the four original lines will drop out, and one of the four will have, to a contrived line that we introduce, the compounded ratio.) We can retain the original lines in the compounded ratio if we wish, but then the terms of the ratio must move up a dimension, to rectangles (whose sides are the original lines). See figure 1.

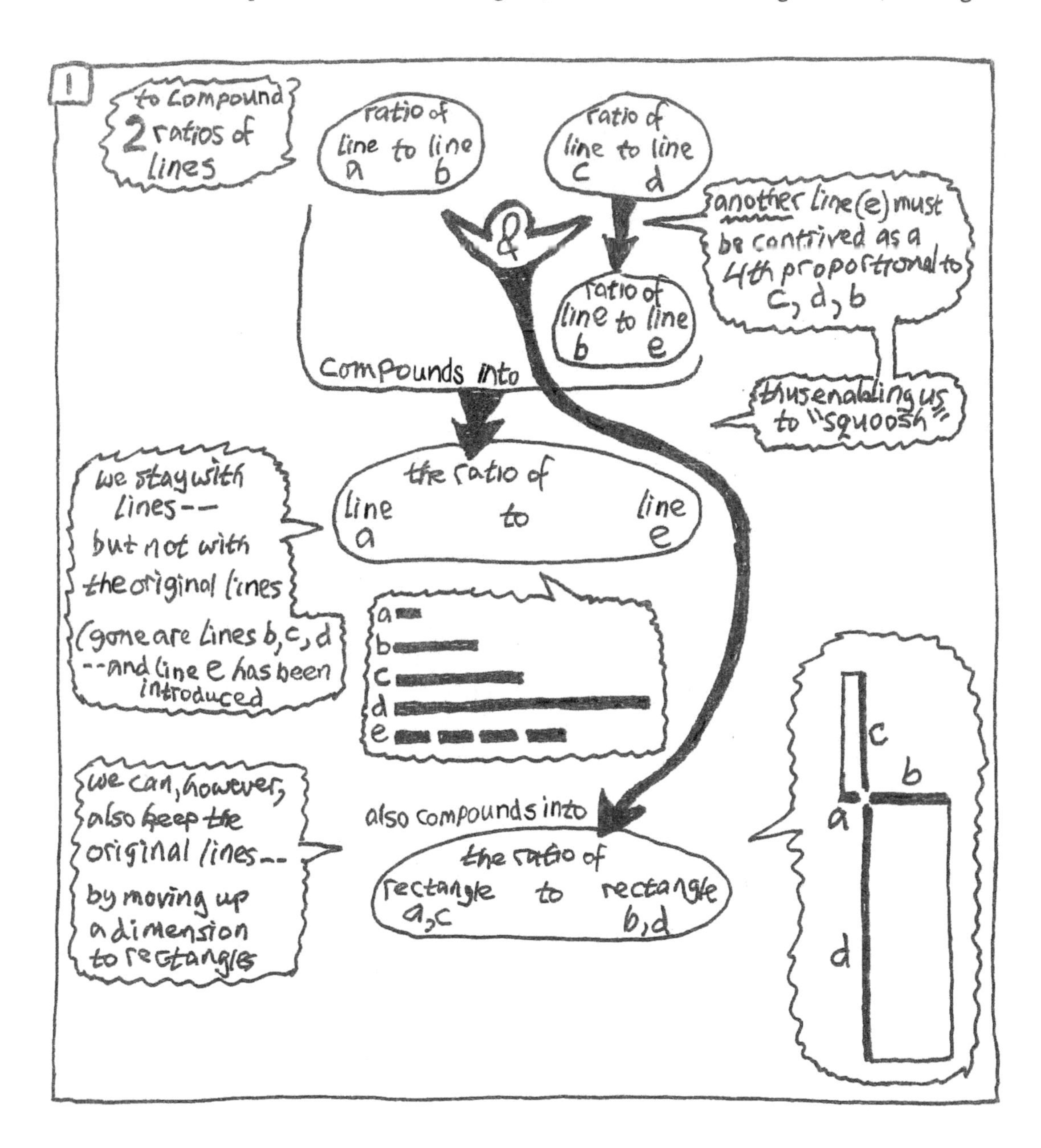

Likewise in compounding *three* ratios of lines, we can get a ratio of lines as our result—but only if we do not stay with the original lines. (Five of the six original lines will drop out, and one of the six will have, to a contrived line that we introduce, the compounded ratio.) Here also, we can retain the original lines in the compounded ratio if we wish, but then the terms of the ratio must move up *two* dimensions, to *parallelepipeds* (whose sides are the original lines.) See figure 2.

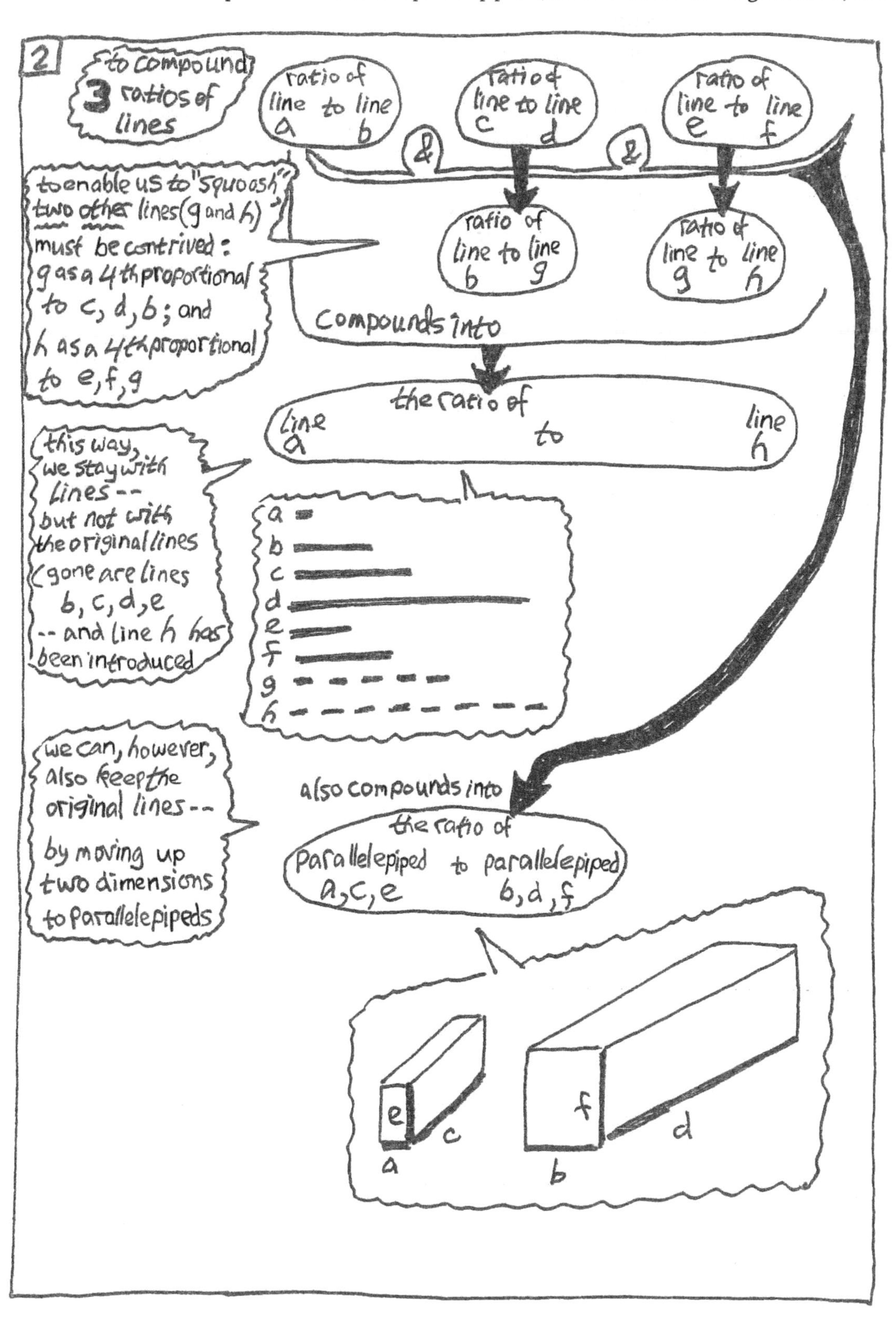

In compounding *four* ratios of lines, however, we have no choice but to contrive lines and give the result as a ratio of one of the original lines to one of the contrived lines. That is because we cannot go up to a box with more dimensions than a parallelepiped. We cannot build something that will have four of the lines as sides at right angles to each other. The compounded ratio cannot be represented as a ratio of two such things.

Unless we let lines drop out in getting to the result, and contrive lines to take their place in expressing the ratios, we will not be able to exhibit the result of compounding as a ratio of two terms. Instead, we will have to speak of "the ratio compounded of the ratios of rectangle [A, C] to rectangle [B, D] and of rectangle [E, G] to rectangle [F, H]," which is no better than, and indeed is even worse than leaving it at speaking of "the ratio compounded of the ratios of line A to line B, and of line C to line D, and of line E to line F, and of line G to line H." If that is too cumbersome, we might—as an *abbreviation*—write this:

the ratio of (A, C, E, G) to (B, D, F, H).

Of course, if A, B, C, D, E, F, G, H were numbers, then there would be no difficulty in speaking like that about the ratio compounded of the ratios of A to B, and of C to D, and of E to F, and of G to H. The result of the compounding would be the ratio of one number to another (the ratio's two numbers being products of multiplication: one of them the product of multiplying A, C, E, G together; and the other, the product of multiplying B, D, F, H together). This would be

the ratio of (A*C*E*G) to (B*D*F*H).

But as we saw in Part Two of this guidebook, relations of magnitudes and relations of multitudes do not completely correspond to each other.

When we name lines as "a" or "b" or "c" or "d" or "e" or "f" or "g" or "h," then we will name a rectangle by using the abbreviation "a,c" or "b,d,"; and a parallelepiped, by using the abbreviation "a,c,e" or "b,d,f." But when we use the abbreviation "a,c,e,g" or "b,d,f,h" we are not naming a box at whose corners meet four lines perpendicular to each other. Rather, we are indicating a process of manipulation whereby we get terms to express the ratio compounded out of ratios in which the antecedents are (in order) a, c, e, and g, and the consequents are (in order) b, d, f, and h. It is tiresome to spell out the cumbersome manipulation all the time, and it eases our mental labor if we speak sometimes as if, for example, the abbreviation "(a,c,e,g)" were a name like the abbreviation "(a,c)" or "(a,e)." Perhaps it would be better for us to feel some inhibition about saying things like "(a,c,e,g)."

Descartes suggests that it was a related inhibition that kept the ancients from doing at all what he can teach us to do quite easily. He also suggests that the ancients were not quite so stiff and formal as they liked to present themselves as being—that they were led by mere pretentiousness to keep off-stage as vulgar what Descartes would like to bring to center-stage. As we shall see when we examine what Descartes wrote, the manipulation required for solving problems will not be eased by using the abbreviation "the ratio of (a,c,e,g) to (b,d,e,h)" as if it referred to something. We can, after all, always say that it does mean something—that it refers to a ratio compounded out of certain other ratios. Rather, what will ease the manipulation is overcoming the distinction that lies behind the seeming duplication of labor in the Fifth and Seventh Books of Euclid's *Elements*. Why and how to overcome that distinction will be our chief concern in looking at Descartes.

Why were locus problems important to the ancients? Locus theorems show that if a line is of sort A, then it is a locus of sort Z. When lines that have already been obtained are shown to have some locus property, then the question arises whether it is *only* they that have the property. If indeed only those lines have that property, the property can claim definitively to characterize those lines. Locus problems find out, if a line is a locus of sort Z, what sort of a line it then is—for example, a line of sort A. To show that it is a line of that sort, you must construct it as a line of sort A.

If you *define* some line as being a locus of sort Z—then you need to *demonstrate* that the line so defined *exists*: that is, you need to *construct* it. Why? Because it is possible that there is *no line* such that *every* point on it satisfies the locus property. The points that have the property may lie on a line and may be infinitely numerous—and yet the collection of points may lack continuity.

Straight and circular lines are the only lines lying in a plane that are given by postulate; all others must be obtained by construction. Though we can characterize them in many ways, we must obtain them by construction—and that may mean by letting the surface of a solid be intersected by the plane in which the curves so obtained are to lie.

But there are many curves that we might not think to construct because we haven't even thought about them. How might we consider in somewhat orderly fashion the various curves that are possible? One way to proceed would be to move a point through different positions while keeping constant certain things about its distances from other points or else from lines in fixed positions.

For some features, we already have the lines that would be traced out by a point that kept such features constant as its position varied.

If what is constant is the point's *distance* from a fixed *point*—
then the locus is a circle.

If its *distance* from a fixed *straight line*—
then another straight line.

If the *sum (or the difference) of its distances* from *two* fixed *points*—
then an ellipse (or an hyperbola).

If the *ratio of its distances* from *two* fixed *straight lines*—
then another straight line.

If the *ratio of its distances* from a fixed *point and* a fixed *straight line*—
then a conic section (whose sort depends on whether the ratio is the same as, or is greater than, or is less than a ratio of equality).

Now what if what is constant is the *ratio compounded of the ratios of its distances* from *three* fixed *straight lines*? Then the locus is a conic section. What if from *four* fixed straight lines? Again, it is a conic section. From *five*, or *six*, or *more*? Who knows?

The n-line locus is a curve such that, while distances to a point on it from the n fixed straight lines are not constant, and the distances are not even in constant ratio with each other, nonetheless each of these non-constant ratios varies in a way that is correlated with the way in which the others vary—so that the ratio compounded of these non-constant ratios is a constant ratio. What is it a ratio of?

Consider the 6-line locus. It is a line each of whose points is at such distances from the sides of a 6-sided polygon that,

> though none of the six distances is constant,
> nor is any of the three ratios that are formed by
> taking them pairwise (in any combination)
> yet the ratio that's *compounded of* those three non-constant ratios
> nonetheless *is* constant.

We can think of it this way:

> if a parallelepiped is contained by the three distances that
> are the antecedents in the three ratios,
> and another parallelepiped is formed by the distances that
> are the consequents in the three ratios,
> then the ratio of the one parallelepiped to the other parallelepiped
> is a constant ratio.

Apollonius is pleased to have figured out things that are the necessary means for solving certain loci. Pappus is displeased by what he regards as Apollonius's boasting. Pappus, in the Seventh Book of his *Collection,* compares Apollonius unfavorably to Euclid, as follows:

He [Apollonius] says in his Third Book that
the locus with respect to three or four lines
had not been completed by Euclid;
but neither he himself nor anyone else could have added,
by means of only the conic things that
had been shown up to the time of Euclid,
even a small thing to the things written by Euclid—
as Apollonius himself bears witness to
when he calls it impossible to complete the locus apart from
the things that he himself was compelled to write up.
Now Euclid accepted Aristaeus as worthy of esteem
for the conic things that he had contributed—
not overtaking or intending to overthrow his treatment of those things,
being equitable and gracious toward any who
were able to augment mathematics however much,
as one should be,
and in no way being disposed to take offense,
but accurate, not boastful like that one [Apollonius],
he wrote as much as was possible to
show of the locus by means of the conics of that one [Aristaeus],
not saying that what had been shown was complete,
for it would have been necessary to blame this,
but it is not so now in any way,
since [Apollonius] himself is not called to account though he
left most things in his conics incomplete,
and he was able to add the remaining things to the locus
having seen beforehand
the things already written by Euclid about the locus,
and having spent much time at leisure with those who had
learned mathematics from Euclid in Alexandria,
which was the source of his becoming
not at all unaccomplished mathematically.

Pappus then goes on to present locus problems that are still unsolved—by Apollonius or anyone else.

Pappus, as we have seen, criticizes the language used by some who speak of the locus problems: these people speak of the ratio obtained by compounding many ratios (say, five of them) almost as if it were the same as a ratio of a multi-dimensional figure to another such figure (say, a five-dimensional figure).

Now a ratio compounded of *two ratios of lines* is the same as a ratio of *two rectangles* (or—if you are willing to contrive fourth proportionals—of two lines; or —if you wish—of two parallelepipeds). And a ratio compounded of *three ratios of lines* is the same as a ratio of *two parallelepipeds* (or—if you are willing to contrive fourth proportionals—of two lines, or of two rectangles). But a ratio compounded of *four ratios of lines* cannot be expressed by two terms *unless* you are willing to contrive fourth proportionals—the two terms in the compounded ratio can be lines or rectangles or parallelepipeds, but nothing of higher dimension. That is why, for people who think like Pappus, to speak of

"the ratio of (A, B, C, D, E) to (A', B', C', D', E')"

can only be an abbreviation—an abbreviation meaning merely this:

"the ratio compounded of the five ratios:

of A to A'
and
of B to B'
and
of C to C'
and
of D to D'
and
of E to E'."

Pappus's complaint may seem like mere pedantry if you believe the heart of things to be the instrumental power derived from manipulating quantities, as in the compounding of ratios—rather than believing the heart of things to be the visualization of form, using the embodiment of ratios in figures for the sake of the insight into form that is thus provided.

But if you condemn Pappus merely for technical pedantry, perhaps you are not being radical enough. Perhaps you must go further and condemn him for proud ignorance.

Is compounding merely a clumsy way to manipulate quantity, and is construction of solid loci merely a clumsy and limited way to handle relations of quantity?

If you reject the classical view of mathematics, then Pappus's complaint about the mathematicians' misleading technical jargon is worse than an expression of technical pedantry. For a man like Descartes, it is the expression of a point of view which overlooks the chief point of mathematics—its power as an instrument for solving problems. To Descartes, mathematics is not of value as a path toward insight into form; it is the means for achieving mastery. Not mastery in this or that narrow field of endeavor, but universal mastery—mastery over nature.

PART VII
ARITHMETICAL PROBLEM-SOLVING, AND THE ANALYTIC ART

Diophantus's *Arithmetica*: selections Viète's *Introduction to the Analytic Art*

CHAPTER 1. SOLVING ARITHMETICAL PROBLEMS: DIOPHANTUS

geometrical and arithmetical problems

We have been considering for some time how to represent the relationships of size that straight lines have to one another. Back in Part Three of this guidebook, when we began to examine the propositions in which Apollonius characterizes the conic sections, we saw how the notion of application of areas enabled us to say things about the relative sizes of straight lines that could not be said by directly comparing the lines themselves. Even before that (way back in Part Two), when we examined some propositions from the Second Book of Euclid's *Elements* (especially the 5th and 6th propositions) we saw how the building of squares and rectangles also enabled us to say things about the relative sizes of straight lines that could not be said by directly comparing the lines themselves.

The propositions in the Second Book of the *Elements* enable us to find straight lines when we do not have them but we do have other straight lines to which they are related by way of squares and rectangles.

For example, suppose that we need to find a straight line (call it x) that is related as follows to two given straight lines (call them a and b) such that half of a is larger than b:

> a rectangle which is thrown alongside the one given line a
> and is equal to a square on the other given line b
> will fall short of the rectangle whose sides are
> the given line a and the sought-for line x
> by a square whose side is the sought-for line x (fig. 1).

The 5th proposition of the Second Book of the *Elements* can help us to find line x, given lines a and b, as follows. First, what is shown by II.5 is depicted in figure 2.

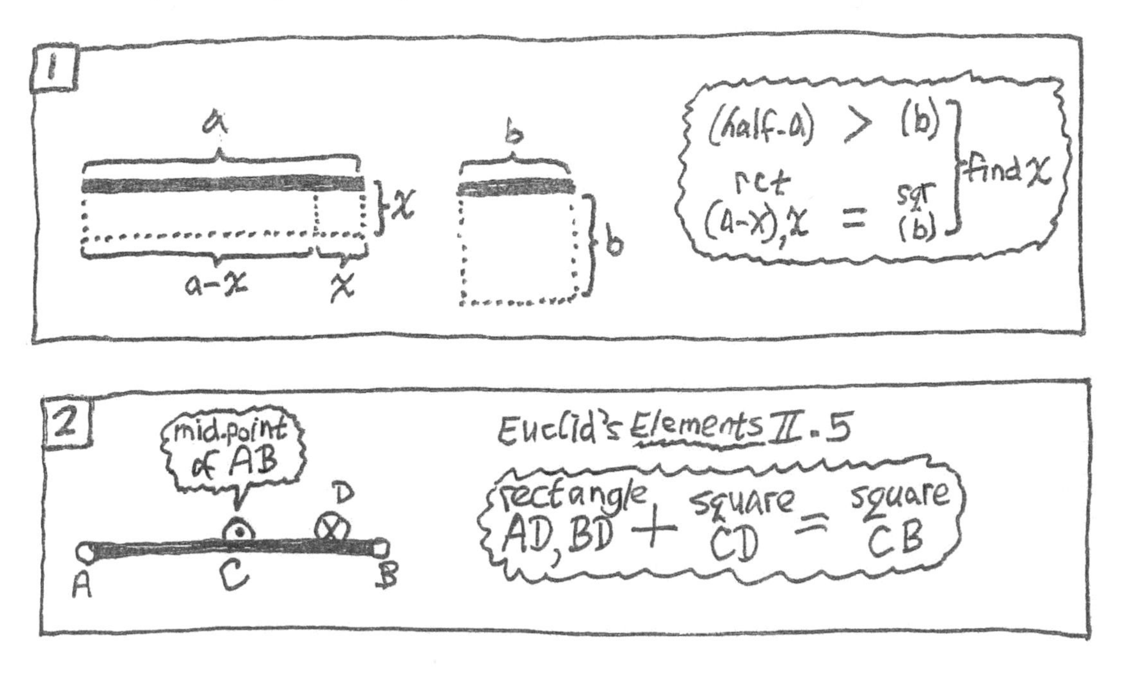

Now, to solve our problem,

> let line AB be equal to the one given line a;
> now, at AB's midpoint C,
> erect a perpendicular CO that is equal to the other given line b;
> now, with point O as center and with half of line a as radius,
> draw a circle;
> this circle will—since half of line a is greater than line b—
> cut AB at two points;
> let one of these points be D.

Now if we let DB be line x, then we can show that

rectangle [(a-x), x]	is equal to	square [b];

they are equal to each other because they are both equal to the same thing (fig. 3).

That problem in geometry has, as its counterpart in arithmetic, the following numerical problem:

> given two numbers, the one being smaller than the double of the other,
> find a number that is
> the "side" of a "square" number, when this "square" number
> is less than the "plane" number which is the product of
> multiplying the unknown number (as one "side")
> by the one given number (as the other "side")—
> and is less, moreover, by a "square" number whose "side" is
> the other given number.

Now you may be tempted immediately to replace this problem in numbers by the corresponding problem in algebra. Resist that temptation. What we are considering may be *like* solving for x when (given that a >2b) we have $ax - x^2 = b^2$; but it is *not* the *same as* what it is like. The difference is the very thing that we are trying to understand.

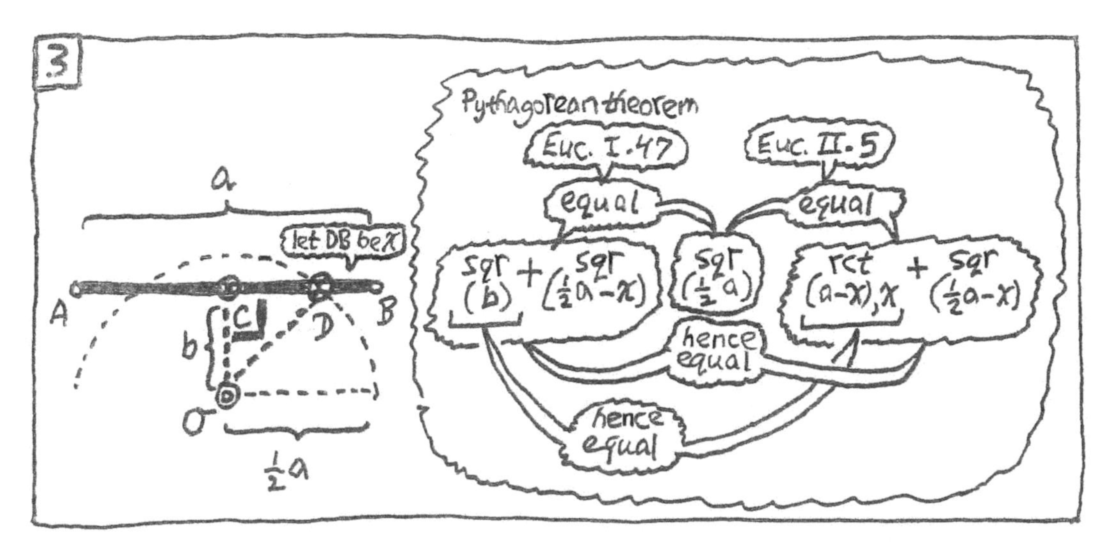

Similarly for the 6th proposition of the Second Book of Euclid's *Elements*. What II.6 shows, is depicted in figure 4. (In II.6, as in II.5, the rectangle contained by AD and BD is equal to the difference of two squares—one of them being raised from CD and the other from CB; but whereas in II.5 it was the square from CB that was the larger, now in II.6 it is the square from CD that is the larger.)

Suppose now that we need to find a straight line (call it *x*) that is related as follows to two given straight lines (call them *a* and *b*) such that half of *a* is larger than *b*:

a rectangle which is thrown alongside the one given line *a* and is equal to a square on the other given line *b* will exceed by a square whose side is the sought-for line x (fig. 5).

The 6th proposition of the Second Book of the *Elements* can help us to find line x, given lines a and b, as follows.

To solve this problem, then,

let AB be equal to the one given line *a*; at AB's endpoint B, erect a perpendicular BP that is equal to the other given line *b*, and join point P to AB's midpoint C; now, with point C as center and with line CP as radius, draw a circle; this circle will—since half of line *a* is greater than line *b*— cut AB's prolongation; let the point where it does so be D.

Now if we let DB be line *x*, then we can show that

rectangle [(*a* + *x*), x]	is equal to	square [*b*];

they are equal to each other because the results will be equal to each other when the same thing is added to both (fig. 6).

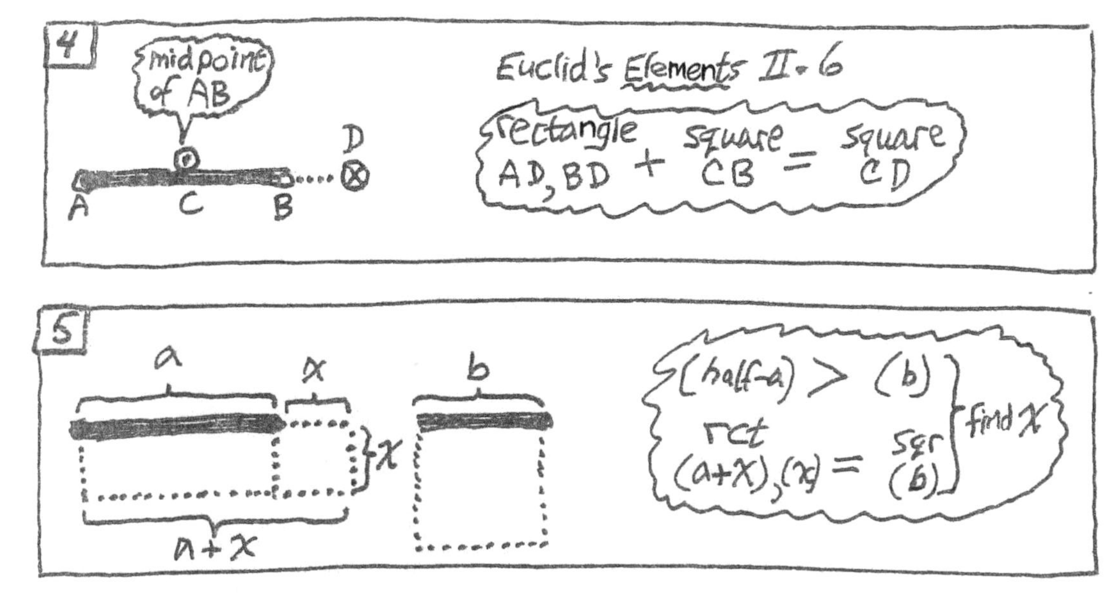

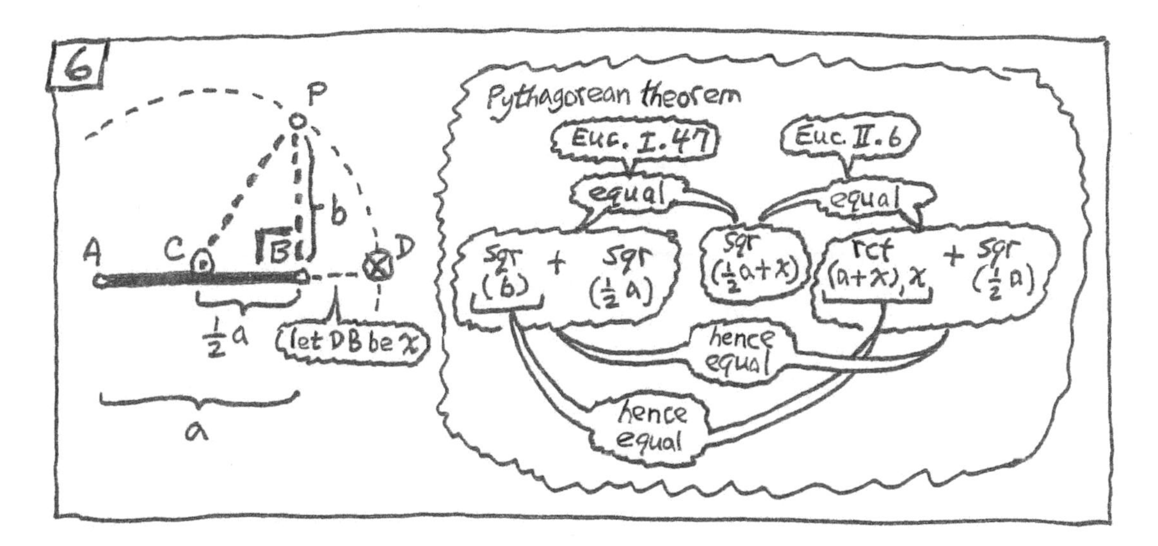

That problem in geometry has, as its counterpart in arithmetic, the following numerical problem:

> given two numbers, the one being smaller than the double of the other,
> find a number that is
> the "side" of a "square" number, when this "square" number will—
> if it is added to the "plane" number which is the product of
> multiplying the unknown number (as one "side")
> by the one given number (as the other "side")—
> be equal to a "square" number whose "side" is
> the other given number.

Again, resist the temptation immediately to replace this by the corresponding problem in algebra—namely, solving for x when (given that $a > 2b$) we have the equation $ax + x^2 = b^2$.

How, then, are arithmetical problems to be solved if not by having recourse to the algebra with which we are so familiar nowadays? To see how, we need to take a look at the work of another Greek writer of antiquity—Diophantus, whose dates are around 250 AD. Our consideration of Diophantus will be our last look at the ancients in our preparation for the radical transformation in mathematics that was the root of Descartes' project for the total transformation of the world in which human beings live.

nomenclature

The treatment of arithmetical matters by Diophantus is, he says, for someone eager to learn the finding of solutions to problems in numbers. That person will know, he says, that the numbers are all composed of some multitude of units. It is clearly established, he continues, that the numbers spring forth boundlessly (*eis apeiron*). Diophantus goes on to name some kinds of them:

"squares"	come from	some number multiplied by ("made manifold by") itself—and this number is called a *"side of the square"*;
"cubes"	come from	squares multiplied by their sides;
"powero-powers" (*dynamodynameis*),	come from	squares multiplied by themselves;
"powero-cubes" (*dynamokubeis*)	come from	squares multiplied by cubes from the same sides;
"cubo-cubes"	come from	cubes multiplied by themselves.

Readers of this guidebook have already been told by Euclid that numbers are multitudes of units, and have already heard from him of the numbers that are called "squares" and "cubes." The number nine, for example, is a "square," since it comes from multiplying the number three by itself (that is to say: three multitudes of three units each make a multitude of nine units); and the "side" of the "square" that is nine is three; and when the "square" nine is multiplied by its "side" three, it makes the "cube" that is twenty-seven.

After we considered what Euclid said, we took note (in Chapter 1 of Part Three of this guidebook) of the Greek usage according to which a side that has the power to give rise to some square is referred to as that square's "power" (*dynamis*). Now we see Diophantus give a name to a number that is itself a square and whose side is also a square. The number that is named the *dynamodynamis* ("powero-power") is formed by taking some number that is itself a square, and using this square number as a power that makes another square. The resulting square is not called a squaro-square. To call it that, would name its components as if they were resultants. Rather, it is called a powero-power. That names its components as if they indeed are components. (Diophantus was not alone in using this terminology of "powero-power." The ancient mathematician Hero used the term "dynamodynamis" to refer to a four-fold product made from ingredients furnished by the measurement of the sides of a triangle.)

So, for example, eighty-one is a "powero-power," since the "square" number nine has, when multiplied by itself, the power to give the "square" number eighty-one; likewise, from the "square" nine, when it is multiplied by twenty-seven (which is the "cube" from the same "side," three), comes the "powero-cube" two hundred forty-three; and from the "cube" twenty-seven, when it is multiplied by itself, comes the "cubo-cube" seven hundred twenty-nine.

These sorts of numbers—"squares," their "sides," "cubes," "powero-powers," "powero-cubes," and "cubo-cubes"—are singled out by Diophantus because, as he says, it is from the adding, subtracting, and multiplying of these, or from the ratio of them to one another, or of each of them to their own "sides," that most arithmetical problems are contrived. These problems are solved, he says, by proceeding along the road (*hodos*) shown below.

Diophantus starts his reader out by noting how each of these numbers, having obtained an abbreviated name, is recognized as an "element" of "arithmetical theory." These abbreviated names indicate the kinds of these numbers. Diophantus refers to each kind as a "form" (*eidos*; the plural of which is *eidê*). Diophantus will present problems, and will solve them, in terms of the numerical *eidê*. An *eidos* represents a characteristic of numbers of a certain kind; these characteristics will allow various rules for calculating with numbers to be used as well for calculating with *eidê*.

The square is called "power" and its sign is "D^{y}"—an abbreviation for the Greek word *dynamis*. (The term "power," it should be noted, as used nowadays refers, not to what has the power to give rise to a square, but rather to the square that is the effect of that power.) The term "power"—alone and in compounds—will be used by Diophantus to refer only to what arises from a number that is as yet *undetermined*. (To refer to what arises from a *known* number that has been multiplied by itself, Diophantus uses the term "square.") The number that is still being sought is, like any number that is already known, a multitude of units. It is merely *as yet* undetermined; the aim is to find that determinate multitude of units (or those determinate multitudes of units) which will solve the problem. To designate a square number that is not composed entirely of already determined numbers, Diophantus draws a little square followed by the final letters that are used in Greek nouns or adjectives to indicate gender, number, and case; but, for convenience, what this little square designates will be designated by us as "Sq."

The "cube" (*kybos* in Greek) has as its sign "K^{y}"; the "dynamo-dynamis" (the "square" multiplied by itself) has "$D^{y}D$"; the "dynamo-cube" (the "square" multiplied by that "cube" which arises from the very same "side" as the "square" does) has "DK^{y}"; the "cubo-cube" (the "cube" multiplied by itself) has "$K^{y}K$." These abbreviations are all used only for a number whose "side" is as yet undetermined.

The number that, while having none of these properties, is a multitude of units that is as yet undetermined is called a "number" (in Greek, *arithmos*) and its sign is "A^{r}." There is another sign for the unchanging unit (*monad*) in multitudes that are determinate—it is "M^{o}." This is usually prefixed to the known number of them, which is indicated in the Greek manner of writing numbers, that is, by a letter of the alphabet with a horizontal line above it. So, putting a bar over an alpha indicates "one," over a beta indicates "two," over a gamma indicates "three," and so on.

As corresponding English abbreviations, what we shall write is this : "Un" for units; "Nu" for the undetermined "number"; "Po" for its "power," which is produced when it is multiplied by itself; "Cu" for its "cube"; and, correspondingly, "PoCu" and "CuCu." Rather than Greek letters capped by a bar, we shall employ letters of our Latin alphabet preceded by the sign "#"; but sometimes, for convenience, we'll employ our usual Hindu-Arabic numerals.

Now in dealing with numbers, it is often necessary (as we see, for example, at the beginning of the 31st problem of Diophantus's Book Four, or at the beginning of the 9th through 12th problems of his Book Five) to partition the original unit and count up some of the fragments into which the original unit is broken. ("Fragment," like "fracture," derives from the Latin word for "break," which also gives us our word "fraction.") Whatever is measured by the original unit—thus yielding some number of units—will, when divided by a smaller unit, yield a larger number of units; by dividing the unit we thus multiply the number of units that will be found in whatever is measured. If, for example, we break the unit in three, then something that was measured as having four of the old units will now have twelve of the new units.

Diophantus reminds the reader that the equal parts into which the old unit is divided are called by names that resemble the names of the numbers—the fragment that is the "third" part being named after the number named "three," for example, and the fragment that is the "fourth" part being named after the number "four"; so also, he says, the numbers of the various forms now being named will have the corresponding parts of the unit named after them:

from this name:	will come this name of a part of the unit:
the "number" (*arithmos*)	a "numberth" (*arithmoston*)
the "power" (*dynamis*)	a "powerth" (*dynamoston*)
the "cube" (*kubos*)	a "cubeth" (*kuboston*)
the "powero-power" (*dynamodynamis*)	a "powero-powerth" (*dynamodynamoston*)
the "powero-cube" (*dynamokubos*)	a "powero-cubeth" (*dynamodynamoston*)
the "cubo-cube" (*kubokubos*)	a "cubo-cubeth" (*kubokuboston*).

Such a part of the unit will have as its sign the sign of what it is named after (this original sign being "A^{r}" when the fragment is "arithmetic" or "numeric," "D^{y}" when the fragment is "dynamic" or "power-ish," "K^{y}" when the fragment is "cubic," and so on)—with an additional mark, however, to distinguish its form (*eidos*). So, in English we might distinguish it by marking it with a preceding double comma like this: ",,"—thus writing " ,,Cu" as an abbreviation to indicate a cubeth (a "cubic" fragment), and " ,,3" as an abbreviation to indicate a 3rd part of a unit (a fragment that is one-third of it).

We need not here go further into the details of Diophantus's notation. We can attend to such details when necessary in considering some of his propositions. The chief thing to bear in mind—a matter of the utmost importance—is this: all of his signs are merely abbreviations for words. That cannot be emphasized too much: all of his signs are merely abbreviations for words.

multiplication

Having set out the nomenclature for each of the numbers, Diophantus passes on to the multiplying of them. He notes that this is something that is clear on account of the very naming of them, but he nonetheless elaborates, showing how rules of calculation ordinarily used for numbers are to be used also with his nomenclature, thus:

A "number" multiplied		
by a "number,"	makes	a "power";
by a "power,"	makes	a "cube";
by a "cube,"	makes	a "powero-power";
by a "powero-power,"	makes	a "powero-cube";
by a "powero-cube,"	makes	a "cubo-cube";
and a "power"		
by a "power,"	makes	a "powero-power";
by a "cube,"	makes	a "powero-cube";
by a "powero-power,"	makes	a "cubo-cube";
and a "cube"		
by a "cube,"	makes	a "cubo-cube."
And every number multiplied		
by its same-named part [of a unit]	makes	a unit.

What Diophantus means by the last point is this: a third part of a unit taken three times makes a unit, as does a fourth part of a unit taken four times, and so on; we say, therefore, that three multiplied by a third part of a unit must make a unit, as does four multiplied by a fourth part of a unit, and so on; and therefore, that a unit is made when, for example, a "number" is multiplied by the "numeric" fragment of a unit, or when a "cube" is multiplied by the "cubic" fragment of a unit. (Note, by the way, that when Diophantus in that last point refers to every number, he does not mean every as yet unknown "number" in contradistinction to its "power" or its "cube" or some other self-multiple of it; rather, he means any number at all, including self-multiples of as yet unknown numbers—"powers," "cubes," and so on.

So: the unit being unchangeable and always constant,
the form (*eidos*) that is multiplied by it will be the same form (*eidos*).

The unit, it must be remembered, since it is not a multitude of units, is therefore not itself a number. The unit does, however, enter into calculations with numbers. When you add a unit to some multitude, or subtract a unit from it, you will get a different multitude; but nonetheless you will not get a different multitude when you "multiply" by a unit, or "divide" by it. The point being made by Diophantus, then, is this: multiplying by the unit will not change the *eidos* any more than it changes the multitude—which is not at all.)

Diophantus continues:

The parts with names that resemble the names of numbers,
when multiplied by one of themselves, will make
parts with names that resemble the names of numbers,
such as the following:
the “numberth”

by the “numberth,”	makes	the “powerth”;
by the “powerth,”	makes	the “cubeth”;
by the “cubeth,”	makes	the “powero-powerth”;
by the “powero-powerth,”	makes	the “powero-cubeth”;
by the “powero-cubeth,”	makes	the “cubo-cubeth.”

Finally—having already dealt with the multiplication

of some form of number,
by some form of number—

and of some form of number,
by its same-named part of a unit—

and of the “numberth” part of a unit,
by some form of number—

Diophantus now will deal with the multiplication

of the unit’s part that is same-named as some form of number,
by some other form of number.

(In this immediate sequel, each subsection will omit to mention the item that is covered by the statement in which he has said that every number will, when multiplied by the same-named part of it, make a unit. For example, in treating the multiplication of a “numberth” part of a unit, Diophantus will omit multiplication of it by the “number.” Likewise, in treating the multiplication of a “powerth” part of a unit, he will omit multiplication of it by the “power.” And so on.) Here, then, is Diophantus’s continuation:

And this will happen name-resemblance-wise—

the "number^th"			
	[—		]
	by the "power,"	makes	the "number";
	by the "cube,"	makes	the "power";
	by the "powero-power,"	makes	the "cube";
	by the "powero-cube,"	makes	the "powero-power";
	by the "cubo-cube,"	makes	the "powero-cube";
and the "power^th"			
	by the "number,"	makes	the "number^th";
	[—		]
	by the "cube,"	makes	the "number";
	by the "powero-power,"	makes	the "power";
	by the "powero-cube,"	makes	the "cube";
	by the "cubo-cube,"	makes	the "powero-power";
and the "cube^th"			
	by the "number,"	makes	the "power^th";
	by the "power,"	makes	the "number^th";
	[—		]
	by the "powero-power,"	makes	the "number";
	by the "powero-cube,"	makes	the "power";
	by the "cubo-cube,"	makes	the "cube";
and the "powero-power^th"			
	by the "number,"	makes	the "cube^th";
	by the "power,"	makes	the "power^th";
	by the "cube,"	makes	the "number^th";
	[—		]
	by the "powero-cube,"	makes	the number";
	by the "cubo-cube,"	makes	the "power";
and the "powero-cube^th"			
	by the "number,"	makes	the "powero-power^th";
	by the "power,"	makes	the "cube^th";
	by the "cube,"	makes	the "power^th";
	by the "powero-power,"	makes	the "number^th";
	[—		]
	by the "cubo-cube,"	makes	the "number";
and the "cubo-cube^th"			
	by the "number,"	makes	the "powero-cube^th";
	by the "power,"	makes	the "powero-power^th";
	by the "cube,"	makes	the "cube^th";
	by the "powero-power,"	makes	the "power^th";
	by the "powero-cube,"	makes	the "number^th"
	[—		].

What Diophantus has presented about the results of multiplication involving the various sorts of numerical *eidê* will now be depicted in overview, with abbreviations, in the following two boxes:

in eidetic abbreviation			*which is to say*	
this	*by this*	*makes this*	*this by this*	*makes this*
\/	\/	\/	\/ \/	\/
Nu	Nu	Po	(n)*(n)	(nn)
Nu	Po	Cu	(n)*(nn)	(nnn)
Nu	Cu	PPo	(n)*(nnn)	(nnnn)
Nu	PPo	PCu	(n)*(nnnn)	(nnnnn)
Nu	PCu	CCu	(n)*(nnnnn)	(nnnnnn)
Po	Po	PPo	(nn)*(nn)	(nnnn)
Po	Cu	PCu	(nn)*(nnn)	(nnnnn)
Po	PPo	CCu	(nn)*(nnnn)	(nnnnnn)
Cu	Cu	CCu	(nnn)*(nnn)	(nnnnnn)
any *eidos*	the part with the same name	Un		
(e.g., Nu	Nuth	Un	(n)*(n)th	unit
or Po	Poth	Un	(nn)*(nn)th	unit
etc.)			etc.	
any *eidos*	Un	same *eidos*		
(for example				
Nu	Un	Nu	(n)*[unit]	(n)
Po	Un	Po	(nn)*[unit]	(nn)
etc.)			etc.	
Nuth	Nuth	Poth	(n)th*(n)th	(nn)th
Nuth	Poth	Cuth	(n)th*(nn)th	(nnn)th
Nuth	Cuth	PPoth	(n)th*(nnn)th	(nnnn)th
Nuth	PPoth	PCuth	(n)th*(nnnn)th	(nnnnn)th
Nuth	PCuth	CCuth	(n)th*(nnnnn)th	(nnnnnn)th

in eidetic abbreviation			*which is to say*		
this	*by this*	*makes this*	*this*	*by this*	*makes this*
\/	\/	\/	\/	\/	\/
[—-		]	[—-		]
Nuth	Po	Nu	(n)^{th*}	(nn)	(n)
Nuth	Cu	Po	(n)^{th*}	(nnn)	(nn)
Nuth	PPo	Cu	(n)^{th*}	(nnnn)	(nnn)
Nuth	PCu	PPo	(n)^{th*}	(nnnnn)	(nnnn)
Nuth	CCu	PCu	(n)^{th*}	(nnnnnn)	(nnnnn)
Poth	Nu	Nuth	(nn)^{th*}(n)		(n)th
[—-		]	[		]
Poth	Cu	Nu	(nn)^{th*}(nnn)		(n)
Poth	PPo	Po	(nn)^{th*}(nnnn)		(nn)
Poth	PCu	Cu	(nn)^{th*}(nnnnn)		(nnn)
Poth	CCu	PPo	(nn)^{th*}(nnnnnn)		(nnnn)
Cuth	Nu	Poth	(nnn)^{th*}(n)		(nn)th
Cuth	Po	Nuth	(nnn)^{th*}(nn)		(n)th
[—-		]	[—-		]
Cuth	PPo	Nu	(nnn)^{th*}(nnnn)		(n)
Cuth	PCu	Po	(nnn)^{th*}(nnnnn)		(nn)
Cuth	CCu	Cu	(nnn)^{th*}(nnnnnn)		(nnn)
PPoth	Nu	Cuth	(nnnn)^{th*}(n)		(nnn)th
PPoth	Po	Poth	(nnnn)^{th*}(nn)		(nn)th
PPoth	Cu	Nuth	(nnnn)^{th*}(nnn)		(n)th
[—-		]	[		]
PPoth	PCu	Nu	(nnnn)^{th*}(nnnnn)		(n)
PPoth	CCu	Po	(nnnn)^{th*}(nnnnnn)		(nn)
PCuth	Nu	PPoth	(nnnnn)^{th*}(n)		(nnnn)th
PCuth	Po	Cuth	(nnnnn)^{th*}(nn)		(nnn)th
PCuth	Cu	Poth	(nnnnn)^{th*}(nnn)		(nn)th
PCuth	PPo	Nuth	(nnnnn)^{th*}(nnnn)		(n)th
[—-		]	[		]
PCuth	CCu	Nu	(nnnnn)^{th*}(nnnnnn)		(n)
CCuth	Nu	PCuth	(nnnnnn)^{th*}(n)		(nnnnn)th
CCuth	Po	PPoth	(nnnnnn)^{th*}(nn)		(nnnn)th
CCuth	Cu	Cuth	(nnnnnn)^{th*}(nnn)		(nnn)th
CCuth	PPo	Poth	(nnnnnn)^{th*}(nnnn)		(nn)th
CCuth	PCu	Nuth	(nnnnnn)^{th*}(nnnnn)		(n)th
[—-		]	[		]

After presenting the rules of calculation that tell what is the *eidos* that results when *eidê* are multiplied by each other, Diophantus turns to multiplications involving not only numbers that are "present" but also those that are "lacking." What he has to say about that will be clearer if we first consider what might lead to it. Consider the following examples, which are followed by depictions of them, in figures 7 and 8:

Fifteen is the result that we get
if we suppose five to be present and we multiply by three;
and since a presence of three results when
there is both a presence of nine and a lacking of six,
we can say that fifteen results when
a present five is multiplied by the result of there being
both a presence of five and a lacking of six.
But when a presence of five is multiplied by
a presence of nine alone,
the result is a present forty-five;
and a presence of forty-five
can be a net presence of fifteen only if
there is also a lacking of thirty.
Hence, when a *presence of five* is multiplied by a *lacking of six*,
the result must be a *lacking of thirty*.

By using the procedure of the previous example,
it can be established that a lacking of thirty results when
a lacking of three is multiplied by a presence of ten;
and since a presence of ten results when
there is both a presence of twenty-two and a lacking of twelve,
we can say that a lacking of thirty results when
a lacking of three is multiplied by the result of
a presence of twenty-two and a lacking of twelve;
and that is to say that
a lacking of thirty is the net result of there being:
both a lacking of three multiplied by a presence of twenty-two,
and a lacking of three multiplied by a lacking of twelve.
But the former multiplication—
namely, of a lacking of three by a presence of twenty-two—
makes a lacking of sixty-six.
Now with a lacking of sixty-six, how can we have as the net result
our original lacking of merely thirty?
We can have it only if there is another thirty-six also present.
Since we have already taken into account
everything but the result when
the *lacking of three* is multiplied by the *lacking of twelve*,
the result of this multiplication must be
a *presence of thirty-six*.

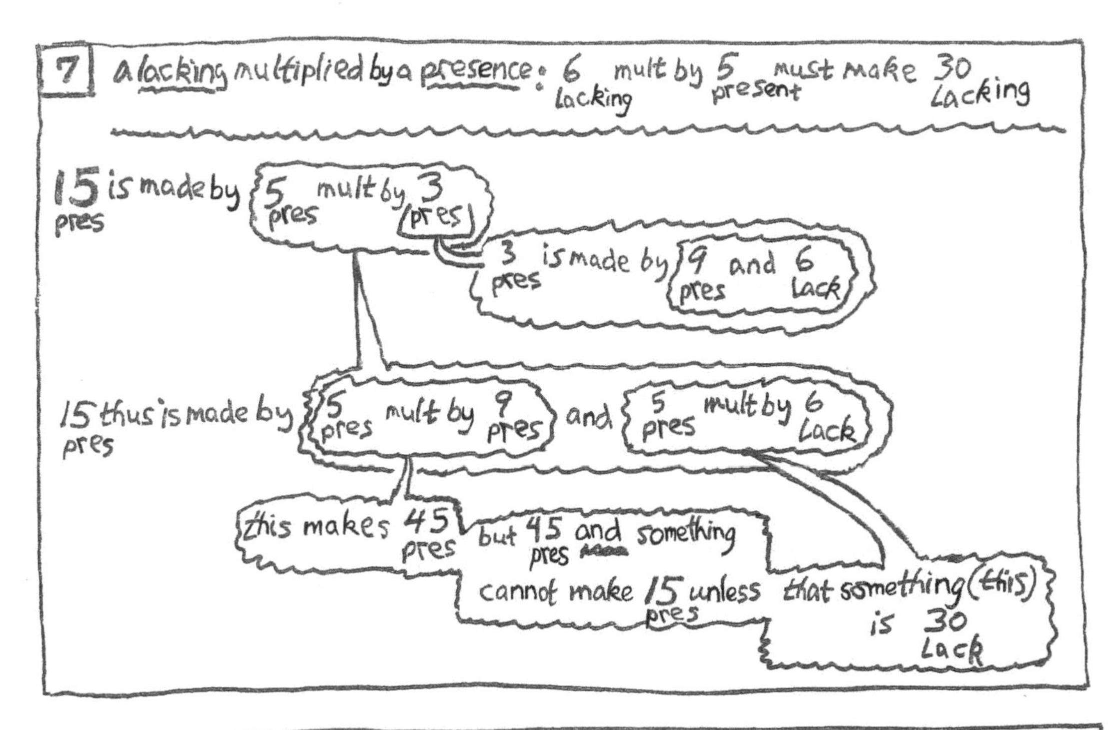

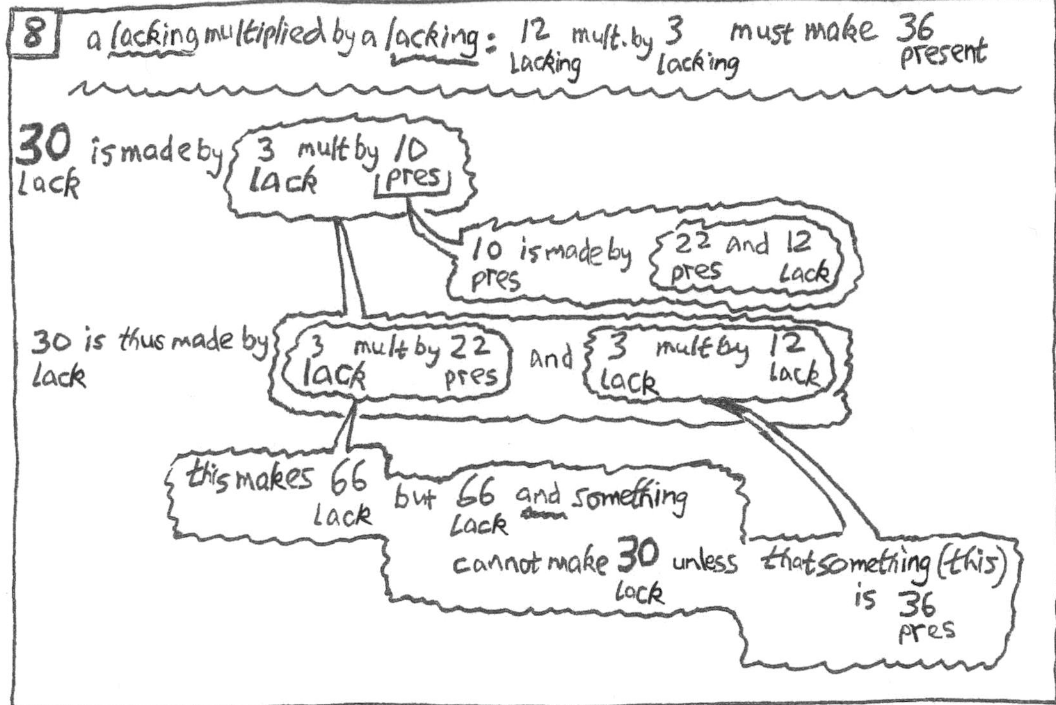

That is the sort of thinking which can lead to what Diophantus says next:

> a lacking multiplied by a lacking makes a presence,
> while a lacking multiplied by a presence makes a lacking.

A lacking is, in Greek, a *leipsis* (from the root for "taking"), and the sign that Diophantus uses to indicate a lacking combines the Greek letters for "L" and "I." Similarly, we shall use here "Li"—which suggests the English "*Lacking*."

more of what's prefatory

Division is a partitioning which multiplies the items to be counted. That is why Diophantus can go on to say this:

> the *multiplications* having been explained,
> the *divisions* of the *eidê* that have been previously laid out for you
> are clear.

And he continues:

> it is well for beginners to get exercise
>
> in doing additions, takings away, and multiplications in
> the things that have to do with the *eidê*—
>
> and in how you add *eidê*, both those that are present and those that are lacking,
> to other *eidê* that are not the same multiple
> [we would say: to other "terms that don't have the same coefficients"] ,
> themselves being either present or both present and lacking likewise—
>
> and in how, from *eidê*, both those that are present and others that are lacking,
> you take away others that are
> either present or both present and lacking likewise.

He goes on to say what that exercise is needed for:

> And then, if from some problem
> some *eidê* become equal to *eidê* that
> are the same but are not the same multitude
> [again, we would say: are "like" terms that "don't have the same coefficients"],
> it will be necessary to take away the like ones from the like ones,
> from each of the [two] parts
> [we would say: from each of the two "sides of the equation"]—
> until one *eidos* becomes equal to one *eidos*.

Here is an example of what Diophantus means:

> Suppose that
>
> (7 "cubes" & 2 "squares") are equal to (4 "cubes" & 2 "squares" & one "number.")
>
> From each of the two parts of the statement of equality, take away 4 "cubes"—so:
>
> (3 "cubes" & 2 "squares") are equal to (2 "squares" & one "number.")
>
> Now, from each of the two parts in the last statement, take away 2 "squares,"
> leaving one *eidos* equal to one *eidos*—so:
>
> (3 "cubes") are equal to (one "number.")

Diophantus continues:

But if somehow in either or both (parts)
there are any *eidê* that are lacking,
it will be necessary:

to add the *eidê* that are lacking in both parts,
until the *eidê* of both parts become present,

and again to take away the like ones from the like ones
until in each of the parts one *eidos* is left.

Here is an example of what Diophantus means:

Suppose that

(7 "cubes" lacking 9 "squares") are equal to (3 "cubes" lacking one "square.")

Add to both parts the *eidê* (9 "squares") that are lacking on the left—so:

(7 "cubes") are equal to (3 "cubes" & 8 "squares.")

Now, from each of the two parts in the last statement, take away 3 "cubes," leaving one *eidos* equal to one *eidos*—so:

(4 "cubes") are equal to (8 "squares.")

or, to simplify,

(one "cube") is equal to (2 "squares.")

And so, says Diophantus,

if the things supposed by the propositions admit of it,
let that be what is done by this technical pursuit—
until one *eidos* be left equal to one *eidos*;
and later we shall show you how to solve when there are
two *eidê* that are left equal to one.

propositions

Diophantus says after that:

And now, let us to the propositions (*protaseis*)—
traveling a road (*hodos*) that has the material gathered together
most in accordance with the *eidê* themselves.

And there being very many in number and very great in mass,
and, on account of this, established slowly by those who take them up,
and there being in them things hard to remember,
it seemed fitting to divide the things that are admitted among them—
mostly the things in the beginning that are elemental—
driving through, as is fitting, from simpler to more tangled,
since thus will the road go well (*euodeuta*) for beginners,
and how they are led will be remembered.

Diophantus concludes his prefatory statement by remarking that what he is doing will be done in thirteen books. (Seven of them, however, have been lost—unlike the books of Euclid's *Elements*, all thirteen of which are extant.)

Some modern readers of Diophantus have complained that his work is a mess. They would prefer his work to have been organized by types of equations, say, and methods of solution. But his work is organized by his own concerns, which are the possible relations to one another of sorts of numbers, particularly the sorts that are "squares" and "cubes" and their "sides." These numbers, it should be borne in mind, are multitudes of units, obtained by counting off items in a collection (or obtained by counting off the fragments that were themselves obtained by partitioning the original items in a collection). They are not what are now called the members of the set of real numbers. To avoid what would now be called "negative" or "irrational" solutions to an equation, Diophantus in many propositions introduces a de-limiting provision. Thus, when an equation is of a sort that nowadays would be said to have a "negative" or "irrational" solution, Diophantus characterizes what he is dealing with as "impossible" or "not enunciable." "Not enunciable" is, in Greek, *ou rhêtê* (which is, in Latin, *ir-rational)*. Diophantus speaks of its being "absurd" (*atopos*: "nowhere," or, more literally, "place-less") for a number to be non-enunciable.

Fractional parts of whatever happens to be the unit of calculation, however, are enunciable; they can be counted off. So, they can (like the unit) be considered as belonging to the company of numbers; they are numbers as-it-were. For example, the 23rd problem of Book Four—which is

> to find three numbers such that
> the solid [that is produced] from them
> [when all three of the numbers are multiplied together],
> less any one of them,
> makes a square—

that has this as its solution (the abbreviations having been replaced by the words that they indicate):

> (first) seventeen eighths;
> and (second) one unit;
> and (third) twenty-five eighths.

Indeed, fragments of a unit are sometimes explicitly called numbers by Diophantus in stating problems. For example, what is required by one of his problems (the 11th of Book Five) is "to separate a unit into three numbers," each of them making a "square" when there is added to it a previously given number (the same number being given for all three).

Now let us look at some sample propositions from the beginning of Diophantus's First Book of things arithmetical. The abbreviations that we shall use, in order to have convenient English counterparts of the Greek abbreviations of Diophantus, are listed in the following box:

unit	Un	"powero-power"	PPo
"number"	Nu	"powero-cube"	PCu
"power"	Po	"cubo-cube"	CCu
"cube"	Cu	"square"	Sq

one	a	ten	j	one hundred	s
two	b	twenty	k	two hundred	t
three	c	thirty	l	three hundred	u
four	d	forty	m	four hundred	v
five	e	fifty	n	five hundred	w
six	f	sixty	o	six hundred	x
seven	g	seventy	p	seven hundred	y
eight	h	eighty	q	eight hundred	z
nine	i	ninety	r		

the number ten (for example)	#j
of units, ten of them	Un #j
of "numbers," ten of them	Nu #j
of "powers," ten of them—less one "number"	Po #j Li Nu #a
the fraction with the same name as ten (namely, a tenth)	,,j
of the tenth parts of the unit, three of them (three-tenths)	j\c

In the following sampling of propositions, continuous text of Diophantus has been broken up and laid out so that the arrangement of text on the page will exhibit more clearly the structure of each proposition.

(I.1)

To separate a set-down (epitakhthenta) *number*
into two numbers
in the excess that is given.

So, let the given number be #s and let the excess be .... Un #m.

To find the numbers—

let the less be ordained .... Nu #a
then the greater will be .... Nu #a Un #m

then both together will be .... Nu #b Un #m
but they have been given as Un #s
then .... Un #s are equal to Nu #b Un #m

and I take like things away from like:

from	and from
the #s	the #b
[I take away]	likewise [I take away]
Un #m	#m units

remaining are.. Nu #b
which are equal to .... Un #o,

therefore, ... each Nu
becomes .... Un #l [that is the letter "el"].

As to what was supposed:
the less will be .... Un #l
and the greater .... Un #p.

And the demonstration (*apodeixis*) is manifest (*phaneron*).

Let us look again at the proposition. This time, however, let us make things more convenient for an overview—by using our own numerals, as well as by expanding the abbreviated terms and adding some words of clarification, while omitting some words that we can do without. Here is a restatement of that proposition (I.1):

To separate a given number into two numbers with a given excess.

The given number 100 is to be separated into two numbers
such that the greater exceeds the lesser by 40 units.

[*THE ANALYSIS* is as follows.]

Let the lesser be .... one "(unknown) number,"
then the greater will be one "(unknown) number" and 40 units;
then
both together will be .... 2 "(unknown) numbers" and 40 units;

and what both are together
has been given as being
100 units.

So: (100 units) are equal to (2 "(unknown) numbers" and 40 units)

and taking away likes from likes (namely, units from units), that is,

from 100 units and also from 40 units
taking away 40 units taking away 40 units

then (60 units) are equal to (2 "(unknown) numbers");
hence (30 units) are equal to (each "(unknown) number)."

As to what was supposed:

the lesser [which is one "(unknown) number"]
will be 30 units,

the greater [which exceeds one "(unknown) number" by 40 units]
will be 70 units.

And *THE DEMONSTRATION* is manifest.

(I.2)

It is required to separate the set-down number into two numbers
in the ratio that is given:

So, to separate—let it be the #o—into
two numbers in the ratio that is #c^pl [3^fold—that is, triple].

Let the less be ordained	Nu #a	
then the greater will be	Nu #c	
—that is, the greater is triple the less—		
it's required, further, that the two		
[the lesser and the greater together] be equal to..		Un #o
but the two together are..	Nu #d	
then..	Nu #d are equal to.............	Un #o
and then..	the Nu........is...........................	Un #je
The less, then, will be..		Un #je;
And the greater,..		Un #me.

And here is a restatement of that proposition (I.2):

The given number 60 is to be separated into two numbers
such that the greater is triple the less.

Let the lesser be..one "(unknown) number"
and then the greater (being its triple) will be...........3 "(unknown) numbers";

therefore both together are.....................................4 "(unknown) numbers,"
and it is required that both together be equal to.................................... 60 units;

4 "(unknown) numbers," then, are equal to... 60 units,
and hence the "(unknown) number" is.. 15 units.

The lesser [which is one "(unknown) number"] will be.................... 15 units, and
the greater [which is triple the "(unknown) number"] will be............. 45 units.

(I.4)

To find two numbers in a ratio that is given,
so that their excess is also given.

So, let it be set down
that the greater is the #e^pl of the lesser
and that their excess makes Un #k.

Let the lesser be ordained.... Nu #a,
then the greater will be......... Nu #e;
further, I wish..................... Nu #e... to exceed...Nu #a...by.......... Un #k;

but their excess.......[that is, Nu #e's excess over Nu #a]..is..Nu #d,
[and] these are equal to.. Un #k.

The lesser number will be.. Un #e;

the greater [will be the lesser together with the excess—that is]....... Un #ke;

and the greater stays #e^pl the lesser.

And here is a restatement of that proposition (I.4):

The greater number is to be 5-fold the lesser,
and is to exceed it by 20 units.

Let the lesser be......................one "(unknown) number";
then the greater will be...............5 "(unknown) numbers."

The excess of greater over less—which should be............................... 20 units—
thus is...4 "(unknown) numbers";

hence these................................4 "(unknown) numbers" are equal to.. 20 units.

Therefore the lesser
[which is..........................one "(unknown) number"] will be......... ..5 units,
and the greater
[which is.............................5 "(unknown) numbers"] will be...... 25 units,

the greater [which exceeds the lesser by 20 units] being 5-fold the lesser.

(I.7)

From the same number, to take away two given numbers,
and make the remainders have to each other a ratio that is given.

So, let it be set down that from the same number
there is taken away the #s and also the #k,
and that this makes the greater the #c^{pl} of the lesser.

Let the [number that is] sought be ordained.........................Nu #a

and if from this I take away
(on the one hand) the #s, the remainder is..........................Nu #a Li Un #s,
and (on the other hand) the #k, the remainder is.................Nu #a Li Un #k;

and it will be required that
the greater things be #c^{pl} of the lesser;
then thrice the lesser things are equal to the greater,
and thrice the lesser things become ....Nu #c Li Un #u;

these [that is, thrice the lesser things—namely, ...Nu #c Li Un #u]
are equal to
Nu #a Li Un #k;

let the lack [in thrice the lesser] be added in common
[that is, let (Un #u) be added to each of the equals—
added both to....... Nu #c Li Un #u
and also to........... Nn #a Li Un #k

—and then]........... Nu #c
become equal to
Nu #a and Un #tq;

and let like things be taken away from like things
[that is,........................both from .... Nu #c
and also from........... Nu #a and Un #tq]
let................ Nu #a be taken away;

what remains is.. Nu #b
are equal to
Un #tq;

and.. Nu
becomes.
Un #sm.

As to what was supposed:
I ordained the sought number to be Nu #a; it will then be...Un #sm;
and if from this [U #sm] I take away
(on the one hand) the #s, what remain are.........................Un #m,
and (on the other) the #k, [what remain are].......................Un #sk,
and the things that are greater stay triple the lesser.

And here (with the numbers 100 and 20 treated in reverse order for convenience) is a restatement of that proposition (I.7):

From the same number, there is to be taken away 20 and also 100,
so that the one remainder is 3-fold the other one.

Let the number that is sought be
one "(unknown) number".

The one remainder is........one "(unknown) number" less 20 units,
the other remainder is.......one "(unknown) number" less 100 units;

and the former is 3-fold the latter—hence:

one "(unknown) number" less 20 units
is equal to
3 "(unknown) numbers" less 300 units;

adding to each of them the same number of units—
namely, the 300 units that the second of them lacks,
the result will be that
the 3 "(unknown) numbers"
are equal to
the one "(unknown) number" and 280 units;

now—taking a like thing (one "unknown number")
away from like things ("unknown numbers" in each part of the statement of equality)
what remains will be:
2 "(unknown) numbers"
are equal to
280 units;

hence ....the one "number"
becomes
140 units.

And the number that is......140 less 20 units
is 3-fold the number that is 140 less 100 units.

Diophantus also solves arithmetical problems that are more complicated—problems involving numbers that are "squares." Here, for example, is the first of the new problems in his Second Book (this problem, because of the way in which the text has come down to us, is called the 8th rather than the 1st item in the Second Book):

(II.8)

To separate a set-down "square" into two "squares."

So let it be set down, to separate the #jf into two "squares."

And let the ast [the 1st] be...................... Po #a,
then the other will be..............Un #jf Li Po #a;

then it will be required that .....Un #jf Li Po #a
are equal to
a Sq;

I mold the Sq from
whatever-many Nu's that there are,
Li as many Un's as
the "side" of the #jf Un has—
let it be [for example]...........Nu #b Li Un #d;

then, [multiplying that by itself,]
the Sq itself will be.............Po #d Un #jf Li Nu #jf;

these
[items that make up the Sq itself]
are equal to......................Un #jf Li P o #a;

let the lack be added in common [that is, to both]
and from like things [let] like things [be taken],

then.....................................Po #e
are equal to
Nu #jf

and.......................................the Nu
becomes
#jf fifths.

They'll be: the one,................ ke \ tnf [of twenty-fifths, two hundred fifty-six] ;
and......... the other,........... ke \ smd [of twenty-fifths, one hundred fifty-four].

And the two put together
make..............................ke\v [of twenty-fifths, four hundred]
—or..............................Un #jf [sixteen units]—
and each is a square.

Here is a restatement of that proposition (II.8):

The square number 16 is to be separated into two square numbers.

Let the first one be one "(unknown) power."
The other one will then be 16 units less this one "power."

Then 16 units less one "power" is required to be equal to some "square."
What "square"? One that is molded from [generated from a "side" that is]
whatever-many "numbers" (let it be, say, 2 "numbers")
less a certain number of units (namely, the number that is
the "side" belonging to 16 units—that is, 4 units).

From that, it follows that 5 "numbers" will be equal to 16 units—
for the following reason:
The square—since it is generated by
the multiplication-by-itself of (2 "numbers" less 4 units)—
will itself be equal to this: (4 "powers" and 16 units less 16 "numbers").
But the square is also equal to this: (16 units less one "power").
Since things equal to the same thing are equal to each other,
it follows that
(4 "powers" and 16 units less 16 "numbers") <u>equal</u> (16 units less one "power").
To both of these, let us add 16 "numbers" and one "power"
(making up for what is lacking on the left and also for what is lacking on the right,
without losing the equality),
and, from both of them, let us take away 16 units;
what we get is that (5 "powers") <u>equal</u> (16 "numbers");
and now, let us divide both of these by one "number,"
and the result is this: (5 "numbers") <u>equal</u> (16 units).

Now, since 5 "numbers" are equal to 16 units,
one "number" is therefore equal to 16 of the unit's 5th parts.
We began by letting the first number sought be one "power"
(a "power" being merely the product when
a "number" is multiplied by itself),
and one "number" has been shown to be equal to 16 / 5ths;
therefore the first number sought must be the product when
that "number" 16 / 5ths is multiplied by itself—namely, 256 / 25ths;

and the second number sought
(since it was 16 units less one "power") must therefore be this: 144 / 25ths.

These (namely: 256 / 25ths, and 144 / 25ths) will,
when put together, make 400 / 25ths—or the given 16 units;
and each of them is a square.

The problems solved by Diophantus involve numbers that are "cubes" as well as those that are "squares." The 18th proposition of the Fourth Book, for example, shows how to find two numbers that will satisfy the following requirement: the "cube" that arises from the first, being added on to the second, will make a "cube"—while the "square" that arises from the second, being added on to the first, will make a "square." We shall not here examine Diophantus's solution to this problem. If what you have seen of Diophantus has made you think that though he wields a very cumbersome instrument, what he does with it is really very simple, then you might try finding two such numbers on your own.

To readers nowadays, Diophantus's treatment of things arithmetical may seem to be a messily arranged presentation, in strange terms, using a cumbersome notation, of familiar things algebraical. We must bear in mind, however, that Diophantus is showing how to solve problems that demand the finding of definite numbers (countable multitudes of units) that have certain characteristics. He is not seeking general solutions of types of equations that relate constant and variable algebraic "quantities," though he does sometimes do more than find a single numerical solution. That is, he does sometimes solve a problem which requires him to find required numbers "indeterminately" (*en tô aoristô*)—a problem, that is to say, which requires him to find a way in which numbers of the sort requested will be obtained in a supply that is boundless (*a+oristos*).

Consider the 19th problem of his Fourth Book:

to find three numbers indeterminately (*en tô aoristô*)
such that the product of any two will,
with one additional unit, make a square.

In treating this problem, Diophantus says that the problem has been solved indeterminately (*en tô aoristô*) when the seeking boundlessly (*aoristôs*) finds that

whatever not-yet-determined "number" you may wish
to be one of the three,
the two other ones will be these:
that "number" increased by two units,
and four of those "numbers" increased by four units.

This notion of indeterminate solution recurs right after the next problem (after the next one, because the next one itself does not need it). The 21st problem of the Fourth Book proceeds as follows.

(IV.21)

To find three numbers in proportion
so that with any two the excess be a "square."

Let the lesser be ....Nu #a
and the middle be ....Nu #a Un #d,
so that the excess be a Sq;
and the c[rd] [3rd] be ....Nu #a Un #jc,
so that also the excess of this
with respect to the middle be a Sq;

and then, if the excess of the greatest and the least were a Sq,
then the excess of any two would have been
solved indeterminately (*en tô aoristô*).

But the greatest exceeds the least by.................................... Un#jc;
and the Un #jc [which make up the excess]
are put together from the Sqs—
that is, from #d and #i
[#d being the excess of the middle over the less,
and #i—which is the difference between #jc and #d—
being the excess of the 3rd over the middle].

In other words: The least is some "number" that is not yet determined. The middle exceeds it by a "square" number. As for this "square" by which the middle exceeds the least, let it be 4 units. Also, the greatest exceeds the middle by a "square." As for this "square" by which the greatest in its turn exceeds the middle, let it be 9 units. Thus we would now seem to have, for any choice of a least number, a solution of the problem: in order to obtain the middle one, we would just add another 4 units onto the one that is chosen to be least; and in order to obtain the third (the greatest one), we would just add onto the middle one another 9 units.

But, alas, what might seem to be a solution is not yet one. The excess of the middle over the least was indeed made to be a "square" (namely, 4 units), and that of the greatest over the middle was likewise indeed made to be a "square" (namely, 9 units); but the consequence is, that the excess of the greatest over the least (the 13 units that result from adding 9 units to 4 units) is *not* a "square." Hence, Diophantus must continue:

Therefore what arises for me is to find
two squares equal to one square.

Suppose we do find two "square" numbers that, when added together, are equal to a number that is also "square." We again let the least of the three numbers that we are seeking be an as yet unknown "number." But now, to obtain the middle one, let us add the one "square" onto that "number"; and to obtain the greatest, let us add the other "square" onto the middle one. Since the excess of the greatest over the least of the three is the sum of those two "squares," this excess will—by our supposition—also be a "square." There's the rub: by our supposition.

But how do we find what we have supposed—namely, two "square" numbers that together are equal to a third number that's also "square"? Diophantus indicates how:

> And this is easy from
> a right-angled triangle.

That is to say: if we construct a right triangle whose one leg is a certain number of units long, and whose other leg is a certain other number of units long, and whose hypotenuse is yet a third number of units long—then those three numbers will be the "sides" of three "squares" such that the first two together equal the third.

Take, for example, a right triangle whose legs have as their lengths, respectively, three units and four units—and whose hypotenuse is five units long. The "square" numbers whose "sides" are the numbers three, four, and five are respectively—the numbers nine, sixteen, and twenty-five. Such a triangle is used by Diophantus:

> Now there are the #i and the #jf;
> and I ordain the least to be.. Nu #a,
> and the middle to be.. Nu #a Un #i,
> and the c[rd] [3rd] to be..Nu #a Un #ke;
> and the excess of any two
> [that is, the excess that any one of the three of them
> has over any other one of the three] is a Sq.

We now know how to find three numbers such that any one of them will differ from any other one of them by a number that is a square. But, Diophantus points out, the three numbers must meet an additional condition for the problem to be solved:

> What remains is for them to be proportional.

Diophantus shows how to get them proportional:

> Now if three numbers be proportional,
> the [number produced] by the extremes is equal to
> the [square number arising] from the middle.
>
> But the [number produced by] the greatest and the least—
> this is the [number produced] by the extremes—is..........................Po #a Nu #ke.
> And the [square number arising] from the middle is........................Po #a Nu #jh Un #qa
> [which is] eq[ual] to
> Po#a Nu #ke.
>
> And the "number" becomes g \ qa.

In other words:

What is the product when you multiply together
the least number (one as yet unknown "number")
and the greatest (one such "number" and 25 units)?
That product is this: (one "power" and 25 "numbers").

And what is the product when you multiply by itself
the middle number (one "number" and 9 units)?
That product is this: (one "power" and 18 "numbers" and 81 units).

Now, the one product is equal to the other—
since multiplying the least by the greatest yields the same number as does
multiplying the middle by itself—
which is to say that
(one "power" & 25 "numbers") equal (one "power" & 18 "numbers" & 81 units).
If we take away, from both of these, (one " power" & 18 "numbers"),
Then what we get is this: (7 "numbers") equal (81 units);
and if we divide both of these by 7,
one "number" turns out to be this:
of 7th's, one "number" turns out to have 81.

And from that, Diophantus can tell how many sevenths constitute not only the first of the three numbers that are sought, but also the second and the third:

As to what was supposed, they will be:
the ath [1st], \qa; and the bth [2nd], \smd; and the cth [3rd], \tnf.

How many sevenths constitute the 1st of the three? As was said, as many as are in one "number": 81 of them. And how many constitute the 2nd? Well, the second of the numbers is this one "number" and 9 units; and hence what is needed are 144 of those sevenths. And the 3rd (which is one "number" and 25 units) needs 256 of them.

We have just examined an arithmetical problem in the course of whose solution a collection of numbers with certain requisite properties is obtained by Diophantus from a well-known right-angled triangle. An entire book of Diophantus's arithmetical things (the Sixth Book) is devoted to problems of finding a right-angled triangle when various suppositions are made about the numbers "in" various of its features: the number of units of length in the legs (that is, in the sides at right angles to each other), or in the hypotenuse, or in the perimeter; or the number of units of area.

In other words, the Sixth Book solves problems involving three numbers that are "sides" such that the "squares" of two of the "sides" are together equal to the "square" of the third "side" (which third "side" is the number "in" the triangle's hypotenuse, while the other "sides" are the numbers "in" the triangle's sides that are at right angles to each other—its legs). Also involved in one or another of the problems are: the number that is the sum of the three "sides" (which is the number "in" the triangle's perimeter), and the number that is half the product of the two "sides" that are the numbers "in" the legs (which is the number that is "in" the area).

Diophantus is not here interested in right triangles as such, or even in right triangles whose sides are commensurable. He is interested in collections of numbers that are "in" the features of such triangles and that have certain properties. The solutions of these problems give the numbers that are "in" the three sides of a triangle whose features (legs, hypotenuse, perimeter, or area) have "in" them numbers as given in the suppositions of the problem. Here are three examples of problems solved by Diophantus in the course of his Sixth Book of arithmetical things:

(1) To find a right-angled triangle such that the [number] in the hypotenuse,
lacking the [number] in each of the right-angle sides, makes a cube.

(3) To find a right-angled triangle such that the [number] in the area,
added to a given number, makes a square.

(17) To find a right-angled triangle such that the [number] in the area,
added to the [number] in the hypotenuse, makes a square —
while the [number] in the perimeter is a cube.

With those enunciations, we complete our examination of Diophantus—and with him, our consideration of ancient authors. Of those authors, Diophantus and Pappus are the ones in whom Descartes found hints of a way to get past the variegated outer husk of mathematics and into its inner kernel. That way, he said, was obscured by the synthetic presentation which dressed up the work of the great geometers of antiquity, Apollonius being pre-eminent among them. Descartes found the obscurity to be both witting and unwitting. It was witting obscurantism, since the analysis that had to precede synthesis was almost always deliberately covered up. The obscurity was unwitting also, since the ancients were unaware of the real worth of mathematics, being so eager for mathematical fruit—for particular mathematical truths to look upon in wonder—that they overlooked the profoundly illuminating and truly fruitful power that lurked within the study of mathematics, constituting its homogeneous hard core. To indicate what he had in mind, Descartes presented his solution of a geometrical problem, a problem that was based upon what Apollonius taught but that no one had been able to solve, despite the many centuries that had passed since Pappus stated it in ancient times.

Now we may have completed our examination of the mathematicians of antiquity; but before we pass on to Descartes, we need to take into consideration a development that intervened between the work of Diophantus and the work of Descartes. Although Diophantus's arithmetical work was part of the mathematical tradition of Greek antiquity, it owed something to mathematical traditions that had been passed down farther east. We cannot here consider the numerical discoveries of the Babylonians and their possible contributions to the work of Diophantus, nor can we consider any of the later Hindu and Arabic contributions to algebra. This algebra from the East was inherited by the West, which studied and advanced it, albeit outside the official curriculum of the academic institutions. What we do need to consider here, is what happened in the West—just before the time of Descartes—when the vigorous pursuit of studies that derived from medieval Arabic algebraic work came together with a revival in the study of Diophantus's ancient Greek arithmetical work, and with a revival as well in the study of the ancient Greek geometrical work presented in the writings of Apollonius and of Pappus.

The outcome of this meeting was the rise of a new and most ambitious mathematical art that dealt with a new and most peculiar object. The object dealt with by the art was called a "species" by the man who discovered or invented the art, which he called the "analytic" art. *Analysis* is a Greek term of which we have already seen a lot; and *species* is merely the Latin equivalent (coined, it seems, by Cicero) of the Greek term *eidos*, of which we have already seen a lot too. The man was Viète, who was a French lawyer of the sixteenth century (1540–1603, to be exact).

Let us consider the ambition of Viète's art and the peculiarity of what it dealt with. We shall then be ready to consider Descartes. Descartes in his correspondence denied a charge that he had not gone beyond Viète; but Descartes' assertion that he had indeed gone beyond Viète was an admission that Viète had indeed gone beyond the ancients in some respects. Although we cannot here explore Viète's various narrowly technical accomplishments, we do need to see how he altered the character of mathematics. The work in which Viète made that great alteration had been made known to others and had been highly regarded for years by the time that Descartes came of age. Descartes in his correspondence went so far as to say that his own great alteration of the character of mathematics did not rely upon any reading of Viète's great innovative work; but even if Descartes' work was not in fact a deliberate innovation upon Viète's prior innovation, Descartes' most radical transformation of the legacy of Apollonius, Diophantus, and Pappus is easier to appreciate if one first takes a look at the great innovation of the previous generation. What, then, was the great accomplishment of Viète?

CHAPTER 2. INTRODUCING VIÈTE'S ANALYTIC ART

Referring to his great accomplishment as "the restored mathematical analysis" or "the new algebra," Viète in 1591 published a general introduction to it, under the title of *Introduction to the Analytic Art*. He wrote the book in Latin, but the title contains, not the Latin word *introduction,* but rather the (transliterated) equivalent Greek word *isagogê*. For terminology throughout the book, Viète draws upon Greek.

(Note that the following translation of Viète owes much to the work of J. Winfree Smith, but with modification in the direction of greater literalness, and in a presentation such that the layout of the text can help to make the structure of the thought easier to see than it would be if the print stretched uniformly across the page from margin to margin.)

Viète's letter of dedication opens with praise for his patroness and then goes on to characterize his enterprise as follows:

Things which are new are wont to be put forth raw and unformed,
and then to be polished up and perfected by succeeding ages.
Behold, the art which I present is new;
or at least it is so ancient and befouled by barbarians and covered with filth,
that in order to introduce into it an entirely new form,
I have held it necessary to think out and publish a new vocabulary
(its pseudo-technical terminology having all been gotten rid of,
lest it retain its filth and stink inveterately)—
by which new vocabulary, since ears be hitherto too little accustomed to it,
it will hardly happen that many will not be deterred and offended
even at the very threshold.
And under their Algebra or Almucabula,
which they praised and called "the great art,"
all mathematicians recognized that incomparable gold was hiding,
though in truth they found very little of it.
They vowed hecatombs and provided sacrifices to the Muses and Apollo
if someone would raise up to solution one or another problem
of the order of a sort such as we exhibit by tens and by twenties beyond,
since our art is the most certain finder of all things mathematical.

Thus introducing his *Introduction to the Analytic Art*, Viète has gone so far as to claim that the analytic art, since it is the most certain finder of all things mathematical, solves freely by the score such problems as classical mathematicians were grateful to solve singly; but he will go even further than that in concluding his *Introduction*. In the last line of the last chapter, he will go so far as to make the claim that the analytic art, as put into proper form by him, "arrogates to itself by right the proud problem of problems, which is TO LEAVE NO PROBLEM UNSOLVED."

Viète's concern for an art that would solve the problem of solving any problem in all of mathematics seems to have been bound up with his work in astronomy and cosmology. That cosmic work may have been his ultimate concern, but it was not transformational and it cannot be our concern here. Rather, we turn directly to the first chapter of his *Introduction to the Analytic Art*. This chapter defines analysis, divides it into parts, and begins to discuss its first part, which Viète calls "zetetics."

the first chapter

> CHAPTER I.
> *About the definition and partitioning of Analysis,*
> *and about those things which help Zetetic*
>
> There is a certain way of seeking into truth in mathematical things—
> which Plato is said to have found—
>
> named *Analysis* by Theon and defined by him as:
> assumption of what is sought as if it were granted,
> and going through what is its consequence,
> to what is granted to be true.
>
> So, contrariwise, *Synthesis*, [defined as:]
> assumption of what is granted,
> and going through what is its consequence,
> to the reaching and the comprehending of what is sought.

Viète in this book is interested not so much in mathematical truth itself as in a certain "way" of "seeking" it. When one takes this way, one assumes as granted the very thing that is sought; and one proceeds—through what follows from that—to arrive at the thing that is granted to be true. The name of this way is "analysis." By contrast, "synthesis" is the name of a way, not of seeking, but rather of presenting mathematical truth that has already been found. Taking this way of synthesis, one assumes the thing that is granted to be true; and one proceeds—through what follows from that—to conclude with, and to comprehend, the thing that is sought. Analysis and synthesis proceed in contrary ways.

The classic text for the elements of mathematics, Euclid's *Elements*, proceeds synthetically. An ancient note to Euclid's *Elements* shows how each of the first five theorems of the Thirteenth Book may be understood as a synthetic presentation, in reverse order, of the steps of a prior analysis. In that ancient note are definitions of analysis and synthesis. The definitions were ascribed to Theon of Alexandria. Pappus gave clearer versions of the definitions at the beginning of the Seventh Book of his *Mathematical Collection* (where he presents, as we have seen, the *Storehouse of Analysis*), and Viète elsewhere refers to Pappus's work. Here, however, Viète refers to Theon's note on Euclid, thus suggesting that gold lies hidden beneath the surface of those very texts (such as the *Elements*) that are accepted as the peaks of classical mathematics. Indeed, Viète indicates that what he will bring to light, and will clean off and renovate, is of more general interest and is even older than the reference to Theon would suggest, for the analytic way is said to have been the discovery of no less a personage than Plato. (What made it possible to say that, was Plato's depiction of Socrates as proceeding by investigating the consequences of taking for granted, as if known, what in fact is not yet known but is rather merely supposed—and is being sought for.) It seems that the classics covered up the analytic way long before it was befouled by barbarians.

Viète goes on to distinguish the parts of the analytical art. The analysis of the ancients was two-fold, but Viète makes the analytical art three-fold. It is fitting, he says, to establish a third kind of analysis.

For the ancients—as we learn from Pappus, and as Theon reports—there were two kind of analysis. The one kind of analysis was *theoretical*, whose Greek root means "to gaze" and is also the root of "theorem" (something to gaze at). The other kind was *problematical*, whose Greek root means "to throw out in front" and is also the root of "problem" (something thrown out before someone as a challenge to do or make something).

The theoretical analysis, which sought a demonstration of a truth which could be gazed at, was also called *zetetic*, whose Greek root means "to seek." The problematic analysis, which provided the solution for the challenge thrown out in front, was also called *poristic*, whose Greek root means "to provide."

For the subdivisions of the renovated analytical art, Viète takes two of the ancient Greek names. The ones that he takes are *zetetic* and *poristic*—rather than the two names that point, respectively, to theorems and to problems. For the third subdivision, the one established by himself, he uses ancient Greek roots to make up two new names (*rhetic* and *exegetic*—we'll consider these in a moment).

By "zetetic art," Viète tells us, "is found the equation or proportion that holds among the magnitude that is sought and those that are given"; and "from the equation set up or the proportion," he says, "there is produced"—by "exegetic art"—"the magnitude itself which is being sought." The three arts of zetetic, poristic, and exegetic make up "the whole" of the "analytical art" which "may be defined as the science or discipline or teaching (the *doctrina*) of finding well in things mathematical."

The *whole* art has to do with *finding*; the *part* of it that is of most interest to us is the one that has to do with *seeking*—namely, zetetics. This is the part that constitutes the bulk of Viète's *Introduction.* On this first (the zetetic) part of Viète's analytic art, we shall now concentrate our attention, with a look first at how it is completed by the last (the exegetic) part.

What is sought may be a number—a tally told off as the outcome of a count—and hence it is a "rhetic" art which *tells* us as found what was sought, when what was sought is arithmetical. Or what was sought may be geometrical (a length, a surface, or a solid)—and hence it is an "ex+egetic" art which *leads out* before us what was sought. But in this introduction to the analytic art the emphasis is on zetetics, and Viète does not keep strictly separate the two ways that the worked-up equation or proportion can be made to yield the sort of thing that was sought. The nomenclature that he uses varies. In this first chapter, the third part of the analytic art is called both "rhetic or exegetic" and "exegetic"; later, the title of the seventh chapter speaks of "rhetic," while the body of the chapter speaks of "rhetic or exegetic"; but the eighth chapter, which is the final chapter, speaks throughout of "exegetic," not only with reference to geometry but also explicitly with reference to arithmetic; and elsewhere, in the title of another of his treatises, the one on the numerical resolution of powers, he calls the work not only "numerical" but also "exegetic." But although the terminology is used indiscriminately, the two ends are clearly distinguished. This third part of the analytical art, Viète will say in the seventh chapter, is most pertinent to the ordering of the art. The whole analytic art, that is to say, is ordered with a view to its end, which is either the telling of a multitude or the setting out of a magnitude (a length, a surface, or a solid). The whole analytic art is an instrument, a mere instrument, for a finding that is either arithmetical or geometrical.

So Viète continues as follows:

> And although the ancients propounded an analysis so-much-as-two-fold
> (the *zetetic* and *poristic* [both in Greek letters],
> to which Theon's definition most applies),
> it is nonetheless suitable for yet a third species to be constituted,
> which is called *rhetic or exegetic* [again in Greek],
> so there be:
>
> *Zetetic*, by which is found an equality or proportion
> of that magnitude about which there is a seeking,
> with those [magnitudes] which are given;
>
> *Poristic*, by which is examined the truth of a theorem about
> the equality or proportion that is set in order;
>
> *Exegetic*, by which, from the equality or proportion that is set in order,
> the magnitude about which there is a seeking is itself exhibited.
> And thus the whole Analytic art,
> claiming for itself that three-fold function,
> may be defined as:
> the science or discipline or teaching (*doctrina*)
> of finding well in things mathematical.

We have not yet considered the second of the parts of Viète's analytic art—namely, poristics. He himself never made much of it, even though he did subsequently make much of the other parts—zetetics and exegetics.

It is not easy to see just what Viète meant by "poristic." His zetetics and poristics do not correspond to, respectively, that ancient zetetic analysis which was theoretical and that ancient poristic analysis which was problematical. Moreover, in the sixth chapter of the *Introduction*, which purports to discuss poristics, Viète is concerned largely with a certain synthesis and its relation to analysis. And it is puzzling to be told by Viète that poristic analysis deals with theorems, rather than with problems; that it goes from the known to the unknown, rather than the reverse; and that it facilitates synthesis.

Nor is it easy to see why, although he went on to write much on zetetics and on exegetics, he neglected poristics.

A restoration of Viète's poristic art, and therewith an explanation for Viète's neglect of it, has been attempted in a doctoral dissertation by Richard D. Ferrier. (See Chapter VIII of his *Two Exegetical Treatises of François Viète Translated, Annotated, and Explained*; Department of History and Philosophy of Science, Indiana University, 1980.) We shall make use of Ferrier's persuasive account.

But before we more closely consider parts of Viète's work, let us briefly consider in overview what sort of thing it is that Viète is presenting in his *Introduction to the Analytic Art*.

Suppose that we are confronted with the following geometrical problem:

> *given*: two straight lines,
>
> *find*: a third straight line, such that
>
> the square raised on one of the given lines
> is equal to one quarter of what remains when
> the square raised on the other given line
> is taken away from the square on the line sought.

We can solve this problem analytically as follows.

First, we proceed "zetetically." We begin, that is to say, with the given relationship of size among the thing sought and the things that are given, and we treat these geometrical things in the same way that Diophantus treated arithmetical things. That is to say, we let the various sizes that are involved be named by *letters*, so that we have a comparison of size among the various *eidê*—in other words, (translating Greek into Latin) a comparison of size among the various *species*. Now, calculating with the species, we transform the equation until we can reduce it to one in which the species that corresponds to the magnitude that is sought is presented in terms of the species that correspond to the magnitudes that are given.

Thus, in the problem before us, we might begin by letting A be the length sought, and K and L be the given lengths. We are given that A's square less K's square is equal to 4 of L's square. Adding the same magnitude (K-squared) to equal magnitudes (namely, to A-squared less K-squared, on the one hand, and, on the other, to quadruple L-squared), the results are equal—namely, A's square is equal to K's square and four of L's square taken together. But the quadruple of the result of squaring L is equal to squaring the double of L. Hence we have the following transformed equality: A's square is equal to K's square and L-doubled's square taken together. Thus, the species (A) that corresponds to the magnitude that is sought is stated in terms of the species (K and L) that correspond to the magnitudes that are given.

But the thing that is sought and the things that are given are lengths. We want to set out the *length* that is sought. Our zetetic proceeding tells us that the line sought is such that the square on it is equal to what we shall get by making a square on a line that is double the one given line, and putting that square together with a square made on the other given line. That has transformed our problem but has not yet resolved it into something that will enable us to set out the length sought. Using the outcome of having proceeded "zetetically," we now want to proceed "exegetically." A further resolution of the outcome of our zetetic proceeding is needed to prepare us for our exegetic proceeding. A poristic proceeding would provide that preparation.

Now here are two geometrical theorems to provide the help that we need in the construction-work that will set out the length sought. They are theorems about givenness. *One* of the theorems is this:

> when a straight line is given,
> then also given is its double—
>
> which is shown by merely taking a line equal to the given line,
> and putting it end-to-end in line with the given line,
> thus making the double line.

The *other* theorem is this:

> when two lines are given,
> then also given is the line whose square is equal to
> the squares (taken together) on the two given lines—
>
> which is shown by taking a line equal to one of the two given lines,
> and putting it end-to-end at right angles to the other given line,
> and then joining the other endpoints, thus making
> a right triangle's hypotenuse, whose square will be equal to
> the squares (taken together) on the sides.

With those theorems about givenness in mind, "exegetical" analysis will show us how to proceed when we seek a straight line such that the square arising from it is equal to the sum of two squares, one of them arising from one given straight line, and the other of them from the double of another given straight line.

Once that straight line has been constructed (following the steps that demonstrate those two theorems about givenness that provide what is needed), then it is easy to demonstrate that this line that has been constructed is the very line required by the problem. We merely have to present—in reverse order—the geometrical statements that correspond to the statements about species that are the steps of the zetetic analysis.

The problem that we have just considered is a *geometrical* problem. In the solution of an *arithmetical* problem, the zetetic work concludes when the species of the sought number is equal to an expression in terms of the species of the given numbers. It is easy to compute the sought number if it is equal to, say, this: the product of the two given numbers multiplied together, taken six times, less the first of the given numbers taken four times, all divided by the second of the given numbers cubed. But what if what is equal to that, is not the sought number itself, but rather is its square? In this latter case, to do the requisite computation would require some learning in the rhetic art.

Now let us consider how the analytic art relates to the work of Pappus and of Diophantus as it bears on the business of analysis and synthesis.

First, let us remind ourselves about the parts of a proposition. The enunciation of a proposition can be restated in the following form:

theorem: ....if (c) is so, then (z) is too,
or
problem: ....given (c), to do (z).

A proposition can be so simple as to have a schematic structure like the following—

ANALYSIS
(the way to a demonstration)

this finds out how, supposing that what is sought were so,
it follows that what is given originally would have to be so too:

supposing that indeed....................(z) were so,
it follows from this that.................. (a) would be so;
and from this in turn that...............(b) would be so;
and from this in turn that...............(c) would be so.

SYNTHESIS
(the demonstration)

this shows how, because what is originally given is in fact so,
therefore what is sought must be so too:

now, it is given as true that............(c) is so,
from which it follows that..............(b) must be so;
from which it follows that..............(a) must be so;
from which it follows that..............(z) must be so.

That simple structure of a proposition is depicted in figure 1.

But what if the link from (b) to (c) is not reversible? Then the way to a demonstration needs to provide a link from (c) to something else—that in its turn will be linked to (b). See figure 2.

This complication is not found in all propositions. When it is found in a problem, the missing link characteristically involves construction. The problem, after having been "transformed," is then "resolved"—the *analysis being thus completed* by showing that if something were given, then something *can* be constructed which *will* be constructed at the *beginning of the synthesis* and which will be shown, at the beginning of the demonstration proper (which then will be the second part of the synthesis), to entail (b). In this case—that is, when the proposition does have a resolution and construction—the structure will be as laid out in the box following.

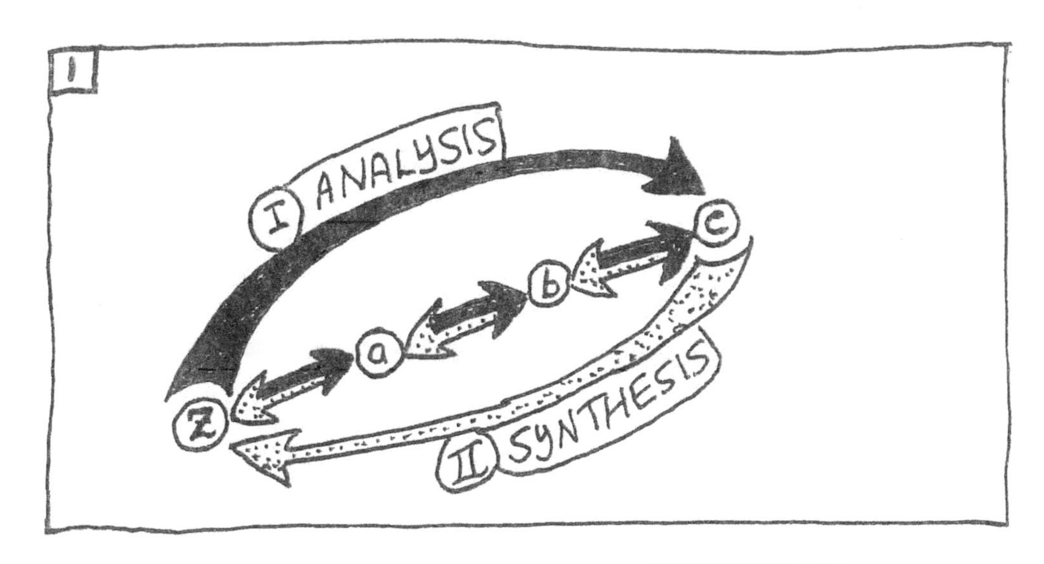

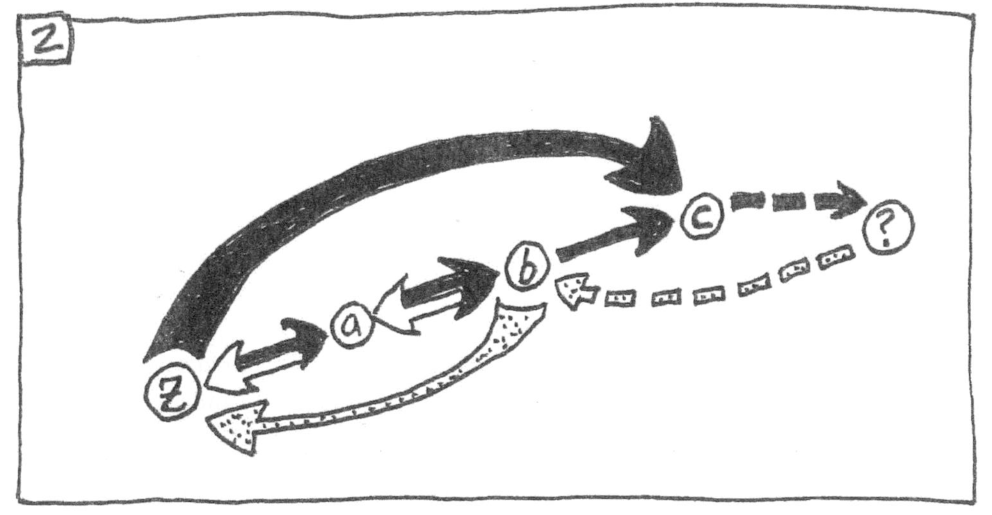

theorem: ....if (c) is so, then (z) is too,
or
problem: ....given (c), to do (z).

[I] ANALYSIS

[1] TRANSFORMATION
supposing that what's sought (z) is so or is done
entails something that was originally given (c):

supposing that..............................what is sought (z) is so, or is done,
it follows that................................(a) is so,
and from this in turn that...............(b) is so,
and from this in turn that...............(c) is so—this (c) being given
either originally
or independently

[2] RESOLUTION
(c)'s being given either originally or independently
entails that
(e) is given by construction (that is, that it can be constructed):

but if...(c) is so,
it follows that................................(d) is so,
and from this in turn
it follows that................................(e) is so—this (e) being some
construct

[II] SYNTHESIS

[1] CONSTRUCTION
the steps in the resolution—that is, the steps in showing (e)'s constructibility—
are repeated in the same order, but now as steps in showing (e)'s construction:

because...(c) is indeed given
either originally
or independently,
therefore.......................................(d),
and therefore..................................(e)

[2] DEMONSTRATION
the construction of (e) from what is given originally or independently
entails that what is sought (z) is so, or is done

from...(e)'s construction,
it follows that................................(b) is so,
and from this that...........................(a) is so,
and from this that...........................(z) is so, or is done

How does that illuminate the relation between Viète's zetetic and poristic? This is shown in figure 3, which depicts logical relations (note the bidirectional arrows). The movement of thought in all that, is afterward depicted by the sequence in figure 4 (where all the arrows are unidirectional and the several stages are separated).

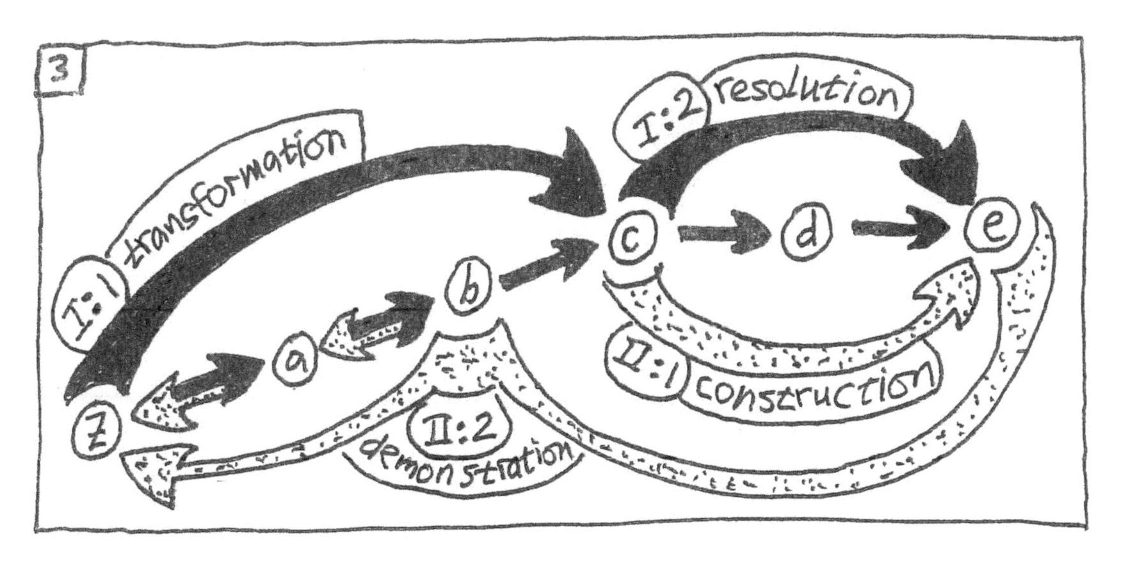

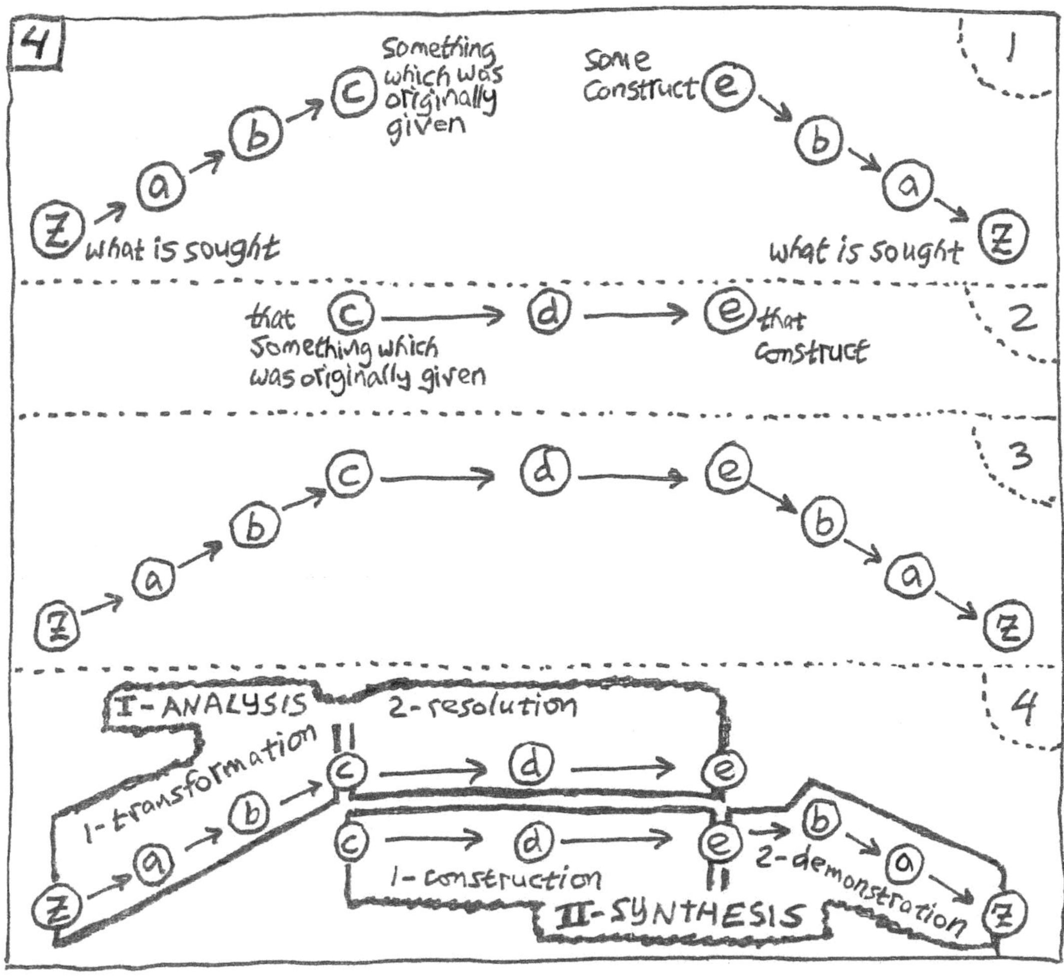

We can now say more about Viète's poristic. It is hard to say what he means by the little that he does say in his *Introduction to the Analytic Art*. The only surviving specimens of his poristic are propositions VI and VII of the *Supplement to Geometry*, which is one of his exegetical treatises. What was Viète's poristic, asks Ferrier in his dissertation, and why did so little become of it? Ferrier tries to give some account of "what the second part of the new analysis was and why it suffered this fate." He says that while there may remain difficulties even with his account, he is "not aware of any other attempt to say what poristic means for Viète that actually takes into account his practice in the *Supplementum geometriae* in any detail." We cannot here examine Ferrier's detailed treatment of Viète's *Supplement*; but the gist of Ferrier's account is as follows (in a summary paraphrase that largely selects and rearranges Ferrier's own words).

The authoritative statements on analysis that came down to Viète from antiquity were by Pappus, Theon, and Proclus. Pappus, who gives analysis a two-fold division, says that the geometer in theoretical analysis supposes the thing sought as existing, and in problematical analysis supposes the thing proposed as known; that is what Pappus says—even though it is theorems that are known, and it is figures that exist. Analysis is not made two-fold by Theon, another ancient authority on analysis; Theon supplied only theoretical analyses. Like Pappus, however, Proclus, the last of the ancient authorities on analysis, suggests that the chief use of analysis is in problems. The ancient teachings on analysis thus seemed confusing.

Now, Viète did not deny or discount the importance of the difference between theorems and problems. But he did confuse Pappus's problematic analysis with the resolution of an analysis. Viète made zetesis and poristic part of the same procedure. His zetesis corresponds to what is now called "the transformation" (or the analysis proper); his poristic corresponds to what is now called "the resolution." See the box following.

ANALYSIS	The *transformation* ends when what is given in consequence of supposing as done what the problem proposes to be done is shown to be what's given truly and originally. From this fact, the *resolution* begins; it ends when the constructibility of the thing sought has been proved.
SYNTHESIS	The *construction*, constructing in fact what has (merely) been proved to be constructible, repeats the resolution step-by-step in the same order. The *demonstration* repeats the transformation step-by-step in reverse order.

While there is a *transformation* in *all* analyses (both theoretical and problematical), there is a *resolution* in *most* analyses *but not* in *all* of them.

The transformation (the first part of the analysis) ends when the argument has led to a proposition (almost always a proportion or an equation) that has been true from the start or can be given independent of the hypothesis.

The resolution (the second part of the analysis) shows that the proposition that terminates the transformation, together with the conditions stated in the original problem, both of which are truly given, can make the remaining parts of the figure (which enter into the subsequent proof) truly given. The argument of the resolution employs a figure that is not given; it may be drawn, but it is not understood to be given until the true synthesis has been accomplished. Resolution, unlike ordinary geometry, deals with objects only in their character as objects that can be given; that is, rather than constructing them (as a problem does), it demonstrates that they *can* be constructed. It demonstrates their constructibility—it demonstrates that what is originally given is sufficient to provide us with what we want. Resolution shows us that what we want is providable (or in Greek, is *poriston*).

An argument like a resolution would be facilitated by a collection of antecedent propositions stating that the givenness of certain objects entails the givenness of others. Such a collection, which would save analysts much labor, is the book by Euclid that received the Latin name *Data* (which means "givens," as does its Greek name: *Dedomena*). This book begins by speaking of givenness as *providability*; it says: "We are able to provide for ourselves (*poristhai*) ...." A datum is a theorem, a theorem that shows that the givenness of the known entails the givenness of the unknown. (For example, the 58th proposition of the *Data* asserts that if a given [parallelogrammic] area be applied to a given straight line so as to be deficient by a [parallelogrammic] figure given in form [a parallelogram is "given in form"—that is, in *eidos* or "looks"—when what is given are its angles and the ratio of its sides], then the breadths [sides] of the deficiency have also been given. This corresponds to the theorem that is the 28th proposition of the Sixth Book of Euclid's *Elements*.) Pappus lists Euclid's *Data* in his *Storehouse of Analysis*, along with another book by Euclid, called *Porisms*—which (like *data* theorems) had to with problems, and dealt with the possibility of the determination of geometrical objects.

Poristics thus appears to embrace the theorems (arising from the proportion or equation) which, though proved as theorems without reference to givenness, serve to prepare the equation or proportion for the construction (or computation) of the determinate solution which is performed on determinate magnitudes (or numbers) by exegetic (or rhetic). Poristics shows that the requisite construct is providable (*poriston*)—it shows that what is originally given will suffice to give us the construction that we need.

From that, we can see why poristics plays so small a role in Viète's analysis. *Poristics* becomes superfluous to the extent that bodies of problems are successfully transformed by *zetetics* into standard forms whose syntheses are regularly provided by *exegetics*. In a problem for which there already exists a standard construction or a regular computation, there is no need to establish the possibility of the subsequent synthesis. Not so for the ancients, however, who never methodized their analyses to that extent. Viète works out a great simplification; he reduces his analyses to equations of a limited number of types, and then systematically arranges and solves those canonical equations. That is why Viète placed such a great value on the third (exegetic) part of his new analysis. The development of exegetics obviates the need for ordinary resolutions, and hence for poristics.

Having thus obtained from Ferrier some help in understanding Viète's division of the analytic art, we can now prepare to consider more closely what is of most particular interest in that art—namely, reckoning by species. To start us off in our consideration of reckoning by species, consider figure 5.

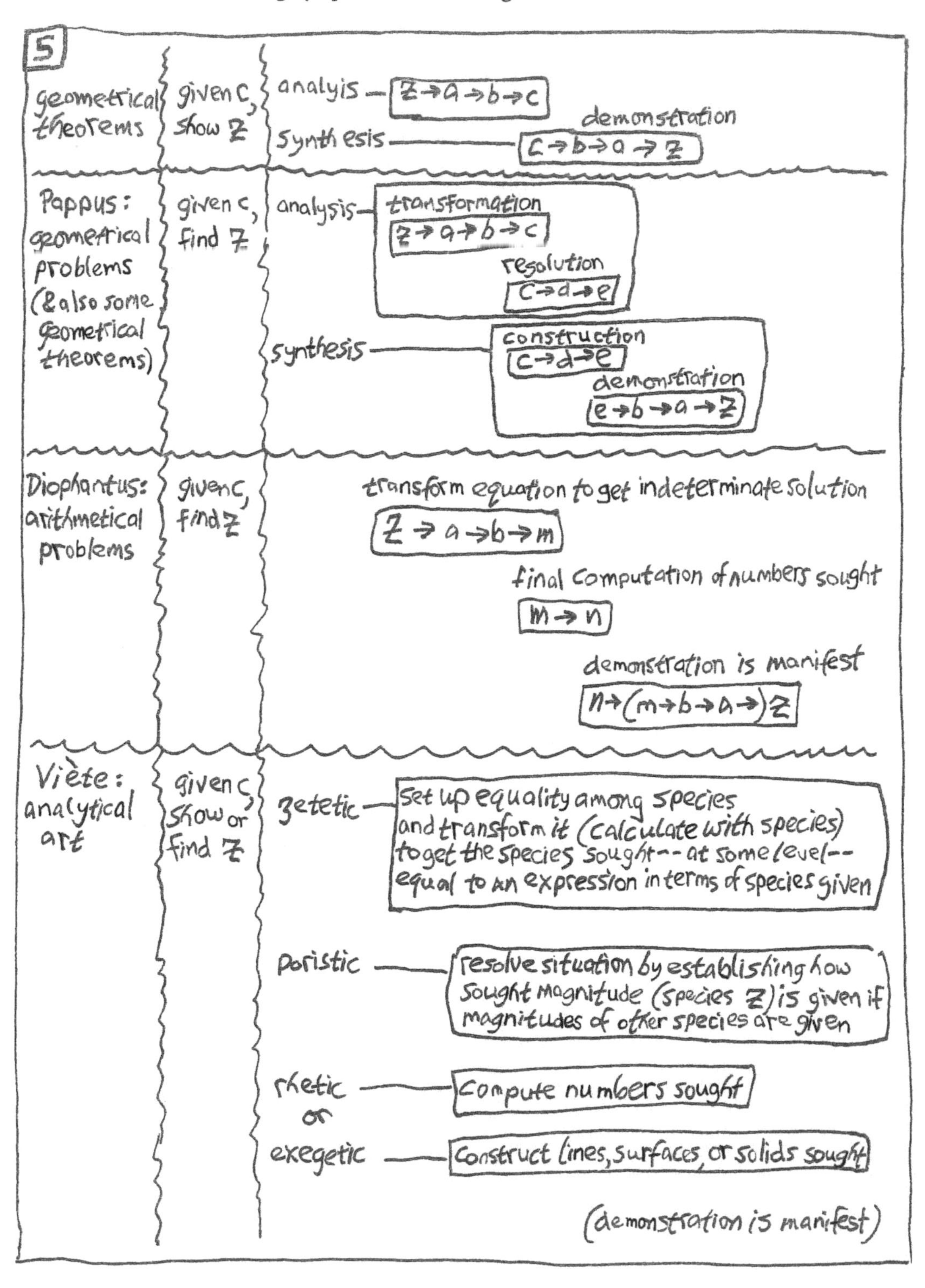

In a geometrical theorem, the synthesis can be (though it may not be) simply a demonstration that repeats the steps of the analysis in reverse order.

In the geometrical problems presented by Pappus, things are different. In these geometrical problems, the synthesis begins with a construction. This construction repeats in the very *same* order the sequence of steps of the resolution with which the analysis *ended*; and while the demonstration with which the synthesis ends does repeat in reverse order *most* of the steps of the transformation with which the analysis began, the demonstration begins with a step that came *before* the very end of the transformation.

In the arithmetical problems presented by Diophantus, there is (with respect to being analytic) a resemblance not to geometrical problems but rather to geometrical theorems. Diophantus ends problems by saying that the demonstration (which he omits) is manifest. Why would he call the omitted demonstration manifest? Because if you insisted on including it, you would simply present in reverse order the steps in what Diophantus has already presented. The synthesis (if included) would not be complicated by the presence of any construction; the calculations are simply reversible.

Still, Diophantus does deal in problems rather than in theorems. He says that his treatment of matters arithmetical is for someone eager to learn to find solutions to problems in numbers. Now there was, in Greek antiquity, another sort of treatment of matters arithmetical; it was for someone eager to learn to see what the different kinds of numbers are. Such a "theoretical" treatment of numbers aimed at insight into what is, rather than at capability for problem-solving. The *Introduction to Arithmetic* of Nicomachus, which became the classic text of that sort, was written in about 100 AD. This was a century before Diophantus—and four centuries after Euclid, who presents theorems of theoretical arithmetic in his *Elements*. Such theorems constitute the Seventh, Eighth, and Ninth Books, which are followed by books which show that numerical ratios will not suffice for treating the ratios of magnitudes. Now as we have seen, the relation in size of one number to another is, in Greek, a *logos*, as is the relation in size of one magnitude to another of the same kind. *Logos* (from a root meaning "lay" or "select" or "collect") can also mean "word" or "speech" or "reason" or "account" or "count." Hence the words *logidzomai*—meaning "to take into account" or "to count" or "to calculate"—and *logistês*—meaning "reasoner" or "calculator" or "auditor of accounts" or "teacher of things arithmetical." Hence *logistikê* is what one learns in order to become a skillful calculator. The aim of Diophantus is to enable his reader to find solutions to problems in numbers by very clever calculation.

Like Diophantus's art of arithmetical problem-solving, Viète's zetetics (the first part of his analytic art) proceeds by calculation. Viète will say soon that it proceeds through a kind of *logistic*.

But Viète's calculative inquiry, unlike Diophantus's, closes without arriving at definite numbers. It is not the work of Viète's zetetics to say what are the countable numbers—or to show what are the constructible figures—which will do what is required. By assigning elsewhere than zetetics the final computation of definite numbers—or the construction of definite figures—Viète makes his analytical art of inquiry resemble the analysis in geometrical problem-solving as presented by Pappus, for whom the construction (which constitutes, together with the demonstration, the synthesis) is not part of the analysis.

Viète's work combines something from Diophantus and something from Pappus: it has the calculational character of a problem-solving that is arithmetical, and the analytical character of a problem-solving that is not.

Viète *calculates* with his *species*, as Diophantus does with his *eidê*; and, like Diophantus, Viète can use what he calculates with, in order to solve problems about numbers. But Viète's *species*—unlike Diophantus's *eidê*—can also be used, and indeed are used just as much, in order to solve problems about lengths, surfaces, and solids; Viète's *species*—unlike Diophantus's *eidê*—admit of geometrical exegesis because they are not abbreviations for numbers.

Viète continues the first chapter of the *Introduction* by saying this:

> Now as for what pertains to Zetetic, it is established by the art of logic
> through syllogisms and enthymemes whose supports are
> those very stipulations (*symbola*) by which
> equalities and proportions are come to as conclusions—
> [these supports being] to be derived as much from common notions
> as from theorems to be set in order by the force of Analysis itself.
>
> The form of entering upon a Zetesis, however,
> is peculiar on account of the art,
> now at last not exercising its Logic in numbers,
> which was the tediousness of the ancient Analysts;
> but in a new way, through a logistic by species—
> [this speci-al logistic that is] to be introduced [being]
> much more fruitful and more powerful than the numerical one
> for comparing magnitudes with one another—
>
> the law of homogeneous [magnitudes] having been put forward first,
> and then the customary series or ladder (*scala*) having been set up,
> a ladder, as it were, of magnitudes that
> ascend and descend from kind to kind (*ex genere ad genus*)
> proportionally by their own force,
> by which the grades (*gradus*) and the kinds (*genera*)
> of those same [magnitudes]
> may be designated and distinguished.

The ancient analysts were tedious, says Viète: they performed calculations on numbers only. At least in public that is all they did. Viète will tell us later, at the end of the fifth chapter, that they engaged in something of a cover-up. For now, what we are told is merely that calculations can be performed on species as well as on numbers. And among species, as among numbers, comparisons can be made in equations. Comparisons of magnitudes—for which calculation is a powerful instrument, and yet for which numbers are unsatisfactory—are what seem to be of concern. The success of the logistic done through species depends upon establishing a certain "law of homogeneous magnitudes"; and to designate and distinguish the "grades and kinds of magnitudes," there has been established a scale or "ladder" of magnitudes. The foundations of zetetic discourse are the "stipulations" by which one arrives at equations and proportions that compare magnitudes.

Those fundamental stipulations of zetetics will be dealt with next—in chapter II; and then the law of homogeneous magnitudes, and the grades and kinds of the magnitudes that are compared, will be dealt with—in chapter III. The rest of Viète's *Introduction to the Analytic Art* is laid out as follows:

Chapter IV will treat the precepts of
the logistic which is done through species;

and chapter V will treat the laws of zetesis,
in which the speci-al logistic is employed.

Following those zetetic chapters will come two chapters
(VI and VII) that treat, respectively, the two remaining
subdivisions of the analytic art—poristic and exegetic.

Chapter VIII, the final chapter of the *Introduction*,
will first deal with the denoting of equations,
and will then present the epilogue of the analytic art.

the second chapter

In the first chapter, Viète said that the supports of zetetic are certain stipulations. The word that he used, *symbola*, is a transliterated Greek legal term referring to contracts. Viète borrowed the term for what that chapter described as those very stipulations which logical argument uses for concluding that there is an equality or proportionality, and which are to be derived from common notions as well as from theorems to be set in order by force of analysis itself.

Now in the second chapter he gives us a numbered list of sixteen stipulations, introduced by the following words:

CHAPTER II
About the stipulations of equalities and proportions

The more noted stipulations about equations and proportions
which are found in the *Elements*,
the Analytical Art assumes as having been shown forth (*demonstrata*)—
such as are, just about, [the following.]

The first stipulations are these:

1. That the whole [*totum*] is equal to its parts [taken all together].
2. That what are equals to the same thing are equals to each other.

Stipulation no. 1 appears as a nineteenth axiom in Clavius's much-read edition of Euclid's *Elements*; stipulation no. 2 is the first common notion in the First Book of the *Elements*.

The next stipulations, like the previous ones, have to do with equality; now, however, they have to do with equality when certain operations are performed upon equals:

3. That if equals are added to equals, ... the wholes are equals.
4. That if equals are taken away from equals,..... the remainders are equals.
5. That if equals are multiplied by equals, ... the products are equals.
6. That if equals are divided by equals, ... the results are equals.

Stipulations nos. 3 and 4 are the second and third common notions in the First Book of the *Elements*. Stipulations nos. 5 and 6 correspond to the interpolations that were labeled the fifth and sixth common notions in the First Book of the *Elements*. The outcomes of multiplication here translated as "products" are *facta* (more literally "things made or done"); the outcomes of division are *orta* (more literally "things arisen"). It might not be easy for one of Viète's readers to answer another reader who, seeing that Viète will speak of multiplying or dividing by a species, asked this question: what might be meant by "multiplication" or "division" if the multiplier or divisor be a species that is not merely standing in for some number (that is, for some multitude of units)?

The next stipulation has to do with proportionality:

> 7. That if anythings be proportionals directly,
> [then] they are [also] proportionals inversely and alternately.

What this means, is depicted in figure 6. It corresponds, for alternation, to the 16th proposition of the Fifth Book of the *Elements*, but no proposition is given there for inverted proportionality. The definitions for that Fifth Book, however, do include alternate ratio and inverse ratio, both of which we have seen employed by Apollonius. As we read Viète's stipulations on proportionality, we must bear in mind that the Fifth Book of Euclid's *Elements* is not about the proportionality of "anythings" whatever; it is about the proportionality of *magnitudes*—of *any* magnitudes, to be sure, but *not* of any *numbers*. Those books of the *Elements* that are about numbers do not contain a proposition that demonstrates for numbers the truth which corresponds to what the 16th proposition of the Fifth Book demonstrates for magnitudes (indeed, which corresponds to what the proposition demonstrates for four magnitudes that are tacitly taken to be all of the same kind).

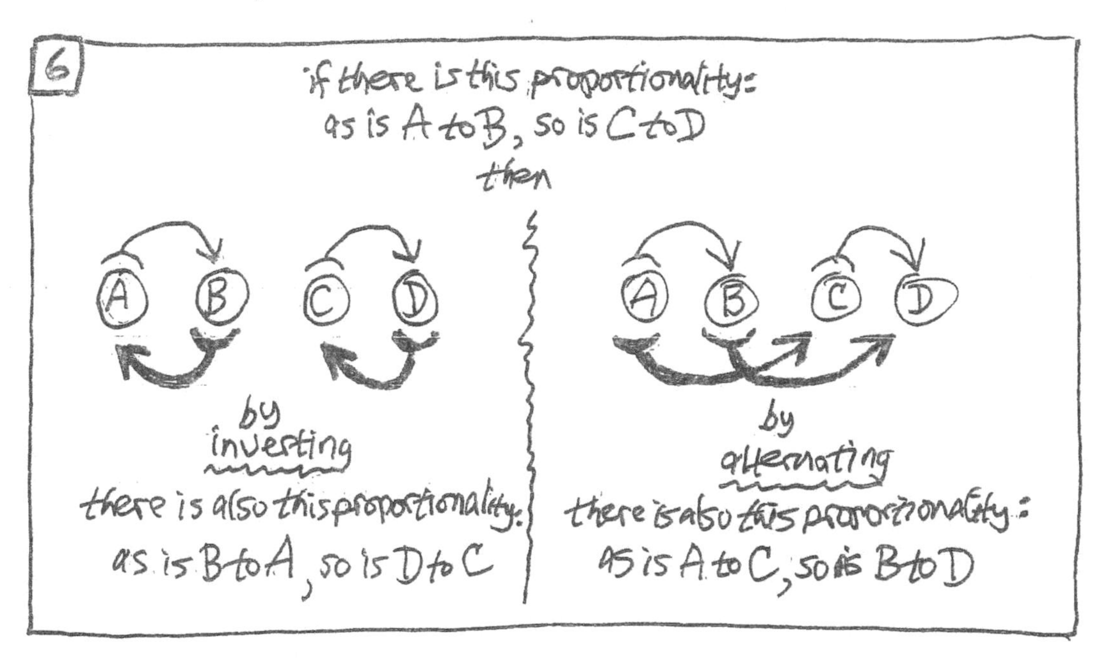

The next stipulations are these:

8. That if like proportionals are	added to like proportionals, the wholes are proportionals.
9. That if like proportionals are	taken away from like proportionals, the remainders are proportionals.
10. That if proportionals are	multiplied by proportionals, the products (*facta*) are proportionals.
11. That if proportionals are	divided by proportionals, the results (*orta*) are proportionals.

That says what is depicted (using the familiar signs now current) as follows:

if there is this proportionality—

as A is to B, ...so........C......is to....................D,

then there are also these proportionalities—

as A is to B, ...	so (A+C)....is to..............	(B+D),	and
	so (A- C)....is to...............	(B-D),	and
	so (A*C)....is to...............	(B*D),	and
	so (A/C)....is to................	(B/D).	

(with the proviso—but only in adding and in taking away—
that C be like A, and D be like B)

Those stipulations, like the one before it, have to do with proportionality—now, however, they do so when certain operations are performed upon proportionals.

The proviso of likeness that is included in stipulations no. 8 and no. 9 (for adding and for taking away) is not included in stipulations no. 10 and no. 11 (for multiplying and for dividing). Again, it is not clear what multiplying or dividing means if the multiplier or divisor be a species that is not taking the place of some number. At any rate, it seems that the products of multiplying proportionals are in the ratio that is compounded from the ratios of the proportionals. (And the results of dividing proportionals are in the ratio that, when compounded with one ratio, would yield the other one.)

Stipulations no. 8 and no. 9 correspond to the 12th and the 19th propositions of the Fifth Book of the *Elements*.

Stipulation no. 10 corresponds (with suitable elaboration) to two propositions of the *Elements*—the 23rd proposition of the Sixth Book (for straight lines) and the 5th proposition of the Eighth Book (for numbers). (Euclid's proposition VI.23 shows that rectangles are to each other in the ratio that is compounded of the ratios of their sides—which are the straight lines that contain the rectangles, or, we might say, make them. Euclid's proposition VIII.5 shows that "plane" numbers are to each other in the ratio that is compounded of the ratios of their "sides"—which are the numbers that "make" the plane numbers, or, we might say, "produce" them.) Stipulation no. 11 is the converse of no. 10.

Earlier, no. 4 and no. 5 said that when equals are multiplied or divided by equals, the products or results are equals. Now, stipulation no. 12 gives, for multiplication and division, a more general consequence from a condition that is more general in one respect and more restricted in another:

> 12. That by a common multiplier or divisor,
> an equality is not to be changed, nor is a ratio.

This says what is depicted (using the familiar signs now current) as follows:

If......A..................is equal to, or is in a certain ratio to................B,

then (A*C)...........is equal to, or is in that same ratio to.................(B*C),
and (A/C)............is equal to, or is in that same ratio to................(B/C).

For multiplication of numbers, this corresponds to the 17th proposition of the Seventh Book of the *Elements*.

Next comes another stipulation about products. This one, like the very first of all the stipulations, has to do with the whole and its parts all taken together:

> 13. That products by the several segments (*facta sub singulis segmentis*)
> [all these several products being taken together]
> are equal to the product by all [the segments taken together] (*sub tota*).

This says what is depicted (using the familiar signs now current) as follows:

(A*B) + (A*C) is equal to A*(B + C).

This stipulation's speaking of "segments" might seem to make it less general than the other stipulations, which seem to apply as much to numbers as to straight lines. If that were so, and "segments" referred merely to the pieces into which a straight line is cut, then "things made" would not apply to products of multiplying numbers but merely to the rectangles that are made by the straight lines that contain them—and then this stipulation would not merely correspond to, but would be identical with the 1st proposition of the Second Book of Euclid's *Elements*. As will become clear, however, what Viète has in mind are segments of species, rather than of straight lines. Species are not numbers—but neither are they straight lines or plane figures or solids. Species may be used to investigate all of those things, but species themselves are not as such any of them.

The next stipulation, which is another one about the products of multiplying or the results of dividing, is the first stipulation to use the term "magnitudes"; the term continues to be used in both of the remaining stipulations that come after this one. What this one stipulates is:

> 14. That products produced by magnitudes "continually,"
> or results resulting from them "continually,"
> are equal—whatever be the order of the magnitudes in the
> production or application (*ductio vel adplicatio*) [that is to say, multiplication or division].

The proposition in the *Elements* that comes closest to this (despite the stipulation's speaking of *ductio* and *adplicatio*, rather than of "multiplication" and "division") is the 16th proposition of the Seventh Book—which says for numbers, that if C is the product made when A multiplies B, and D is the product made when B multiplies A, then C is equal to D. Stipulation no. 14 says what is depicted (using the familiar signs now current) as follows:

if........................C is equal to (A*B),
and....................D is equal to (B*A),

then...................C
is equal to
D.

(In the *Elements* , there is no proposition corresponding to this for straight lines—since it goes without saying that the rectangle contained by one straight line and another one is equal to the rectangle contained by the other straight line and the first one.) Although proposition VII.16 is the closest that the *Elements* comes to what this stipulation says, that is nonetheless not very close, since this stipulation says what is depicted (using the familiar signs now current) thus:

(A*B)*C)...........is equal to...........A*(B*C).

(A/B)/C.............is equal to...........(A/C)/B.

In presenting the next stipulation, Viète does what he does not do for any of the other stipulations: he introduces it. His introduction uses, for emphasis, the word for "sovereign" in Greek.

> However, the sovereign (*kyrion*) stipulation of equalities and proportions,
> and one that is all-momentous in analyses, is:
>
> 15. If there be three or four magnitudes,
> but what is [produced] by (*sub*) the extreme terms
> is equal to what is [produced]
> by (*sub*) the [single] middle term with itself,
> or by the [two] middle terms [together],
> then they [the three or four magnitudes] are proportional.

That says what is depicted (using the familiar signs now current) as follows:

> if A, B, and C are three magnitudes such that.......(A*C) is equal to (B*B),
> then..as A is to B, so B is to C;
>
> or if A, B, C, and D are four magnitudes such that (A*D) is equal to (B*C),
> then..as A is to B, so C is to D).

After saying that such equality entails the proportionality of the magnitudes involved, Viète then says also that the proportionality entails the equality:

> And conversely,
> 16. If there be three or four magnitudes,
> and as the first is to the second, so that second, or else some third, is to another,
> then what would be [produced] by (*sub*) the extreme terms
> will be equal to what would be [produced] by (*sub*) the middle ones.

That says what is depicted (using the familiar signs now current) as follows:

> if A, B, and C are three magnitudes such that.......as A is to B, so B is to C,
> then...(A*C) is equal to (B*B);
>
> or if A, B, C, and D are four magnitudes such that as A is to B, so C is to D,
> then...(A*D) is equal to (B*C).

In the stipulations prior to this last pair, the form of statement was indirect: the verb was in the infinitive. The translation might have read as follows: "for X to be Y [is stipulated]"—or (less literally) "[it is stipulated] that X is Y." Now, however, in the stipulation that Viète says is sovereign or all-momentous in analyses (and also in the converse of this stipulation), the form of statement is direct: the verb is in the indicative. The change in grammatical form lends emphasis to the importance of this pair of stipulations, which correlate equality and proportionality.

Where is there, in the *Elements*, a counterpart of the statement that equality between one product of two magnitudes and another product of two magnitudes is correlated with proportionality among the magnitudes? For numbers (for multitudes, that is to say, but not for magnitudes) such a correlation is shown in the 19th proposition of the Seventh Book. It is not shown in the Fifth Book for magnitudes generally (how, after all, could it be?), but in the Sixth Book it is shown for straight lines (if, that is, we may say that straight lines "produce" or "make" the rectangles that they are said to "contain"). That proposition of the Sixth Book in which it is shown is the 16th; and it is shown in the 17th proposition for a special case (namely, when three rather than four straight lines are proportional, and one of the rectangles is that special rectangle called a square).

The significance of this last pair of stipulations (nos. 15 and 16) is indicated in Viète's very next sentence after he has presented them, which is the only epilogue he gives to any of the stipulations. This sentence, which is the concluding paragraph of the chapter, says:

And so, a proportion can be called what is
put together [or constituted] from an equality
(*constitutio aequalitatis*);
an equality can be called what is
unloosened [or resolved] from a proportion
(*resolutio proportionis*).

That is to say: you never have only an equation or only a proportion; having either one of them will always give you the other. From an equality, a proportion is built up; and from that structure when it is dissolved again, there will come an equality.

The items in these equalities and proportions are called by Viète "magnitudes." In some ways they are like what Euclid calls "magnitudes," but in some ways they are different. Like the "numbers" of Euclid (which are "multitudes" not "magnitudes"), these "magnitudes" of Viète admit of calculation together—they admit not only of the adding of one of them to another one, and of the taking away of one of them from another one that is bigger, but also of the multiplying of one of them by another one, and of the dividing of one of them by another one! In this second chapter of his *Introduction*, Viète has not spelled out the complications that arise from the heterogeneity of the things he has been treating. He will turn to doing so now, in the third chapter.

the third chapter

Heterogeneity is a source of complication because it restricts the comparing of magnitudes. Two magnitudes are "compared" with a view to their being equal to each other or not. More than two of them can enter into a comparison if some of them "affect" each other; there then turn out to be two items which can be compared.

> CHAPTER III
> *About the law of homogeneous [magnitudes]*
> *and the grades and kinds (gradibus ac generibus] of compared magnitudes*
>
> The prime and perpetual law of equalities or proportions—
> which, because it is conceived concerning homogeneous [magnitudes],
> is called the law of homogeneous [magnitudes]—
> is this:
>
> Homogeneous [magnitudes]
> are to be compared to
> homogeneous [magnitudes].
>
> For, as Adrastus said, one cannot know
> how those which are heterogeneous
> may be affected among each other (*inter se adfecta*).

The remark by Adrastus to which Viète refers was made in the course of explicating a definition from the Fifth Book of Euclid's *Elements*. What Euclid himself says, in that definition, is this: a ratio is a sort of relation in size between two homogeneous magnitudes.

What Adrastus says in that connection, as reported by Theon, is this: it is impossible to know what relations non-homogeneous [magnitudes] have to one another. This does not mean that a ratio of one straight line to another straight line, for example, cannot be the same as the ratio of one square to another square (or even of one weight to another weight); proportions like that are often found in the writings of the ancient mathematicians. Rather, what Adrastus means is that a straight line, for example, cannot have a ratio to a square (or to a weight). Viète takes what was said in the context of the proportionality of magnitudes (whether measured or not) and uses it in the context of comparisons that will admit of calculation—that is, in the context of the equality of magnitudes that are "speci-al." Viète presents the following laws that govern in this context:

And so,

> If a magnitude is added to a magnitude,
> this [latter] one is homogeneous with that [former] one.
>
> If a magnitude is subtracted from a magnitude,
> this [latter] one is homogeneous with that [former] one.
>
> If a magnitude is led onto a magnitude (*in magnitudem ducitur*),
> what it produces (*fit*) is heterogeneous
> [both] with this [latter] one and with that [former] one.
>
> If a magnitude is applied to a magnitude,
> this [latter] one is heterogeneous with that [former] one.

Not having attended to which things was the cause of
the fogginess and blindness of the ancient analysts.

Before considering that cryptic last remark by Viète, let us look at an example of what he is talking about, and consider his terminology. A straight line that is 5 meters long is *homogeneous* with a straight line that is 7 meters long to which it is *added*, or from which it is *subtracted*. But when a straight line that is 5 meters long is *led onto* a line that is 7 meters long, then the magnitude that it makes or produces is not a straight line but a rectangular surface (one that is 35 square meters in area)—this surface being *heterogeneous* with each of the lines that made it. And a rectangular surface that is 35 square meters in area is *heterogeneous* with a straight line that is 7 meters long to which it is *applied* (and is also heterogeneous with what results from the application—this result being a straight line that's 5 meters long).

The terminology having to do with products is a Latin version of ancient Greek terminology. To refer to drawing a straight line, Euclid used the Greek verb *agô*, which is "to lead." In Latin, "to lead" is *ducô*, which also has the sense "to draw (forth)"; the direction is more pronounced in our "pro+duct" (from the Latin "led+forth"). Now, a straight line that is "led onto" another straight line will make a rectangle. (When, on the other hand, a surface is "applied to" a straight line, there "arises" a straight line that is the other side of a rectangle which is equal in area to the applied surface and which has the first straight line as a side.) A rectangle is contained "by" (*hypo*) its sides. (This Greek preposition *hypo*, which can mean "by" when used to refer to an agent or a means, has as its primary sense "under"—which can be seen in the word "hypo+tenuse," for the side that "stretches+under" a triangle's right-angled vertex.) Euclid followed an earlier tradition of naming kinds of numbers by using the names of figures formed with the same number of points; he called some numbers "sides," he called some numbers "squares" or "planes," and he called some numbers "cubes" or "solids." Diophantus followed Euclid in this, and he extended the terminology. Diophantus also spoke of a number as being "made by" the numbers "from" which it comes into being when you "multiply" one of those numbers "onto" the other; here, the word "by" translates the Greek preposition *hypo* used with the genitive case, and the word "onto" translates the Greek preposition *epi* used with the accusative case. In Viète, the Greek *hypo* becomes the Latin *sub* ; and the Greek *epi* becomes the Latin *in*. The Greek verb for doing or making (*poieô*) becomes the Latin *facio*, so that the things made are called by Viète *facta*.

Now what exactly is Viète referring to when he speaks (at the very end of the passage last quoted) about the failing of the ancient analysts? After all, those ancients did know that a magnitude has a ratio to another magnitude only if the two magnitudes are of the same kind; and they did know that a ratio of magnitudes of some one kind is always the same or is greater or less than another ratio of magnitudes, even when the magnitudes of the other ratio are of another kind than the magnitudes of the first ratio. They did know also that a proportion of certain magnitudes is sometimes equivalent to an equality of certain other magnitudes; and they did know that in the case of multitudes, which easily admit of calculations of many sorts, a proportion is always equivalent to an equality.

Yes, they did know all that. But (Viète is saying) they did not attend to the ways in which magnitudes can admit of calculations of all sorts despite their differences of kind—that's why the ancient analysts saw many things only very dimly or even not at all. They had a notion of comparison that was too restricted. A less restricted notion of comparison would have enabled them to see certain differences in kinds of magnitudes as gradations or steps arranged in a sort of vertical sequence. Magnitudes of different kinds can be taken as different rungs on the same ladder, so that the *variety* of magnitudes becomes a *numerical order of rank or degree*.

"Ladder" in Latin is *scala*. The ladderly arrangement or scale of magnitudes is now presented by Viète, who does not suffer from what he describes as "the fogginess and blindness of the ancient analysts":

2. Magnitudes which ascend or descend
from kind to kind (*ex genere ad genus*)
proportionally by their own force
are called "ladder [magnitudes]" (*scalares*).

3. Of ladder magnitudes,

the first is "Side," or "Root."	(*Latus, seu Radix*)
2. "Square."	(*Quadratum*)
3. "Cube."	(*Cubus*)
4. "Squaro-square."	(*Quadrato-quadratum*)
5. "Squaro-cube."	(*Quadrato-cubus*)
6. "Cubo-cube."	(*Cubo-cubus*)
7. "Squaro-squaro-cube."	(*Quadrato-quadrato-cubus*)
8. "Squaro-cubo-cube."	(*Quadrato-cubo-cubus*)
9. "Cubo-cubo-cube."	(*Cubo-cubo-cubus*)

And successively in that series and method are to be denominated
the ones that are left.

Magnitudes are said to "ascend or descend proportionally by their own force" when they are related to each other as follows: the first magnitude is to the second as the second is to the third, which is as the third is to the fourth, which is as the fourth is to the fifth, and so on. It will turn out that these "ladder magnitudes" correspond to what we now call "powers of the unknown quantity." That is to say: if the unknown quantity is called "x," then the ladder magnitudes would correspond to what is depicted in the following box. (Again—and it cannot be repeated too often—keep in mind that "corresponds to" is not the same as "is the same as." As we shall see, it is significant that the names of the ladder magnitudes on different rungs do not differ in a merely numerical way.)

1. x^1
2. x^2
3. x^3
4. x^4
5. x^5
6. x^6
7. x^7
8. x^8
9. x^9

etc.

For magnitudes that are unknown, Viète has just given names to their "grades" (that is, to what we now would call their "degrees"). Now, for magnitudes that are "compared," he goes on to give names to their corresponding "kinds." (What in English is called a "rung" of a "ladder" is in Latin a *gradus* of a *scala*. What in English is called a "kind" is in Latin a *genus*, the plural being *genera*.)

> 4. Of compared magnitudes, the kinds are
> (as they were declared in order concerning ladder [magnitudes]):
>
> 1. Length or breadth.
> 2. Plane.
> 3. Solid.
> 4. Plano-plane.
> 5. Plano-solid.
> 6. Solido-solid.
> 7. Plano-plano-solid.
> 8. Plano-solido-solid.
> 9. Solido-solido-solid.
>
> And successively in that series and method
> are to be denominated the ones that are left.

It will turn out that *compared* magnitudes of the nth level can, by the law of homogeneous magnitudes, be equated with a *ladder* magnitude of the nth grade but not with one of any other grade. Now our 4th power of an unknown quantity (our x^4) corresponds to Viète's ladder magnitude of the 4th grade. To be capable of being equated with such a ladder magnitude, which is Viète's "squaro-square," a compared magnitude would have to be what Viète calls a "plano-plane," which would correspond to an item of ours like "a^2b^2" or like "ax^3" (if "a" and "b" correspond to "lengths or breadths") or like "ax^2"(if "a" corresponds to a "plane"). Whereas Viète's various ladder magnitudes correspond to our powers of an unknown quantity, his kinds of compared magnitudes are like our terms that contain some unknown quantity or quantities (and that do so either exclusively or along with a known quantity).

Note that Viète's terminology of levels (both of ladder magnitudes and of compared magnitudes) is—like the terminology of Diophantus—not multiplicative but additive. Consider the following illustrative example. The 3rd rung of the ladder (or the 3rd kind of compared magnitude) is named "cube" (or "solid"), but what about "cubo-cube" (or "solido-solid")—what does it name? Not the 3-*times*-3 (not, that is to say, the 9th level) but rather the 3-*plus*-3 (that is to say, the 6th level). The "cubo-cube" is *not* a cubing *of* a cubing—which would correspond to our "$(x^3)^3 = x^{3*3} = x^9$"; what it is, rather, is a cubing *and* a cubing—which corresponds to our "$x^3*x^3 = x^{3+3} = x^6$." The number of Viète's rung on the ladder corresponds to what we now call "the exponent or power of the unknown quantity."

Viète borrows the terminology of Diophantus, and supplements it. He also modifies it. Let us compare the two terminologies. First, let us briefly look back at Diophantus. For an *as yet unknown* number and the products of multiplying it by itself one or more times, Diophantus used the following terminology:

> (1) the unknown number,
> he called the "number" (*arithmos*);
>
> (2) the product of multiplying it by itself,
> he called the "power" (*dynamis*);
>
> (3) the product of multiplying it by itself and then again by itself,
> the "cube";
>
> (4) and then again,
> the "powero-power";
>
> (5) and then again,
> the "powero-cube";
>
> (6) and then again, finally,
> the "cubo-cube."

Those names were sometimes used in full by Diophantus; sometimes he used abbreviations for them. For a *determinate* number, Diophantus sometimes uttered its proper name and sometimes he used the Greek alphabetic notation for it, or else he used—without abbreviation—the names that he gave to the products of multiplying a determinate number by itself one or more times. The names that he gave were these:

> the product of multiplying a determinate number by itself,
> he called a "square" (*tetragonos*);
>
> and it itself,
> he called the square's "side" (*pleura*);
>
> next, as with numbers that are indeterminate,
> came "cube" and the rest—unabbreviated, however.

Now what happens to that terminology in the hands of Viète is this:

> The terminology that
> was used by Diophantus for the *products of a determinate number*
> is used by Viète for the *ladder-rungs of an unknown magnitude*.
>
> A terminology which is
> like that, but does not suggest the equality of the component "sides,"
> is then used by Viète for the *kinds of compared magnitudes*.

See the box following.

Diophantus		Viète	
number that is determinate	number that is as yet unknown	ladder magnitudes —rungs of a magnitude which is unknown	compared magnitudes —equated with ladder magnitudes which are known, at least partly
side	number (& abbrev.)	side or root	length or breadth [of square]
square	power (& abbrev.)	square	plane
cube	cube (& abbrev.)	cube	solid
	powero-power (& abbrev)	squaro-square	plano-plane
	powero-cube (& abbrev.)	squaro-cube	plano-solid
	cubo-cube (& abbrev.)	cubo-cube	solido-solid
		etc.	etc.

Both for magnitudes that are given and for magnitudes that are unknown, Viète will use letters of the alphabet—consonants for magnitudes that are given, and vowels for those that are unknown.

Viète continues as follows:

5. Out of the series of ladder [magnitudes] (*ex serie scalarium*)
the higher grade at which (*in quo*)
the compared magnitude takes its stand (*consistit*)
going up from (*exinde à*) the "side"
is called the "power."
The remaining lower ladder [magnitudes]
are grades "along-the-way (*parodici*)" to the "power"

Think of yourself as going up the series of steps that are ladder magnitudes. As you go from the bottom step, which is the grade called the "side," you eventually reach that higher step with which some compared magnitude may be equated (the step at which that compared magnitude can be said to take its stand—the step in which or with which it can be said to consist). This higher grade is called the side's "power." The remaining steps, those that are lower than the "power," are grades that are called "along-the-way" to the "power." *Par-od-ici* is a Latinate adjective coined from Greek words: "along (*para*) the way (*hodos*)." Like many of Viète's coinages, this one failed as currency. Failure was not, however, the fate of another neologism, which we shall see in his next paragraph; we still use his term "coefficient."

6. A power is "pure" when it is free of any "affecting."
[And a power is] "affected," into which there is mixed
a [grade that is] homogeneous [with it, but is produced] by
a grade along-the-way to the power and [with that grade]
an adopted (*adscita*) co-productive (*coefficiente*) magnitude.

7. Adopted magnitudes
by (*sub*) [one of] which—
and [along with it,] a grade that's along-the-way—
there is made [a product] homogeneous with the power that is affected,
shall be called "sub-graduals."

Such an adopted magnitude is called by that Latin term "sub-gradual" not so much because any such factor is at some grade "below" (*sub*) the powe, as because it is that "by" (*sub*) which the factor that is a grade-along-the-way is multiplied, in order to make the product that is homogeneous with the power that is affected.

That may all be clearer if we consider an example. As our "compared magnitude," let us take this: a "squaro-square" along with a "plano-plane"—the plano-plane being taken as the product of a "cube" by a "length." Where in that, might there be a "pure power," and what grades might there be "along-the-way" to it? What might be the "affected power," and what magnitudes might be "adopted"; what might be "coefficient"; and what might be "sub-gradual"? For help with that, see the box following.

> Now, a "squaro-square" by itself is
> a *pure power* of the 4th grade.
>
> *Along-the-way* from a "side or root" to such a pure power would be
> a "square" (2nd grade) or a "cube" (3rd grade).
>
> The compared magnitude that we are considering
> is an *affected power*;
> mixed in with the pure power (the "squaro-square") there is
> a magnitude homogeneous with it (a "plano-plane")—
> a "plano-plane" is homogeneous with a "squaro-square" because
> it is on the same 4th level.
>
> The "plano-plane" is produced by
> a particular grade that is along-the-way to the power,
> namely, a "cube" (which is at the 3rd level),
> and, along with it, a magnitude that is *co-efficient* in the producing,
> namely, a "length" (which is at the 1st level, like the power's "side or root").
>
> A "length" is the "*subgradual*" when
> the grade "along-the-way" to a power at the 4th level is a "cube."
> (What would be the subgradual when
> the grade along-the-way to a power at the 4th level is a "square"?
> —In this case, the subgradual would be a "plane.")

At the risk of misleading, we might say that the pure power "squaro-square" corresponds to our "x^4" nowadays. The affected power that we considered as an example corresponds to our "x^4+ax^3" when the quantity "a" has the same dimensions that "x" has. (In the case of the affected power that corresponds to our "x^4+ax^2" [that is, when the grade along-the-way is not "x^3" but "x^2"] then the quantity "a" has the same dimensions that "x^2" has; and so the product will be at the 4th level like the pure power "x^4.") Of course, what we do nowadays is not the same as what Viète is doing. What we do nowadays merely corresponds to what he is doing. Viète emphasizes the law of homogeneous magnitudes and uses a complex terminology of grades and kinds, whereas we nowadays use exponents in an enterprise of quantifying homogenization. In our notation, and in much more than our notation, we are—as we shall see later—followers of Descartes.

CHAPTER 3. RECKONING BY "SPECIES"

In Greek, technical knowledge about numbers (*arithmoi*) was called *arithmetikê*. Technical knowledge about the ratios and relations of numbers (their *logoi*) was called *logistikê*—a term that was used also for the technical know-how about numerical relations which enables one to calculate or reckon through numbers. That logistic art—that art of reckoning through numbers—does not suffice, however, as an instrument for the comprehensive art of finding in things mathematical. To acquire the art of finding the sizes not only of unknown numbers but also of unknown lines, surfaces, and solids, one must be able to reckon through species. Now, in his fourth chapter, Viète tells how to do so. He begins as follows:

> CHAPTER IV. *About the precepts of the speci-al reckoning-art (logistices speciosae)*
>
> The numerical reckoning-art (*logistice numerosa*) is one that
> is exhibited through numbers;
> the speci-al (*speciosa*) reckoning-art is one that
> is exhibited through species or forms of things (*species seu rerum formas*),
> as perhaps through letters of the alphabet (*Alphabetica elementa*).
> Of the reckoning-art that is speci-al, the canonical precepts are four,
> as are those of the one that is numerical.

The four canonical precepts of the reckoning-art that is numerical are these:

(1) to *add* one number to another;

(2) to *subtract* one number from another number that is greater;

(3) to *multiply* one number by another;

(4) to *divide* one number by another.

While all numbers are homogeneous, all magnitudes are not; so, while the canonical precepts of numerical reckoning do not need elaboration, those of speci-al reckoning do. Viète's elaborations on the precepts were printed in smaller type in his published *Introduction*.

PRECEPT I. *Magnitude to magnitude—to add [one to the other]*

Let there be two magnitudes A and B. It is required to add one to the other.
Since, therefore, a magnitude is to be added to a magnitude,
but homogeneous ones do not affect heterogeneous ones,
the two magnitudes which are put forward (*proponuntur*) to be added are homogeneous.
But being more (*plùs*) or being less (*minùs*) do not constitute diverse kinds (*genera*).
Wherefore a mark of coupling or joining together is fittingly added;
and, aggregated, they will be: "A plus B"—if they be simple "lengths" or "breadths."
But if they ascend [that is, are higher] upon the ladder that has been set out,
or they share in kind (*genere*) with ones that ascend,
they will be designated by their own denomination which is suitable—
"A square plus B plane" is said, or "A cube plus B solid," and similarly in the rest.
Analysts are accustomed, however, to indicate by the symbol (*symbolo*) "+" the affecting by a joining.

That is to say: in designating "A" and "B" to indicate their sum when they are at the first level, what will be said is simply "A plus B" without any mention of their level; but when they are taken at a level higher than the first, then their level will be mentioned. For example: if the unknown "A" and the given "B" are each taken at the second level (that is, if each is multiplied by itself), then what indicates their sum will explicitly designate "A" as a "square" and "B" as a "plane": what will be said is this: "A square plus B plane."

Now we turn from addition to subtraction. Viète's presentation of the next precept begins as follows:

PRECEPT II. *Magnitude from magnitude—to subtract [one from the other]*

Let there be two magnitudes A and B—
that former one being greater, this latter one being lesser.
It is required to subtract the lesser from the greater.
Since, therefore, a magnitude is to be subtracted from a magnitude,
but homogeneous magnitudes don't affect heterogeneous ones,
the two magnitudes which are put forward (*proponuntur*) to be subtracted
are homogeneous.
But being more (*plùs*) or being less (*minùs*) do not constitute diverse kinds (*genera*).
Wherefore by a mark of disjoining or mulcting
the subtraction of the lesser from the greater will be fittingly made;
and, disjoined, they will be: "A minus B"—if they be simple "lengths" or "breadths."
But if they ascend upon the ladder that has been set out,
or they share in kind (*genere*) with ones that ascend,
they will be designated by their own denomination which is suitable—
as "A square minus B plane" is said, or "A cube minus B solid,"
and similarly in the remaining ones.

In the elaboration of this precept, subtraction so far has resembled addition except for this: either of two magnitudes can be added to the other, but the magnitude which is the one that is taken away cannot be the one that is greater. In the sequel, there is more about what distinguishes subtraction.

Viète continues his elaboration on the precept for subtraction as follows:

> Nor would the work (*opus*) be otherwise
> if the magnitude that is to be subtracted is itself already affected,
> since the whole and the parts should not be evaluated by a different law (*jure*);
> as: if from A is to be subtracted "B plus D,"
> the remainder will be "A minus B, minus D,"
> with the magnitudes B and D having been subtracted piece-by-piece.
> But if D is already denied to B itself,
> and [after that, the] "B minus D" be subtracted from A,
> the remainder will be "A minus B plus D,"
> since in subtracting magnitude B there is subtracted
> what is, by [the amount of] magnitude D,
> more than equal [to what has been subtracted];
> for that reason, it is to be compensated by the addition of that [magnitude D].

Viète's argument for a change of sign in subtraction—namely, that

(A minus [B *minus* D]) is equal to ([A minus B] plus D)—

may be clearer if you consider the following numerical example:

15 *minus* (7 *minus* 2)	What you are *supposed* to take away from 15 is less than 7; indeed it is 2 less than 7.
is equal to	
(15 *minus* 7) *plus* 2.	But if what you *do* take away is 7 itself (rather than being 2 less than 7), then what you will have taken away from 15 is greater by 2 than it should have been. So the remainder that will be left (when 7 is taken away from 15) will be too small by 2— and so, in order to *compensate* for that, what you will need to do is *add* 2.

The elaboration of the precept for subtraction concludes as follows:

> Analysts are, however, accustomed to indicate
> by the symbol "—" the affecting by a mulct.
> And this [affecting by a mulct] is, in Diophantus, *leipsis* ("lacking") [in Greek letters],
> as the affecting by joining-onto is *hyparxis* ("being present") [in Greek letters].
> When, however, it is not put forward (*proponitur*) whether
> a magnitude be the greater or the lesser,
> and nonetheless a subtraction is to be done,
> [then] the mark of the difference is "====="
> —that is, when the lesser is unknown (*incerto*) [undetermined]—
> as: "A square" and "B plane" being put forward, the difference will be
> "A square ===== B plane," or "B plane ===== A square."

What looks like an elongated equal-sign in Viète's work is *not* one. In Viète's work, there is no sign of equality at all. Not until later was a sign of equality introduced into mathematics. The elongated side-by-side double-dash is used by Viète as a sign for the difference between the two magnitudes between which the sign stands. Thus, "B ===== A" means the very same thing that "A ===== B" does. What it means is "A – B" when A is the greater; but when B is the greater, then what it means is "B – A." To say "A minus B" (that is, to say "A less B") makes sense only if B is less than A.

After finishing with subtraction, Viète goes on to the next of the canonical precepts. Its presentation begins as follows:

PRECEPT III. *Magnitude onto magnitude—to lead [one onto the other]*

Let there be two magnitudes A and B.
It is required to lead one onto the other [thus making a product].
Since, therefore, a magnitude is to be led onto a magnitude,
they will effect, by that leading of theirs,
[a product which is] a magnitude heterogeneous to their very selves;
and what may be [effected] by (*sub*) them will be fittingly designated by
that same word "onto" (*in*) or "by" (*sub*)—
as "A onto B," by which is signified
that this has been "led onto" that,
or as others [might say],
that [a magnitude] has been "produced by" (*factam sub*) A and B—
and that [said] simply, if indeed A and B be simple "lengths" or "breadths."
But if they ascend [that is, are higher] upon the ladder,
or they share in kind (*genere*) with those [that do] ,
it is suitable to employ the denominations of
the ladder [magnitudes] themselves or of those which share in kind with them—
as perhaps "A square onto B," or "A square onto B plane or solid,"
and similarly in the ones that are left.

So far, the elaboration of this precept for making products resembles that earlier elaboration for addition except for this: the sum of two magnitudes is homogeneous with each of the magnitudes, but the product and any factor that enters into its production are heterogeneous. In the sequel, there is more about what distinguishes the making of products.

The elaboration of this precept continues as follows:

> But if [both of the] magnitudes to be led, or one or the other of them,
> be of two or more names,
> nothing different happens in the work (*opere*)—
> since the whole is equal to its parts [taken all together], and so
> [taken all together] the products by the segments of some magnitude
> are equal to the product by the whole [magnitude].

To see why Viète speaks of the "segments" of a magnitude, you may imagine a line L that is cut up into segments P, Q, and R. The rectangle produced by that whole line L along with another line M will be equal to the rectangles, taken all together, produced by this line M along with the several segments P, Q, R. See figure 1. Viète, of course, is not speaking about lines and rectangles. If he were, then this would merely be the enunciation of the 1st proposition of the Second Book of Euclid's *Elements*. Rather, Viète is speaking of "magnitudes" and their products when one of them is "led onto" another. He has in mind something that resembles what we do when we say nowadays that if "l" and "m" are quantities, and if "l" is equal to the sum of quantities "p+q+r," then

$$m*l = m*(p+q+r) = (m*p) + (m*q) + (m*r).$$

Our quantities of two terms and of many terms (what we call "binomial" and "polynomial" quantities, thus employing both Greek and Latin roots) are like what Viète calls "magnitudes of two (*duorum*) or of many (*plurium*) names (*nominum*)."

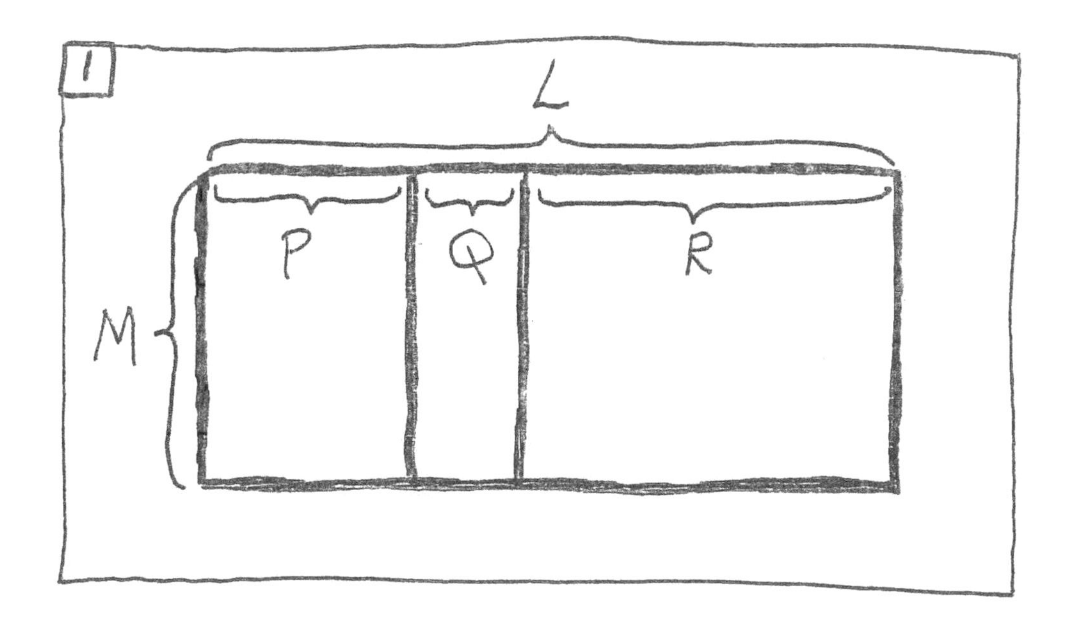

Viète continues with the making of products—as follows:

And if the affirmed (*affirmatum*) name of one magnitude is
led onto another magnitude's also affirmed name,
what is produced (*quod fiet*) will be affirmed;
and when [it is led] onto [another magnitude's name that is] denied (*negatum*),
[what is produced] will be denied (*negatum*).

In naming his magnitudes, Viète uses letters of the alphabet. Their "looks" (*species*) put us in mind of them. But what is on our mind when we name them? A name that is "affirmed" (*affirmatum*) is assertive; it adds firmness. A name that is "denied" (*negatum*) detracts from the strength of what it is denied to. A magnitude may be added on to any other one with which it is homogeneous, or may be taken away from a greater one with which it is homogeneous. In speaking of a taking away, the name of the lesser magnitude is *denied* to the name of the greater one from which it is taken away. Names that are not denied are *affirmed*. This is all worth dwelling upon because Viète's "affirmed" and "denied" are not our "positive" and "negative"—although there are some similarities. If they were the same—if Viète had recognized any such thing as "negative" magnitudes—he would not have said that we may write "A – B" only when the magnitude A is greater than the magnitude B. In other words, Viète's "A ===== B" is not the same as our current "$|A - B|$."

Now it is clear that when A is led onto "B+C," the product must be equal to "AB+AC"; and that when A is led onto "B ===== C," the product must be equal to "AB ===== AC." But it is more difficult to deal with what is produced when both factors involve denials, as when "A ===== B" is led onto "D ===== G." With that, Viète deals next, as follows:

For which precept, moreover, a consequence is that
in the leading of one denied name onto another
what is produced be affirmed,
as when A ===== B is led onto D ===== G;
since what is produced from an affirmed A [led] onto a G denied
remains denied, which is to deny or diminish too much,
inasmuch as A is led forward—not accurately—as
the magnitude to be led [onto another].
And similarly, what is produced from a denied B [led] onto a D affirmed
remains denied, which is again to deny too much,
inasmuch as D is led forward—not accurately—as
the magnitude to be led [onto another] ;
therefore, in compensation, when a B denied is led onto a G denied,
what is produced is affirmed.

This argument for "compensation" in the making of products resembles the earlier argument for "compensation" in subtraction. It is a bit more complicated now, however. Let us spell it out.

As in the earlier situation, so here too a numerical example can clarify Viète's argument. How is it, then, that the product of two denied names must be affirmed?

Suppose we have the following two denials: to 5, is denied 3; and to 11, is denied 7. When one denial's outcome (which is 2) is multiplied by the other denial's outcome (which is 4), the product must be 8.

Now, let us put aside, for a moment, the denials of the 3 and of the 7. What would be the product of the 5 alone by the 11 alone? It would be 55. That is our first outcome, but it cannot be our final outcome. It is too big, since in getting it nothing has been denied to either factor. What we want is the product when 3 is denied to 5 in one factor, and 7 is denied to 11 in the other factor.

Consider the last-named factor—namely, the 11 to which should be denied 7. If we do make that denial, what then would the 5 alone produce when multiplied by an 11 to which was indeed denied 7? The product would be this: (5 by 11) to which is denied (5 by 7). And since this number is "55 to which is denied 35," a net affirmation of 20 is the outcome that we have now. That is our second outcome, but it cannot be our final outcome. It is too small, since in getting it we denied too much. In getting it, we denied 35, which came from the multiplication by 7 of 5—that is: from the multiplication, by 7, of the *full* 5—even though to the 5 should be denied 3.

If now we do take some account of that 3 which should be denied to the 5, then, when we multiply by 11, we make an additional denial of 33. When we make this denial of 33, in addition to the previous affirmation of 20 that was our second outcome, we now get as our third outcome a net denial of 13. However, in getting this new outcome, we again have denied too much: the denied 33 was the product of the multiplication by 3 of a full 11, even though the to the 11 should be denied 7.

So we have denied too much yet again. That is why we do not yet have a final outcome. What we should get as our final outcome is that affirmation of 8 which we saw at the outset is the product of the two factors that include denials. Instead, our latest outcome is a denial of 13. In order to get from this denial of 13 to an affirmation of 8, we would need to make an additional affirmation of 21.

That 21 which we still need to affirm—where can it come from? It is the product of multiplying the two numbers (3 and 7) that were originally denied. We shall therefore obtain the product of

(5 denied 3) by (11 denied 7)

if we do this:

affirm (5 by 11) & deny (5 by 7) & deny (11 by 3) & *affirm* (3 by 7).

That is to say: when the one *denied* number (3) is *multiplied by* the other *denied* number (7), then the product is a number that must be *affirmed*.

Now reverting to large print, Viète's treatment of the making of products concludes by specifically laying out, as follows, what was mentioned at the beginning of the elaboration of the precept: the heterogeneity of the product and any of its factors:

The denominations of what's produced (*factorum*) by magnitudes ascending proportionally from kind to kind are related to each other in this straightforward manner:		
Side onto itself	produces (*facit*) ...	Square.
Side onto Square	produces	Cube
Side onto Cube	produces	Squaro-square.
Side onto Squaro-square	produces	Squaro-cube.
Side onto Squaro-cube	produces	Cubo-cube.
And [also] when changed around; that is:		
Square onto Side	produces	Cube;
Cube onto Side	produces	Squaro-square, etc.
Again,		
Square onto itself	produces	Squaro-square.
Square onto Cube	produces	Squaro-cube.
Square onto Squaro-square	produces	Cubo-cube.
And [also] when changed around.		
Again,		
Cube onto itself	produces	Cubo-cube.
Cube onto Squaro-square	produces	Squaro-squaro-cube.
Cube onto Squaro-cube	produces	Squaro-cubo-cube.
Cube onto Cubo-cube	produces	Cubo-cubo-cube.
And [also] when changed around.		
And in that order successively onward.		
Equally in the [magnitudes that are] homogeneous,		
Breadth onto Length	produces	Plane.
Breadth onto Plane	produces	Solid.
Breadth onto Solid	produces	Plano-plane.
Breadth onto Plano-plane	produces	Plano-solid.
Breadth onto Plano-solid	produces	Solido-solid.
And [also] when changed around		
Plane onto Plane	produces	Plano-plane.
Plane onto Solid	produces	Plano-solid.
Plane onto Plano-solid	produces	Solido-solid.
And [also] when changed around.		
Solid onto Solid	produces	Solido-solid.
Solid onto Plano-plane	produces	Plano-plano-solid.
Solid onto Plano-solid	produces	Plano-solido-solid.
Solid onto Solido-solid	produces	Solido-solido-solid.
And [also] when changed around.		
And in that order successively onward.		

Viète now presents the last of the canonical precepts of the reckoning that is speci-al. The presentation begins as follows:

> PRECEPT IV. *Magnitude to magnitude—to apply [one to the other]*
>
> Let there be two magnitudes A and B. It is required to apply one to the other.
>
> Since, therefore, a magnitude is to be applied to a magnitude,
> but higher [magnitudes] are applied to lower [magnitudes]
> [and] homogeneous [magnitudes] to heterogeneous [magnitudes],
> [that is to say: since a higher magnitude of one kind
> is to be applied to a lower magnitude of another kind,]
> those magnitudes which are put forward (*proponuntur*) are heterogeneous.
> Let A, if you will, be a "length"; [and] B, a "plane."
> And so, a rod-like mark will fittingly go between
> the higher B, which is applied, and the lower A, to which the application is made.

Note that when Viète writes "$\frac{\text{B plane}}{\text{A}}$" where we might say that the higher (the plane called B) is "divided by" the lower (the length called A), Viète is saying that the higher (the plane called B) is "applied to" the lower (the length called A).

The elaboration of the precept continues as follows:

> But the magnitudes themselves
> [namely, the ones that result from application]
> will be denominated by their own grades—
> [that is, by the grades] at which they stick or to which they have been carried down
> in the ladder of [magnitudes that are] proportional
> or [in the ladder of the magnitudes that are] homogeneous [with them]—
> as: "$\frac{\text{B plane}}{\text{A}}$."
> By which symbol (*symbolo*) may be signified
> the "breadth" which "B plane" makes (*facit*) when applied to "length" A.
>
> And if B be given to be "cube," and A to be "plane,"
> [the result] will be displayed "$\frac{\text{B cube}}{\text{A plane}}$."
> By which symbol may be signified
> the "breadth" which "B cube" makes when applied to "A plane."
>
> And if B be put [forward to be] "cube," [and] A [to be] "length,"
> [the result] will be displayed "$\frac{\text{B cube}}{\text{A}}$."
> By which symbol may be signified
> the "plane" which results (*oritur*) [arises] from the application of "B cube" to A;
> and in that order endlessly (*in infinitum*).
>
> Nor will anything different be observed in binomial or polynomial [magnitudes].

"Binomial or polynomial" magnitudes are what Viète a short while back referred to as magnitudes "of two or more names" (*duorum vel pluorum nominum*).

Now reverting to large print, Viète lays out specifically what was mentioned at the beginning of the elaboration of the precept—namely, the heterogeneity of the magnitude that is applied and the magnitude that it is applied to—along with the results that arise from this, as follows.

The denominations of results (*ortorum*) arising from application
by magnitudes ascending proportionally from kind to kind
are related to each other in this straightforward manner:

Square	applied to Side	returns Side.
Cube	applied to Side	returns Square.
Squaro-square	applied to Side	returns Cube.
Squaro-cube	applied to Side	returns Squaro-square.
Cubo-cube	applied to Side	returns Squaro-cube.

And [also] when changed around—that is:

Cube	applied to Square	returns Side.
Squaro-square	[applied] to Cube	[returns] Side.

Again,

Squaro-square	applied to Square	returns Square.
Squaro-cube	applied to Square	returns Cube.
Cubo-cube	applied to Square	returns Squaro-square.

And [also] when changed around.

Again,

Cubo-cube	applied to Cube	returns Cube [corrected].
Squaro-cubo-cube	applied to Cube	returns Squaro-cube.
Cubo-cubo-cube	applied to Cube	returns Cubo-cube.

And [also] when changed around.

And in that order successively onward.

Equally in the [magnitudes that are] homogeneous,

Plane	applied to Breadth	returns Length.
Solid	applied to Breadth	returns Plane.
Plano-plane	applied to Breadth	returns Solid.
Plano-solid	applied to Breadth	returns Plano-plane.
Solido-solid	applied to Breadth	returns Plano-solid.

And [also] when changed around.

Plano-plane	applied to Plane	returns Plane.
Plano-solid	applied to Plane	returns Solid.
Solido-solid	applied to Plane	returns Plano-plane.

And [also] when changed around.

Solido-solid	applied to Solid	returns Solid.
Plano-plano-solid	applied to Solid	returns Plano-plane.
Plano-solido-solid	applied to Solid	returns Plano-solid.
Solido-solido-solid	applied to Solid	returns Solido-solid.

And [also] when changed around.

And in that order successively onward.

The elaboration of the precept then continues as follows:

Moreover, whether in additions and subtractions of magnitudes
or in multiplications and divisions,
application does not get in the way of the precepts that have been set out—
this being observed—
that when the higher as well as the lower magnitude in application
is led onto the same magnitude,
nothing of the magnitude resultant from the application
is added or is taken away, in kind or in value —
since what the multiplication effects above, the same thing the division dissolves, as:

$\frac{\text{B onto A}}{\text{B}}$ is A, and $\frac{\text{B onto A plane}}{\text{B}}$ is A plane.

And so, in *additions*—
let it be required: to $\frac{\text{A plane}}{\text{B}}$, to add Z.

The sum will be: $\frac{\text{A plane + Z onto B.}}{\text{B}}$

Or let it be required: to $\frac{\text{A plane}}{\text{B}}$, to add $\frac{\text{Z square}}{\text{G}}$.

The sum will be: $\frac{\text{G onto A plane + B onto Z square}}{\text{B onto G.}}$

In *subtractions*—
let it be required: from $\frac{\text{A plane}}{\text{B}}$, to subtract Z.

The remainder will be: $\frac{\text{A plane − Z onto B}}{\text{B.}}$

Or let it be required: from $\frac{\text{A plane}}{\text{B}}$, to subtract $\frac{\text{Z square}}{\text{G}}$.

The remainder will be: $\frac{\text{A plane onto G − Z square onto B}}{\text{B onto G.}}$

In *multiplications*—

let it be required: to lead $\frac{\text{A plane}}{\text{B}}$ onto B.

The effect will be A plane.

Or let it be required: to lead $\frac{\text{A plane}}{\text{B}}$ onto Z.

The effect will be: $\frac{\text{A plane onto Z}}{\text{B}}$.

Or, finally: let it be required to lead $\frac{\text{A plane}}{\text{B}}$ onto $\frac{\text{Z square}}{\text{G}}$.

The effect will be: $\frac{\text{A plane onto Z square}}{\text{B onto G}}$.

In *applications*—

let it be required: to apply $\frac{\text{A cube}}{\text{B}}$ to D.

Having led each magnitude onto B,
the result will be: $\frac{\text{A cube}}{\text{B onto D}}$.

Or let it be required: to apply (B onto G) to $\frac{\text{A plane}}{\text{D}}$.

Having led each magnitude onto D,
the result will be: $\frac{\text{B onto G onto D}}{\text{A plane}}$.

Or, finally, let it be required: to apply $\frac{\text{B cube}}{\text{Z}}$ to $\frac{\text{A cube}}{\text{D plane}}$.

The result will be: $\frac{\text{B cube onto D plane}}{\text{Z onto A cube}}$.

We have now been prepared for the core activity of Viète's art. In his *Chapter II*, Viète has presented the stipulational foundations that support the arguments employed in the art of seeking magnitudes of unknown sizes; in his *Chapter III*, he has presented the law that requires homogeneity in making comparisons of size, and the scale that will enable magnitudinal heterogeneity to become a numerical order of rank, thus turning variety into degree; and in *Chapter IV*, he has presented the precepts in accordance with which the art of reckoning through species operates in four ways, corresponding to the addition, subtraction, multiplication, and division of the art of reckoning through numbers. Now, in *Chapter V*, he will turn to presenting the laws in accordance with which the art of seeking the solution to mathematical problems performs its task: first, it sets up an equality that compares magnitudes, and thus relates the magnitudes that are given and the magnitude that is sought; then, performing the operations of the art of reckoning with species, it changes the statement of equality into another more manageable statement that nonetheless remains a statement of equality; finally, it obtains an orderly equality that expresses clearly the relationship among the given magnitudes and the magnitude that is sought or its "power" or the "grades-along-the-way" to its power.

CHAPTER 4. FINDING UNKNOWN SIZES BY ZETETIC

Now, with Viète's fifth chapter, we have arrived at the core of his *Introduction to the Analytic Art.* It begins as follows.

> CHAPTER V *About the zetetic laws*
>
> The form of a zetesis that is to be accomplished
> is contained in just about these laws:

Of these zetetic laws, there are fourteen. The first three of them tell how to gain access to what sets you on the road to a solution of your problem:

> 1 If there's a seeking about length, but the equality or proportion is
> hidden under the wrappings of the things that are put forward (*proponuntur*),
> let the *length sought* be a *Side*.
>
> 2 If there's a seeking about planarity, but the equality or proportion is
> hidden under the wrappings of the things that are put forward,
> let the *planarity sought* be a *Square*.
>
> 3 If there's a seeking about solidity, but the equality or proportion is
> hidden under the wrappings of the things that are put forward,
> let the *solidity sought* be a *Cube*.
>
> That [magnitude] about which there's a seeking will, therefore,
> ascend or descend by its own force through whatever are
> the grades of the compared magnitudes.

The proposition that states the problem covers over the statement about the relationship of size (the equality or proportion) that will set you on the road to the solution. What affords access to the road—what makes the relationship manifest—is a determination of *the level of what is being sought*. (Viète now employs the ladder terminology which he presented earlier, in sections 2-3-4 of his Chapter III, and which he then employed in Chapter IV for laying out what it is that will be produced when species are multiplied and what it is that will result when species are applied.)

The unknown magnitude of the zetesis will be a "side" by itself if the problem is about "length," but it will be the "side" (or "root") of a "square" if the problem is about "planarity," or of a "cube" if it is about "solidity."

The levels are related proportionally. For example: "side" is to "square" as "square" is to "cube"; and, more generally, the unknown taken at one level is, to the one at the next higher level, as the one at this next higher level is, to the one at the level which is next higher after that. What determines the level at which the unknown is to be taken—whether it is to be higher or lower—is the consideration that *what is sought* must be at the *same* level as *what it is compared with*, since only homogeneous magnitudes can be compared.

Once the level has been determined, you are ready to be set on the road to solution by a statement of equality or proportion that relates, with respect to size, what is being sought and what (at the same level) it is being compared with.

The next law now tells you that you must set up a proper statement of equality in order to be set upon the road to solution:

4 Let the magnitudes that are given as well as those that are sought
be assimilated and compared—
in accordance with the condition stated by the search (*questioni*)—
by adding, subtracting, multiplying, and dividing,
with the constant law of homogeneous [magnitudes] being observed everywhere.
It is therefore manifest
that something is going to be found at last that
is equal to the magnitude about which there is a seeking,
or is equal to the power of itself to which it [this sought magnitude] ascends;
and [it is also manifest] that it [what is found] will be produced (*factum*)
[either] entirely by magnitudes that are given,
or partly by magnitudes that are given, and
partly by the unknown (*incerta*)
[the undetermined magnitude itself] about which there is a seeking
or by a grade of it that is along-the-way (*parodico*) to the power.

At last you may have a statement in which *the unknown magnitude taken at some level* (either it itself at the first level, as "side," or its higher level "power," that is, its "square" or its "cube" or higher) *is equal to what is made up entirely of magnitudes that are given*. For example, it may be equal to the product made by leading one given magnitude onto another; or equal to such a product plus quadruple a third given magnitude; or equal to the remainder when, from the result of applying one given magnitude to a second one, there is then taken away half of a third one.

Corresponding to such statements of equality, what we would write nowadays is that x^n (where n is 1 or 2 or 3 etc.) is equal to ($b*c$), or equal to ($b*c + 4c$), or equal to ($b/c — d/z$).

Or *what the "power" of the unknown magnitude is equal to, may be made up only partly of magnitudes that are given*; that is, *it may partly be made up also of the unknown magnitude taken at some level lower than the power*. For example, the "power" may be a "cube" that is equal to the following magnitude: the product made by leading a given magnitude onto the "square," less the product made by leading another given magnitude onto quadruple the "side."

Corresponding to such a statement of equality, what we would write nowadays is that x^3 is equal to ($bx^2 — 4cx$).

Of course, what we would write nowadays obliterates the reason for insisting on a law of homogeneous magnitudes. The use of exponents, which we owe to Descartes, sweeps away the elaborate apparatus of Diophantus-like names for the items on the ladder of magnitudes. The various grades and kinds of powers of the unknown magnitude and of the compared magnitudes—these are all homogenized under the influence of Descartes. But more of that later.

Now Viète turns to a notational aid for the analytic work of dealing with statements of equality.

> 5 So as to aid this work (*opus*) by some art,
> let the magnitudes that are given be distinguished from
> the unknown ones (*incertis*) that are sought,
> by a convention (*symbolo*) that is constant, perpetual, and clearly seen—
> as perhaps by designating the magnitudes that are sought out (*quaesititias*)
> by the letter (*elemento*) A or another vowel-letter (*litera*) E, I, O, U, Y;
> [and by designating the magnitudes that are] given,
> by the letters B, G, D, or other consonants.

Note that although he uses the Latin alphabet for the purpose, Viète selects and arranges his consonant examples to follow the Greek alphabetical order (with "B, G, D, and others" corresponding to Beta, Gamma, Delta, and others). It was Diophantus's use of Greek letters and of signs made from modified forms of those letters, as abbreviations for numbers of various kinds, that prepared for Viète's use of letters of the alphabet for species in his analytic art—consonants for magnitudes that are given, vowels for those that are sought. This allocation of consonants and vowels seems to be governed by the expectation that in any problem the given magnitudes will be more numerous than the ones that are sought. (Two modifications turned Viète's usage into our own: first a man named Harriott used lower-case letters instead of capitals; then Descartes used letters from the beginning of the alphabet instead of Viète's consonants, and letters from the end of the alphabet instead of Viète's vowels.)

Now Viète turns to laws that say how to set up the equation. (As we consider them, remember that the "power" is the unknown at the highest level at which it is found in the equation, and that a "grade along-the-way to the power" is the unknown at a level lower than that of the "power.")

The 8th law will deal with the part of the equality that contains the unknown's "power." But that law is preceded by two laws dealing with the part of the equality that contains what is to be equated with the unknown's power. More precisely, these two preceding laws deal with that part containing what is to be equated, when this is constituted by the addition or subtraction of products. The 6th law applies when the products are made entirely by given magnitudes; and the 7th law applies when the products are made by given magnitudes cooperatively with the same grade that is along-the-way to the power.

It will be easier to understand Viète's statement of these two laws if we consider some examples before we read Viète.

First, consider the part of the equality that is to be equated with the unknown's "power." It is to be constituted by the addition or subtraction of products. So, let the given magnitudes be B and C, and also D and F. They'll make the products "B onto C" and "D onto F." Put each of them right next to the mark of affecting by addition. They will add up to *"B onto C + D onto F."* (Or subtract them; all you need to do is use a "minus" sign instead of a "plus" sign.) That sum or that difference constitutes one product, a "plane," like each of the two products (namely, "B onto C" and "D onto F") that coalesce to make it; and this single magnitude of the second level that is made by coalescence of known magnitudes may be compared to some magnitude that is at the same level (the second level) and is made by a magnitude that is sought.

Now let the sought magnitude be A; then its "power" of the second level will be "A square." So: "B onto C + D onto F" and "B onto C — D onto F" are each homogeneous with "A square," and their sum (or difference) is thus "homogeneous in the comparing." And since B, C, D, F are all magnitudes that are given, therefore also given is the unit that is their measure; this is some "length." Therefore also given is the "unit that is the measure" of their products ("B onto C," and "D onto F") and the measure also "of the coalesced sum or difference of these products." This unit that is also given is the "plane" whose "length" is the unit for the given magnitudes B, C, D, F. The coalesced sum (or difference) of products will appear as *one* side or "part of an equation" whose *other* "part" will be the unknown magnitude taken at the same level, as in the following:

(A square) is equal to (B onto C + D onto F).

That is an example of the *6th law*, where the products are made entirely by magnitudes that are given. Now let us consider an example of the *7th law*, where those products are made by given magnitudes in cooperation with the same "grade that is along-the-way to the power." (In both products, that is to say, there will appear as a factor the unknown at the same level—a level that is lower than the level of the "power".) So, let B and C be magnitudes that are given, and let each make a product that is at, say, the third level. Now the grade at the third level of the ladder of the sought magnitude A would be "A cube"; but we are considering not that power of A, but rather a grade that is along-the-way to that power—so take "A square." Then the products we are talking about are these: "A square onto B" and "A square onto C." Put each of them right next to the mark of affecting by addition. They will add up to "A square onto B + A square onto C." (Or subtract them; all you need to do is use a "minus" sign instead of a "plus" sign.) That sum or that difference constitutes one product that is a "solid," like each of the two products that coalesce to make it (namely, "A square onto B" and "A square onto C"). The two products coalesce, through the affecting, into a single product that is homogeneous with each of them (is "homogeneous in the affecting"). This single magnitude of coalescence, since it is a "solid" (that is, a compared magnitude of the third level), can be equated with a power where the unknown is taken at the third level. It can be equated, that is to say, with a "cube." It is produced, however, by a magnitude that is a "square"—namely, by "A square." It is produced, that is to say, by that "grade along-the-way" (to the cube power) which is the same as the grade by which are produced each of the products that coalesce in the affecting (namely, "A square onto B" and "A square onto C"). The coalesced sum of products will appear as *one* part of an equality whose *other* part will be the unknown magnitude taken at the same level, as in the following:

(A cube) is equal to (A square onto B + A square onto C).

Here, then, are Viète's 6th and 7th zetetic laws:

6 Products made by (*facta sub*) given magnitudes
are added to or are subtracted from one another
in accordance with the juxtaposed mark of their affecting,
and they coalesce into one product (*factum*)—
as for which, let it be [a magnitude that is]
"homogeneous in the comparing,"
or [a magnitude that is produced] "by a given measure:"
and let it itself make one part of the equation (*unam aequationis partem facito*).

7 Equally, products [made] by given magnitudes
and by the same grade that is along-the-way to the power
are added to or are subtracted from one another,
in accordance with the juxtaposed mark of their affecting,
and they coalesce into one product—
as for which, let it be
[a magnitude that is] "homogeneous in the affecting,"
or [a magnitude that is produced] "by the grade."

The previous two laws dealt with that part of the equality which says what is to be equated with the unknown's "power" when what is to be equated is constituted by products that are added to or subtracted from one another. Now, in the 8th law, Viète goes on to deal with the other part of the equality—the part which contains the unknown's "power." He says this:

8 Homogeneous [magnitudes that are products]
by grades [that are along-the-way] to a power which
they [that is, those magnitudes] affect or by which they are affected—
let them [namely, those magnitudes so produced]
be accompanied [by that power],
and let them make, with the power itself, [as] one [magnitude],
the other part of an equality.

And so, therefore, a homogeneous [magnitude that is a product] by a given measure—
let it be expressed (*enuncietur*) by the power that is
designated at its own kind or order (*à suo genere vel ordine*):

purely—if that [power] is indeed pure of affecting;

but if accompanying it [namely, the power]
there are homogeneous [magnitudes] of affecting,
indicated both by the symbol of affecting
and by the symbol of the grade [along-the-way to the power]—
[then let the homogeneous magnitude that is a product by a given measure]
[be expressed by] one [magnitude that is put together as follows:]
with [the power] itself [as one ingredient,]
[and] with [as another ingredient] a grade [that] co-effects
[a product homogeneous with the power]
by means of an adopted (*adscititia*) magnitude.

That too will be clearer if we consider some examples. So, let A be a magnitude that is sought, and take the third-level power of it: this is "A cube." Any compared magnitude that is homogeneous with the cube must be at the third level too: it must be a solid, and this may be the product of a plane and a length. Now, let B, C, D be the magnitudes that are given; then a homogeneous magnitude that is a product "by a given measure" would be, say, this: "B plane onto C + D solid." This magnitude is "homogeneous" with "A cube" because it too is at the third level; and it is "produced by a given measure" because its factors (the given magnitudes B, C, D) are all measured by the same unit. Now, we might let the homogeneous product by a given measure (namely, the aforementioned "B plane onto C + D solid") be "expressed purely by the power that is designated at its own kind or order" (namely, "A cube"). In that case, our equality would be this:

(A cube) is equal to (B plane onto C + D solid).

However, the homogeneous magnitude that is produced by a given measure may be expressed by the power *not* "purely" (that is to say: *not* free from any affecting) but rather "accompanied" in its part of the equality by homogeneous magnitudes that affect it—these homogeneous magnitudes being products whose factors are "grades along-the-way to the power." The grades, along with an adopted magnitude, co-produce the homogeneous magnitudes that affect the power. (In other words: together, they make such a homogeneous magnitude; that is, they co-effect it.) Now since the power that we've taken is "A cube," the grade that is along-the-way to the power will be "A square." And adopted magnitudes that will be co-productive of the power that we have taken (in other words, magnitudes that function like what we now—following Viète's lead—call "coefficients") will be "lengths": let them be G and H. And so there will be two homogeneous products that accompany the power that we have taken (the power "A cube"). These two accompanying powers will be: "A square onto G" and "A square onto H." Now let us admit both of them into the affecting of "A cube"—letting one of the affectings be additive, and the other one be subtractive. And what we shall get from that, is this: "A cube + A square onto G — A square onto H." Our equality in that case would be:

(A cube + A square onto G — A square onto H) is equal to
[B plane onto C + D solid).

In the *earlier* case that we considered for this 8th law, the magnitude "B plane onto C + D solid" was equated with the "power" (namely, "A cube"). It was "expressed by" the cube *"purely."* That is to say: the magnitude was equated with a cube that was not affected at all. *Now*, however, the magnitude is expressed by a power (namely, "A cube") that is indeed *"affected."* What "B plane onto C + D solid" is equal to now, is not "A-cube" purely, but rather is "A cube + A square onto G — A square onto H." Now, "A cube" is affected by an affirmed magnitude ("A square onto G") that is homogeneous with it (since it is at the third level too), and that is made by a square ("A square") along with an adopted magnitude (G); and "A cube" is affected also by a denied magnitude ("A square onto H") that is homogeneous with it (since it is at the third level too) and that is made by a square ("A square") along with an adopted magnitude (H).

Now, having given laws that tell the analyst how to set up an equation in order to solve the problem, Viète in the next three laws will give the analyst ways to change such a statement of equality into a more manageable statement that remains a statement of equality.

As usual, the terminology introduced by Viète is derived from ancient Greek. His names for the three processes are: *anti+thesis* (putting opposite); *hypo+bibasm* (making go down under); *para+bolism* (throwing alongside).

In *antithesis* (law 9), there is contraposing: an affecting or affected magnitude is put on the opposite side of an equation, and is changed to its contrary—that is, from being affirmed to being denied, or vice versa.

In *hypobibasm* (law 10), there is a depressing: the level of the things that are equal is lowered by applying each of them to the sought magnitude; that is, both parts of the equality are divided by the unknown magnitude.

In *parabolism* (law 11), there is an applying: each of the things that are equal is applied to the same given magnitude; that is, both parts of the equality are divided by a given magnitude.

Viète presents *antithesis* first. He explains that it is employed when two homogeneous products that are mixed together on the same side of the equality need to be unmixed and put on opposite sides. The two homogeneous products that need to be put on opposite sides of the equality are these: the one that is produced by the given measure, and the one that is produced by the grade that is along-the-way to the power. He explains what antithesis does—how it takes any magnitude that we obtain entirely from given magnitudes, and puts it on the opposite side of the equation from any magnitude that has as a factor the unknown taken at any level. He then demonstrates that the process works properly.

> 9 And so, for that reason, if it happens that a homogeneous [magnitude that is]
> [produced] by a given measure
> is mixed in with a homogeneous [magnitude that is]
> [produced] by a grade [that is on-the-way to the power] ,
> let there be an Antithesis.
>
> Antithesis is when affecting or affected magnitudes transit,
> [going across] from one part of an equation into the other
> under the contrary mark of affecting—
> by which work (*opere*), equality is not changed.
> That, however, is to be demonstrated [here] along the way.

Before we examine Viète's demonstration, however, let us consider an example that displays what happens in antithesis. Consider the following equality:

(A cube....................			
plus	A square onto G		
minus	A square onto H		
minus	B plane onto C		
plus	F plane onto K)	is equal to	(D solid).

How antithesis changes that equality into another one, is displayed as follows:

(A cube..................		is equal to	(D solid..................
plus	A square onto G		\/
minus	A square onto H		\/
~~minus~~	~~B plane onto C~~ >>>>>>>>>>>>>>>>>		*plus* B plane onto C
~~plus~~	~~F plane onto K~~ >>>>>>>>>>>>>>>>>		*minus* F plane onto K).

The result of that antithesis is this: all the solids that are produced entirely by given magnitudes are placed together in the right-hand part of the equality, so as to affect only each other.

That resembles what we do nowadays to an equation when we transpose so as to combine all known quantities. For example—

if we have this equation:	$x^3 + gx^2 - hx^2 - b^2c + f^2k = d^3$,	
then if we		
combine all the constant terms,		
we get this equation:	$x^3 + gx^2 - hx^2$	$= (d^3 + b^2c - f^2k)$.

Now, here is Viète's demonstration. (The demonstrations for the propositions appropriate to the three laws of which this is the first, were published in smaller print, like the elaborations of the four precepts in the previous chapter.)

PROPOSITION I. *That by antithesis, equality is not changed.*

Let it be put forward (*proponantur*) that
"A square minus D plane" is equal to "G square minus B onto A."
I say that
"A square plus B onto A" is equal to "G square minus D plane,"
and that through this transposition under the contrary mark of affecting,
the equality is not changed.

Since indeed
"A square minus D plane" is equal to "G square minus B onto A,"
let there be added to both [sides] "D plane plus B onto A."
Therefore, by the common notion [equals are the outcome when equals are added to equals],
"A square minus D plane" plus "D plane plus B onto A" is equal to
"G square minus B onto A" plus "D plane plus B onto A."
Now, a denied affecting in the same part [that is, on the same side] of an equation
strikes out an affirmed [affecting]:
there [on the first side], the affecting by "D plane" vanishes;
here [on the second side], the affecting by "B onto A" vanishes;
and "A square plus B onto A" will survive equal to "G square plus D plane."

Viète next presents *hypobibasm*:

10 And if it happens that all the given magnitudes
are led onto a grade [that is along-the-way to the power],
and for that reason a homogeneous [magnitude which is]
[produced] by a measure given overall (*omnino*)
does not immediately offer itself,
let there be a Hypobibasm.
Hypobibasm is like a depressing
of the power and of the grades along-the-way [to it],
with the order of the ladder being observed,
until the homogeneous [magnitude that is produced] by the more depressed grade
falls into [being] a homogeneous [magnitude which is] given overall and
to which the rest [of the magnitudes] are compared—
by which work [of hypobibasm], equality is not changed.
That, however, is to be demonstrated [here] along the way:

PROPOSITION II. *That by hypobibasm, equality is not changed*

Let it be put forward that

"A cube	plus B onto A square"	is equal to	"Z plane onto A."

I say that

"A square	plus B onto A"	is equal to	"A plane."

[To have done] that, indeed, is to have divided all the solids by a common divisor,
by which it is determined that an equality is not changed.

What is here put forward is an equality in which every one of the magnitudes that are given has coefficient with it (has, that is, making a product with it) some grade along-the-way to the sought magnitude's power. Thus, in Viète's example, what is coefficient with the given magnitude B is "A square," and what is coefficient with the given magnitude " Z plane" is A. Therefore antithesis cannot bring about what we would like—namely, a situation in which a solid produced entirely by given magnitudes is set apart on one side of an equality, or several such solids are so placed together as to affect only each other.

What can bring about that situation, however, is hypobibasm. The solid in the right-hand part of the equality (namely, "Z plane onto A") will (if it is divided by "the side A") be lowered a level, and it will, being thus depressed, become "Z plane"—which is not produced by any magnitudes but ones that are given. And if the other part of the equality is divided by that same magnitude (namely, by "side A") and thus is lowered a level too—then, by being thus depressed, it will become "A square plus B"; then equality is maintained between the (depressed) parts on the right and on the left. That resembles what we do nowadays when we lower the degree of an equation through division by the unknown of the highest degree that is found in every term. For example:

if we have this equation:	$x^7 + bx^3 = c^2x^2$,
then we divide throughout by x^2,	
to get an equation in which there is a term (here it is c^2) that	
contains no quantities but ones that are given—namely, ...	$x^5 + bx = c^2$.

Finally, Viète presents *parabolism*:

11 And if it happens that
the higher grade to which the sought magnitude ascends
does not subsist of itself but is led onto some magnitude [that is] given,
let there be a Parabolism.
Parabolism is the common application,
to a given magnitude which is led onto a higher grade of the sought [magnitude],
of homogeneous [magnitudes] by which an equation is constituted;
so that that [higher] grade will claim for itself the name of "the power,"
and by that [namely, that grade which claims the name of "the power"]
the equation at last subsists—
by which work [of parabolism], equality is not changed.
That, however, is to be demonstrated [here] along the way:

PROPOSITION III. *That by parabolism, equality is not changed.*

Let it be put forward that
"B onto A square .... plus D plane onto A" .... is equal to........... "Z solid."
I say that through a parabolism
"A square .... plus $\frac{\text{D plane}}{\text{B}}$ onto A" .... is equal to .... "$\frac{\text{Z solid}}{\text{B}}$"

[To have done] that, indeed, is
to have divided all the solids by a common divisor,
by which it is determined that an equality is not changed.

That resembles what we do nowadays when we change into unity the constant quantity [what we call a coefficient) in the term containing the unknown of highest degree in the equation—doing so through dividing every term in the equation by the coefficient that we want to change into unity in that term containing the unknown of highest degree. For example—

if we have this equation: .... $bx^7 + c^2x^3 = d^3x^2$,
then we divide all the terms by b, so that
the coefficient of the term that contains x^7
changes into unity: .... $x^7 + c^2x^3/b = d^3x^2/b$.

Now Viète presents the last three of his zetetic laws. They are about the setting in order that is the completion of the zetetic work. The first of these three laws begins as follows:

12 And [when antithesis, hypobibasm, and parabolism have done their work,]
then the equality is esteemed (*censetor*) to be "clearly expressed" (*exprimi*)
and is said to be "set in order";

and here is an example of an equality that is clearly expressed and set in order:

(A square + B onto A) is equal to (C onto D + C onto F).

That equality is said to be clearly expressed and set in order because whatever work was needed has been done on it; there is no need for antithesis, or for hypobibasm, or for parabolism:

there is not any need for *antithesis*—
since all of the terms involving the magnitude A that is sought
are in one part of the equality (the left side),
and all of the magnitudes in the other part (the right side) are given;

nor is there any need for *hypobibasm*—
since there is in the equality a term that
does not involve the sought magnitude;

nor, finally, is there any need for *parabolism*—
since the term that involves the sought magnitude at
its highest level in the equality
is not a product that the sought magnitude at this level
makes with any given magnitude.

Equivalent to that equality is a proportionalism, but caution is required in making the restatement:

[that equality is something] to be restated (*revocanda*), if you please,
as a proportionalism (*ad Analogismum*),
with such a precaution (*tali praesertim cautione*) [as the following:]

that [products made] by the *extremes* correspond (*respondeant*) to
the power [taken] together with the [products that]
[are] homogeneous [with the power]
[and are part] of the affectings [of it];

and indeed
that [products made] by the *means* correspond to
a [product that]
[is] homogeneous [with the power]
[and is produced] by the given measure.

The proportionalism into which that equality can be restated is this:

as (A) is to (C), so (D + F) is to (A + B).

That proportionalism can be such a restatement for the following reason:

When we lead one extreme (A + B) onto the other extreme (A),
the product is this—"A square + B onto A";
and when we lead one mean (C) onto the other mean (D + F),
the product is this—"C onto D + C onto F";
and each of these products is one part of that equality which we had:
the one product (of the extremes) is equal to
the other product (of the means).

Now, that product of the *extremes* (namely, "A square + B onto A")
corresponds to this:
the power (namely, "A square") taken together with
a product (namely, "B onto A") that is homogeneous with the power
(since both it and the power are at the 2nd level);

and this product (namely, "B onto A")
affects the power (namely, "A square").

That product of the *means* corresponds to
a product (namely, "C onto D + C onto F") that is homogeneous with
the product (namely, "A square + B onto A") of the extremes
(since all the terms of both products are at the 2nd level—
namely, "A square," "B onto A," "C onto D," and "C onto F");

and this product (namely "C onto D + C onto F")
is produced by the given measure
(since all the terms in it—namely, "C onto D" and "C onto F"—
contain only magnitudes that are given—namely, C, D, F—
and therefore have a common unit).

Note that although Viète back in his Chapter II used the Latin term *proportio*, what he uses now is a term Latinized from the Greek. However, the Greek term that he draws upon is not the ordinary term for proportion (*analogia*). He speaks, rather, of restating the equality *ad Analogismum*. That is why here it is translated as "proportionalism," rather than as "proportion." The locution is used again in the next law.

13 Whence, moreover,
let "a proportionalism that is set in order" be defined as:
a series of three or four magnitudes
so expressed (*effata*) in terms either pure or affected
that all are given except that one about which there's a seeking
or else its power and the grades along-the-way to it [that is, to the power].

In the discussion of Viète's 12th section, a few paragraphs back, we looked at the following example of an equality that is "clearly expressed and set in order":

(A square + B onto A) is equal to (C onto D + C onto F);

and we examined also its restatement into a proportionalism:

as (A) is to (C), so (D plus F) is to (A plus B).

Now we have been told that the proportionalism is, like the equality itself, "set in order."

Likewise set in order is the proportionalism into which one can restate the following more complicated equality:

(A cube + A square onto D + A onto D onto F + A square onto F) is equal to (B onto C).

The proportionalism into which one can restate that equality is this:

as (A square + A onto F) is to (*B*), so (*C*) is to (*A* + *D*).

All of the magnitudes are expressed in terms (either pure or affected) that are given (namely, B, C, D, F)—or that are sought (namely, A) or else its power (namely, "A square") and grades along-the-way to the power (here, of course, since the power is merely at the second level, there is no grade along-the-way to it that is above the first level at which A—the magnitude that is sought—itself subsists).

And now the concluding law of zetetic tells us when to consider the zetesis complete:

14 At last, when an equality has thus been set in order,
or when a proportionalism has been set in order,
let Zetetic be esteemed to have fulfilled its functions.

The chapter is not yet quite complete, however. At its very end, Viète returns to the question of the newness of seeking the size of what is unknown by reckoning with species:

> Now zetetic was practiced most subtly of all by Diophantus,
> in those books which he wrote about what is arithmetical.
> In truth, however, he exhibited it just like something that is
> undertaken through numbers but not through species—
> which [species], nonetheless, he used—
> by which his subtlety and skill might be the more admired;
> inasmuch as those things which
> appear subtler and more abstruse
> in the logistic that is numerical (*numeroso*)
> are, in the one that is speci-al (*specioso*),
> quite familiar and immediately obvious.

Viète suggests that reckoning with species (rather than with numbers) originated among the ancients, long before Viète's own time; and so, the originality of Viète—his innovation—would seem to be this: what the ancients did covertly, he did openly.

Before we continue on our way through the *Introduction to the Analytic Art*, let us therefore examine some examples of Viète's zetetic work alongside the corresponding items from Diophantus's numerical work. The examples come from another one of Viète's writings, his *Five Books of Zetetical Things*, where he sets the zetetic art to work solving numerical problems. We shall look at the first four problems, along with the corresponding problems from Diophantus's *Thirteen Books of Arithmetical Things*. Here is a tabulation of the correspondence.

Viète, *Zetetica*, Book One:	(1)	(2)	(3)	(4)
	|	|	|	|
	\/	\/	\/	\/
Diophantus, *Arithmetica*, Book One:	(1)	(4)	(2)	(7).

In reading these problems you'll avoid some puzzlement if you bear in mind that Viète, like Diophantus earlier, writes the amount of something after (rather than before) the something of which it is the amount. So Diophantus writes, for instance, Nu#b"—meaning "of the [unknown) number, two of them," and he writes "Un#m"—meaning "of units, thirty of them." And Viète writes "A2"—meaning "of A, two of them," and he writes "D 1/2 "—meaning "of D, one-half of it."

Note that letters (and modified letters) are used by Diophantus as mere *abbreviations* for *words* that name numbers which can be counted off, if not at the outset, then at least eventually—whereas the letters used by Viète are not mere abbreviations for such words. Although it is true that numerical values can be assigned to the letters used in Viète's zetesis, it is also true that what is assigned to them could just as well be such magnitudes as lines, surfaces, or solids.

VIETE I.1

Given the difference of two "sides" and their sum, to find the "sides."

Let B be given as the difference of the two "sides," and also D be given as their sum.
It is required to find the "sides."

Let the lesser "side" be A; therefore the greater will be A+B.
On that account, the sum of the "sides" will be A2+B.
But that same [sum] is given as D. Wherefore A2+B is equated to D.
And by antithesis, A2 will be equated to D — B;
and if they are all halved, A will be equated to D 1/2 — B 1/2.

Or let the greater "side" be E. Therefore the lesser will be E — B.
On that account the sum of the "sides" will be E2 — B.
But the same [sum] is given as D. Wherefore E2 — B will be equated to D.
And by antithesis, E2 will be equated to D+B;
and if they are all halved, E will be equated to D 1/2 + B 1/2.

Therefore, given the difference of the two sides and their sum,
the "sides" are found. For, indeed,
half the sum of the "sides" minus half their difference is equal to the lesser "side";
and [half the sum of the "sides"] plus [half their difference is equal] to the greater—
which very thing the zetesis shows.

Let B be 40; D, 100. A is made 30; E, 70.

That corresponds to this:

DIOPHANTUS I.1

To separate a given number into two numbers in the excess that is given.

So let the given number be #s [number100], and let the excess be Un#m [40 units].
To find the numbers.

Let the lesser be ordained .... Nu#a [one (unknown) "number"];
then the greater will be .... Nu#a Un#m [one "number" and 40 units].
Then both together will be .... Nu#b Un#m [2 "numbers" and 40 units].
But they've been given as .... Un#s [100 units].

Then Un#s [100 units] are equal to Nu#b Un#m [2 "numbers" and 40 units].
And I take like things away from like:
from #s [100], I take away Un#m [40 units];
and from the Nu#b Un#m [2 "numbers" and 40 units] likewise I take away Un#m [40 units].
Nu#b [2 "numbers"] are left equal to Un#o [60 units].
Therefore each Nu ["number"] becomes Un#l [30 units].

As to what was supposed:
the lesser will be Un#l [30 units]; and the greater, Un#p [70 units].
And the demonstration is manifest.

VIETE I.2

Given the difference of two "sides" and their ratio, to find the "sides."

Let B be given as the difference of the two "sides,"
and also let the ratio of the lesser "side" to the greater be given as R to S.
It is required to find the "sides."

Let the lesser "side" be A. Therefore, the greater "side" will be A+B.
Wherefore, A is to A+B, as R is to S.
Which proportionalism having been resolved,
S onto A will be equated to R onto A + R onto B.
And by transposition under the opposite mark of affecting,
S onto A — R onto A will be equated to R onto B,
and, when all are divided by S — R, $\frac{\text{R onto B}}{\text{S} - \text{R}}$ will be equated to A.
Whence, as S — R is to R, so B is to A.

Or, let the greater "side" be E. Therefore, the lesser "side" will be E — B.
Wherefore, E is to E — B, as S is to R.
Which proportionalism having been resolved,
R onto E will be equated to S onto E — S onto B.
And by a suitable transposition,
S onto E — R onto E will be equated to S onto B. Whence, as S — R is to S, so B is to E.

Therefore, given the difference of the two "sides," and their ratio,
the "sides" are found. For indeed,
as the difference of the two "sides" that are in the same ratio
is to the greater or to the lesser of them,
so the difference of the "sides" that are true [that is, are the ones being sought]
is to the greater or to the lesser of them.

Let B be 12; R, 2; S, 3. A is made 24; E, 36.

That corresponds to this:

DIOPHANTUS I.4

To find two numbers in a ratio that is given, so that their excess is also given.

So let it be set down that:
the greater is the #e^{pl} [5-fold] of the lesser, and their excess makes Un#k [20 units].

Let the lesser be ordained Nu#a [one "number"].Then the greater will be Nu#e [5 "numbers"].
Further, I wish the Nu#e [5 "numbers"] to exceed the Nu#a [1 "number"] by Un#k [20 units];
but their excess is Nu#d [4 "numbers"]—these [therefore] are equal to Un#k [20 units].

The lesser number will be Un#e [5 units], [and] the greater will be Un#ke [25 units];
and the greater stays #e^{pl} [5-fold] the less.

VIÈTE I.3
Given the sum of two "sides and their ratio, to find the "sides."

Let G be given as the sum of the two "sides,"
and the ratio of the lesser to the greater be given as R to S.
It is required to find the "sides."

Let the lesser "side" be A. Therefore, the greater side will be G — A.
Wherefore, A is to G — A, as R is to S.
Which proportionalism having been resolved,
S onto A will be equated to R onto G — R onto A.
And transposition having been done according to the art,
S onto A +R onto A will be equated to R onto G. Whence, as S+R is to R, so G will be to A.

Or, let the greater "side" be E. Therefore, the lesser "side" will be G — E.
Wherefore, as E to G — E, so S to R.
Which proportionalism having been resolved,
R onto E is equated to S onto G — S onto E.
And transposition having been done according to the art,
S onto E + R onto E will be equated to S onto G. Whence, as S + R to S, so G to E.

Therefore, given the sum of the two "sides," and their ratio,
the "sides" are given. Indeed,
as the sum of the two "sides" that are in the same ratio
is to the greater or to the lesser of them,
so the sum of the "sides" that are true [that is, are the ones being sought]
is to the greater or to the lesser of them.

Let G be 60; R, 2; S, 3. A will be 24; E, 36.

That corresponds to this:

DIOPHANTUS I.2
It is required to separate a set-down number into two numbers in the ratio that is given.

So, to separate—let it be the #o [number 60]—into two numbers
in the ratio that's #c^{pl} [3-fold].
It is required to find the "sides."

Let the lesser be ordained Nu#a [one "number"].
Then the greater will be Nu#a [3 "numbers"]—that is, the greater is #c^{pl} [3-fold] the lesser.
Further, it is required that the two [together] be equal to Un#o [60 units].
But the two together are Nu#d [4 "numbers"].
Then Nu#d [4 "numbers"] are equal to Un#o [60 units];
and then the "number" is Un#je [15 units].

The lesser, then, will be Un#je [15 units]; and the greater, Un#me [45 units].

VIÈTE I.4

Given two "sides" that are deficient with respect to one that is just right, along with the ratio of the deficits, to find the one that is just right.

The two given "sides" that are deficient with respect to one that is just right—let them be: first, B; second, D.
And let there also be the given ratio of the first's defect to the second's defect—as R to S.
It is required to find the "side" that is just right.

Let A be the first [side's] defect. Therefore B+A will be the "side" that is just right. Since, however,
as R is to S, so A is to $\frac{\text{S onto A}}{\text{R}}$, consequently $\frac{\text{S onto A}}{\text{R}}$ will be the second [side's] defect.
Whereby D + $\frac{\text{S onto A}}{\text{R}}$ will be the "side" that is just right too [like B+A], and
on that account D + $\frac{\text{S onto A}}{\text{R}}$ will be equated to B+A [both being the side that is just right].
[In the last-mentioned equality, now let] all [of the terms be led] onto R:
Therefore D onto R + S onto A will be equal to B onto R + A onto R.
And when the equality has been set in order,
D onto R ===== B onto R will be equated to R onto A ===== S onto A.
Whence, as R=====S to R, so D=====B will be to A.

Or, let E be the second [side's] defect. Therefore D+E will be that "side" that is just right.
Since, however,
as S is to R, so E is to $\frac{\text{R onto E}}{\text{S}}$, consequently $\frac{\text{R onto E}}{\text{S}}$ will be the first [side's] defect.
Whereby B + $\frac{\text{R onto E}}{\text{S}}$ will be the "side" that is just right too [like D+E], and
on that account will be equated to D+E [both being the side that is just right].
[In the last-mentioned equality, now let] all [of the terms be led] onto S:
Therefore B onto S + R onto E will be equated to D onto S + S onto E.
And when the equality has been set in order,
D onto S ===== B onto S will be equated to R onto E ===== S onto E.
Whence, as R=====S to S, so D=====B will be to E.

Therefore, two "sides" that are deficient with respect to one that is just right
having been given, along with the ratio of the deficits, the one that is just right is found.
For, indeed, *as the difference between the defects is to*
the defect of the first or of the second of the "sides" that are in the same ratio
[this is the ratio that the difference between R and S has to one of them, to R or to S]
so the true difference between the deficient "sides"—
that which is between the defects [themselves]—is
to the true defect of the first or of the second "side"
[this is the ratio that the difference between D and B has to E or to A].
By suitably restoring which defect [that is, by adding the D or B]
to a deficient "side" [to E or to A, which are obtained above as fourth proportionals],
there is made the "side" that is just right [which is what is to be found].

Let B be 76; D, 4; R, 1, S, 4. A is made 24; E, 96. [And—]

ALTERNATIVELY

Given two "sides" that are deficient with respect to one that is just right,
along with the ratio of the deficits, to find the one that is just right.

The two given "sides" that are deficient with respect to one that is just right—
let them again be: first, B; second, D.
And let there also be the given ratio of the first's defect to the second's defect—as R to S.
It is required to find the "side" that is just right.

Let A be that [just-right "side"]
Therefore A — B will be the defect of the first "side,"
and A — D will be the defect of the second.

Whereby, as A — B to A — D, so R to S.
By which proportionalism having been resolved,
R onto A — R onto D will be equated to S onto A — S onto B.
And transposition having been done according to the art,
S onto A ===== R onto A will be equated to S onto B ===== R onto D.

And so $\dfrac{\text{S onto B ===== R onto D}}{\text{S ===== R}}$ will be equated to A.

Therefore, there having been given
two "sides" that are deficient with respect to the true one, along with the ratio of the deficits,
the one that is just right is found.
For, indeed,
when the difference between the "rectangle" [produced] by
the first deficient "side" and what in the same ratio corresponds to the second's defect
[namely, B and S]
and the "rectangle" [produced] by
the second deficient "side" and what in the same ratio corresponds to the first's defect
[namely, D and R]
will be applied to the difference between the corresponding defects [namely, S and R]
there will arise the just-right "side" about which there is a seeking.

Let B be 76; D, 4; R, 1; S, 4.
A is made 100.

That corresponds to this:

DIOPHANTUS I.7
From the same number, to take away two given numbers—
and make the remainders have to each other a ratio that is given.

So, let it be set down that
from the same number, there is taken away the #s and also the #k [numbers 100 and 20],
and that this makes the greater [remainder] the #c^{pl} [3-fold] of the lesser.

Let the [number] being sought be ordained Nu#a [one "number"].
And if from this I take away [on the one hand] the #s [100],
the remainder is Nu#a Li Un#s [one "number" lacking 100 units],
and [I take away also on the other hand] the #k [20],
the remainder is Nu#a Li Un#k [one "number" lacking 20 units].

And it will be required that
the greater [remainder] be #c^{pl} [3-fold] the lesser [remainder];
then thrice the lesser things are equal to the greater,
and thrice the lesser things become .... Nu#c Li U#u [3 "numbers" lacking 300 units];
these are equal to .... Nu#a Li Un#k [one "number" lacking 20 units].

Let the lack be added in common [that is, let 300 units be added to each of the equals—
added to the 3 "numbers" lacking 300 units,
and also to the one "number" lacking 20 units];

[and then:] Nu#c [3 "numbers"] become equal to
Nu#a and U#tq [one "number" and 280 units];

and let like things be taken away from like—
[that is, from both of those two equals just mentioned]
let N#a [one "number"] be taken away—what remains is:

Nu#b [2 "numbers"] are equal to
Un#tq [280 units];

and Nu#a [one "number"] becomes
Un#sm [140 units]

As to what was supposed:
I ordained the sought "number" to be N#a; it will then be U#sm [140 units];
and if from this
[on the one hand] I take away the #s [100], what remain are U#m [40 units],
[and on the other, I take away] the #k [20], [what remain are] U#sk [120 units];
and the greater [that is, these 120 units] stay triple the lesser [that is, those 40 units].

So much for Viète's *Five Books of Zetetical Things*. Here we cannot go through more than that bit of its beginning. We should take note, however, of its very end. What is placed at the end of Viète's fifth zetetical book is that very problem which corresponds to the one that is placed at the end of Diophantus's fifth arithmetical book. Viète's analytical art looks back as well as looking forward.

CHAPTER 5. PORISTIC

Having finished treating zetetic, Viète moves on to a brief discussion of the other two parts of the analytic art—poristic and rhetic. First comes his sixth chapter—on poristic. Be patient reading through it if you have trouble seeing what Viète is getting at; help is available in the comments that follow what he himself says. Here, then, is what Viète's chapter says about poristic:

CHAPTER VI. *About the examination of theorems through [the] Poristic [art]*

Zetesis having been accomplished (*perfecta*), the Analyst
moves from supposing (*ab hypothesi*) to positive statement (*ad thesin*)
[*hypothesis* and *thesis* being transliterated Greek],
and exhibits theorems gotten hold of by his own finding,
in accordance with the ordaining of the art, subject to the laws
kata pantos, *kath'auto*, *kath'olou prôton* [in Greek letters].

Which [theorems],
although from Zetesis they have their own demonstration and firmness,
are nonetheless subject to the law of synthesis,
which is deemed (*censetur*) to be a way of demonstration that
has more to do with the art of logic [is more *logikôterê* (in Greek letters)];
and if at any time there is [demonstrative] work (*opus*) [to be done],
they [the theorems] are proved through it [namely, through synthesis]
by the great miracle of the finding-out art (*artis inventricis*).

And so for that reason (*idcirco*),
the footsteps of the Analysis are retraced (*repetuntur*).
Which [retracing] is itself Analytic;
nor—since [it has been] introduced by speci-al Logistic—
[does it] by this time involve busyness (*jam negociosum*).

But if something that has been found that is unfamiliar (*alienum*)
proposes itself [for demonstration],
or there is something offered fortuitously
whose truth is to be weighed out and sought into,
then the way of Poristic is to be tried first,
from which a return to the synthesis,
one-step-after-another (*deinceps*), would be easy—
as those examples about the matter (*de re*) have been brought forth (*prolata*)
by Theon in the *Elements*, by Apollonius of Perga in the *Conics*,
and also by Archimedes himself in various books.

The zetesis is complete when the equation that was formulated at the outset has been transformed into a final formulation in which it can be solved. That zetetic transformation has the following form:

> since the thing that is sought satisfies the opening formulation,
> it will therefore satisfy the final formulation.

The thing that is sought *would*, then, be a thing that is found—*if* in fact the final formulation that has been obtained by zetetic *will* in the end give what is sought—*if*, in other words, one *givenness* (of those numbers or of those magnitudes which correspond to the species that are *consonant*-letters in the final formulation) *guarantees* another *givenness* (of that number or of that magnitude which corresponds to the species that is the *vowel*-letter in the final formulation).

But will the latter givenness follow indeed from the former? Without a guarantee that it does, the problem—even though it has been transformed—is still unresolved. Until the guarantee is provided, the solution is still "iffy"—is still based on a supposition—is still hypothetical. To resolve the problem, hypothesis must be superseded by demonstrated thesis. That is the work of poristic.

Poristic resolves the problem that zetetic has transformed. Poristic does that by providing a theorem, a positive statement found by it that shows how the one givenness follows from the other. Then, after poristic has accomplished a resolution of the problem, the very same steps that were taken in the course of the resolution are retraced—in the very same order—as either the number sought is computed by rhetic, or the magnitude sought is constructed by exegetic. After that, the demonstration is synthesized. The very same steps that were taken in the course of the analytic transformation are retraced—this time, however, in reverse order—by the synthetic demonstration.

A theorem of poristic needs to be demonstrated; that is to say: poristic at work needs to do some synthesizing of its own. To do that synthesizing, poristic needs to do some zetesis of its own. Such zetesis (making use of the special logistic) will facilitate the finding of a demonstration of that poristic theorem. The synthesis to demonstrate the poristic theorem will be an easy retracing, in reverse, of the footsteps in the zetesis, since the steps of the zetesis involve predications of attributes that are predicable of *every* instance of their subjects, and are *essential* to their subjects (since they observe the law of homogeneous magnitudes), and are *completely convertible* with their subjects (since they manipulate equations).

That is why Viète refers to those laws that are named by Greek terms. (They derive from Aristotle's *Posterior Analytics* I.4, and were emphasized by Peter Ramus, an influential author in Viète's time.) The terms concern predication: a predicate (P) that is predicated of a subject (S) is said to be predicated

> *kata pantos*, when P is predicated of every instance of an S;
>
> *kath'auto*, when being P is essential to being an S at all,
> rather than merely incidental;
>
> *kath'olou prôton*, when not only is P predicated of every S,
> but also S is predicated of every P
> (that is, P is completely convertible with S).

We previously looked at the place of poristic in Viète's enterprise, when we examined his first chapter.

CHAPTER 6. RHETIC

Having treated zetetic and poristic, Viète now, in his seventh chapter, comes to the last of the three parts of the analytic art: rhetic.

> CHAPTER VII *About the function of [the] Rhetic [art]*
>
> The equation of the magnitude about which there is a seeking
> having been set in order,
> *rhetic* or *exegetic* [in Greek letters—including the “oi,” which is [illegible]]
> exercises its function—
> which is to be deemed to be the remaining part of Analytic,
> and, indeed, to pertain most to the setting in order of the [Analytic] art
> (since the remaining two [namely, zetetic and poristic] be
> more [a matter of] examples than of precepts,
> as is to be conceded by right to the [Ramist] logicians) —
> [with this remaining part exercising its function]
>
> as much about *numbers*
> if the seeking (*questio*) is about a magnitude that is to
> be expressed (*explicanda*) by a number,
>
> as about *lengths, surfaces, or solids* (*corpora*) [literally, “bodies”]
> if it be appropriate for the magnitude to
> be exhibited by the thing itself (*re ipsa*).
>
> And here [in the latter case],
> the Analyst presents himself as a Geometer
> by effecting a true work (*opus verum efficiendo*)
> [that is: by performing a true construction]
> after [doing] another one similar to the true one;
> [but] there [in the former case—of seeking a number],
> [he presents himself as] a reckoner (*Logistam*) [a calculator]
> by resolving whatever powers [have been] exhibited in a number,
> the powers being either pure or affected.
> [To resolve such powers is to get a numerical solution to an equation.]
>
> And either in things Arithmetical or in things Geometrical,
> he does not fail to bring forth
> some specimen of his own artful work (*artificii sui*),
> according to the condition set by the equality that was found
> or by the proportionalism (*Analogismi*) that
> was taken from it when it had been set in order.

As a geometer, the analyst solves a geometrical problem that resembles his analytic problem, but differs from it. The geometrical construction retraces the steps in his analytical resolution of his problem. The analyst as geometer hides his analytical capability: he presents his solution as if done merely by a process of synthesis, and only after that (as if merely to be helpful to those who are interested in the calculations) does he present the solution as a consequence of the equation that he recognized in the problem.

And in truth, not every Geometrical effecting (*effectio*) [construction]
is neatly put together (*concinna*),
indeed problems singly have their own elegances;
the truth is this:
the effecting that is preferred to others
is that which lays out (*arguit*) and demonstrates,
not the putting together of the work (*compositionem operis*) from the equality,
but [rather] the equality from the putting together—
the very putting together itself [thus laying out] itself.

And so, the Geometer—[being] an artful worker (*artifax*)—
although educated [in] Analytic, dissimulates that,
and just like someone who is thinking [only] about
[how to go about] effecting the work,
presents and explicates his problem synthetically,
[and] then, to be helpful to Logisticians [calculators],
gets the theorem together (*concipit*) and demonstrates it
from the proportion or equation that is
recognized (*adgnita*) in it [that is, in the problem].

CHAPTER 7. COMPLETING VIÈTE'S INTRODUCTION

Viète's previous chapters have treated the analytic art in its several parts—the zetetic most extensively, the task of which part is to set up and handle equations. Nomenclature for equations will now be treated in his eighth and final chapter, which will end as an epilogue about what the analytical art is to accomplish.

CHAPTER VIII *The denoting of equations, and the epilogue of the art*

"Equation" pronounced simply (*simpliciter prolata*) [that is, without a qualification]
is the name (*vox*) received in Analytic by
an equality that is properly set in order through Zetesis.
2 And so an equation is a comparing of
a magnitude that is unknown (*incerta*) [undetermined]
with [one that is] known (*certa*) [determined].
3 The magnitude that is unknown (*incerta*)
is either a root or a power (*radix vel potestas*).

Viète has used "root" to mean "side" (a "side" being an unknown magnitude at the very first level of the ladder magnitudes). He introduced that usage back in section 3 of his Chapter III. Now we shall see him extend his use of the term. In his new usage, a magnitude at the first level can be called a "root" not only when it is unknown but also when it is not. Not only a "side" but also a "length" (section 12 of Chapter VIII will tell us) can be called a root. But so can compared magnitudes at a level above the first: for example, a "plane" (section 13) can be a root, and so can a "solid" (section 14). Indeed, a given magnitude at any level (call it the "n^{th}" level) is the "root" of a series of ladder magnitudes at the level 2n, at the level 3n, and so on. So, for example, the "solid" will be the root of the powers that are called "cubo-cube," "cubo-cubo-cube," "cubo-cubo-cubo-cube," etc.

4 Again, a power is either pure or affected.
5 Affecting is through either denial or affirmation.
6 When [a magnitude that is] homogeneous [with and] affecting [a power]
is denied to the power,
[then] the denial is "direct."
7 When, on the contrary, the power is denied to [a magnitude that]
[is] homogeneous [with and] affecting [the power]
and [is a product of multiplication] by a grade [along-the-way to the power],
[then] the denial is "inverse."

Consider, for example, this power: "A cube." And this magnitude that is homogeneous with it: "A square onto G plus A square onto H." And this affecting of the power by the homogeneous magnitude: a denial. Then the denial will be *direct* when we take this: *the power* minus *the magnitude that is denied to the power*. But the denial will be *inverse* when that order of terms is inverted—that is, when we take this: *what the power is denied to* minus *the power*. So our example of the difference between the two sorts of denial is this:

direct: ....*(A cube) minus* (A square onto G plus A square onto H);
inverse: ....(A square onto G plus A square onto H) *minus (A cube)*.

Viète continues:

> 8 A "measuring subgradual [magnitude]" (*subgradualis metiens*)
> is [what] itself is a measure of the grade [that is] homogeneous and affecting.

Consider, for example, this unknown magnitude at the third level: "A cube." Suppose that it is a power that is affected by the addition to it of this third-grade magnitude: "A square onto G." This third-grade magnitude ("A square onto G") which affects "A cube," and is homogeneous with it, will be measured by the magnitude "G." This magnitude "G" is adopted by the lower-level (second-grade) magnitude "A square," which co-effects—by means of "G"—the product "A square onto G"; and this product affects that magnitude ("A cube") to which "A square" is a grade-along-the-way, and with which the product "A square onto G" is homogeneous. The magnitude "G" is thus a measuring magnitude that is "subgradual." ("Subgraduals" were defined, in section 7 of Viète's Chapter III, as "adopted magnitudes by which, together with a grade that is along-the-way, there is made a product homogeneous with the power that is affected.")

> 9 It is required, however,
> in the unknown (*incerta*) [undetermined] part of the equation,
> that [the following] be designated:
> the order [that is to say, the level] of the power and of the grades [along-the-way],
> and also the quality or sign of the affecting.
> [It is required,] moreover,
> that the adopted subgradual magnitudes themselves be given.

So, for example, if some solid made entirely by given magnitudes is one part of an equation (that is: is one part of an equality that has been set in order by zetesis), then the other part of the equation must say what is the highest level of the unknown magnitude, and what are the other (lower) levels of the unknown magnitude, and also whether they are added or subtracted; and, as for the magnitudes adopted by the lower levels of the unknown magnitude in order to co-effect the homogeneous products that affect the power—these adopted magnitudes must be given. Consider the following equation:

> (A cube + A square onto B — A onto C plane) is equal to
> (B plane onto C + D cube).

That equation meets the requirements. The one part of the equation (the right side) is a third-level magnitude (it is a "solid") that is entirely made up of given magnitudes. What about the other part, the left side? In it, the third is the highest level of the unknown: the power is of the order "cube." The first of the homogeneous affecting magnitudes is designated as "affirmed"; and the second one, as "denied." In the first of the affecting magnitudes, the grade that is along-the-way to the power is designated as being of the order "square," and the measuring subgradual magnitude that it adopts is a given "length"; and in the second of the affecting magnitudes, the grade that is along-the-way to the power is designated as being of the order "side" or "root," and the measuring subgradual magnitude that it adopts is a given "plane" (given because its "length" is given). The equation thus says what it must say in order to fulfill the requirements.

10 The first grade that's along-the-way to the power
is the root about which there's a seeking.

The last one (*extremus*) is that one which is
lower than the power by one grade of the ladder.
It is usual moreover (*solet autem*) for this [last one] to be called (*exaudiri*)
by the name (*voce*) of "epanaphora."

Ep+ana+phora is (transliterated) Greek for "what is carried+up+onto." The grade of the ladder that is along-the-way to the power, and is just one level below the power, is "carried up onto the power" through its being multiplied by the root that is the first grade of the ladder. So a square is the epanaphora of a cube, which is the epanaphora of a squaro-square, which is the epanaphora of a squaro-cube, and so on.

11 A grade that's along-the-way to a power
is the "reciprocal" of [another grade that is] along-the-way,
when the power is [the product] of the one's being led onto the other.
Thus an adopted [magnitude] is the reciprocal of the grade which it sustains.

What, for example, is the "reciprocal" of a "square" that is a grade along-the-way to a "cube"? The reciprocal is a "side"—since the product that will be made when a "square" is led onto a "side" is a "cube." Thus an adopted "plane" that is led onto the "side" to produce a "solid" will be the reciprocal of that "side."

12 From a *length* as the root, the grades along-the-way to a power are
those very ones which are designated on the ladder.

13 From a *plane* as the root, the grades-along- the-way are:

the Square	or	the PLANE
the Squaro-square	or	the PLANE'S Square
the Cubo-cube	or	the PLANE'S Cube

and in that order successively onward.

14 From a *solid* as the root, the grades-along-the-way are:

the Cube	or	the SOLID
the Cubo-cube	or	the SOLID'S Square
the Cubo-cubo-cube	or	the SOLID'S Cube.

That is to say: If you take as your root the item at the level (n), then the successive steps on the way to the power will be at the following levels: the level (n), the level (2n), the level (3n), and so on. That is because when the item at the level (n) is multiplied by itself, the product is at the level (n + n). And when this item at the level (2n) is itself in turn multiplied by the root item that is lower down at the level (n), the product is at the level (2n + n). And so on.

15 Square, Squaro-square, Squaro-cubo-cube, and those [powers] which
are produced continuously in that order from these very same ones,
are "powers of the simple intermediate" (*potestates simplicis medii*);
the rest [of the powers] are
"[powers] of the multiple [intermediate] (*reliquae multiplicis*)."

The word here translated "intermediate" is the Latin word that is used for what we call a "mean." Any item in a list of "powers of the simple intermediate" is at the level whose number is the mean proportional between the level-number of the item that it comes right after, and the level-number of the item that it comes right before. Let us first set out the list of items given here:

item	*level*	*produced by*
Square	2	(A onto A)
Squaro-square	4	(A onto A) onto (A onto A)
Squaro-cubo-cube	8	[(A onto A) onto (A onto A)] onto [(A onto A) onto (A onto A)] .

Now as 2 is to 4, so 4 is to 8; and, because of that proportionality, we say that 4 is the mean proportional between 2 and 8. Generally, a "mean" is something that is middling—in other words, it is in the middle; it is in-between; it is intermediate. What is intermediate in a proportion is what is intermediate between the terms at the two extremes. When these two terms that are intermediate (that is to say, are "means") are the same, each of them is "*the* mean"—and, since it is the mean in a proportion, it is "the mean proportional." So the level-numbers of the "powers of the simple intermediate" constitute a series—a series in which any of the numbers is the mean proportional between the number that it follows and the number that it precedes. Hence the name "powers of the simple intermediate." The rest of the powers—the powers at the many different levels that are located on the ladder amid these levels of the powers of the simple intermediate—are the "powers of the *multiple* intermediate."

16 The known (*certa*) [determined] magnitude with which
the rest [of the magnitudes] are compared
is the "homogenous [term] of comparison."

Here is an example:

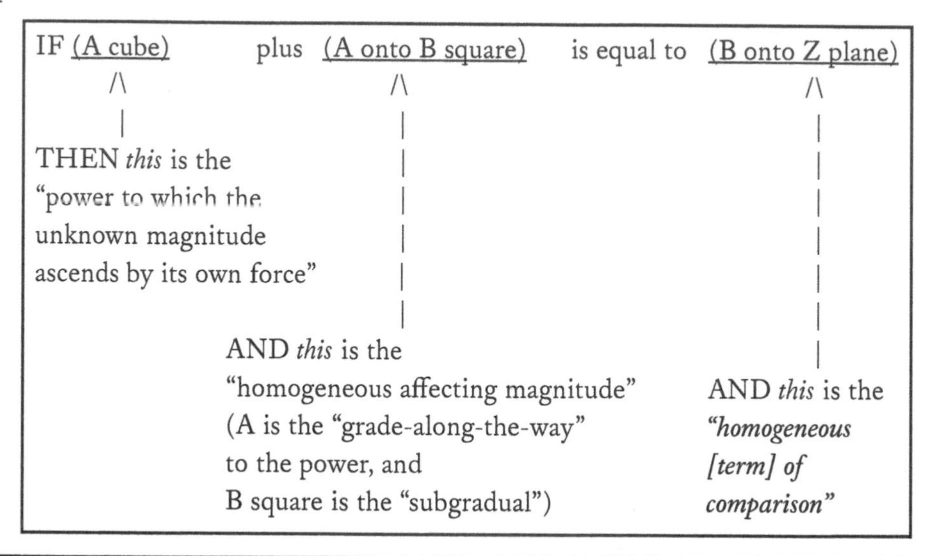

17 In [the case of] numbers,
the homogeneous [terms] of comparison are units.

Although there is some resemblance, calculation with species is not calculation with numbers. In the case of *non*-numerical equality, the homogeneous term of comparison in *one* equation is *not necessarily* homogeneous with the homogeneous term of comparison in *another* equation. Consider, for example, this equation:

(A cube) plus (A square onto B) is equal to (C plane onto D).

Its homogeneous term of comparison is the magnitude "C plane onto D"—which is a "solid." It would be impossible for the homogeneous term of comparison in that equation to be "C plane" rather than "C plane onto D." Why? Because the rest of the magnitudes in that equation are at the third level (as "C plane onto D" is), whereas "C plane" is at the second level. But the magnitude "C plane," would indeed be the homogeneous term of comparison in the following equation:

(A square) plus (A onto B) is equal to (C plane).

In the case of *numerical* equality, by contrast, the homogeneous term of comparison for *any* equation *must* be homogeneous with the homogeneous term of comparison for *every other*. If "C" is some number, then also "C plane" is a number, and so too is "C cube." Indeed, the very same number may be both a "square" and a "cube." (For example, 64 is both the square of 8 and the cube of 4.) Any number can be compared with any other number—because every number is a number of units.

Next in Viète's discussion of nomenclature, is the subject of the kinds of equations. When the equality has been set in order, one part will contain all the instances of the unknown magnitude. This part is equal to a homogeneous term of comparison (a magnitude which involves no magnitudes but given ones, and which is at the same level as the part that contains every instance of the unknown magnitude). It is the character of the *part containing every instance of the unknown magnitude* that characterizes the whole *equation*. Here are some examples:

if part containing every instance of the unknown is:	*then equation is:*
(A)	"simple absolutely"
(A cube)	"simple ladderwise"
(A cube) plus (A square onto B) minus (A onto C)	"polynomial"

In an equation that is "simple absolutely," a "side" (A) might be equal to such a given "length" as "B," or "B minus C," or "B onto C/ D." In an equation that is "simple ladderwise," a magnitude "A cube" might be equal to any one of a variety of "solids"; so might, in an equation that is "polynomial," a polynomial like the last example above. A polynomial containing a "power" at the level (n) might have as many as (n — 1) affectings, but it may have fewer affectings than that. How many fewer? As many as are the levels (lower than the power) at which the unknown is not to be found in the polynomial. Here is Viète:

18 When that root about which there is a seeking—
[while] staying at its own base [that is, at the first level]—
is compared to a homogeneous magnitude that is given,
the equation is "simple absolutely."

19 When a power of that root about which there is a seeking —
[while being] pure of affecting—
is compared to a homogeneous [magnitude that is] given,
the equation is "simple climactically" [simple "ladder"-wise (from Greek)].

20 When a power of that root about which there's a seeking—
[while being] affected by
a designated grade [that is along-the-way to the power]
and a coefficient [magnitude that is] given—
is compared to a homogeneous magnitude that is given,
the equation is "polynomial" [from Greek for "many" and Latin for "name"]
in accordance with (*pro*) [in proportion to] the affecting's multitude and variety.

21 The power can be enfolded with as many affectings as there are
grades along-the-way to the power. And so:

Square can be affected	by Side.
Cube	by Side and Square.
Squaro-square	by Side, Square, and Cube.
Squaro-cube	by Side, Square, and Cube [and Squaro-square] ,

and endlessly onward in that series.

22 Proportionalisms are distinguished and receive their nomenclature from
the kinds (*generibus*) of the Equations into which they fall when resolved.

Now Viète turns from "the denoting of equations" to the other item that he promised in the title of this final chapter of his *Introduction to the Analytic Art*: he turns to "the epilogue of the art." This deals with what the art is to accomplish. First, arithmetically—

> 23 For Exegetics in things Arithmetical,
> an educated Analyst is instructed:
>
> to add a number to a number;
> to subtract a number from a number;
> to lead a number onto a number;
> to divide a number by a number.
>
> Further, the Art hands over the resolution of any powers whatever,
> either ones that are pure
> or—what the ancients and also the moderns (*novi*) were ignorant of—
> ones that are affected.

That further business of the art is the program of Viète's *Treatise on the Numerical Resolution (for Exegetics) of Powers, Pure and also Affected*, which was published in 1600 (and, with this expanded title, in 1646). The "numerical resolution of a power" is the finding of a number that solves the equation which contains that power (the power being the magnitude where the unknown magnitude is at its highest level in the equation). The first section of Viète's treatise treats powers that are pure; the second section treats powers that are affected. Here, for example, is the equation treated in the first problem of each section of that treatise:

> *1st section*
> From a pure "square" given in numbers, to elicit the side analytically.
> Let it be proposed that 1 Q [one "square" (*Quadratum*)]
> be equal to 2916.
>
> *2nd section*
> From an affected "square" given in numbers—
> affected [that is to say] by the adjoining of
> a "plane" [that's produced] by a "side" and a given coefficient "length"—
> to elicit the "side" analytically.
> Let it be proposed that 1Q + 7N
> [one "square" and seven of the number that is the "side"] be equal to 60750.

Nowadays, we would state the two problems as the finding of a suitable x such that:

In the one problem,	x^2	=	2916.
In the other problem,	$x^2 + 7x$	=	60750.

The epilogue now goes on to deal with what the art is to accomplish geometrically:

> 24 For Exegetic in things Geometrical,
> [the analytic art] selects and recounts
> the more canonical effectings (*effectiones*) by which
> the equations of "sides" and "squares" may be interpreted (*explicentur*).

That is the program of Viète's *Canonical Recounting of Geometrical Effectings*, which was published in 1646. In this work, Viète provides "canonically" (that is, in a form that is standard and orderly) "effectings" (that is, constructions) which furnish geometrical counterparts to the magnitudes and relations or operations that appear in equations that contain no terms above the second level. Above the second level, geometrical "explication" (that is, interpretation) becomes trickier, as Viète goes on to say in the *Introduction*:

> 25 For Cubes and Squaro-squares, [the analytic art] demands (*postulat*) —
> in order that the deficiency of Geometry
> may be supplied as if (*quasi*) by Geometry—[this postulate:]
>
> From any point whatever, to draw a straight line that
> intercepts any two lines whatever,
> in such a way that the segment between the intercepts
> will be of any possible predetermined size whatever.
>
> This [demand] being conceded—
> indeed, it is an *aitêma* [(in Greek letters) a demand or postulate]
> not *dysmêkhanon* [(in Greek letters) something difficult to devise]—
> it solves *eutekhnôs* [(in Greek letters) with well-employed art]
> the more famous problems which have
> hitherto been said to be *aloga* [(in Greek letters) irrational],
> [namely]:
>
> the mesographicum;
> the cutting of an angle into three equal parts;
> the finding of the side of the heptagon;
> and as many others as fall into the formulations of equations in which
> Cubes are compared with Solids, Squaro-squares with Plano-planes—
> either purely or with affecting.

Viète's *Supplement of Geometry*, which was published in 1593, begins by restating the postulate about the intercept and goes on to present his solution to the three problems that he names here. The first of these is the problem of finding two mean proportionals to two given lines (this is the "*meso+graphicum*" or—to translate the Greek roots—the "mean+drawing" problem), whose solution will yield the solution to the celebrated problem of doubling any given cube. The other two problems are the problem of trisecting any given angle, and of finding the side of the regular heptagon that can be inscribed in any given circle. Conceding the postulate which Viète demands makes it possible to solve the problems while avoiding the higher-level equations to which the problems would otherwise lead.

This avoidance of higher-level equations by using an additional geometric postulate, which thus simplifies the task of the analytical art in solving those geometrical problems, raises the question whether higher level equations should have any part to play in geometrical problem-solving at all. Equations that resolve into proportions in which more than three ratios of sides or lengths are compounded would seem to have no geometric interpretation—since a ratio compounded of three ratios of lengths is the same as a ratio of solids, and figures beyond that are unimaginable. Viète raises the question and gives an answer.

26 Indeed, since all magnitudes are
lines, surfaces, or solids (*corpora*) [literally: "bodies"],
what truly can be the use in human affairs (*in rebus humanis*)
of proportions beyond the triplicate ratio,
or [even] as far as the quadruplicate ratio—
if not perchance in the cuttings-up (*sectionibus*) of angles,
so that we might obtain in consequence (*consequamur*)
the angles from the sides of figures, or the sides from the angles?

27 Therefore [the analytic art] reveals the
mystery of the cuttings-up (*sectionum*) of angles—known to no one hitherto—
[with a view] either to things Arithmetical or to things Geometrical;
and [the analytic art] educates in how [to do the following]:

Given the ratio of angles, to give the ratio of the sides.

To make an angle be to an angle so, as a number is to a number.

Viète treated such cuttings-up of angles in *The more general theorems for the analytic of angular cuttings-up*, which was published in 1615. (The title uses a Greek word to characterize the theorems —*katholikôtera*. The reference to the analytic art in the title was dropped in the republication of 1646.)

Having spoken of the analytic art's ability to treat angles, relating their sizes and the sizes of straight lines and of numbers, Viète goes on to say that nonetheless the art cannot relate the sizes of straight lines and curves—precisely because of the kind of a thing an angle is:

28 [The analytic art] does not compare a straight line to a curve,
because an angle is something [that is]
intermediate between (*medium quiddam inter*) a straight line and a plane figure;
and so the law of homogeneous [magnitudes] seems to be opposed.

Having mentioned, by the way, something that the analytic art will not do, Viète concludes with a general statement about how much it will do; he concludes by declaring the immense ambition of his art:

> 29 And so, the Analytic art—at last having been put into the threefold form of Zetetic, Poristic, and Exegetic—arrogates to itself by right the proud problem of problems, which is:
>
> TO LEAVE NO PROBLEM UNSOLVED
> *(NULLUM NON PROBLEMA SOLVERE)*.

PART VIII PROBLEM-SOLVING, EQUATIONS, AND METHODICAL MASTERY OF CURVES

Descartes' *Geometry*

CHAPTER 1. TURNING TO DESCARTES' *GEOMETRY*

The *Geometry* is the final item in the volume by Descartes which contains it. The first of the items is called a "discourse," and the other three are called "essays." The title of the volume is "Discourse about the method for well conducting one's reason and seeking the truth in the sciences; plus the Dioptrics, the Meteors, and the Geometry—which are try-outs ["essays"] of this method."

The initial essay consists of a mixture of mathematical and physical considerations; the middle essay consists of physical considerations unmixed with mathematical ones; and the final essay consists of mathematical considerations unmixed with physical ones. (See the letter that Descartes wrote to an unknown correspondent at the end of 1637 [Adam & Tannery: I, 370].)

The *Discourse*, which opens the volume that is closed by the *Geometry*, twice speaks of "this volume" (first, in its 3rd part's 6th paragraph; after that, in its 6th part's 4th paragraph). The *Discourse* reports that this volume is the author's initial public offering because he suppressed an earlier writing which he mentions as his "treatise" (*traité*) five times (twice in the 5th part, in its 2nd paragraph and again in its 9th; and thrice in the 6th part, in the 1st and 3rd and 8th paragraphs); he also uses forms of the related verb five times (*traitant*, at I.6; *traitent*, at I.7 and I.10; *traiter*, at V.1; *traitais*, at V.5). The *Discourse* says, of the author's earlier treatise, that he was prevented from publishing it by "certain considerations"; however, after speaking of these, the *Discourse* goes on to tell us (in the 6th part's 8th paragraph) that while all these considerations together caused him to wish not at all to divulge the treatise which he had in his hands—and even caused him to resolve against letting, during his lifetime, any other one be seen "which would be so general, or from which one could understand the foundations of my physics"—nonetheless, there have since then been some other reasons which have obliged him to "set down here some particular essays," and to "render to the public some account of my actions and of my designs." For a full understanding of the author's designs, we must study the entire volume in which his *Geometry* appears. To understand the entire volume, however, we must work our way through its several items, the one that appears to be most formidable being the one that we shall examine now—namely, the *Geometry*.

(Note that the translations from the French of Descartes—like those presented earlier in this guidebook from the Greek of Pappus and of Diophantus, and from the Latin of Viète—will be so presented that the layout of the text can help to make the structure of the thought easier to see than it would be if the print stretched uniformly across the page from margin to margin.)

The *Geometry* is an item set apart in its form. Divided into three "books," it is the volume's only item that is divided into "books": the *Discourse* is divided into six "parts"; and the two other essays of the method—the *Dioptrics* and the *Meteors*—are each divided into ten "discourses." While the discourses of the *Dioptrics* are all referred to numerically (from "first" to "tenth"), only the first nine of the discourses of the *Meteors* are referred to in that way: the concluding discourse of the *Meteors* is referred to not as the "tenth" discourse but as the "last" one. The *Dioptrics* refers to itself as "this treatise," in its last paragraph; the *Meteors* likewise so refers to itself in its last paragraph. The *Geometry*, unlike the other two essays, does not so refer to itself in its last paragraph.

The *Geometry* is unlike the other essays also in its being preceded by a "Notice" (*Advertissement*); and this prefatory notice is the place where the *Geometry* refers to itself as "this treatise." This notice that precedes the *Geometry* is printed in italics. Only one other item in the volume is preceded by an italicized statement— the *Discourse*, the item with which the volume opens. The only other "Notice" (*Advertissement*) in the volume comes after the *Geometry*, the item with which the volume closes. This notice in italics precedes three appended tables, one for each of the essays: first comes a "table of the principal difficulties which are explained in (*en*) the Dioptrics," and then a "table of the principal difficulties which are explained in (*aux*) the Meteors"; then comes a "table of the matters of (*de*) the Geometry." The notice that follows the *Geometry* tells the reader not to try to avoid labor by selecting matters from an essay and looking at them in isolation from the other matters that are placed before and after. Here is the notice:

Notice (*Advertissement*)

Those who do not visit the tables of books but for the sake of
choosing there the matters which they wish to see and exempting themselves from
the trouble (peine) *of reading the rest*
will not draw any satisfaction from this one here;
for the explanation of the questions which are noted therein depends
almost always so expressly on what precedes them,
and often also on what follows them,
that one would not know how to understand it perfectly unless
one reads with attention the whole book.
But for those who will have already read it,
and who will know well enough the most general things that it contains,
this table could serve them as much
to make them remember places where it has spoken of more particular [things] which will
have escaped from their memory,
as often also to make them be on the alert (prendre garde) *for those [things]*
which they will perhaps have passed without taking note (remarquer) *of them.*

The *Geometry* is set apart from the other essays in its not being mentioned by name anywhere in the prefatory *Discourse*. (The *Meteors* is mentioned once, together with the *Dioptrics*, in the 6th part's 10th paragraph; the very next paragraph contains the only other mention of the *Dioptrics* by name in the *Discourse*.) Yet, while the *Geometry* is not mentioned by name in the *Discourse*, its subject matter is pervasive there. Some indication why, is suggested by the notice that precedes the *Geometry*. What it says is the following:

Notice (*Advertissement*)

Up to here, I have tried to render myself intelligible to everybody;
but as for this treatise (traité)*, I fear that it will not be read but by*
those who already know what is in the books of Geometry—
for inasmuch as they [those books] contain
quite a few (plusieurs) *very well demonstrated truths,*
I believed that it would be superfluous to repeat them;
and [yet,] for all that, I did not refrain from making use of them (m'en servir)*.*

Like the other two essays, this one makes use of assertions that it does not demonstrate. However, the assertions in the *Dioptrics* and in the *Meteors* are suppositions for which, although the author could present an argument, he does not, since to do so would be premature, and the instructive consequences of his supposing them for now will themselves constitute a kind of argument—whereas the mere assertions upon which he relies here in The *Geometry* need no argument. Arguments would be superfluous. The arguments that he might present for what he relies upon in his *Geometry* would merely repeat the very good demonstrations in geometry books which his readers may be expected to have read already. Geometry differs from the other sciences—the sciences where there are physical considerations—in that a body of reliable geometrical knowledge already exists. Later he will several times refer to it as "the common Geometry." The special geometry that is presented here, however, is far advanced beyond the elementary common geometry. And it plays a unique part in Descartes' initial public offering. In a letter occasioned by his friend Mersenne's report of another author's adverse judgment on Descartes' writings, Descartes wrote back to his friend at the end of 1637 [Adam & Tannery: I, 478–80] as follows:

I am not comfortable being obliged to speak favorably of myself;
but because there are few people who would be able to understand my Geometry
and you desire that I send you what is the opinion that I have of it,
I believe it is appropriate that I tell you
that it is such that I would wish nothing more of it,
and that through the Dioptrics and through the Meteors I have
only tried to persuade that my method is better than the ordinary [one]—
but through my Geometry I claim to have demonstrated it.

For from the beginning onwards I solve a problem (*question*) which,
by the witness of Pappus, could not be solved by any of the ancients;
and, one can say, could no more be solved by any of the moderns, since
none of them has written of it, and nonetheless the most able [of them] have tried to
find the other things that Pappus in the same place says were sought by the ancients …
[here Descartes mentions the authors of books like
"The Renovated Apollonius" and "The Dutch Apollonius"];
but none of those have known how to do what the ancients were ignorant of.

After that, what I give in the second book, touching
the nature and properties of curved lines and the fashion of examining them, is,
it seems to me, as much beyond the ordinary geometry as
the rhetoric of Cicero is beyond the ABC's of children.
[Here Descartes goes on to say that he finds ridiculous the promise—
made by the author whose judgment on Descartes' writings occasioned this letter—
that in a preface he will give means that will be better than Descartes' means
(presented in the second book of his *Geometry*, after his general discussion of curves)
for finding the tangents of all curved lines.]

And so much fault is there in saying that
the things I have written could be easily drawn from Viète, that—on the contrary—
the cause that my treatise is difficult to understand is that I tried to put nothing in it but
what I believed had not been known by him or anyone else.
As one can see if what I wrote about the number of roots which are in each equation, ...
—[in] the place [in the third book] where I begin to give the rules of my Algebra—
is compared with what Viète wrote about it,
at the very end of his book On the Emendation of Equations;
for one will see that I determine it generally in all equations,
whereas he gave naught but some particular examples,
about which he nonetheless makes so great a fuss that
he wanted to conclude his book with that—
by which he has shown that he could not determine it in general.
And so, I began where he had finished up his achievement;
which I nonetheless did without thinking of it, for since I received your last [letter] I have
leafed through Viète more than I had ever done previously,
having found it here by chance in the hands of one of my friends;
and, between us, I do not find that he knew so much of it as I had thought,
notwithstanding that he was very able.

It remains [to be said, that], having
determined as I did in each genre of problems (*questions*) all that can be done there,
and shown the means of doing it,
I claim not only that one should believe
that I have done something more than those who have preceded me,
but also that one should be persuaded that
our posterity will never find anything in this matter which
I could not have found as well as they, if I had wanted to take the trouble to seek it.

Let us turn now to the *Geometry* of Descartes. For ease of reference, we shall number Descartes' paragraphs, though he himself did not number them. Unfortunately, previous translations of Descartes into English have assumed that there is no compelling reason to preserve Descartes' paragraphing. To blur Descartes' paragraphing, even with the best of intentions, is, however, to obscure the design of his work. One or another translator may have thought that by conflating paragraphs or splitting them apart he was making clearer the intention of an indifferent or careless or incompetent author, but there is reason to believe that Descartes gave careful attention to his paragraphing and knew quite well what he was doing with it. That he cared a lot about his paragraphing is attested by his letter to Mersenne of 23 June 1641; in it, transmitting the Fifth Objections to his *Meditations* (the objections by Gassendi) along with his own Replies, Descartes tells Mersenne to make sure that the printer corrects the alterations that he had made in the paragraphing, alterations made either by putting in new breaks where Descartes did not have them or by omitting breaks where Descartes did have them (Adam & Tannery: III, 386). It is true that Descartes' paragraphing here in the *Geometry* may at times seem strange. Consider, for example, his devoting a whole paragraph to what is in the First Book's 20th paragraph (not to speak of his making the 29th paragraph so brief), while giving merely one paragraph to what is in the 30th paragraph. We shall see, however, as we proceed, that numbering Descartes' paragraphs is a help in figuring out the design of Descartes' work.

Descartes' translators not only tamper with his paragraphing (as his printers tried to do); they go so far as to transpose the parts of his work. With no apparent reason and with no mention of the fact, the only English translation of the complete *Discourse and Essays* places the *Geometry* between the *Dioptrics* and the *Meteors*, rather than where it belongs—at the end of the volume, following both of them. And then within this Englished version of the *Geometry*, important terms that Descartes takes over from the traditional study of conic sections are rendered unrecognizable—the upright side, for example, becomes "the right side," and the lines drawn ordinatewise are said to be "in order," while tangents become "contingents." In both of the major translations of the *Geometry*, little attention is given to whether Descartes in some place speaks in fact of method or speaks merely of the fashion in which something is done: both promiscuously become "method"—this, in translating an author for whom method is a theme of the utmost importance. (Even in little details that might be significant, we are betrayed by the translation of the *Discourse and Essays* just mentioned—which labels the "last" discourse of the *Meteors*, like that of the *Dioptrics*, its "tenth.") There is no need now to heap up more examples of how existing translations of the *Geometry* lead the reader astray. In the next chapter of this guidebook, we shall take note of one more example of mistranslation, the most egregious such example, which will serve to emphasize a matter upon which everything else depends.

Besides numbering Descartes' paragraphs, this translation will impose upon him in another respect. As previously, in this guidebook's translations of Diophantus, Pappus, and Viète, the text of Descartes is here laid out not as it originally appeared—word after word, line after line, so as to fill the page—but rather so broken up and so spaced out as to give visual display to the structure of thought. (Note, however, that the figures you will see within the shaded heavy-bordered boxes that contain the translation are mere reproductions of the figures that appeared on the pages of Descartes' *Geometry* as it was published.)

CHAPTER 2. INTRODUCING "THE TRUE METHOD" IN GEOMETRY

THE
GEOMETRY.

FIRST BOOK.

*Of problems that one can construct without
employing therein aught but circles and straight lines.*

[1] All the problems of Geometry can
easily be reduced to such terms that,
for constructing them [that is, for constructing the problems] after that,
there is no need of aught but
knowing the length of some straight lines.

Descartes has referred the reader to the previously available books of geometry for demonstrations of several truths already known. For him to repeat them in this book would be superfluous. Indeed, this will not be a book of geometrical demonstrations in the classic fashion; it will be an innovative treatment of geometrical problem-solving. While the previously available books of geometry may contain constructions for several problems already solved, this book will treat, not several of the problems of geometry, but all of them. It will treat *all* of the construction-problems that geometry *can* propose. It will treat them, not by producing a construction for *every* one of them, but by showing an easy *way* to reduce *any* one of them to such terms that the construction will require nothing more than knowledge of this: how long are certain straight lines.

[2] And as all Arithmetic is composed of
naught but four or five operations—which are
Addition, Subtraction, Multiplication, Division,
and the Extraction of roots, which one can take for a species of Division—
so one has no other thing to do in Geometry,
touching the lines that one is seeking (*cherche*),
for preparing them to be known (*connuës*),
than [this:]

with respect to them, to add others of them,

or to take away [others] of them—

or else,
having one [line]
which, for relating it (*pour la rapporter*) to numbers
as well as possible (*d'autant mieux*),
I shall name "the unit,"
and which can ordinarily be taken at discretion,
then, having two others of them besides,
to find a fourth [one] of them, which is
to one of these two as the other is to the unit—
which is the same as Multiplication;

or else to find a fourth [one] of them, which is
to the one of these two as the unit is to the other—
which is the same as Division;

or finally, to find one or two or several mean proportionals
between the unit and some other line—
which is the same as to take the square root or cube root, etc.

And I am not afraid to introduce these terms of Arithmetic into Geometry,
to the end of rendering myself more intelligible.

To render himself "more intelligible" in presenting a way for problems in geometry to be "easily" reduced to finding out how long are certain straight lines, the author is not afraid to introduce terms of arithmetic into geometry. In this, as we can see from what he will say later (in the 23rd paragraph), he differs from those who wrote the old geometry books to which the prefatory notice referred. "In the fashion in which they explained themselves," he will say, there was much "obscurity" and "difficulty" (*embaras*). (Of his own "fashion," he will speak elsewhere in the *Geometry*.) What caused the ancients' fashion was the "scruple which the ancients had about using terms of Arithmetic in Geometry"—"which could proceed from naught but the fact that they did not see clearly enough their relation (*rapport*)." Descartes can proceed in his own fashion, fearlessly introducing certain terms of arithmetic into geometry, because he himself sees clearly their relation. The differences in kind of the items studied should not be allowed to obscure the essential matter—which is how the items are related.

Among the many misleading things in previous translations of Descartes' *Geometry*, perhaps the worst is the mistranslation which makes Descartes say the opposite of what he means to say about this fundamental matter. Where the French of the 23rd paragraph says *le scrupule, que faisoient les anciens d'user des termes*, the translators take the relative pronoun *que* as the subject of the relative clause rather than as its direct object, and so they take the *d'* as a kind of "to" rather than as a kind of "from." In the mistranslation, the scruple made the ancients (to) use the terms—whereas, in the correct translation, the ancients made a scruple against using the terms (that is, they kept "away from" using the terms). Unless emended, the translation makes it impossible to understand Descartes' transformation of mathematics. It is impossible to understand what Descartes is doing if he is not introducing arithmetic terms into geometry, where the ancients, on principle, were kept from using them despite what their use might contribute as a labor-saving device or power-source.

Descartes' *Geometry* is devoted to problem-solving, not to the display of theorems in the fashion of Euclid and Apollonius. The titles of its opening First Book and its closing Third Book both speak of problems; and he says at several places in the *Geometry* that its chief benefit is not in its demonstrating things for readers to look at, but rather in its affording them an opportunity to exercise themselves in this mode of finding out.

The ancients, by contrast, did not find out how finding out was the heart of the matter. They exhausted themselves, and their audience, by heaping up displays of what they happened to find. Right before quoting Pappus as evidence of the weakness of the ancients in solving problems, Descartes will criticize the ancients for putting so much labor into so many big books in which the mere order of their propositions lets us know that they did not have the true method of finding all the things that can be found, but rather only amassed those of them that they happened to encounter. After quoting Pappus, he will exhibit his own power by solving what tied the ancients up in knots, thus exhibiting his own power, while encouraging his readers to adopt his mode of operation.

He begins by saying that all the problems of geometry can be constructed by knowing the lengths of certain straight lines. After showing how to operate on lines without constraint, he will emphasize the sufficiency of this for all the problems of ordinary geometry, something that he believes was not noticed by the ancients.

The ancients could not notice what their scruple kept them from seeing. Descartes claims to lack the scruple of the ancients, not because he is unscrupulous, but because he is more clear-sighted. To judge his claim, we must answer two questions: First, why was it that the ancients had a scruple against introducing the terms of arithmetic into geometry, and what is it that Descartes sees clearly enough to be free of the scruple? And second, what is it that Descartes is free *for* by being free *from* the scruple?

What Descartes does first in the *Geometry* can help us to think about the first question. He then goes on to do what can help us to think about the second. What he does first will be clearer to us if we have some notion of where he is headed—that is, if we have some notion of the answer to the second question. For now, let it suffice to say that Descartes wishes to make problem-solving in geometry largely a manipulation of *equations*—setting them up, solving them, and interpreting the solution. For that, it is necessary to be able to compare quantities of all sorts with each other. The relations among heterogeneous items must be represented by operations upon homogeneous items.

The numbers studied by arithmetic are homogeneous, being multitudes of nondescript units—unlike the numbers of grapes that are bunches of grapes, or of cows that are herds of cows. And the figures studied by geometry are heterogeneous—for one cannot compare the magnitude of a cube with that of a straight line by saying, for example, that the former is triple the latter. But while the bare numbers counted off by arithmetic possess homogeneity, they are discrete: they lack the continuity possessed by the figures imagined by geometry, which are heterogeneous. So Descartes will engage in calculative operations on items that combine the advantages of arithmetical homogeneity and geometrical continuity. An item of that sort, which lack the discreteness of numbers and the heterogeneity of figures might be given the composite name "number-line." The yield from Cartesian calculations with such "number-lines" will continue to be just as manageable "number-lines"; these will, however, facilitate the solution of geometrical problems that involve lines, surfaces, and solids of very many different sorts.

What are the terms of arithmetic which Descartes introduces into geometry? He gives certain geometrical constructions for introducing them, but he does not explain the rationale behind what he does. He does not show how geometrical constructions are suitable counterparts of numerical operations. Let us consider the matter somewhat more explicitly than he does.

In the *first* place: when, to some A, is put some B, we write "A + B." We do that with numbers—as when to 5 units, we put 2 units, getting 7 units as the sum. And we also do it with lines, as when, as follows,

to line A:	!_____!
we put line B:	!__!
getting line A + B:	!_____!__! .

In the *second* place: when, from some A, is taken away some B, then we write "A – B." We do that with numbers—as when from 5 units, we take away 2 units, getting 3 units as the difference. And we also do it with lines, as when, as follows,

from line A:	!________!
we take away line B:	!__!
getting line A – B:	!_____!.

Since, for Descartes, it goes without saying that those are the geometrical counterparts of arithmetical addition and subtraction, what he himself begins with is not those two, which he does not even mention. He begins, rather, with the geometrical counterparts of arithmetical multiplication and division.

[3] For example, let AB be the unit,
and [let] multiplying BD by BC be [what is] needed:
I have naught [to do] but
to join the points A and C, then to draw DE parallel to CA—
and BE is the product of this Multiplication.

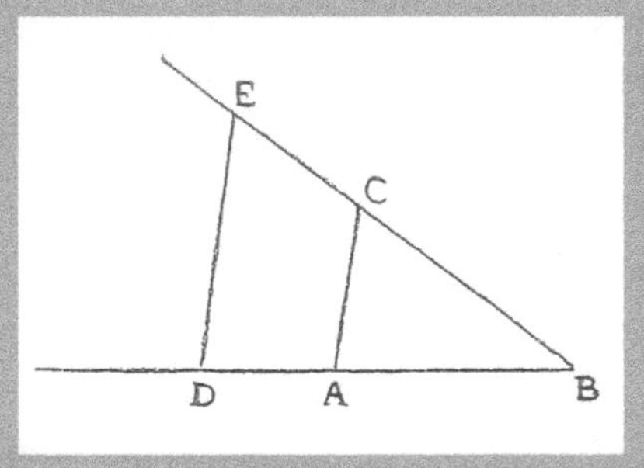

[4] Or else, if dividing BE by BD is [what is] needed:
having joined the points E and D, I draw AC parallel to DE—
and BC is the product of this division.

Thus there comes, in this *third* place, the immensely innovative step with which Descartes himself begins his introduction of the terms of arithmetic into geometry. Let us consider more closely what might lie behind this step of multiplication. (After that, we shall consider what is in the fourth place—namely, the step of division, which follows closely upon that of multiplication.)

When some A is taken some B times, we write "A*B." We do that with numbers, as when we take 4 units thrice—that is, when 4 units are taken 3 times—thus producing 12 units, that is, getting 12 units as the product. And we also do it with a line, as when, as follows,

```
line A:                      !_____!
is taken thrice (3 times),
getting line 3A:             !_____!______!______!.
```

Now, in speaking of numbers, we may say hurriedly "4 times 3 is 12," but what is meant is merely repeated addition, in which the first number is added to itself, and the other number is the number of times we take the first one. That is to say: "4*3" means merely "4 units taken 3 times." Since, however, when 4 units are taken 3 times, the yield is the same number of units as when 3 units are taken 4 times, we may speak briefly, without paying attention, and say "4 times 3." When we write "A*B" and say "A times B," both A and B are numbers. If A is a line, however, then "A*B" can only mean "line A taken B times"—where B is a number. Or if B is a line, then "A*B" can only mean "line B taken A times"—where A is a number. What, after all could "A*B" mean if A and B were *both* lines? But suppose we insist on introducing into geometry the term "product" from arithmetic. How could we do it?

We might try the easiest answer first. We might say that what is meant by the expression "line A * line B" is merely "that line whose number of units is equal to this: (the number of units in line A) * (the number of units in line B)." That way, the line that is the product of a 4-unit line and a 3-unit line would be a 12-unit line. But the trouble with that is this: not all lines are commensurable. If we say what we mean by speaking of the line that is the product of multiplying two lines together, then it has to work for any two lines whatever. If we were to say that the product-line is the line whose measure is the product of the measures of the two factor-lines, we would have a product-line only when we were lucky enough to have two factor-lines that can be measured together—that is, by the same unit.

So next we might try this instead: "line A * line B" means "the rectangle contained by the lines A and B." That way, two lines will have a product whatever may be the ratio of the one to the other. Even incommensurable lines will have a product. But the trouble with that is this: the product will be a magnitude of a different kind from the factors. And what about the product of more than two factors? With three of them, we would go from rectangles to parallelepipeds—but what could we do after that?

We are having trouble. Why? Because of the difference between multitudes and magnitudes. According to Descartes' unfinished early presentation of "rules for the direction of the native wit," neither multitudes nor magnitudes are the object of mathematics as such; its object, rather, is quantity. Quantity does have two aspects—looked at one way it is order, looked at the other way it is measure—but in dealing with quantity as such we are not hampered as we are in dealing as the ancients did with arithmetic and geometry.

Arithmetic deals with counted-up multitudes of units as such; geometry, with figures that we visualize. Multitudes as such are homogeneous (every number of units can be compared in size with every other), but multitude is restricted to the commensurable (since it is constituted of discrete, countable units); on the other hand, magnitude is not restricted to the commensurable (since it is continuous), but magnitudes are heterogeneous (a magnitude can be compared in size only with another magnitude of the same kind). By contrast, pure quantity—the object of universal mathematics—is both homogeneous and continuous. In handling it, we can compare size without hindrance. We are not hindered by any need to take into account differences in kind, nor by any restriction to what is countable.

What has just been said might make it seem as if the object of Descartes' mathematics is a composite that draws together, in no order, one feature from each of the two sorts of objects of elemental mathematical study. But Descartes did not write an *Arithmetic* that incorporated geometrical terms; rather, he wrote a *Geometry* that incorporated arithmetical terms.

What he sought was not an understanding of "bare numbers" or of "visualizable figures." His rules disparaged those who devoted themselves merely to a study of either or both. What he sought was a means to render the world intelligible as a material universe of mere extension, different portions of which are distinguishable only by their different relative motions. His geometrical work was an ingredient in his project of mastering nature. The link between the viewer and the visible world was the visualizable; the means to make the study of the visualizable the means of mastering the visible was the introduction of the terms of arithmetic into geometry.

But let us return to our consideration of the arithmetic operations and our quest for geometric counterparts of them. We were considering multiplication, and the difficulties of introducing into geometry (which deals with heterogeneous magnitudes) that ease of operation which we have when using the term "multiplication" in arithmetic (which deals with homogeneous multitudes). Whereas here with the operation of multiplication, the relevant contrast is between heterogeneity and homogeneity, another contrast—the contrast between the discrete or commensurable, on the one hand, and the continuous or incommensurable, on the other—will appear shortly when we discuss the operation of extracting roots.

We have been trying to get a correspondence between the operations of arithmetic and geometry. Our first difficulty has been in finding out how to get a product "A*B" when both factors A and B are lines. We have seen that if we try this:

(line * line) ——> (number * number) ——> (number) ——> (line),

then the product is nice (the lines behave like numbers, since their product is of the same kind as they), but the process is not useable unless the two factor-lines are commensurable. And if we try this:

(line * line) ——> (a box with the lines as sides situated at right angles),

then, although the process is useable for any two lines even if they are incommensurable, the product is not nice (the lines do not behave like numbers, since their product is not of the same kind as they), and (because of that difference in kind) the process is not useable for more than three lines. What, then, to do?

What we *can* do is this: multiply any two numbers and produce a number;	what we *wish* to do is this: "multiply" any two lines and produce a line.
The operation makes sense for *numbers*;	but how can we define it for *lines*?

We seek something that is
common to both
number relations and relations of *lines*.

What is that?

Ratio.

Numbers and lines
both enter into
ratios!

As Descartes says, in Part II of his *Discourse*: although the several mathematical studies may have had different objects (numbers, or figures, or magnitudes of some other sort), what they have had in common is that every one of them considers relationships of size (ratios), and especially the sameness of those relations (proportions), among its objects. He, in his work, employs lines to bear those relations, he says—lines with which he associates numbers.

Consider this equation: "A * B = P." It seems to have meaning for numbers but not for lines. How can it get meaning for lines? By making use of what has meaning for both numbers and lines—namely, proportionality. Here is how.

We want to get a definition of multiplication for two straight lines (A and B) that will give as their product, when the lines are commensurable, a line (P) the number of whose units is the product of the number of units in the one line and the number of units in the other. But we also want the definition to be useable when the lines are incommensurable, and when there are more than two or three of them.

To get the definition we want, it helps to keep in mind that two statements are said to be "equivalent" when neither one of them can be true unless the other one is true too. Now for numbers, there is a proportion that is equivalent to that equation which we are considering (namely, "P = A* B"), and that proportion for numbers is this: "as a unit is to A, so B is to P." But a proportion like that has meaning for lines too. So, we say that, for lines, the equation "P = A * B" will mean merely that, for the lines, this proportion holds: "as a unit is to A, so B is to P." Figure 1 depicts this process of definition.

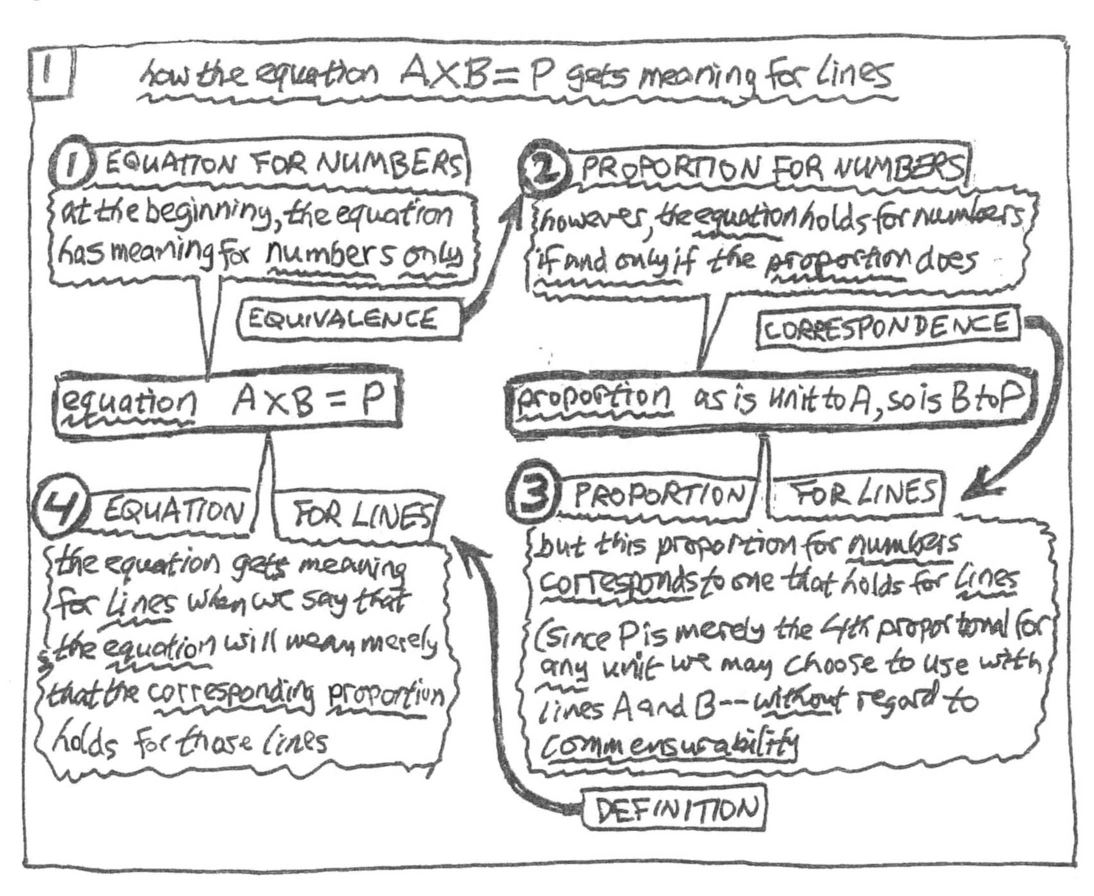

Thus we now have a *definition* for the product of two lines that are to be its factors:

> the product is that fourth proportional which is obtained when
> some unit-line is taken as the first term,
> and the two factor-lines are made the second and third terms.

All that we need now is a *construction* that will give us such a product-line. Descartes simply borrows from the 12th proposition of the Sixth Book of Euclid's *Elements* the procedure for finding a fourth proportional when three straight lines are given. It works for Descartes like this. Given lines A and B, we have defined their product P when a line of any size has been chosen as a unit. Line P is the fourth proportional to these three items: the unit-line, and line B, and line A. Since lines are proportional if they are corresponding sides of similar triangles, we take the three lines that we already have and we use them to make similar triangles that will give us the line we seek. Referring to figure 2, consider this:

> We lay out lines A and B so as to make an angle of any size;
> we choose a line of any size to be the unit-line, and we lay it along B.
> We join the endpoint of the unit-line to the endpoint of line A,
> thus getting a triangle whose sides are the unit-line and line A.
> We prolong line A,
> and from the endpoint of B we draw a line parallel to
> the line that we drew to get the first triangle,
> thus getting another triangle, which
> has B as one side, and is similar to the first triangle.
> Now (by the similarity of the triangles)
> as the unit-line is to corresponding side B,
> so is A to its corresponding side—which therefore
> (by the definition) is the product-line P.

Note that for the given lines A and B the length of the line that is their product will depend on how long a unit-line we may have happened to choose.

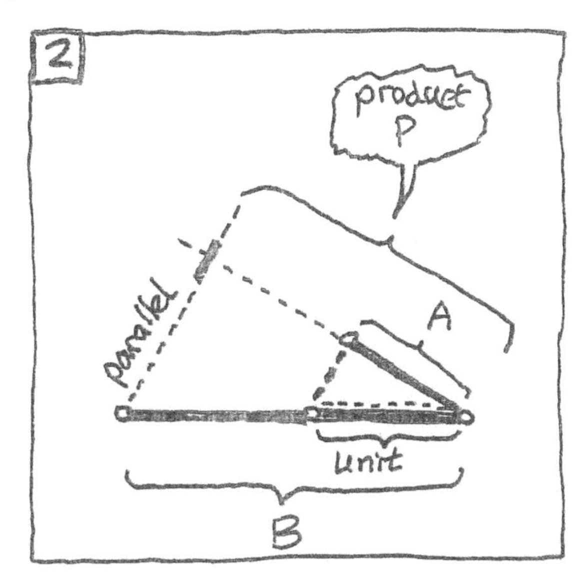

Now, in the *fourth* place (having defined addition, subtraction, and multiplication), that procedure which we have used to define multiplication for two lines will supply the definition of another operation as well: division for two lines.

For numbers, division is the inverse of multiplication (in other words, the equation "A/B = Q" means that "A = B * Q"). And so we say that for lines division will be defined as the inverse of multiplication. In accordance with the way in which the product of two lines has already been defined, therefore, the quotient of two lines must be this: the fourth proportional of the following three items—the divisor, and some unit, and the dividend. Referring to figure 3, consider this:

Whereas ....A*B = P	.........meant that as the unit is to A, so B is to P—	
now ....A/B = Q	........means that as B is to the unit, so A is to Q.	

The construction of the quotient-line uses the same proposition from Euclid as was used for the construction of the product-line; the only change is in the order of the lines, so that the proper lines will be used for corresponding sides of similar triangles.

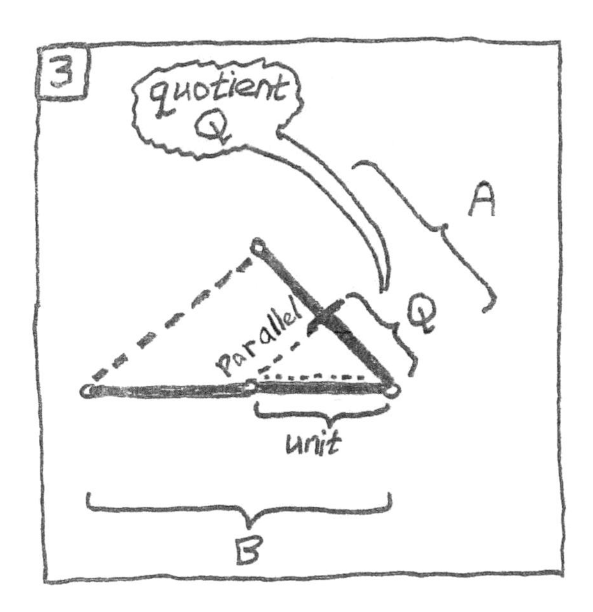

We have now seen how to define lines that are the outcomes of four arithmetic operations on a pair of lines. Given two lines, we have constructions for getting their sum, their difference, their product and their quotient.

Descartes does not stop with that, but he does not go on in quite the same way. He has said that there are "four or five" operations of arithmetic. Why the "or"? Extraction of roots, he has said, can be taken for a species of division. Why, then, not say that there are *only* four operations of arithmetic but that besides them there is a sub-category of the fourth that is of special interest? Or, since exponentiation can be taken for a species of multiplication, why not say that there are four or *six* operations of arithmetic?

Let us confine our discussion to the *square* root for now (about other roots, Descartes says, he will speak later on). Let us ask: is the extraction of square roots primarily an operation on *lines* that is now to be defined for numbers, or is it primarily an operation on *numbers* that is now to be defined for lines?

Having raised that question about the *definition*, let us put it aside for a moment, while we consider Descartes' *construction* for extracting a square root. Given a line A, how are we to find another line R that is its square root? What we need is a line that, once a line of any size has been chosen as a unit, will be the mean proportional between the line chosen as unit and the given line A. In other words, the following proportion holds of line R: as the unit-line is to R, so R is to the given line A. For this, Descartes simply borrows, from the 13th proposition of the Sixth Book of Euclid's *Elements,* the procedure for finding a mean proportional when two straight lines are given.

That is a proposition on which we have much relied in our study of the conic sections. It gives us the equality of a certain square (the one that arises from a semi-chord perpendicular to the diameter of a circle) and a certain rectangle (the one that is contained by the segments into which that semi-chord splits the diameter).

Descartes' construction of the square root works as follows (for which, see figure 4):

> Prolong A by the length of the unit,
> and use that composite line as the diameter of a semicircle.
> At the common endpoint of line A and the unit-line,
> erect a perpendicular and extend it until it meets the circumference.
> That perpendicular semi-chord will be
> the mean proportional between the segments into which it splits the diameter.
> Because of the proportion (namely,
> as the unit-line is to line R, so line R is to line A),
> it follows from the definition of multiplication for lines
> that A is the product of multiplying R by itself.
> That is to say: A = R*R.

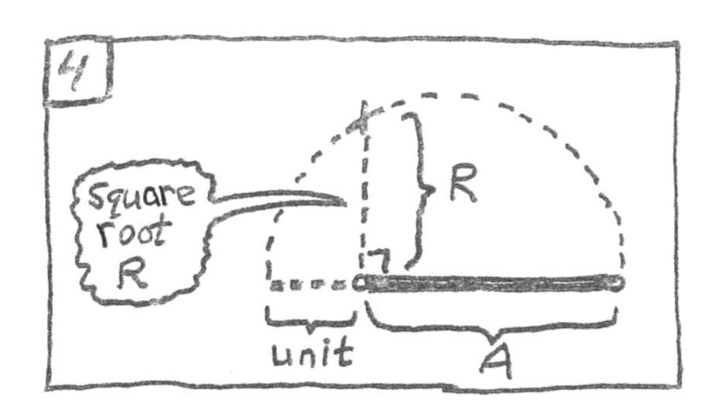

Now here is how Descartes says it:

> [5] Or else, if [what is] needed is extracting the square root of GH:
> I add it [GH] onto straight line FG, which is the unit,
> and, dividing FH into two equal parts at point K,
> I draw from center K the circle FIH,
> then, raising from point G, at right angles upon FH, a straight line as far as I,
> it is GI [that is] the root sought.
>
> I say nothing here about the cube root, nor about other [roots],
> because I shall speak of it more conveniently hereafter.

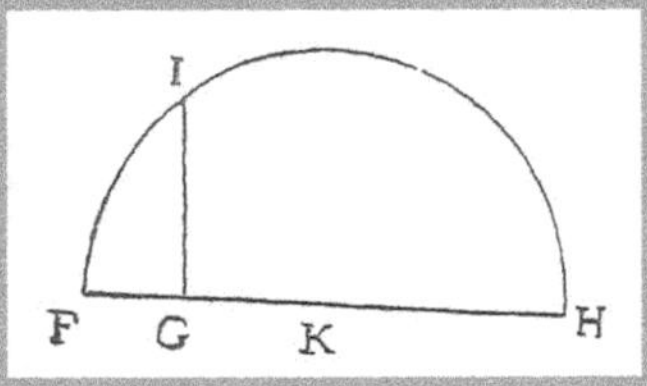

In a sense, "root" is a *numerical* term: "R is the *n*-th root of P" means, when R and *n* are numbers, "R*R*R*R* . . . (with R being taken *n* times) = P." But *not every number* has a square root, or a cube root, or some *n*-th root, and the difficulty cannot be handled as division was handled—namely, by fractions. ("Fractiones" is Latin for what in French means "broken" numbers.) Statements about *fractions* can be translated *directly* into statements about *ratios of numbers*, but irrationals cannot be gotten by "breaking" the unit into smaller units. (As we saw earlier, in Part Two of this guidebook, considering the Euclidean foundation of our study, the definition of ratio for incommensurable magnitudes depends upon the fact that such a ratio uniquely separates all ratios of commensurable magnitudes—every one of these ratios being either greater than it or less than it.) That is to say: the difficulty is not merely that numbers are discrete, since the difficulty with discreteness can be handled merely by "breaking" the unit—that is, by taking a new unit that is smaller. The difficulty, rather, is incommensurability.

But if we accept multiplication of lines, and we choose a line as unit, then every *line* will have an *n*-th root. That is, we can say that R is the *n*-th root of P if what we mean is this: when R is taken as a factor *n* times, the product of the Rs is P.

The extraction of the root of a line, like the multiplication of lines, requires that we choose a unit-line. Unlike with the multiplication of lines, however, the difficulty with the extraction of roots is not in making the operation universally applicable to lines, but in making it universally applicable to numbers. And that is so even though this process (root-extraction) which is universally applicable to lines is derived from a process (multiplication) which originally was universally applicable to numbers but not to lines.

Extraction of a root is strange. In a way, the root is universally defined first for lines (which by their nature do not have any unit), and only afterwards for numbers (which by their nature do have a unit); but in a way, that is not so, for its size in Descartes' construction is not universal, for it depends on the size of the line chosen as the unit.

For considering roots other than the square root, the more convenient place will be the beginning of the Third Book, where Descartes refers to the so-called proportional compass, a device for root-extraction that he will introduce at the beginning of the Second Book.

Now that we have seen how Descartes, being free from the scruple of the ancients, introduces the terms of arithmetic into geometry, let us consider what he does next. We shall thus be considering what it was, that freedom from the scruple freed him *for*. From considering how Descartes introduces the terms of arithmetic into geometry, we now turn to considering why he wants to do it.

Why, then, does Descartes want to introduce the terms of arithmetic into geometry? Because by making it possible to compare quantities freely, he makes it possible to turn geometrical problem-solving into equation-solving. The homogenizing device that makes it all possible is the replacement of multitudes and magnitudes of all sorts by quantities that are all represented by straight lines. Often the lines do not even need to be drawn. The lines, that is to say, are symbols. When convenient, a mere letter will do as a symbol. But keep a list, Descartes will say.

Also, he will introduce various simplifications in notation in order to facilitate the labor that is involved in solving problems.

Descartes tells us how to solve a problem: suppose the problem to have been solved; run through the difficulty, treating unknowns no differently from knowns, according to the order that shows most naturally how they depend upon one another, until you get an equation that relates the unknown to the known. The question is determined if you can get as many equations as there are unknown lines; that is to say (speaking as we would nowadays): if you solve that many equations simultaneously, the unknown quantity will eventually be found in a single equation that equates that unknown quantity with the outcome of operations involving no other quantities but ones that are known.

The method of solution thus proceeds in the ancient mode of geometrical analysis. It employs a novelty, however: equations—equations relating quantities by arithmetical operations, even though the quantities are not so much arithmetical as they are linear. (Again, see the second part of Descartes' *Discourse*.) The mode of proceeding is a kind of calculation.

For a relatively simple example that shows what Descartes means, consider figure 5.

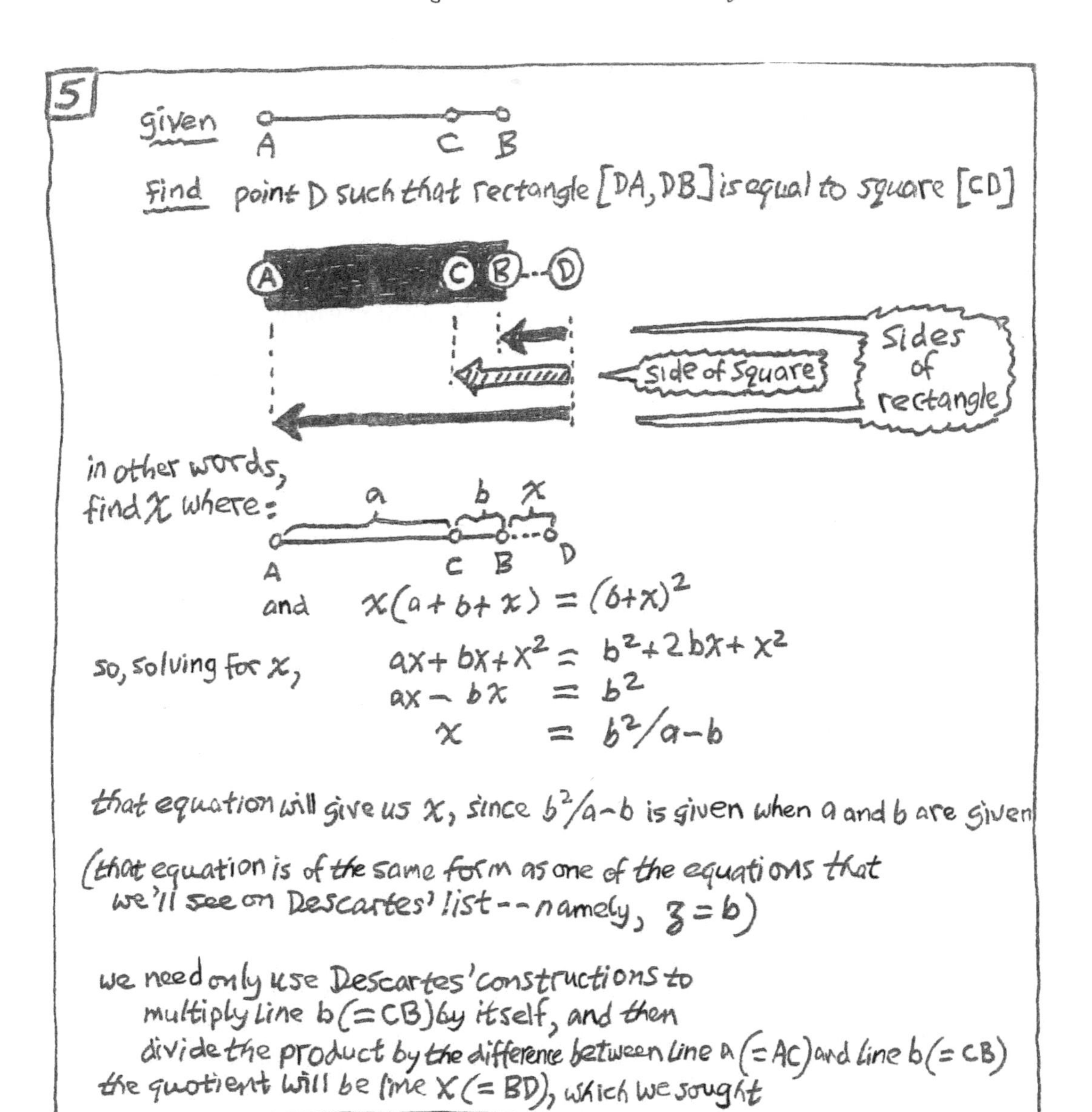
5
given
A
C B
find point D such that rectangle [DA, DB] is equal to square [CD]
A
C
B
D
side of square
sides of rectangle
in other words, find x where:
a
b
x
A
C
B
D
and $x(a+b+x) = (b+x)^2$
so, solving for x,
$ax + bx + x^2 = b^2 + 2bx + x^2$
$ax - bx = b^2$
$x = b^2/a-b$
that equation will give us x, since $b^2/a-b$ is given when a and b are given
(that equation is of the same form as one of the equations that we'll see on Descartes' list -- namely, $z = b$)
we need only use Descartes' constructions to
multiply line b (= CB) by itself, and then
divide the product by the difference between line a (= AC) and line b (= CB)
the quotient will be line x (= BD), which we sought

Descartes will go on to say that if the problem is constructible by straight lines, by circles, by conic sections, or by curves no more than a degree or two more composite than conic sections, then all the unknowns are reducible to a single quantity by combining the equations.

Descartes says that he won't go into more detail. The principal thing to be drawn from this science is not any particular results that can be laid before the reader's eye. Rather, it is the cultivation of the mind in exercising it in this science. More than once he makes this point that the science he is teaching is a science of empowerment. His book is a methodology, not a thesaurus of theorems.

So instead of presenting a heap of particular results, Descartes says he will only give some words of advice: divide whenever possible in solving these equations, so that you will reach the simplest terms to which the problem can be reduced. In other words: don't do any more labor than you must. The labor-saving is made possible by the homogenizing breakthrough that carries Descartes beyond Viète. Dimensionality replaces rank. Difference in degree replaces difference in kind.

If a problem can be solved by ordinary geometry (that is: by lines that are straight or circular, traced on a plane surface), then when the last equation has been entirely solved, there will remain at most only

$z^2 = az \pm b^2$, where z is an unknown quantity and a and b are known quantities.

In solving a geometrical problem, you would like eventually to get an equation that contains a single unknown and no other quantities but known ones. Then you have to construct the unknown line by combining in some way the simple constructions that Descartes has presented at the beginning of the book. If all you have is that desired equation, then you have not yet solved the geometrical problem: you need to translate the algebraic relation into a construction that will give you the required geometrical entity. Descartes will give some examples of composite constructions from which you will get the unknown line once you have obtained an equation of the sort desired.

He then will go on to spend most of the First Book getting set up to solve the four-line locus problem that Pappus presents. He will obtain algebraic expressions of the distances to the fixed lines from a point on the curve that satisfies what the locus problem calls for. The Second Book will open, not with the solution to the problem, but with a discussion of how to distinguish geometrical curves comprehensively, and then it will solve the locus problem, showing that the locus is a conic section of one sort or another, depending on the form of the algebraic expressions for the geometrical relations involved. There is a tie between the conic sections and certain equations of what we nowadays call the second degree. (Note, however, that in the terminology of Descartes, it is not equations but curves that have "degrees"; and the terms in equations have "dimensions" not "degrees.") Descartes will then go on, in the rest of the Second Book, and in the Third, to show the power of his method in geometry.

But all that is some way ahead of us. Let us now get back to following Descartes step-by-step. We have seen him provide geometrical counterparts to the arithmetical operations. He has defined the operations on lines. Let us see what he goes on to say right after doing so.

[6] But often one has not any need to trace these lines on paper thus,
and it suffices to designate them by some letters,
each [line] by one [letter] alone.
[The marginal title reads: use of *chiffres* in geometry.]

As, for adding the line BD to GH:
I name the one a and the other b, and write $a + b$;
And $a—b$, for subtracting b from a ;
And ab, for multiplying them, the one by the other;
And a/b , for dividing a by b ;
And aa or a^2, for multiplying a by its own self;
And a^3, for multiplying it yet one more time by a,
and so on to infinity;
And Ö($a^2 + b^2$), for extracting the square root of $a^2 + b^2$;
And Ö C.($a^3—b^3 + abb$), for extracting the cube root of $a^3—b^3 + abb$,
and so on of others.

[7] Where[in] it is to be noted (*remarquer*) that
by a^2 or b^3 or the like (*semblables*)
I ordinarily conceive naught but lines, all simple—
although, for names used in Algebra to serve me,
I name them squares or cubes, etc.

[8] It is also to be noted
that all the parts of one line itself
should ordinarily be expressed by as many dimensions—
the one [part by] as [many as] the other—
when the unit is not determined in the problem (*question*),
as here a^3 contains as many of them [dimensions] as [does] abb or b^3,
of which is composed the line that I have named Ö C.($a^3—b^3 + abb$);
but it is not the same when the unit is determined,
because it [the unit] can be understood [to be] everywhere
throughout all [of it] (*par tout*)
whether there are too many or too few dimensions: as,
if there is need of extracting the cube root of $aabb—b$,
there is need of thinking that the quantity $aabb$ is divided one time by the unit,
and that the other quantity b is multiplied two times by the same [unit].

Descartes wants to make geometric use of algebraic "*chiffres.*" (This term, which is introduced in the headings of the printed pages of the book, is not used here in the text of *The Geometry* itself, though it is used in the text of the *Discourse* that precedes the essays in the volume.) Descartes wants to calculate with letters. With letters, the steps in analysis will be expressed more compactly; by calculating, the analyst turns the steps into equation-manipulations whose sequence is reversible. Descartes retains some of the algebraic nomenclature, but he greatly simplifies the notation. He introduces what came to be called "exponents." His simplification makes dimensionality a trivial affair. The unit chosen at the discretion of the investigator enables him to homogenize "quantity." The operations of arithmetic, which work upon numbers, can also work upon the "lines, all simple" that represent one or another "quantity." Just as operations upon numbers yield numbers, so operations upon quantity-lines yield quantity-lines. Whatever difficulty might be felt about, for example, subtracting a quantity-line from the product of multiplying four quantity-lines together, the difficulty can be overcome by doing the following before subtracting:

take the one quantity-line and multiply it by the unit, twice,
and
take the product of the four quantity-lines and divide it by the unit, once;

thus you will obtain, for the subtraction, two terms, both of which are
products of multiplying three quantity-lines together.

Descartes can subtract his one-dimensional quantity-line, *b* , from his four-dimensional quantity-line, *aabb* , as follows:

he conceives of the four-dimensional quantity-line *aabb*
as being divided by the unit-line once,
which leaves its size the same but reduces its dimensions to three;

and he conceives of the one-dimensional quantity-line *b*
as being multiplied by the unit-line,
which leaves its size the same also but increases its dimensions to two,
and then, multiplication by the unit-line a second time will
leave the size the same but now increase the dimensions to three;

thus, two straight lines that are equal, respectively, to *b* and to *aabb*
will both have three dimensions.

The number of "dimensions" here does not refer to what distinguishes straight lines from planes, or either of them from solids; rather, it refers to what distinguishes one straight line from another, depending on how the straight line is conceived of as being generated. If *a* and *b* are straight lines that each have one dimension, then *ab* is a straight line having two dimensions, and *aabb* is a straight line having four dimensions. In the common geometry, by contrast, one magnitude cannot be added to another if the two magnitudes are of different kinds: a straight line (which has one dimension in the non-Cartesian sense) can be added to another straight line; and a plane figure (which has two dimensions) can be added to another plane figure; but a straight line cannot be added to a plane figure.

In a way, it therefore does not much matter how many dimensions a term in an equation has. The choice of a unit that makes possible the employment of equations in place of proportions by importing arithmetical operations without scruple into geometry makes it possible to raise or lower the number of dimensions in a term without altering the value of the quantity. One simply multiplies or divides a line by a unit line. For example, consider the terms *ab* and x^2 and x^3. The term *ab* has the same number of dimensions as does the term x^2, but it has one fewer dimensions than does the term x^3. Unless x is equal to the unit, x^3 is not equal to x^2; but it (x^3) is equal to x^3/unit, which has the same number of dimensions as does x^2 and as does *ab*. (As we shall see, however, the number of dimensions of the unknown quantity *will* turn out to matter in the following way: The higher the number of dimensions of the highest power of the unknown quantity that appears in some equation, the more composite is its curve—since higher also is the number of mean proportionals between the unit and that power of the unknown quantity. For example, x is the first of *four* mean proportionals between the unit and x^5, but is the first of only three of them between the unit and x^4.)

We come now to the central paragraph of what is—when we count by paragraphs—the first half of the First Book of the *Geometry*. This First Book contains 34 paragraphs, the first 17 of which introduce the author's new "method." This introduction will culminate, at the end of the 17th paragraph, in a critique of the ancients: the prolixity of the ancient geometers (whose books were so big and so many), and the very order in which they presented what they had found, suffice to let us know that the ancients did not have "the true method for finding them all." After that critique, the remaining 17 paragraphs of this First Book will take as their point of departure the as-yet unsolved locus problem of Pappus, and will begin to show how far beyond the ancients one can go by doing geometry in the Cartesian way. Here, then, is the central and by far the longest paragraph in the author's introduction of his innovative geometrical method:

[9] What is left to say is this (*Au reste*):
to the end of not failing to remember the names of these lines,
there is need always to make a separate register,
in proportion (*à mesure*) as one assigns them or changes them—
writing, for example:
AB æ 1, that is to say, AB is equal to 1
GH æ *a*
BD æ *b*, etc.
So, wishing to solve (*resoudre*) some problem,
one should at the outset (*d'abord*) consider it as already done,
and give names to all the lines which seem necessary for constructing it—
to those which are unknown as well as to the others.

Then, without considering any difference between
these lines known and unknown,
one should run through the difficulty,
according to the order which shows, most naturally of all,
how (*en quelle sorte*) they mutually depend, the ones on the others—
until one has found means of
expressing one and the same quantity in two fashions, which is called an Equation,
since the terms of the one of these two fashions are equal to those of the other.

And one should find as many of such Equations
as one has supposed of lines which were unknown.
Or else if there are not so many [equations] found
[as there are unknown lines supposed],
notwithstanding that one has omitted nothing of
what is desired in the problem (*question*),
[then] this witnesses that it [the problem] is not entirely determined.
And then one can at discretion take known lines for
all the unknown [lines] to which no Equation corresponds.

After this, if there yet remain several [unknown lines],
There is need to make use, in order, of each of the Equations which also remain—
be it in considering it all alone, be it in comparing it with the others—
for explicating each of these unknown lines, and
making [things] so, in unmixing them, that
naught but one alone of them remains equal to some other which is known,
or else the square,
or the cube,
or the square of the square,
or the supersolid,
or the square of the cube, etc.
be equal to what is produced by the addition or subtraction of

two or several other quantities, of which the one be known and the others be
composed of several mean proportionals between the unit and
this square, or cube, or square of the square, etc.
multiplied by other known [lines].

Which I write thus (*en cete sorte*):

z æ b ;
or
z^2 æ $-az + bb$;
or
z^3 æ $+az^2 + bbz - c^3$;
or
z^4 æ $az^3 - c^3z + d^4$;
etc.

That is to say:

z, which I take for the unknown quantity, is equal to b ;
or
the square of z is equal to the square of b, less a multiplied by z ;
or
the cube of z is equal to a multiplied by the square of z,
plus the square of b multiplied by z,
less the cube of c ;
and so the others.

The straight lines that are named in the ordinary geometry by the two upper-case letters that name their endpoints have been given new names that are more compact, consisting of a single lower-case letter. The points they may stretch between in some configuration do not matter so much as the quantities they represent in some equation. The author here introduces the practice, retained in the algebra that is familiar to us, of assigning to quantities that are known the letters at the beginning of the alphabet, and assigning to those that are unknown the letters at the end of it. These algebraic names replace the more cumbersome names during the solving of the problem. During the solving, the unknowns are employed in calculations no differently than are the knowns.

[10] And one always can so reduce all the unknown quantities to one alone—
when the Problem can be constructed through circles and straight lines,
or also [when it is constructible] by conic sections,
or even by some other line that is
naught but one or two degrees more complex (*composée*).

But I am not stopping to explain this in more detail,
because I would deprive you of
the pleasure of learning it by your own self—
and the utility of cultivating your mind (*esprit*) in exercising yourself in it,
which is, in my opinion (*a mon avis*),
the principal [utility] which one can draw from this science.

Also because I note nothing in it so difficult that
those who will be a little versed in the common Geometry and in Algebra,
and who will be on the alert (*prendront garde a*) about all that is in this treatise,
will not be able to find [it].

[11] That is why I shall be content here to notify you that,
provided that in unmixing these Equations
one does not fail to make use of (*a se servir de*)
all the divisions that will be possible,
one will infallibly have the simplest terms to which
the problem (*question*) can be reduced.

As was mentioned earlier Descartes speaks throughout of how many "degrees" a curve has. He does not speak, as we do nowadays, of the degree of a term in an equation. What a term in an equation does have, for him, is not a degree but "dimensions." What a curve's degrees are, we have not yet been told. Later, we shall be told that conic sections are curves of the first degree. We shall also be told what would make a curve of the second degree, and of the third, fourth, fifth, and so on. Curves of a higher number degree are said to be "more composite"; but what the put-together ingredients are, we have not yet been told.

The first 11 paragraphs of this 17-paragraph first half of the First Book have shown a new way to go about resolving any problem in geometry into the solving of an equation that is as simple as can be. In the central (6th) paragraph of these 11, the lines that represented quantities were replaced by letters, thus allowing quantitative relationships of the ordinary geometry to be thoroughly enciphered in an equation. Now the remaining 6 paragraphs of this First Part will show how the equation will yield the construction sought in any problem of the ordinary geometry.

> [12] And [I shall be content to give you notice also] that
> if it can be solved (*resolue*) by the ordinary Geometry—
> that is to say, [by] making use of (*se servant*) naught but
> lines straight and circular traced on a plane surface—
> [then] when the last Equation will have been entirely unmixed,
> there will remain at the very most (*tout au plus*) naught but
> an unknown square equal to
> what is produced by the addition or subtraction of
> its root multiplied by some known quantity
> and some other quantity [that is] also known.

The last equation, that is to say, when entirely unmixed will look like this:

x^2 æ $ax + c$;
or
x^2 æ $c - ax$;
or
x^2 æ $ax - c$.

The next several paragraphs show how to obtain, from these three unmixed equations, a construction for the line represented by the unknown quantity x. Each of the next three paragraphs (13–15) shows a construction for one of the three possible equations, and the paragraph immediately following them (16) shows the case when the last equation is such as to render construction impossible.

[13] And then this root, or unknown line, is found easily.

For if I have, for example, z^2 æ $az + bb$:

I make the right-angled triangle NLM,
of which the side LM is equal to b, the square root of the known quantity bb,
and the other [side] LN is $(1/2)a$, the half of the other known quantity, which was multiplied by z, which I suppose to be the unknown line;

then, prolonging MN, the "base" [now called the "hypotenuse"] of this triangle, to O, in such a way that (*en sorte que*) NO is equal to NL—

[then] the whole OM is z, the line sought.
And it is expressed in this way: z æ $(1/2)a + \sqrt{[(1/4)aa + bb]}$.

When z^2 is equal to $az + b^2$, then the equation's "root" (z itself) will be equal to:

$$(1/2)a + \sqrt{[(1/4)a^2 + b^2]}.$$

That result can be obtained by employing the algebraic technique of "completing the square," as in figure 6. So "the unknown line" (z) can be obtained, since line a and line b are given, by employing the constructions that have already been presented—from which we get, step-by-step, the lines

a^2 and $(1/4)a^2$ and b^2 and $[(1/4)a^2 + b^2]$ and
$\sqrt{[(1/4)a^2 + b^2]}$ and $(1/2)a$ and $\{(1/2)a + \sqrt{[(1/4)a^2 + b^2]}\}$, which last is line z itself.

Descartes takes a shortcut (see figure 7):

he obtains the line $\sqrt{[(1/4)a^2 + b^2]}$ in only one step, and then
in one more step, he obtains the line equal to the sum of it and the line $(1/2)a$.

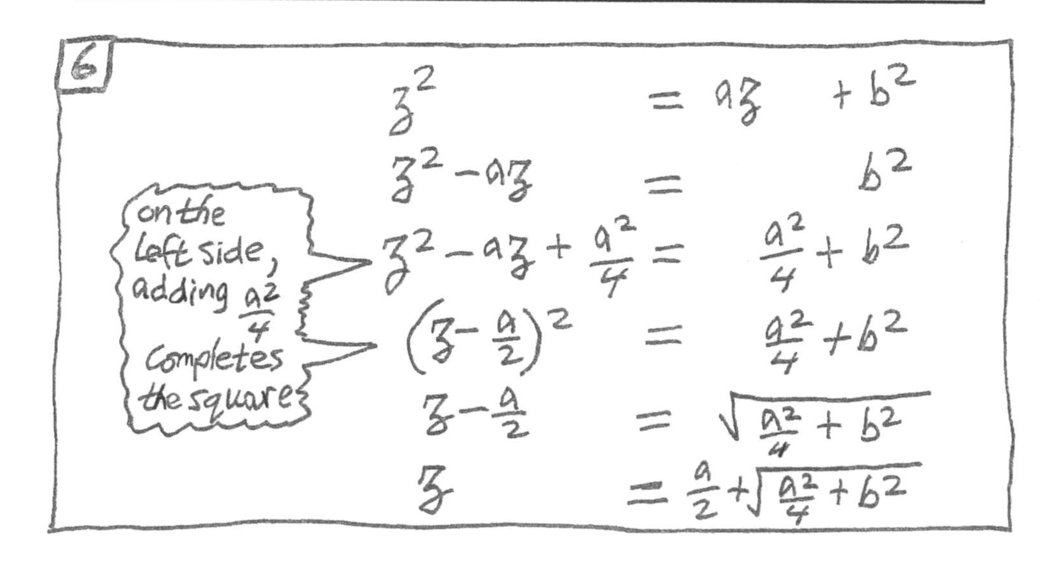

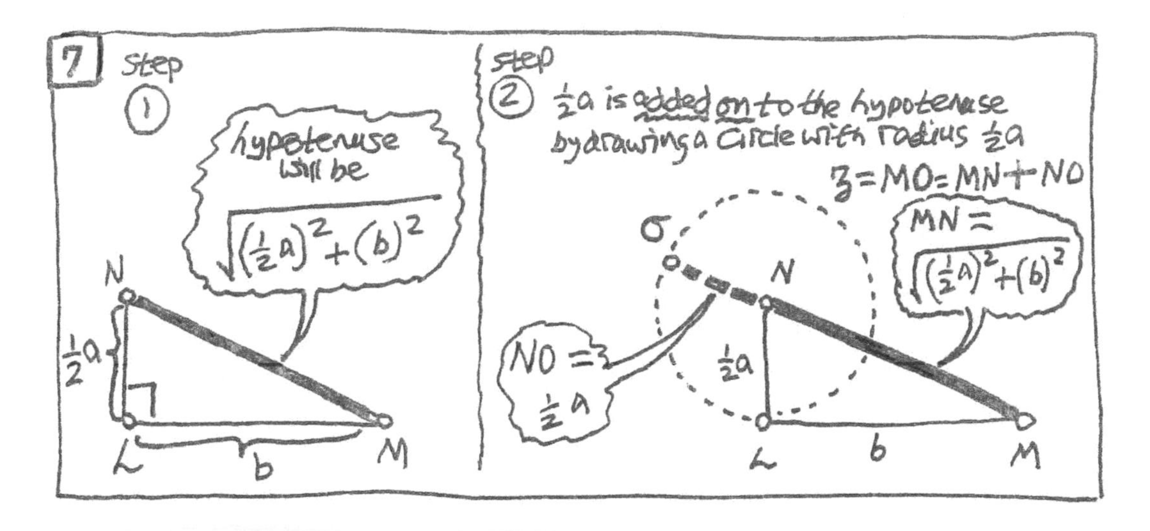

[14] But if I have yy æ $-ay + bb$
and y be the quantity that there is need to find:

I make the same right-angled triangle NLM;
and from its "base" [that is, its hypotenuse] MN, I take away NP equal to NL—

and [then] the remainder PM is y, the root sought.
In [such] fashion that (*De façon que*) I have: ...... y ae $-(1/2)a + \sqrt{[(1/4)aa + bb]}$

And it is all the same if I had x^4 æ $-ax^2 + b^2$.
PM would be x^2, and I would have: .... x ae $\sqrt{\{-(1/2)a + \sqrt{[(1/4)aa + bb]}\}}$
and so others.

This situation resembles the previous one, except that now the line $(1/2)a$ is taken away from the hypotenuse rather than added on to it. See figure 8. In neither case does Descartes take what we would call the "negative" root: in the problems that he is solving, there is no straight line that would correspond to the negative root.

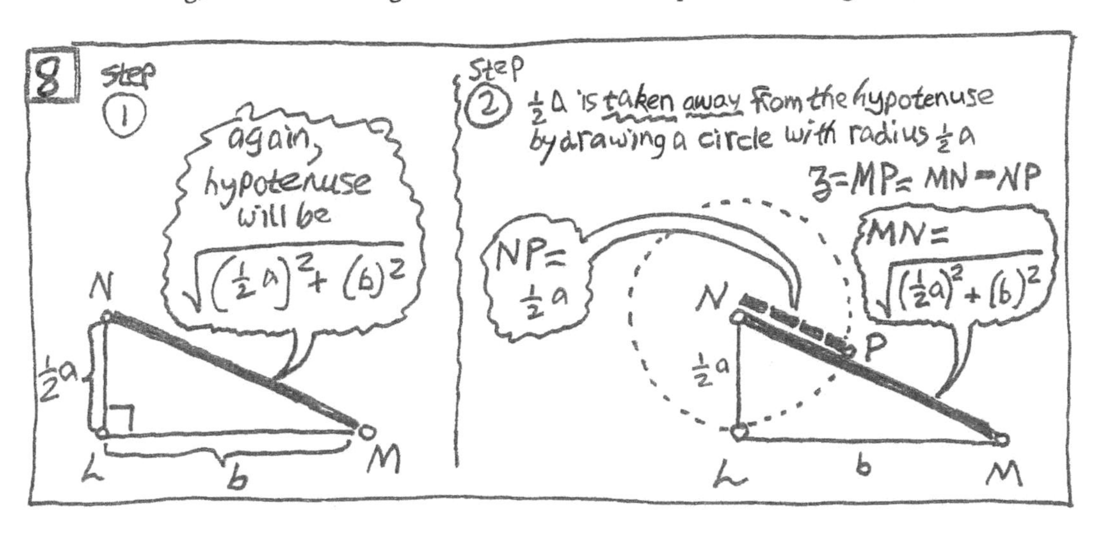

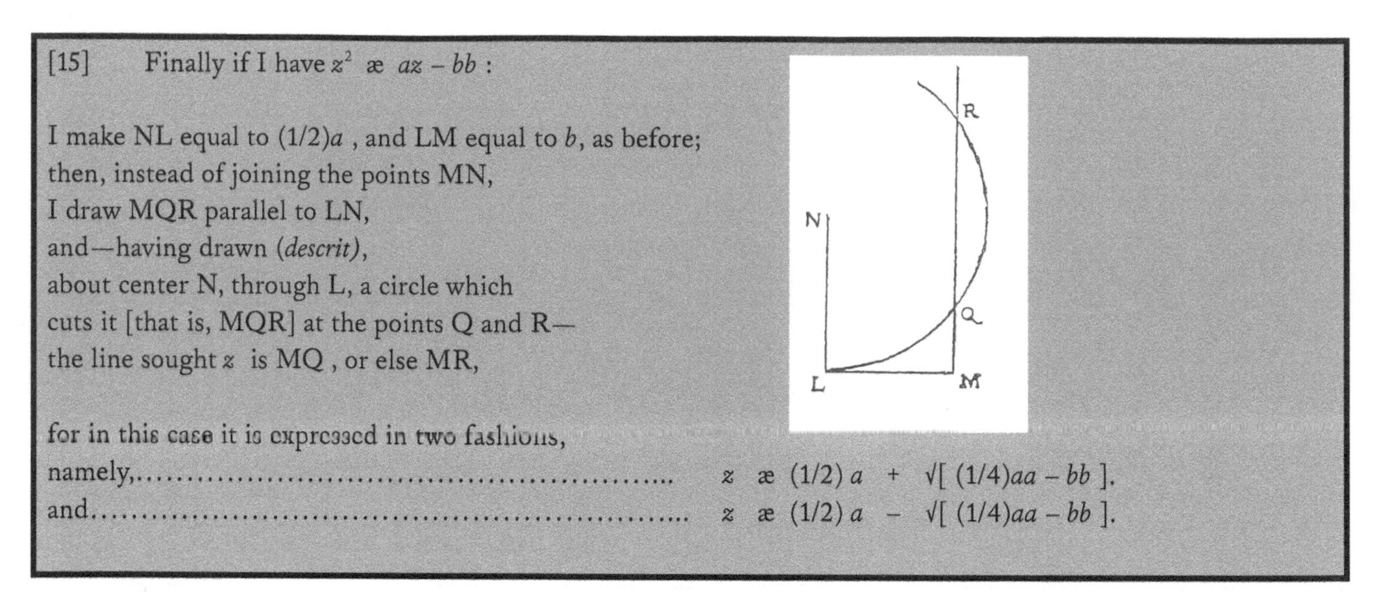

[15] Finally if I have z^2 æ $az - bb$:

I make NL equal to (1/2)a , and LM equal to b, as before;
then, instead of joining the points MN,
I draw MQR parallel to LN,
and—having drawn (*descrit*),
about center N, through L, a circle which
cuts it [that is, MQR] at the points Q and R—
the line sought z is MQ , or else MR,

for in this case it is expressed in two fashions,
namely,.. z æ (1/2) a + √[(1/4)aa – bb].
and.. z æ (1/2) a – √[(1/4)aa – bb].

Both values of z are given by Descartes because both roots are positive. That is so for the second root because [(1/4)a^2 – b^2] must be less than (1/4)a^2. When what z is equal to is

$$(1/2)a + \sqrt{[(1/4)aa - bb]},$$

then the unknown line will be MR; and when what z is equal to is

$$(1/2)a - \sqrt{[(1/4)aa - bb]},$$

then the unknown line will be MQ. Again, this is a shortcut construction. Why it works, is shown in figure 9.

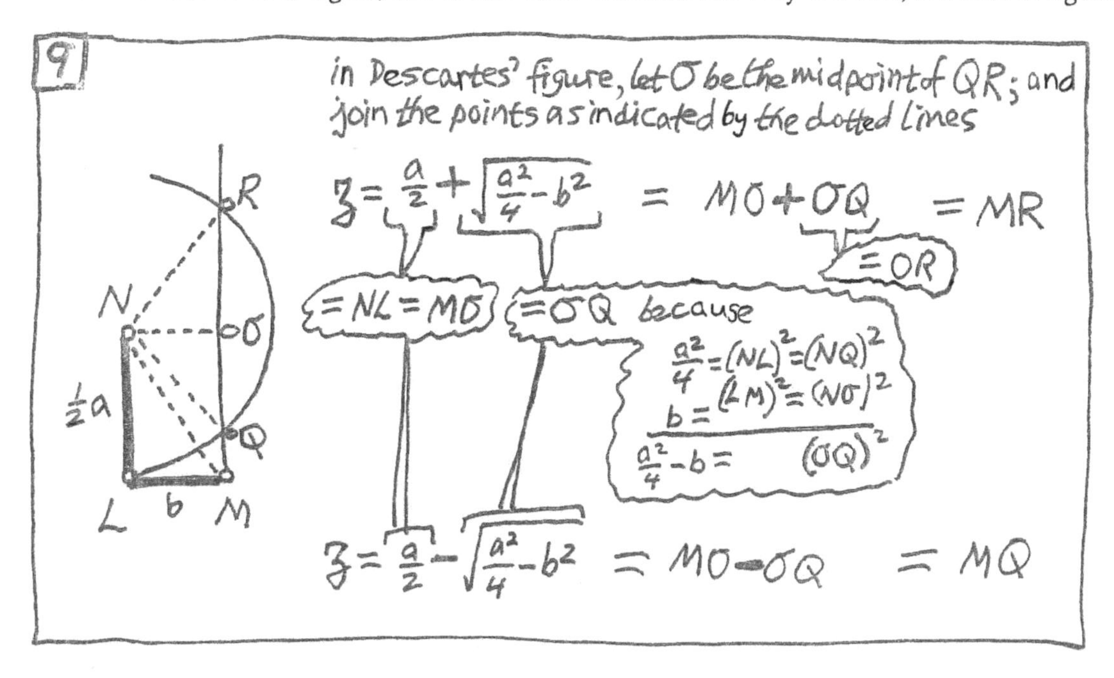

> [16] And if the circle
> which, having its center at point N, passes through the point L
> does not cut or touch the straight line MQR,
> [then] there is not any root in the Equation,
> so that (*de façon que*) one can give assurance that
> the construction of the proposed problem is impossible.

The circle will be tangent to the line MQR if (1/2) a is equal to b ; and the roots will then be equal. The circle will not cut or even touch the line MQR if (1/2) a is less than b ; the roots would then be square roots of negatives. In the problems that Descartes is solving, there is no straight line that would correspond to such a root.

There is no mention at all of an equation like this: x^2 æ $-ax-c$. Any root of this equation would have to be negative, but in the problems that Descartes is solving, there is no straight line that would correspond to such a root.

Four figures have now been explicated. The first two were part of the introduction of arithmetical terms into geometry: they illustrated the operations of multiplying (or dividing) two lines and of extracting the square root of a line. The last two of the figures gave constructions for finding an unknown line that is the root of certain equations involving the square of the unknown line.

> [17] What is left to say is this (*Au reste*):
> these same roots can be found by an infinity of other means,
> and I only want to put these here, as very simple,
> to the end of making you see that
> one can construct all the Problems of the ordinary Geometry
> without doing another thing than the little which is comprised in
> the four figures that I have explicated.
>
> Which is what I do not believe the ancients noted,
> for otherwise they would not have taken the trouble (*pris la peine*) to
> write so many big books of them,
> where the order alone of their propositions makes us know that
> they did not have the true method for finding them all,
> but only gathered together those which they had come upon (*rencontrées*).

The author of *The Geometry*, by contrast with the ancient geometers, does not proceed by the troublesome gathering together of the outcomes of lucky encounters: he thinks he has the true method for finding all that can be found in geometry. Having introduced that method in these first 17 paragraphs of the First Book, he will now, in the remaining 17 paragraphs, begin to show how far beyond the ancients one can go by doing geometry in the methodical, Cartesian fashion. His point of departure is the locus problem that according to Pappus had not been entirely solved by Euclid, or by Apollonius, or by any other geometer (and that also had not been solved by anyone else in all of the many centuries that had passed since Pappus passed away). If the mere order of the propositions presented by the prolix ancient geometers does not suffice to make one see the ancients' lack of method, Descartes expects that one will be able to see it very clearly by examining Pappus's problem under the guidance of Descartes.

CHAPTER 3. SURPASSING THE ANCIENTS: THE SOLUTION OF PAPPUS'S PROBLEM

[18] And one can very clearly see it also from what
Pappus put at the beginning of his Seventh Book,
where, after having stopped several times to enumerate
all that had been written in Geometry by those who preceded him,
he speaks finally of a problem (*question*) which he says that
neither Euclid, nor Apollonius, nor any other
had known how to solve entirely (*sceu entierement resoudre*),
and here are his words:

[marginal title reads: I cite the Latin version rather than the Greek text,
to the end that everyone may understand it more easily]

[19] *Which locus as to three and four lines, however,*
he (Apollonius) in the Third Book says was not completed by Euclid;
nor was he himself able to complete it; nor was anyone else,
but neither [was anyone able] to add anything, [even] a very little,
to those [things] which Euclid had written so much by means of those conics,
*[and] which had been shown forth (*praemonstrata*) by Euclid's times, etc.*

[20] And a little later, he thus explains what this problem (*question*) is:

[21] *And the locus as to three and four lines—*
in connection with which he (Apollonius) throws himself around so grandly
*and shows not a bit of gratitude (*ostentat nulla habita gratia*) to him who wrote earlier—*
*is of this sort (*hujusmodi*):*

If, three lines being given in position,
from one and the same point straight lines are drawn to the three lines,
[meeting these lines] in given angles,
*and there be given the ratio (*proportio*)*
of the rectangle contained by two drawn [lines] to the square of the remaining [line]—
[then] the point falls on a solid locus given in position,
that is, on one of the three conic sections.

And if the lines are drawn to four straight lines given in position,
[meeting those lines] in given angles,
*and there be given the ratio (*proportio*)*
of the rectangle contained by two drawn [lines]
to the [rectangle] contained by the remaining [lines]—
likewise, the given point falls upon a conic section [given] in position.

Indeed, if [drawn] to merely two lines, well then, a plane locus is shown.
But if to more than four, [then] the point falls on loci [that are]
not yet known but merely called lines;
what sort they are, moreover, or what property they have, is not established.
One of them, not the first one but the one that is seen most manifestly,
they have put together, showing it to be useful.

On reading that, one might well ask: what is the one that has been "put together"—and how is it "not first" but "seen most manifestly"—and how is it "shown to be useful"? The difficulty is dispelled when one checks, against Pappus's Greek, the Latin translation of Pappus that Descartes was using here. The Latin translation is misleading, for what the Greek of Pappus says is that they did *not* do the job—the job of synthesizing any of these lines, having shown it to serve for these loci—not even for that one which would seem to be the first and most manifest of them.

The quotation from Pappus continues as follows—

Their propositions, however, are these:

[22] *If, from any point, straight lines are drawn*
to five lines given in position, [meeting these lines] in given angles,
and there be given the ratio (proportio)
of the rectangular parallelepiped solid which is contained by
three [of the] drawn lines
to the rectangular parallelepiped solid which is contained by
the remaining two [drawn lines] and any given line whatever—
[then] the point falls on a line given in position.

If, however, [the lines are drawn] to six [lines given in position],
and there be given the ratio (proportio)
of the solid contained by three [of the] lines [drawn to the lines given in position]
to the solid which is contained by the three remaining [drawn] lines—
[then] again the point falls on a line given in position.

But if to more than six,
until now they have not been able to say
whether there be given a ratio (proportio)
of whatever is contained by four lines
to what is contained by the remaining [lines],
since there is not anything contained by more than three dimensions.

That reference by Pappus to dimensionality, which involves the *kinds* of things that geometers have the ability to *visualize*, provokes Descartes to interrupt with a remark on the blindness and the weakness of the ancients. By his clear insight into relationship, Descartes overcomes the blindness and the weakness of those whose vision is delimited by attending to superficial kinds of things:

[23] Wherein I pray you to note (*remarquer*) in passing that
the scruple which the ancients had (*faisoient*)
about using terms from Arithmetic in Geometry—
which could proceed from naught but that
they did not see clearly enough their relation (*rapport*)—
caused much obscurity and difficulty (*embaras*) in
the fashion in which they [the ancients] explained themselves.
For Pappus follows up in this way (*sorte*):

[24] *Now (*autem*), they go along with those who, a bit before, interpreted*
such things as not signifying a single [thing] in any way comprehensible—
[namely, such things as] what is contained by these [lines].
But it is legitimate
*both to say this by means of compound ratios (*per conjunctas proportiones*)*
*and generally to demonstrate it in the aforesaid ratios (*in dictis proportionibus*),*
and, in these, in this mode:

If, from any point, straight lines are drawn
to straight lines given in position, [meeting these lines] in given angles,
*and there be given a ratio compounded (*proportio conjuncta*) of those which*
one of the drawn [lines] has to one,
and another has to another,
and yet another to yet another,
and the remaining one to a given line, if there be seven,
and the remaining one to the [other] remaining one, if indeed there be eight—
[then] the point will fall on lines given in position.

And likewise however many they be,
[the two sets of them being] unequal or equal in multitude
[that is, whether the number of them be odd or even],
though these correspond to the locus in respect to four lines,
they put nothing, then, in such a way that the line may be denoted, etc.

[25] The problem (*question*), therefore, which had been
begun to be solved (*resoudre*) by Euclid, and followed up by Apollonius,
without having been finished up (*achevée*) by anyone,
was such:

Having three or four or a greater number of straight lines given in position,
one calls for (*demande*),
first, a point from which one can draw as many other straight lines,
one onto each of the given [lines], which make with them given angles,
and [second],
that the rectangles contained within two of these [lines] which
will be thus drawn from one and the same point
have the given ratio (*proportion*)
with the square of the third,
if there are naught but three of them;

or else with the rectangle of the two others,
if there are four of them;

or else, if there are five of them,
that the parallelepiped composed of three
have the given ratio (*proportion*)
with the parallelepiped composed of
the two which remain and one other given line.

Or if there are six of them,
that the parallelepiped composed of three
have the given ratio (*proportion*)
with the parallelepiped [composed] of the three others.

Or if there are seven of them,
that what is produced when one multiplies four of them by one another
have the given ratio (*raison*)
with what is produced by the multiplication of
the three others and yet another line, a given [one].

Or if there are eight of them,
that the product of the multiplication of four
have the given ratio (*proportion*)
with the product of the four others.

And so this problem (*question*) can be extended to every other number of lines.

It is manifest here that Descartes is free of the scruple that the ancients had: when he has four lines from which he wants to produce something that will bear a ratio to something produced by four other lines—that is, when he has too many lines to compose a two-dimensional or a three-dimensional box—he forms a product by just "multiplying" the *lines* together. As arithmetical multiplication operates to get a product from two numbers, or from as many numbers as you please, so Cartesian "multiplication" operates to get a product from two lines, or from as many lines as you please. His paragraph goes on as follows:

Then, because there is always an infinity of
different points which can satisfy what is here called for (*demandé*),
it is also required to know and to trace
the line in which they should all be found,

and Pappus says that
when there are naught but three or four given lines,
it [namely, this line in which all the satisfying points are to be found]
is one of the conic sections,
but he does not undertake to determine it, nor to draw (*descrire*) it—

no more than [he undertakes]
to explicate those [lines] where all these points should be found
when the problem (*question*) is proposed in [reference to]
a greater number of lines.

He only adds that the ancients had imagined [that is, visualized] one of them—
[one] which they showed to be useful,
but which seemed the most manifest and which all the same was not the first.

Which gave me occasion to try (*essayer*)
if through the method of which I make use (*la methode dont je me sers*)
one can go as far as they did.

Pappus's problem gave Descartes occasion to try the method of which he makes use. The thing that is responsible for the prominence of Pappus's problem in Descartes' *Geometry* appears to be merely the chance encounter of an innovative writer with an ancient book that he happened to read. But considering that our innovative writer's 17th paragraph did criticize the ancient writers for their dependence upon lucky encounters, we might ask ourselves whether the appearance is deceptive. Perhaps the particular problem of Pappus provides the most appropriate access to the most general geometrical apprehension of curves that one can get. Perhaps it is our author who is deceptive, delighting in setting us up for instructive astonishment. Reporting that his reading of Pappus on the 3- and 4- and many-line locus problem gave him occasion to make trial whether through his method one can go as far as they did, he says merely "as far as they did" even though we shall see later that he is convinced that he has gone, not merely as far as they did, but much, much farther—indeed, that he has gone, as it were, all the way. Much, to be sure, remains to be explored by his readers and their successors; but he has discovered the territory and he has provided them with a general description of it, along with reliable equipment for discovering further particulars, and also with comprehensive instructions concerning how to set the particular discoveries in order.

[26] And first I knew that,
this problem (*question*) being proposed in naught but three or four or five lines,
one can [in this case] always find the sought points by means of the simple Geometry,
that is to say, not making use (*se servant*) of aught but ruler and compass,
nor doing anything but what has already been said—
except only when there are five lines given, if they are all parallel.

In which case [namely, when the problem is proposed in
five lines if they are all parallel],
as also when the problem (*question*) is proposed in six, or 7, or 8, or 9 lines,
one can always find the sought points by means of the Geometry of solids,
that is to say, by employing in it some of the three conic sections—
except only when there are nine lines given, if they are all parallel.

In which case again (*derechef*) [namely, when the problem is proposed in
nine lines if they are all parallel],
and also (*encore*) in 10, 11, 12, or 13 lines,
one can find the sought points by means of a curved line which should be
one degree more composite (*composée*) than the conic sections—
except in thirteen, if they are all parallel,

in which case [namely, when the problem is proposed in
thirteen lines if they are all parallel],
and in fourteen, 15, 16, and 17,
there will be need to employ a curved line [which will be]
still more composite (*composée*), by one degree, than the preceding—

and so on, to infinity. [See box following]

number of given lines proposed in the locus-problem	*the curve that is employed along with a straight line to find points on the line that will be the locus*
3 or 4 or (not all parallel) 5	a circle
5 all parallel or 6 or 7 or 8 or (not all parallel) 9	a conic section
9 all parallel or 10 or 11 or 12 or (not all parallel) 13	a curve one degree more composite than a conic section
13 all parallel or 14 or 15 or 16 or (not all parallel) 17	a curve one degree more composite than a curve one degree more composite than a conic section
and so on, to infinity	

That 26th paragraph is the central paragraph of the First Book's second half. Perhaps this would be a good place to pause for an overview of the structure of the whole First Book. Here it is:

The 17-paragraph first half of this First Book (as we saw earlier)
consisted of two parts:

an 11-paragraph first part (1–11), in which
any problem of geometry was algebraicized
(the central [6th] paragraph of this part
being the replacement—
by letters— of the lines that were taken to represent quantities),

followed by

a 6-paragraph second part (12–17), in which
the equation that results from algebraicization yielded
the construction that solved any problem of the ordinary geometry.

This First Book's 17-paragraph second half (18–34)
also consists of two parts:

an 11-paragraph first part (18–28), in which
the author successfully tries his method on an ancient unsolved problem
(the central 6th paragraph of this part [the 23rd of the entire First Book]
being his interruption of Pappus to take note of
the blindness and the weakness of the ancients, who lacked his method),

followed by

a 6-paragraph second part (29–34), in which
he starts off the demonstration of his superiority to the ancients.

The central paragraph (9) of the entire first half of the First Book
(that is, of 1–17, which is the half that introduces the author's new method)
tells the first steps that he takes in solving a geometrical problem.

The central paragraph (26) of the entire second half of the First Book
(that is, of 18–34, which is the half that begins to show
how far beyond the ancients one can go
by using the new method that the author has introduced)
tells the first steps that he takes in finding the solution of
the ancient unsolved problem that was the occasion for trying out his method.

That last paragraph by Descartes himself, on the first steps that he takes in solving Pappus's problem, spoke of what it takes to find *points* that will satisfy what is called for in the problem. Descartes now goes on to speak of finding the *line* that is the place (the locus) where all such points are encountered.

[27] Then I found also that
when there is naught but three or four lines given,
the points sought are encountered
not only in one of the three conic sections
but sometimes also in the circumference of a circle or in a straight line.

And that
when there are five, or six, or seven, or eight of them,
all these points are encountered in some one of the lines which are
more composite (*composée*) by one degree than the conic sections,
and it is impossible to imagine [that is, to visualize] any of them that
may not be useful to this problem (*question*),
but they can also again (*derechef*) be encountered
in a conic section or in a circle or in a straight line.

And if there be nine, or 10, or 11, or 12 of them,
these points will be encountered in a line which can be
more composite (*composée*) than the preceding ones by naught but one degree,
but all those which are more composite (*composée*) by one degree can be used for it
(*y peuvent servir*),

etc. so on, to infinity. [See box following.]

the number of given lines that are proposed in the locus-problem	*the line that will be the locus of points that satisfy what is called for by the locus problem*
3 or 4	a conic section (but sometimes also a circle or a straight line)
5 or 6 or 7 or 8	a curve that is one degree more composite than a conic section (but sometimes also a conic section or a circle or a straight line)
9 or 10 or 11 or 12	a curve that is one degree more composite than a curve that is one degree more composite than a conic section (etc.)
and so on, to infinity	

It would be hard to exaggerate the importance of what has just been said in Descartes' last two paragraphs. Some locus problems of five or six or seven or eight lines will be special cases which yield straight lines or circles or conic sections, but for the most part the locus problems of five or six or seven or eight lines will yield curves that are more composite than the conics, and more so by exactly one "degree"; moreover, every curve that is more composite than the conics by exactly one degree will solve a locus problem of five or six or seven or eight lines. And so on up a ladder of kinds of curves, in which the compositeness of the curves increases by exactly one degree with the addition of one more given line or two or three or four more of them—one kind of curve coming after another, on and on, endlessly. The implication is this: the n-line locus problem of Pappus is not just any problem; to investigate *that* problem, is to investigate *all* of the infinitely many kinds of curves that can be located on Descartes' scale of infinitely increasing compositeness.

Now we might ask ourselves whether all such curves are indeed *all* the curves that there are. Are there any curves that will not be found among the curves located on Descartes' scale of compositeness? To answer that, we might ask ourselves why curves that precede other curves on the scale are said to be less composite. One way to think about the matter is to consider that in the locus problem corresponding to the *curve* that is less composite, the statement of what is called for involves a *ratio* that is less composite—that is, it involves a ratio that is compounded out of fewer ratios. The determining ratio for a more composite curve is compounded out of a greater number of ratios than is the determining ratio for a less composite curve.

Another way to think about the matter is to consider how one might *draw* curves that come after the conic sections on the scale of compositeness. Descartes' next paragraph says something about drawing the very next curves on the scale of compositeness—the curves, that is to say, which come immediately after the conic sections.

[28] What is left to say is this (*Au reste*):
the first, and the most simple, of all [the lines] after the conic sections
is that which one can draw (*descrire*) by
the intersection of a parabola and a straight line,
in the fashion which will be explained soon.

So (*En sorte que*) I think I have entirely satisfied
what Pappus tells us had been sought in this by the ancients;
and I shall try to put the demonstration of it in a few words,
for I am already bored from writing about it so much.

At a number of places in his *Geometry*, Descartes indicates that he could say more than he does. Later on, he will say that by leaving it to his readers to follow up on his suggestions, he will teach them more than they would learn if he were less thrifty with his words. Here, however, his thrift with words is attributed merely to his boredom. He waits until later to replace that haughty reason with one that is much more generous. He waits also to disclose the extent of his generosity: while he says here that he will soon explain the fashion in which one can draw the curves that come immediately after the conic sections on the scale of compositeness, he does not say here that he will explain also how to draw the curves that come after those (that is, after the ones that come immediately after the conic sections).

The treatment of the problem of Pappus will occupy the rest of the *Geometry*'s First Book, which is entitled "Of problems that one can construct without employing aught but circles and straight lines." The treatment of that problem of Pappus will be finished up in the Second Book. That Second Book, which is entitled "On the nature of curved lines," opens with seven paragraphs that comprehensively distinguish the geometric curves; and the central paragraph of those seven begins to show how any of the geometrical curves can be drawn by the intersection of two lines: one of these intersecting lines is another curve that is simpler by just one degree than the curve that is to be drawn; and the other intersecting line is straight. The curve that, here in the First Book, he says he will soon explain how to draw, will later turn out to be only the first step up the never-ending ladder that Descartes will teach his readers to climb.

Descartes has told us of his solution of the problem of Pappus. But although he has determined what sort of line will be the locus of points that satisfy the condition of the problem, and indeed has done so for any number of lines, he has not yet shown us how to draw the curves of higher degrees than the conic sections. He has not yet even demonstrated what he has determined.

Now he begins the demonstration of his solution to the problem of Pappus.

[29] Let AB, AD, EF, GH, etc. be several lines given in position,
and let it be necessary (*qu'il faille*) to find a point, as C,
having drawn from which [point C], onto the given [lines],
other straight lines, as CB, CD, CF, and CH,
in such a way that (*en sorte que*) the angles CBA, CDA, CFE, CHG, etc. be given,
and that what is produced by the multiplication of one part of these lines
be equal to what is produced by the multiplication of the others,
or else that they have some other given ratio (*proportion*)
[that is, some ratio other than a ratio of equality],
for this does not render the problem (*question*) more difficult.

[See figure 1.]

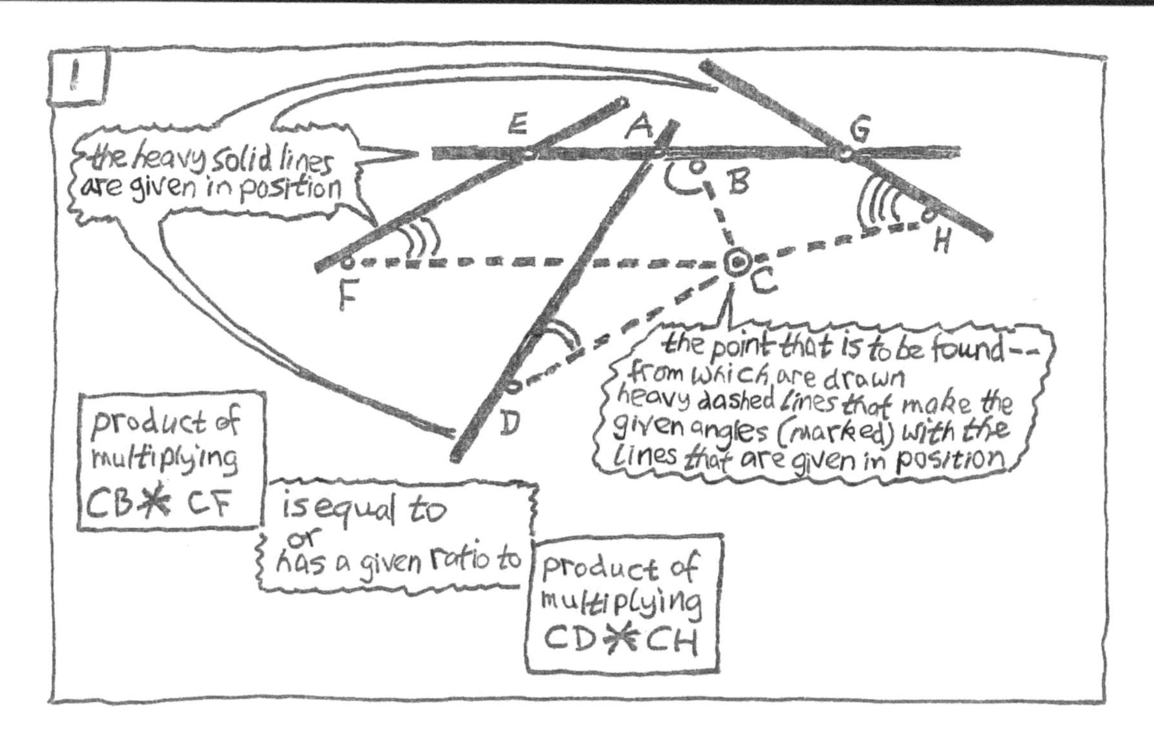

We shall begin by supposing that we already have such a point C as we are seeking. The condition called for in order for C to be such a point will be stated in terms of products whose factors are lines drawn from point C to the given lines. Those lines the sizes of which we need to find in order to form the products are: BC, CD, CF, and CH.

To obtain those sizes, we shall relate (or refer—the verb is *rapporter*) those lines to two "principal" lines.

The first of the lines whose size we need to find (BC),
we shall consider to be one of the principal lines;

the line (AB) whose position is given and which
meets BC (or whose prolongation meets it),
we shall consider to be the other one of the principal lines.

And we shall associate size-names with
each of those lines (AB and BC) that
we are considering to be our
lines of reference or "principal lines":

with regard to one of the lines (whose position is given),
the part which is bounded on one end (A)
by the line AD,
and on the other end by point B (to be found),
we shall call x;

and the other line (BC),
we shall call y. (See figure 2.)

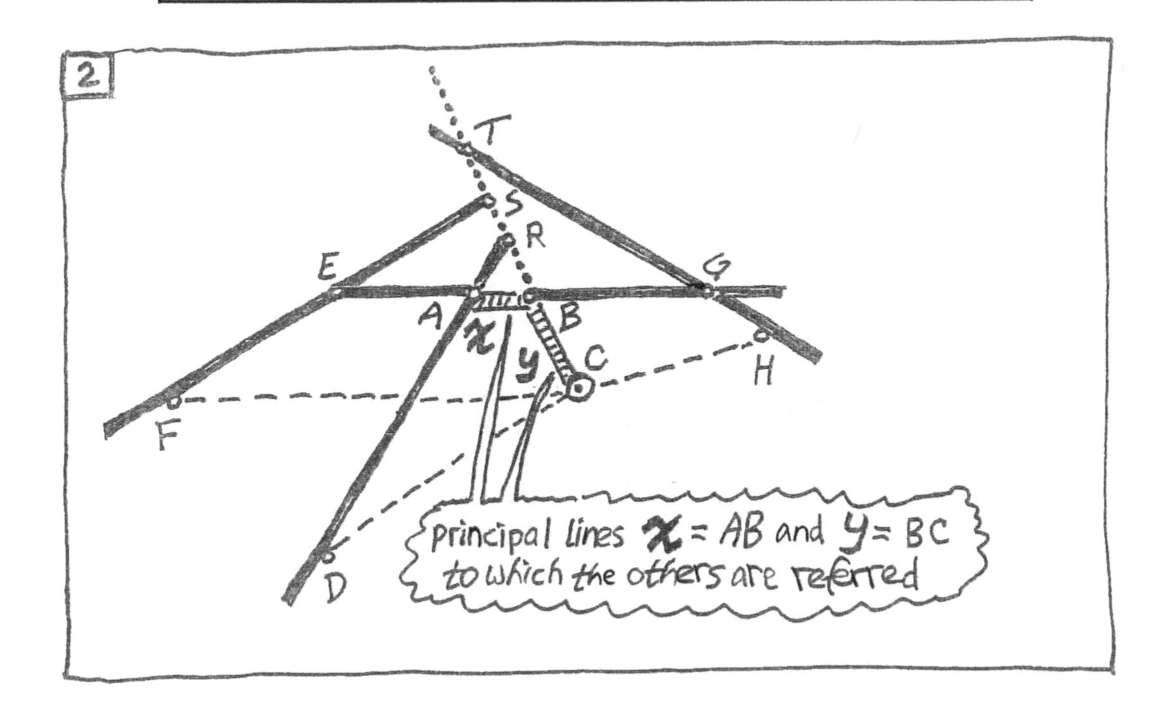

[30] First I suppose the thing as already done;
and for unmixing myself from the confusion of all these lines,
I consider one of the given [lines]
and one of those which there is need to find —
for example, AB and CB—
as the [the lines which are] principal, and
to which I try to relate [or: refer] (*rapporter*) all the others, thus:

Let the segment of the line AB which is between the points A and B be named x,
and let BC be named y;

and let all the other given lines be prolonged until they cut these two,
[these two being] also prolonged if there is need,
and if they are not parallel to them;
as you see here that they cut the line AB at the points A, E, G,
and [cut the line] BC at the points R, S, T. [Again, see figure 2.]

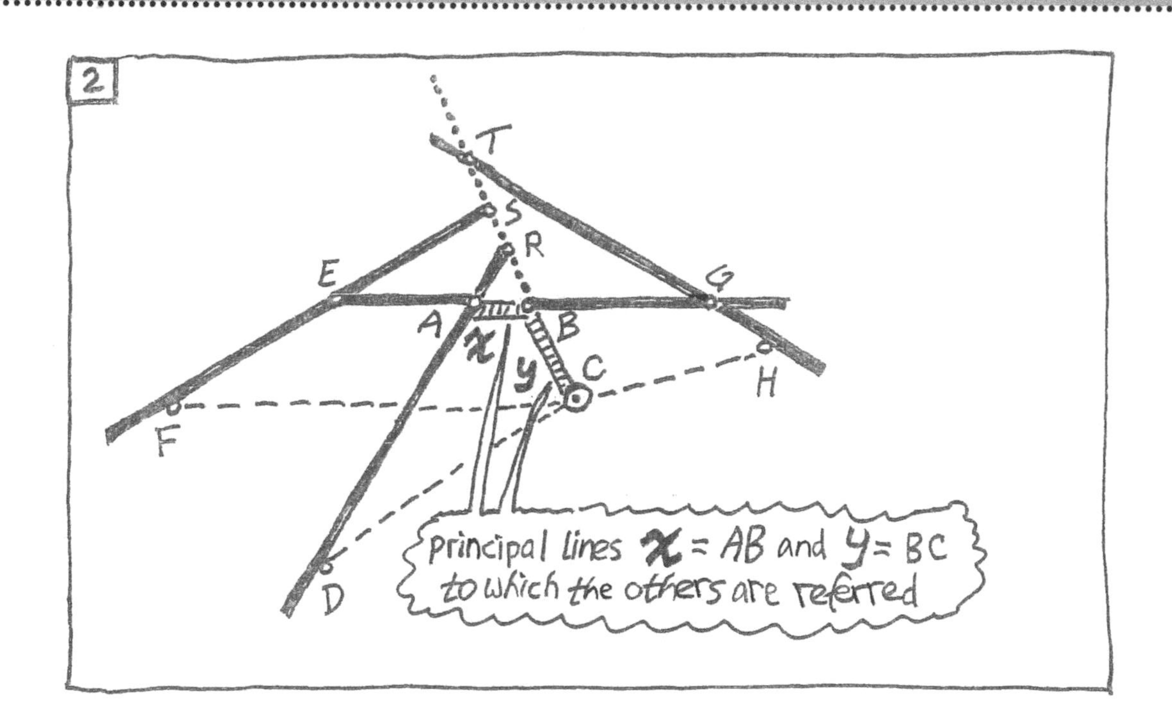

Just why are we doing this—why do we choose *two* lines as reference lines, and why *these* two? We are not told here. What we are told here, is only that we are doing it for the sake of "unmixing" ourselves from "the confusion of all these lines."

Why might we consider lines AB and BC to be our principal lines, the lines to which we shall try to relate all others? Well, every point C on that curve that we seek is determined by an equation that relates the sizes of the four lines that are drawn from C at given angles to the four given lines. These four drawn lines that determine any given point C on the curve thus stretch from C to points B and D and F and H. Now consider any one of these lines—say, CB. If we take some such CB, then B will be at some distance, along the given line, from given point A. That is, if we could figure out how to describe the relationship between distance AB (which we call quantity x) and distance BC (which we call quantity y) then we could obtain as many points C as we please. In other words, every point C on the curve would be determined by an equation relating BC to only one quantity, rather than being determined by an equation relating it to three quantities.

What we need to try to do, then, is to obtain an equation that will relate quantity x to quantity y—that is, will relate AB to BC—by using the equation that relates this line BC (drawn to given line AB from point C) to the other three lines drawn to given lines from that point C. We begin by finding expressions that relate each of these three lines (namely, CD and CF and CH) to x and y.

First, we obtain CD. We do so by obtaining CR, which we obtain by first obtaining BR. We take a line z of any size we please; it will determine some other length b such that the ratio of z to b will be the same as the given ratio of AB to BR. (The ratio of AB to BR is given because they are sides of triangle ABR, whose angles are given. Whenever the angles of a triangle are given, that fixes its shape—that is, fixes the ratio of any one of its sides to any other.) That enables us to obtain an expression for line CD, as follows.

Then, because all the angles of the triangle ARB are given,
the ratio (*proportion*) that is between the sides AB and BR is also given,
and I put it as that of z to b,
in [such a] fashion that, AB being x ,..... RB will be.......... (bx/z),
and............................ the whole [line] CR will be........$y + (bx/z)$,
because the point B falls between C and R—
for if R should fall between C and B,
[then ....that line] CR would be ..$y - (bx/z)$,
and if C should fall between B and R,
[then...............................that line] CR would be $-y + (bx/z)$,

Just the same (*tout de meme*) are given the three angles of the triangle DRC,
and in consequence also [given is]
the ratio (*proportion*) that is between the sides CR and CD,
which I put as [that] of z to c :
in [such a] fashion that, ...CR being......... $y + (bx/z)$,
[then]...CD will be..... $(cy/z) + bcx/zz$. [See figure 3.]

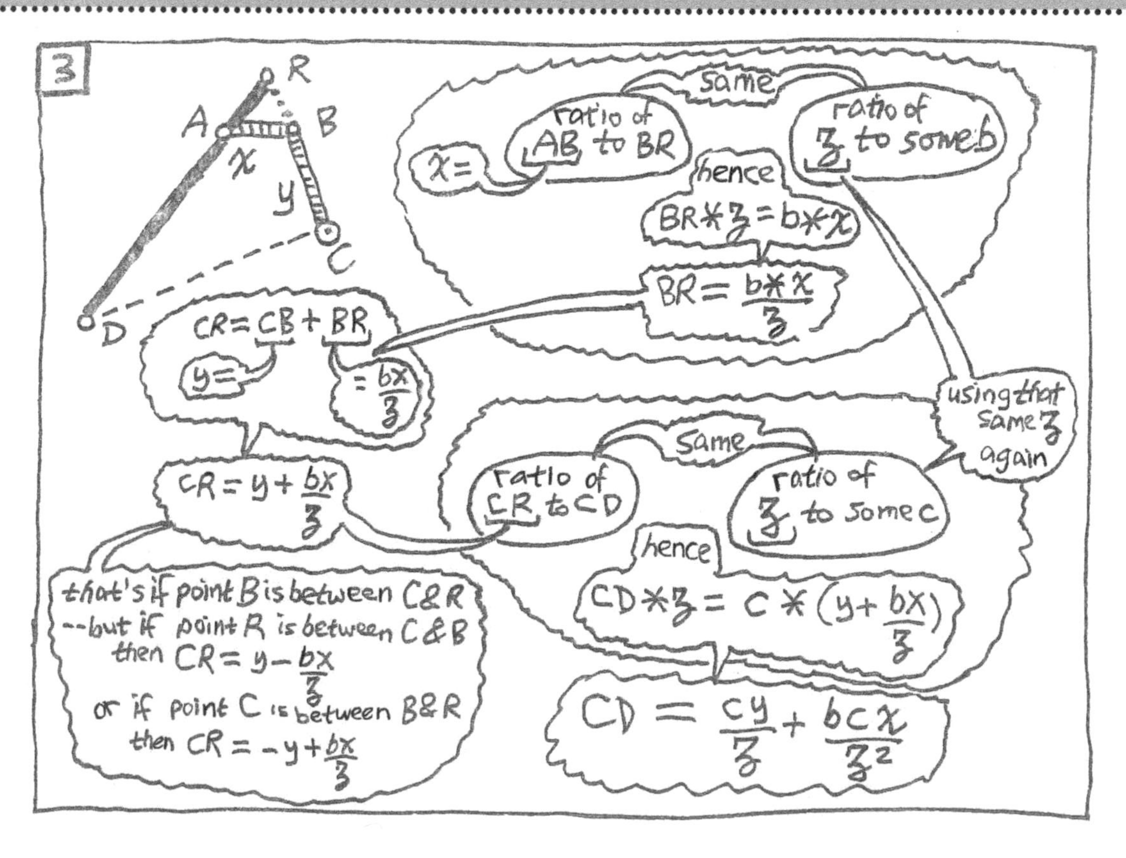

Next, we obtain CF. We do so by obtaining CS, which we obtain by first obtaining BS, as follows.

After that, because the lines AB, AD, and EF are given in position,
the distance that is between the points A and E is also given,
and if one calls it k,
[then] one will have EB equal to.......... $k + x$;
but this would be............................ $k - x$ if the point B should fall between E and A;
and [it would be]............................... $-k + x$ if E should fall between A and B.

And because the angles of the triangle ESB are all given,
also given is the ratio (*proportion*) of BE to BS,
and I put it as [that of] z to d ,
with the result that (*sibienque*) BS is....... $(dk + dx)/z$,
and.............. the whole [line] CS is $(zy + dk + dx)/z$;
but this would be.............................. $(zy - dk - dx)/z$
if the point S should fall between B and C,
and this would be......................... $(-zy + dk + dx)/z$
if C should fall between B and S.
Moreover, the three angles of the triangle FSC are given,
and consequently (*en suite*) [also given] is the ratio (*proportion*) of CS to CF,
which [ratio] let be as [that] of z to e ,
and..... the whole [line] CF will be $(ezy + dek + dex)/zz$. [See figure 4.]

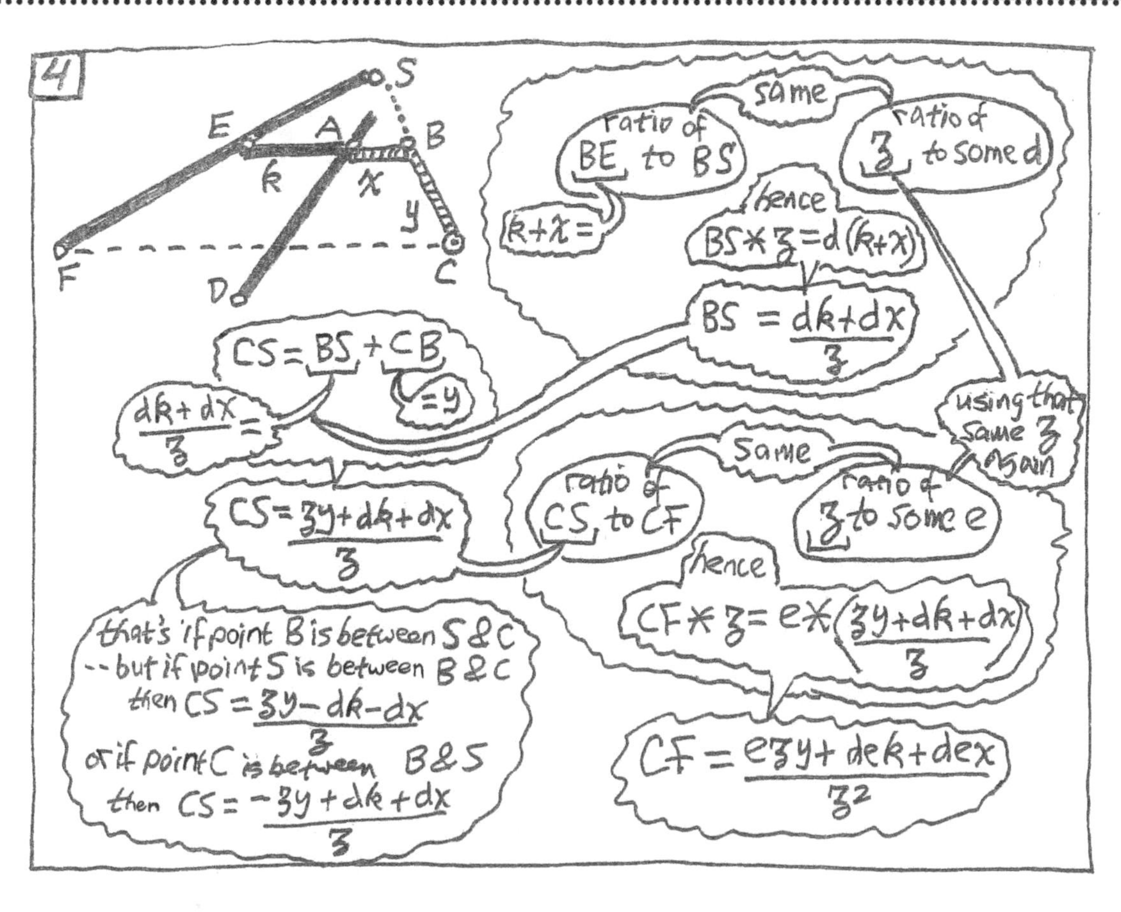

And, finally, we obtain CH. We do so by obtaining CT, which we obtain by first obtaining BG, as follows.

In the same fashion, AG, which I name l, is given, and BG is $l - x$,
and because, from the triangle BGT, also given is the ratio (*proportion*) of BG to BT,
which [ratio] let be as [that] of z to f,
and ....BT will be $(fl - fx)/z$,
and [one will have] ....CT æ $(zy + fl - fx)/z$,
Then again (*derechef*)
the ratio (*proportion*) of TC to CH is given, because of the triangle TCH,
and, putting it as [that] of z to g ,
one will have... CH æ $(gzy + fgl - fgx)/z^2$. [See figure 5.]

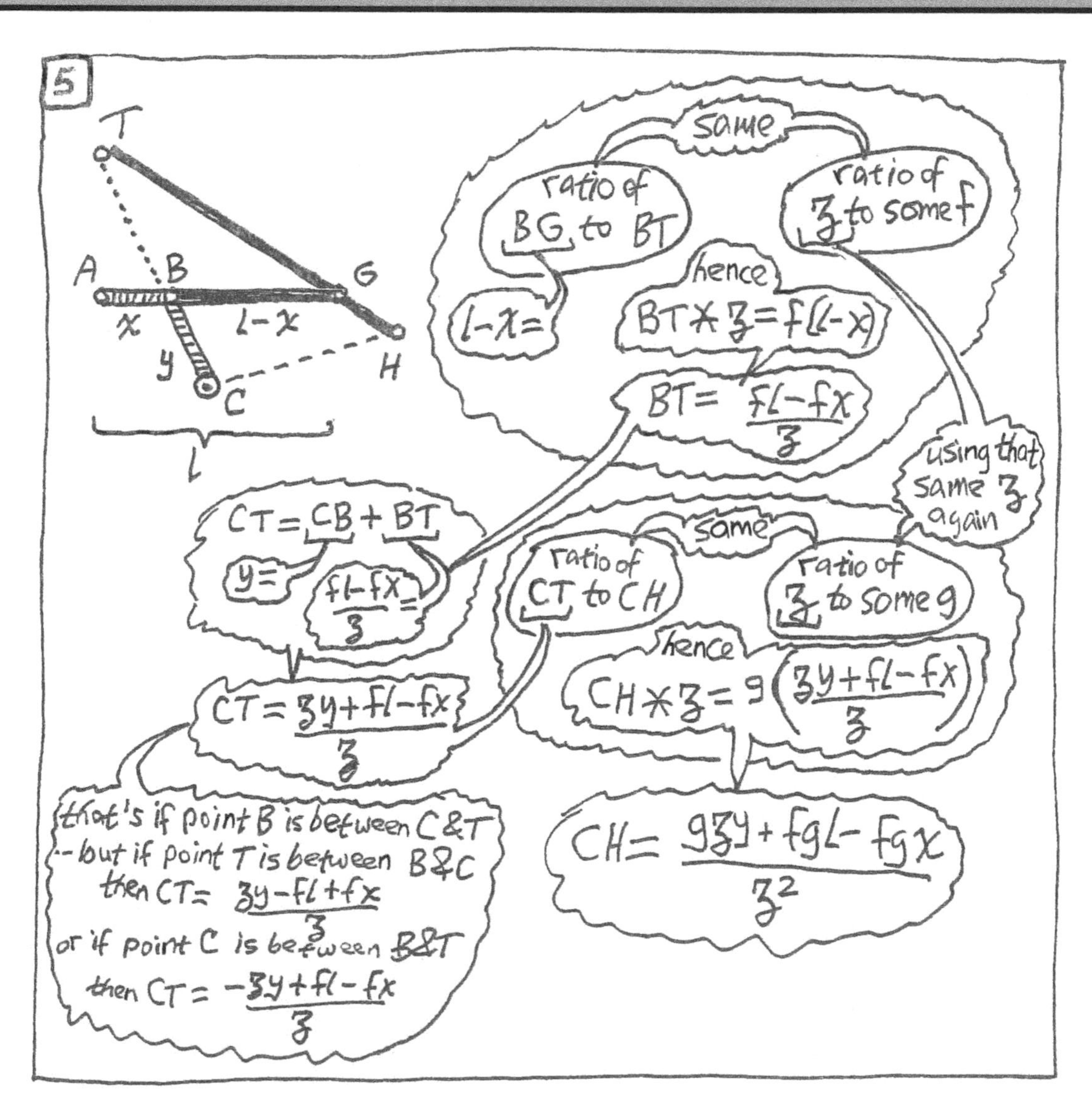

Paragraph 30 has yielded terms expressing the distances along lines drawn, at given angles, from the same point to four given lines. We need those distances (namely: CB, CF, CD, CH) to find the point C, which is determined by the condition that the products CB*CF and CD*CH are in a given ratio (which may be a ratio of equality). To avoid getting mixed up by all those lines, we referred all four of the distances to two lines. As these two reference lines, we took one of the lines which we need to find (CB, say, which we call y) and another line (AB, say, which we call x) which lies along a line whose position is given and whose size we shall have once we have that line CB (which we call y). Using them, what we have just obtained in Descartes' example is displayed as follows:

Take any length you please (calling it z); then:

$$CD = (cy)/z + (bcx)/z^2$$

with b being determined by $AB/BR = z/b$
and c being determined by $CR/CD = z/c$

$$CF = (ezy + dek + dex)/z^2$$

with d being determined by $BE/BS = z/d$
and e being determined by $CS/CF = z/e$

$$CH = (gzy + fgl — fgx)/z^2$$

with f being determined by $BG/BT = z/f$
and g being determined by $CT/CH = z/g$

(the arrangements of + and - signs will be different for different relative positions of the points)

And now paragraphs 31 and 32 will say what "you thus see" about the terms. First (in paragraph 31), you see that however many given lines there may be, each distance will be expressed by three terms (some of them possibly being null terms).

[31] And so (ainsi) you see that:
whatever the number of lines given in position that one could have,
all the lines drawn onto them from the point C at given angles
following the tenor of the problem (*question*)
can always be expressed by three terms each—

of which the one [term] is composed of
the unknown quantity y
multiplied or divided by some other [quantity that is] known,

and the other [term is composed] of
the unknown quantity x
also multiplied or divided by some other [quantity that is] known,

and the third [term is composed] of
a quantity [that is] totally (*toute*) known;

except only if they [the lines given in position] are parallel—

either to the line AB,
in which case the term composed from the quantity x will be null,

or else to the line CB,
in which case that which is composed from the quantity y will be null—

which is thus too manifest for me to stop to explain it.

And, as for the signs + and — which are joined to these terms,
they can be changed in all the fashions imaginable.

What is said about the terms in the expressions for the lengths of those lines which are drawn from the point to those lines which are given in position—that is true no matter how many there may be of these lines which are given in position. We have seen a 4-line example. In that case, the drawn lines can be expressed as follows:

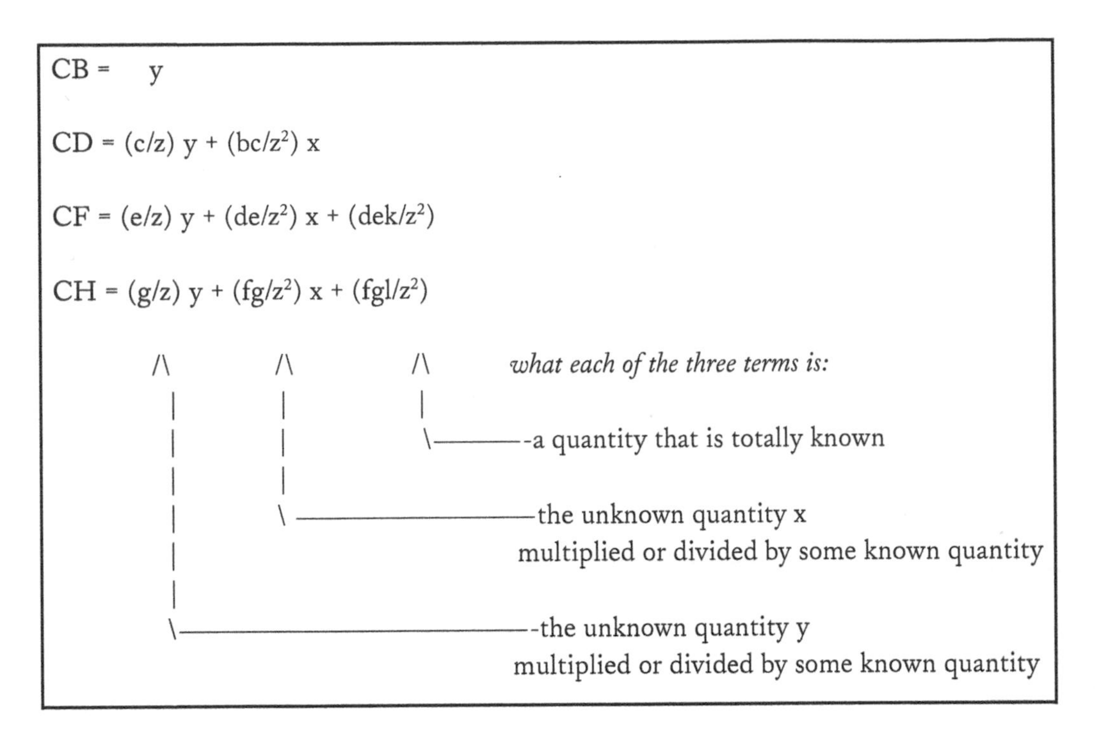

CB = y
CD = (c/z) y + (bc/z²) x
CF = (e/z) y + (de/z²) x + (dek/z²)
CH = (g/z) y + (fg/z²) x + (fgl/z²)
what each of the three terms is:
-a quantity that is totally known
the unknown quantity x
multiplied or divided by some known quantity
-the unknown quantity y
multiplied or divided by some known quantity

Displayed in the following box is a 4-line example where the lines given are parallel, for which see figure 6. (To form triangles of determinate shape, we draw line UC in the most convenient manner, that is, perpendicular to the parallels.)

Take any length you please (calling it z); then there will be lengths (b,c,d,e,f,g,h) such that:

$AS = b/z \qquad CB/CA = e/z$

$ST = c/z \qquad CE/CS = f/z$

$TU = d/z \qquad CH/CT = g/z$

$\qquad\qquad\qquad CM/CU = h/z.$

Only one line of reference (y) is needed for expressing the drawn lines:

$$
\begin{aligned}
CB &= (e/z)\ y && = (e/z)\ y \\
CE &= (f/z)\ [y + b/z] && = (f/z)\ y + [(fb) \quad /z^2] \\
CH &= (g/z)\ [y + (b+c)/z] && = (g/z)\ y + [(gb+gc) \quad /z^2] \\
CM &= (h/z)\ [y + (b+c+d)/z] && = (h/z)\ y + [(hb+hc+hd) \quad /z^2]
\end{aligned}
$$

—and none of the expressions contains an x-term.

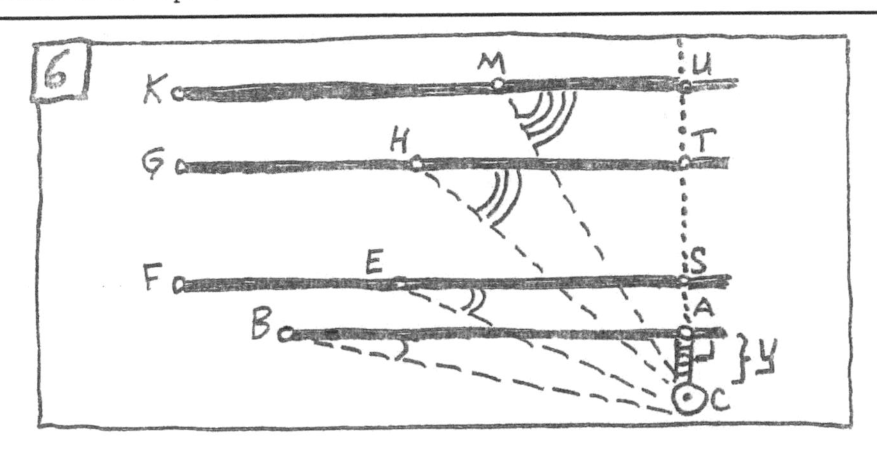

As we saw in Descartes' own example, the signs that are joined to the terms (namely, + and –) will depend upon the relative positions of the points.

Note that *z* does not name a given length; that is why the letter is taken from the end of the alphabet. But the length that it names is not unknown. It is chosen as you please, to be employed throughout the solution of the problem, something like a unit.

Note also that *x* and *y* are called "quantities," and that they are called "unknown" quantities. In the next paragraph (32), Descartes will speak of "multiplying several of these lines, one by the other" and of the dimensions that "the quantities *x* and *y* which are found in the product can each have."

[32] Then you see also that
multiplying several of these lines, one by the other,
the quantities *x* and *y* which are found in the product
can each have there [in the product] naught but
as many dimensions as there are lines which have been thus multiplied
for whose explication they [*x* and *y*] serve,

so that (*en sorte que*)
they will never have more than two dimensions
in what will be produced by the multiplication of naught but two lines,
nor [will they have] more than three [dimensions]
in what will be produced by the multiplication of naught but three [lines],
and so on, to infinity.

Thus we have seen first (31) that no matter how many lines are given in position, the length of any line drawn from point C making a given angle with those lines can always be expressed by three terms: one term consisting of the unknown quantity *y* multiplied or divided by a known quantity; another term consisting of the unknown quantity *x* multiplied or divided by a known quantity; and a third term consisting of a totally known quantity. (In the exceptionally simple case when all the lines given in position are parallel, one or the other term containing the unknown quantity—either *x* or *y*—will come to nothing.)

Because of that, we have just seen also (32) something about the terms in the expressions that result when we multiply several of those lines together. For neither of the quantities *x* and *y* which are in the product, will the number of dimensions be greater than the number of lines that are multiplied together to make the product. (Multiplying 2 lines together will not yield an expression that has a term containing *x* or *y* with an exponent greater than 2; nor will multiplying 3 lines yield an exponent greater than 3; and so on, endlessly.)

The next two paragraphs (33, for the case when there are no more than 5 given lines; and 34, for the case when there are 6 or more) will say how those expressions will yield points on the curved line that is called for—points by means of which we can try to draw the curved line upon which the points fall—and will also say how to begin to characterize such curves.

Since the product of some of the lines drawn from point C to the given lines will have a given ratio (possibly a ratio of equality) to the product of the other lines so drawn, we can obtain an equation that will enable us, for any size we take one of the unknowns as having (say, y), to find the size that the other unknown must then have (say, x).

Now from what has been said, it is evident that if we are not given more than 5 lines, then, in the equation, x cannot have an exponent greater than 2. (Here is why: The expression for the first line, CB, does not contain any x. When this line is multiplied by one of the other lines drawn from C, in which expression the exponent of x is 1, then also in the expression for the product the exponent of x will be 1. Since in the expression for each of the two other drawn lines also the exponent of x is 1, in the expression for their product the exponent of x will be 2.) So if we choose some size for y, then we shall get x by setting x^2 equal to this: $+px + q$ (where p and q are quantities that are known, and there might be a minus-sign rather than a plus-sign). Consider the following:

```
Suppose, for example, that  CB*CD = CF*CH
                             |   |    |   |
            y = ------------/    |    |   \------------ = (g/z) y – (fg/z^2) x + (fgl/z^2)
                                 |    |
   (c/z) y + (bc/z^2) x = -------/    \------------- = (e/z) y + (de/z^2) x + (dek/z^2)
```

Then

$(c/z)\, y^2 + (bc/z^2)\, xy =$

$(eg/z^2)\, y2 - (efg/z^3)\, xy + efgl/z^3)\, y - (deg/z^3)\, xy - (defg/z^4)\, x^2$
$+ (defgl/z^4)\, x + (degk/z^3)\, y - (defgk/z^4)\, x + (defgkl/z^4) =$

$-(defg/z^4)x2 - (efg/z^3)\, xy + (eg/z^2)\, y^2 + (defgl/z^4)\, x + (efgl/z^3)\, y + (defgkl/z^4)$
$+ (deg/z^3)\, xy \qquad - (defgk/z^4)\, x + (degk/z^3)\, y$

Hence,

$(defg/z^4)\, x^2 =$

$(eg/z^2 - c/z)\, y^2 + \{[(deg - efg)/z^3 - (bc/z^2)]\}\, xy + [(defgl - defgk)/z^4]\, x$
$+ [(efgl + degk)/z^3)]\, y + (defgkl/z^4)$

Hence,

$x^2 =$

$(z^4/defg)\, \{[(deg - efg - bcz)/z^3]\, y + [(defgl - defgk)/z^4]\}\, x$
$+ (z^4/defg)\, \{[(eg - cz)/z^3]\, y^2 + [(efgl + degk)/z^3]\, y + (defgkl/z^3)\}$

That is,

$x^2 = +\ (\text{or} -)\ px + (\text{or} -)\ q$

We can, therefore, find the size of x , for any size y that we choose, by making use merely of ruler and compass. And so, by endlessly taking different sizes for y, we endlessly obtain different corresponding sizes for x—thus endlessly finding different points on the curve, by means of which we draw the curve that is called for.

The letters a and b have been used in getting expressions for the lines drawn from point C to the given lines. Now in presenting the expression for what is equal to the square of x, Descartes will use those letters differently: he will now use the letter a for the known quantity by which the quantity x is multiplied, and the letter b for the known quantity that is combined with ax additively or subtractively.

[33] Moreover, because for determining the point C,
there is naught but a single condition that is required—namely, that
what is produced by the multiplication of a certain number of these lines
be equal to, or, what is no less easy (*est de rien plus mal aysé*),
have the given ratio (*proportion*) to
what is produced by the multiplication of the others—
one can [therefore] take at discretion
one of the two unknown quantities x or y,
and seek the other by means of that Equation;

in which it is evident that
when the problem (*question*) is not proposed in more than five lines,
the quantity x which does not serve for the expression of the first [line]
can always have naught but two dimensions,
in [such] fashion that, taking a known quantity for y,
there remains naught but

xx æ +(or—) ax + (or—) bb ,

and so one will be able to find the quantity x with ruler and compass,
in the fashion explained just a little while ago.
Even thus, taking successively
infinite different sizes [magnitudes] (*grandeurs*) for the line y,
one will find also their infinite [corresponding sizes] for the line x,
and so one will have
an infinity of different points such as the one that's marked C,
by means of which one will describe the curved line called for (*demandée*).

We seek to find the curved line that is called for—a line whose points satisfy a certain condition. We began by finding that if we had a point on the curved line, then its relationship to the given lines would be expressed by a certain equation. This equation yields as many points on the curve as we please. Note that what we have found out is how to find as many points lying on the called-for curved line as we please—that is to say, endlessly many different points on the curved line, but nonetheless not all of them. Descartes' procedure may help us to draw the line, but it does not quite yield the line.

Having just considered the problem when no more than 5 lines are given, Descartes in the next paragraph moves on to consider the problem when the number of given lines is 6 or more.

[34] It can be done also—the problem (*question*) having been proposed in
six or a greater number of lines—
if there are among the given [lines] some which are parallel to BA or BC,
[so] that one of the two quantities x or y has
naught but two dimensions in the Equation,
and so one could find the point C with ruler and compass.

But if, in the contrary, they are all parallel,
even though the problem (*question*) be proposed in naught but five lines,
this point C could not be so found,
because, the quantity x not being found in the Equation at all,
to take a known quantity for that [one] that is named y
will be permitted no more,
but this [namely, y] is what it will be necessary to seek.
And because it will have three dimensions,
one could not find it but in extracting the root of a cubic Equation,
which cannot be done generally without employing at least one conic section.

And even though there be up to nine lines given,
provided that they not be all parallel,
one can always make [things so] that
the Equation does not mount but as far as the square's square,
by means of which
one can also always solve (*resoudre*) it through conic sections,
in the fashion which I'll explain hereafter.

And even though there be up to thirteen [lines given],
one can always make [things so] that
it [the Equation] does not mount but as far as the cube's square;
in consequence of which one can solve (*resoudre*) it by means of a line which
is by naught but one degree more complex (*composée*) than the conic section,
in the fashion which I shall explain hereafter.

Even though the problem be proposed in more than 5 lines, the point C can still be found by ruler and compass if some of the lines are parallel to either AB or BC. Why? Because in this case one of the quantities *x* or *y* may still have merely 2 dimensions (that is, have an exponent smaller than 3). But the point C cannot be found by ruler and compass if all the given lines are parallel, even though there will not be any quantity *x* in the equation; and so, instead of getting an *x* from the equation by taking some known quantity for *y*, we would have to solve the equation for *y*; but *y* will have 3 dimensions (that is, have an exponent of 3), so the equation to be solved will be a cubic equation, which generally cannot be solved without employing a conic section.

Even though the given lines be as many as 9, still, if they are not all parallel, the equation can be so expressed as to be no higher than the square's square (that is, expressed with the unknown quantity's having an exponent no greater than 4). This can always be solved by employing conic sections, in the fashion that will be explained hereafter. But if the given lines be no more than 13 in number, then the equation can be so expressed as to be no higher than the cube's square (that is, expressed with the unknown quantity's having an exponent no greater than 6). This can always be solved by employing a curve that is more composite by just one degree than the conic sections. (It appears that the compositeness of the curve that will solve the problem increases by exactly one degree for every 2 dimensions by which the dimensions of the unknown increase, and that the unknown increases by exactly one dimension for every 2 additional lines that are given.) Here is a tabulation:

number of given lines proposed in the locus-problem	*number of dimensions in the equation that the points have*	*curve which is employed to solve the equation that the points have*	*curve which is the locus—the line where the points are located*
3 or 4	2 in x and y	a circle	a conic section
5 not all parallel	2 in x or y	a circle	(special case)
5 all parallel or 6 or 7 or 8	3 or 4 in x and y	a conic section	a curve that is one degree more composite than a conic section
9 not all parallel	4 in x or y	a conic section	(special case)
9 all parallel or 10 or 11 or 12	5 or 6 in x and y	a curve that is one degree more composite than a conic section	a curve that is two degrees more composite than a conic section
13 not all parallel	6 in x or y	a curve that is one degree more composite than a conic section	(special case)

We come now to the concluding sentence of the concluding 34th paragraph of the First Book:

> And this here is the first part of what I had to demonstrate;
> but before I pass to the second, there is need that I say
> something in general about the nature of curved lines.

The demonstration has been treated in these last 7 paragraphs of this First Book. The last 6 of these paragraphs have constituted the first part of the demonstration. Before passing on to the second part, which will constitute the remainder of the demonstration, Descartes for some reason needs to say something in general about the nature of curved lines. He does not say it here at the end of the First Book; rather, he will say it at the beginning of the Second Book, whose title (which greatly contrasts with the titles of the First Book and the Third) is "Of the nature of curved lines." What he says in general about the nature of curved lines will occupy only the first 7 paragraphs of the Second Book; immediately after that, will come 17 paragraphs which will constitute the second part, the concluding part, of the demonstration, after which will come 40 paragraphs more before the Second Book is done.

Why does Descartes split the demonstration into two parts which are separated by a general statement about the nature of curved lines which begins a second book the conclusion of which does not come until long after the demonstration is done?

CHAPTER 4. COMPREHENSIVELY DISTINGUISHING THE CURVES THAT ARE GEOMETRICAL

Descartes' *Geometry* consists of three books, from whose titles it appears that the *Geometry* is predominantly about constructions for solving problems. Two of the three books have titles declaring them to be about that. The First Book treats problems that require constructions in which merely "plane" resources are sufficient—problems for whose solution one needs to do no more than draw lines that are straight or are circumferences of circles; and the Third Book treats problems that require constructions in which "solid" or "supersolid" resources are necessary—problems for whose solution one needs to draw lines that are cut into conic surfaces (which are surfaces of solids) or lines that are "over beyond" such "solid" lines (that is, are super-solid). The First Book and the Third together would thus seem to exhaust all the kinds of geometrical problems that there can be. Why, then, the Second Book?

This central book might seem to be a bridge—from problems that call for constructions in which the lines of elementary geometry are sufficient, to problems that call for constructions in which curves treated in higher study are necessary. But is it merely that? The title of the Second Book speaks not of construction-work but of nature; it declares that it is about the nature of curved lines. The Second Book opens with what the First Book in closing said was needed before finishing up the problem of Pappus—namely, a general discussion of the nature of curved lines. This discussion, which comprehensively distinguishes the kinds of curves, occupies merely the first 7 paragraphs of this Second Book, which then devotes 17 paragraphs to the final second part of the demonstration of Descartes' solution of Pappus's problem. What the title declares the Second Book to be about, thus indeed might seem to be a bridge—a bridge, however, between the two parts of the demonstration of Descartes' solution of Pappus's problem. Nonetheless, the Second Book does not finish up with that finishing up of Pappus's problem: it goes on for fully 40 paragraphs more. The organization of these 40 paragraphs is as follows. First there comes a single paragraph followed by a 7-paragraph stretch that concludes the Second Book's first half; and then there comes a single paragraph followed by a 7-paragraph stretch that opens the Second Book's second half. After that, comes a 24-paragraph stretch that constitutes the rest of the Second Book. These 24 paragraphs have a very important place in the design of the entire *Geometry*. They are preceded by 74 paragraphs in all (the 34 paragraphs that constitute the entire First Book, together with the first 40 paragraphs of the Second Book) and they are also followed by 74 paragraphs (the 74 paragraphs that constitute the entire Third Book). The final 24 paragraphs of the central book are thus the central paragraphs of the entire *Geometry*. What those paragraphs contain would seem to be the heart of the matter. Let us work our way toward it, beginning at the beginning of the Second Book.

THE
GEOMETRY.

SECOND BOOK.

Of the nature of curved lines.

[1] The ancients noted very well that among the Problems of Geometry,
some are plane, others [are] solid, and others [are] linear—that is to say, that

some can be constructed in tracing naught but straight lines and circles,

whereas others cannot be [constructed] unless one employs as means
some conic section,

nor finally [can] others [be constructed] unless one employs
some other line [that is] more composite (*composée*) [than a conic section].

But I am astonished that among these more composite (*composée*) lines
they did not distinguish different degrees beyond that [degree of a conic section],
and I would not know how to comprehend why they named them [that is,
named the lines that are more composite than conics] "mechanical" rather than "Geometrical."

For to say that this was done because there is need to make use (*se servir*) of
some mechanical instrument (*machine*) for drawing (*descrire*) them—
that very reason would necessitate rejecting circles and straight lines,
seeing that on paper one does not draw (*descrit*) them [that is, circles and straight lines]
but with compass and ruler, which one can also name "mechanical instruments" (*machines*).

No more is it because the instruments (*instruments*) which serve (*servent*) to trace them,
being more composite (*composés*) than the ruler and compass, cannot be so exact (*justes*)—
for that reason would necessitate rejecting them from Mechanics,
where the exactness (*justesse*) of the works which issue from the hand
is desired more than [it is desired] from Geometry,
where it is only the exactness (*justesse*) of reasoning that one seeks (*recherche*), which can,
without doubt, be as perfect touching these [more composite] lines as touching the others.

I also would not say that it would be because they [namely, the ancients]
did not want to augment the number of their postulates (*demandes*),
and were content that one accord them [those postulates alone which allow] that
they could join two given points by a straight line, and [could]
draw (*descrire*) about a given center a circle which would pass through a given point—
for they did not have (*fait*) a scruple about supposing, beyond that,
for treating conic sections, that one could cut every given cone by a given plane;
and, for tracing all the curved lines which I intend to introduce here,
there is need of supposing nothing but that two or several lines can
be moved, the one by [or upon, or through] (*par*) the other,
and their intersections mark out others of them—
which would seem to me no more difficult [to suppose] at all
[than to suppose, as the ancients did, that one could cut every given cone by a given plane].

It is true that they [the ancients] did not so entirely
receive the conic sections into their Geometry,
and I do not wish to undertake to change the names which have been approved by usage,
but it is, it seems to me, very clear that,
taking, as one does,
for "Geometrical" what is precise and exact (*precis et exact*), and for "Mechanical" what is not,
and considering
Geometry as a science which teaches generally to know the measures of all bodies,
one should
no more exclude the more composite (*composées*) lines than the more simple ones,
provided that one could imagine [that is, visualize] them to be
drawn (*descrites*) by one continuous movement, or
by several [movements] which succeed each other and the last of which are
entirely ruled by those which precede them—
for by this means one can always have an exact [*exacte*] knowledge of their measure.

But perhaps what hindered the ancient Geometers from receiving those which
were more composite (*composées*) than the conic sections
is that the first ones which they considered
were by chance the Spiral, the Quadratrix, and the like,
which truly do not belong to aught but the Mechanical ones,
and are not of the number of those which I think ought to be received here
[that is, they are not among the ones that are Geometrical],
because one imagines [that is, visualizes] them as
drawn (*descrites*) by two separate movements which do not have between them any
relation (*rapport*) that one could measure exactly (*exactement*);
although afterward they [the ancient Geometers] did examine
the Conchoid, the Cissoid, and some little bit of others which are of them
[that is, are among those that should be received as Geometrical]—
all the same, because they perhaps did not sufficiently note their properties,
they did not make more of them (*en ... fait plus d'estat*) than they did of the first ones
[namely, the spiral, the quadratrix, and the like—
which should be received as Mechanical].

Or else—
seeing that as yet they knew naught but little touching the conic sections,
and that, even touching what could be done with ruler and compass,
there remained to them much of which they were ignorant—
it is that they believed they should not cut into more difficult material.

But because I hope that from now on those who have the cleverness [or skill] (*addresse*)
to make use (*se servir*) of the Geometrical calculation
[or reckoning or "calculus"] (*du calcul Geometrique*) here proposed
will not [in matters] touching plane problems or solid ones
find enough (*assé dequoy*) to be stopped,
I believe that it is appropriate that I invite them to other researches (*recherches*), where
they will never lack exercise.

In extending the scope of geometry to include innumerably many curves beyond the conic sections (though not to include all curves beyond the conic sections), Descartes presents his notion of what the science of geometry teaches. According to him, its teaching is not about figures (as his predecessors might say) nor about space (as his successors might say); rather, it teaches, generally, knowledge of the measures of all bodies. That is to say, it deals with those features of bodies that can be put precisely and exactly into ratios and equations. Geometry is the universal science that treats bodies quantitatively.

Descartes names four curves that are beyond the conic sections in degree: the spiral, the quadratrix, the conchoid, and the cissoid. This is an echo of something that Pappus says in the fourth book of his *Collection*.

Speaking of why the ancient geometers were at an impasse in seeking to trisect a given angle, Pappus says they could not solve the problem because the conic sections had not yet been synthesized by them—and so they made the error of seeking to solve through "planes" a problem that by nature is "solid." Pappus adds that, on the other hand, it seems to be no small error when geometers solve a "plane" problem through conics (that is, through "solid" means), let alone through some of the lines (that is, through the "linear" means) that have been studied by the newer geometers. It is in explaining all this, that Pappus says that we say there are three kinds (*genê*) of problems in geometry: we call some of them "plane," some "solid," and some "linear." Those should be called "plane" which can be solved through straight lines and circumferences of circles, since these lines have their genesis in a plane. Those are called "solid" whose finding out takes one or even several of the conic sections, since for their construction (*kataskeuê*) it is necessary to make use of certain surfaces of solid figures (namely, surfaces of cones). There remains the third kind (*genos*) of problems, the one called "linear," since for their construction (*kataskeuê*) lines are taken other than the ones aforementioned—that is, lines are taken which have a more complex (*poikilôtera*) and forced (*bebiasmenê*) genesis, being generated from more disorderly (*atakterai*) surfaces and from interwoven (*epipeplegmenai*) movements. Mentioning several geometers who have made discoveries in the course of investigating such lines, Pappus concludes his account of the third genre as follows—"others of the same kind (*genos*) are: spirals and quadratrices and cochloids and cissoids."

These four curves that Pappus mentioned together are the four that Descartes names here. The first two of them, Descartes takes to be "mechanical"; and the last two of them, he takes to be "geometrical." (What Pappus called "cochloids" later came to be called "conchoids." The names suggest that the curve looks like some shellfish—that it has the *eidos* either of the *kokhlos* or of the *konkhê*.) Of the two curves that Descartes takes to be mechanical, the quadratrix is the one about which Pappus transmitted material that is particularly relevant for appreciating Descartes' denial of geometrical status to the curve; and of the two curves that Descartes takes to be geometrical, the conchoid is the one that Descartes mentions again (in the 5th paragraph) and again (in the 7th). Let us therefore leave to one side the first and the last of the curves—the spiral and the cissoid—and concentrate on the two that are central—the quadratrix and the conchoid.

First, the "quadratrix." The name taken over into English is Latin. It means "square-er." Presenting the curve in his *Collection*, Pappus says that for the squaring of the circle earlier geometers took a certain line that they named from its property; they called it the *tetragônizousa* (Greek for "square-er"). To square the circle—in other words, to construct a square that will be equal in area to a circle whose radius is given—that was one of the great problems of antiquity. This curve that was used to do it, the quadratrix, is traced out by the intersection of two straight lines—a horizontal straight line descending down a square, and a radius rotating round a quadrant at the corner of the square—each of the straight lines moving equably (that is, never getting any faster or slower).

Pappus says that the curve has a genesis of the following sort (see figure 1):

Let ABCD be a square, and let a circular arc be described with center A.

Let AB be so moved that
point A remains fixed while point B is carried along arc BED;
and let BC be so moved that
BC, while remaining always parallel to AD, follows point B in its motion along BA;

and in equal times, moving uniformly,
AB passes through angle BAD (that is, point B passes along the arc BD)
and BC passes by the straight line BA (that is, point B traverses the length of BA)
so AB and BC will coincide at the same time with straight line AD;

and while the motion is in progress,
the straight lines BC and BA as they move
will cut one another at a certain point which continually changes place as they do,
and that point of intersection (F) will describe—
in the space between the straight lines BA, AD and the circular arc BED—
a concave curve BFG.

The principal property of that curve, says Pappus, is this:

if any straight line (such as AFE) be drawn to the circumference,
the ratio of the whole circular arc BED to the arc ED will be
the same as the ratio of the straight line BA to straight line FH;

and that property, which is evident from the genesis of the curve, is what is used to show how to square the circle.

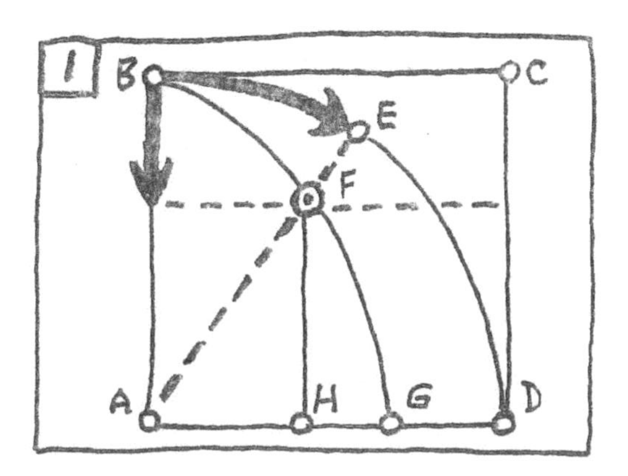

Pappus reports that two aspects of the matter bothered a certain man named Sporus—and with good reason (*eu-logôs*).

In the first place, the very making of the curve presupposes something that the curve (once it has already been made) is to be used to get. The two points that begin to move equably from point B, one of them along a straight line and the other along a circumference of a circle, are brought down together in equal time—the one onto point A, and the other onto point D—and therefore the speeds of the two movements must be in just that ratio which straight line AB has to circumference BED; but how could the whole business be possible unless that ratio (*logos*) is known? If speeds lacking just that fit were used, then things would not come together unless by chance—and, Pappus asks, would that not be irrational (*alogon*)?

After that, there is the other bother. This one has to do with the fact that in using the curve to square the circle, a certain point is needed that is not found in generating the curve. That bothersome point is the point that bounds the curve—namely, point G, where the curve would cut the straight line AD. When CB and BA have been carried down together as was said, they will coincide at last with line AD and no longer make a cut into each other. When they coincide with line AD, their intersection ceases to be—and will have to be replaced by the bounding-point of the curved line where it meets straight line AD. The points of the curve are found by getting the intersections, but the bounding-point is not so found. Someone might say that the curved line is to be thought of as being prolonged as far as AD—as we suppose straight lines to be prolonged. This, however, does not follow from the original suppositions. Rather, point G would have to be obtained by obtaining in advance the ratio of the circumference to the straight line; and since that ratio is not given, one must not (relying on the men who found the line) accept its making, on the ground of its being more mechanical and, for many problems, useful in respect to mechanical affairs.

That's what Pappus, following Sporus, says about the quadratrix. Nowadays, mathematicians can prove not only that the speeds of the movements are not in the ratio that an integer has to an integer, but that they are not even in any of the following ratios:

> the ratio that the mean proportional has to
> either of two lines that have to one another the ratio of two integers,
>
> or the ratio that one of the two mean proportionals has to
> either of two lines that have to one another the ratio of two integers,
>
> or the ratio that one of the three mean proportionals has to
> either of two lines that have to one another the ratio of two integers,
>
> or ... etc.

The relationship involved, that is to say, transcends expression in the algebra of polynomials. That is why nowadays the quadratrix is said to be a curve that is "transcendental" rather than "algebraic."

Now, having looked at the generative definition of the quadratrix, let us take a look at how this curve is used to square the circle.

Here is how the quadratrix provides a square equal to a given circle (see figures 2 and 3):

> If ABCD is a square, and BED is the arc of a circle with center C,
> and BGH is a quadratrix generated as said before,
>
> then (and this will be shown in figure 3)
> the ratio of arc DEB to straight line BC is not the same as
> the ratio of BC
> to a straight line that is greater than CH or to a straight line that is less than CH;
>
> hence that ratio of arc DEB to BC will be the very same as that ratio of BC to CH;
>
> and, therefore, if, a straight line be taken that is a
> third proportional to straight lines HC and CB,
> then that straight line will be equal to the arc BED—
> and a line that is its quadruple will be equal to the perimeter of the circle.
>
> But, as Archimedes has demonstrated,
> the rectangle contained by a circle's perimeter and its radius
> is the double of the circle—
> and, since a straight line equal to the circle's perimeter has been found,
> a square can easily be constructed equal to the circle itself.

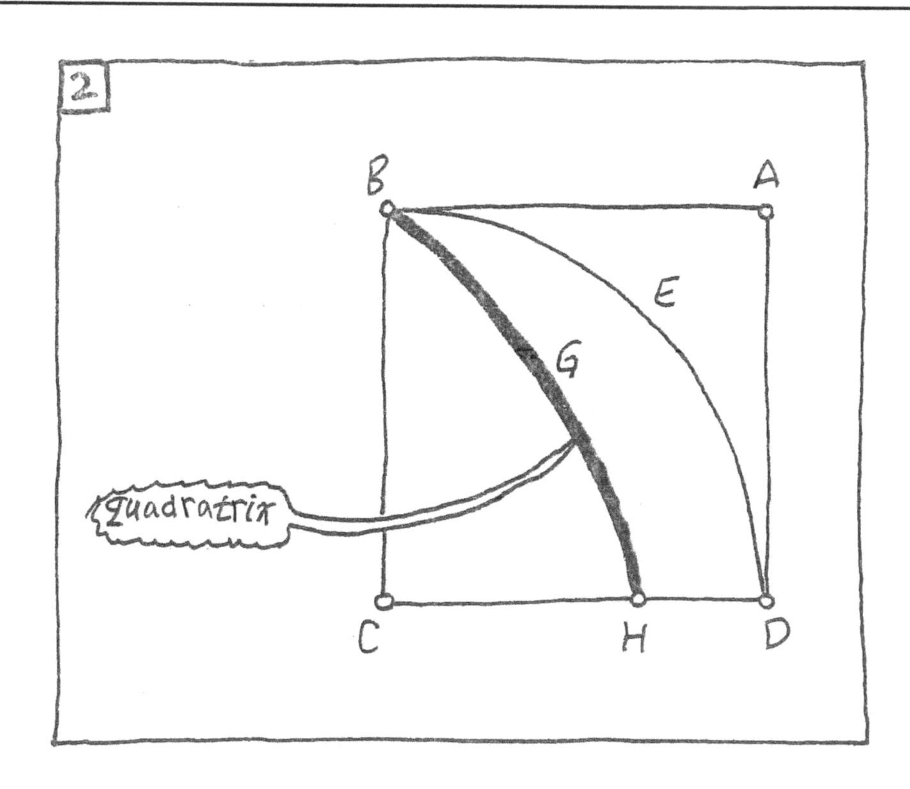

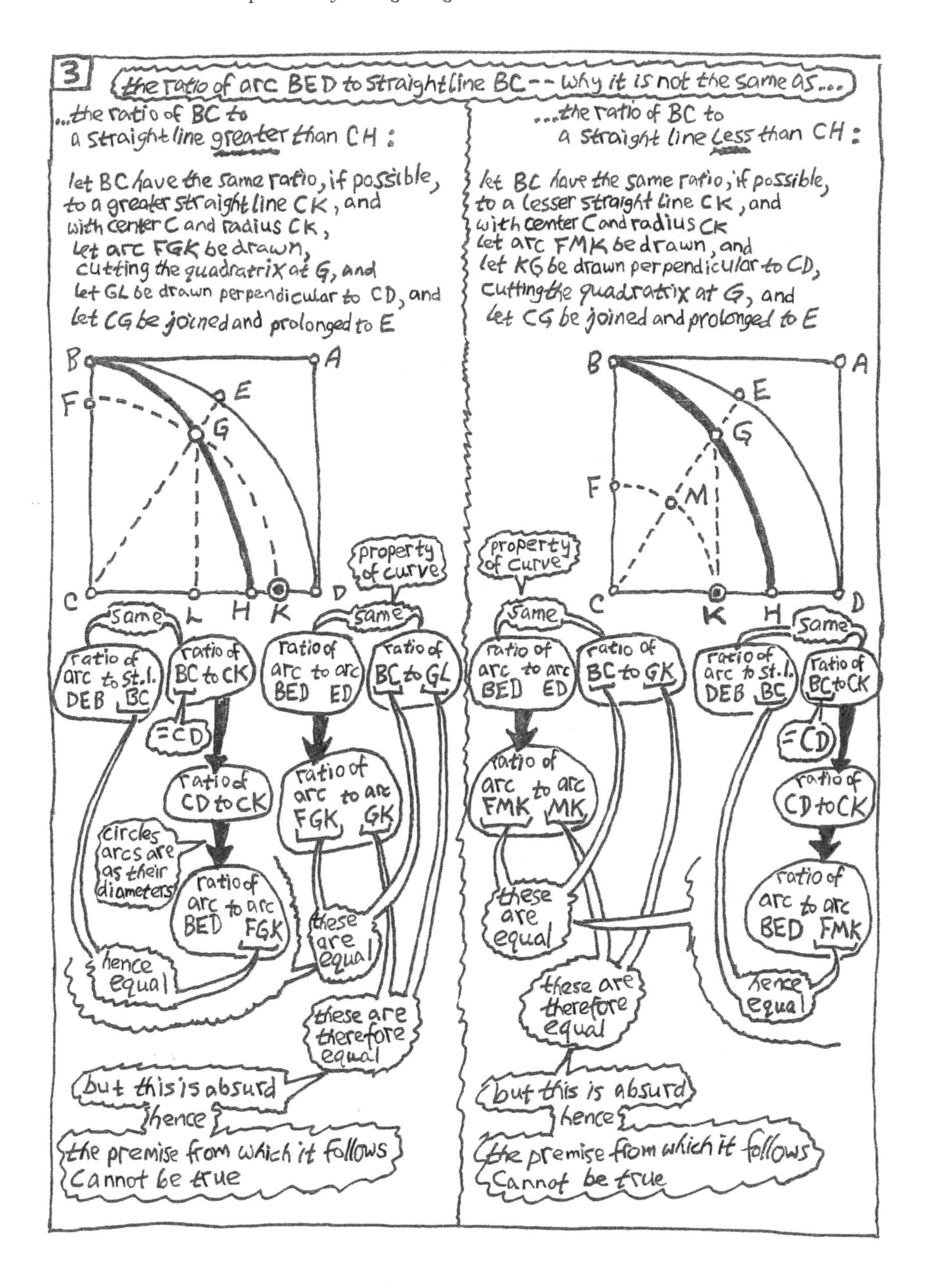
3
the ratio of arc BED to straight line BC -- why it is not the same as...
...the ratio of BC to a straight line greater than CH:
let BC have the same ratio, if possible, to a greater straight line CK, and with center C and radius CK, let arc FGK be drawn, cutting the quadratrix at G, and let GL be drawn perpendicular to CD, and let CG be joined and prolonged to E
B
A
F
E
G
C
L
H
K
D
property of curve
same
same
ratio of arc DEB to st. l. BC
ratio of BC to CK
ratio of arc BED to arc ED
ratio of BC to GL
=CD
ratio of CD to CK
ratio of arc FGK to arc GK
circles arcs are as their diameters
ratio of arc BED to arc FGK
these are equal
hence equal
these are therefore equal
but this is absurd
hence
the premise from which it follows cannot be true
...the ratio of BC to a straight line less than CH:
let BC have the same ratio, if possible, to a lesser straight line CK, and with center C and radius CK let arc FMK be drawn, and let KG be drawn perpendicular to CD, cutting the quadratrix at G, and let CG be joined and prolonged to E
B
A
E
G
F
M
C
K
H
D
property of curve
same
same
ratio of arc BED to arc ED
ratio of BC to GK
ratio of arc DEB to st. l. BC
ratio of BC to CK
=CD
ratio of arc FMK to arc MK
ratio of CD to CK
ratio of arc BED to arc FMK
these are equal
these are therefore equal
hence equal
but this is absurd
hence
the premise from which it follows cannot be true

So much for the quadratrix and the problem it was used to solve, the squaring of the circle. Another one of the great problems of antiquity was the doubling of the cube—the problem of finding the side of a cube that will have double the volume of a cube whose side is given. Presenting the cochloid (that is, the conchoid) in his *Collection*, Pappus says that for the doubling of the cube a certain line is brought up by Nicomedes, and that it has a genesis of the following sort (see figure 4):

> Let there be a straight line AB, with CDF at right angles to it;
> and on CDF let there be taken a certain given point (E);
> while point E remains in the same position,
> let straight line CDF be so rotated about E that D always
> moves along the straight line AB and does not fall beyond it.
> (Since the rotation will make EG bigger than ED,
> the line CDF will slide upward through E.)
>
> The motion being after this fashion on either side,
> it is clear that point C will describe a curve such as LCM,
> and its property is such:
>
> when any straight line drawn from point E falls upon the curve,
> the portion cut off between the straight line AB and the curve ACM
> is equal to the straight line CD—
>
> for AB is stationary and point E is fixed,
> and when D goes to G,
> then straight line CD will coincide with GH, and point C will fall upon H,
> and hence GH is equal to CD—
>
> similarly, if any other straight line from point E falls upon the curve,
> then the portion cut off by the curve and the straight line AB
> will make a straight line equal to CD.
>
> The straight line AB is called the ruler, the point E is called the pole,
> portion CD is called the interval,
> and the curve LCM is called the first cochloidal line—since there are
> second and third and fourth cochloids which are useful for other theorems.

Pappus reports that Nicomedes demonstrated that the line can be drawn instrumentally (*organikôs*), and that it continually approaches closer to the ruler—which is to say that of all the perpendiculars drawn to the straight line AB from points on the line LCH, the greatest is the perpendicular CD, while any perpendicular drawn near CD is always greater than another that is more remote; and also that any straight line in the space between the ruler and the cochloid will, when prolonged, be cut by the cochloid.

It is clear from what has been said, Pappus goes on, that if there is an angle, such as GAB, and a point C outside the angle, then it is possible so to draw CG as to make KG, between the line and AB, equal to a given straight line, as follows (see figure 5):

> From point C, let CH be drawn perpendicular to AB, and
> prolonged to D so that DH is equal to the given straight line;
> and, with C for pole, the given straight line DH for interval, and AB for ruler,
> let the first cochloid be drawn;
>
> then, by what's been said, it will meet AG;
> let it meet it in G, and let CG be joined.
>
> Therefore, KG will be equal to the given line.

Pappus shows how one can, by doing that, obtain two means in continuous proportion between two given straight lines—which will solve the problem of doubling the cube. But why is the old problem of doubling a cube solved if you can find two mean proportionals between one given line and another? Because, as we learn from Euclid (*Elements* XI.33), cubes are to one another in the triplicate ratio of their sides.

That is to say, if s is the smaller cube's side and d is the side of the cube whose volume is double the smaller ones, then the double ratio (namely, the ratio of 2 to 1) will be the ratio that is the triplicate of the ratio of d to s. So, if we find x and d such that

> as $2s$ is to x, so also x is to d, and so also d is to s,

then— since those three ratios are the same, and the ratio compounded of those three same ratios is the ratio of $2s$ to s, which is the ratio of 2 to 1—we shall have the line d whose ratio to s, when triplicated, is the double ratio. In other words, if we double the side of the cube that is given, and find two mean proportionals between the given side and its double, then the smaller of the two means will be the side of a cube that has double the volume of the given cube.

Now, how does Pappus demonstrate that two mean proportionals can be found by using the cochloid? This will be shown in figures 6 and 7.

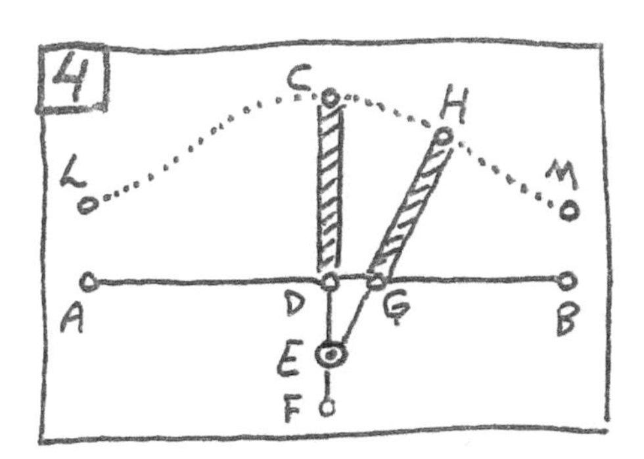

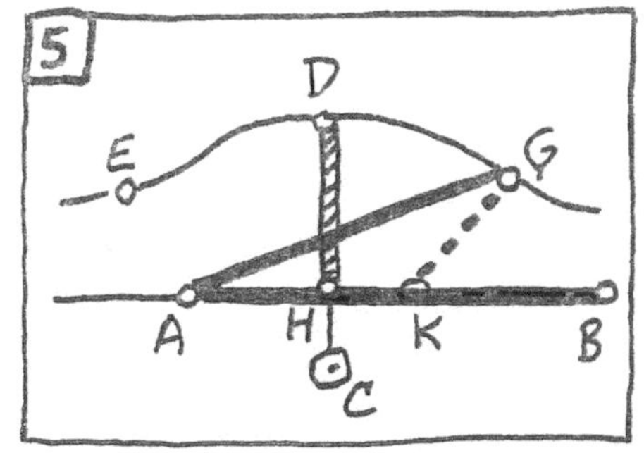

First, the necessary construction is provided, as follows (see figure 6)—

Given two straight lines CL and LA
(placed for convenience at right angles to one another),
to find between them two lines that are means in continuous proportion:

let parallelogram ABCL be completed;

let each of the straight lines AB, BC be bisected, at points D and E respectively;

let DL be joined and prolonged, meeting CB prolonged in G;

let EF be so drawn that CF is equal to AD;

let FG be joined, and parallel to it, let CH be drawn;

and, since angle KCH is given,
from given point F let FHK be so drawn as to make HK equal to AD or CF—
that this is possible is shown by the cochloidal line
(note, however, that in this case, unlike the previous one,
the curve is between the pole and its ruler, and therefore the interval must be marked off on EF below the ruler);

let KL be joined and prolonged, meeting AB prolonged in M.

And now it can be shown that LC is to KC, as KC is to MA, as MA is to AL.

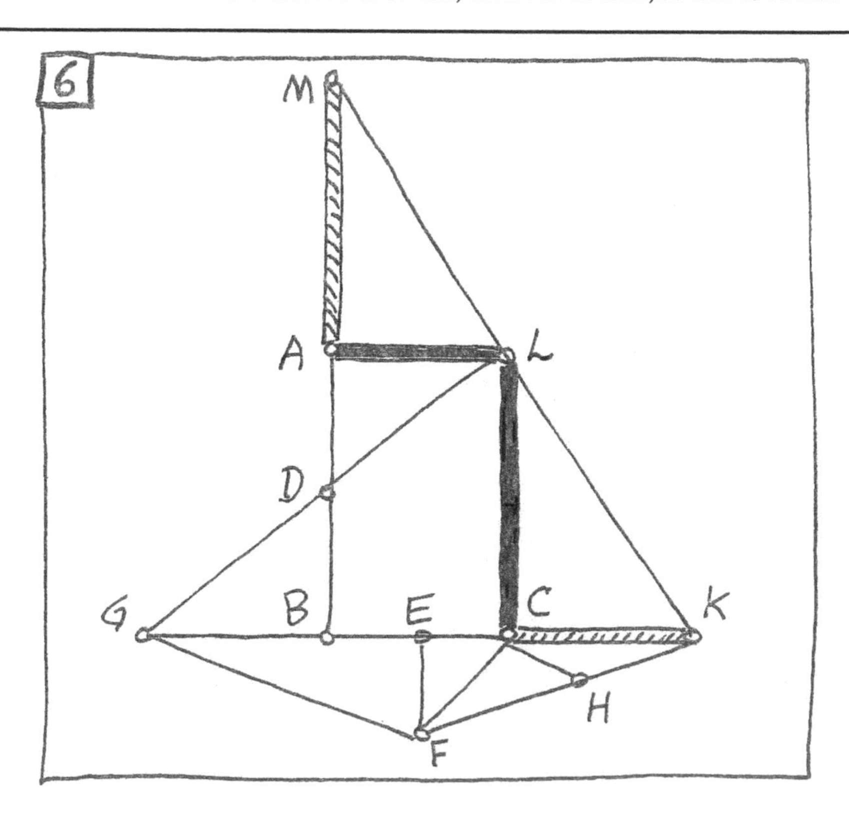

Next, figure 7 displays the demonstration that KC and MA are two means in continuous proportion with LC and AL.

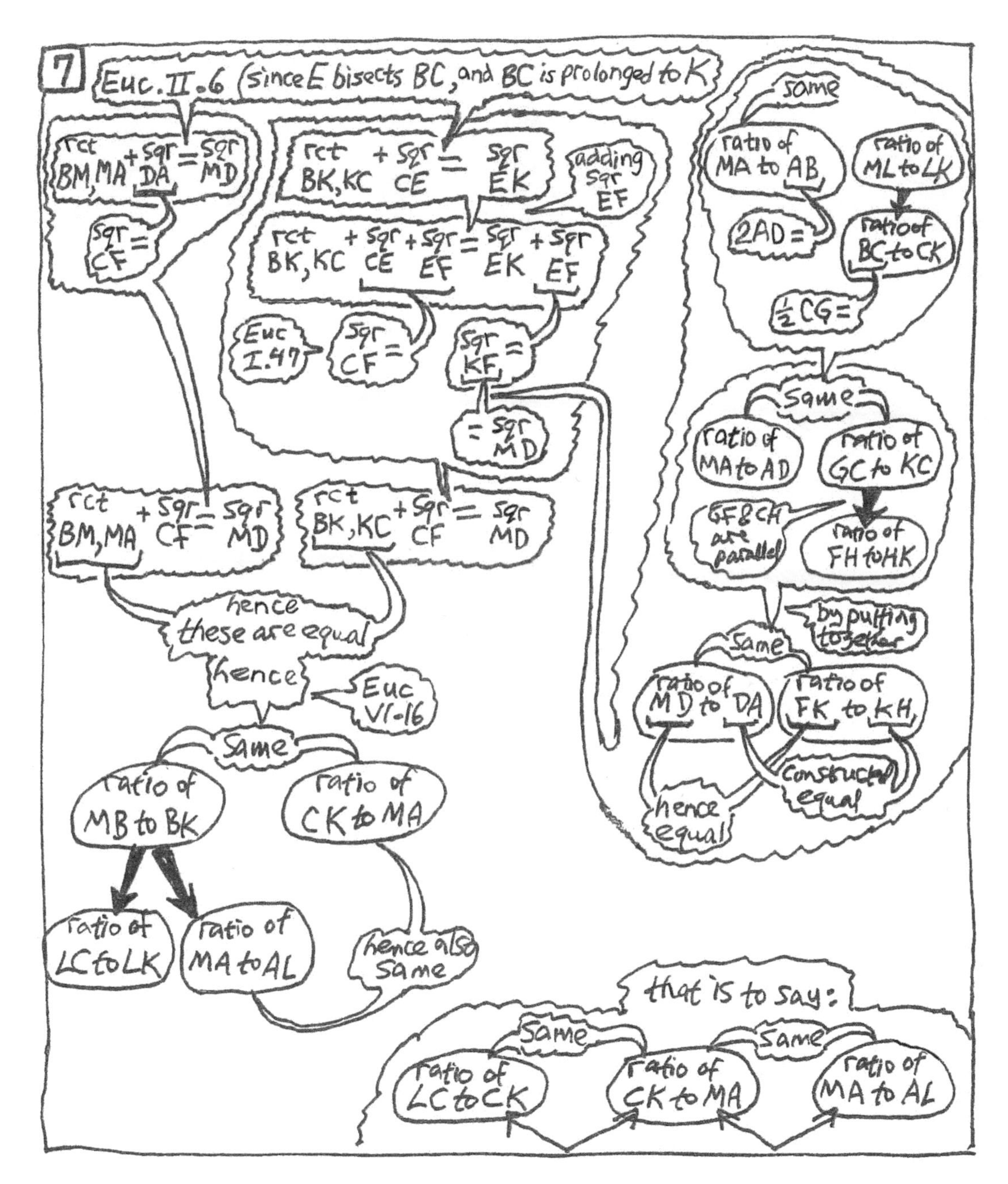

Now let us return to the text of Descartes.

As the first step in the researches beyond conics to which he invites those who have what it takes to make good use of his geometrical "calculus," Descartes tells them to look at certain curves which he supposes to have been drawn through the aid of a certain instrument.

[2] See the lines AB, AD, AF, and the like
[all of them but BA being dotted curves in the accompanying figure],
which I suppose to have been drawn (*descrites*) through the aid of
the instrument YZ, which is composed (*composée*) of several rulers so joined that,
that [one] which is marked YZ being stopped on the line AN,
one can open and close the angle XYZ;

and when it is totally closed, the points B, C, D, E, F, G, H are all assembled at point A;

but in the measure that one opens it,
the ruler BC, which is joined at right angles with XY at the point B,
pushes, toward Z, the ruler CD,
which slides on YZ, while always making right angles with it;
and CD pushes DE,
which slides on YX all the same while remaining parallel to BC;
DE pushes EF; EF pushes FG, this last [one] pushing GH;
and one can conceive an infinity of others
which are pushed consecutively in the same fashion, and of which
the ones always make the same angles with YX, and the others with YZ.

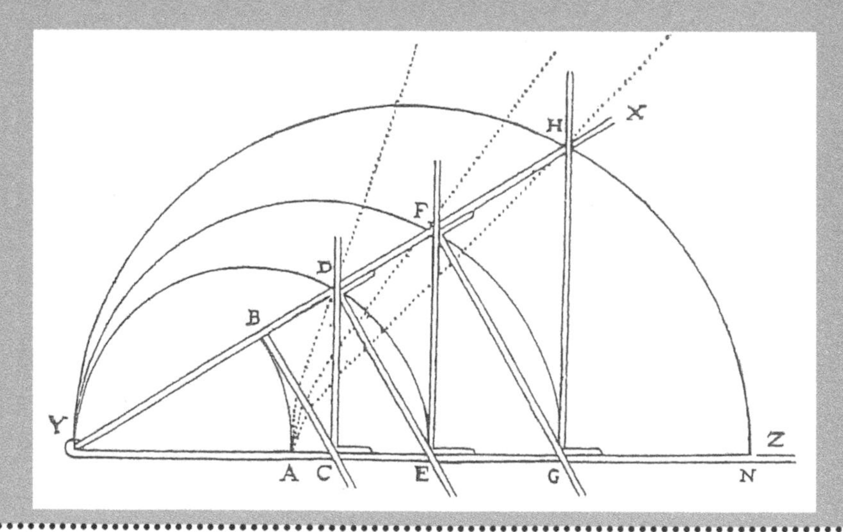

The movement of the parts of the instrument may be easier to visualize by means of figure 8.

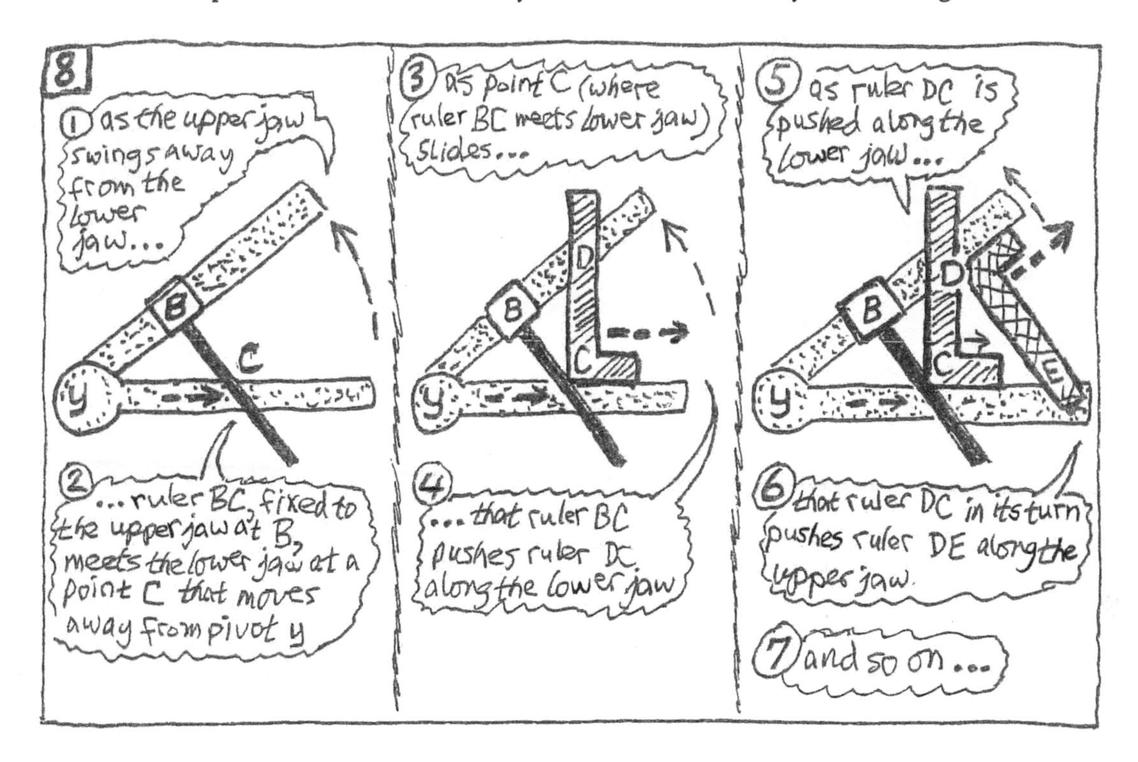

Now while one thus opens the angle XYZ,
the point B describes the line AB, which is a circle;
and the other points D, F, H, where the intersections of the other rulers are made,
describe the other curved lines AD, AF, AH,
of which the later ones are,
in order (*par ordre*), more composite (*composé*) than the first one,
and this [first one is more composite] than the circle;

but I do not see what can hinder that one conceive the drawing (*description*) of this first one
as clearly and distinctly
as [one conceives the drawing] of the circle, or at least of the conic sections,

nor what can impede that one conceive
the second, and the third, and all the others that one can draw (*descrire*) as well as the first;

nor, consequently, that one receive all [of them] in the same fashion
for serving (*pour servir*) in the speculations of Geometry.

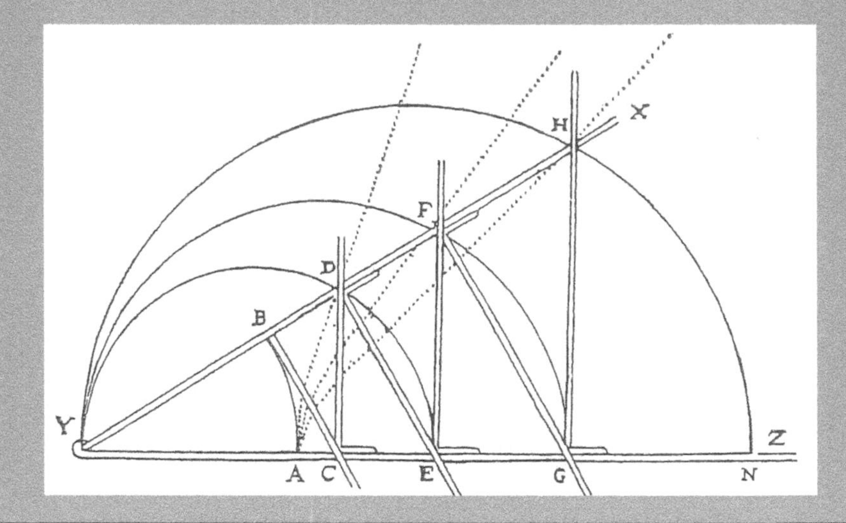

Each curve traced out by a point (namely, curve AD by point D, and curve AF by point F, and curve AH by point H) is the result of a succession of movements, but each later motion is ruled entirely by the movement that precedes it—so you know the point's exact location resulting from the preceding movement; and so these curves drawn by Descartes' instrument are to be called "geometrical," whereas such curves as the spiral and the quadratrix are not.

Such curves as the spiral and the quadratrix do not seem to come in any order, whereas the curves that are successively drawn by Descartes' instrument do come in an order; they come in an order of increasing compositeness. The increasing compositeness comes to sight as an increase in the number of rulers that are put together to draw the curve. Can anything more be said about the matter?

Here, Descartes himself does not say. Here, there does not seem to be much that is of general significance about the instrument for drawing curves that he presents. Only later, at the beginning of the Third Book, will Descartes make it clear that his instrument is a means of finding, between a unit-line and a given line, as many mean proportionals as we please. We could use it, for example, to double a cube by finding two mean proportionals between a pair of lines. (That is, we could find two mean proportionals between the given cube's side and a line that is its double, and use the smaller mean proportional as a side for the cube that is to be the double.) But does the finding of multiple mean proportionals have any general significance? It sure does: to find n-many mean proportionals between the unit and any given quantity is, as we shall see, to solve what we nowadays would call an equation of the n-th degree. For now, however, Descartes leaves off considering the work of that instrument (which, by the way, is called a "proportional compass"). Descartes is not in fact very interested in this or that particular instrument for tracing curves. He is, however, very interested in the orderly and comprehensive general treatment of curves by means of equations that express relations among straight lines.

[3] I could put here several other means
for tracing and conceiving curved lines which would be
more and more composite (*composée*) by degrees to the infinite;

but, for comprehending together all those which are in nature (*sont en la nature*)
and distinguishing them by order (*par ordre*) in certain kinds (*genres*),

I know nothing better to say than

that all the points of those [curves] which one can name Geometrical,
that is to say, which fall under some precise and exact measure,
necessarily have some relation (*rapport*) to all the points of a straight line,
which [relation] can be expressed by some equation—
in all [the points], by one and the same [equation].

And that when this equation does not mount but as far as
to the rectangle [produced] from two indeterminate quantities,
or else to the square from one and the same [indeterminate quantity],
[then] the curved line is of the first and most simple kind (*genre*),
in which there is naught that is comprised but
the circle, the parabola, the hyperbola, and the ellipse;

but that when the equation mounts as far as
to the third or fourth dimension of two [indeterminate quantities] or of one of two [of them]—
for there is need of two of them [that is, two indeterminate quantities are needed]
to explicate here the relation (*rapport*) of one point to another—
[then] it [the curved line] is of the second [kind];

and that when the equation mounts as far as
the 5th or sixth dimension,
it [the curve] is of the third [kind];

and so of the others on to the infinite.

In determining a curve's genre, we are to consider how far its equation mounts—that is, we are to consider the number of dimensions in the term where the number of dimensions is greatest. (Note, however, that what counts is the *total* number of dimensions in the term, since the dimensions can belong to both of the two indeterminate quantities together, and not just to one of them by itself. For example: the term ax^5 has 5 dimensions, as has the term by^5, but so also does the term ax^4y, or the term bx^2y^3.)

The distinguishing of the kinds of curves, ordered according to rank as determined by how composite they are, is not primarily a matter of seeing how many parts are put together to make up the contraption that Descartes has employed for tracing and conceiving them. He shows this by providing next an example of other means for tracing and conceiving curves.

[4] What if (*comme si*) I wish to know of what kind (*genre*)
is the line EC, which I imagine [that is, visualize] to be drawn (*descrite*) by
the intersection of the ruler GL and
the rectilinear plane [figure] CNKL, whose side KN is indefinitely prolonged toward C,
and which, being moved upon the plane downward in a straight line—

> that is to say, [being moved] in such a way (*en telle sorte*) that
> its [side, or in a later example its] diameter KL is found always applied onto
> some place (*endroit*) on the line BA [which line is] prolonged
> from one part and from the other [that is, in both directions]—

makes this ruler GL move circularly around point G,
because it [the ruler GL] is so joined to it [namely, to figure CNKL] that
it [namely, the ruler] always passes through point L.

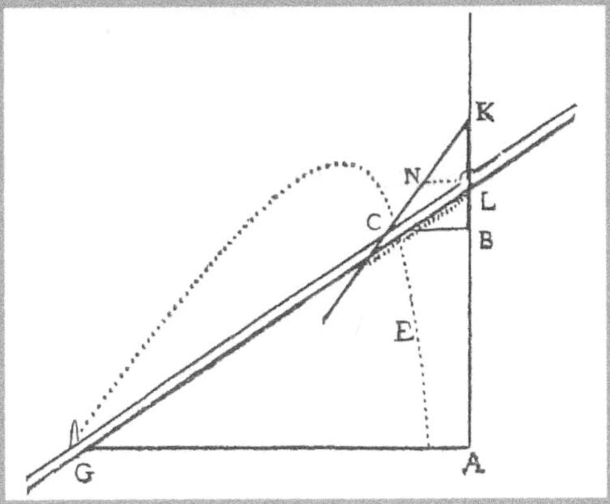

To visualize more easily how that instrument for drawing Descartes' curve EC is put together, see figure 9.

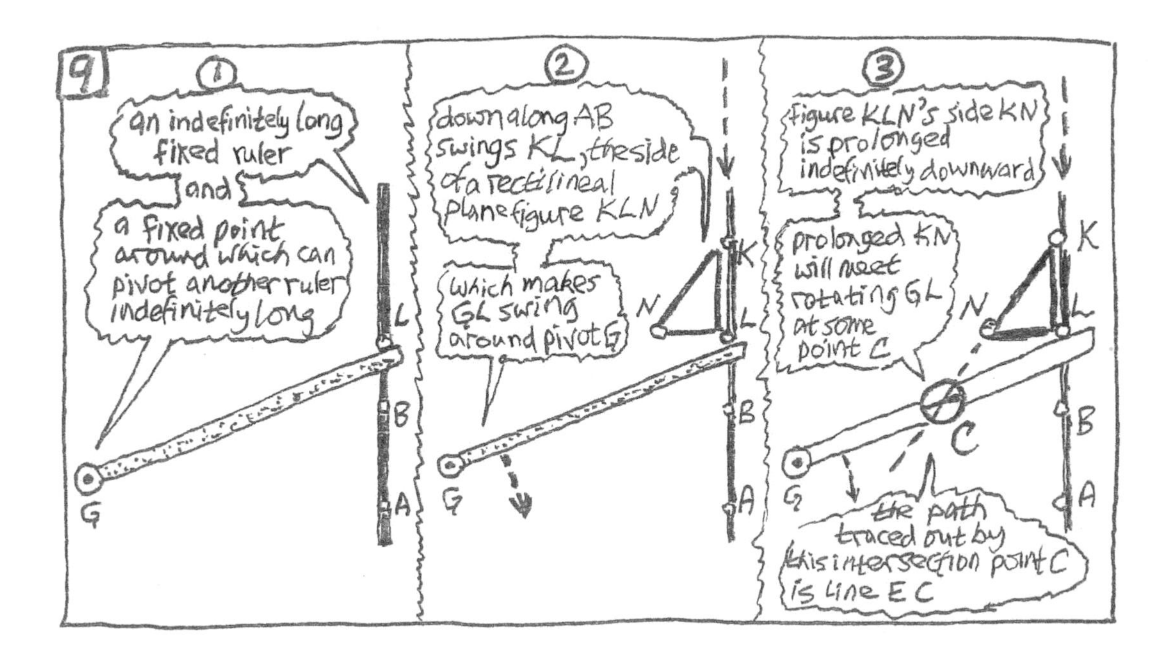

I choose a straight line, as AB,
for relating (*rapporter*), to its different points, all those [points] of this curved line EC;
and in this line AB, I choose a point, as A, for beginning this calculation (*calcul*) through it.

I say that I choose the one [the straight line] and the other [the point],
because it is [a matter of one's being] free to take them such as one wishes;
for although there is many a choice for rendering the equation shorter and easier—
all the same, in whatever fashion one takes them, one can always make [things to be so]
that the [curved] line may appear [as being] of the same kind (*genre*),
as it is easy to demonstrate.

After that, taking a point at discretion in the curve, as C, on which
I suppose that the instrument which serves for drawing (*descrire*) it [the curve] is applied,
I draw from this point C the line CB parallel to GA,
and because CB and BA are two quantities [that are] indeterminate and unknown,
I name one of them y and the other x ;
but, to the end of finding the relation (*rapport*) of the one to the other,
I consider also the known quantities which determine the describing of this curved line—as

GA, ...	which I name a ,
KL, ...	which I name b ,
and NL parallel to GA	which I name c ;

Note that even though the two reference lines may appear to be drawn perpendicular to each other, they do not have to be drawn that way. (Line GA could be drawn so that it does not make a right angle with line AB.) See figure 10.

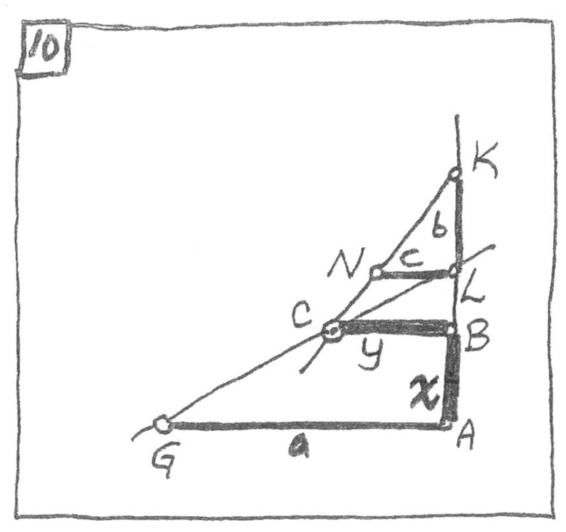

then I say, as NL is to LK, or c [is] to b,
so CB or y is to BK, which [BK] is consequently........... $(b/c)\,y$,

and [hence] BL is..$(b/c)\,y - b$,
and [hence] AL is ....$x + (b/c)\,y - b$,

moreover, as CB is to LB, ...or y [is] to.............................. $(b/c)\,y - b$
so a, or GA, is to..................... LA, or............... $x + (b/c)\,y - b$,

in [such] fashion that (*de façon que*),
multiplying the second [term in the proportion] by the third,
one produces... $(ab/c)\,y - ab$,
which is equal to ....$xy + (b/c)\,yy - by$,
which is produced in
multiplying the first [term in the proportion] by the last,

and so the equation which it was necessary to find is ....yy æ $cy - (c/b)\,xy + ay - ac$,
from which one knows that the line EC is
of the first kind (*genre*)—as in fact it is not other than an hyperbola.

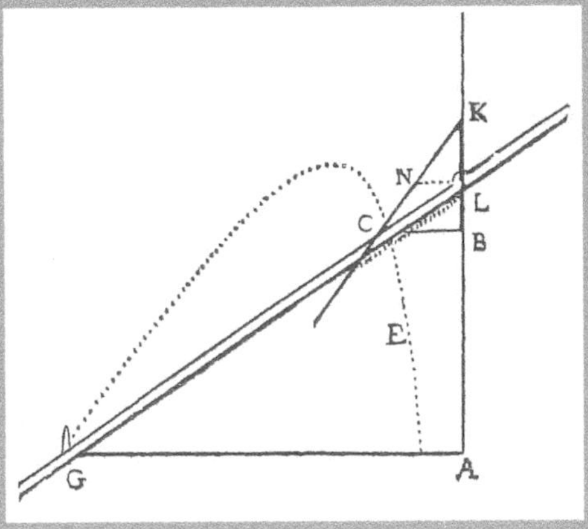

The derivation of the equation for curve EC is depicted in the following figure, after which will come a figure depicting a demonstration that the curve is an hyperbola. (See figures 11 and 12.)

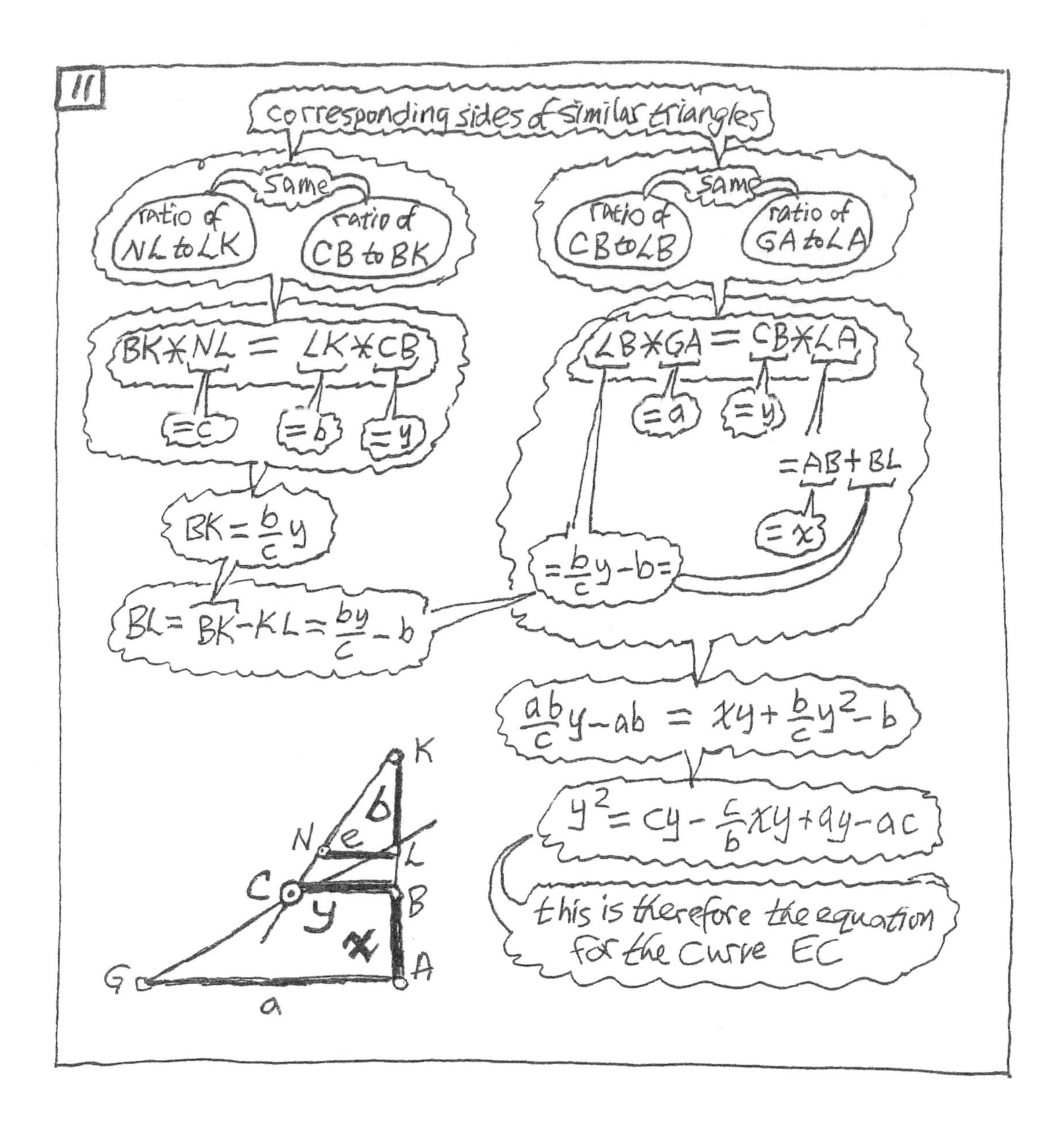
11
corresponding sides of similar triangles
same
ratio of NL to LK
ratio of CB to BK
same
ratio of CB to LB
ratio of GA to LA
BK*NL = LK*CB
=c
=b
=y
BK = b/c y
BL = BK − KL = by/c − b
LB*GA = CB*LA
=a
=y
=AB+BL
=x
= b/c y − b =
ab/c y − ab = xy + b/c y² − b
y² = cy − c/b xy + ay − ac
this is therefore the equation for the curve EC
K
b
N
e
L
C
B
y
x
G
A
a

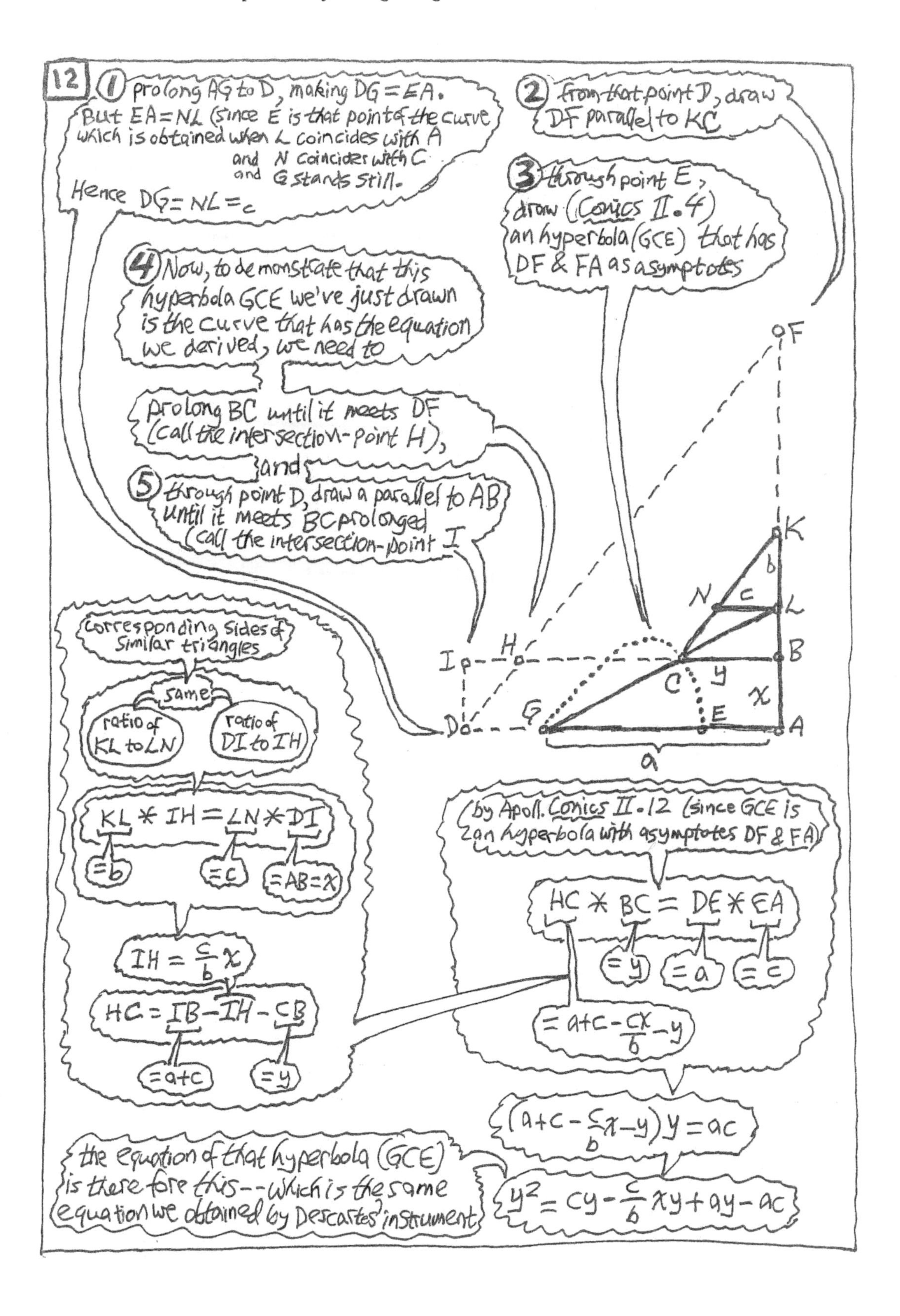
12
① prolong AG to D, making DG = EA.
But EA = NL (since E is that point of the curve which is obtained when L coincides with A and N coincides with C and G stands still.
Hence DG = NL = c
② from that point D, draw DF parallel to KC
③ through point E, draw (Conics II.4) an hyperbola (GCE) that has DF & FA as asymptotes
④ Now, to demonstrate that this hyperbola GCE we've just drawn is the curve that has the equation we derived, we need to
prolong BC until it meets DF (call the intersection-point H),
and
⑤ through point D, draw a parallel to AB until it meets BC prolonged (call the intersection-point I
Corresponding sides of similar triangles
same
ratio of KL to LN
ratio of DI to IH
KL * IH = LN * DI
= b
= c
= AB = x
IH = c/b x
HC = IB − IH − CB
= a+c
= y
(by Apoll. Conics II.12 (since GCE is an hyperbola with asymptotes DF & FA)
HC * BC = DE * EA
= y
= a
= c
= a+c − cx/b − y
(a+c − c/b x − y) y = ac
y² = cy − c/b xy + ay − ac
the equation of that hyperbola (GCE) is therefore this -- which is the same equation we obtained by Descartes' instrument
F
K
b
N
c
L
B
I
H
C
y
x
D
G
E
A
a

In the 4th paragraph, the central and longest of the 7 paragraphs that comprehensively distinguish the geometrical curves, Descartes has used an equation to identify a curve, more complex than a circle, that he has generated by means other than the multi-ruler instrument that he first introduced for the generation of endlessly many different geometrical curves of increasing complexity. The curve that he has examined in that 4th paragraph turns out to be merely one of the conic sections. He therefore now goes on to show how one can, by means of this second device for tracing curves, ascend from the conic sections, step-by-step, to higher curves.

[5] But if (*Que si*) in the instrument which serves for drawing (*descrire*) it [the curve],
one makes [things so] that, instead of the straight line CNK,
it be this Hyperbola, or some other curved line of the first kind (*genre*),
which terminates the plane [figure] CNKL,
[then] the intersection of this line and the ruler GL will describe,
instead of the Hyperbola EC, another curved line, which will be of the second kind (*genre*).

As, if CNK is a circle of which L be the center,
[then] one will describe the first Conchoid of the ancients;

and if it is a Parabola of which the diameter be KB,
[then] one will describe the curved line which a little while ago I said to be
the first and simplest for Pappus's problem (*question*) when there are
naught but five straight lines given in position.

But if instead of one of these curved lines of the first kind (*genre*),
it is one of the second [kind] that terminates the plane [figure] CNKL,
[then] one will describe by its means one of the third [kind] of them,
or if it is one of the third [kind] of them [that terminates the plane figure],
[then] one will describe one of the fourth [kind] of them,
and so on to the infinite,
as it is very easy to know by calculation (*par le calcul*).

And in whatever other fashion that one imagines [visualizes]
the drawing (*description*) of a curved line,
provided that it be of the number of those which I call Geometrical,
one will always be able to find an equation for determining all its points
in this way (*en cete sorte*).

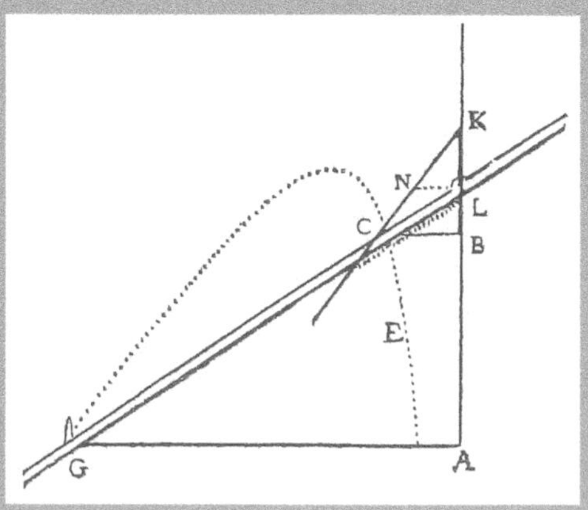

The solution to the 5-line locus problem—according to which the curve which is traced by point C if line CNK is a parabola having KB as its diameter—that is something which Descartes does not show until the 20th paragraph, which is right after he returns to Pappus's problem (in the 19th paragraph).

But if we know what a conchoid is, then Descartes expects us to see that the conchoid is the curve which is traced out by point C if CNK is a circumference of a circle having L as its center. Here is why the conchoid is that curve (see figure 13):

> For every position of point L on fixed straight line AB,
> there is a point C such that the distance from C to L
> (taken along a straight line that is drawn from fixed point G to that L)
> is the same as it is for every other position of L
> (since LC is always a radius of the same circle).
>
> That is, C lies on a conchoid—whose pole is G, ruler is AB, and interval is LC.
>
> Note, however, that with this conchoid,
> unlike the one that we examined earlier, the curve is between its pole and its ruler.
> (In the one that we examined earlier, the ruler was between the pole and the curve.)

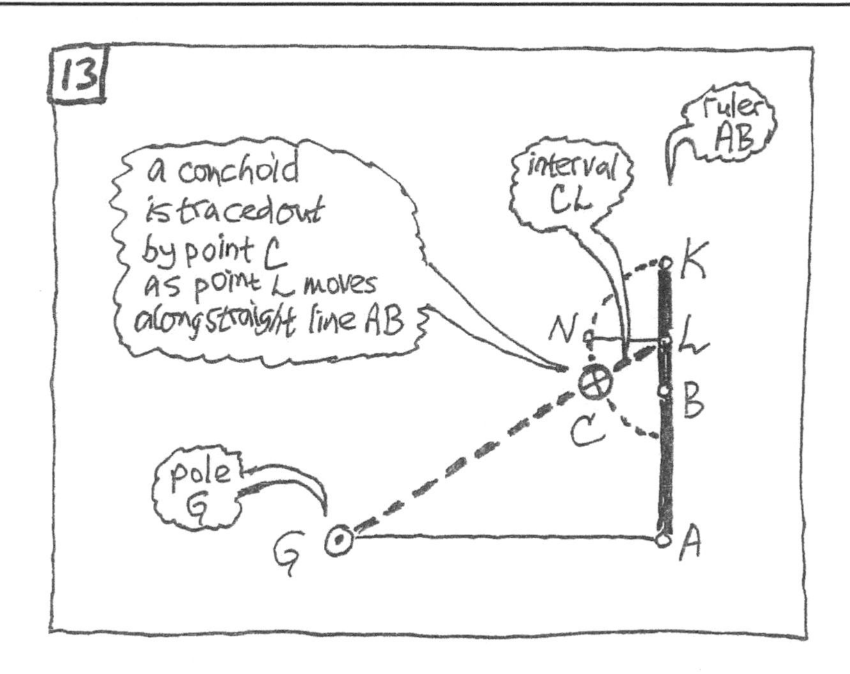

If we wanted to, we could now figure out an equation for the conchoid. Descartes has shown us how to figure out an equation for the curve that is traced out by the intersection point C whatever may be the figure that is used for line KN. Here is how to do the job (see figure 14):

Let	GA	$= a$
	KL	$= b$
and	AB	$= x$
	CB	$= y$
	KB	$= z$.
Then	LB	$= z - b$
	AL	$= x + z - b$.

Now, because parallel lines make similar triangles,

the ratio of	is the same as	the ratio of
GA to AL		CB to BL ;

hence $a/(x + z - b) = y/(z - b)$;

and hence $z = (xy - by + ab)/(a - y)$.

That is true whatever may be the figure CNKL; and now,
given the particular figure used for line KN, we can get a second equation for z,

and then, equating to each other the two expressions for what is equal to z,
we shall get an equation for that curve which is traced out by
C, the point where straight line GL intersects line KN.

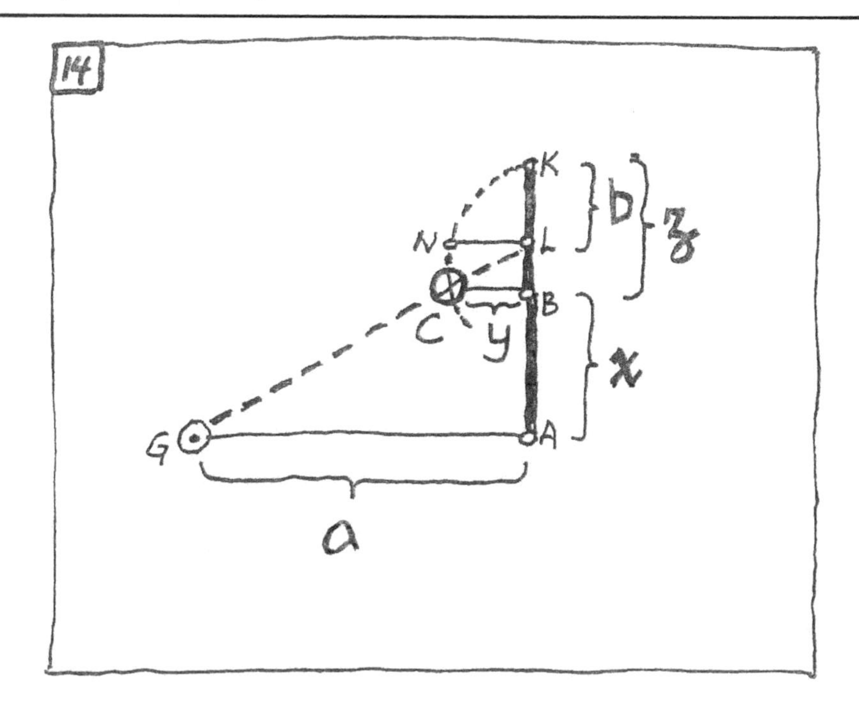

We could also get equations for those curves which Descartes said would be traced by that multi-ruler contraption which he trotted out as he began this Second Book but quickly put aside. First, for curve AD, traced out by point D, see figure 15.

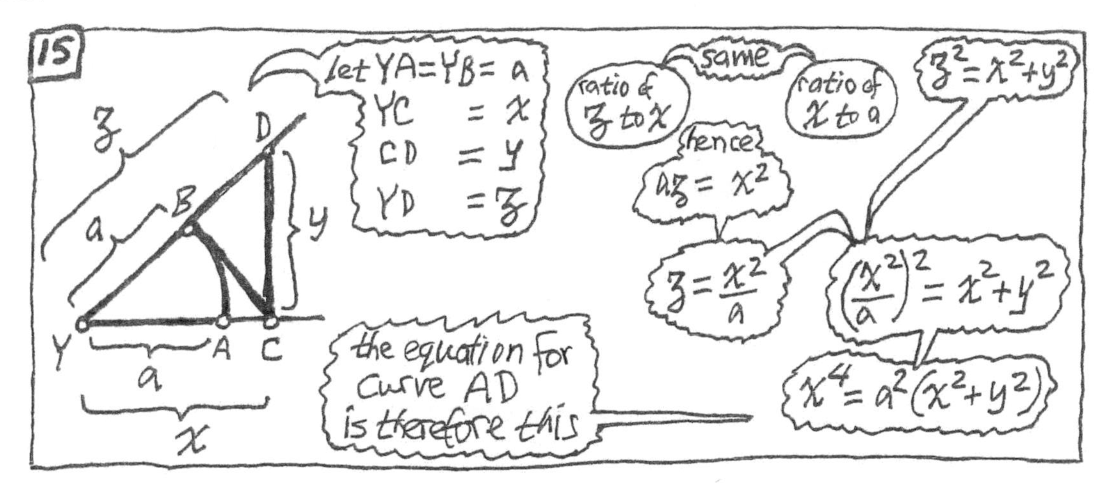

Now for curve AF, traced out by point F, see figure 16.

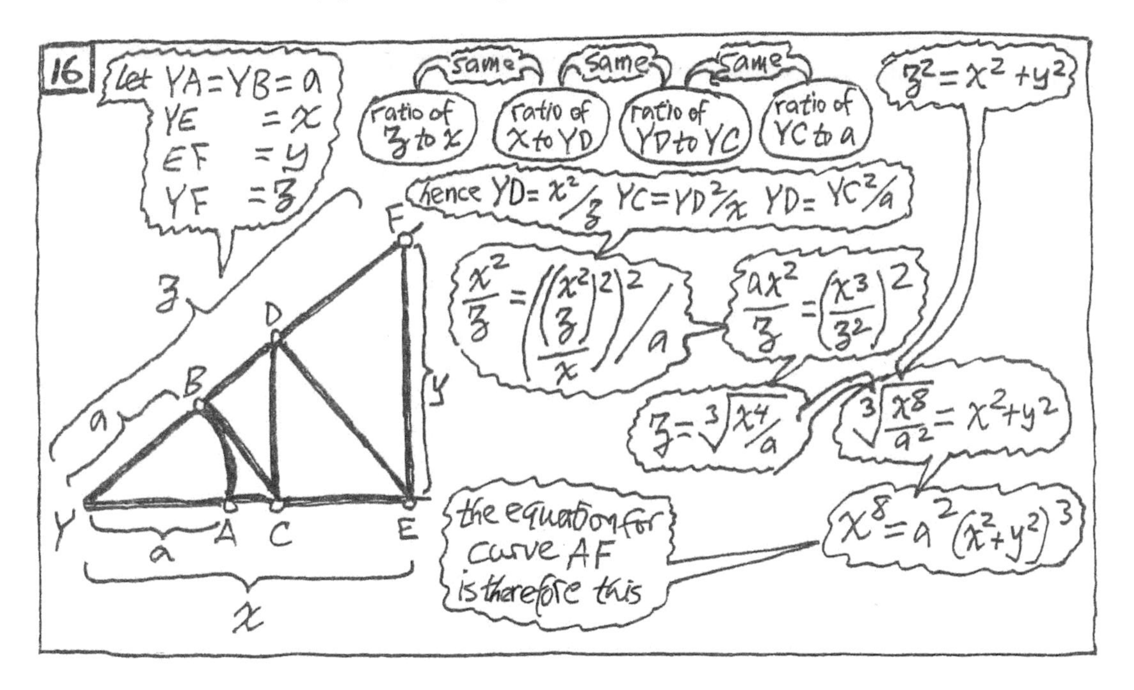

And finally, proceeding in the same fashion, we would get an equation for curve AH, traced by point H. For all three curves, our results are as follows:

curve	*equation*
AD	$x^4 = a^2 (x^2 + y^2)$
AF	$x^8 = a^2 (x^2 + y^2)^3$
AH	$x^{12} = a^2 (x^2 + y^2)^5$

[6] What is left to say is this (*Au reste*):
I put the curved lines which make this equation mount up to the square of a square
into the same kind (*genre*) as those which make it mount up to the cube;
and [I put] those whose equation mounts up to the square of a cube
into the same kind (*genre*) as those which mount up but to the supersolid;
and so on, of the others.

The reason of which is that there is a general rule for reducing
to the cube all the difficulties which go with the square of a square,
and to the supersolid all those which go with the square of a cube—
in [such] fashion that one should not give consideration to
the [ones that are] more composite (*composées*).

[7] But it is to be noted that among the lines of each kind (*genre*),
even though the greater part be equally composite (*composées*),
so that (*en sorte que*) they can serve
to determine the same points and to construct the same problems,
there are at the same time also some of them which are more simple,
and which do not have so much extent in their power—

as, among those of the first kind (*genre*),
besides the Ellipse, the Hyperbola, and the Parabola,
which are equally composite (*composées*),
there is also comprised within it the circle,
which is manifestly simpler [than are the ellipse, the hyperbola, and the parabola],

and among those of the second kind (*genre*),
there is the common Conchoid (*Conchoide vulgaire*), which has its origin from the circle,
and there are others of them which,
although they do not have so much extent as [do]
the greater part of those of the same [namely, the second] kind (*genre*),
at the same time cannot be put in the first [kind (*genre*)].

Now it is clear why, for every *two* additional *dimensions* in the *equation*, Descartes counts only *one* additional *genre* of *curve*. The reason is this: when a curve's equation goes up to the square of a square, then its difficulties can be reduced to those of a curve whose equation goes up to a cube; and when the equation goes up to the square of a cube, then its difficulties can be reduced to those of a supersolid; and so on up the ladder. Nowadays, we would say that when a curve's equation is of the 4th degree, the difficulties can be reduced to the difficulties of a curve whose equation is of the 3rd degree; and when the equation is of the 6th degree, the difficulties can be reduced to those of the 5th; and so on. There is still a vestige of the older terminology of kinds in Descartes' discussion—he speaks, for example, of equations mounting up to "the square of a square" or to "the square of a cube"—but he nonetheless distinguishes the infinitely many genres of curves numerically, and he does so with a view to how their equations can be solved. For him, the parabola, the hyperbola, and the ellipse are curves of the same genre, but that is not because all of them have their genesis in the cutting of a cone. Nor for Descartes are they to be studied primarily with a view to their distinctive *eidê*. Access to the nature of things will come by way of equations. Descartes will improve our ability to handle and employ equations. He says here that there is a general rule for reducing those difficulties to which he has referred. Not until the Third Book, however, does he present the rule for performing the reductions. We are now only at the end of the beginning of the Second Book.

Here in the Second Book's beginning, what has been accomplished? These last 7 paragraphs have extended the scope of geometry to curves beyond the conic sections. The curves that are to be taken as "geometrical" include more than circles and conic sections, but they do not include all curves. The ones that are not included are, however, excluded for a reason. Those which are of a geometrical nature comprise a system; they can be treated methodically; they can be comprehended all together and distinguished by genres which line up in an order of compositeness. The succession in the degrees of compositeness of the geometrical curves corresponds to a rise in the number of dimensions in that unknown term which has the most dimensions in the equation that expresses how the points on the geometrical curve are related to certain straight lines. The geometrical "calculus" makes it possible to investigate the curves by means of equations. As the term with the most dimensions in the equation mounts higher and higher, the algebraic relations correspond to the putting together of more and more ratios between the unit chosen at discretion and the unknown quantity marked off along one of the reference lines. The successive movements which trace out any curve that is to be taken as geometrical must fall under some precise and exact measure by which the earlier movement entirely rules the later one. The systematic character of the treatment of curves depends upon the restricted character of the expansive definition of those curves which are to be taken as "geometrical." If the totality of geometrical curves did not exclude curves that are beyond expressibility in algebraic equations, then the geometrical curves could not be treated "methodically"; not every one of them could be placed at some step in a successive order of genres. The manner in which the geometrical curves are distinguished in their totality from those which are not geometrical is what makes possible the manner in which they are distinguished from each other, as falling into particular genres arranged in an order of succession that is endless but exhaustive.

Where in the argument of Descartes' *Geometry* does that leave us now?

Earlier, the first 17 paragraphs of the First Book introduced the terms of arithmetic into geometry, so that the solution of geometrical problems could become a geometrical calculus, thus vastly increasing human constructive power; and the last 17 paragraphs of the First Book gave an answer to the ancient unsolved question put by Pappus, and presented the first part of a demonstration that Descartes' answer does in fact solve the problem.

Now, before going on to the second part of the demonstration, the Second Book's first 7 paragraphs have expanded the scope of geometry to include curves more composite than the conic sections. Indeed, these paragraphs have laid out the totality of geometrical curves as a system: these are methodically examinable as a succession of ever more composite geometrical entities, grouped into particular genres of ever higher degree, succeeding one another in an order that goes along with the order of ever-mounting equations that express them.

And now, this 7-paragraph presentation of Descartes' new way to distinguish the geometrical curves—in their totality from others, and from each other in the order of their particular genres—will be followed by 17 paragraphs which will pursue to completion the demonstration that the author's geometrical calculus along with his system of geometrical curves will solve Pappus's problem.

CHAPTER 5. THE SOLUTION OF PAPPUS'S PROBLEM: FINISHING UP

The next 17 paragraphs (that is, paragraphs 8–24 of Book Two) finish up the solution of Pappus's problem.

[8] Now after having thus reduced all the curved lines to certain kinds (*genres*),
it is easy for me to follow through on the demonstration of
the response which I made a little while ago to Pappus's problem (*question*).

For first, having made [one] see, above, that when there is naught but three or 4 straight lines,
the equation which serves to determine the points [that are] sought
does not mount up but to the square,
it is evident that the curved line where these points are found
is necessarily some one of those [curved lines] of the first kind (*genre*),
because this same equation explicates the relation which all the points
of the [curved] lines of the first kind (*genre*) have to those of a straight line.

And [it is evident] that when there are no more than 8 straight lines given,
this equation does not mount up but to the square of a square at the very most,
and that consequently the line [that is] sought can be naught but
of the second kind (*genre*) or below.

And that when there are no more than 12 lines given,
the equation does not mount up but to the square of a cube,
and that consequently the line [that is] sought is of naught but
the third kind (*genre*) or below.
And so on, of the others.

And, indeed (*mesme*), because the position of the given straight lines
can vary in all sorts (*sortes*) [of ways],
and consequently [can] make the known quantities change
in all the fashions imaginable, as many as the signs + and — [make possible],
it is evident that there is not any curved line of the first kind (*genre*) which
may not be useful for this problem (*question*) when it is proposed in 4 straight lines,
nor any [curve line] of the second [kind] which may not be useful when it is proposed in eight,
nor [any curved line] of the third [kind] when it is proposed in twelve,
and so on, of the others.

So that (*ensorte que*) there is not a curved line that
falls under calculation [or: under "the calculus"] (*le calcul*) and could be received in Geometry
which may not be useful there for some number of lines.

This 8th paragraph has thus returned to Pappus's problem by saying why the discussion of the kinds (*genres*) of curves was not a digression but is in fact relevant to the solution of Pappus's problem when more than a few straight lines are given.

The next 8 paragraphs now treat the various cases when merely a few, that is, when merely 3 or 4 straight lines are given. The 8 paragraphs that will come right after those 8 have the following contents:

> there will come, first, 2 paragraphs that
> will consider what has been accomplished;
>
> then, there will come 4 paragraphs that
> will treat some of the cases when 5 straight lines are given
> and will conclude by saying that it is
> unnecessary to treat any more cases
> because the fashion of finding points that has been explained
> gives means of drawing the curves;
>
> and, finally, there will come 2 paragraphs that
> will further discuss the means of drawing curves.

Thus, the entire 17-paragraph second part of the treatment of the problem of Pappus, which we have just begun, will have the following structure:

```
paragraph number        how many paragraphs

[8]                     1
                                                    \
[9] to [16]             8                           |
                                                    |
[17] to [18]            2                 \         |——14
                                          |         |
[19] to [22]            4        \        |——8      /
                                 |——6     |
[23] to [24]            2        /        /
```

when 3 or 4 straight lines are given (*paragraphs 9–18*)

[9] But it is necessary here more particularly that
I determine and give the fashion of finding the line sought
which serves in each case when there is naught but 3 or 4 straight lines given;

and one will see by [this] very means
that the first kind (*genre*) of curved lines does not contain any others of them but
the three conic sections and the circle.

[10] Let us again take up the 4 lines AB, AD, EF, and GH given above,
and let it be necessary to find another line, in which are encountered an infinity of points such as C, having drawn from which [point C]
the 4 lines CB, CD, CF, and CH at given angles onto the given [lines],
CB multiplied by CF produces a total equal to [that produced by] CD multiplied by CH;

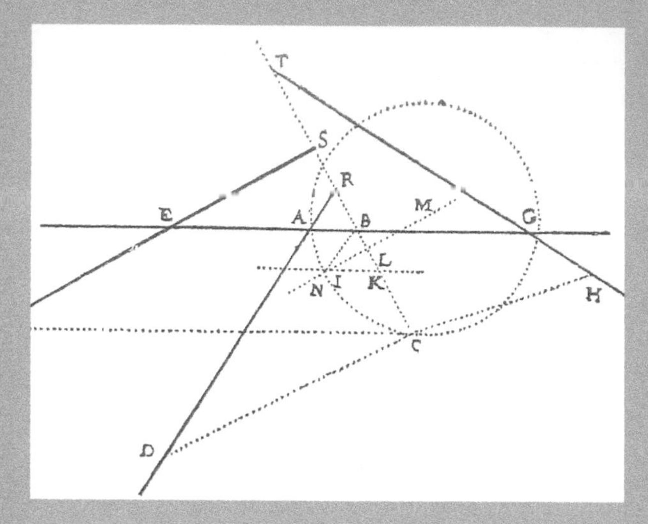

[see figure 1]

that is to say, having made [as was done earlier]

CB æ y, CD æ $(czy + bcx)/zz$,

CF æ $(ezy + dek + dex)/zz$, and CH æ $(gzy + fgl + fgx)/zz$,

[we obtain the following equation by equating the products and transposing terms:]

yy æ
$[(dekzz + cfglz)\,y + (-dezzx - cfgzx + bcgzx)\,y + (bcfglx - bcfgxx)\,]/(ezzz - cgzz)$,

at least [that is the equation] in [the case of our] supposing ez [to be] greater than eg,
for if it were less, [then] it would be necessary to change
all the lines [that is, the signs before them] + [to —] and — [to +].

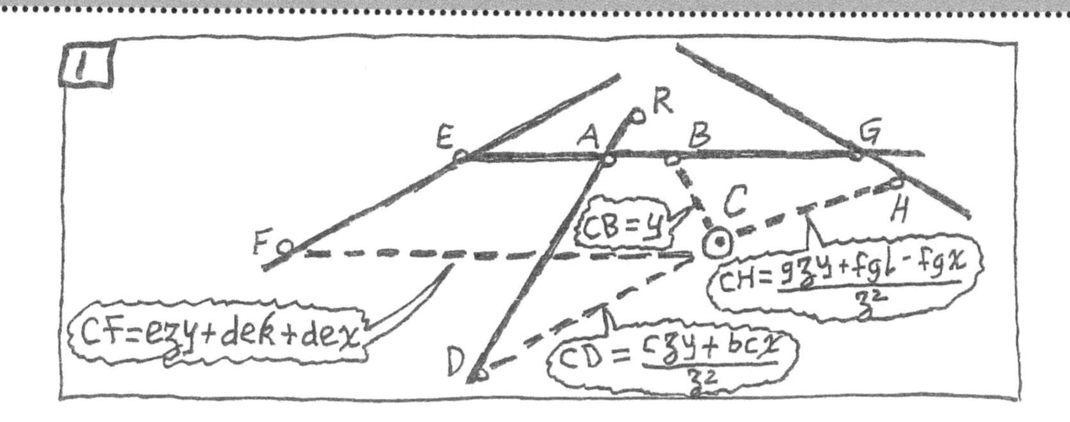

Note that Descartes is taking the case where the constant ratio of the two products is a ratio of equality. This will make the resulting equation a bit simpler than it would be otherwise. But even so, things will not be as simple as we might like. We shall have to deal with the consequences of the following fact: while the equation contains terms having x without y and y without x, the rest of the complicated side of the equation has x and y together, rather than having neither of them. There would have been a term without either of them—that is to say, there would have been a term entirely constituted of quantities that are given—if the products that were taken to be equal were CB*CD and CF*CH, rather than CB*CF and CD*CH.

In any case, Descartes sets before us only some of his steps. In a letter to Mersenne, he spoke about his handling of the problem of Pappus: he said that he had given only the construction and demonstration without including the whole analysis; that is, that he had done the work as architects build—giving specifications and leaving the manual labor to carpenters and masons.

Nowadays we would write what is in the box following—

Expressions for the distances from point C to the four given lines as obtained 14 paragraphs earlier (in the 30th paragraph of the First Book) are:

$CB = y$ $\qquad$ $CD = (czy + bcx)/z^2$

$CF = (ezy + dek + dex)/ z^2$ $\qquad$ $CH = (gzy + fgl - fgx)/ z^2$

The product (CB * CF) is now set equal to the product (CD * CH), yielding the following:

$$(ezy^2 + deky + dexy/ z^2 = \{[c\,(zy + bx)]\ [g\,(zy + fl - fx)]\}/ z^4$$

$$(ez^3y^2 + dekyz^2 + dexyz^2)/z^4 = [cg\,(z^2y^2 + bxzy + zyfl + bxfl - zyfx - bxfx)]/ z^4$$

$$ez^3y^2 + dekyz^2 + dexyz^2 = cgz^2y^2 + cgbxzy + cgzyfl + cgbxfl - cgzyfx - cgbxfx$$

$$y^2\,(ez^3 - cgz^2) = y\,(-dekz^2 - dexz^2 + cgbxz + cgflz - cgzfx) + cgbxfl - cgbxfx.$$

Finally, we obtain what y^2 is equal to—expressing it by grouping together

terms with y and also with some z but without any x	terms with y and also with some z and with some x too	terms without any y and also without any z but with some x :
$[y(-dekz^2 + cfglz) +$	$y(-dez^2x - cfgzx + bcgzx) +$	$(bcfglx - bcfgx^2)/\ ]\ /\ (ez^3\ cgz^2).$

Now since the quantities all represent some length of line, the smaller ones are taken away from the larger; this accounts for which of the terms are preceded by a plus-sign and which by a minus. If the quantity *ez* were less than the quantity *cg*, then the denominator of the expression to which y^2 is equal would have to be, not $ez^3 - cgz^2$, but rather $-ez^3 + cgz^2$, and then the signs in the numerator would have to be changed too.

Why is it, that if we suppose a configuration in which CR is greater than CB, and CT is greater than CS, then ez^3 must be greater than cgz^2? Figure 2 shows why.

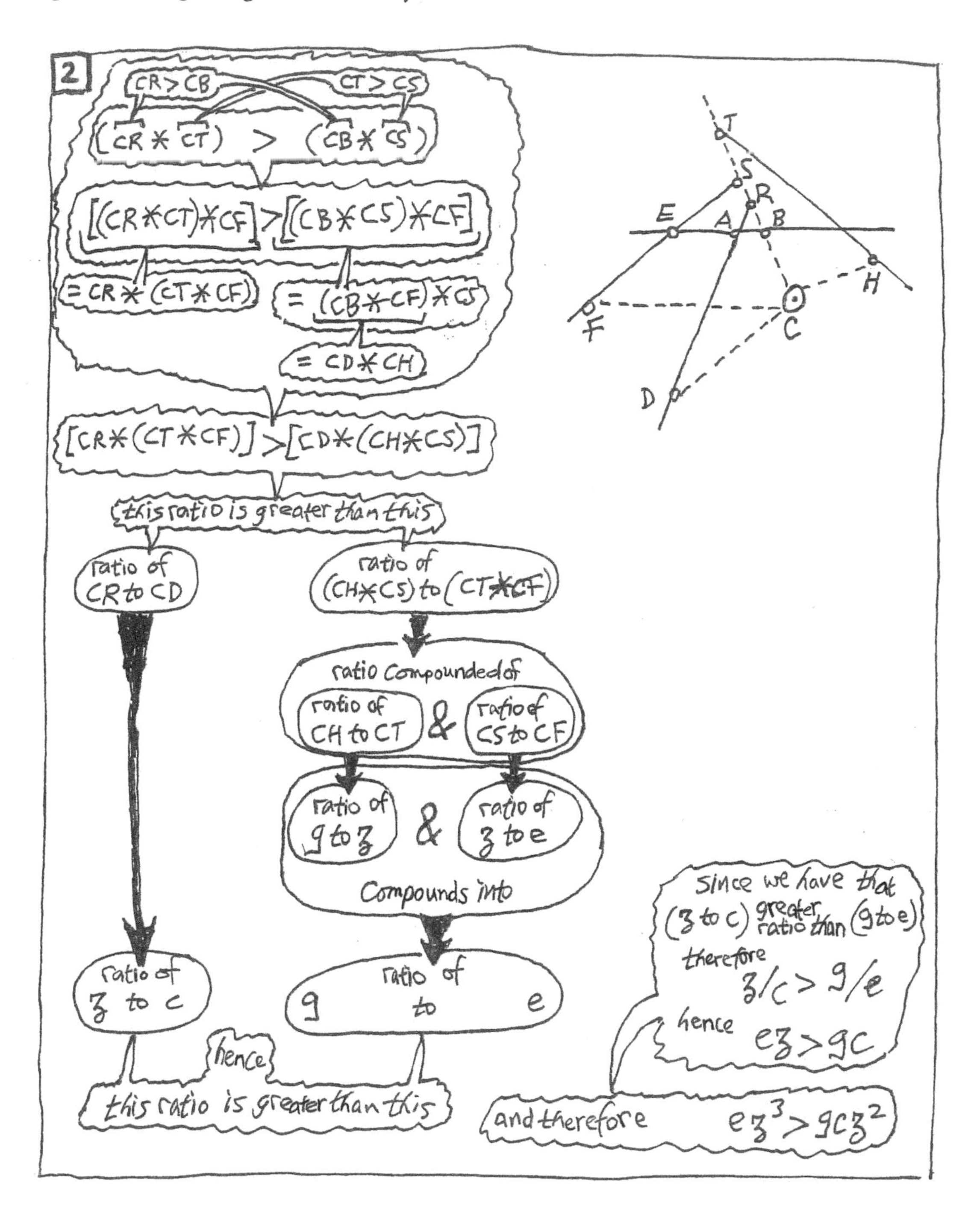

And if the quantity y be found [to be] null or [to be] less than nothing in this equation,
when one supposed the point C [to lie] in the angle DAG,
it would be necessary also to suppose it [to lie] in the angle DAE, or EAR, or RAG,
in changing the line + and – according as would be requisite for this effect.

And if in all these 4 positions the value of y is found to be null,
[then] the problem (*question*) would be impossible in the case proposed.

All of Descartes' letters represent quantities which refer to, or are represented by, lengths of straight lines. If a letter represents an outcome of combining arithmetical operations upon pairs of lines of some length (that is, upon pairs of lines that are greater than nothing), then that letter will represent a quantity which refers to, or is represented by, either a line (which is greater than nothing) or the nothing that results when one line is subtracted from another line to which it is equal. The minus-sign is a sign of subtraction, and a line cannot be taken away from a smaller line or from nothing at all. (That is why the equations that were treated in paragraphs 13–15 of the First Book, while they did include the three equations of the form where y^2 equals $ay + c$ or equals $c - ay$ or equals $ay - c$, did not include the one where it equals $-y - c$.)

The point C must be supposed to lie in that region (labeled, in figure 3, as I or II or III or IV) for which the quantity y will be positive in the equation, and the signs of terms accordingly will either be plus and minus, or else be minus and plus.

What Descartes says is misleading, though. It is true that the signs in the equation will have to be changed (as Descartes says) if the point is in region II, III, or IV. But it is also true (although Descartes neglects to say so) that some modification of sign will be necessary also in part of region I—namely, in that part contained by AD and AQ, the latter being a straight line from point A that is drawn parallel to line BC. In the rest of region I (namely, in the part contained by AQ and AB) the signs remain unmodified from what they are in Descartes' equation. The part of region I where some modification of sign would be needed (which is the part contained by angle DAQ) is indicated by stripes in figure 4.

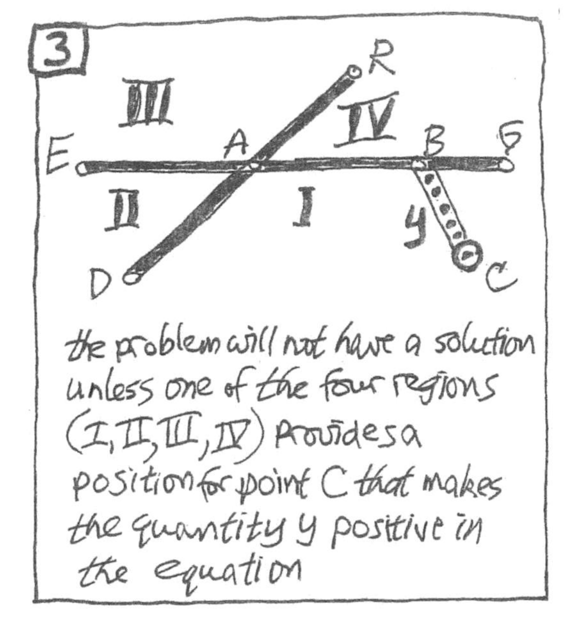

3
R
III
IV
E
A
B
G
II
I
y
D
C
the problem will not have a solution unless one of the four regions (I, II, III, IV) provides a position for point C that makes the quantity y positive in the equation

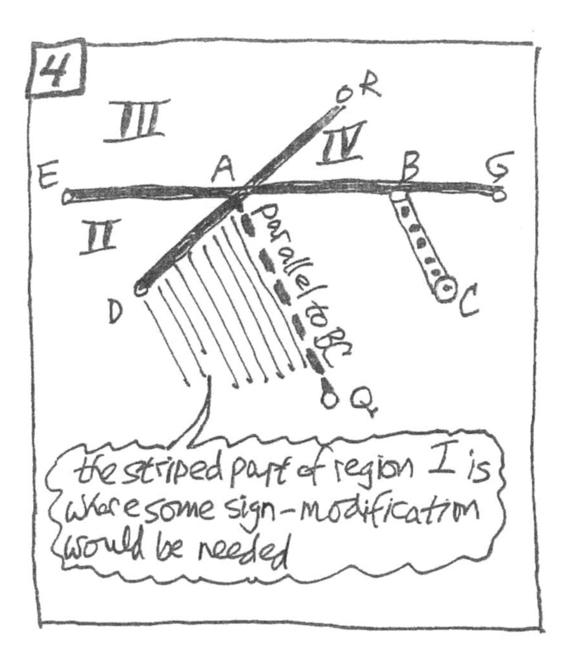

But here let us suppose it is possible
[to get a solution for the problem in the case proposed]

and—for abbreviating the terms of it—
instead of the quantities $(cfglz - dekzz) / (ez^3 - cgzz)$, let us write $2m$,
and instead of $(dezz + cfgz - bcgz) / (ez^3 - cgzz)$, let us write $2n/z$;

and so we shall have

$$yy \;\text{æ}\; 2my - (2n\,xy/z) + [(bcfglx - bcfgxx) / (ez^3 - cgzz)],$$

of which the root is

$$y \;\text{æ}\; m - (nx/z) + \sqrt{\{mm - (2mnx/z) + (nnxx/zz) + [(bcfglx - bcfgxx) / (ez^3 - cgzz)]\}};$$

and again for abbreviating,
instead of $-(2mn/z) + [(bcfgl) / (ez^3 - cgzz)]$, let us write o,
and instead of $(nn/zz) - [(bcfg) / (ez^3 - cgzz)]$, let us write p/m,
for, these quantities being all given, we can name them as we please.

And thus we have $y \;\text{ae}\; m - (n/z)x + \sqrt{[mm - ox + (p/m)xx]}$,
which should be the length of the line BC,
leaving AB or [to use the name that we assigned it] x undetermined.

And it is evident that,
the problem (*question*) being proposed in naught but three or four lines,
one can always have such terms, except
that some of them can be null,
and that the signs + and − can be changed in different ways.

Descartes does not consider the case where he would get an equation like this:

$$y^2 = ay - b^2;$$

rather, he considers only the case where he would get an equation like this:

$$y^2 = ay + b^2.$$

He therefore does not consider this solution:

$$y = m - (n/z)\,x - \sqrt{[\,m^2 + ox + (p/m)\,x^2]};$$

rather, he considers only this solution:

$$y = m - (n/z)\,x + \sqrt{[\,m^2 + ox + (p/m)\,x^2]}.$$

He also does not consider the possibility that he would have something more complicated than the m^2 that is the first of the terms inside the radical. The m^2 arises from the squaring of $\{(1/2)\,[2m - 2\,(n/z)]\}$. But if, instead of taking CB*CF and CD*CH, what he had taken had been CB*CD and CF*CH, then what would be inside the radical would be not simply m^2; rather, it would be m^2 + something (and then, too, the coefficient of x^2 would not be p/m). The outcome is simplest when the first term inside the radical (namely, m^2) is the square of the first term outside (namely, m), but other cases are possible. Descartes has taken the terms CB, CD, CF, CH in the particular way that will yield a simpler outcome than he would get otherwise.

Having obtained a value for BC algebraically in the preceding paragraph, we now proceed to construct it geometrically in the next paragraph. By constructing line BC we are preparing what we need in order to find point C—that is, what we need in order to find the locus of points that satisfy the condition for C. The line BC has been named y, so the lines that are the ingredients in constructing BC will be expressed by the terms found in the expression that is equal to y , namely:

> m , from which we take away $(n/z)x$, and then,
> to the difference, add $\sqrt{[mm + ox - (p/m)x]}$,
> where x is what we named AB.

So Descartes goes on:

> [11] After this, I make KI equal and parallel to BA,
> in such a way that (*en sorte que*) it cuts from BC the part BK equal to m ,
> because here [in the expression for BC] there is $+m$;
>
> and I would have added it [that is, added BK onto BC]
> in drawing this line IK on the other side [of AB]
> if [in the expression for BC] there had been $-m$;
>
> and I would not have drawn it at all
> if the quantity m had been null.

In other words: if, in the expression for y, we had had m with a minus instead of a plus sign, we would then prolong CB to the other side of AB, and then take point K on it at a distance m from B; in that case, BK would make KC in a different way—not by being cut away from BC, but rather by being added onto BC. But if m had counted for nothing in the expression for y, then there would have been no need for any BK at all. See figure 5.

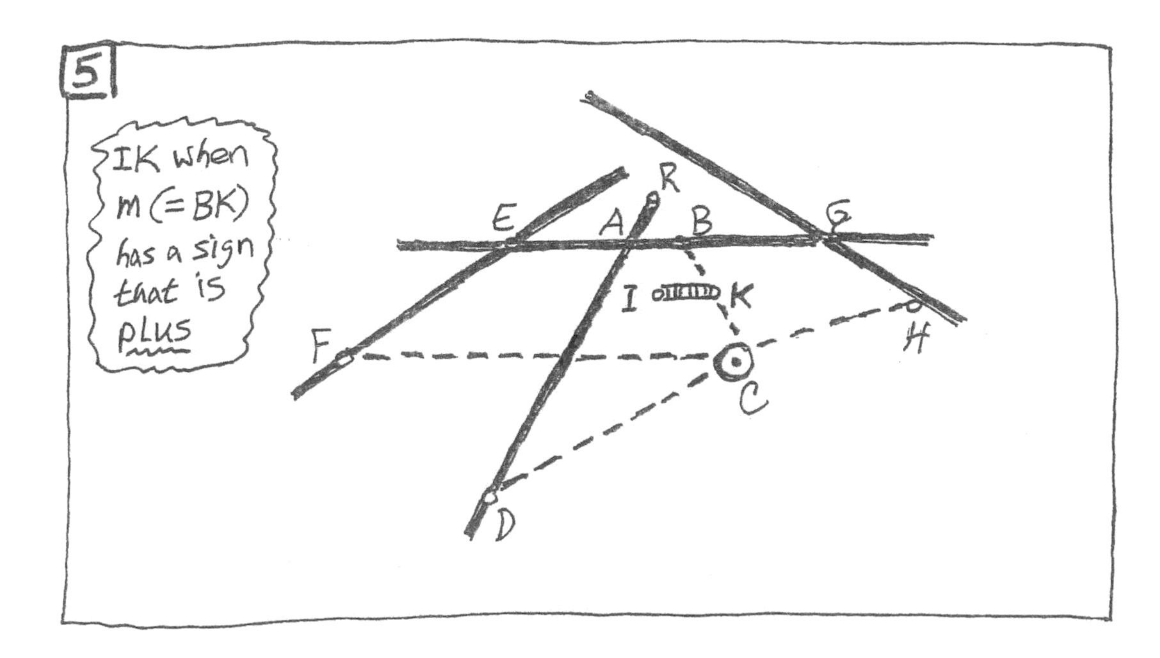

Then I draw also IL,
in such a way that (*en sorte que*)
the line IK is to KL as z is to n ;
that is to say that, IK being x , KL is $(n/z)x$.

And by the same reasons I know also
the ratio (*proportion*) which is between KL and IL,
which I put as [the same as the ratio] between n and a ,
so that (*si bien que*), KL being $(n/z)x$, IL is $(a/z)x$.

And I make the point K be between L and C,
because here [in the expression] there is $-(n/z)x$,
instead of which I would put L between K and C if I had had $+(n/z)x$;
and I would not have drawn this line IL at all if $(n/z)x$ had been null. [See figure 6.]

If $(n/z)x$ is negative, then wherever point C is located, point L will be on the other side of K from it. If $(n/z)x$ is positive, then point L will be on the same side of K as C. If $(n/z)x$ is zero, then points K and L will coincide.

For triangle ILK, the ratio of two sides (of IK to KL) is given (it is the ratio of z to n), and also given is the angle between the two sides—this angle IKL is the supplement of given angle KBA (since IK is parallel to AB). The shape of that triangle ILK is therefore given, and hence so is the ratio of any two of its sides (not just the ratio of the two sides first mentioned, of IK to KL, but also that of KL to IL). But n is a given quantity—it was introduced as an abbreviation:

for $[(dezz + cfgz—bcgz) / (ez^3—cgzz)]$, we wrote $2n/z$.

Since the ratio of KL to IL is given, and so is n, therefore also given is the fourth proportional—we shall call it a—such that as KL is to IL, so n is to a. And now we can say what is the quantity to which IL will be equal when IK is taken to be x. When IK is taken to be x , then IL will be $(a/z)\,x$. See figure 7.

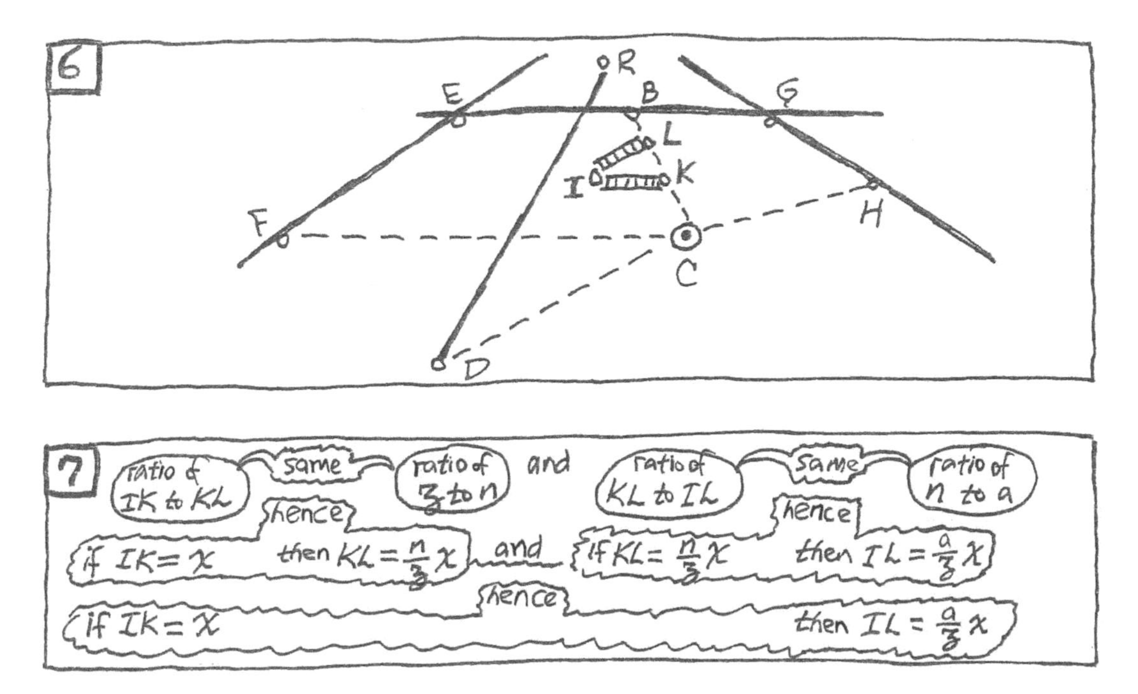

But just why is it, that we should draw those straight lines KI and IL? And once we have seen that the part of BC which is KL is $(n/z)\,x$—well, just why is it, that we should want to know that IL is $(a/z)\,x$? Those things are not at all clear. We may suppose that they will come in handy; but just how it is that they will come in handy, we do not yet know.

Be that as it may, line BC is equal to (BK – KL + LC), and we have constructed lines BK and KL, thus getting the easiest part of BC, namely, the line (BK – KL), which is merely $m - (n/z)\,x$. All that is left, therefore, is to construct line LC; and then we shall be able to determine where point C is located. See figure 8.

[12] Now this [being] done, there does not remain to me any more than,
for the line LC, these terms: LC æ $\sqrt{[\ mm + ox + (p/m)xx\ ]}$,

whence I see that if they were null,
[then] this point C would be found in the straight line IL;

and that were they such that the root could be extracted (*tirer*) from them,
that is to say, [were they such]

> that, *mm* and *(p/m)xx* being marked by one and the same sign, + or – ,
> *oo* were equal to 4*pm*,
>
> or else that
> the terms *mm* and *ox*, or *ox* and *(p/m)xx* were null,

[then] this point C would be found in another straight line which
would not be any less easy to find than IL.

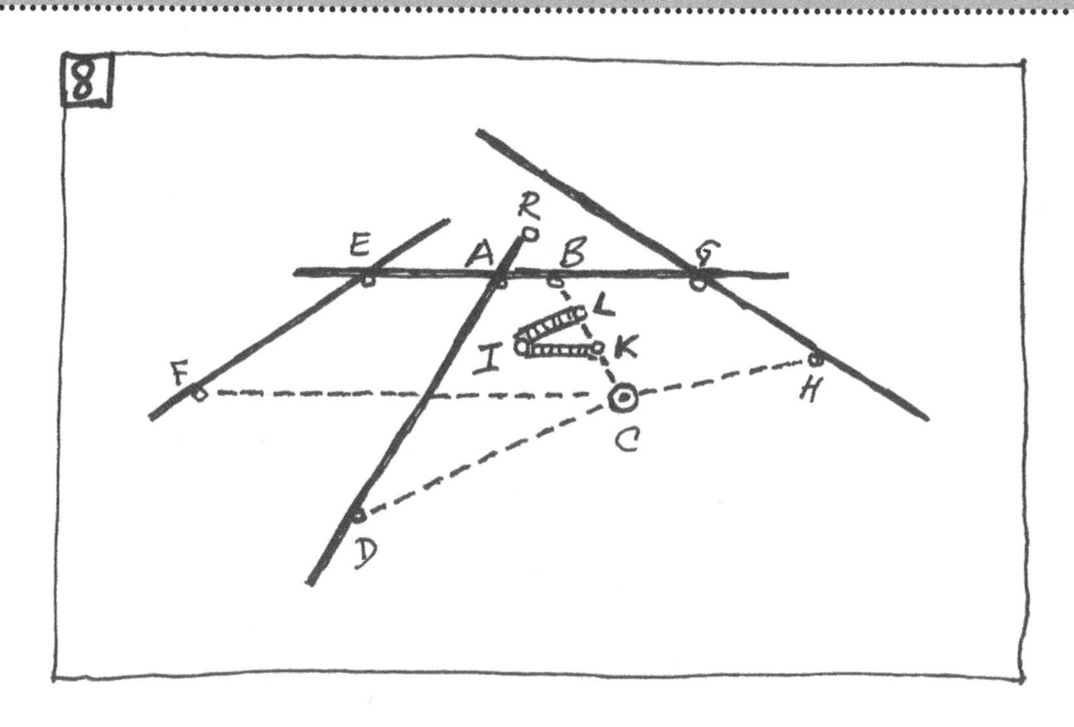

Once we get LC ,we proceed to characterize the locus of point C. (Consider the box following.)

```
y  =  [m – (n/z)x] + √[ m² + ox + (p/m)x² ]
!—!   !————————-!  !————————————————————!
/\        /\                 /\
|         |                  |
[BC]   [BK – BL]           [LC]
```

Examining the expression for LC, Descartes first discusses three cases when the locus of point C will be a straight line.

If LC amounts to nothing, so that BC is simply equal to (BK – KL), then

$$y = m - (n/z)\,x\ ;$$

and then point C is found to be located on the straight line IL.

Also in some other cases, point C will be found to be located on a straight line, but in these two cases the straight line will not be IL.

The first such case is this: when the expression for LC—namely, the expression $\sqrt{[\,m^2 + ox + (p/m)x^2\,]}$, can be rewritten so that no term containing x is inside a radical sign. (Nowadays we would call such an expression a perfect square.) That will be possible if the term m^2 and $(p/m)\,x^2$ are marked by the same sign (either plus or minus) and $o^2 = 4pm$. Why? Because if so, then LC will be equal to

$$\sqrt{\{\,m^2 + 2\,[\,\sqrt{(pm)}\,]\,x + (p/m)\,x^2\,\}}\text{ , which is }\{m + [\,\sqrt{(p/m)}\,]\,x\}.$$

But if LC is equal to that quantity, then BC = y must be equal to

$$m - (n/z)\,x + \{m + [\,\sqrt{(p/m)}\,]\,x\};$$

that is to say:

$$y = 2m + \{[\,\sqrt{(p/m)}\,] - (n/z)\}\,x.$$

In that case, point C is found to be located on a straight line (other than IL).

The second case in which C will be so located is the case when ox is nothing and one of the following alternatives is true:

either	*or*
m^2 is also nothing	$(p/m)\,x^2$ is also nothing
and then LC $= [\,\sqrt{(p/m)}\,]\,x$	and then LC $= m$
and then $y = \{[\,\sqrt{(p/m)}\,] - (n/z)\}\,x;$	and then $y = 2m - (n/z)\,x.$

The 12th paragraph began by setting forth an expression for LC, which allowed Descartes to begin to find the locus of point C. He began by identifying cases where the locus will be merely a straight line. Now, in the rest of the 12th paragraph, Descartes will turn to cases other than those that he has just discussed. What line will be the locus of point C if it is not a straight line? First comes the answer: one of the three conic sections or a circle. Then comes an identification of the orientation of its diameter and ordinate. And then comes an identification of what determines whether the locus will be one or another particular sort of curve among those that he has first characterized more generically as conic sections and circles.

But when this is not [so],
[then] this point C is always
in one of the three conic sections or in a circle,

of which one of the diameters is in the line IL,
and the line LC is one of those [lines] which
is applied ordinatewise (*par ordre*) to this diameter,
or, to the contrary,
LC is parallel to the diameter, to which
what is in the line IL is applied ordinatewise (*par ordre*).

To wit:

if the term [$(p/m)\ xx$] is null,
[then] this conic section is a Parabola;

and if it is marked by the sign +,
[then] it is an Hyperbola;

and finally if it is marked by the sign −,
[then] it is an Ellipse, except only
if the quantity *aam* is equal to *pzz* and the angle ILC be right,
in which case one has a circle instead of an Ellipse.

And now the paragraph will conclude by treating the case of the first of the conic sections, the parabola—discussing the size of the upright side and the position of the axis and its vertex. The next paragraph, the 13th, will treat those of these curves that have a center—namely, the circle, the ellipse, and the hyperbola—discussing the positions of the center of the various curves, as well as the sizes of the upright side and transverse side, and the positions of the axis and its vertex. Right after that, the 14th paragraph will further treat the case of the hyperbola.

In all of these three paragraphs, except for one case of the hyperbola, Descartes states conclusions without presenting demonstrations. Indeed, having given us a general indication of how to figure out problems, Descartes sees no need to give us a particular account of how one might arrive at the conclusions that he here asserts.

Perhaps, then, we would do well to pause to try to figure out just how one might figure out what Descartes here is satisfied merely to assert. Unless we do so, Descartes' findings will not be truly ours.

Let us start with the easiest case. Descartes has told us that if certain things are true of the algebraic expression for line BC, then the locus of point C will be a straight line. What if we turned the matter about, and asked: what if the locus were a straight line—what things would then be true of the algebraic expression for line BC? Let us do it. Let us ask this question: to what would BC (which is the quantity y) then be equal? Well, we figured out a while ago that, whatever the locus might be, y must be equal to

$$y = [m - (n/z)x] + \sqrt{[\, m^2 + ox + (p/m)x^2 \,]}.$$

Now if the locus of point C were a straight line, then would not that expression for the quantity y have to have the form that results when

> the product of multiplying the quantity x by some constant quantity
> is added to, or subtracted from,
> some (perhaps another) constant quantity?

In other words, would not the expression have to be without any x-quantity of two dimensions: would there not have to be no x^2 in it? If so, then all we need to do is consider what conditions will give that form to the expression for y. The easiest such condition to see is this: the term inside the radical would amount to nothing; that alone would suffice for the expression to have the needed form; y will be equal to just this: $[m—(n/z)\, x]$. And then the locus will be the straight line IL. The quantity (n/z) will tell us how much IL slants, and the minus-sign will tell us which way it slants, while the quantity m will tell us how far the straight-line locus is from one of the given straight lines. Now what other conditions would give to the expression for y the needed form? If we think that question through, we should arrive at Descartes' cases for a locus that is a straight line other than IL.

Why, then, didn't Descartes himself make things easier for us by telling us how to get to his conclusions?

Early on in the First Book, Descartes had said that he would not stop to explain in more detail something that is, after all, not so difficult that it cannot be found by those who, versed a little in the common geometry and in algebra, will be on the alert—will be watchful about (*prendront garde a*) all that is in this treatise. They could find it if, in order to get it, they had to seek it. He makes them seek it, he says, because if he merely gave it away "I would deprive you of the pleasure of learning it by your own self—and of the utility of cultivating your own mind (*esprit*) in exercising yourself in it, which is, in my opinion (*a mon avis*), the principal [utility] which one can draw from this science."

That first mention of utility was followed by the last 24 paragraphs of that First Book. Then—after, first, the 24 opening paragraphs of the Second Book that take the reader through to the final solution of the Pappus problem, and then, 16 paragraphs more that act as mediation—the last 24 paragraphs of this Second Book will be devoted to utility. The utility that was first mentioned, which was the principal utility of this science, has to do with cultivating one's own *esprit*, as we have seen; we shall consider the discussion of the other utility in its proper place. For now, we would do well to consider only the part of it where Descartes says more about the reasons for his deliberate (or should we say willful?) omissions.

We shall look, that is, at what he says in the next-to-last paragraph of the next-to-last book of the *Geometry* (II.63). After that, we shall look at what he says in the last paragraph of the preparatory part of the last book (III.44), and then at what he says in the last paragraph of that last book.

In the first justification of his withholding of some knowledge that he could impart (I.10), the author coupled pleasure and utility for the reader; in the first of the two central justifications (II.63), the author speaks of pain as a means of producing esteem in the reader; Descartes says that it would be possible to pass beyond what he has just said and say something further: "For this is no more difficult than what I have just explained—or rather it is a much easier thing, because the road (*chemin*) of it is opened. But I like better that others seek it, to the end that if they have a bit more trouble (*peine*) to find it, that would cause them to esteem so much more the finding (*invention*) of the things which are demonstrated." Perhaps what is meant is that readers will not properly weigh what they obtain too lightly—or perhaps it is that they will esteem too lightly the giver of a gift that is given too lightly. Descartes has opened the road. Will an explorer earn more gratitude if he acts as a tour-guide too?

In the second of the two central justifications (III.44), the author also speaks of the reader's pain, but now he couples pain and what is useful for learning. After giving rules for reducing the number of dimensions in an equation, in preparation for constructing the solution of problems that might otherwise seem to be of greater complexity than in fact they are, he says: "I have omitted here the demonstrations of most of what I have said, because they seemed to me so easy (*faciles*) that, provided you take the trouble (*peine*) to examine methodically if I have been faulty (*si jay failly*), they will present themselves to you of themselves; and it will be more useful to learn them in this fashion than in reading them."

The final justification of his withholding of some knowledge that he could impart (III.74) is the final sentence of the final paragraph of the final book of the *Geometry*, which is the final item in the whole volume that opens with Descartes' *Discourse on the Method of Conducting Well One's Reason*. That final paragraph comes right after the discussion of supersolid problems reaches its concluding paragraph; this concluding paragraph on supersolid problems (the next-to-last paragraph of all) says that a certain inconvenience for practice could be remedied by composing other rules in imitation of those presented here, as one could do in a thousand sorts of ways. Now this merely suggests that the author has left many things for his readers to do which he himself could do for them but won't; in the next, the very last paragraph of the *Geometry*, he goes on to speak more explicitly about his designing policy of omission. He speaks of the endless vista of problems and solutions that he has opened up—an infinite but orderly and gradual way into the construction of solutions for problems that are ever, and ever, and ever more complex. The paragraph begins by saying "my design is not to make a big book, and I try rather to comprehend much in few words—as one will perhaps judge that I have done, if one considers that, having reduced to one and the same construction all the Problems of one and the same kind (*genre*), I have given, all together, the fashion of reducing them to an infinity of different others, and so of solving each of them in an infinity of fashions.

"Then, beyond that," the paragraph continues, "having constructed all those which are plane, by a circle's cutting a straight line—and [having constructed] all those which are solid, by a circle's cutting also a Parabola—and finally [having constructed] all those which are more composite (*composés*) by one degree, by a circle's cutting entirely the same a line which is more composite (*composée*) than the Parabola by one degree—there is naught needed (*il ne faut*) but to follow the same way (*voye*) for constructing all those which are more composite (*composés*) to the infinite (*a l'infini*). For in the matter of Mathematical progressions, when one has the first two or three terms, it is not difficult (*malaysé*) to find the others."

It seems that some effort will be required if posterity is to make further progress, but that the way will not be painfully arduous. Posterity's pleasures in finding for themselves what Descartes deliberately refrained from going after and bringing back for them, while telling them not only that it was there to be found but also how to prepare to go after it themselves, will only increase their gratitude to Descartes. The final paragraph's next—and final—sentence is this: "And I hope that our posterity (*neveux*) will be grateful by recognizing me (*me sçauront gré*) [literally: will recognize me willingly], not only for the things which I have here explained, but also for those which I have omitted with a will (*voluntairment*) [intentionally, or perhaps willfully], to the end of leaving them the pleasure of finding them (*les inventer*)."

Let us return to the 12th paragraph of the Second Book. The paragraph concludes by treating the case of the parabola. We have already been told that if $(p/m)\,x^2$ is null, then the locus is a parabola. Now we are told what will be the size of its axis's upright side, and what will be the position of its axis and its vertex.

But (*Que*) if this section is a Parabola,

[then] its upright side (*costé droit*) is equal to *oz/a*;

and its diameter [that is, "the" diameter that is the axis] is always in line IL;

and for finding the point N which is its vertex (*sommet*),
it is necessary to make IN equal to *amm/oz*

[and it is necessary] that...... the point I be between L and N...if the terms are $+mm \ + ox$,
or else that........................ the point L be between I and N...if they are....... $+ mm \ - ox$,
or else it would be necessary that N be between I and L...if there were.... $-mm + ox$.

But, in the fashion in which the terms have been posed here, there can never be $-mm$.

And finally the point N would be the same as the point I if the quantity *mm* were null.

By means of which it is easy to find this parabola
through the 1st Problem of the 1st Book of Apollonius.

That "1st problem of the 1st Book of Apollonius" is the *Conics*' 52nd proposition: to draw a parabola in a plane when given the position of the axis and its vertex, and the size of its upright side.

Again let us see how to arrive at Descartes' conclusions. Let us ask: if the locus were a parabola, then what would BC (which is y) be equal to? In other words, what conditions would constrain that expression which we obtained for y whatever the locus might be? That expression was this:

$$y = [m - (n/z)x] + \sqrt{[\, m^2 + ox + (p/m)x^2 \,]}.$$

Now let us make things easier by asking not just what if the locus were a parabola, but what if it were a parabola whose axis lies along that straight line IL which was the locus in the simplest straight-line case. Then LC (which is at right angles to IL, and is bounded by it and by point C of the curve) would be an ordinate; its abscissa would be the straight line that lies along IL and runs from point L to the vertex—which vertex let us call point N. And the quantity that is equal to the upright side, let us call u.

So, by the characteristic feature of the parabola according to Apollonius (*Conics* I.11), we shall get $LC^2 = NL*u$.

For each point C, we shall have to draw a point C' that will be located just as far from L as point C is, but opposite C, along the prolongation of CL through IL. But where on IL is point N to be located? We have been supposing that point I is to the left and L to the right; now IN (the distance from I to vertex point N) is a fixed length (let us call it v), while NL (the abscissa) is different for each C (and so, is different for each x). There are three possible locations for point N relative to points I and L: that point N could be on the far side of either one relative to the other, or it could be between them. So, considering only the half of the parabola that is below the axis, we shall get what is depicted in figure 9.

What we have now, is this—that if the locus were a parabola, then LC^2 would be equal to one of the following three possible expressions:

either $uv + (ua/z)\,x$,
or $uv - (ua/z)\,x$,
or $-uv + (ua/z)\,x$.

But, because of what we saw earlier, LC^2 must be also be equal to the following expression: $m^2 + ox + (p/m)\,x^2$. Why? Because we saw that, whatever the locus might be, LC is equal to the expression $\sqrt{[\,m^2 + ox + (p/m)x^2\,]}$.

So, since things equal to the same thing are equal to each other, and, taking for convenience only the first of the possibilities above (the possibility, that is, where neither term in the expression, neither $uv\,x$ nor $(ua/z)\,x$, has a negative-sign), we get the following equation—

this expression:	uv	+	$(ua/z)\,x$		
must be equal to					
this expression:	m^2	+	ox	+	$(p/m)\,x^2$.

Now for this x^2 term in this latter member of the equation to be utterly absent from that former member of the equation, the quantity p/m (which is the coefficient of the x^2 term) must amount to nothing. That is to say: if the locus is to be a parabola, then the quantity p/m must amount to nothing.

Also, equating the coefficients of the other terms that correspond to each other in the equation, we get the following two equations:

uv	(ua/z)
is equal to	is equal to
m^2	o.

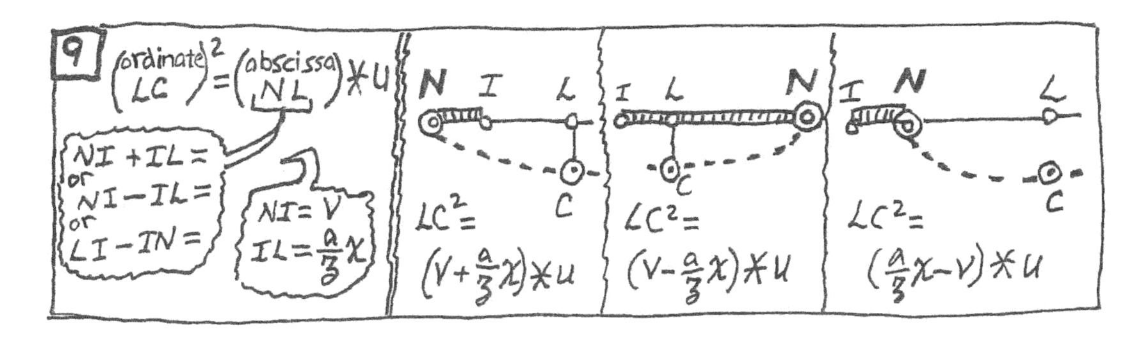

From the latter equation ($ua/z = o$), it follows that u (which is the upright side) is equal to oz/a.

And from that, combined with the former equation ($uv = m^2$), it follows that v (which is the distance of the vertex N from point I) is equal to m^2/u ; in other words, IN is equal to am^2/oz, N lying on the far side of I from L. See figure 10.

It should be easy to see what would follow from a like consideration of the other two possibilities for the location, relative to points I and L, of the vertex point N.

And now that we have done a bit of analysis for the cases of the straight line and the parabola, it may be clearer just why it was helpful for Descartes in the previous paragraph to draw straight lines KI and KL, and for him to mention, once we had seen that the part of BC which is KL is $(n/z)\,x$, that IL is $(a/z)\,x$.

What still remains to be accounted for in the present paragraph is his saying that in the case of the parabola there can never be, in the fashion in which the terms have been posed here, a quantity m^2 with a minus-sign attached to it. Why does he say that? Because it matters which distances are combined in the equality (or in the given ratio) that is part of the posing of the locus problem. Descartes has supposed that the locus problem here requires that the lines which are drawn to the given lines from point C—that is, the drawn lines CB, CD, CF, CH—are to be combined in such fashion as to make the products CB*CF and CD*CH the terms in the equality. Because of that, it can be shown that no term entirely constituted of known quantities is available to combine with the m^2 term that arises from solving the quadratic equation. The m^2 cannot have a minus-sign. And here there is no possibility that point N can be located between points I and L. Things would have been different if the products that Descartes took had been, say, CB*CD and CF*CH.

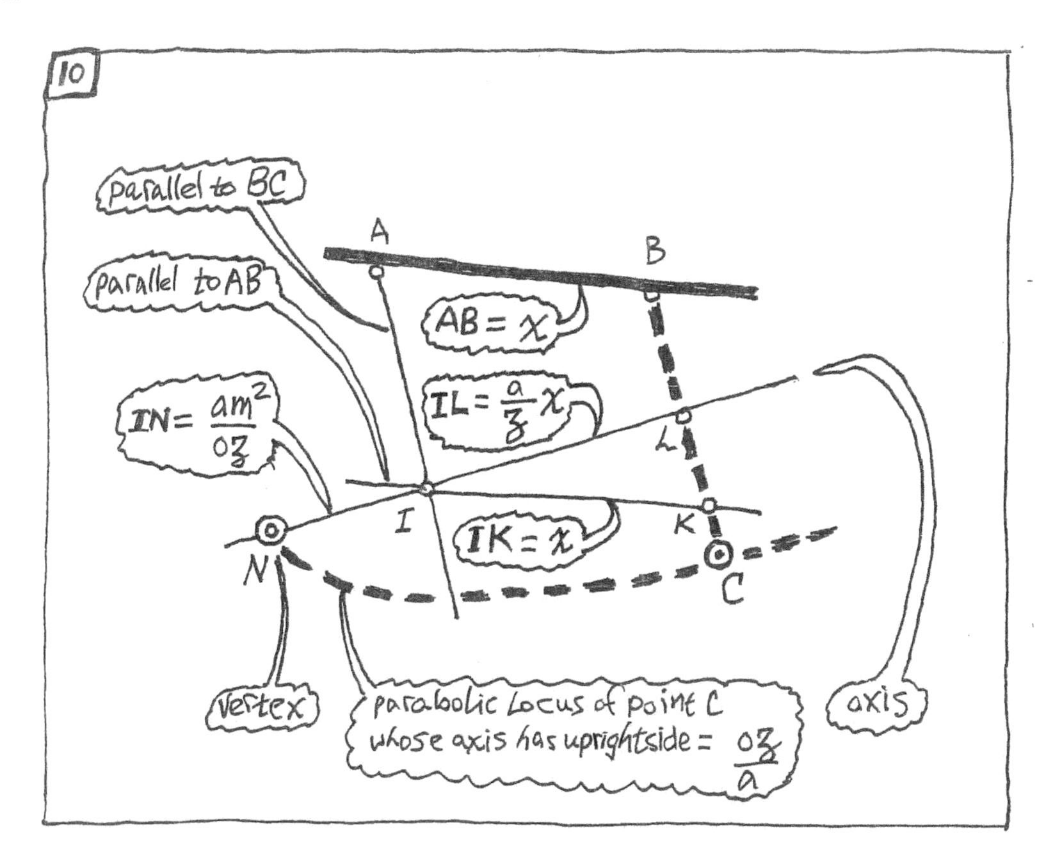

Descartes now considers more particularly the situation when the curve is one of the conic sections that (unlike the parabola) has a center, and hence has not only an upright side but also a transverse side.

As an exercise, you might wish to do now—for the circle, the ellipse, and the hyperbola—what we have done already for the straight line and the parabola namely, figure out how to get as conclusions the statements that Descartes will now merely assert. In other words, before going on to read what Descartes says next, you might ask yourself the following questions: if the locus were a circle—or if it were an ellipse—or if it were an hyperbola—what then would BC (which is y) be equal to? In other words, what conditions would constrain that expression which we obtained for y whatever the locus might be? Again, that expression is this:

$$y = [m - (n/z)x] + \sqrt{[m^2 + ox + (p/m)x^2]}.$$

Again, as with the parabola, you will make things easier by asking not just what if the locus were a circle, or an ellipse, or an hyperbola, but by asking the question more specifically: what if it were one of those curves and its axis lay along that straight line IL which was the locus in the simplest straight-line case?

Again, as with the parabola, LC (which is at right angles to IL, and is bounded by it and by point C of the curve) would be an ordinate; its abscissa would be the straight line that lies along IL and runs from point L to the vertex—which vertex let us call point N. And the quantity that is equal to the upright side, could be called u. Then, if the center were called point M, the size of the transverse side would be twice MN.

Now, by the characteristic feature of the hyperbola (or the ellipse), you would get the following statement about the square on the ordinate:

LC^2 = NL*u plus (or, in the ellipse, minus) the quantity NL*[NL*(u/ 2MN)].

Pause for a moment to make sure it is clear that the above equation is in fact the algebraic counterpart of Apollonius's enunciations for the 12th and 13th propositions of the First Book of the Conics.

And again, as with the parabola, a number of possible locations would have to be considered for the points relative to each other, each yielding an expression for the abscissa NL in terms of the following distances: from the point I to the vertex point N (that is, the distance which, in considering the parabola, we called v), and from point I to point L, which distance we were told by Descartes was equal to $[(a/z)x]$.

So, what you would have now is this: if the locus were a parabola, then LC^2 would be equal to some expression involving u and v and a and z and MN. But earlier we saw that, whatever the locus might be,

LC is equal to the expression $\sqrt{[m^2 + ox + (p/m)x^2]}$, which means that LC^2 must be equal to the expression $m^2 + ox + (p/m)x^2$.

(Remember, though, that the plus and minus signs must be appropriately chosen for the particular arrangement of points in the figure.)

So, you can equate the two expressions for the same quantity LC^2, and then you can equate the coefficients of terms that correspond to each other on one side of the equation and the other. From that you can go on to figure out what must be the sizes of the upright side and the transverse side, and the distances to point I from the center and from the vertex. All that will prepare you for what Descartes says in his next paragraph, the 13th.

You can then prepare yourself for what he says right after that, in the 14th paragraph, by asking some questions about the case of the hyperbola when the axis lies not along the straight line IL, but rather at right angles to it. Letting point M continue to be the center, can you find an expression that will tell you where, on a line that is perpendicular to straight line ILM at point M, to find the vertex (call it point O)? Can you find expressions that will tell you the sizes of the upright side and the transverse side?

[13] But (*Quo*) if the line called for (*demandee*) is
a circle or an Ellipse or an Hyperbola,

it is necessary first to seek the point M which is the center,
and which is always in the straight line IL, where one finds it in taking (*aom/2pz)* for IM,

so that (*en suite que*) if *o* is null, [then] this center is exactly (*justement*) at the point I.

And if the line sought is a circle or an Ellipse,
one should take point M
on the same side—with respect to point I—as point L when one has $+ox$;
and when one has $-ox$, one should take it on the other [side].

But totally to the contrary in the Hyperbola:
if one has $-ox$, this center M should be toward L [from I];
and if one has $+ox$, it should be on the other side.

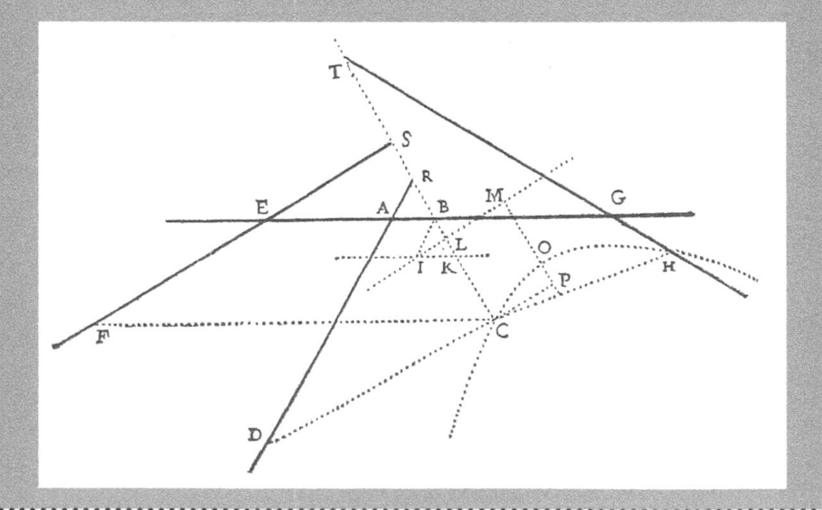

After that, the upright side (*costé droit*) of the figure
should be $\sqrt{[(oozz/aa) + (4mpzz/aa)]}$
when one has $+mm$ and the line sought is a circle or an Ellipse,
or else when one has $-mm$, and it is an Hyperbola.

And it should be $\sqrt{[(oozz/aa) - (4mpzz/aa)]}$
if, the line sought being a circle or an Ellipse, one has $-mm$,
or else if,
being an Hyperbola, and the quantity oo being greater than $4mp$, one has $+mm$.

But (*que*) if the quantity mm is null, [then] this upright side (*costé droit*) is oz/a;
and if oz is null, [then] it is $\sqrt{(4mpzz/aa)}$.

Then for the transverse side (*costé transversant*)
it is necessary to find a line which is to this upright side as aam is to pzz,
to wit, if this upright side is $\sqrt{[(oozz/aa) + (4mpzz/aa)]}$,
[then] the transverse [side] is $\sqrt{[(aaoomm/ppzz) + (4aam^3/pzz)]}$,

And in all these cases
the diameter of the section is in the line IM,
and LC is the one of those [lines] which is applied ordinatewise (*par ordre*).

So that (*Si bien que*), making MN equal to the half of the transverse side,
and taking it on the same side of point M as is the point L,
one has the point N for the vertex (*sommet*) of the diameter;

Note that if one has $+mm$ and $+nx$, then the problem is possible only if nn is greater than $4mp$.

after which (*ensuite dequoy*),
it is easy to find the section through the second and 3rd problem of the 1st bk. of Apollonius.

The 2nd problem of the 1st Book of Apollonius begins with the *Conics*' 54th proposition: to draw an hyperbola in a plane when given the position of the axis and its vertex and the sizes of its transverse side and upright side. The 3rd problem begins with the 56th proposition: to draw an ellipse in a plane when given the sizes and positions of its two axes.

Two statements made about the ellipse in this 13th paragraph—

that the diameter is in line IL, and LC is drawn ordinatewise, and

that the transverse side is to the upright side as a^2m is to pz^2

—can explain why it was said, in the previous (12th) paragraph, that there is an exception to the statement that if the term p/m is marked by a minus-sign, then the equation indicates the locus to be an ellipse. The exception is that if ILC is an angle that is right, and if a^2m is equal to pz^2, then the locus will be a circle. Why is it, that in this case, the ellipse will turn into a circle? Here is why. Since angle ILC is right, the diameter is an axis; and since a^2m is equal to pz^2, the transverse side is equal to the upright side. Now the conjugate diameter of an ellipse's axis is the ellipse's other axis, and we learn from Apollonius (*Conics* I.15) that any diameter of an ellipse is the mean proportional between its conjugate diameter and this conjugate diameter's upright side. It follows that since the axis of what would here seem to be an ellipse is equal to its upright side, then this axis must be equal to its conjugate diameter, namely, the other axis, and thus what would seem to be an ellipse is, in fact, a circle—since its axes are equal. See figure 11.

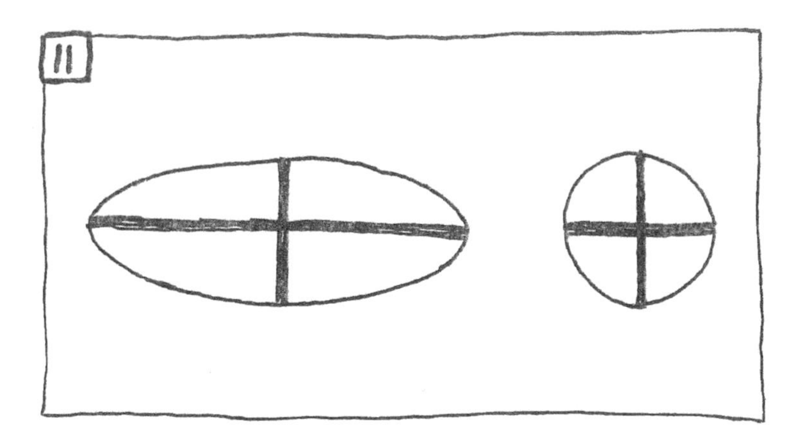

[14] But when (*quand*), this section being an Hyperbola,
one has $+mm$, and the quantity oo is null or less than $4pm$,
[then] one should
draw from the center M the line MOP parallel to LC, and [draw] CP parallel to LM,
and
make MO equal to $\sqrt{[\ mm - (oom/4p)\ ]}$, or else make it equal to m if the quantity ox is null.

Then [it is necessary] to consider the point O as the vertex (*sommet*) of this Hyperbola,
of which the diameter is OP,
and CP [is] the line which is applied to it ordinatewise (*par ordre*),
and its upright side (*costé droit*) is $\sqrt{[\ (4a^4m^4/ppz^4) + (a^4oom^3/p^3z^4)\ ]}$,
and its transverse side (*costé transversant*) is $\sqrt{[\ 4mm - (oom/p)\ ]}$.

Except when ox is null, for then
the upright side is $2aamm/pzz$,
and the transverse [side] is $2m$.

And so it is easy to find them through the 3rd problem of the 1st book of Apollonius.

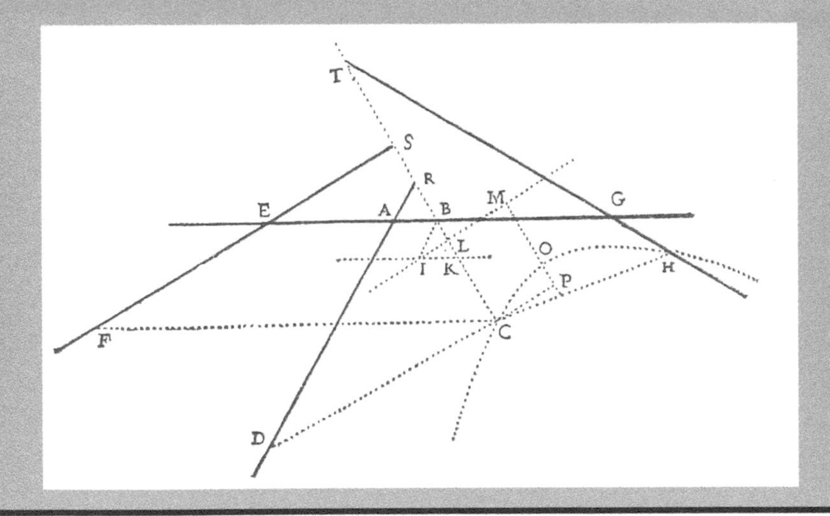

Having reported the results of analyzing what would be required for the locus to be a straight line or one of the conic sections, a parabola, or an hyperbola, or an ellipse, or a circle, Descartes now comes to the demonstrations of what he has said.

[15] And the demonstrations of all this are evident; for,
composing a space [in other words, forming a product] from
the quantities which I have assigned
for the upright side and the transverse [side]
and for the segment of the diameter NL or OP,
following the tenor of the 11th, of the 12th, and of the 13th theorems of
the 1st book of Apollonius,
one will find all the same terms from which is composed
the square of the line CP or CL which is applied ordinatewise to this diameter.

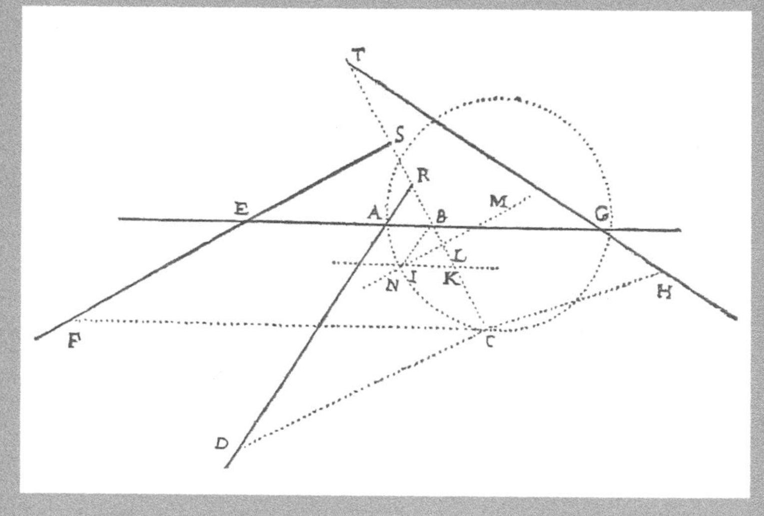

The demonstrations, he says, are evident: just take what he has said will have to be the axis and its vertex, and what will have to be the abscissa and its ordinate, and also take the quantities that he has said will have to be the upright side and the transverse side, and then merely employ the tenor of the theorems in which Apollonius characterized the three conic sections—*Conics* I. 11 (parabola), I.12 (hyperbola), and I.13 (ellipse). But, although Descartes says that the demonstrations are evident, he gives an example. He treats the locus when it is an ellipse (or, rather, an ellipse or a circle: the circle here would seem to be a special case of the ellipse). Readers who would like some exercise might try treating the locus when it is not an ellipse. In this guidebook, we shall look only at Descartes' example. Here it is.

As, in this example,
taking IM, which is $(aom/2pz)$, away from NM, which is $\{(am/2pz)\ \sqrt{(oo + 4mp)}\}$,
[then] I have IN;

to which, adding IL, which is $\{(a/z)\ x\}$, I have
NL, which is...... [IL − IM + NM,
that is:]........... $\{(a/z)x\} - \{aom/2pz\} + \{(am/2pz)\ \sqrt{(oo + 4mp)}\}$

and this
being multiplied by $\{(z/a)\ \sqrt{(oo + 4mp)}\}$, which is the upright side of the figure,
there comes $\{x\sqrt{(oo + 4mp)} - (om/2p)\ \sqrt{(oo + 4mp)} + (moo/2p) + 2mm\}$
for the rectangle [produced by the abscissa-to-be NL and the upright side],

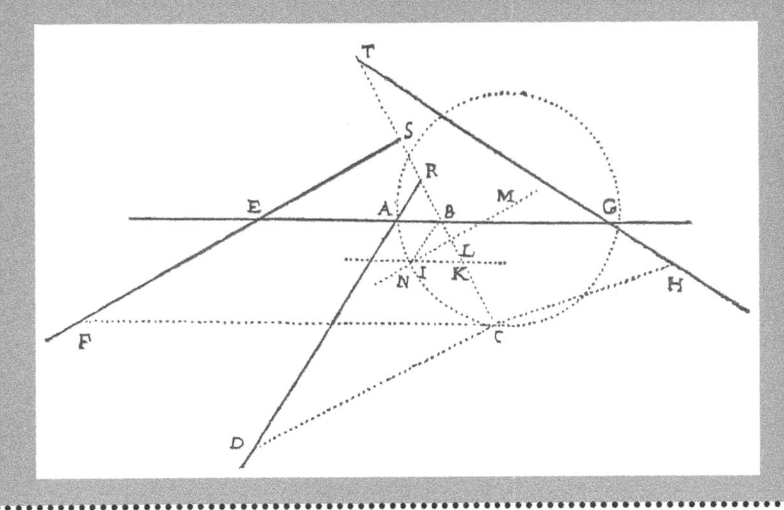

from which [rectangle (NL*upright side)] it is necessary to take away a space which is to the square of NL as the upright side is to the transverse [side],

and this square of NL is.... $\{(aa/zz)xx - (aaom/pzz)x + (aam/pzz)x \sqrt{(oo + 4mp)} + (aaoomm/2ppzz) + (aam^3/pzz) - (aaomm/2ppzz) \sqrt{(oo + 4mp)}\}$,
which it is necessary to
divide by *aam* and to multiply by *pzz*, because these terms express (*expliquent*) the ratio (*proportion*) which is between the transverse side and the upright [side],

and there comes $\{(p/m)xx - ox + x \sqrt{(oo + 4mp)} + (oom/2p) - (om/2p) \sqrt{(oo + 4mp)} + mm\}$,
which it is necessary to take away from the previous rectangle,

and one finds $\{mm + ox - (p/m)xx\}$ *for the square of CL*,
which consequently is a line applied ordinatewise,
in an ellipse or in a circle, to the segment of the diameter NL.

Descartes' example shows that the locus whose equation is

$$y = m - (n/z)x + \sqrt{[m^2 + ox - (p/m)x^2]}$$

will be an ellipse (or circle). That is shown by showing that for point C which is determined by that equation, the lines NL and LC have the relation that belongs to an abscissa and ordinate in an ellipse—namely, that

$$(NL^*u) - (\text{a certain space that we shall call S}) = LC^2.$$

How does Descartes show that [(NL*u) – (S)] is equal to LC^2? He relies on what we already know about LC, namely, that LC^2 is equal to $[m^2 + ox - (p/m)x^2]$; and so, he merely needs to show that [(NL*u)—(S)] is equal to that same thing to which LC^2 is equal.

How, then, does Descartes show what needs to be shown—that

$$\boxed{(\mathrm{NL}^*u) - (\mathrm{S}) \;=\; m^2 + ox - (p/m)\,x^2\ ?}$$

He shows it as follows. Making use of the results (reported earlier) of analyzing what would be required for the locus to be an ellipse, he obtains an expression for each of the terms (NL^*u) and (S); then, from the former, he takes away the latter. The particulars are depicted in figure 12.

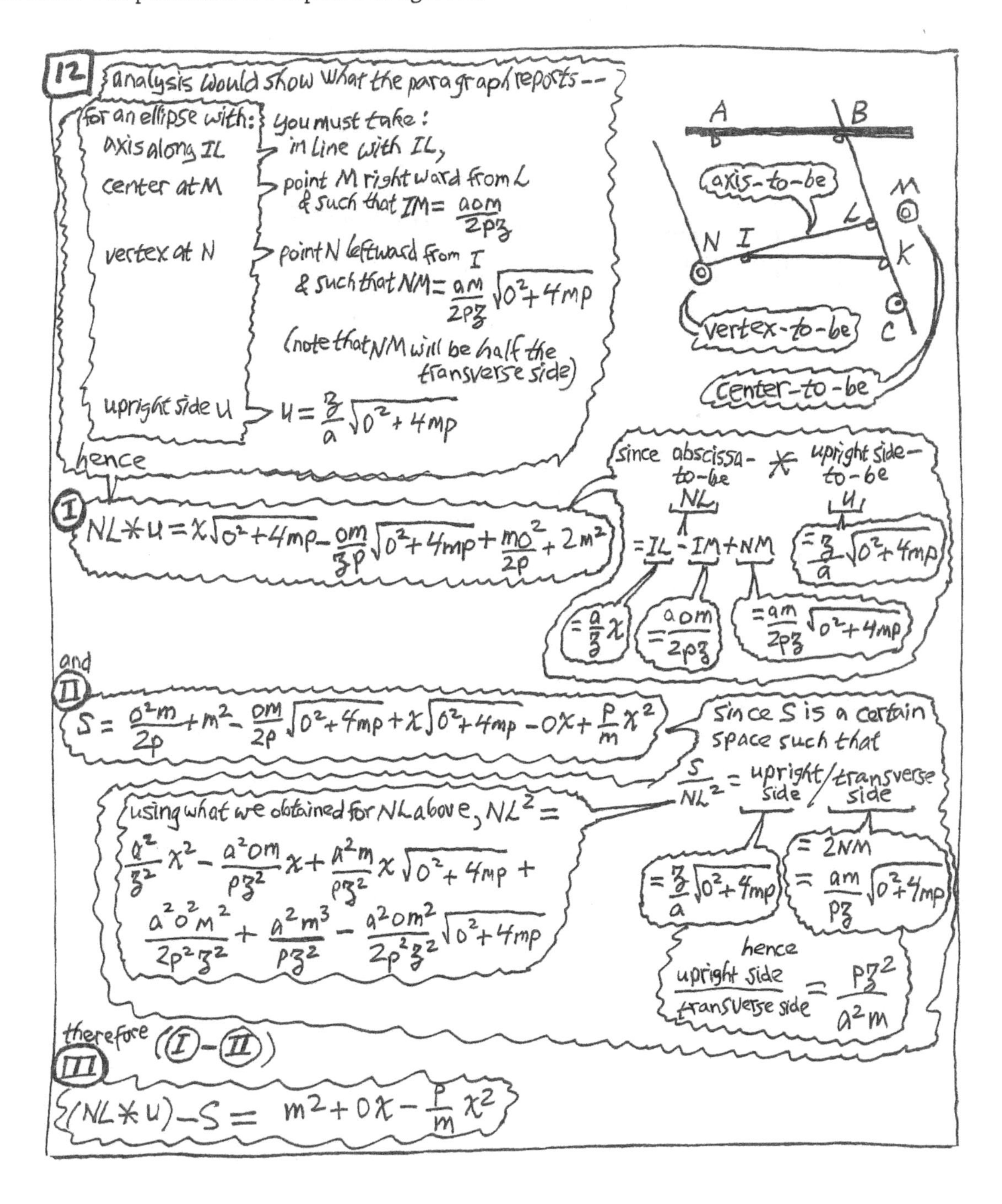

That is where Descartes' example stops. To make the demonstration more explicit, we might go on to draw some lines that would show how to bring into the matter what Descartes refers to at the beginning of the paragraph—namely, the 13th theorem of the First Book of Apollonius's *Conics*. Here is how to make the drawing (see figure 13):

let NH = ...upright side *u;*
let MQ = MN,
so that NQ = ... transverse side;
complete......... rectangle[NH, NQ];
draw..............diagonal HQ;
complete......... rectangle [HT, TD].

Now what needs to be shown is that square [LC] is equal to rectangle [NT, NL]. Why? Because if that equality is shown, then

square [LC] has been applied to NH in such a way that
it falls short by a rectangle [HT, TD] which
has breadth NL and is similar to rectangle [NH, NQ];

and then, by Apollonius, *Conics* I.13,

LC acts as ordinate,
NL acts as abscissa,
NH acts as upright side, and
NQ acts as transverse side—

in an ellipse. But how is the equality shown? As depicted in figure 14.

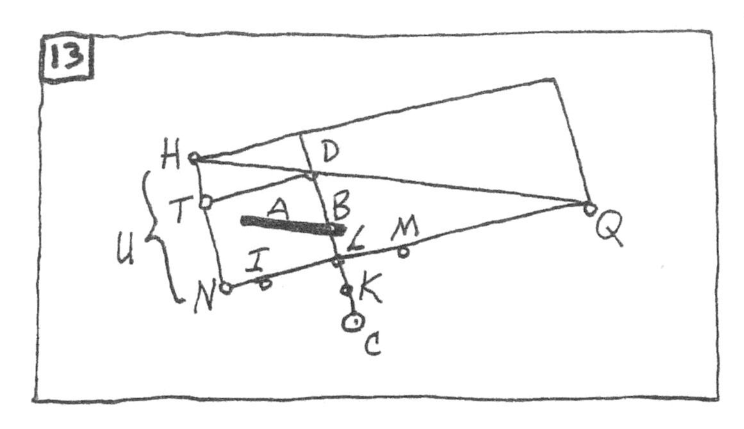

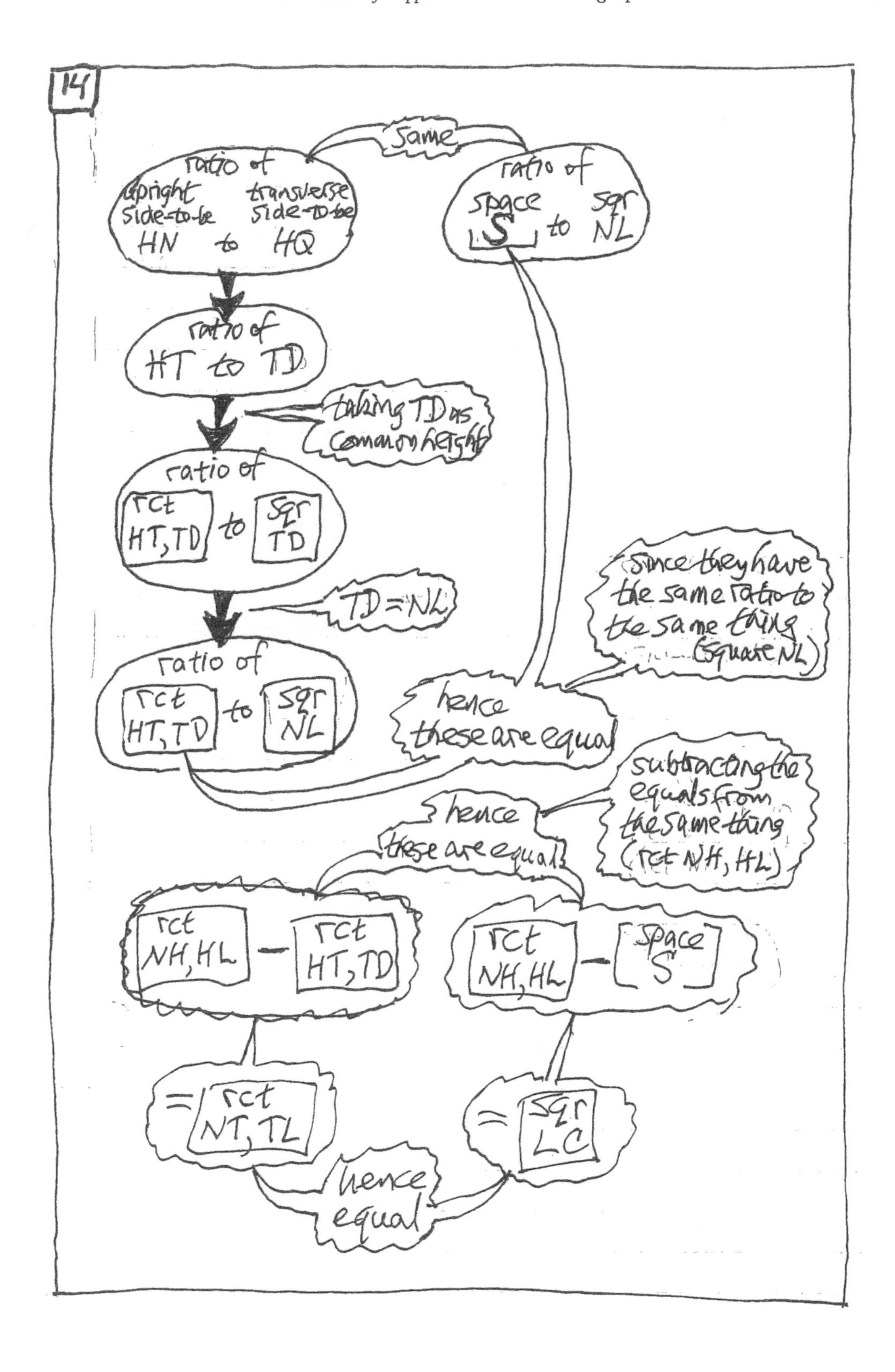
14
same
ratio of upright side-to-be HN to transverse side-to-be HQ
ratio of space S to sqr NL
ratio of HT to TD
taking TD as common height
ratio of rct HT, TD to sqr TD
TD = NL
since they have the same ratio to the same thing (square NL)
ratio of rct HT, TD to sqr NL
hence these are equal
subtracting the equals from the same thing (rct NH, HL)
hence these are equal
rct NH, HL − rct HT, TD
rct NH, HL − space S
= rct NT, TL
= sqr LC
hence equal

Now that the 15th paragraph has sketched the demonstration that would complete the synthesis that would be required for the classical solution of a geometrical problem, the 16th paragraph goes on to provide a numerical explication in the manner of Diophantus. That is to say: having introduced the terms of arithmetic into geometry, Descartes blurs the difference in manner of treatment between geometrical problems and arithmetical problems.

[16] And what if one wishes to express (*expliquent*) all the quantities by means of numbers, making, for example, EA æ 3,
AG æ 5,
AB æ BR,
BS æ (1/2) BE,
GB æ BT,
CD æ (3/2) CR,
CF æ 2 CS,
CH æ (2/3) CT, and
the angle ABR be of 60 degrees, and finally
the rectangle from the two [lines] CB and CF be equal to
the rectangle from the two others CD and CH,
for it is necessary to have
all these things, to the end that the problem (*question*) be entirely determined.

The entirely determined problem, with all the quantities expressed by means of numbers, is depicted in figure 15. Descartes is now ready to show its solution.

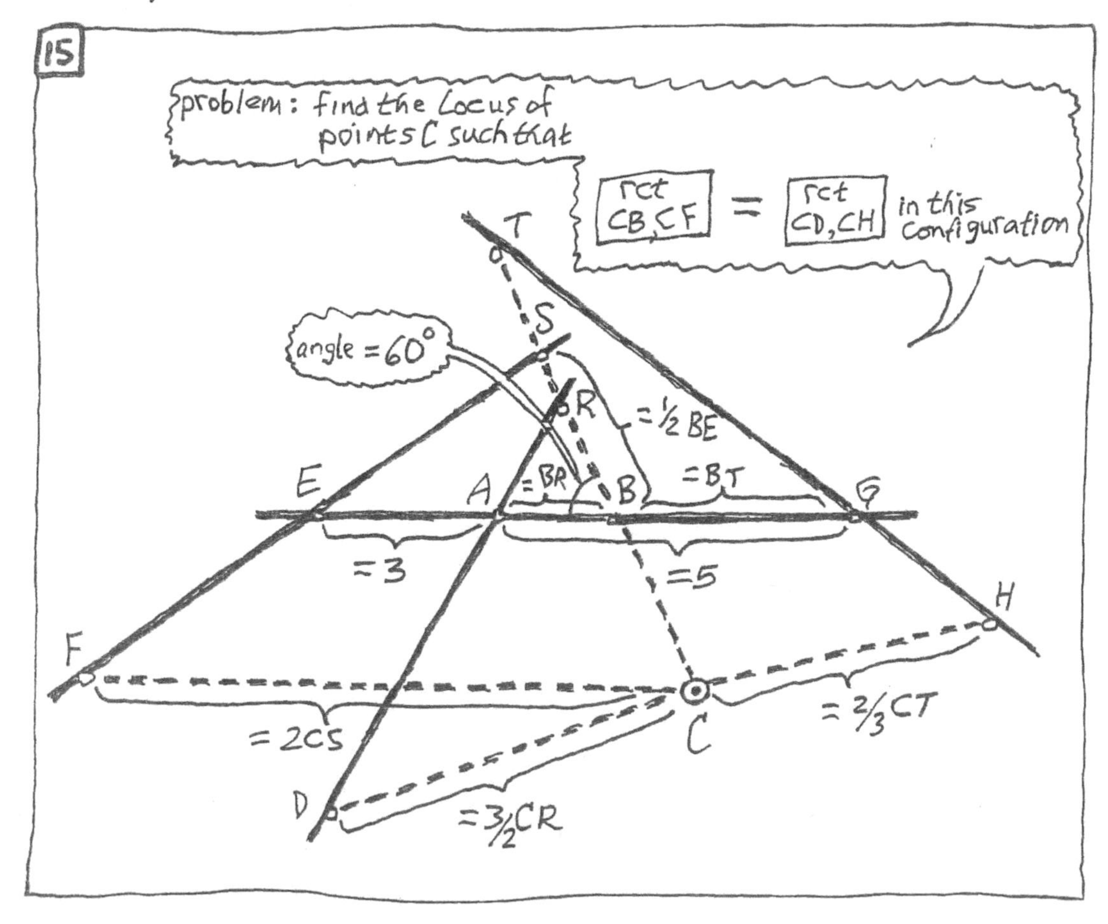

> And with this, supposing AB æ *x* and CB æ *y*,
> one finds, by [doing things in] the fashion explained here above,
>
> [that] *yy* æ 2*y* – *xy* + 5*x* – *xx*
> and *y* æ 1 – (1/2)*x* + √{ 1 + 4*x* – (3/4)*xx* };

We get the equation for *y* as follows. We let AB be equal to *x* as before; and, also as before, we let CB be equal to *y*, and then (using the determinations that we were given in the posing of the problem) we get expressions for the other lines (namely, CD, CF, CH) that are drawn from point C—expressions in terms of *x* and *y*. See figure 16.

And now we substitute those expressions into the statement that, by determining the relation between the lines that are drawn from C, determines the point C:

CB	*	CF	=	CD	*	CH
|		|		|		|
\/		\/		\/		\/
!—!		!————!		!————!		!————!
y		2[*y* + (1/2)(3 + *x*)]		[(3/2)(*y* + *x*)]		[(2/3)(*y* + 5 – *x*)]

And that—if you do the calculations—will all work out to the following equation:

$$y^2 = 2y - xy + 5x - x^2.$$

And solving the equation for *y*, we shall get Descartes' equation:

$$y = 1 - (1/2)\,x + \sqrt{[\,1 + 4x - (3/4)\,x^2\,]}.$$

That equation, by the way, has the form of the following:

$$y = m - (n/z)\,x + \sqrt{[\,m + ox - (p/m)\,x^2\,]}.$$

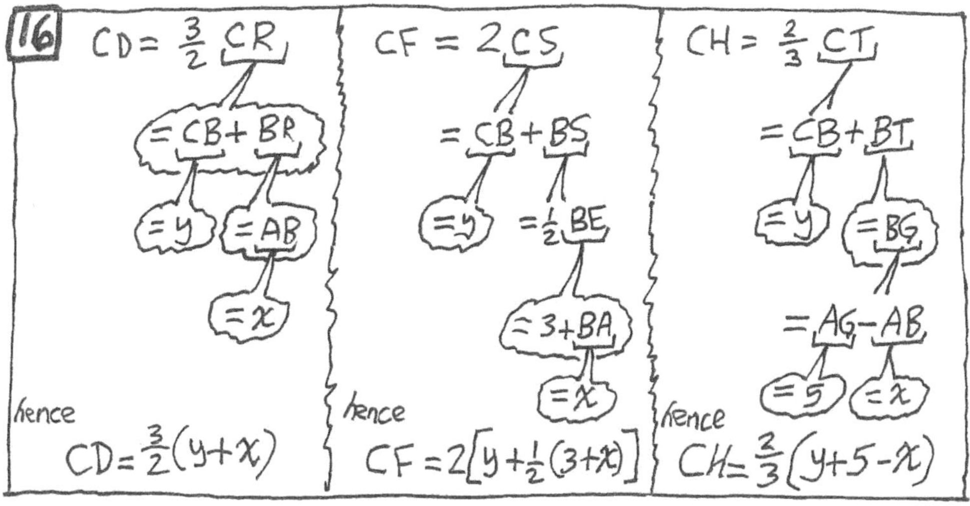

16
CD = 3/2 CR
= CB + BR
= y
= AB
= x
hence
CD = 3/2 (y + x)
CF = 2 CS
= CB + BS
= y
= 1/2 BE
= 3 + BA
= x
hence
CF = 2[y + 1/2 (3 + x)]
CH = 2/3 CT
= CB + BT
= y
= BG
= AG – AB
= 5
= x
hence
CH = 2/3 (y + 5 – x)

Now, before we can continue with Descartes, we need to put lines IK and IL back into the figure. See figure 17.

so that (*si bienque*) BK should be 1, and KL should be the half of KI,
and ILK is [a] right [angle]
because the angle IKL or ABR is of 60 degrees, and
[the angle] KIL, which is the half of KIB or of IKL, [is] of 30 [degrees].

And because IK or AB is named x,
KL is $(1/2)\,x$ and IL is $x\,\sqrt{(3/4)}$.

That is to say, BK (which corresponds to the first term in that expression which we found to be equal to y) should be 1. And KL should be half of KI. (Why? Because KL is to KI as BS is to BE, and BS is given as being half of BE.)

And ILK is a right angle. (Why? Because it is the third angle in a triangle—namely, triangle IKL—whose other two angles are, respectively, of 60° and of 30°. Angle IKL is of 60°, since it is equal to the angle ABR, which is given to be of 60°; and angle KIL is of 30°, since it is half of angle KIB, or of angle IKL, which is of 60°.)

Now we have already said that KL should be half of KI; but, since KI is equal to AB, which is x, we should go on to say that KL is half of x. And what about the other line that was put into the picture along with KL? That is, what about IL? Well, consider again the right triangle IKL. One of its legs is KL, and its other leg is IL; its hypotenuse is KI. So $KL^2 + IL^2 = KI^2$. But, as we have just seen, leg KL is half of hypotenuse KI. Hence, the other leg squared (in other words, IL^2) will be equal to KI^2 minus $[(1/2)\,KI]^2$. That is, it will be equal to $(3/4)KI^2$. That is to say, IL will be equal to $[KI\,\sqrt{(3/4)}]$. But, as we said, KI is equal to AB, which is x. So IL is equal to $[x\,\sqrt{(3/4)}]$.

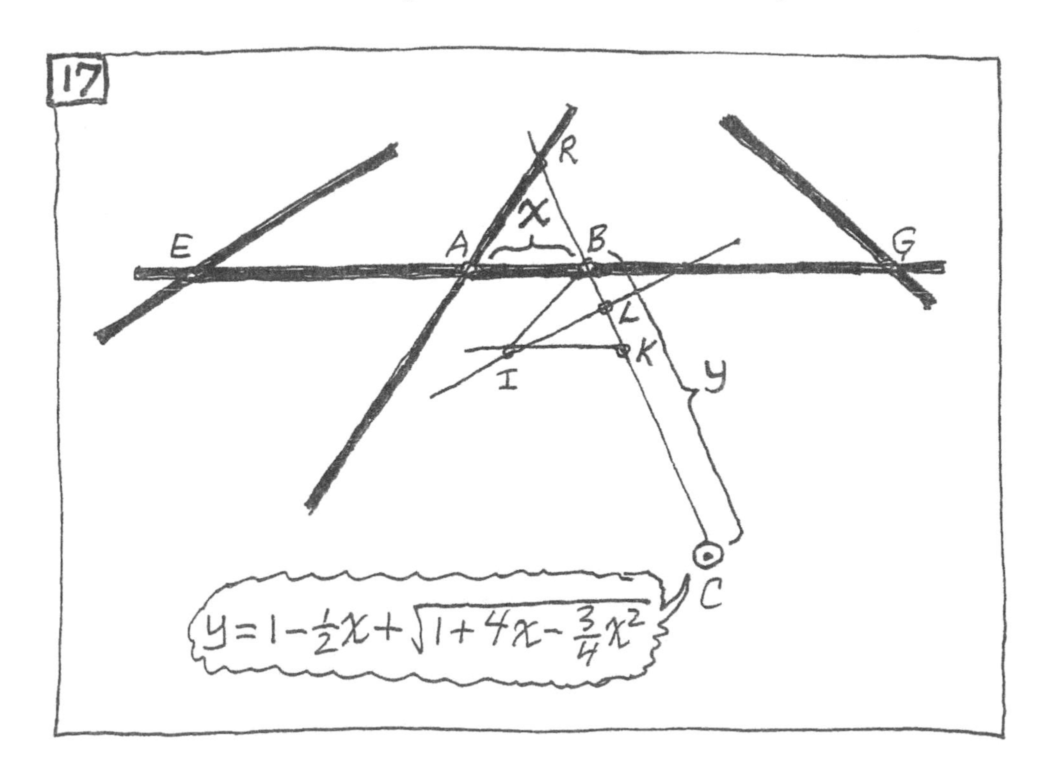

So we have now taken care of both of the lines (KL and IL) that we just drew back in. When Descartes himself first put those lines into the picture a little while ago, he drew IL in such a way that IK/KL = z/n and thus IK/IL = z/a. We shall use that in a minute.

Now when Descartes then went on to write the equation (back before expressing the quantities by means of numbers), BK was expressed by the term m, and KL by the term $(n/z)\,x$. The equation then had the following form:

$$y = m - (n/z)\,x + \sqrt{[\,m + ox - (p/m)\,x^2\,]}.$$

The quantity that we were then calling z was like a unit. Now Descartes will tell us that it is one.

And since our equation for y, now that the quantities are expressed by numbers, is

$$y = 1 - (1/2)\,x + \sqrt{[\,1 + 4x - (3/4)\,x^2\,]},$$

it follows that 1 and 4 and 3/4 are the quantities that a little while ago we were calling, respectively, m and o and p (since m is 1, we can speak of p by itself, rather than of p/m).

It goes without saying that (z being the unit) the quantity that a little while ago we were calling n is just 1/2. And

since, as we saw a minute ago,.................	IK/KL = z/n and IK/IL = z/a,
it follows that....................................	(z/n) KL = IK = (z/a) IL;
and from that in turn it follows that...........	$a = n$ (IL/KL).
But (as we have seen) n is just 1/2,	
while IL is $[x\,\sqrt{(3/4)}]$ and KL is $[(1/2)\,x]$.	

We conclude that $\sqrt{(3/4)}$ is the quantity that we were then calling a.

That is why Descartes goes on to say:

and the quantity which a little while ago was called z is 1,
what was a is $\sqrt{(3/4)}$, what was m is 1, what was o is 4, and what was p is 3/4,

in [such] fashion (*de façon*) that one has $\sqrt{(16/3)}$ for IM and $\sqrt{(19/3)}$ for NM,

and because aam, which is 3/4, is here equal to pzz, and the angle ILC is [a] right [angle],
[therefore] one finds that the curved line NC is a circle.

And one can easily examine all the other cases in the same way (*sorte*).

But why does one have √(16/3) for IM, and why √(19/3) for NM? Well, as was said in the 12th paragraph, what is indicated by the minus-sign attached to the x^2 term inside the radical of the equation for y, is that the curve will be an ellipse (or a circle); and, as was shown in the 13th paragraph, when the x-term inside the radical has a plus-sign attached to it, the axis of the ellipse will lie along line IL, with the center M lying to the right of point I at such distance from it that

$$\text{IM} = aom/2pz,$$

and with the vertex N lying to the left of point I at such distance from center M that

$$\text{NM} = 1/2 \text{ the transverse side, the transverse side being } \sqrt{[\,(a^2o^2m^2/p^2z^2) + (4a^2m^3/(pz^2)\,]}\,.$$

To get the numbers that Descartes has for IM and NM here, we need only substitute the numbers that he has here for the quantities that a little while ago were called a, m, o, p, and z.

The 12th paragraph also said that instead of an ellipse we shall have a circle when the angle ILC is right and the quantity a^2m is equal to the quantity pz^2. Two statements made in the 13th paragraph can explain why what would seem to be an ellipse is, in fact, a circle. Here, angle ILC is a right angle and quantities a^2m and pz^2 are indeed equal to each other (since each of them is 3/4).

We can draw the circle that is the locus, then, if we locate its center at point M on straight line L to the right of point I, at a distance from it of IM = √(16/3), employing as radius NM = √(19/3). (You might note the ease with which we have come to speak of "numbers" which are the square roots of fractions and which themselves cannot even be fractions, let alone be numbers with which we could count.) See figure 18.

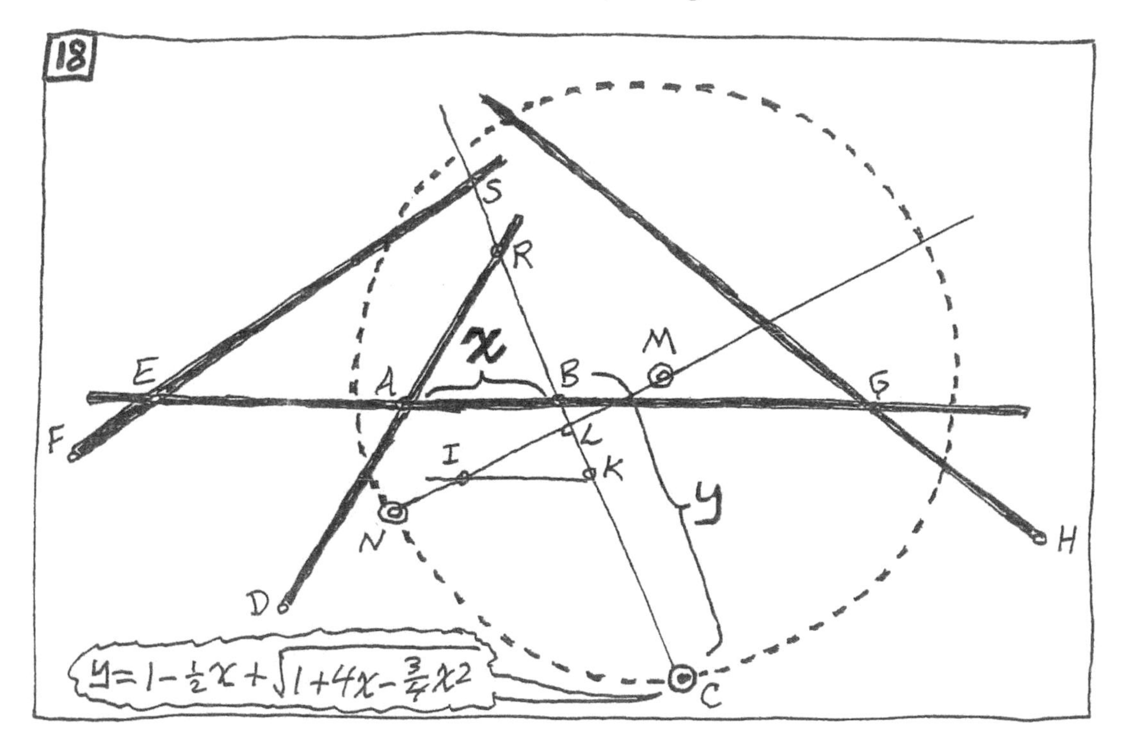

what has been accomplished (*paragraphs 17–18*)

[17] What is left to say is this (*Au reste*):
because the equations which mount up to naught but the square
are all comprised in what I have just explained (*expliquer*),
not only is the problem of the ancients in 3 or 4 lines here entirely finished up (*achevé*),
but [so] also [is] all that which belongs to what they
call the composition of the solid loci (*lieux solides*)
and consequently also [all that which belongs] to
that of the plane loci (*lieux plans*), because they are comprised in the solid [loci].

For these loci are not anything unless when it is a problem (*question*) of finding some point to
which there is lacking one condition for being entirely determined,
so that, as happens in this example,
all the points of one and the same line can be taken for that which is called for (*demandé*).

And if this line is straight or circular, one names it a plane locus.
But if it is a parabola or an hyperbola or an ellipse, one names it a solid locus.

And all the same and whatever it is (*quantes que cela est*),
one can come to an Equation which contains two unknown quantities and
is like (*pareille a*) some one of those which I have just solved (*viens de resoudre*).

But (*Que*) if the line which so determines the point sought
is more composite (*composée*) by one degree than the conic sections,
one can name it, in the same fashion, a supersolid (*sursolide*);
and so on, of the others.

And if there are lacking two conditions for the determination of this point,
the locus (*lieu*) where it is found is a surface which, all the same,
can be either plane, or spherical, or more composite (*composée*).

But the highest goal (*but*) which the ancients had in this matter
was to arrive (*parvenir*) at the composition of the solid loci;
and it seems that all of what Apollonius wrote about the conic sections
was with naught but the design (*dessein*) to seek it [namely, the composition of the solid loci].

[18] Moreover, one sees here that
what I have taken for the first kind (*genre*) of curved lines
does not comprehend [that is, include] any other [curves] than
the circle, the parabola, the hyperbola, and the ellipse,
which is all that I have undertaken (*entrepris*) to prove.

That last (18th) paragraph, which tells of the completion of all that the author undertook to prove, is the 11th paragraph of the stretch that finishes up the solution of Pappus's problem. The finishing up goes further in the remaining 6 paragraphs of the stretch. Thus, this third 17-paragraph stretch is divided like the other two that have preceded it—into 11 paragraphs followed by 6 paragraphs.

some cases when 5 straight lines are given (*paragraphs 19-22*)

[19] But if (*Que si*) the problem (*question*) of the ancients
is proposed in five lines which are all parallel,
it is evident that the point sought will always be in a straight line.

But if it is proposed in five lines of which
there are four of them that are parallel
and [it is proposed] that the fifth cuts them at right angles,
and indeed that all the lines drawn from the point sought meet them also at right angles,
and finally [it is proposed] that the parallelepiped composed from
three of the lines so drawn onto three of those which are parallel
be equal to the parallelepiped composed from [the following three lines, namely]
the two lines drawn [from the point sought],
the one [of which is drawn] onto the fourth of those lines which are parallel
and the other [of which is drawn] onto that [line] which cuts them at right angles,
and from a third line, [one that is] given—
which is, it seems, the simplest case that one can imagine after the preceding [one]—

[then] the point sought will be in a curved line which is described by
the movement of a parabola in the fashion explained below.

[20] Let it be, for example,
that the lines sought [*sic*; correction: "proposed"] are AB, IH, ED, GF, and GA;
and that one calls for (*demande*) the point C such that (*en sorte que*),
drawing CB, CF, CD, CH, and CM at right angles onto the given [lines],
the parallelepiped [which is produced] from the three [lines] CF, CD, and CH
be equal to that [parallelepiped which is produced]
from the two other [lines] CB and CM and from a third [line], let which [third one] be AI.

I put CB æ y. CM æ x. AI or AE or GE æ a,
in [such] fashion that, the point C being between the lines AB and DE,
I have CF æ $2a - y$, CD æ $a - y$, and CH æ $y + a$;
and, multiplying these three [CF, CD, CH] by one another,
I have $y^3 - 2ayy - aay + 2a^3$
equal to the product from the three others [CB, CM, AI], which is axy.
[See figure 19.]

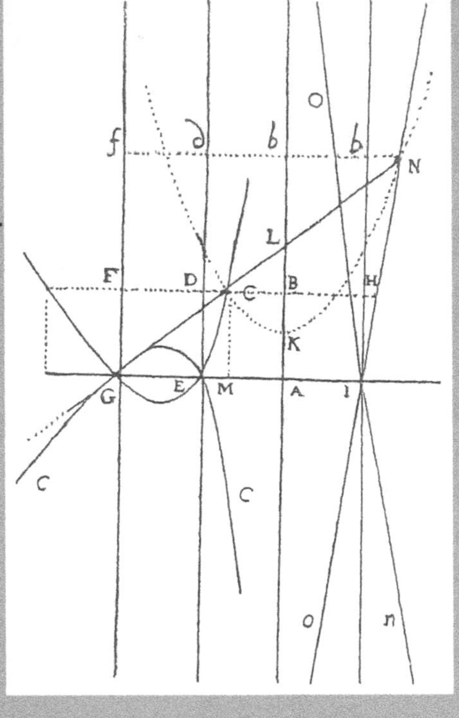

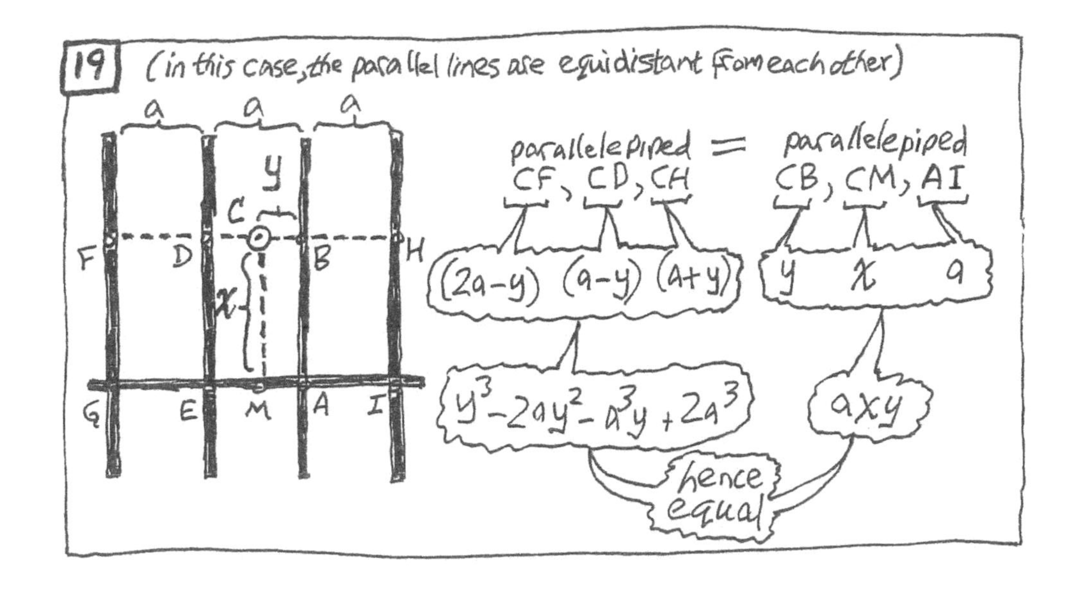

Having obtained an equation for any point on that curve which is sought, Descartes now tells how to draw a certain curve whose points will turn out to have that equation:

> After that, I consider the
> curved line CEG which I imagine [or visualize] to be drawn (*descrite*) by the intersection of
> the parabola CKN, which one makes move in such a way (*sorte*) that
> its diameter KL is always on the straight line AB,
> and the ruler GL, which turns meanwhile around point G in such a way (*sorte*) that
> it passes always in the plane of this parabola through the point L.
> and I make KL æ a,
> and I also make equal to a the principal upright side, that is to say, that [upright side] which
> is related [*se rapporte*] to the axis of this parabola,
> and GA æ $2a$,
> and CB or MA æ y,
> and CM or AB æ x. [See figure 20.]

Now Descartes goes on to show that the points of that curve which he has told us how to draw will have the very equation that was shown to belong to the points that the locus problem called for:

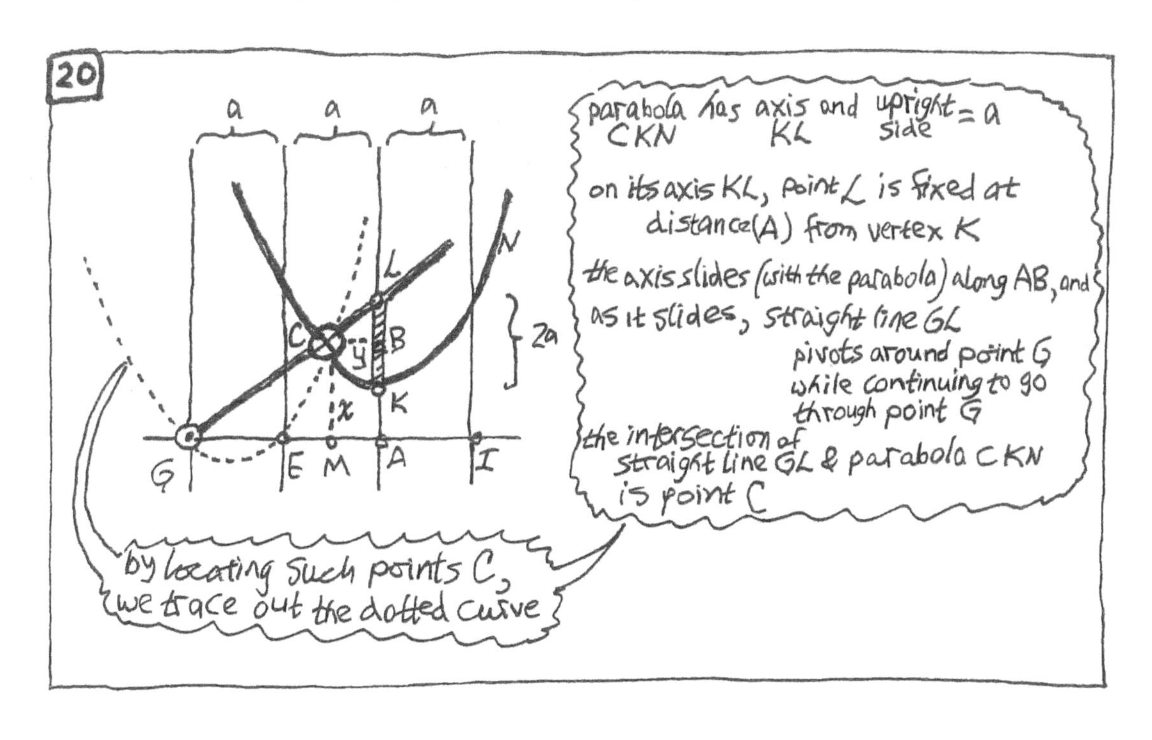

Then, because of the similar triangles GMC and CBL,
[line] GM, which is $2a - y$, is to MC, which is x , as CB, which is y, is to BL,
which [BL] is consequently $xy/(2a - y)$.

And because LK is a, BK is $\{a - [xy/(2a - y)]\}$

And finally,
because this same BK, being a segment of the diameter of the Parabola, is
to BC, which is applied to it ordinatewise,
as this [same BC] is to the upright side, which is a,
calculation (*le calcul*) shows that $y^3 - 2ayy - aay + 2a^3$ is equal to axy;

and consequently that the point C is that which was called for (*demandé*). [See figure 21.]

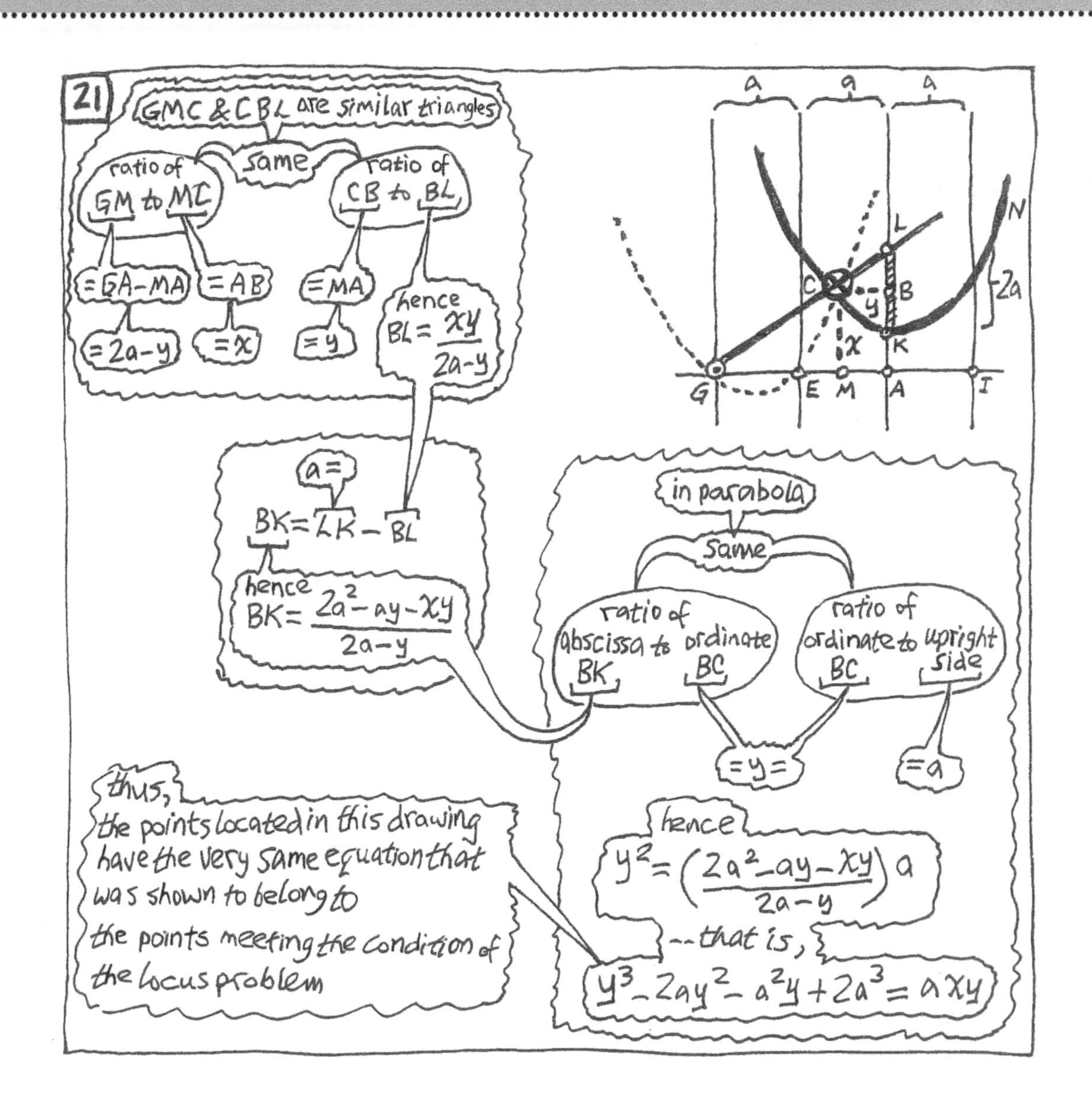

And it [the point C] can be taken

in such spot (*endroit*) of the line CEG as one wishes to choose,

or also
in its adjoined line cEGc which is
described in the same fashion, except that the vertex (*sommet*) of the Parabola is turned toward the other side [that is, in the other direction],

or finally
in their contraposed NIo, nIO, which are drawn (*descrites*) by
the intersection which the line GL makes on the other side of the Parabola KN.

[See figure 22.]

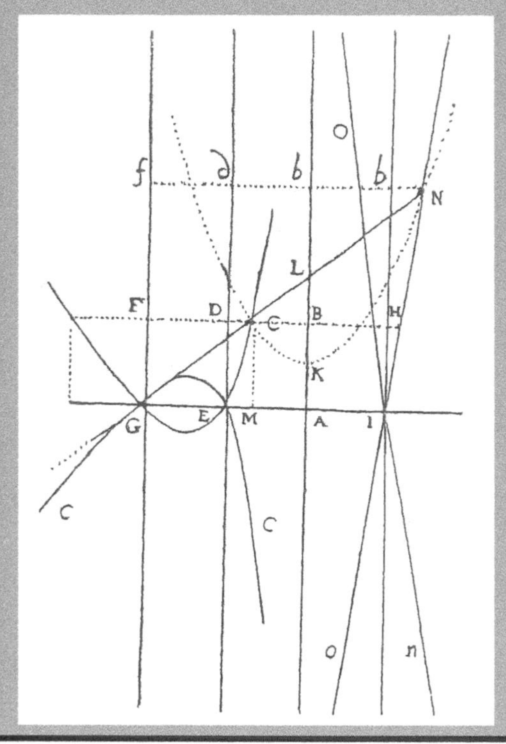

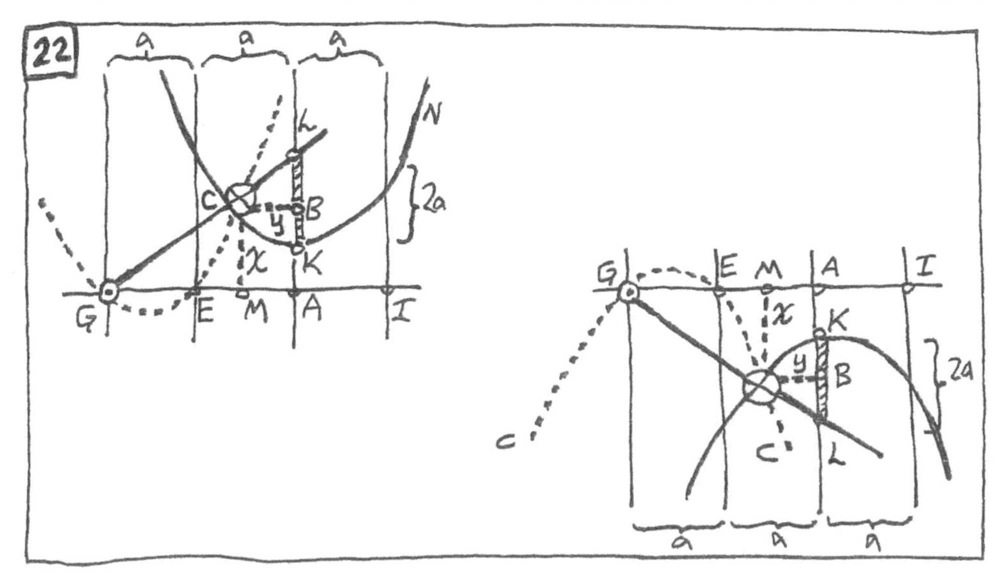

[21] Now again (*Or encore*)
[if] the parallel given [lines] AB, IH, ED, and GF were not equally distant [from each other],
and GA did not cut them at right angles,
nor also [did] the lines drawn toward them from the point C,
[then] this point C would not always be found in a curved line which is of this same nature.

And sometimes it can also be found [in such a curve] even though
not any of the given lines be parallel.

Descartes has been considering the five-line locus when four of the straight lines are parallels and the fifth is a transversal. He will continue to do so; but now he will change the condition that is imposed on the lines drawn from the point whose locus is sought. Now, although these drawn lines will continue to form the equal parallelepipeds, they will be differently combined to do so, as follows:

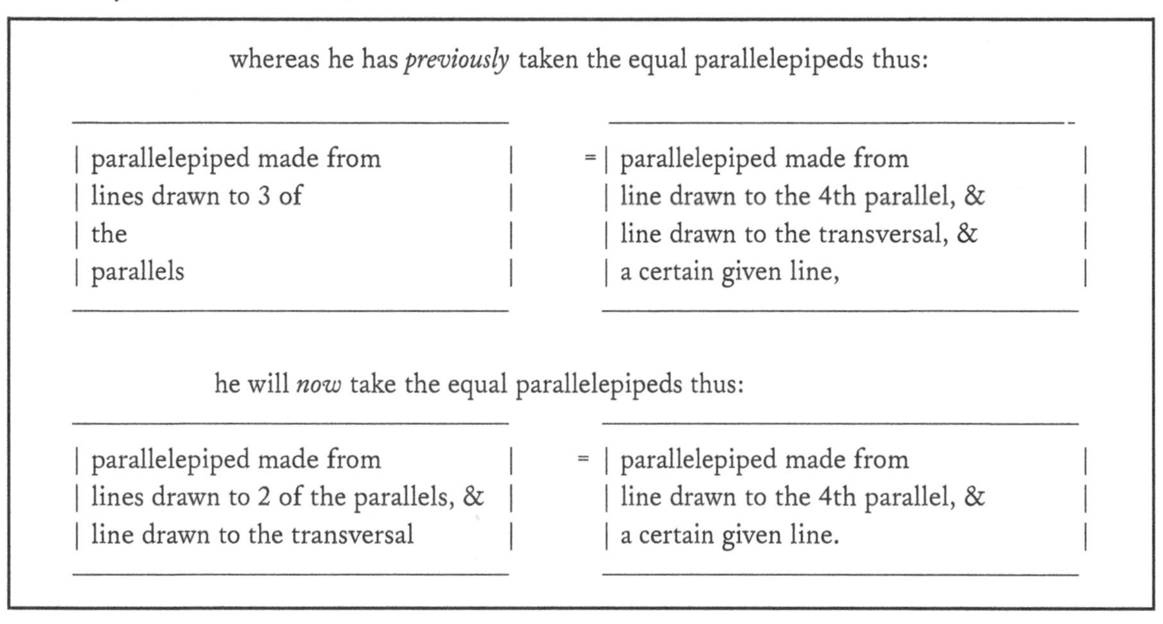

whereas he has *previously* taken the equal parallelepipeds thus:
parallelepiped made from
lines drawn to 3 of
the
parallels
=
parallelepiped made from
line drawn to the 4th parallel, &
line drawn to the transversal, &
a certain given line,
he will *now* take the equal parallelepipeds thus:
parallelepiped made from
lines drawn to 2 of the parallels, &
line drawn to the transversal
=
parallelepiped made from
line drawn to the 4th parallel, &
a certain given line.

In other words: if the earlier equation is taken, but one of the ingredient parallel lines on the left side of the equation is interchanged with the ingredient transversal line on the right side, then we shall have the equation that states the condition that will now determine the point whose locus is sought.

But [what] if, when there are 4 of them thus parallel
and a fifth which crosses them,
and the parallelepiped [produced]
from three of the lines drawn from the point sought,
the one [being drawn] onto the fifth
and the two others onto 2 of those which are parallel,
be equal to that [parallelepiped produced]
from the two [lines] drawn onto the two other parallels
and from one other line [that is] given.

[Then] this point sought is in a curved line of another nature,
to wit, in one which is such that,
all the straight lines applied ordinatewise to its diameter
being equal to those of a conic section,
the segments of this diameter which are between the vertex and these lines
[that is, the abscissas of the curve of another nature]
have the same ratio (*proportion*) to a certain given line as this given line has to
[the abscissas of the conic sections, that is, to]
the segments of the diameter of the conic section to which
lines that are equal (*pareilles*) are applied ordinatewise

That is to say, the other curve is of this sort (see figure 23):

the abscissa of any point on the other curve
has the same ratio to a given line as this given line has to
the abscissa of a point on a conic section whose ordinate is the same as
the ordinate of the given point on the other curve.

And I would not know how to say truly that this line be less simple than the preceding [one],
which I believed all the same (*toutefois*) should be taken for the first,
because the drawing (*description*) and the calculating of it are in some fashion easier.

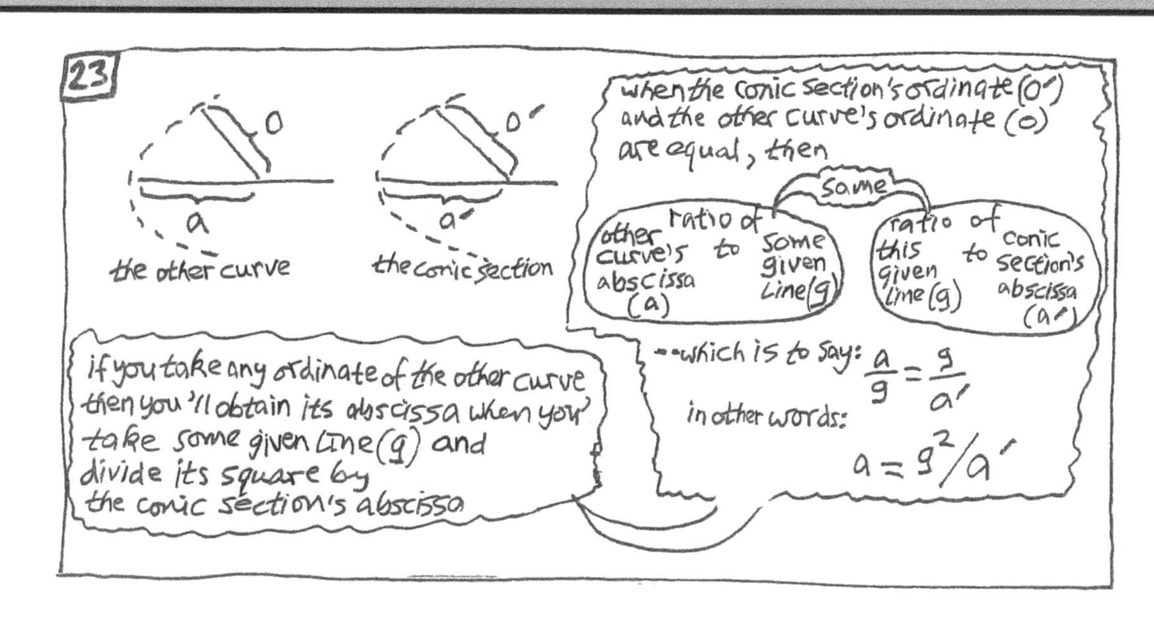

Whereas the equation that was yielded by the calculation for the preceding curve is this:

$y^3 - 2ayy - aay + 2a^3$	is equal to	axy,

the equation that would be yielded by a calculation for the present curve is this:

$- ayy + aay$	is equal to	$axy - xyy + 2aax$.

Why cannot Descartes say that the latter is less simple than the former? Because for both of the equations there is in the highest term the same number of dimensions (namely, 3). In the former equation, the highest term is y^3; in the latter, it is xyy. But he does treat the former curve first—because it is somehow easier to draw, easier to get an equation for. That is to say: curves in the same genre may have the same degree of compositeness but nonetheless be of species which differ somehow in the ease with which they can be drawn and with which an equation can be obtained for them.

[22] As for the lines of which use is made for
the other cases (*se servent aux autres cas*),
I will not stop to distinguish them by species (*par especes*),
for I did not undertake (*entrepris*) to say all,
and, having explained the fashion of finding an infinity of points through which they pass,
I think I have sufficiently given the means of drawing (*descrire*) them.

The author's enterprise is not exhaustive in all respects. In some respects, however, it is comprehensive. He says that he has given a means sufficient for drawing all "geometrical" curves. Descartes does not seem to distinguish between having the means for drawing a continuous curve and having the means for calculating where to find an infinity of distinct points on it. The statement we have just read leads him now to conclude this discussion of Pappus's problem by resuming his earlier discussion of the tracing of geometrical curves.

That earlier discussion of the tracing of geometrical curves took place in a stretch of 7 paragraphs that came right in the middle of the discussion of Pappus's problem. In that stretch of 7 paragraphs between the two 17-paragraph parts of the discussion of Pappus's problem, only the paragraph right in the middle of the stretch, that is, the 4th one, contains an exercise in what elsewhere he calls "the geometrical calculus here proposed," saying that he hoped the calculus would serve those who have the "*addresse*" for exerting themselves further in "other researches" to which he invited them. Before moving on to advanced studies, however, they need to complete their basic training. And before taking care of that, the instructor now needs to say a few words more about the fashion of drawing curves that he has shown them.

more on drawing curves (*paragraphs 23-24*)

[23] But (*Mesme*) it is appropriate to note that
there is a great difference between
this fashion of finding several (*plusieurs*) points for tracing a curved line
and that [fashion] of which one makes use (*dont se sert*) for [such curves as]
the spiral and its like;

for by this latter [fashion] one does not find
indifferently all the points of the line which one seeks,
but only those which can be determined by
some measure simpler than that which is requisite for composing it,
and so, to speak properly, one does not find a single one of its points,
that is to say, a single one of those which are so properly its own that
they could not be found but through it [the line];
instead of that, there is not any point
in the lines which serve for the proposed problem (*question*)
that could not be encountered among those which are
determined by the fashion explained a little while ago.
And because this fashion of tracing a curved line
in finding indifferently several (*plusieurs*) of its points
extends to naught but those which can also be described by
a movement [that is] regular and continuous,
one should not reject it entirely from Geometry.

That is to say: It is true that the Cartesian fashion of tracing curves does not yield a line, but merely yields many points that lie on it. However, not only does it yield as many points of the line as you please; it also yields any of the points you please. The curves that can be traced by locating their points in the Cartesian fashion should therefore not be excluded from geometry. The fashion in which Descartes traces them differs from the fashion in which one traces the curves that Descartes does exclude as non-geometrical. In these "non-geometrical" curves, there are located as many points as you please, to be sure—but the fashion in which such a curve is traced yields only some of its points rather than any of them you please. Only those of the curve's points can be located which are accessible to determination by a quantification simpler than is that property which is peculiar to the curve.

Now, having argued for permitting in geometry a certain fashion of tracing curves by locating points determined by quantitative relations among straight lines, Descartes will go on to argue for also permitting in geometry a certain fashion of tracing curves by locating points determined by quantitative relations among straight lines that are embodied in wires or cords. He refers in particular to the use of a cord coiled around pins and stretched taut, by means of which it is possible to draw ellipses and hyperbolas. We have looked at such uses of a cord earlier, in Chapter 3 of Part VI of this guidebook, toward the end of our study of conic sections.

Descartes delimits geometry as a field of study by including within it curves that had not traditionally been included, while continuing to exclude curves whose treatment would require unattainable exactness in knowing ratios that would bring straight lines together with lines that are curved.

> [24] And one should no more reject that [fashion of tracing a curve] where
> one makes use of (*on se sert d'*) a wire or a coiled cord
> for determining the equality or the difference of
> two or several straight lines which can be drawn
>   from each point of the curve that one seeks,
>   to certain other points, or onto certain other lines at certain angles,
> as we did in the Dioptrics to explain the Ellipse and the Hyperbola;
>
> for even though one could not receive there [in Geometry] any lines which resemble cords—
> that is to say, which become now straight, now curved—
> because the ratio (*proportion*) which is between the straight [line] and the curved,
> not being known, and even, I believe, not [being] able to be known by men,
> one is not able to conclude anything about that which would be exact and sure.
>
> All the same, because one makes use of (*se sert de*) cords
> in these constructions [performed in the Dioptrics]
> for naught but determining straight lines whose length one knows perfectly,
> that [use of cords] should not make one reject them.

Here, for the first time since the notice that immediately preceded the *Geometry*, Descartes refers to the fact that this treatise comes late in a volume that discusses other matters first. More particularly, he here refers to the way that the ellipse and hyperbola were explicated in the *Dioptrics*. The *Dioptrics* will again be mentioned 17 paragraphs hence (in the 41st paragraph). Later (in the 45th paragraph) it will be said that the act of drawing certain ovals (ovals that he is there discussing) imitates the act of drawing the ellipse and hyperbola that was presented in the *Dioptrics*.

Descartes introduces the ellipse and hyperbola at the beginning of the Eighth Discourse of the *Dioptrics*, as follows:

Eighth Discourse

Of the Figures that Transparent Bodies Should Have in Order to Divert Rays by Refraction in All the Fashions that Serve Sight

Now, to the end that I can soon tell you more exactly in what sort we should make these artificial organs for rendering them the most perfect they can be, there is need that I explain beforehand the figures that the surfaces of transparent bodies should have for bending and diverting rays of light in all the fashions that can serve my design. In which if I cannot render myself sufficiently clear and intelligible to all the world—because it is a matter of Geometry that's a little difficult—I will at least try to be sufficiently [clear] for those who will have learned only the first Elements of that science. And at first, to the end of not keeping them in suspense, I shall tell them that all the figures of which I have to speak to them here will not be composed but of Ellipses or Hyperbolas, and of circles or straight lines.

The Ellipse, or the Oval, is a curved line which the Mathematicians are accustomed to setting out for us by cutting through a cone or cylinder, and which I have also sometimes seen employed by gardeners in partitionings of their flower beds, where they draw it in a manner which is truly very coarse and little exact, but which makes for, it seems to me, better comprehending its nature than does the cutting (*section*) of a cylinder or a cone. They plant in the earth two pickets—as, for example, the one at point H, the other at point I—and having knotted together the two ends of a cord, they pass it around them in the fashion that you see here, [namely,] BHI. Then, placing the end of their finger in this cord, they conduct it all around these two pickets, while always pulling it to them with equal force, to the end of keeping it extended evenly, and so they draw on the earth the curved line DBK, which is an Ellipse. And if, without changing the length of this cord BHI, they only plant their pickets H and I a little closer, the one to the other, they will yet again draw an Ellipse, but which will be of another species than the preceding one; and if they plant them a little closer yet, they will draw yet another; and finally, if they join them entirely together, it will be a circle that they will draw. Whereas, if they diminish the length of the cord in the same proportion as [they diminish] the distance of these pickets, they will draw a good lot of Ellipses which will be different in size, but which will be all of the same species. And so you see that there can be an infinity of all different species of them, so that they differ not less, the one from the other, than the last does from the circle; and that of each species there can be all sizes of them; and that if from a point, as B, taken as you please in some one of these Ellipses, one draws two straight lines toward the two points H and I, where the two pickets should be planted for drawing it, these two lines BH and BI, joined together, will be equal to its greatest diameter DK, as can easily be proven through construction. For the portion of the cord which extends from I toward B and is from there bent back as far as to H, is the same which extends from I toward K or toward D and from there also bends back as far as to H: so that DH is equal to IK, and HD plus DI, which has the value amounting to HB plus BI, is equal to the whole line DK. And finally, the Ellipses that one draws while always putting the same proportion between their greatest diameter DK and the distance from the points H and I, are all of the same species. And because of a certain property of these point H and I, of which you will hear afterwards, we will name them the burning points, the one interior and the other exterior: to wit, if one relates them to the half of the ellipse which is toward D, I will be the exterior; and if one relates them to the other half which is toward K, it will be the interior; and when we shall speak without distinction of the burning point, we shall always intend to be speaking of the exterior one. ....

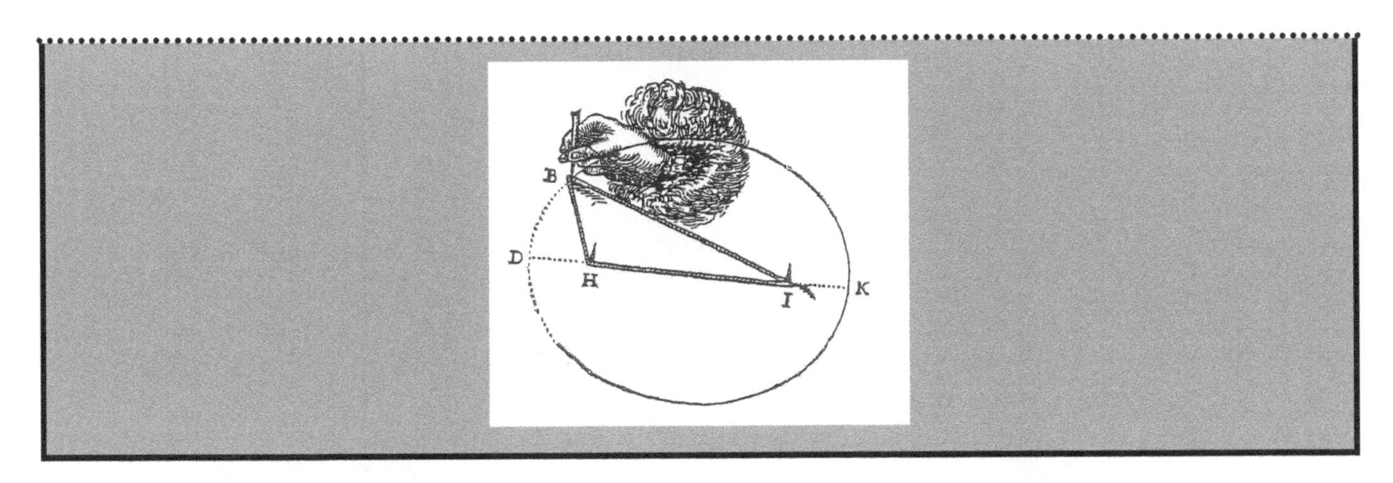

That having been said, Descartes goes on to discuss the use of the ellipse in a lens (a transparent body made to focus the rays of light that pass through it). He tells how, given the points H and I, to get the tangent to the ellipse at any point B on it, but says "I do not furnish a demonstration of this, because the geometers know it well enough, and others would only be bored to hear it." (What Descartes will not furnish can be found in the presentation of the 48th proposition of *Conics* III, in Chapter 3 of Part VI of this guidebook.) Descartes does, however, demonstrate what directly concerns the determination of the desired focal arrangement—and then he turns to the hyperbola.

The Hyperbola is also a curved line which the mathematicians explain by the cutting (*section*) of a cone, as with the Ellipse. But to the end of making you conceive it better, I shall here again introduce a gardener who puts it into service in compassing the embellishment of some flower-bed. He yet again plants his two pickets at the points H and I; and having attached to the end of a long ruler the end of a cord which is a little shorter, he makes a round hole at the other end of this ruler, into which he makes the picket I enter, and he makes a loop at the other end of this cord, which loop he passes around the picket H. Then, putting his finger at point X, where they are attached, the one to the other, he runs it from there down as far as D, nonetheless always keeping the cord entirely joined to the ruler as if glued against it from the point X as far as the place where he is touching it, and with this cord stretched entirely taut: by means of which, constraining this ruler to turn around the picket I in the measure that he lowers his finger, he draws on the earth the curved line XBD, which is part of an Hyperbola. And after this, turning his ruler in the other direction toward Y, he draws in the same fashion another part, YD. And, moreover, if he passes the loop of his cord around the picket I, and the other end of his ruler around the picket H, he will draw another Hyperbola SKT, entirely similar and opposite to the preceding. But if, without changing his pickets or his ruler, he only makes his cord a little longer, he will draw an Hyperbola of another species; and if he makes it yet a little longer, he will draw one of yet another species, until, making it entirely equal to the ruler, he will draw, in place of an Hyperbola, a straight line. Then, if he changes the distance of his two pickets in the same proportion as the difference between the lengths of the ruler and the string, he will draw Hyperbolas which will all be of the same species but whose similar parts will be different in size. And finally, if he equally increases the lengths of the cord and the ruler, without changing either their difference or the distance of the pickets, he will always not draw any but the same Hyperbola, but he will draw a greater part of it. For this line is of such a nature that, although it always curves more and more toward the same side, it can all the same extend to infinity without its extremities ever meeting. And so you see that it has in several fashions the same relation to the straight line that the Ellipse has to the circular. And you see also that it has an infinity of different species of them, and that in each species there is an infinity of them whose similar parts are different in size. And [you see], moreover, that if from a point, as B, taken as you please in one of them, one draws two straight lines toward the two points, as H and I, where the two pickets should be planted for drawing it, and which we shall again name the burning points, the difference of these two lines, HB and IB, will always be equal to the line DK, which marks the distance between the opposite hyperbolas. This happens because BI is longer than BH by just so much as the ruler has been taken longer than the cord, and also because DI is that much longer than DH. For, if one shortened this one, DI, by [the length of] KI, which is equal to DH, we would have DK for their difference. And, finally, you see that the Hyperbolas which we draw while always putting the same proportion between DK and HI are all of the same species. …

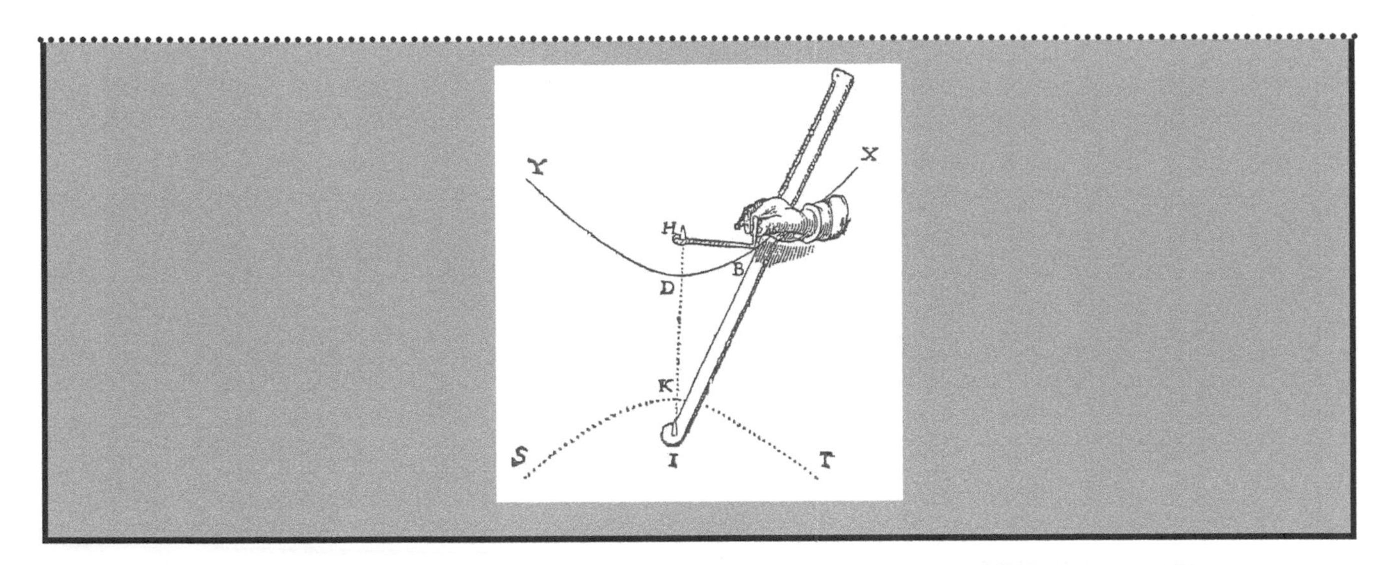

Having said that, Descartes goes on to discuss the use of the hyperbola in a lens, a transparent body made to focus the rays of light that bend when they pass through it. He tells how, using the given points H and I, one can get the tangent to the hyperbola at any point B on it, and says "the geometers know sufficiently well the demonstration of this." (For this, you can look again at the presentation of the 48th proposition of *Conics* III, in Chapter 3 of Part VI of this guidebook.) Descartes does, however, demonstrate what directly concerns the determination of the desired focal arrangement.

In this Eighth Discourse of the *Dioptrics*, Descartes also shows how, by means of such lenses, one can "make all the rays which come from a certain point, or tend toward a point, or are parallel, to be exactly changed from one of these three dispositions to another." He goes on to say "we can imagine yet an infinity of other lenses which, like those mentioned above," can be used to cause such changes in the dispositions of rays of light. "But I do not think I have any need to speak of them here, because I shall be able to explain them more conveniently hereafter in the *Geometry*, and because those that I have drawn are the most proper of all for my design, as I now wish to try proving...." And so, this Eighth Discourse concludes by saying that "we must conclude from all this that the hyperbolical and elliptical lenses are preferable to all the others which can be imagined, and even that the hyperbolical are in almost everything preferable to the elliptical. Following which I shall now say in what fashion it seems to me we must compose each species of telescope, in order to make them as perfect as possible."

In the first of the reasons given by this Eighth Discourse of the *Dioptrics* for the preferability of lenses that are hyperbolical or elliptical, Descartes refers to lines less simple and less easy to trace than those that are straight, those that are circular, and those that the geometers treat as conic sections:

> ...the figures of some are much easier to trace than those of others; and, it is certain, that after the straight line, the circular line, and the parabola—which alone cannot suffice for tracing any of these lenses, as everyone can easily see if he examines it—there are none simpler than the ellipse and the hyperbola. So that, the straight line being easier to trace than the circular, and the hyperbola no less easy than the ellipse, those lenses whose figures are composed of hyperbolas and straight lines are the easiest which can be cut, and then there follow those whose figures are composed of ellipses and circles, so that all the others, which I have not explained, are less so.

The *Dioptrics* is the first of three "try-outs" (*essais*) in a single volume introduced by a *Discourse on the Method for Well Conducting One's Reason and Seeking Truth in the Sciences*. That *Discourse* is followed by the ten "discourses" that constitute the *Dioptrics*, which itself is followed by the ten "discourses" that constitute the *Meteorology*. Only then come the three "books" that constitute the *Geometry*. With the paragraph of the *Geometry* that calls attention to the fact that the *Geometry* is merely one item in the volume in which it appears, the solution of Pappus's problem is finished up, and we have come to the end of the *Geometry*'s beginning.

The treatise called the *Geometry* is the volume's only item that is divided into "books"; but, like all the other items, it is also divided into paragraphs. These paragraphs, however, are not equally distributed among the three books of the *Geometry*: the First Book has 34, the Second has 64, and the Third has 74. If the 172 paragraphs of the whole were equally distributed among the three books, then each would get 57 paragraphs, with one paragraph left over after that distribution. As it turns out, the 58th paragraph of the *Geometry* is the Second Book's 24th paragraph—the last paragraph of the finishing up of Pappus's problem. It is preceded by 57 paragraphs of the *Geometry* and is followed by the *Geometry*'s remaining twice-57 paragraphs. (What marks the completion of the first 57 of these latter twice-57 paragraphs is, as it turns out, the Third Book's 17th paragraph.) The *Geometry* thus provides the means of finishing up an item on the agenda of the ancients, and, after doing so, devotes twice as much of itself to putting those means to use dealing with its own agenda.

Of those 57 paragraphs constituting that initial one-third of the *Geometry* which deals with the ancient agenda in the new way, the 28th paragraph of the First Book is the one that is central. It is also central with respect to substructure, in the following way. Taking all 58 paragraphs (the 57 plus the one that is disregarded when we divide 172 into thirds), the initial one-third of the *Geometry* is, as we have seen, divided into three stretches of 17 paragraphs each, with a 7-paragraph insertion between the central stretch of 17 and the last stretch of 17, both of which treat Pappus's problem; and each of the three 17-paragraph stretches is composed of an 11-paragraph stretch followed by a 6-paragraph stretch. Now, the 28th paragraph of the First Book, which is the center of the entire initial third of the *Geometry*, is also located within the central stretch of its three 17-paragraph stretches; indeed, it is located at the structural midpoint of the stretch: it is the last paragraph in the first of the two sub-sections (the one of 11 paragraphs); there the author sums up what he has accomplished in that sub-section, and he tells what he is going on to do in the second of the two sub-sections (the one of 6 paragraphs).

Those formal considerations would seem to reinforce the view that the end of the first 24 paragraphs of the Second Book of the *Geometry* marks the end of the beginning of this treatise. The structure of this beginning of Descartes' *Geometry* is laid out in figure 24.

24 the beginning of Descartes' *Geometry* -- its structure

- **17** introducing "the true method" in geometry
  - **11** a new way to resolve any problem in geometry into the solving of an equation — First Book 1–11
  - **6** how the equation can yield the construction that's sought in any problem of the ordinary geometry — 12-17
- **17** surpassing the ancients: the solution of Pappus's problem
  - **11** the method successfully tried on an ancient unsolved problem — 18–28
  - **6** first part of the demonstration that in this the ancients have in fact been surpassed — 29–34
- **7** comprehensively distinguishing the geometrical curves as an ordered system — Second Book 1–7
- **17** the solution of Pappus's problem: finishing up
  - **11** completion of all that he undertook to prove — 8–18
  - **6** going further — 19-24

58 total in the beginning = 57+1 paragraphs
total remaining in the *Geometry* = 57 × 2 paragraphs
total remaining in the Second Book = 40 paragraphs

Contrary to the ancient Greek proverb, however, here the beginning is not more than half of the whole, but is merely the least bit more than half of what remains of the whole. The beginning, as we have seen, consists of one-third (plus one) of the 172 paragraphs of the whole *Geometry*. Now, having made a beginning in the study of the *Geometry*, we should consider briefly how that beginning fits into what Descartes does later, in the rest of the treatise, and earlier, in the volume of which the treatise is the final item. (For the text of the rest of the *Geometry*, however, as well as for the text of the rest of the volume of which it is the final item, readers will have to look outside this already very lengthy guided study.)

CHAPTER 6. THE MOST USEFUL PROBLEM IN GEOMETRY, AND THE USEFULNESS OF THE GEOMETRY

The very beginning of the *Geometry* was a 17-paragraph stretch that introduced "the true method in geometry." The rest of the beginning presented the solution, by that method, of the ancient unsolved problem of Pappus. This solution of the problem was presented in two 17-paragraph stretches that surrounded a 7-paragraph stretch where the field of geometry was vastly expanded in such a way as to make it methodically surveyable by means of the geometrical calculus introduced in the *Geometry*'s very first 17-paragraph stretch.

At the end of the first 24 paragraphs of the Second Book, the solution of Pappus's problem has been finished up. The remaining 40 paragraphs of the Second Book will show how the solution of Pappus's problem is merely the first example of the power of this *Geometry*, which far surpasses the "common" geometry. This 40-paragraph stretch will fall into a 16-paragraph stretch and a 24-paragraph stretch. Whereas the first 24 paragraphs of the Second Book have treated a problem that might seem to be of concern only to those who are preoccupied with geometry, the final 24 paragraphs will treat geometry as a tool for making tools of service for physics and the mastery of nature. Between these two 24-paragraph stretches which begin and end the Second Book, there lies a 16-paragraph stretch which shows how—in matters of great importance to the common geometry—this *Geometry* is much more powerful than the common geometry. That 16-paragraph stretch on matters of great importance to the common geometry is also a preparation for the 24-paragraph stretch that follows it and concludes the Second Book.

This 24-paragraph stretch that concludes the 40-paragraph stretch at the end of the Second Book is preceded by the entire 34 paragraphs of the First Book and the first 40 paragraphs of the Second Book. That is to say, it follows 74 paragraphs of the *Geometry*. It also precedes 74 paragraphs of the *Geometry*. Thus this 24-paragraph stretch is the very center of the *Geometry*. See figure 1.

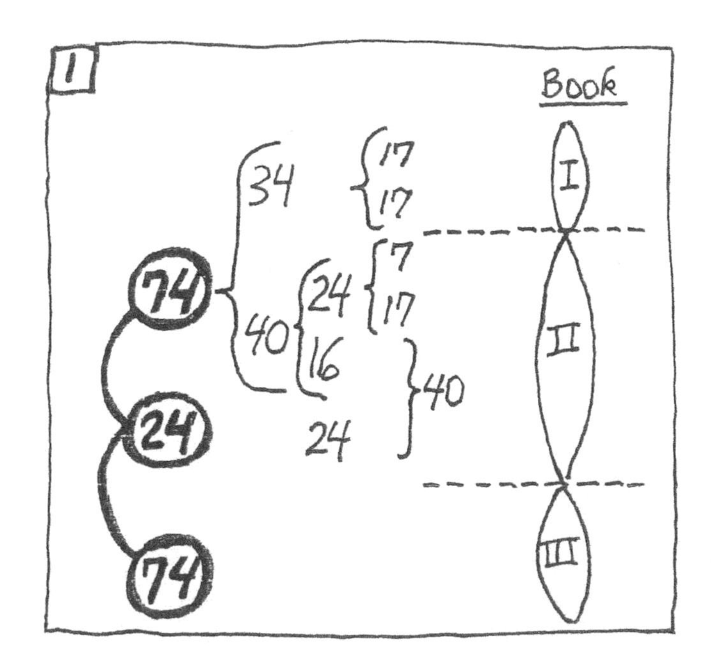

At the very heart of the *Geometry* is a discussion of certain ovals. That discussion begins by speaking of its own usefulness for the *Dioptrics*, and it ends by saying that although there is more to the geometry of those ovals, that part of geometry has nonetheless been excluded from the *Geometry* because it was not useful for the *Dioptrics*. The last paragraph of that central stretch of 24 paragraphs which concludes the Second Book (the book on the nature of curved lines) speaks of the extendibility of the whole affair (not merely the affair of those ovals) to three dimensions. That is to say: the core section treats of the utility of this geometry—its role in a physics applied to the mastery and possession of nature—and concludes with a reminder that what is laid out upon a flat sheet of paper is only the first step in dealing with the world that extends outward from the paper indefinitely in all directions.

As has been said, the initial 24 paragraphs of the *Geometry*'s Second Book constitute the end of the beginning of the entire *Geometry*, and the final 24 paragraphs of the Second Book constitute the center of the entire *Geometry*. Between the initial 24 paragraphs and the final 24 paragraphs, stretch the middle 16 paragraphs of the Second Book. See figure 2.

The very opening of the middle book of the *Geometry* (its opening 7-paragraph stretch) made it possible to move on from the 17-paragraph stretch which preceded it (and which concluded the First Book) to the 17-paragraph stretch which followed it; likewise, the middle of the middle book of the *Geometry* (its middle 16-paragraph stretch) makes it possible to move on from the 24-paragraph stretch which precedes it (and which opened the Second Book) to the 24-paragraph stretch which follows it (and which will conclude the Second Book). See figure 3.

The middle of the middle book of the *Geometry* mediates between the end of the beginning of the treatise and what lies at the heart of the treatise. These middle paragraphs of the Second Book will lead us from the presentation of "the true method in Geometry," which has finished up the construction of what easily solves the ancient hitherto unfinished problem, to a presentation of the usefulness of that method in finding the means for figuring out how to construct the artificial means of vastly increasing humanity's natural power of vision. Let us now see how.

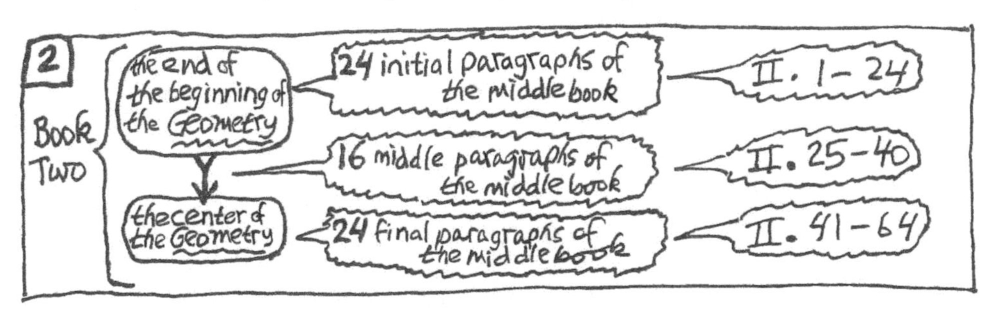

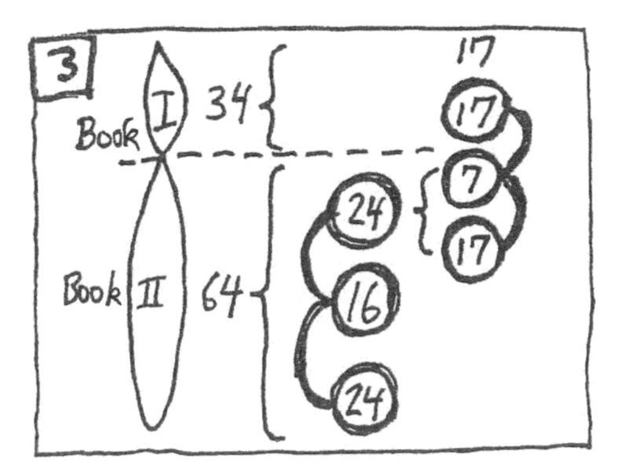

normals to curves (*paragraphs 25–40*)

The middle 16 paragraphs of the Second Book open as follows:

[25] Now (*Or*) from that alone—that one knows the relation (*rapport*)
which all the points of a curved line have to all those of a straight line
in the fashion that I have explained—
it is easy to find also
the relation which they have to all the other points and lines [that are] given;
and in consequence (*en suite*) to know
the diameters, the axes, the centers, and other lines or points to which
each curved line will have some relation (*rapport*) [that is]
more particular or more simple than [the one that the curve has] to the other [lines or points],
and thus
to imagine [that is, to visualize] different means for drawing (*descrire*) them [the curves],
and to choose the easiest (*les plus faciles*) of them [those means].

And one can even also by that alone (*par cela seul*) find
almost (*quasi*) all that can be determined touching the
magnitude of the space which they [that is, curved lines] comprehend,
without there being need that I give more of [a] disclosure of it (*en donne plus d'ouverture*).

And finally, as for all the other properties that one can attribute to lines [that are] curved,
they depend on naught but
the magnitude of the angles that they make with certain other lines.
But when one can draw straight lines which cut them at right angles,
at points where they are met by those [straight lines] with which they
make the angles that one wishes to measure,
or—which I take for the same thing here—which cut their tangents,
[then] the magnitude of these angles is no less easy (*n'est pas plus malaysée*) to find
than if they were contained between two lines [that are] straight.
This is why I believe myself to have put here
all that is required for the elements of curved lines
when I have given generally the fashion of drawing straight lines which fall at right angles
upon such of their points [that is, such points of curved lines] as one will wish to choose.
And I dare say that this here is the problem [which is] the most useful and the most general
not only that I know, but even that I have ever desired to know in Geometry.

That is to say, the equation that relates the points on the curve to the points on straight lines can also be used to obtain lines such as diameters and axes and to obtain points such as centers—as well as to obtain lines that are tangent to curves at any given points (and hence to obtain the lines called normals—that is, straight lines perpendicular to those tangents at those points of tangency) and to obtain points that are points of application (in other words, foci).

Why might the problem of drawing a tangent to a curve be an important problem in geometry? Well, if the key to geometrical problem-solving is the relation between lines that are curved and lines that are straight, then an interest in the curviness of a curve at a point might lead us to investigate how it departs from a straight line which meets it but does not cross it. But the general usefulness of the tangent-problem in geometry also has something to do with the general usefulness of geometry in physics and in usefully applied physics, as we shall see in the final 24 paragraphs of the Second Book.

The solution of the most generally useful problem in geometry—namely, how to draw, at any point on a curve, a straight line perpendicular to the tangent at that point—is begun in the next paragraph, the 26th, which is accompanied by a figure. Later, in the final 24-paragraph stretch, in the first of the 3 paragraphs included so that he will not "omit the demonstration of what I have said" about optical properties of certain ovals, Descartes draws a line striking one of those ovals at right angles, at a point taken "at discretion"—which, he says, is "easy by the preceding problem"; the demonstration there in the 54th paragraph is accompanied by the figure that appears here for the 26th paragraph, which treats the first step of the problem that occupies the middle 16 paragraphs of the Second Book.

Now, before looking at the Second Book's final stretch of 24 paragraphs, let us consider briefly the remaining paragraphs of this 16-paragraph road that takes us from the initial 24 paragraphs of the Second Book to its final 24. For our present purposes, it will suffice merely to take note of the organization of this mediating section, which treats this most useful and most general problem.

If one makes an effort to detect the structural principle behind the organization of this middle stretch of paragraphs in the middle Book of the *Geometry*, one's attention is drawn to the simplest arrangement of items that can have a middle. That is to say, this portion of the text emphasizes organization by triples: subdivisions into an initial item and a final item—and an item in the middle.

If any readers of the present guided study wish to go beyond that portion of Descartes' text which has already been covered here, the following figure will be helpful as they read on through these middle 16 paragraphs of the Second Book of the *Geometry*. (Later, there will also be such figures to help them read on through the concluding 24 paragraphs of the Second Book, and through the 74 paragraphs of the Third Book.) See figure 4.

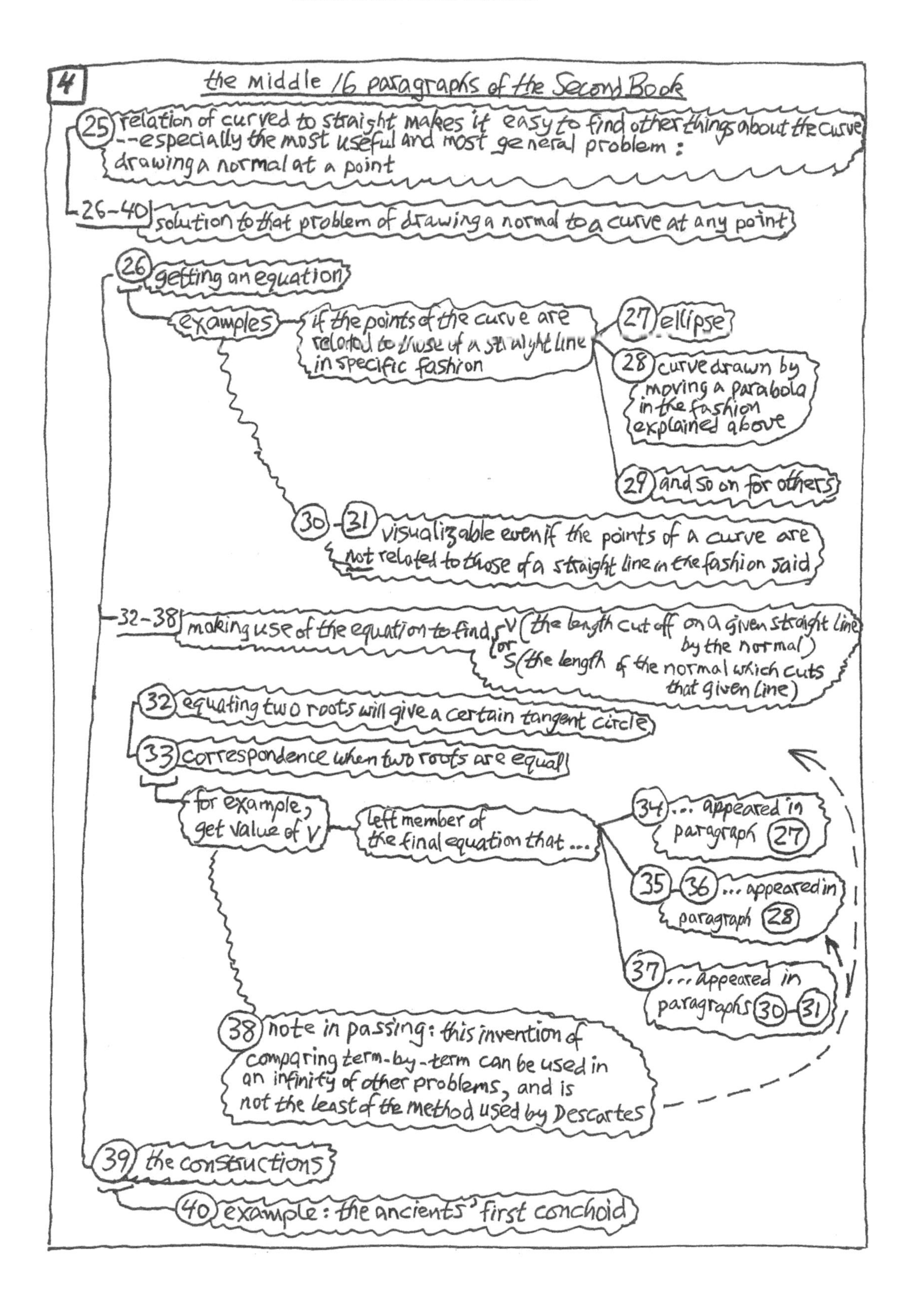
4
the middle 16 paragraphs of the Second Book
25 relation of curved to straight makes it easy to find other things about the curve --especially the most useful and most general problem : drawing a normal at a point
26–40 solution to that problem of drawing a normal to a curve at any point
26 getting an equation
examples
if the points of the curve are related to those of a straight line in specific fashion
27 ellipse
28 curve drawn by moving a parabola in the fashion explained above
29 and so on for others
30 - 31 visualizable even if the points of a curve are not related to those of a straight line in the fashion said
32–38 making use of the equation to find V (the length cut off on a given straight line by the normal) or S (the length of the normal which cuts that given line)
32 equating two roots will give a certain tangent circle
33 correspondence when two roots are equal
for example, get value of V
Left member of the final equation that ...
34 ... appeared in paragraph 27
35 - 36 ... appeared in paragraph 28
37 ... appeared in paragraphs 30 - 31
38 note in passing: this invention of comparing term-by-term can be used in an infinity of other problems, and is not the least of the method used by Descartes
39 the constructions
40 example: the ancients' first conchoid

In this mediating 16-paragraph stretch, the opening paragraph (25) generalizes what has come before, and speaks of that problem in particular (of all those whose solution is indicated by what has come before) which is the most useful and the most general in geometry. (This problem is the drawing of a straight line that makes right angles with a given curve at any given point on the curve—that is, the straight line perpendicular to the tangent at the point of tangency—the line that came to be called the "normal.") The rest of the paragraphs in this portion present the solution of that problem of drawing the normal to a curve at any one of its points. The solution has three parts: the initial step, obtaining a certain equation in x or y, is treated in the 26th paragraph, and the final step, doing the construction, is treated in the 39th paragraph; the middle step, making use of the equation so as to find a certain quantity v or s that determines a certain point required for the construction, is treated in the 32nd and 33rd paragraphs. These two paragraphs are the middle paragraphs of this portion, which comprises the 16 middle paragraphs of the middle book of the *Geometry*; the rest of the portion consists of the 7 paragraphs which precede them, and the 7 paragraphs which follow them. The first step is treated in a single paragraph, as is the final step; only the middle step is treated in two paragraphs.

The treatment of the middle step is peculiar also in the structure of its sequel. Each treatment of a step is followed by an exemplification.

The treatment of the initial step (26th paragraph) is followed by exemplification that occupies paragraphs 27–31. The last two of these (30–31) deal with the same example, one where the first step is possible even though the points of the curve are related to those of a straight line not "in the fashion that I said" but rather in some totally other fashion that one can visualize (*sçauroit imaginer*). The trio of paragraphs preceding this pair deal with the other member of this binary division. This member has a triple structure—first, (in the 27th paragraph) a curve familiar from Apollonius—the ellipse; then, (in the 28th paragraph) the curve described by the movement of a parabola "in the fashion explained here, above"; and finally (in the 29th paragraph) only this: "And so [on], of the others." So the first step has the structure, not of a triple, but of a binary division whose first member is a triple—unless we are to leave it as a string of four items whose presentation in five paragraphs serves to make central the one-sentence paragraph of indefinite extension (the 29th) which emphasizes the author's systematization of curves and helps to situate in the middle of this portion those paragraphs that treat the middle step (the 32nd and 33rd).

This pair of paragraphs that treat the middle step (32–33), as has been said, makes use of a certain equation in x or y, so as to find a certain quantity v or s that determines a certain point required for the construction of the line that is normal to the curve. First, the 32nd paragraph speaks of equating two roots of an equation in order to have a certain circle that is tangent to the curve at the point in question; then the 33rd paragraph speaks of a correspondence of terms when the two roots are equal. The sequel to the treatment of the middle step occupies paragraphs 34–38. The first four of these sequel-paragraphs has the structure of a triple: they present three examples, the very examples that were presented in the first step. (What was the fourth item in the first step's exemplification is now left out, since it was not truly an example: it merely said "And so on, of the others.") The initial member of the triple is the left-hand side of the concluding equation from paragraph 27; now, for this step, it is taken up in the 34th paragraph. The final member of the triple is the left-hand side of the concluding equation from paragraphs 30–31; now, for this step, it is taken up in a single paragraph (rather than a pair), the 37th. The middle member of the triple is the left-hand side of the concluding equation from paragraph 28; now, for this step, it is taken up in the 35th and 36th paragraphs.

The treatment of the final step (39) says that he does not add the construction by which the tangents or the perpendiculars that are sought can be drawn in consequence of the calculating that he has just explained—because it is always easy to find them, although often one has need of a little "*addresse*" for rendering them short and simple. The sequel consists of a single paragraph (the 40th), presenting a simple example: the first conchoid of the ancients.

We have still not dealt with the last paragraph in the sequel to the treatment of the middle step. This is the 14th of the paragraphs in the whole 16-paragraph middle portion of the Second Book. This paragraph (the 38th of the Second Book) takes note (*remarquer*) of something about an expression that the middle step makes use of; the example that it employs is the middle example, not the one that has just been presented (namely, the final one). That is to say, this paragraph (38) directs attention to the middle example of the middle step. The middle step is the only one whose treatment occupies a pair of paragraphs (32–33) rather than a single one. It is also the only step whose sequel (34–38) contains the exemplification organized as a triple. It is also the only step whose middle example occupies a pair of paragraphs. It is also the only step whose sequel (34–38) contains more than exemplification. The final paragraph (38) of the sequel to the treatment of the middle step refers not only to the middle example; it concludes by referring to the treatment of the middle step itself. Things have been omitted from the treatment, Descartes admits.

We must admit that this passage was omitted back when (in the course of examining the 12th paragraph of the Second Book) we considered what Descartes says about the reasons for his deliberate omissions. Here (38), concerning the expression used in the middle example, he says that what he takes note of "you could easily see by experience, for if it were necessary that I stop to demonstrate all the theorems of which I made some mention, I would be constrained to write a volume much bigger than I desire." We omitted that passage from our earlier consideration because, while Descartes here calls attention to his omissions, he says little by way of justifying them; rather, as if by way of compensation for what he has omitted to do for his readers, he puts them on notice about the importance of what he has in fact done for them. These readers, who have not been put on notice since the paragraph of "*advertissement*" that preceded the *Geometry*, are now (38) told this concerning the expression that he makes use of in the middle example: "But I do indeed (*bien*) wish in passing to give you notice (*vous avertir*) that the contrivance (*invention*) of supposing two equations of the same form, for comparing separately all the terms of the one to those of the other, and so making several of them [that is, several equations] to be born from a single one, can serve for an infinity of other Problems, and is not one of the least [of the contrivances] of the method of which I make use (*dont je me sers*)."

Descartes can forgo an exhaustive presentation of what he has come upon in the course of making use of his method. What he says here suggests that it is best if he uses a limited space to teach his readers to follow him in making use of his method. The rest can be left to them.

Descartes' ovals: the *Geometry* makes itself useful (*paragraphs 41-64*)

These 24 paragraphs to which we now come—in which the *Geometry* makes itself useful—not only constitute the end of the treatise's central book; they also constitute the center of the whole treatise.

Because of the central location of this 24-paragraph stretch on utility, its own central paragraphs are the central paragraphs of the entire treatise. These paragraphs—the 52nd and 53rd of the Second Book—are the concluding paragraphs of 8 paragraphs (46–53) on the genres and properties of certain ovals. In the 5 paragraphs that precede those 8, the ovals have their dioptric usefulness mentioned (41) and have the fashion in which they are drawn presented (42–45); and in the 3 paragraphs that follow those 8, the ovals have their convergence properties demonstrated (54–56). The 52nd paragraph completes the consideration of the different properties of the two parts of the ovals in each genre, and the 53rd paragraph names certain points "the burning points" of these ovals, after the example of those points of the ellipses and hyperbolas which were so named in the *Dioptrics*.

The 8 paragraphs that present the burning points of various genres of certain ovals are thus surrounded by a total of 8 paragraphs that prepare what will be presented and then afterward demonstrate what has been presented. These 16 paragraphs (41–56) are followed by an addendum of 8 more paragraphs (57–64).

This addendum consists mostly of another 7-paragraph stretch; this treats the ovals and presents matters omitted from the *Dioptrics*; its next-to-last paragraph says what will be omitted because it is not useful for the *Dioptrics*, and its last paragraph says what will be omitted so that readers will do it for themselves. The concluding paragraph of the Second Book, which comes right after the 7-paragraph stretch, speaks of something else that the reader can do—namely, extend all this, from the plane to the world of three dimensions—and it speaks also of what has not been omitted from the *Geometry*. See figure 5.

5 the center of the entire Geometry: the Second Book's concluding paragraphs
74 paragraphs precede it (= First Book entire & Second Book's opening 40 paragraphs)
74 paragraphs follow it (= Third Book entire)
paragraph numbers in Second Book
Geom.'s central 24
16 certain ovals
5 getting them
1 why: their dioptric usefulness #41
4 how: fashion of drawing them #42-45
8 the ovals themselves
1 their genres #46
7 their properties #47-53
3 consequences: demonstration of their convergence properties #54-56
addendum 8
7 what has been omitted from the Dioptrics #57-63
1 extendability to three dimensions & what has not been omitted from the Geometry #64

In this 24-paragraph stretch where organization by 8-paragraph groupings is so important, there is another octet of interest: after an opening paragraph that speaks of the theory of dioptrics as well as of catoptrics, a total of 8 of the paragraphs speak of the *Dioptrics* (*la Dioptrique*).

The single mention of the theory of dioptrics takes place in paragraph 41. This opening paragraph of the stretch says: "certain ovals are very useful for the theory of catoptrics and dioptrics (*tres utiles pour la Theorie de la Catoptrique et de la Dioptrique*)." The 8 mentions of the *Dioptrics* ([also] *la Dioptrique*) are as follows.

The first sub-stretch tells the fashion in which Descartes draws the ovals; its opening and closing paragraphs (42 and 45) say: certain things are drawn according to the proportion which measures the refractions, "if one wishes to make use of them [the ovals] for the *Dioptrics* (*si on s'en veut servir pour la Dioptrique*)"; a certain movement of a point will draw this oval in imitation of what has been "said in the *Dioptrics* (*dit en la Dioptrique*)" about the ellipse and the hyperbola. After a paragraph on the various kinds of such ovals, the next sub-stretch treats their properties; its first two paragraphs (47 and 48) and its last paragraph (53) mention the *Dioptrics*: certain refractions can all be measured by a certain proportion, following what has been "said in the *Dioptrics* (*dit en la Dioptrique*)"; it is evident from what has been "demonstrated in *the Dioptrics* (*demonstré en la Dioptrique*)" that, things being posed in a certain way, certain angles are unequal—those of reflection as well as those of refraction; one can name certain points the burning points of the ovals, after the example of those of ellipses and hyperbolas, which have been "so named in the *Dioptrics* (*ainsi nommé en la Dioptrique*)." The next sub-section demonstrates the properties; its opening paragraph (54) says: it is very evident from what has been "said in the *Dioptrics* (*dit en la Dioptrique*)" that a certain ray entering a lens is curved in a certain way. The next sub-stretch is the final one on the usefulness of the ovals in the *Dioptrics*. Its opening paragraph (57) says that it is now necessary for Descartes to give satisfaction with respect to what he omitted in the *Dioptrics* (*omis en la Dioptrique*) when noting certain things; and, after treating two cases, he says in the sub-stretch's penultimate paragraph (62) that one could extend these two problems to an infinity of other cases, which he doesn't stop to deduce, because they did not have any use in the *Dioptrics* (*ils n'ont eu aucun usage en la Dioptrique*).

The reader should by now have seen some reasons to believe that Descartes is a far more careful writer than he is usually taken to be. It would seem to be more prudent to risk reading him over-attentively, rather than to risk treating his scientific essays as if they constituted a grab-bag of things that he happened to come across and then hastily threw together to get some attention as a rising young *savant*. The fact is, that Descartes was a man with a plan—a very designing man with a plan.

The volume that was the initial public offering in Descartes' enterprise consists of a discourse about the method for conducting one's reason well and seeking the truth in the sciences, followed by three essays of this method. We are examining the last of these essays, whose mention of the first essay we have just considered. Now let us consider the first essay's mention of the last.

The first essay (the *Dioptrics*) mentions the last essay (the *Geometry*) only once. This is in the 63rd paragraph of the *Dioptrics* (the 131st paragraph of the entire volume), which is the 22nd paragraph of the *Dioptrics*' 8th Discourse.

The paragraph occupies an important place in the structure of the *Dioptrics*. The *Dioptrics* consists of ten Discourses. The first six of them present a theory of vision; this opening of the essay occupies 29 paragraphs in all. Its closing too occupies 29 paragraphs in all; these treat practical aspects of the perfecting of vision by the contriving of lenses, the discussion of which occupies the second part of the 8th discourse together with the entirety of the *Dioptrics*' two remaining discourses. Between the opening block of 29 paragraphs and the closing block of 29 paragraphs, the theory of the practice of the perfecting of vision by the contriving of lenses is treated in a block of 34 paragraphs, which occupies the 7th discourse together with the first part of the 8th. The 8th Discourse, by far the largest one in the *Dioptrics*, contains 29 paragraphs in all, and the beginning of the second of the two parts into which this Discourse falls is its 22nd paragraph, the paragraph with the single mention of the *Geometry*.

(To observe the existence of structural features of that sort is not to deny that there are also structural features which simply coincide with the manifest division of the *Dioptrics* into Discourses. For example: the first two Discourses in their entirety form a chunk of 14 paragraphs, as does the last Discourse by itself; and the remaining 54 paragraphs of the *Dioptrics* that fall between them divide evenly at the break between the 7th and the 8th Discourses—the last paragraph of the 7th Discourse being the 41st of the *Dioptrics*, while the 7th and 8th Discourses in their entirety form a chunk of 41 paragraphs.)

The context of the *Dioptrics*' single mention of the *Geometry* is Descartes' assertion about the rays of light which (1) come from one point or (2) tend to one point or (3) are parallel—that they can all be made to change exactly from one to another of those three dispositions, as the glass through which the rays pass is taken to have various shapes. One can visualize (*imaginer*) endlessly many more such lenses that do what is done by the lenses that he has already drawn (*descrits*), says Descartes (who has concluded his depiction by saying, in the preceding one-sentence paragraph: "And all this, it seems to me, is so clear that to understand it there is need only to open the eyes and consider the figures.")

What he then says about the *Geometry* is this: "But I do not think [myself] to have any need to speak about it [the matter] here, because I could more conveniently explain it hereafter in the *Geometry*." (Also, he adds, "because those that I have drawn [*descrit*] are the most appropriate to my design; so that I wish to try now to prove, and make you see by the same means, which among them are the most appropriate, by making you consider all the principal things in which they differ.")

The place in the *Geometry* that is more convenient for explaining that dioptrical material is, of course, these 24 paragraphs which we have been considering at the end of the *Geometry*'s Second Book.

The preceding 16 paragraphs of the Second Book of the *Geometry* have treated the problem of drawing the tangent to a curve at any point, which is the "problem that is the most useful ... in Geometry." Inverse to that, is the dioptric problem: the problem of drawing a curve whose every tangent will have a certain relation to a certain pair of rays that emanate from the point of tangency. The 16 central paragraphs of the *Geometry*'s Second book, which treat the tangent problem, thus prepare for the final 24 paragraphs of the *Geometry*'s Second Book, which treat the dioptrical problem. These final 24 paragraphs of the *Geometry*'s Second Book provide a general treatment of the dioptrical problem. These paragraphs present certain ovals which give shape to lenses; in certain cases, these ovals take on hyperbolic or elliptical shapes, which are the easier shapes to handle that are treated in the *Dioptrics*. Descartes' ovals thus give a general solution to a problem that is most useful for the whole conduct of life. The solution of the dioptrical problem, Descartes says in the *Dioptrics*, will make it possible to perfect one of the most useful inventions that could be—one which seems to have opened the road for us to come to a knowledge of nature much greater and more perfect than our forefathers had. Appropriately enough, these final 24 paragraphs of the *Geometry*'s central book are the central paragraphs of the whole *Geometry*.

CHAPTER 7. COMPLETING THE *GEOMETRY*

The title of the Third Book corresponds to that of the First:

> THE
> GEOMETRY.
>
> THIRD BOOK.
>
> *Of the construction of Problems that are*
> *Solid or more than Solid (*Solides ou plusques Solides*).*

According to its title, the Third Book is about the construction of problems that are more than plane, but there is a bit of a wait to reach the treatment proper of such problems—for, although the book is 74 paragraphs long, the construction of problems that are more than plane is not treated until its final 30 paragraphs. In the book's initial 4 paragraphs, we learn why there must be a discussion preliminary to the treatment proper of those problems which are more than plane; so, besides the 34 paragraphs that together constitute the beginning and the end of the Third Book, there are 40 paragraphs of preparation for the end. Those other 34 paragraphs—the ones by which the 40 paragraphs of preparation are surrounded—have the following arrangement: a stretch of 17 paragraphs (preceded by a preparation of 40 paragraphs) surrounded by another 17 (= 4 + 13) paragraphs. See figure 1.

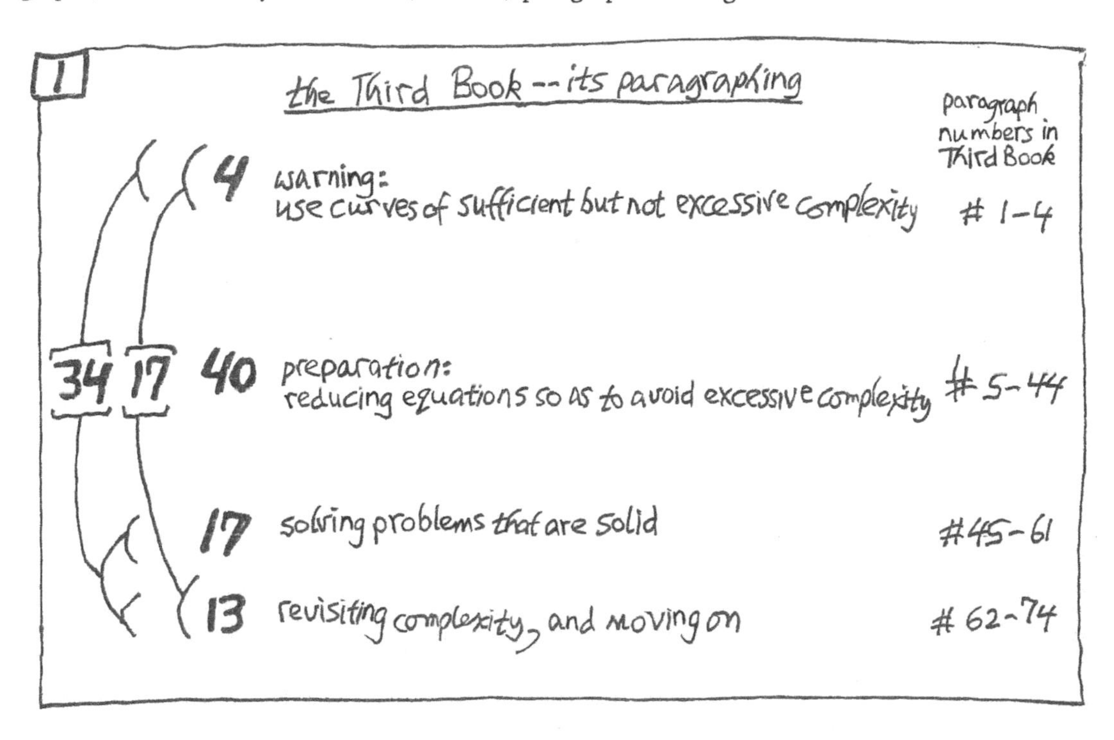

We have seen 17-paragraph building blocks used by Descartes before. Now we see them used by him again.

The First Book consisted of two such blocks. The second of them was also the first of two such blocks which treated the locus problem of Pappus. The last paragraph of the First Book, its 34th—which was also the last paragraph of the first part of the treatment of Pappus's problem—said that before finishing up the problem, something in general needed to be said about the nature of curves. The Second Book opened with some paragraphs that said it, and then turned to presenting the second part of the treatment of Pappus's problem, in the form of another block of 17 paragraphs, which was followed by the 40 paragraphs of the rest of the Second Book.

Now the Third Book—before reaching the solution of problems that are solid, which also occupies a 17-paragraph block—opens with a few paragraphs of warning to use curves of sufficient but not excessive complexity, which is followed by a 40-paragraph discussion of how to do so by reducing the number of dimensions in certain equations. The Third Book closes—following the solution of problems that are solid—by revisiting complexity and moving on beyond problems that are solid. These closing paragraphs, taken together with the Third Book's opening paragraphs (the ones that warn against solving a problem by using a curve of more than merely sufficient complexity), amount to 17 in number. (The *Geometry*, by the way, is not the only one of Descartes' writings in which 17-paragraph chunks are employed as building blocks. He makes some use of them in his *Dioptrics*, as was pointed out above, and, in addition, he makes some use of them elsewhere than in that volume which contains the *Geometry*—for example, at the very end of his career, in his *Passions of the Soul.* Descartes is not the only author who has made formal use of the number 17. There was an old tradition that went back, through medieval writings in Arabic, to still older writings in ancient Greek, in which the number 17 was associated with nature. And, by the way, there is also an old tradition, going back at least to the Hebrew scriptures, in which the number 40 was associated with preparation.)

The 40 paragraphs of intervening preparation for solving problems that are solid (and more than solid) begin with a paragraph saying that something in general needs to be said about the nature of equations. Descartes' earlier statement of the need to say something in general about the nature of *curves* took place right before the opening of the Second Book; the statement of the need now to say something in general about the nature of *equations* takes place right after the opening of the Third Book. The first statement was made in order to complete, in the Second Book, the solution of the problem that arose in the First Book; the second statement is made in order to present, in the Third Book, the solution of problems that involve greater complexity than the earlier problem did. The first statement was made in the 34th paragraph of the First Book; the second, in paragraph III.5, which is 68 paragraphs later—that is to say, twice-34 paragraphs after it. The paragraph which is just-34 paragraphs after that 34th paragraph of the First Book—namely, the 34th paragraph of the Second Book—was the first of the paragraphs in the Second Book that solved an equation to yield a construction for the problem that Descartes called the most useful and the most general of geometrical problems that he knows or ever desired to know.

Now let us look more closely at how Descartes puts together the contents of this Third Book of his *Geometry*.

initial statement (*paragraphs 1-4*)

Although the Second Book has expanded geometry by receiving into it curves in an endless order of increasingly composite kinds (*genres*), this Third Book's initial 4 paragraphs issue a warning about the use of composite curves to solve problems in geometry: one should be careful to choose a curve of the appropriate kind—a curve of just the right degree of compositeness—a curve that is composite enough, but not more composite than would serve to determine the quantity that is sought. In constructing the solution to a problem (says the first paragraph) it is a fault to make use of a curve that is not as simple as possible; on the other hand (says the last of these paragraphs) it is also a fault to toil uselessly under the wish to construct the solution to some problem by means of a line of a kind that is more simple than is permitted by the nature of the problem. Exemplification of the first fault is provided in the 2nd paragraph, and in the 3rd, and at the beginning of the 4th. By employing curves drawn by means of the instrument explained earlier, the central paragraphs of the initial statement show how to obtain, between a pair of given lines, 2 mean proportionals (this is treated in the 2nd paragraph) and 4 or 6 mean proportionals (this is treated in the 3rd paragraph). But (says the 4th paragraph) those 2 or 4 or 6 mean proportionals can be obtained by employing curves of kinds that are less composite.

Here is Descartes' figure for the 2nd paragraph of the Third Book. It is, in fact, the same one that he presented earlier for the 2nd paragraph of the Second Book. Now, however, we shall see why there are semicircles drawn in it.

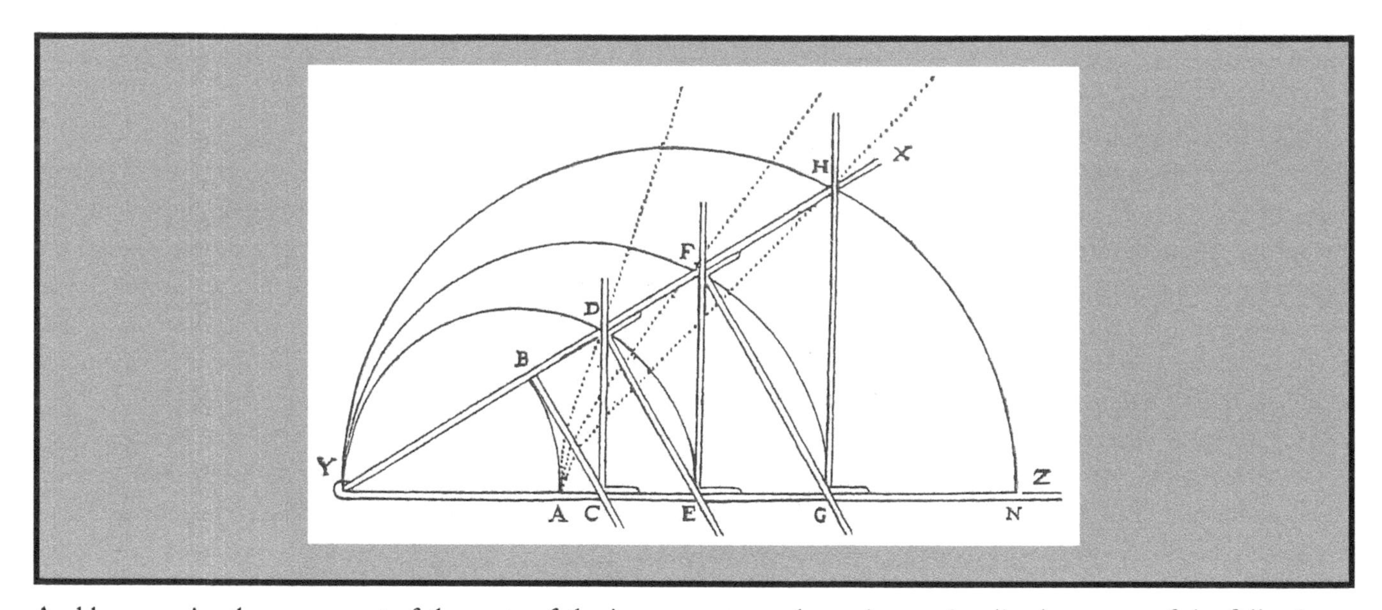

And here, again, the movement of the parts of the instrument may be easier to visualize by means of the following figure (fig. 2):

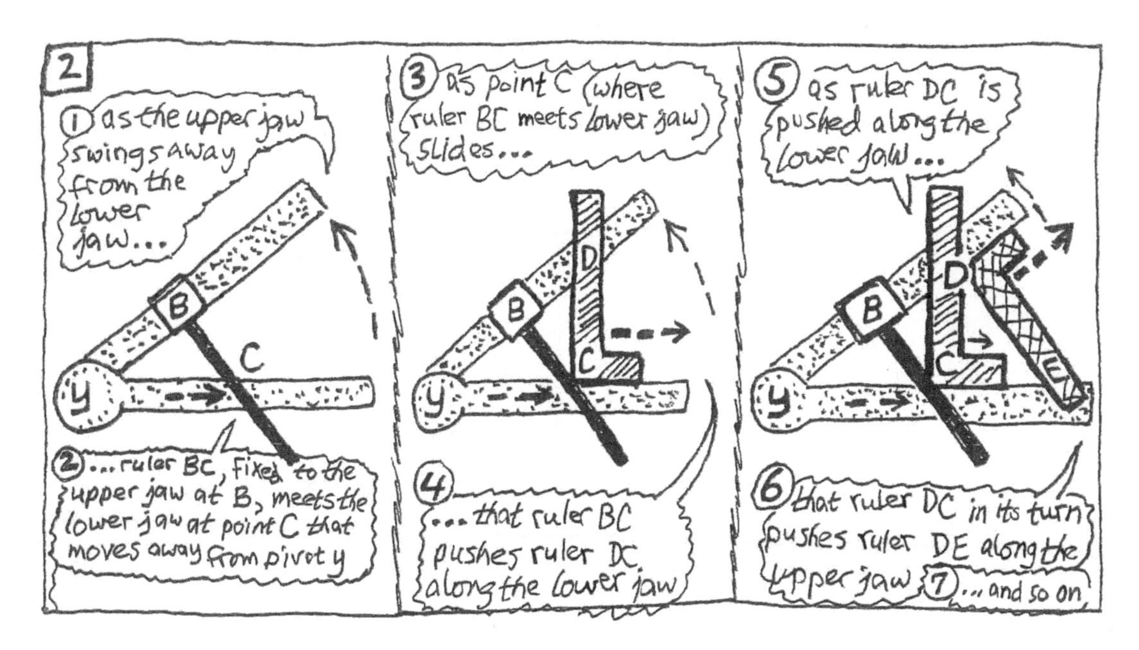

The curve traced out by point D, or by point F, or by point H, and so on, will yield 2, or 4, or 6—or as many mean proportionals as you please between the pair of given lines, in the following fashion.

Suppose you are required to obtain 2 mean proportionals between a pair of given lines. Adjust the instrument so that YA is equal to the first of the given lines, and YE equal to the second. Now draw a circle whose diameter is this second line YE. The circle will cut the dotted curve AD at some point D, thus determining (on the upper jaw of the instrument) the straight line YD. This straight line YD will be one of the 2 mean proportionals that are required. (The other one will be YC.) Why is that? Because

YA (which is equal to YB) is to YC, as YC is to YD, as YD is to YE.

And here is the reason for that: the instrument is so built that each of the sliding rulers makes right angles with one of the jaws (that is, with YX or YZ), and hence the triangles YBC, YCD, YDE are similar (since each of them has a right angle, and all of them share the angle made by the joined jaws); and hence the corresponding sides are proportional.

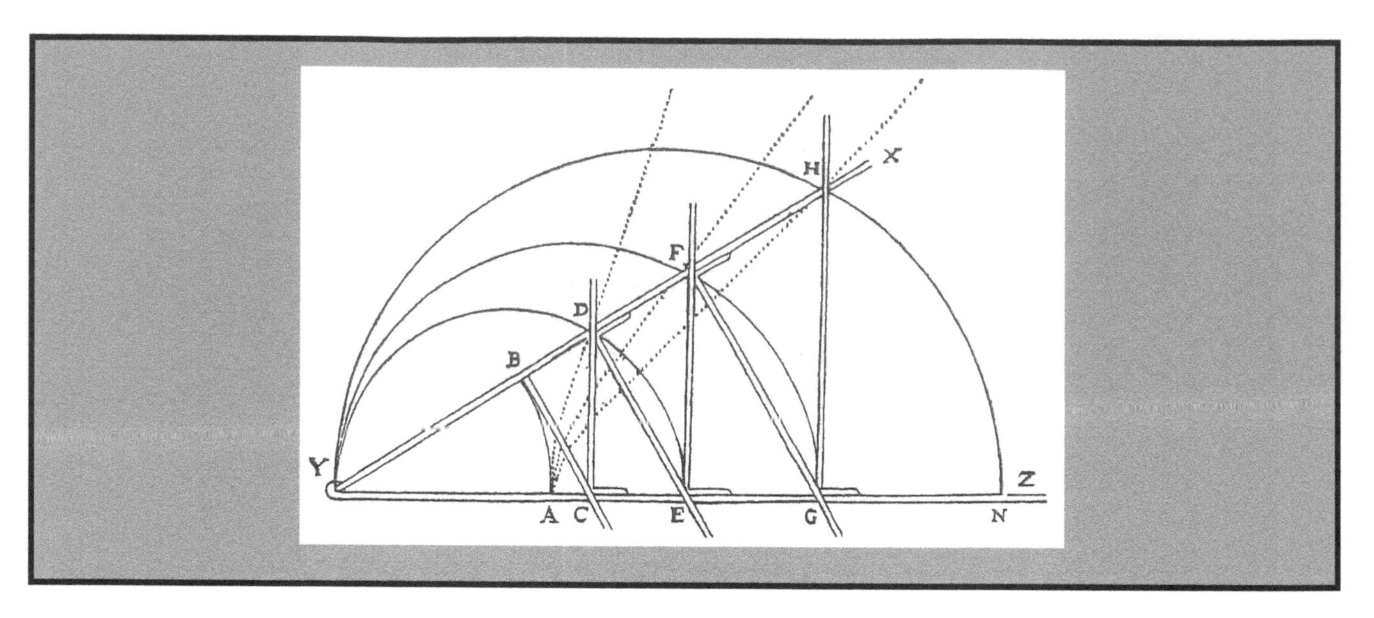

Likewise, suppose that between a pair of given lines you are required to obtain not 2 but 4 mean proportionals. Again adjust the instrument so that YA is equal to the first of the given lines; but now it is YG that is to be equal to the second. Again draw a circle whose diameter is this second line, which is now YG. The circle will cut the (next) dotted curve AF at some point F, thus now determining (on the upper jaw of the instrument) the straight line YF. This straight line YF will be one of the 4 mean proportionals that are required.

And likewise for 6 mean proportionals: the line on the lower jaw that is to be equal to the second given line will now be YN, which will also be the circle's diameter; the dotted curve that it will intersect will now be AH, and now the straight line YH will be one of the 6 mean proportionals that are required.

In that fashion, the instrument that Descartes has explained is employed for finding as many mean proportionals as you please. Descartes goes on to say that he does not believe that there would be any easier fashion of finding them, nor any whose demonstration would be more evident. Still, he says, when one says that one should choose the simplest curve that is possible for constructing the solution of a problem, what one should mean by "simplest" is not the one that can be drawn most easily, or the one that makes the construction or the demonstration easiest, but principally this: a curve of the simplest kind (*genre*) that would serve to determine the quantity that is sought. For example, he says, 2 mean proportionals can be found by means of conic sections, which are of the first genre of curves—rather than by means of the dotted curve AD, which is of the second genre. Or 4 mean proportionals can be found by means of curves of a genre that is not so composite (*composeé*) as the one to which the dotted curve AF belongs; or 6 mean proportionals can be found by means of curves of a genre that is not so composite as the one to which the dotted curve AH belongs. (Back in the first volume of this guidebook, at the end of Chapter 2 of Part V, we considered an instance of the employment of conic sections to find 2 mean proportionals between a pair of given lines. The job was done by means of parabolas, in order to solve the problem of getting a cube that would be double the size of a given cube.)

The drawing of the semicircles indicates how far we have come. Our way to a single mean proportional is via the simplest of the curves, the circle. A half-chord that is perpendicular to a diameter is the single mean proportional between the segments into which that half-chord splits the diameter. But now we are after a multiplicity of mean proportionals. For these, the circle-drawing compass is not helpful; what is helpful is the proportional compass. The proportional compass is the technological embodiment of Descartes' geometrical methodology. It guarantees the exhaustive productivity of his methodical process.

Descartes' example of finding a number of mean proportionals might seem to be of less general significance than it truly is. Why is it, that his example is more significant than might appear at first glance? Because the finding of mean proportionals and the solving of certain equations go hand in hand. How so? Well, consider what is meant by speaking of the n-th root of some number a. That root is the number x such that if n of these x's are multiplied together, then their product is a; or, to put it into an equation, $x^n = a$. Now the proportion that corresponds to that equation will need a unit in it, so let us first rewrite the equation. And because $x^n = x*x^{(n-1)}$, we shall, calling the unit u, obtain the equation $x*x^{(n-1)} = a*u$. This means that $u/x = x^{(n-1)}/a$, or, in other words, that the following proportion holds:

u is to... x, as .... x^{n-1} is to a.

But then all of the ratios in the following box are the same:

u is to.....x,

as ....x is to x^2,

as.................... x^2 is to x^3,

as............................ x^3 is to x^4,

etc. ...

as.. x^{n-3} is to x^{n-2},

as.. x^{n-2} is to x^{n-1},

as .... x^{n-1} is to a.

In other words:

x is the first of many mean proportionals —indeed, to be precise, it is the first of $n - 1$ of them— between the unit and a.

Now, between the unit and any given *number* there will *not* be *any* given number of mean proportionals. There are, for example, only two mean proportionals between the unit and 8; and there is no mean proportional at all between the unit and 2. But there *will* be *any* given number of mean proportionals between any line taken as a unit and any given *line*. Descartes' instrument (his so-called proportional compass) enables you, having taken some line to be the unit, to obtain the n-th root of any given length a. That is to say, the instrument enables you to construct the solution to a problem that gives rise to the equation $x^{(n-1)} = a$.

It is true, however, that the equation obtained in the course of solving a problem is likely to involve more than the equality of (on the one hand) an unknown of some number of dimensions and (on the other hand) a given quantity. In other words, the equation will probably not have a form so simple as $x^{(n-1)} = a$. Nonetheless, the example suggests that there is a correlation between (on the one hand) the number of dimensions to which the equation mounts and (on the other hand) the degree of compositeness of the curve which will serve to solve it. That is to say, if some general knowledge about the nature of equations could be put together with what we have already learned about the nature of curves, then we would be better able to determine what level of compositeness in a curve would be high enough, while being no higher than is necessary, to find the root of any particular equation—and thereby to construct the solution of that problem from which the equation arose.

So it makes sense that the very next paragraph (III.5), the first of the 40 paragraphs that prepare for the treatment of the construction for solid problems, should go on to make the following announcement: to the end of giving some rules for avoiding each of the two faults that have been mentioned, something in general needs to be said about the nature of equations.

That need is satisfied in the 18 paragraphs which follow the announcement; these paragraphs show how to see whether an equation can be reduced to another one with a root whose value can be found by a curve that is less composite than the curves used to solve other equations whose unknown has the same number of dimensions as the unknown has in the original equation.

Those 18 paragraphs on rules for reducing equations are followed by 18 paragraphs which put the rules to work in the sort of affair one might expect to be treated first. And what would that be? Well, Descartes is going to treat the construction of solutions to solid and supersolid problems. Problems of the first of these kinds, the solid problems, are those whose construction requires making use of a conic section. One would expect a problem to be solid when, in the course of finding its construction, we come to an equation in which the unknown has 3 dimensions. Yet, in seeking the construction that will solve some problem, it might be possible that although we would come to such an equation, this equation could be reduced, so that the construction could be accomplished without making use of any conic section, but merely making use of ruler and compass. That is to say: when we come to an equation in which the unknown has 3 dimensions, we should find out whether the problem is plane—before we press into service for our construction-work a curve more composite than a circle, such as a parabola.

After the 18 paragraphs in which that is shown, a 3-paragraph addendum—on what does not need to be shown in this connection—completes the 40-paragraph preparation for the treatment of problems whose construction requires curves of more than plane compositeness.

In the treatment proper of these problems of a higher order—the treatment, that is to say, which occupies the final 30 paragraphs of the final Third Book of the *Geometry*—the first 18 paragraphs present rules for constructing solutions to problems that are solid, and the remaining 12 paragraphs do it for problems that are more than solid.

The structure of the Third Book is laid out in the box following.

LAY-OUT OF THE THIRD BOOK

(warning to use curves of sufficient but not excessive complexity to
solve problems—
INITIAL STATEMENT (4 paragraphs: #1–4):
GENRES OF CURVES AND NATURE OF PROBLEMS

(1 par.) Genres of curves (degrees of compositeness) and
nature of problems (what is required for constructing solutions)

(1 par.) Example: using the curves drawn by the instrument described above, for
constructing mean proportionals—
2 of them
(1 par.) 4 or 6 of them

(1 par.) In addition to that one fault (when those mean proportionals can be obtained
by means of simpler curves), a fault on the other side is
uselessly toiling with curves of a kind that is simpler than is necessary
===

(reducing equations to avoid excessive compositeness—)
PREPARATION (40 paragraphs: #5–44):
NATURE OF EQUATIONS AND THEIR REDUCTION

/— (1par.) announcement that, to avoid the faults, need to say something in general
| about the nature of equations
|
\— (18 par.) nature of equations and rules for reducing them (general)

/— (18 par.) reducing equations that have unknowns in 3 dimensions or 4
| to find out whether the problem is not solid but merely plane
| (i.e., construction merely with ruler and compass)
|
\—(3 par.) addendum on what need not be shown
===
FINAL TREATMENT (30 paragraphs: #45–74):
RULES FOR CONSTRUCTING SOLUTIONS OF PROBLEMS BEYOND THE PLANAR

/— (18 par.) solid
| (17 par.) problems that are solid
| (1 par.) transition: revisiting compositeness—what is needed for
| the solution to be possible, especially for solids
|
\— (12 par.) more than solid
(1 par.) continuation of possibility, for supersolids
(10 par.) problems that are supersolid
(1 par.) and so on

Now let us look more closely at the contents of the parts of the Third Book.

preparation (*paragraphs 5–44*)

The 40-paragraph preparation that follows the 4-paragraph initial statement has as its core two stretches of 18 paragraphs each:

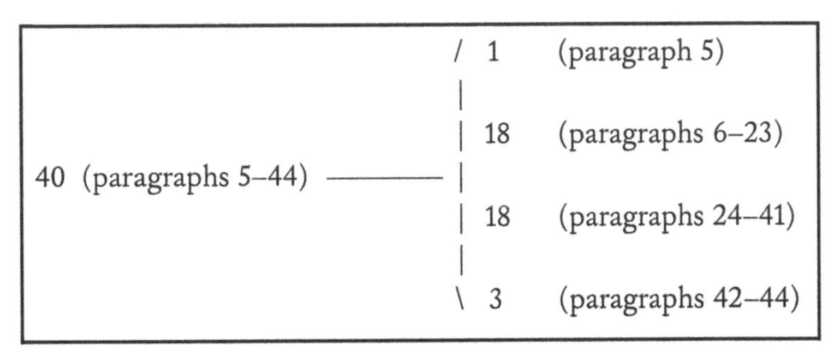

The preparation opens with a one-paragraph announcement of a need that its 18-paragraph sequel satisfies: by saying something in general about the nature of equations, it is possible to give some rules for avoiding the faults, mentioned just previously, of trying to construct the solution of a problem by means of a curve that is too composite or is not composite enough.

After its first 18-paragraph stretch, which is general, the preparation goes on to its second one, which is more particular: it gives rules for telling from the equation whether a problem whose equation might make it seem to be more than planar will turn out to be merely planar after all; this is followed by a 3-paragraph addendum to what is said in giving those rules.

What is said in general about the nature of equations makes it possible to reduce some equations to other equations whose unknowns have fewer dimensions. When one comes to an equation in the course of seeking the construction of some problem, it is useful to find out whether, instead of finding the root of this equation by employing a curve of as much compositeness as is characteristic of equations whose unknowns have that many dimensions, one might rather employ a less composite curve to find the root of an equation to which the original equation might be reduced. The preparation's opening part thus shows how to reduce an equation to another equation whose unknown has fewer dimensions.

More particularly, the contents of those paragraphs (5–23) which open the preparation are as follows.

To the end of giving some rules to avoid each of the two faults—the first of the paragraphs (5) announces—something in general needs to be said about the nature of equation; the paragraph then says what an equation is.

An equation can have as many roots as the unknown has dimensions, says the next paragraph (6); roots can be false—that is, less than nothing—says the next paragraph (7), referring to what we would call "negative" roots. The next paragraph (8) tells how one can diminish the number of dimensions of an equation when one knows one of its roots; and the next (9) tells how one can examine whether some given quantity is the value of a root.

The next (10) tells how many of an equation's roots can be true; the next (11) tells how one can change certain signs in an equation so that false roots become true and true roots become false.

The next (12) tells how one can change an equation by substituting for the unknown quantity an expression that adds to or subtracts from the unknown (for example, substitutes $x + 5$ for x) so that one can augment or diminish an equation's roots without having to know them. After that comes one example of augmenting (13) and one of diminishing (14); the next (15) adds that in augmenting the true roots, one diminishes the false, and vice versa.

Now (*Or*), says the next paragraph (16), one can—in this changing of the value of roots without having to know them—remove the second term of an equation, as the last example showed, and as the next paragraph (17) shows by another example; in such changing, one can also make all the false roots of an equation become true without the true becoming false, as the next two paragraphs (18–19) show.

The next paragraph (20) tells how one can make an equation's places all be filled—that is, how one can get an equation with a certain number of dimensions that has a term for every smaller number of dimensions. For example, if we want to go from an equation with its term of highest dimensions being x^5 to an equation with its term of highest dimensions being y^6, how to get an equation that will contain terms with y^5, and y^4, and y^3, and y^2, and y.

The next (21) shows how one can multiply or divide roots by any known quantity without having to know the roots, which can serve to reduce broken numbers (that is, fractions) to whole numbers, or surd numbers (that is, irrational numbers) to rationals; this multiplying or dividing can also make the known quantity of some term of the equation be equal to some other given quantity, as the next paragraph (22) shows.

In the next paragraph (23) what remains to be said is this (*Au reste*): roots, both true and false, can be real or imaginary. Roots that are imaginary, Descartes does not pursue, however, for his interest lies not in relations simply, but rather in geometry with a view to physics and its applications. While his *Geometry* may take the decisive step toward conceiving mathematics as a formal study, his own work is not purely formal.

Now, in the second 18-paragraph stretch, Descartes gets more particular: he presents rules on what to do when the equation that one comes to, in the course of seeking the construction for a problem, has an unknown quantity whose dimensions number 3 or 4. He shows how to deal with the fact that while the problem might well be solid, it might merely be plane: in other words, the construction for the problem, while it might require employment of a conic section, might require a curve no more composite than a circle.

Of the 18 paragraphs of rules for trying to effect a reduction, half a dozen (24–29) treat the case when the equation's unknown has 3 dimensions, and a dozen (30–41) treat the case when it has 4 dimensions.

More particularly, the contents of those half dozen paragraphs are as follows. Now (*Or*) when, for finding the construction of the solution to some problem, one comes to an equation in which the unknown quantity has 3 dimensions, the first of the paragraphs (24) says how to reduce the equation in order to find out if the problem is a plane (that is, if the construction can be done merely by means of ruler and compass). Four paragraphs of exemplification follow (25–28). But (*Mais*), says the next paragraph (29), if the equation is such that the problem must be solid, it is as much a fault to try to construct the solution by employing only circles and straight lines as it would be to employ conic sections to construct one that needs naught but circles.

The contents of the next dozen paragraphs, saying what to do if the unknown quantity has 4 dimensions, are as follows. The first three of the paragraphs (30–32) tell how to reduce such an equation. Two paragraphs of exemplification follow (33–34). The next two paragraphs (35–36) tell what to do afterward if the equation is reduced to 3 dimensions by the method already used. Two paragraphs of exemplification follow (37–38). But (*Mais*), says the next paragraph (39), to the end of better knowing the utility of this rule, let us apply it to some problem; the problem is treated in the next paragraph (40), and also in the next one (41), where Descartes puts readers on notice (*avertir*) that if one first comes by a road (*chemin*) that leads to a very composite equation, one can ordinarily come by another road to a simpler equation.

In the 3-paragraph addendum to what he has said in giving those rules, Descartes goes on to say the following things: why he won't add what he could; what he prefers to say instead; and what he omits and why he omits it.

The addendum's initial paragraph (42) says that he could add different rules for the situation for which he has presented the rules—namely, when the equation goes to the cube (3 dimensions), or to the square of a square (4 dimensions), and the problem turns out to be plane rather than solid; but different rules would be superfluous—for when the problems are plane, one can always find the construction by the rules presented. The addendum's central paragraph (43) says that he could also add rules for the situation that would come next in order—when the equation goes as far as to the supersolid or square of a cube (5 dimensions) or beyond; but he prefers to comprehend them all in one rule, and to say in general what to do: "when one has tried to reduce them [namely, the equations] to the same form as those of as many dimensions which come from the multiplication of two others which have less of them [namely, dimensions], and, having enumerated (*dénombré*) all the means by which this multiplication is possible, the thing was not able to succeed by any, one should assure himself that they cannot (*elles ne sçauroient*) be reduced to simpler [ones]. So that (*En sorte que*) if the unknown quantity has 3 or 4 dimensions, the problem for which one seeks is solid; and if it has 5 of them, or 6, it is more composite (*composé*) by one degree; and so [on] of the others." The addendum's final paragraph (44) says what he omits and why he omits it. Note that 44 paragraphs have intervened since he gave (in the 63rd paragraph of the Second Book) a reason for omission different from the reason given here in the 44th paragraph of the Third Book. In the earlier paragraph, pointing out that he had not gone on to do something that would have been "no more difficult than what I have just explained, or, rather, something much more easy, because the way (*chemin*) is open," he said: "I have preferred that others seek it, to the end that if they have yet a little trouble (*peine*) finding it, that may make them esteem so much more the finding out (*invention*) of the things which are here demonstrated." Now, in the Third Book, he says: "What is left to say is this (*Au reste*): here I have omitted the demonstrations of most of what I have said, because they seemed to me so easy (*faciles*) that, provided that you take the trouble (*peine*) to examine methodically if I have been faulty (*si jay failly*), they will present themselves to you of themselves; and it will be more useful to learn them in this fashion than in reading them." Descartes speaks here of examining "methodically"; he spoke of "method" 9 paragraphs earlier (35) and he will speak of it also 18 paragraphs later (62).

The gist of the addendum that ends the preparation for the subject proper of the Third Book is laid out in the box following.

if equation has an unknown, the number of whose dimensions is:	and the equation cannot be reduced so as to make the problem a:	then the problem will be a:
3 (cube), or 4 (square of square)	plane	solid
5 (square of cube)	solid	supersolid
and so on—for every increase of a dimension or two in the equation's unknown quantity		the problem will go up one step

final treatment (*paragraphs 45-74*)

The 30 final paragraphs of the Third Book treat this final book's subject proper—the construction of solutions to problems that are solid or more than solid. The treatment of this final subject opens with 18 paragraphs on problems that are solid; it closes with 12 paragraphs on problems that are more composite by one degree than solid problems are, and (in the final paragraph) on what he has done in the book and what he has omitted from it.

The 18-paragraph treatment of problems that are solid consists of a 17-paragraph treatment proper, followed by a methodological paragraph that revisits the question of compositeness.

The 17-paragraph treatment proper opens with five paragraphs (45–49) on rules for finding—once one has made sure that the problem is solid (and not merely plane)—the roots of the equation by means of a conic section, indeed by means of a parabola, which is simplest. How to proceed is shown in three of these paragraphs (45–47), and then two paragraphs demonstrate that the job will thus indeed be done (48–49). What is presented in those five paragraphs is then exemplified in the twelve paragraphs which follow. These twelve paragraphs of exemplification open with three paragraphs on certain particular problems—the finding of 2 mean proportionals between a given pair of lines (50), and the division of a given angle into 3 equal parts (51–52); then come six paragraphs showing, by an examination of all the alternative possible equations, that all other solid problems reduce to those two problems (53–58); then come three concluding paragraphs on the significance of this (59–61).

The concluding paragraph of those last 17 paragraphs makes the following claim: "And one can also, in consequence of (*en suite de*) this here, express the roots of all equations which mount up as far as to the square of a square, by the rules explained here above. So that (*En suite que*) I do not know anything more to desire in this matter. For in short [or: finally] (*enfin*) the nature of these roots does not permit one to express them in terms more simple, nor to determine them by any construction which is both (*ensemble*) more general and more easy (*facile*)." Descartes' reflection on that claim occasions a final methodological paragraph before he moves on from problems that are solid. In this methodological paragraph (62), he makes an admission: "It is true that I have not yet said upon what reasons I make for myself a foundation for daring to be so sure if a thing is possible or it is not"; "But," he continues, "if one is on the alert (*prent garde*) how, by the method of which I make use (*par la methode dont je me sers*), all that falls under the consideration of the Geometers reduces itself to one and the same kind (*genre*) of Problems, which is to seek the value of roots of some Equation"—then, if one is thus on the alert, "one will judge well that it is not hard (*malaysé*) to make an enumeration (*dénombrement*) of the ways (*voyes*) by which one can find them, which would be sufficient for demonstrating that one has chosen [the way that is] the most general and the most simple." The remainder of the paragraph speaks "particularly" of solid problems, which, he reminds readers he has said, cannot be constructed without employing some line more composite than one that is circular—a thing that is easy to find from the fact that all of those problems reduce to two constructions. He then shows why some conic section not a circle is what is needed for finding 2 points which will provide 2 mean proportionals between a given pair of lines, or 2 points which will divide a given arc into 3 equal parts.

After that conclusion to the treatment of problems that are solid, readers are ready to move right into the treatment of problems of the very next kind, the kind that is more composite by just one degree. The very next paragraph (63) begins this treatment by saying the following: "But for this same reason it is impossible that any of the Problems which are more composite by one degree (*d'un degré plus composé*) than the solid [ones], and which presuppose the finding (*invention*) of four mean proportionals or the division of an angle into five equal parts, could be constructed by any of the conic sections. This is why I shall believe [myself] to have done everything in this as best could be done, if I give a general rule for constructing them, employing therein the curved line which is described by the intersection of a Parabola and a straight line, in the fashion explained here, above; for I dare to give assurance (*assurer*) that there is not any of them more simple in [its] nature (*la nature*) which could serve to this same effect; and you have seen how it [that is, the curve so described] immediately follows the conic sections in that problem (*question*), so much sought by the ancients, the solution of which teaches, in order [*par ordre*], all the curved lines which should be received in Geometry."

That is the opening paragraph in the 12-paragraph treatment of problems that are more than solid (63–74). Except for the last of these 12, the treatment deals with problems that are only one degree more complex than solid problems.

First come ten paragraphs on the construction of such problems. These open with two paragraphs (63–64) on the need to employ a parabola to get the curve you need, and on rules for obtaining the equation that you need for obtaining the curve that you need, and for obtaining the roots of the equation by cutting that curve with a circle. Next come three paragraphs which show how to obtain the curve in another way—namely, without employing the parabola (65)—and which demonstrate that the same curve is thus obtained differently (66–67). Then come another three paragraphs, these demonstrating that what were claimed to be the roots are indeed the roots sought (68–70). Finally, the ten paragraphs conclude with two paragraphs of exemplification: finding 4 mean proportionals between a given pair of lines (71), and dividing a given angle into 5 equal parts (72)—or (also paragraph 72) other examples such as inscribing in a circle a polygon of 11 or 13 sides.

An 11th paragraph (73) is, as it were, a note appended to the treatment of the construction of solutions to problems that are strictly supersolid (that is, are one degree more complex than solid problems): "the fact is to be noted (*il est remarquer*) that in several of these examples it can happen that the parabola—which is the curve of the second kind (*genre*) that is used—is cut by the circle so obliquely that the point of their intersection is difficult to recognize, and hence the construction may not be convenient for practice (*commode pour la pratique*)." The remainder of this paragraph says that nonetheless the inconvenience would be "easy (*aisé*) to remedy by composing other rules, in imitation of this rule here, as one can compose them in ways of a thousand sorts (*de mille sortes*)."

By thus concluding the discussion of supersolid problems with a paragraph saying that a certain inconvenience for practice could be remedied by composing other rules in imitation of those presented here, as one could do "in ways of a thousand sorts," the author merely suggests that he has left many things for his readers to do that he himself could do for them but won't. In the next, the very last paragraph of the *Geometry*, he goes on to speak more explicitly about his designing policy of omission. Before we go on to look at it, we should recall what he has already said about his policy of omission.

Descartes has told us from time to time that he wouldn't do something or other. His longest such statement was the 3-paragraph addendum to the paragraphs in which he told how to determine whether a problem is merely plane, when the equation that one comes to in the course of seeking its solution might seem to indicate that it is solid. After the first two of those paragraphs spoke of rules that he could have added but wouldn't, the last one—the 44th paragraph of the Third Book of the *Geometry*—said this: "here I have omitted the demonstrations of the greater part of what I have said, because they seemed to me so easy that, provided that you take the trouble (*peine*) to examine methodically if I have been faulty, they will present themselves to you of themselves; and it will be more useful to learn them in this fashion than in reading them." Thus in the last of the 40 paragraphs that prepared for the treatment of the proper subject of the final book of the *Geometry*, he made his only statement of omission that mentioned examining things "methodically," while speaking of a fashion of learning that is more "useful" than mere reading. His great theme throughout the entire volume that includes the *Geometry* is the usefulness, for true learning, of being methodical—true learning being what is most useful for other things, whose promotion is protected by the effect of promoting Cartesian "method."

Now, in the very last paragraph of the *Geometry*, Descartes will speak his final words about what he has omitted to do—and why. This was considered earlier in this guidebook, when the 12th paragraph of the *Geometry*'s Second Book was under examination and we considered passages, located throughout the *Geometry*, where Descartes speaks of his omissions. What he says here in the last paragraph, is the following:

[III.74] But (*Mais*) my design is not to make a big book,
and I try rather to comprehend much in few words—

as one will perhaps judge that I have done, if one considers that,
having reduced to one and the same construction
all the Problems of one and the same kind (*genre*),
I have given, all together, the fashion of
reducing them to an infinity of different others,
and so of solving each of them in an infinity of fashions.

Then, beyond that—having constructed

all those which are plane,
in cutting by a circle
a straight line,

and all those which are solid,
in cutting also by a circle
a Parabola,

and finally all those which are more composite (*composés*) by one degree,
in cutting—entirely the same—by a circle
a line which is more composite than the Parabola by one degree—

there is naught needed (*il ne faut*) but to follow the same way (*voye*)
for constructing all those which are more composite (*composés*) to the infinite (*a l'infini*).

For in the matter of Mathematical progressions,
when one has the first two or three terms, it is not difficult (*malaysé*) to find the others.

And I hope that our posterity (*neveux*) will be grateful by recognizing me (*me sçauront gré*)
[literally: will recognize me willingly],
not only for the things which I have here explained,
but also for those which I have omitted with a will (*voluntairment*)
[that is, on purpose—or willfully],
to the end of leaving them the pleasure of finding them (*les inventer*).

Descartes tells us that he has provided for his successors to have the pleasure of coming upon, by themselves, things that he has not already explained to them in so many words. Because of what he has already explained to them, however, as he now explains, any exploration in which they might engage hereafter will take place on territory that has already been surveyed by him. How so?

In considering that question, we would do well to keep in mind some words from the letter written by Descartes to Mersenne at the end of the year in which he published the volume that contained the *Discourse* and its three companion essays. (Here we are interested only in the end and in the beginning of what was quoted from the letter we considered earlier in this guidebook, in the first chapter of this eighth part, right after looking at the notice which Descartes placed right before the *Geometry* [Adam & Tannery: I, 480, 478].) The letter speaks about what Descartes has accomplished in the *Geometry*: "Having determined as I did in each kind of problem (*genre de questions*) all that can be done there, and shown (*monstré*) the means of doing it," he says in the letter, "I claim not only that one should believe that I have done something more than those who have preceded me, but also that one should be persuaded that our posterity (*neveux*) will never find anything in this matter which I could not have found as well as they if I had wanted to take the trouble (*peine*) to seek it." But before making that claim about what he has accomplished mathematically in his *Geometry*, the letter makes a claim about what he has accomplished by means of what he has accomplished mathematically in his *Geometry*: "through the *Dioptrics* and through the *Meteors*, I have only tried to persuade that my method is better than the ordinary, but I claim to have demonstrated (*demonstré*) it through my *Geometry*."

Now, those two Cartesian claims—the one that is merely mathematical and the one that is methodological—how are they related? And what has that to do with the final paragraph of the *Geometry*, at which we have just looked?

Consider the "proportional compass." Descartes brought the instrument forward in the 2nd paragraph of the Second Book in order to obtain, in order, "geometrical" curves, as many as you please, of ever-increasing complexity. In the very next paragraph, the 3rd paragraph of the Second Book, Descartes moves on, however, to another means of tracing and conceiving a series of curves of ever-increasing complexity: beginning with the simplest curve, a circle, as many points as you please are obtained on a more complex curve by cutting the simpler curve with a straight line, as follows:

Cut this:	*to get a curve that is:*
a circle,	a parabola;
a parabola,	a curve that is one degree more composite than a parabola;
this last curve,	a curve that is one degree more composite than the last curve;
and so on, endlessly.	

The proportional compass reappeared in the 2nd paragraph of the Third Book, where it was shown to be useful for obtaining mean proportionals by finding the intersection of a straight line and a curve drawn by that compass. The more complex the make-up of the proportional compass, the greater the number of mean proportionals—and the greater is the complexity of the curve used to get them. When the proportional compass is used to find mean proportionals, in order to find roots of the equation used for solving a problem, then we cannot be sure that the curve which that compass draws to do it is no more complex than necessary. That is why the curves employed to solve problems in the Third Book are traced by other means, the means presented in the 3rd paragraph of the Second Book. These curves, which are gradually and endlessly more complex, can be used to solve problems by obtaining the roots of equations that mount ever higher—with those roots being obtained by making a circle cut a line (a curve except in the simplest case, which is a straight line) as follows:

Cut this:	*to solve a problem that is:*
straight line	plane;
parabola,	solid;
curve that is one degree more composite than a parabola;	supersolid:
and so on, endlessly	

By considering the proportional compass, Descartes has taken the matter of the only study that he held to be truly solid hitherto, and has laid it out methodically, with a view to its utility in a grand transformational project.

Conclusion: Descartes' project

Descartes believes himself to have solved the problem of finding solutions to all geometrical problems that it is possible to solve by constructing geometrical curves. And how has he solved that grand problem of problem-solving? By proceeding *methodically*: by taking knowledge to be a progression from what is clearly seen as simple, through an order of kinds distinguished by degrees of compositeness, to results that are not only reliable but also comprehensive.

Alert readers of Descartes will be reminded of the rules of method that he presents in what we might call his "methodology"—the *Discourse about the Method for Well Conducting One's Reason and Seeking the Truth in the Sciences*, with which he chooses to open the volume which he chooses to close with the treatise called the *Geometry*.

The *Geometry* concludes the initial phase of Descartes' very ambitious project. It is the culmination of the volume with which he took his project public. In the course of the four centuries since then, there has been a radical transformation in the world around us, in the way we live within it, and in the thoughts we think about that world and about the way we live within it. To understand that transformation, we need to understand the transformation in mathematics that has been central to it. The mathematical transformation got underway with Descartes' *Geometry*, a work that John Stuart Mill went so far as to call the greatest single step in the progress of the exact sciences. Descartes' work has made our technology possible, but his was not a merely technical enterprise. Indeed, Descartes is most widely known as the father of modern philosophy. We would do well to consider who might be the grandfather, and even the great-grandfather, of modern thought, but Descartes needs special attention, since it is with him that the transformative mathematical work of modernity gets underway.

Descartes presents his entire project most comprehensively in this anonymously released first publication. In the introductory part of it, the *Discourse about the Method for Well Conducting One's Reason and Seeking the Truth in the Sciences*, he says that all the received bodies of learning are useless—including mathematics, the two branches of which, arithmetic and geometry, deal merely with, respectively, bare numbers and imaginary figures. But, he says, mathematics—unlike all the other received bodies of learning—is certain. In traditional learning, that is to say, he finds nothing to be fruitful, and only one part of it to be solid. He sees a need to make what is solid also fruitful.

The fruits of knowledge, says Descartes in his later book that he called the *Principles of Philosophy*, are to be gathered from the three branches of the tree of knowledge—namely, mechanics (in which we learn how to master our environment), medicine (in which we learn how to master our bodies), and morals (in which we learn how to master our psyches)—and these three branches grow out from the tree's trunk, which is natural philosophy, that is, physical science.

And what does physical science look like in the Cartesian enterprise?

Physical science, in the Cartesian enterprise, is to be thoroughly geometricized—the world in all its variety and change being merely an endless homogeneous stuff stretching outward in all directions, the portions of which differ only in how the stuff is moving. And geometry, in the Cartesian enterprise, is to be thoroughly arithmeticized—a universalization of calculation being the means of overcoming obstacles to quantification, to an assimilation of how much and how many which facilitates the solving of problems by the manipulation of equations.

Descartes' program for the geometricization of the world is most easily seen in the book of his that he called *The World*. This was the book with which he had planned to go public, but which he decided not to publish when he heard of Galileo's troubles with the authorities. Descartes took his project public instead with the book that contained not only the cunning *Discourse about the Method of Well Conducting One's Reason and Seeking the Truth in the Sciences* but also the three companion pieces that he called tryouts (*essays*), testing whether by means of them he could direct science onto the right road.

In the *Discourse*, Descartes suggests briefly (what he had set out at some length in *The World*) how the world could come into being out of chaos by the merely natural effects of matter in motion but lacking any animation; and he also suggests that what animated human nature, namely, the human heart, is merely a furnace which heats up animal spirits that are just as bodily—just as explicable as merely natural effects of matter in motion but lacking any animation—as are the fine spirits produced when brandy is distilled.

The first two tryouts exhibit prime examples of the benefits of his project. In the central tryout (the *Meteorology*), things that appear in the sky, evoking human fear and prayer, wonder and worship, are shown to be explicable as merely natural effects of matter in motion—some of which, men could learn to manipulate, and thus create wonders that would inspire awe in other men; and in the first of the tryouts (the *Dioptrics*), mental tools developed in the later-placed *Geometry* are applied to the problem of devising bodily tools for overcoming the natural limitations of the human eye, so that by manufacturing microscopes and telescopes we can behold the parts of nature that are nearby but very small or large but very far away.

The project promoted by Descartes is mastery of nature. That is the end which he sets before us in the *Discourse*. What he does not tell us there, is where it came from—there he does not tell us that he learned from Francis Bacon to look upon the world as a quarry rather than a spectacle. Bacon himself confessed to being much beholden to an earlier author, and it was in the spirit of that earlier proponent of rational control that Bacon set Descartes and many others on the way to teaching mankind that beholding is subordinate to mastery, that wonder should give way to problem-solving. What Descartes brought to this modern project which took mastery of nature as its end was an emphasis on mathematics as means to the end. Descartes' take on things was to transform the world, and human life within it, by transforming mathematics.

The program for transforming mathematics is most easily seen in the *Discourse*'s accompanying tryout called the *Geometry*. Descartes' transformation of mathematics combined two ingredients, as we have seen—one of which we might call "geometrical," the other, "arithmetical." In this methodology of mastery, problem-solving analysis was to be given primacy over synthesizing demonstration of theorems; and all kinds of problems were to be solved by calculation, employing equations in place of proportions. Progress would not have to wait upon good luck; insight would abound upon the path of methodical manipulation.

Humanity after Descartes did his work, like humanity before the transformation that Descartes wrought, could neither learn nor act without employing its capabilities both for insight and for manipulation, but it makes a difference just what is understood to be the proper relation between viewing being and technological doing. There is no way simply to separate the question of what mathematics is from questions about what learning is, and about who the learners are, and about what it is for which they act. Our study of how Descartes transformed mathematics, and of what it was that he transformed, should help us to a better understanding of what we are and what we seek.

APPENDIX TO PART VI:

Diagrams for Selected Propositions:

for Chapter 1—from *Conics* II: #1–9, 28–29, 10, 12–15, 17

for Chapter 2—from *Conics* II: #44–49

for Chapter 3—from *Conics* III: #42, 45–52

for Chapter 6—from *Conics* III: #1–2, 16, 54

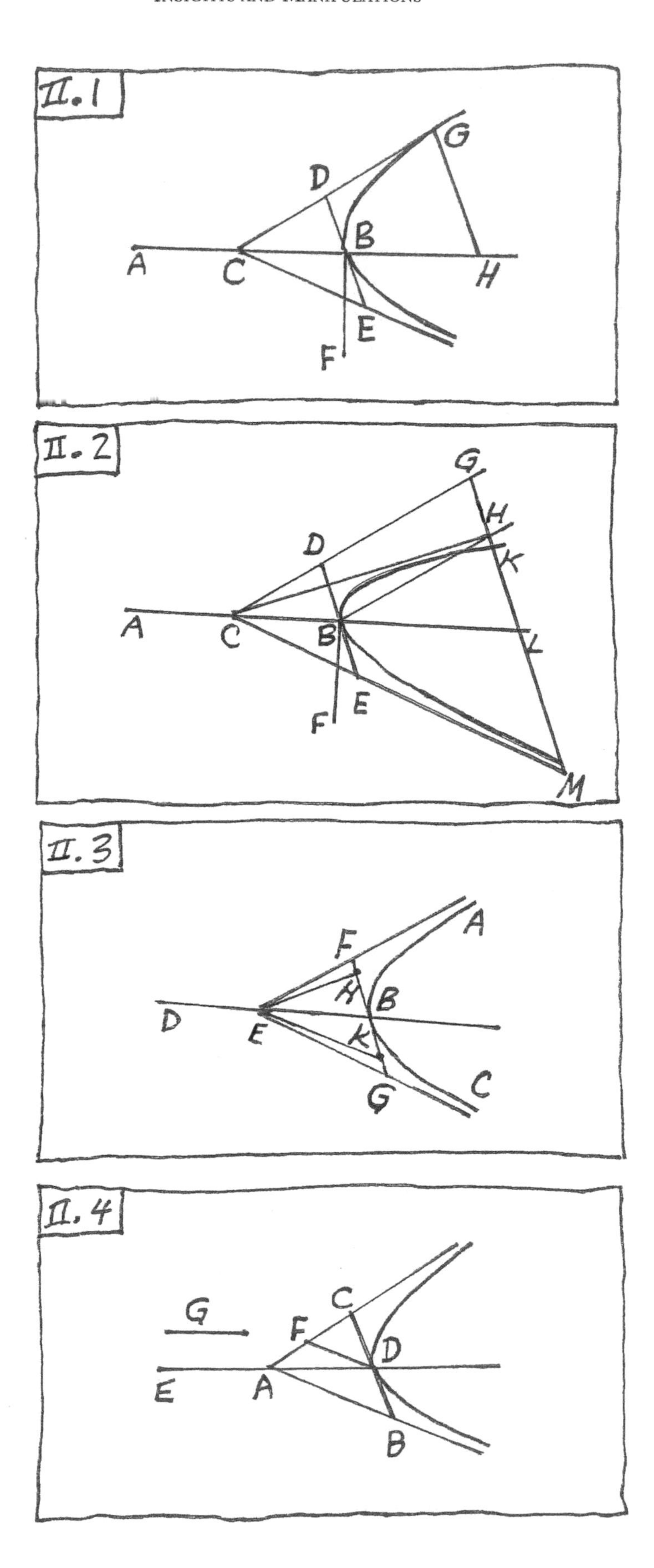
II.1
D
G
B
A
C
H
E
F
II.2
G
H
D
K
A
C
B
L
E
F
M
II.3
A
F
H
B
D
E
K
C
G
II.4
C
G
F
D
E
A
B

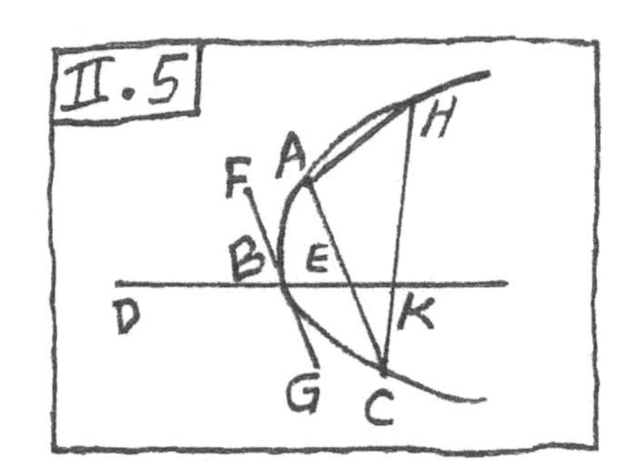
II.5
F
A
H
B
E
D
K
G
C

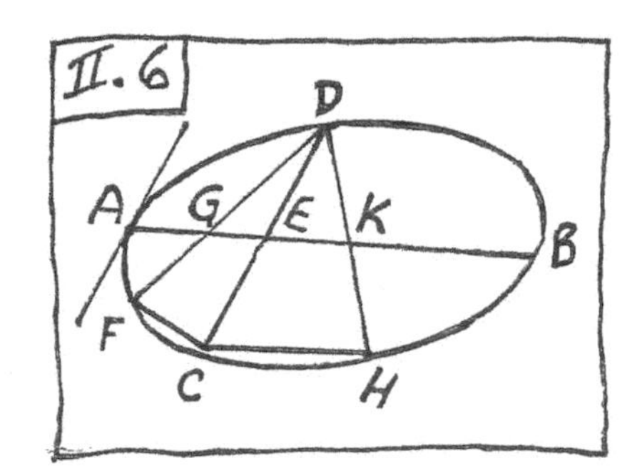
II.6
D
A
G
E
K
B
F
C
H

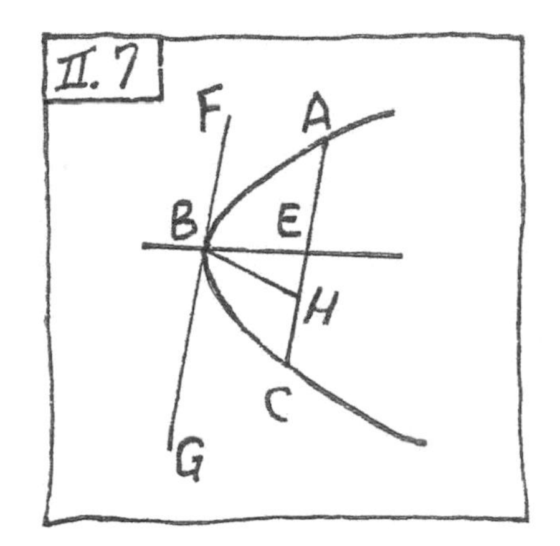
II.7
F
A
B
E
H
C
G

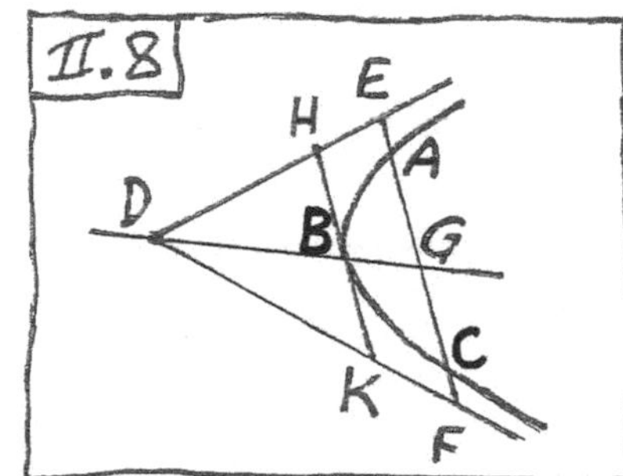
II.8
E
H
A
D
B
G
C
K
F

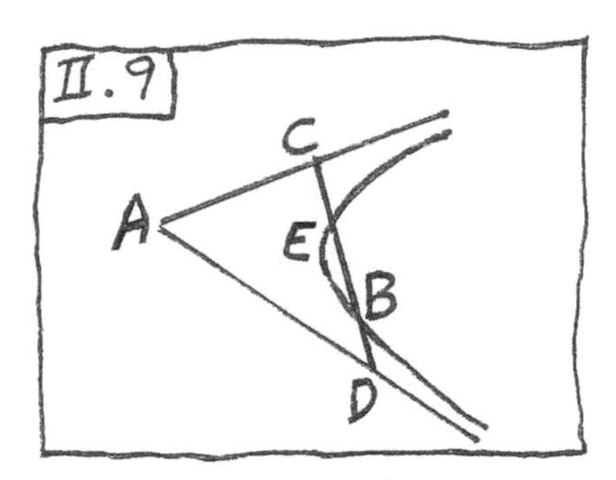
II.9
C
A
E
B
D

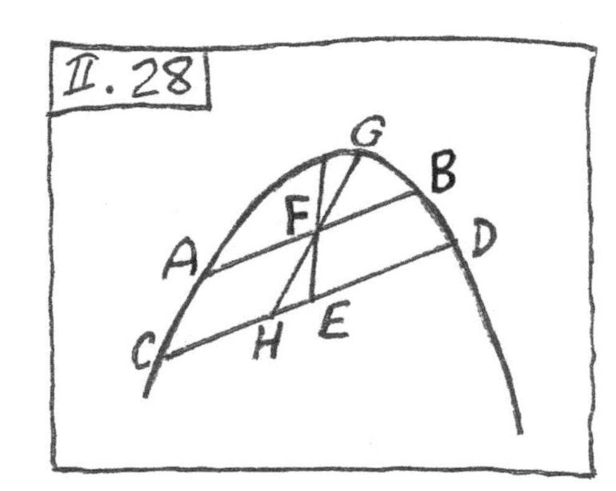
II.28
G
B
F
D
A
C
H
E

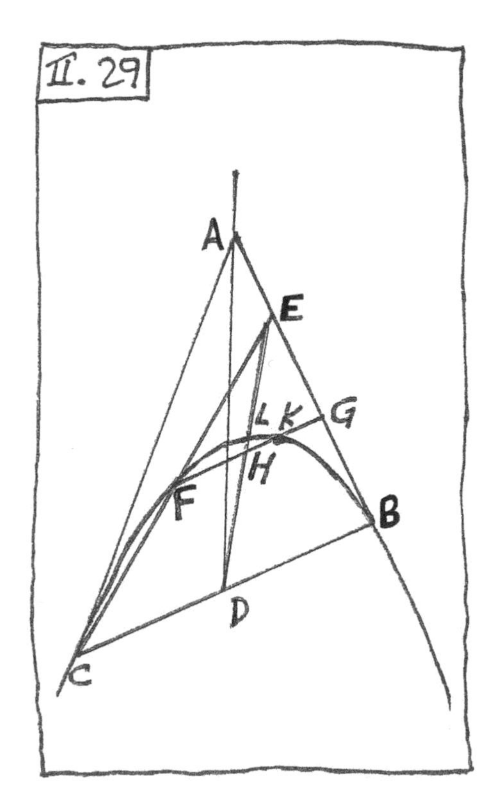
II.29
A
E
L
K
G
H
F
B
D
C

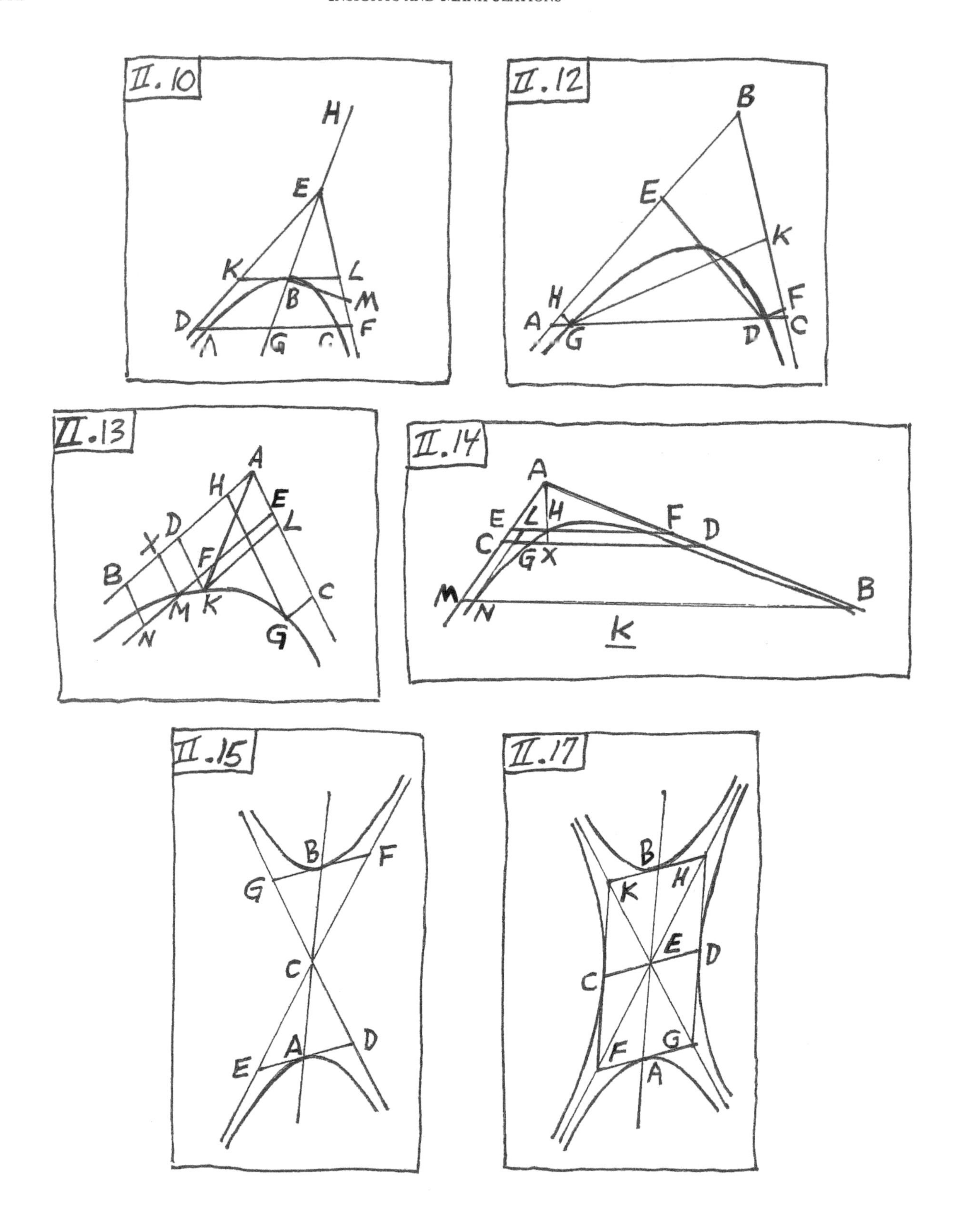
II.10
H
E
K
L
B
M
D
F
G
II.12
B
E
K
H
F
A
G
D
C
II.13
A
H
E
D
L
F
B
C
M
K
N
G
II.14
A
E
L
H
F
D
C
G
X
B
M
N
K
II.15
B
F
G
C
A
D
E
II.17
B
K
H
E
D
C
F
G
A

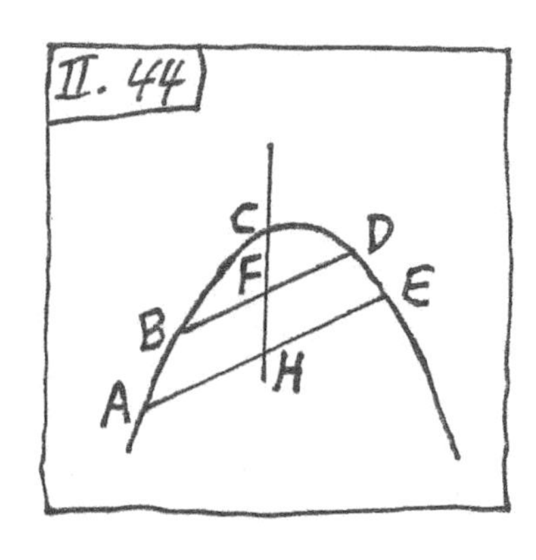
II. 44
C
D
F
E
B
A
H

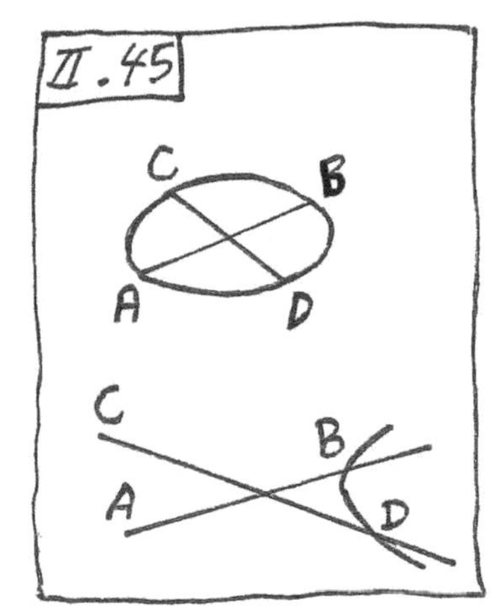
II. 45
C
B
A
D
C
B
A
D

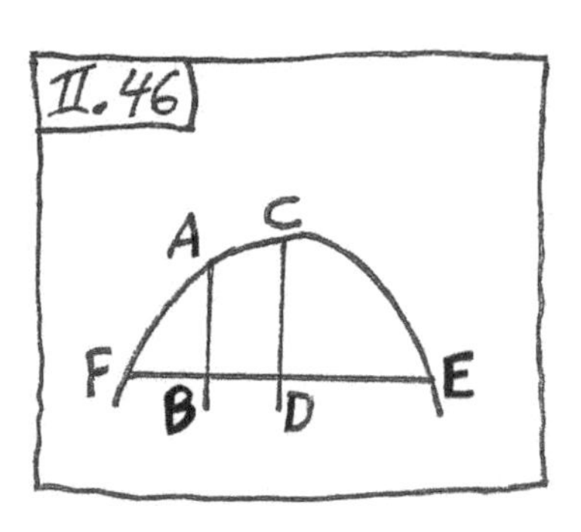
II. 46
A
C
F
B
D
E

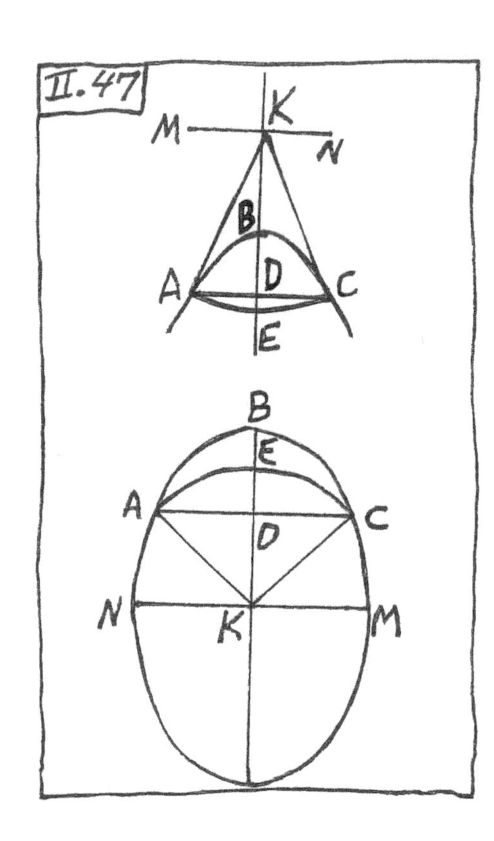
II. 47
M
K
N
B
A
D
C
E
B
E
A
C
D
N
K
M

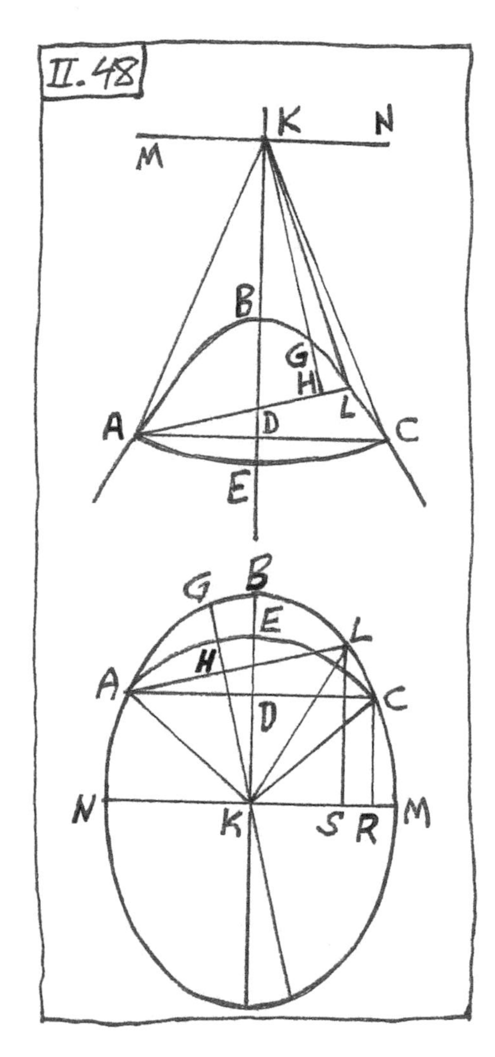
II. 48
K
N
M
B
G
H
L
A
D
C
E
G
B
E
L
H
A
C
D
N
K
S
R
M

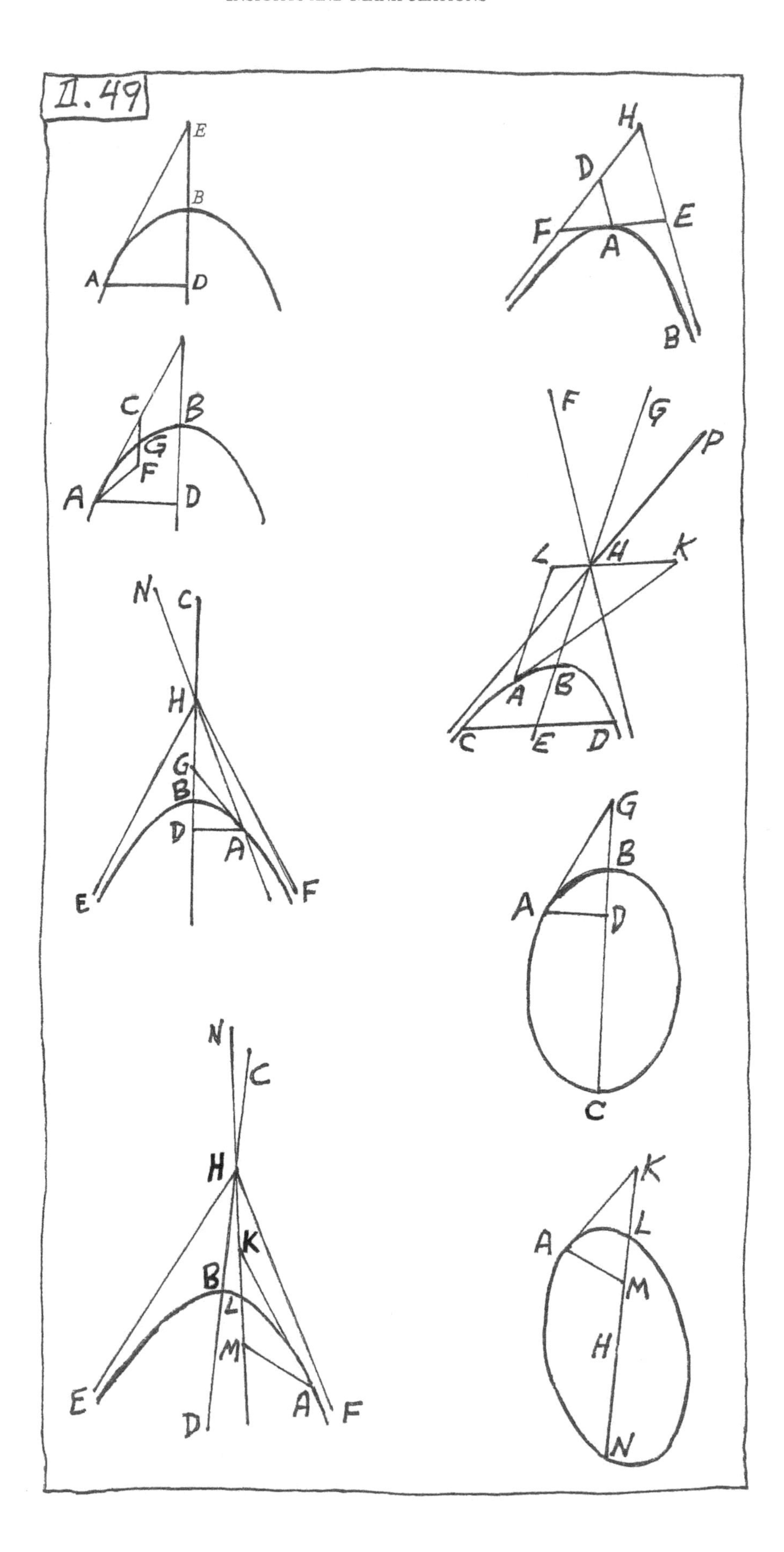
II.49
E
B
A
D
H
D
F
E
A
B
C
B
G
F
A
D
F
G
P
L
H
K
A
B
C
E
D
N
C
H
G
B
D
A
E
F
G
B
A
D
C
N
C
H
K
B
L
M
E
D
A
F
K
L
A
M
H
N

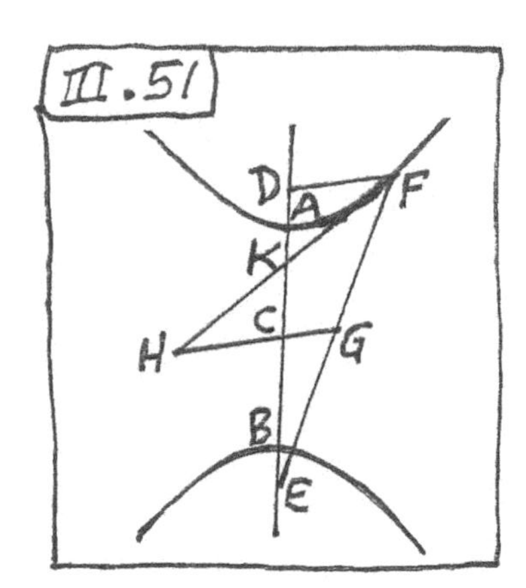
III.51
D
A
F
K
C
H
G
B
E

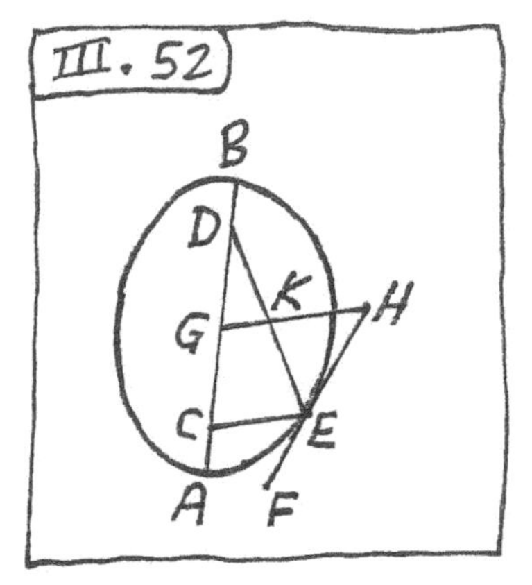
III.52
B
D
K
H
G
C
E
A
F

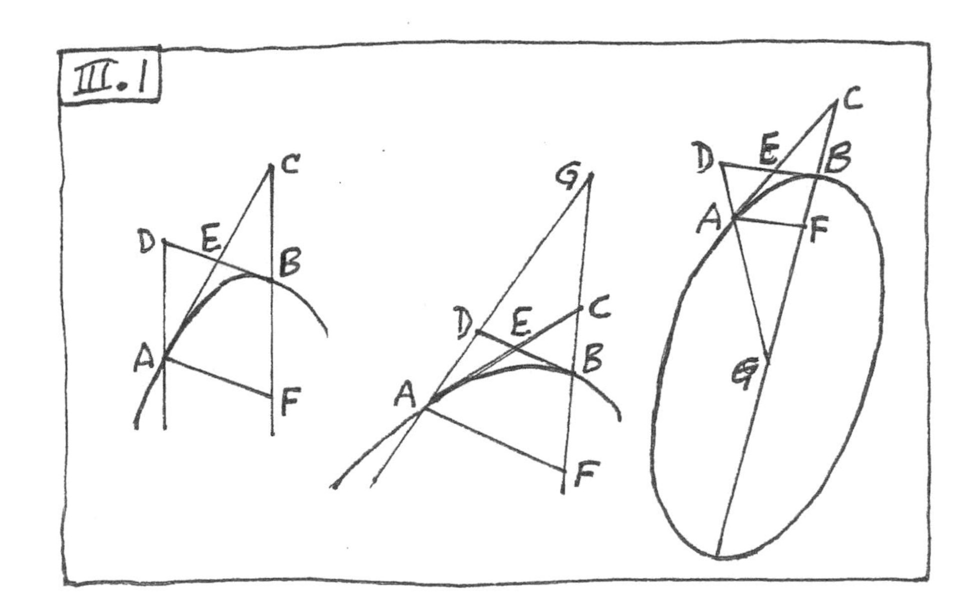
III.1
C
D
E
B
A
F
G
D
E
C
B
A
F
C
D
E
B
A
F
G

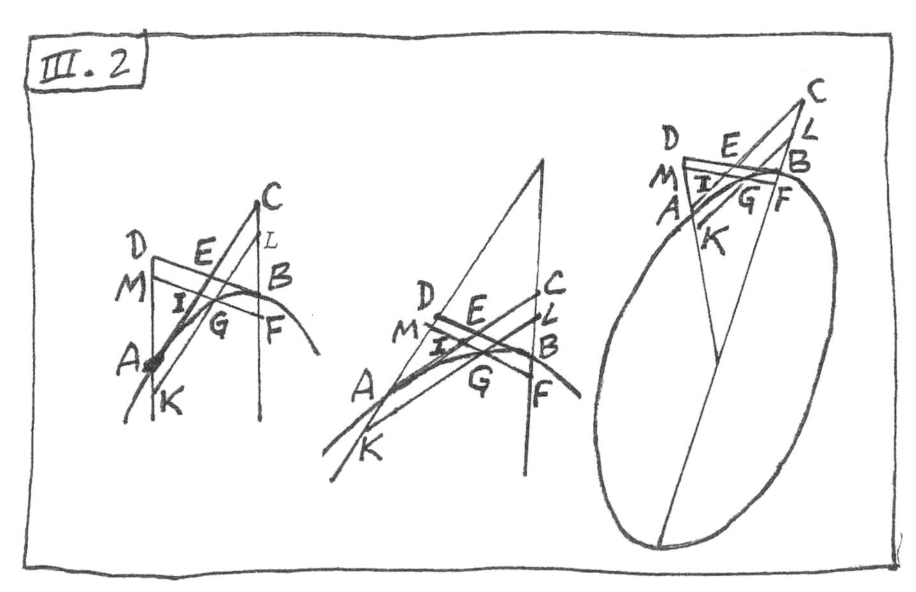
III.2
C
L
D
E
B
M
I
G
F
A
K
D
E
C
L
M
I
B
A
G
F
K
C
L
D
E
B
M
I
G
F
A
K

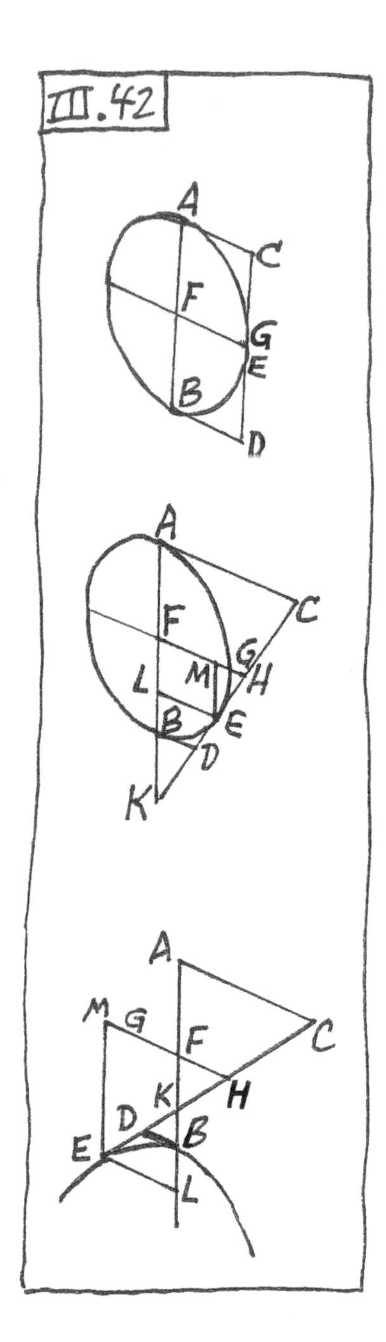
III.42
A
C
F
G
E
B
D
A
C
F
G
H
L
M
B
E
D
K
A
M
G
F
C
K
H
D
B
E
L

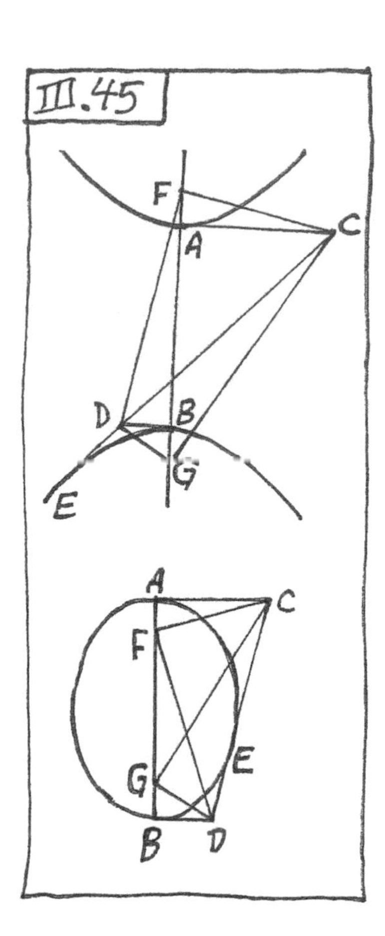
III.45
F
C
A
D
B
G
E
A
C
F
E
G
B
D

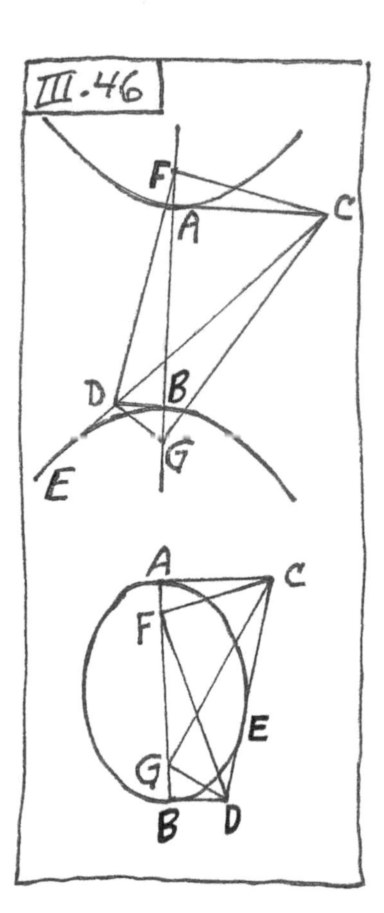
III.46
F
C
A
D
B
G
E
A
C
F
E
G
B
D

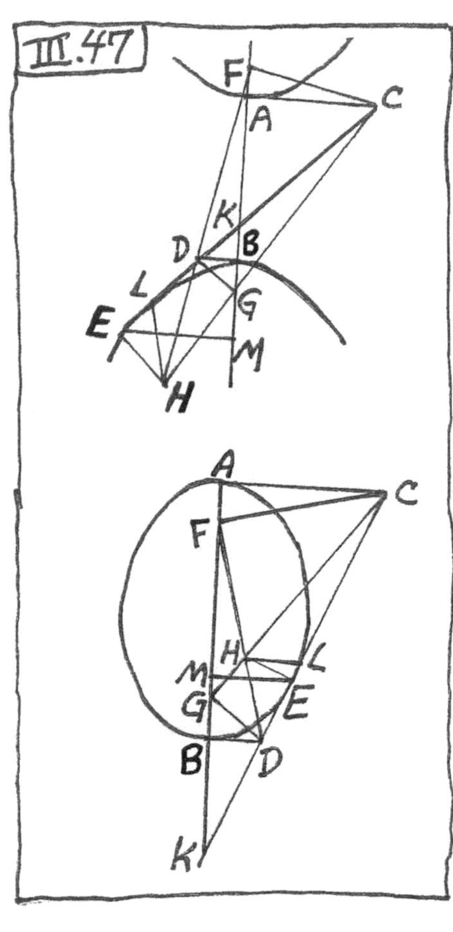
III.47
F
C
A
K
D
B
L
G
E
M
H
A
C
F
H
L
M
G
E
B
D
K

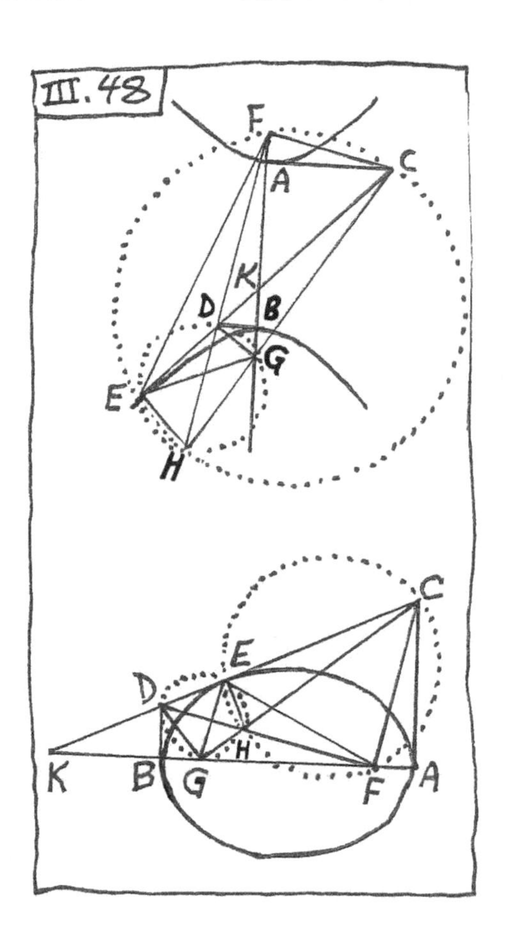
III.48
F
C
A
K
D
B
G
E
H
C
E
D
H
K
B
G
F
A

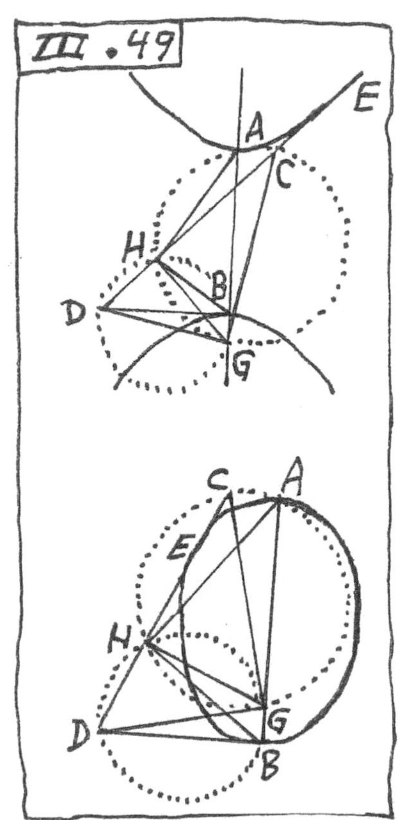
III.49
E
A
C
H
B
D
G
C
A
E
H
G
D
B

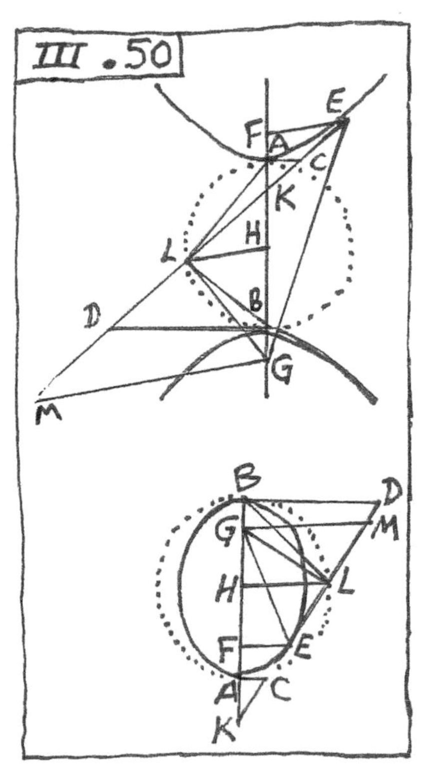
III.50
E
F
A
C
K
H
L
D
B
G
M
B
D
G
M
H
L
F
E
A
C
K

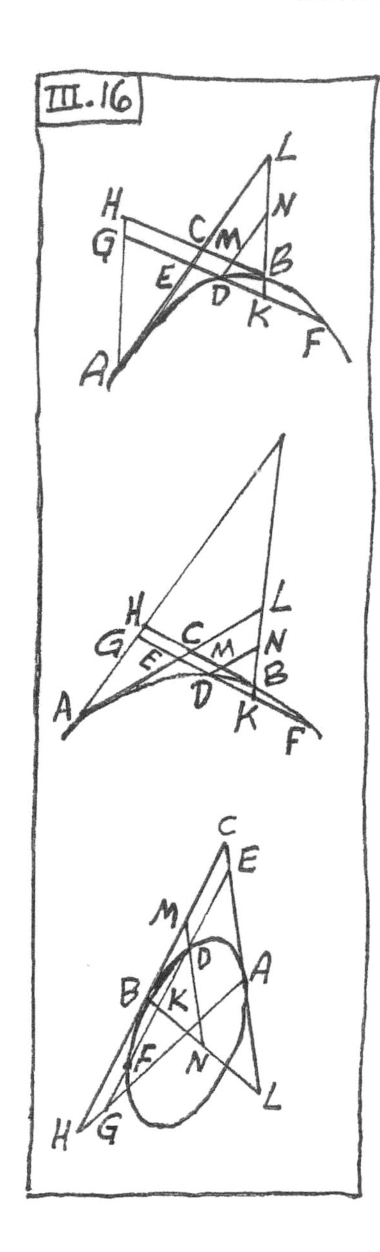
III.16
L
H
N
C
M
G
B
E
D
K
A
F
H
L
C
G
M
N
E
B
A
D
K
F
C
E
M
D
A
B
K
F
N
L
H
G

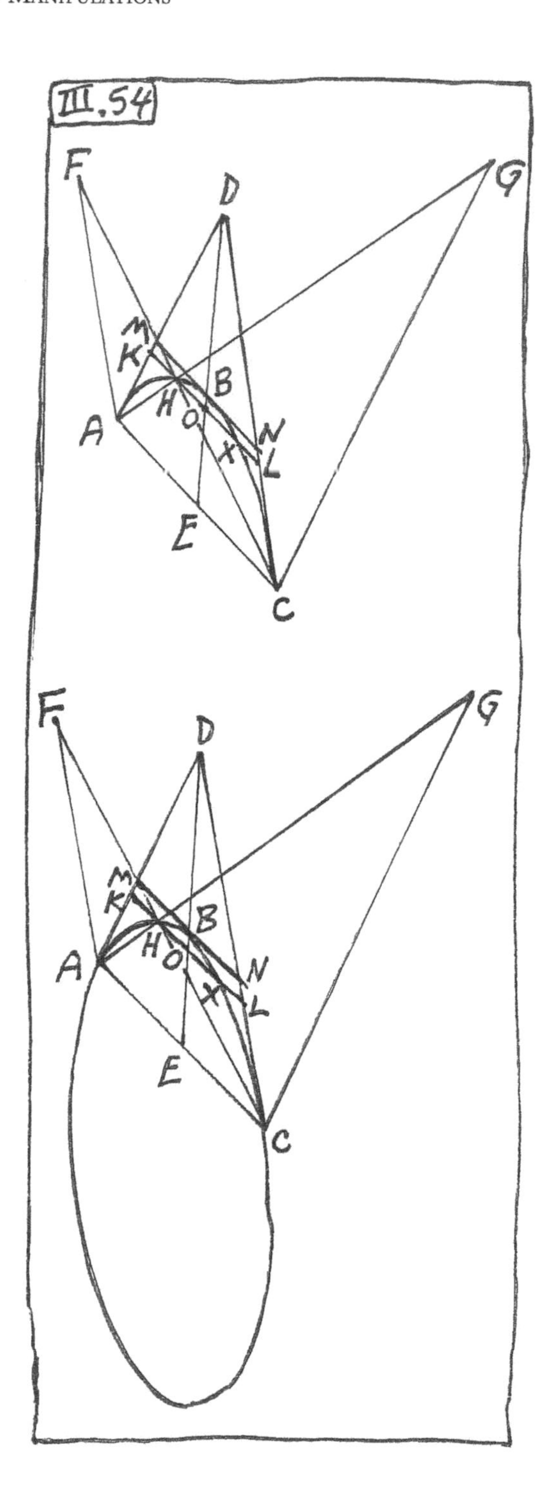
III.54
F
G
D
M
K
B
A
H
O
N
X
L
E
C
F
G
D
M
K
B
A
H
O
N
X
L
E
C